AQA A Level Further Maths

Year 1 / AS Level

Series Editor
David Baker

Authors
Brian Jefferson, David Bowles, Eddie Mullan, Garry Wiseman, John Rayneau, Katie Wood, Mike Heylings, Rob Wagner

Powered by MyMaths.co.uk

OXFORD
UNIVERSITY PRESS

OXFORD
UNIVERSITY PRESS

Great Clarendon Street, Oxford, OX2 6DP, United Kingdom

Oxford University Press is a department of the University of Oxford.

It furthers the University's objective of excellence in research, scholarship, and education by publishing worldwide. Oxford is a registered trade mark of Oxford University Press in the UK and in certain other countries.

First published in 2018

British Library Cataloguing in Publication Data
Data available

978 0 19 841292 2

10 9 8 7 6 5 4 3

Paper used in the production of this book is a natural, recyclable product made from wood grown in sustainable forests.
The manufacturing process conforms to the environmental regulations of the country of origin.

Printed and bound by CPI Group (UK) Ltd, Croydon, CR0 4YY

Acknowledgements

Authors
Brian Jefferson, David Bowles, Eddie Mullan, Garry Wiseman, John Rayneau, Katie Wood, Mike Heylings, Rob Wagner

Editorial team
Dom Holdsworth, Ian Knowles, Matteo Orsini Jones

With thanks also to Geoff Wake, Matt Woodford, Brian Brooks, Jon Stone, Deb Dobson, Katherine Bird, Linnet Bruce, Keith Gallick, Laurie Luscombe, Susan Lyons and Amy Ekins-Coward for their contribution.

Index compiled by Milla Hills

Although we have made every effort to trace and contact all copyright holders before publication, this has not been possible in all cases. If notified, the publisher will rectify any errors or omissions at the earliest opportunity.

p3: Shutterstock; **p30:** Shutterstock; **p35:** josemoraes/iStockphoto; **p71:** Shutterstock; **p75:** Atosan/Shutterstock; **p122:** Manzotte Photography/Shutterstock; **p122:** Soleil Nordic/Shutterstock; **p125:** BEST-BACKGROUNDS/Shutterstock; **p143:** MarkLG/Shutterstock; **p176:** Shutterstock; **p179:** science photo/Shutterstock; **p202:** Shutterstock; **p205:** Dolomites-image/iStockphoto; **p230:** Andyd/iStockphoto; **p230:** Steve Mann/Dreamstime; **p235:** FatCamera/iStockphoto; **p256:** selensergen/iStockphoto; **p256:** Tatiana Shepeleva/Shutterstock; **p261:** choja/iStockphoto; **p272:** Roman023_photography/Shutterstock; **p272:** Annette Shaff/Shutterstock; **p275:** Wafuefotodesign/Dreamstime.com; **p315:** Alphotographic/iStockphoto; **p341:** El Nariz/Shutterstock; **p341:** El Nariz/Shutterstock; **p345:** kynny/iStockphoto; **p409:** Pavel L Photo and Video/Shutterstock; **p426:** turtix/Shutterstock; **p431:** xieyuliang/Shutterstock; **p467:** Monkey Business Images/Shutterstock; **p479:** Rvector/Shutterstock; **p479:** Zvonimir Atletic/Shutterstock.com

Message from AQA

This student book has been approved by AQA for use with our qualification. This means that we have checked that it broadly covers the specification and we are satisfied with the overall quality. We have not, however, reviewed the MyMaths and InvisiPen links, and have therefore not approved this content.

We approve books because we know how important it is for teachers and students to have the right resources to support their teaching and learning. However, the publisher is ultimately responsible for the editorial control and quality of this book.

Please note that mark allocations given in assessment questions are to be used as guidelines only: AQA have not reviewed or approved these marks. Please also note that when teaching the AQA A Level Further Maths course, you must refer to AQA's specification as your definitive source of information. While the book has been written to match the specification, it cannot provide complete coverage of every aspect of the course.

Full details of our approval process can be found on our website: www.aqa.org.uk

Contents

STATISTICS

Chapter 10: Discrete and continuous random variables

Chapter 11: Hypothesis testing and contingency tables

DISCRETE

Chapter 12: Graphs and networks 1

Chapter 13: Critical path analysis 1

Chapter 14: Linear programming and game theory 1

Chapter 15: Abstract algebra

About this book

This book has been specifically created for those studying the AQA 2017 Further Mathematics AS and A Level. It has been written by a team of experienced authors and teachers, and it's packed with questions, explanation and extra features to help you get the most out of your course.

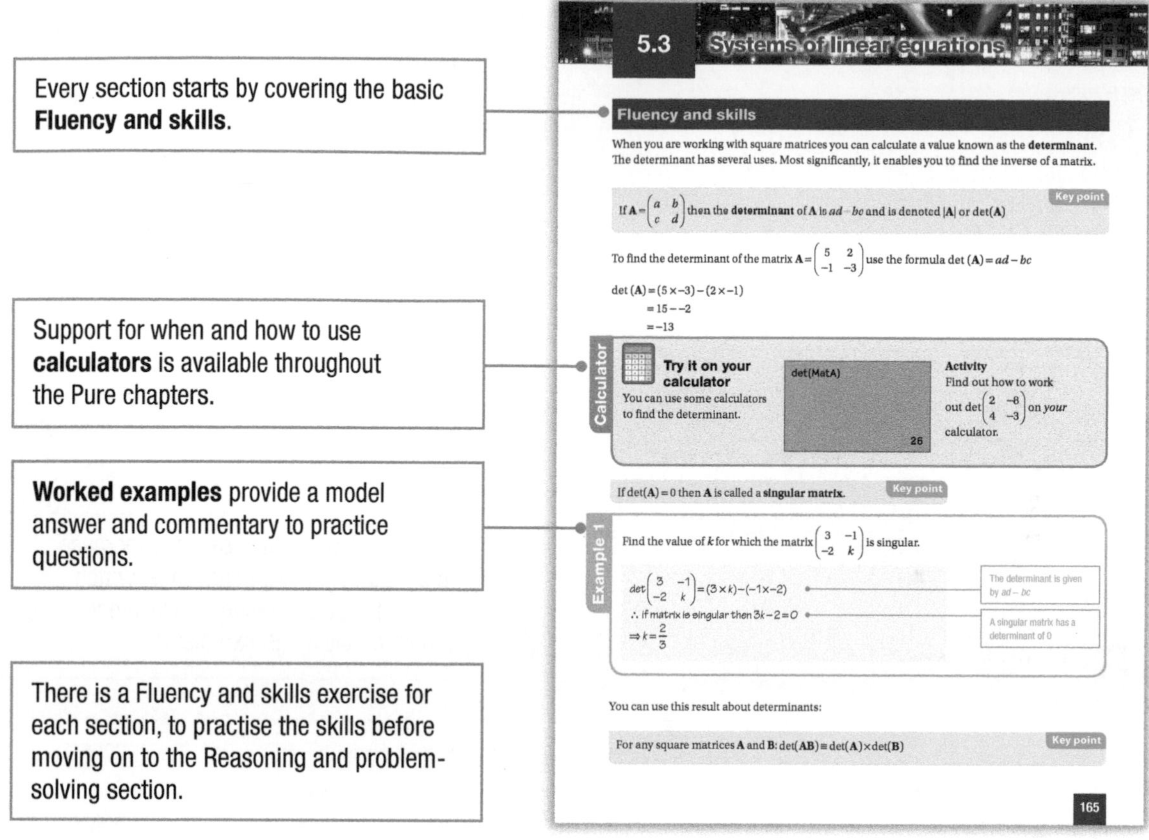

Every section starts by covering the basic **Fluency and skills**.

Support for when and how to use **calculators** is available throughout the Pure chapters.

Worked examples provide a model answer and commentary to practice questions.

There is a Fluency and skills exercise for each section, to practise the skills before moving on to the Reasoning and problem-solving section.

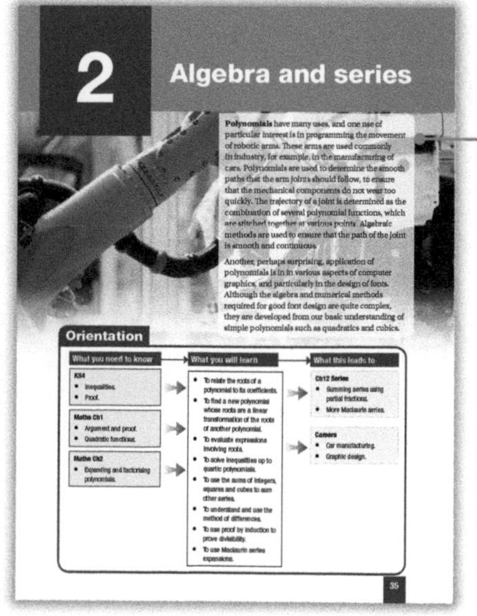

On the chapter **Introduction page**, the **Orientation box** explains what you should already know, what you will learn, and what this leads to.

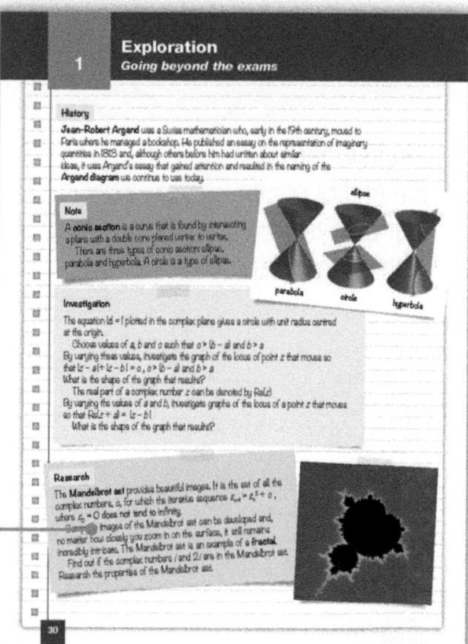

At the end of every chapter, an **Exploration page** gives you an opportunity to explore the subject beyond the specification.

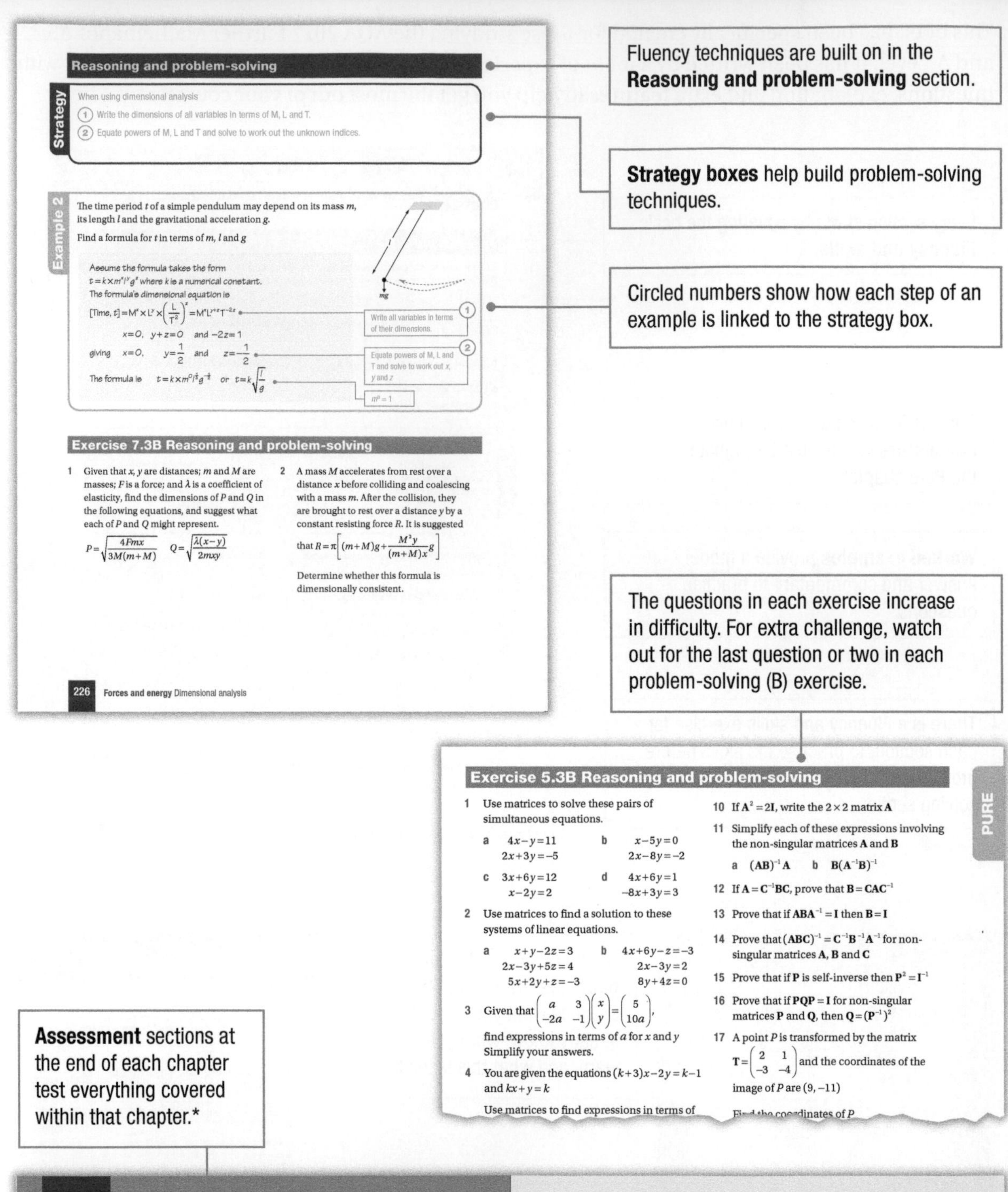

Fluency techniques are built on in the **Reasoning and problem-solving** section.
Strategy boxes help build problem-solving techniques.
Circled numbers show how each step of an example is linked to the strategy box.
The questions in each exercise increase in difficulty. For extra challenge, watch out for the last question or two in each problem-solving (B) exercise.
Assessment sections at the end of each chapter test everything covered within that chapter.*

3 Assessment

1 a Sketch the graph of $y=\frac{x+5}{x-7}$, clearly labelling any intercepts with the coordinate axes. **[4 marks]**

b Write down the equations of the asymptotes to the curve. [2]

c Find the coordinates of the points where the curve $y=\frac{x+5}{x-7}$ meets the line $y=-x$ [3]

2 a Sketch the curve $y=\frac{5x^2}{x^2+1}$ [3]

b Find the points of intersection between the curve $y=\frac{5x^2}{x^2+1}$ and the line $y=1$ [3]

3 a Express the Cartesian coordinates $(\sqrt{3},-3)$ as polar coordinates, giving the exact value of r and writing θ in terms of π where $-\pi<\theta\le\pi$ [4]

11 a Sketch the curve $y=\frac{x^2-5x+4}{x^2+1}$ [3] b Solve the inequality $\frac{x^2-5x+4}{x^2+1}>1-x$ [1]

12 A rational function is defined by $f(x)=\frac{3x^2+5x-8}{x^2+2x-6}$

A sketch of the curve $y=f(x)$ is shown.

a For what values of x is the function undefined? Leave your answer as a surd. [2]

b Where does the curve of the function cut the coordinate axes? [3]

c For what values of x is $f(x)>1$? [5]

13 a Show that the equation $xy-5y+x-9=0$ can be written as $(x-5)(y+1)=k$ where k is a constant to be found. [2]

* Please note that mark allocations given in assessment questions are to be used as guidelines only: AQA have not reviewed or approved these marks.

1 Complex numbers 1

Complex numbers are a crucial tool in the design of modern electrical components, such as motherboards. Electrical engineers use them to simplify many of their calculations. This helps them to analyse varying currents and voltages in electrical circuits. When the current flows in one direction constantly, the resistance can often be calculated using a simple formula. However, when the direction of the current is alternating, this formula doesn't work. Therefore, engineers express the quantities as complex numbers, which makes it easier to understand the processes involved and perform the necessary calculations.

Complex numbers are used in many other fields such as chemistry, economics, and statistics. Their unique properties provide powerful ways of solving and interpreting complicated equations that can be applied in a wide variety of contexts. From some of the deepest mysteries in number theory to the signal processing used in playing digital music, these so-called imaginary numbers have led to a surprising array of advances in the real world.

Orientation

What you need to know

KS4
- Use the quadratic formula.

Maths Ch1
- Solve simultaneous equations.

Maths Ch3
- Work with sine and cosine functions.

What you will learn

- To calculate with complex numbers in the form $a + bi$
- To understand and use the complex conjugate.
- To solve quadratic, cubic and quartic equations with real coefficients.
- To convert between modulus-argument form and the form $a + bi$ and calculate with numbers in modulus-argument form.
- To sketch and interpret Argand diagrams.

What this leads to

Ch16 Complex numbers 2
- Exponential form.
- The Euler formula $e^{i\theta} = \cos\theta + i\sin\theta$
- De Moivre's formula.
- Roots of unity.

Careers
- Electrical engineering.

1.1 Properties and arithmetic

Fluency and skills

Some equations, including those deriving from real-world situations, have no real solutions. For example, up to this point, you have been unable to solve equations such as $x^2=-1$. You know that the solutions are $x=\pm\sqrt{-1}$, but this is not a real number and is difficult to manipulate. In order to solve this problem, mathematicians denoted the square root of negative one by i. This means that the solutions of the equation $x^2=-1$ can be written as $\pm i$. Although this number is known as 'imaginary', it means that all polynomial equations do indeed have solutions.

A good analogy is negative numbers: these can be hard to grasp in isolation. For example, the concept of 'minus one apple' is a difficult one, but it's useful in sums such as 3 apples – 1 apple = 2 apples. In exactly the same way, imaginary numbers are essential in many calculations and have many real-world applications.

Key point

The **imaginary number** i is defined as $i=\sqrt{-1}$

You can solve the equation $x^2=-9$ by square-rooting both sides to give $x=\pm\sqrt{-9}$

Then, because $\sqrt{-9}$ can be written as $\sqrt{9}\sqrt{-1}$, you can use the fact that $i=\sqrt{-1}$ to give $x=\pm 3i$

Notice that once you have defined the square root of minus one, the square root of all other negative numbers can be written in terms of i

For example, to solve the equation $(x-3)^2=-5$, you first square-root both sides to give $x-3=\pm\sqrt{-5}$, and since $\sqrt{-5}=\sqrt{5}i$, this gives the solutions $x=3\pm\sqrt{5}i$

Numbers with both real and imaginary parts are called complex numbers.

Key point

Complex numbers can be written in the form $a+bi$ where $a, b\in\mathbb{R}$. The set of complex numbers is denoted $\mathbb{C}$

Complex numbers can be added, subtracted and multiplied by a constant in the same way as algebraic expressions.

Example 1

Simplify the expression $3(4-7i)-2(3-2i)$

$3(4-7i)-2(3-2i)=(12-21i)-(6-4i)$ — Multiply real and imaginary parts by the constant.

$=6-17i$ — Simplify real parts: $12-6=6$. Simplify imaginary parts: $-21i+4i=-17i$

When multiplying complex numbers together, you will use the fact:

Key point

$i^2 = -1$ (since $i^2 = (\sqrt{-1})^2$)

Example 2

Solve the equation $(x+7)^2 = -16$

$x + 7 = \pm\sqrt{-16}$ — Square-root both sides of equation.

$x = -7 \pm \sqrt{-16}$

$= -7 \pm \sqrt{16}\sqrt{-1}$

$= -7 \pm 4i$ — Since $\sqrt{-1} = i$

Calculator

Try it on your calculator

Some calculators enable you to manipulate complex numbers.

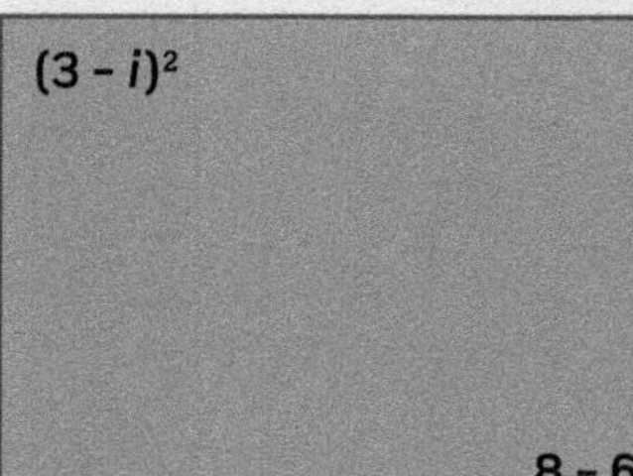

Activity

Find out how to work out $(3-i)^2$ on *your* calculator.

Use your calculator to find these expressions.

a $(2+i)(5-7i)$ **b** $4i(2-8i)$

c $(30+5i) \div (2-i)$

To rationalise the denominator in surd form such as $\frac{1}{a+\sqrt{b}}$, you multiply the numerator and the denominator by $a-\sqrt{b}$

You can use a similar method to simplify fractions with complex denominators.

Doing this will always change the denominator into a positive real number.

Key point

To simplify $\frac{1}{a+bi}$, multiply the numerator and denominator by $a - bi$ (this is called the **complex conjugate**).

Example 3

Simplify $\frac{1+3i}{1-2i}$

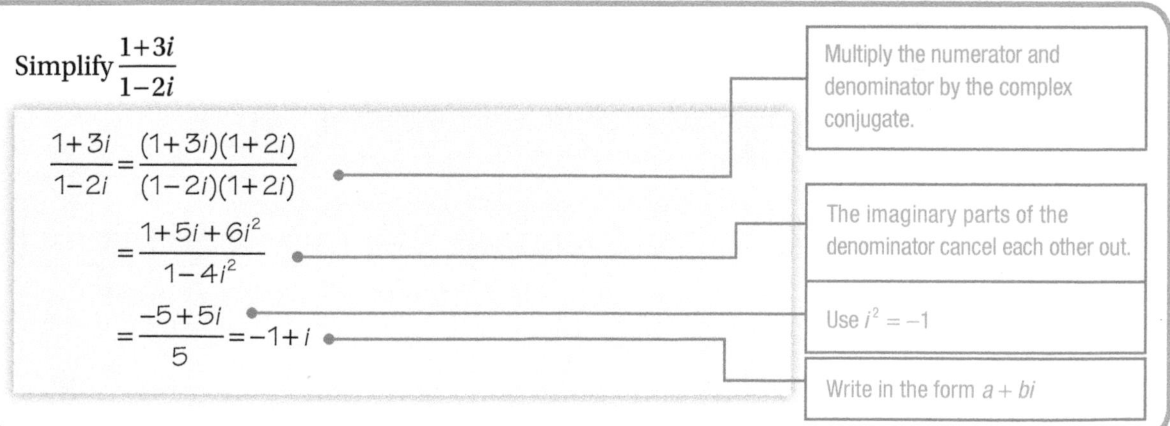

$$\frac{1+3i}{1-2i} = \frac{(1+3i)(1+2i)}{(1-2i)(1+2i)}$$

Multiply the numerator and denominator by the complex conjugate.

$$= \frac{1+5i+6i^2}{1-4i^2}$$

The imaginary parts of the denominator cancel each other out.

$$= \frac{-5+5i}{5} = -1+i$$

Use $i^2 = -1$

Write in the form $a + bi$

Exercise 1.1A Fluency and skills

1 Solve these equations.

a $x^2=-25$ b $x^2=-121$

c $x^2=-20$ d $x^2+8=0$

e $z^2=-9$ f $z^2+12=0$

2 Simplify these expressions, giving your answers in the form $a+bi$ where $a, b\in\mathbb{R}$

a $(2+3i)+(5-9i)$

b $(5-7i)-(12+3i)$

c $3(6-9i)$ d $3(2+10i)+5(4-i)$

e $4-9(7i+5)$ f $2(6-2i)-3(2i-5)$

3 Write each of these expressions in the form $a+bi$ where $a, b\in\mathbb{R}$

a $(2+3i)(i+5)$ b $(7-i)(6-3i)$

c $i(8-3i)$ d $(9-4i)^2$

4 Fully simplify each of these expressions.

a i^3 b i^4 c i^5

d $(2i)^3$ e $(3i)^4$ f $2i^2(5i-9)^2$

5 Simplify these fractions, giving your answers in the form $a+bi$ where $a, b\in\mathbb{R}$

a $\frac{3}{2+i}$ b $\frac{2i}{1-5i}$ c $\frac{1+7i}{3-i}$

d $\frac{i+3}{2i-1}$ e $\frac{6+3i}{i-\sqrt{2}}$ f $\frac{\sqrt{2}i-\sqrt{6}}{\sqrt{3}-i}$

6 You are given that $z_1=3i-2$, $z_2=4+i$

Calculate these expressions, fully simplifying your answers.

a z_1+z_2 b z_1z_2

c $\frac{z_1}{z_2}$ d $\frac{z_2}{z_1}$

Reasoning and problem-solving

Strategy

To solve equations involving imaginary numbers

1. Write all the numbers and expressions in the form $a+bi$
2. Equate real parts and imaginary parts on both sides of the equation or identity.
3. Solve the equations simultaneously.

Example 4

Find real numbers a and b such that $(a+5i)(2-i)=9+bi$

$(a+5i)(2-i)=9+bi$

$2a-ai+10i-5i^2=9+bi$ — Expand the brackets. (1)

$(2a+5)+(10-a)i=9+bi$ — Simplify using $i^2=-1$ and collect together real and imaginary terms.

Re: $2a+5=9$

$a=2$ — Equate real parts. (2)

Im: $10-a=b$

$b=8$ — Equate imaginary parts. (2)

Example 5

Find the complex numbers z such that $z^2 = 5+12i$

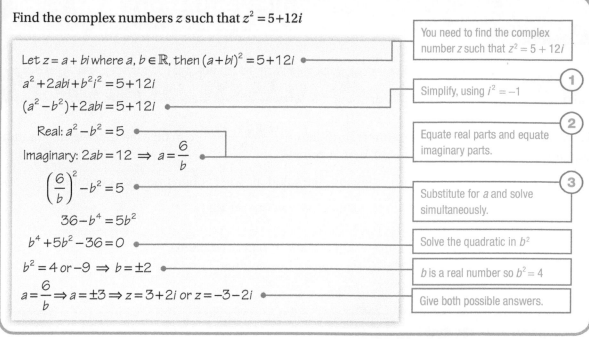

Let $z = a + bi$ where $a, b \in \mathbb{R}$, then $(a+bi)^2 = 5+12i$ — You need to find the complex number z such that $z^2 = 5 + 12i$

$a^2 + 2abi + b^2i^2 = 5+12i$

$(a^2 - b^2) + 2abi = 5+12i$ — Simplify, using $i^2 = -1$ (1)

Real: $a^2 - b^2 = 5$

Imaginary: $2ab = 12 \Rightarrow a = \frac{6}{b}$ — Equate real parts and equate imaginary parts. (2)

$\left(\frac{6}{b}\right)^2 - b^2 = 5$ — Substitute for a and solve simultaneously. (3)

$36 - b^4 = 5b^2$

$b^4 + 5b^2 - 36 = 0$ — Solve the quadratic in b^2

$b^2 = 4$ or $-9 \Rightarrow b = \pm 2$ — b is a real number so $b^2 = 4$

$a = \frac{6}{b} \Rightarrow a = \pm 3 \Rightarrow z = 3+2i$ or $z = -3-2i$ — Give both possible answers.

Exercise 1.1B Reasoning and problem-solving

1 Find the real numbers a and b such that $(2a+i)^2 = 35 - bi$

2 Find the values $a, b \in \mathbb{R}$ that satisfy the equation $(2-bi)(3+4i) = a - 13i$

3 Solve the equation

$z(2-5i) = 13 - 18i$ for $z \in \mathbb{C}$

4 Solve the equation

$(x+3i)(10-7i) = 34+6i$

for a complex number x

5 Find the complex numbers z_1 and z_2 such that $z_1 + z_2 = 11 - 3i$ and $z_2 - z_1 = 5 + 7i$

6 Find the complex numbers z and w that satisfy the equations

$z + 2w = 6i$ and $3z - 4w = 20 + 23i$

7 Find $z_1, z_2 \in \mathbb{C}$ such that

$2z_1 - 3z_2 = 10 + 8i$ and $5z_1 - \frac{1}{2}z_2 = 4 + 6i$

8 Find the complex numbers w in each of these cases.

a $w^2 = 30i - 16$ b $w^2 = -3 - 4i$

c $w^2 - 1 = 20(1-i)$

9 Calculate the square roots of $2 - 4\sqrt{2}i$

10 Solve these simultaneous equations, where w and z are complex numbers.

$w^2 + z^2 = 0$

$z - 3w = 10$

11 Solve each of these equations to find $z \in \mathbb{C}$

a $z^2 - 2z = -50$ b $z^2(1+i) = 7 - 17i$

Full A Level

12 Simplify each of these expressions, giving your answer in the form $a + bi$ in each case.

a $(1+2i)^4$ b $(2-5i)^5$

c $(3i-1)^3(1+i)$

13 Find $a, b \in \mathbb{R}$ such that

$(a+i)^4 = 28 + bi$

1.2 Solving polynomial equations

Fluency and skills

Using complex numbers, the quadratic formula $x=\frac{-b\pm\sqrt{b^2-4ac}}{2a}$ can be used to solve all quadratic equations of the form $ax^2+bx+c=0$, with $a, b, c\in\mathbb{R}$

When the discriminant (b^2-4ac) is negative the solutions will contain imaginary numbers.

Because the $\pm$ in the formula gives two solutions, you can see that:

Key point

If $z=a+bi$ is a solution of a quadratic equation with **real coefficients** then the complex conjugate, $z^*=a-bi$, will also be a solution.

Example 1

Given that $z=5-3i$, find

a z^* **b** zz^* **c** $z+z^*$

a $z^*=5+3i$ — Write the complex conjugate of z

b $zz^*=(5-3i)(5+3i)$

$=25+15i-15i-9i^2$

$=25+9$ — Since $i^2=-1$

$=34$ — zz^* will always be a real number.

c $z+z^*=5-3i+5+3i$

$=10$ — $z+z^*$ will always be a real number.

If you know the roots of a quadratic equation, then you can find the original equation by multiplying the factors together.

Example 2

Find a quadratic equation with real coefficients that has one root of $2+7i$

Roots are $2+7i$ and $2-7i$ — The complex conjugate will also be a root.

So the factors are $x-(2+7i)$ and $x-(2-7i)$ — Ensure you know the difference between a *root* and a *factor*.

Equation is $(x-(2+7i))(x-(2-7i))=0$

$x^2-x(2-7i)-x(2+7i)+(2+7i)(2-7i)=0$

$x^2-4x+53=0$ — Since $-49i^2=49$

Complex roots for any real polynomial equation will always occur in **complex conjugate pairs**. Therefore, a polynomial equation will always have an even number of complex solutions. This implies that a cubic equation will always have at least one real root (as there cannot be three complex roots). You can use these facts to help you solve cubic and quartic equations.

PURE

Example 3

The quartic equation $3x^4+ax^3+bx^2+cx+d=0$ has solutions $-3, \frac{2}{3}, 5+i$ and $5-i$. Find the values of the real constants a, b, c and d

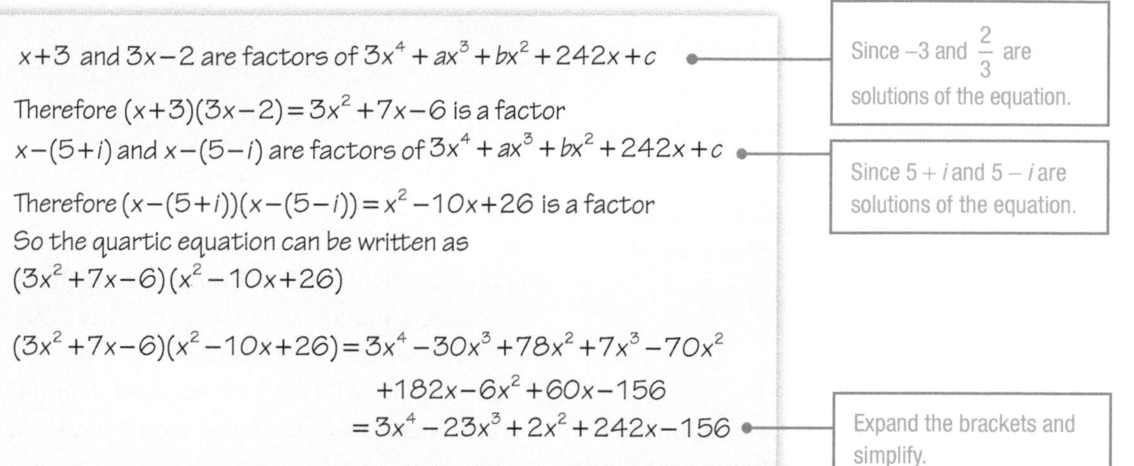

$x+3$ and $3x-2$ are factors of $3x^4+ax^3+bx^2+242x+c$
Therefore $(x+3)(3x-2)=3x^2+7x-6$ is a factor
$x-(5+i)$ and $x-(5-i)$ are factors of $3x^4+ax^3+bx^2+242x+c$
Therefore $(x-(5+i))(x-(5-i))=x^2-10x+26$ is a factor
So the quartic equation can be written as
$(3x^2+7x-6)(x^2-10x+26)$
$(3x^2+7x-6)(x^2-10x+26)=3x^4-30x^3+78x^2+7x^3-70x^2$
$+182x-6x^2+60x-156$
$=3x^4-23x^3+2x^2+242x-156$
Therefore, $a=-23$, $b=2$, $c=242$, $d=-156$

Since -3 and $\frac{2}{3}$ are solutions of the equation.

Since $5+i$ and $5-i$ are solutions of the equation.

Expand the brackets and simplify.

Exercise 1.2A Fluency and skills

1 Write the complex conjugate of z in each case.

- a $z=5-2i$
- b $z=8+i$
- c $z=5i-6$
- d $z=\sqrt{2}-i\sqrt{3}$
- e $z=\frac{1}{3}+4i$
- f $z=\frac{2}{3}i-5$

2 Given that $z=9-2i$, calculate

- a zz^*
- b $z+z^*$
- c $z-z^*$
- d $\frac{z}{z^*}$
- e $(z^*)^*$
- f $\frac{z^*}{z}$

3 Given that $w=-\sqrt{6}+\sqrt{2}i$, calculate

- a ww^*
- b $w+w^*$
- c $w-w^*$
- d $\frac{w}{w^*}$
- e $w^2+(w^*)^2$
- f $(w+w^*)^2$

4 Solve each of these quadratic equations.

- a $x^2+5x+7=0$
- b $x^2-3x+5=0$
- c $2x^2+7x+7=0$
- d $3x^2-10x+9=0$

5 Find a quadratic equation with solutions

- a $x=7$ and $x=-4$
- b $x=3+5i$ and $x=3-5i$
- c $x=-1-9i$ and $x=-1+9i$
- d $x=-5+4i$ and $x=-5-4i$

6 A quadratic equation has a solution of $z=\sqrt{3}+i$

- a Write down the other solution of the equation.
- b Find a possible equation.

7 Find a quadratic equation where one solution is given as

- a $2+i$
- b $4-3i$
- c $7i-1$
- d $-5-2i$
- e $a+3i$
- f $5-bi$

8 Find a cubic equation with solutions

a $x=-5, x=2+3i$ and $x=2-3i$

b $x=2, x=i-1$ and $x=-i-1$

c $x=0, x=\sqrt{3}+2i$ and $x=\sqrt{3}-2i$

d $x=3, x=-\sqrt{2}-i$ and $x=-\sqrt{2}+i$

9 A cubic equation has solutions of $z=-\frac{1}{2}$ and $z=6-2i$

a Write down the other solution of the equation.

b Find a possible equation that has these solutions.

10 You are given that $f(x)=x^3+9x^2+25x+25$ and $f(-5)=0$

a Show that the other solutions of the equation $f(x)=0$ satisfy $x^2+4x+5=0$

b Solve the equation $f(x)=0$

11 Given that $1+8i$ is a root of $x^4+4x^3+66x^2+364x+845=0$

a Show that two of the other solutions to the quartic equation satisfy $x^2+6x+13=0$

b Find all the solutions of the equation $x^4+4x^3+66x^2+364x+845=0$

12 Use the solutions of these cubic equations to find the values of the real constants a, b and c

a $x^3+ax^2+bx+c=0$ with solutions $x=-3, x=1+3i$

b $ax^3+bx^2+27x+c=0$ with solutions $x=-6i-1, x=5$

13 Use the solutions of these quartic equations to find the values of the real constants a, b and c

a $x^4+ax^3+bx^2+8x+c=0$ with solutions $x=1+2i, x=2$

b $x^4-8x^3+ax^2+bx+c=0$ with solutions $x=1+i, x=3-2i$

14 Use the fact that $7i$ is a root to show that the quartic equation $x^4-x^3+43x^2-49x-294=0$ can be written in the form $(x^2+A)(x+B)(x+C)=0$ where A, B and C are constants to be found. You must show your working.

Reasoning and problem-solving

You can find a polynomial equation when given its roots.

Strategy 1

To derive a polynomial equation when given its roots

1. Use the fact that if z is a root of $f(z)=0$ then z^* will also be a root.
2. Multiply the factors together.
3. Simplify and write the equation in descending powers of x or z

You can apply this result to other problems that involve quadratics.

Example 4

A quadratic equation has roots α and β

Show that the equation is $x^2-(\alpha+\beta)x+\alpha\beta=0$

Roots are α and β so equation is $(x-\alpha)(x-\beta)=0$ — (1) If α and β are roots then $x-\alpha$ and $x-\beta$ must be factors, so multiply these together.

which becomes $x^2-\alpha x-\beta x+\alpha\beta=0$ — (3) Expand brackets then simplify.

$\therefore x^2-(\alpha+\beta)x+\alpha\beta=0$ — You can also use this equation and the factor theorem to prove its factors are α and β

Example 5

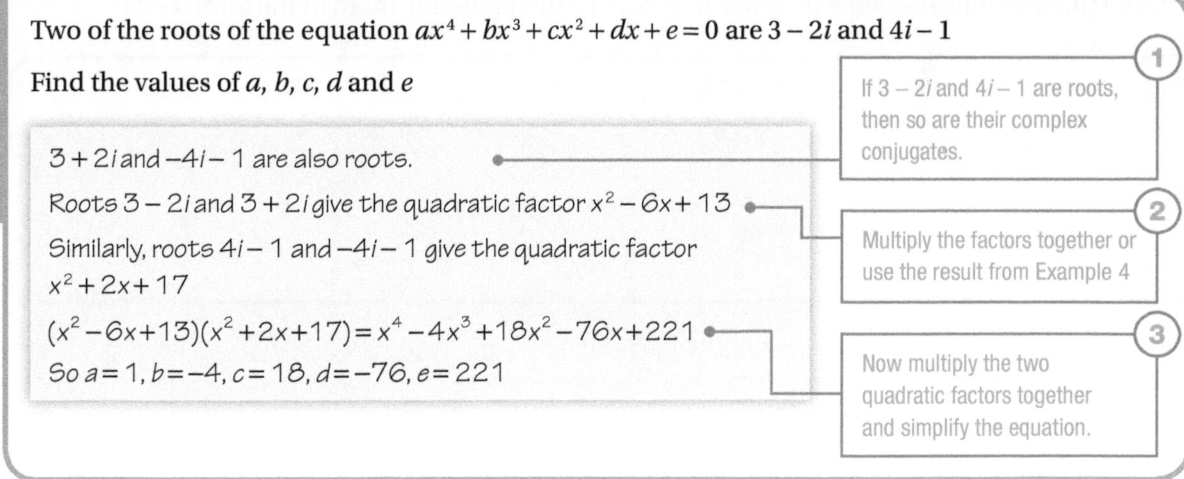

Two of the roots of the equation $ax^4 + bx^3 + cx^2 + dx + e = 0$ are $3 - 2i$ and $4i - 1$

Find the values of a, b, c, d and e

$3 + 2i$ and $-4i - 1$ are also roots.

Roots $3 - 2i$ and $3 + 2i$ give the quadratic factor $x^2 - 6x + 13$

Similarly, roots $4i - 1$ and $-4i - 1$ give the quadratic factor $x^2 + 2x + 17$

$(x^2 - 6x + 13)(x^2 + 2x + 17) = x^4 - 4x^3 + 18x^2 - 76x + 221$

So $a = 1, b = -4, c = 18, d = -76, e = 221$

1 If $3 - 2i$ and $4i - 1$ are roots, then so are their complex conjugates.

2 Multiply the factors together or use the result from Example 4

3 Now multiply the two quadratic factors together and simplify the equation.

You can solve a cubic equation by using the factor theorem to find the first solution, then dividing by the factor found and solving the remaining quadratic equation.

Strategy 2

To find solutions of polynomial equations with real coefficients

1. Use the fact that if z is a root of $f(z) = 0$ then z^* will also be a root.
2. Factorise $f(z)$ using the factor you know.
3. Equate coefficients to find the value of unknown constants.

Example 6

$f(z) = z^4 + az^3 - 15z^2 + bz - 18 = 0$ where a and b are integers.

Use the fact that $f(1-i) = 0$ to solve the equation $f(z) = 0$ and find the values of a and b

$f(1-i) = 0$ implies that $z = 1 - i$ is a solution of $f(z) = 0$

$z = 1 - i$ and $z = 1 + i$ are both solutions

So $z - (1 - i)$ and $z - (1 + i)$ are factors of $f(z)$

Therefore, $z^2 - 2z + 2$ is a factor of $z^4 + az^3 - 2z^2 + bz - 18 = 0$

$z^4 + az^3 - 15z^2 + bz - 18 = (z^2 - 2z + 2)(z^2 + kz - 9)$ for some real k

$= z^4 + kz^3 - 9z^2 - 2z^3 - 2kz^2 + 18z + 2z^2 + 2kz - 18$

$= z^4 + (k-2)z^3 + (-7 - 2k)z^2 + (18 + 2k)z - 18$

$z^2 : -15 = -7 - 2k \Rightarrow 2k = 8$

$\Rightarrow k = 4$

So $(z^2 - 2z + 2)(z^2 + 4z - 9) = 0$

The solutions of $z^2 + 4z - 9 = 0$ are $z = -2 \pm \sqrt{13}$

$z : b = 18 + 2k = 18 + 2(4) = 26$

$z^3 : a = k - 2 = 4 - 2 = 2$

So the quartic equation is $z^4 + 2z^3 - 15z^2 + 26z - 18 = 0$

It has two real solutions, $z = -2 \pm \sqrt{13}$ and two complex solutions, $z = 1 \pm i$

1 Since $i - 1$ is a solution, $i + 1$ will also be a solution.

Multiply the factors or use the result that $z^2 - (\alpha + \beta)z + \alpha\beta = 0$ has solutions α and β

2 Factorise $f(z)$, use the fact that the constant term is -18 and $2 \times (-9) = -18$

3 Equate coefficients of z^2 to find the value of k

3 Equate coefficients of z to find the value of b

3 Equate coefficients of z^3 to find the value of a

You can prove results involving complex numbers by manipulating them in the form $a + bi$

Strategy 3

To prove results involving complex numbers

1. Write complex numbers in the form $a + bi$
2. Use the fact that if $z = a + bi$ then $z^* = a - bi$
3. Manipulate in the usual way to prove the result.

Example 7

Prove that $(z + w)^* \equiv z^* + w^*$ for all $z, w \in \mathbb{C}$

Let $z = a + bi$ and $w = c + di$ with $a, b, c, d \in \mathbb{R}$ — 1 Write in the form $a + bi$

$(z + w)^* = ((a + c) + (b + d)i)^*$

$= (a + c) - (b + d)i$ — 2 Find complex conjugate.

$= (a - bi) + (c - di)$ — 3 Rearrange equation.

$= z^* + w^*$ as required.

Exercise 1.2B Reasoning and problem-solving

1 Prove these equations for all $z, w \in \mathbb{C}$

a $(zw)^* = z^*w^*$

b $(z^*)^* = z$

c $\left(\dfrac{z}{w}\right)^* = \dfrac{z^*}{w^*}$

2 Prove that, for all $z \in \mathbb{C}$

a $z + z^*$ is real,

b $z - z^*$ is imaginary,

c zz^* is real.

3 Find the possible complex numbers z such that $z + z^* = 6$ and $zz^* = 58$

4 Find the possible complex numbers w such that $w - w^* = 18i$ and $ww^* = 85$

5 Find the complex number z such that

$z + z^* = 2\sqrt{3}$ and $\dfrac{z}{z^*} = \dfrac{1}{2} - \dfrac{1}{2}\sqrt{3}i$

6 Find the complex number w such that

$w - w^* = 4i$ and $\dfrac{w}{w^*} = \dfrac{3}{5} + \dfrac{4}{5}i$

7 Given that $x = 7$ is a solution of the equation $x^3 - 9x^2 + 31x + k = 0$

a Find the value of k

b Solve the equation fully.

8 Given that $x - 1$ is a factor of the equation $x^3 + 2x^2 + 5x + k = 0$, find the value of k and solve the equation.

9 Find all the solutions of the equation $x^4 - 4x^3 + 4x^2 - 4x + 3 = 0$ given that $x^2 - 4x + 3$ is a factor of $x^4 - 4x^3 + 4x^2 - 4x + 3$
You must show your working.

10 The curve of $y=x^4-6x^3+ax^2+bx+c$ is shown where a, b, and c are real constants.

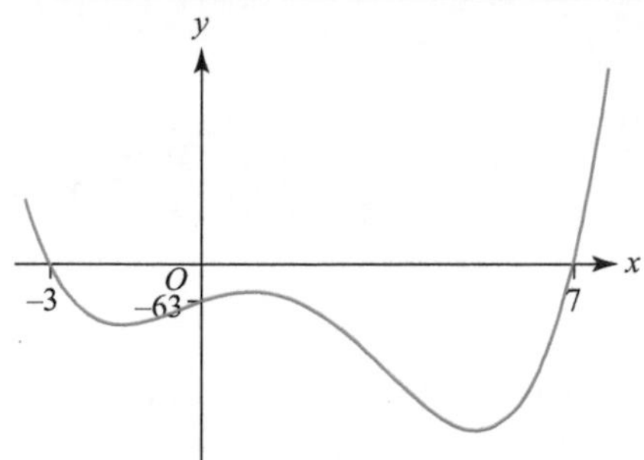

Find all the solutions of the equation $x^4-6x^3+ax^2+bx+c=0$

11 The equation $x^4+kx^3+kx^2-110x-111=0$ has a root of $i-6$

a Find the value of k

b Solve the equation.

12 Given that $5-i$ is a root of the equation $x^4+Ax^3+28x^2-20x+52=0$, find the value of A and solve the equation.

13 Find a quartic equation with real coefficients and repeated root

a $1+3i$

b $2i-3$

14 $f(x)=x^3-19x^2+89x+109$

a Write $f(x)$ as the product of a linear and a quadratic factor.

b Sketch the graph of $y=f(x)$

15 a Write the quartic expression

$4x^4+12x^3-35x^2-300x+625$

as the product of two quadratic expressions with real coefficients.

b Sketch the graph of

$y=4x^4+12x^3-35x^2-300x+625$

16 The curve of a quartic equation $y=p(x)$ is shown.

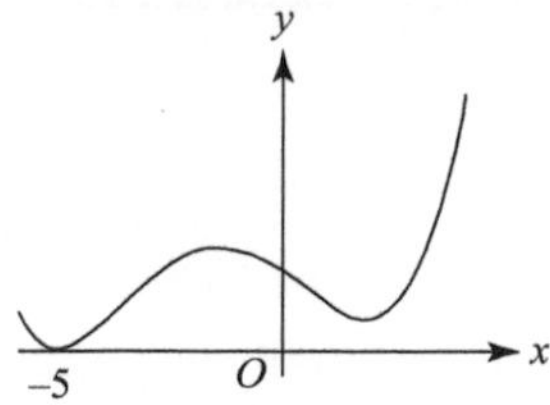

a What does the graph tell you about the nature of the roots of $p(x)=0$?

Given that $p(x)=x^4+6x^3+ax^2+bx+125$ where a and b are integers

b Solve the equation $p(x)=0$

17 If the roots of a cubic equation are a, b and c, and the coefficient of x^3 is 1, write the equation in terms of a, b and c

18 The solutions of the equation $x^3+5x^2+Ax+12=0$ are α, β and γ

Write the value of

a $\alpha+\beta+\gamma$ b $\alpha\beta\gamma$

19 A quartic equation has roots α, β, γ and δ

Given that the coefficient of x^4 is 1, write an expression in terms of α, β, γ and δ for

a The coefficient of x^3

b The constant term.

20 $f(z)=az^5+bz^4+cz^3+dz^2+ez-841$ where a, b, c, d, and e are real numbers.

The complex number $5-2i$ is a repeated root of the equation $f(z)=0$

Given that $f\left(\frac{1}{2}\right)=0$, find the values of a, b, c, d and e

21 Three solutions of a polynomial equation with real coefficients are -2, $4-i$ and $3i$

a State the lowest possible order of the equation.

b Find an equation of this order with these solutions.

1.3 Argand diagrams

Fluency and skills

You represent real numbers visually as points on a number line, but imaginary numbers will not fit onto the real number line. Therefore, we need a new number line of imaginary numbers. Combining these two number lines together gives a plane. So complex numbers can be represented visually as points on a plane.

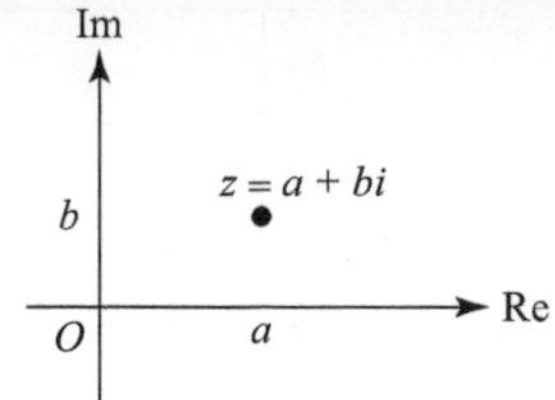

This plane is called an **Argand diagram**. Argand diagrams are used to represent complex numbers graphically.

Key point

The complex number $z = a + bi$ can be represented as the point (a, b) on an Argand diagram.

In an Argand diagram, the horizontal axis is the **real axis** and the vertical axis is the **imaginary axis**.

Example 1

Given that $z = 1 + 3i$, show z and z^* on an Argand diagram.

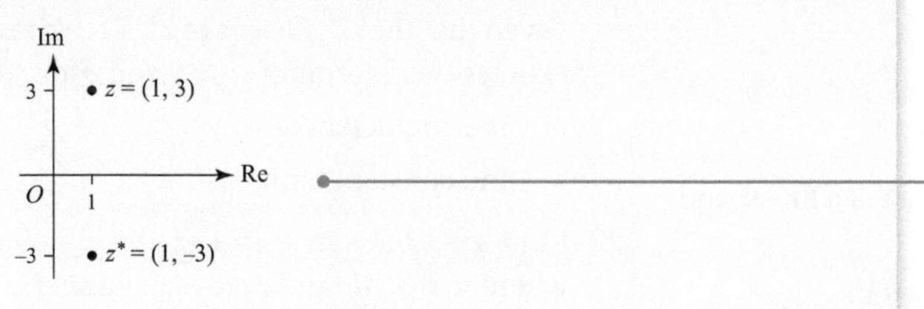

The real axis is labelled 'Re' and the imaginary axis is labelled 'Im'.

z^* is the point z reflected in the real axis.

Example 2

Complex numbers can also be shown as position vectors.

Given that $z = 4 + i$ and $w = -2 + i$, show z, w and $z + w$ on an Argand diagram.

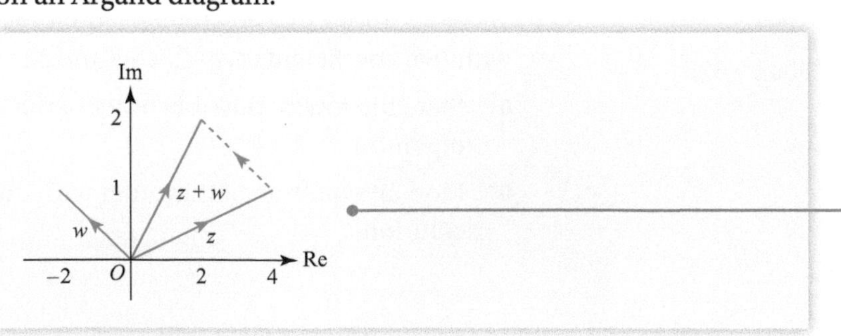

The sum can be shown as a vector addition on the Argand diagram.

Ensure that you label each vector clearly.

Example 3

Given that $z = 1 + 2i$, show z, $3z$ and $-2z$ on an Argand diagram.

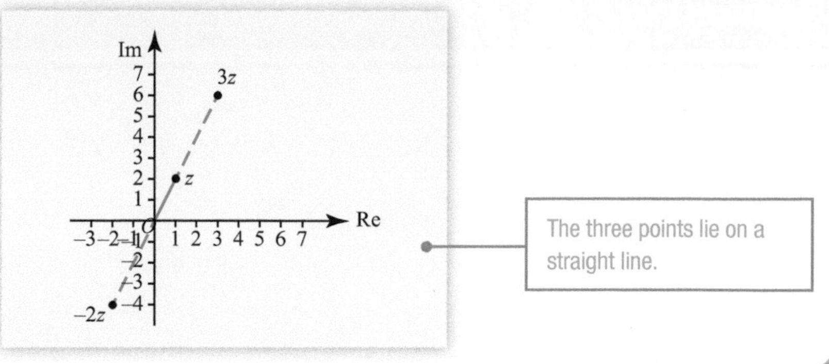

Exercise 1.3A Fluency and skills

1 Write the complex numbers represented by the vectors $\overrightarrow{OA}$, $\overrightarrow{OB}$, $\overrightarrow{OC}$, $\overrightarrow{OD}$, $\overrightarrow{OE}$ and $\overrightarrow{OF}$, as shown in the Argand diagram, in the form $a + bi$

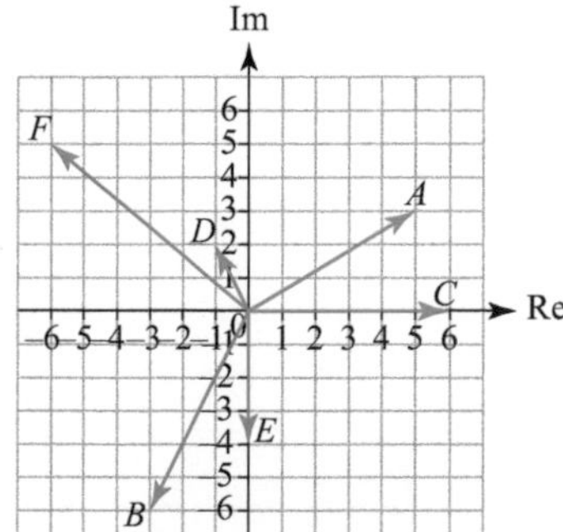

2 The Argand diagram shows the complex numbers u, v, w and z

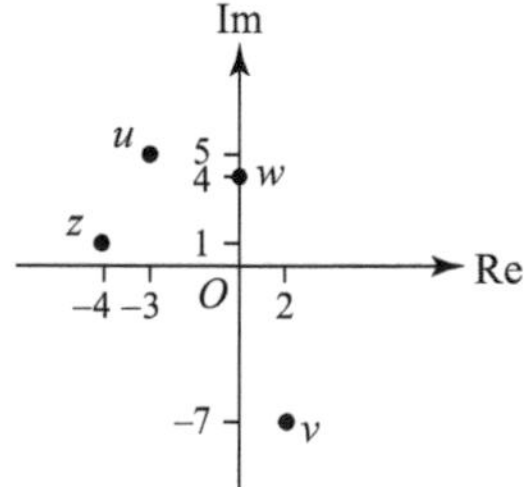

Write the complex numbers u, v, w and z in Cartesian form.

3 $z_1 = 5 - 8i$, $z_2 = 2 + 4i$

Show z_1, z_2 and $z_1 + z_2$ on an Argand diagram.

4 $z = 2 - 7i$, $w = 6i - 4$

Show z, w and $z - w$ on an Argand diagram.

5 Show the vector addition $z = (3 + 5i) + (2 - 7i)$ on an Argand diagram.

6 Show the vector subtraction $z = (6 - 2i) - (2 - 5i)$ on an Argand diagram.

7 Given that $w = \sqrt{3} + 2i$, draw w and w^* on an Argand diagram and describe the geometric relationship between them.

8 Given that $z^* = -7 + i$, draw z and z^* on the same Argand diagram and describe the geometric relationship between them.

9 Solve each of these equations and plot the solutions on an Argand diagram. Describe the geometric relationship between the two solutions in each case.

a $x^2 = -16$ b $x^2 = -80$

10 a Solve the equation $(z + 1)^2 = -3$

b Plot the solutions on an Argand diagram.

c Describe the geometric relationship between the two solutions.

11 a Solve the equation $(1 - z)^2 = -25$

b Plot the solutions on an Argand diagram.

c Describe the geometric relationship between the two solutions.

12 For each value of z, show z and iz on an Argand diagram. What transformation maps z to iz in each case?

a $z = -3 - 4i$ b $z = 2 - 6i$

13 For the complex numbers given, draw w and i^2w on the same Argand diagram and describe the geometrical relationship between them.

a $w = 5 - 7i$ b $w = -2 + 3i$

Reasoning and problem-solving

Strategy

To solve geometric problems involving complex numbers

1. Draw an Argand diagram.
2. Use rules for calculating area, lengths and angles.
3. Use the fact that the product of gradients of perpendicular lines is –1
4. Fully define transformations.

Example 4

$f(x)=2x^4+x^3-15x^2+100x-250$

a Find all the roots of the equation $f(x)=0$ given that $x+5$ is a factor of $f(x)$

b Find the area of the quadrilateral formed by the points representing the four roots.

c Prove that the quadrilateral contains two right angles.

a $2x^4+x^3-15x^2+100x-250=(x+5)(2x^3-9x^2+30x-50)=0$ — Use long division or an inspection method to factorise.

$\therefore 2x^3-9x^2+30x-50=0$

Solve to give roots $x=\frac{5}{2}, 1\pm 3i$ — Some calculators have an equation solver. This is the easiest method to use to find the three remaining roots.

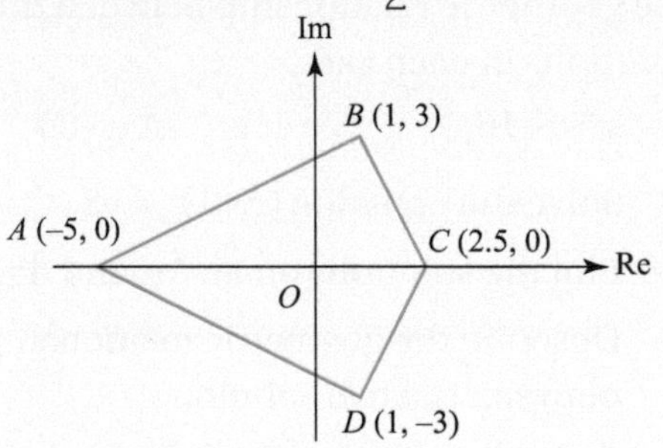

(1) Show the roots on an Argand diagram.

b Area $=2\left(\frac{1}{2}\times 7.5\times 3\right)=22.5$ square units — (2) The quadrilateral formed is a kite which we can split into two triangles.

c Gradient of $AB=\frac{3}{6}=\frac{1}{2}$

Gradient of $BC=\frac{-3}{1.5}=-2$

$\frac{1}{2}\times -2=-1$ so $\angle ABC$ is a right angle. — (3) The product of the gradients is –1

Similarly for $\angle ADC$

Example 5

You are given $z = 1 - 3i$

The points A, B and C are represented by the complex numbers z, z^2 and $z^2 - 4z$ respectively.

a Find the complex number $z^2 - 4z$ in the form $a + bi$

b Describe the transformation that maps line segment OA to CB

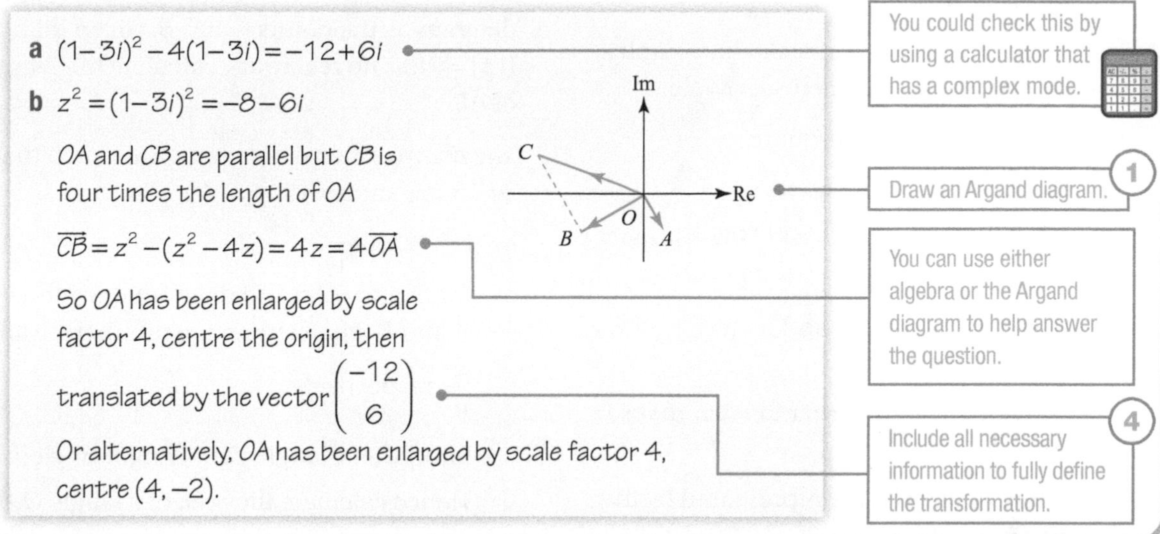

a $(1-3i)^2 - 4(1-3i) = -12 + 6i$

You could check this by using a calculator that has a complex mode.

b $z^2 = (1-3i)^2 = -8 - 6i$

OA and CB are parallel but CB is four times the length of OA

Draw an Argand diagram.

$\overrightarrow{CB} = z^2 - (z^2 - 4z) = 4z = 4\overrightarrow{OA}$

You can use either algebra or the Argand diagram to help answer the question.

So OA has been enlarged by scale factor 4, centre the origin, then translated by the vector $\begin{pmatrix} -12 \\ 6 \end{pmatrix}$

Or alternatively, OA has been enlarged by scale factor 4, centre $(4, -2)$.

Include all necessary information to fully define the transformation.

Exercise 1.3B Reasoning and problem-solving

1 $z_1 = 5 - 2i$, $z_2 = \dfrac{20 + 21i}{z_1}$

a Find z_2 in the form $a + bi$

b Show the points A and B representing z_1 and z_2 respectively on an Argand diagram.

c Show that AOB is a right angle.

2 **a** Show the three roots of the equation $x^3 + x^2 + 6x - 8 = 0$ on an Argand diagram.

b What type of triangle is formed by the points representing the three roots?

c Find the exact area of the triangle.

3 The points A, B and C on an Argand diagram represent the solutions of the cubic equation $x^3 - 9x^2 + 16x + 26 = 0$

Calculate the area of triangle ABC

4 **a** Given that the expression $x^4 + x^3 + 3x + 9$ can be written in the form $(x^2 + 3x + 3)(x^2 + Ax + B)$, find the values of A and B

b Solve the equation $x^4 + x^3 + 3x + 9 = 0$

c Show the roots of $x^4 + x^3 + 3x + 9 = 0$ on an Argand diagram.

d What type of quadrilateral is formed by the points representing the roots?

e Find the area of the quadrilateral.

5 The quartic equation $x^4 + 8x^3 + 40x^2 + 96x + 80 = 0$ has a repeated real root.

a Show that the repeated root is $x = -2$

b Calculate the other solutions of the equation.

c Show all the solutions on an Argand diagram.

6 The points A, B, C and D represent the solutions of the quartic equation $x^4 - 3x^3 + 10x^2 - 6x - 20 = 0$

a Use the factor theorem to find two real solutions.

b Calculate the area of quadrilateral $ABCD$

7 The solutions of the cubic equation $x^3 - x^2 + 9x - 9 = 0$ are represented by the points A, B and C on an Argand diagram.

a Prove that triangle ABC is isosceles.

b Calculate the area of triangle ABC

8 You are given that $w = 5 - 2i$

The points A, B and C represent the complex numbers w, $w + iw$ and iw respectively.

a Prove that $OABC$ is a square.

b Find the area of the square.

9 The points P, Q and R represent the complex numbers z, z^* and $2z + z^*$, where $z = 2 - 4i$

a Show the points P, Q and R on an Argand diagram.

b Describe the transformation that maps OP to QR

10 The points A, B and C are represented by the complex numbers z, z^3 and $z^3 - 2z$ where $z = 2 + i$

a Find the complex number $z^3 - 2z$ in the form $a + bi$

b Describe the transformation that maps line segment OA to CB

11 The solutions of a quartic equation are $z = a \pm bi$ and $w = c \pm di$, where $a, b, c, d \in \mathbb{C}, b \neq 0, d \neq 0$

Find the area of the quadrilateral formed by the roots on an Argand diagram.

12 Given that $z = 8 + 12i$, show the complex numbers z, zi, zi^2 and zi^3 on an Argand diagram.

Describe the transformations that map z to each of the other points.

13 The points A and B represent the complex numbers z and iz respectively. Prove that OA is perpendicular to OB

14 The points P and Q represent the complex numbers $4 + 6i$ and $-3 + 4i$ respectively. Find the area of the triangle OPQ

15 Find the area of the shape formed by the solutions of the equation $x^5 - x^4 + 18x^3 - 18x^2 + 81x - 81 = 0$

16 $f(z) = z^4 - 4z^3 + 56z^2 - 104z + 676$

The equation $f(z) = 0$ has complex roots z_1 and z_2 which are represented on an Argand diagram by the points A and B. Given that $f(z) = 0$ has no real roots, calculate the length of AB

17 The complex numbers w and z are such that $w = 3 + 2i$ and $z = 12 - 5i$

a Show that $\frac{z}{w} = 2 - 3i$

b Mark on an Argand diagram the points A and B representing the numbers w and $\frac{z}{w}$, respectively.

c Show that triangle OAB is right-angled.

d Hence calculate the area of triangle OAB

Full A Level

18 $f(z) = z^4 - az^3 + 47z^2 - bz + 290$ where $a, b \in \mathbb{R}$ are constants.

Given that $z = 2 - 5i$ is a root of the equation $f(z) = 0$, show all the roots of $f(z) = 0$ on an Argand diagram.

19 $f(z) = az^4 + bz^3 + cz^2 + dz + e$ where $a, b, c, d, e \in \mathbb{R}$ are constants.

The solutions of $f(z) = 0$ are plotted on an Argand diagram.

a Describe the shape formed by the points representing the solutions of $f(z) = 0$ when the equation has

i Precisely two (distinct) real roots,

ii Precisely one real root,

iii No real roots.

Given that the points representing the solutions of $f(z) = 0$ form a square,

b Calculate the values of the constants a, b, c, d and e when

i $z = 2i$ is a root of $f(z) = 0$

ii $z = 1 + 3i$ is a root of $f(z) = 0$

1.4 Modulus-argument form and loci

Fluency and skills

You have seen how the complex number $z = a + bi$ can be represented on an Argand diagram.

The length of the vector representing z is known as the **modulus** of z and written $|z|$

The angle between the positive real axis and the vector representing z is known as the **(principal) argument** of z and is written $\arg(z)$

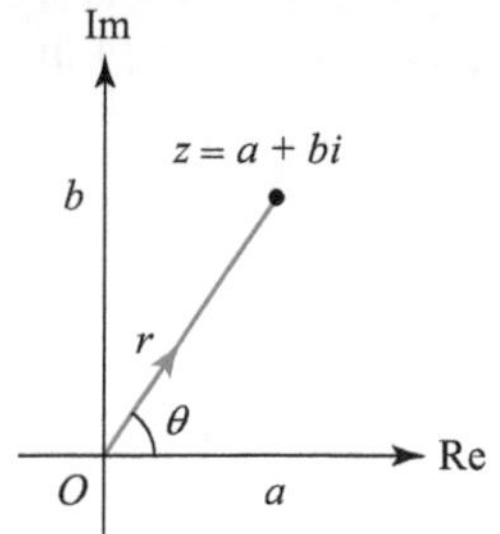

Key point

The modulus of the complex number $z = a + bi$ is given by $|z| = \sqrt{a^2 + b^2}$

Key point

Write $\arg(z) = \theta$ where $-\pi < \theta \leq \pi$

Example 1

Find the argument and modulus of the complex number $w = -3 + 3i$

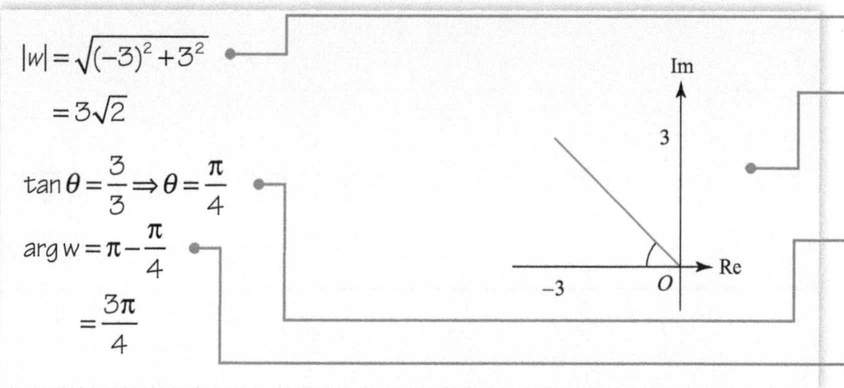

Use $|w| = \sqrt{a^2 + b^2}$

An Argand diagram will help you to find the correct angle.

We always use radians on Argand diagrams: π radians is equivalent to $180°$, so $\frac{\pi}{4}$ radians is $45°$

You need the angle with the positive real axis.

Using $w = -3 + 3i$ from the previous example, if you wished to find the modulus of $-2w$ then it is clear from the diagram that $|-2w| = |-2||w|$

The argument of $-2w$ can also be seen to be given by $\arg(-2w) = \arg(w) + \pi$ which is $\arg w + \arg(-2)$

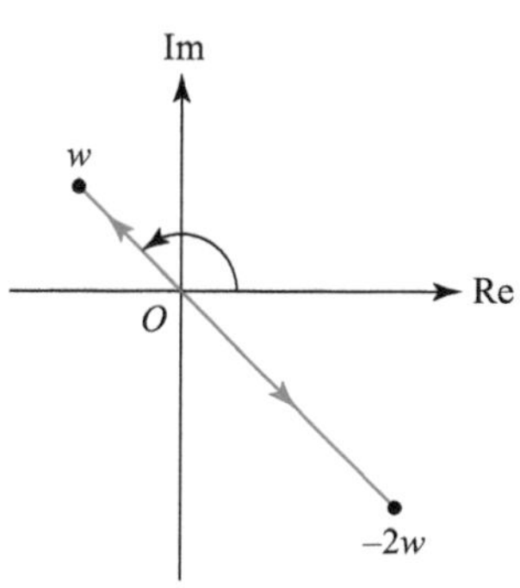

In fact, these results can be generalised for any two complex numbers.

Key point

$|z_1 z_2| = |z_1||z_2|$ and $\left|\frac{z_1}{z_2}\right| = \frac{|z_1|}{|z_2|}$ for all $z_1, z_2 \in \mathbb{C}$

Key point

$\arg(z_1 z_2) = \arg(z_1) + \arg(z_2)$ and

$\arg\left(\frac{z_1}{z_2}\right) = \arg(z_1) - \arg(z_2)$ for all $z_1, z_2 \in \mathbb{C}$

You can quote these results and will learn how to prove them in the next section.

Calculator

Try it on your calculator

Some calculators can be used to find the argument of a complex number. Work out how to find the argument of $5 - 6i$ on *your* calculator.

arg(5 – 6*i*)

–0.8760580506

Example 2

Given that $z_1 = \sqrt{3} + i$ and $z_1 z_2 = -4 - 4i$, find the argument and modulus of z_2

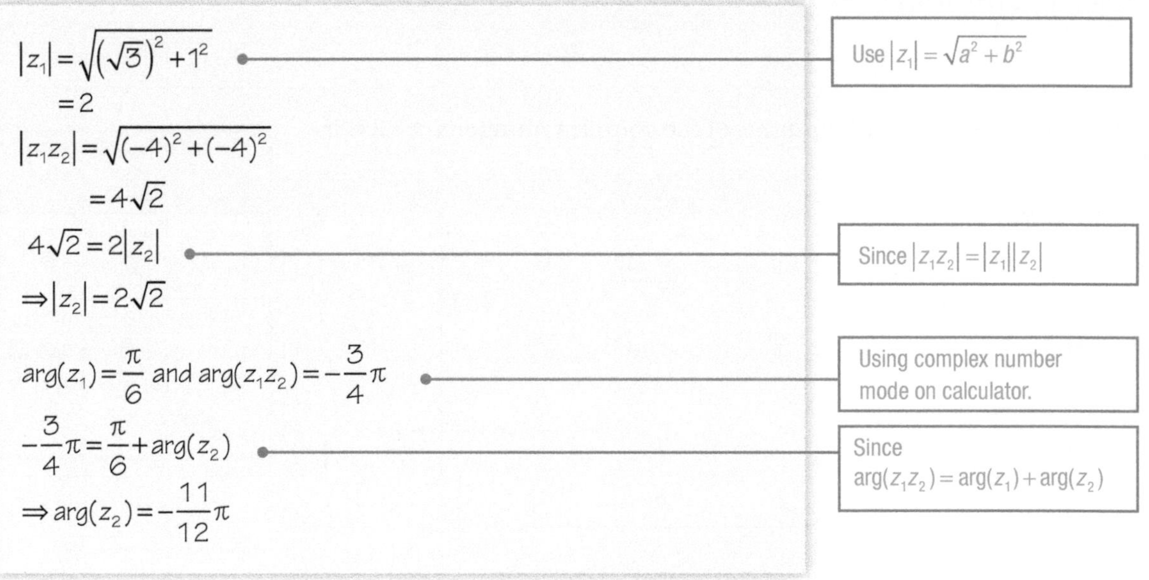

$|z_1| = \sqrt{(\sqrt{3})^2 + 1^2}$ — Use $|z_1| = \sqrt{a^2 + b^2}$

$= 2$

$|z_1 z_2| = \sqrt{(-4)^2 + (-4)^2}$

$= 4\sqrt{2}$

$4\sqrt{2} = 2|z_2|$ — Since $|z_1 z_2| = |z_1||z_2|$

$\Rightarrow |z_2| = 2\sqrt{2}$

$\arg(z_1) = \frac{\pi}{6}$ and $\arg(z_1 z_2) = -\frac{3}{4}\pi$ — Using complex number mode on calculator.

$-\frac{3}{4}\pi = \frac{\pi}{6} + \arg(z_2)$ — Since $\arg(z_1 z_2) = \arg(z_1) + \arg(z_2)$

$\Rightarrow \arg(z_2) = -\frac{11}{12}\pi$

Instead of the form $z = a + bi$, sometimes known as Cartesian form, you can write complex numbers in **modulus** and **argument** form.

You can see in the diagram that the real component of r is $r\cos\theta$ and the imaginary component is $r\sin\theta$

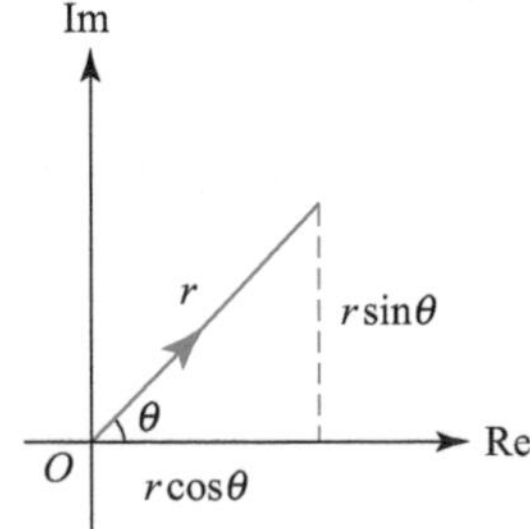

Key point

The **modulus-argument form** of the complex number $z = a + bi$ is given by $z = r(\cos\theta + i\sin\theta)$ where r is the modulus of z and θ is the argument.

The argument is measured in radians, where π radians is equal to 180°

Example 3

Write the number $z = 7 - i$ in modulus-argument form.

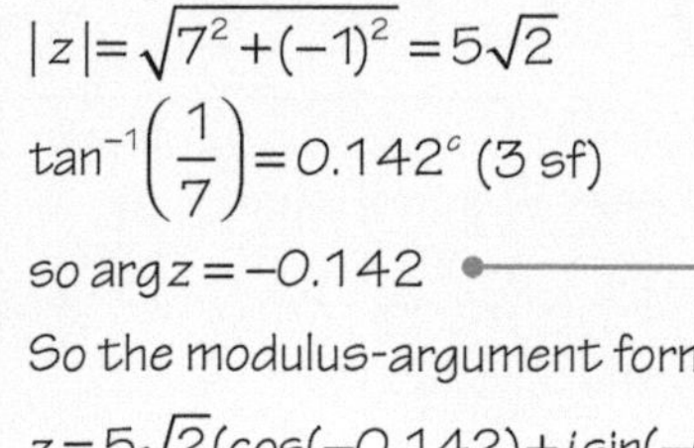

$|z| = \sqrt{7^2 + (-1)^2} = 5\sqrt{2}$

$\tan^{-1}\left(\frac{1}{7}\right) = 0.142^c$ (3 sf)

so $\arg z = -0.142$

So the modulus-argument form is

$z = 5\sqrt{2}(\cos(-0.142) + i\sin(-0.142))$

Im, Re, O, θ, 7, −1

Ensure you find the angle with the positive x-axis.

Remember that $-\pi < \theta \leq \pi$ and that an angle measured clockwise will be negative.

Calculator

Try it on your calculator

Some calculators can be used to convert to and from modulus-argument form. Find out how to convert $\sqrt{3} + i$ to modulus-argument form on *your* calculator. Also find out how to enter a number in modulus-argument form and convert back to the form $a + bi$

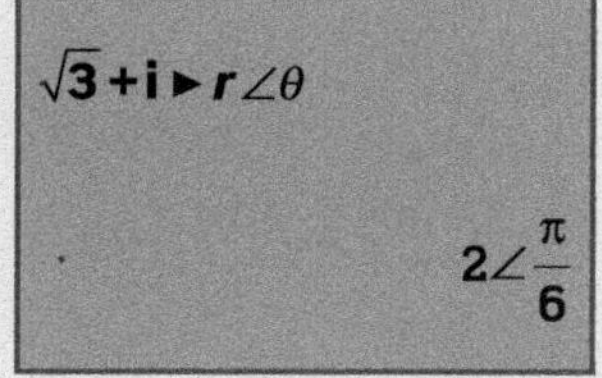

You can draw **loci** in an Argand diagram. These are the set of points that obey a given rule.

Example 4

Sketch the locus of points that satisfy $|z| = 4$

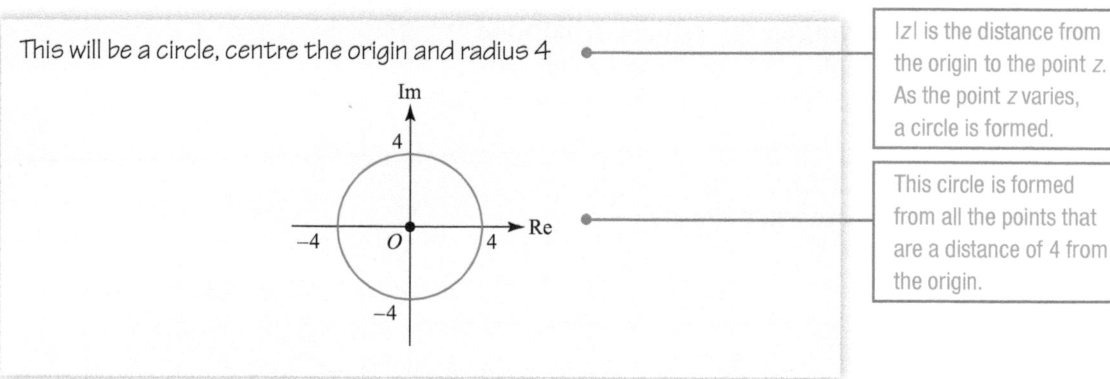

$|z|$ is the distance from the origin to the point z. As the point z varies, a circle is formed.

This circle is formed from all the points that are a distance of 4 from the origin.

If we have any fixed point z_1, then

Key point

The locus of points satisfying $|z - z_1| = r$ will be a circle centre z_1 and radius r

Example 5

Sketch the locus of points that satisfy $|z-2+3i|=2$

$|z-(2-3i)|=2$

First write in the form $|z-z_1|=r$

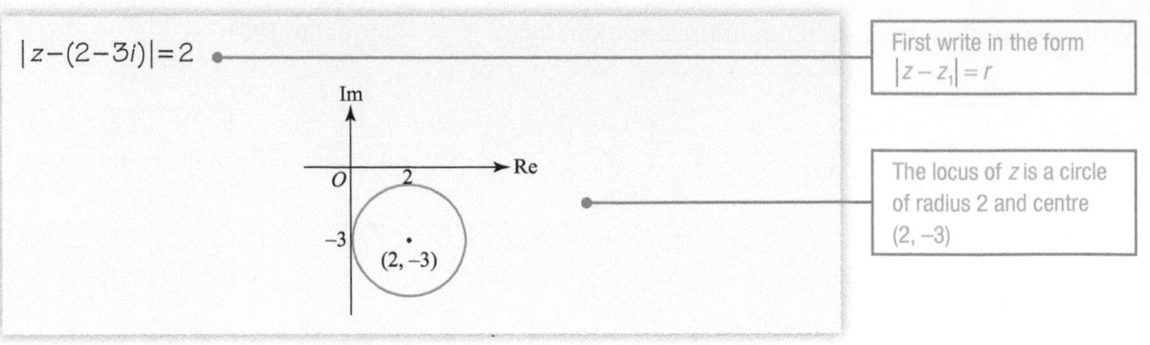

The locus of z is a circle of radius 2 and centre (2, –3)

Example 6

Sketch the locus of points satisfying $\arg(z)=\frac{2\pi}{3}$

This will be a line that makes an angle of $\frac{2\pi}{3}$ with the positive real axis.

arg(z) is the angle between the positive real axis and the line representing z

Im

$\frac{2\pi}{3}$

Re

O

Notice that the line ends at the origin.

Key point

The locus of points satisfying $\arg(z-z_1)=\theta$ is a **half-line** from the point z_1 at an angle of θ to the positive real axis.

Example 7

Sketch the locus of z where $\arg(z+2+3i)=\frac{\pi}{6}$

$\arg(z-(-2-3i))=\frac{\pi}{6}$

First write in the form $\arg(z-z_1)=\theta$

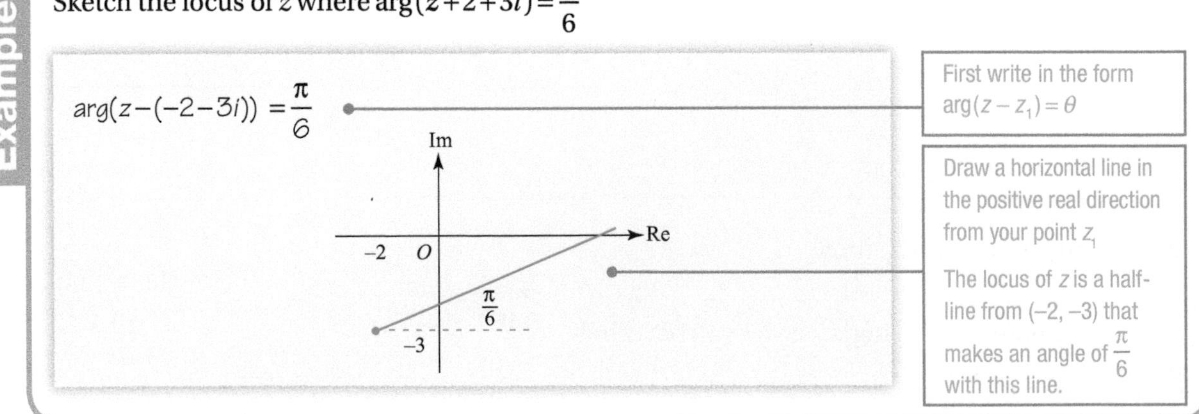

Draw a horizontal line in the positive real direction from your point z_1

The locus of z is a half-line from (–2, –3) that makes an angle of $\frac{\pi}{6}$ with this line.

Key point

The locus of points satisfying $|z-z_1|=|z-z_2|$ is the perpendicular bisector of the line joining z_1 and z_2

This is because the locus includes all points that are equidistant from the fixed points z_1 and z_2

Example 8

Sketch the locus of points satisfying $|z-2|=|z+3i|$

We have $|z-2|=|z-(-3i)|$

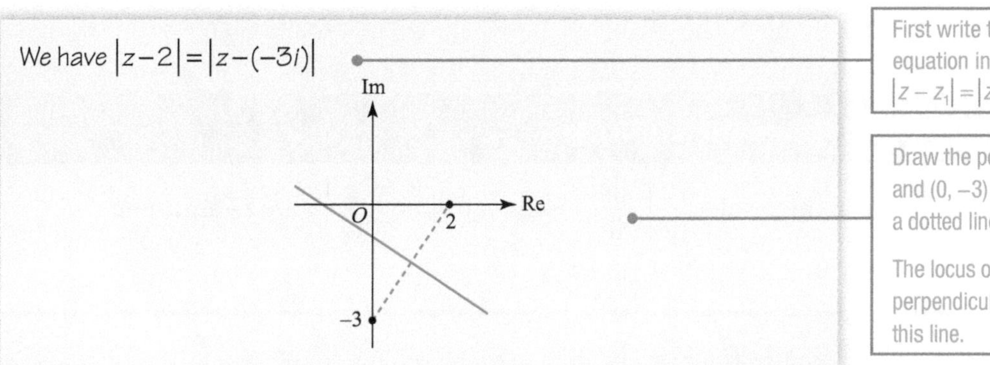

First write the equation in the form $|z-z_1|=|z-z_2|$

Draw the points (2, 0) and (0, −3) and join with a dotted line.

The locus of z is the perpendicular bisector of this line.

Exercise 1.4A Fluency and skills

1 Find the modulus and argument of each complex number.

a $12+5i$ b $4-3i$

c $-3i$ d $-6-8i$

e $-1+7i$ f $-2-i$

g $-\sqrt{2}-\sqrt{2}i$ h $\sqrt{3}-\sqrt{6}i$

2 Verify in each case that $|zw|=|z||w|$ and $\left|\frac{z}{w}\right|=\frac{|z|}{|w|}$

a $z=1+3i, w=-5+2i$

b $z=-2-i, w=\sqrt{5}i$

c $z=-\sqrt{3}+6i, w=1-\sqrt{3}i$

3 In each case, verify that $\arg(zw)=\arg z+\arg w$ and $\arg\left(\frac{z}{w}\right)=\arg z-\arg w$

a $z=1+i, w=3+\sqrt{3}i$

b $z=i, w=2-2i$

c $z=\sqrt{3}-3i, w=-2i$

4 Given that $z=1-\sqrt{3}i$ and $zw=\sqrt{3}+3i$, find the modulus and the argument of w

5 Given that $z_1=-1-i$ and $\frac{z_1}{z_2}=3\sqrt{2}+\sqrt{6}i$, find the modulus and the argument of z_2

6 Given that $z=-2\sqrt{3}+6i$ and $\frac{w}{z}=\sqrt{3}-i$, find the modulus and the argument of w

7 Write each of these complex numbers in Cartesian form.

a $3\left(\cos\left(\frac{\pi}{2}\right)+i\sin\left(\frac{\pi}{2}\right)\right)$

b $5(\cos(-\pi)+i\sin(-\pi))$

c $10\left(\cos\left(\frac{5\pi}{6}\right)+i\sin\left(\frac{5\pi}{6}\right)\right)$

d $\sqrt{3}\left(\cos\left(-\frac{2\pi}{3}\right)+i\sin\left(-\frac{2\pi}{3}\right)\right)$

8 Write each of these numbers in modulus-argument form.

a $z=3+3i$ b $z=1-\sqrt{3}i$

c $z=-2\sqrt{3}-2i$ d $z=-4+9i$

9 Sketch the locus of z in each case.

a $|z|=7$ b $|z-2|=5$

c $|z-i|=3$ d $|z-(1+2i)|=2$

e $|z-3+5i|=5$ f $|z+4-2i|=4$

10 Sketch the locus of z in each case.

a $\arg z = -\frac{\pi}{4}$ **b** $\arg(z-3) = \frac{\pi}{2}$

c $\arg(z+i) = \frac{3\pi}{4}$ **d** $\arg(z-2i) = -\frac{\pi}{6}$

e $\arg(z-2+i) = \frac{5\pi}{6}$ **f** $\arg(z-4-i) = -\frac{2\pi}{3}$

g $\arg(z+5-7i) = -\frac{\pi}{3}$

11 Sketch the locus of z in each case.

a $|z| = |z+4|$ **b** $|z-2i| = |z|$

c $|z-2i| = |z+2|$ **d** $|z+6+2i| = |z+6|$

e $|z+4-i| = |z-5+2i|$

Reasoning and problem-solving

You need to be able to prove the results: $\left|\frac{z}{w}\right| = \frac{|z|}{|w|}$, $|zw| = |z||w|$, $\arg\left(\frac{z}{w}\right) = \arg z - \arg w$ and $\arg(zw) = \arg z + \arg w$

Strategy 1

To prove results about modulus and argument

1. Write complex numbers in modulus-argument form.
2. Simplify powers of i
3. Split into real and imaginary parts.
4. Use the addition formulae for sine and for cosine.

Example 9

Prove that $\left|\frac{z}{w}\right| = \frac{|z|}{|w|}$ and $\arg\left(\frac{z}{w}\right) = \arg z - \arg w$ for all $z, w \in \mathbb{C}$

Let $z = |z|(\cos A + i\sin A)$ and $w = |w|(\cos B + i\sin B)$ — Write both numbers in modulus-argument form. (1)

Then $\frac{z}{w} = \frac{|z|(\cos A + i\sin A)}{|w|(\cos B + i\sin B)}$

$= \frac{|z|(\cos A + i\sin A)(\cos B - i\sin B)}{|w|(\cos B + i\sin B)(\cos B - i\sin B)}$

$= \frac{|z|(\cos A\cos B - i\cos A\sin B + i\cos B\sin A - i^2\sin A\sin B)}{|w|(\cos^2 B - i^2\sin^2 B)}$ — Expand the brackets.

$= \frac{|z|(\cos A\cos B + \sin A\sin B + i(\cos B\sin A - \cos A\sin B))}{|w|(\cos^2 B + \sin^2 B)}$ — Use the fact that $i^2 = -1$ (2). Separate real and imaginary parts. (3)

$= \frac{|z|(\cos(A-B) + i\sin(A-B))}{|w|(1)}$

$= \frac{|z|}{|w|}(\cos(A-B) + i\sin(A-B))$ — Use $\cos(A-B) = \cos A\cos B + \sin A\sin B$ and $\sin(A-B) = \cos A\sin B - \sin A\cos B$ (4)

Therefore the number $\frac{z}{w}$ has modulus $\frac{|z|}{|w|}$ and argument $A - B$

So $\left|\frac{z}{w}\right| = \frac{|z|}{|w|}$ and $\arg\left(\frac{z}{w}\right) = A - B = \arg z - \arg w$

You know how to find the Cartesian equation of certain loci by drawing the graph and using known equations of circles or lines. However, it is possible to find the Cartesian equation of any locus by setting $z = x + iy$ and finding a relationship between x and y

PURE

Strategy 2

To find the Cartesian equation of a locus

1. Write z as $x + iy$
2. Calculate the modulus.
3. Use tan to form an equation from the argument.
4. Rearrange to the required form.

See Maths Ch1.5

For a reminder of the equations of lines and circles.

Example 10

Find the Cartesian equations of these loci.

a $|z-3+4i| = \sqrt{5}$

b $\arg(z+i) = -\dfrac{\pi}{6}$

a Let $z = x + iy$ — (1) Write in Cartesian form.

$$|x+iy-3+4i| = \sqrt{5}$$

$$\sqrt{(x-3)^2+(y+4)^2} = \sqrt{5}$$ — (2) Find the modulus of the left-hand side.

$$(x-3)^2+(y+4)^2 = 5$$

So the locus is a circle with centre $(3, -4)$ and radius $\sqrt{5}$

b Let $z = x + iy$ — (1) Write in Cartesian form.

$$\arg(x+iy+i) = -\frac{\pi}{6} \text{ so } \arg(x+(y+1)i) = -\frac{\pi}{6}$$

$$\frac{y+1}{x} = \tan\left(-\frac{\pi}{6}\right) \text{ so } \frac{y+1}{x} = -\frac{\sqrt{3}}{3}$$ — (3) Since $\tan\theta = \dfrac{\text{opposite}}{\text{adjacent}}$

$$\therefore \sqrt{3}x + 3y + 3 = 0$$ — (4) When rearranged, you can see this is the equation of a line.

As the locus is a half-line, we only need the part where $x > 0$ and $y < -1$

Strategy 3

To find a region bounded by a locus

1. Sketch the locus of the boundary of the region.
2. Test a point to see if it is inside the region or not.
3. Shade the correct area.

Example 11

Shade these sets of points.

a $\{z \in \mathbb{C} : |z-5| \geq 3\}$

b $\left\{z \in \mathbb{C} : 0 < \arg(z-4i) < \frac{3\pi}{4}\right\}$

$\{z \in \mathbb{C} : |z-5| \geq 3\}$ is the set of complex numbers z such that $|z-5| \geq 3$

a

Im, Re, O, 2, 5

1 Sketch the locus of z

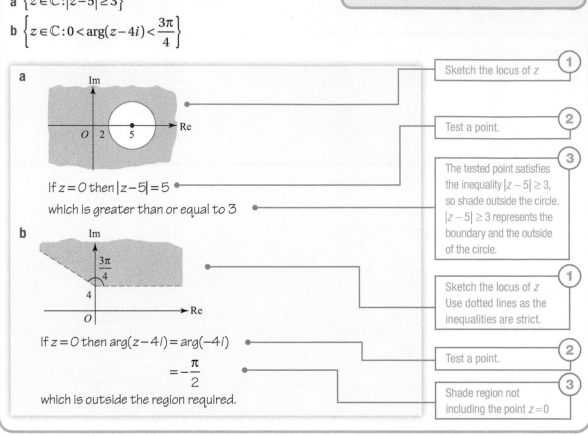

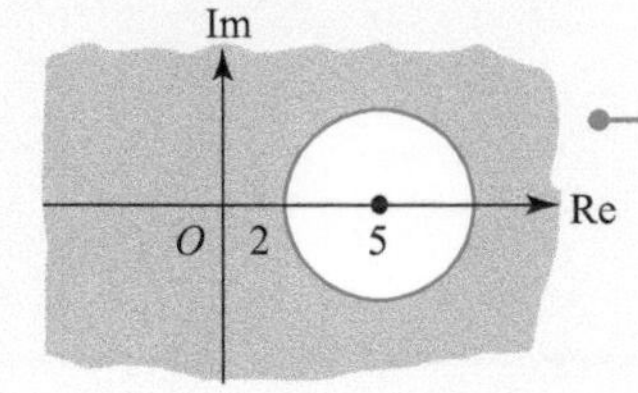

If $z = 0$ then $|z-5| = 5$

2 Test a point.

which is greater than or equal to 3

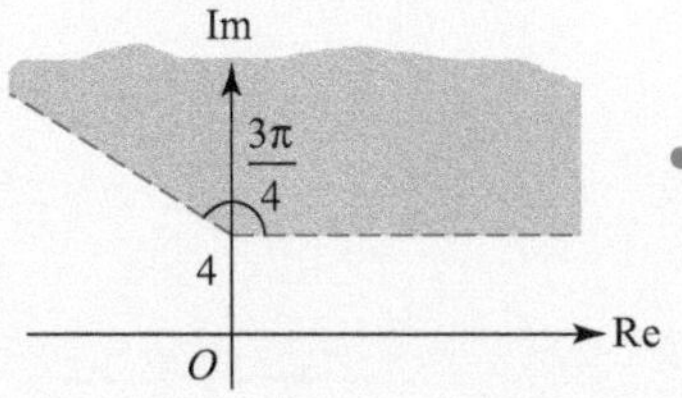

3 The tested point satisfies the inequality $|z-5| \geq 3$, so shade outside the circle. $|z-5| \geq 3$ represents the boundary and the outside of the circle.

b

Im, Re, O, 4, $\frac{3\pi}{4}$

1 Sketch the locus of z. Use dotted lines as the inequalities are strict.

If $z = 0$ then $\arg(z-4i) = \arg(-4i)$

$= -\frac{\pi}{2}$

2 Test a point.

which is outside the region required.

3 Shade region not including the point $z = 0$

Exercise 1.4B Reasoning and problem-solving

1 Prove that $|zw| = |z||w|$ and $\arg(zw) = \arg z + \arg w$ for all $z, w \in \mathbb{C}$

2 Show that the locus of points satisfying $|z+3-2i| = 4$ is a circle and sketch it.

3 a Show that the locus of points satisfying $|z-2-i| = 2$ is a circle.

The circle touches the imaginary axis at the point A and crosses the real axis at B and C

b Calculate the exact area of triangle ABC

4 Find the Cartesian equation of each locus.

a $|z-5| = |z|$ b $|z+2| = |z-2i|$

c $|z-4i| = |z+2|$ d $|z+1-i| = |z-3|$

e $|z-5+i| = |z+i-2|$

f $|z-7+4i| = |z+6-3i|$

5 In each case, find the Cartesian equation of the line on which the half-line lies.

a $\arg(z-3i) = \frac{\pi}{4}$ b $\arg(z+5) = \frac{\pi}{2}$

c $\arg(z+2-i) = \frac{\pi}{3}$ d $\arg(z-4+i) = \frac{2\pi}{3}$

6 Find the Cartesian equation of the locus of points satisfying $|z-2|=|z+3-i|$

7 Find the Cartesian equation of the locus of $|z-\sqrt{2}+i|=|z-1-\sqrt{2}|$

8 Find the Cartesian equation of the locus of $|z-2|=|4i-z|$

9 Shade the region that satisfies

a $|z-4i|\leq 3$ b $|z+2-i|\geq 1$

c $4\leq|z|\leq 10$ d $2<|z-5+2i|<5$

10 Find the area of the region that satisfies $\sqrt{7}\leq|z+3-7i|\leq 7$

11 Shade the region represented by each inequality.

a $0\leq\arg z\leq\frac{\pi}{3}$ b $-\frac{5\pi}{6}<\arg(z-3i)<0$

c $\frac{\pi}{2}\leq\arg(z-2-i)<\frac{3\pi}{4}$

12 Find the area of the region that satisfies $|z-5+2i|<8$ and $0\leq\arg(z-5+2i)\leq\frac{\pi}{2}$

13 Shade the region that satisfies

a $|z-3|\geq|z+5|$ b $|z-i|<|z-3i|$

c $|z-2-4i|\leq|z+8-4i|$

d $|z-3-i|>|z-5+3i|$

e $|z-1+2i|\leq|z-3-2i|$ and $|z|>|z-2|$

14 Sketch and shade the region that satisfies both $|z-3+3i|\leq 3$ and $-\frac{\pi}{4}\leq\arg z\leq 0$

15 Sketch and shade the region that satisfies both $|z|\leq|z-8i|$ and $|z-2i|\geq 8$

16 Sketch and shade the region that satisfies both $\frac{\pi}{3}<\arg(z-2)<\frac{2\pi}{3}$ and $|z-2i|<|z-4i|$

17 Shade on an Argand diagram the set of points $\{z\in\mathbb{C}:|z+i-2|\leq 1\}$

18 Shade on an Argand diagram the set of points $\{z\in\mathbb{C}:|z+2i|>|z-2|\}$

19 Shade on an Argand diagram the set of points $\{z\in\mathbb{C}:|z|\geq 4\}\cap\left\{z\in\mathbb{C}:-\frac{\pi}{3}<\arg(z)<\frac{\pi}{3}\right\}$

20 Shade on an Argand diagram the set of points $\{z\in\mathbb{C}:|z-3i|>|z+5i|\}\cap\left\{z\in\mathbb{C}:-\pi\leq\arg(z)\leq-\frac{\pi}{2}\right\}$

PURE

Full A Level

21 a Use algebra to show that the locus of points satisfying $|z+3|=2|z-6i|$ is a circle, then sketch it.

b Shade the region that satisfies $|z+3|\leq 2|z-6i|$ and $|z-1-3i|\leq 20$

22 The point P represents a complex number z on an Argand diagram such that $|z-3+i|=3$

a State the Cartesian equation of the locus of P

The point Q represents a complex number z on an Argand diagram such that $|z+2-i|=|z-1+2i|$

b Find the Cartesian equation of the locus of Q

c Find the complex number that satisfies both $|z-3+i|=3$ and $|z+2-i|=|z-1+2i|$, giving your answer in surd form.

23 Find the complex number that satisfies both $|z-3i|=4$ and $\arg(z-3i)=-\frac{\pi}{4}$

MyMaths 2054, 2055, 2057 SEARCH

1 Summary and review

Chapter summary

- The imaginary number i is defined as $i=\sqrt{-1}$
- Complex numbers written in the form $a+bi$ can be added, subtracted and multiplied in the same way as algebraic expressions.
- Powers of i should be simplified: $i^2=-1$, $i^3=-i$ and so on.
- The complex conjugate of the number $z=a+bi$ is $z^*=a-bi$
- Fractions with a complex number in the denominator can be simplified by multiplying the numerator and the denominator by the complex conjugate of the denominator.
- Complex roots of polynomial equations with real coefficients occur in conjugate pairs.
- The complex number $z=a+bi$ can be represented by the point (a, b) on an Argand diagram.
- The modulus of a complex number $z=a+bi$ is given by $|z|=\sqrt{a^2+b^2}$
- The (principal) argument of a complex number is the angle between the vector representing it and the positive real axis. Write $\arg z=\theta$ where $-\pi<\theta\le\pi$
- $|z_1z_2|=|z_1||z_2|$ and $\left|\frac{z_1}{z_2}\right|=\frac{|z_1|}{|z_2|}$ for all $z_1, z_2\in\mathbb{C}$
- $\arg(z_1z_2)=\arg(z_1)+\arg(z_2)$ for all $z_1, z_2\in\mathbb{C}$
- The modulus-argument form of the complex number $z=a+bi$ is given by $z=r(\cos\theta+i\sin\theta)$ where r is the modulus of z and θ is the argument.
- The locus of points satisfying $|z-z_1|=r$ will be a circle, centre z_1 and radius r
- The locus of points satisfying $\arg(z-z_1)=\theta$ is a half-line from the point z_1 at an angle of θ to the positive real axis.
- The locus of points satisfying $|z-z_1|=|z-z_2|$ is the perpendicular bisector of the line joining z_1 and z_2

Check and review

You should now be able to...	Try Questions
✔ Add, subtract, multiply and divide complex numbers in the form $a+bi$	1
✔ Understand and use the complex conjugate.	1–3
✔ Solve quadratic, cubic and quartic equations with real coefficients.	2–4
✔ Calculate the modulus and the argument of a complex number.	5
✔ Convert between modulus-argument form and the form $a+bi$	6, 7
✔ Multiply and divide numbers in modulus-argument form.	8
✔ Sketch and interpret Argand diagrams.	9
✔ Construct and interpret loci in the Argand diagram.	10–14

PURE

1 Given that $z=5-4i$ and $w=-2-3i$, find each of these in the form $a+bi$

a $z+w$ b $3w$

c $2z-w$ d zw

e z^2+w^2 f $(z+w)^2$

g z^* h w^*

i $\frac{z}{w}$ j $\frac{3w}{2z}$

k $2\div z$ l $w^*\div 3i$

2 $f(x)=ax^2+bx+c=0$ where a, b and c are real numbers.

Given that $f(x)$ has a root $x=-7+2i$

find possible values of a, b and c

3 $f(z)=z^3+az^2+93z-130$ where a is a real number.

Given that $f(2)=0$,

a Find the value of a

b Solve the equation $f(z)=0$

4 Solve the quartic equation

$x^4+ax^3-ax^2+bx+169=0$

given that a and b are real constants and that $3+2i$ is one solution of the equation.

5 Calculate the modulus and the argument of these complex numbers.

a $2+9i$ b $3-3i$

c $7i$ d $-2i$

e $-1+4i$ f $-3-4i$

6 Write these complex numbers in modulus-argument form.

a $8+6i$ b $-12+5i$

c $-2-2i$ d $\sqrt{3}-i$

e $\sqrt{5}\left(\cos\left(\frac{\pi}{3}\right)-i\sin\left(\frac{\pi}{3}\right)\right)$

7 Write these in the form $a+bi$

a $2\left(\cos\left(\frac{\pi}{6}\right)+i\sin\left(\frac{\pi}{6}\right)\right)$

b $\sqrt{3}\left(\cos\left(-\frac{\pi}{4}\right)+i\sin\left(-\frac{\pi}{4}\right)\right)$

8 Given that

$$z=\sqrt{6}\left(\cos\left(-\frac{\pi}{3}\right)+i\sin\left(-\frac{\pi}{3}\right)\right)$$

and

$$w=\sqrt{3}\left(\cos\left(\frac{\pi}{6}\right)+i\sin\left(\frac{\pi}{6}\right)\right)$$

find these in modulus-argument form.

a zw b $\frac{z}{w}$

c $\frac{w}{z}$

9 Given that $z_1=5-2i$ and $z_2=-2+3i$, draw these complex numbers on the same Argand diagram.

a z_1 b z_2

c z_1+z_2 d z_1-z_2

10 Sketch these loci on separate Argand diagrams and give the Cartesian equation of each.

a $|z|=7$ b $|z-8|=5$

c $|z+3-i|=3$ d $|z-2-3i|=2$

11 Sketch these loci on separate Argand diagrams.

a $\arg z=\frac{\pi}{6}$ b $\arg(z-3)=-\frac{\pi}{3}$

c $\arg(z+2i)=\frac{2\pi}{3}$ d $\arg(z+1-i)=-\frac{\pi}{4}$

12 Sketch these loci on separate Argand diagrams and give the Cartesian equation of each.

a $|z|=|z-6|$ b $|z+2i|=|z-8i|$

c $|z-2i|=|z+4|$ d $|z-1-i|=|z+1+i|$

13 Sketch and shade the region satisfying each inequality.

a $|z-5+2i|\leq 2$ b $\frac{\pi}{2}\leq\arg(z-i)<\frac{5\pi}{6}$

c $3\leq|z+3-5i|\leq 5$ d $|z-6i|<|z+6|$

14 Sketch and shade the region satisfying both of these inequalities.

$|z-7|\geq 7$ and $-\frac{\pi}{2}\leq\arg(z-7)\leq 0$

Exploration

Going beyond the exams

History

Jean-Robert Argand was a Swiss mathematician who, early in the 19th century, moved to Paris where he managed a bookshop. He published an essay on the representation of imaginary quantities in 1813 and, although others before him had written about similar ideas, it was Argand's essay that gained attention and resulted in the naming of the **Argand diagram** we continue to use today.

Note

A **conic section** is a curve that is found by intersecting a plane with a double cone placed vertex to vertex.

There are three types of conic section: ellipse, parabola and hyperbola. A circle is a type of ellipse.

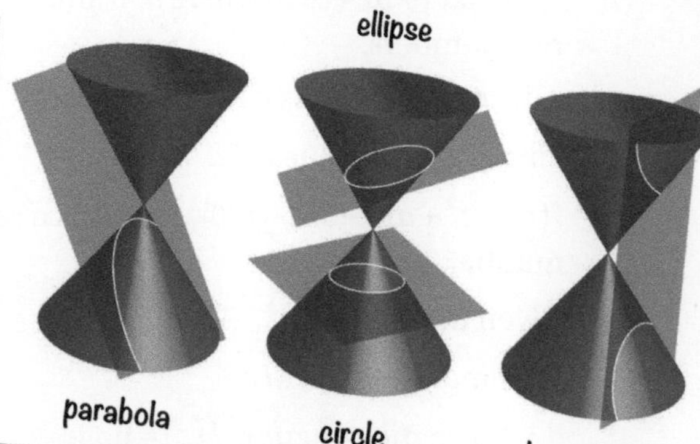

Investigation

The equation $|z| = 1$ plotted in the complex plane gives a circle with unit radius centred at the origin.

Choose values of a, b and c such that $c > (b - a)$ and $b > a$

By varying these values, investigate the graph of the locus of point z that moves so that $|z - a| + |z - b| = c$, $c > (b - a)$ and $b > a$

What is the shape of the graph that results?

The real part of a complex number z can be denoted by $\text{Re}(z)$

By varying the values of a and b, investigate graphs of the locus of a point z that moves so that $\text{Re}(z + a) = |z - b|$

What is the shape of the graph that results?

Research

The **Mandelbrot set** provides beautiful images. It is the set of all the complex numbers, c, for which the iterative sequence $z_{n+1} = z_n^2 + c$, where $z_0 = 0$ does not tend to infinity.

Computer images of the Mandelbrot set can be developed and, no matter how closely you zoom in on the surface, it still remains incredibly intricate. The Mandelbrot set is an example of a **fractal**.

Find out if the complex numbers i and $2i$ are in the Mandelbrot set.

Research the properties of the Mandelbrot set.

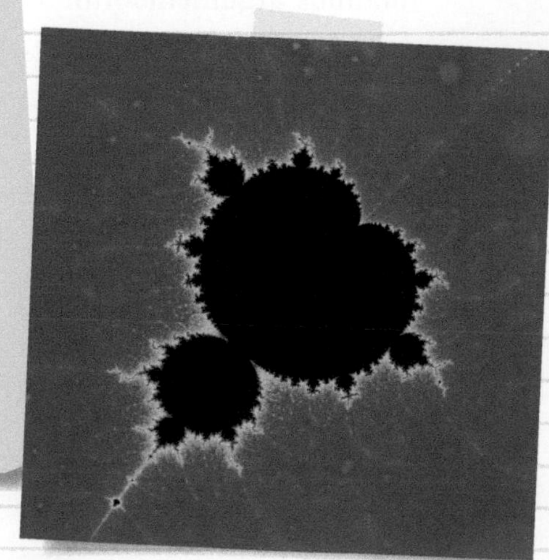

1 Assessment

1 The complex numbers z and w are given as $z = 9 - 11i$ and $w = 7 + 3i$

a Calculate these expressions, giving the answers in the form $a + bi$

i $w - z$ ii wz iii $\frac{w}{z}$ **[5 marks]**

b Calculate the modulus and the argument of z, giving your answers to 3 significant figures. **[3]**

2 The quadratic equation $x^2 + 2x + 5 = 0$ has complex roots α and β

a Find α and β **[2]**

b Show α and β on an Argand diagram. **[2]**

c Describe the transformation that maps α to β **[1]**

3 $z = \frac{5+2i}{3+i}$

a Show that $z = a + bi$ where a and b are constants to be found. **[3]**

b Calculate the value of

i $z + z^*$ ii $z - z^*$ iii zz^* **[4]**

c i Draw z and iz on the same Argand diagram.

ii Describe the geometric relationship between z and iz **[3]**

4 Find the values of a and b such that $(a + bi)^2 = -15 + 8i$ **[6]**

5 Find the values of a and b such that $(a + bi)^2 = 1 + 2\sqrt{2}i$ **[6]**

6 $f(x) = x^3 - 2x^2 + kx + 26$ for some real constant k

a Given that $x = 2 - 3i$ is a solution of $f(x) = 0$, find all the roots of $f(x)$ **[3]**

b Find the area of the triangle formed by the three roots of $f(x)$ **[2]**

7 Given that $5 - i$ is a root of the equation $x^3 + ax^2 + bx - 182 = 0$

a Find the other two roots, **[4]**

b Calculate the values of a and b **[3]**

8 Solve the equation $z - 2z^* = -6 + 27i$ **[3]**

9 Solve the simultaneous equations

$$z - 2w = 8 + 5i$$
$$2z - 3w = 13 + 3i$$ **[4]**

10 Given that $z = 3 + 3\sqrt{3}i$ and $w = -\sqrt{2} + \sqrt{2}i$

a Calculate the modulus and argument of z **[3]**

b Calculate the modulus and argument of w **[3]**

c Hence, or otherwise, find the value of

i $|zw|$ **ii** $\left|\frac{z}{w}\right|$

iii $\arg(zw)$ **iv** $\arg\left(\frac{z}{w}\right)$ **[6]**

11 Given that $z = 4 + \sqrt{2}i$ and $zw = 5\sqrt{2} - 2i$

a Find w in modulus-argument form. **[6]**

The points A and B represent w and zw on an Argand diagram.

b Calculate the distance AB, giving your answer as a surd. **[2]**

12 a Write the complex number $z = \sqrt{3} - i$ in modulus-argument form. **[3]**

The complex number w has argument $-\frac{\pi}{12}$

b Find the argument of

i zw **ii** $\frac{z}{w}$ **[4]**

c Find $|w|$ given that $|zw| = 10$ **[2]**

13 a Sketch the locus of points that satisfy

i $|z - 4i| = 2$ **ii** $|z + 3| = |z - 3i|$ **[4]**

b Find the Cartesian equation of the locus of points drawn in part **a**. **[3]**

14 a Show that the locus of points satisfying $|z + 3 - 4i| = 4$ is a circle. **[3]**

b Sketch the locus of points satisfying $|z + 3 - 4i| = 4$ **[2]**

c Shade in the region that satisfies $|z + 3 - 4i| \leq 4$ **[2]**

15 For the locus of points satisfying $|z + 5| = |z - i|$

a Sketch the locus, **[2]**

b Find the Cartesian equation. **[3]**

16 Find the square roots of the complex number $39 - 80i$ **[6]**

17 Solve the equation $z^2 = 4 - 2\sqrt{5}i$ **[6]**

18 The solutions of the quartic equation $x^4 - 14x^3 + ax^2 - bx + 58 = 0$ are represented on an Argand diagram by the points P, Q, R and S

Given that a and b are real constants and that one solution of the equation is $w = 7 + 3i$

calculate the area of the quadrilateral $PQRS$ **[6]**

19 Two solutions of a cubic equation with real coefficients are $x = -3$ and $x = i - 4$

a State the third solution of the equation. **[1]**

b Find a possible equation with these solutions. **[3]**

20 Two solutions of a quartic equation with real coefficients are $z = -2 + i$ and $z = 3 - 5i$

a State the two other solutions of the equation. **[2]**

b Find a possible equation with these solutions. **[3]**

21 A quartic equation with real coefficients $ax^4+bx^3+cx^2+dx+e=0$ has exactly two distinct solutions, one of which is $1+5i$

Find the values of a, b, c, d and e **[4]**

22 Solve the simultaneous equations

$$z^2-w^2=-6$$

$$z+2w=3$$ **[6]**

23 Solve the simultaneous equations.

$$z^2+w^2=30$$

$$z+3w=20$$ **[4]**

24 The equation $2x^5+x^4+36x^3+18x^2+162x+81=0$ has a repeated quadratic root.

a Show that $x=3i$ is a solution of the equation. **[2]**

b Fully factorise the equation. **[3]**

25 The curve of $y=x^4-4x^3-x^2+6x+18$ is shown.

Find all the solutions of the equation $x^4-4x^3-x^2+6x+18=0$ **[4]**

26 A quadratic equation has roots α and β, where α has argument $\frac{5\pi}{6}$ and modulus 3

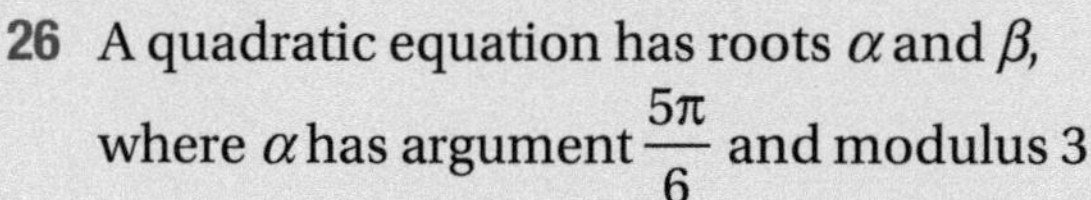

a Find β in modulus-argument form. **[2]**

b Calculate

i $|\alpha\beta|$ **ii** $\left|\frac{\alpha}{\beta}\right|$ **iii** $\arg(\alpha\beta)$ **[6]**

c Find the quadratic equation with roots α and β **[3]**

27 The complex numbers z_1 and z_2 are given by $z_1=3+ai$ and $z_2=2-i$ where a is an integer.

a Find $\frac{z_1}{z_2}$ in terms of a **[3]**

Given that $\left|\frac{z_1}{z_2}\right|=\sqrt{18}$

b Find the possible values of a **[5]**

28 Given that $w=\frac{1}{2}\left(\cos\frac{\pi}{3}-i\sin\frac{\pi}{3}\right)$

a Write w in modulus-argument form, **[1]**

b Find the complex number z such that $\arg(wz)=-\frac{\pi}{6}$ and $\left|\frac{w^2}{z}\right|=3$

Write your answer in exact Cartesian form. **[3]**

29 Describe the locus shown in terms of z **[2]**

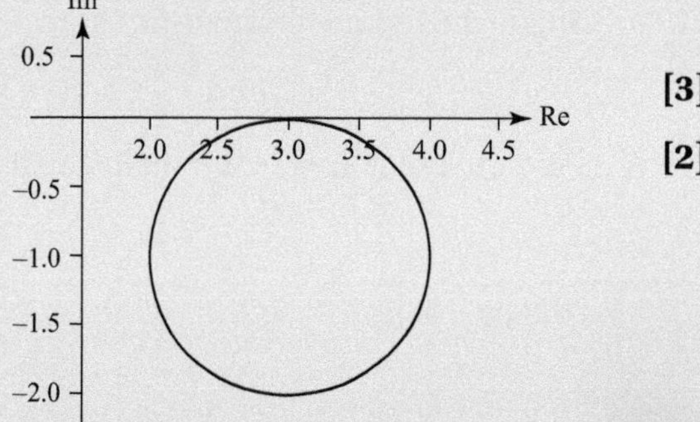

30 a Sketch the locus of points that satisfy $\arg(z-2i)=\frac{\pi}{4}$ **[3]**

b Find the Cartesian equation of the locus drawn in part **a**. **[3]**

31 Sketch and shade the region satisfying
$0<\arg(z+1+3i)\leq\frac{2\pi}{3}$ **[5]**

32 Sketch and shade the region satisfying
$|z-i|\geq 1$ and $|z+i|\leq 2$ **[4]**

33 The complex numbers $4-3i$ and $-2-i$ represent the points A and B respectively.

Find the area of the triangle OAB **[5]**

34 The locus of the complex number z is a half-line as shown.

Describe the locus in terms of z **[3]**

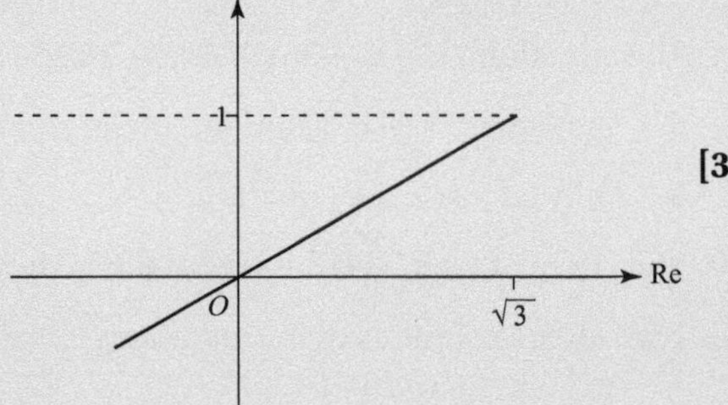

35 The point P represents a complex number z on an Argand diagram such that

$|z+2|=3|z-2i|$

Show that, as z varies, the locus of P is a circle and state the radius of the circle and the coordinates of the centre. **[6]**

36 The point P represents a complex number z on an Argand diagram such that

$$|z-1|=1$$

The point Q represents a complex number z on an Argand diagram such that

$$\arg(z-i)=-\frac{\pi}{4}$$

a Sketch the loci of P and Q on the same axes. **[4]**

b Find the complex number that satisfies both

$|z-1|=1$ and $\arg(z-i)=-\frac{\pi}{4}$ **[6]**

37 a Find the possible complex numbers, z, that satisfy both $|z+3i|=|z-i|$ and $|z-2+i|=4$ **[4]**

b Sketch the region that satisfies both $|z+3i|>|z-i|$ and $|z-2+i|\geq 4$ **[5]**

38 a Shade the region that satisfies $1\leq|z-3i|\leq 3$ **[4]**

b Find the exact area of the shaded region. **[2]**

39 a Shade the region that satisfies both

$|z+i-3|\leq 2$ and $-\frac{\pi}{2}\leq\arg(z+i)\leq 0$ **[4]**

b Find the exact area of the shaded region. **[2]**

2 Algebra and series

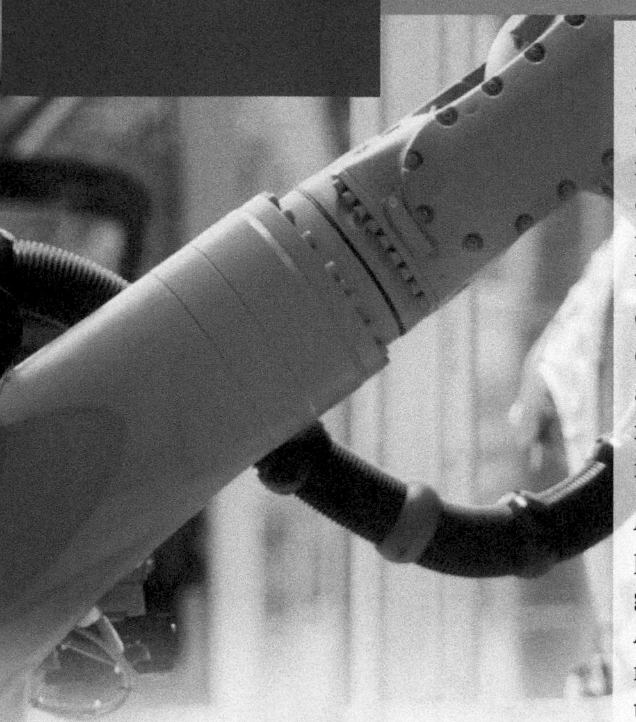

Polynomials have many uses, and one use of particular interest is in programming the movement of robotic arms. These arms are used commonly in industry, for example, in the manufacturing of cars. Polynomials are used to determine the smooth paths that the arm joints should follow, to ensure that the mechanical components do not wear too quickly. The trajectory of a joint is determined as the combination of several polynomial functions, which are stitched together at various points. Algebraic methods are used to ensure that the path of the joint is smooth and continuous.

Another, perhaps surprising, application of polynomials is in in various aspects of computer graphics, and particularly in the design of fonts. Although the algebra and numerical methods required for good font design are quite complex, they are developed from our basic understanding of simple polynomials such as quadratics and cubics.

Orientation

What you need to know

KS4

- Inequalities.
- Proof.

Maths Ch1

- Argument and proof.
- Quadratic functions.

Maths Ch2

- Expanding and factorising polynomials.

What you will learn

- To relate the roots of a polynomial to its coefficients.
- To find a new polynomial whose roots are a linear transformation of the roots of another polynomial.
- To evaluate expressions involving roots.
- To solve inequalities up to quartic polynomials.
- To use the sums of integers, squares and cubes to sum other series.
- To understand and use the method of differences.
- To use proof by induction to prove divisibility.
- To use Maclaurin series expansions.

What this leads to

Ch12 Series

- Summing series using partial fractions.
- More Maclaurin series.

Careers

- Car manufacturing.
- Graphic design.

2.1 Roots of polynomials

Fluency and skills

See Maths Ch 2.3
For a reminder of the factor theorem.

The roots of polynomials can be found using the factor theorem and long division. You are now going to learn how to transform one polynomial into another polynomial with roots that are related in some way.

Quadratic equations

Let the quadratic equation $ax^2 + bx + c = 0$ have roots $x = \alpha$ and $x = \beta$

Dividing through by a gives

$$x^2 + \frac{b}{a}x + \frac{c}{a} = 0$$

Since $x = \alpha$ and $x = \beta$ are the roots of this quadratic, you can write the equation in the form

$$(x - \alpha)(x - \beta) = 0$$

Expanding the brackets gives

$$x^2 - (\alpha + \beta)x + \alpha\beta = 0$$

Comparing the two versions of the quadratic equation gives

$$x^2 + \frac{b}{a}x + \frac{c}{a} \equiv x^2 - (\alpha + \beta)x + \alpha\beta = 0$$

So, comparing the coefficients for x and the constant gives $(\alpha + \beta) = -\frac{b}{a}$ and $\alpha\beta = \frac{c}{a}$

For a quadratic equation:

Key point

The sum of the roots $= \alpha + \beta = -\frac{b}{a}$ and the product of the roots $= \alpha\beta = \frac{c}{a}$

This also shows that all quadratics can be written in the form
$x^2 - (\text{sum of the roots})x + (\text{product of the roots}) = 0$

Example 1

The roots of $x^2 - 7x + 12 = 0$ are $x = \alpha$ and $x = \beta$. Without finding the values of α and β separately,

a Write down the values of $\alpha + \beta$ and $\alpha\beta$

b Hence find the quadratic equations whose roots are **i** α^2 and β^2 **ii** $\frac{1}{\alpha}$ and $\frac{1}{\beta}$

a $a = 1, b = -7$ and $c = 12$, so

$\alpha + \beta = -(-7) = 7$ and $\alpha\beta = 12$ — $(\alpha + \beta) = -\frac{b}{a}$ and $\alpha\beta = \frac{c}{a}$

(Continued on the next page)

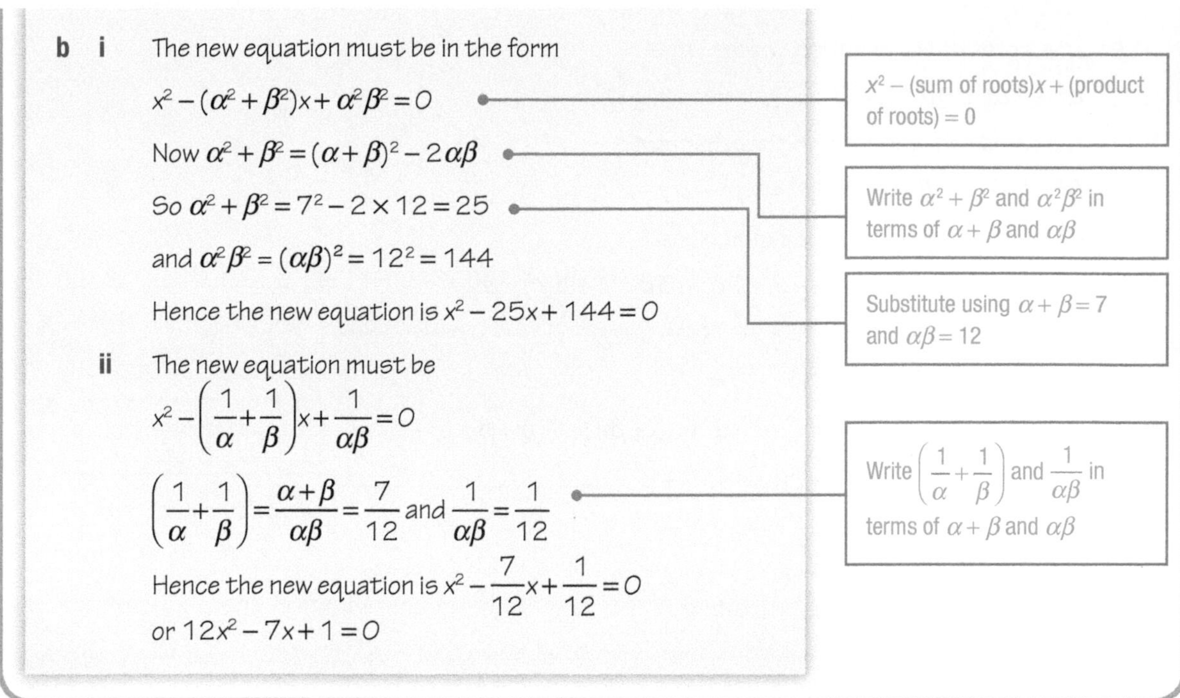

b i The new equation must be in the form

$x^2 - (\alpha^2 + \beta^2)x + \alpha^2\beta^2 = 0$

Now $\alpha^2 + \beta^2 = (\alpha + \beta)^2 - 2\alpha\beta$

So $\alpha^2 + \beta^2 = 7^2 - 2 \times 12 = 25$

and $\alpha^2\beta^2 = (\alpha\beta)^2 = 12^2 = 144$

Hence the new equation is $x^2 - 25x + 144 = 0$

ii The new equation must be

$x^2 - \left(\frac{1}{\alpha} + \frac{1}{\beta}\right)x + \frac{1}{\alpha\beta} = 0$

$\left(\frac{1}{\alpha} + \frac{1}{\beta}\right) = \frac{\alpha + \beta}{\alpha\beta} = \frac{7}{12}$ and $\frac{1}{\alpha\beta} = \frac{1}{12}$

Hence the new equation is $x^2 - \frac{7}{12}x + \frac{1}{12} = 0$

or $12x^2 - 7x + 1 = 0$

x^2 – (sum of roots)x + (product of roots) = 0

Write $\alpha^2 + \beta^2$ and $\alpha^2\beta^2$ in terms of $\alpha + \beta$ and $\alpha\beta$

Substitute using $\alpha + \beta = 7$ and $\alpha\beta = 12$

Write $\left(\frac{1}{\alpha} + \frac{1}{\beta}\right)$ and $\frac{1}{\alpha\beta}$ in terms of $\alpha + \beta$ and $\alpha\beta$

Cubic equations

There are similar relationships between the coefficients of x and the roots of the equation for higher-order equations (cubics, quartics, and so on).

Let the cubic equation

$ax^3 + bx^2 + cx + d = 0$ have roots $x = \alpha$, $x = \beta$ and $x = \gamma$

Let $ax^3 + bx^2 + cx + d = 0 \equiv (x - \alpha)(x - \beta)(x - \gamma) = 0$

$$\equiv (x^2 - \alpha x - \beta x + \alpha\beta)(x - \gamma) = 0$$

$$\equiv x^3 - \alpha x^2 - \beta x^2 + \alpha\beta x - x^2\gamma + \alpha\gamma x + \beta\gamma x - \alpha\beta\gamma = 0$$

$$\equiv x^3 - (\alpha + \beta + \gamma)x^2 + (\alpha\beta + \beta\gamma + \gamma\alpha)x - \alpha\beta\gamma = 0$$

Dividing $ax^3 + bx^2 + cx + d = 0$ by a gives $x^3 + \frac{b}{a}x^2 + \frac{c}{a}x + \frac{d}{a} = 0$

Hence

$(\alpha + \beta + \gamma) = -\frac{b}{a}$ $\quad(\alpha\beta + \beta\gamma + \gamma\alpha) = \frac{c}{a}$ $\quad$ and $\alpha\beta\gamma = -\frac{d}{a}$

Example 2

The roots of the equation $x^3 - 7x^2 + 3x + 2 = 0$ are α, β and γ. Find the values of

a $\alpha^2 + \beta^2 + \gamma^2$ **b** $\alpha^3 + \beta^3 + \gamma^3$ **c** $(\alpha+2)(\beta+2)(\delta+2)$

a $(\alpha + \beta + \gamma)^2 = \alpha^2 + 2\alpha\beta + 2\alpha\gamma + \beta^2 + 2\beta\gamma + \gamma^2$

$= (\alpha^2 + \beta^2 + \gamma^2) + 2(\alpha\beta + \beta\gamma + \alpha\gamma)$

$7^2 = (\alpha^2 + \beta^2 + \gamma^2) + 2(3)$

$\alpha^2 + \beta^2 + \gamma^2 = 49 - 6 = 43$

Consider the expression $(\alpha + \beta + \gamma)^2$ to find a way to write $\alpha^2 + \beta^2 + \gamma^2$

Substitute $\alpha + \beta + \gamma = 7$ and $\alpha\beta + \beta\gamma + \alpha\gamma = 3$ and rearrange.

(Continued on the next page)

b α is a root of the equation, which means

$\alpha^3 - 7\alpha^2 + 3\alpha + 2 = 0$, so by rearranging $\alpha^3 = 7\alpha^2 - 3\alpha - 2$

The same can be done with the other roots.

$\beta^3 = 7\beta^2 - 3\beta - 2 \qquad \gamma^3 = 7\gamma^2 - 3\gamma - 2$

Now find the sum of these expressions.

$$\alpha^3 + \beta^3 + \gamma^3 = 7\alpha^2 - 3\alpha - 2 + 7\beta^2 - 3\beta - 2 + 7\gamma^2 - 3\gamma - 2$$

$$= 7(\alpha^2 + \beta^2 + \gamma^2) - 3(\alpha + \beta + \gamma) - 6$$

Substitute $\alpha^2 + \beta^2 + \gamma^2 = 43$ and $\alpha + \beta + \gamma = 7$

$$= 7(43) - 3(7) - 6 = 274$$

c $(\alpha+2)(\beta+2)(\delta+2) = \alpha\beta\delta + 2(\alpha\beta + \alpha\delta + \beta\delta) + 4(\alpha + \beta + \delta) + 8$

Expand the brackets.

$$= -2 + 2(3) + 4(7) + 8$$

Use $\alpha\beta\delta = -2$, $\alpha\beta + \alpha\delta + \beta\delta = 3$ and $\alpha + \beta + \delta = 7$

$$= 40$$

Quartic equations

The same method can also be used to find the relationships between the roots, α, β, γ and δ, and the coefficients of the quartic equation $ax^4 + bx^3 + cx^2 + dx + e = 0$

Key point

$(\alpha + \beta + \gamma + \delta) = -\frac{b}{a}$ i.e. $\Sigma\alpha = -\frac{b}{a}$

$(\alpha\beta + \alpha\gamma + \alpha\delta + \beta\gamma + \beta\delta + \gamma\delta) = \frac{c}{a}$ i.e. $\Sigma\alpha\beta = \frac{c}{a}$

$(\alpha\beta\gamma + \beta\gamma\delta + \gamma\delta\alpha + \delta\alpha\beta) = -\frac{d}{a}$ i.e. $\Sigma\alpha\beta\delta = -\frac{d}{a}$

$\alpha\beta\gamma\delta = \frac{e}{a}$

$\alpha + \beta + \gamma + \delta$ is often abbreviated to $\Sigma\alpha$

$\alpha\beta + \beta\gamma + \gamma\delta + \delta\alpha$ is often abbreviated to $\Sigma\alpha\beta$

$\alpha\beta\gamma + \beta\gamma\delta + \gamma\delta\alpha + \delta\alpha\beta$ is often abbreviated to $\Sigma\alpha\beta\gamma$

Exercise 2.1A Fluency and skills

1 Find the sum and product of the roots of these equations.

a $x^2 - 5x + 9 = 0$ b $x^2 + 6x + 7 = 0$

c $x^2 - 8x - 12 = 0$ d $x^2 + 10x - 5 = 0$

e $3x^2 - 12x + 8 = 0$ f $4x^2 + x + 6 = 0$

2 Write the sum, the sum of the products in pairs and the product of the roots of these equations.

a $x^3 + 4x^2 - 9x - 14 = 0$

b $x^3 - 7x^2 - 11x + 12 = 0$

c $x^3 - 13x^2 + 22x - 26 = 0$

d $2x^3 + 5x^2 + 17x - 21 = 0$

e $4x^3 - x^2 + 3x + 8 = 0$

f $\frac{1}{2}x^3 + \frac{3}{8}x^2 - \frac{3}{4}x + \frac{5}{16} = 0$

3 A quadratic equation $3x^2 + kx - 4 = 0$ has roots α and β

Find an expression in terms of k for

a $\alpha + \beta$ b $(\alpha+1)(\beta+1)$

c $\alpha^2 + \beta^2$ d $(\alpha-4)(\beta-4)$

e $\alpha^3 + \beta^3$ f $\frac{1}{\alpha} + \frac{1}{\beta}$

4 A cubic equation $p(x)=0$ has roots α, β and γ

Use the facts that $\alpha+\beta+\gamma=5$, $\alpha\beta\gamma=-2$ and $\alpha\beta+\alpha\gamma+\beta\gamma=4$ to write down a possible expression for $p(x)$

5 These equations have roots α, β, γ and δ

a $x^4-13x^2+36=0$

b $x^4-4x^3-7x^2+22x+24=0$

Find the values of

i $\alpha^2+\beta^2+\gamma^2+\delta^2$ ii $\frac{1}{\alpha}+\frac{1}{\beta}+\frac{1}{\gamma}+\frac{1}{\delta}$

6 The cubic equation $2x^3+ax^2+bx+c=0$ has roots α, β and γ

Find an expression in terms of *a*, *b* and *c* for

a $\alpha+\beta+\gamma$ b $\alpha\beta+\alpha\gamma+\beta\gamma$

c $\alpha\beta\gamma$ d $\alpha^2+\beta^2+\gamma^2$

e $\alpha^3+\beta^3+\gamma^3$ f $(1+\alpha)(1+\beta)(1+\gamma)$

g $\frac{1}{\alpha}+\frac{1}{\beta}+\frac{1}{\gamma}$ h $(\alpha\beta)^2+(\alpha\gamma)^2+(\beta\gamma)^2$

7 The quartic equation $ax^4+bx^3+cx^2+dx+e=0$ has roots α, β, γ and δ. Show that

a $\alpha+\beta+\gamma+\delta=-\frac{b}{a}$

b $\alpha\beta+\alpha\gamma+\alpha\delta+\beta\gamma+\beta\delta+\gamma\delta=\frac{c}{a}$

c $\alpha\beta\gamma+\alpha\beta\delta+\alpha\gamma\delta+\beta\gamma\delta=-\frac{d}{a}$

d $\alpha\beta\gamma\delta=\frac{e}{a}$

8 The quartic equation $3x^4+kx^3+7x^2-2x+k=0$ has roots α, β, γ and δ

Find an expression in terms of *k* for

a $\alpha^2+\beta^2+\gamma^2+\delta^2$ b $\frac{1}{\alpha}+\frac{1}{\beta}+\frac{1}{\gamma}+\frac{1}{\delta}$

PURE

Reasoning and problem-solving

Sometimes it is useful to transform an equation into another whose roots are related in a simple way to the roots of the original equation.

Key point

If the roots are transformed in a linear way, so that $y=mx+c$, then you transform the equation by substituting $x=\frac{y-c}{m}$

If the new roots are reciprocals, so that $y=\frac{1}{x}$, then you transform the equation by substituting $x=\frac{1}{y}$

Strategy 1

To solve a question involving the transformation of one polynomial into another

1. Rewrite the transformation $y=mx+c$ as $x=\frac{y-c}{m}$
2. Substitute $\frac{y-c}{m}$ for x in the original polynomial and simplify to produce the transformed equation.

Example 3

The roots of the equation $x^3 - 2x^2 - x + 2 = 0$ are α, β and γ

a Write the value of $\alpha\beta\gamma$

b **i** Write down a cubic equation with roots $\alpha - 2$, $\beta - 2$ and $\gamma - 2$

ii By considering the product of the roots of this new equation, find the value of $\alpha\beta\gamma(1+\beta\gamma)(1+\alpha\gamma)(1+\alpha\beta)$

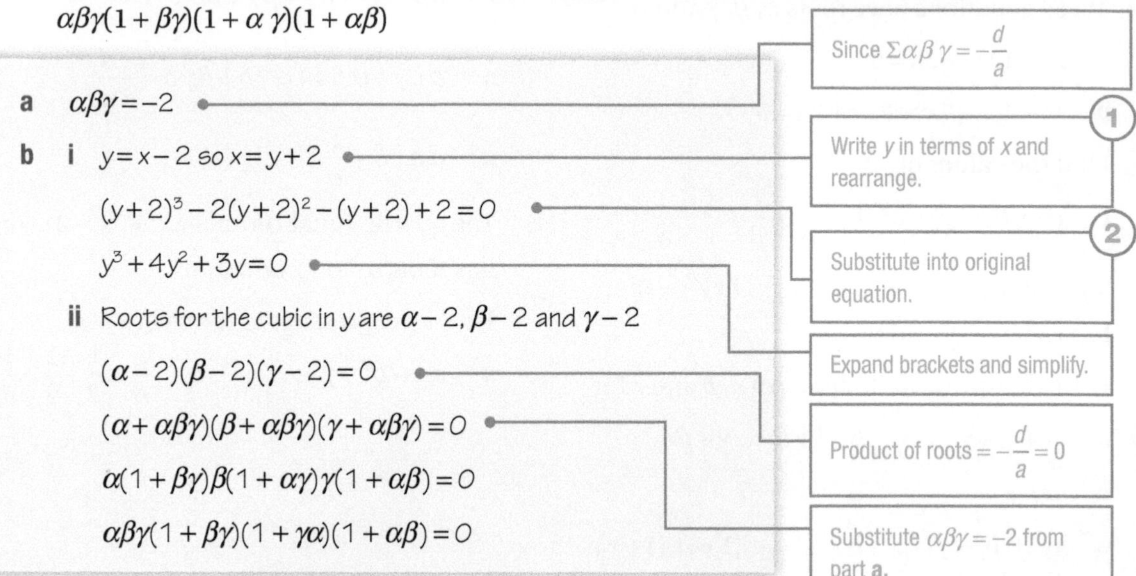

a $\alpha\beta\gamma = -2$ — Since $\Sigma\alpha\beta\gamma = -\frac{d}{a}$

b **i** $y = x - 2$ so $x = y + 2$ — (1) Write y in terms of x and rearrange.

$(y+2)^3 - 2(y+2)^2 - (y+2) + 2 = 0$ — (2) Substitute into original equation.

$y^3 + 4y^2 + 3y = 0$ — Expand brackets and simplify.

ii Roots for the cubic in y are $\alpha - 2$, $\beta - 2$ and $\gamma - 2$

$(\alpha-2)(\beta-2)(\gamma-2) = 0$ — Product of roots $= -\frac{d}{a} = 0$

$(\alpha + \alpha\beta\gamma)(\beta + \alpha\beta\gamma)(\gamma + \alpha\beta\gamma) = 0$ — Substitute $\alpha\beta\gamma = -2$ from part **a**.

$\alpha(1+\beta\gamma)\beta(1+\alpha\gamma)\gamma(1+\alpha\beta) = 0$

$\alpha\beta\gamma(1+\beta\gamma)(1+\gamma\alpha)(1+\alpha\beta) = 0$

Strategy 2

To solve problems with unknown roots and coefficients

1. Use a coefficient you know the value of to form an equation.
2. Solve this equation to find the roots.
3. Use this root to find the value of the others.
4. Use the roots with the rules learnt to find the other coefficients in the equation.

Example 4

The quartic equation $2x^3 + px^2 + qx - 50 = 0$ has roots α, $\frac{50}{\alpha}$, $\alpha + \frac{50}{\alpha} + \frac{5}{2}$

Solve the equation and find the values of p and q

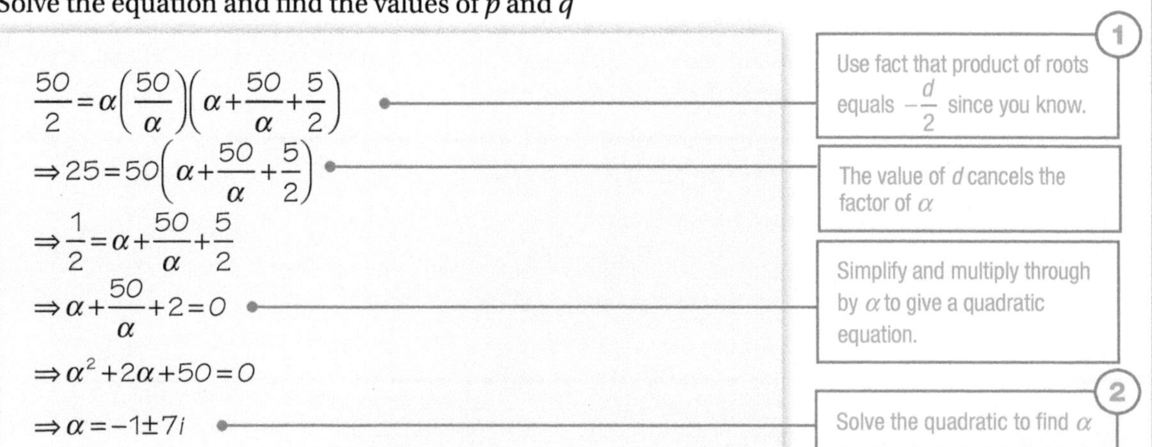

$\frac{50}{2} = \alpha\left(\frac{50}{\alpha}\right)\left(\alpha + \frac{50}{\alpha} + \frac{5}{2}\right)$ — (1) Use fact that product of roots equals $-\frac{d}{2}$ since you know.

$\Rightarrow 25 = 50\left(\alpha + \frac{50}{\alpha} + \frac{5}{2}\right)$ — The value of d cancels the factor of α

$\Rightarrow \frac{1}{2} = \alpha + \frac{50}{\alpha} + \frac{5}{2}$

$\Rightarrow \alpha + \frac{50}{\alpha} + 2 = 0$ — Simplify and multiply through by α to give a quadratic equation.

$\Rightarrow \alpha^2 + 2\alpha + 50 = 0$

$\Rightarrow \alpha = -1 \pm 7i$ — (2) Solve the quadratic to find α

(*Continued on the next page*)

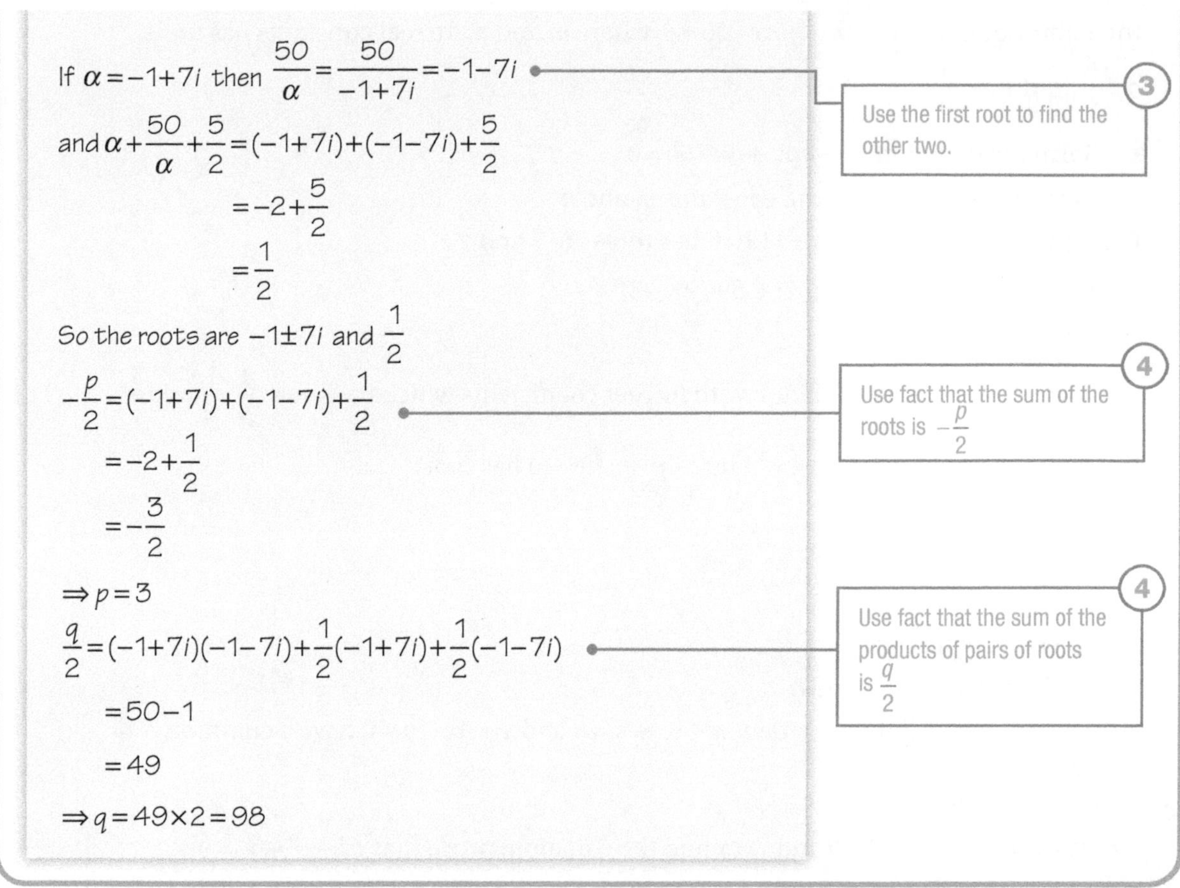

If $\alpha = -1+7i$ then $\frac{50}{\alpha} = \frac{50}{-1+7i} = -1-7i$

and $\alpha + \frac{50}{\alpha} + \frac{5}{2} = (-1+7i)+(-1-7i)+\frac{5}{2}$

$= -2+\frac{5}{2}$

$= \frac{1}{2}$

So the roots are $-1 \pm 7i$ and $\frac{1}{2}$

$-\frac{p}{2} = (-1+7i)+(-1-7i)+\frac{1}{2}$

$= -2+\frac{1}{2}$

$= -\frac{3}{2}$

$\Rightarrow p = 3$

$\frac{q}{2} = (-1+7i)(-1-7i)+\frac{1}{2}(-1+7i)+\frac{1}{2}(-1-7i)$

$= 50-1$

$= 49$

$\Rightarrow q = 49 \times 2 = 98$

3 Use the first root to find the other two.

4 Use fact that the sum of the roots is $-\frac{p}{2}$

4 Use fact that the sum of the products of pairs of roots is $\frac{q}{2}$

Exercise 2.1B Reasoning and problem-solving

1 Find the polynomial whose roots are

  a Reciprocals of, b Triple

  the roots of $x^3+3x^2+5x+1=0$

2 The equation $x^3-2x^2-x+2=0$ has roots α, β and γ

  a Find a cubic equation with roots of $\alpha-2$, $\beta-2$ and $\gamma-2$

  b Solve this cubic equation.

  c Use the substitution $y=x-2$ to find the roots of the original equation in x

3 The equation $x^3-7x^2+9x+11=0$ has roots α, β and γ

  Without calculating the roots, find an equation with roots of $\frac{\alpha}{2}-1, \frac{\beta}{2}-1$ and $\frac{\gamma}{2}-1$

4 The equation $x^4+6x^3+7x+8=0$ has roots α, β, γ and δ

  Without calculating the roots, find an equation with roots of $2\alpha-1$, $2\beta-1$, $2\gamma-1$ and $2\delta-1$

5 The equation $x^3-9x^2+6x+k=0$ has roots α, β, γ and δ

  a Find an equation with roots of $\alpha-4$, $\beta-4$, $\gamma-4$ and $\delta-4$

  b If $k=56$, demonstrate how you can use your solution to part **a** to find the solutions to $x^3-9x^2+6x+k=0$

6 The cubic equation $x^3+mx^2+nx-50=0$ where m and n are real constants has roots $\alpha, \frac{10}{\alpha}$ and $\alpha+\frac{10}{\alpha}-1$

a Solve the equation $x^3+mx^2+nx-50=0$

b Calculate the values of the constants m and n

7 The equation $4x^3+kx^2-11x+119=0$ has roots α, β and γ

a Find the value of k when $\alpha+\beta+\gamma=-6$

For this same value of k

b Write down a cubic equation with integer coefficients which has roots $\frac{1}{\alpha}, \frac{1}{\beta}$ and $\frac{1}{\gamma}$

8 The quartic equation $4x^4-44x^3+px^2+qx+4165=0$ has roots α, β, γ and δ where $\alpha<0$

Given that $\alpha=\beta$ and $\gamma=\frac{85}{\delta}$

a Solve the equation,

b Find the values of p and q

9 Find the value of b if the equations $x^2+6x+b=0$ and $x^2+4x-b=0$ have a common root that is not equal to 0

10 α, β and γ are the roots of $x^3=4x+3$
By substituting $x=\alpha$, $x=\beta$ and $x=\gamma$ into the equation, prove that $\alpha^3+\beta^3+\gamma^3=9$

11 α, β and γ are the roots of $2x^3-2x^2-6x-3=0$
By considering the sum of the roots, find a suitable substitution to transform this equation into a polynomial with roots $\alpha+\beta$, $\beta+\gamma$ and $\gamma+\alpha$

12 The equation $x^3-4x^2-x+7=0$ has roots α, β and γ

a Write down the value of $\alpha+\beta+\gamma$

b Use the substitution $x=4-y$ to find a cubic equation in y

c By considering the product of the roots for this equation, find the value of $(\beta+\gamma)(\alpha+\gamma)(\alpha+\beta)$

Full A Level

2.2 Inequalities

Fluency and skills

The rules for solving inequalities are very similar to those for solving equations, but there are two differences.

See Maths Ch 1.7
For a reminder about solving inequalities.

Key point

- Answers to inequalities are a range of values rather than individual values.
- When you multiply or divide by a negative number you reverse the inequality sign.

You need to be particularly careful to avoid incorrect assumptions when working with inequalities that involve negative numbers.

For example, to solve the inequality $\frac{2x-1}{-5} < 3+x$ you can multiply both sides of the equation by −5 but you must remember to reverse the direction of the inequality sign:

so $\frac{2x-1}{-5} < 3+x$ becomes $2x-1 > -15-5x$

rearrange to give $7x > -14$

so the solution is $x > -2$ (a range of values)

Example 1

If $a < b$, is $a^2 < b^2$? Explain your answer.

Not always.

If $a = 2$ and $b = 3$ then $2^2 < 3^2$ — Since $4 < 9$

but if $a = -3$ and $b = 2$ then $(-3)^2 > 2^2$ even though $-3 < 2$ — Since $9 > 4$

If an expression can be positive or negative, you can consider each case separately.

For example, to solve the inequality $\frac{5}{x+2} > 1$ you could consider the cases when $x + 2$ is positive and $x + 2$ is negative separately.

If $x + 2$ is positive then $x > -2$ and $5 > x+2 \Rightarrow x < 3$ so $-2 < x < 3$

If $x + 2$ is negative then $5 < x+2 \Rightarrow x > 3$, but this cannot be true since we know that $x + 2$ is negative so $x < -2$

Hence the solution is $-2<x<3$

An alternative method is to multiply both sides of the equation by $(x+2)^2$ since you know this will be a positive number.

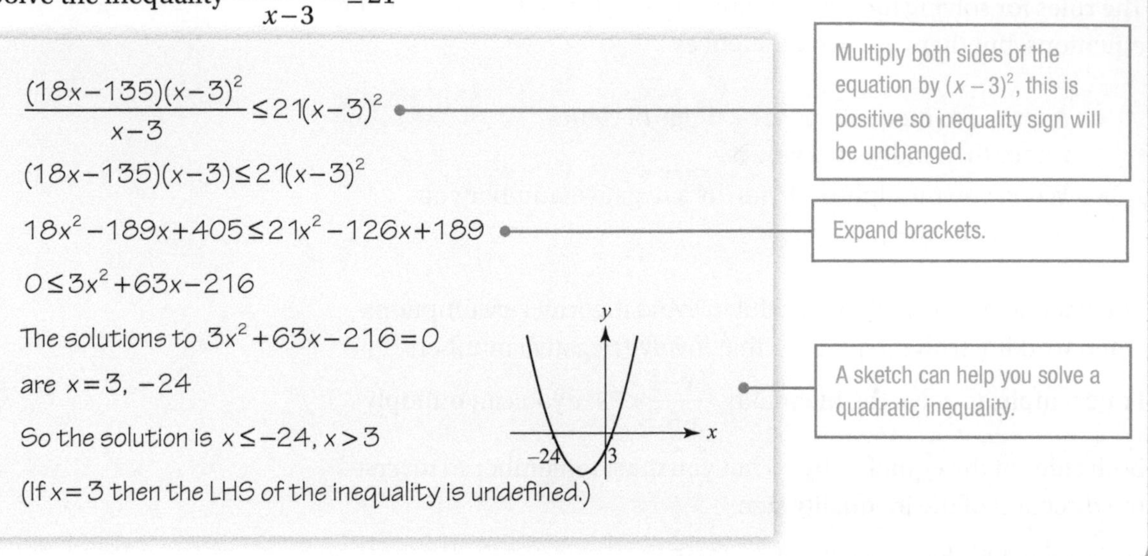

Example 2
Solve the inequality $\frac{18x-135}{x-3} \le 21$
$\frac{(18x-135)(x-3)^2}{x-3} \le 21(x-3)^2$
$(18x-135)(x-3) \le 21(x-3)^2$
$18x^2-189x+405 \le 21x^2-126x+189$
$0 \le 3x^2+63x-216$
The solutions to $3x^2+63x-216=0$
are $x=3, -24$
So the solution is $x \le -24, x>3$
(If $x=3$ then the LHS of the inequality is undefined.)
y
x
−24
3
Multiply both sides of the equation by $(x-3)^2$, this is positive so inequality sign will be unchanged.
Expand brackets.
A sketch can help you solve a quadratic inequality.

If the inequality in Example 2 had been $\frac{18x-135}{x-3}<0$, you can multiply both sides by $(x-3)$, and use the same argument as in Example 2

Case 1: $x>3$

Solve $18x-135<0$ to get $x<7.5$

Case 2: $x<3$

Solve $18x-135>0$ to get $x>7.5$, which isn't possible.

So, $\frac{18x-135}{x-3}<0$ when $3<x<7.5$

Alternatively, to work out when $\frac{18x-135}{x-3}$ is negative, you could consider the signs of the numerator and denominator for different values of x

	$x<3$	$3<x<7.5$	$x>7.5$
$18x-135$	−ve	−ve	+ve
$x-3$	−ve	+ve	+ve
$\frac{18x-135}{x-3}$	+ve	−ve	+ve

If you need to solve an inequality of the form $f(x) \ge 0$ or $f(x) \le 0$, remember to include values where $f(x)=0$ (where the graph intersects the x-axis).

So, $\frac{18x-135}{x-3}<0$ when $3<x<7.5$

You could also sketch a graph of the function $y = \frac{18x - 135}{x - 3}$ and find the values of x where $y < 0$ and the graph lies below the x-axis.

PURE

Example 3

a Sketch the graph of $f(x) = (x+3)(x+1)(x-4)$

b Use your sketch to solve **i** $f(x) = 0$ **ii** $f(x) > 0$ **iii** $f(x) < 0$

You can use a graphical calculator to help you sketch this graph.

a

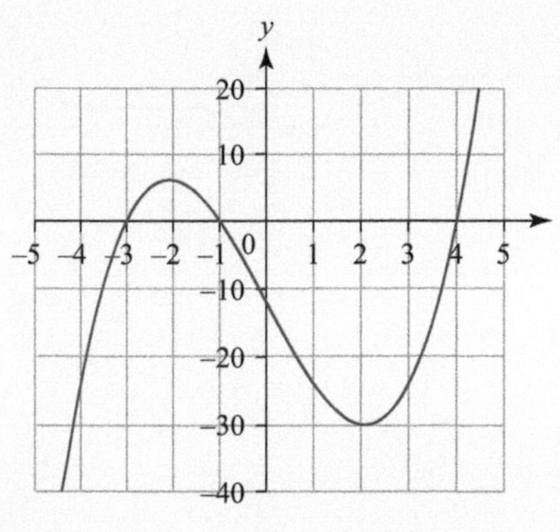

Use where the graph intersects, is above and is below the x-axis to solve the inequalities.

b i $(x+3)(x+1)(x-4) = 0$ when $x = -3$ or $x = -1$ or $x = 4$

ii $(x+3)(x+1)(x-4) > 0$ when $-3 < x < -1$ or $x > 4$

iii $(x+3)(x+1)(x-4) < 0$ when $x < -3$ or $-1 < x < 4$

Example 4

a For the function $f(x) = \frac{5x+15}{x-1}$, use an algebraic method to find the values of x where

i $f(x) = 0$ **ii** $f(x) > 0$ **iii** $f(x) < 0$

b Using graph plotting software, sketch a graph of $y = f(x)$ to confirm your answers.

You can use a graphical calculator to help you sketch this graph.

a i $\frac{5x+15}{x-1} = 0$ when $5x + 15 = 0$ so when $x = -3$

$f(x) = 0$ when numerator $= 0$

ii $\frac{5x+15}{x-1} > 0$ when numerator and denominator have the same sign.

If $x > -3$, numerator is positive.

For denominator to also be positive, $x > 1$

So $x > 1$

x must satisfy $x > -3$ and $x > 1$

If $x < -3$, numerator is negative.

For denominator to also be negative $x < 1$

So $x < -3$

x must satisfy $x < -3$ and $x < 1$

$\frac{5x+15}{x-1} > 0$ when $x > 1$ or $x < -3$

Combine the results.

(*Continued on the next page*)

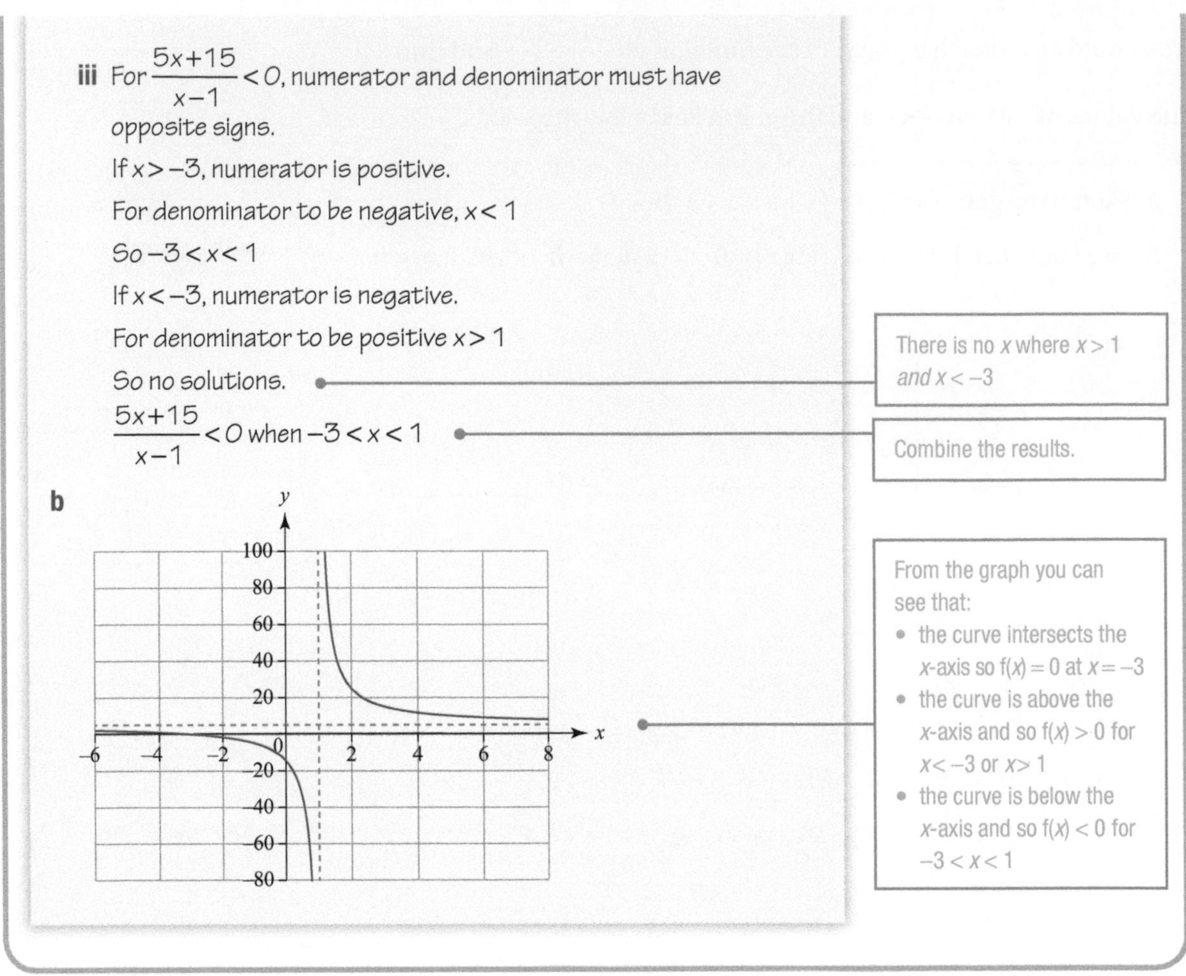

iii For $\frac{5x+15}{x-1}<0$, numerator and denominator must have opposite signs.

If $x>-3$, numerator is positive.

For denominator to be negative, $x<1$

So $-3<x<1$

If $x<-3$, numerator is negative.

For denominator to be positive $x>1$

So no solutions.

There is no x where $x>1$ and $x<-3$

$\frac{5x+15}{x-1}<0$ when $-3<x<1$

Combine the results.

b

From the graph you can see that:

- the curve intersects the x-axis so $f(x)=0$ at $x=-3$
- the curve is above the x-axis and so $f(x)>0$ for $x<-3$ or $x>1$
- the curve is below the x-axis and so $f(x)<0$ for $-3<x<1$

In Example 4, for $x<-3$ the numerator is negative and for $x>-3$ it is positive so $x=-3$ is known as a **critical value**. Similarly, $x=1$ is the critical value for the denominator.

Exercise 2.2A Fluency and skills

1 By trying values for a, b, x and y, show that each of these statements are sometimes, but not always, true.

a If $a<b$ then $\frac{1}{a}<\frac{1}{b}$

b If $a<x$ and $b<y$ then $a-b<x-y$

c If $a<x$ and $b<y$ then $ab<xy$

2 Solve these inequalities algebraically.

a $\frac{x-1}{x}<1$ b $\frac{2x-3}{3x}\le 1$

c $\frac{x-1}{x+2}\ge -1$ d $\frac{2x+5}{x-2}>3$

3 Solve these inequalities by sketching a graph in each case.

a $x(x+4)(x-5)>0$

b $(x+5)(x+4)(x-5)<0$

c $(2x+3)(3x-4)(5x-9)\le 0$

d $(x+4)(x-2)^2\ge 0$

e $(3x-5)(2x+3)^2>0$

f $(x^2-25)(3x+7)<0$

g $(x-3)(x^2-x-12)\le 0$

4 Solve these inequalities.

a $x(x+1)(x-2)(x-9)>0$

b $(x+2)(x+1)(x-3)(x-7)<0$

c $(2x+1)(4x-5)(3x-4)(x-11)\leq 0$

d $(x-1)^2(x-3)^2\geq 0$

e $(2x-7)^2(3x+2)^2>0$

f $(x^2-49)(x^2-5)<0$

g $(x^2-3x-10)(x^2-x-12)\leq 0$

5 For what values of x are the following functions positive?

a $\frac{x-2}{x}$ b $\frac{x-2}{x+3}$

c $\frac{x-2}{x-5}$ d $\frac{(x+1)(x+2)}{x}$

e $\frac{(x+3)(x-4)}{(x+5)}$ f $\frac{(2x-3)(3x+4)}{(x-7)}$

g $\frac{(4-x)(5+2x)}{(2x+3)}$ h $\frac{(3x-2)(2x+4)}{2x-3}$

6 For what values of x are the following functions negative?

a $\frac{x+1}{x}$ b $\frac{x+3}{x+4}$

c $\frac{x-3}{x+2}$ d $\frac{(x+3)(x-5)}{x}$

e $\frac{(x+7)(x+4)}{(x+6)}$ f $\frac{(4x-3)(2x+5)}{(x+6)}$

g $\frac{(9-2x)(6+5x)}{(2x-1)}$ h $\frac{(2x+5)(x-2)}{x+1}$

PURE

Reasoning and problem-solving

Strategy

To solve an inequality algebraically

1. Factorise and simplify using the same rules as for equations, but remember to change the direction of the inequality sign if you multiply or divide by a negative value.
2. Use graphs or critical values to investigate where functions are positive or negative.
3. Solve the inequality by considering all possibilities and cases.

Example 5

a Solve the inequality $x^4-37x^2+36\geq 0$ algebraically.

b Using a graphical calculator or graph sketching software on your computer, sketch a graph to confirm your answer.

a $f(x)\equiv x^4-37x^2+36$ is a quadratic in x^2

$f(x)\equiv(x^2-1)(x^2-36)$ — Factorise the polynomial. (1)

$\equiv(x-1)(x+1)(x-6)(x+6)$

(*Continued on the next page*)

Solve $(x-1)(x+1)(x-6)(x+6) \geq 0$

$(x+6)(x+1)(x-1)(x-6) \geq 0$

The critical values are $x = -6, -1, 1$ and 6

When $x < -6$, f(x) is positive.

When $-6 < x < -1$, f(x) is negative.

When $-1 < x < 1$, f(x) is positive.

When $1 < x < 6$, f(x) is negative.

When $x > 6$, f(x) is positive.

Hence $x^4 - 37x^2 + 36 \geq 0$ when

$x \leq -6$ or $-1 \leq x \leq 1$ or $x \geq 6$

b

It might be helpful to re-order the brackets into order of size of the root.

2 At these values one factor = 0
This means $f(x) = 0$ and the graph crosses the x-axis.

All four factors are negative.

Three negative and one positive factor.

Two negative and two positive factors.

One negative and three positive factors.

All four factors are positive.

3 Include the critical points as the inequality is ≥ 0 and so includes 0

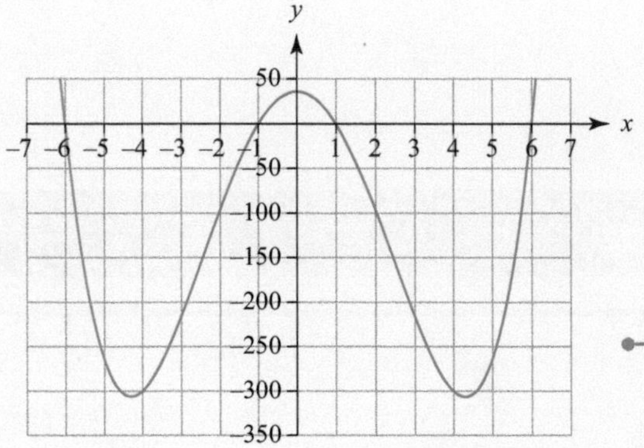

2 $x^4 - 37x^2 + 36 \geq 0$ when the curve intersects or is above the x-axis.
You can see that this is when $x \leq -6$ or $-1 \leq x \leq 1$ or $x \geq 6$

It is sometimes necessary to find inequalities for functions with real roots.

Example 6

The function $f(x) = \dfrac{x^2 - 4x + 4}{x}$ intersects the straight line $y = k$

a Form a quadratic equation in x and k

b Hence find the values of k for which $f(x)$ has real roots.

c Use a graphical calculator or graph sketching software on your computer to confirm your answer.

a $k = \dfrac{x^2 - 4x + 4}{x}$

$kx = x^2 - 4x + 4$

$x^2 - (k+4)x + 4 = 0$

1 Multiply both sides by x and simplify to form a quadratic equation in x

(*Continued on the next page*)

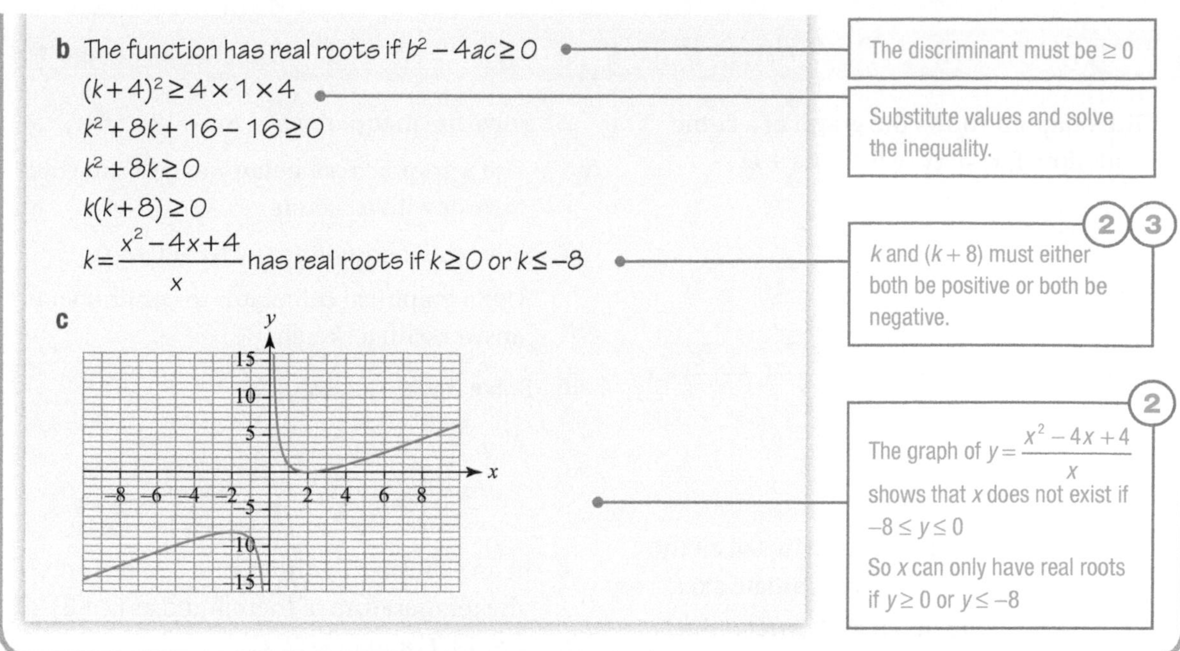

b The function has real roots if $b^2 - 4ac \geq 0$

$(k+4)^2 \geq 4 \times 1 \times 4$

$k^2 + 8k + 16 - 16 \geq 0$

$k^2 + 8k \geq 0$

$k(k+8) \geq 0$

$k = \dfrac{x^2 - 4x + 4}{x}$ has real roots if $k \geq 0$ or $k \leq -8$

The discriminant must be ≥ 0

Substitute values and solve the inequality.

k and $(k+8)$ must either both be positive or both be negative. (2) (3)

c

(2) The graph of $y = \dfrac{x^2 - 4x + 4}{x}$ shows that x does not exist if $-8 \leq y \leq 0$

So x can only have real roots if $y \geq 0$ or $y \leq -8$

You can solve double inequalities using the same methods.

Example 7

Find the values of x for which $0 \leq \dfrac{4+x}{3-x} < 1$

Consider $0 \leq \dfrac{4+x}{3-x}$

Critical values are $x = -4$ and $x = 3$

For $x < -4$, $\dfrac{4+x}{3-x}$ is negative.

For $-4 < x < 3$, $\dfrac{4+x}{3-x}$ is positive.

For $x > 3$, $\dfrac{4+x}{3-x}$ is negative.

(2) Consider the signs of the numerator and denominator for each set of values.

So $\dfrac{4+x}{3-x} \geq 0$ when $-4 \leq x < 3$

Do not include 3 as $\dfrac{4+x}{3-x}$ is not defined at $x = 3$

Now consider $\dfrac{4+x}{3-x} < 1$

When $x < 3$

$4 + x < 3 - x$

$x < -\dfrac{1}{2}$

$(3 - x)$ is positive.

When $x > 3$

$4 + x > 3 - x$

$x > -\dfrac{1}{2}$, but we have the more restrictive condition $x > 3$

(1) $(3 - x)$ is negative so reverse the inequality.

So $\dfrac{4+x}{3-x} < 1$ when $x < -\dfrac{1}{2}$ or $x > 3$

Full solution: $0 \leq \dfrac{4+x}{3-x} < 1$ when $-4 \leq x < -\dfrac{1}{2}$

(3) Find the values which satisfy both inequalities.

Exercise 2.2B Reasoning and problem-solving

1 The diagram shows the graph of a cubic function, $f(x) = Ax^3 + Bx^2 + Cx + D$

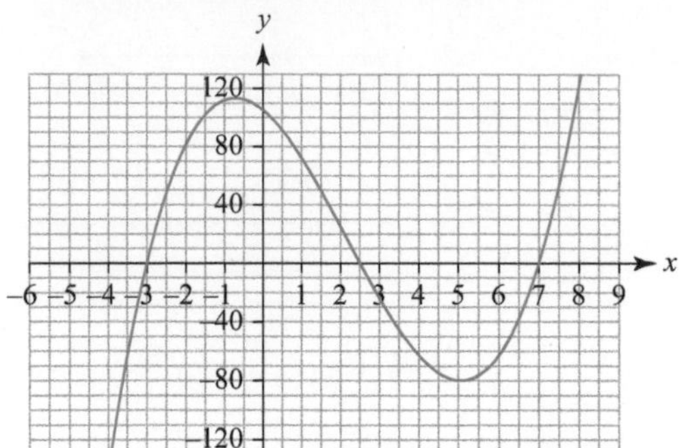

a Write down the coordinates of all the intercepts with the coordinate axes.

b Hence write down and expand the equation of the function.

c Write down the values of A, B, C and D

d Estimate and write down the coordinates of the turning points.

e Write down the ranges of values of x where $f(x) < 0$

2 The diagram shows the graph of a quartic function, $f(x) = Ax^4 + Bx^3 + Cx^2 + Dx + E$

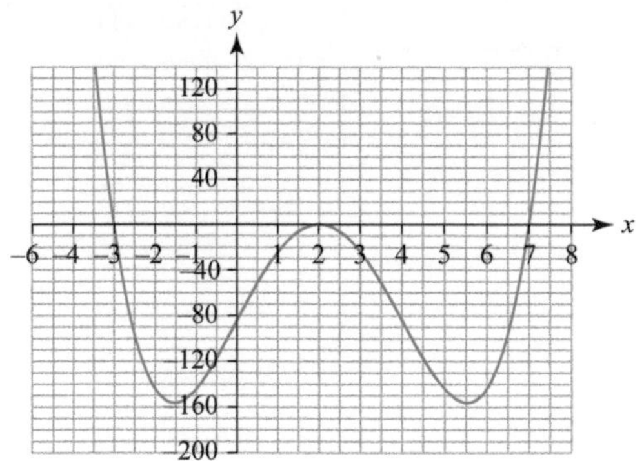

a Write down the coordinates of all the intercepts with the coordinate axes. Explain why there are only three x-intercepts even though this quartic has four real roots.

b Hence write down and expand the equation of the function.

c Write down the values of A, B, C, D and E

d Estimate and write down the coordinates of the turning points.

e Write down the ranges of values of x where $f(x) > 0$

3 Solve the inequality $x^3 - 8x^2 + 12x \geq 0$

Use a graphical calculator to confirm your answer with a sketch.

4 Solve the inequality $x^4 - 20x^2 + 64 \leq 0$

Use a graphical calculator to confirm your answer with a sketch.

5 Solve these inequalities.

a $\frac{x+3}{x-2} < x+3$ b $\frac{x-1}{x+4} > x - \frac{1}{4}$

c $\frac{x-5}{x-4} \geq 5 - x$ d $\frac{2x-3}{x+1} \leq x+9$

6 In an experiment, three students recorded the temperature of their liquid as $(x+3)$ °C, $(x-2)$ °C and $(x-7)$ °C.

For what values of x is the product of these temperatures positive?

7 $f(x) \equiv x^4 - 34x^2 + 225$. Find the range of values of x for which $f(x) \leq 0$

8 Find the range of values for which

$10 - 2x < \frac{8}{x}$

9 a Find the range of values of y for which the function $y = \frac{x^2 - 3x + 2}{x}$ has real roots.

b Use a graphical calculator or graph sketching software on your computer to confirm your results.

10 a Find the range of values of x for which the function $y = \frac{2x^2 - 5x - 12}{x+2}$ is positive.

b Use a graphical calculator or graph sketching software to confirm your answer.

11 a For all values of θ, $-1 \leq \sin\theta \leq 1$
Hence find the values of x such that $\sin\theta = \frac{x-3}{x+3}$ has a solution in θ

b Confirm your answer using a sketch.

12 If x is any positive number, prove that the sum of the number and its reciprocal cannot be less than 2

Full A Level

2.3 Summing series and the method of differences

Fluency and skills

If you want to write the sum of a series of n terms $u_1, u_2, \ldots, u_r \ldots u_{n-1}, u_n$, you can use sigma notation, Σ

Σ means 'find the sum of all these terms.'

$\sum_{r=1}^{r=n} u_r$ or $\sum_1^n u_r$ means 'find the sum of all the terms from u_1 to u_n.'

u_r is the general term and, if all terms are defined algebraically, u_r is a function of the variable.

Key point

If $S_n = u_1 + u_2 + \ldots + u_r + \ldots + u_{n-1} + u_n$, then $S_n = \sum_1^n u_r$

S_n means 'the sum of these n terms.'

If $u_r = r^2$ then $S_n = 1^2 + 2^2 + 3^2 + \ldots + r^2 \ldots + n^2 = \sum_1^n r^2$

You need to know the formulae for the sums of integers, squares and cubes.

Finding the sum of integers from 1 to n

$$\sum_1^n r = 1 \quad + \quad 2 \quad + \quad 3 \quad + \ldots + \quad (n-2) \quad + \quad (n-1) \quad + \quad n$$

Write the same sequence in reverse:

$$\sum_1^n r = n \quad + (n-1) + (n-2) + \ldots + \quad 3 \quad + \quad 2 \quad + \quad 1$$

Add the two together:

$$2\sum_1^n r = (n+1) + (n+1) + (n+1) + \ldots + (n+1) \quad + \quad (n+1) \quad + (n+1)$$

Every term is equal to $n + 1$. Try it with some numbers.

There are n lots of $(n + 1)$ so $2\sum_1^n r = n(n+1)$ and so $\sum_1^n r = \frac{n(n+1)}{2}$

You will be expected to remember and be able to quote without proof, the formulae for Σr

Key point

$$\sum_1^n r = 1 + 2 + \ldots + n = \frac{n(n+1)}{2}$$

Example 1

Find the value of $2 + 4 + 6 + 8 + \ldots + 40$

$2+4+6+8+\ldots+40 = 2(1+2+3+4+\ldots+20)$ — Rewrite as a sum of integers.

$= 2\sum_1^{20} r$

$= 2 \times \frac{20(20+1)}{2} = 2 \times \frac{20 \times 21}{2}$ — Substitute $n = 20$ into the formula.

$= 420$

Sum of squares

You can derive the formula for the sum of squares using the method of differences. You will see how to do this later in this section.

Key point

$$\sum_{1}^{n} r^2 \equiv \frac{n(n+1)(2n+1)}{6}$$

Sum of cubes

Compare sums of r and r^3 for different values of n

n	Σr	Σr^3
1	1	1
2	$1+2=3$	$1+8=9$
3	$1+2+3=6$	$1+8+27=36$
4	$1+2+3+4=10$	$1+8+27+64=100$
5	$1+2+3+4+5=15$	$1+8+27+64+125=225$

For each value of n, $\Sigma r^3 = (\Sigma r)^2$

Hence $\Sigma n^3 = (\Sigma n)^2 = \left(\frac{n(n+1)}{2}\right)^2 = \frac{n^2(n+1)^2}{4}$

Key point

$$\sum_{1}^{n} r^3 = \frac{n^2(n+1)^2}{4}$$

Example 2

Evaluate $\sum_{5}^{n} r(r+4)$

$$\sum_{5}^{n} r(r+4) = 5(9)+6(10)+\ldots+n(n+4)$$

Write out some of the sequence to help you understand it.

$$= \sum_{1}^{n} r(r+4) - \sum_{1}^{4} r(r+4)$$

This is the same as $\sum_{1}^{n} r(r+4)$ without the first four terms.

$$= \sum_{1}^{n} r^2 + 4\sum_{1}^{n} r - (1(5)+2(6)+3(7)+4(8))$$

$$= \left[\frac{n}{6}(n+1)(2n+1)\right] + 4\frac{n(n+1)}{2} - 70$$

Use the standard formulae.

$$= n(n+1)\left[\frac{1}{6}(2n+1)+\frac{4}{2}\right] - 70$$

$$= n(n+1)\left[\frac{2n+1+12}{6}\right] - 70$$

$$= \frac{n(n+1)(2n+13)-420}{6}$$

$$= \frac{(n-4)(2n^2+23n+105)}{6}$$

Simplify.

Example 3

a Find a formula for the sum to n terms of the series

$1\times2\times4+2\times3\times5+3\times4\times6+\ldots$

b Find the sum of the first 10 terms and check your formula.

a The general term is $r(r+1)(r+3)$

$$\sum_1^n r(r+1)(r+3) \equiv \sum_1^n (r^3+4r^2+3r)$$

Rewrite in terms of sums you know.

$$\equiv \sum_1^n r^3 + 4\sum_1^n r^2 + 3\sum_1^n r$$

$\Sigma 4n^2 \equiv 4\Sigma n^2$ and $\Sigma 3n \equiv 3\Sigma n$

$$\equiv \frac{n^2(n+1)^2}{4} + 4\left[\frac{n}{6}(n+1)(2n+1)\right] + 3\frac{n(n+1)}{2}$$

Substitute formulae for Σn^3, Σn^2 and Σn, and simplify.

$$\equiv \frac{n(n+1)}{12}[3n(n+1)+8(2n+1)+18]$$

$$\equiv \frac{n(n+1)}{12}[3n^2+19n+26]$$

$$\equiv \frac{n(n+1)(n+2)(3n+13)}{12}$$

b $$\sum_1^{10} r(r+1)(r+3) \equiv \frac{10\times11}{12}(3\times10^2+19\times10+26)$$

Find the sum using the formula.

$$=4730$$

$$(1\times2\times4)+(2\times3\times5)+\ldots+(10\times11\times13)$$

$$=8+30+\ldots+1430=4730$$

Calculate the first 10 terms and add to check your formula.

If a series has alternate positive and negative terms, $\Sigma f(x)$ can be written with a $(-1)^n$ multiple in it.

For example, $-1+2-3+4-\ldots$ for n terms, is expressed as $\sum_1^n (-1)^n r$

The method of differences

There is no general method for summing series.

Sometimes it is not possible to express Σn as an algebraic expression.

For example $1+\frac{1}{2}+\frac{1}{3}+\frac{1}{4}+\frac{1}{5}+\ldots+\frac{1}{n}$

However, if the general term of a function can be expressed as $f(r+1)-f(r)$, you can find the sum of the series using the **method of differences**.

Example 4

Find the sum to n terms of the series $\frac{1}{r} - \frac{1}{r+1}$

$$\sum_1^n \frac{1}{r(r+1)} = \frac{1}{1} - \frac{1}{2}$$
$$+\frac{1}{2} - \frac{1}{3}$$
$$+\frac{1}{3} - \frac{1}{4}$$
$$\downarrow \quad \downarrow$$
$$+\frac{1}{n} - \frac{1}{(n+1)}$$

Write the 'differences' for successive terms vertically and eliminate terms wherever possible.

$$\text{Thus } \sum_1^n \frac{1}{r(r+1)} = 1 - \frac{1}{(n+1)} = \frac{n}{(n+1)}$$

Collect up the remaining terms and simplify.

Example 5

The expression $\frac{18}{(3r-2)(3r+1)(3r+4)}$ can be written as $\frac{1}{(3r-2)} - \frac{2}{(3r+1)} + \frac{1}{(3r+4)}$

Find the sum to n terms of the series $\frac{18}{(3r-2)(3r+1)(3r+4)}$

$$\sum_1^n \frac{18}{(3r-2)(3r+1)(3r+4)} \equiv \sum_1^n \left[\frac{1}{(3r-2)} - \frac{2}{(3r+1)} + \frac{1}{(3r+4)}\right]$$
$$\equiv \frac{1}{1} - \frac{2}{4} + \frac{1}{7}$$
$$= \frac{1}{4} - \frac{2}{7} + \frac{1}{10}$$
$$= \frac{1}{7} - \frac{2}{10} + \frac{1}{13}$$
$$= \frac{1}{10} - \frac{2}{13} + \frac{1}{16}$$
$$\downarrow \quad \downarrow \quad \downarrow$$
$$= \frac{1}{3n-8} - \frac{2}{3n-5} + \frac{1}{3n-2}$$
$$= \frac{1}{3n-5} - \frac{2}{3n-2} + \frac{1}{3n+1}$$
$$= \frac{1}{3n-2} - \frac{2}{3n+1} + \frac{1}{3n+4}$$

Write the 'differences' for successive terms vertically and eliminate terms wherever possible.

$$\text{Hence } \sum_1^n \frac{18}{(3r-2)(3r+1)(3r+4)} \equiv \frac{1}{1} - \frac{2}{4} + \frac{1}{4} + \frac{1}{3n+1} - \frac{2}{3n+1} + \frac{1}{3n+4}$$
$$\equiv \frac{3}{4} - \frac{1}{3n+1} + \frac{1}{3n+4}$$

Collect up the remaining terms and simplify.

Exercise 2.3A Fluency and skills

PURE

1 Write out these series.

a $\sum_{1}^{5} r^3$ b $\sum_{4}^{n} r^2$

c $\sum_{1}^{n}(r^2-2r)$ d $\sum_{1}^{n}\frac{1}{m+2}$

e $\sum_{1}^{6}(-1)^r r^3$ f $\sum_{n-3}^{n} r(r+1)$

2 Use Σ notation to write these sums.

a $1+3+5+\ldots+31$

b $1^5+2^5+3^5+\ldots+n^5$

c $1+\frac{1}{2}+\frac{1}{3}+\frac{1}{4}+\ldots+\frac{1}{n+1}$

d $3-6+9-12+\ldots-42$

e $1\times3+2\times4+3\times5+\ldots$ for n terms

f $\frac{1\times4}{3}-\frac{3\times5}{4}+\frac{5\times6}{5}-\frac{7\times7}{6}+\ldots$ for n terms

3 Use standard results to prove each of these results

a $\sum_{r=1}^{n}(r^2+6r-3)=\frac{1}{6}n(2n^2+21n+1)$

b $\sum_{r=1}^{n}(2r^2-7r)=\frac{1}{6}n(n+1)(4n-19)$

c $\sum_{r=1}^{n}(8-5r^2)=-\frac{1}{6}n(10n^2+15n-43)$

d $\sum_{r=1}^{n}(r+1)^2=\frac{1}{6}n(2n^2+9n+13)$

e $\sum_{r=1}^{n}(r^3+2r+3)=\frac{1}{4}n(n^3+2n^2+5n+16)$

f $\sum_{r=1}^{n}(2r^3+3r^2)=\frac{1}{2}n(n+1)(n^2+3n+1)$

g $\sum_{r=1}^{n}(r^3-2r^2+3r)=\frac{1}{12}n(n+1)(3n^2-5n+14)$

h $\sum_{r=1}^{n}(r+2)^3=\frac{1}{4}n(n^3+10n^2+37n+60)$

4 a How many terms are there in the series $\sum_{1}^{3n}(r^2-r+3)$?

b Write and simplify the $(2n+1)$th term.

5 Find the sums of these series.

a $1+2+3+\ldots+3n$

b $1^2+2^2+3^2+\ldots+(2n-1)^2$

c $1^3+2^3+3^3+\ldots+(2n-1)^3$

d $1+3+5+\ldots+(2n-1)$

e $1\times2+2\times3+3\times4+\ldots+(n)(n+1)$

f $0+2+6+12+\ldots+(n^2-n)$

6 a Find a formula for the sum to n terms of the series

$0\times1+1\times2+2\times3+\ldots+(n-1)n$

b Find the sum of the first 10 terms and check your formula.

7 Find the rth term and the sum to n terms of these series.

a $1\times2\times3+2\times3\times4+3\times4\times5+\ldots$

b $1\times4+3\times5+5\times6+\ldots$

c $1\times3\times5+2\times4\times6+3\times5\times7+\ldots$

d $2+6+10+16+\ldots$

e $2+5+10+17+\ldots$

8 Evaluate

a $\sum_{5}^{n}(2r+3)$ b $\sum_{3}^{n-2}(r^2-r)$

c $\sum_{n}^{2n+1}(r^3+3r)$

9 Find the sum of each series, using the method of differences where appropriate.

a $\sum_{1}^{n}(2-3r)$

b $\sum_{1}^{n}\left(\frac{1}{r}-\frac{1}{r+1}\right)$

c $\sum_{1}^{n}\left(\frac{1}{r}-\frac{1}{r+2}\right)$

d $\sum_{1}^{n}\left(\frac{1}{r}-\frac{1}{r+3}\right)$

e $\sum_{1}^{n}\left(\frac{1}{2r-1}-\frac{1}{2r+1}\right)$

f $\sum_{1}^{n}\left(\frac{1}{r+2}-\frac{2}{r+3}+\frac{1}{r+4}\right)$

Reasoning and problem-solving

Strategy

To solve problems involving sums of series

1. Rearrange the expression into multiples of integers, squares and cubes, or into the form $f(x+1) - f(x)$
2. Substitute standard formulae or terms as required.
3. Simplify the resulting expression for the sum.

Example 6

a Simplify the expression $(2r+1)^3 - (2r-1)^3$

b Hence, use the method of differences to show that $\sum_{1}^{n} r^2 \equiv \frac{n(n+1)(2n+1)}{6}$

a $(2r+1)^3 - (2r-1)^3 \equiv (8r^3 + 12r^2 + 6r + 1) - (8r^3 - 12r^2 + 6r - 1)$

$\equiv 24r^2 + 2$

b $24r^2 + 2 \equiv (2r+1)^3 - (2r-1)^3$

$$r^2 \equiv \frac{1}{24}[(2r+1)^3 - (2r-1)^3 - 2]$$

$$\sum_{1}^{n} r^2 \equiv \frac{1}{24}(3^3 - 1^3 - 2)$$
$$+ \frac{1}{24}(5^3 - 3^3 - 2)$$
$$+ \frac{1}{24}(7^3 - 5^3 - 2)$$
$$+ \frac{1}{24}(9^3 - 7^3 - 2)$$
$$\downarrow \quad \downarrow \quad \downarrow$$
$$+ \frac{1}{24}[(2n-1)^3 - (2n-3)^3 - 2]$$
$$= \frac{1}{24}[(2n+1)^3 - (2n-1)^3 - 2]$$

Write the 'differences' for successive series of terms vertically and eliminate terms wherever possible. (2)

Hence $\sum_{1}^{n} r^2 \equiv \frac{1}{24}[(2n+1)^3 - 1^3 + n \times (-2)]$

Collect up the remaining terms and simplify. (3)

$$\equiv \frac{1}{24}[8n^3 + 12n^2 + 6n + 1 - 1 - 2n]$$
$$\equiv \frac{1}{24}[8n^3 + 12n^2 + 4n]$$
$$\equiv \frac{n}{6}[2n^2 + 3n + 1]$$
$$\equiv \frac{n}{6}(n+1)(2n+1)$$

Example 7

a Show that $\dfrac{2}{r(r+1)(r+2)}$ can be written as $\dfrac{1}{r}-\dfrac{2}{r+1}+\dfrac{1}{r+2}$

b Hence or otherwise find the sum to n terms of the series

$$\frac{1}{3}+\frac{1}{12}+\frac{1}{30}+\frac{1}{60}+\ldots+\frac{1}{n\times(n+1)\times(n+2)}$$

c Use your result to deduce $\displaystyle\sum_{1}^{\infty}\frac{2}{n(n+1)(n+2)}$

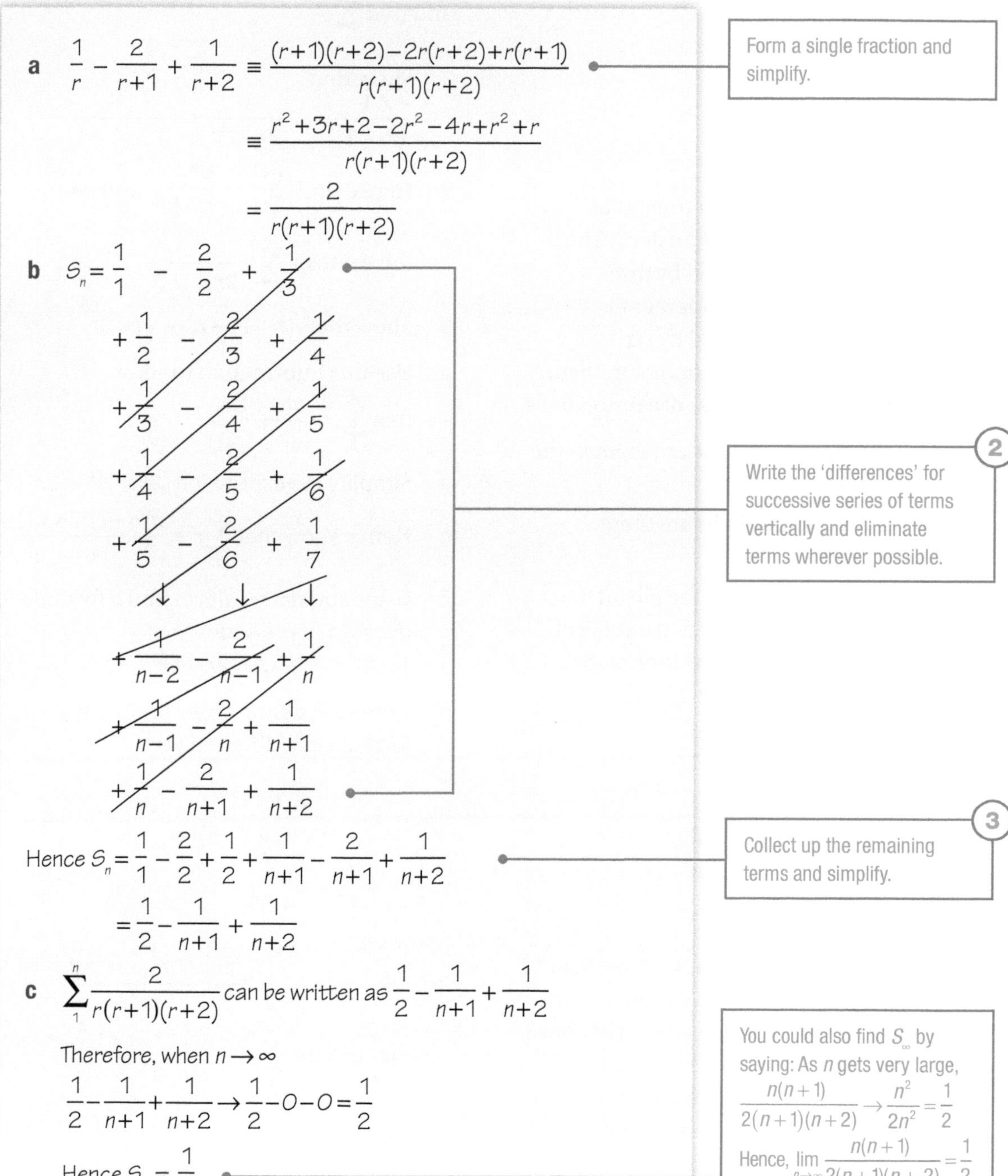

a $$\frac{1}{r}-\frac{2}{r+1}+\frac{1}{r+2}\equiv\frac{(r+1)(r+2)-2r(r+2)+r(r+1)}{r(r+1)(r+2)}$$

$$\equiv\frac{r^2+3r+2-2r^2-4r+r^2+r}{r(r+1)(r+2)}$$

$$=\frac{2}{r(r+1)(r+2)}$$

Form a single fraction and simplify.

b $$S_n=\frac{1}{1}-\frac{2}{2}+\frac{1}{3}$$
$$+\frac{1}{2}-\frac{2}{3}+\frac{1}{4}$$
$$+\frac{1}{3}-\frac{2}{4}+\frac{1}{5}$$
$$+\frac{1}{4}-\frac{2}{5}+\frac{1}{6}$$
$$+\frac{1}{5}-\frac{2}{6}+\frac{1}{7}$$
$$\downarrow\quad\downarrow\quad\downarrow$$
$$+\frac{1}{n-2}-\frac{2}{n-1}+\frac{1}{n}$$
$$+\frac{1}{n-1}-\frac{2}{n}+\frac{1}{n+1}$$
$$+\frac{1}{n}-\frac{2}{n+1}+\frac{1}{n+2}$$

Write the 'differences' for successive series of terms vertically and eliminate terms wherever possible.

Hence $$S_n=\frac{1}{1}-\frac{2}{2}+\frac{1}{2}+\frac{1}{n+1}-\frac{2}{n+1}+\frac{1}{n+2}$$

$$=\frac{1}{2}-\frac{1}{n+1}+\frac{1}{n+2}$$

Collect up the remaining terms and simplify.

c $\displaystyle\sum_{1}^{n}\frac{2}{r(r+1)(r+2)}$ can be written as $\dfrac{1}{2}-\dfrac{1}{n+1}+\dfrac{1}{n+2}$

Therefore, when $n\to\infty$

$$\frac{1}{2}-\frac{1}{n+1}+\frac{1}{n+2}\to\frac{1}{2}-0-0=\frac{1}{2}$$

Hence $S_\infty=\dfrac{1}{2}$

You could also find S_∞ by saying: As n gets very large, $\dfrac{n(n+1)}{2(n+1)(n+2)}\to\dfrac{n^2}{2n^2}=\dfrac{1}{2}$

Hence, $\displaystyle\lim_{n\to\infty}\frac{n(n+1)}{2(n+1)(n+2)}=\frac{1}{2}$

PURE

Exercise 2.3B Reasoning and problem-solving

1 The sum to n terms of a series is $2n^2 + 4n$

a By considering S_1, S_2 and S_3, find the first three terms of the series.

b By considering S_{n-1} and S_n, find the nth term.

2

Admiral Nelson's ship has a triangular pile of cannonballs stacked on deck. The top cannon ball is supported by three cannonballs in the layer underneath it. These three are supported by six cannonballs in the layer underneath them, and so on. There are n layers of cannon balls.

a How many cannon balls are there in the bottom layer?

b How many cannonballs are there altogether?

c Calculate the weight of the pile of 'six pounder' balls contained in a stack of 10 layers. A 'six pounder' cannon ball weighs approximately 2.722 kg.

3

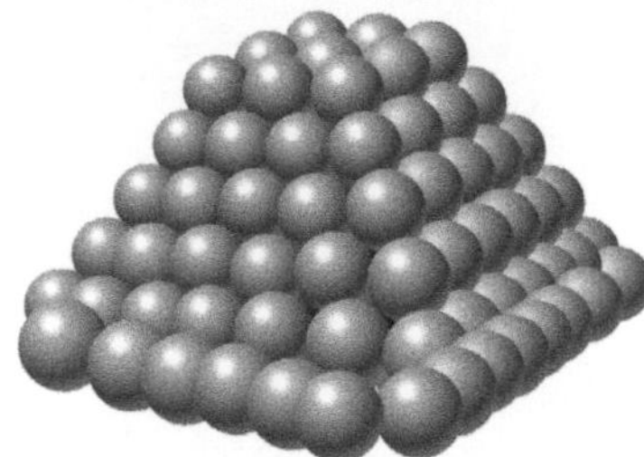

The Spanish Armada's cannon balls are in square pyramids. The base is a square with $2n$ cannonballs in each side, the layer above has $(2n - 1)$ cannonballs in each side, and so on. There are n layers in total. How many cannon balls are there in this pyramid?

4 a Write down the first four terms and the nth term of the series whose general term is $\left(\frac{1}{r^2} - \frac{1}{(r+1)^2}\right)$

b Find $\sum_1^n \left(\frac{1}{r^2} - \frac{1}{(r+1)^2}\right)$ and hence find the sum to infinity.

5 Write down the general term in the series $S = 1(m) + 2(m-1) + 3(m-2) + \ldots + (m)(1)$ and find $\sum_1^m S$

6 a Show that
$$\frac{1}{(2r-1)^2} - \frac{1}{(2r+1)^2} \equiv \frac{8r}{((2r)^2 - 1)^2}$$

b Hence find $\sum_1^n \frac{8r}{((2r)^2-1)^2}$ and deduce the value of $\sum_1^n \frac{r}{((2r)^2-1)^2}$

7 a Show that $r(r+1) - r(r-1) \equiv 2r$

b Use this information to show that $\sum_1^n r \equiv \frac{n(n+1)}{2}$

8 a Simplify the expression $(2r+1)^3 - (2r-1)^3$

b Hence show that $\sum_1^n r^2 \equiv \frac{n(n+1)(2n+1)}{6}$

9 a Use standard results to find a formula for the sum of the series $1^2 + 3^2 + 5^2 + \ldots + (2r-1)^2$

b Confirm your result by finding the formula in a different way.

10 Show that $\frac{2(x-4)}{x(x+2)(x+4)}$ can be written as $-\frac{1}{x} + \frac{3}{x+2} - \frac{2}{x+4}$

11 Show that
$$2r^2 + 3r + \frac{1}{r} - \frac{1}{r+1} \equiv \frac{2r^4 + 5r^3 + 3r^2 + 1}{r(r+1)}$$
and hence find $\sum_1^n \frac{2r^4 + 5r^3 + 3r + 1}{r(r+1)}$

Full A Level

2.4 Proof by induction

Fluency and skills

e Maths 1.1

or a minder of ethods of oof.

There are a number of different methods of directly proving a mathematical statement. **Proof by induction** is a method that is generally used to prove a mathematical statement for all natural numbers (positive integers).

Other methods of direct proof are **proof by deduction** and **proof by exhaustion.**

It is a powerful method that can be used in many different contexts. The principle behind proof by induction is that if you prove a statement is true when $n = 1$, and if you can prove it is true for $n = k + 1$ by assuming it is true for any $n = k$, then you can deduce that the statement must be true for all $n \in \mathbb{N}$

$\mathbb{N}$ means the set of natural numbers: these are the positive integers 1, 2, 3, ...

Key point

The three key steps to a proof by induction are:

1 Prove the statement is true for $n = 1$

2 Assume the statement is true for $n = k$ and use this to prove the statement is true for $n = k + 1$

3 Write a conclusion.

You must always explain the steps fully and end your proof with a conclusion.

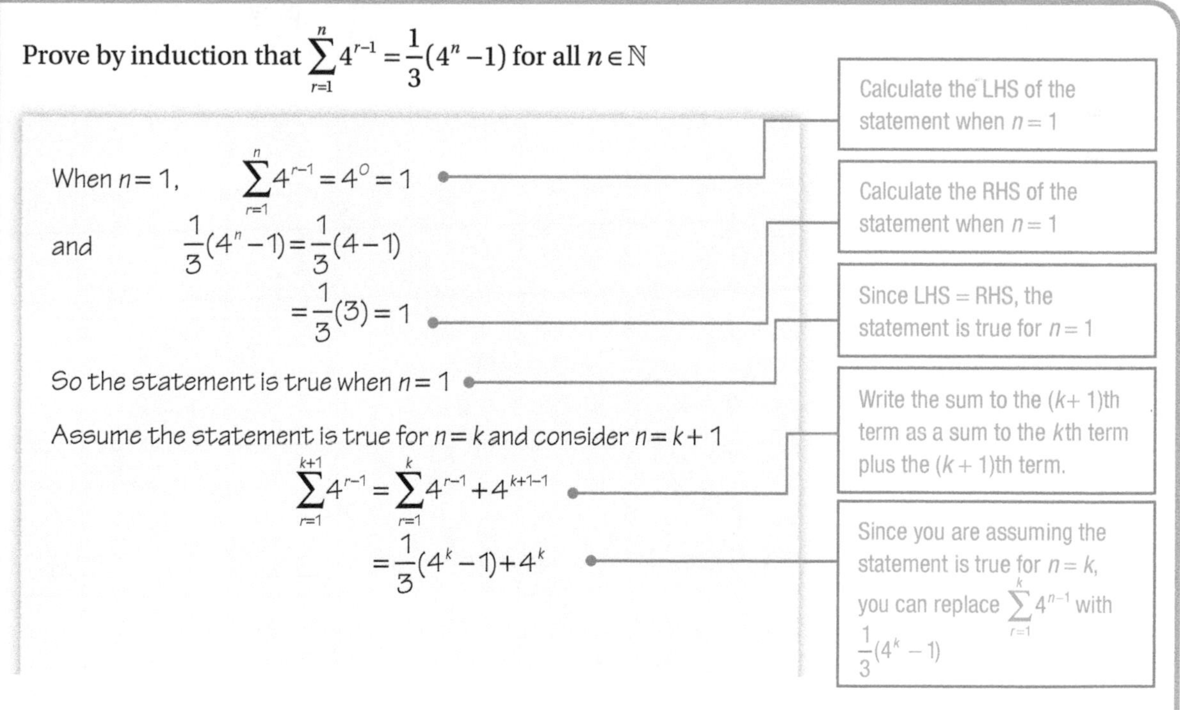

Example 1

Prove by induction that $\sum_{r=1}^{n} 4^{r-1} = \frac{1}{3}(4^n - 1)$ for all $n \in \mathbb{N}$

When $n = 1$, $\sum_{r=1}^{n} 4^{r-1} = 4^0 = 1$ — Calculate the LHS of the statement when $n = 1$

and $\frac{1}{3}(4^n - 1) = \frac{1}{3}(4 - 1)$

$= \frac{1}{3}(3) = 1$ — Calculate the RHS of the statement when $n = 1$

So the statement is true when $n = 1$ — Since LHS = RHS, the statement is true for $n = 1$

Assume the statement is true for $n = k$ and consider $n = k + 1$

$\sum_{r=1}^{k+1} 4^{r-1} = \sum_{r=1}^{k} 4^{r-1} + 4^{k+1-1}$ — Write the sum to the $(k+1)$th term as a sum to the kth term plus the $(k+1)$th term.

$= \frac{1}{3}(4^k - 1) + 4^k$ — Since you are assuming the statement is true for $n = k$, you can replace $\sum_{r=1}^{k} 4^{n-1}$ with $\frac{1}{3}(4^k - 1)$

(*Continued on the next page*)

$$=\frac{1}{3}(4^k-1+3(4^k))=\frac{1}{3}(4(4^k)-1)$$

Collect like terms and then use index laws.

$$=\frac{1}{3}(4^{k+1}-1)$$

This is $\frac{1}{3}(4^n-1)$ with n replaced by $k+1$

So the statement is true when $n=k+1$

The statement is true for $n=1$ and by assuming it is true for $n=k$ it is shown to be true for $n=k+1$

Therefore, by mathematical induction, it is true for all $n\in\mathbb{N}$

You must always write a conclusion.

Example 2

Prove by induction that $\sum_{r=1}^{n} r^2=\frac{1}{6}n(n+1)(2n+1)$ for all $n\in\mathbb{N}$

When $n=1$, $\sum_{r=1}^{n} r^2=1^2=1$

Calculate the LHS of the statement when $n=1$

and $\frac{1}{6}n(n+1)(2n+1)=\frac{1}{6}\times1\times(1+1)(2\times1+1)=\frac{1}{6}(2)(3)$
$=1$

Calculate the RHS of the statement when $n=1$

So the statement is true when $n=1$

Since LHS = RHS, the statement is true for $n=1$

Assume statement is true for $n=k$ and substitute $n=k+1$ into the formula:

$$\sum_{r=1}^{k+1} r^2=\sum_{r=1}^{k} r^2+(k+1)^2$$

Write the sum to the $(k+1)$th term as sum to the kth term plus the $(k+1)$th term.

$$=\frac{1}{6}k(k+1)(2k+1)+(k+1)^2$$

Since you are assuming the statement is true for $n=k$, you can replace $\sum_{r=1}^{k} r^2$ with $\frac{1}{6}k(k+1)(2k+1)$

$$=\frac{1}{6}(k+1)[k(2k+1)+6(k+1)]$$

$$=\frac{1}{6}(k+1)(2k^2+k+6k+6)$$

$$=\frac{1}{6}(k+1)(2k^2+7k+6)$$

$$=\frac{1}{6}(k+1)(k+2)(2k+3)$$

Look for common factors – avoid multiplying out brackets unless necessary.

$$=\frac{1}{6}(k+1)((k+1)+1)(2(k+1)+1)$$

This is $\frac{1}{6}n(n+1)(2n+1)$ with n replaced by $k+1$

So statement is true when $n=k+1$

The statement is true for $n=1$ and by assuming it is true for $n=k$ it is shown to be true for $n=k+1$

Therefore, by mathematical induction, it is true for all $n\in\mathbb{N}$

You must always write a conclusion.

Exercise 2.4A Fluency and skills

PURE

1 Use proof by induction to prove these statements for all $n \in \mathbb{N}$

a $\sum_{r=1}^{n} 1 = n$

b $\sum_{r=1}^{n} r = \frac{1}{2}n(n+1)$

c $\sum_{r=1}^{n} (2r+3) = n(n+4)$

d $\sum_{r=1}^{n} r(r+1) = \frac{1}{3}n(n+1)(n+2)$

e $\sum_{r=1}^{n} (r-1)^2 = \frac{1}{6}n(n-1)(2n-1)$

f $\sum_{r=1}^{n} (r+1)(r-1) = \frac{1}{6}n(2n+5)(n-1)$

2 Use induction to prove these sums for all natural numbers, n

a $4 + 9 + 16 + 25 + \ldots + (n+1)^2 = \frac{1}{6}n(2n^2 + 9n + 13)$

b $1 + 5 + 25 + 125 + \ldots + 5^{(n-1)} = \frac{1}{4}(5^n - 1)$

3 Prove by induction that $\sum_{r=1}^{n} r^3 = \frac{1}{4}n^2(n+1)^2$ for all $n \in \mathbb{N}$

4 Prove by induction that $\sum_{r=1}^{2n} r = n(2n+1)$ for all $n \in \mathbb{N}$

5 Prove by induction that $\sum_{r=1}^{2n} r^2 = \frac{1}{3}n(2n+1)(4n+1)$ for all $n \in \mathbb{N}$

6 Use induction to prove these statements for all $n \in \mathbb{N}$

a $\sum_{r=1}^{n} 2^r = 2(2^n - 1)$

b $\sum_{r=1}^{n} 3^r = \frac{3}{2}(3^n - 1)$

c $\sum_{r=1}^{n} 4^r = \frac{4}{3}(4^n - 1)$

d $\sum_{r=1}^{n} 2^{r-1} = 2^n - 1$

e $\sum_{r=1}^{n} 3^{r-1} = \frac{1}{2}(3^n - 1)$

f $\sum_{r=1}^{n} \left(\frac{1}{2}\right)^r = 1 - \left(\frac{1}{2}\right)^n$

7 Prove by induction that $\sum_{r=1}^{n} \frac{1}{r(r+1)} = \frac{n}{n+1}$ for all $n \in \mathbb{N}$

8 Prove by induction that $\sum_{r=2}^{n} \frac{1}{r(r-1)} = \frac{n-1}{n}$ for all $n \in \mathbb{N}$

9 Prove by induction that $\sum_{r=1}^{n} \frac{1}{r^2+2r} = \frac{n(3n+5)}{4(n+1)(n+2)}$ for all $n \in \mathbb{N}$

Reasoning and problem-solving

See Ch5.1

To see how to apply induction to matrix proofs.

As well as proving sums of series, you can use proof by induction in other contexts such as to prove an algebraic expression is divisible by a given integer.

Strategy

To prove an expression is divisible by a particular integer

1. Substitute $n = 1$ into the expression and show the result is divisible by the integer you are using.
2. Assume the expression is divisible by the integer for $n = k$ and substitute in $n = k + 1$
3. Separate the part you know to be divisible by the integer given.
4. Write the conclusion.

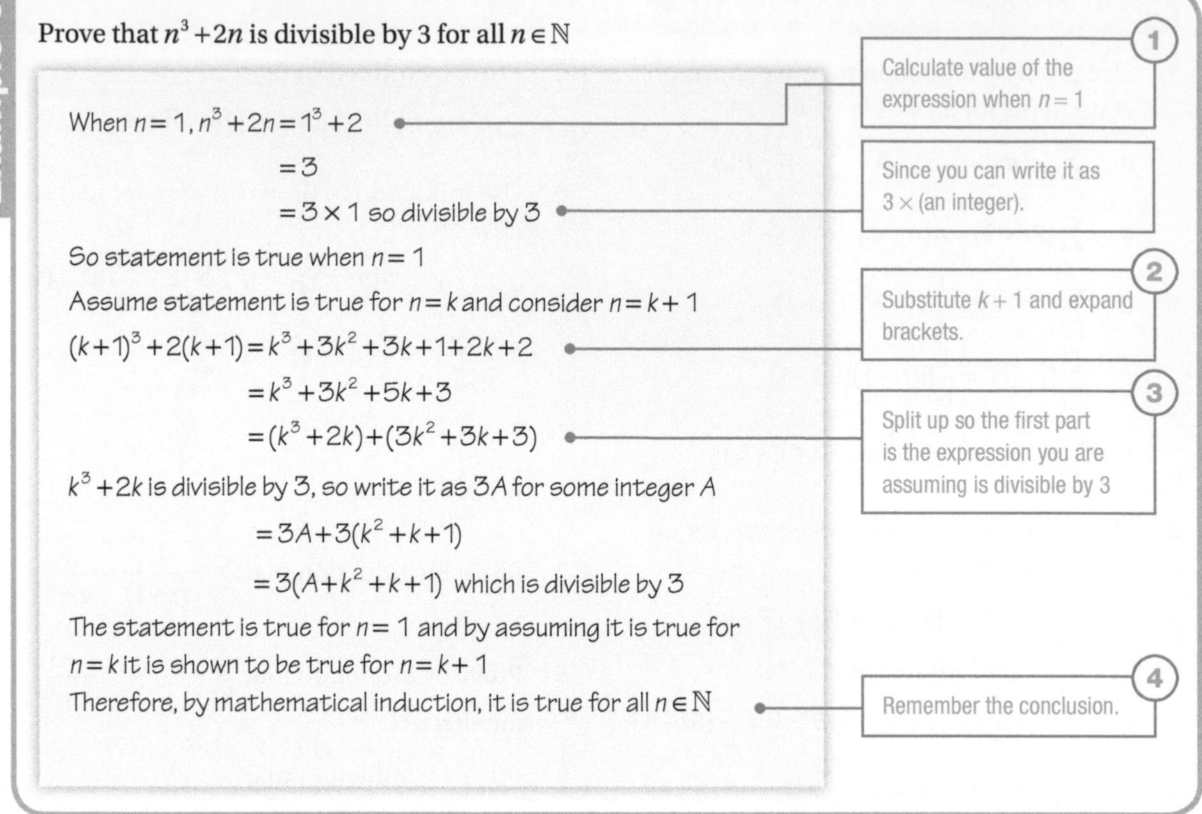

Example 3

Prove that n^3+2n is divisible by 3 for all $n \in \mathbb{N}$

When $n=1$, $n^3+2n=1^3+2$

$=3$

$=3\times 1$ so divisible by 3

So statement is true when $n=1$

Assume statement is true for $n=k$ and consider $n=k+1$

$(k+1)^3+2(k+1)=k^3+3k^2+3k+1+2k+2$

$=k^3+3k^2+5k+3$

$=(k^3+2k)+(3k^2+3k+3)$

k^3+2k is divisible by 3, so write it as $3A$ for some integer A

$=3A+3(k^2+k+1)$

$=3(A+k^2+k+1)$ which is divisible by 3

The statement is true for $n=1$ and by assuming it is true for $n=k$ it is shown to be true for $n=k+1$

Therefore, by mathematical induction, it is true for all $n \in \mathbb{N}$

1. Calculate value of the expression when $n=1$

Since you can write it as $3\times$ (an integer).

2. Substitute $k+1$ and expand brackets.

3. Split up so the first part is the expression you are assuming is divisible by 3

4. Remember the conclusion.

Example 4

Given that $f(n)=2^{n+1}+3^{2n-1}$

prove by induction that $f(n)$ is a multiple of 7 for all positive integers n

When $f(1)=2^{n+1}+3^{2n-1}=2^2+3^1$

$=4+3$

$=7$

$=7\times 1$ so a multiple of 7

Assume $f(k)$ is a multiple of 7 and consider $f(k+1)$

$f(k+1)=2^{k+1+1}+3^{2(k+1)-1}$

$=2^{k+2}+3^{2k+1}$

$=2(2^{k+1})+3^2(3^{2k-1})$

$=2(2^{k+1})+9(3^{2k-1})$

$=2(2^{k+1}+3^{2k-1})+7(3^{2k-1})$

$=2f(k)+7(3^{2k-1})$

1. Calculate the value of the expression when $n=1$

2. Substitute $k+1$ into expression.

Use index laws to write in terms of 2^{n+1} and 3^{2n-1}

3. Separate out $2\times f(k)$

(Continued on the next page)

which is a multiple of 7 since $f(k)$ and $7(3^{2k-1})$ are both multiples of 7

$f(1)$ is a multiple of 7 and by assuming that $f(k)$ is a multiple of 7 it is shown that $f(k+1)$ is a multiple of 7

Therefore, by mathematical induction, $f(n)$ is a multiple of 7 for all positive integers n

Write the conclusion. 4

PURE

Exercise 2.4B Reasoning and problem-solving

1 Use proof by induction to prove these statements for all $n \in \mathbb{N}$

a $n^2 + 3n$ is divisible by 2

b $5n^2 - n$ is divisible by 2

c $8n^3 + 4n$ is divisible by 12

d $11n^3 + 4n$ is divisible by 3

2 Prove by induction that $7n^2 + 25n - 4$ is divisible by 2 for all $n \in \mathbb{N}$

3 Prove by induction that $n^3 - n$ is divisible by 3 for all $n \geq 2$, $n \in \mathbb{N}$

4 Prove by induction that $10n^3 + 3n^2 + 5n - 6$ is divisible by 6 for all $n \in \mathbb{N}$

5 Use proof by induction to prove these statements for all $n \in \mathbb{N}$

a $6^n + 9$ is divisible by 5

b $3^{2n} - 1$ is divisible by 8

c $2^{3n+1} - 2$ is divisible by 7

6 Prove by induction that $5^n - 4n + 3$ is a multiple of 4 for all $n \in \mathbb{N}$

7 Prove by induction that $3^n + 2n + 7$ is a multiple of 4 for all $n \in \mathbb{N}$

8 Prove by induction that $7^n - 3n + 5$ is a multiple of 3 for all $n \in \mathbb{N}$

9 Use proof by induction to prove these statements for all $n \in \mathbb{N}$

a $\sum_{r=n+1}^{2n} r^2 = \frac{1}{6}n(2n+1)(7n+1)$

b $\sum_{r=n}^{2n} r^3 = \frac{3}{4}n^2(5n+1)(n+1)$

c $8^n - 5^n$ is divisible by 3 for all $n \in \mathbb{N}$

2.5 Maclaurin series 1

Fluency and skills

It is sometimes useful to be able to express a function as a series of terms. The **Maclaurin series** for a function f(x) is based on three assumptions:

- f(x) can be expanded as a **convergent** infinite series of terms
- each of the terms in f(x) can be differentiated
- each of the differentiated terms has a finite value when $x = 0$

A convergent series is one where an infinite number of terms has a finite sum.

The proof of these assumptions is outside the scope of this chapter.

With these assumptions, you can write

$\mathrm{f}(x) \equiv a_0 + a_1x + a_2x^2 + a_3x^3 + a_4x^4 + a_5x^5 + a_6x^6 + \ldots + a_rx^r + \ldots$ where $a_0, a_1, a_2, \ldots$ are constants.

Differentiating repeatedly and substituting $x = 0$, gives

$$\mathrm{f}(x) \equiv a_0 + a_1x + a_2x^2 + a_3x^3 + a_4x^4 + a_5x^5 + \ldots + a_rx^r + \ldots$$

So $\mathrm{f}(0) \equiv a_0$

$$\mathrm{f}'(x) \equiv a_1 + 2a_2x + 3a_3x^2 + 4a_4x^3 + 5a_5x^4 + \ldots + ra_rx^{r-1} + \ldots$$

So $\mathrm{f}'(0) \equiv a_1$

$$\mathrm{f}''(x) \equiv 2 \times 1a_2 + 3 \times 2a_3x + 4 \times 3a_4x^2 + 5 \times 4a_5x^3 + \ldots + r(r-1)a_rx^{r-2} + \ldots$$

So $\mathrm{f}''(0) \equiv 2!a_2$

$$\mathrm{f}'''(x) \equiv 3 \times 2 \times 1a_3 + 4 \times 3 \times 2a_4x + 5 \times 4 \times 3a_5x^2 + \ldots + r(r-1)(r-2)a_rx^{r-3} + \ldots$$

So $\mathrm{f}'''(0) \equiv 3!a_3$

Continuing this process gives the definition of the Maclaurin series.

Key point

$$\mathrm{f}(x) \equiv \mathrm{f}(0) + x\mathrm{f}'(0) + \frac{x^2}{2!}\mathrm{f}''(0) + \frac{x^3}{3!}\mathrm{f}'''(0) + \ldots + \frac{x^r}{r!}\mathrm{f}^r(0) + \ldots$$

This is the definition of a Maclaurin series.

A Maclaurin series can be found for any function that meets the conditions given above. You should be able to recognise and use the Maclaurin series for the following functions and know the values of x for which they are valid.

You do not need to be able to prove the conditions for validity of these Maclaurin series.

Key point

$$(1+x)^n = 1 + nx + \frac{n(n-1)}{2!}x^2 + \ldots + \frac{n(n-1)\ldots(n-1+r)}{r!}x^r + \ldots \text{ for } -1 < x < 1,\ n \in \mathbb{R}$$

$$\mathrm{e}^x = 1 + x + \frac{x^2}{2!} + \ldots + \frac{x^r}{r!} + \ldots \text{ for all } x$$

$$\ln(1+x) = x - \frac{x^2}{2} + \frac{x^3}{3} - \ldots + (-1)^{r+1}\frac{x^r}{r} + \ldots \text{ for } -1 < x \le 1$$

$$\sin x = x - \frac{x^3}{3!} + \frac{x^5}{5!} - \ldots + (-1)^r\frac{x^{2r+1}}{(2r+1)!} + \ldots \text{ for all } x$$

$$\cos x = 1 - \frac{x^2}{2!} + \frac{x^4}{4!} - \ldots + (-1)^r\frac{x^{2r}}{(2r)!} + \ldots \text{ for all } x$$

Example 1

Use the Maclaurin expansion of $(1+x)^n$ to find the first five terms of the series for $f(x)=\dfrac{1}{(1-x)^2}$

$$(1+x)^n = 1 + nx + \frac{n(n-1)}{2!}x^2 + \ldots + \frac{n(n-1)\ldots(n-1+r)}{r!}x^r + \ldots$$

This expansion will be given in your formulae booklet.

$$\frac{1}{(1-x)^2} \equiv 1 + 2x + \frac{(-2)(-3)}{2}x^2 + \frac{(-2)(-3)(-4)}{6}(-x)^3 + \frac{(-2)(-3)(-4)(-5)}{24}x^4 + \ldots$$

$$\equiv 1 + 2x + 3x^2 + 4x^3 + 5x^4 + \ldots$$

For the expansion of $\dfrac{1}{(1-x)^2}$, $n=-2$ and x is replaced by $-x$

If n is not a positive integer, the expansion of $(1+x)^n$ is valid for $-1<x<1$

This means that in Example 1, the expansion for $(1-x)^{-2}$ is valid when $-1<-x<1$

This can be rearranged to $-1<x<1$

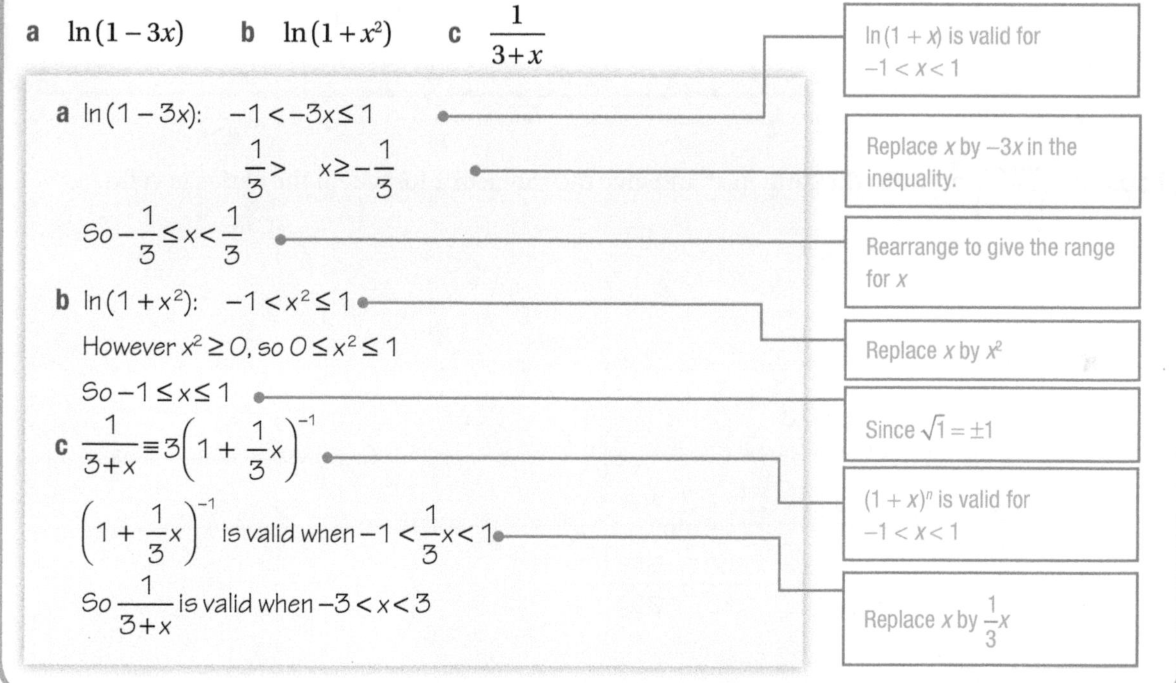

Example 2

Write down the range of values of x that make the Maclaurin series for these functions valid.

a $\ln(1-3x)$ **b** $\ln(1+x^2)$ **c** $\dfrac{1}{3+x}$

a $\ln(1-3x)$: $-1<-3x\le 1$

$\dfrac{1}{3} > x \ge -\dfrac{1}{3}$

So $-\dfrac{1}{3}\le x<\dfrac{1}{3}$

b $\ln(1+x^2)$: $-1<x^2\le 1$

However $x^2\ge 0$, so $0\le x^2\le 1$

So $-1\le x\le 1$

c $\dfrac{1}{3+x}\equiv 3\left(1+\dfrac{1}{3}x\right)^{-1}$

$\left(1+\dfrac{1}{3}x\right)^{-1}$ is valid when $-1<\dfrac{1}{3}x<1$

So $\dfrac{1}{3+x}$ is valid when $-3<x<3$

$\ln(1+x)$ is valid for $-1<x\le 1$

Replace x by $-3x$ in the inequality.

Rearrange to give the range for x

Replace x by x^2

Since $\sqrt{1}=\pm 1$

$(1+x)^n$ is valid for $-1<x<1$

Replace x by $\dfrac{1}{3}x$

Exercise 2.5A Fluency and skills

1 Use the Maclaurin expansion to find the first five terms of these functions.

a e^{2x} **b** e^{-x} **c** e^{-3x}

d $e^{\frac{x}{2}}$ **e** $e^{\frac{-x}{3}}$

2 Use the Maclaurin expansion to

i Find the first five terms of these functions,

ii Find the range of values of x for which each expansion is valid.

a $\ln(1-x)$ **b** $\ln(1+2x)$ **c** $\ln(1-3x)$

d $\ln\left(1+\dfrac{x}{2}\right)$ **e** $\ln\left(1-\dfrac{x}{3}\right)$

3 Use the Maclaurin expansion to find the first four terms of these functions.

a $\sin 2x$ **b** $\sin\dfrac{x}{2}$ **c** $\sin(-x)$

d $\sin(-3x)$ **e** $\sin\dfrac{3x}{2}$

4 Use the Maclaurin expansion to find the first four terms of these functions.

a $\cos 4x$ b $\cos\frac{x}{3}$ c $\cos(-x)$

d $\cos(-2x)$ e $\cos\frac{-x}{2}$

5 For each function

i Find the first four terms of the Maclaurin expansion,

ii Find the range of values of x for which the expansion is valid.

a $(1+x)^{\frac{1}{2}}$ b $\sqrt{(1-x)}$

c $\sqrt[3]{(1+x)}$ d $(1+2x)^{\frac{3}{2}}$

e $(1-3x)^{\frac{5}{2}}$ f $\frac{1}{1-x}$ g $\frac{3}{1+x}$

6 Use the Maclaurin series to write down the first four terms in these expansions.

a $\sin(x^2)$ b $\sin(\sqrt{x})$

c $\cos(x^2)$ d $\cos(\sqrt{x})$

Reasoning and problem-solving

Strategy

To find series expansions for compound functions

1. Use standard series to expand each part of the function.
2. Combine the series to give the expansion of the compound function.
3. Consider the validity of the values of x

Example 3

Expand $\frac{\sin x}{\sqrt{(1+x)}}$ as far as the term in x^5 and give the range of x for which the series is valid.

$$\frac{1}{\sqrt{(1+x)}} \equiv (1+x)^{\frac{-1}{2}}$$

$$(1+x)^n = 1 + nx + \frac{n(n-1)}{2!}x^2 + \frac{n(n-1)(n-2)}{3!}x^3 + \ldots$$

$$(1+x)^{-\frac{1}{2}} = 1 + \left(-\frac{1}{2}\right)x + \left(\frac{3}{4}\right)\left(\frac{x^2}{2!}\right) + \left(-\frac{15}{8}\right)\left(\frac{x^3}{3!}\right) + \left(\frac{105}{16}\right)\left(\frac{x^4}{4!}\right) + \left(\frac{945}{32}\right)\left(\frac{x^5}{5!}\right) + \ldots$$

$$= 1 - \frac{1}{2}x + \frac{3x^2}{8} - \frac{5x^3}{16} + \frac{35x^4}{128} - \frac{63x^5}{256} + \ldots$$

(1) Replace n by $-\frac{1}{2}$ in the expansion for $(1+x)^n$

$$\sin x = x - \frac{x^3}{3!} + \frac{x^5}{5!} - \ldots$$

(1) Use the standard expansion for $\sin x$

$$\frac{\sin x}{\sqrt{(1+x)}} \equiv (\sin x)(1+x)^{-\frac{1}{2}}$$

$$= \left(x - \frac{x^3}{6} + \frac{x^5}{120}\right)\left(1 - \frac{1}{2}x + \frac{3x^2}{8} - \frac{5x^3}{16} + \frac{35x^4}{128} - \frac{63x^5}{256} + \right)$$

$$= x - \frac{1}{2}x^2 + \frac{3x^3}{8} - \frac{5x^4}{16} + \frac{35x^5}{128} + \ldots - \frac{x^3}{6} + \frac{x^4}{12} - \frac{x^5}{16} + \ldots + \frac{x^5}{120} + \ldots$$

(2) Multiply the expansions to find the expansion for the compound function. Only multiply out the terms that give you powers up to x^5

(Continued on the next page)

Hence $\frac{\sin x}{\sqrt{(1+x)}} \equiv x - \frac{1}{2}x^2 + \frac{5x^3}{24} - \frac{11x^4}{48} + \frac{421x^5}{1920} + \ldots$

The expansion for $(1+x)^{-\frac{1}{2}}$ is valid for $-1 < x < 1$

The expansion for $\sin x$ is valid for all values of x

So, the expansion for $\frac{\sin x}{\sqrt{(1+x)}}$ is valid for $-1 < x < 1$

3 Choose the range that is valid for both series.

PURE

Exercise 2.5B Reasoning and problem-solving

1 Use the Maclaurin series to find the first three non-zero terms of these functions.

a $e^x \sin x$ b $e^{\sin x}$

c $\sin x + \cos x$ d $\cos^2 x$

2 Find the first three non-zero terms in the Maclaurin expansion of $x\cos x$. Give the values of x for which the expansion is valid.

3 Find the first three non-zero terms in these Maclaurin expansions. Find the range of values of x for which each expansion is valid.

a $(1+3x)^{\frac{1}{2}}$ b $\ln\left(\frac{1-x}{1+x}\right)$

c $\frac{x}{1-x}$ d $\frac{\cos 2x}{1+x}$

4 a Find the first four non-zero terms in the expansion of

i $\sqrt{1+\frac{1}{x}}$ ii $\ln\left(1+\frac{1}{x}\right)$

b Find the range of values of x for which the expansion of $\ln\left(1+\frac{1}{x}\right)$ is valid.

5 a Make use of known series expansions to obtain these expansions.

i $\frac{1}{2}(e^x - e^{-x})$ (up to the term in x^5)

ii $\cos x$ (up to the term in x^4)

iii $\ln(1+\cos x)$ (use your answer to part **ii** and expand up to the term in x^{12})

b Find the range of values of x for which the expansion of $\ln(1+\cos x)$ is valid.

6 a Use the Maclaurin series to approximate $e^{-0.6}$ (to 4 sf).

b i By writing 3.5 as $3\left(1+\frac{1}{6}\right)$, using $\ln 3 \approx 1.09861$ and the expansion of $\ln(1+x)$, find the value of $\ln(3.5)$ correct to 5 sf.

ii Explain why just substituting $x = 2.5$ does not give a valid answer.

Full A Level

7 At any point x, the gradient of exponential function e^{kx} is ke^{kx}

This means that the derivative of e^{kx} is ke^{kx}

a Use this to derive the Maclaurin series for e^{2x} from first principles using the definition of the Maclaurin series.

b Check your answer by substituting into the standard formula for the expansion of e^x

c Substitute $x = \frac{1}{4}$ in your first five terms and hence obtain an approximation for $\sqrt{e}$

8 a The differential of $\sin x$ is $\cos x$ and the differential of $\cos x$ is $-\sin x$
Use these facts and the definition of the Maclaurin series to find the first four terms of the function $f(x) = \sin x$

b Substitute $x = \frac{\pi}{2}$ into your expansion and find the percentage error between that and the exact value of $\sin\frac{\pi}{2}$

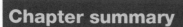

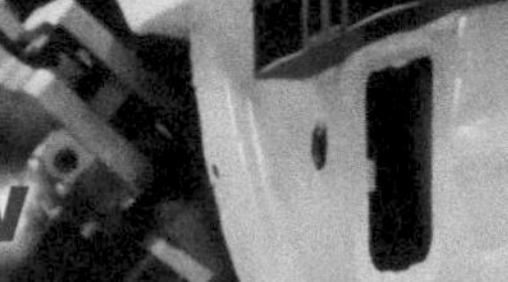

2 Summary and review

Chapter summary

- For the quadratic equation $ax^2+bx+c=0$, the sum of the roots $=-\frac{b}{a}$ and the product of the roots $=\frac{c}{a}$
- If the roots of the cubic equation $ax^3+bx^2+cx+d=0$ are α, β and γ, then
  $(\alpha+\beta+\gamma)=-\frac{b}{a}$ $\quad(\alpha\beta+\beta\gamma+\gamma\alpha)=\frac{c}{a}$ $\quad$ and $\quad\alpha\beta\gamma=-\frac{d}{a}$
- If the roots of the quartic equation $ax^4+bx^3+cx^2+dx+e=0$ are α, β, γ and δ, then
  $(\alpha+\beta+\gamma+\delta)=-\frac{b}{a}$ $\quad(\alpha\beta+\beta\gamma+\gamma\delta+\delta\alpha)=\frac{c}{a}$ $\quad(\alpha\beta\gamma+\beta\gamma\delta+\gamma\delta\alpha+\delta\alpha\beta)=-\frac{d}{a}$ $\quad$ and $\quad\alpha\beta\gamma\delta=\frac{e}{a}$
- If roots are transformed by a linear amount so $y=mx+c$, then substitute $x=\frac{y-c}{m}$
- To solve inequalities, follow the same rules as in solving equations, but remember that
  - answers to inequalities come as a range of values
  - when you multiply or divide by a negative number you reverse the inequality sign.
- $\sum_{1}^{n}u_r$ means 'find the sum of all the terms $u_1, u_2, \ldots, u_r, \ldots u_{n-1}, u_n$'
- $\sum_{1}^{n}r\equiv 1+2+3+\ldots+n=\frac{n(n+1)}{2}$
- $\sum_{1}^{n}r^2\equiv\frac{n}{6}(n+1)(2n+1)$
- $\sum_{1}^{n}r^3\equiv\frac{n^2(n+1)^2}{4}$
- If a series has alternate positive and negative terms, include $(-1)^{r+1}$ in $\Sigma f(r)$
  For example, $1-2+3-4+\ldots+$ for n terms, can be written as $\sum_{1}^{n}(-1)^{r+1}$
- If you can find a function f(x) such that the general term, u_r can be expressed as f($r+1$) – f(r), you can find the sum of the series using the method of differences.
- To prove a statement by induction you must first prove the result for $n=1$ then assume the statement is true for $n=k$ and use this fact to prove the statement is true for $n=k+1$. Always remember to write a conclusion.
- You should recognise the following Maclaurin series and their range of validity:
  - $(1+x)^n=1+nx+\frac{n(n-1)}{2!}x^2+\ldots+\frac{n(n-1)\ldots(n-1+r)}{r!}x^r+\ldots$ for all x if n is a positive integer, otherwise $-1<x<1,\ n\in\mathbb{R}$
  - $e^x=1+x+\frac{x^2}{2!}+\ldots+\frac{x^r}{r!}+\ldots$ for all x
  - $\ln(1+x)=x-\frac{x^2}{2}+\frac{x^3}{3}-\ldots+(-1)^{r+1}\frac{x^r}{r}+\ldots$ for $-1<x\le 1$
  - $\sin x=x-\frac{x^3}{3!}+\frac{x^5}{5!}-\ldots+(-1)^r\frac{x^{2r+1}}{(2r+1)!}+\ldots$ for all x
  - $\cos x=1-\frac{x^2}{2!}+\frac{x^4}{4!}-\ldots+(-1)^r\frac{x^{2r}}{(2r)!}+\ldots$ for all x

Check and review

You should now be able to...	Try Questions
✔ Relate the roots of a polynomial to its coefficients.	3, 4
✔ Find a new polynomial whose roots are a linear transformation of the roots of another polynomial.	2–5
✔ Evaluate expressions involving roots.	6
✔ Solve inequalities up to quartic polynomials.	7–9
✔ Use the formulae for the sums of integers, squares and cubes to sum other series.	10–12
✔ Understand and use the method of differences.	13–16
✔ Use proof by induction to prove divisibility.	17–19
✔ Use Maclaurin series expansions.	20–22

1 Write down the sum, the sum of the products in pairs and the product of the roots of these equations.

a $x^3+5x^2+2x+1=0$

b $x^3+9x^2-11x-8=0$

2 Find the polynomial whose roots are double the roots of $x^3-4x^2+6x-2=0$

3 The cubic equation $2x^3-35x^2+kx+510=0$ has solutions α, β and γ

a Use the fact that $\alpha\beta+\alpha\gamma+\beta\gamma=51$ to find the value of k

b Find a cubic equation with roots $\frac{1}{\alpha}+1$, $\frac{1}{\beta}+1$ and $\frac{1}{\gamma}+1$

4 The cubic equation $x^3+ax^2+bx+2=0$ has roots $\alpha, \frac{1}{\alpha}$ and $\alpha+\frac{1}{\alpha}-2$

a Solve the equation,

b Find the values of a and b

5 The roots of the equation $x^3-x^2-3x+1=0$ are α, β and γ

a Write down the value of $\alpha\beta\gamma$

b i Find a cubic equation with roots $\alpha-1, \beta-1, \gamma-1$

ii By considering the product of the roots of this new equation, find the value of $\alpha\beta\gamma(1+\beta\gamma)(1+\gamma\alpha)(1+\alpha\beta)$

6 The equation $2x^3-6x^2+3x-1=0$ has roots α, β and γ

Find the values of

a $\frac{1}{\alpha}+\frac{1}{\beta}+\frac{1}{\gamma}$

b $(2-\alpha)(2-\beta)(2-\gamma)$

c $\alpha^2+\beta^2+\gamma^2$

7 Solve these inequalities algebraically.

a $\frac{3x-15}{8x}\geq 0$ b $\frac{x}{x-4}<0$

c $\frac{x-1}{x+2}>4$ d $\frac{4x}{2x-5}\geq -4$

e $\frac{x-3}{x-2}<3-x$ f $\frac{2x-3}{x+4}\geq 3$

8 Solve these inequalities.

a $x(2x+3)(3x-2)<0$

b $(x+6)(x+1)(x-8)>0$

c $(3x+8)(x-1)(2x-5)(x-8)\leq 0$

d $(x^2+6x+5)(x^2-10x+21)\leq 0$

9 Solve these inequalities.

a $1 \le \frac{2x-5}{x-2} \le 3$

b $-8 \le \frac{3x+1}{x+4} \le 2$

10 Write out these sequences.

a $\sum_{3}^{10} r$

b $\sum_{n}^{n+4} r^2$

c $\sum_{1}^{n} (r^2 + 5r)$

d $\sum_{1}^{n} \frac{1}{m-3}$

e $\sum_{2}^{10} (-1)^r r^2$

f $\sum_{n-1}^{n+3} 2r(3r-2)$

11 Find the sums of these series to n terms.

a $3 - 2r$

b $2r^2 - 4r$

c $5r + r^3$

12 a Find $\sum_{n}^{2n+1} (r^3 - 3r + 2)$

b Verify your solution using $n = 6$

13 Find the sums of these series to n terms.

a $\frac{1}{r+1} - \frac{1}{r+2}$

b $\frac{1}{r+2} - \frac{1}{r+3}$

14 Use the method of differences to find the sum to n terms of the series

$\frac{1}{2} + \frac{1}{6} + ... + \left(\frac{1}{n} - \frac{1}{n+1}\right)$

Hence write down the sum to infinity.

15 Show that $t^2(t+1)^2 - t^2(t-1)^2 \equiv 4t^3$

Hence show that $\sum_{1}^{n} t^3 = \frac{n^2(n+1)^2}{4}$

16 The expression $\frac{2n+1}{n^2(n+1)^2}$ can be written as $\frac{1}{n^2} - \frac{1}{(n+1)^2}$. Use the method of differences on $\left(\frac{1}{n^2} - \frac{1}{(n+1)^2}\right)$ to show that

$\sum_{1}^{k} \left(\frac{1}{n^2} - \frac{1}{(n+1)^2}\right) = \frac{k(k+2)}{(k+1)^2}$

17 Prove by induction that $3^{2n+1} + 1$ is divisible by 4 for all $n \in \mathbb{N}$

18 Prove by induction that $2^{2n} - 3n + 2$ is divisible by 3 for all $n \in \mathbb{N}$

19 Prove by induction that $6^n - 1$ is divisible by 5 for all $n \in \mathbb{N}$

20 Find the validity of each of these functions.

a $(1+x^2)^{-\frac{1}{2}}$

b $\ln\left(1 - \frac{x}{2}\right)$

21 Find the first three non-zero terms in these Maclaurin expansions.

a $\ln(1 + x^2)$

b $(1+x)^{-\frac{1}{3}} \sin 2x$

22 Use the Maclaurin expansion of e^x to find the first four non-zero terms of $e^{\frac{x}{3}}$ and hence find an approximate value of $\sqrt[3]{e}$ to 4 dp.

Exploration

Going beyond the exams

Did you know?

The **Riemann zeta function** is an infinite summation formula.
It is given by the formula

$$\zeta(s)=\sum_{n=1}^{\infty}\frac{1}{n^s}$$

where s is any complex number with $\mathrm{Re}(s) > 1$

The Riemann zeta function is a famous function in mathematics because it has a link to prime numbers. The **Riemann hypothesis**, proposed by Bernhard Riemann, is a conjecture that makes statements about the zeros of the zeta function. The hypothesis implies results about how the prime numbers may be distributed.

The Riemann hypothesis is yet to be proved, but there is $1 million for the first person who manages it.

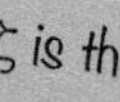

Note

ζ is the Greek letter "zeta".

Research

The **Basel problem** was first posed by **Pietro Mengoli** in 1644. The Basel problem simply asks for the precise summation of the reciprocals of the squares of the natural numbers.
In other words, $\zeta(2)$

Many famous mathematicians failed to solve the problem until **Leonhard Euler**, at the age of 28, solved it in 1734.
Euler found the solution to be

$$\sum_{n=1}^{\infty}\frac{1}{n^2}=\frac{\pi^2}{6}$$

Find out how Euler solved the problem.

After Euler, **Augustin-Louis Cauchy** came up with an alternative proof for the result.
Find out Cauchy's approach to the problem.

Leonhard Euler

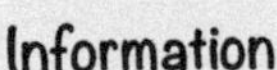

Information

The sum of the natural numbers is a divergent sequence. This can be shown by considering increasingly longer finite sums of the natural numbers:

$$\sum_{n=1}^{100}\frac{1}{n}=5050 \quad \sum_{n=1}^{1000}\frac{1}{n}=500\,500 \quad \sum_{n=1}^{10000}\frac{1}{n}=50\,005\,000$$

As the length of the sequence increases, so does the sum. Therefore as the length tends to infinity, so does the sum.

Research

In 1913, the Indian mathematician **Ramanujan** claimed to be able to show that

$$\sum_{n=1}^{\infty}\frac{1}{n}=-\frac{1}{12}$$

Find out about the proof he offered, and how it related to the Riemann zeta function.

2 Assessment

1 The quadratic equation $3x^2 + 4x + 2 = 0$ has roots α and β. Work out the value of $\alpha + \beta$
Select the correct answer.

a $\frac{2}{3}$ b $-\frac{4}{3}$ c $\frac{4}{3}$ d $\frac{2}{3}$ **[1 mark]**

2 The quartic equation $x^4 - 4x^2 + x - 2 = 0$ has roots α, β, γ and δ. Work out the value of $\sum \alpha\beta\delta$

Select the correct answer.

a -3 b 3 c -1 d 1 **[1]**

3 Solve the inequality $(x-1)(x+3) > 0$
Select the correct solution.

a $0 < x < -3$ or $x < 1$ b $x < -3$ or $0 < x < 1$ c $-3 < x < 1$ or $x > 0$ d $-3 < x < 0$ or $x > 1$ **[1]**

4 Given that $(1+x)^n \approx 1 + nx + \frac{n(n-1)}{2!}x^2 + \ldots + \frac{n(n-1)\ldots(n-1+r)}{r!}x^r + \ldots$

a Find the expansion of $\frac{1}{(1+5x)^3}$ in ascending order up to the term in x^3 **[4]**

b State the range of values of x for which this expansion is valid. **[1]**

5 a Use the Maclaurin expansion of e^x to find the first four terms in the expansion of $e^{-\frac{x}{2}}$ **[3]**

b Use your expansion to estimate the value of $e^{0.05}$ **[3]**

6 Express the sum $\sum_{r=1}^{n}(6r^2 + 2r)$ as a polynomial in n **[3]**

7 A cubic equation is given by $4x^3 + 3x = 1 + 6x^2$

a Use the substitution $u = 2x - 1$ to form a cubic in u **[3]**

b Solve this cubic equation in u **[2]**

c Hence, or otherwise, solve the original equation in x **[2]**

8 Solve the inequality $1 + \frac{x+2}{-5} > x - \frac{x-2}{3}$ **[3]**

9 a Express $\sum_{r=1}^{n}(3r^2 - 2r - 1)$ as a factorised polynomial in n **[4]**

b Hence evaluate the sum to $n = 16$ **[1]**

10 The nth term of a sequence is defined by $u_n = (2n-1)(n+3)$

a Express the sum to n terms of the sequence as a function of n **[4]**

b Evaluate the sum of the first 15 terms. **[2]**

c Find the sum of the 16th to the 20th term. **[3]**

11 a Form a polynomial function, $f(x)$, whose roots are 2, 3 and −4 and whose curve passes through the point (1, 5) **[3]**

b The function $y = f(x)$ undergoes the transformation described by $y = \frac{1}{2}x + 1$
Find the equation of the transformed polynomial, $g(x)$ **[3]**

12 The roots of the cubic $2x^3 - x^2 - 3x + 2 = 0$ are α, β and γ

a Write down the values of $\alpha\beta\gamma$, $\sum\alpha\beta$ and $\sum\alpha$ **[3]**

b Find the cubic equation with roots $2\alpha - 1$, $2\beta - 1$ and $2\gamma - 1$ **[4]**

13 a Use the Maclaurin expansion of $\ln(1+x)$ to find the expansion of $\ln(1-x^2)$ up to the term in x^4 **[3]**

b State the range of values of x for which this series is valid. **[2]**

14 Find the first four non-zero terms in the series expansion of $\dfrac{\sin\left(\dfrac{x}{3}\right)}{e^x}$ **[5]**

15 A cubic function is given by $f(x) = 2x^3 + 3x^2 - 8x + 3$

a Evaluate $f(1)$ and write down a linear factor of $f(x)$ **[1]**

b Factorise $f(x)$ fully. **[3]**

c Hence solve the inequality $f(x) < 0$ **[2]**

16 Prove by induction that $\sum_{r=1}^{n} 4^{r-1} = \frac{1}{3}(4^n - 1)$ for all $n \in \mathbb{N}$ **[6]**

17 Prove by induction that $4n^3 + 8n$ is divisible by 12 for all $n \in \mathbb{N}$ **[6]**

18 Prove by induction that $4^{2n+1} - 1$ is a multiple of 3 for all $n \in \mathbb{N}$ **[6]**

19 A function is given by $f(x) = \dfrac{(x-3)(x+4)}{x-5}$

a Sketch the curve $y = f(x)$ and find the set of values for x for which the function is positive. **[3]**

b Find the set of values for which $f(x) \le -8$ **[4]**

20 a Show that $\dfrac{1}{x} - \dfrac{1}{x+1} = \dfrac{1}{x(x+1)}$ **[1]**

b Find an expression in n for $\sum_{r=1}^{n} \dfrac{1}{r(r+1)}$ **[3]**

c Hence evaluate the sum $\dfrac{1}{1\times2} + \dfrac{1}{2\times3} + \dfrac{1}{3\times4} + \dfrac{1}{4\times5} + \ldots + \dfrac{1}{99\times100}$ **[2]**

d Evaluate $\sum_{r=50}^{99} \dfrac{1}{r(r+1)}$ **[2]**

21 a Prove that $\dfrac{1}{x^2-1} = \dfrac{1}{2}\left(\dfrac{1}{x-1} - \dfrac{1}{x+1}\right)$; $x > 1$ **[2]**

b Find an expression in n for $\sum_{r=2}^{n} \dfrac{1}{r^2-1}$ **[4]**

c Evaluate $\dfrac{1}{3} + \dfrac{1}{8} + \dfrac{1}{15} + \dfrac{1}{24} + \ldots + \dfrac{1}{399}$ **[3]**

d Find $\sum_{r=2}^{\infty} \dfrac{1}{r^2-1}$ **[2]**

22 Given that $\sum_{r=1}^{n} 1 = n$, that $\sum_{r=1}^{n} r = \frac{n}{2}(n+1)$ and that $(r+1)^3 - r^3 = 3r^2 + 3r + 1$

find an expression for $\sum_{r=1}^{n} r^2$ in terms of n **[6]**

23 The quartic equation $x^4 + x + 1 = 0$ has four roots α, β, γ and δ

a Find the quartic equation with integer coefficients which has roots 3α, 3β, 3γ and 3δ **[3]**

b By considering the substitution $u = x^2$, or otherwise, find the quartic equation with roots α^2, β^2, γ^2 and δ^2 **[3]**

24 Solve the inequality $\frac{x+1}{2x-2} \geq 2x+2$ **[7]**

25 The function $\mathrm{f}(x) = \frac{x^2+x+4}{x}$ intersects the straight line $y = k$

a Form a quadratic equation in x and k and hence find the set of values of k for which $\mathrm{f}(x)$ has real roots. **[4]**

b Hence show that the equation $\sin\theta = \frac{x^2+x+4}{x}$ has no solutions. **[3]**

26 a Expand $\ln\left(\frac{1+4x}{\sqrt{1-2x}}\right)$ up to the term in x^3 **[5]**

b State the range of values of x for which this expansion is valid. **[2]**

27 a Prove that $\frac{1}{x(x+1)(x+2)} = \frac{1}{2}\left(\frac{1}{x} - \frac{2}{x+1} + \frac{1}{x+2}\right)$ **[2]**

b Hence find an expression in n for $\sum_{r=1}^{n} \frac{1}{r(r+1)(r+2)}$ **[4]**

c Evaluate $\frac{1}{11\times12\times13} + \frac{1}{12\times13\times14} + \frac{1}{13\times14\times15} + \ldots + \frac{1}{20\times21\times22}$ **[3]**

28 a Show that $2x + \frac{3}{x} - \frac{3}{x+2} = \frac{2x^3+4x^2+6}{x(x+2)}$ **[2]**

b Hence find, as a single fraction, an expression for $\sum_{r=1}^{n} \frac{2r^3+4r^2+6}{r(r+2)}$ **[6]**

29 a Prove by induction that $\sum_{r=2}^{n} \frac{1}{r^2-1} = \frac{3n^2-n-2}{4n(n+1)}$ for all $n \geq 2 \in \mathbb{N}$ **[8]**

b Explain why $\sum_{r=1}^{n} \frac{1}{r^2-1}$ does not exist. **[1]**

3 Curve sketching 1

It is important that aircraft pilots and air traffic controllers are able to locate the position of aircraft precisely, as well as their direction of travel and speed. To do so, they use a modified system of **polar coordinates**. The system takes the direction of magnetic north as 360°, and measures angles clockwise from this. Just like in Cartesian coordinates, in polar coordinates two values can be used to determine the precise location of a point accurately: distance from the origin and the angle measured from magnetic north.

Polar coordinates are useful in many situations when working with phenomena on or near the Earth's surface. For example, meteorologists use them to model existing weather systems and predict future weather, pilots of ocean-going tankers use them for navigation, and cartographers use them to map the Earth.

Orientation

What you need to know

KS4
- Quadratic functions.
- Cartesian coordinates.

Maths Ch2
- Curve sketching.

What you will learn

- To sketch graphs of linear and quadratic rational functions, finding asymptotes, turning points, limits, etc.
- To solve inequalities using graph sketches.
- To use polar coordinates and to convert between Cartesian and polar coordinates.
- To use the equations of parabolas, ellipses and hyperbolae.
- To use the definitions and graphs of hyperbolic functions and their inverses.

What this leads to

Ch18 Curve sketching 2
- Relationships between zeros and asymptotes.
- Combined transformations.
- Hyperbolic functons.

Ch19 Integration 2
- Integrating hyperbolic functions.

Careers
- Aircraft pilot.
- Cartography.

3.1 Linear rational functions

Fluency and skills

When solving problems involving rational functions, you may find it helpful to sketch a graph.

When sketching a graph of a function of the form $\frac{ax+b}{cx+d}$

- find the intercepts where the curve crosses the axes,
- find any asymptotes,
- consider what happens when x and y become very large and approach infinity.

When sketching a graph you might not need all of these techniques. Just use the relevant ones for each question.

For $y = f(x)$, vertical asymptotes occur at values where the denominator is zero.

To find horizontal asymptotes, rearrange to $x = f(y)$ and find the values of y which make the denominator zero.

Example 1

Find the key features (intercepts and asymptotes) and sketch the graph of $y = \frac{(x+1)}{(x-4)}$

$y = \frac{(x+1)}{(x-4)}$

When $x = 0$, $y = -\frac{1}{4}$ and when $y = 0$, $x = -1$ — Find the intercepts by substituting $x = 0$ and $y = 0$

So the curve crosses the axes at $\left(0, -\frac{1}{4}\right)$ and $(-1, 0)$

When $x - 4 = 0$, $x = 4$ — The vertical asymptote is where the denominator is zero.

So $x = 4$ is the vertical asymptote.

As x becomes very large $y \to \frac{x}{x} \to \frac{1}{1} \to 1$ — Consider large values for x

So $y = 1$ is the horizontal asymptote.

Putting $x = 1000$ gives a value just greater than 1, so the curve is above the asymptote, and tends to 1 as x tends to ∞. As x tends to 4 from the right, y tends to ∞. As x tends to ∞, y tends to 1 from above. — Looking at specific large and small values tells you on which side of the asymptote the curve lies.

Putting $x = -1000$ gives a value just less than 1, so the curve is below the asymptote, and tends to 1 as x tends to $-\infty$. Since the curve passes through $\left(0, -\frac{1}{4}\right)$, y must therefore tend to $-\infty$ as x tends to 4 from the left.

As y becomes very large $x \to \frac{4}{1} = 4$ — Consider large values for y

(This is another indication that $x = 4$ is an asymptote.)

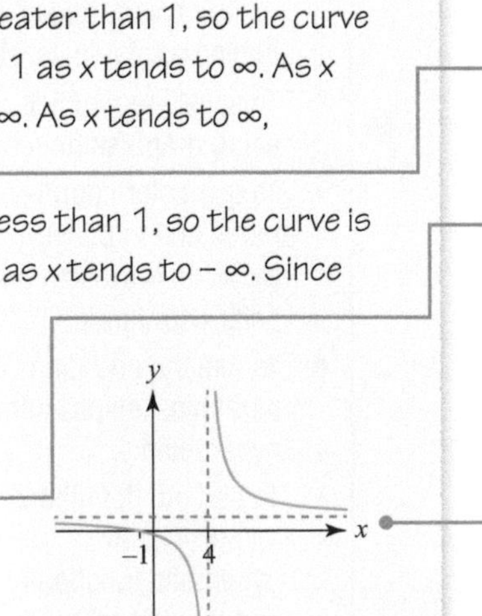

Sketch the graph using the key features to help you.

Asymptotes are usually marked with dashed or dotted lines.

Check your sketch by plotting the curve using a graphical calculator or graph plotting software.

You can of course draw the graph using a graphical calculator and use this to read off the key features directly. However, you should be able to work them out algebraically because you may not be able to plot all curves using a calculator.

PURE

Example 2

Find the key features and sketch the graph of $y=\dfrac{(x+a)}{(x-2)}$ where $a>0$

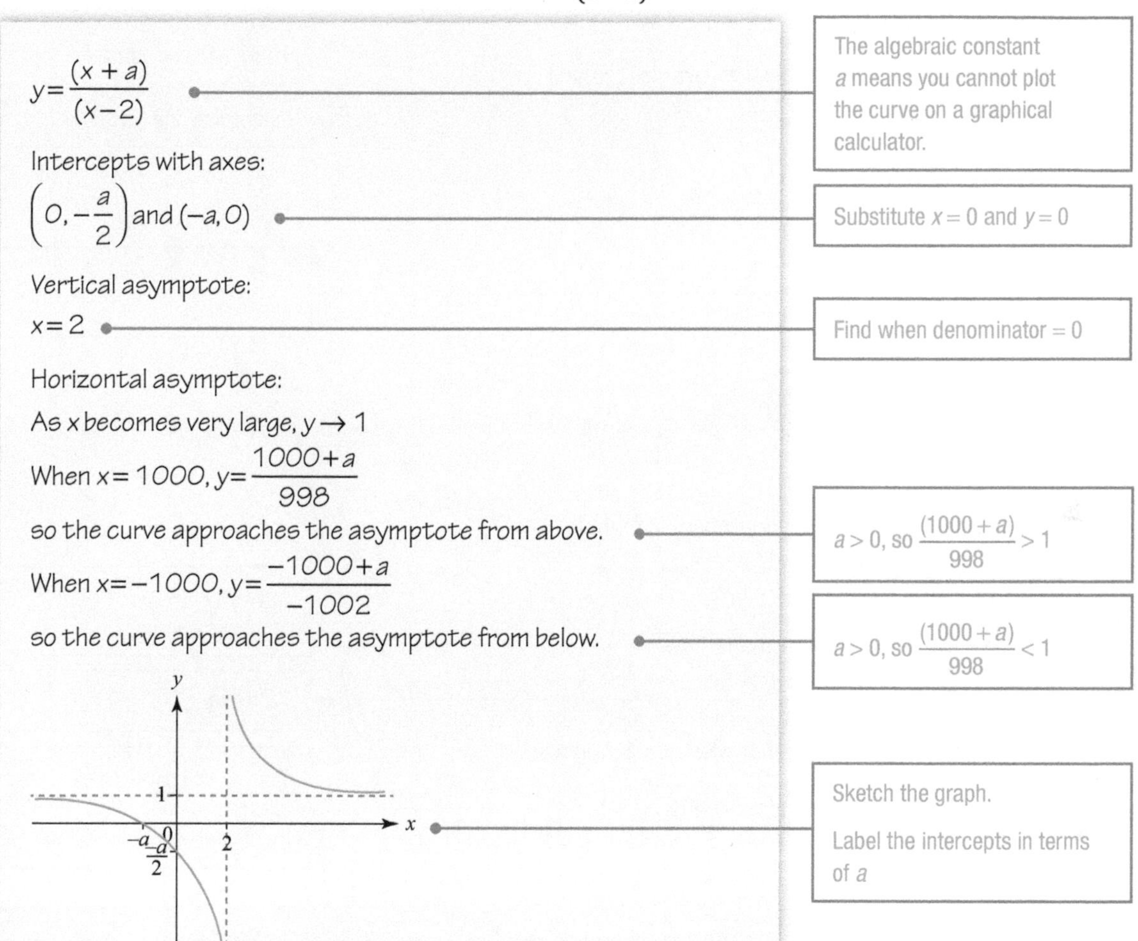

$y=\dfrac{(x+a)}{(x-2)}$

Intercepts with axes:

$\left(0,-\dfrac{a}{2}\right)$ and $(-a, 0)$

Vertical asymptote:

$x=2$

Horizontal asymptote:

As x becomes very large, $y \to 1$

When $x=1000$, $y=\dfrac{1000+a}{998}$

so the curve approaches the asymptote from above.

When $x=-1000$, $y=\dfrac{-1000+a}{-1002}$

so the curve approaches the asymptote from below.

The algebraic constant a means you cannot plot the curve on a graphical calculator.

Substitute $x=0$ and $y=0$

Find when denominator $=0$

$a>0$, so $\dfrac{(1000+a)}{998}>1$

$a>0$, so $\dfrac{(1000+a)}{998}<1$

Sketch the graph.

Label the intercepts in terms of a

Example 3

a Identify the key features and sketch the graphs of $y=\dfrac{x}{(x-5)}$ and $y=5x-24$

b Find the points of intersection.

a For $y=\dfrac{x}{(x-5)}$:

when $x=0$, $y=0$ and when $y=0$, $x=0$

When $x-5=0$, $x=5$, so $x=5$ is a vertical asymptote

Substitute $x=0$ and $y=0$

Find the vertical asymptote.

(Continued on the next page)

$y=\frac{x}{(x-5)} \rightarrow y(x-5)=x$

So, $x=\frac{5y}{y-1}$

Rearrange to the form $x=\dots$ to find the horizontal asymptote.

When $y-1=0$, $y=1$, so $y=1$ is a horizontal asymptote.

As x becomes very large $y\rightarrow\frac{1}{1}=1$

You can substitute large values of x and y to see from which side the curve approaches the asymptote.

$y=5x-24$ is a straight line with gradient 5 and y-intercept -24

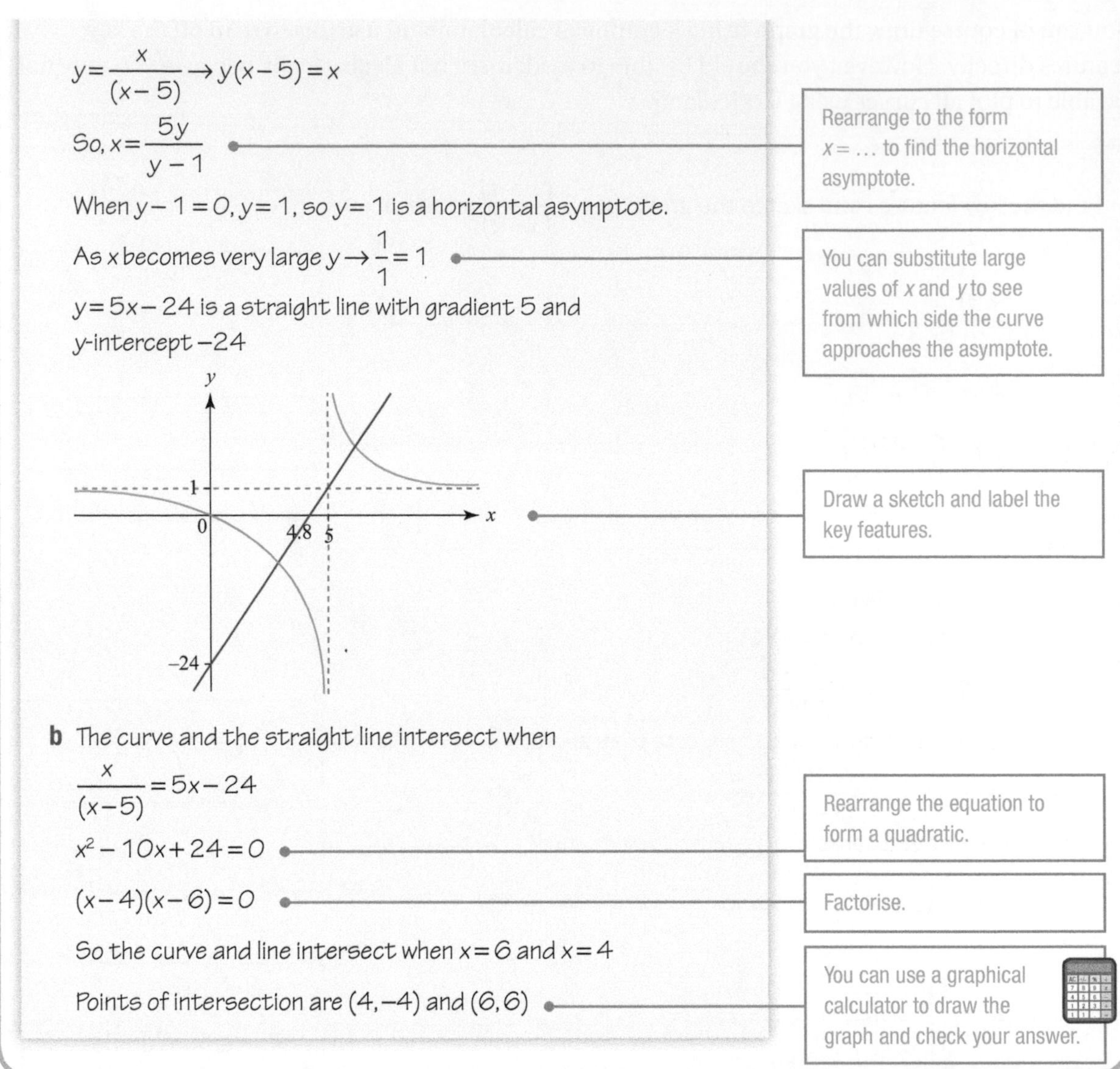

Draw a sketch and label the key features.

b The curve and the straight line intersect when

$\frac{x}{(x-5)}=5x-24$

$x^2-10x+24=0$

Rearrange the equation to form a quadratic.

$(x-4)(x-6)=0$

Factorise.

So the curve and line intersect when $x=6$ and $x=4$

Points of intersection are $(4,-4)$ and $(6,6)$

You can use a graphical calculator to draw the graph and check your answer.

Exercise 3.1A Fluency and skills

Wherever possible, you should use a graphical calculator or graph sketching software on your computer to check your answers to the questions in this exercise.

1 For each of these graphs

i Determine any intercepts with the axes,

ii Find any vertical and horizontal asymptotes,

iii Draw a sketch.

a $y=\frac{x+1}{x-2}$ b $y=\frac{x-1}{x+2}$

c $y=\frac{x+5}{x-3}$ d $y=\frac{6x-5}{8x+3}$

e $y=\frac{x+12}{5x-12}$ f $y=\frac{7x-10}{3x-20}$

2 Find any asymptotes and intercepts with the axes for each of following graphs.

a $y=\frac{x+a}{2x-1}$ b $y=\frac{3x+4}{x+b}$

c $y=\frac{2x-a}{3x-5}$ d $y=\frac{x-a}{2x+b}$

3 Identify the key features, sketch each pair of equations and find the points of intersection. Show all your working.

a $y=\frac{x+3}{x-4}$ and $y=2$

b $y=\frac{x+6}{x-4}$ and $y=6$

c $y=\dfrac{x+12}{5x-12}$ and $y=0$

d $y=\dfrac{4-3x}{2x-3}$ and $y=-8$

e $y=\dfrac{x+5}{x+1}$ and $y=2x+1$

f $y=\dfrac{2x-6}{x+5}$ and $y=3-x$

g $y=\dfrac{10-6x}{x+3}$ and $y=7x+15$

h $y=\dfrac{8x+48}{3x-8}$ and $y=24-x$

4 By sketching a curve and a line, solve these inequalities graphically. Show all your working and label all the key features.

a $\dfrac{x-2}{x+5}<8$ **b** $\dfrac{x-15}{x+6}\le 4$

c $\dfrac{4x-1}{3x+4}<2$ **d** $\dfrac{3x+5}{2x-3}\ge 11$

Reasoning and problem-solving

Sometimes, solving an inequality algebraically can be quite complicated. You can use a graphical method to solve the inequality or to check your algebraic result.

Strategy

When solving inequalities graphically

1. Investigate the key features to find the shape of the graph of the functions.
2. Sketch the graphs and find the points of intersection.
3. Use the graphs to write down the ranges of the variable which solve the inequality.

Example 4

Solve the inequality $\dfrac{18x-135}{x-3}>21-x$ graphically.

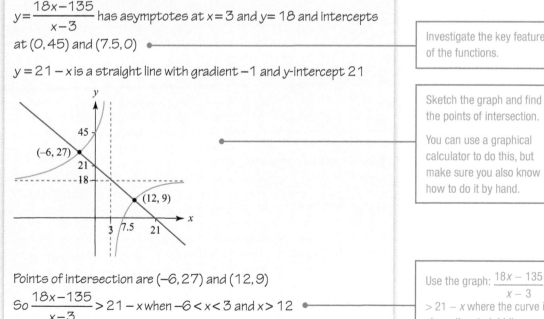

$y=\dfrac{18x-135}{x-3}$ has asymptotes at $x=3$ and $y=18$ and intercepts at $(0, 45)$ and $(7.5, 0)$

Investigate the key features of the functions. (1)

$y=21-x$ is a straight line with gradient -1 and y-intercept 21

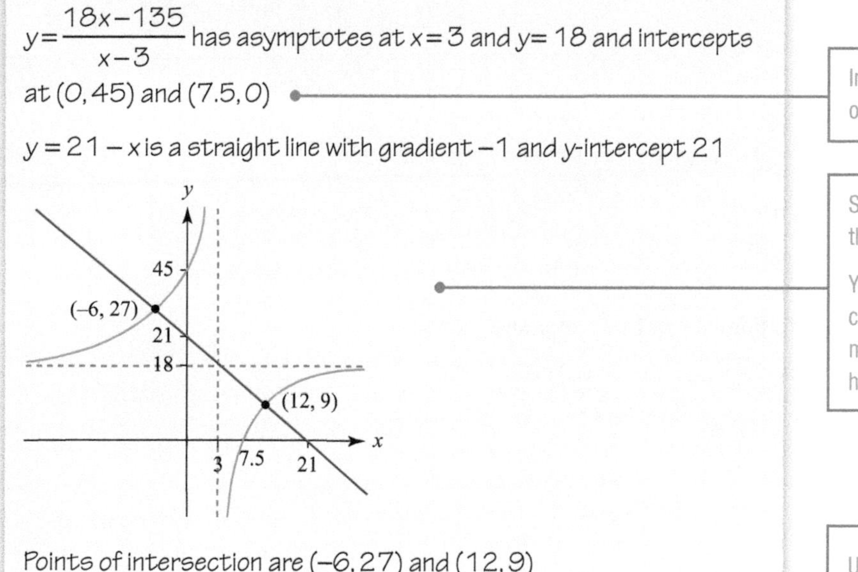

Sketch the graph and find the points of intersection. (2)

You can use a graphical calculator to do this, but make sure you also know how to do it by hand.

Points of intersection are $(-6, 27)$ and $(12, 9)$

So $\dfrac{18x-135}{x-3}>21-x$ when $-6<x<3$ and $x>12$

Use the graph: $\dfrac{18x-135}{x-3}>21-x$ where the curve is above the straight line. (3)

Example 5

Following an experiment, student A's results fit the graph of $y(x-3)=2$, whilst student B's results fit with $y(x+2)=3$

The students plotted their graphs of y against x on the same grid.

a Find the point of intersection of their graphs.

b Using a graphical calculator or graph sketching software on your computer, find the range of values of x where the graph of student A gives higher results than that of student B.

c Confirm your results algebraically.

a $y=\dfrac{2}{x-3}=\dfrac{3}{x+2}\rightarrow 2(x+2)=3(x-3)$

$2x+4=3x-9$

$x=13$

The point of intersection is $(13, 0.2)$

b $y=\dfrac{2}{x-3}$ has asymptotes at $x=3$ and $y=0$

1 Investigate the key features.

$y=\dfrac{3}{x+2}$ has asymptotes at $x=-2$ and $y=0$

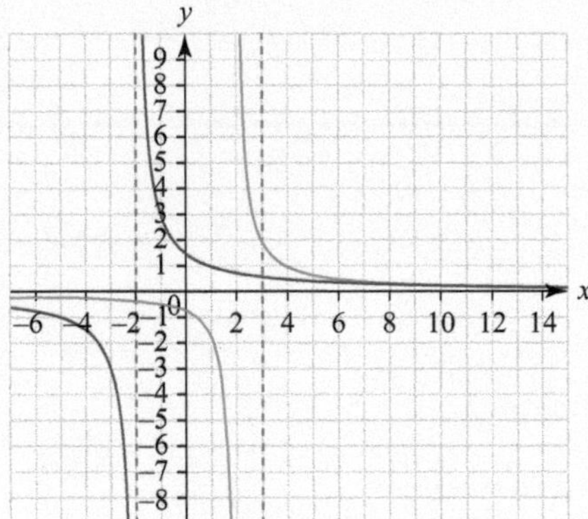

2 Sketch the graphs.
The blue graph is $y=\dfrac{2}{x-3}$
The red graph is $y=\dfrac{3}{x+2}$

The results of student A are higher than student B when the blue graph is above the red graph.

$\dfrac{2}{x-3}>\dfrac{3}{x+2}$ when $x<-2$ or when $3<x<13$

c Case 1: both denominators positive or both negative

See Ch2.2 For a reminder on solving inequalities algebraically.

$2(x+2)>3(x-3)$ if $(x+2)(x-3)>0$

i.e. if $x>3$ or $x<-2$

$2x+4>3x-9$

$13>x$

So, $x<-2$ or $3<x<13$

(*Continued on the next page*)

Case 2: one denominator positive and one negative

$2(x+2)<3(x-3)$ if $(x+2)(x-3)<0$

i.e. if $-2<x<3$

$2x+4<3x-9$

$13<x$

So no solution since $-2<x<3$

Remember to reverse the inequality sign when you multiply by a negative number.

PURE

Exercise 3.1B Reasoning and problem-solving

Use a graphical calculator or graph plotting software to answer the questions in this exercise.

1 Solve these inequalities graphically, labelling all the intercepts, asymptotes and points of intersection.

a $\frac{x+5}{x+1}>2x+1$ b $\frac{2x-6}{x+5}<3-x$

c $\frac{x+8}{x-4}\geq 2x-9$ d $\frac{2x+6}{2x-3}\geq 14-2x$

e $\frac{10-6x}{x+3}\leq 7x+15$ f $\frac{8x+48}{3x-8}>24-x$

2 Solve these inequalities graphically, labelling all the intercepts, asymptotes and points of intersection.

a $-2<\frac{x+2}{x-4}<3$ b $-9\geq\frac{2x+5}{2x-5}\geq 0$

c $-1\leq\frac{3x-5}{2x-5}\leq 1$ d $2-x<\frac{x-12}{x-6}\leq 6-\frac{x}{2}$

e $11-3x\leq\frac{2x-18}{x-3}\leq\frac{9-x}{3}$

f $\frac{2}{x+3}>\frac{3}{x-1}$ g $\frac{5}{x+4}\leq\frac{3}{x-2}$

3 John says that the ratio of $2x$ to $x+2$ is always less than 1. Use graphs to explain why John is wrong.

4 Sketch two appropriate graphs to solve the inequality $\frac{3}{x-1}-x-1\geq 0$

5 For all real values of θ, $-1\leq\sin\theta\leq 1$
Find the values of x for which $\sin\theta=\frac{x-5}{x+5}$ is satisfied by real values of θ

6 For all real values of θ, $-1\leq\cos\theta\leq 1$
Find the values of x for which $\cos\theta=\frac{x+1}{1-x}$ is satisfied by real values of θ

7 a Use a graphical calculator or graph sketching software to sketch the graph of $y=\frac{2x}{x+1}-\frac{x-3}{x-5}$

b Show that the graph crosses the y-axis at $y=-0.6$

c Show that the asymptotes of the graph are $x=-1$, $x=5$ and $y=1$

8 When is $\frac{x}{x+1}\geq\frac{x+1}{x}$?

9 You can solve $x^2-2x-15=0$ by sketching, on the same grid, the graph of $y=x-2$ and an appropriate curve.

a Give the equation of an appropriate curve and draw a sketch to solve the equation.

b Verify your solution by solving $x^2-2x-15=0$ algebraically.

10 Solve the equation $x^3-x^2-6x=0$ by using a graphical calculator or graph sketching software to sketch, on the same grid, the graph of $y=x^2$ and an appropriate curve of the form $\frac{Ax}{x+B}$
State the values of A and B

11 Three fractions are $\frac{1}{x}$, $\frac{x}{2}$ and $\frac{3}{x+2}$
Sketch suitable graphs on the same grid and calculate the exact ranges of values of x when each fraction is the largest of the three.

Full A Level

3.2 Quadratic rational functions

Fluency and skills

You can sketch graphs of quadratic rational functions using similar techniques to those you learned in Section 3.1.

When sketching a graph of a function of the form $\frac{ax^2+bx+c}{dx^2+ex+f}$

- find any asymptotes,
- find the intercepts where the curve crosses the axes,
- consider what happens when x and y approach infinity,
- find any limitations on the possible ranges of y by solving the function for y and/or x and use this to find any maximum or minimum points on the curve.

See Ch18.4
To learn about oblique asymptotes

You need to be able to find the key features of a graph using algebraic techniques, but make sure you also know how to plot graphs on a graphical calculator to check your answers.

Example 1

For the graph of $y=\frac{x^2-16}{x^2+4}$

a Find the intercepts where the curve crosses the axes,

b Show that there is no vertical asymptote, but that there is a horizontal asymptote at $y=1$

a When $x=0$, $y=\frac{-16}{4}=-4$

When $y=0$, $x^2-16=0$ (provided $x^2+4\neq 0$)

So the curve crosses the axes at $(0,-4)$ $(4,0)$ and $(-4,0)$

Find the intercepts by substituting $x=0$ and $y=0$

b $x^2+4\neq 0$ for any real value of x

so there is no vertical asymptote.

A vertical asymptote occurs when the denominator is zero.

Rearranging: $y(x^2+4)=x^2-16$

$$4y+16=x^2(1-y)$$

$$x=\sqrt{\frac{4y+16}{1-y}}$$

Rearrange to $x=\ldots$ and find when the denominator is zero to find any horizontal asymptote.

There is a horizontal asymptote at $y=1$

Alternatively,

As x gets very large $y\to\frac{x^2}{x^2}=1$

So $y=1$ is a horizontal asymptote.

You can also find horizontal asymptotes by considering large values of x

Let $y=k$ so $\frac{x^2-16}{x^2+4}=k\Rightarrow x^2-16=kx^2+4k$

$\Rightarrow(1-k)x^2=4k+16\Rightarrow x^2=\frac{4k+16}{1-k}$

Since $x\geq 0$, $k\leq 1$. However, as $x\to\infty$, $\frac{x^2}{x^2}\to 1$

so $y=1$ is a horizontal asymptote.

Hence $y\leq 1$

Sketch the graph using a graphical calculator to confirm your values for the intercepts and asymptote.

Example 2

Find the range of x and y values for which the graph of $y=\dfrac{x^2+6x+9}{x^2+3x+3}$ exists.

Use this information to find any turning points. Check your answers using a graphical calculator.

x values

The graph exists for all values of x except for any vertical asymptotes.

Vertical asymptotes occur when the denominator $= 0$

Since x^2+3x+3 has no real solutions, there is no vertical asymptote.

This means the graph exists for all real values of x

y values

Consider where the graph crosses the line $y=k$ and deduce allowable values for k

Let $y=k$

Rearrange the equation

Solve for x to find the limits on the range of y

$$k(x^2+3x+3)=x^2+6x+9$$

$$(k-1)x^2+3(k-2)x+3(k-3)=0$$

x can only take real values when $b^2-4ac\geq 0$

$$9(k-2)^2-4(k-1)(3)(k-3)\geq 0$$

$$-3k^2+12k\geq 0$$

$$3k(k-4)\leq 0$$

Change the direction of the inequality since you have multiplied by -1

So the curve only exists for $0\leq k\leq 4$

since $y=k$, $0\leq y\leq 4$

Turning points

As x gets very large $y\rightarrow \dfrac{x^2}{x^2}=1$, so $y=1$ is a horizontal asymptote.

$y=1$ is the only horizontal asymptote and $0\leq y\leq 4$

So $y=0$ and $y=4$ must be either minimum or maximum points.

When $y=4$

Substitute $y=4$ into the equation and solve.

$$3x^2+6x+3=0 \text{ so } 3(x+1)^2=0$$

Hence the maximum point is $(-1, 4)$

When $y=0$

$$x^2+6x+9=0 \text{ (provided } x^2+3x+3\neq 0)$$

So, $(x+3)^2=0$ and the minimum point is $(-3, 0)$

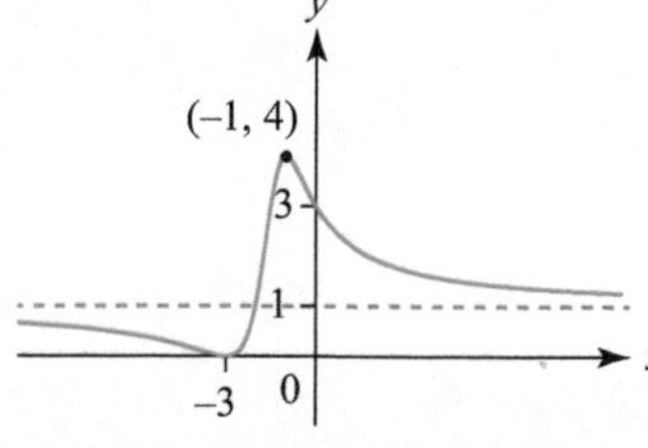

Check your answers with a graphical calculator.

Notice that, in some circumstances, a curve can cross an asymptote.

Example 3

Identify the key features and sketch the graph of $y=\dfrac{x^2-x-6}{x^2-x-12}$

Intercepts

When $x=0$, $y=\dfrac{-6}{-12}=\dfrac{1}{2}$

When $y=0$, $x^2-x-6=0$ (provided $x^2-x-12\neq 0$)

$(x-3)(x+2)=0$

$x=3$ or $x=-2$

(Check: $(3)^2-(3)-12\neq 0$ and $(-2)^2-(-2)-12\neq 0$)

So, the curve crosses the axes at $\left(0, \dfrac{1}{2}\right)$, $(-2, 0)$ and $(3, 0)$

Substitute $x=0$ and $y=0$

Asymptotes

$x^2-x-12=0$

$(x+3)(x-4)=0$

So, $x=-3$ and $x=4$ are vertical asymptotes.

The denominator is zero at a vertical asymptote.

As $x\to\infty$, $\dfrac{x^2}{x^2}\to 1$, so $y=1$ is a horizontal asymptote.

Consider large values of x

Substituting $x=\pm 1000$, say, shows the graph approaches the asymptote from above on both sides of the graph.

Range

Consider where the graph crosses the line $y=k$ and deduce allowable values for k

Let $y=k$

Rearranging the equation:

Solve for x to find the limits on the range of y

$k(x^2-x-12)=x^2-x-6$

$\Rightarrow (k-1)x^2-(k-1)x+(6-12k)=0$

x can only take real values when

$(k-1)^2-4(k-1)(6-12k)\geq 0$

$49k^2-74k+25\geq 0$

$(49k-25)(k-1)\geq 0$

$49k-25\geq 0$ or $k-1\geq 0$

So

$k\geq \dfrac{25}{49}$ or $k\geq 1$

So the curve exists for $k\leq \dfrac{25}{49}$ or $k\geq 0$

However, because of the asymptote, y does not exist when $\dfrac{25}{49}<y\leq 1$

Since $k=y$

$y\leq \dfrac{25}{49}$ or $y\geq 1$

(*Continued on the next page*)

Turning points

When $y = 1$

$y = 1$ is an asymptote as x tends to $\pm\infty$

When $y = \frac{25}{49}$

$$\left(\frac{25}{49}-1\right)x^2+\left(1-\frac{25}{49}\right)x+\left(6-12\times\frac{25}{49}\right)=0$$

$$\frac{-24}{49}x^2+\frac{24}{49}x-\frac{6}{49}=0$$

$$4x^2-4x+1=0$$

$$(2x-1)^2=0$$

So there is a turning point at $x = \frac{1}{2}$

This shows that $\left(\frac{1}{2}, \frac{25}{49}\right)$ is a maximum point.

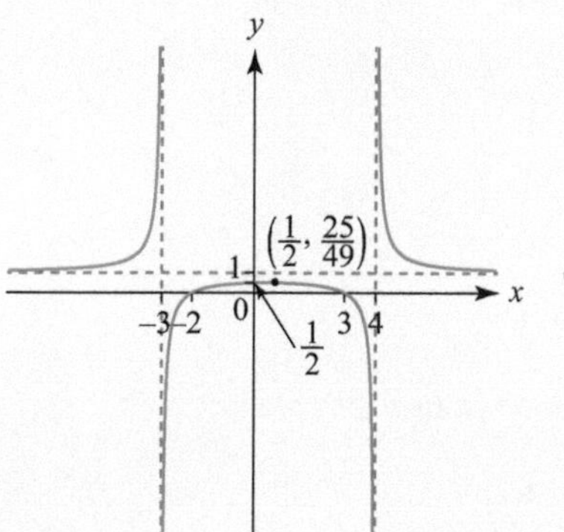

Use these features to sketch the graph.

You should use a graphical calculator to check your sketch.

Exercise 3.2A Fluency and skills

Use a graphical calculator or graph sketching software to check your answers in this exercise.

1 Find any asymptotes and points of intersection with the axes for each graph.

a $y=\frac{x^2-9}{x^2+9}$ b $y=\frac{x^2+9}{x^2-9}$

c $y=\frac{x^2}{x^2+1}$ d $y=\frac{x^2+5x}{x^2+5}$

e $y=\frac{x^2-3x+4}{x^2+4x}$ f $y=\frac{x^2-3x-4}{x^2+4x}$

2 For each of these functions, determine the ranges of x and y values for which the graph exists.

a $y=\frac{x^2}{x^2+4x-3}$ b $y=\frac{x^2-3x}{x^2-3x+2}$

c $y=\frac{x^2-3x-4}{x^2-3x+1}$ d $y=\frac{2-x}{2+x^2}$

3 For each of these curves, find any local maximum and minimum points.

a $y=\frac{x^2}{x+1}$ b $y=\frac{12x}{x^2+4}$

c $y=\frac{6}{x^2+4}$ d $y=\frac{5}{x^2+4x}$

4 Sketch these curves.

a $y=\frac{x^2-2x}{x^2-4x+8}$ b $y=\frac{x^2-8x}{x^2-5x-6}$

c $y=\frac{x^2-4x}{x^2-5x+6}$ d $y=\frac{x^2-4x+4}{x^2-5x-6}$

Reasoning and problem-solving

Strategy

To sketch a curve of the form $y=\frac{ax^2+bx+c}{dx^2+ex+f}$

1. Find any points of intersection with the axes and any asymptotes.
2. Investigate the areas where y and x have real values and determine any maxima/minima.
3. Sketch the curve and use a graphical calculator to check your answer.

Example 4

Sketch the curve $y=\frac{x^2+x-6}{x^2-x-6}$

Consider the function $y=\frac{x^2+x-6}{x^2-x-6}$

This can be written as $y=\frac{(x+3)(x-2)}{(x-3)(x+2)}$

Therefore there are critical values at $x=-3, -2, 2$ and 3

Considering the sign of y between and on either side of these values we obtain:

	$x<-3$	$-3<x<-2$	$-2<x<2$	$2<x<3$	$x>3$
y	negative	positive	negative	positive	negative

Hence, every time a critical value of x is passed, y changes from one side of the x-axis to the other.

Also, when $x=0$, $y=1$ so the curve passes through $(0, 1)$

There are vertical asymptotes at $x=-2$ and 3 (you can see this by considering the denominator of the curve).

There is a horizontal asymptote because $y\to 1$ as $x\to\infty$

And the graph crosses this horizontal asymptote once as can be seen by solving

$$\frac{x^2+x-6}{x^2-x-6}=1$$

$$\Rightarrow x^2+x-6=x^2-x-6$$

$$\Rightarrow x=0$$

(*Continued on the next page*)

Although we cannot say that y is continuous, the curve actually passes through every value of y

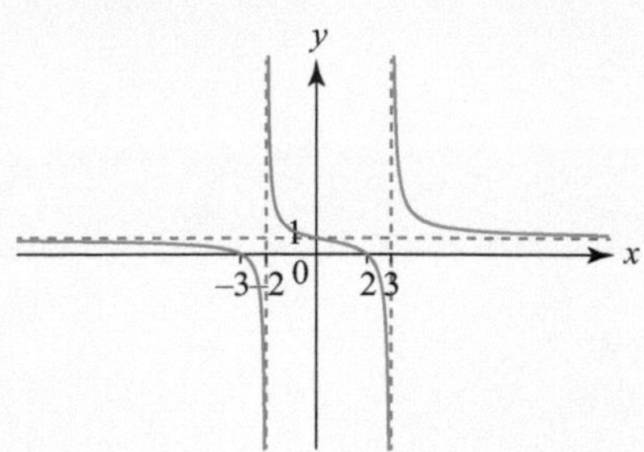

Exercise 3.2B Reasoning and problem-solving

Use a graphical calculator or graph sketching software to answer the questions in this exercise where appropriate.

1 Prove that $-\frac{1}{2} \le \frac{x}{x^2+1} \le \frac{1}{2}$ and hence sketch the graph of the function.

2 Prove that $0 \le \frac{x^2}{x^2+1} \le 1$ and use this to sketch the graph of the function.

3 Prove that $y = \frac{x^2}{x^2-1}$ cannot lie between 0 and 1 and use this to sketch the graph.

4 Prove that $y = \frac{x^2+5}{x-2}$ has a minimum value of 10 and a maximum value of −2, and explain this apparent contradiction using a sketch graph.

5 Sketch the graph of the function $\frac{2x^2+3x-2}{x^2-2x+2}$

6 a Find the greatest value of c for which the value of the expression $\frac{x^2-4x+4}{x^2+c}$ is 5

b Sketch the graph of the expression for this value of c

7 Calculate the range of values of C for which the equation $\frac{1-3x}{x^2+x} = C$ has real roots. Hence find the coordinates of the local maximum and minimum points on the curve.

8 Show that $y = \frac{x^2+A}{x^2-B^2}$ has real solutions except for the range $-\frac{A}{B^2} < y < 1$. Hence sketch the graph of $y = \frac{x^2+4}{x^2-3^2}$ and find the coordinates of the maximum turning point.

9 a Sketch on the same grid the graphs of $y = \frac{4}{x^2-x}$ and $y^2 = \frac{4}{x^2-x}$

b Find the exact values of x and y at the points of intersection of the graphs and hence solve the inequality $\frac{4}{x^2-x} > \sqrt{\frac{4}{x^2-x}}$

c Explain why $y^2 = \frac{4}{x^2-x}$ has no value between $x = 0$ and $x = 1$

10 a Prove that $y = \frac{x^2-A}{x^2-4x-A}$ has real roots if $A \ge 0$

b i Sketch the graph when $A = 4$ and show there is no turning point.

ii Solve the equation $\frac{x^2-4}{x^2-4x-4} = 0$

11 If $y = \frac{kx}{x^2-x+1}$, find the minimum and maximum values of y in terms of k
Hence sketch the curve when $k = 6$

Full A Level

3.3 Polar coordinates

Fluency and skills

A point can be defined by its (x, y) coordinates, but you could also define it by a distance from the origin and an angle as shown in the diagram.

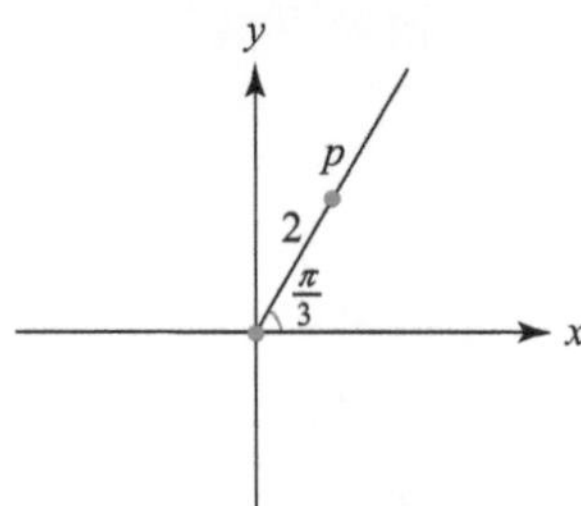

The distance of P from the origin is 2, and the angle it makes with the positive x axis is $\frac{\pi}{3}$

These two numbers describe the point perfectly, and they are called the polar coordinates of the point P, written $\left(2, \frac{\pi}{3}\right)$.

However, the angle is not unique: there is an infinite number of ways to represent a point in polar coordinates. For example,

$$\left(2, \frac{\pi}{3}\right) = \left(2, -\frac{5\pi}{3}\right) = \left(2, \frac{7\pi}{3}\right) = \left(2, -\frac{11\pi}{3}\right)$$

The letters r and θ are usual for the distance and the angle.

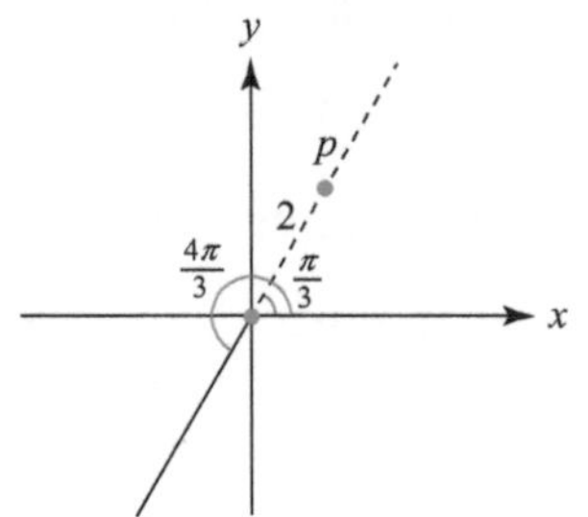

It is possible to use a negative value of r, by going backwards from the origin:

$$\left(2, \frac{\pi}{3}\right) = \left(-2, \frac{4\pi}{3}\right) = \left(-2, -\frac{2\pi}{3}\right)$$

Every point, however, can only be written in one way if we always choose $r \geq 0$ and $-\pi < \theta \leq \pi$ or $0 \leq \theta < 2\pi$

The point (0, 0) is called the pole, and the positive x axis, which is where the angle is measured from, is called the initial line.

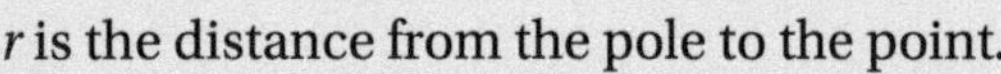

Key point

r is the distance from the pole to the point.

θ is the angle between the initial line and the line segment from the pole to the point, measured counterclockwise.

You can convert polar (r, θ) coordinates into Cartesian (x, y) coordinates. In the example above, the x coordinate is $2\cos\frac{\pi}{3}$ and the y coordinate is $2\sin\frac{\pi}{3}$

PURE

The same method will always work, so

Key point

To convert from polar to Cartesian, use

$x = r\cos\theta$

$y = r\sin\theta$

To convert from Cartesian to polar, use

$r^2 = x^2 + y^2$

$\tan\theta = \dfrac{y}{x}$

but you should draw a diagram to ensure that you choose the right value of θ

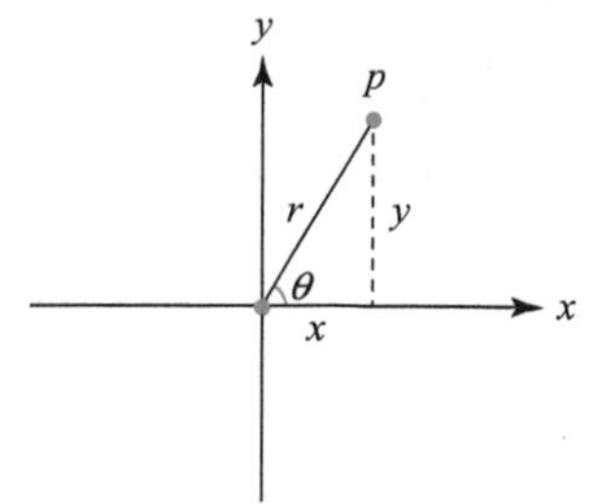

Example 1

Convert the polar coordinates $\left(5, -\dfrac{3\pi}{4}\right)$ into Cartesian coordinates.

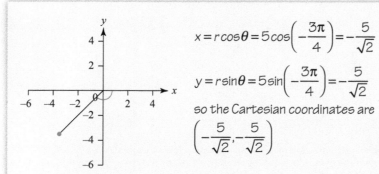

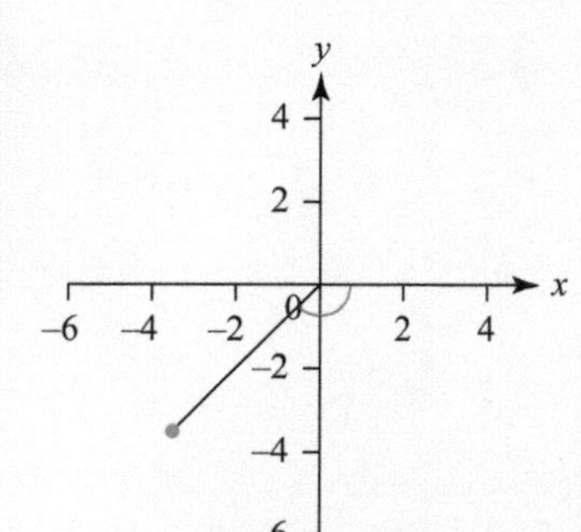

$x = r\cos\theta = 5\cos\left(-\dfrac{3\pi}{4}\right) = -\dfrac{5}{\sqrt{2}}$

$y = r\sin\theta = 5\sin\left(-\dfrac{3\pi}{4}\right) = -\dfrac{5}{\sqrt{2}}$

so the Cartesian coordinates are

$\left(-\dfrac{5}{\sqrt{2}}, -\dfrac{5}{\sqrt{2}}\right)$

Example 2

Convert the Cartesian coordinates $(-2\sqrt{3}, 2)$ into polar coordinates.

First, draw a sketch.

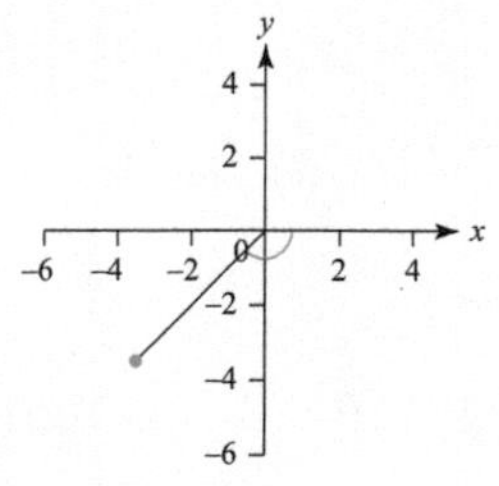

$r^2 = x^2 + y^2 = 12 + 4 = 16$

$\Rightarrow r = 4$

$\tan\theta = \dfrac{y}{x} = -\dfrac{1}{\sqrt{3}}$

$\Rightarrow \theta = -\dfrac{\pi}{6}, \dfrac{5\pi}{6}$

but the sketch shows which value to use, so polar coordinates are

$\left(4, \dfrac{5\pi}{6}\right)$

An equation involving x and y can be represented as a graph. So can an equation involving r and θ

Example 3

Draw the graph

$r = 3(1 - \sin\theta),\ 0 \le \theta < 2\pi$

This means that for any value of θ in the range 0 to 2π, you can calculate the value of r and plot that point. It is a good idea to make a table of values using easy values of θ

Then you can join the points up to make a smooth curve.

θ	0	$\frac{\pi}{6}$	$\frac{\pi}{3}$	$\frac{\pi}{2}$	$\frac{2\pi}{3}$	$\frac{5\pi}{6}$	π	$\frac{7\pi}{6}$	$\frac{4\pi}{3}$	$\frac{3\pi}{2}$	$\frac{5\pi}{3}$	$\frac{11\pi}{6}$	2π
r	3	1.5	0.40	0	0.40	1.5	3	4.5	5.6	6	5.6	4.5	3

Polar graph paper can be useful: the lines represent points with the same θ coordinate, and the circles represent points with the same r coordinate.

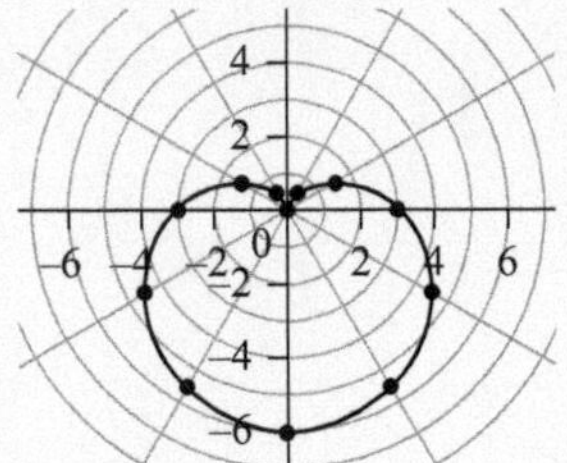

Example 4

a Draw the graph
$r = 5\sin 3\theta,\ 0 \le \theta < 2\pi$

b Draw the graph
$r = 5\sin 3\theta,\ 0 \le \theta < 2\pi,\ r \ge 0$

a You could draw up a table of values as before, but you do not usually do this for (x, y) graphs, and you don't always need to for (r, θ) graphs.

First, notice that $-1 \le \sin 3\theta \le 1$, so $-5 \le r \le 5$

It is useful to find the values of θ that give r these maximal and minimal values:

$r = 5 \Rightarrow \sin 3\theta = 1 \Rightarrow 3\theta = \frac{\pi}{2}, \frac{5\pi}{2}, \frac{9\pi}{2}, \ldots$

$\Rightarrow \theta = \frac{\pi}{6}, \frac{5\pi}{6}, \frac{3\pi}{2}, \ldots$

Also,

$r = -5 \Rightarrow \sin 3\theta = -1 \Rightarrow 3\theta = \frac{3\pi}{2}, \frac{7\pi}{2}, \frac{11\pi}{2}, \ldots$

$\Rightarrow \theta = \frac{\pi}{2}, \frac{7\pi}{6}, \frac{11\pi}{6}, \ldots$

(Continued on the next page)

And it would also be a good idea to find where r is zero:

$r = 0 \Rightarrow \sin 3\theta = 0 \Rightarrow 3\theta = 0, \pi, 2\pi, 3\pi, 4\pi, 5\pi, 6\pi, \ldots$

$\Rightarrow \theta = 0, \frac{\pi}{3}, \frac{2\pi}{3}, \pi, \frac{4\pi}{3}, \frac{5\pi}{3}, 2\pi, \ldots$

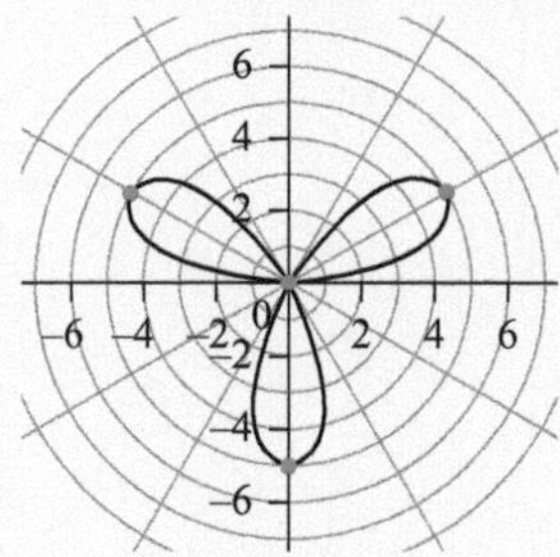

As θ increases from 0, r increases from 0 until it reaches 5, when $\theta = \frac{\pi}{6}$

Then r decreases until it reaches 0 when $\theta = \frac{\pi}{3}$

This makes the top right loop on the curve.

When θ is between $\frac{\pi}{3}$ and $\frac{2\pi}{3}$, r is negative, going from 0 to -5 and back to 0 again. This makes the bottom loop.

Then when θ is between $\frac{2\pi}{3}$ and π, r goes from 0 to 5 and back to 0 again, making the top left loop.

When θ goes above π, the curve traces over the same path that it has already completed.

b Here, you are only interested in the positive values of r (or zero), so you only need values of θ for which $5\sin 3\theta$ is positive or zero. This gives you the top right loop when $0 \leq \theta \leq \frac{\pi}{3}$, the top left loop when $\frac{2\pi}{3} \leq \theta \leq \pi$, and the bottom loop when $\frac{4\pi}{3} \leq \theta \leq \frac{5\pi}{3}$.

However, if the equation had been

$$r = 5\sin 3\theta,\ 0 \leq \theta < \pi,\ r \geq 0$$

with θ only going as far as π, then the bottom loop would be missing.

Key point

Pay careful attention to the range of values of θ and r specified in the question.

PURE

Example 5

a Draw the graph
$r=1-2\sin\theta,\ 0\le\theta<2\pi$

b Draw the graph
$r=1-2\sin\theta,\ 0\le\theta<2\pi,\ r\ge0$

a First notice that

$-1\le 1-2\sin\theta\le 3$

$\Rightarrow -1\le r\le 3$

When $r=0$, $1-2\sin\theta=0 \Rightarrow \theta=\frac{\pi}{6},\frac{5\pi}{6}$

When $r=-1$, $1-2\sin\theta=-1 \Rightarrow \theta=\frac{\pi}{2}$

When $r=3$, $1-2\sin\theta=3 \Rightarrow \theta=\frac{3\pi}{2}$

When $r=1$, $1-2\sin\theta=1 \Rightarrow \theta=0,\ \pi$

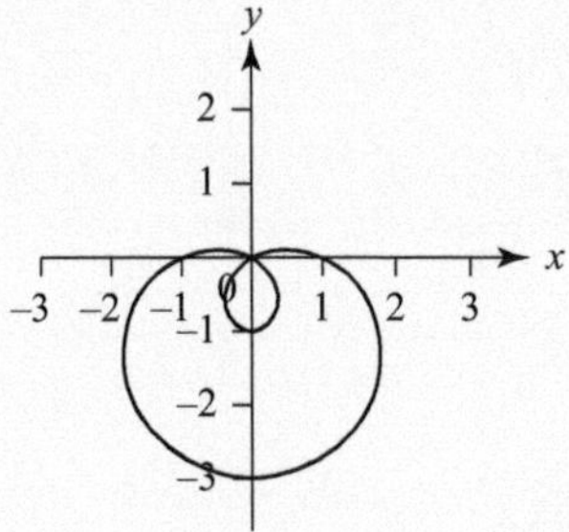

When $0<\theta<\frac{\pi}{6}$, r decreases from 1 to 0

When $\frac{\pi}{6}<\theta<\frac{5\pi}{6}$, r decreases from 0 to -1 and increases back to 0

When $\frac{5\pi}{6}<\theta<2\pi$, r increases from 0 to 3 and decreases back to 1

b As in example 4b, you are not interested in the values of θ for which r is negative, so the smaller inner loop is missing.

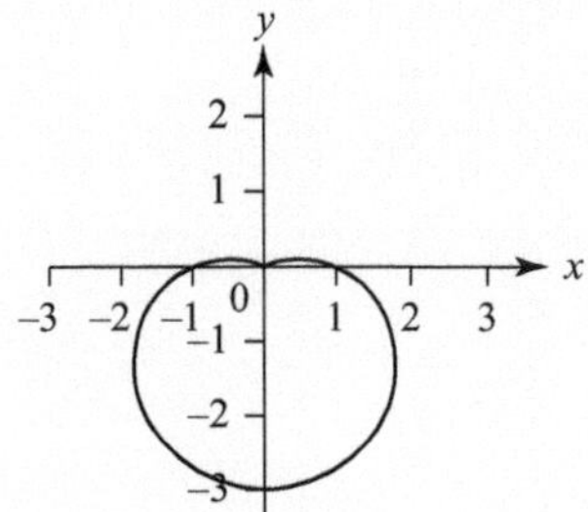

Example 6

Draw these graphs:

a $r = 4$ **b** $\theta = \frac{5\pi}{6}$ **c** $\theta = \frac{5\pi}{6}, r \geq 0$ **d** $r = \theta, r \geq 0$

a This graph includes all points that are 4 units away from the pole, so it is a circle radius 4 with centre at the pole.

b This graph includes all points on the line that forms an angle $\frac{5\pi}{6}$ with the positive x axis, so it is the line $y = -\frac{\sqrt{3}}{3}x$

c In this case, r cannot take negative values, so this graph is only the half line.

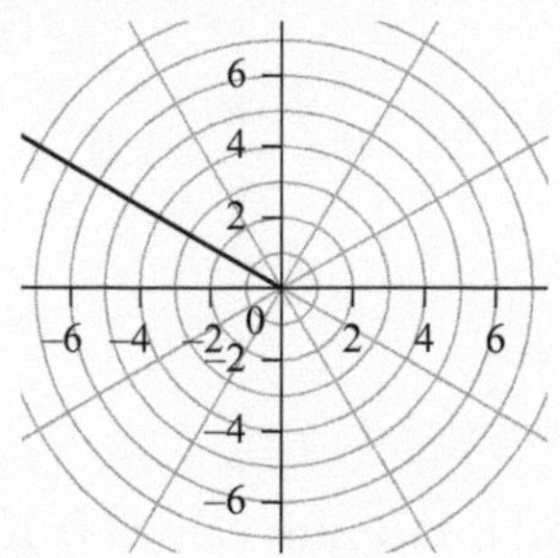

d As θ gets bigger, so does r

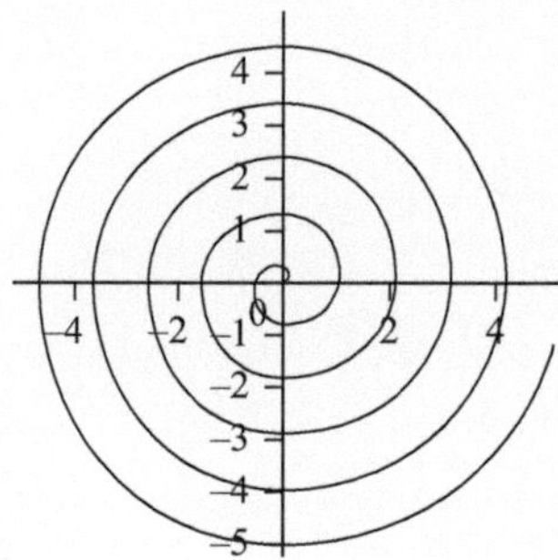

Exercise 3.3A Fluency and skills

1 Draw a polar chart similar to the one shown and plot the given points.

Convert each set of polar coordinates to their Cartesian equivalent.

a $\left(2, \frac{\pi}{3}\right)$ **b** $\left(4, \frac{4\pi}{3}\right)$ **c** $\left(2, \frac{5\pi}{3}\right)$

d $\left(3, \frac{2\pi}{3}\right)$ **e** $\left(3, -\frac{5\pi}{6}\right)$ **f** $\left(3, -\frac{2\pi}{3}\right)$

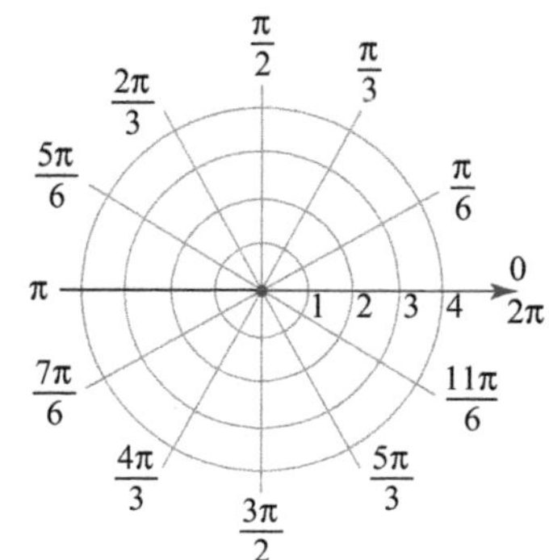

2 Convert these Cartesian coordinates to polar form (r, θ) where $0 < \theta \leq 2\pi$. Give your answers to 3 sf where appropriate.

a $(2\sqrt{3}, 2)$ b $(-3, 3\sqrt{3})$ c $(4, -4\sqrt{3})$ d $(-5, -12)$

3 Convert these Cartesian coordinates to polar form (r, θ) where $-\pi < \theta \leq \pi$

Give r as a surd and angles to 3 sf where appropriate.

a $(-2\sqrt{3}, -2)$ b $(5, -5\sqrt{3})$ c $(-3, 4)$ d $(1, 6)$

4 State the polar equation of the curve $x^2 + y^2 = 5$ and describe the curve.

5 Describe the curve $r = 4$ and state its Cartesian equation.

6 Sketch the graphs for each of these polar equations and find the Cartesian equation of the line they lie on.

a $\theta = \frac{\pi}{4}, r \geq 0$ b $\theta = \frac{\pi}{2}, r \geq 0$ c $\theta = \frac{3\pi}{4}, r \geq 0$ d $\theta = -\frac{\pi}{6}, r \geq 0$

7 a Plot the curve $r = \theta$ in the domain $0 \leq \theta \leq 6\pi$ using step-sizes of $\frac{\pi}{4}$

b Plot $r = \frac{10}{2 + \cos\theta}$ in the domain $0 \leq \theta \leq 2\pi$

8 Sketch these polar curves for $0 \leq \theta \leq 2\pi$ and state the maximum and minimum value of r in each case.

a i $r = \cos 2\theta, r \geq 0$ ii $r = \cos 2\theta, r \in \mathbb{R}$

b $r = \sin 4\theta, r \geq 0$

c i $r = 2\cos 3\theta, r \geq 0$ ii $r = 2\cos 3\theta, r \in \mathbb{R}$

d i $r = 4\sin 2\theta, r \geq 0$ ii $r = 4\sin 2\theta, r \in \mathbb{R}$

e $r = 1 + \cos\theta$

f $r = 4 + \sin\theta$

g $r = 3 - 2\cos\theta$

h i $r = 3 - 5\cos\theta, r \geq 0$ ii $r = 3 - 5\cos\theta, r \in \mathbb{R}$

i $r = 2\theta$

j i $r^2 = 4\sin 2\theta, r \geq 0$ ii $r^2 = 4\sin 2\theta, r \in \mathbb{R}$

PURE

Reasoning and problem-solving

An equation that is simple in (r, θ) coordinates can be complicated in (x, y) coordinates, and vice versa. You can always convert one into the other, though.

Key point

To convert from polar to Cartesian, use

$$r^2 = x^2 + y^2$$

$$\tan\theta = \frac{y}{x}$$

$$\cos\theta = \frac{x}{r} = \pm\frac{x}{\sqrt{x^2+y^2}}$$

$$\sin\theta = \frac{y}{r} = \pm\frac{y}{\sqrt{x^2+y^2}}$$

[Choose + or – depending on whether r is positive or negative]

To convert from Cartesian to polar, use

$$x = r\cos\theta$$

$$y = r\sin\theta$$

The difficulty sometimes lies in simplifying the answer.

Example 7

Convert the equation $\frac{x^2}{3} + \frac{y^2}{4} = 1$ to a polar equation.

$$\frac{(r\cos\theta)^2}{3} + \frac{(r\sin\theta)^2}{4} = 1$$

$$\Rightarrow 4r^2\cos^2\theta + 3r^2\sin^2\theta = 12$$

$$\Rightarrow r^2 = \frac{12}{4\cos^2\theta + 3\sin^2\theta}$$

This is a perfectly good answer, although we could also write:

$$r^2 = \frac{12}{3(\cos^2\theta + \sin^2\theta) + \cos^2\theta}$$

$$\Rightarrow r^2 = \frac{12}{3 + \cos^2\theta}$$

Notice that this implies that r could be positive or negative. So long as the range of values of θ is big enough, however, it makes no difference whether we include the negative values of r or not.

Example 8

Convert the equation $r=\dfrac{\tan\theta}{\cos\theta}$ to a Cartesian equation.

When $r>0$

$$\sqrt{x^2+y^2}=\frac{y}{x}\div\frac{x}{\sqrt{x^2+y^2}}$$

$$=\frac{y\sqrt{x^2+y^2}}{x^2}$$

$$\Rightarrow 1=\frac{y}{x^2}$$

$$\Rightarrow y=x^2$$

When $r<0$

$$-\sqrt{x^2+y^2}=\frac{y}{x}\div\frac{x}{-\sqrt{x^2+y^2}}$$

$$\Rightarrow y=x^2$$

So this is the Cartesian equation of the graph.

We can also find the points of intersection of two curves given their polar equations without having to convert into Cartesian equations.

Example 9

a Find the points of intersection of the curves

$r=\sin 2\theta\ \ r\geq 0$ and $r=\cos 2\theta\ \ r\geq 0$

b Find the points of intersection of the curves $r=\sin 2\theta$ and $r=\cos 2\theta$, where r can be any real number.

In this question, only the values of θ for which r is positive are considered for each curve.

At the points of intersection, $\sin 2\theta=\cos 2\theta$

$$\Rightarrow \tan 2\theta=1$$

$$\Rightarrow 2\theta=\frac{\pi}{4},\frac{5\pi}{4},\frac{9\pi}{4},\frac{13\pi}{4}\ldots$$

$$\Rightarrow \theta=\frac{\pi}{8},\frac{5\pi}{8},\frac{9\pi}{8},\frac{13\pi}{8}\ldots$$

However, r is not positive on both curves for all these values of θ

When $\theta=\dfrac{\pi}{8}$ or $\theta=\dfrac{9\pi}{8}$, $r=\sin 2\theta=\cos 2\theta=\dfrac{\sqrt{2}}{2}$

When $\theta=\dfrac{5\pi}{8}$ or $\theta=\dfrac{13\pi}{8}$, $r=\sin 2\theta=\cos 2\theta=-\dfrac{\sqrt{2}}{2}<0$

So the only solutions are

$$\left(\frac{\sqrt{2}}{2},\frac{\pi}{8}\right),\left(\frac{\sqrt{2}}{2},\frac{9\pi}{8}\right)$$

(Continued on the next page)

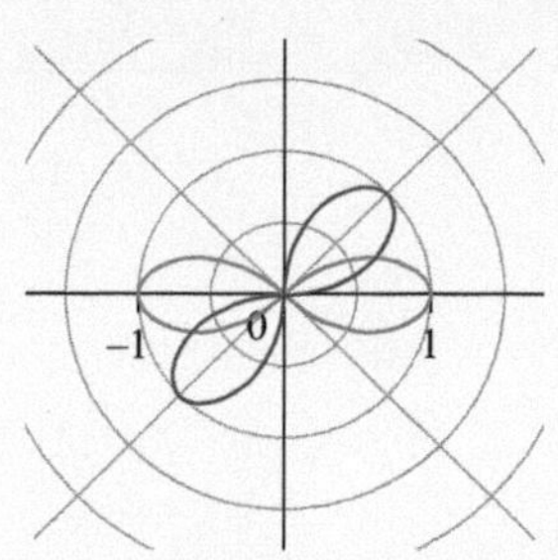

Drawing the curves will help you to visualise what is going on, but you need to choose values of θ that make r positive on each curve:

$0 \le \theta \le \frac{\pi}{2}$ or $\pi \le \theta \le \frac{3\pi}{2} \Rightarrow r = \sin 2\theta \ge 0$

$-\frac{\pi}{4} \le \theta \le \frac{\pi}{4}$ or $\frac{3\pi}{4} \le \theta \le \frac{5\pi}{4} \Rightarrow r = \cos 2\theta \ge 0$

b At the points of intersection, $\sin 2\theta = \cos 2\theta$

$\Rightarrow \tan 2\theta = 1$

$\Rightarrow 2\theta = \frac{\pi}{4}, \frac{5\pi}{4}, \frac{9\pi}{4}, \frac{13\pi}{4} \dots$

$\Rightarrow \theta = \frac{\pi}{8}, \frac{5\pi}{8}, \frac{9\pi}{8}, \frac{13\pi}{8} \dots$

This gives 4 points of intersection. However, drawing the curves shows 8 points of intersection. Where do the other 4 come from? The problem is that polar coordinates are not unique. In this case, notice that the point $\left(\frac{\sqrt{2}}{2}, \frac{3\pi}{8}\right)$ is on the curve $r = \sin 2\theta$

But this point is the same as $\left(-\frac{\sqrt{2}}{2}, \frac{11\pi}{8}\right)$, which is on the curve $r = \cos 2\theta$, and at this point, $\tan 2\theta = -1$

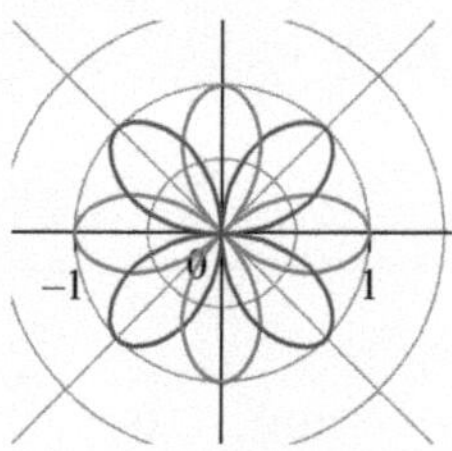

This means you also need to find the solutions to

$\tan 2\theta = -1$

$\Rightarrow 2\theta = \frac{3\pi}{4}, \frac{7\pi}{4}, \frac{11\pi}{4}, \frac{15\pi}{4} \dots$

$\Rightarrow \theta = \frac{3\pi}{8}, \frac{7\pi}{8}, \frac{11\pi}{8}, \frac{15\pi}{8} \dots$

giving 8 points of intersection:

$\left(\frac{\sqrt{2}}{2}, \frac{\pi}{8}\right), \left(\frac{\sqrt{2}}{2}, \frac{3\pi}{8}\right), \left(\frac{\sqrt{2}}{2}, \frac{5\pi}{8}\right), \left(\frac{\sqrt{2}}{2}, \frac{7\pi}{8}\right), \left(\frac{\sqrt{2}}{2}, \frac{9\pi}{8}\right), \left(\frac{\sqrt{2}}{2}, \frac{11\pi}{8}\right), \left(\frac{\sqrt{2}}{2}, \frac{13\pi}{8}\right), \left(\frac{\sqrt{2}}{2}, \frac{15\pi}{8}\right)$

Perhaps the easiest way to deal with this is to sketch the two curves, and use the idea of symmetry to see where all the solutions lie.

Exercise 3.3B Reasoning and problem-solving

1 Find the polar equation for

a The line passing through the origin with gradient 1

b The line passing through the origin with gradient 2

c The positive y-axis

d The positive x-axis

e $y=5$

f $x=3$

g $y=x+1$

h $y=mx+c$

i The circle centre the origin, radius 5

j The circle with centre (5, 12) and radius 13

k The circle with centre (3, 0) passing through the origin

l The circle with centre (0, 4) passing through the origin

m The circle with centre (5, 5) passing through the origin.

n $y=x^2$

o $x^2+y^2+2xy=5$

2 Convert these polar equations into Cartesian form and describe the graphs.

a $r=\dfrac{1}{\sin\theta+\cos\theta}$ b $r=\dfrac{5}{\sin\theta-2\cos\theta}$

c $r=\dfrac{2}{\cos\theta}$ d $r=\dfrac{-2}{3\sin\theta}$

e $r=4\sin\theta$ f $r=8\cos\theta+6\sin\theta$

3 Find the points of intersection between these pairs of curves for $-\pi<\theta\leq\pi$ and $r\geq 0$
Give angles in terms of π

a $r=2$, $r=3-2\sin\theta$

b $r=\cos 4\theta$, $r=\dfrac{1}{2}$

c $r=\sin 2\theta$, $r=\cos 2\theta$

4 Show algebraically that any spiral of the form $r=a\theta$ will intersect any circle centred on the origin precisely once.

5 For each pair of equations, the domain is $0 \le \theta \le 2\pi$, $r \ge 0$ unless otherwise stated.

i Sketch the curves on the same graph,

ii Find their points of intersection.

a $r=\sqrt{3}\sin\theta$, $r=\cos\theta$

b $r=\theta$, $r=\theta-\sin\theta$

c $r=2+\cos\theta$, $r=\cos^2\theta$

d $r=\sin^2\theta$, $r=\cos\theta$

e $r=-\sin 3\theta$, $r=\sqrt{3}\cos 3\theta$, $r \ge 0$

f $r=-\sin 3\theta$, $r=\sqrt{3}\cos 3\theta$, $r \in \mathbb{R}$

6 The family of curves $r=a+4\cos\theta$, $-\pi<0\le\pi$, $r\in\mathbb{R}$ produces different shapes depending on the choice of value for a

a Show that if a is greater than 4 then the curve never passes through the pole.

b Prove that the curve is symmetrical about the initial line for any value of a

c Hence, sketch the curve when

i $a=0$ ii $0<a\le 4$ iii $4<a<8$ iv $a\ge 8$

d State, in terms of a, the maximum and minimum values of r for the curves in part **c**.

7 Show that the graphs of $r=\sec\theta$ and $r^2=2\tan\theta$ intersect only once whether or not negative values of r are included, and find their point of intersection.

8 Find the points of intersection between these pairs of curves for $-\pi<\theta\le\pi$ and $r\ge 0$
Give your coordinates to 3 sf.

a $r=2\cos 2\theta$, $r=1+\sin\theta$

b $r^2=4\cos\theta$, $r=2-\cos\theta$

What is the difference if r can take any value, including negatives?

9 Show that the points of intersection between the curves $r=2\sin^2\theta$ and $r=\cos 2\theta$ form a rectangle and find its area.

3.4 Parabolas, ellipses and hyperbolas

Fluency and skills

Conic sections are a special category of curve. They are formed by the intersection of a plane and a cone. You can think of this as slicing through the cone. The angle at which the plane intersects the cone determines the type of curve which is formed. For example, a plane parallel to the base of the cone will form a circle and a plane parallel to the curved face will cut through the base and the curved face and form a **parabola**.

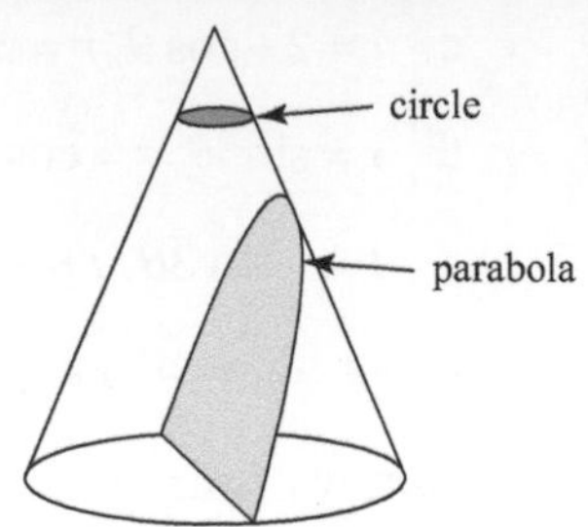

You are already familiar with the parabola $y = Ax^2$ and with transformations of this curve.

Parabolas can also be of the form $x = Ay^2$ which can be rearranged to make y^2 the subject.

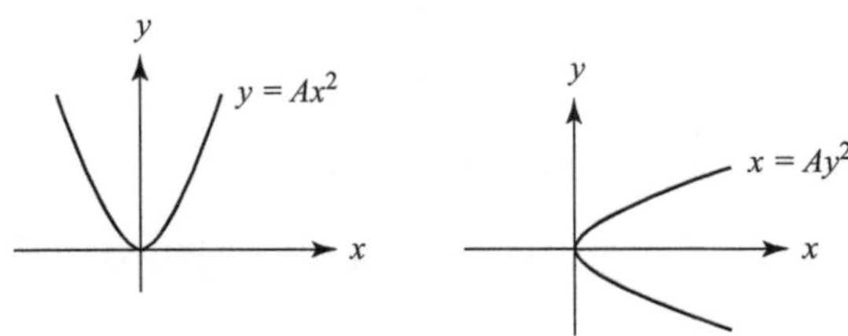

Key point

An equation of the form $y^2 = kx$ describes a parabola with its vertex at the origin.

You may also see the standard equation of a parabola written as $y^2 = 4ax$

Example 1

A curve has equation $y^2 = 24x$

a Sketch the curve, giving the coordinates of any intercepts with the coordinate axes.

b Find the points of intersection between the curve and the line with equation $y = 9 - 2x$

a

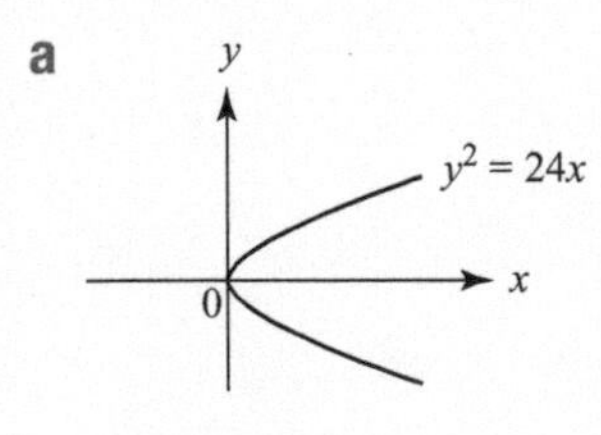

b $(9-2x)^2 = 24x$ — Solve the equations simultaneously.

$81 - 36x + 4x^2 = 24x$

$4x^2 - 60x + 81 = 0$

(*Continued on the next page*)

$x = \frac{27}{2}, \frac{3}{2}$

$y = -18, 6$

So the curve and line intersect at $\left(\frac{27}{2}, -18\right)$ and $\left(\frac{3}{2}, 6\right)$

Factorise or use your calculator to solve the quadratic.

Another type of conic section is an **ellipse**. The ellipse is formed by the intersection of a cone with a plane that cuts through only the curved surface of the cone. If this plane is parallel to the base of the cone then a circle if formed, but in general an ellipse will be formed.

circle

ellipse

You know that the equation of a circle with centre the origin is $x^2 + y^2 = r^2$

This can be written $\frac{x^2}{r^2} + \frac{y^2}{r^2} = 1$

In contrast, an ellipse does not have a constant radius.

Its equation can be written $\frac{x^2}{a^2} + \frac{y^2}{b^2} = 1$ where $a, b > 0$

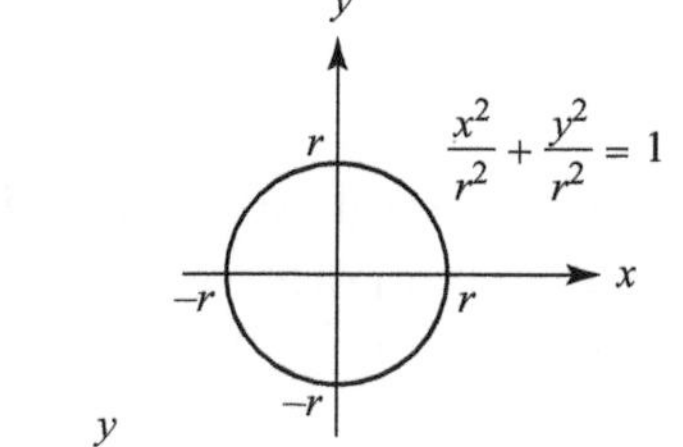

Key point

An equation of the form $\frac{x^2}{a^2} + \frac{y^2}{b^2} = 1$ describes an ellipse centred on the origin. The ellipse will pass through the points $(\pm a, 0)$ and $(0, \pm b)$

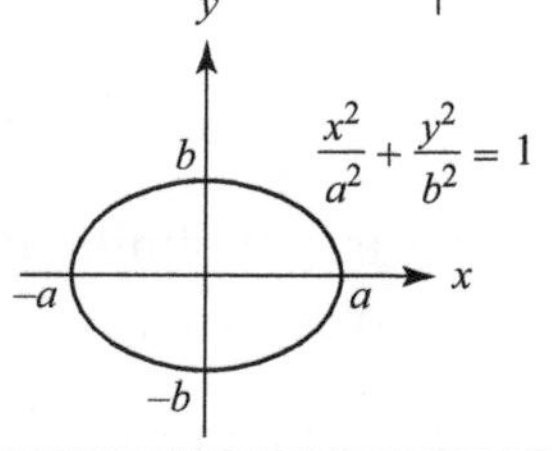

Example 2

Sketch each of these ellipses, clearly labelling their points of intersection with the coordinate axes.

a $\frac{x^2}{4} + \frac{y^2}{9} = 1$ **b** $x^2 + 3y^2 = 3$

a You can see that $a^2 = 4$ so $a = 2$

Check by letting $y = 0$, this gives $x^2 = 4 \Rightarrow x = \pm 2$

and $b^2 = 9$ so $b = 3$

Check by letting $x = 0$, this gives $y^2 = 9 \Rightarrow y = \pm 9$

So the graph looks like this

y
3
-2
2
x
-3

Try to make your graph approximately to scale.

b Rearrange the equation to $\frac{x^2}{3} + y^2 = 1$

So $a = \sqrt{3}$ and $b = 1$

Writing the equation in the usual form makes it easier to identify the values of a and b

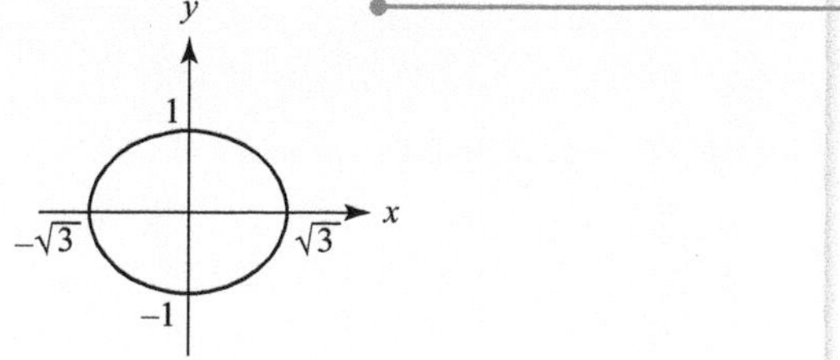

Example 3

A circle with equation $x^2 + y^2 = 9$ is stretched by scale factor 2 in the y-direction to form an ellipse.

Sketch the ellipse and state its equation.

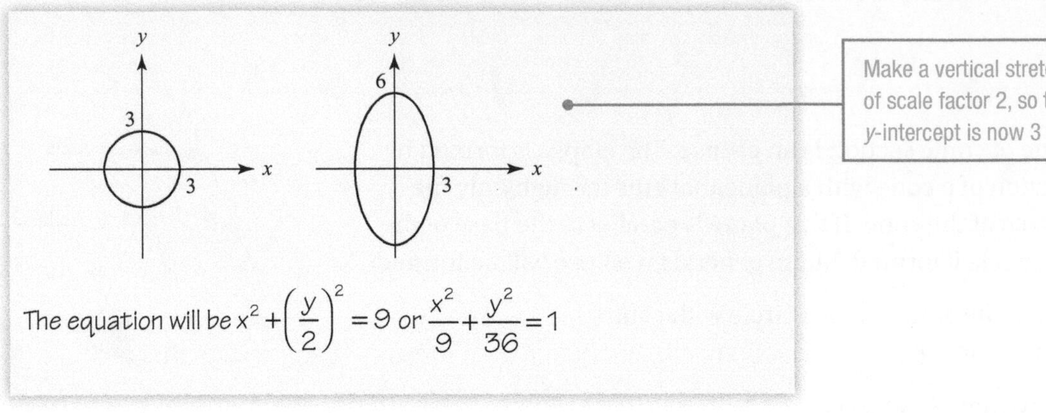

Make a vertical stretch of scale factor 2, so the y-intercept is now $3 \times 2 = 6$

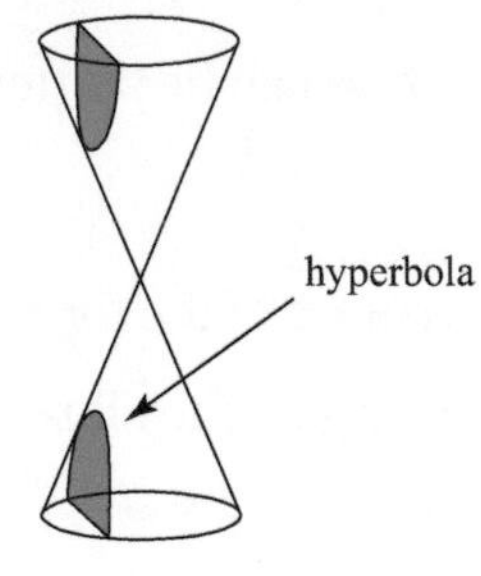

The final conic section is a **hyperbola**, this can be formed by the intersection of a cone with a plane that is steeper than the curved face of the cone. A double-cone is shown as the plane will intersect both cones.

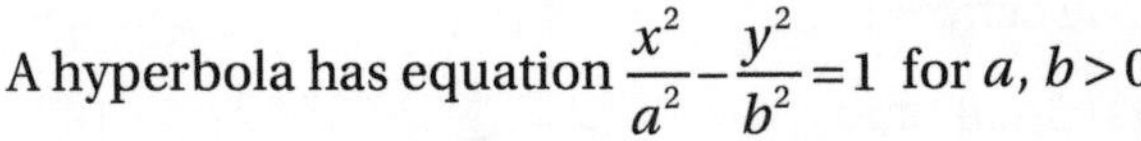

A hyperbola has equation $\frac{x^2}{a^2} - \frac{y^2}{b^2} = 1$ for $a, b > 0$

Notice that this is very similar to the equation of the ellipse. Only the minus sign is different.

You can see that when $y = 0$, $x^2 = a^2$ so the curve crosses the x-axis at $(\pm a, 0)$. However, if you substitute $x = 0$ into the equation you get

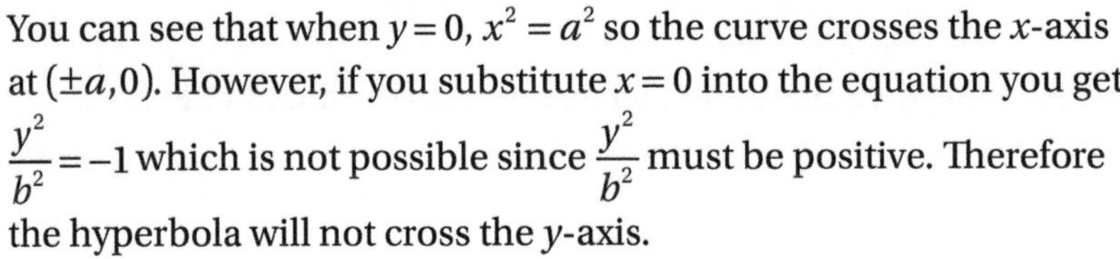

$\frac{y^2}{b^2} = -1$ which is not possible since $\frac{y^2}{b^2}$ must be positive. Therefore the hyperbola will not cross the y-axis.

In fact, since the equation can be rearranged to give $\frac{x^2}{a^2} = 1 + \frac{y^2}{b^2}$,

you can see that $\frac{x^2}{a^2} \geq 1$, therefore x is only defined when $|x| \geq a$

As x and y become large, $\frac{x^2}{a^2} = 1 + \frac{y^2}{b^2}$ becomes $\frac{x^2}{a^2} \approx \frac{y^2}{b^2}$

So the graph has asymptotes at $y^2 = \frac{b^2}{a^2}x^2$, i.e. $y = \pm\frac{b}{a}x$

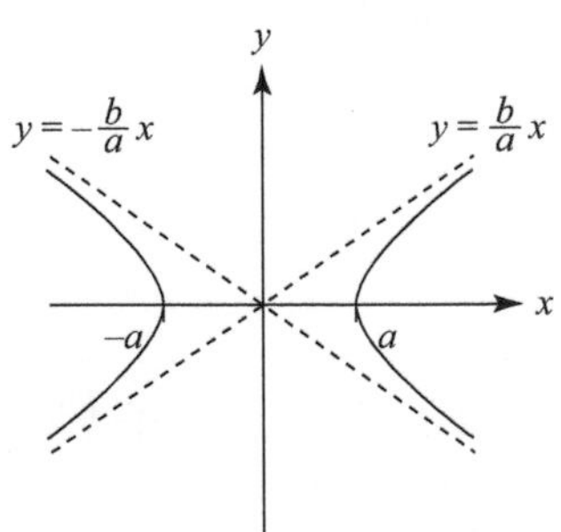

Key point

An equation of the form $\frac{x^2}{a^2} - \frac{y^2}{b^2} = 1$ describes a hyperbola centred on the origin with asymptotes $y = \pm\frac{b}{a}x$

Example 4

Sketch the hyperbola $\frac{x^2}{4}-\frac{y^2}{12}=1$ and state the equations of the asymptotes.

In this equation, $a=\sqrt{4}=2$ and $b=\sqrt{12}=2\sqrt{3}$

$\frac{b}{a}=\frac{2\sqrt{3}}{2}$, so the equations of the asymptotes are $y=\pm\sqrt{3}x$

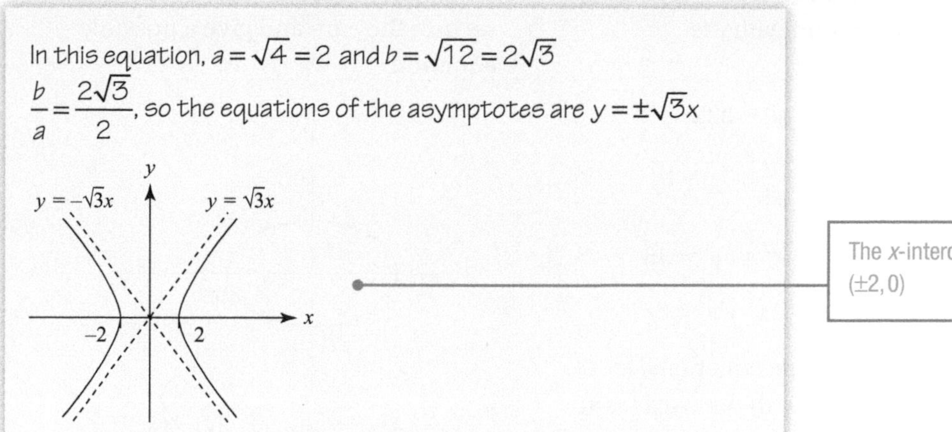

The x-intercepts are at $(\pm2, 0)$

A hyperbola whose asymptotes are the coordinate axes is called a **rectangular hyperbola**. The equation of a rectangular hyperbola is $xy=c^2$

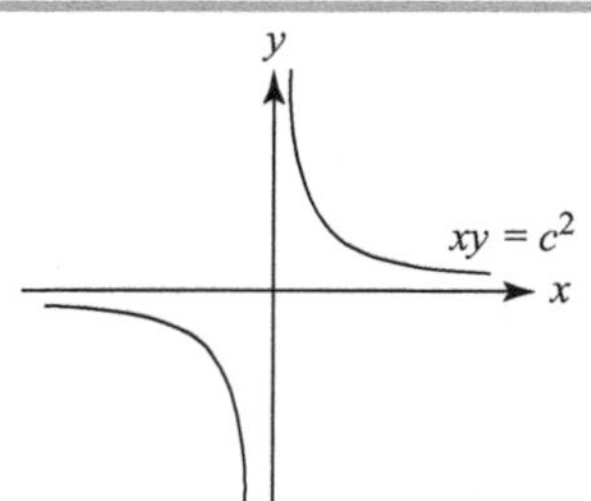

Key point

An equation of the form $xy=c^2$ describes a rectangular hyperbola centred on the origin with asymptotes $y=0$ and $x=0$

Example 5

A curve, C, has equation $xy=25$

a Sketch C, giving the equations of the asymptotes.

The line with equation $y=2x+5$ intersects C at the points A and B

b Calculate the length of the line segment AB

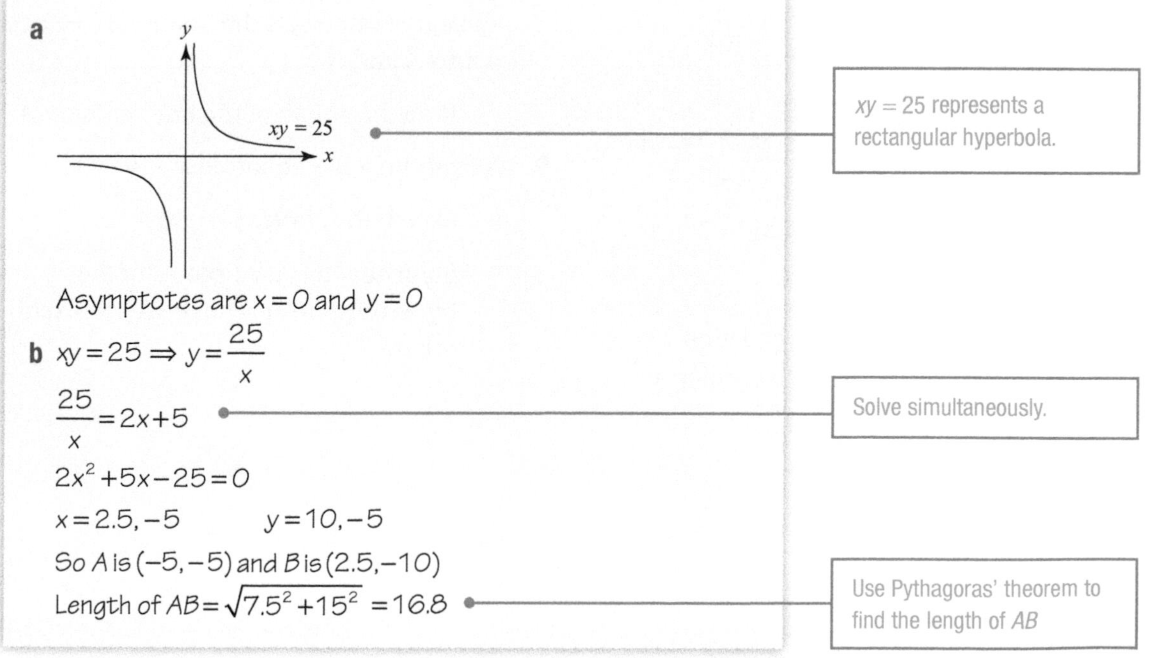

a

Asymptotes are $x=0$ and $y=0$

$xy=25$ represents a rectangular hyperbola.

b $xy=25 \Rightarrow y=\frac{25}{x}$

$\frac{25}{x}=2x+5$

Solve simultaneously.

$2x^2+5x-25=0$

$x=2.5, -5 \qquad y=10, -5$

So A is $(-5, -5)$ and B is $(2.5, -10)$

Length of $AB=\sqrt{7.5^2+15^2}=16.8$

Use Pythagoras' theorem to find the length of AB

Exercise 3.4A Fluency and skills

1 Sketch each of these curves, clearly labelling their points of intersection with the coordinate axes.

a $y=2x^2$

b $y^2=32x$

c $\frac{x^2}{9}+\frac{y^2}{4}=1$

d $\frac{x^2}{16}+\frac{y^2}{25}=1$

e $x^2+\frac{y^2}{7}=1$

f $x^2+9y^2=36$

g $9x^2+4y^2=36$

h $4x^2+y^2=8$

2 Sketch the curves with these equations, label the points of intersection with the x-axis and state the equation of the asymptotes in each case.

a $\frac{x^2}{4}-\frac{y^2}{9}=1$

b $\frac{x^2}{5}-\frac{y^2}{3}=1$

c $xy=14$

d $x=\frac{3}{y}$

e $x^2-\frac{y^2}{2}=1$

f $3x^2-y^2=6$

g $x^2-5y^2=5$

h $2x^2-5y^2=20$

3 A parabola has equation $y^2=20x$

a Sketch the curve.

b Find the points of intersection between the parabola and the line with equation $y=2x$

4 An ellipse has equation $\frac{x^2}{4}+\frac{y^2}{5}=1$

a Sketch the curve.

b Find the points of intersection between the ellipse and the line $y=x+2$

5 An ellipse has equation $x^2+3y^2=7$

a Sketch the curve.

b Show that the line with equation $2x+3y=7$ is a tangent to the ellipse $x^2+3y^2=7$

6 For each graph, state what type of conic section they are and give a possible equation.

a

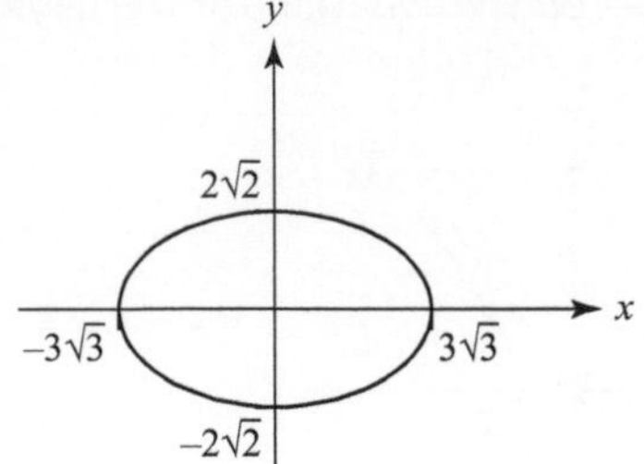

b

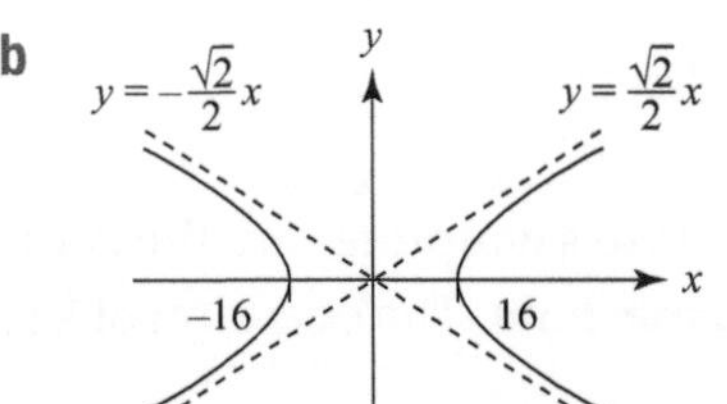

7 The asymptotes of the hyperbola $\frac{x^2}{a^2}-\frac{y^2}{b^2}=1$ are $y=\pm 3x$
Find the equation of the hyperbola given that it passes through the point $(1, 1)$

8 A curve has equation $xy=16$

a What name is given to this type of curve?

b Sketch the curve, giving the equations of the asymptotes.

The curve intersects the line $y=3x+2$ at the points A and B

c Find the length of the line segment AB

9 A hyperbola has equation $x^2-y^2=k^2$

a Sketch the curve.

b Show that the asymptotes of the hyperbola are at right angles to each other.

PURE

Reasoning and problem-solving

Replacing $f(x)$ by $f(x-c)$ in an equation translates the curve c units in the x-direction. Similarly, replacing $g(y)$ by $g(y-c)$ translates the curve c units in the y-direction.

To sketch translations of conic sections, identify the type of curve involved and find the vertex (for a parabola) or centre (for a hyperbola or an ellipse) of the transformed curve.

Strategy 1

To sketch a translation of a conic section

1. Identify the curve by converting the equation to a standard form.
2. Find the centre or vertex of the transformed curve.
3. Sketch the transformed curve, stating the values of any intercepts and the equations of any asymptotes.

Example 6

Sketch the curve of $y^2 = 2x+10$

The equation can be written as $y^2 = 2(x+5)$

The vertex of the curve has moved from the origin to $(-5, 0)$

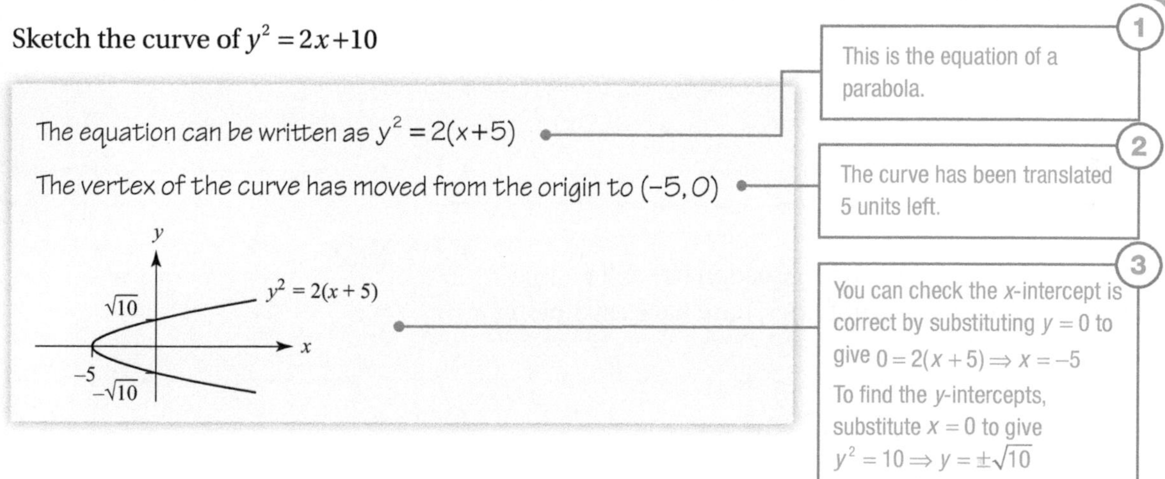

1. This is the equation of a parabola.
2. The curve has been translated 5 units left.
3. You can check the x-intercept is correct by substituting $y = 0$ to give $0 = 2(x+5) \Rightarrow x = -5$
   To find the y-intercepts, substitute $x = 0$ to give $y^2 = 10 \Rightarrow y = \pm\sqrt{10}$

Key point

A parabola with equation $(y-y_1)^2 = 4a(x-x_1)$ will have its vertex on the point (x_1, y_1)

Example 7

Sketch the curve of $\frac{x^2}{4}+(y-5)^2 = 25$

The equation can be written as $\frac{x^2}{100}+\frac{(y-5)^2}{25} = 1$

The centre of the curve has moved from the origin to $(0, 5)$

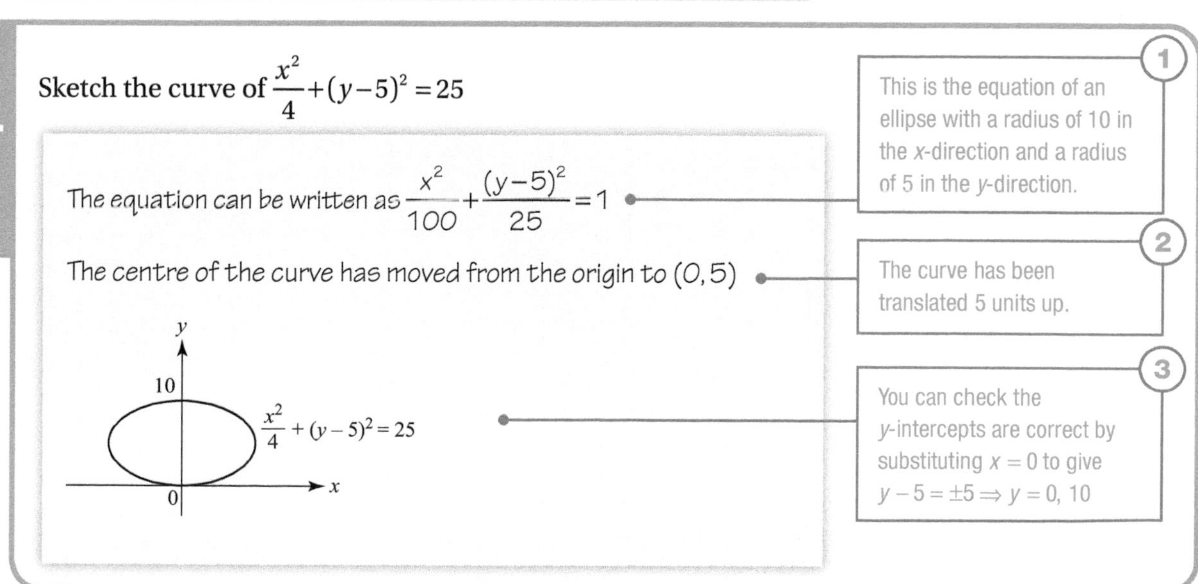

1. This is the equation of an ellipse with a radius of 10 in the x-direction and a radius of 5 in the y-direction.
2. The curve has been translated 5 units up.
3. You can check the y-intercepts are correct by substituting $x = 0$ to give $y - 5 = \pm 5 \Rightarrow y = 0, 10$

Key point

An ellipse with equation $\frac{(x-x_1)^2}{a^2}+\frac{(y-y_1)^2}{b^2}=1$ will be centred on the point (x_1, y_1) and have radius of a in the x-direction and b in the y-direction.

Example 8

Sketch the curve of $x+xy=4$

The equation can be written as $x(y+1)=4$

The centre of the curve has moved from the origin to $(0, -1)$

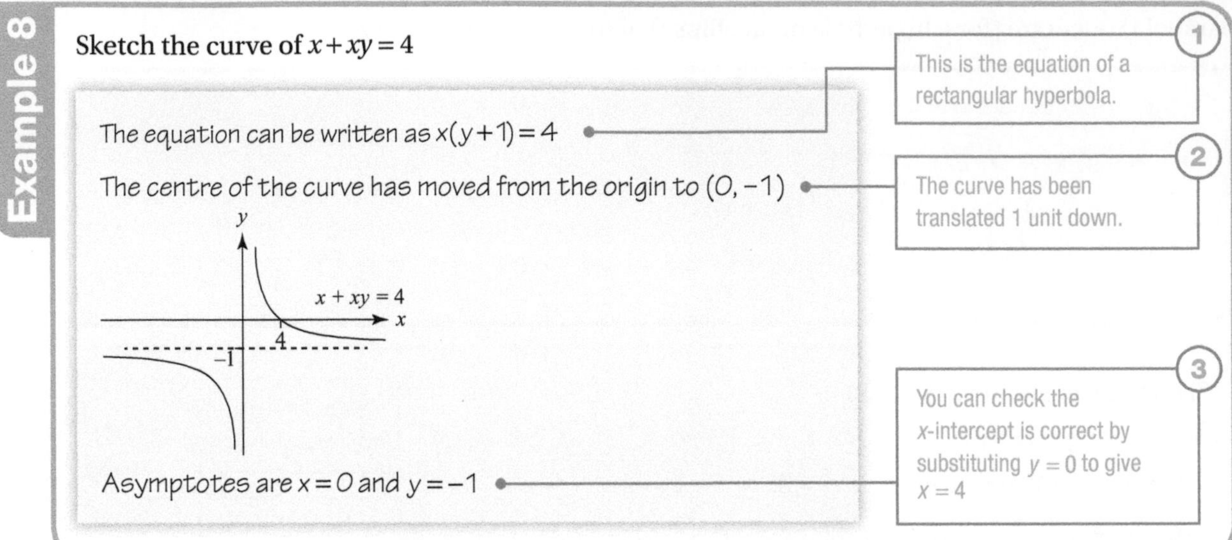

Asymptotes are $x=0$ and $y=-1$

1. This is the equation of a rectangular hyperbola.
2. The curve has been translated 1 unit down.
3. You can check the x-intercept is correct by substituting $y=0$ to give $x=4$

Key point

A rectangular hyperbola with equation $(x-x_1)(y-y_1)=c^2$ will be centred on the point (x_1, y_1) and have asymptotes $x=x_1$, $y=y_1$

Example 9

Sketch the curve of $12(x-3)^2-3y^2=48$

The equation can be written as $\frac{(x-3)^2}{4}-\frac{y^2}{16}=1$

The centre of the curve has moved from the origin to $(3, 0)$

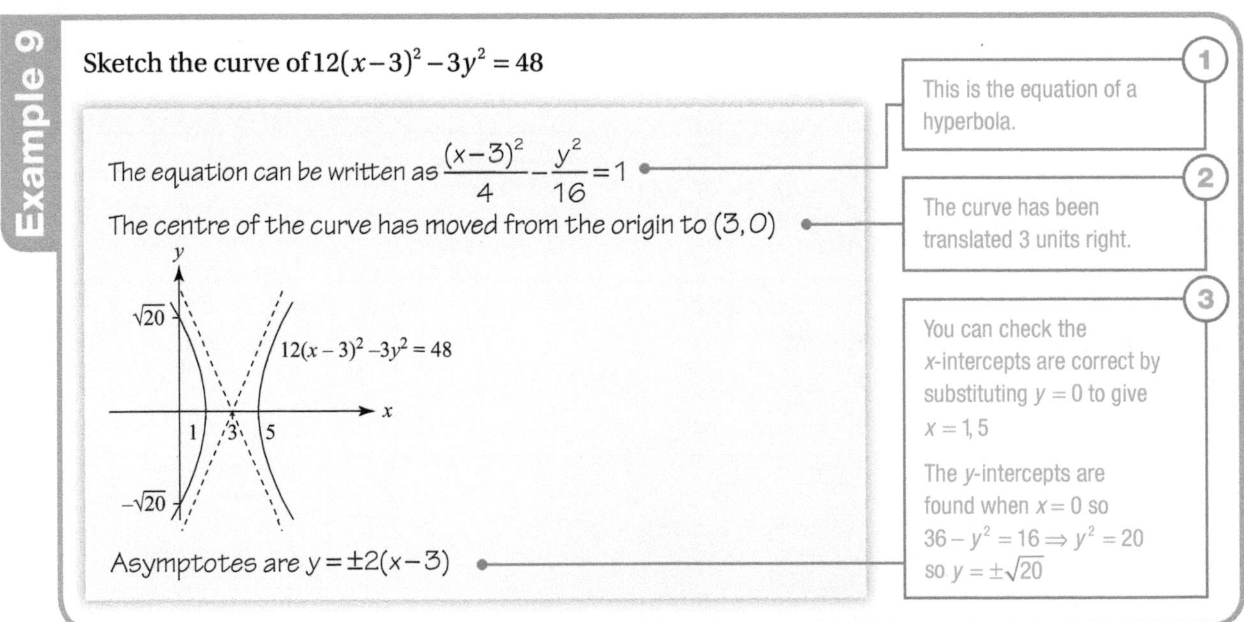

Asymptotes are $y=\pm 2(x-3)$

1. This is the equation of a hyperbola.
2. The curve has been translated 3 units right.
3. You can check the x-intercepts are correct by substituting $y=0$ to give $x=1, 5$

   The y-intercepts are found when $x=0$ so $36-y^2=16 \Rightarrow y^2=20$ so $y=\pm\sqrt{20}$

Key point

A hyperbola with equation $\frac{(x-x_1)^2}{a^2}-\frac{(y-y_1)^2}{b^2}=1$ will be centred on the point (x_1, y_1) and have asymptotes $y-y_1=\pm\frac{b}{a}(x-x_1)$

Replacing $f(x)$ by $f\left(\frac{x}{k}\right)$ in an equation will stretch the curve by scale factor k in the x-direction. Similarly, replacing $g(y)$ by $g\left(\frac{y}{k}\right)$ will stretch the curve by scale factor k in the y-direction.

PURE

Strategy 2

To sketch a stretch or a reflection in a coordinate axis of a conic section

1. Identify the curve by converting the equation to a standard form.
2. Consider the direction to stretch in or the axis to reflect in.
3. Sketch the transformed curve, stating the values of any intercepts and the equations of any asymptotes.

Example 10

Sketch the curve of $y^2 = -2x$

Notice that a stretch or a reflection in a coordinate axis will not affect the vertex.

The equation can be written as $y^2 = 2(-x)$

The curve has been reflected in the line $x = 0$

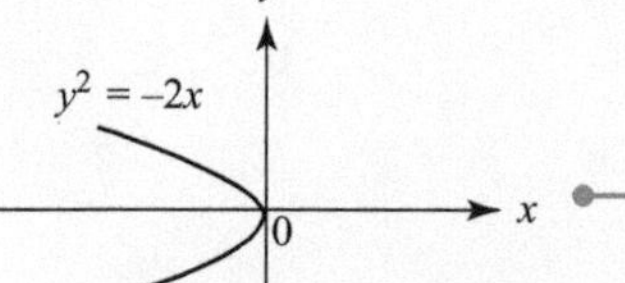

1. This is the equation of a parabola.
2. Note that in this case a reflection in the line $y = 0$ would have no effect since $(-y)^2 = y^2$
3. The x-intercept remains at the origin.

Example 11

The curve with equation $x^2 + 5y^2 = 5$ is stretched by scale factor 3 in the y-direction. Write the equation of the transformed curve and state its points of intersection with the coordinate axes.

Equation of original curve is $\frac{x^2}{5} + y^2 = 1$

A stretch of scale factor 3 in the y-direction means equation becomes $\frac{x^2}{5} + \left(\frac{y}{3}\right)^2 = 1$

Which simplifies to $\frac{x^2}{5} + \frac{y^2}{9} = 1$

This is an ellipse which intersects the coordinate axes at $(\pm\sqrt{5}, 0), (0, \pm 3)$

1. This is the equation of an ellipse.

Example 12

The curve with equation $x^2 - 2y^2 = 8$ is stretched by scale factor $\frac{1}{2}$ in the x-direction.
Sketch the transformed curve and state the equations of its asymptotes.

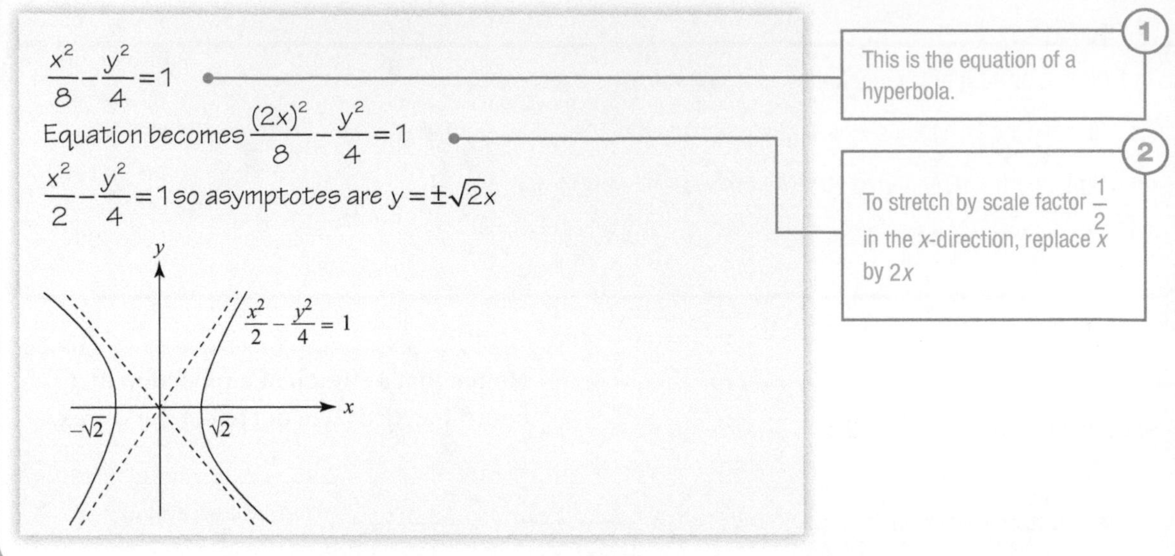

$\frac{x^2}{8} - \frac{y^2}{4} = 1$
Equation becomes $\frac{(2x)^2}{8} - \frac{y^2}{4} = 1$
$\frac{x^2}{2} - \frac{y^2}{4} = 1$ so asymptotes are $y = \pm\sqrt{2}x$
y
$\frac{x^2}{2} - \frac{y^2}{4} = 1$
x
$-\sqrt{2}$
$\sqrt{2}$
1 This is the equation of a hyperbola.
2 To stretch by scale factor $\frac{1}{2}$ in the x-direction, replace x by $2x$

You can reflect in the line $y = x$ by reversing the roles of x and y in the equation.

- A parabola of the form $y^2 = kx$ will become $x^2 = ky$
- An ellipse of the form $\frac{x^2}{a^2} + \frac{y^2}{b^2} = 1$ will become $\frac{y^2}{a^2} + \frac{x^2}{b^2} = 1$
- A hyperbola of the form $\frac{x^2}{a^2} - \frac{y^2}{b^2} = 1$ will become $\frac{y^2}{a^2} - \frac{x^2}{b^2} = 1$ with asymptotes at $x = \pm\frac{b}{a}y$
- A rectangular hyperbola of the form $xy = c^2$ will become $yx = c^2$ so no change!

To reflect in the line $y = -x$, you need to replace x with $-y$ and y with $-x$ in the equation.

Example 13

The curve with equation $(x-1)^2 + 36y^2 = 9$ is reflected in the line $y = -x$

Sketch the transformed curve and state its equation.

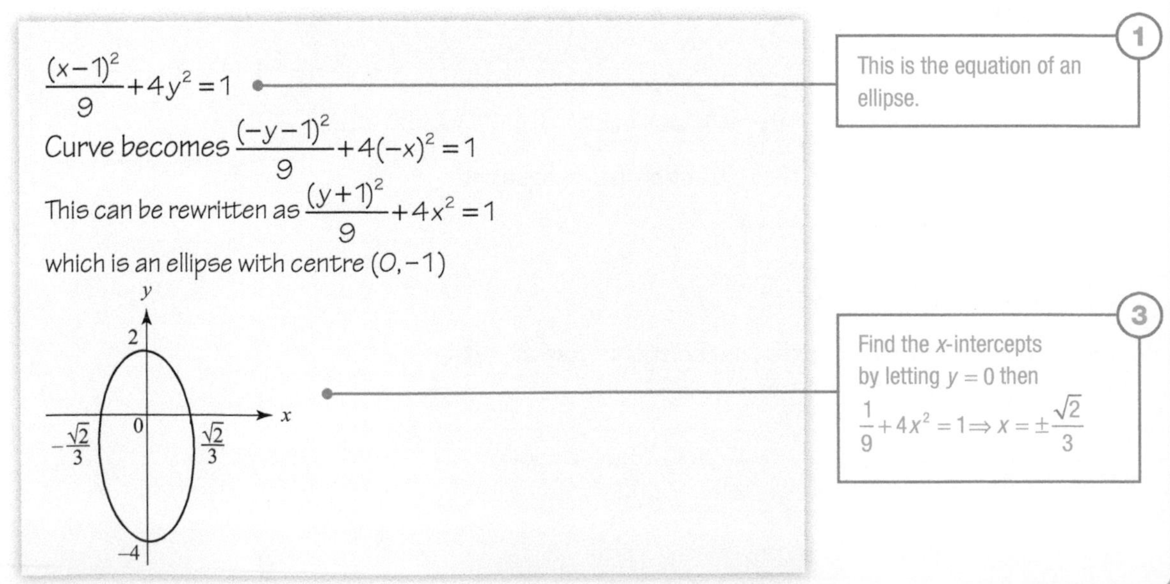

$\frac{(x-1)^2}{9} + 4y^2 = 1$
Curve becomes $\frac{(-y-1)^2}{9} + 4(-x)^2 = 1$
This can be rewritten as $\frac{(y+1)^2}{9} + 4x^2 = 1$
which is an ellipse with centre $(0, -1)$
y
2
x
0
$-\frac{\sqrt{2}}{3}$
$\frac{\sqrt{2}}{3}$
-4
1 This is the equation of an ellipse.
3 Find the x-intercepts by letting $y = 0$ then $\frac{1}{9} + 4x^2 = 1 \Rightarrow x = \pm\frac{\sqrt{2}}{3}$

Exercise 3.4B Reasoning and problem-solving

1 Sketch each of these curves, clearly labelling the points of intersection with the coordinate axes in each case.

a $y^2=3(x-2)$ b $(y-3)^2=3x$

c $(y+2)^2=8(x-1)$ d $y^2=4x+8$

e $y^2+2y=x$ f $\frac{x^2}{5}+\frac{(y+3)^2}{4}=1$

g $\frac{(3+x)^2}{9}+\frac{y^2}{4}=1$ h $x^2+9(y-2)^2=36$

i $(x-4)^2+4(y-2)^2=16$

j $4x^2-8x+y^2-12y+4=0$

2 Sketch each of these curves and state the equations of the asymptotes in each case.

a $x(y+3)=9$ b $(x-4)y=25$

c $(x+3)(y-3)=10$ d $xy-2x=4$

e $xy-5y=2$ f $\frac{x^2}{2}-\frac{(y+5)^2}{9}=1$

g $\frac{(x+2)^2}{16}-\frac{y^2}{25}=1$ h $x^2-4(y-7)^2=36$

i $(x-5)^2-2(y+2)^2=16$

j $x^2+6x-3y^2+30y=147$

3 Sketch each of these curves, clearly labelling the points of intersection with the coordinate axes in each case.

a $y^2=-5x$ b $-xy=7$

4 Write down the equation of the image of each of these transformations of the curve $y^2=8x$ in the form $y^2=\mathrm{f}(x)$

a Stretch by scale factor 3 in the x-direction

b Stretch by scale factor $\frac{1}{4}$ in the y-direction

c Reflection in the x-axis

d Reflection in the y-axis.

5 Write down the equation of the image of each transformation of the curve $y^2=-4x$

a Translation by the vector $\begin{pmatrix}2\\-3\end{pmatrix}$

b Reflection in the line $y=x$

6 Write the equation of the image of each transformation of the curve $\frac{x^2}{12}+\frac{y^2}{4}=1$

Write the equations in the form $\frac{x^2}{a^2}+\frac{y^2}{b^2}=1$

a Stretch vertically by scale factor 2

b Stretch horizontally by scale factor -5

c Reflection in the line $y=0$

d Reflection in the line $y=-x$

7 Write the equation of the curve formed by translating the ellipse $x^2+2y^2=8$ by the vector $\begin{pmatrix}3\\-1\end{pmatrix}$

8 For each of these transformations of the curve $x(y-1)=4$

i Write down the equation of the image,

ii State the equations of the asymptotes.

a Reflection in the line $x=0$

b Translation by the vector $\begin{pmatrix}5\\2\end{pmatrix}$

c Reflection in the line $y=x$

d Reflection in the line $y=-x$

9 For each of these transformations of the curve $9x^2-y^2=81$

i Write the equation of the image,

ii State the equations of the asymptotes.

a Stretch vertically by scale factor -3

b Reflection in the x-axis

c Translation by the vector $\begin{pmatrix}-1\\4\end{pmatrix}$

d Reflection in the line $y=x$

10 A curve has equation $\frac{(x-k)^2}{a^2}-(y+k)^2=a^2$ where a and k are constants, $a, k>0$

a State the equations of the asymptotes to the curve.

b Sketch the reflection of the curve in the line $y=-x$ and state the equations of the asymptotes of the reflected curve.

3.5 Hyperbolic functions 1

Fluency and skills

If you plot all the points of the form $(\sin\theta, \cos\theta)$ where $0 \le \theta \le 2\pi$, you produce a circle (shown by the red curve).

There exists another pair of functions, denoted by $\sinh\theta$ and $\cosh\theta$. If you plot the points $(\sinh\theta, \cosh\theta)$, you produce the curve shown in blue.

This curve is called a hyperbola and so the functions are referred to as **hyperbolic** functions and defined using exponentials.

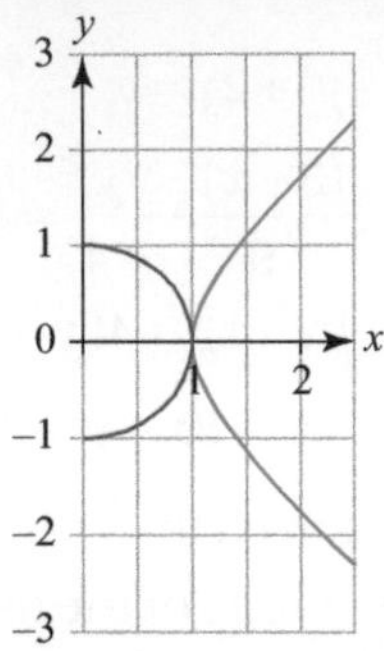

Key point

$$\sinh x = \frac{e^x - e^{-x}}{2} \quad \cosh x = \frac{e^x + e^{-x}}{2}$$

$$\tanh x = \frac{\sinh x}{\cosh x} = \frac{e^x - e^{-x}}{e^x + e^{-x}}$$

These functions are commonly read as 'shine', 'cosh' and 'tanch' or 'than'.

Example 1

Find the exact value of

a $\tanh 0$ **b** $\sinh(\ln 3)$

a $\tanh 0 = \frac{e^0 - e^{-0}}{e^0 + e^{-0}}$ — Use the definition of $\tanh x$

$= \frac{1-1}{1+1}$

$= 0$

b $\sinh(\ln 3) = \frac{e^{\ln 3} - e^{-\ln 3}}{2}$

$= \frac{e^{\ln 3} - e^{\ln 3^{-1}}}{2}$ — Use the fact that $a \ln b = \ln b^a$

$= \frac{3 - \frac{1}{3}}{2}$

$= \frac{4}{3}$

Calculator

Try it on your calculator

You can use a calculator to evaluate hyperbolic functions and inverse hyperbolic functions.

sinh (ln5)

$\frac{12}{5}$

Activity

Find out how to evaluate sinh (ln 5) on your calculator.

The graph $y = \cosh x$

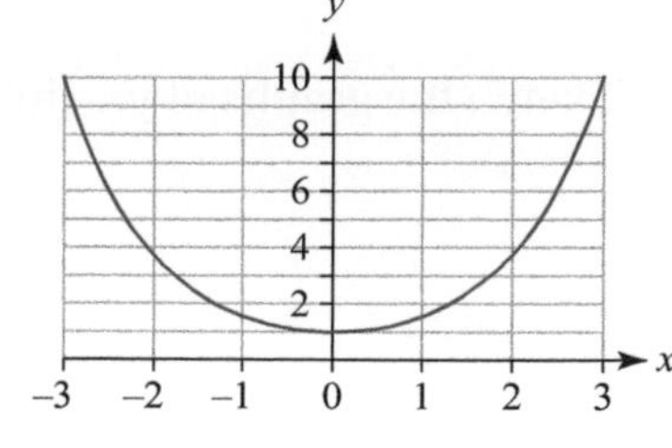

The curve has a minimum point at (0, 1)

So $y \geq 1$

The curve is symmetrical about the y-axis.

The graph $y = \sinh x$

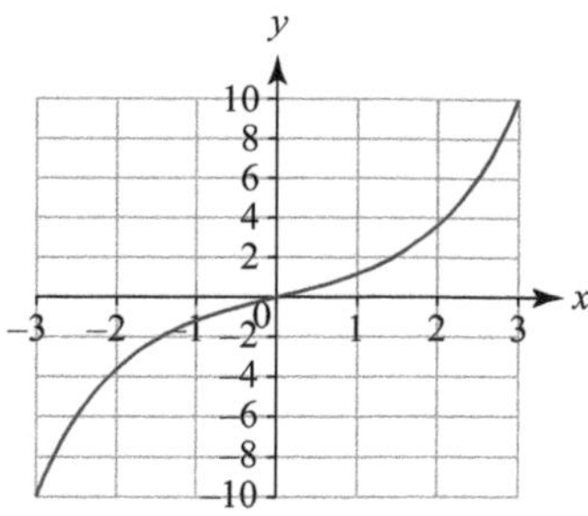

The curve has rotation symmetry about the origin.

The graph $y = \tanh x$

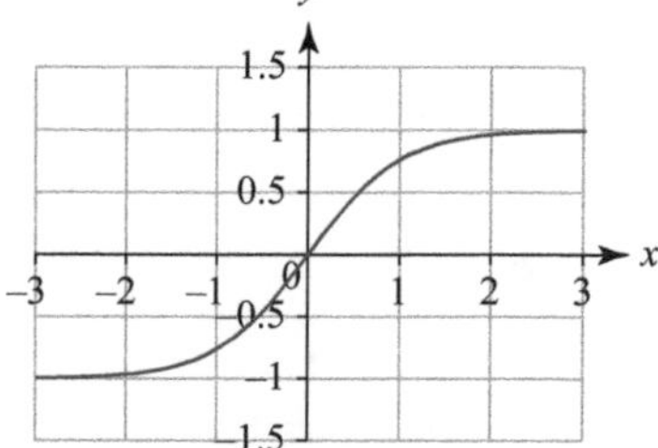

As $x \to +\infty \Rightarrow e^{-x} \to 0 \Rightarrow \tanh x \to \pm 1$

The curve has asymptotes at $y = 1$ and $y = -1$

So $-1 < y < 1$

The curve has rotational symmetry about the origin.

Key point

Graphs of hyperbolic functions can be transformed in the same way as other functions so for $y = f(x)$:

- $y = af(x)$ is $y = f(x)$ stretched vertically by scale factor a
- $y = f(ax)$ is $y = f(x)$ stretch horizonally by scale factor $\frac{1}{a}$
- $y = a + f(x)$ is $y = f(x)$ translated vertically by a units
- $y = f(x + a)$ is $y = f(x)$ translated horizontally by $-a$ units.

Example 2

Sketch the graph of $y = 2 + \cosh x$

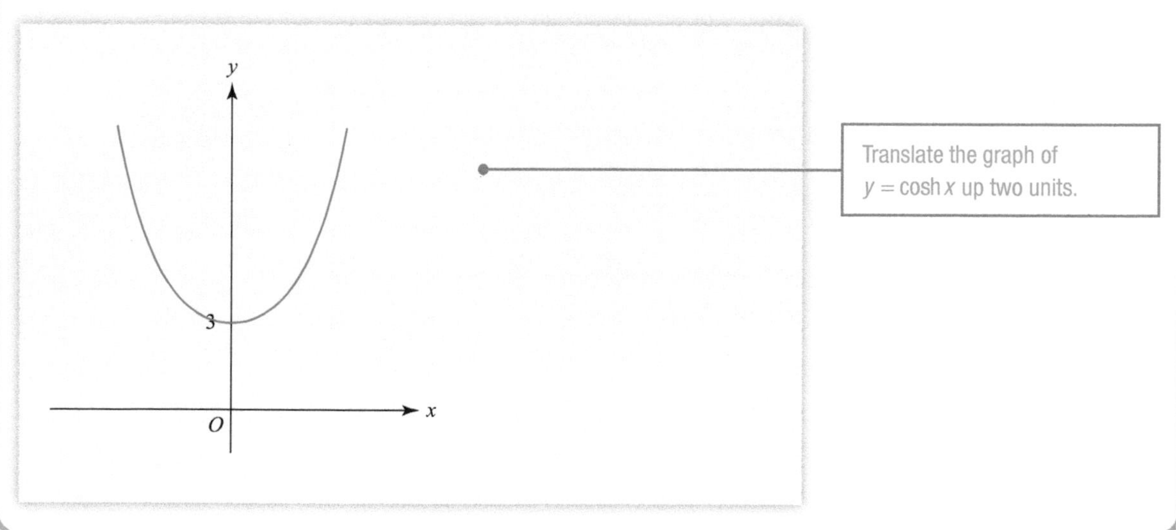

Translate the graph of $y = \cosh x$ up two units.

Example 3

Sketch the graphs of $y = \sinh x$ and $y = \sinh\dfrac{x}{2}$ on the same axes.

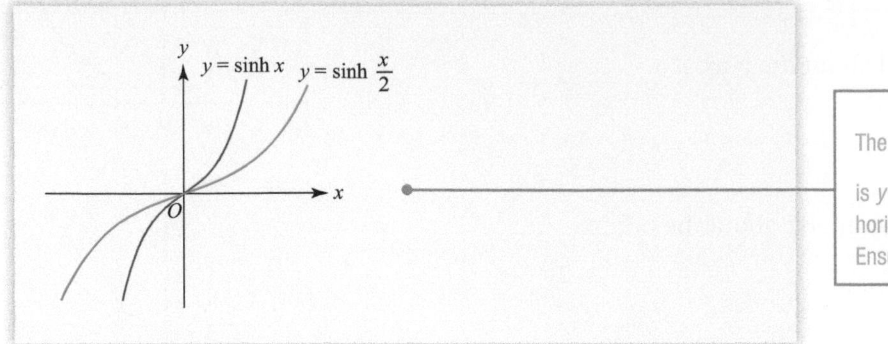

The graph of $y = \sinh\dfrac{x}{2}$ is $y = \sinh x$ stretched horizontally by scale factor 2. Ensure you label each curve.

If you are given the value of $\sinh x$, $\cosh x$ or $\tanh x$ you can use their definitions to find the value of x

Example 4

Solve the equation $\tanh x = \dfrac{1}{2}$

$$\frac{e^x - e^{-x}}{e^x + e^{-x}} = \frac{1}{2}$$

Use the definition of $\tanh x$

$$2(e^x - e^{-x}) = e^x + e^{-x} \Rightarrow e^x = 3e^{-x}$$

Multiply both sides by e^x

$$e^{2x} = 3$$

$$2x = \ln 3 \Rightarrow x = \frac{1}{2}\ln 3$$

Take logarithms on both sides.

This example could also be done on a calculator using $\tanh^{-1}\dfrac{1}{2}$, however your calculator will give a decimal, not an exact answer.

Example 5

Solve the equation $\sinh x + 2\cosh x = 2$

y = sinh x/2

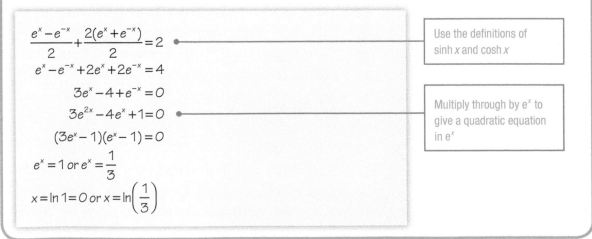

$\frac{e^x - e^{-x}}{2} + \frac{2(e^x + e^{-x})}{2} = 2$
Use the definitions of sinh x and cosh x
$e^x - e^{-x} + 2e^x + 2e^{-x} = 4$
$3e^x - 4 + e^{-x} = 0$
$3e^{2x} - 4e^x + 1 = 0$
Multiply through by e^x to give a quadratic equation in e^x
$(3e^x - 1)(e^x - 1) = 0$
$e^x = 1$ or $e^x = \frac{1}{3}$
$x = \ln 1 = 0$ or $x = \ln\left(\frac{1}{3}\right)$

Exercise 3.5A Fluency and skills

1 Work out the value of

a $\sinh 0$ b $\cosh(-1)$ c $\sinh(\ln 2)$

d $\cosh(-\ln 3)$ e $\cosh 0$ f $\tanh(\ln 2)$

2 Sketch these graphs.

a $y=\sinh(x-2)$ b $y=10+\sinh x$ c $y=\tanh(x-1)$

d $y=\cosh(x+2)$ e $y=2+\tanh x$ f $y=1-\cosh x$

3 The function f is defined as $f(x)=\sinh(x)$

a Sketch the graphs of $y=f(x)$ and $y=f(2x)$ on the same set of axes.

b Solve the equation $f(2x)=2$, giving your answer to 3 significant figures.

4 Sketch the graphs of $y=\cosh\left(\frac{x}{3}\right)$ and $y=\cosh x$ on the same set of axes.

5 The function f is defined as $f(x)=\tanh x$

a Sketch the graph $y=-2f(x)$

b Solve the equation $-2f(x)=1$, giving your answer to 3 significant figures.

6 Given that $f(x)=\tanh(x+a)$, $a>0$

a Sketch the graph of $y=f(x)$

b Write down the equations of the asymptotes.

7 Given that $g(x)=\tanh x$, $a>0$

a Sketch the graph of $y=a+g(x)$

b Write down the equations of the asymptotes.

8 Given that $f(x)=\cosh x$, $x\in\mathbb{R}$, sketch the graphs of

a $y=f(x-a)$, for $a>0$ b $y=f(x)+a$, for $a>0$ c $y=af(x)$, for $a<0$

9 Solve each of these equations, giving your answers to 3 sf.

a $\sinh x=5$ b $\cosh x=2$ c $\tanh x=-\frac{1}{2}$

d $\cosh(x+1)=3$ e $\sinh(3x)=4$ f $2\tanh(x)+1=2$

10 Solve these equations, leaving your answer as a logarithm in its simplest form where appropriate. Show your working.

a $\sinh x=0$ b $\cosh x=1$ c $\sinh x=\frac{3}{4}$

d $\cosh x=\frac{17}{8}$ e $\tanh x=\frac{3}{5}$ f $\tanh x=\frac{40}{41}$

11 Solve these equations, leaving your answer as a logarithm in its simplest form where appropriate.

a $\cosh x-\sinh x=2$ b $2\sinh x+3\cosh x=3$

c $3-\sinh x=2\cosh x$ d $\frac{1}{\sinh x}-\frac{1}{\cosh x}=4e^x$

Reasoning and problem-solving

Hyperbolic functions have identities which are very similar to the trigonometric ones you already know. You can prove these using the definitions.

Key point

Learn this important identity: $\cosh^2 x - \sinh^2 x \equiv 1$

Notice how the sign is different to the trigonometric identity $\cos^2 x + \sin^2 x \equiv 1$

Strategy 1

To solve problems using hyperbolic functions

1. Use the definitions of $\sinh x$, $\cosh x$ and $\tanh x$
2. Use laws of indices and logarithms as necessary to simplify expressions.

Example 6

Prove that $\cosh(A+B) \equiv \cosh(A)\cosh(B) + \sinh(A)\sinh(B)$

$$\cosh(A)\cosh(B)+\sinh(A)\sinh(B)$$

$$\equiv \left(\frac{e^A+e^{-A}}{2}\right)\left(\frac{e^B+e^{-B}}{2}\right)+\left(\frac{e^A-e^{-A}}{2}\right)\left(\frac{e^B-e^{-B}}{2}\right)$$

(1) Use the definitions of $\cosh x$ and $\sinh x$

$$\equiv \frac{e^{A+B}+e^{(A-B)}+e^{-(A-B)}+e^{-(A+B)}}{4}+\frac{e^{A+B}-e^{(A-B)}-e^{-(A-B)}+e^{-(A+B)}}{4}$$

(2) Expand the brackets, using laws of indices.

$$\equiv \frac{2e^{A+B}+2e^{-(A+B)}}{4}$$

$$\equiv \frac{e^{A+B}+e^{-(A+B)}}{2}$$

$\equiv \cosh(A+B)$ as required

Example 7

Prove that $\cosh^2 x - \sinh^2 x \equiv 1$

$$\cosh^2 x - \sinh^2 x \equiv \left(\frac{e^x+e^{-x}}{2}\right)^2-\left(\frac{e^x-e^{-x}}{2}\right)^2$$

(1) Use the exponential definitions.

$$\equiv \frac{e^{2x}+2+e^{-2x}}{4}-\frac{e^{2x}-2+e^{-2x}}{4}$$

(2) Simplify the numerator.

$$\equiv \frac{4}{4}$$

$\equiv 1$ as required

Example 8

Derive the logarithmic form of the inverse function $\cosh^{-1} x$

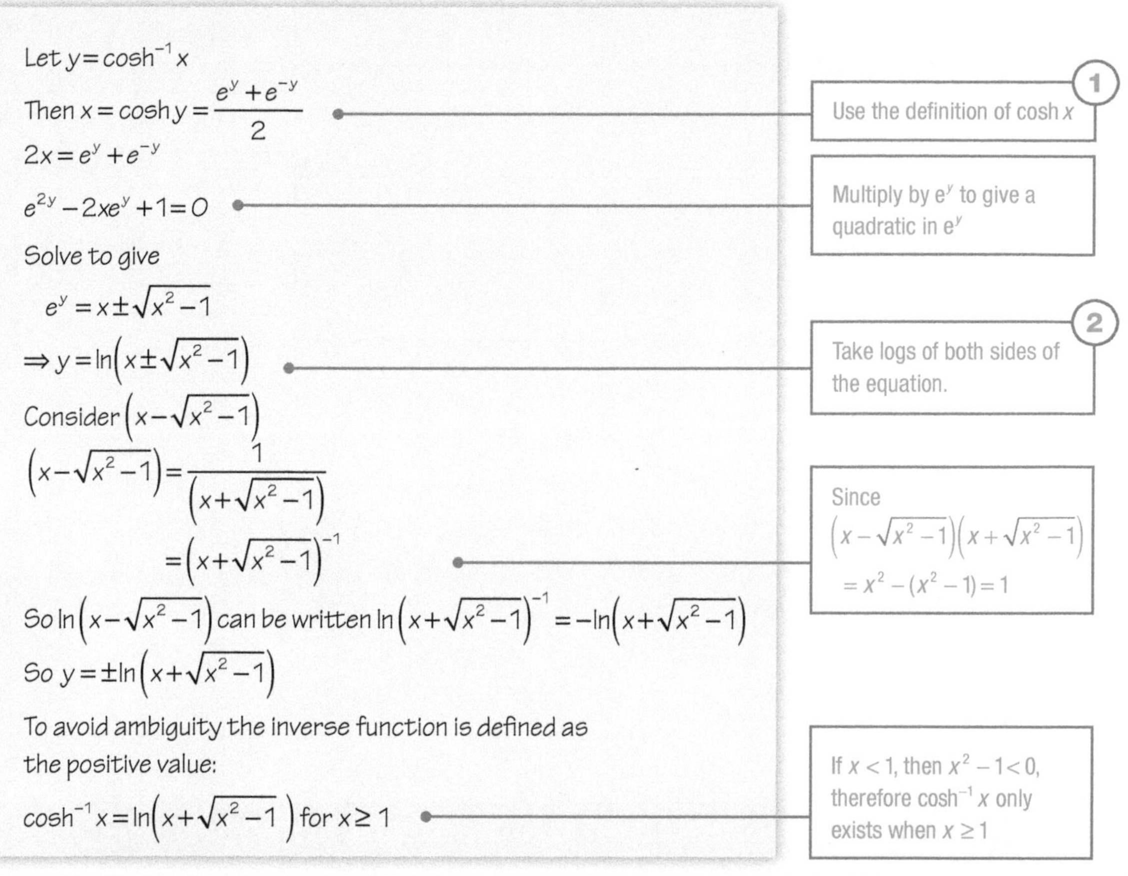

Let $y = \cosh^{-1} x$

Then $x = \cosh y = \frac{e^y + e^{-y}}{2}$ — Use the definition of $\cosh x$ (1)

$2x = e^y + e^{-y}$

$e^{2y} - 2xe^y + 1 = 0$ — Multiply by e^y to give a quadratic in e^y

Solve to give

$e^y = x \pm \sqrt{x^2 - 1}$

$\Rightarrow y = \ln\left(x \pm \sqrt{x^2 - 1}\right)$ — Take logs of both sides of the equation. (2)

Consider $\left(x - \sqrt{x^2 - 1}\right)$

$\left(x - \sqrt{x^2 - 1}\right) = \frac{1}{\left(x + \sqrt{x^2 - 1}\right)}$

$= \left(x + \sqrt{x^2 - 1}\right)^{-1}$ — Since $\left(x - \sqrt{x^2 - 1}\right)\left(x + \sqrt{x^2 - 1}\right) = x^2 - (x^2 - 1) = 1$

So $\ln\left(x - \sqrt{x^2 - 1}\right)$ can be written $\ln\left(x + \sqrt{x^2 - 1}\right)^{-1} = -\ln\left(x + \sqrt{x^2 - 1}\right)$

So $y = \pm \ln\left(x + \sqrt{x^2 - 1}\right)$

To avoid ambiguity the inverse function is defined as the positive value:

$\cosh^{-1} x = \ln\left(x + \sqrt{x^2 - 1}\right)$ for $x \geq 1$ — If $x < 1$, then $x^2 - 1 < 0$, therefore $\cosh^{-1} x$ only exists when $x \geq 1$

The other inverse hyperbolic functions can be derived in a similar way.

Key point

$$\sinh^{-1} x = \ln\left(x + \sqrt{x^2 + 1}\right)$$

$$\cosh^{-1} x = \ln\left(x + \sqrt{x^2 - 1}\right);\ x \geq 1$$

$$\tanh^{-1} x = \frac{1}{2}\ln\left(\frac{1 + x}{1 - x}\right);\ -1 < x < 1$$

You can also refer to the inverse functions as arsinh x or arcsinh x, arcosh x or arccosh x and artanh x or arctanh x

Strategy 2

To solve quadratic equations involving hyperbolic functions

1. Use identities to write the equation in terms of a single hyperbolic function.
2. Solve the quadratic equation to find the value of $\sinh x$, $\cosh x$ or $\tanh x$
3. Use the definitions of the inverse hyperbolic functions to find the exact values of x

Example 9

Solve the equation $\sinh x + 2\cosh^2 x = 3$

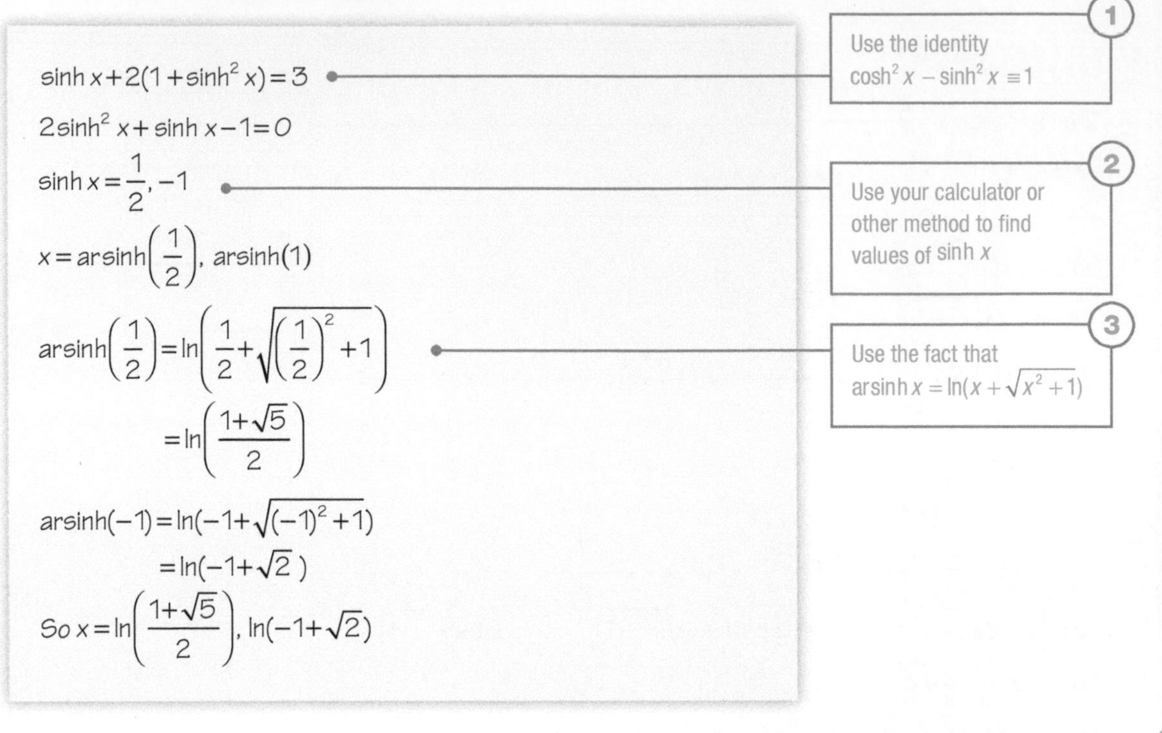

$\sinh x + 2(1+\sinh^2 x) = 3$
$2\sinh^2 x + \sinh x - 1 = 0$
$\sinh x = \frac{1}{2}, -1$
$x = \text{arsinh}\left(\frac{1}{2}\right), \text{arsinh}(1)$
$\text{arsinh}\left(\frac{1}{2}\right) = \ln\left(\frac{1}{2} + \sqrt{\left(\frac{1}{2}\right)^2 + 1}\right)$
$= \ln\left(\frac{1+\sqrt{5}}{2}\right)$
$\text{arsinh}(-1) = \ln(-1 + \sqrt{(-1)^2 + 1})$
$= \ln(-1+\sqrt{2})$
So $x = \ln\left(\frac{1+\sqrt{5}}{2}\right), \ln(-1+\sqrt{2})$
1
Use the identity $\cosh^2 x - \sinh^2 x \equiv 1$
2
Use your calculator or other method to find values of $\sinh x$
3
Use the fact that $\text{arsinh}\, x = \ln(x + \sqrt{x^2+1})$

Exercise 3.5B Reasoning and problem-solving

1 Solve the equations, giving your answers as exact logarithms where appropriate.

a $\sinh^2 x + \sinh x = 6$

b $2\cosh^2 x - 5\cosh x + 3 = 0$

c $3\tanh^2 x - 2 = 0$

d $\cosh^2(2x) - 3\cosh(2x) = 0$

2 Solve the equations. Give your answers as exact logarithms where appropriate.

a $\sinh^2 x + \cosh x = 1$

b $\cosh^2 x + 3\sinh x = 5$

c $2\sinh x + 3\cosh^2 x = 3$

d $2\tanh^2 x + 3\tanh x = 2$

e $\sinh 2x = 3$

f $\cosh 3x = 4$

g $2\tanh 3x = 1$

3 Use the definitions of $\sinh x$ and $\cosh x$ to prove that $\cosh^2 x - \sinh^2 x \equiv 1$

4 Use the definitions of $\cosh x$ and $\sinh x$ to prove these identities.

a $\sinh x(A+B) = \sinh(A)\cosh(B) + \sinh(B)\cosh(A)$

b $\sinh(A-B) = \sinh(A)\cosh(B) - \sinh(B)\cosh(A)$

c $\sinh(2x) = 2\sinh(x)\cosh(x)$

5 a Use the definitions of sinh x and cosh x to prove that $\tanh x = \frac{e^{2x}-1}{e^{2x}+1}$

b Hence show that $\tanh(2x) = \frac{2\tanh x}{1+\tanh^2 x}$

6 a Use the definition of cosh x to prove that $\cosh(2x) \equiv 2\cosh^2 x - 1$

b Hence find the exact solutions to the equation $\cosh(2x) + \cosh x = 5$

7 Use a similar method as in Example 8 to prove that $\text{arsinh}\, x = \ln\left(x+\sqrt{x^2+1}\right)$

Explain clearly why $\text{arsinh}\, x \neq \ln\left(x-\sqrt{x^2+1}\right)$

8 a Prove that $\text{artanh}\, x = \frac{1}{2}\ln\left(\frac{1+x}{1-x}\right)$

b Explain why the formula in part **a** is only valid for $-1 < x < 1$

9 Solve the simultaneous equations

$3\cosh x + 2\sinh x = 1$ and $\cosh^2 x + \sinh x = 3$

10 Solve the simultaneous equations

$\cosh x - 2\sinh^2 x = 3$ and $\cosh x + \sinh x = 4$

11 Solve the equation $\cosh 2x + 7 = 7\cosh x$

12 a Show that $3\sinh\left(\frac{x}{2}\right) - \sinh(x) \equiv \sinh\left(\frac{x}{2}\right)\left(3-2\cosh\left(\frac{x}{2}\right)\right)$

b Hence, find the exact solutions to the equation $3\sinh\left(\frac{x}{2}\right) - \sinh(x) = 0$

13 a Show that $\frac{1}{\cosh(x)+1} + \frac{1}{1-\cosh(x)} \equiv -\frac{2}{\sinh^2 x}$

b Hence solve the equation $\frac{1}{\cosh(x)+1} + \frac{1}{1-\cosh(x)} + 8 = 0$

14 a Given that $\sinh(2x) = \cosh^2 x$, show that $\tanh x = \frac{1}{2}$

b Hence solve the equation $\sinh(2x) - \cosh^2 x = 0$

15 Solve the equation $\text{arcosh}\, x = 2\ln x$

16 Given that $\sinh x = 2$, $x > 0$, calculate the exact value of

a $\cosh x$ b $\tanh x$

17 Given that $\cosh x = 3$, $x > 0$, calculate the exact value of

a $\sinh x$ b $\tanh x$

18 Given that $\tanh x = \frac{1}{2}$, $x > 0$, calculate the exact value of

a $\sinh x$ b $\cosh x$

19 a Use the definitions to find the derivatives of $\sinh x$ and $\cosh x$

b Find the gradient of both curves when $x = 0$

c Show that for $x > 0$ the curve $y = \sinh x$ is always steeper than $y = \cosh x$

d Find the gradient of the tangent to the curve $y = \tanh x$ at $x = 0$

3 Summary and review

Chapter summary

- When sketching the graph of a function of the form $\dfrac{ax+b}{cx+d}$
  - find the intercepts and any asymptotes,
  - consider what happens when x and y become very large and approach infinity.
- When sketching the graph of a function of the form $\dfrac{ax^2+bx+c}{dx^2+ex+f}$
  - use the above techniques and also find any limitations on the possible ranges of y
  - use the results of these techniques to find any turning points.
- You can use sketches of graphs to solve inequalities.
- r is the distance from the pole to the point.
- θ is the angle between the initial line and the line segment from the pole to the point, measured counterclockwise.
- To convert from polar to Cartesian, use

  $x = r\cos\theta$

  $y = r\sin\theta$
- To convert from Cartesian to polar, use

  $r^2 = x^2 + y^2$

  $\tan\theta = \dfrac{y}{x}$

  but you should draw a diagram to ensure that you choose the right value of θ
- To convert from polar to Cartesian, use

  $r^2 = x^2 + y^2$

  $\tan\theta = \dfrac{y}{x}$

  $\cos\theta = \dfrac{x}{r} = \pm\dfrac{x}{\sqrt{x^2+y^2}}$

  $\sin\theta = \dfrac{y}{r} = \pm\dfrac{y}{\sqrt{x^2+y^2}}$

  [Choose + or – depending on whether r is positive or negative]
- To convert from Cartesian to polar, use

  $x = r\cos\theta$

  $y = r\sin\theta$
- The equation of a parabola centred at the origin is $y^2 = 4ax$
- The equation of an ellipse centred at the origin is $\dfrac{x^2}{a^2}+\dfrac{y^2}{b^2}=1$
- The equation of a hyperbola centred at the origin with asymptotes at $y=\pm\dfrac{b}{a}x$ is $\dfrac{x^2}{a^2}-\dfrac{y^2}{b^2}=1$
- A rectangular hyperbola centred at the origin with asymptotes $y=0$ and $x=0$ has equation $xy=c^2$

- The hyperbolic functions are defined using exponentials and have graphs as shown.
    - 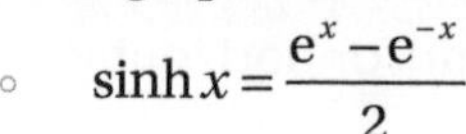
      $\sinh x = \dfrac{e^x - e^{-x}}{2}$
    - $\cosh x = \dfrac{e^x + e^{-x}}{2}$
    - $\tanh = \dfrac{e^x - e^{-x}}{e^x + e^{-x}}$

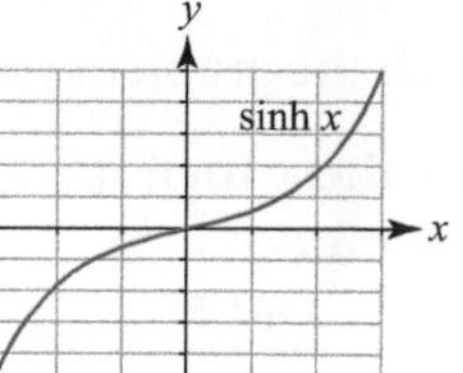

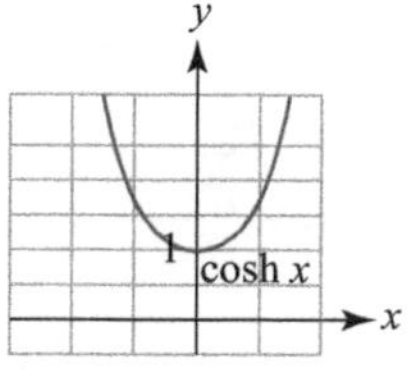

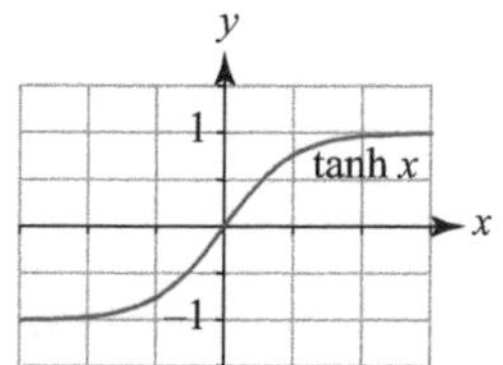

- $\cosh^2 x - \sinh^2 x \equiv 1$
- The inverse hyperbolic functions are
    - $\sinh^{-1} x = \ln\left(x + \sqrt{x^2 + 1}\right)$
    - $\cosh^{-1} x = \ln\left(x + \sqrt{x^2 - 1}\right),\ x \geq 1$
    - $\tanh^{-1} x = \dfrac{1}{2}\ln\left(\dfrac{1+x}{1-x}\right),\ -1 < x < 1$

Check and review

You should now be able to...	Try Questions
✔ Graph functions of the form $\dfrac{ax+b}{cx+d}$ and $\dfrac{ax^2+bx+c}{dx^2+ex+f}$, and solve associated inequalities.	1–3
✔ Use quadratic theory to find the range of possible values of a function and find stationary points.	4, 5
✔ Convert between polar and Cartesian coordinates and equations.	6–9
✔ Sketch polar curves and find points of intersection.	10–13
✔ Sketch a parabola with vertex at the origin and an ellipse centred at the origin, giving the coordinates of the points of intersection with the coordinate axes.	14, 15
✔ Sketch a hyperbola centred at the origin, giving the equations of the asymptotes.	16, 17
✔ Sketch single transformations of conic sections.	18–20
✔ Find the equations of conic sections reflected in the lines $y = x$ or $y = -x$	21
✔ Know the definition of $\sinh x$, $\cosh x$ and $\tanh x$ in terms of exponentials; calculate exact values and solve equations.	22, 25, 26
✔ Sketch hyperbolic graphs and transformations.	23
✔ Recall and use the identity $\cosh^2 x - \sinh^2 x \equiv 1$	24
✔ Derive the logarithmic form of the inverse hyperbolic functions and calculate exact values.	26–27

1 Sketch the graph of $y=\frac{(x-1)}{(x+4)}$

2 Solve these inequalities graphically, naming any intercepts and vertical and/or horizontal asymptotes.

a $-3<\frac{x+12}{x-4}<9$ b $7\geq\frac{x-6}{x-12}\geq-5$

3 Sketch the graph of

a $y=\frac{x^2+12x+24}{x^2+2x+2}$ b $y=\frac{x^2+12x+24}{x^2+2x-3}$

4 Find the values of y for which the graph of the function $y=\frac{x^2-3x-4}{x^2-3x+1}$ exists.

5 Without using calculus, find the coordinates of the turning points of the curve with equation $y=\frac{x^2-5x}{x^2-5x+5}$ and find the maximum value.

6 Convert the sets of polar coordinates to Cartesian coordinates.

a $\left(5,\frac{\pi}{3}\right)$ b $\left(5,-\frac{\pi}{3}\right)$

7 Convert the sets of Cartesian coordinates to polar coordinates.

a $\left(2\sqrt{3},2\right)$ b $\left(-2\sqrt{3},2\right)$

c $\left(-2\sqrt{3},-2\right)$ d $\left(2\sqrt{3},-2\right)$

8 Express these Cartesian equations in polar form.

a $y=3x$ b $y=2x+1$

c $x^2+y^2=16$ d $(x-1)^2+(y-1)^2=2$

9 Express these polar equations in Cartesian form.

a The circle $r=4\sin\theta$, $r\geq0$, stating its centre and radius,

b The graph $r=2-4\sin\theta$, $r\geq0$

10 For each polar equation

i Sketch the half-line it represents,

ii Give the Cartesian equation of the line it lies on.

a $\theta=-\frac{3\pi}{4}, r\geq0$ b $\theta=\frac{\pi}{3}, r\geq0$

11 For each of the polar equations

i Sketch the graph for $r\geq0$ and $0\leq\theta<2\pi$

ii State the maximum and minimum values of r

a $r=3$ b $r=7\cos\theta$

c $r=4\cos2\theta$ d $r=2+\sin\theta$

e $r=4-3\cos\theta$ f $r=3\theta$

12 Find the points of intersection between the polar curves $r=2$ and $r=3-2\cos\theta$

13 Find the points of intersection between the polar curves $r = \sin 2\theta$ and $r = \sqrt{3}\cos 2\theta$ where $r \geq 0$. What difference does it make if r can take any value, including negatives?

14 a Sketch the parabola $y^2 = 4x$

b Find the points of intersection between this curve and the line $y = x - 3$

15 Sketch each of these ellipses, clearly labelling where they cross the coordinate axes.

a $\frac{x^2}{36} + \frac{y^2}{49} = 1$ b $\frac{x^2}{12} + y^2 = 2$

16 Sketch the hyperbola $\frac{x^2}{8} - \frac{y^2}{4} = 1$, giving the equations of the asymptotes.

17 A rectangular hyperbola has equation $xy = 5$

a Sketch the curve and state the equations of the asymptotes.

b Show that the hyperbola does not intersect the line with equation $x + y = 1$

18 Sketch these parabolas, clearly labelling where they cross the coordinate axes.

a $y^2 = -2x$ b $(y-1)^2 = 3x$

19 Sketch these ellipses, clearly labelling where they cross the coordinate axes.

a $\frac{x^2}{3} + \frac{(y-2)^2}{4} = 1$ b $\frac{(x-5)^2}{25} + (y+1)^2 = 1$

20 Sketch these hyperbolas and state the equations of their asymptotes.

a $\frac{x^2}{3} - \frac{(y+5)^2}{4} = 1$ b $\frac{(x+3)^2}{9} - \frac{(y-2)^2}{4} = 1$

c $-xy = 7$ d $(x+3)(y-4) = 2$

21 a The parabola $\frac{x^2}{8} + y^2 = 2$ is reflected in the line $y = x$. Sketch the curve and write down its equation.

b The hyperbola $6x^2 - 2y^2 = 6$ is reflected in the line $y = -x$. Sketch the curve and write down the equations of its asymptotes.

22 Calculate the exact value of

a $\cosh(\ln 5)$ b $\tanh(-\ln 2)$

23 Sketch these graphs for $a \geq 1$

a $y = a\cosh x$ b $y = \sinh(x - a)$

c $y = a\tanh x$ d $y = \cosh x - a$

24 Solve the equation $\cosh^2 x + \sinh x = 3$

25 Solve these equations, giving your answers as logarithms.

a $\sinh x + \cosh x = 3$ b $\cosh x + 1 = e^x$

26 Use the definition of $\cosh x$ to prove that $\text{arcosh}(2x) = \ln\left(2x + \sqrt{4x^2 - 1}\right)$

27 Find the exact solution to the equations

a $\sinh x = 2$ b $\tanh x = -\frac{1}{2}$

Assume $r \geq 0$ in each case.

Exploration

Going beyond the exams

History

Two mathematicians, the Flemish Gregoire de Saint-Vincent and the Italian Bonaventura Cavalieri, introduced the concepts of the polar coordinate system independently in the mid-seventeenth century. Cavalieri used them to solve a problem involving the area within an Archimedian spiral. The French mathematician Blaise Pascal used polar coordinates to calculate the lengths of parabolic arcs like the ones on this bridge.

Did you know?

Although many bridges are known as suspension bridges, they are actually suspended deck bridges. The cables follow a parabolic curve.

Using a photograph of a famous bridge, such as the Forth Road Bridge, together with graph plotting technology, explore how well you can model the suspension cables using quadratic and hyperbolic functions.

Research

Coordinate systems are used to locate the position of a point in space. Explore coordinate systems that can be used effectively in three-dimensions, for example, the spherical coordinate system commonly used by mathematicians. You should also research cylindrical coordinate systems. These are often used by engineers. Explore when, and why, these systems are used.

Different coordinate systems are used in other subject areas. For example, geographers use a geographical coordinate system, and space scientists and astronomers use a celestial coordinate system. Explore the advantages and disadvantages of these systems.

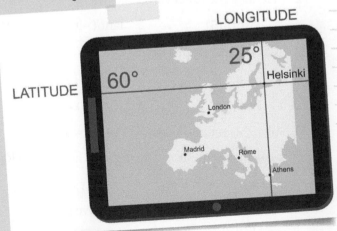

3 Assessment

1 a Sketch the graph of $y=\frac{x+5}{x-7}$, clearly labelling any intercepts with the coordinate axes. **[4 marks]**

b Write down the equations of the asymptotes to the curve. **[2]**

c Find the coordinates of the points where the curve $y=\frac{x+5}{x-7}$ meets the line $y=-x$ **[3]**

2 a Sketch the curve $y=\frac{5x^2}{x^2+1}$ **[3]**

b Find the points of intersection between the curve $y=\frac{5x^2}{x^2+1}$ and the line $y=1$ **[3]**

3 a Express the Cartesian coordinates $(\sqrt{3},-3)$ as polar coordinates, giving the exact value of r and writing θ in terms of π where $-\pi<\theta\leq\pi$ **[4]**

b Sketch the polar graph with equation

i $r=9$ **[2]** ii $\theta=\frac{\pi}{6}, r\geq 0$ **[1]**

c Find the Cartesian equations of the curves in part **b**. **[5]**

4 a Express the polar coordinates $\left(4,-\frac{5\pi}{6}\right)$ as exact Cartesian coordinates. **[3]**

b Find the Cartesian equation of the polar curve

$r=2\sin\theta$ **[3]**

c Sketch the graph of the curve in part **b**. **[3]**

5 a Sketch the graph of $y^2=3x$ and state the name given to this type of curve. **[3]**

The line with equation $y=x+c$ is a tangent to $y^2=3x$

b Calculate the value of c **[4]**

6 Given that $k>3$, sketch the curve with these equations. Label the points of intersection with the coordinate axes and give the equations of any asymptotes.

a $\frac{x^2}{9}+\frac{y^2}{k^2}=1$ **[3]** b $\frac{x^2}{9}-\frac{y^2}{k^2}=1$ **[4]**

7 Sketch these polar curves.

a $r=4\theta$ for $0\leq\theta\leq 3\pi$ **[2]** b $r=\cos 3\theta$ **[2]**

c $r=1+\sin\theta$ **[2]** d $r=8+3\cos\theta$ **[2]**

8 a Use the exponential definition of $\sinh x$ to show that $\sinh(\ln 2)=\frac{3}{4}$ **[3]**

b Solve the equation $\sinh x=3$, giving your solution to 1 decimal place. **[3]**

9 a Sketch the graph of $y=\tanh(x+1)$ **[2]**

b Write down the equations of the asymptotes to the curve in part **a**. **[2]**

c Use the exponential definitions of $\sinh x$ and $\cosh x$ to show that $\tanh x=\frac{e^{2x}-1}{e^{2x}+1}$ **[4]**

d Hence solve the equation $\tanh x=\frac{1}{2}$. Give your answer as a logarithm. **[3]**

10 a Sketch the graph of $y=\frac{2+x}{2x-1}$ **[3]** b Solve the inequality $\frac{2+x}{2x-1}\geq\frac{x}{x+4}$ **[5]**

11 a Sketch the curve $y=\dfrac{x^2-5x+4}{x^2+1}$ **[3]** **b** Solve the inequality $\dfrac{x^2-5x+4}{x^2+1}>1-x$ **[1]**

12 A rational function is defined by $f(x)=\dfrac{3x^2+5x-8}{x^2+2x-6}$

A sketch of the curve $y=f(x)$ is shown.

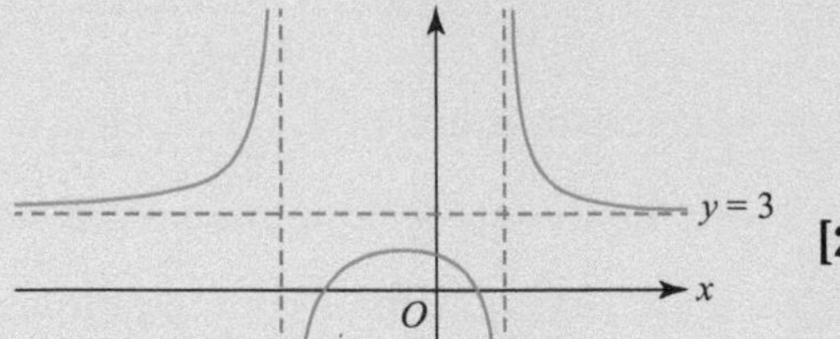

a For what values of x is the function undefined? Leave your answer as a surd. **[2]**

b Where does the curve of the function cut the coordinate axes? **[3]**

c For what values of x is $f(x) > 1$? **[5]**

13 a Show that the equation $xy-5y+x-9=0$ can be written as $(x-5)(y+1)=k$ where k is a constant to be found. **[2]**

b Hence sketch the curve of $xy-5y+x-9=0$, giving the equations of the asymptotes. **[3]**

14 Sketch the curve with equation $\dfrac{(x+2)^2}{4}+(y-2)^2=1$, labelling any points of intersection with the coordinate axes. **[3]**

15 Show that the curve $y=\dfrac{1}{x}$ has polar equation $r^2=2\operatorname{cosec}2\theta$ **[4]**

16 a Give the Cartesian equation of the curve with polar equation $r=3\sin 2\theta$ for $r\geq 0$ and $0\leq\theta\leq 2\pi$ **[4]**

b Sketch the polar curve found in part **a**. **[2]**

c State the maximum value of $r=3\sin 2\theta$ and the values of θ at which it occurs. **[3]**

17 a State the maximum and minimum values of $r=4+3\cos\theta$ where $0\leq\theta<2\pi$ **[2]**

b Sketch the curve $r=4+3\cos\theta$ **[2]**

18 a Sketch the graphs of $y=\sinh x$ and $y=\sinh 2x$ on the same diagram. **[3]**

b If $y=\sinh x$, use the exponential definition of $\sinh x$ to show that $x=\ln\left(y+\sqrt{y^2+1}\right)$ **[5]**

19 The hyperbola with equation $\dfrac{x^2}{10}-\dfrac{y^2}{5}=1$ is reflected in the line $y=x$

a Write down the equation of the image. **[1]**

b Sketch the image and state the equations of the asymptotes. **[3]**

20 Sketch each transformation of the parabola with equation $y^2=12x$ and write down its equation.

a A stretch of scale factor 2 in the y-direction **[2]**

b A translation by the vector $\begin{pmatrix}3\\-8\end{pmatrix}$ **[2]**

c A reflection in the line $y=-x$ **[2]**

21 a Sketch these curves on the same diagram, where $r\geq 0$

i $r=\sin 4\theta$ **[2]** ii $r=2\sin 3\theta$ **[2]**

b What difference does it make to the diagram if r can take any value? **[3]**

22 a Use the definitions of $\cosh x$ and $\sinh x$ to prove that $\sinh(2x)\equiv 2\sinh x\cosh x$ **[3]**

b Hence find the exact solutions to these equations.

i $\sinh(2x)-3\sinh^2 x=0$ ii $\sinh x\cosh x - \cosh^2(2x)+4=0$ **[10]**

4 Integration 1

In economics many factors, such as currency rates, vary with time. Changes are often not sudden. Economists may model them using functions that, although complex, are developed using fundamental functions with which we are familiar, such as trigonometric functions. An average value of the factor might then be found over a given period by using integration to find the area under the function. This is then equated to the area of a rectangle with the period as base and mean value as height.

This method of finding a mean value using integral calculus has applications in other very different areas. For example, it is used extensively in electrical and electronic engineering in the design of essential circuits that are used everywhere, from delivering electricity to our homes, through to ensuring that all electrical appliances, transport, computers, and so on, work as intended.

Orientation

What you need to know	What you will learn	What this leads to
Maths Ch1 • Finding solutions of simultaneous equations. **Maths Ch4** • Finding definite integrals.	• To calculate the mean value of a function. • To find the area enclosed by a curve and lines. • To calculate the volume of revolution when rotated around the x-axis. • To calculate the volume of revolution when rotated around the y-axis. • To calculate more complicated volumes of revolution by adding or subtracting volumes.	**Ch19 Integration 2** • Surface area and volumes of revolution. **Careers** • Architecture.

4.1 Mean values

Fluency and skills

You already know that you can use integration to find the area, A, enclosed by the curve $y=\mathrm{f}(x)$, the x-axis and the lines $x=a$ and $x=b$: $A=\int_a^b \mathrm{f}(x)\,\mathrm{d}x$

The Mean Value Theorem is based on the idea that for any continuous function there is always a rectangle with exactly the same area, A. The base of this rectangle will be between a and b. The height, h, of the rectangle is known as the mean value of the function f(x)

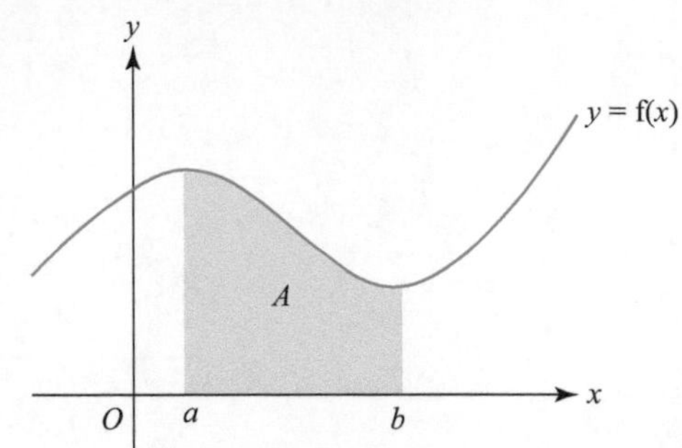

You can calculate the height of the rectangle by dividing its area, found by integration, by its width. So $h=\frac{1}{b-a}\int_a^b \mathrm{f}(x)\mathrm{d}x$

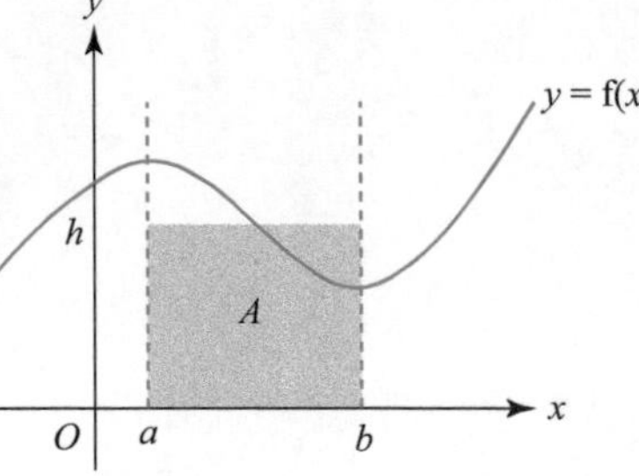

Key point

The mean value of a function $\mathrm{f}(x)$ in the range $a \le x \le b$ is given by $\frac{1}{b-a}\int_a^b \mathrm{f}(x)\mathrm{d}x$

Example 1

Calculate the mean value of the function $\mathrm{f}(x)=2x^3+6x-1$ in the interval [2, 5]

Mean value $=\frac{1}{5-2}\int_2^5 (2x^3+6x-1)dx$ — Use the definition of the mean value as $\frac{1}{b-a}\int_a^b \mathrm{f}(x)\mathrm{d}x$

$=\frac{1}{3}\left[\frac{x^4}{2}+3x^2-x\right]_2^5$

$=\frac{1}{3}\left(\frac{625}{2}+75-5\right)-\frac{1}{3}(8+12-2)$ — Substitute in limits.

$=\frac{243}{2}$

So the mean value of $2x^3+6x-1$ in the interval [2, 5] is $\frac{243}{2}$ (or 121.5)

Example 2

Calculate the mean value with respect to t of the function $t^2(3t+1)(t-2)$ for $1 \le t \le 3$

$t^2(3t+1)(t-2) = t^2(3t^2-5t-2)$

$= 3t^4 - 5t^3 - 2t^2$ — Expand the brackets and use index rules to simplify.

Mean value $= \frac{1}{3-1}\int_1^3 (3t^4-5t^3-2t^2)\,dt$ — Use the definition of the mean value as $\frac{1}{b-a}\int_a^b f(t)\,dt$ since our function is in terms of the variable t

$= \frac{1}{2}\left[\frac{3}{5}t^5 - \frac{5t^4}{4} - \frac{2}{3}t^3\right]_1^3$

$= \frac{1}{2}\left(\frac{729}{5} - \frac{405}{4} - 18\right) - \frac{1}{2}\left(\frac{3}{5} - \frac{5}{4} - \frac{2}{3}\right)$ — Substitute in limits.

$= \frac{209}{15}$

So the mean value of $t^2(3t+1)(t-2)$ for $1 \le t \le 3$ is $\frac{209}{15}$ (or 13.9)

Calculator

Try it on your calculator

Definite integrals can be worked out on your calculator.

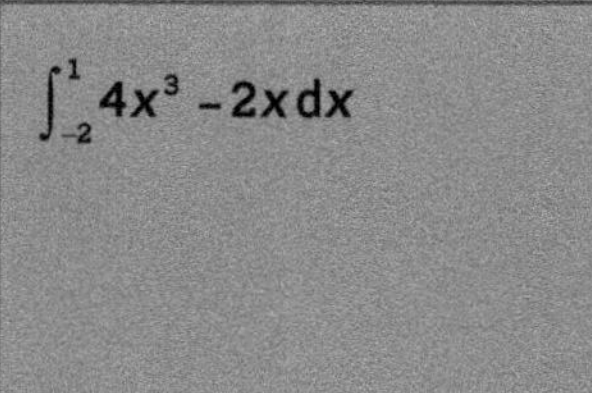

Activity

Find out how to work out $\int_{-2}^{1} 4x^3 - 2x\,dx$ on *your* calculator.

Exercise 4.1A Fluency and skills

1 The graph of $y = x^3 - 3x^2 + 6$ is shown.

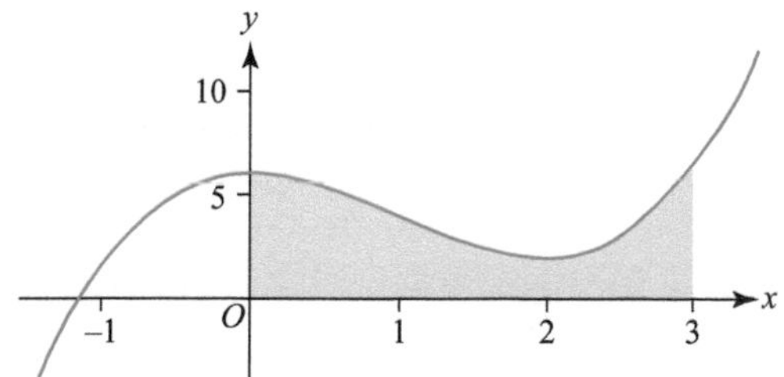

a Calculate the area bounded by the curve, the coordinate axes and the line $x = 3$

b Work out the mean value of $x^3 - 3x^2 + 6$ for $0 \le x \le 3$

2 $f(x) = x^2 - 3x + 2$

a Sketch the graph of $y = f(x)$, labelling where the curve crosses the coordinate axes.

b Calculate the area bounded by the curve $y = f(x)$ and the x-axis.

c Work out the mean value of $f(x)$ for each of these intervals.

i $1 \le x \le 2$ ii $0 \le x \le 1$

3 $g(x) = x^3 + 5x^2$

a Sketch the graph of $y = g(x)$, labelling where the curve crosses the coordinate axes.

b Calculate the area bounded by the curve $y=g(x)$ and the x-axis.

c Work out the mean value of $g(x)$ for each of these intervals.

i $-5 \le x \le 0$ **ii** $0 \le x \le 2$

4 Calculate the mean value of the function $f(x)=5x^4$ in each of these intervals.

a $[0, 4]$ **b** $[1, 3]$

c $[-1, 2]$ **d** $\left[-\frac{1}{2}, \frac{1}{2}\right]$

5 Calculate the mean value of $2x^2+3$ for each of these ranges of values of x

a $0 \le x \le 3$ **b** $2 \le x \le 6$

c $-1 \le x \le 2$ **d** $-2 \le x \le -1$

6 Calculate the mean value of $3x-8x^3$ for x between

a 1 and 4 **b** 0 and 3

c −1 and 1 **d** −3 and −2

7 Show that the mean value of $\frac{1}{x^2}$ for $2 \le x \le 5$ is $\frac{1}{10}$

8 Show that the mean value of $\sqrt{x}$ for $1 \le x \le 4$ is $\frac{14}{9}$

9 Show that the mean value of $\frac{1}{\sqrt{x}}$ for $4 \le x \le 9$ is 0.4

10 Given that $f(x)=\frac{2\sqrt{x}+3x}{2x}$

a Write $f(x)$ in the form Ax^c+B, for constants A, B and c and state their values.

b Show that the mean value of $f(x)$ in the interval $[1, 9]$ is 2

11 Given that $g(x)=\frac{3x-x^2}{5\sqrt{x}}$

a Write $g(x)$ in the form Ax^c+Bx^d for constants A, B, c and d and state their values.

b Calculate the exact mean value of $g(x)$ for $2 \le x \le 4$

12 Find the mean value of each of these functions of t for the range given.

a $2t(5t^3-8t+1)$ for $-1 \le t \le 0$

b $3t\sqrt{t}-2t^2$ for $\frac{1}{4} \le t \le 4$

c $\frac{1}{2t\sqrt{t}}$ for $1 \le t \le 9$

d $\frac{(t^3)^{\frac{1}{2}}}{t}$ for $4 \le t \le 64$

13 Find an expression for the mean value of $(1-2t)^2$ across the interval

a $0 \le t \le T$ **b** $T \le t \le T+2$

Give your answers as polynomials in T

14 Calculate the mean value of the function $\frac{4+x}{\sqrt{x}}$ for $2 \le x \le 8$

Give your answer in the form $A\sqrt{2}$

15 Derive an expression in terms of a for the mean value of the function $3x^3-\frac{1}{x^3}$ for x in the interval $[a, 2a]$. Give your answer in its simplest form.

16 Find an expression in terms of X for the mean value of the function $\frac{1}{2x^4}$, for x in the interval $\left[0, \frac{1}{X}\right]$. Simplify your answer.

Reasoning and problem-solving

Strategy

To solve problems on mean values

1. Use the formula for the mean value of a function.
2. Form one or more equations using the information provided.
3. Solve the quadratic or simultaneous equations.

Example 3

The velocity of a particle after t seconds is given by $v = \frac{1}{8}t\left(9 - \frac{4}{\sqrt{t}}\right)$ m s^{-1}

a Show that the mean velocity for $1 \le t \le 4$ is 2.03 m s^{-1} to 3 significant figures.

b Calculate the mean acceleration of the particle over the same time period.

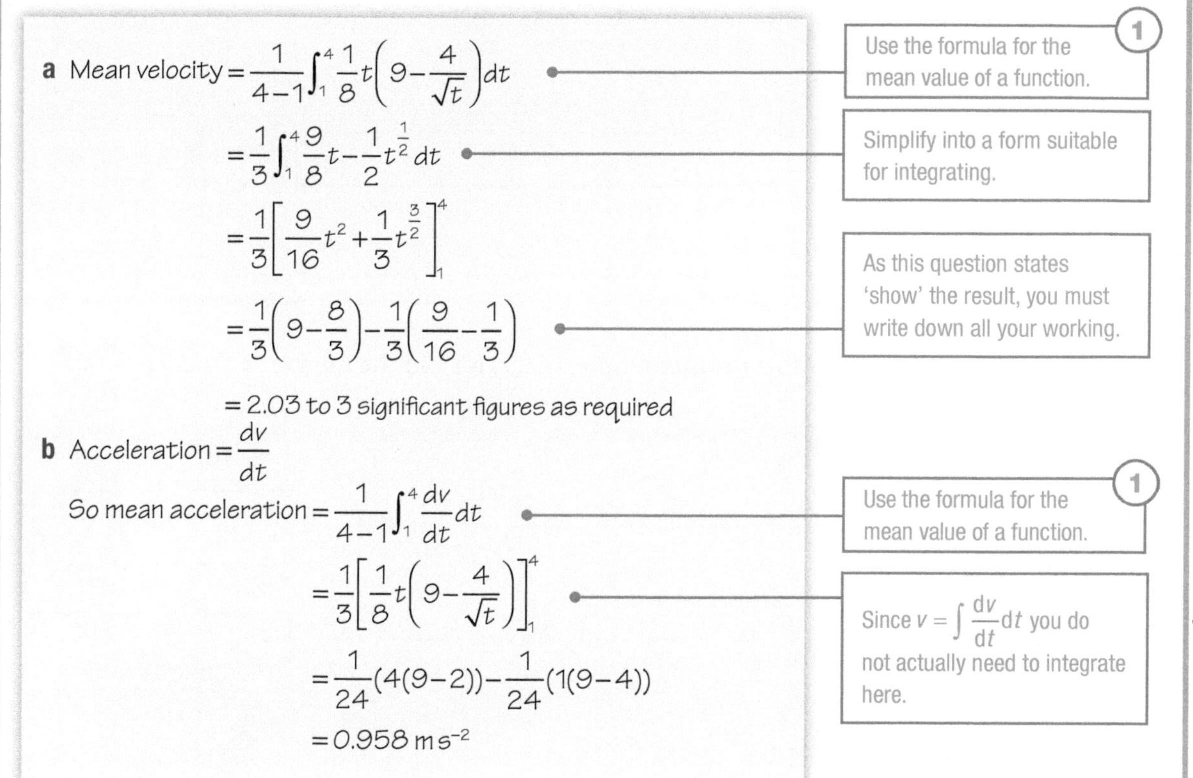

a Mean velocity $= \frac{1}{4-1}\int_1^4 \frac{1}{8}t\left(9-\frac{4}{\sqrt{t}}\right)dt$
$= \frac{1}{3}\int_1^4 \frac{9}{8}t - \frac{1}{2}t^{\frac{1}{2}}\,dt$
$= \frac{1}{3}\left[\frac{9}{16}t^2 + \frac{1}{3}t^{\frac{3}{2}}\right]_1^4$
$= \frac{1}{3}\left(9-\frac{8}{3}\right) - \frac{1}{3}\left(\frac{9}{16} - \frac{1}{3}\right)$
= 2.03 to 3 significant figures as required
b Acceleration $= \frac{dv}{dt}$
So mean acceleration $= \frac{1}{4-1}\int_1^4 \frac{dv}{dt}dt$
$= \frac{1}{3}\left[\frac{1}{8}t\left(9-\frac{4}{\sqrt{t}}\right)\right]_1^4$
$= \frac{1}{24}(4(9-2)) - \frac{1}{24}(1(9-4))$
$= 0.958$ m s^{-2}
1 Use the formula for the mean value of a function.
Simplify into a form suitable for integrating.
As this question states 'show' the result, you must write down all your working.
1 Use the formula for the mean value of a function.
Since $v = \int \frac{dv}{dt}dt$ you do not actually need to integrate here.

See Maths Ch7

For a reminder on the link between differentiation and kinematics

Example 4

The mean value of the function $4x + 7$ in the interval $[a, b]$ is 2 and in the interval $[a, 2b]$ is 3
Calculate the values of a and b

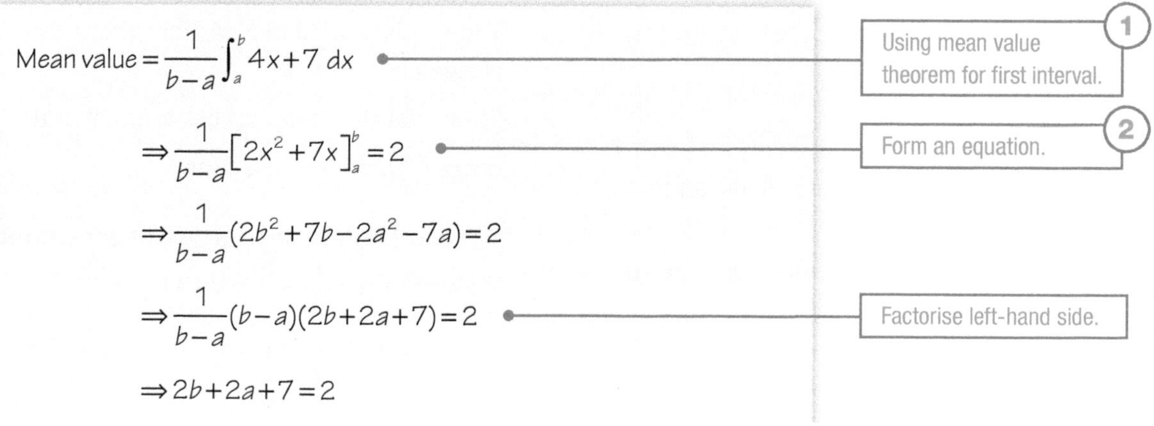

Mean value $= \frac{1}{b-a}\int_a^b 4x+7\,dx$
$\Rightarrow \frac{1}{b-a}\left[2x^2+7x\right]_a^b = 2$
$\Rightarrow \frac{1}{b-a}(2b^2+7b-2a^2-7a) = 2$
$\Rightarrow \frac{1}{b-a}(b-a)(2b+2a+7) = 2$
$\Rightarrow 2b+2a+7 = 2$
1 Using mean value theorem for first interval.
2 Form an equation.
Factorise left-hand side.

(*Continued on the next page*)

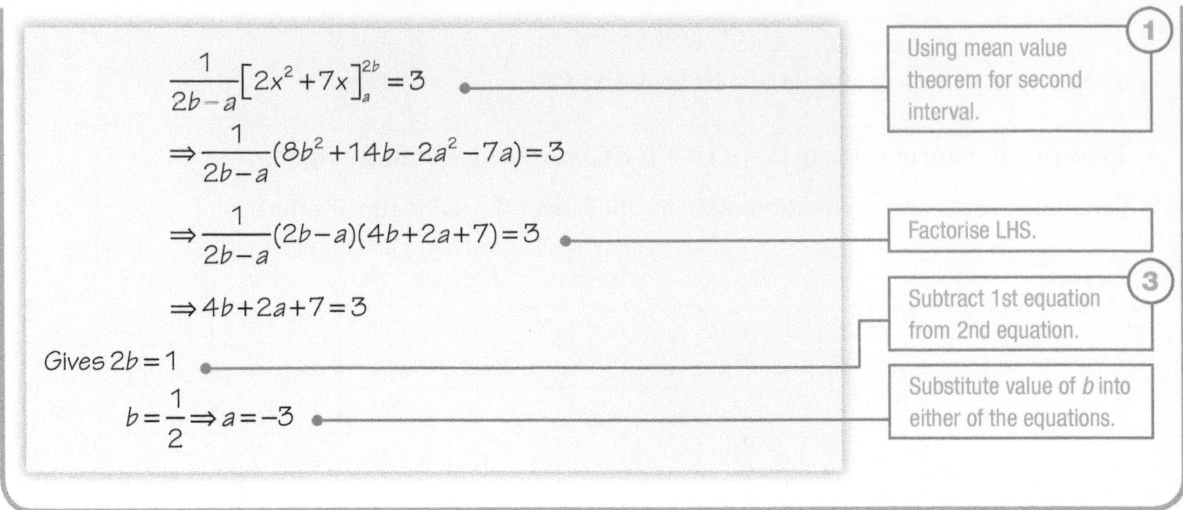

$$\frac{1}{2b-a}\left[2x^2+7x\right]_a^{2b}=3$$

Using mean value theorem for second interval.

$$\Rightarrow \frac{1}{2b-a}(8b^2+14b-2a^2-7a)=3$$

$$\Rightarrow \frac{1}{2b-a}(2b-a)(4b+2a+7)=3$$

Factorise LHS.

$$\Rightarrow 4b+2a+7=3$$

Gives $2b=1$

Subtract 1st equation from 2nd equation.

$$b=\frac{1}{2}\Rightarrow a=-3$$

Substitute value of b into either of the equations.

Example 5

Given that $f(x)=\dfrac{2x+1}{\sqrt{x}}$, show that the mean value of $f(x)$ for x in the interval $2\le x\le 8$ is $\dfrac{31}{9}\sqrt{2}$

$$f(x)=\frac{2x+1}{x^{\frac{1}{2}}}$$

$$=2x^{\frac{1}{2}}+x^{-\frac{1}{2}}$$

Write in simplified index form.

$$\text{Mean value }=\frac{1}{8-2}\int_2^8\left(2x^{\frac{1}{2}}+x^{-\frac{1}{2}}\right)dx$$

Use the definition of the mean value.

$$=\frac{1}{6}\left[\frac{4}{3}x^{\frac{3}{2}}+2x^{\frac{1}{2}}\right]_2^8$$

$$=\frac{1}{6}\left(\frac{64}{3}\sqrt{2}+4\sqrt{2}\right)-\frac{1}{6}\left(\frac{8}{3}\sqrt{2}+2\sqrt{2}\right)$$

$$=\frac{31}{9}\sqrt{2}\text{ as required}$$

Exercise 4.1B Reasoning and problem-solving

1 The velocity of a particle after t seconds is given by $v=\frac{2}{5}t^3-\frac{1}{5}t\ \text{m s}^{-1}$

  a Calculate the mean velocity for $1\le t\le 3$

  b Show that the mean acceleration for $1\le t\le 3$ is $5\ \text{m s}^{-2}$

2 The velocity of a particle after t seconds is given by $v=\frac{3t^2-5}{5}\ \text{ms}^{-1}$

  a Show that the mean velocity over the first 5 seconds is $4\ \text{ms}^{-1}$

  b Calculate the mean acceleration over the first 5 seconds.

3 The displacement of a particle after t seconds is given by $s=2t\sqrt{t}$

Show that the mean acceleration over the range $\frac{1}{2}\le t\le 1$ is $6-3\sqrt{2}\ \text{ms}^{-2}$

4 The acceleration of a particle after t seconds is given by $a=\frac{t}{10}-\frac{10}{t^3}$ for $t>0$

  a Show that the mean acceleration for $1\le t\le 3$ is $-\frac{91}{45}\ \text{m s}^{-2}$

  b Given that after 1 second the particle is travelling at $5\ \text{m s}^{-1}$, calculate the mean velocity for $1\le t\le 3$

5 Show that the mean value of a straight line $y = mx + c$ in the interval $[a, b]$ is given by $\frac{m(a+b)}{2} + c$

6 Show that the mean value of the curve $y = x^2$ for $0 \le x \le a$ is given by $\frac{a^2}{3}$

7 $g(x) = x^3 + 4x^2 - 5x$

a Sketch the graph of $y = g(x)$, labelling where the curve crosses the coordinate axes.

b Calculate the area bounded by the curve $y = g(x)$ and the x-axis.

c Find the mean value of $g(x)$ for values of x in each of these intervals

i $[-5, 0]$ ii $[0, 1]$

8 $g(x) = \sqrt{x}$

The mean value of $g(x)$ in the interval $[1, 4]$ is double the mean value of $g(x)$ in the interval $[0, k]$ for $k > 0$

Find the exact value of k

9 $f(x) = 2x^2 + a$

The mean value of $f(x)$ in the interval $[-3, 3]$ is -2. Calculate the value of a

10 The mean value of the function $f(x) = x - 5$ for $0 \le x \le a$ is -1.5

Calculate the value of a

11 The mean value of the function $g(x) = x^2 + 2x - 1$ for $0 \le x \le X$ is 17

Calculate the possible values of X

12 Given $h(x) = 2x + 1$

The mean value of $h(x)$ for $a \le x \le b$ is 4 and the mean value of $h(x)$ for $a \le x \le 2b$ is 8

Calculate the values of a and b

13 Is the mean value of the function $x^4 - 2x^3 + 3x - 5$ greater in the range $[1, 3]$ or in the range $[-2, -1]$? Show your working and state how much bigger the greater mean value is.

14 Particle A has a speed at time t seconds given by $t^2 + t$ and particle B has a speed at time t seconds given by $\sqrt{t}$

Which particle has the fastest average speed over the range $0 \le t \le \frac{1}{2}$? Show all your working and write down precisely how much quicker this particle's average speed is.

PURE

Full A Level

15 Show that the mean value of the function $\ln x$ in the interval $\frac{1}{2} \le x \le 3$ is $a\ln b + c$ where a, b and c are constants to be found.

16 $f(x) = \frac{x}{4x^2 - 3}$

Find an expression for the mean value of $f(x)$ in the interval $[k, k+1]$. Give your answer as a single logarithm.

17 The graph of $y = A + \frac{1}{x - B}$ is shown.

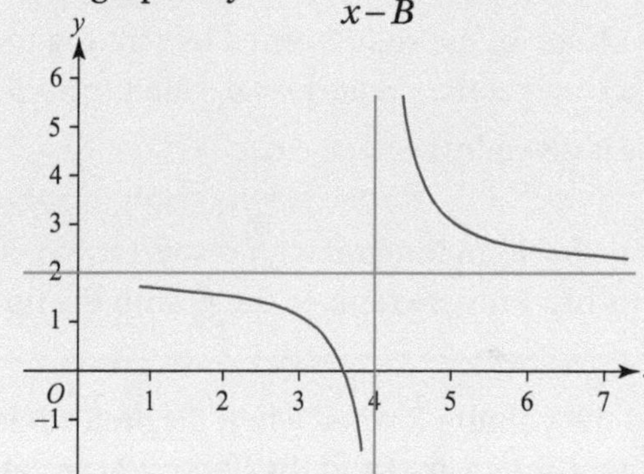

a Calculate the mean value of the function for $5 \le x \le 7$

b Is it possible to calculate the mean value of the function for $0 \le x \le 7$?

Explain your answer.

18 Calculate the mean value of each of these functions for the interval given.

a $\sin x$ for $\left[0, \frac{\pi}{3}\right]$ b $\cos 2x$ for $\left[\frac{\pi}{6}, \frac{\pi}{3}\right]$

c e^{3x} for $[0, 1]$ d $2xe^x$ for $[0, 3]$

4.2 Volumes of revolution

Fluency and skills

You can use calculus to find the volumes of 3D shapes. You can also generate the formulae for shapes like cones and spheres.

Look at this graph of $y = \mathrm{f}(x)$. If the section of this curve between $x = a$ and $x = b$ is rotated 360° around the x-axis it will form a solid.

Since the curve has rotated through 360°, the cross-section of this solid will always be a circle with radius y.

Now consider thin strips of thickness δx. Each of these strips is approximately a cylinder so it will have a volume of $\delta V = \pi y^2 \delta x$

The volume of the whole solid is the sum of these thin strips. Using integration to sum these strips gives the formula

$$V = \int_a^b \pi y^2 \, \mathrm{d}x$$

Key point

The volume of the solid formed by rotating the curve $y = \mathrm{f}(x)$ between $x = a$ and $x = b$ a full turn around the x-axis is given by $V = \int_a^b \pi y^2 \, \mathrm{d}x$

Because this is an integral with respect to x, you need the expression in terms of x only, so replace the y^2 with the function in terms of x

Example 1

Find the volume formed when the area enclosed by the curve with equation $y = x^3 + 1$, the coordinate axes and the line $x = 2$ is rotated

a 360° around the x-axis, **b** 180° around the x-axis.

a $V = \pi \int_0^2 (x^3 + 1)^2 \, dx$ — Using $V = \int_a^b \pi y^2 \, dx$ where $y = x^3 + 1$

$= \pi \int_0^2 x^6 + 2x^3 + 1 \, dx$ — Expand and simplify before attempting to integrate.

$= \pi \left[\frac{x^7}{7} + \frac{x^4}{2} + x \right]_0^2$

$= \pi \left(\frac{128}{7} + 8 + 2 - 0 \right)$

$= \frac{198}{7}\pi$ cubic units

b Rotating through 180° produces exactly $\frac{1}{2}$ of the shape in part **a** so the volume in this case is $\frac{1}{2}\left(\frac{198}{7}\pi\right) = \frac{99}{7}\pi$ cubic units.

PURE

You could also be asked to calculate the volume when a curve is rotated around the y-axis. In this case, the roles of x and y are swapped.

Therefore you need the function to be $x = \mathrm{f}(y)$ and the limits to be $y = a$ and $y = b$. Then the volume is $V = \int_a^b \pi x^2 \, \mathrm{d}y$

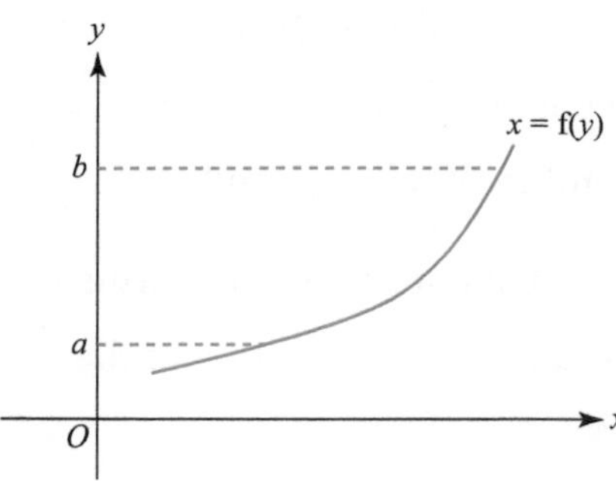

Key point

The volume of the solid formed by rotating the curve $x = \mathrm{f}(y)$ between $y = a$ and $y = b$ a full turn around the y-axis is given by $V = \int_a^b \pi x^2 \, \mathrm{d}y$

Example 2

The region A is enclosed by the curve with equation $y = x^5$, the y-axis and the lines $y = 0.5$ and $y = 1$

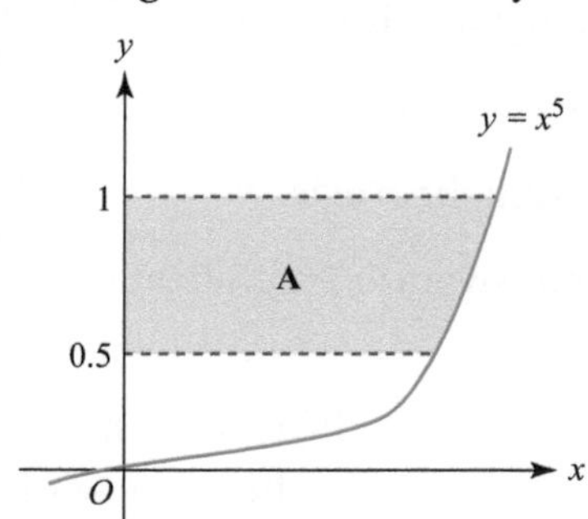

Calculate the volume when region A is rotated 2π radians around the y-axis.

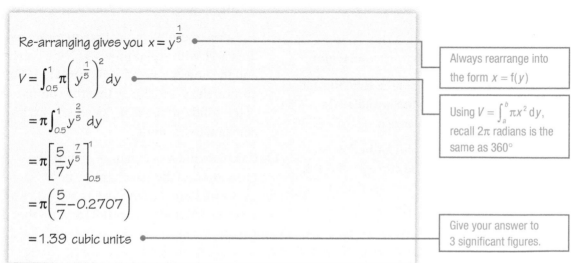

Re-arranging gives you $x = y^{\frac{1}{5}}$ — Always rearrange into the form $x = \mathrm{f}(y)$

$V = \int_{0.5}^{1} \pi \left(y^{\frac{1}{5}}\right)^2 \mathrm{d}y$ — Using $V = \int_a^b \pi x^2 \, \mathrm{d}y$, recall 2π radians is the same as 360°

$= \pi \int_{0.5}^{1} y^{\frac{2}{5}} \, \mathrm{d}y$

$= \pi \left[\frac{5}{7} y^{\frac{7}{5}}\right]_{0.5}^{1}$

$= \pi \left(\frac{5}{7} - 0.2707\right)$

$= 1.39$ cubic units — Give your answer to 3 significant figures.

Exercise 4.2A Fluency and skills

1 Calculate the volume formed when the region bounded by the curve $y=3x^2$, the x-axis and the lines $x=1$ and $x=4$ is rotated 360° around the x-axis.

2 The region R is enclosed by the curve $y=8-x^3$ and the coordinate axes.

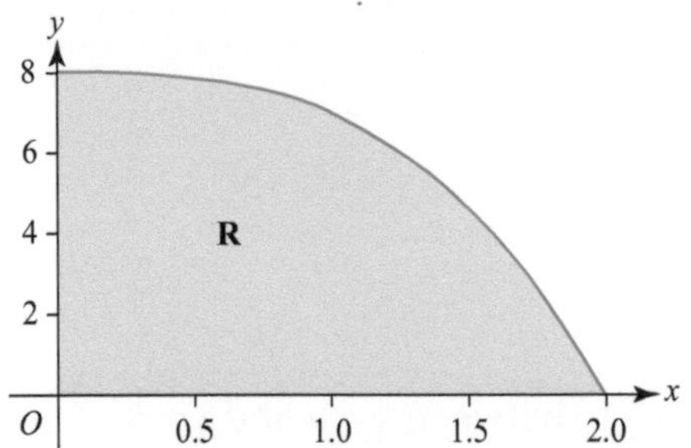

a Show that the area of R is 12 square units.

The region R is rotated 360° around the x-axis.

b Calculate the volume of the solid formed.

3 The shaded region is enclosed by the curve $y=x^4-x^3$ and the x-axis.

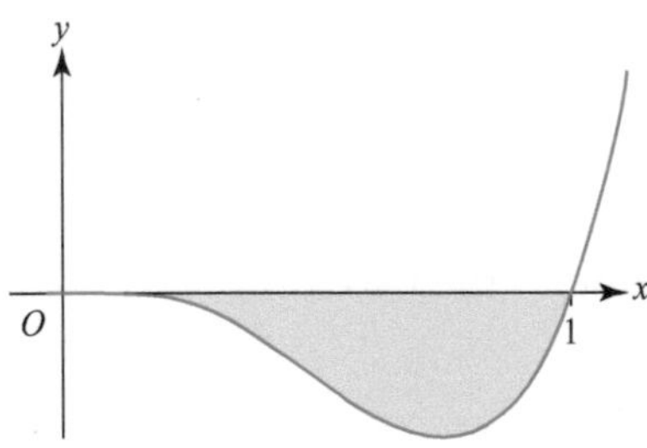

a Calculate the area of the shaded region.

b Calculate the volume of the solid formed when the shaded region is rotated 2π radians around the x-axis.

4 Find the volume of the solid formed when the region bounded by the curve $y=4x^2-5$, the x-axis and the lines $x=0.5$ and $x=1$ is rotated 180° around the x-axis.

5 Calculate the volume of the solid formed when the region bounded by the curve $x=2y^3-y$ and the lines $y=1$ and $y=2$ is rotated 2π radians around the y-axis.

6 The region R is enclosed by the curve $y=\frac{1}{x^2}$, the x-axis and the lines $x=\frac{1}{2}$ and $x=1$

Show that the volume of the solid formed when R is rotated 2π radians around the x-axis is $\frac{7}{3}\pi$

7 The region A is enclosed by the curve $y=\frac{2+\sqrt{x}}{5x^2}$, the x-axis and the lines $x=1$ and $x=3$

a Show that the area of A is $\frac{10-2\sqrt{3}}{15}$

b Calculate the volume of the solid formed when the region A is rotated

i 360° around the x-axis,

ii 90° around the x-axis.

Give your answers to 3 significant figures.

8 Calculate the volume of the solid formed when the area bounded by the curve with equation $y=4-x^2$, the positive x-axis and the positive y-axis is rotated 360° around

a The x-axis, b The y-axis.

9 The area bounded in the first quadrant by the curve with equation $y=4x^4$, the y-axis and the lines $y=2$ and $y=8$ is rotated 360° around the y-axis. Show that the volume of the solid formed is $A\sqrt{2}\pi$ where A is a constant to be found.

10 Calculate the exact volume of the solid formed when the region in the first quadrant bounded by the curve $y=3x-x^3$ and the x-axis is rotated 180° around the x-axis.

Reasoning and problem-solving

Strategy

To solve problems involving volumes of revolution

1. Sketch a graph. Note carefully which axis you are rotating the graph about.
2. Choose the correct version of the formula for a volume of revolution, either $V = \int_a^b \pi x^2 \, dy$ or $V = \int_a^b \pi y^2 \, dx$
3. Add or subtract another volume where necessary.

Example 3

The region R is enclosed by the curve $y = \sqrt{x}$ and the lines $y = x - 6$, $x = 9$ and $y = 0$

Calculate the volume of the solid formed when R is rotated 360° around the x-axis.

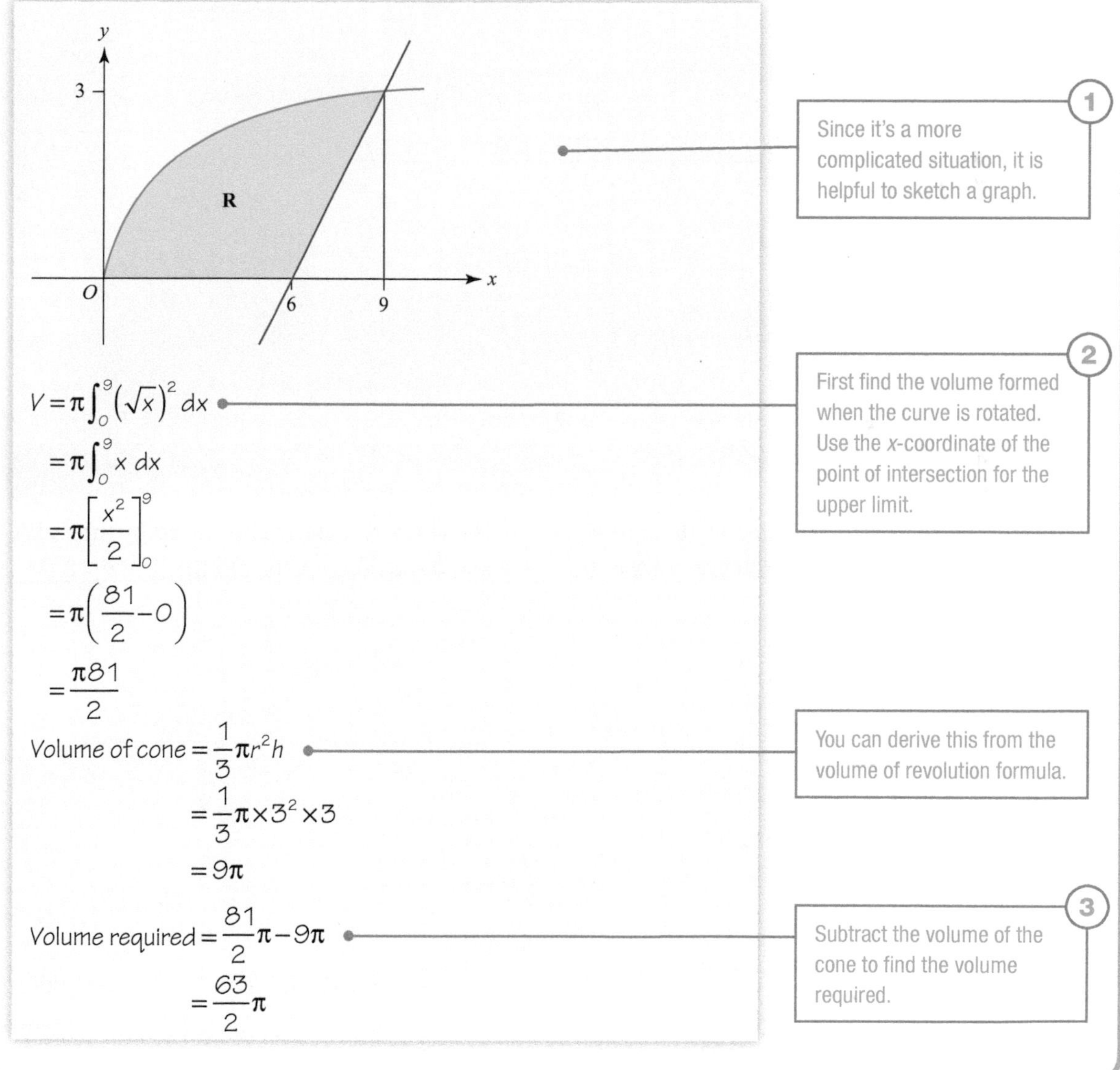

(1) Since it's a more complicated situation, it is helpful to sketch a graph.

$$V = \pi \int_0^9 \left(\sqrt{x}\right)^2 dx$$
$$= \pi \int_0^9 x \, dx$$
$$= \pi \left[\frac{x^2}{2}\right]_0^9$$
$$= \pi \left(\frac{81}{2} - 0\right)$$
$$= \frac{\pi 81}{2}$$

(2) First find the volume formed when the curve is rotated. Use the x-coordinate of the point of intersection for the upper limit.

$$\text{Volume of cone} = \frac{1}{3}\pi r^2 h$$
$$= \frac{1}{3}\pi \times 3^2 \times 3$$
$$= 9\pi$$

You can derive this from the volume of revolution formula.

$$\text{Volume required} = \frac{81}{2}\pi - 9\pi$$
$$= \frac{63}{2}\pi$$

(3) Subtract the volume of the cone to find the volume required.

You can use volumes of revolution to prove some standard volume formulae.

Example 4

By rotating the curve $y=\sqrt{r^2-x^2}$ between $-r$ and r around the x-axis, show that the volume of a sphere is $\frac{4}{3}\pi r^3$

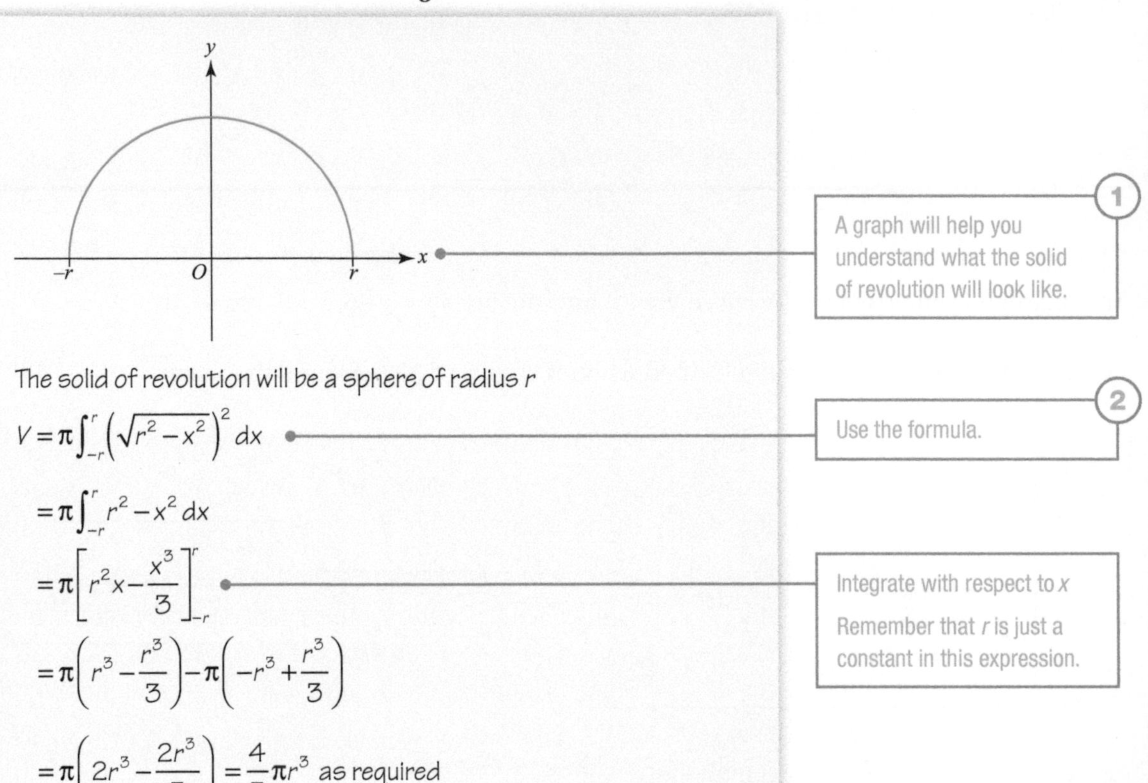

y
x
$-r$
O
r
The solid of revolution will be a sphere of radius r
$V=\pi\int_{-r}^{r}\left(\sqrt{r^2-x^2}\right)^2 dx$
$=\pi\int_{-r}^{r} r^2-x^2\,dx$
$=\pi\left[r^2x-\frac{x^3}{3}\right]_{-r}^{r}$
$=\pi\left(r^3-\frac{r^3}{3}\right)-\pi\left(-r^3+\frac{r^3}{3}\right)$
$=\pi\left(2r^3-\frac{2r^3}{3}\right)=\frac{4}{3}\pi r^3$ as required
1
A graph will help you understand what the solid of revolution will look like.
2
Use the formula.
Integrate with respect to x
Remember that r is just a constant in this expression.

Exercise 4.2B Reasoning and problem-solving

1 The shaded region is bounded by the curve $y=9x-8x^2$, the x-axis and the line $y=2x-1$

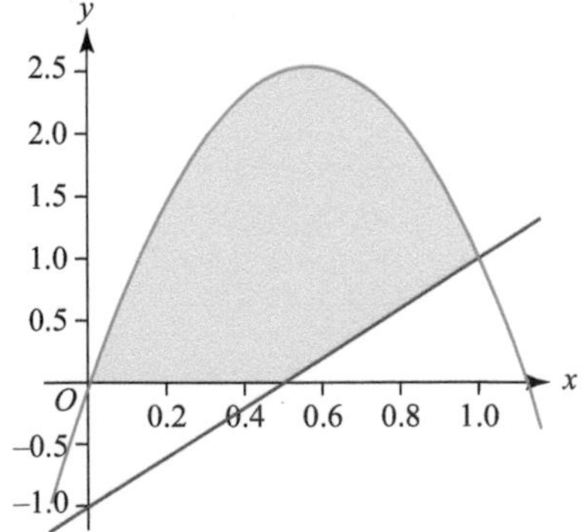

a Calculate the area of the shaded region.

b Calculate the volume of the solid formed when the shaded region is rotated 360° around the x-axis. Give your answer in terms of π

2 The shaded region is bounded by the curve $y=10-x^3$, the y-axis and the line $y=2$

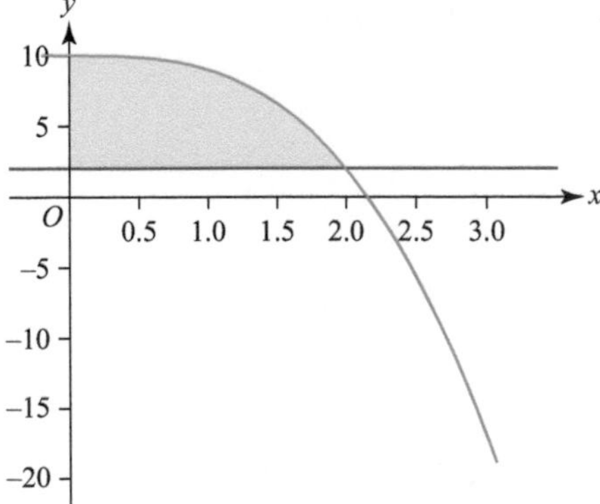

a Calculate the area of the shaded region.

b Calculate the volume of the solid formed when the shaded region is rotated 360° around the x-axis. Give your answer in terms of π

3 The region R is bounded in the first quadrant by the curve with equation $y=16-x^4$, the line $y=7$ and the y-axis.

a Show that the area of R is $\frac{36}{5}\sqrt{3}$ square units.

b Calculate the volume of the solid formed when R is rotated 360° around the x-axis.

4 The region A, to the right of the y-axis, is bounded by the curve with equation $y=8+2x-x^2$, the line $y=x-4$ and the y-axis.

a Show that the area of A is $\frac{104}{3}$ square units.

b Calculate the volume of the solid formed when A is rotated 180° around the x-axis.

5 The region R is bounded by the curve $y=\sqrt{x}$, the line $y=x-2$ and the x-axis.

a Calculate the area of R.

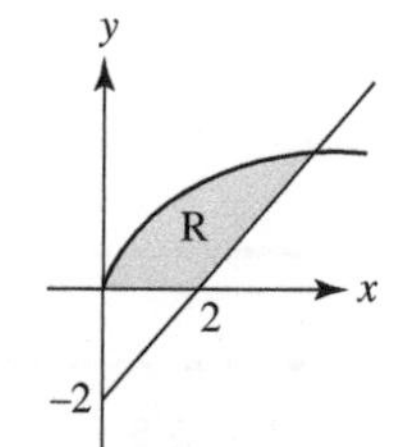

R is rotated 180° around the x-axis.

b Calculate the volume of the solid formed.

6 By rotating the line $y=ax$ between $x=0$ and $x=b$ 360° around the x-axis, show that the volume of a cone is $\frac{1}{3}\pi r^2 h$, where r is the radius of the base and h is the height.

7 Use a volume of revolution to show that the volume of a cylinder is $\pi r^2 h$, where r is the radius and h is the height.

8 Use a volume of revolution to derive a formula for the volume of a hemisphere of radius r

9 The section above the x-axis between $x=1$ and $x=3$ of the curve with equation $y=\frac{Ax}{\sqrt{x}}$ is rotated 2π radians around the x-axis and the volume of the solid formed is $\frac{8}{3}\pi$

a Calculate the value of A

b Calculate the volume when the same section of the curve is rotated 2π radians around the y-axis.

10 The parabola $y^2=28x$ intersects the line $x=k$ at the points A and B

The area bounded by the parabola and the line segment AB is $130\frac{2}{3}$ square units.

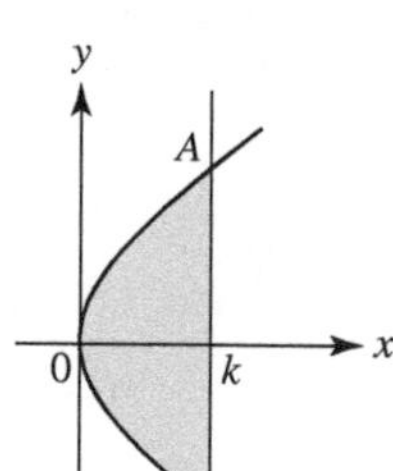

a Find the coordinates of A and B

The shaded region is rotated 180° around the x-axis.

b Calculate the volume of the solid formed.

11 Calculate the volume of the solid formed when each of these curves is rotated 360° around the x-axis between the limits shown. Give each of your answers in its simplest form.

a $y=\frac{1}{2\sqrt{x}}$, $x=2$ and $x=5$

b $y=\sec x$, $x=\frac{\pi}{6}$ and $x=\frac{\pi}{3}$

c $y=(2x-5)^3$, $x=3$ and $x=3.5$

d $y=3e^{2x}$, $x=-\frac{1}{4}$ and $x=0$

12 Find the volume of the solid formed when the region in question **2** is rotated 180° around the y-axis. Give your answer in terms of π

Full A Level

PURE

4 Summary and review

Chapter summary

- The mean value of a function $f(x)$ in the range $a \le x \le b$ is given by $\frac{1}{b-a}\int_a^b f(x)dx$
- The area enclosed between the curve with equation $y = f(x)$, the x-axis and the lines $x = a$ and $x = b$ is given by $\int_a^b y\ dx$
- The volume of the solid formed by rotating the curve $y = f(x)$ between $x = a$ and $x = b$ a full turn around the x-axis is given by $V = \int_a^b \pi y^2\ dx$
- The volume of the solid formed by rotating the curve $x = f(y)$ between $y = a$ and $y = b$ a full turn around the y-axis is given by $V = \int_a^b \pi x^2\ dy$

Check and review

You should now be able to...	Try Questions
✔ Calculate the mean value of a function.	1–5
✔ Find the area enclosed by a curve and lines.	7, 10, 12
✔ Calculate the volume of revolution when rotated around the *x*-axis.	6–8, 11, 12
✔ Calculate the volume of revolution when rotated around the *y*-axis.	9–11
✔ Calculate more complicated volumes of revolution by adding or subtracting volumes.	11–13

1 Find the mean value of the function $f(x) = 8x^3 + x^2 - 4$ for $1 \le x \le 3$

2 Show that the mean value of the function $g(x) = \frac{1+x}{\sqrt{x}}$ in the interval [2, 8] is $A\sqrt{2}$ where A is a constant to be found.

3 Work out the mean value of $4x^3 - \frac{16}{x^3}$ in the interval $\frac{1}{2} \le x \le 2$

4 Show that the mean value of the function $\frac{2-x}{2x^3}$ for x in the interval $[a, 1]$ is $\frac{1}{2a^2}$ and $a > 0$

5 The velocity of a particle after t seconds is given by $v = \frac{1}{20}(5t^4 - 3t^2)\ \mathrm{m\,s^{-1}}$

a Calculate the mean velocity over the first 4 seconds.

b Calculate the mean acceleration over the first 4 seconds.

6 Calculate the volume of the solid formed when the line with equation $y = 1 - 3x$ between $x = 2$ and $x = 5$ is rotated 2π radians around the x-axis.

7 The region R is enclosed by the curve with equation $y = 3 - \frac{1}{x^2}$, the x-axis and the line $x = 2$

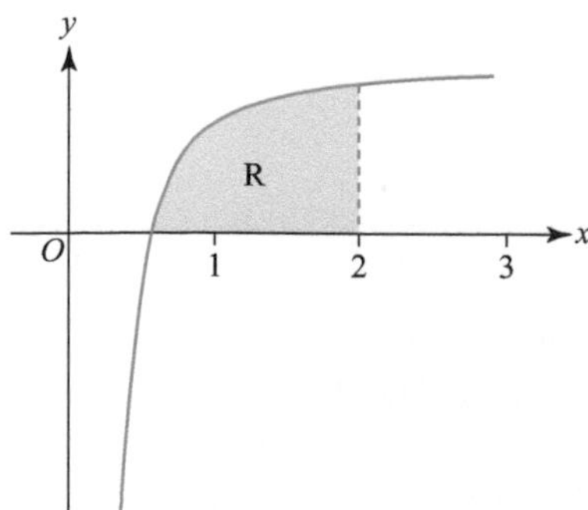

a Show that the area of R is $\frac{13 - 4\sqrt{3}}{2}$ square units.

The region R is rotated 360° around the x-axis.

b Calculate the volume of the solid formed.

8 Calculate the volume of the solid formed when the curve with equation $y = 5\sqrt{x}$ between $x = 1$ and $x = 3$ is rotated 180° around the x-axis.

9 Show that the volume of revolution when $x = 4y^2$ between $y = 0$ and $y = \frac{1}{2}$ is rotated 2π radians around the y-axis is $\frac{\pi}{10}$ cubic units.

10 The shaded region is bounded by the curve with equation $x = y^4 + 1$, $y \geq 0$, the coordinate axes and the line with equation $y = 2$

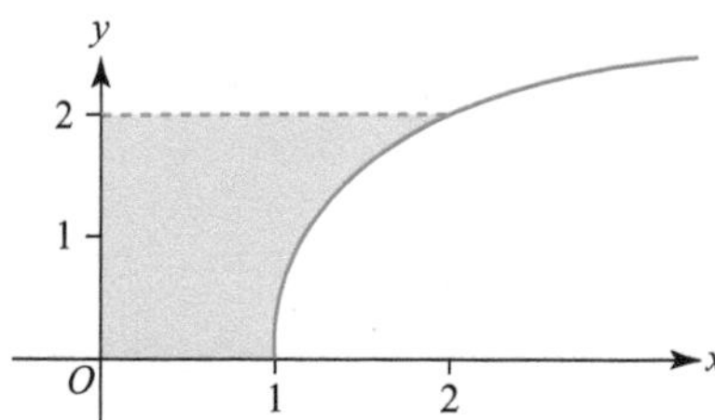

a Show that the area of the shaded region is $\frac{42}{5}$ square units.

The region is rotated 360° around the y-axis.

b Calculate the volume of the solid formed.

11 The region A, bounded by the curve with equation $y = 2 - \sqrt{x}$, the line $y = 2 - \frac{x}{2}$ and the y-axis, is rotated 360° around the x-axis.

a Sketch a graph showing the region A.

b Calculate the volume of revolution when region A is rotated 2π radians around the x-axis.

c Calculate the volume of revolution when region A is rotated 2π radians around the y-axis.

12 The shaded region is bounded by the curve with equation $y = 5x - 4x^2$, the line $y = 1$ and the x-axis.

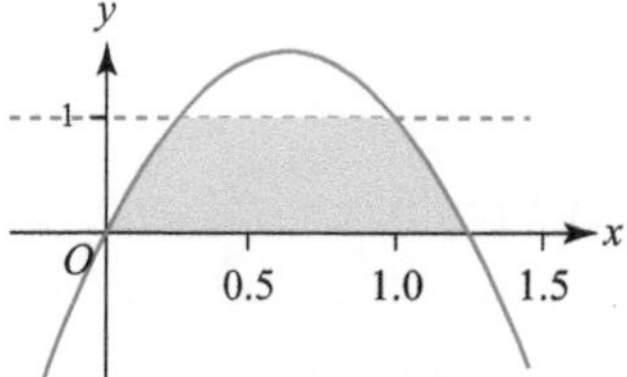

a Calculate the area of the shaded region.

b Calculate the volume of the solid formed when the shaded region is rotated 180° around the x-axis.

13 The region A is bounded by the curve with equation $y = 2\sqrt{x}$ and the line $y = x$

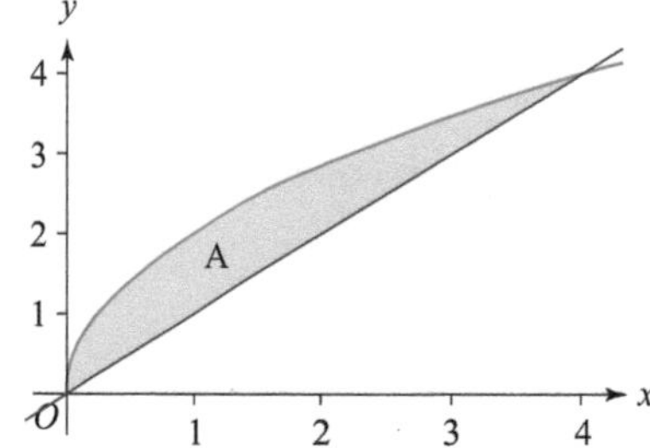

Calculate the volume of revolution when the shaded region is rotated 2π radians around

a The x-axis, **b** The y-axis.

PURE

Investigation

A different method of finding the volume of a solid by considering the revolution of a function, relies on considering the solid to be composed of thin cylindrical shells.

Each cylindrical shell is formed by rotating a vertical strip about the y-axis. The inner surface area of this cylinder is a rectangle with width $2\pi x$ and height y, so it has area $2\pi xy$

This means that the volume, δv, of a cylinder of width δx is given by $\delta v = 2\pi xy\delta x$

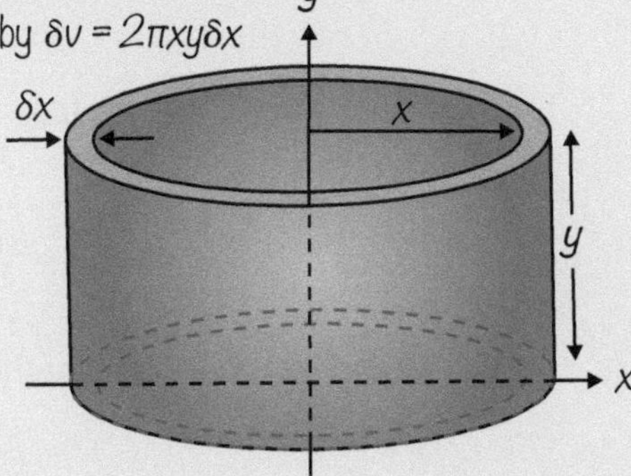

Summing these cylinders between $x = x_1$ and $x = x_2$, and letting δx tend to zero, results in the formula $v = 2\pi \int_{x_1}^{x_2} xy\,dx$

Convince yourself that this formula is correct by using it to find

a The volume of a cylinder of radius a and height h

b The volume of a cone of radius a and height h

Information

Gabriel's horn is a geometric solid that has a finite volume but an infinite surface area. This is sometimes known as the painter's paradox as although the horn would contain a finite volume of paint you would never have enough paint to paint its outside.

The solid is formed by rotating the function $f(x) = \frac{1}{x}$ about the x-axis for $x > 1$

Have a go

When carrying out an integral with no limit on the upper bound, it is usual practice to take an arbitrary value, say $x = a$, and then consider what happens as a tends to infinity.

Try the method described above to calculate the finite volume of Gabriel's horn.

Challenge

The formula for a **surface area of revolution** that is found by rotating $y = f(x)$ from $x = a$ to $x = b$ about the x-axis is given by

$$A = 2\pi \int_a^b y\sqrt{1 + \left(\frac{dy}{dx}\right)^2}\,dx$$

Use this formula to show that Gabriel's horn has an infinite surface area.

4 Assessment

1 Calculate the mean value of these functions for the limits shown.

a x^2+3x+1 for $0 \le x \le 2$ b $x^3(2-x^3)$ for $-1 \le x \le 1$ **[8 marks]**

2 You are given that $f(x)=\dfrac{x^3-3\sqrt{x}}{2x^2}$

a Write $f(x)$ in the form Ax^m+Bx^n where A, B, m and n are constants to be found. **[2]**

b Calculate the mean value of $f(x)$ in the interval [1, 4]. **[4]**

3 Find the mean value of the function $g(x)=x\sqrt{x}+\dfrac{3}{x^2}$ for $1 \le x \le 3$

Give your answer in the form $a+b\sqrt{3}$ where a and b are constants to be found. **[4]**

4 The velocity of a particle after t seconds is given by $v=5+3t^2$ m s^{-1}

a Calculate the mean velocity for $1 \le t \le 5$ **[4]**

b Calculate the mean acceleration for $1 \le t \le 5$ **[4]**

5 The mean value of the function $1+2x^3$ in the interval $[0, a]$ is 109

Evaluate a **[5]**

6 The shaded region is bounded by the curve $y=x^2-2x$, the x-axis and the line $x=4$

a Find the value of a **[2]**

b Calculate the area of the shaded region. **[3]**

The shaded region is rotated one full turn around the x-axis.

c Calculate the volume of the solid formed. **[4]**

7 Calculate the volume of revolution when the region bounded by the curve $y=2\sqrt{x}$, the x-axis and the line $x=3$ is rotated 360° about the x-axis. Give your answer in terms of π **[4]**

8 The region R is bounded by the curve $y=\dfrac{2}{x}+x^3$, the x-axis and the lines $x=1$ and $x=2$

Calculate the volume of revolution when R is rotated 360° around the x-axis. **[4]**

9 Calculate the volume of revolution when the region bounded by the curve $y=x^2(5-x)$ and the x-axis is rotated 180° around the x-axis. Give your answer to 3 significant figures. **[5]**

10 The acceleration of a particle is given by $a=\dfrac{3}{2}t^2-\dfrac{2}{t^3}$

Given that after 1 second the particle is travelling at 3 m s^{-1}, calculate the mean velocity between 1 and 4 seconds after starting. **[5]**

11 $f(x)=ax^2+bx^5$

The mean value of the function $f(x)$ in the interval [1, 2] is 35 and the mean value of $f(x)$ in the interval [0, 3] is 99

Calculate the values of a and b **[5]**

12 The curve with equation $y=x-k\sqrt{x}$ between $x=0$ and $x=1$ is rotated 2π radians around the x-axis and the volume of revolution is $\dfrac{11}{15}\pi$. Calculate the possible values of k **[5]**

13 The mean value of $f(x)=Ax(2x^2-1)^3$ in the interval $\left[\frac{1}{2}, 1\right]$ is 15

Calculate the value of A **[4]**

14 The shaded region is part of an ellipse with equation $\frac{x^2}{9}+\frac{y^2}{4}=1$

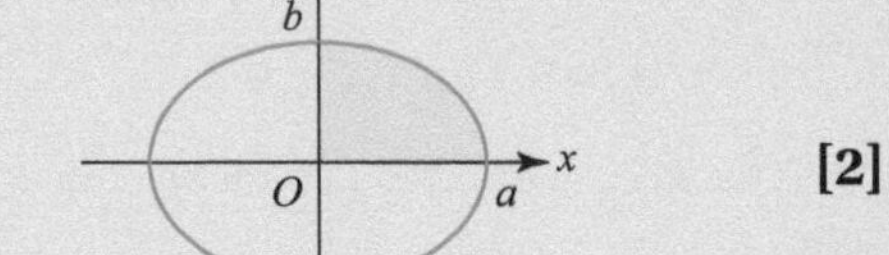

a Write the values of a and b in the diagram. **[2]**

b Calculate the volume when the shaded area is rotated π radians about the x-axis. **[5]**

15 a Sketch the graphs of $y=5-|x|$ and $y=|x^2-25|$ on the same axes. **[4]**

b The enclosed region bounded by $y=5-|x|$ and $y=|x^2-25|$ is rotated 360° around the x-axis. Calculate the volume of the solid formed. **[5]**

16 $g(x)=e^{2x}+x^3+2$

Show that the mean value of the function $g(x)$ in the interval [0, 2] is $\frac{1}{4}e^4+k$, where k is a constant to be found. **[4]**

17 Calculate the mean value of the function $f(x)=\frac{1}{2x-3}$ for x in the range $a \le x \le 3$

Give your answer in its simplest form in terms of a **[4]**

18 Calculate the mean value of each of these functions for the limits shown

a $\sin 2x$ for $0 \le x \le \frac{\pi}{6}$ b $\sin^2 x$ for $\frac{\pi}{3} \le x \le \frac{\pi}{2}$ **[9]**

19 Calculate the mean value of the function $f(t)=\frac{t}{2}\ln t$ for t in the region $2 \le t \le 4$

Give your answer in the form $A+B\ln 2$ where A and B are constants to be found. **[6]**

20 The region R is bounded by the curve with equation $= xe^{\frac{x}{2}}$, the x-axis and the line $x=2$

a Calculate the area of R **[4]**

b Calculate the volume of revolution when R is rotated 360° around the x-axis. **[6]**

21 The curve C is defined by the parametric equations

$y=3t^2+1, x=5t$

The region bounded by C, the x-axis and the lines $x=5$ and $x=10$ is rotated π radians around the x-axis. Calculate the volume of revolution of the solid formed. **[6]**

22 a Sketch the graph of $y=\cosh 2x$ **[2]**

The region R is bounded by the curve $y=\cosh 2x$ and the lines $x=\pm 1$

b Calculate the volume of the solid formed when R is rotated 90° about the x-axis. **[6]**

23 The region R is bounded by the curve with equation $y=\tan x$, the x-axis and the line $x=\frac{\pi}{4}$

R is rotated $k\pi$ radians around the x-axis and the volume of the solid formed is $\frac{\pi}{3}-\frac{\pi^2}{12}$

Calculate the value of k **[6]**

24 The curve C is defined by the parametric equations

$x=2\sin\theta, y=\sin 2\theta, 0 \le \theta \le \frac{\pi}{2}$

Calculate, in terms of π, the volume of the solid formed when the region bounded by C and the y-axis is rotated 360° around the y-axis. **[8]**

5 Matrices 1

The mathematics of matrices is used in a diverse range of applications. For example, civil and structural engine engineers find them useful because they can be used to find solutions to **systems of linear simultaneous** equations with multiple unknowns. When analysing the forces in, for example, the design of a new bridge, the mathematics can involve systems of equations where many forces need to be found. The use of matrices, together with computer software, provides a powerful method that ensures that the equations can be solved and the forces involved calculated. This allows more complex structures to be built than was previously the case.

Matrix algebra methods can be used in other branches of engineering. For example, to find the current flowing in various parts of electrical circuits, to determine traffic flow in models of road networks, the production of different chemicals in an industrial process, and so on.

Orientation

What you need to know

KS4

- Transforming objects by reflection, rotation and enlargement.

Maths Ch1

- Solving simultaneous equations.

Maths Ch6

- Using vectors.

What you will learn

- To identify the order of a matrix.
- To add, subtract and multiply matrices by a scalar or a conformable matrix.
- To apply linear transformations given as matrices, describe transformations given as matrices and write linear transformations as a matrix.
- To find invariant points and lines of linear transformations.
- To calculate determinants and inverses of matrices.
- To use matrices to solve systems of linear equations.

What this leads to

Ch22 Further Matrices

- Eigenvalues and eigenvectors.
- 3×3 systems.
- Matrix diagonalisation.

Ch12 Graphs and networks

- Matrix formulation of Prim's algorithm.

Ch14 Linear programming and game theory

- Zero sum games.

5.1 Properties and arithmetic

Fluency and skills

See Maths Ch6
For a reminder of vectors.

Matrices are a way of representing information in a form that can be manipulated mathematically. You will have already used vectors, which are a particular sort of matrix.

Key point

A **matrix** with n rows and m columns has **order** $n \times m$

For example, the matrix $\begin{pmatrix} 5 & 3 & -1 \\ -4 & 0 & 5 \end{pmatrix}$ is said to have order 2×3

The matrix $\begin{pmatrix} 2 \\ 7 \\ -5 \end{pmatrix}$ is said to have order 3×1

Computers use matrices to carry out an operation, such as addition or subtraction, and to carry it out on multiple numbers simultaneously. This is particularly used in the processing of computer graphics.

Key point

If two matrices have the same order then they can be added or subtracted by adding or subtracting their corresponding **elements**.

You can think of putting the two matrices on top of each other and adding or subtracting the elements that match.

Key point

To multiply a matrix by a constant, you should multiply each of its elements by that constant.

Example 1

You are given that $\mathbf{A} = \begin{pmatrix} 2 & 5 \\ 0 & 4 \\ -1 & -3 \end{pmatrix}$, $\mathbf{B} = \begin{pmatrix} 2 & -3 \\ 0 & 5 \end{pmatrix}$ and $\mathbf{C} = \begin{pmatrix} 6 & 15 \\ 0 & 12 \\ -3 & -9 \end{pmatrix}$. Find, if possible,

a $\mathbf{A}+\mathbf{B}$ **b** $\mathbf{A}+\mathbf{C}$ **c** $2\mathbf{B}$ **d** $\mathbf{C}-3\mathbf{A}$

a Not possible as **A** and **B** are not of the same order.

b $\mathbf{A}+\mathbf{C} = \begin{pmatrix} 2 & 5 \\ 0 & 4 \\ -1 & -3 \end{pmatrix} + \begin{pmatrix} 6 & 15 \\ 0 & 12 \\ -3 & -9 \end{pmatrix}$

$= \begin{pmatrix} 2+6 & 5+15 \\ 0+0 & 4+12 \\ -1+-3 & -3+-9 \end{pmatrix}$ — Add corresponding elements of **A** and **C**

$= \begin{pmatrix} 8 & 20 \\ 0 & 16 \\ -4 & -12 \end{pmatrix}$

(*Continued on the next page*)

c $2\mathbf{B} = 2\begin{pmatrix} 2 & -3 \\ 0 & 5 \end{pmatrix}$

$= \begin{pmatrix} 2\times 2 & 2\times -3 \\ 2\times 0 & 2\times 5 \end{pmatrix}$

Multiply every element of **B** by 2

$= \begin{pmatrix} 4 & -6 \\ 0 & 10 \end{pmatrix}$

d $\mathbf{C} - 3\mathbf{A} = \begin{pmatrix} 6 & 15 \\ 0 & 12 \\ -3 & -9 \end{pmatrix} - 3\begin{pmatrix} 2 & 5 \\ 0 & 4 \\ -1 & -3 \end{pmatrix}$

$= \begin{pmatrix} 6-3\times 2 & 15-3\times 5 \\ 0-3\times 0 & 12-3\times 4 \\ -3-3\times -1 & -9-3\times -3 \end{pmatrix}$

Multiply every element of **A** by 3 then subtract from the corresponding element in **C**

$= \begin{pmatrix} 0 & 0 \\ 0 & 0 \\ 0 & 0 \end{pmatrix}$

This is called a zero matrix as all its elements are 0

Key point

The **zero matrix, 0**, is a matrix, of any order, with all elements equal to zero.

Matrices can only be multiplied together if the number of columns in the first matrix is the same as the number of rows in the second matrix.

Key point

If two matrices can be multiplied, then it is said that they are **conformable for multiplication**.

When multiplying matrices, to find the element in the nth row and mth column, you must multiply the first term of the nth row in the first matrix by the first term of the mth column in the second matrix, then the second term of the nth row by the second term of the mth column and so on, then add these terms together.

Key point

The product of an $n \times m$ matrix and an $m \times p$ matrix has **order** $n \times p$

To find the first term in a matrix multiplication, look at the first row of the first matrix and the first column of the second matrix.

$\begin{pmatrix} a & d \\ b & e \\ c & f \end{pmatrix}\begin{pmatrix} w & y \\ x & z \end{pmatrix} = \begin{pmatrix} aw+dx & \cdots \\ \cdots & \cdots \\ \cdots & \cdots \end{pmatrix}$

Then consider the first row and second column.

$\begin{pmatrix} a & d \\ b & e \\ c & f \end{pmatrix}\begin{pmatrix} w & y \\ x & z \end{pmatrix} = \begin{pmatrix} aw+dx & ay+dz \\ \cdots & \cdots \\ \cdots & \cdots \end{pmatrix}$

Then move on to using the second row of the first matrix. Continue in this way until the last term which is found by considering the final row of the first matrix and the final column of the second.

$\begin{pmatrix} a & d \\ b & e \\ c & f \end{pmatrix}\begin{pmatrix} w & y \\ x & z \end{pmatrix} = \begin{pmatrix} aw+dx & ay+dz \\ bw+bx & by+ez \\ cw+fx & cy+fz \end{pmatrix}$

Example 2

If $\mathbf{A} = \begin{pmatrix} 3 & 0 \\ -1 & 2 \\ 7 & -4 \end{pmatrix}$, $\mathbf{B} = \begin{pmatrix} 5 & 1 \\ -3 & 0 \end{pmatrix}$ find, if possible, the products

a **AB** **b** **BA** **c** $\mathbf{B}^2$

a $\mathbf{AB} = \begin{pmatrix} 3 & 0 \\ -1 & 2 \\ 7 & -4 \end{pmatrix}\begin{pmatrix} 5 & 1 \\ -3 & 0 \end{pmatrix}$

$= \begin{pmatrix} (3\times5)+(0\times-3) & (3\times1)+(0\times0) \\ (-1\times5)+(2\times-3) & (-1\times1)+(2\times0) \\ (7\times5)+(-4\times-3) & (7\times1)+(-4\times0) \end{pmatrix}$

To find the term in the 3rd row and the 2nd column: multiply the first term from the 3rd row of matrix **A** by the first term from the 2nd column of matrix **B**. Then multiply the second term from the 3rd row of matrix **A** by the second term in the 2nd column of matrix **B**. Finally, add these values together.

$= \begin{pmatrix} 15 & 3 \\ -11 & -1 \\ 47 & 7 \end{pmatrix}$

Notice how the product of a 3×2 matrix and a 2×2 matrix is a 3×2 matrix.

b Not possible as **B** has 2 columns but **A** has 3 rows.

c $\mathbf{B}^2 = \begin{pmatrix} 5 & 1 \\ -3 & 0 \end{pmatrix}\begin{pmatrix} 5 & 1 \\ -3 & 0 \end{pmatrix}$

$\mathbf{B}^2$ means $\mathbf{B}\times\mathbf{B}$
It does not mean square each element of the matrix.

$= \begin{pmatrix} (5\times5)+(1\times-3) & (5\times1)+(1\times0) \\ (-3\times5)+(0\times-3) & (-3\times1)+(0\times0) \end{pmatrix} = \begin{pmatrix} 22 & 5 \\ -15 & -3 \end{pmatrix}$

Key point

It is only possible to find $\mathbf{A}^2$ if **A** is a **square matrix**, that is, it has the same number of rows as columns.

Calculator

Try it on your calculator

Some calculators can be used to add, subtract and multiply matrices.

MatA × MatB

$\begin{bmatrix} 15 & 3 \\ -11 & -1 \\ 47 & 7 \end{bmatrix}$

Activity

Find out how to work out $\begin{pmatrix} 3 & 0 \\ -1 & 2 \\ 7 & -4 \end{pmatrix}\begin{pmatrix} 5 & 1 \\ -3 & 0 \end{pmatrix}$ on *your* calculator.

Key point

The **transpose** of a matrix is formed by swapping the rows and columns.

For example, the transpose of the matrix $\begin{pmatrix} 1 & -2 & 4 \\ 0 & 5 & -6 \end{pmatrix}$ is $\begin{pmatrix} 1 & 0 \\ -2 & 5 \\ 4 & -6 \end{pmatrix}$: the first row of the matrix became the first column of the transpose and the second row of the matrix became the second row of the transpose.

You can use this result about transposes:

Key point

Give two $n\times n$ matrices **A** and **B**: $(\mathbf{AB})^{\mathrm{T}} \equiv \mathbf{B}^{\mathrm{T}}\mathbf{A}^{\mathrm{T}}$

Example 3

The matrix $\mathbf{A}=\begin{pmatrix}1 & 0 & -3\\ k & 2 & 4\\ 5 & -6 & 0\end{pmatrix}$ and $\mathbf{B}=\begin{pmatrix}2 & k & 0\\ -3 & 1 & -5\\ 0 & 1 & -1\end{pmatrix}$

Without finding **AB**, find $(\mathbf{AB})^{\mathrm{T}}$ in terms of k

$$\mathbf{A}^{\mathrm{T}}=\begin{pmatrix}1 & k & 5\\ 0 & 2 & -6\\ -3 & 4 & 0\end{pmatrix}$$

$$\mathbf{B}^{\mathrm{T}}=\begin{pmatrix}2 & -3 & 0\\ k & 1 & 1\\ 0 & -5 & -1\end{pmatrix}$$

Write down the transpose of matrices **A** and **B**

$$(\mathbf{AB})^{\mathrm{T}}=\begin{pmatrix}2 & -3 & 0\\ k & 1 & 1\\ 0 & -5 & -1\end{pmatrix}\begin{pmatrix}1 & k & 5\\ 0 & 2 & -6\\ -3 & 4 & 0\end{pmatrix}$$

Since $(\mathbf{AB})^{\mathrm{T}}\equiv\mathbf{B}^{\mathrm{T}}\mathbf{A}^{\mathrm{T}}$

$$=\begin{pmatrix}2 & 2k-6 & 28\\ k-3 & k^2+6 & 5k-6\\ 3 & -14 & 30\end{pmatrix}$$

Exercise 5.1A Fluency and skills

1 a Write down the order of these matrices.

i $\begin{pmatrix}5 & -3\\ 1 & 0\\ 9 & 11\end{pmatrix}$ ii $\begin{pmatrix}9\\ 3\\ -2\end{pmatrix}$

iii $\begin{pmatrix}3 & -7\\ -2 & 0\end{pmatrix}$ iv $\begin{pmatrix}-3 & 0 & 5 & 8\\ 2 & -1 & 4 & 0\end{pmatrix}$

b Which of the matrices in part **a** is a square matrix?

2 Calculate

a $\begin{pmatrix}5 & -1 & 2\\ 0 & 6 & 4\end{pmatrix}+\begin{pmatrix}-1 & 4 & 0\\ 5 & -7 & 2\end{pmatrix}$

b $\begin{pmatrix}4 & -2\\ 8 & -5\end{pmatrix}-\begin{pmatrix}8 & 4\\ 7 & -2\end{pmatrix}$

c $4\begin{pmatrix}-1 & 0 & 4\\ 2 & 5 & -3\\ 8 & -6 & 0\end{pmatrix}$

d $\begin{pmatrix}-6 & 14\\ 3 & 9\\ 2 & -4\end{pmatrix}+\frac{1}{2}\begin{pmatrix}8 & -4\\ 6 & 3\\ -2 & 0\end{pmatrix}$

3 If $\mathbf{A}=\begin{pmatrix}9 & -4\\ 0 & -2\end{pmatrix}$, $\mathbf{B}=\begin{pmatrix}8 & 4\\ 2 & -6\\ -2 & 0\end{pmatrix}$,

$\mathbf{C}=\begin{pmatrix}-5 & -2\\ 7 & 0\\ 3 & 1\end{pmatrix}$, $\mathbf{D}=\begin{pmatrix}0 & 1\\ -5 & 4\end{pmatrix}$

a Calculate if possible or, if not, explain why.

i $\mathbf{A}+\mathbf{D}$ ii $\mathbf{A}+\mathbf{B}$

iii $\mathbf{B}-\mathbf{C}$ iv $3\mathbf{B}$

b Show that

i $5\mathbf{A}-\mathbf{D}=\begin{pmatrix}45 & -21\\ 5 & -14\end{pmatrix}$

ii $2\mathbf{C}+7\mathbf{B}=2\begin{pmatrix}23 & 12\\ 14 & -21\\ -4 & 1\end{pmatrix}$

4 Calculate these matrix products.

a $\begin{pmatrix}-3 & 0\\ 1 & 4\end{pmatrix}\begin{pmatrix}2 & 5 & -1 & 6\\ -3 & 1 & 0 & -4\end{pmatrix}$

b $\begin{pmatrix} -2 & 1 \\ 0 & 9 \\ -5 & 0 \end{pmatrix}\begin{pmatrix} -6 & 3 \\ 4 & 8 \end{pmatrix}$

c $\begin{pmatrix} 5 & -3 & 0 \\ 2 & -4 & 1 \end{pmatrix}\begin{pmatrix} -3 \\ -5 \\ -1 \end{pmatrix}$

d $\begin{pmatrix} 9 & -2 \end{pmatrix}\begin{pmatrix} 0 & 1 & -2 \\ 4 & 7 & -4 \end{pmatrix}$

e $\begin{pmatrix} 4 \\ 3 \end{pmatrix}\begin{pmatrix} -1 & 2 & 5 \end{pmatrix}$

f $\begin{pmatrix} 1 & 3 & -1 \\ -2 & 0 & 2 \\ 0 & -3 & 1 \end{pmatrix}^2$

5 Show that $\begin{pmatrix} 3 & 2 & 0 \\ -1 & 0 & -2 \end{pmatrix}\begin{pmatrix} 4 & 1 \\ 0 & 3 \\ -3 & 0 \end{pmatrix} = \begin{pmatrix} 12 & 9 \\ 2 & -1 \end{pmatrix}$

6 If $\mathbf{A} = \begin{pmatrix} 3 \\ 7 \end{pmatrix}$, $\mathbf{B} = \begin{pmatrix} 0 & 2 \\ -1 & 3 \end{pmatrix}$,

$\mathbf{C} = \begin{pmatrix} 2 & 9 & 4 \\ -3 & 0 & -5 \end{pmatrix}$, $\mathbf{D} = \begin{pmatrix} -1 \\ 5 \\ 2 \end{pmatrix}$,

find, if possible, or if the calculation is not possible, explain why.

a **AB** b **CD** c **BCD**

d **CB** e $\mathbf{B}^2\mathbf{A}$ f $\mathbf{C}^2$

7 Calculate these matrix products, simplifying each element where possible.

a $\begin{pmatrix} 2a & 3 \\ 1 & -1 \end{pmatrix}\begin{pmatrix} 2 & 4 & 3 \\ a & 2 & -a \end{pmatrix}$

b $\begin{pmatrix} a \\ 2 \\ -a \end{pmatrix}\begin{pmatrix} 3 & -a \end{pmatrix}$

c $\begin{pmatrix} -1 & a & 0 \end{pmatrix}\begin{pmatrix} 3a & a \\ 4 & 2 \\ -1 & 5 \end{pmatrix}$

d $\begin{pmatrix} a & 3 \\ -2 & a \end{pmatrix}^2$

8 Given $\mathbf{A} = \begin{pmatrix} 2 & 5 \\ -1 & 2 \end{pmatrix}$, show that

$\mathbf{A}^3 = \begin{pmatrix} -22 & k \\ -7 & -22 \end{pmatrix}$, stating the value of k

9 The matrix $\mathbf{A} = \begin{pmatrix} 3 & k \\ 2 & 0 \end{pmatrix}$ and $\mathbf{B} = \begin{pmatrix} k & 4 \\ 0 & -2 \end{pmatrix}$

Find $(\mathbf{AB})^{\mathrm{T}}$ in terms of k

10 The matrix $\mathbf{A} = \begin{pmatrix} 4 & 0 & k \\ -1 & 3 & 0 \\ 0 & 2 & 1 \end{pmatrix}$ and

$\mathbf{B} = \begin{pmatrix} 5 & 1 & 3 \\ 3 & 0 & -1 \\ 2 & k & 2 \end{pmatrix}$

Find $(\mathbf{AB})^{\mathrm{T}}$ in terms of k

Reasoning and problem-solving

Strategy 1

To solve problems involving matrix arithmetic

1. If two matrices are of equal order, then equate corresponding elements.
2. Solve the equations simultaneously to find the values of unknowns.
3. Use subscript notation for the elements of general matrices, for example, a_1, a_2 etc.

You can solve problems involving unknown elements using basic algebra.

Example 4

Given that $\begin{pmatrix} x & 5 \\ 3 & 3y \end{pmatrix}\begin{pmatrix} 1 \\ 5 \end{pmatrix} = \begin{pmatrix} y \\ x \end{pmatrix}$, find the values of x and y

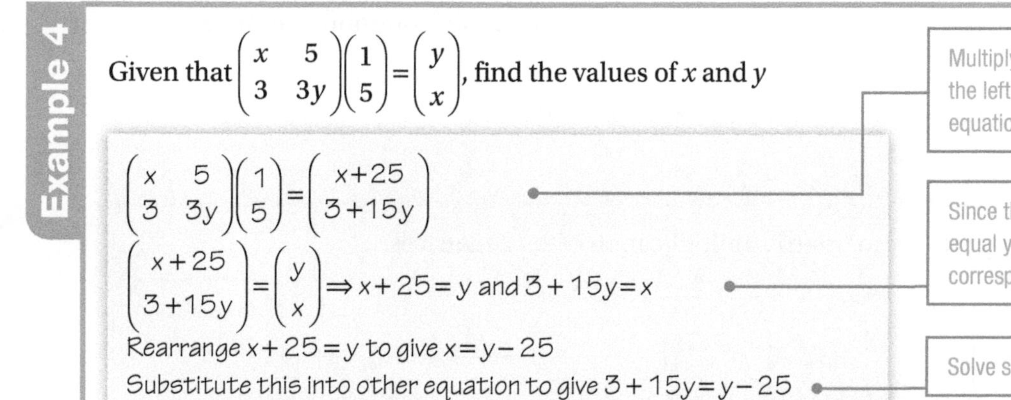

$\begin{pmatrix} x & 5 \\ 3 & 3y \end{pmatrix}\begin{pmatrix} 1 \\ 5 \end{pmatrix} = \begin{pmatrix} x+25 \\ 3+15y \end{pmatrix}$ — Multiply the matrices on the left-hand side of the equation.

$\begin{pmatrix} x+25 \\ 3+15y \end{pmatrix} = \begin{pmatrix} y \\ x \end{pmatrix} \Rightarrow x+25=y$ and $3+15y=x$ — (1) Since the matrices are equal you can equate corresponding elements.

Rearrange $x+25=y$ to give $x=y-25$

Substitute this into other equation to give $3+15y=y-25$ — (2) Solve simultaneously.

$14y=-28 \Rightarrow y=-2$

$\therefore x=-2-25=-27$ — You could now use matrix multiplication on your calculator to check the answer.

Alternatively, you may be given a system of linear equations and have to write it in matrix form.

You will be shown later how to use matrices to solve 2×2 systems of equations.

Example 5

Write a matrix equation for each of these systems of simultaneous equations.

a $x+7y=-5$
$3x-y=29$

b $3x-2y-5z=-3$
$x+7y+4z=30$
$5x-9z=13$

a $\begin{pmatrix} 1 & 7 \\ 3 & -1 \end{pmatrix}\begin{pmatrix} x \\ y \end{pmatrix} = \begin{pmatrix} -5 \\ 29 \end{pmatrix}$ — (1) Check by equating the individual elements.

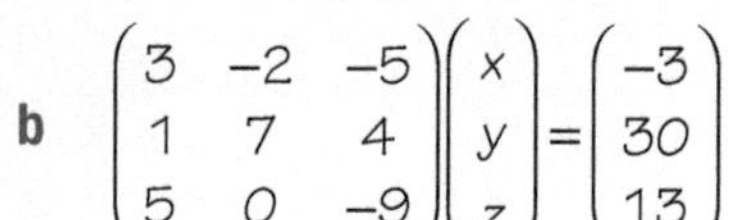

b $\begin{pmatrix} 3 & -2 & -5 \\ 1 & 7 & 4 \\ 5 & 0 & -9 \end{pmatrix}\begin{pmatrix} x \\ y \\ z \end{pmatrix} = \begin{pmatrix} -3 \\ 30 \\ 13 \end{pmatrix}$ — Notice that the number of variables determines the number of columns in the matrix, and the number of equations determines the number of rows.

Example 6

Use the result $(\mathbf{AB})^{\mathrm{T}} \equiv \mathbf{B}^{\mathrm{T}}\mathbf{A}^{\mathrm{T}}$ to prove that $(\mathbf{MPR})^{\mathrm{T}} \equiv \mathbf{M}^{\mathrm{T}}\mathbf{P}^{\mathrm{T}}\mathbf{R}^{\mathrm{T}}$ for all matrices **M**, **P** and **R** that are conformable for multiplication.

$$(\mathbf{MPR})^{\mathrm{T}} \equiv ((\mathbf{MP})\mathbf{R})^{\mathrm{T}}$$

Since matrix multiplication is associative.

$$\equiv \mathbf{R}^{\mathrm{T}}(\mathbf{MP})^{\mathrm{T}}$$

Use result $(\mathbf{AB})^{\mathrm{T}} \equiv \mathbf{B}^{\mathrm{T}}\mathbf{A}^{\mathrm{T}}$ with $\mathbf{A} = \mathbf{MP}$ and $\mathbf{B} = \mathbf{R}$

$$\equiv \mathbf{R}^{\mathrm{T}}\mathbf{P}^{\mathrm{T}}\mathbf{M}^{\mathrm{T}} \text{ as required}$$

Since $(\mathbf{MP})^{\mathrm{T}} \equiv \mathbf{P}^{\mathrm{T}}\mathbf{M}^{\mathrm{T}}$

Matrix addition is both **associative**, so $\mathbf{A}+(\mathbf{B}+\mathbf{C}) = (\mathbf{A}+\mathbf{B})+\mathbf{C}$, and **commutative**, so $\mathbf{A}+\mathbf{B} = \mathbf{B}+\mathbf{A}$

Matrix multiplication is associative, that is, $(\mathbf{AB})\mathbf{C} = \mathbf{A}(\mathbf{BC})$, but it is not commutative since, in general, $\mathbf{AB} \neq \mathbf{BA}$

Matrix multiplication is also **distributive**, that is, $\mathbf{A}(\mathbf{B}+\mathbf{C}) = \mathbf{AB}+\mathbf{AC}$

Example 7

Prove the associative property for matrix multiplication of 2×2 matrices.

3

Let $A = \begin{pmatrix} a_1 & a_2 \\ a_3 & a_4 \end{pmatrix}$, $B = \begin{pmatrix} b_1 & b_2 \\ b_3 & b_4 \end{pmatrix}$ and $C = \begin{pmatrix} c_1 & c_2 \\ c_3 & c_4 \end{pmatrix}$

Use subscript notation for the elements of general 2×2 matrices.

See Ch2.4 For a reminder of proof by induction.

$$\text{Then } (AB)C = \left[\begin{pmatrix} a_1 & a_2 \\ a_3 & a_4 \end{pmatrix}\begin{pmatrix} b_1 & b_2 \\ b_3 & b_4 \end{pmatrix}\right]\begin{pmatrix} c_1 & c_2 \\ c_3 & c_4 \end{pmatrix}$$

$$= \begin{pmatrix} a_1b_1 + a_2b_3 & a_1b_2 + a_2b_4 \\ a_3b_1 + a_4b_3 & a_3b_2 + a_4b_4 \end{pmatrix}\begin{pmatrix} c_1 & c_2 \\ c_3 & c_4 \end{pmatrix}$$

First find **AB**

$$= \begin{pmatrix} (a_1b_1 + a_2b_3)c_1 + (a_1b_2 + a_2b_4)c_3 & (a_1b_1 + a_2b_3)c_2 + (a_1b_2 + a_2b_4)c_4 \\ (a_3b_1 + a_4b_3)c_1 + (a_3b_2 + a_4b_4)c_3 & (a_3b_1 + a_4b_3)c_2 + (a_3b_2 + a_4b_4)c_4 \end{pmatrix}$$

Now multiply by **C**

$$= \begin{pmatrix} a_1b_1c_1 + a_2b_3c_1 + a_1b_2c_3 + a_2b_4c_3 & a_1b_1c_2 + a_2b_3c_2 + a_1b_2c_4 + a_2b_4c_4 \\ a_3b_1c_1 + a_4b_3c_1 + a_3b_2c_3 + a_4b_4c_3 & a_3b_1c_2 + a_4b_3c_2 + a_3b_2c_4 + a_4b_4c_4 \end{pmatrix}$$

Expand the brackets.

$$= \begin{pmatrix} a_1(b_1c_1 + b_2c_3) + a_2(b_3c_1 + b_4c_3) & a_1(b_1c_2 + b_2c_4) + a_2(b_3c_2 + b_4c_4) \\ a_3(b_1c_1 + b_2c_3) + a_4(b_3c_1 + b_4c_3) & a_3(b_1c_2 + b_2c_4) + a_4(b_3c_2 + b_4c_4) \end{pmatrix}$$

Factorise out terms from matrix **A**

$$= \begin{pmatrix} a_1 & a_2 \\ a_3 & a_4 \end{pmatrix}\begin{pmatrix} b_1c_1 + b_2c_3 & b_1c_2 + b_2c_4 \\ b_3c_1 + b_4c_3 & b_3c_2 + b_4c_4 \end{pmatrix}$$

$$= \begin{pmatrix} a_1 & a_2 \\ a_3 & a_4 \end{pmatrix}\left[\begin{pmatrix} b_1 & b_2 \\ b_3 & b_4 \end{pmatrix}\begin{pmatrix} c_1 & c_2 \\ c_3 & c_4 \end{pmatrix}\right]$$

$$= A(BC) \text{ as required}$$

You can apply the method of proof by induction to matrices.

PURE

Strategy 2

To solve induction problems involving matrices

1. Check the base case.
2. Assume the statement is true for $n = k$
3. Use $A^{k+1} = A \times A^k$ to show the statement is true for $n = k + 1$
4. Write a conclusion.

Example 8

Prove by induction that $\begin{pmatrix} 1 & 0 \\ 2 & 1 \end{pmatrix}^n = \begin{pmatrix} 1 & 0 \\ 2n & 1 \end{pmatrix}$ for all positive integers n

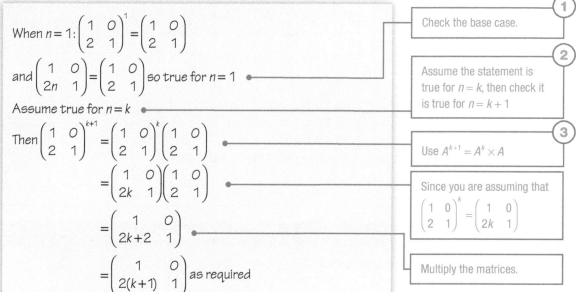

When $n = 1$: $\begin{pmatrix} 1 & 0 \\ 2 & 1 \end{pmatrix}^1 = \begin{pmatrix} 1 & 0 \\ 2 & 1 \end{pmatrix}$

and $\begin{pmatrix} 1 & 0 \\ 2n & 1 \end{pmatrix} = \begin{pmatrix} 1 & 0 \\ 2 & 1 \end{pmatrix}$ so true for $n = 1$ — (1) Check the base case.

Assume true for $n = k$ — (2) Assume the statement is true for $n = k$, then check it is true for $n = k + 1$

Then $\begin{pmatrix} 1 & 0 \\ 2 & 1 \end{pmatrix}^{k+1} = \begin{pmatrix} 1 & 0 \\ 2 & 1 \end{pmatrix}^k \begin{pmatrix} 1 & 0 \\ 2 & 1 \end{pmatrix}$ — (3) Use $A^{k+1} = A^k \times A$

$= \begin{pmatrix} 1 & 0 \\ 2k & 1 \end{pmatrix}\begin{pmatrix} 1 & 0 \\ 2 & 1 \end{pmatrix}$ — Since you are assuming that $\begin{pmatrix} 1 & 0 \\ 2 & 1 \end{pmatrix}^k = \begin{pmatrix} 1 & 0 \\ 2k & 1 \end{pmatrix}$

$= \begin{pmatrix} 1 & 0 \\ 2k+2 & 1 \end{pmatrix}$ — Multiply the matrices.

$= \begin{pmatrix} 1 & 0 \\ 2(k+1) & 1 \end{pmatrix}$ as required

Since true for $n = 1$ and assuming true for $n = k$ implies true for $n = k + 1$, therefore true for all positive integers n — (4) Write the conclusion.

Strategy 3

To solve problems involving tables by using matrices

1. Convert from tabular form into a matrix of suitable order.
2. Identify any relevant vectors.
3. Multiply the matrix by the vector to perform the necessary data analysis.
4. Write a conclusion, if required.

Example 9

The table shows the probabilities of a spring day being rainy in four UK cities.

	March	**April**	**May**
London	0.32	0.3	0.29
Edinburgh	0.38	0.33	0.37
Cardiff	0.42	0.37	0.36
Belfast	0.45	0.38	0.38

(Continued on the next page)

a Write a matrix, **A**, to represent the table of information and a vector, **x**, to represent the number of days in each month.

b Show how you can use your matrices to calculate the total number of rainy days expected in each city over the three months.

a $A=\begin{pmatrix}0.32 & 0.3 & 0.29\\0.38 & 0.33 & 0.37\\0.42 & 0.37 & 0.36\\0.45 & 0.38 & 0.38\end{pmatrix}$ $x=\begin{pmatrix}31\\30\\31\end{pmatrix}$

The columns of the matrix represent the months and the rows represent the cities. March and May have 31 days but April has 30 (1)

b $\begin{pmatrix}0.32 & 0.3 & 0.29\\0.38 & 0.33 & 0.37\\0.42 & 0.37 & 0.36\\0.45 & 0.38 & 0.38\end{pmatrix}\begin{pmatrix}31\\30\\31\end{pmatrix}=\begin{pmatrix}0.32\times31+0.3\times30+0.29\times31\\0.38\times31+0.33\times30+0.37\times31\\0.42\times31+0.37\times30+0.36\times31\\0.45\times31+0.38\times30+0.38\times31\end{pmatrix}$

Multiply the vector by the matrix. (3)

$=\begin{pmatrix}27.91\\33.15\\35.28\\37.13\end{pmatrix}$

So we expect a total of 28 rainy days in London, 33 in Edinburgh, 35 in Cardiff and 37 in Belfast.

Write a conclusion. Give answers to the nearest day. (4)

Exercise 5.1B Reasoning and problem-solving

1 Find the values of a, b, c and d such that

$\begin{pmatrix}1 & a\\3 & 7\end{pmatrix}+\begin{pmatrix}-2 & 4\\b & -5\end{pmatrix}=\begin{pmatrix}c & 1\\-2 & d\end{pmatrix}$

2 Find the values of a, b and c such that

$\begin{pmatrix}3 & a\\1 & -2\end{pmatrix}\begin{pmatrix}5 & b\\4 & 1\end{pmatrix}=\begin{pmatrix}3 & 9\\c & 2\end{pmatrix}$

3 Find the values of x and y in each case.

a $\begin{pmatrix}x & 3\\2 & 0\end{pmatrix}\begin{pmatrix}2 & 5\\y & -y\end{pmatrix}=\begin{pmatrix}10 & -17\\4 & 10\end{pmatrix}$

b $\begin{pmatrix}x & -1\\y & 2\end{pmatrix}^2=\mathbf{0}$

4 Solve each of these equations to find the possible values of x

a $\begin{pmatrix}3 & x & -2\\4 & 0 & 0\end{pmatrix}\begin{pmatrix}1 & 1\\5x & 0\\x & 5\end{pmatrix}=\begin{pmatrix}6 & -7\\4 & 4\end{pmatrix}$

b $\begin{pmatrix}7 & x & 4\end{pmatrix}\begin{pmatrix}x & 3 & 0\\x & 0 & 2\\5 & -1 & x\end{pmatrix}=\begin{pmatrix}10 & 17 & 6x\end{pmatrix}$

c $\begin{pmatrix}5 & x\\-x & -5\end{pmatrix}^2=16\begin{pmatrix}1 & 0\\0 & 1\end{pmatrix}$

d $\begin{pmatrix}0 & x\\x & 1\end{pmatrix}^3=\begin{pmatrix}4 & -10\\-10 & 9\end{pmatrix}$

5 Calculate the values of a, b and c in this matrix equation.

$\begin{pmatrix}-1 & 2 & 5\\4 & 1 & 0\\0 & 3 & 0\end{pmatrix}\begin{pmatrix}a\\b\\c\end{pmatrix}=\begin{pmatrix}-3\\25\\-9\end{pmatrix}$

6 Calculate the values of x, y and z in this matrix equation.

$\begin{pmatrix}2 & 0 & 0\\0 & 4 & 6\\-1 & 3 & 1\end{pmatrix}\begin{pmatrix}x\\y\\z\end{pmatrix}=\begin{pmatrix}6\\10\\8\end{pmatrix}$

7 Calculate the values of a, b and c in this matrix equation.

$\begin{pmatrix}1 & 3 & 2\\5 & 0 & 0\\7 & -2 & 1\end{pmatrix}\begin{pmatrix}a\\b\\c\end{pmatrix}=\begin{pmatrix}5\\-5\\3\end{pmatrix}$

8 Write a matrix equation for each of these systems of simultaneous equations.

a $2x+5y=6$
$6x-y=3$

b $x+y-z=-4$
$2x+y+z=4$
$3x+2y+2z=10$

9 Prove by induction that $\begin{pmatrix}1 & 5\\0 & 1\end{pmatrix}^n=\begin{pmatrix}1 & 5n\\0 & 1\end{pmatrix}$ for all positive integers n

10 Prove by induction that
$\begin{pmatrix}5 & 4\\0 & 1\end{pmatrix}^n=\begin{pmatrix}5^n & 5^n-1\\0 & 1\end{pmatrix}$
for all positive integers n

11 Prove by induction that
$\begin{pmatrix}-2 & -1\\9 & 4\end{pmatrix}^n=\begin{pmatrix}1-3n & -n\\9n & 3n+1\end{pmatrix}$
for all positive integers n

12 Prove by induction that
$\begin{pmatrix}1 & 4\\0 & 2\end{pmatrix}^n=\begin{pmatrix}1 & 4(2^n-1)\\0 & 2^n\end{pmatrix}$
for all positive integers n

13 Prove that $\mathbf{A}+(\mathbf{B}+\mathbf{C})=(\mathbf{A}+\mathbf{B})+\mathbf{C}$ for any 3×2 matrices **A**, **B** and **C**

14 Prove that matrix addition is commutative for any 2×2 matrices.

15 Prove by counter-example that matrix multiplication is not commutative.

16 Prove that $\mathbf{A}(\mathbf{B}+\mathbf{C})=\mathbf{AB}+\mathbf{AC}$ for any 2×2 matrices **A**, **B** and **C**

17 Prove that $\mathbf{A}\begin{pmatrix}1 & 0\\0 & 1\end{pmatrix}=\begin{pmatrix}1 & 0\\0 & 1\end{pmatrix}\mathbf{A}$ for any 2×2 matrix **A**

18 Prove that $\mathbf{B}\begin{pmatrix}1 & 0 & 0\\0 & 1 & 0\\0 & 0 & 1\end{pmatrix}=\begin{pmatrix}1 & 0 & 0\\0 & 1 & 0\\0 & 0 & 1\end{pmatrix}\mathbf{B}$ for any 3×3 matrix **B**

19 The profit (in £1000) made by three employees for a company over the four quarters of a year is shown in the table.

	Q1	Q2	Q3	Q4
Employee A	8	14	15	17
Employee B	6	11	7	9
Employee C	9	18	19	12

The company plans to pay bonuses of 5% of profit from Q1, 2% from Q2, 3% from Q3 and 1% from Q4

a Write a matrix to represent the table of information and a vector to represent the percentage paid in bonuses.

b Show how you can use your matrices to calculate the amount of bonus paid to each of the employees.

c Write down the total amount of bonuses paid to all three employees.

20 The stationery requirements of a group of Maths teachers is given in the table.

	Red pens	Pencils	Rulers	Board pens	Paper clips
Teacher 1	5	20	7	6	50
Teacher 2	4	15	15	8	100
Teacher 3	12	4	2	30	50

Red pens cost 12p each, pencils cost 8p, rulers 18p, board pens 30p and paper clips 1p

a Show how you can use matrices to calculate the expenditure on each type of stationery by each teacher.

b Write down the total cost of all the stationery required.

Full A Level

21 Given that
$\begin{pmatrix}a & b\\c & d\end{pmatrix}\begin{pmatrix}e & f\\g & h\end{pmatrix}=\begin{pmatrix}1 & 0\\0 & 1\end{pmatrix}$,

a Verify that $e=\dfrac{d}{ad-bc}$ and $g=-\dfrac{c}{ad-bc}$

b Find similar expressions for f and h in terms of a, b, c and d

22 Prove by induction that
$\begin{pmatrix}0 & 1\\2 & 0\end{pmatrix}^{2n}=2^n\begin{pmatrix}1 & 0\\0 & 1\end{pmatrix}$ for all $n\in\mathbb{N}$

5.2 Transformations

Fluency and skills

Matrices can be used to represent certain transformations such as some rotations, reflections and enlargements. Transformations that can be described in this way are known as **linear**.

Key point

In order to find the image of a vector $\begin{pmatrix} x \\ y \end{pmatrix}$ under a transformation $\mathbf{T} = \begin{pmatrix} a & b \\ c & d \end{pmatrix}$, we **pre-multiply** the vector by the matrix, so $\begin{pmatrix} x' \\ y' \end{pmatrix} = \begin{pmatrix} a & b \\ c & d \end{pmatrix}\begin{pmatrix} x \\ y \end{pmatrix}$

A point (x, y) can also be represented by the vector $\begin{pmatrix} x \\ y \end{pmatrix}$

To find the image of a point $A(1, -3)$ under the transformation $\mathbf{T} = \begin{pmatrix} 1 & 0 \\ 0 & -1 \end{pmatrix}$,

multiply the vector $\begin{pmatrix} 1 \\ -3 \end{pmatrix}$ by $\mathbf{T}$

$$\begin{pmatrix} x' \\ y' \end{pmatrix} = \begin{pmatrix} 1 & 0 \\ 0 & -1 \end{pmatrix}\begin{pmatrix} 1 \\ -3 \end{pmatrix} = \begin{pmatrix} 1 \\ 3 \end{pmatrix}$$

This is a reflection in the x-axis.

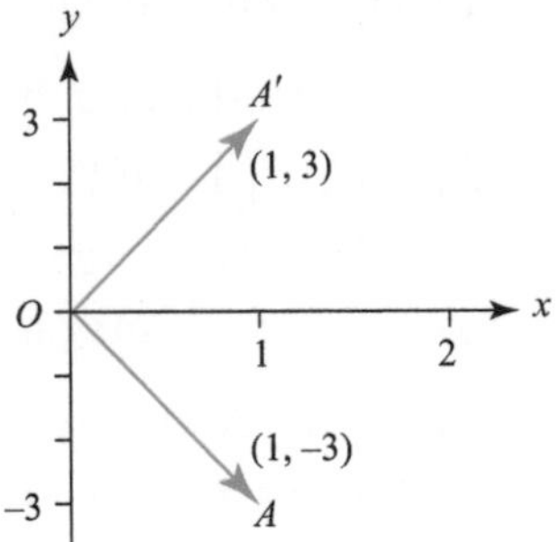

Key point

The matrix $\begin{pmatrix} 1 & 0 \\ 0 & -1 \end{pmatrix}$ represents a reflection in the x-axis.

The matrix $\begin{pmatrix} -1 & 0 \\ 0 & 1 \end{pmatrix}$ represents a reflection in the y-axis.

The matrix $\begin{pmatrix} 0 & 1 \\ 1 & 0 \end{pmatrix}$ represents a reflection in the line $y = x$

The matrix $\begin{pmatrix} 0 & -1 \\ -1 & 0 \end{pmatrix}$ represents a reflection in the line $y = -x$

You can find matrices that represent particular transformations by considering the effect the transformation will have on a pair of vectors.

Example 1

Find the 2×2 matrix that represents a rotation of 90° anticlockwise about the origin.

Let the matrix be given by $\begin{pmatrix} a & b \\ c & d \end{pmatrix}$

Choose two points and their images after the rotation.

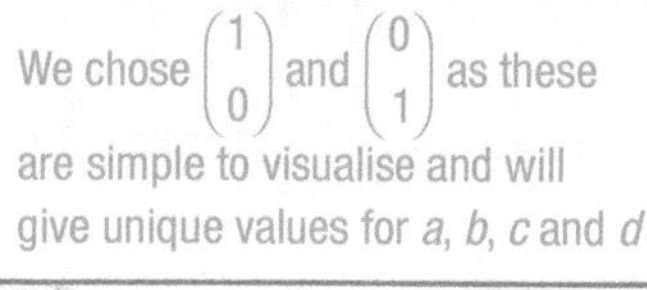

We chose $\begin{pmatrix} 1 \\ 0 \end{pmatrix}$ and $\begin{pmatrix} 0 \\ 1 \end{pmatrix}$ as these are simple to visualise and will give unique values for a, b, c and d

(1, 0) moves to (0, 1) and (0, 1) moves to (−1, 0)

So $\begin{pmatrix} a & b \\ c & d \end{pmatrix}\begin{pmatrix} 1 \\ 0 \end{pmatrix} = \begin{pmatrix} 0 \\ 1 \end{pmatrix}$ and $\begin{pmatrix} a & b \\ c & d \end{pmatrix}\begin{pmatrix} 0 \\ 1 \end{pmatrix} = \begin{pmatrix} -1 \\ 0 \end{pmatrix}$

So from the first equation: $a = 0$, $c = 1$

From the second equation: $b = -1$ and $d = 0$

Therefore the transformation matrix is $\begin{pmatrix} 0 & -1 \\ 1 & 0 \end{pmatrix}$

y
(0, 1)
(−1, 0)
O
(1, 0)
x

Notice that the first column of the matrix is the image of $\begin{pmatrix} 1 \\ 0 \end{pmatrix}$ and the second column is the image of $\begin{pmatrix} 0 \\ 1 \end{pmatrix}$

Key point

The matrix $\begin{pmatrix} a & b \\ c & d \end{pmatrix}$ represents the transformation that maps $\begin{pmatrix} 1 \\ 0 \end{pmatrix}$ to $\begin{pmatrix} a \\ c \end{pmatrix}$ and $\begin{pmatrix} 0 \\ 1 \end{pmatrix}$ to $\begin{pmatrix} b \\ d \end{pmatrix}$

It follows from this fact that the matrix $\begin{pmatrix} 1 & 0 \\ 0 & 1 \end{pmatrix}$ maps all points to themselves.

In number multiplication the identity is 1 since $1 \times a = a \times 1 = a$

Key point

The matrix $\mathbf{I} = \begin{pmatrix} 1 & 0 \\ 0 & 1 \end{pmatrix}$ is known as the **identity matrix** and has the property that $\mathbf{AI} = \mathbf{IA} = \mathbf{A}$ for all 2×2 matrices

This can be generalised for any square matrix, in particular $\mathbf{I}_3 = \begin{pmatrix} 1 & 0 & 0 \\ 0 & 1 & 0 \\ 0 & 0 & 1 \end{pmatrix}$

To find the 2×2 matrix that represents a stretch of scale factor 3 parallel to the x-axis, look at what happens to the vectors $\begin{pmatrix} 1 \\ 0 \end{pmatrix}$ and $\begin{pmatrix} 0 \\ 1 \end{pmatrix}$

A stretch 'parallel to the x-axis' can be thought of a stretch 'along the x-axis' so only the x-coordinates are affected.

The image of $\begin{pmatrix} 1 \\ 0 \end{pmatrix}$ under this transformation is $\begin{pmatrix} 3 \\ 0 \end{pmatrix}$

The image of $\begin{pmatrix} 0 \\ 1 \end{pmatrix}$ under this transformation is $\begin{pmatrix} 0 \\ 1 \end{pmatrix}$

Therefore the transformation matrix is $\begin{pmatrix} 3 & 0 \\ 0 & 1 \end{pmatrix}$

Key point

The matrix $\begin{pmatrix} k & 0 \\ 0 & 1 \end{pmatrix}$ represents a stretch of scale factor k parallel to the x-axis.

The matrix $\begin{pmatrix} 1 & 0 \\ 0 & k \end{pmatrix}$ represents a stretch of scale factor k parallel to the y-axis.

Transformations can be applied to several points at once by forming a matrix from their position vectors.

Example 2

Find the image of the points (2, –1), (–4, 5) and (–3, 0) under the transformation described by the matrix $\begin{pmatrix} 2 & 0 \\ 0 & 2 \end{pmatrix}$. Describe the transformation geometrically.

$$\begin{pmatrix} 2 & 0 \\ 0 & 2 \end{pmatrix}\begin{pmatrix} 2 & -4 & -3 \\ -1 & 5 & 0 \end{pmatrix} = \begin{pmatrix} 4 & -8 & -6 \\ -2 & 10 & 0 \end{pmatrix}$$

Ensure you always pre-multiply by the transformation matrix.

This is an enlargement of scale factor 2, centre the origin.

Key point

The matrix $\begin{pmatrix} k & 0 \\ 0 & k \end{pmatrix}$ represents an enlargement of scale factor k, centre the origin.

See Maths Ch3.1 For a reminder of exact values of trig functions.

To find the matrix that represents a 60° rotation anticlockwise about the origin, look at what happens to the vectors $\begin{pmatrix} 1 \\ 0 \end{pmatrix}$ and $\begin{pmatrix} 0 \\ 1 \end{pmatrix}$

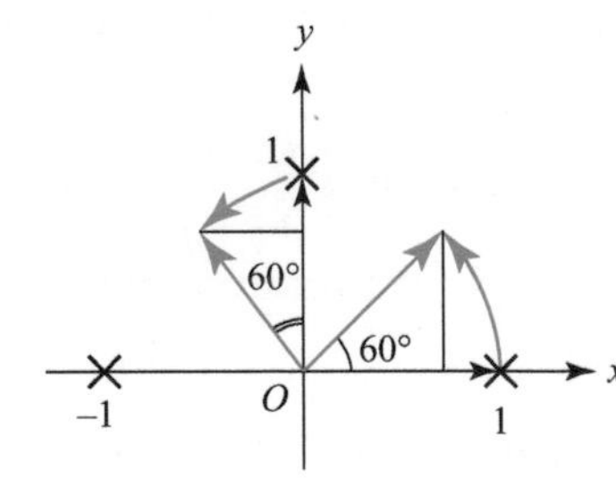

The image of $\begin{pmatrix} 1 \\ 0 \end{pmatrix}$ under this transformation is $\begin{pmatrix} \cos 60° \\ \sin 60° \end{pmatrix}$

and the image of $\begin{pmatrix} 0 \\ 1 \end{pmatrix}$ is $\begin{pmatrix} -\sin 60° \\ \cos 60° \end{pmatrix}$

Therefore the matrix is $\begin{pmatrix} \cos 60° & -\sin 60° \\ \sin 60° & \cos 60° \end{pmatrix} = \begin{pmatrix} \frac{1}{2} & -\frac{\sqrt{3}}{2} \\ \frac{\sqrt{3}}{2} & \frac{1}{2} \end{pmatrix}$

This result can be generalised for any angle.

Key point

The matrix $\begin{pmatrix} \cos\theta & -\sin\theta \\ \sin\theta & \cos\theta \end{pmatrix}$ represents an anticlockwise rotation by angle θ about the origin. Positive angles always give anticlockwise rotations.

Transformations can also be combined and you can use the product of the matrices to represent the combined transformation.

> **Key point**
>
> If the transformation represented by matrix **A** is followed by the transformation represented by matrix **B**, then **BA** is the combined transformation.

Notice that if you wish to apply transformation **A** to a vector first, you should multiply by **A** then by **B** hence **BA**

Example 3

Find the single matrix that represents a rotation of 270° anticlockwise followed by a reflection in the y-axis.

Rotation of 270°: $\begin{pmatrix}1\\0\end{pmatrix}$ is transformed to $\begin{pmatrix}0\\-1\end{pmatrix}$ and $\begin{pmatrix}0\\1\end{pmatrix}$ is transformed to $\begin{pmatrix}1\\0\end{pmatrix}$

Therefore $\mathbf{A}=\begin{pmatrix}0 & 1\\-1 & 0\end{pmatrix}$

Reflection in the y-axis: $\begin{pmatrix}1\\0\end{pmatrix}$ is transformed to $\begin{pmatrix}-1\\0\end{pmatrix}$ and $\begin{pmatrix}0\\1\end{pmatrix}$ is transformed to $\begin{pmatrix}0\\1\end{pmatrix}$

Therefore $\mathbf{B}=\begin{pmatrix}-1 & 0\\0 & 1\end{pmatrix}$

So combined transformation $=\mathbf{BA}$

$$=\begin{pmatrix}-1 & 0\\0 & 1\end{pmatrix}\begin{pmatrix}0 & 1\\-1 & 0\end{pmatrix}$$

$$=\begin{pmatrix}0 & -1\\-1 & 0\end{pmatrix}$$

The order of the matrices is important.

This is a reflection in the line $y=-x$

You may recall learning about 3D coordinates in GCSE Maths. Points in 3D space have 3D position vectors, and you will learn more about these in your Further Maths course.

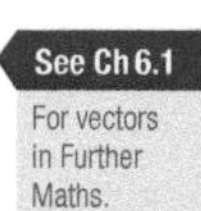
See Ch 6.1
For vectors in Further Maths.

You can use 3×3 matrices to represent transformations in 3D.

To reflect in the plane $x=0$, the 'mirror' is a plane through the y-axis and extended in the positive and negative z-directions.

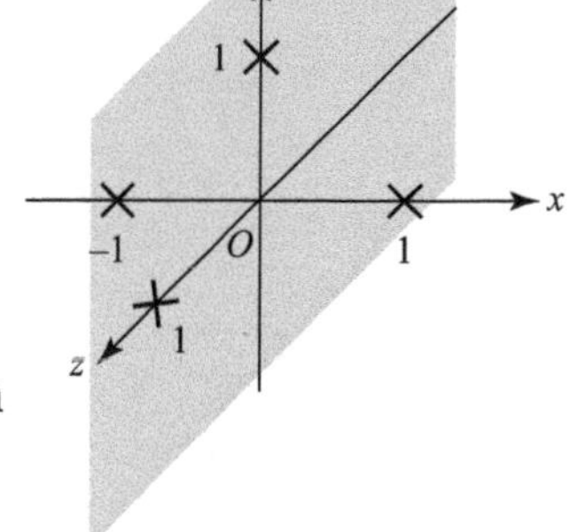

The image of the vector $\begin{pmatrix}1\\0\\0\end{pmatrix}$ will be $\begin{pmatrix}-1\\0\\0\end{pmatrix}$

The vectors $\begin{pmatrix}0\\1\\0\end{pmatrix}$ and $\begin{pmatrix}0\\0\\1\end{pmatrix}$ are in the 'mirror' plane so are unchanged when reflected in x

Therefore the transformation matrix is $\begin{pmatrix}-1 & 0 & 0\\0 & 1 & 0\\0 & 0 & 1\end{pmatrix}$

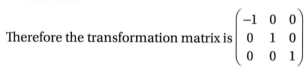

The matrices for reflection in the planes $y=0$ and $z=0$ can be found using the same method.

Key point

The 3D reflection matrices you need to use are as follows.

Reflection in $x = 0$: $\begin{pmatrix} -1 & 0 & 0 \\ 0 & 1 & 0 \\ 0 & 0 & 1 \end{pmatrix}$

Reflection in $y = 0$: $\begin{pmatrix} 1 & 0 & 0 \\ 0 & -1 & 0 \\ 0 & 0 & 1 \end{pmatrix}$

Reflection in $z = 0$: $\begin{pmatrix} 1 & 0 & 0 \\ 0 & 1 & 0 \\ 0 & 0 & -1 \end{pmatrix}$

You also need to know about 3D rotation around one of the coordinate axes.

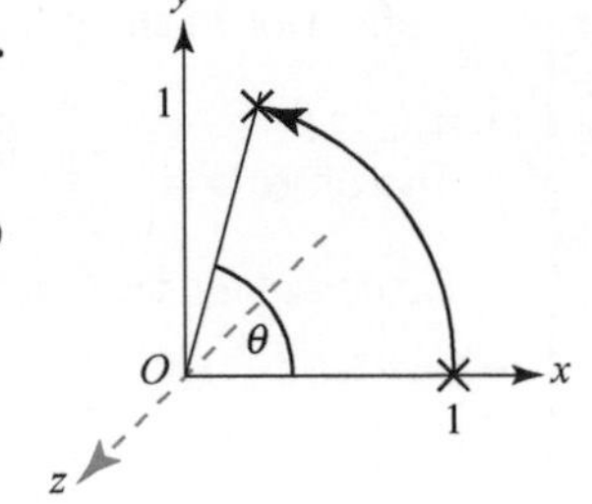

This diagram illustrates rotating the point (1, 0, 0) by angle θ anticlockwise around the z-axis. Notice how the z-coordinate of the point is unchanged so this is actually the same as rotating around the origin in the 2D case.

Therefore you can represent anticlockwise rotation around the z-axis by the matrix

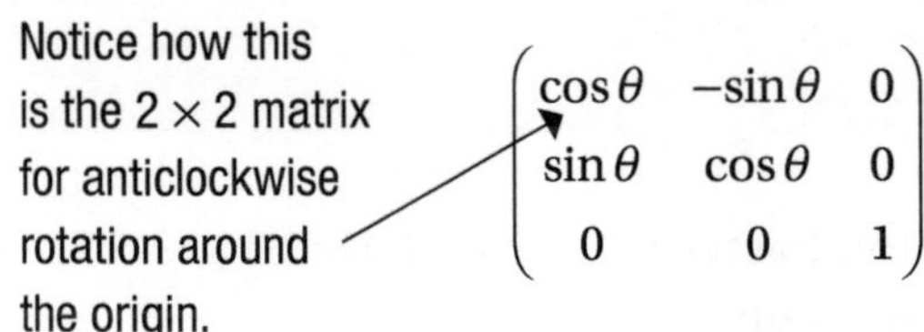

$$\begin{pmatrix} \cos\theta & -\sin\theta & 0 \\ \sin\theta & \cos\theta & 0 \\ 0 & 0 & 1 \end{pmatrix}$$

Consider an anticlockwise rotation around the x-axis or the y-axis in a similar way.

Key point

The 3D rotation matrices you need to use are as follows.

Rotation around the x-axis: $\begin{pmatrix} 1 & 0 & 0 \\ 0 & \cos\theta & -\sin\theta \\ 0 & \sin\theta & \cos\theta \end{pmatrix}$

Rotation around the y-axis: $\begin{pmatrix} \cos\theta & 0 & \sin\theta \\ 0 & 1 & 0 \\ -\sin\theta & 0 & \cos\theta \end{pmatrix}$

Rotation around the z-axis: $\begin{pmatrix} \cos\theta & -\sin\theta & 0 \\ \sin\theta & \cos\theta & 0 \\ 0 & 0 & 1 \end{pmatrix}$

Example 4

Find the 3×3 matrix, **T**, that represents a rotation of 45° anticlockwise around the x-axis and find the image of the point $P(1, 4, -2)$ under **T**

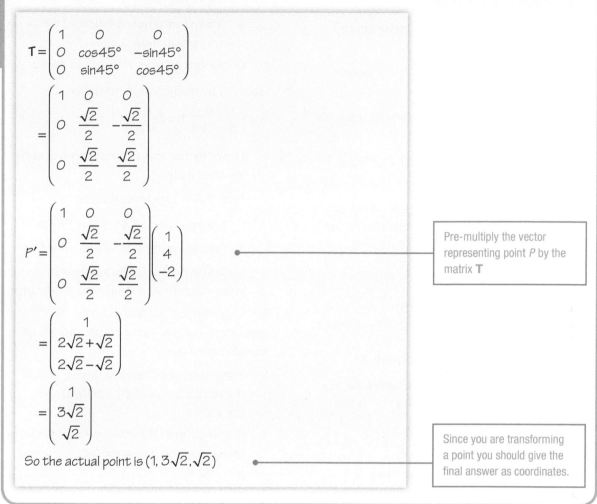

$$T = \begin{pmatrix} 1 & 0 & 0 \\ 0 & \cos 45° & -\sin 45° \\ 0 & \sin 45° & \cos 45° \end{pmatrix}$$

$$= \begin{pmatrix} 1 & 0 & 0 \\ 0 & \frac{\sqrt{2}}{2} & -\frac{\sqrt{2}}{2} \\ 0 & \frac{\sqrt{2}}{2} & \frac{\sqrt{2}}{2} \end{pmatrix}$$

$$P' = \begin{pmatrix} 1 & 0 & 0 \\ 0 & \frac{\sqrt{2}}{2} & -\frac{\sqrt{2}}{2} \\ 0 & \frac{\sqrt{2}}{2} & \frac{\sqrt{2}}{2} \end{pmatrix} \begin{pmatrix} 1 \\ 4 \\ -2 \end{pmatrix}$$

Pre-multiply the vector representing point P by the matrix **T**

$$= \begin{pmatrix} 1 \\ 2\sqrt{2} + \sqrt{2} \\ 2\sqrt{2} - \sqrt{2} \end{pmatrix}$$

$$= \begin{pmatrix} 1 \\ 3\sqrt{2} \\ \sqrt{2} \end{pmatrix}$$

So the actual point is $(1, 3\sqrt{2}, \sqrt{2})$

Since you are transforming a point you should give the final answer as coordinates.

Exercise 5.2A Fluency and skills

1 The matrix **B** represents a stretch of scale factor 5 parallel to the x-axis and a stretch of scale factor -2 parallel to the y-axis.

a Use a diagram to show the transformation of the vectors $\begin{pmatrix} 0 \\ 1 \end{pmatrix}$ and $\begin{pmatrix} 1 \\ 0 \end{pmatrix}$ under matrix **B**

b Write down matrix **B**

2 The matrix **A** represents a rotation of 180°

a Use a diagram to show the transformation of the vector $\begin{pmatrix} 1 \\ 0 \end{pmatrix}$ under this transformation.

b Write down the image of the vector $\begin{pmatrix} 0 \\ 1 \end{pmatrix}$ under this transformation.

c Write down the matrix **A**

3 Find the 2×2 matrix which represents each of these transformations.

a Reflection in the line $y = x$

b Rotation of 30° anticlockwise about the origin

c Enlargement of scale factor 5, centre the origin

d Stretch parallel to the y-axis of scale factor 4

4 A triangle has vertices (0, –2), (2, 3) and (–4, 3)

a Find the image of these points under the transformation represented by the matrix

$\mathbf{T} = \begin{pmatrix} 2 & 0 \\ 0 & 1 \end{pmatrix}$

b Describe geometrically the transformation represented by matrix **T**

5 The matrix $\frac{\sqrt{2}}{2}\begin{pmatrix} 1 & 1 \\ -1 & 1 \end{pmatrix}$ represents an anticlockwise rotation of θ degrees, centre the origin. Find the value of θ

6 A transformation is given by $\mathbf{A} = \begin{pmatrix} 1 & 0 \\ 0 & 0.5 \end{pmatrix}$

a Find the image of the vectors $\begin{pmatrix} 6 \\ -2 \end{pmatrix}$ and $\begin{pmatrix} -1 \\ 4 \end{pmatrix}$ under the transformation represented by matrix **A**

b Describe geometrically the transformation represented by matrix **A**

7 Write the transformation represented by each of these matrices.

a $\begin{pmatrix} 3 & 0 \\ 0 & 3 \end{pmatrix}$ b $\begin{pmatrix} 1 & 0 \\ 0 & 2 \end{pmatrix}$

c $\frac{\sqrt{2}}{2}\begin{pmatrix} -1 & -1 \\ 1 & -1 \end{pmatrix}$ d $\begin{pmatrix} 0 & 1 \\ -1 & 0 \end{pmatrix}$

e $\begin{pmatrix} -1 & 0 \\ 0 & 1 \end{pmatrix}$ f $\begin{pmatrix} \cos 20° & \sin 20° \\ -\sin 20° & \cos 20° \end{pmatrix}$

8 Given the transformation matrices

$\mathbf{A} = \begin{pmatrix} 1 & 1 \\ 0 & 1 \end{pmatrix}$ and $\mathbf{B} = \begin{pmatrix} 0 & -1 \\ -1 & 1 \end{pmatrix}$,

a Find the matrix representing

i Transformation **A** followed by transformation **B**

ii Transformation **B** followed by transformation **A**

b Comment on your answers to part **a**.

9 Give the transformation matrices

$\mathbf{A} = \begin{pmatrix} 1 & 0 \\ 0 & -1 \end{pmatrix}$, $\mathbf{B} = \begin{pmatrix} 0 & 1 \\ -1 & 0 \end{pmatrix}$

a Describe the transformations **A** and **B** geometrically.

b Find the matrix representing

i Transformation **A** followed by **B**

ii Transformation **B** followed by **A**

c Describe both transformations in part **b** geometrically as a single transformation.

10 Find the single 2×2 matrix that represents each of these combinations of transformations.

a A rotation of 45° anticlockwise about the origin followed by a reflection in the y-axis

b A reflection in the line $y = -x$ followed by an enlargement of scale factor 2 about the origin

c A stretch parallel to the x-axis of scale factor –2 followed by a clockwise rotation of 90° about the origin.

11 A square has vertices at (2, 5), (2, 1), (–3, 1) and (–3, 5)

The square is rotated 225° anticlockwise about the origin and then stretched by a scale factor of –2 in the y-direction.

Find the vertices of the image of the square following these transformations.

12 a Write a matrix **A** that represents a rotation of 135° anticlockwise about the origin.

b Show that $\mathbf{A}^2 = \begin{pmatrix} 0 & 1 \\ -1 & 0 \end{pmatrix}$

c Show that $\mathbf{A}^2$ represents a clockwise rotation of 90° about the origin.

d Describe the transformation represented by $\mathbf{A}^3$

13 The rotation represented by the 2×2 matrix $\mathbf{A}$ is such that $\mathbf{A}^2 = \mathbf{I}$ and $\mathbf{A} \neq \mathbf{I}$

a Write down a possible matrix **A** and describe the transformation fully.

The rotation represented by the 2×2 matrix $\mathbf{B}$ is such that $\mathbf{B}^3 = \mathbf{I}$ and $\mathbf{B} \neq \mathbf{I}$

b Write down a possible matrix **B** and describe the transformation fully.

14 Matrix **P** represents a reflection in the x-axis and matrix **Q** represents a stretch of scale factor 3 parallel to the y-axis.

a Find the single matrix that represents transformation **P** followed by transformation **Q**

b Describe geometrically the single transformation that could replace this combination of transformations.

15 The 2×2 matrix **M** represents an enlargement around the origin of scale factor k, $k > 0$, followed by an anticlockwise rotation of θ, $0 \leq \theta \leq 180$ around the origin.

a Write **M** in terms of k and θ

The image of the point $(1, -2)$ under **M** is $(2, 6)$

b Find the values of k and θ

16 A transformation is represented by the matrix $\mathbf{T} = \begin{pmatrix} 1 & 0 & 0 \\ 0 & -1 & 0 \\ 0 & 0 & 1 \end{pmatrix}$

a Find the image of the points $(4, -1, 0)$ and $(2, 5, 3)$ under **T**

b Describe this transformation geometrically.

17 Find the 3×3 matrix that represents each of these transformations.

a A rotation of $30°$ anticlockwise around the x-axis

b A reflection in $z = 0$

c A rotation of $45°$ anticlockwise around the y-axis.

18 The matrix $\begin{pmatrix} 1 & 0 & 0 \\ 0 & \frac{1}{2} & \frac{\sqrt{3}}{2} \\ 0 & -\frac{\sqrt{3}}{2} & \frac{1}{2} \end{pmatrix}$ represents an anticlockwise rotation of θ degrees around one of the coordinate axes.

a Find the value of θ and state which axis the rotation is around.

b Find the image of the vector $3\mathbf{i} + 4\mathbf{j} - \mathbf{k}$ under this transformation.

19 The matrix $\frac{1}{2}\begin{pmatrix} 1 & \sqrt{3} & 0 \\ -\sqrt{3} & 1 & 0 \\ 0 & 0 & 2 \end{pmatrix}$ represents an anticlockwise rotation of θ around one of the coordinate axes. Find the value of θ and state which axis the rotation is around.

20 The point $A(3, 7, -2)$ is transformed by a transformation matrix **M** to the point $A'(3, 7, 2)$

a Describe this transformation geometrically.

b Write down the transformation matrix **M**

21 Describe the transformation represented by each of these matrices.

a $\begin{pmatrix} 1 & 0 & 0 \\ 0 & 0 & -1 \\ 0 & 1 & 0 \end{pmatrix}$ b $\begin{pmatrix} -1 & 0 & 0 \\ 0 & 1 & 0 \\ 0 & 0 & 1 \end{pmatrix}$

c $\begin{pmatrix} -1 & 0 & 0 \\ 0 & -1 & 0 \\ 0 & 0 & 1 \end{pmatrix}$ d $\begin{pmatrix} -\frac{1}{2} & 0 & \frac{\sqrt{3}}{2} \\ 0 & 1 & 0 \\ -\frac{\sqrt{3}}{2} & 0 & -\frac{1}{2} \end{pmatrix}$

22 The 3×3 matrix **P** represents a clockwise rotation of $135°$ around the y-axis.

a Find **P** in its simplest form.

b Find the image of the point $(\sqrt{2}, 0, 1)$ under the transformation represented by **P**

23 The 3×3 matrix **T** represents an anticlockwise rotation of $330°$ around the z-axis.

a Find **T** in its simplest form.

b Find the image of the point $(\sqrt{3}, 0,1)$ under the transformation **T**

Reasoning and problem-solving

A point which is unaffected by a transformation is known as an invariant point.

Key point

Given a transformation matrix **T** and a position vector **x**, if **Tx** = **x** then **x** represents an **invariant point**.

You can be asked to find invariant points for a specific transformation.

Strategy

To find invariant points and lines

1. Write vectors in a general form.
2. Pre-multiply a vector by a transformation matrix to form a system of simultaneous equations.
3. Solve simultaneous equations to find either an invariant point or a line of invariant points.
4. Equate coefficients to find equations of invariant lines.

Example 5

Find any invariant points under the transformation given by $\begin{pmatrix} 3 & 1 \\ -1 & 2 \end{pmatrix}$

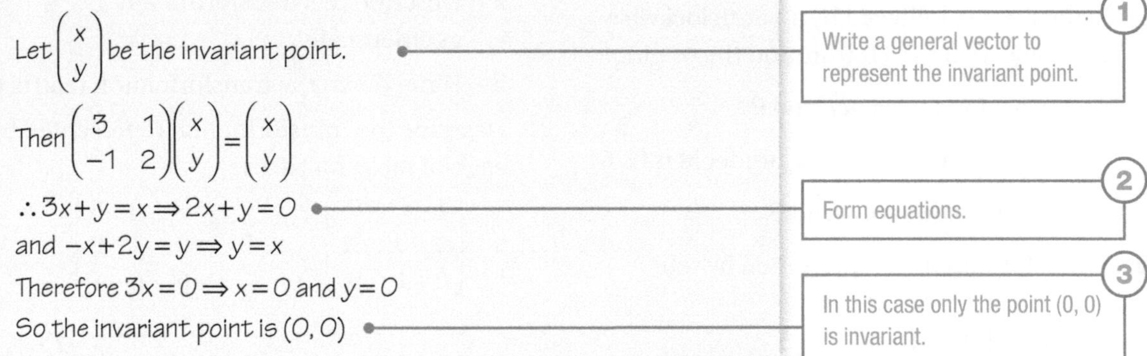

Let $\begin{pmatrix} x \\ y \end{pmatrix}$ be the invariant point.

Then $\begin{pmatrix} 3 & 1 \\ -1 & 2 \end{pmatrix}\begin{pmatrix} x \\ y \end{pmatrix} = \begin{pmatrix} x \\ y \end{pmatrix}$

$\therefore 3x + y = x \Rightarrow 2x + y = 0$

and $-x + 2y = y \Rightarrow y = x$

Therefore $3x = 0 \Rightarrow x = 0$ and $y = 0$

So the invariant point is $(0, 0)$

1 Write a general vector to represent the invariant point.

2 Form equations.

3 In this case only the point (0, 0) is invariant.

Key point

For any linear transformation, (0, 0) is an invariant point.

Some transformations have a **line of invariant points** where every point satisfying a certain property is invariant.

For example, consider a reflection in the line $y = x$

All points on the line will be unaffected by the transformation.

Therefore $y = x$ is a line of invariant points.

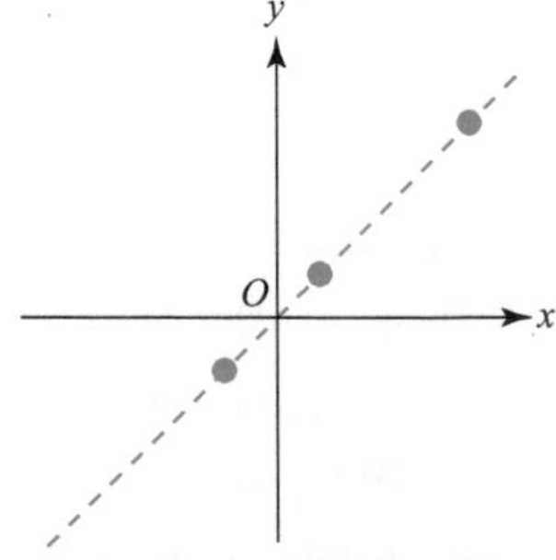

Key point

If every point on a line is mapped to itself under a transformation then it is known as a **line of invariant points**.

Example 6

Find the invariant points under the transformation given by $\begin{pmatrix} 2 & 3 \\ -1 & -2 \end{pmatrix}$

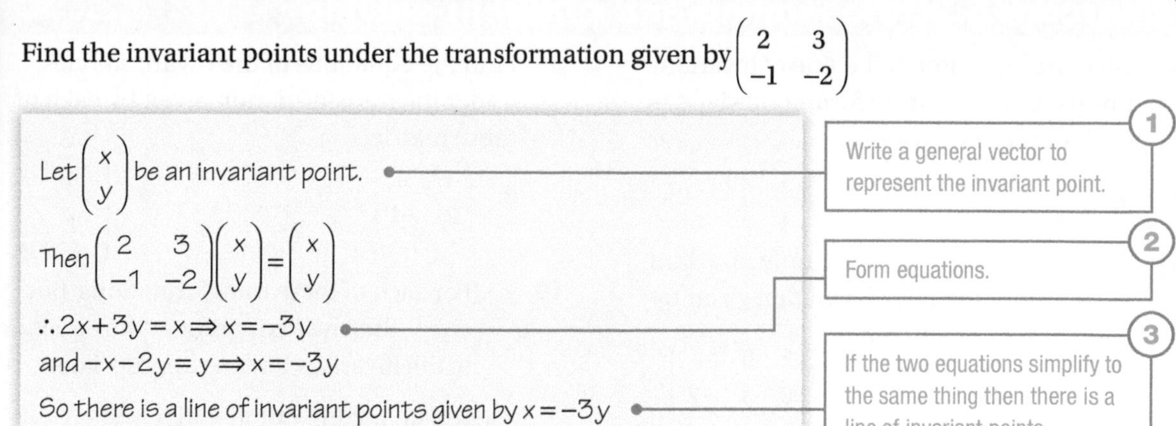

Let $\begin{pmatrix} x \\ y \end{pmatrix}$ be an invariant point.

Then $\begin{pmatrix} 2 & 3 \\ -1 & -2 \end{pmatrix}\begin{pmatrix} x \\ y \end{pmatrix}=\begin{pmatrix} x \\ y \end{pmatrix}$

$\therefore 2x+3y=x \Rightarrow x=-3y$

and $-x-2y=y \Rightarrow x=-3y$

So there is a line of invariant points given by $x=-3y$

1. Write a general vector to represent the invariant point.
2. Form equations.
3. If the two equations simplify to the same thing then there is a line of invariant points.

In some cases, every point on a line will map to another point on the same line.

For example, consider a rotation of 180° around the origin. The point (0, 0) is invariant but if you take any line through the origin then every other point on that line will map to a different point on the same line. So any line through the origin will be an invariant line.

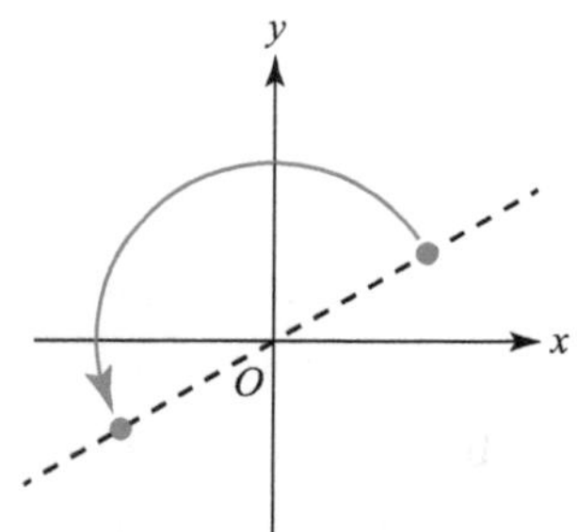

Key point

If every point on a line is mapped to another point on the same line then it is known as an **invariant line**. Lines of invariant points are a subset of invariant lines.

To find the equations of invariant lines use $y=mx+c$ to rewrite your general vector and $\begin{pmatrix} x' \\ y' \end{pmatrix}$ for the image of that vector. You should then use $y'=mx'+c$ to find the possible values of m and c

Example 7

Find the equations of the invariant lines of the transformation given by $\begin{pmatrix} 1 & 1 \\ 0 & 3 \end{pmatrix}$

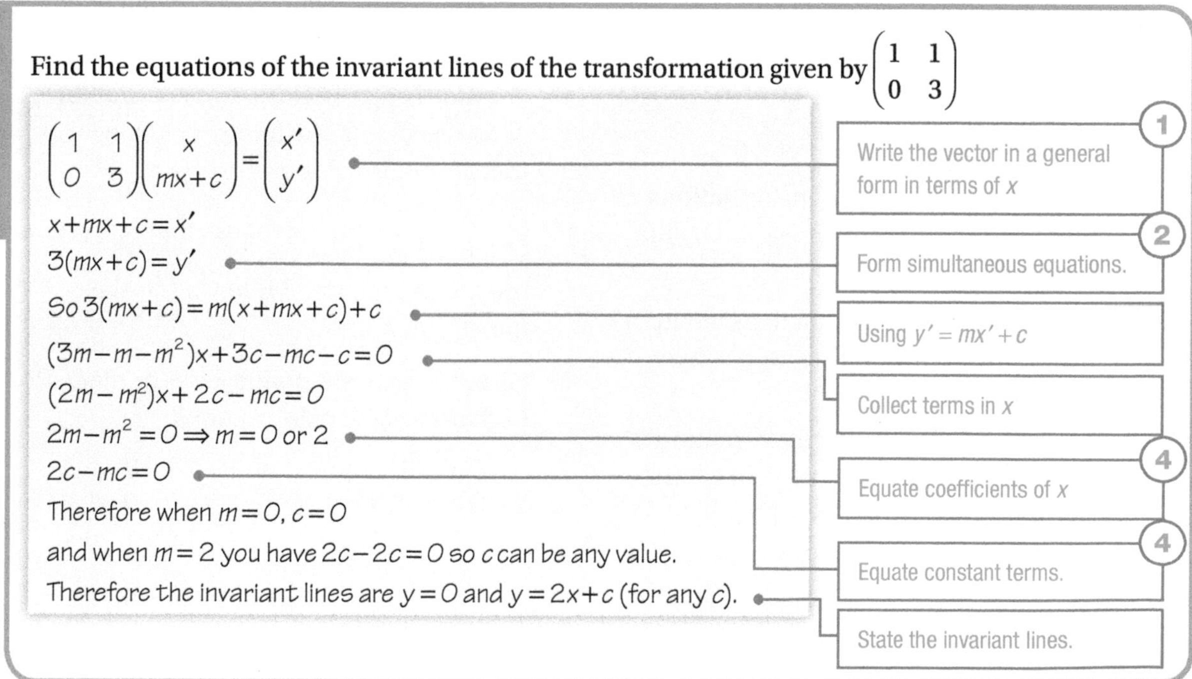

$\begin{pmatrix} 1 & 1 \\ 0 & 3 \end{pmatrix}\begin{pmatrix} x \\ mx+c \end{pmatrix}=\begin{pmatrix} x' \\ y' \end{pmatrix}$

$x+mx+c=x'$

$3(mx+c)=y'$

So $3(mx+c)=m(x+mx+c)+c$

$(3m-m-m^2)x+3c-mc-c=0$

$(2m-m^2)x+2c-mc=0$

$2m-m^2=0 \Rightarrow m=0$ or 2

$2c-mc=0$

Therefore when $m=0$, $c=0$

and when $m=2$ you have $2c-2c=0$ so c can be any value.

Therefore the invariant lines are $y=0$ and $y=2x+c$ (for any c).

1. Write the vector in a general form in terms of x
2. Form simultaneous equations.
   Using $y'=mx'+c$
   Collect terms in x
4. Equate coefficients of x
4. Equate constant terms.
   State the invariant lines.

Exercise 5.2B Reasoning and problem-solving

1 Give the equation of the line of invariant points under the transformation given by each of these matrices.

a $\begin{pmatrix} 1 & -2 \\ 0 & 3 \end{pmatrix}$ b $\begin{pmatrix} 5 & 2 \\ 4 & 3 \end{pmatrix}$

2 Show that the origin is the only invariant point under the transformation given by each of these matrices.

a $\begin{pmatrix} 3 & -2 \\ 2 & 3 \end{pmatrix}$ b $\begin{pmatrix} 2 & 0 & 1 \\ 0 & 3 & -2 \\ 1 & 0 & -4 \end{pmatrix}$

3 In each of these cases in 2D, decide whether or not the order the transformations are applied affects the final image. Justify your answers.

a A reflection in the y-axis and a stretch parallel to the y-axis

b A rotation about the origin and an enlargement with centre the origin

c A reflection in the line $y=x$ and a stretch along the x-axis

4 Describe the combination of transformations that will be represented by each of these transformation matrices.

a $\begin{pmatrix} 3 & 0 \\ 0 & -2 \end{pmatrix}$ b $\begin{pmatrix} 0 & 4 \\ 4 & 0 \end{pmatrix}$

5 Find the equations of the invariant lines under each of these transformations.

a $\begin{pmatrix} 2 & 0 \\ 1 & -3 \end{pmatrix}$ b $\begin{pmatrix} 2 & -1 \\ -3 & 0 \end{pmatrix}$

6 a Suggest two different transformations that have an invariant line given by the equation $x=0$

b Which of your transformations in part **a** has a line of invariant points given by $x=0$?

7 The invariant lines under transformation **A** are given by the equations $x+y=c$ where c can be any value. Describe a possible transformation **A** geometrically.

8 The invariant lines under transformation **B** are given by the equations $y=kx$ where k can be any value. Describe transformation **B** geometrically.

9 Find the equations of the invariant lines under the transformation given by each of these matrices.

a $\begin{pmatrix} 1 & 2 \\ 2 & -1 \end{pmatrix}$ b $\begin{pmatrix} 3 & 0 \\ 0 & 3 \end{pmatrix}$ c $\begin{pmatrix} -\frac{3}{5} & \frac{4}{5} \\ \frac{4}{5} & \frac{3}{5} \end{pmatrix}$

10 a For each of these transformations, find either the invariant point or the equations of the invariant lines as appropriate.

i Reflection in the x-axis

ii Rotation of 90° around the origin

iii Stretch of scale factor 2 parallel to the x-axis

iv Reflection in the line $y=-x$

b For each of the invariant lines in part **a**, explain whether or not it is a line of invariant points.

11 A transformation is represented by the matrix $\mathbf{T}=\begin{pmatrix} -2 & 3 \\ -3 & 4 \end{pmatrix}$

a Find the line of invariant points under transformation **T**

b Show that all lines of the form $y=x+c$ are invariant lines of the transformation **T**

12 The matrix representing transformation **A**, a reflection in the line $x=y$ followed by transformation **B** is given by $\begin{pmatrix} -1 & 0 \\ 0 & 1 \end{pmatrix}$. Find the matrix that represents the transformation **B** followed by the transformation **A**

13 Give the invariant points, lines or planes for each of these transformations.

a $\begin{pmatrix} -1 & 0 & 0 \\ 0 & 1 & 0 \\ 0 & 0 & 1 \end{pmatrix}$ b $\begin{pmatrix} 1 & 0 & 0 \\ 0 & 0 & 1 \\ 0 & 1 & 0 \end{pmatrix}$

c $\begin{pmatrix} 0 & 1 & 0 \\ 0 & -1 & 0 \\ 0 & 0 & 1 \end{pmatrix}$ d $\begin{pmatrix} 1 & 0 & 0 \\ 1 & 0 & 0 \\ 0 & 0 & 1 \end{pmatrix}$

5.3 Systems of linear equations

Fluency and skills

When you are working with square matrices you can calculate a value known as the **determinant**. The determinant has several uses. Most significantly, it enables you to find the inverse of a matrix.

Key point

If $\mathbf{A}=\begin{pmatrix} a & b \\ c & d \end{pmatrix}$ then the **determinant** of **A** is $ad-bc$ and is denoted $|\mathbf{A}|$ or $\det(\mathbf{A})$

To find the determinant of the matrix $\mathbf{A}=\begin{pmatrix} 5 & 2 \\ -1 & -3 \end{pmatrix}$ use the formula $\det(\mathbf{A})=ad-bc$

$$\det(\mathbf{A})=(5\times-3)-(2\times-1)$$
$$=15--2$$
$$=-13$$

Calculator

Try it on your calculator

You can use some calculators to find the determinant.

det(MatA)

26

Activity

Find out how to work out $\det\begin{pmatrix} 2 & -8 \\ 4 & -3 \end{pmatrix}$ on *your* calculator.

Key point

If $\det(\mathbf{A})=0$ then **A** is called a **singular matrix.**

Example 1

Find the value of k for which the matrix $\begin{pmatrix} 3 & -1 \\ -2 & k \end{pmatrix}$ is singular.

$\det\begin{pmatrix} 3 & -1 \\ -2 & k \end{pmatrix}=(3\times k)-(-1\times-2)$ — The determinant is given by $ad-bc$

$\therefore$ if matrix is singular then $3k-2=0$ — A singular matrix has a determinant of 0

$\Rightarrow k=\frac{2}{3}$

You can use this result about determinants:

Key point

For any square matrices **A** and **B**: $\det(\mathbf{AB})\equiv\det(\mathbf{A})\times\det(\mathbf{B})$

Example 2

The 2 × 2 matrices **A, B** and **C** are such that **C** = **AB**

Given that $\mathbf{C}=\begin{pmatrix}2 & -4\\ 3 & -5\end{pmatrix}$ and $\det(\mathbf{A})=\frac{1}{4}$, calculate the determinant of **B**

$\det(\mathbf{C})=2\times(-5)-(-4)\times 3$

$=-10+12$

$=2$

Use $ad-bc$ to calculate the determinant of **C**

So $\frac{1}{4}\times\det(\mathbf{B})=2$

$\Rightarrow \det(\mathbf{B})=2\times 4$

$=8$

Use the result $\det(\mathbf{AB})\equiv\det(\mathbf{A})\times\det(\mathbf{B})$

Recall that $\mathbf{I}=\begin{pmatrix}1 & 0\\ 0 & 1\end{pmatrix}$ is the identity matrix.

Key point

The inverse of a matrix, **A**, is $\mathbf{A}^{-1}$ where $\mathbf{AA}^{-1}=\mathbf{A}^{-1}\mathbf{A}=\mathbf{I}$

Key point

If $\mathbf{A}=\begin{pmatrix}a & b\\ c & d\end{pmatrix}$ then the inverse matrix is given by $\mathbf{A}^{-1}=\frac{1}{\det(\mathbf{A})}\begin{pmatrix}d & -b\\ -c & a\end{pmatrix}$

Notice how this definition will not work for a matrix which is singular. Singular matrices do not have an inverse. Therefore, only non-singular matrices will have an inverse.

Example 3

Find the inverse of the matrix $\mathbf{B}=\begin{pmatrix}2 & 5\\ -1 & 4\end{pmatrix}$

$\det(\mathbf{B})=(2\times 4)-(5\times -1)$

$=8--5$

$=13$

First calculate the determinant using $ad-bc$

$\mathbf{B}^{-1}=\frac{1}{13}\begin{pmatrix}4 & -5\\ 1 & 2\end{pmatrix}$

Use $\mathbf{B}^{-1}=\frac{1}{\det(\mathbf{B})}\begin{pmatrix}d & -b\\ -c & a\end{pmatrix}$

Checking answer:

$\mathbf{BB}^{-1}=\frac{1}{13}\begin{pmatrix}2 & 5\\ -1 & 4\end{pmatrix}\begin{pmatrix}4 & -5\\ 1 & 2\end{pmatrix}$

$=\frac{1}{13}\begin{pmatrix}13 & 0\\ 0 & 13\end{pmatrix}$

$=\begin{pmatrix}1 & 0\\ 0 & 1\end{pmatrix}$

$=\mathbf{I}$

You can check your answer by verifying that $\mathbf{BB}^{-1}=\mathbf{I}$

Calculator

Try it on your calculator

Some calculators can be used to find the inverse of a matrix.

MatA^{-1}

$$\begin{bmatrix} -4 & 3 \\ -3 & 2 \end{bmatrix}$$

Activity

Find out how to work out the inverse of $\begin{pmatrix} 2 & -3 \\ 3 & -4 \end{pmatrix}$ on *your* calculator.

Using the result $\det(\mathbf{AB}) \equiv \det(\mathbf{A}) \times \det(\mathbf{B})$ with $\mathbf{B} = \mathbf{A}^{-1}$ you can see that

$\det(\mathbf{AA}^{-1}) \equiv \det(\mathbf{A}) \times \det(\mathbf{A}^{-1})$

Then, since $\mathbf{AA}^{-1} = \mathbf{I}$ (the identity matrix), you have

$\det(\mathbf{I}) \equiv \det(\mathbf{A}) \times \det(\mathbf{A}^{-1})$

Think about the 2×2 identity matrix, $\mathbf{I}_2 = \begin{pmatrix} 1 & 0 \\ 0 & 1 \end{pmatrix}$, the determinant of this is $\det(\mathbf{I}_2) = 1 - 0 = 1$

The same will be true for an identity matrix of any size (you do not need to prove this fact).

Therefore, $1 \equiv \det(\mathbf{A}) \times \det(\mathbf{A}^{-1})$

$\Rightarrow \det(\mathbf{A}^{-1}) \equiv \dfrac{1}{\det(\mathbf{A})}$

You can use this result about the determinant of an inverse:

Key point

For any square, non-singular matrix **A**: $\det(\mathbf{A}^{-1}) \equiv \dfrac{1}{\det(\mathbf{A})}$

Example 4

Given that the matrix **C** is self-inverse, deduce the possible values of $\det(\mathbf{C})$

C is self-inverse so $\mathbf{C}^{-1} = \mathbf{C}$

Therefore $\det(\mathbf{C}^{-1}) = \det(\mathbf{C})$

$\Rightarrow \det(\mathbf{C}) = \dfrac{1}{\det(\mathbf{C})}$

Use the result $\det(\mathbf{C}^{-1}) \equiv \dfrac{1}{\det(\mathbf{C})}$

$\Rightarrow [\det(\mathbf{C})]^2 = 1$

$\Rightarrow \det(\mathbf{C}) = \pm 1$

Exercise 5.3A Fluency and skills

1 Show that $\det\begin{pmatrix} -7 & 3 \\ 5 & -4 \end{pmatrix} = 13$

2 Calculate the determinants of each of these matrices.

a $\begin{pmatrix} 2 & 4 \\ 5 & 1 \end{pmatrix}$ b $\begin{pmatrix} -3 & 2 \\ 0 & 1 \end{pmatrix}$

c $\begin{pmatrix} 4a & 3 \\ -a & -2 \end{pmatrix}$ d $\begin{pmatrix} 3\sqrt{2} & -\sqrt{2} \\ -2\sqrt{2} & \sqrt{2} \end{pmatrix}$

3 Show that $\det\begin{pmatrix} a+b & 2a \\ 2b & a+b \end{pmatrix} = (a-b)^2$

4 Find the possible values of k given that $\det\begin{pmatrix} 7 & k \\ -k & 2-k \end{pmatrix} = 2$

5 Decide whether each of these matrices is singular or non-singular. You must show your working.

a $\begin{pmatrix} 3 & 1 \\ -6 & 2 \end{pmatrix}$ b $\begin{pmatrix} 2 & 3 \\ 4 & 6 \end{pmatrix}$

c $\begin{pmatrix} -a & -2 \\ -2a & 4 \end{pmatrix}$ d $\begin{pmatrix} \sqrt{3} & 3 \\ -2\sqrt{3} & -6 \end{pmatrix}$

6 Find the values of x for which each of these matrices is singular.

a $\begin{pmatrix} x & 1 \\ 5 & 2 \end{pmatrix}$ b $\begin{pmatrix} x & -3 \\ -1 & x \end{pmatrix}$

c $\begin{pmatrix} 2x & x \\ -3 & x \end{pmatrix}$ d $\begin{pmatrix} 4 & -x \\ -x & x \end{pmatrix}$

7 Given that $\mathbf{A} = \begin{pmatrix} y+3 & y+5 \\ -1 & y-1 \end{pmatrix}$, calculate the possible values of y for which the matrix **A** has no inverse.

8 Given that $\mathbf{A} = \begin{pmatrix} -1 & 2 \\ 0 & 5 \end{pmatrix}$ and $\mathbf{B} = \begin{pmatrix} 0 & -3 \\ 2 & 4 \end{pmatrix}$

a Calculate the values of

i $\det(\mathbf{A}) \times \det(\mathbf{B})$ ii $\det(\mathbf{A}) + \det(\mathbf{B})$

b Show that

i $\det(\mathbf{AB}) = -30$ ii $\det(\mathbf{A}+\mathbf{B}) = -7$

9 The square matrices **A** and **B** have determinants $\frac{1}{3}$ and -6 respectively.

Calculate the determinant of

a **AB** b $\mathbf{A}^2$

10 The 2 × 2 matrices **A**, **B** and **C** are such that **C** = **AB**

Given that $\mathbf{C} = \begin{pmatrix} 3 & -2 \\ 4 & 8 \end{pmatrix}$, and $\det(A) = -4$, calculate

a $\det(\mathbf{B})$ b $\det(\mathbf{A}^2)$

11 The non-singular 2 × 2 matrix, **M**, has determinant 7

Calculate $\det(\mathbf{M}^{-1})$

12 The matrix $\mathbf{M} = \begin{pmatrix} 2 & a \\ 3 & 5 \end{pmatrix}$ has determinant k

Given that $\det(\mathbf{M}^{-1}) = \frac{1}{5}$

a State the value of k

b Calculate the value of a

13 Find the inverse of each of these matrices.

a $\begin{pmatrix} 7 & 2 \\ 4 & 5 \end{pmatrix}$ b $\begin{pmatrix} -4 & 3 \\ 8 & -1 \end{pmatrix}$

c $\begin{pmatrix} 2x & 5 \\ -x & -3 \end{pmatrix}$ d $\begin{pmatrix} 4\sqrt{5} & -\sqrt{5} \\ 2\sqrt{5} & \sqrt{20} \end{pmatrix}$

14 Show that the inverse of $\begin{pmatrix} 6 & 1 \\ -9 & -1 \end{pmatrix}$ is $\begin{pmatrix} -\frac{1}{3} & -\frac{1}{3} \\ 3 & 2 \end{pmatrix}$

PURE

15 $\mathbf{A}=\begin{pmatrix} b & 2a \\ b & 3a \end{pmatrix}$

a Find the inverse of **A**, writing each element in its simplest form.

b Show that $\det(\mathbf{A}^{-1})=\frac{1}{\det(\mathbf{A})}$ for the matrix **A** given.

16 Show that the matrix $\begin{pmatrix} 3 & -2 \\ 4 & -3 \end{pmatrix}$ is self-inverse.

17 Find the values of a such that the matrix $\begin{pmatrix} a & -3 \\ 2 & -a \end{pmatrix}$ is self-inverse.

18 Find the value of a such that the matrix $\begin{pmatrix} 1 & 3-a \\ a-5 & 2-a \end{pmatrix}$ is self-inverse.

19 $\mathbf{B}=\begin{pmatrix} x+1 & -x \\ x-1 & 2x \end{pmatrix}$

a Given that the determinant of **B** is 10, find the possible values of x

b Write the inverse of **B** in terms of x

20 $\mathbf{T}=\begin{pmatrix} 2 & 0 \\ 0 & 2 \end{pmatrix}$

a Show that $\mathbf{T}^{-1}=\frac{1}{2}\mathbf{I}$

b Describe the transformations represented by **T** and $\mathbf{T}^{-1}$ geometrically.

21 a Write the matrix representing rotation of 135° anticlockwise about the origin.

b Show that the inverse of this matrix is $\frac{1}{\sqrt{2}}\begin{pmatrix} -1 & 1 \\ -1 & -1 \end{pmatrix}$

22 For each of these transformations, explain whether it is always, sometimes or never self-inverse. For those which are sometimes self-inverse, state when this occurs.

a Reflection in a coordinate axis

b Enlargement centre the origin

c Rotation about the origin

d Reflection in the line $y=\pm x$

Reasoning and problem-solving

Matrices can be used to solve systems of linear equations.

Take, for example, the system of equations $ax+by=x'$
$cx+dy=y'$

If $\mathbf{M}=\begin{pmatrix} a & b \\ c & d \end{pmatrix}$ then $\mathbf{M}\begin{pmatrix} x \\ y \end{pmatrix}=\begin{pmatrix} a & b \\ c & d \end{pmatrix}\begin{pmatrix} x \\ y \end{pmatrix}=\begin{pmatrix} ax+by \\ cx+dy \end{pmatrix}$ so the equations can be written using matrices as $\mathbf{M}\begin{pmatrix} x \\ y \end{pmatrix}=\begin{pmatrix} x' \\ y' \end{pmatrix}$

Then you can pre-multiply both sides by $\mathbf{M}^{-1}$ to give $\mathbf{M}^{-1}\mathbf{M}\begin{pmatrix} x \\ y \end{pmatrix}=\mathbf{M}^{-1}\begin{pmatrix} x' \\ y' \end{pmatrix}$

As $\mathbf{M}^{-1}\mathbf{M}=\mathbf{I}$, this can be rewritten as $\begin{pmatrix} x \\ y \end{pmatrix}=\mathbf{M}^{-1}\begin{pmatrix} x' \\ y' \end{pmatrix}$ and you can calculate the values of x and y

Strategy

To solve problems involving systems of linear equations

1. Rewrite a system of linear equations using matrices.
2. Pre-multiply or post-multiply a matrix by its inverse.
3. Use the fact that $\mathbf{AA}^{-1}=\mathbf{A}^{-1}\mathbf{A}=\mathbf{I}$

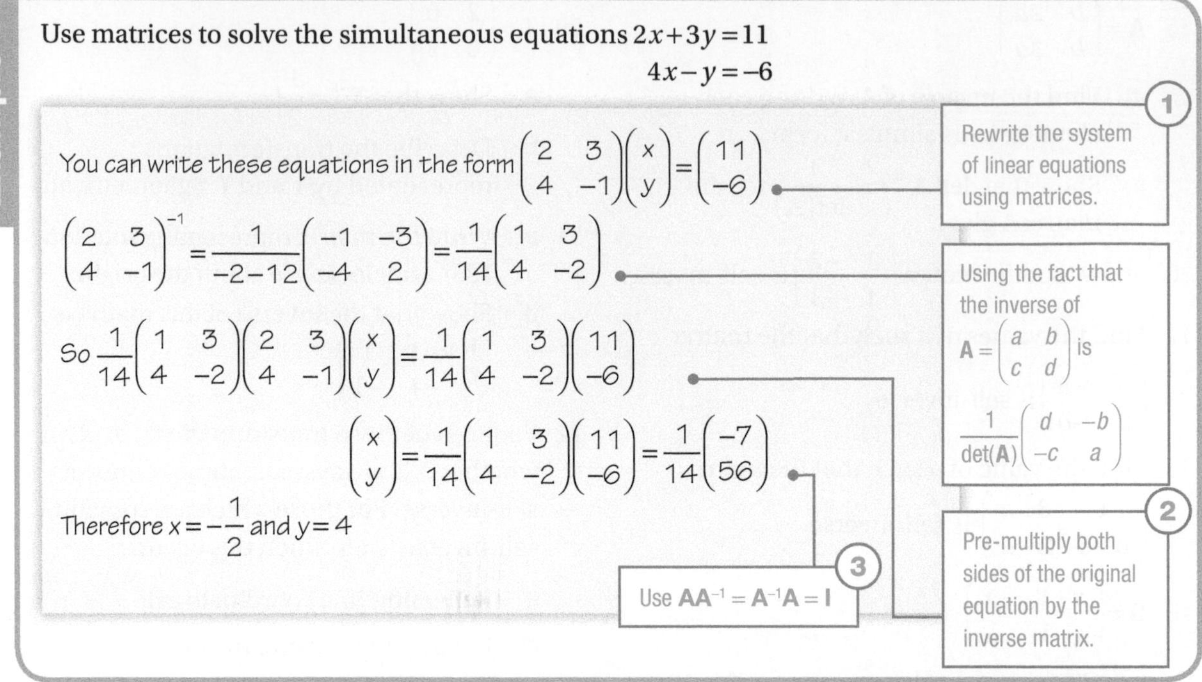

Example 5

Use matrices to solve the simultaneous equations $2x+3y=11$
$4x-y=-6$

You can write these equations in the form $\begin{pmatrix} 2 & 3 \\ 4 & -1 \end{pmatrix}\begin{pmatrix} x \\ y \end{pmatrix}=\begin{pmatrix} 11 \\ -6 \end{pmatrix}$

$\begin{pmatrix} 2 & 3 \\ 4 & -1 \end{pmatrix}^{-1}=\frac{1}{-2-12}\begin{pmatrix} -1 & -3 \\ -4 & 2 \end{pmatrix}=\frac{1}{14}\begin{pmatrix} 1 & 3 \\ 4 & -2 \end{pmatrix}$

So $\frac{1}{14}\begin{pmatrix} 1 & 3 \\ 4 & -2 \end{pmatrix}\begin{pmatrix} 2 & 3 \\ 4 & -1 \end{pmatrix}\begin{pmatrix} x \\ y \end{pmatrix}=\frac{1}{14}\begin{pmatrix} 1 & 3 \\ 4 & -2 \end{pmatrix}\begin{pmatrix} 11 \\ -6 \end{pmatrix}$

$\begin{pmatrix} x \\ y \end{pmatrix}=\frac{1}{14}\begin{pmatrix} 1 & 3 \\ 4 & -2 \end{pmatrix}\begin{pmatrix} 11 \\ -6 \end{pmatrix}=\frac{1}{14}\begin{pmatrix} -7 \\ 56 \end{pmatrix}$

Therefore $x=-\frac{1}{2}$ and $y=4$

1 Rewrite the system of linear equations using matrices.

Using the fact that the inverse of $\mathbf{A}=\begin{pmatrix} a & b \\ c & d \end{pmatrix}$ is $\frac{1}{\det(\mathbf{A})}\begin{pmatrix} d & -b \\ -c & a \end{pmatrix}$

2 Pre-multiply both sides of the original equation by the inverse matrix.

3 Use $\mathbf{AA}^{-1}=\mathbf{A}^{-1}\mathbf{A}=\mathbf{I}$

You can construct matrix proofs by pre-multiplying or post-multiplying both sides of an equation.

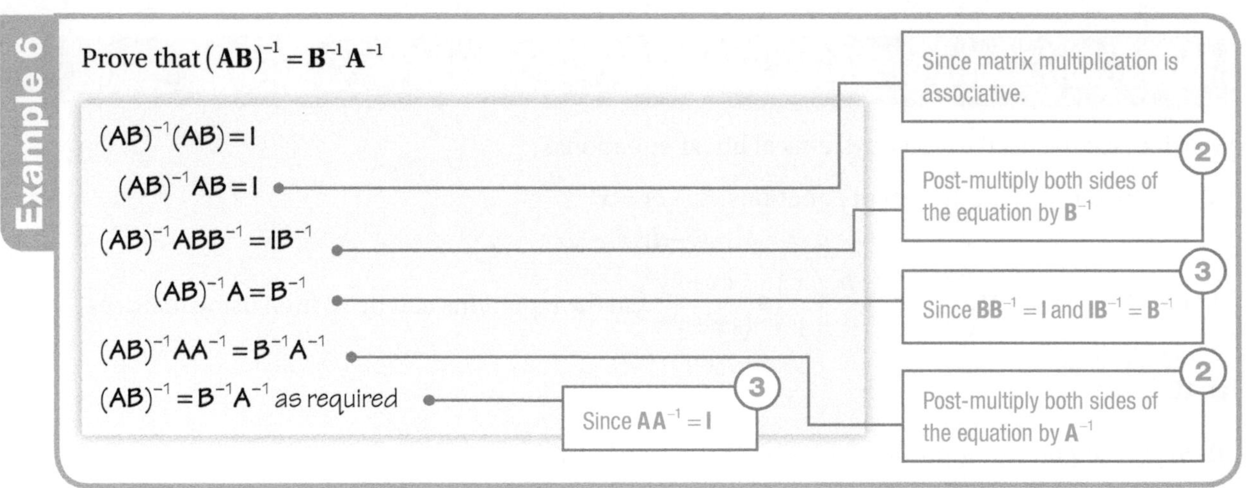

Example 6

Prove that $(\mathbf{AB})^{-1}=\mathbf{B}^{-1}\mathbf{A}^{-1}$

$(\mathbf{AB})^{-1}(\mathbf{AB})=\mathbf{I}$

$(\mathbf{AB})^{-1}\mathbf{AB}=\mathbf{I}$ — Since matrix multiplication is associative.

$(\mathbf{AB})^{-1}\mathbf{ABB}^{-1}=\mathbf{IB}^{-1}$ — 2 Post-multiply both sides of the equation by $\mathbf{B}^{-1}$

$(\mathbf{AB})^{-1}\mathbf{A}=\mathbf{B}^{-1}$ — 3 Since $\mathbf{BB}^{-1}=\mathbf{I}$ and $\mathbf{IB}^{-1}=\mathbf{B}^{-1}$

$(\mathbf{AB})^{-1}\mathbf{AA}^{-1}=\mathbf{B}^{-1}\mathbf{A}^{-1}$ — 2 Post-multiply both sides of the equation by $\mathbf{A}^{-1}$

$(\mathbf{AB})^{-1}=\mathbf{B}^{-1}\mathbf{A}^{-1}$ as required — 3 Since $\mathbf{AA}^{-1}=\mathbf{I}$

Key point

If **A** and **B** are non-singular matrices then $(\mathbf{AB})^{-1}=\mathbf{B}^{-1}\mathbf{A}^{-1}$

Exercise 5.3B Reasoning and problem-solving

PURE

1 Use matrices to solve these pairs of simultaneous equations.

a $4x-y=11$
$2x+3y=-5$

b $x-5y=0$
$2x-8y=-2$

c $3x+6y=12$
$x-2y=2$

d $4x+6y=1$
$-8x+3y=3$

Full A Level

2 Use matrices to find a solution to these systems of linear equations.

a $x+y-2z=3$
$2x-3y+5z=4$
$5x+2y+z=-3$

b $4x+6y-z=-3$
$2x-3y=2$
$8y+4z=0$

3 Given that $\begin{pmatrix} a & 3 \\ -2a & -1 \end{pmatrix}\begin{pmatrix} x \\ y \end{pmatrix}=\begin{pmatrix} 5 \\ 10a \end{pmatrix}$,

find expressions in terms of a for x and y
Simplify your answers.

4 You are given the equations $(k+3)x-2y=k-1$ and $kx+y=k$

Use matrices to find expressions in terms of k for x and y

5 Given that $\mathbf{A}\begin{pmatrix} -7 & -2 \\ 0 & 1 \end{pmatrix}=\begin{pmatrix} 14 & 8 \\ 7 & 0 \end{pmatrix}$, find the matrix $\mathbf{A}$

6 Given that $\begin{pmatrix} 3 & -5 \\ -1 & 2 \end{pmatrix}\mathbf{B}=\begin{pmatrix} 12 & -1 \\ -4 & 0 \end{pmatrix}$, find the matrix $\mathbf{B}$

7 Given that $\mathbf{P}\begin{pmatrix} 2a & b \\ -a & -b \end{pmatrix}=\begin{pmatrix} a & 2a \\ a & -b \end{pmatrix}$, find an expression for $\det(\mathbf{P})$ in terms of a and b
Fully simplify your answer.

8 If $\mathbf{A}=\begin{pmatrix} 3 & 5 \\ 2 & 4 \end{pmatrix}$ and $\mathbf{AB}=\begin{pmatrix} 6 & 8 \\ -2 & 0 \end{pmatrix}$, find $\mathbf{B}$

9 If $\mathbf{B}=\begin{pmatrix} 2 & 3 \\ -3 & -1 \end{pmatrix}$ and $\mathbf{AB}=\begin{pmatrix} 10 & 15 \\ 11 & 13 \end{pmatrix}$, find the matrix $\mathbf{A}$

10 If $\mathbf{A}^2=2\mathbf{I}$, write a possible 2×2 matrix $\mathbf{A}$

11 Simplify each of these expressions involving the non-singular matrices $\mathbf{A}$ and $\mathbf{B}$

a $(\mathbf{AB})^{-1}\mathbf{A}$ b $\mathbf{B}(\mathbf{A}^{-1}\mathbf{B})^{-1}$

12 If $\mathbf{A}=\mathbf{C}^{-1}\mathbf{BC}$, prove that $\mathbf{B}=\mathbf{CAC}^{-1}$

13 Prove that if $\mathbf{ABA}^{-1}=\mathbf{I}$ then $\mathbf{B}=\mathbf{I}$

14 Prove that $(\mathbf{ABC})^{-1}=\mathbf{C}^{-1}\mathbf{B}^{-1}\mathbf{A}^{-1}$ for non-singular matrices $\mathbf{A}$, $\mathbf{B}$ and $\mathbf{C}$

15 Prove that if $\mathbf{P}$ is self-inverse then $\mathbf{P}^2=\mathbf{I}^{-1}$

16 Prove that if $\mathbf{PQP}=\mathbf{I}$ for non-singular matrices $\mathbf{P}$ and $\mathbf{Q}$, then $\mathbf{Q}=(\mathbf{P}^{-1})^2$

17 A point P is transformed by the matrix $\mathbf{T}=\begin{pmatrix} 2 & 1 \\ -3 & -4 \end{pmatrix}$ and the coordinates of the image of P are (9, –11)

Find the coordinates of P

18 Prove that $\det(\mathbf{AB})=\det(\mathbf{A})\det(\mathbf{B})$ for any non-singular 2×2 matrices $\mathbf{A}$ and $\mathbf{B}$

19 Prove that $\det(\mathbf{A}^{-1})=[\det(\mathbf{A})]^{-1}$ for any non-singular 2×2 matrix $\mathbf{A}$

20 a Write the matrix that represents

i a rotation of angle θ anticlockwise around the origin,

ii a rotation of angle θ clockwise around the origin.

b Hence, deduce the identity $\sin^2\theta+\cos^2\theta\equiv 1$

21 By considering a square with one vertex at the origin, show that, under any stretch, **T**, in the x- or y-direction (or both),

a The image of the square will be a rectangle,

b The area of the image will be the area of the original square multiplied by the determinant of **T**

Full A Level

5 Summary and review

Chapter summary

- A matrix with m rows and n columns has order $m \times n$
- Matrices of the same order can be added by adding corresponding elements.
- A matrix can be multiplied by a constant by multiplying each of its elements by that constant.
- Two matrices are conformable for multiplication if the number of columns in the first matrix is the same as the number of rows in the second matrix.
- The product of an $n \times m$ matrix and an $m \times p$ matrix has order $n \times p$
- Under the transformation given by the matrix $\begin{pmatrix} a & b \\ c & d \end{pmatrix}$ the vector $\begin{pmatrix} 1 \\ 0 \end{pmatrix}$ is transformed to $\begin{pmatrix} a \\ c \end{pmatrix}$ and the vector $\begin{pmatrix} 0 \\ 1 \end{pmatrix}$ is transformed to $\begin{pmatrix} b \\ d \end{pmatrix}$
- The matrix representing transformation **A** followed by transformation **B** is given by the product **BA**
- The identity matrix is a matrix with ones along the leading diagonal and zeros elsewhere, so in 2D, $\mathbf{I} = \begin{pmatrix} 1 & 0 \\ 0 & 1 \end{pmatrix}$
- The zero matrix is a matrix with 0 as every element, so in 2D, $\mathbf{0} = \begin{pmatrix} 0 & 0 \\ 0 & 0 \end{pmatrix}$
- The invariant points of the matrix **T** can be found by solving $\mathbf{T}\begin{pmatrix} x \\ y \end{pmatrix} = \begin{pmatrix} x \\ y \end{pmatrix}$
- The invariant lines of the matrix **T** can be found by solving $\mathbf{T}\begin{pmatrix} x \\ mx+c \end{pmatrix} = \begin{pmatrix} x' \\ mx'+c \end{pmatrix}$ to find the values of m and c
- The determinant of a 2×2 matrix is $\det\begin{pmatrix} a & b \\ c & d \end{pmatrix} = ad - bc$
- A matrix is singular if its determinant is zero.
- Under a transformation **T**, area of image = area of original $\times |\det(\mathbf{T})|$
- If $\det(\mathbf{T}) > 0$ then the transformation represented by **T** preserves orientation.
- If $\mathbf{A} = \begin{pmatrix} a & b \\ c & d \end{pmatrix}$ then its inverse is $\mathbf{A}^{-1} = \frac{1}{\det(\mathbf{A})}\begin{pmatrix} d & -b \\ -c & a \end{pmatrix}$
- For any non-singular matrix **A**, $\mathbf{A}\mathbf{A}^{-1} = \mathbf{A}^{-1}\mathbf{A} = \mathbf{I}$

Check and review

You should now be able to...	Try Questions
✔ Identify the order of a matrix.	1
✔ Add and subtract matrices of the same order and multiply matrices by a constant.	1, 2
✔ Multiply conformable matrices.	1, 2
✔ Apply a linear transformation given as a matrix to a point or vector.	3, 4
✔ Describe transformations given as matrices geometrically.	3, 4
✔ Write a linear transformation given geometrically as a matrix.	5, 6
✔ Find the matrix that represents a combination of two transformations.	7
✔ Find invariant points and lines.	8, 9
✔ Calculate the determinant of a 2×2 matrix.	10
✔ Understand what is meant by a singular matrix.	11
✔ Find the inverse of a 2×2 matrix.	12–14
✔ Use matrices to solve systems of linear equations.	15
✔ Describe the geometrical significance of solutions.	3, 4

1 $\mathbf{A}=\begin{pmatrix} 6 & 2 \\ -4 & 12 \end{pmatrix}$, $\mathbf{B}=\begin{pmatrix} 0 & 2 & -1 \\ 5 & 4 & 0 \end{pmatrix}$, $\mathbf{C}=\begin{pmatrix} -1 & -5 \\ 4 & 0 \\ 0 & 2 \end{pmatrix}$ and $\mathbf{D}=\begin{pmatrix} 8 & -2 & 3 \\ 0 & -1 & 0 \end{pmatrix}$

- a State the order of each of the matrices.
- b Find each of these matrices.
  - i $\mathbf{B}+\mathbf{D}$
  - ii $3\mathbf{C}$
  - iii $\mathbf{D}-2\mathbf{B}$
  - iv $\frac{1}{2}\mathbf{A}$
- c Calculate these matrix products where possible. If it is not possible, explain why not.
  - i $\mathbf{AB}$
  - ii $\mathbf{BA}$
  - iii $\mathbf{BC}$
  - iv $\mathbf{CA}$
  - v $\mathbf{AC}$
  - vi $\mathbf{CD}$
  - vii $\mathbf{A}^2$
  - viii $\mathbf{C}^2$

2 Given that $\mathbf{A}=\begin{pmatrix} a & 3 \\ a & 2 \end{pmatrix}$, $\mathbf{B}=\begin{pmatrix} -1 & a & 3 \\ 0 & 1 & a \\ 2a & -3 & 0 \end{pmatrix}$ and $\mathbf{C}=\begin{pmatrix} a & -1 \\ 0 & 3 \\ -2 & 2a \end{pmatrix}$,

find each of these matrices in terms of a

a **CA**

b **BC**

c $\mathbf{A}^2$

d $\mathbf{B}^2$

3 a Apply each of these transformations to the point (–1, 2) and describe the effect of the transformation geometrically.

i $\begin{pmatrix} 0 & 1 \\ 1 & 0 \end{pmatrix}$ ii $\begin{pmatrix} 7 & 0 \\ 0 & 1 \end{pmatrix}$ iii $\begin{pmatrix} 0 & -1 \\ 1 & 0 \end{pmatrix}$ iv $\begin{pmatrix} -\frac{\sqrt{3}}{2} & \frac{1}{2} \\ -\frac{1}{2} & -\frac{\sqrt{3}}{2} \end{pmatrix}$

b A triangle has area 5 square units. Give the area of the image of the triangle under each of the transformations given in part **a.**

c Explain which of the transformations in part **a** preserves orientation.

4 Apply each of these transformations to the point (3, –2, 4) and describe the effect of the transformation geometrically.

a $\begin{pmatrix} 1 & 0 & 0 \\ 0 & \frac{\sqrt{2}}{2} & -\frac{\sqrt{2}}{2} \\ 0 & \frac{\sqrt{2}}{2} & \frac{\sqrt{2}}{2} \end{pmatrix}$

b $\begin{pmatrix} 1 & 0 & 0 \\ 0 & 1 & 0 \\ 0 & 0 & -1 \end{pmatrix}$

5 Write down a 2×2 matrix to represent each of these transformations.

a Reflection in the line $x=0$

b Rotation of 315° anticlockwise around the origin

c Enlargement of scale factor –3, centre the origin.

6 Write down a 3×3 matrix to represent each of these 3D transformations.

a Rotation of 45° anticlockwise around the y-axis

b Reflection in the plane $y=0$

7 Write down the single 2×2 matrix that represents each of these transformations.

a Rotation of 270° anticlockwise around the origin followed by a stretch of scale factor 2 parallel to the y-axis

b Reflection in the line $y=-x$ followed by reflection in the x-axis

c Reflection in the x-axis followed by rotation of 150° anticlockwise about the origin.

8 Find the invariant points under each of these transformations.

a $\begin{pmatrix} -1 & 0 \\ 4 & -1 \end{pmatrix}$

b $\begin{pmatrix} 2 & -3 \\ -1 & 4 \end{pmatrix}$

9 Find all of the invariant lines under each of these transformations.

a $\begin{pmatrix} 0 & 2 \\ 3 & 1 \end{pmatrix}$

b $\frac{1}{5}\begin{pmatrix} -4 & 3 \\ 3 & 4 \end{pmatrix}$

10 Work out the determinant of each of these matrices.

a $\begin{pmatrix} 2 & -1 \\ 4 & -3 \end{pmatrix}$

b $\begin{pmatrix} 3a & b \\ -2a & b \end{pmatrix}$

11 Calculate the values of x for which these matrices are singular.

a $\begin{pmatrix} 3x & 4 \\ 5 & 5 \end{pmatrix}$

b $\begin{pmatrix} x & -1 \\ 4 & -2x \end{pmatrix}$

12 Find the inverse, if it exists, of each of these matrices. If it does not exist, explain why not.

a $\begin{pmatrix} 6 & 3 \\ -1 & -2 \end{pmatrix}$

b $\begin{pmatrix} 0 & 1 \\ 3 & -2 \end{pmatrix}$

c $\begin{pmatrix} 3 & -6 \\ -4 & 8 \end{pmatrix}$

d $\begin{pmatrix} 5a & 2b \\ 7a & 3b \end{pmatrix}$

13 Given that $\mathbf{A} = \begin{pmatrix} 5 & 7 \\ 1 & -4 \end{pmatrix}$ and $\mathbf{AB} = \begin{pmatrix} 10 & 22 \\ 2 & -1 \end{pmatrix}$, work out matrix **B**

14 Given that $\mathbf{C} = \begin{pmatrix} 9 & -2 \\ 5 & 1 \end{pmatrix}$ and $\mathbf{BC} = \begin{pmatrix} -17 & -11 \\ -60 & 7 \end{pmatrix}$, work out matrix **B**

15 Use matrices to solve each pair of simultaneous equations.

a $x + 6y = 7$
$3x - y = -17$

b $5x - 7y = 8$
$10x + 3y = -1$

PURE

5 Exploration

Going beyond the exams

History

The Russian mathematician **Andrey Andreyevich Markov** (1856 – 1922) used matrices in his work on **stochastic processes**. Stochastic processes are collections of random variables. They can be used to show how a system might change over time.

There are now many applications of **Markov processes** including random walks, the Gambler's ruin problem, modelling queues arriving at an airport, exchange rates and the PageRank algorithm.

ICT

Here is an example of a Markov process that helps to predict long-term weather probabilities.

In a simplified model the weather on any one day can either be dry or wet.

Assume that the weather today is dry so the probabilities of dry and wet are given by $\mathbf{W}_0 = [1 \;\; 0]$

Let $\mathbf{P} = \begin{bmatrix} 0.75 & 0.25 \\ 0.6 & 0.4 \end{bmatrix}$ where $\mathbf{P} = \begin{bmatrix} P(\text{dry tomorrow} \mid \text{dry today}) & P(\text{wet tomorrow} \mid \text{dry today}) \\ P(\text{dry tomorrow} \mid \text{wet today}) & P(\text{wet tomorrow} \mid \text{wet today}) \end{bmatrix}$

It follows that the weather tomorrow is given by

$$\mathbf{W}_1 = \mathbf{W}_0\mathbf{P} = [1 \;\; 0]\begin{bmatrix} 0.75 & 0.25 \\ 0.6 & 0.4 \end{bmatrix} = [0.75 \;\; 0.25]$$

The weather the next day would be given by

$$\mathbf{W}_2 = \mathbf{W}_1\mathbf{P} = [0.75 \;\; 0.25]\begin{bmatrix} 0.75 & 0.25 \\ 0.6 & 0.4 \end{bmatrix} = [0.7125 \;\; 0.2875]$$

Using a spreadsheet, find (to 4 decimal places) the 1×2 matrix that the weather probabilities tend towards.

0.75
0.25
0.4
Dry
Wet
0.6

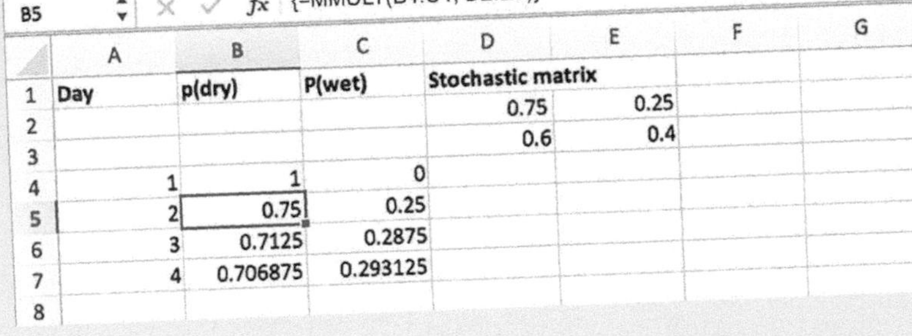

B5 | {=MMULT(B4:C4, D2:E3)}

	A	B	C	D	E	F	G
1	Day	p(dry)	P(wet)	Stochastic matrix			
2				0.75	0.25		
3				0.6	0.4		
4	1	1	0				
5	2	0.75	0.25				
6	3	0.7125	0.2875				
7	4	0.706875	0.293125				
8							

Research

Matrices can be used to solve systems of linear equations using a method now known as **Gaussian elimination**. The method was first invented in China before being reinvented in Europe in the 1700s.

It is named after **Carl Friedrich Gauss** after the adoption, by professional computers, of a specialised notation that Gauss devised.

Find out about Gaussian elimination.

Karl Friedrich Gauß.

5 Assessment

1 Given that $\mathbf{M}=\begin{pmatrix}5 & 2\\-3 & 1\end{pmatrix}$ and $\mathbf{N}=\begin{pmatrix}-4 & 3\\0 & 2\end{pmatrix}$, calculate

a $3\mathbf{M}$ b $-2\mathbf{N}$ c $\mathbf{M}-\mathbf{N}$ d $\mathbf{MN}$ **[4 marks]**

2 You are told that $\mathbf{M}=\begin{pmatrix}1 & 2\\4 & -3\end{pmatrix}$ and $\mathbf{N}=\begin{pmatrix}2 & 7\\5 & 1\end{pmatrix}$

a Calculate **MN** b Calculate **NM** c What does this say about **M** and **N**? **[3]**

3 Given $\mathbf{M}=\begin{pmatrix}2 & 7\\9 & 11\end{pmatrix}$, calculate

a $\mathbf{MI}$ b $\mathbf{0}+2\mathbf{M}$ c $\mathbf{M}^2$ **[3]**

4 You are told that $\begin{pmatrix}a & -1\\2 & 3\end{pmatrix}\begin{pmatrix}b & 5\\6 & 2\end{pmatrix}\equiv\begin{pmatrix}-2 & 18\\20 & c\end{pmatrix}$ is an identity. Find a, b and c **[2]**

5 Each of these matrices represents a transformation. Fill in the blanks in each sentence.

a The matrix $\begin{pmatrix}6 & 0\\0 & 6\end{pmatrix}$ describes an _______ with scale factor _______.

b The matrix $\begin{pmatrix}0 & -1\\1 & 0\end{pmatrix}$ describes a rotation about _______ by _______ in the anticlockwise direction.

c The matrix $\begin{pmatrix}1 & 0\\0 & -1\end{pmatrix}$ describes a _______ in _______. **[6]**

6 **M** is the matrix which describes a rotation by 135° anticlockwise about the origin. **N** is the matrix which describes a rotation by 45° clockwise about the origin.

a Write matrices **M** and **N**

b Calculate **MN**

c What effect does **MN** have and how does it relate to **M** and **N**? **[5]**

7 Calculate any invariant points for these matrices.

a $\begin{pmatrix}2 & -1\\2 & 3\end{pmatrix}$

b $\begin{pmatrix}5 & 2\\-4 & -1\end{pmatrix}$

c $\begin{pmatrix}1 & 0\\2 & 7\end{pmatrix}$ **[6]**

8 Find the invariant lines under these transformations.

a Reflection in the y-axis in 2D space

b Rotation by 180° clockwise about the z-axis in 3D space. **[2]**

9 The matrix $\mathbf{M}=\begin{pmatrix}4 & 1\\3 & 7\end{pmatrix}$ has the inverse $\mathbf{M}^{-1}=\frac{1}{25}\begin{pmatrix}7 & -1\\-3 & 4\end{pmatrix}$

Find the solution to the equation $\begin{pmatrix}4 & 1\\3 & 7\end{pmatrix}\begin{pmatrix}x\\y\end{pmatrix}=\begin{pmatrix}10\\13\end{pmatrix}$ **[2]**

10 The matrix $\mathbf{M}=\begin{pmatrix}3 & 10\\7 & 5\end{pmatrix}$ has the inverse $\mathbf{M}^{-1}=\frac{1}{55}\begin{pmatrix}-5 & 10\\7 & -3\end{pmatrix}$

Find the solution to the equation $\begin{pmatrix}3 & 10\\7 & 5\end{pmatrix}\begin{pmatrix}x\\y\end{pmatrix}=\begin{pmatrix}42\\43\end{pmatrix}$ **[2]**

11 The determinant of the matrix $\begin{pmatrix}3 & 1\\6 & a\end{pmatrix}$ is $3a-6$

a For what value of a does the matrix $\begin{pmatrix}3 & 1\\6 & a\end{pmatrix}$ have no inverse?

b For that value of a find any invariant points under $\begin{pmatrix}3 & 1\\6 & a\end{pmatrix}$ **[4]**

12 Calculate the determinants of these matrices.

a $\begin{pmatrix}6 & 0\\3 & 2\end{pmatrix}$

b $\begin{pmatrix}1 & 5\\2 & 10\end{pmatrix}$ **[4]**

13 Find the inverses of these matrices.

a $\begin{pmatrix}9 & 16\\5 & 9\end{pmatrix}$

b $\begin{pmatrix}4 & -2\\5 & 3\end{pmatrix}$ **[8]**

14 **a** Find the inverses of $\mathbf{A}=\begin{pmatrix}3 & 1\\5 & 2\end{pmatrix}$ and $\mathbf{B}=\begin{pmatrix}-2 & 4\\1 & -1\end{pmatrix}$

b Calculate $\mathbf{B}^{-1}\mathbf{A}^{-1}$

c Find the inverse of $\mathbf{B}^{-1}\mathbf{A}^{-1}$

d How does this relate to **A** and **B**? **[6]**

6 Vectors 1

Vectors are used extensively in engineering. One area of engineering where they are particularly important is in the design of electrical circuits. The size of circuit varies from the design of heavy engineering plants through to the design of miniature electronic circuits. Vectors can be used to represent current and voltage in an alternating current (A. C.) circuit. These vectors are considered to be rotating to provide different sinusoidal waves and, when two or more are brought together, they can be added using vector addition that takes into account not only the magnitude of the voltage, but also the phase angle(s) between them.

The applications of vector methods are also used in all branches of mechanical, civil and structural engineering. A particularly important use is to model forces. They are applied to calculate forces throughout buildings at the design stage to ensure that buildings are structurally sound and can withstand extreme weather, earthquakes, etc.

Orientation

What you need to know

KS4
- The equation of a line.

Maths Ch1
- Parallel and perpendicular lines.
- Points of intersection.
- Solving simultaneous equations.

Maths Ch6
- Position vector and displacements.
- Magnitude and direction.
- **i**, **j** form.

What you will learn

- To write and use the equation of a line in Cartesian and vector form.
- To decide whether lines are intersecting, parallel or skew and determine any points of intersection.
- To calculate the scalar product and use it to find angles and show lines are perpendicular.
- To find points of intersection, angles and calculate distances between points and lines.

What this leads to

Ch23 Further vectors
- The vector product.
- The scalar triple product.
- Lines and planes.

Ch8 Momentum
- Momentum and impulse in two dimensions.

6.1 The vector equation of a line

Fluency and skills

See Maths Ch6.1
For a reminder of magnitude and direction of vectors.

Vectors are used to describe magnitude and direction in 2D. In this section, you will learn how to express the equation of a 2D line and a 3D line in terms of vectors, as well as how to convert these vector equations to Cartesian form by using parametric equations.

In 3D, the vector $\begin{pmatrix} x \\ y \\ z \end{pmatrix}$ can also be written $x\mathbf{i}+y\mathbf{j}+z\mathbf{k}$ where **i**, **j** and **k** are perpendicular unit vectors.

Vectors can also be used to describe points, lines and planes in 2D or 3D. Suppose you have a line that passes through the point with position vector **a** and is parallel to vector **b**

Then the general point on this line will have position vector **r** where $\mathbf{r}=\mathbf{a}+\lambda\mathbf{b}$. You can see this from the triangle in the diagram. λ is a scalar quantity. It can take any value and determines how far along the line you move.

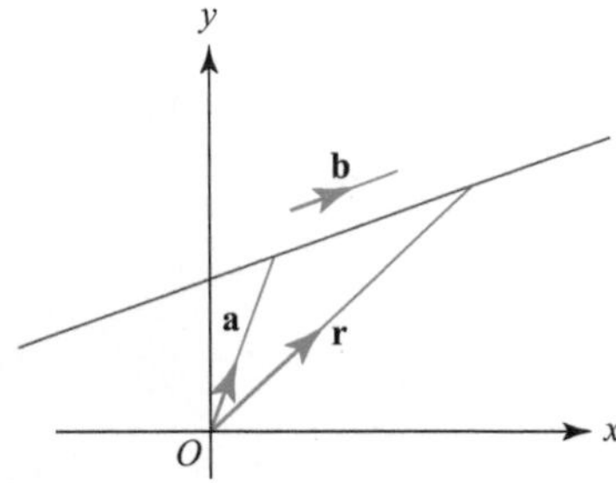

See Maths Ch6.1
For a reminder of finding parallel vectors.

Therefore, in either 2D or 3D:

Key point

$\mathbf{r}=\mathbf{a}+\lambda\mathbf{b}$ is the vector equation of a line which is parallel to the vector **b** and which passes through the point with position vector **a**

Example 1

Work out the vector equation of the line passing through the points with position vectors $\begin{pmatrix} 1 \\ -3 \\ 2 \end{pmatrix}$ and $\begin{pmatrix} 3 \\ 0 \\ -1 \end{pmatrix}$

First, find a vector in the direction of the line.

$\begin{pmatrix} 3 \\ 0 \\ -1 \end{pmatrix}-\begin{pmatrix} 1 \\ -3 \\ 2 \end{pmatrix}=\begin{pmatrix} 2 \\ 3 \\ -3 \end{pmatrix}$ is parallel to the line

To find a vector parallel to the line, subtract the position vector of a point on the line from the position vector of another point on the line.

Therefore the equation of the line is $\mathbf{r}=\begin{pmatrix} 1 \\ -3 \\ 2 \end{pmatrix}+\lambda\begin{pmatrix} 2 \\ 3 \\ -3 \end{pmatrix}$

You can use either of the position vectors given.

Or, alternatively, $\mathbf{r}=\begin{pmatrix} 3 \\ 0 \\ -1 \end{pmatrix}+\lambda\begin{pmatrix} 2 \\ 3 \\ -3 \end{pmatrix}$

Suppose you have a line which is parallel to vector $\mathbf{b}=(b_1, b_2, b_3)$ and passes through the point with position vector $\mathbf{a}=(a_1, a_2, a_3)$

Let the general point on this line be $\mathbf{r}=(x, y, z)$. Then the equation of the line is $(x, y, z)=(a_1, a_1, a_1)+\lambda(b_1, b_2, b_3)$

See Maths Ch12.3

For a reminder of parametric equations.

Therefore the line can be written using **parametric equations:** $x = a_1 + \lambda b_1, y = a_2 + \lambda b_2, z = a_3 + \lambda b_3$

These can be rearranged to give $\frac{x-a_1}{b_1} = \frac{y-a_2}{b_2} = \frac{z-a_3}{b_3}(=\lambda)$, which are the **Cartesian equations of the line.**

PURE

Key point

The **Cartesian equations of a line** parallel to the vector $\mathbf{b} = (b_1, b_2, b_3)$ and passing through the point with position vector $\mathbf{a} = (a_1, a_2, a_3)$ are $\frac{x-a_1}{b_1} = \frac{y-a_2}{b_2} = \frac{z-a_3}{b_3}$

Example 2

The Cartesian equations of a line are $\frac{x-3}{2} = \frac{y+1}{4} = \frac{z-2}{-5}$

Give the vector equation of the line.

The direction vector is $\mathbf{b} = \begin{pmatrix} 2 \\ 4 \\ -5 \end{pmatrix}$

Since the Cartesian equations are of the form $\frac{x-a_1}{b_1} = \frac{y-a_2}{b_2} = \frac{z-a_3}{b_3}$

The position vector is $\mathbf{a} = \begin{pmatrix} 3 \\ -1 \\ 2 \end{pmatrix}$

So the vector equation is $\mathbf{r} = \begin{pmatrix} 3 \\ -1 \\ 2 \end{pmatrix} + t\begin{pmatrix} 2 \\ 4 \\ -5 \end{pmatrix}$

Use $\mathbf{r} = \mathbf{a} + t\mathbf{b}$

Example 3

Give the Cartesian equations of the line with vector equation $\mathbf{r} = \mathbf{i} + 2\mathbf{j} - \mathbf{k} + s(3\mathbf{i} - \mathbf{k})$

The direction vector is $\begin{pmatrix} 3 \\ 0 \\ -1 \end{pmatrix}$

The position vector is $\begin{pmatrix} 1 \\ 2 \\ -1 \end{pmatrix}$

Use $\mathbf{r} = \mathbf{a} + t\mathbf{b}$

The line will lie on the plane $y = 2$

Since the **j**-component in the direction vector is zero, don't write $\frac{y-2}{0}$

Therefore the Cartesian equations of the line are

$\frac{x-1}{3} = \frac{z+1}{-1}, y = 2$

You may have a line that lies on one of the planes $x = a_1$, $y = a_2$ or $z = a_3$.

For example, $\frac{x-a_1}{b_1} = \frac{y-a_2}{b_2}$, $z = a_3$ lies on the plane $z = a_3$, so the **k** component of the direction vector will be zero so the vector equation is $\mathbf{r} = (a_1, a_2, a_3) + s(b_1, b_2, b_3)$.

To determine whether or not a point lies on a line, you can make an equation by using $\mathbf{r} = \mathbf{a} + \lambda\mathbf{b}$ and then substitute the position vector of the given point for **r**. Then, by considering the components of **i**, **j** and **k** separately, you can determine whether or not there is a unique value of λ

Example 4

Establish whether or not the point (−5, 5, 6) lies on the line $\mathbf{r}=\begin{pmatrix}3\\1\\0\end{pmatrix}+\lambda\begin{pmatrix}4\\-2\\-3\end{pmatrix}$

$$\begin{pmatrix}-5\\5\\6\end{pmatrix}=\begin{pmatrix}3\\1\\0\end{pmatrix}+\lambda\begin{pmatrix}4\\-2\\-3\end{pmatrix}$$

Form an equation and substitute in the position vector of the point in place of **r**

i components: $-5=3+4\lambda \Rightarrow \lambda=-2$

j components: $5=1-2\lambda \Rightarrow \lambda=-2$

k components: $6=-3\lambda \Rightarrow \lambda=-2$

Find the value of λ arising from each of the components.

Since you get the same value of λ for each component, (−5, 5, 6) does lie on the line.

If λ was not the same for all three equations, then you would conclude that the point is not on the line.

Exercise 6.1A Fluency and skills

1 Write a vector equation of the line that is parallel to the vector $\begin{pmatrix}1\\-2\\0\end{pmatrix}$ and passes through the point with position vector $\begin{pmatrix}0\\4\\3\end{pmatrix}$

2 Write a vector equation of the line that is parallel to the vector $5\mathbf{i}-3\mathbf{j}+\mathbf{k}$ and passes through the point with position vector $7\mathbf{i}-\mathbf{j}+2\mathbf{k}$

3 Write a vector equation of the line that passes through the points with position vectors $\begin{pmatrix}-7\\2\\-5\end{pmatrix}$ and $\begin{pmatrix}4\\-9\\-2\end{pmatrix}$

4 Write a vector equation of the line that passes through the points with position vectors $2\mathbf{i}+3\mathbf{k}$ and $2\mathbf{i}+\mathbf{j}-\mathbf{k}$

5 Write Cartesian equations of the line that is parallel to the vector $\begin{pmatrix}0\\4\\-1\end{pmatrix}$ and passes through the point with position vector $\begin{pmatrix}5\\2\\-7\end{pmatrix}$

6 Write the Cartesian equations of the line that is parallel to the vector $8\mathbf{i}-2\mathbf{j}+3\mathbf{k}$ and passes through the point with position vector $4\mathbf{i}-2\mathbf{j}$

7 Write the Cartesian equations of the line that passes through the points with position vectors $\begin{pmatrix}9\\4\\-3\end{pmatrix}$ and $\begin{pmatrix}1\\0\\5\end{pmatrix}$

8 Write the Cartesian equations of the line that passes through the points with position vectors $5\mathbf{j}+\mathbf{k}$ and $4\mathbf{i}-\mathbf{j}-\mathbf{k}$

9 Convert each of these vector equations to Cartesian form.

a $\mathbf{r}=\begin{pmatrix}0\\1\\-2\end{pmatrix}+\lambda\begin{pmatrix}-4\\0\\8\end{pmatrix}$ b $\mathbf{r}=\begin{pmatrix}-3\\0\\7\end{pmatrix}+\lambda\begin{pmatrix}5\\-1\\3\end{pmatrix}$

10 Convert each of these Cartesian equations into the form $\mathbf{r}=\mathbf{a}+\lambda\mathbf{b}$

a $\frac{x-5}{2}=\frac{y-1}{-3}=\frac{z+3}{1}$ b $\frac{x+2}{4}=\frac{y}{-2}=\frac{5-z}{3}$

11 Establish whether the point (3, 2, 1) lies on each of these lines.

a $\mathbf{r}=\begin{pmatrix}1\\-4\\3\end{pmatrix}+\lambda\begin{pmatrix}1\\3\\-1\end{pmatrix}$ b $\mathbf{r}=\begin{pmatrix}5\\2\\0\end{pmatrix}+\lambda\begin{pmatrix}2\\0\\1\end{pmatrix}$

12 Establish whether the point (−2, 0, 4) lies on each of these lines.

a $\frac{x-1}{3}=\frac{y+3}{-3}=\frac{z-0}{4}$ b $\frac{x+8}{3}=\frac{y+2}{1}=\frac{z-6}{-1}$

Reasoning and problem-solving

A vector equation of a line is not unique. There are lots of points you can use to set the vector from the origin and you can use any vector which is parallel to the line to give its direction, so there are many correct vector equations for any one line. You need to be able to decide whether or not two equations represent the same line.

Strategy 1

To check whether two equations represent the same line

1. Check whether their direction vectors are equivalent.
2. Check whether they have a point in common.
3. If they have the same direction and a point in common then they must represent the same line.

Example 5

Decide which of these equations represent the same line.

$L_1: \mathbf{r} = \begin{pmatrix} 9 \\ 1 \\ -2 \end{pmatrix} + \lambda \begin{pmatrix} -2 \\ -1 \\ 1 \end{pmatrix}$ $\quad L_2: \mathbf{r} = \begin{pmatrix} 3 \\ 0 \\ 4 \end{pmatrix} + \lambda \begin{pmatrix} 4 \\ 2 \\ -2 \end{pmatrix}$ $\quad L_3: \frac{x-1}{2} = \frac{y+3}{1} = \frac{z-2}{-1}$

All three lines have the same direction since $\begin{pmatrix} -2 \\ -1 \\ 1 \end{pmatrix} = -\begin{pmatrix} 2 \\ 1 \\ -1 \end{pmatrix}$ and $\begin{pmatrix} 4 \\ 2 \\ -2 \end{pmatrix} = 2\begin{pmatrix} 2 \\ 1 \\ -1 \end{pmatrix}$

(1) If a vector is a multiple of another vector then they are parallel.

Take the point $(1, -3, 2)$ which you know lies on line L_3

(2) Alternatively, you could pick a point you know lies on line L_1 or on line L_2

Check if this point lies on L_1: $\begin{pmatrix} 1 \\ -3 \\ 2 \end{pmatrix} = \begin{pmatrix} 9 \\ 1 \\ -2 \end{pmatrix} + \lambda \begin{pmatrix} -2 \\ -1 \\ 1 \end{pmatrix}$

$9 - 2\lambda = 1 \Rightarrow \lambda = 4$

$1 - \lambda = -3 \Rightarrow \lambda = 4$

$-2 + \lambda = 2 \Rightarrow \lambda = 4$

Therefore $(1, -3, 2)$ lies on line L_1

(2) Since the same value of λ is found for each of the components.

So L_1 and L_3 are equations of the same line.

(3) Since they have the same direction and have a point in common.

Check if this point lies on L_2: $\begin{pmatrix} 1 \\ -3 \\ 2 \end{pmatrix} = \begin{pmatrix} 3 \\ 0 \\ 4 \end{pmatrix} + \lambda \begin{pmatrix} 4 \\ 2 \\ -2 \end{pmatrix}$

$3 + 4\lambda = 1 \Rightarrow \lambda = -\frac{1}{2}$

$2\lambda = -3 \Rightarrow \lambda = -\frac{3}{2}$

$4 - 2\lambda = 2 \Rightarrow \lambda = 1$

So $(1, -3, 2)$ does not lie on L_2

Therefore L_2 does not represent the same line as L_1 and L_3

(3) L_2 is parallel to the other lines but does not have a point in common.

If you have two lines in 2D then they must either intersect or be parallel. However, with two lines in 3D they could also be **skew**, that is, they pass one another on different planes so do not intersect but are also not parallel.

For example, in the cuboid shown:

- AB and DC are parallel
- AB and AD intersect
- AB and GH are skew (they will never intersect however far they are extended, but they are not parallel)

F G A B E H D C

You already know that lines are parallel if their direction vectors are multiples of each other. To decide whether or not two lines intersect, you must attempt to solve their equations simultaneously and see if they are consistent.

Strategy 2

To establish if lines are parallel, intersecting or skew

1. Check whether their direction vectors are multiples of each other, in which case they are parallel. You can then use strategy 1 above to determine whether or not they are the same line.
2. If the lines are not parallel, attempt to solve their equations simultaneously.
3. If the solution is consistent then the lines intersect and you can find the point of intersection.
4. If the solutions leads to an inconsistency then the lines are skew.

Example 6

For each pair of lines, decide whether they are parallel, intersecting or skew.

If they intersect, find their point of intersection.

a $L_1: \mathbf{r} = 2\mathbf{i} - \mathbf{j} + \lambda(3\mathbf{i} + \mathbf{j} - 2\mathbf{k}), \quad L_2: 3\mathbf{i} + 2\mathbf{j} - \mathbf{k} + \mu(2\mathbf{i} - \mathbf{k})$

b $L_1: \mathbf{r} = \begin{pmatrix} 1 \\ 0 \\ -8 \end{pmatrix} + \lambda\begin{pmatrix} 2 \\ -2 \\ -3 \end{pmatrix}, \quad L_2: \mathbf{r} = \begin{pmatrix} -4 \\ 7 \\ -1 \end{pmatrix} + \mu\begin{pmatrix} 3 \\ -1 \\ -5 \end{pmatrix}$

a They are not parallel. — (1) Since their direction vectors are not multiples of each other.

$\begin{pmatrix} 2 \\ -1 \\ 0 \end{pmatrix} + \lambda\begin{pmatrix} 3 \\ 1 \\ -2 \end{pmatrix} = \begin{pmatrix} 3 \\ 2 \\ -1 \end{pmatrix} + \mu\begin{pmatrix} 2 \\ 0 \\ -1 \end{pmatrix}$ — You may find it easier to work with column vectors.

i: $2 + 3\lambda = 3 + 2\mu \Rightarrow \mu = \dfrac{3\lambda - 1}{2}$ — (2) Equating the components of **i**

j: $-1 + \lambda = 2 \Rightarrow \lambda = 3$ — (2) Equating the components of **j**

k: $-2\lambda = -1 - \mu \Rightarrow \mu = -1 + 2\lambda$ — (2) Equating the components of **k**

Therefore, from the **i** components, $\mu = \dfrac{3 \times 3 - 1}{2} = 4$

But from the k components, $\mu = -1 + 2 \times 3 = 5$ — (2) Solve simultaneously.

The equations are inconsistent so there is no point of intersection. Therefore the lines are skew. — (4) An inconsistency as μ cannot be both 4 and –5

(*Continued on the next page*)

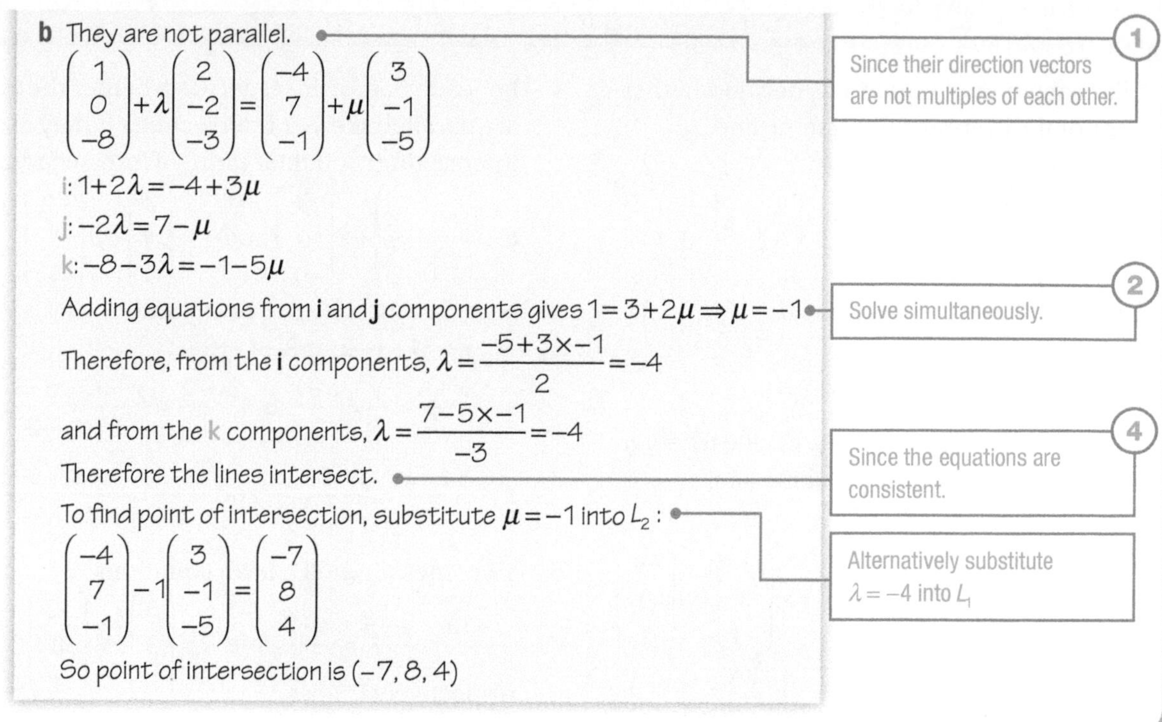

b They are not parallel.

$$\begin{pmatrix}1\\0\\-8\end{pmatrix}+\lambda\begin{pmatrix}2\\-2\\-3\end{pmatrix}=\begin{pmatrix}-4\\7\\-1\end{pmatrix}+\mu\begin{pmatrix}3\\-1\\-5\end{pmatrix}$$

i: $1+2\lambda=-4+3\mu$

j: $-2\lambda=7-\mu$

k: $-8-3\lambda=-1-5\mu$

Adding equations from **i** and **j** components gives $1=3+2\mu \Rightarrow \mu=-1$

Therefore, from the **i** components, $\lambda=\frac{-5+3\times-1}{2}=-4$

and from the **k** components, $\lambda=\frac{7-5\times-1}{-3}=-4$

Therefore the lines intersect.

To find point of intersection, substitute $\mu=-1$ into L_2:

$$\begin{pmatrix}-4\\7\\-1\end{pmatrix}-1\begin{pmatrix}3\\-1\\-5\end{pmatrix}=\begin{pmatrix}-7\\8\\4\end{pmatrix}$$

So point of intersection is $(-7, 8, 4)$

1. Since their direction vectors are not multiples of each other.
2. Solve simultaneously.
4. Since the equations are consistent.

Alternatively substitute $\lambda=-4$ into L_1

PURE

Strategy 3

To apply a matrix transformation to a line

1. Form a vector for a general point on the line.
2. Pre-multiply this vector by the transformation matrix.
3. Write the equation of the image.

See Ch5.2 For a reminder of using matrices for linear transformations.

A line can be transformed by a linear transformation to another line or to a point.

Example 7

The line with equation $\mathbf{r}=\mathbf{i}-\mathbf{j}+\lambda(\mathbf{i}+3\mathbf{j})$ is transformed by the matrix $\mathbf{T}=\begin{pmatrix}1 & -2\\2 & 0\end{pmatrix}$ Work out the equation of the image of the line.

A general point on the line is $\begin{pmatrix}1+\lambda\\-1+3\lambda\end{pmatrix}$

$$\begin{pmatrix}1 & -2\\2 & 0\end{pmatrix}\begin{pmatrix}1+\lambda\\-1+3\lambda\end{pmatrix}=\begin{pmatrix}1+\lambda-2(-1+3\lambda)\\2(1+\lambda)+0\end{pmatrix}$$

$$=\begin{pmatrix}3-5\lambda\\2+2\lambda\end{pmatrix}$$

So the equation of the image is $\mathbf{r}=\begin{pmatrix}3\\2\end{pmatrix}+\lambda\begin{pmatrix}-5\\2\end{pmatrix}$ or alternatively write $\mathbf{r}=3\mathbf{i}+2\mathbf{j}+\lambda(-5\mathbf{i}+2\mathbf{j})$

1. Combine to a single column vector. From the line equation you know that the x component is $1+\lambda$ and the y component is $-1+3\lambda$
2. Pre-multiply by **T**
3. Write the equation of the image in the form required.

Exercise 6.1B Reasoning and problem-solving

1 For each pair of equations, decide whether or not they represent the same line. Show your working clearly.

a $\mathbf{r}=\begin{pmatrix}1\\5\end{pmatrix}+\lambda\begin{pmatrix}4\\-3\end{pmatrix}$ and $\mathbf{r}=\begin{pmatrix}2\\3\end{pmatrix}+\mu\begin{pmatrix}-2\\6\end{pmatrix}$

b $\mathbf{r}=\begin{pmatrix}4\\-1\end{pmatrix}+\lambda\begin{pmatrix}1\\-1\end{pmatrix}$ and $\mathbf{r}=\begin{pmatrix}1\\2\end{pmatrix}+\mu\begin{pmatrix}-1\\1\end{pmatrix}$

c $\mathbf{r}=2\mathbf{i}+7\mathbf{j}+\lambda(\mathbf{i}-\mathbf{j})$ and $y=9-x$

2 For each pair of equations, decide whether or not they represent the same line. Show your working clearly.

a $\mathbf{r}=\begin{pmatrix}-3\\3\\3\end{pmatrix}+\lambda\begin{pmatrix}2\\0\\-1\end{pmatrix}$ and $\mathbf{r}=\begin{pmatrix}1\\3\\-1\end{pmatrix}+\mu\begin{pmatrix}-4\\0\\2\end{pmatrix}$

b $\mathbf{r}=\begin{pmatrix}2\\3\\0\end{pmatrix}+\lambda\begin{pmatrix}1\\1\\-1\end{pmatrix}$ and $\mathbf{r}=\begin{pmatrix}2\\5\\-2\end{pmatrix}+\mu\begin{pmatrix}1\\-1\\1\end{pmatrix}$

c $\mathbf{r}=2\mathbf{j}-\mathbf{k}+\lambda(\mathbf{i}+2\mathbf{j}-3\mathbf{k})$ and $\mathbf{r}=5\mathbf{i}+6\mathbf{j}-10\mathbf{k}+\mu(2\mathbf{i}+4\mathbf{j}-6\mathbf{k})$

See Ch5.2
For a reminder of writing transformations as matrices.

d $\frac{x-2}{1}=\frac{y-3}{3}=\frac{z+1}{-2}$ and $\frac{x+3}{-1}=\frac{y+12}{-3}=\frac{z-9}{2}$

e $\frac{x+5}{-1}=\frac{y-0}{2}=\frac{z-1}{2}$ and $\mathbf{r}=\begin{pmatrix}1\\-12\\11\end{pmatrix}+\lambda\begin{pmatrix}2\\-4\\-4\end{pmatrix}$

3 Find the point of intersection between these pairs of lines.

a $\mathbf{r}=\begin{pmatrix}-1\\1\end{pmatrix}+\lambda\begin{pmatrix}-2\\3\end{pmatrix}$ and $\mathbf{r}=\begin{pmatrix}1\\7\end{pmatrix}+\mu\begin{pmatrix}4\\-3\end{pmatrix}$

b $\mathbf{r}=-4\mathbf{i}+14\mathbf{j}+\lambda(-\mathbf{i}+2\mathbf{j})$ and $\mathbf{r}=\begin{pmatrix}4\\1\end{pmatrix}+\mu\begin{pmatrix}-2\\3\end{pmatrix}$

c $\mathbf{r}=\begin{pmatrix}0\\2\end{pmatrix}+\lambda\begin{pmatrix}-1\\3\end{pmatrix}$ and $y=2x-3$

4 For each pair of lines, decide whether they are parallel, skew or intersecting. If they are intersecting, find their point of intersection.

a $\mathbf{r}=\begin{pmatrix}7\\-6\\-2\end{pmatrix}+\lambda\begin{pmatrix}3\\0\\-1\end{pmatrix}$ and $\mathbf{r}=\begin{pmatrix}3\\4\\0\end{pmatrix}+\mu\begin{pmatrix}-2\\5\\1\end{pmatrix}$

b $\mathbf{r}=\mathbf{i}+2\mathbf{k}+\lambda(3\mathbf{i}+4\mathbf{j}-2\mathbf{k})$ and $\mathbf{r}=3\mathbf{i}+\mathbf{j}-\mathbf{k}+\mu(-6\mathbf{i}-8\mathbf{j}+4\mathbf{k})$

c $\frac{x+5}{-1}=\frac{y}{2}=\frac{z-1}{2}$ and $\frac{x-2}{1}=\frac{y+9}{3}=\frac{z-5}{-2}$

d $\frac{-4-x}{3}=\frac{y-2}{-1}=\frac{z}{5}$ and $\frac{x-6}{2}=\frac{y+8}{-2}=\frac{5-z}{-1}$

5 Two lines, L_1 and L_2, have equations $\frac{x+10}{5}=\frac{y-5}{-1}=\frac{z-5}{2}$ and $\mathbf{r}=\begin{pmatrix}3\\3\\5\end{pmatrix}+\lambda\begin{pmatrix}-3\\0\\4\end{pmatrix}$ respectively.

Given that L_1 and L_2 intersect at the point A, calculate the length of $\overrightarrow{OA}$

6 Find the image of the line $\mathbf{r}=\begin{pmatrix}1\\3\end{pmatrix}+\lambda\begin{pmatrix}-2\\1\end{pmatrix}$ under the transformation given by the matrix $\begin{pmatrix}2&1\\0&-2\end{pmatrix}$

7 Find the image of the line $\mathbf{r}=\begin{pmatrix}2\\-1\end{pmatrix}+\lambda\begin{pmatrix}2\\1\end{pmatrix}$ under a rotation of 90° anticlockwise about the origin.

See Ch5.2
For a reminder of writing transformations as matrices.

8 Find the image of the line $\mathbf{r}=\begin{pmatrix}3\\0\\-1\end{pmatrix}+\lambda\begin{pmatrix}1\\-1\\-2\end{pmatrix}$ under each of these transformations.

a Rotation 90° anticlockwise around the x-axis.

b Rotation 180° around the y-axis.

c Reflection in the plane $z=0$

9 Find the image of the line $\frac{x-1}{2}=\frac{y+3}{-1}=\frac{5-z}{-4}$ after a reflection in the plane $y=0$

6.2 The scalar product

Fluency and skills

A very important concept in the application of vectors is the **scalar product**. As its name suggests, the result of finding the scalar product of two vectors will be a scalar quantity.

Key point

The scalar product $\mathbf{a}\cdot\mathbf{b}$ of vectors $\mathbf{a}$ and $\mathbf{b}$ is defined as $\mathbf{a}\cdot\mathbf{b}=|\mathbf{a}||\mathbf{b}|\cos\theta$ where θ is the angle between the vectors $\mathbf{a}$ and $\mathbf{b}$
The vectors $\mathbf{a}$ and $\mathbf{b}$ are both directed away from the angle (or both directed towards the angle) and $0\le\theta\le180°$

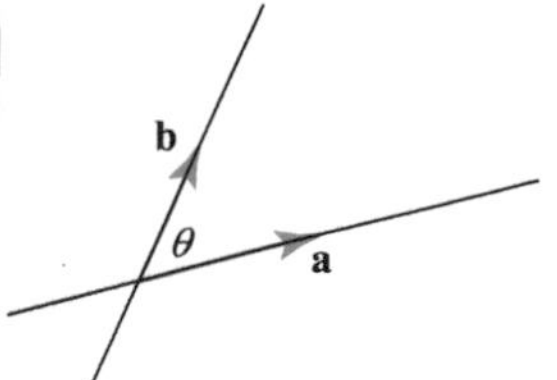

See Maths Ch6.1
For a reminder of magnitude notation.

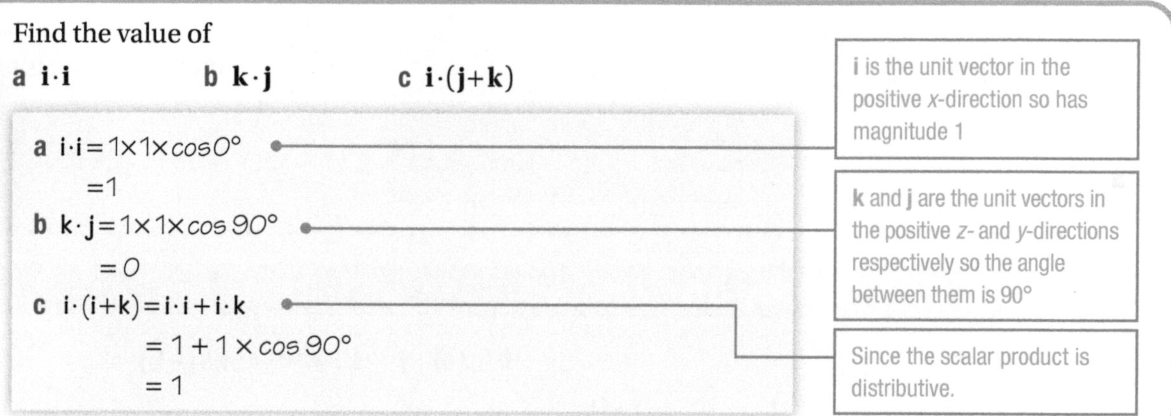

Example 1

Find the value of

a $\mathbf{i}\cdot\mathbf{i}$ b $\mathbf{k}\cdot\mathbf{j}$ c $\mathbf{i}\cdot(\mathbf{j}+\mathbf{k})$

a $\mathbf{i}\cdot\mathbf{i}=1\times1\times\cos0°$
$=1$

i is the unit vector in the positive x-direction so has magnitude 1

b $\mathbf{k}\cdot\mathbf{j}=1\times1\times\cos90°$
$=0$

k and j are the unit vectors in the positive z- and y-directions respectively so the angle between them is 90°

c $\mathbf{i}\cdot(\mathbf{i}+\mathbf{k})=\mathbf{i}\cdot\mathbf{i}+\mathbf{i}\cdot\mathbf{k}$
$=1+1\times\cos90°$
$=1$

Since the scalar product is distributive.

The scalar product is sometimes called the **dot product**. The dot product is a distributive operation. If vectors $\mathbf{a}$ and $\mathbf{b}$ are perpendicular then $\mathbf{a}\cdot\mathbf{b}=0$. This is because $\cos 90°=0$

Key point

If you have two vectors $\mathbf{a}=a_1\mathbf{i}+a_2\mathbf{j}+a_3\mathbf{k}$ and $\mathbf{b}=b_1\mathbf{i}+b_2\mathbf{j}+b_3\mathbf{k}$, then the scalar product is $\mathbf{a}\cdot\mathbf{b}=a_1b_1+a_2b_2+a_3b_3$

Example 2

Given $\mathbf{a}=\begin{pmatrix}1\\-3\\2\end{pmatrix}$ and $\mathbf{b}=\begin{pmatrix}-4\\-1\\5\end{pmatrix}$, calculate

a The scalar product $\mathbf{a}\cdot\mathbf{b}$ b The angle between vectors $\mathbf{a}$ and $\mathbf{b}$

a $\mathbf{a}\cdot\mathbf{b}=\begin{pmatrix}1\\-3\\2\end{pmatrix}\cdot\begin{pmatrix}-4\\-1\\5\end{pmatrix}$
$=(1\times-4)+(-3\times-1)+(2\times5)$
$=9$

Using $\mathbf{a}\cdot\mathbf{b}=a_1b_1+a_2b_2+a_3b_3$

(*Continued on the next page*)

b You need $|\mathbf{a}| = \sqrt{1^2 + 3^2 + 2^2} = \sqrt{14}$

and $|\mathbf{b}| = \sqrt{4^1 + 1^2 + 5^2} = \sqrt{42}$

$\cos\theta = \dfrac{\mathbf{a}\cdot\mathbf{b}}{|\mathbf{a}||\mathbf{b}|}$

$= \dfrac{9}{\sqrt{14}\sqrt{42}}$

$= \dfrac{3\sqrt{3}}{14}$

$\theta = 68.2°$ (1 decimal place)

To find the magnitude of a 3D vector find the sum of the squares of each component and take the square root.

From rearranging $\mathbf{a}\cdot\mathbf{b} = |\mathbf{a}||\mathbf{b}|\cos\theta$

Using $\cos^{-1}\left(\dfrac{3\sqrt{3}}{14}\right)$

Calculator

Try it on your calculator

You can use a calculator to find the scalar product.

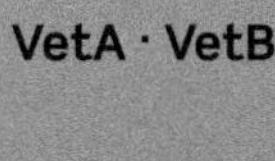

Activity

Find out how to work out $\begin{pmatrix}2\\3\\-1\end{pmatrix}\cdot\begin{pmatrix}-1\\0\\4\end{pmatrix}$ on *your* calculator.

Exercise 6.2A Fluency and skills

1 Find the value of

a $\mathbf{j}\cdot\mathbf{j}$ b $\mathbf{i}\cdot\mathbf{k}$ c $2\mathbf{i}\cdot\mathbf{j}$ d $\mathbf{j}\cdot(\mathbf{i}+\mathbf{j})$

e $\mathbf{i}\cdot(2\mathbf{i})$ f $(\mathbf{i}+\mathbf{j})\cdot(\mathbf{i}-\mathbf{j})$

g $(\mathbf{j}+\mathbf{k})\cdot(\mathbf{i}+\mathbf{j})$ h $(\mathbf{i}+\mathbf{j}+\mathbf{k})\cdot(\mathbf{i}+\mathbf{j}+\mathbf{k})$

2 Calculate these scalar products.

a $\begin{pmatrix}2\\4\\-1\end{pmatrix}\cdot\begin{pmatrix}5\\-3\\-6\end{pmatrix}$ b $(2\mathbf{i}-\mathbf{j}+3\mathbf{k})\cdot(\mathbf{i}+4\mathbf{j}-\mathbf{k})$

c $\begin{pmatrix}\sqrt{2}\\1\\3\end{pmatrix}\cdot\begin{pmatrix}2\\0\\-\sqrt{2}\end{pmatrix}$ d $(5\mathbf{i}+\mathbf{k})\cdot(2\mathbf{j}-3\mathbf{k})$

3 Find an expression in terms of a for the dot product of $(2a\mathbf{i}-\mathbf{j}+a\mathbf{k})$ and $(3a\mathbf{i}-a\mathbf{j}-5\mathbf{k})$

4 Find an expression in terms of k of the dot product of $\begin{pmatrix}2\\k\\-3\end{pmatrix}$ and $\begin{pmatrix}k\\-1\\2\end{pmatrix}$

5 Calculate the angle between the vectors

a $\begin{pmatrix}5\\1\\3\end{pmatrix}$ and $\begin{pmatrix}2\\-2\\4\end{pmatrix}$

b $(5\mathbf{i}+\mathbf{j}-2\mathbf{k})$ and $(4\mathbf{i}+3\mathbf{j}-\mathbf{k})$

c $\begin{pmatrix}2\sqrt{3}\\4\\\sqrt{3}\end{pmatrix}$ and $\begin{pmatrix}-1\\\sqrt{3}\\2\end{pmatrix}$ d $(3\mathbf{j}+2\mathbf{k})$ and $(4\mathbf{i}-\mathbf{j})$

6 Calculate the cosine of the acute angle between each pair of vectors.

Give your answers in terms of a where $a > 0$

a $\begin{pmatrix}a\\1\\0\end{pmatrix}$ and $\begin{pmatrix}1\\0\\a\end{pmatrix}$

b $3a\mathbf{i}-4a\mathbf{j}+5\mathbf{k}$ and $5a\mathbf{j}+5\mathbf{k}$

7 Show that the vectors $\begin{pmatrix}6\\-4\end{pmatrix}$ and $\begin{pmatrix}-8\\-12\end{pmatrix}$ are perpendicular.

8 Show that the vectors $\begin{pmatrix}4\\-1\\5\end{pmatrix}$ and $\begin{pmatrix}-2\\-3\\1\end{pmatrix}$ are perpendicular.

9 Show that the vectors $3\mathbf{i}-5\mathbf{j}-2\mathbf{k}$ and $2\mathbf{i}+4\mathbf{j}-7\mathbf{k}$ are perpendicular.

10 Show that the vectors $2\mathbf{i}+\mathbf{k}$ and $4\mathbf{i}-\mathbf{j}-3\mathbf{k}$ are not perpendicular.

11 Find the value of a for which the vectors $\begin{pmatrix} 3a \\ 2 \\ -2 \end{pmatrix}$ and $\begin{pmatrix} 2 \\ a \\ 4 \end{pmatrix}$ are perpendicular.

12 Find the value of b for which the vectors $(2b\mathbf{i}+\mathbf{j}-\mathbf{k})$ and $(3\mathbf{i}-b\mathbf{j}+10\mathbf{k})$ are perpendicular.

13 Find the values of c for which the vectors $\begin{pmatrix} c \\ 3 \\ -1 \end{pmatrix}$ and $\begin{pmatrix} -c \\ c \\ 2 \end{pmatrix}$ are perpendicular.

PURE

Reasoning and problem-solving

Strategy 1

To find the obtuse angle between two intersecting lines

1. Identify the direction vector of each line, **a** and **b**
2. Use $\cos\theta = \left|\frac{\mathbf{a}\cdot\mathbf{b}}{|\mathbf{a}||\mathbf{b}|}\right|$ to find the acute angle between the lines.
3. Subtract the acute angle from 180° to find the obtuse angle between the lines.

Example 3

Given that L_1 and L_2 have equations $L_1: \mathbf{r} = \begin{pmatrix} 1 \\ -1 \\ 2 \end{pmatrix} + \lambda\begin{pmatrix} 1 \\ -2 \\ 1 \end{pmatrix}$ and $L_2: \frac{x+4}{2} = \frac{y+3}{-1} = \frac{z-1}{1}$ and that they intersect at the point A, find the obtuse angle between L_1 and L_2

The direction vectors are $\begin{pmatrix} 1 \\ -2 \\ 1 \end{pmatrix}$ and $\begin{pmatrix} 2 \\ -1 \\ 1 \end{pmatrix}$ — Identify direction vectors. (1)

$\begin{pmatrix} 1 \\ -2 \\ 1 \end{pmatrix} \cdot \begin{pmatrix} 2 \\ -1 \\ 1 \end{pmatrix} = 2+2+1=5$

$\left|\begin{pmatrix} 1 \\ -2 \\ 1 \end{pmatrix}\right| = \sqrt{1^2+2^2+1^2} = \sqrt{6}$ — Find the magnitude of the two direction vectors. (2)

$\left|\begin{pmatrix} 2 \\ -1 \\ 1 \end{pmatrix}\right| = \sqrt{2^2+1^2+1^2} = \sqrt{6}$

$\cos\theta = \frac{5}{\sqrt{6}\sqrt{6}}$ — First find the acute angle. (2)

$\theta = 33.6°$

So obtuse angle is $180-33.6=146.4°$ (1 decimal place) — Then, subtract from 180° to find the obtuse angle. (3)

Strategy 2

To prove properties of the scalar product

1. Use general vectors, for example $\mathbf{a} = a_1\mathbf{i} + a_2\mathbf{j} + a_3\mathbf{k}$
2. Use the fact that if **a**, **b** are perpendicular vectors then $\mathbf{a}\cdot\mathbf{b} = 0$
3. Use the fact that $\mathbf{i}\cdot\mathbf{i} = \mathbf{j}\cdot\mathbf{j} = \mathbf{k}\cdot\mathbf{k} = 1$

Example 4

Show that $\mathbf{a}\cdot\mathbf{b} = \mathbf{b}\cdot\mathbf{a}$ for any 3D vectors **a** and **b**

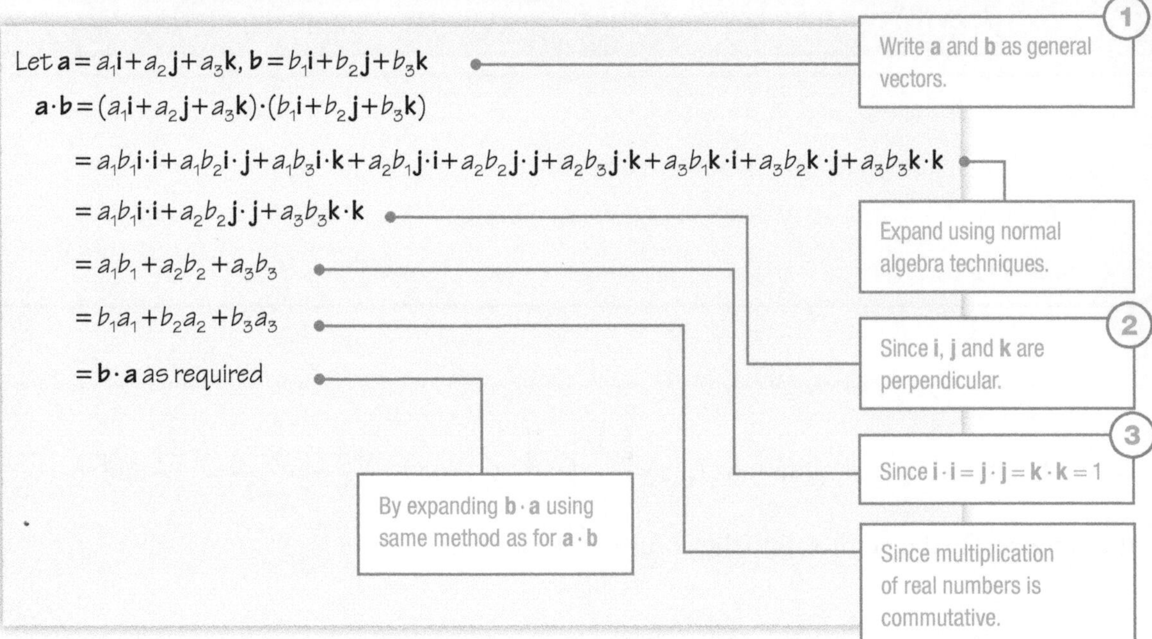

Let $\mathbf{a} = a_1\mathbf{i} + a_2\mathbf{j} + a_3\mathbf{k}$, $\mathbf{b} = b_1\mathbf{i} + b_2\mathbf{j} + b_3\mathbf{k}$
$\mathbf{a}\cdot\mathbf{b} = (a_1\mathbf{i} + a_2\mathbf{j} + a_3\mathbf{k})\cdot(b_1\mathbf{i} + b_2\mathbf{j} + b_3\mathbf{k})$
$= a_1b_1\mathbf{i}\cdot\mathbf{i} + a_1b_2\mathbf{i}\cdot\mathbf{j} + a_1b_3\mathbf{i}\cdot\mathbf{k} + a_2b_1\mathbf{j}\cdot\mathbf{i} + a_2b_2\mathbf{j}\cdot\mathbf{j} + a_2b_3\mathbf{j}\cdot\mathbf{k} + a_3b_1\mathbf{k}\cdot\mathbf{i} + a_3b_2\mathbf{k}\cdot\mathbf{j} + a_3b_3\mathbf{k}\cdot\mathbf{k}$
$= a_1b_1\mathbf{i}\cdot\mathbf{i} + a_2b_2\mathbf{j}\cdot\mathbf{j} + a_3b_3\mathbf{k}\cdot\mathbf{k}$
$= a_1b_1 + a_2b_2 + a_3b_3$
$= b_1a_1 + b_2a_2 + b_3a_3$
$= \mathbf{b}\cdot\mathbf{a}$ as required
1 Write **a** and **b** as general vectors.
Expand using normal algebra techniques.
2 Since **i**, **j** and **k** are perpendicular.
3 Since $\mathbf{i}\cdot\mathbf{i} = \mathbf{j}\cdot\mathbf{j} = \mathbf{k}\cdot\mathbf{k} = 1$
By expanding $\mathbf{b}\cdot\mathbf{a}$ using same method as for $\mathbf{a}\cdot\mathbf{b}$
Since multiplication of real numbers is commutative.

Exercise 6.2B Reasoning and problem-solving

1 Show that the lines $= \begin{pmatrix} 3 \\ -2 \end{pmatrix} + \lambda\begin{pmatrix} -1 \\ 3 \end{pmatrix}$ and $\mathbf{r} = 2\mathbf{i} - \mathbf{j} + \mu(6\mathbf{i} + 2\mathbf{j})$ are perpendicular.

2 Show that the lines $\mathbf{r} = \begin{pmatrix} 5 \\ -2 \\ 4 \end{pmatrix} + \lambda\begin{pmatrix} 6 \\ 7 \\ -5 \end{pmatrix}$ and $\mathbf{r} = \begin{pmatrix} 1 \\ 0 \\ -3 \end{pmatrix} + \lambda\begin{pmatrix} 2 \\ -1 \\ 1 \end{pmatrix}$ are perpendicular.

3 Use the scalar product to show that the lines $\mathbf{r} = \begin{pmatrix} 1 \\ 1 \\ 2 \end{pmatrix} + \lambda\begin{pmatrix} 3 \\ -2 \\ 6 \end{pmatrix}$ and $\frac{x-5}{2} = \frac{y+7}{6} = \frac{z}{1}$ are perpendicular.

4 Show that the lines with equations $\mathbf{r} = \begin{pmatrix} 6 \\ 6 \end{pmatrix} + \lambda\begin{pmatrix} 2 \\ 1 \end{pmatrix}$ and $\mathbf{r} = \begin{pmatrix} 4 \\ 0 \end{pmatrix} + \mu\begin{pmatrix} 3 \\ -1 \end{pmatrix}$ intersect and find the acute angle between them.

5 Show that the lines with equations $\mathbf{r} = \begin{pmatrix} 2 \\ 1 \\ -3 \end{pmatrix} + s\begin{pmatrix} 0 \\ 2 \\ 1 \end{pmatrix}$ and $\mathbf{r} = \begin{pmatrix} 5 \\ 11 \\ 3 \end{pmatrix} + t\begin{pmatrix} 3 \\ -4 \\ -1 \end{pmatrix}$ intersect and find the acute angle between them to the nearest degree.

6 Given that they intersect, find the obtuse angle between the lines with equations

$$\mathbf{r}=\begin{pmatrix}4\\0\\12\end{pmatrix}+\lambda\begin{pmatrix}0\\2\\5\end{pmatrix} \text{ and } \frac{x-1}{3}=\frac{y+2}{-6}=\frac{z+10}{2}$$

7 Show that the lines with equations $\mathbf{r}=(2\mathbf{i}-\mathbf{j}+3\mathbf{k})+\lambda(\mathbf{i}+\mathbf{j}-\mathbf{k})$ and $\mathbf{r}=(2\mathbf{i}-2\mathbf{j}+\mathbf{k})+\mu(-\mathbf{i}+3\mathbf{k})$ intersect and find the acute angle between them to 3 significant figures.

8 Find the values of a for which the lines

$$\mathbf{r}=\begin{pmatrix}3\\-1\\2\end{pmatrix}+\lambda\begin{pmatrix}2\\a\\0\end{pmatrix} \text{ and } \mathbf{r}=\mu\begin{pmatrix}3\\0\\\sqrt{3}\end{pmatrix} \text{ meet at a } 60° \text{ angle.}$$

9 Find the values of a for which the lines

$$\mathbf{r}=\begin{pmatrix}1\\4\\4\end{pmatrix}+\lambda\begin{pmatrix}-2\\a\sqrt{2}\\2\end{pmatrix} \text{ and } \frac{x-3}{1}=\frac{z-1}{-1},\ y=-2$$

meet at a 135° angle.

10 The lines with equations $\frac{x}{4}=\frac{y+1}{8}=\frac{z-3}{1}$ and $\frac{x-1}{2}=\frac{y}{6}=\frac{z+2}{11}$ intersect at the point A

a Find the coordinates of the point A

b Calculate the exact cosine of the acute angle between the two lines at the point A

11 The acute angle between the vectors $\mathbf{a}=\mathbf{i}-k\mathbf{j}$ and $\mathbf{b}=\mathbf{i}+\mathbf{j}$ is 60°

Calculate the possible values of k

12 The acute angle between the vectors $\begin{pmatrix}1\\k\\2\end{pmatrix}$ and $\begin{pmatrix}0\\-1\\1\end{pmatrix}$ is 45°

Calculate the possible values of k

13 You are given that $\mathbf{a}=3\mathbf{i}-\mathbf{k}$ and $\mathbf{b}=\mathbf{i}-2\mathbf{j}+2\mathbf{k}$

Calculate the exact area of triangle ABC

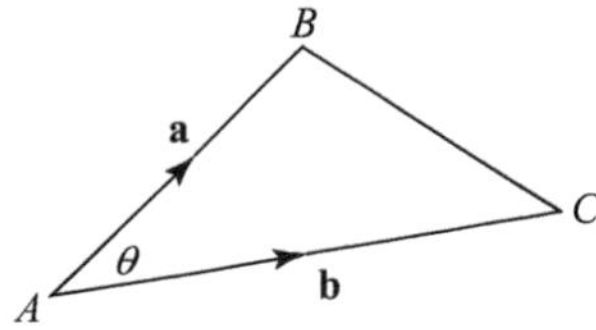

14 The points A, B and C have position vectors $\begin{pmatrix}4\\0\\1\end{pmatrix}$, $\begin{pmatrix}5\\-2\\0\end{pmatrix}$ and $\begin{pmatrix}3\\-1\\1\end{pmatrix}$ respectively.

Calculate the exact area of triangle ABC

15 Given that $\mathbf{a}=a_1\mathbf{i}+a_2\mathbf{j}+a_3\mathbf{k}$ and $\mathbf{b}=b_1\mathbf{i}+b_2\mathbf{j}+b_3\mathbf{k}$, show that $\mathbf{a}\cdot\mathbf{b}=a_1b_1+a_2b_2+a_3b_3$

16 Show that the dot product is distributive, that is, $\mathbf{a}\cdot(\mathbf{b}+\mathbf{c})=\mathbf{a}\cdot\mathbf{b}+\mathbf{a}\cdot\mathbf{c}$ for any 3D vectors **a**, **b** and **c**

17 a Prove that $\mathbf{a}\cdot\mathbf{a}=|\mathbf{a}|^2$ for any vector **a**

b Hence prove the cosine rule: $|\mathbf{a}|^2=|\mathbf{b}|^2+|\mathbf{c}|^2-2|\mathbf{b}||\mathbf{c}|\cos\theta$ for the triangle shown.

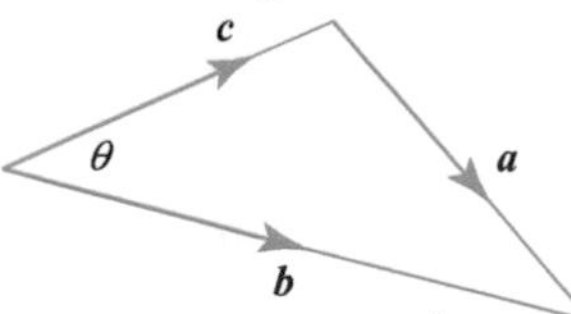

18 a Show that the line $\frac{x+2}{-4}=\frac{y-5}{5}=\frac{z-3}{3}$ cuts the x-axis and find the intercept's coordinates.

b Calculate the acute angle between the x-axis and the line.

19 Three lines have equations as follows: L_1: $\mathbf{r}=6\mathbf{i}-3\mathbf{j}+\lambda(\mathbf{i}+\mathbf{k})$, L_2: $\mathbf{r}=s(3\mathbf{i}-\mathbf{j}+\mathbf{k})$ and L_3: $\mathbf{r}=t(\mathbf{j}+2\mathbf{k})$

The lines L_1 and L_2 intersect at the point A, L_1 and L_2 intersect at the point B, and L_2 and L_3 intersect at the point C as shown.

Calculate the exact area of triangle ABC

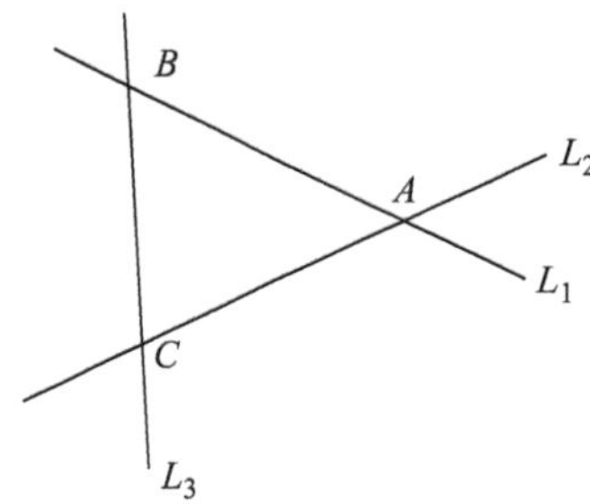

PURE

6.3 Finding distances 1

Fluency and skills

Suppose you wish to find the shortest distance from a line $\mathbf{r}=\mathbf{a}+\lambda\mathbf{b}$ to a given point, C, with position vector $\mathbf{c}$

The shortest possible distance will be CD, which is perpendicular to the original line.

Therefore, the scalar product $(\mathbf{c}-(\mathbf{a}+\lambda\mathbf{b}))\cdot\mathbf{b}=0$ since $\overrightarrow{DC}=\mathbf{c}-(\mathbf{a}+\lambda\mathbf{b})$.

This allows you to find the point of intersection between $\overrightarrow{DC}$ and the line $\mathbf{r}=\mathbf{a}+\lambda\mathbf{b}$. You will then be able to find the length of $\overrightarrow{DC}$ and hence the shortest distance from the point to the line.

Example 1

Calculate the shortest distance from the point (3, 3, 1) to the line with equation $\mathbf{r}=\begin{pmatrix}2\\0\\-6\end{pmatrix}+\lambda\begin{pmatrix}-3\\0\\-1\end{pmatrix}$

$$\begin{pmatrix}3\\3\\1\end{pmatrix}-\left[\begin{pmatrix}2\\0\\-6\end{pmatrix}+\lambda\begin{pmatrix}-3\\0\\-1\end{pmatrix}\right]=\begin{pmatrix}1+3\lambda\\3\\7+\lambda\end{pmatrix}$$

Find a vector that passes through the point and the line.

$$\begin{pmatrix}1+3\lambda\\3\\7+\lambda\end{pmatrix}\cdot\begin{pmatrix}-3\\0\\-1\end{pmatrix}=0$$

Since the vectors are perpendicular, their scalar product is zero.

$$\Rightarrow -3-9\lambda-7-\lambda=0 \quad \Rightarrow \lambda=-1$$

$$\begin{pmatrix}2\\0\\-6\end{pmatrix}-1\begin{pmatrix}-3\\0\\-1\end{pmatrix}=\begin{pmatrix}5\\0\\-5\end{pmatrix}$$

This is the position vector of the point of intersection of the line with the perpendicular through (3, 3, 1)

$$\text{Distance}=\sqrt{(5-3)^2+(0-3)^2+(-5-1)^2}$$
$$=7$$

Find the distance from (3, 3, 1) to the point of intersection.

You already know how to find the point of intersection between two lines, if it exists. If two lines do not intersect, you can instead find the shortest possible distance between them. In the case of parallel lines, this is straightforward as the distance between them is constant.

If two parallel lines have equations $\mathbf{r}=\mathbf{a}+\lambda\mathbf{b}$ and $\mathbf{r}=\mathbf{c}+\mu\mathbf{b}$ then, since their separation is always the same, you can chose any point on either one of the lines (for example, the point with position vector $\mathbf{a}$ on the line $\mathbf{r}=\mathbf{a}+\lambda\mathbf{b}$) and use the method above to find the shortest distance from a point to a line.

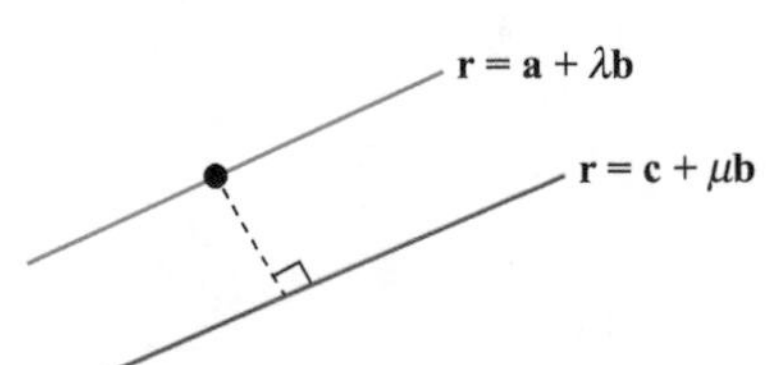

PURE

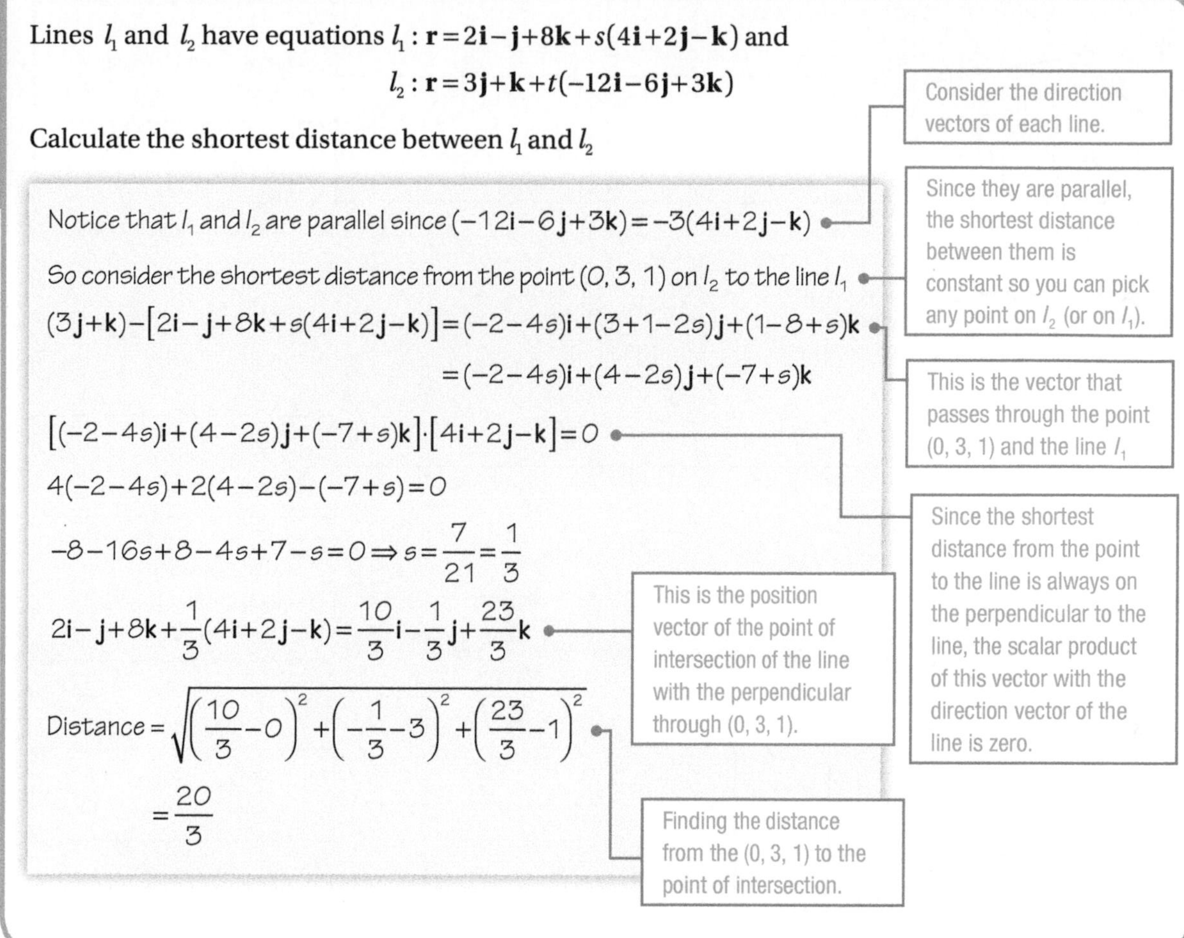

Example 2

Lines l_1 and l_2 have equations $l_1: \mathbf{r} = 2\mathbf{i} - \mathbf{j} + 8\mathbf{k} + s(4\mathbf{i} + 2\mathbf{j} - \mathbf{k})$ and

$l_2: \mathbf{r} = 3\mathbf{j} + \mathbf{k} + t(-12\mathbf{i} - 6\mathbf{j} + 3\mathbf{k})$

Calculate the shortest distance between l_1 and l_2

Notice that l_1 and l_2 are parallel since $(-12\mathbf{i} - 6\mathbf{j} + 3\mathbf{k}) = -3(4\mathbf{i} + 2\mathbf{j} - \mathbf{k})$

Consider the direction vectors of each line.

So consider the shortest distance from the point (0, 3, 1) on l_2 to the line l_1

Since they are parallel, the shortest distance between them is constant so you can pick any point on l_2 (or on l_1).

$$(3\mathbf{j} + \mathbf{k}) - [2\mathbf{i} - \mathbf{j} + 8\mathbf{k} + s(4\mathbf{i} + 2\mathbf{j} - \mathbf{k})] = (-2 - 4s)\mathbf{i} + (3 + 1 - 2s)\mathbf{j} + (1 - 8 + s)\mathbf{k}$$

$$= (-2 - 4s)\mathbf{i} + (4 - 2s)\mathbf{j} + (-7 + s)\mathbf{k}$$

This is the vector that passes through the point (0, 3, 1) and the line l_1

$$[(-2 - 4s)\mathbf{i} + (4 - 2s)\mathbf{j} + (-7 + s)\mathbf{k}] \cdot [4\mathbf{i} + 2\mathbf{j} - \mathbf{k}] = 0$$

Since the shortest distance from the point to the line is always on the perpendicular to the line, the scalar product of this vector with the direction vector of the line is zero.

$$4(-2 - 4s) + 2(4 - 2s) - (-7 + s) = 0$$

$$-8 - 16s + 8 - 4s + 7 - s = 0 \Rightarrow s = \frac{7}{21} = \frac{1}{3}$$

$$2\mathbf{i} - \mathbf{j} + 8\mathbf{k} + \frac{1}{3}(4\mathbf{i} + 2\mathbf{j} - \mathbf{k}) = \frac{10}{3}\mathbf{i} - \frac{1}{3}\mathbf{j} + \frac{23}{3}\mathbf{k}$$

This is the position vector of the point of intersection of the line with the perpendicular through (0, 3, 1).

$$\text{Distance} = \sqrt{\left(\frac{10}{3} - 0\right)^2 + \left(-\frac{1}{3} - 3\right)^2 + \left(\frac{23}{3} - 1\right)^2}$$

$$= \frac{20}{3}$$

Finding the distance from the (0, 3, 1) to the point of intersection.

You can also find the perpendicular distance between skew lines by solving simultaneous equations.

Example 3

Find the minimum distance between the skew lines with equations

$\mathbf{r} = 2\mathbf{i} + \mathbf{j} + \lambda(-\mathbf{j} + 2\mathbf{k})$ and $\mathbf{r} = \mathbf{j} - 2\mathbf{k} + \mu(\mathbf{i} + 2\mathbf{j})$

A general point, P, on the first line has position vector, $\overrightarrow{OP} = \begin{pmatrix} 2 \\ 1-\lambda \\ 2\lambda \end{pmatrix}$.

A general point, Q, on the second line has position vector

$\overrightarrow{OQ} = \begin{pmatrix} \mu \\ 1+2\mu \\ -2 \end{pmatrix}$.

Therefore, the vector joining these two points is

$$\overrightarrow{PQ} = \begin{pmatrix} \mu \\ 1+2\mu \\ -2 \end{pmatrix} - \begin{pmatrix} 2 \\ 1-\lambda \\ 2\lambda \end{pmatrix} = \begin{pmatrix} \mu-2 \\ 2\mu+\lambda \\ -2-2\lambda \end{pmatrix}$$

Use $\overrightarrow{PQ} = \overrightarrow{OQ} - \overrightarrow{OP}$

(*Continued on the next page*)

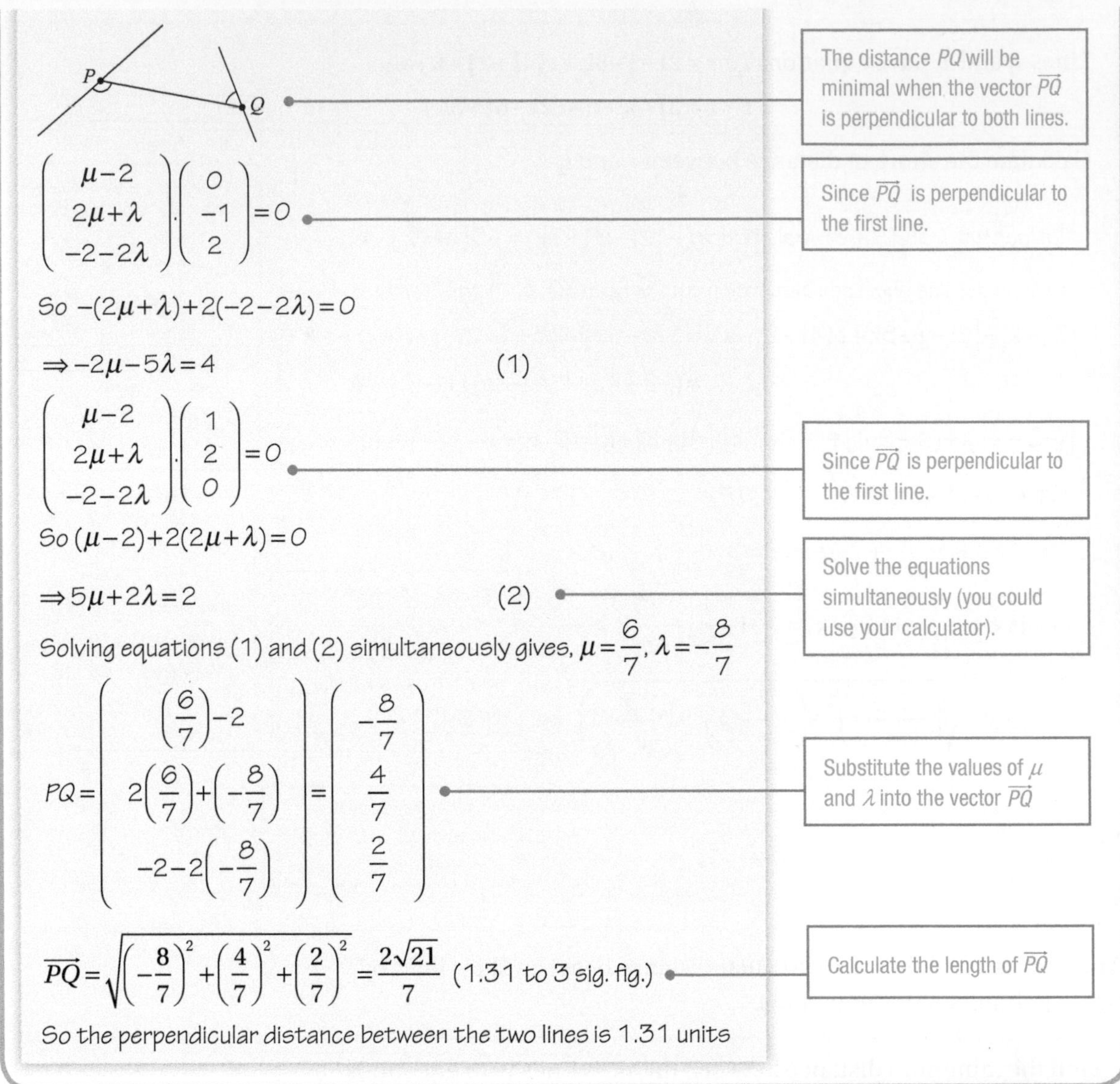

The distance PQ will be minimal when the vector $\overrightarrow{PQ}$ is perpendicular to both lines.

$$\begin{pmatrix} \mu-2 \\ 2\mu+\lambda \\ -2-2\lambda \end{pmatrix} \cdot \begin{pmatrix} 0 \\ -1 \\ 2 \end{pmatrix} = 0$$

Since $\overrightarrow{PQ}$ is perpendicular to the first line.

So $-(2\mu+\lambda)+2(-2-2\lambda)=0$

$\Rightarrow -2\mu-5\lambda=4 \qquad (1)$

$$\begin{pmatrix} \mu-2 \\ 2\mu+\lambda \\ -2-2\lambda \end{pmatrix} \cdot \begin{pmatrix} 1 \\ 2 \\ 0 \end{pmatrix} = 0$$

Since $\overrightarrow{PQ}$ is perpendicular to the first line.

So $(\mu-2)+2(2\mu+\lambda)=0$

$\Rightarrow 5\mu+2\lambda=2 \qquad (2)$

Solve the equations simultaneously (you could use your calculator).

Solving equations (1) and (2) simultaneously gives, $\mu=\frac{6}{7}$, $\lambda=-\frac{8}{7}$

$$PQ = \begin{pmatrix} \left(\frac{6}{7}\right)-2 \\ 2\left(\frac{6}{7}\right)+\left(-\frac{8}{7}\right) \\ -2-2\left(-\frac{8}{7}\right) \end{pmatrix} = \begin{pmatrix} -\frac{8}{7} \\ \frac{4}{7} \\ \frac{2}{7} \end{pmatrix}$$

Substitute the values of μ and λ into the vector $\overrightarrow{PQ}$

$$\overrightarrow{PQ} = \sqrt{\left(-\frac{8}{7}\right)^2+\left(\frac{4}{7}\right)^2+\left(\frac{2}{7}\right)^2} = \frac{2\sqrt{21}}{7} \text{ (1.31 to 3 sig. fig.)}$$

Calculate the length of $\overrightarrow{PQ}$

So the perpendicular distance between the two lines is 1.31 units

Exercise 6.3A Fluency and skills

1 Calculate the shortest distance between the line with equation $\mathbf{r}=6\mathbf{i}-4\mathbf{j}+\mathbf{k}+s(2\mathbf{i}-\mathbf{k})$ and the point $(-4, 2, 6)$

2 Find the shortest distance between the line with equation $\mathbf{r}=\begin{pmatrix} 1 \\ 3 \\ -2 \end{pmatrix}+\lambda\begin{pmatrix} 1 \\ -2 \\ 0 \end{pmatrix}$ and the point $(0, 5, 1)$

3 Find the shortest distance between the line with equations $\frac{x-3}{2}$, $y=1$, $z=-1$ and the point $(0, 3, 1)$

4 Calculate the shortest distance between the line with equation $\mathbf{r}=\mu(\mathbf{i}+\mathbf{j}-\mathbf{k})$ and the point with position vector $2\mathbf{i}-3\mathbf{j}+\mathbf{k}$

PURE

5 Calculate the shortest distance between the point (2, 1, 0) and the line $x = y = 2z$

6 Calculate the shortest distance between these pairs of parallel lines.

a $\mathbf{r} = 5\mathbf{i} + \mathbf{k} + s(\mathbf{i} - \mathbf{j})$ and $\mathbf{r} = 2\mathbf{i} + \mathbf{j} - 7\mathbf{k} + t(\mathbf{i} - \mathbf{j})$

b $\mathbf{r} = \begin{pmatrix} 2 \\ 3 \\ 5 \end{pmatrix} + \lambda \begin{pmatrix} 3 \\ 1 \\ -2 \end{pmatrix}$ and $\mathbf{r} = \begin{pmatrix} -6 \\ 1 \\ -1 \end{pmatrix} + \mu \begin{pmatrix} 9 \\ 3 \\ -6 \end{pmatrix}$

c $x = y + 3 = \frac{z-6}{-2}$ and $x + 1 = y - 4 = \frac{z-8}{-2}$

d $\mathbf{r} = \begin{pmatrix} 0 \\ 4 \\ -1 \end{pmatrix} + \lambda \begin{pmatrix} -3 \\ 5 \\ 1 \end{pmatrix}$ and $\frac{x}{3} = \frac{y+3}{-5} = -z - 1$

e $\mathbf{r} = (2\mathbf{i} - 3\mathbf{j} + 11\mathbf{k}) + s(\mathbf{i} - 5\mathbf{j} - \mathbf{k})$ and $2 - x = \frac{y-1}{5} = z$

7 Calculate the length of the perpendicular between these pairs of skew lines

a $\mathbf{r} = \begin{pmatrix} 3 \\ 0 \\ -1 \end{pmatrix} + \lambda \begin{pmatrix} 1 \\ 1 \\ -1 \end{pmatrix}$ and $\mathbf{r} = \begin{pmatrix} 0 \\ 2 \\ 0 \end{pmatrix} + \mu \begin{pmatrix} 0 \\ 0 \\ 1 \end{pmatrix}$

b $\mathbf{r} = \mathbf{i} - \mathbf{j} + s(\mathbf{i} - \mathbf{k})$ and $\mathbf{r} = \mathbf{j} + 2\mathbf{k} + t(2\mathbf{i} - \mathbf{j})$

c $\mathbf{r} = \mathbf{i} - 4\mathbf{j} + 2\mathbf{k} + s\mathbf{k}$ and $\frac{x-3}{2} = \frac{y+1}{1} = \frac{z-2}{-1}$

Reasoning and problem-solving

An alternative approach to finding the minimum distance between a line and a point is to use calculus. Suppose you have the line $\mathbf{r} = \mathbf{a} + t\mathbf{b}$ and a point with position vector $\mathbf{c}$. Then you need to find the minimum value of $x = |\mathbf{a} + t\mathbf{b} - \mathbf{c}|$ which can be found by using $\frac{dx}{dt} = 0$

Strategy 1

To find the minimum distance between a line and a point using calculus

1. Find the vector joining the point given to a general point on the line.
2. Find an expression for x, the length of this vector.
3. Use $\frac{dx}{dt} = 0$ to calculate the value of t which minimises x
4. Substitute this value of t into your expression for the length.

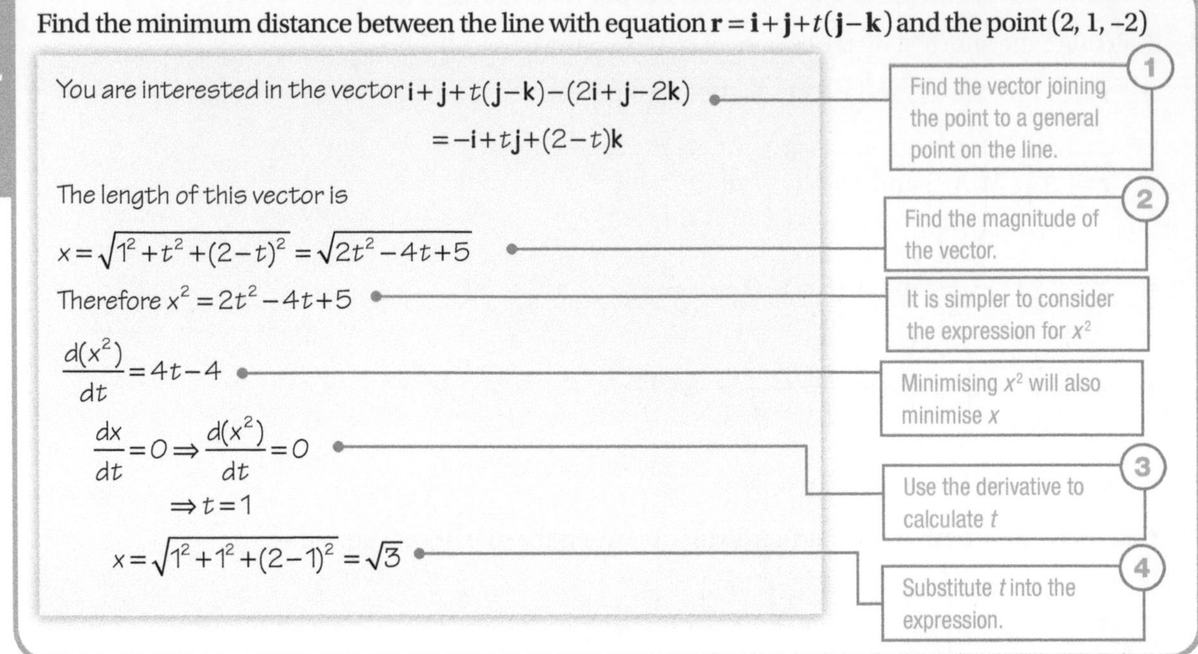

Example 4

Find the minimum distance between the line with equation $\mathbf{r}=\mathbf{i}+\mathbf{j}+t(\mathbf{j}-\mathbf{k})$ and the point (2, 1, −2)

You are interested in the vector $\mathbf{i}+\mathbf{j}+t(\mathbf{j}-\mathbf{k})-(2\mathbf{i}+\mathbf{j}-2\mathbf{k})$

$$=-\mathbf{i}+t\mathbf{j}+(2-t)\mathbf{k}$$

The length of this vector is

$$x=\sqrt{1^2+t^2+(2-t)^2}=\sqrt{2t^2-4t+5}$$

Therefore $x^2=2t^2-4t+5$

$$\frac{d(x^2)}{dt}=4t-4$$

$$\frac{dx}{dt}=0\Rightarrow\frac{d(x^2)}{dt}=0$$

$$\Rightarrow t=1$$

$$x=\sqrt{1^2+1^2+(2-1)^2}=\sqrt{3}$$

1. Find the vector joining the point to a general point on the line.
2. Find the magnitude of the vector.

It is simpler to consider the expression for x^2

Minimising x^2 will also minimise x

3. Use the derivative to calculate t
4. Substitute t into the expression.

A similar process can be used to find the perpendicular distance between a pair of parallel lines.

Example 5

Lines l_1 and l_2 have equations $l_1:\frac{x-1}{4}=\frac{y+1}{2}=\frac{z+2}{-1}$ and $l_2:\mathbf{r}=\begin{pmatrix}1\\3\\0\end{pmatrix}+\lambda\begin{pmatrix}-2\\-1\\0.5\end{pmatrix}$

Use calculus to calculate the shortest distance between l_1 and l_2
Give your answer to 3 significant figures.

The vector equation of l_1 is $\mathbf{r}=\begin{pmatrix}1\\-1\\-2\end{pmatrix}+\lambda\begin{pmatrix}4\\2\\-1\end{pmatrix}$

Since a line with the Cartesian equations $\frac{x-a_1}{b_1}=\frac{y-a_2}{b_2}=\frac{z-a_3}{b_3}$ has vector equation $\mathbf{r}=\begin{pmatrix}a_1\\a_2\\a_3\end{pmatrix}+\lambda\begin{pmatrix}b_1\\b_2\\b_3\end{pmatrix}$.

So l_1 and l_2 are parallel since $\begin{pmatrix}4\\2\\-1\end{pmatrix}=-2\begin{pmatrix}-2\\-1\\0.5\end{pmatrix}$

This is important since the method only works if they are parallel.

So consider the shortest distance from the point (1, 3, 0) on l_2 to the line l_1

Since they are parallel, the shortest distance between them is constant so you can pick any point on l_2 (or on l_1).

(*Continued on the next page*)

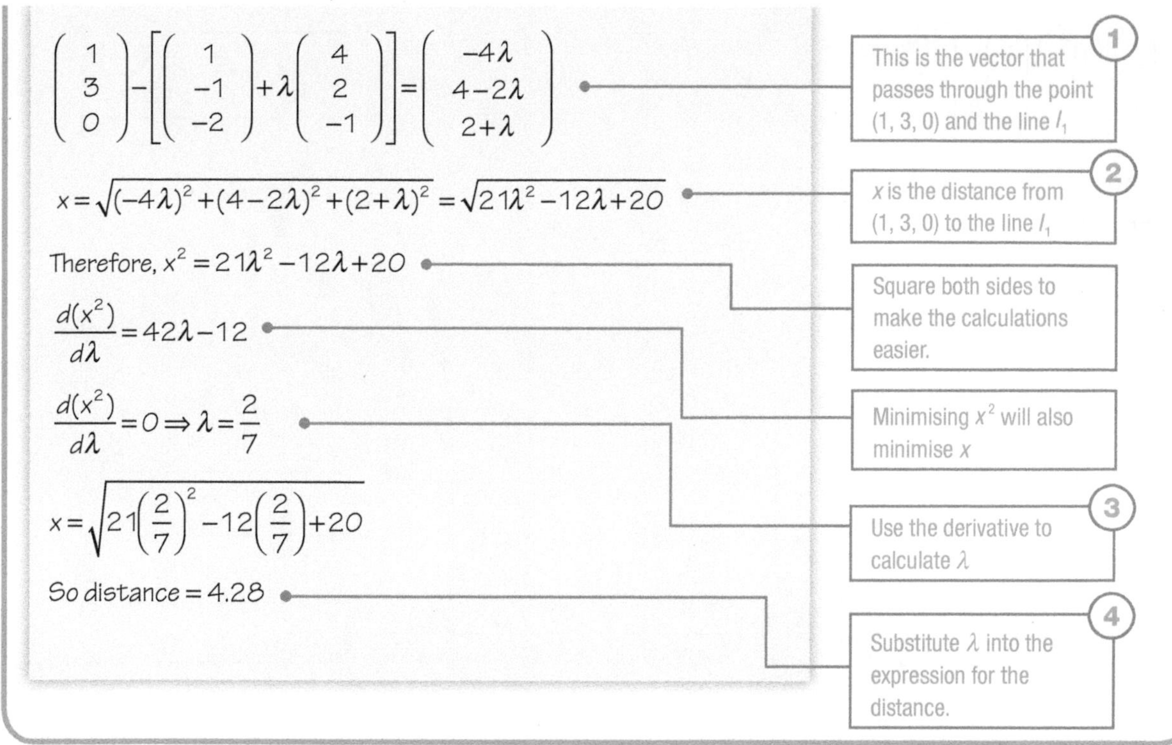

$\begin{pmatrix}1\\3\\0\end{pmatrix}-\left[\begin{pmatrix}1\\-1\\-2\end{pmatrix}+\lambda\begin{pmatrix}4\\2\\-1\end{pmatrix}\right]=\begin{pmatrix}-4\lambda\\4-2\lambda\\2+\lambda\end{pmatrix}$
1 This is the vector that passes through the point (1, 3, 0) and the line l_1
$x=\sqrt{(-4\lambda)^2+(4-2\lambda)^2+(2+\lambda)^2}=\sqrt{21\lambda^2-12\lambda+20}$
2 x is the distance from (1, 3, 0) to the line l_1
Therefore, $x^2=21\lambda^2-12\lambda+20$
Square both sides to make the calculations easier.
$\frac{d(x^2)}{d\lambda}=42\lambda-12$
Minimising x^2 will also minimise x
$\frac{d(x^2)}{d\lambda}=0\Rightarrow\lambda=\frac{2}{7}$
3 Use the derivative to calculate λ
$x=\sqrt{21\left(\frac{2}{7}\right)^2-12\left(\frac{2}{7}\right)+20}$
So distance = 4.28
4 Substitute λ into the expression for the distance.

You can use the perpendicular vector to find the reflection of a point in a line. When a point A is reflected in a line l to give the image A', then l will be the perpendicular bisector of the line segment AA'. You can use this fact to find the coordinates of A'.

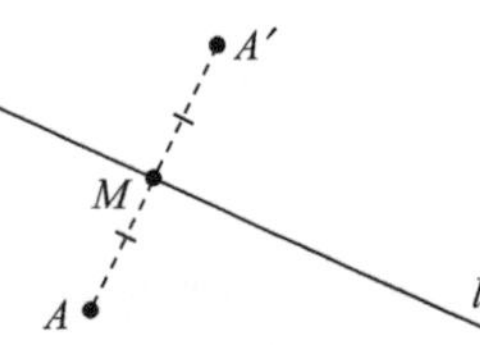

Strategy 2

To find the image of a point reflected in a line

1. Write the vector $\overrightarrow{AM}$
2. Use the fact that $\overrightarrow{AM}$ is perpendicular to the mirror line so the scalar product of $\overrightarrow{AM}$ with the direction vector of the line is zero
3. Use the fact that $\overrightarrow{AA'}=2\overrightarrow{AM}$

Example 6

Find the coordinates of the image of the point $A(-1, 5, 2)$ after it is reflected in the line with equation $\mathbf{r}=\begin{pmatrix}0\\3\\-2\end{pmatrix}+t\begin{pmatrix}1\\1\\-1\end{pmatrix}$

A general point on the line has position vector $\overrightarrow{OM}=\begin{pmatrix}t\\t-3\\2-t\end{pmatrix}$

So the vector $\overrightarrow{AM}=\begin{pmatrix}t\\t+3\\-2-t\end{pmatrix}-\begin{pmatrix}-1\\5\\2\end{pmatrix}=\begin{pmatrix}t+1\\t-2\\-4-t\end{pmatrix}$

1 Using $\overrightarrow{AM}=\overrightarrow{OM}-\overrightarrow{OA}$

(*Continued on the next page*)

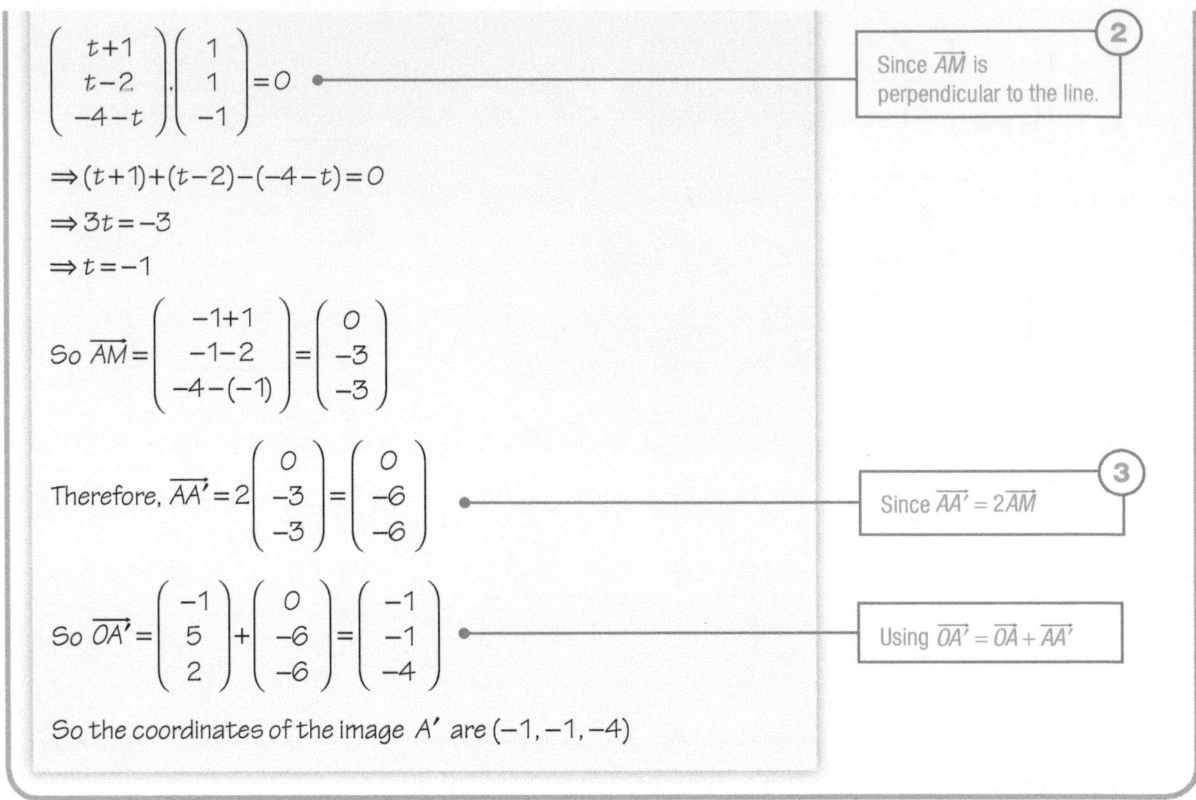

$\begin{pmatrix} t+1 \\ t-2 \\ -4-t \end{pmatrix} \cdot \begin{pmatrix} 1 \\ 1 \\ -1 \end{pmatrix} = 0$
Since $\overline{AM}$ is perpendicular to the line.
$\Rightarrow (t+1)+(t-2)-(-4-t)=0$
$\Rightarrow 3t=-3$
$\Rightarrow t=-1$
So $\overrightarrow{AM} = \begin{pmatrix} -1+1 \\ -1-2 \\ -4-(-1) \end{pmatrix} = \begin{pmatrix} 0 \\ -3 \\ -3 \end{pmatrix}$
Therefore, $\overrightarrow{AA'} = 2\begin{pmatrix} 0 \\ -3 \\ -3 \end{pmatrix} = \begin{pmatrix} 0 \\ -6 \\ -6 \end{pmatrix}$
Since $\overline{AA'} = 2\overline{AM}$
So $\overrightarrow{OA'} = \begin{pmatrix} -1 \\ 5 \\ 2 \end{pmatrix} + \begin{pmatrix} 0 \\ -6 \\ -6 \end{pmatrix} = \begin{pmatrix} -1 \\ -1 \\ -4 \end{pmatrix}$
Using $\overline{OA'} = \overline{OA} + \overline{AA'}$
So the coordinates of the image A' are $(-1, -1, -4)$

Exercise 6.3B Reasoning and problem-solving

1 Use calculus to find the shortest distance between the point $(1, 5, -7)$ and the line with equation

$$\mathbf{r} = \begin{pmatrix} -1 \\ 9 \\ -5 \end{pmatrix} + \lambda \begin{pmatrix} 0 \\ 3 \\ 1 \end{pmatrix}$$

2 Use calculus to find the shortest distance between the point $(-4, 0, 2)$ and the line with equation

$$\frac{x+2}{-1} = \frac{y+2}{2} = \frac{z-1}{1}$$

3 Use calculus to find the shortest distance between these lines.

$$\frac{y+1}{6} = \frac{z}{2}, x = 3 \text{ and } \mathbf{r} = \begin{pmatrix} -1 \\ 1 \\ 5 \end{pmatrix} + \lambda \begin{pmatrix} 0 \\ 3 \\ 1 \end{pmatrix}$$

PURE

4 Determine whether each of these pairs of lines intersect.

If they intersect, find their point of intersection. If they do not intersect, find the minimum distance between them.

a $\mathbf{r}=2\mathbf{i}-\mathbf{k}+s(\mathbf{i}+\mathbf{j}-2\mathbf{k})$ and $\frac{(x-1)}{5}=\frac{(y+5)}{5}=\frac{z}{-10}$

b $\mathbf{r}=\mathbf{i}+5\mathbf{j}+4\mathbf{k}+s(-2\mathbf{i}+\mathbf{j}+4\mathbf{k})$ and $\mathbf{r}=\mathbf{i}+2\mathbf{k}+t(3\mathbf{i}+\mathbf{j}-5\mathbf{k})$

c $\mathbf{r}=2\mathbf{i}+\mathbf{k}+\lambda(\mathbf{i}+2\mathbf{j}-\mathbf{k})$ and $\mathbf{r}=\mathbf{i}-8\mathbf{j}+\mu(\mathbf{i}+3\mathbf{k})$

5 Find the reflection of the point $A(-4, 3, -2)$ in the line $\mathbf{r}=-2\mathbf{i}+5\mathbf{j}-\mathbf{k}+t(\mathbf{i}+\mathbf{j}-\mathbf{k})$

6 Find the reflection of the point $A(3, 0, -4)$ in the line $\mathbf{r}=\mathbf{i}+3\mathbf{j}-2\mathbf{k}+t(\mathbf{i}-2\mathbf{j}-2\mathbf{k})$

7 The points A, B and C are $(2, 6, -4)$, $(0, -2, 3)$ and $(-5, -17, 13)$ respectively.

The line L_1 passes through the origin and point A

Line L_2 passes through the points B and C

Calculate the shortest distance between L_1 and L_2

8 Which of these lines comes closest to the origin?

L_1: $\mathbf{r}=\begin{pmatrix}5\\-2\\-4\end{pmatrix}+\lambda\begin{pmatrix}1\\2\\1\end{pmatrix}$ or L_2: $\mathbf{r}=\begin{pmatrix}-1\\3\\2\end{pmatrix}+\lambda\begin{pmatrix}1\\1\\0\end{pmatrix}$.

Fully explain your reasoning.

9 Which of these two points lie closest to the line $x-4=\frac{y+3}{3}=\frac{7-z}{2}$?

A: $(4, 1, 6)$ or B: $(6, -2, 6)$

Fully explain your reasoning.

10 L_1 and L_2 have equations $\boldsymbol{r}=\boldsymbol{i}+t\boldsymbol{j}$ and $\boldsymbol{r}=\boldsymbol{j}-\boldsymbol{k}+s(\boldsymbol{i}+\boldsymbol{k})$ respectively and intersect at the point A

The line L_3 is parallel to the vector $\boldsymbol{i}+\boldsymbol{j}$ and passes through the point $(2, 1, 0)$.

Calculate the shortest distance from A to the line L_3

Full A Level

6 Summary and review

Chapter summary

- The vector equation of a line parallel to the vector **b** and passing through the point with position vector **a** is $\mathbf{r} = \mathbf{a} + \lambda\mathbf{b}$
- The Cartesian equations of a line parallel to the vector $\mathbf{b} = b_1\mathbf{i} + b_2\mathbf{j} + b_3\mathbf{k}$ and passing through the point with position vector $\mathbf{a} = a_1\mathbf{i} + a_2\mathbf{j} + a_3\mathbf{k}$ are $\frac{x - a_1}{b_1} = \frac{y - a_2}{b_2} = \frac{z - a_3}{b_3}$
- Two lines in 3D can be parallel, intersecting or skew
- The scalar product is defined as $\mathbf{a} \cdot \mathbf{b} = |\mathbf{a}||\mathbf{b}|\cos\theta$ where θ is the angle between the vectors **a** and **b**
- If you have two vectors $\mathbf{a} = a_1\mathbf{i} + a_2\mathbf{j} + a_3\mathbf{k}$ and $\mathbf{b} = b_1\mathbf{i} + b_2\mathbf{j} + b_3\mathbf{k}$, then the scalar product is defined as $\mathbf{a} \cdot \mathbf{b} = a_1b_1 + a_2b_2 + a_3b_3$
- If vectors **a** and **b** are perpendicular then the scalar product, $\mathbf{a} \cdot \mathbf{b} = 0$
- The acute angle, θ, between two lines $\mathbf{r} = \mathbf{a} + \lambda\mathbf{b}$ and $\mathbf{r} = \mathbf{c} + \lambda\mathbf{d}$ can be found using $\cos\theta = \left|\frac{\boldsymbol{b} \cdot \boldsymbol{d}}{|\boldsymbol{b}|\,|\boldsymbol{b}|}\right|$
- The shortest distance between a point and a line can be found by using the fact that it will be perpendicular to the line.
- The shortest distance between two lines can be found by using the fact that it will be perpendicular to both lines.

Check and review

You should now be able to...	Try Questions
✔ Write and use the equation of a line in Cartesian and vector form.	1, 2
✔ Decide whether two lines are intersecting, parallel of skew.	3
✔ Find the coordinates of the point of intersection of two lines.	3
✔ Calculate the scalar product of two vectors.	4, 5
✔ Use the scalar product to find the angle between two vectors.	5, 6
✔ Use the scalar product to show that two lines are perpendicular.	7
✔ Calculate the shortest distance from a point to a line.	8
✔ Calculate the shortest distance between two parallel lines.	9, 10
✔ Calculate the shortest distance between two skew lines.	11

1 A line is parallel to the vector $2\mathbf{i}-3\mathbf{j}+\mathbf{k}$ and passes through the point (1, –1, 4)

Write the equation of the line in

a Cartesian form,

b Vector form.

2 A line passes through the points (3, 0, 5) and (–2, 4, 7)

Write the equation of the line in

a Cartesian form,

b Vector form.

3 For each pair of lines, decide whether they are parallel, skew or intersecting. If they are intersecting, find their point of intersection.

a $\mathbf{r}=\begin{pmatrix}1\\3\\-2\end{pmatrix}+s\begin{pmatrix}-2\\4\\-1\end{pmatrix}$ anc

$\mathbf{r}=\begin{pmatrix}-4\\0\\2\end{pmatrix}+t\begin{pmatrix}5\\-3\\3\end{pmatrix}$

b $\mathbf{r}=4\mathbf{i}-\mathbf{j}+\lambda(2\mathbf{i}+3\mathbf{j}-\mathbf{k})$ and

$\frac{1-x}{10}=\frac{1-y}{15}=\frac{z+3}{5}$

c $\frac{x-2}{3}=\frac{y+1}{2}=\frac{z-3}{-1}$ and

$\frac{x-1}{-4}=\frac{y+3}{-3}=\frac{z-6}{2}$

d $\mathbf{r}=-5\mathbf{i}+2\mathbf{j}+4\mathbf{k}+\lambda(5\mathbf{i}-3\mathbf{j}+\mathbf{k})$ and

$\mathbf{r}=12\mathbf{i}-3\mathbf{j}+6\mathbf{k}+\mu(2\mathbf{i}+4\mathbf{j}-\mathbf{k})$

4 Calculate the value of $\begin{pmatrix}-3\\7\\4\end{pmatrix}\cdot\begin{pmatrix}-2\\5\\-1\end{pmatrix}$

5 For the vectors $\mathbf{a}=2\mathbf{i}-\mathbf{j}+3\mathbf{k}$ and $\mathbf{b}=3\mathbf{i}+4\mathbf{j}-5\mathbf{k}$, calculate

a $\mathbf{a}\cdot\mathbf{b}$

b The acute angle between **a** and **b**

6 Find the obtuse angle between the line $\mathbf{r}=5\mathbf{i}+3\mathbf{j}+2\mathbf{k}+\mathbf{s}(\mathbf{i}+\mathbf{j}-2\mathbf{k})$ and the line $\frac{x-1}{2}=\frac{y+3}{3}=z$

7 Show that the lines $\frac{x-7}{-2}=\frac{y+2}{8}=\frac{z-3}{-4}$ and $\mathbf{r}=\begin{pmatrix}5\\0\\-6\end{pmatrix}+\lambda\begin{pmatrix}-4\\2\\6\end{pmatrix}$ are perpendicular.

8 Calculate the shortest distance from the point (5, -2, 4) to the line with equation $\mathbf{r}=\begin{pmatrix}5\\0\\-1\end{pmatrix}+s\begin{pmatrix}1\\3\\-2\end{pmatrix}$

9 The lines l_1 and l_2 have equations $l_1:\mathbf{r}=(-5\mathbf{i}-\mathbf{j}+2\mathbf{k})+s(3\mathbf{i}-6\mathbf{k})$ and $l_2:\mathbf{r}=(-4\mathbf{i}+3\mathbf{k})+t(-2\mathbf{i}+4\mathbf{k})$.

a Show that l_1 and l_2 are parallel.

b Find the distance between l_1 and l_2

Give your answer to 3 significant figures.

10 Use calculus to find the distance between the lines l_1 and l_2 where l_1 has equation

$\mathbf{r}=\begin{pmatrix}0\\2\\-4\end{pmatrix}+t\begin{pmatrix}1\\5\\-2\end{pmatrix}$

and l_2 is the line through the origin which is parallel to l_1

Give your answer to 3 significant figures.

11 The lines l_1 and l_2 have equations

$\mathbf{r}=\begin{pmatrix}3\\-2\\4\end{pmatrix}+s\begin{pmatrix}2\\1\\-1\end{pmatrix}$ and

$\mathbf{r}=\begin{pmatrix}4\\-1\\5\end{pmatrix}+t\begin{pmatrix}1\\1\\-2\end{pmatrix}$ respectively.

a Show that l_1 and l_2 are skew lines.

a Calculate the minimum distance between l_1 and l_2

6 Exploration

Going beyond the exams

Investigation

A **median** of a triangle is the line joining one of the vertices to the midpoint of the opposite side. The scalar product of vectors can be used to calculate the length of a median.

Consider a triangle defined by the vectors **a**, **b** and **c**, and the median that cuts side **c** is vector **p**

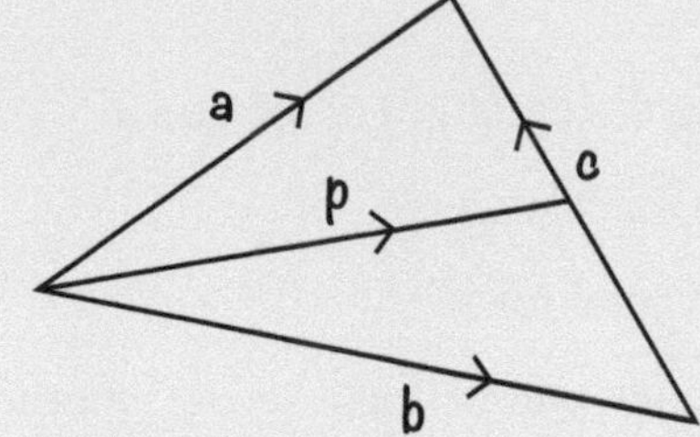

Vector **a** can be expressed as $\mathbf{p} + \frac{\mathbf{c}}{2}$ and

b can be expressed as $\mathbf{p} - \frac{\mathbf{c}}{2}$

Find $\mathbf{a} \cdot \mathbf{a} + \mathbf{b} \cdot \mathbf{b}$ in terms of **p** and **c**

Hence, find an expression for the median in terms of the lengths of the sides of a triangle. Use it to find the lengths of the medians of a triangle with side lengths 3, 4 and 5

Did you know?

The word vector is derived from the Latin word vector, meaning "one who carries or conveys".

Have a go

When **R** is a vector representing a constant resultant force acting on a body which is moved thorough a displacement represented by the vector **r**, the work done is given by $\mathbf{R} \cdot \mathbf{r}$

Using this vector expression, explain why it is a waste of energy to push a heavy object in a direction near perpendicular to the direction in which you wish the body to travel.

To ensure that more than half of a person's efforts are of use, what would you advise about the direction in which they should push?

History

Isaac Newton

The historical development of vectors relies on many underpinning ideas in both pure and applied mathematics.

One important development was that of complex numbers and their geometrical representation. This led mathematicians to wonder how to devise a system that allowed analysis of three-dimensional space.

Alongside such developments, Newton and others were busy considering how to understand and analyse forces and motion. In his famous Principia Mathematica, Newton wrote, "A body, acted on by two forces simultaneously, will describe the diagonal of a parallelogram in the same time as it would describe the sides by those forces separately".

As you will now see, such thinking relating to a parallelogram of forces was very close to what we might now represent by vectors.

6 Assessment

1 a Write down a vector equation of the line that passes through the point with position vector $2\mathbf{i}+4\mathbf{j}-\mathbf{k}$ and is parallel to the vector $\mathbf{i}+\mathbf{j}-2\mathbf{k}$ **[1 mark]**

b Show that the point $(-1, 1, 5)$ lies on the line. **[3]**

2 a Find the vector equation of the line that passes through the points with position vectors $2\mathbf{i}+\mathbf{j}$ and $3\mathbf{i}+\mathbf{k}$ **[2]**

b Show that the point $(8, -5, 7)$ does not lie on the line. **[3]**

3 Find the vector equation of the line that passes through the point with position vector $\begin{pmatrix}2\\1\\0\end{pmatrix}$ and is parallel to the line with equation $\mathbf{r}=\begin{pmatrix}2\\4\\-3\end{pmatrix}+\lambda\begin{pmatrix}8\\-5\\3\end{pmatrix}$ **[2]**

4 The position vectors of points A and B are $\begin{pmatrix}7\\1\\-3\end{pmatrix}$ and $\begin{pmatrix}-2\\8\\-1\end{pmatrix}$ respectively.

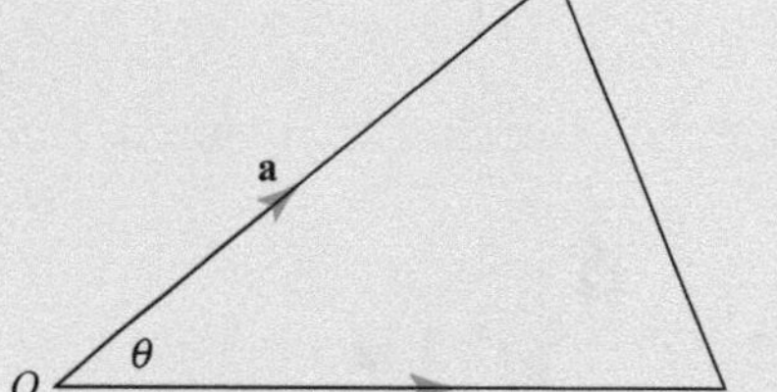

a Calculate $\overrightarrow{OA}\cdot\overrightarrow{OB}$ **[2]**

b Find the size, in degrees, of the acute angle AOB **[3]**

5 Calculate the cosine of the acute angle between the vectors $\begin{pmatrix}4\\3\\5\end{pmatrix}$ and $\begin{pmatrix}-1\\-2\\1\end{pmatrix}$

Give your answer in the form $a\sqrt{3}$ where a is a constant to be found. **[4]**

6 Show that the vectors $-4\mathbf{i}+\mathbf{k}$ and $2\mathbf{i}-\mathbf{j}+8\mathbf{k}$ are perpendicular. **[3]**

7 Given that $\mathbf{a}=\begin{pmatrix}4\\-3\\-7\end{pmatrix}$ and $\mathbf{b}=\begin{pmatrix}-1\\2\\-2\end{pmatrix}$, calculate $\mathbf{a}\cdot\mathbf{b}$ **[2]**

8 Find a vector which is perpendicular to both $\begin{pmatrix}6\\0\\4\end{pmatrix}$ and $\begin{pmatrix}-2\\3\\1\end{pmatrix}$ **[2]**

9 Find the Cartesian equation of this line

$\mathbf{r}=2\mathbf{i}-3\mathbf{k}+s(-\mathbf{i}+4\mathbf{j}+3\mathbf{k})$ **[2]**

10 The point $(2, 7, -3)$ lies on the line with equation $\dfrac{x-a}{2}=\dfrac{y-b}{-3}=\dfrac{z-c}{1}$

a Write down the values of a, b and c **[1]**

b Find the equation of the line in vector form. **[2]**

11 Calculate the perpendicular distance between the lines with equations $\mathbf{r}=3\mathbf{i}-\mathbf{j}+\mu(-2\mathbf{i}+3\mathbf{j}-\mathbf{k})$ and $\mathbf{r}=2\mathbf{i}+\mathbf{j}-\mathbf{k}+\lambda(-2\mathbf{i}+3\mathbf{j}-\mathbf{k})$ **[5]**

12 Is the point $(-2, 4, 7)$ closer to the line with equation $\mathbf{r}=5\mathbf{i}+2\mathbf{k}+\lambda(\mathbf{i}-2\mathbf{j}+4\mathbf{k})$ or the line parallel to the vector $2\mathbf{i}+\mathbf{j}$ which passes through the origin? **[10]**

13 For these pairs of lines, state and justify whether they are skew, parallel or intersecting.

If they are intersecting find their point of intersection.

a $\mathbf{r}=\begin{pmatrix}3\\-1\\5\end{pmatrix}+\lambda\begin{pmatrix}2\\1\\-1\end{pmatrix}$ and $\mathbf{r}=\begin{pmatrix}13\\1\\6\end{pmatrix}+\mu\begin{pmatrix}4\\1\\0\end{pmatrix}$ [5]

b $\mathbf{r}=3\mathbf{i}+\mathbf{j}-\mathbf{k}+s(\mathbf{i}-2\mathbf{j})$ and $\mathbf{r}=5\mathbf{i}+2\mathbf{j}-3\mathbf{k}+t(-2\mathbf{i}+4\mathbf{j})$ [2]

c $\mathbf{r}=\mathbf{i}+2\mathbf{j}+3\mathbf{k}+\lambda(2\mathbf{i}+\mathbf{j}-\mathbf{k})$ and $\mathbf{r}=6\mathbf{i}-\mathbf{k}+\mu(3\mathbf{i}+\mathbf{k})$ [4]

14 a Find the Cartesian equation of the line with vector equation $\mathbf{r}=2\mathbf{i}-3\mathbf{k}+s(-\mathbf{i}+4\mathbf{j}+3\mathbf{k})$ [2]

b Does this represent the same line as $\frac{x-1}{-1}=\frac{y-4}{4}=\frac{z+6}{-3}$? Explain how you know. [3]

15 Which of these equations, if any, represent the same line?

A: $\frac{x-4}{2}=\frac{y-7}{-1}=\frac{z-2}{5}$ B: $\mathbf{r}=\begin{pmatrix}8\\5\\8\end{pmatrix}-\lambda\begin{pmatrix}2\\-1\\5\end{pmatrix}$ C: $\mathbf{r}=\begin{pmatrix}10\\4\\13\end{pmatrix}+\lambda\begin{pmatrix}2\\-1\\5\end{pmatrix}$ [3]

16 Calculate the shortest distance from the point (5, –1, 6) to the line with equation

$\mathbf{r}=\begin{pmatrix}7\\-3\\9\end{pmatrix}+t\begin{pmatrix}2\\1\\-2\end{pmatrix}$ [4]

17 Calculate the perpendicular distance between the lines with equations $\mathbf{r}=3\mathbf{i}-\mathbf{j}+\mu(-2\mathbf{i}+3\mathbf{j}-\mathbf{k})$ and $\mathbf{r}=2\mathbf{i}+\mathbf{j}-\mathbf{k}+\lambda(-2\mathbf{i}+3\mathbf{j}-\mathbf{k})$ [5]

18 The lines l_1 and l_2 have Cartesian equations $\frac{x+1}{2}=\frac{y-4}{7}=\frac{1+z}{-1}$ and $\mathbf{r}=\begin{pmatrix}-5\\4\\1\end{pmatrix}+t\begin{pmatrix}-4\\-2\\2\end{pmatrix}$

a Calculate the acute angle of l_1 and l_2 [4]

b Find the point of intersection of l_1 and l_2 [4]

19 The line l_1 is given by $\mathbf{r}=6\mathbf{i}-3\mathbf{j}+\mathbf{k}+s(-\mathbf{i}+\mathbf{j}-2\mathbf{k})$ and the line l_2 is given by $\frac{x-4}{-5}=\frac{y-2}{2}=\frac{z+5}{-8}$

The point A is the intersection of l_1 and l_2 and the point B has coordinates (1, –2, 7)

Calculate the area of triangle OAB [8]

20 The line l_1 has equation $\mathbf{r}=\lambda(2\mathbf{i}+\mathbf{j}-\mathbf{k})$ and the line l_2 has equation $\mathbf{r}=\mathbf{j}-3\mathbf{k}+\lambda(\mathbf{i}-\mathbf{k})$

a Show that the lines l_1 and l_2 are skew, [3]

b Calculate the perpendicular distance between l_1 and l_2 [7]

The image of the point $A(2, -3, 4)$ when reflected in l_1 is A'.

c Find the coordinates of A' [7]

d Calculate the area of the triangle OAA' [4]

7 Forces and energy

If you've ever ridden a bike down a hill, you'll have noticed that you gain speed without needing to use any energy. Cycling uphill, on the other hand, is hard work and needs a lot of energy to gain any amount of speed and height. The exchange of energy, the work that needs to be done, and the power necessary to do this are all important considerations for vehicle designers who strive for fuel efficiency.

This exchange of energy is a key principle in **regenerative braking,** where kinetic energy is converted to electrical energy, by essentially reversing the drive motor during braking so that it becomes a generator. With electronic technology on the rise, the uses of regenerative braking are becoming ever-more widespread: in all kinds of electric trains, buses and cars.

Orientation

What you need to know

Maths Ch7

- Equations of motion.

Maths Ch8

- Systems of forces.
- Resolving forces.
- Motion under gravity.

Maths Ch4

- Integration.

What you will learn

- To analyse problems using formulae for work, energy and power.
- To use Hooke's law to solve problems involving strings and springs.

What this leads to

Mechanics Ch5

- Simple harmonic motion for strings and springs.
- Architecture.

Careers

- Mechanical engineering.
- Architecture.

7.1 Work, energy and power

Fluency and skills

If you want to move an object, you have to push or pull it and so do some work. How much work you do depends on the force F you exert and the distance x you move the object *in the direction of the force.*

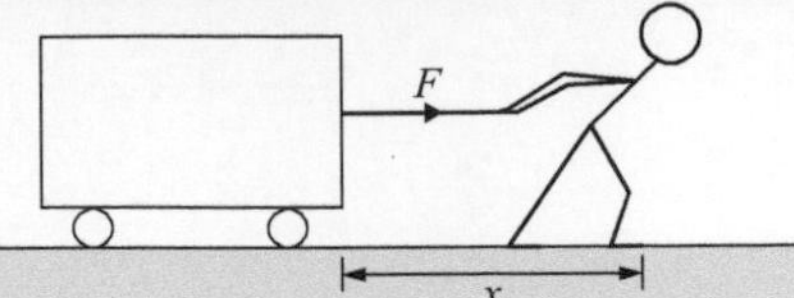

Key point

Work done by a constant force = force × distance

$$= F \times x$$

when F and x are in the same direction.

The units of work are **joules** (J) when force is in N and distance in m.

Joules (J) are the same as N m.

When the force F is not constant, but varies over the distance x moved in the direction of F, the work done by F over a small distance δx is given by $F \times \delta x$

Key point

Summing over the total distance x gives:

the total work done by a variable force $= \int_0^x F\,dx$

Full A Level

A third scenario is when a constant force F acts at an angle θ to the horizontal forward motion. The object does not move vertically, so the vertical component of the force, $F\sin\theta$, does no work. Its horizontal component $F\cos\theta$ moves a distance x in the direction of motion and so does work equal to $F\cos\theta \times x$

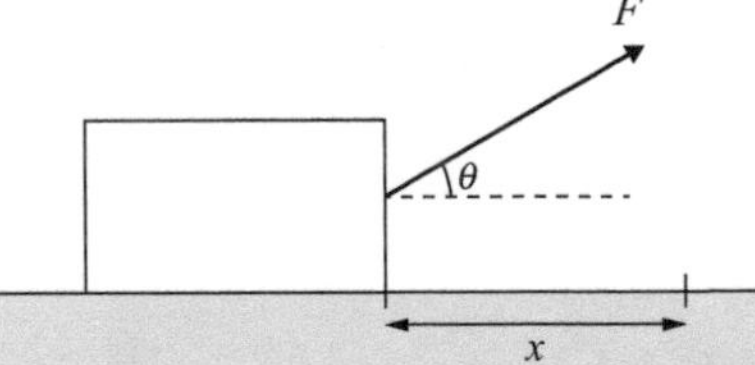

Key point (Full A Level)

Work done by a constant force F is $Fx\cos\theta$ when the angle between F and x is θ.

Example 1

A force F pulls a trolley a horizontal distance x. Find the work done by F if

a F acts horizontally over 4 metres with a constant magnitude of 10 N

b F acts horizontally and varies so that $F = (10 - 2x)$ N for $0 \le x \le 4$ metres

c F acts at 60° to the horizontal over 4 metres with a constant magnitude of 10 N (Full A Level)

(*Continued on next page*)

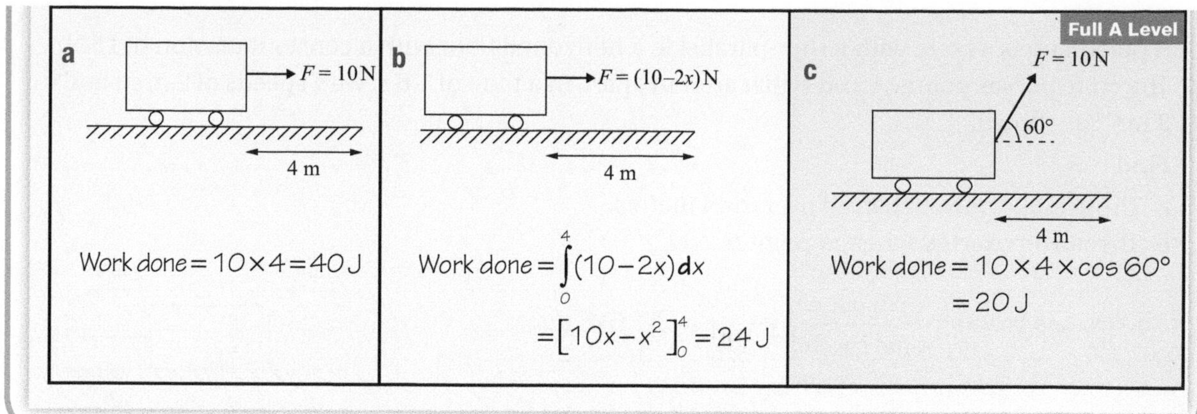

MECH

If you drag an object along a rough floor for a distance x, four forces are acting on the object.

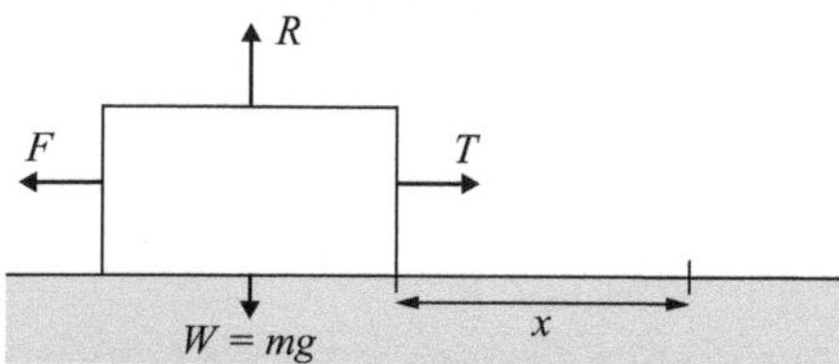

The letter F is often used for any general force and also for a frictional force.

The letter R is often used for a reaction and also for a resisting force.

The weight W and the reaction R between object and floor do no work because they act at right-angles to the motion. The object is dragged by a force T (for example, the tension in a rope or the tractive force of an engine) against a frictional force F. These forces act in the same straight line as the motion of the object, so the work done *by* the dragging force over distance x is $T \times x$, and the work done *against* the frictional force is $F \times x$ (if T and F are constant). Use the word *by* when the force does work and the word *against* when work is done against a resisting force such as friction.

The pulling force of an engine is usually called the **tractive force** or **driving force**. It is often represented by the letter T or P

Power is the rate at which work is done.

If an engine is set to work for a period of time, then the average power of the engine is equal to $\frac{\text{work done}}{\text{time taken}}$

The units of power are **watts** (W) when work is in J and time in s.

If a constant force F N pulls an object in its direction at a constant speed v m s^{-1}, the object moves a distance of v metres every second. For example, consider a train where an engine is pulling coaches at a steady speed. The work done every second by the engine is $F \times v$ joules, so the power of the engine is $F \times v$ watts.

Key point

The power of a constant force F moving at a steady speed v is equal to $F \times v$, where power is in W, force in N and speed in m s^{-1}

If F or v are variable, $F \times v$ gives the power at that particular instant.

Example 2

A person pulls a crate with a rope parallel to a horizontal floor with a constant tension of 15 N. The crate passes points A and B that are 4 m apart, in a time of 1.6 s, with speeds of $2\ \text{m s}^{-1}$ and $3\ \text{m s}^{-1}$ as shown.

Find

a The average power exerted over the 4 metres,

b The power exerted when at point B

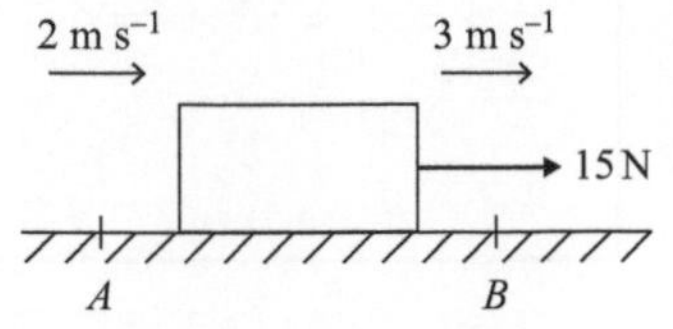

a Average power $= \dfrac{\text{work done}}{\text{time taken}} = \dfrac{15 \times 4}{1.6} = 37.5\ \text{W}$

b Power exerted at $B = F \times v = 15 \times 3 = 45\ \text{W}$.

Example 3

A metal ingot of weight 80 N is pulled at a constant speed of $0.2\ \text{m s}^{-1}$ for 3 metres across a rough horizontal floor by a rope parallel to the floor. There is a constant frictional force F of 30 N. Find

a The work done by each force acting on the ingot

b The power exerted by the tension T in the rope.

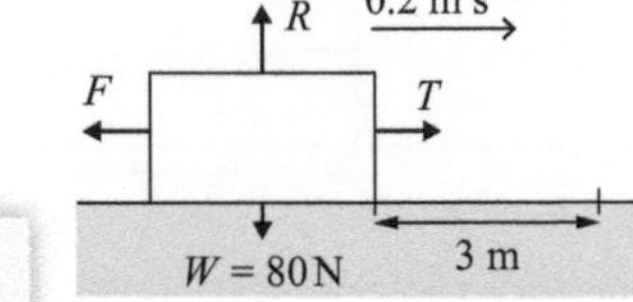

There is no vertical acceleration because there is no vertical motion, and there is no horizontal acceleration because the horizontal motion is at constant speed. So there is no resultant force, and hence,

resolving vertically, reaction $R =$ weight $W = 80\ \text{N}$

resolving horizontally, tension $T =$ friction $F = 30\ \text{N}$

a There is no vertical distance moved, so the work done by R and W is zero.

The horizontal distance moved $= 3\ \text{m}$

The work done *by* tension $T = 30 \times 3 = 90\ \text{N}$

The work done *against* friction $F = 30 \times 3 = 90\ \text{N}$

b Power generated by tension $= T \times v = 30 \times 0.2 = 6\ \text{W}$

Exercise 7.1A Fluency and skills

1 A constant force F pulls an object a distance x in its own direction. Find the work done by the force when

  a $F = 20\ \text{N}, x = 4\ \text{m}$

  b $F = 50\ \text{N}, x = 12\ \text{m}$

  c $F = 0.3\ \text{kN}, x = 0.5\ \text{km}$

2 Each of the three contestants A, B and C, in a strong-man competition drags a weight a distance x along level ground by exerting a constant tension T in a rope which is parallel to the ground. Use this data to find which man does the most work.

	A	B	C
Tension, T	500 N	300 N	800 N
Distance, x	20 m	35 m	15 m

3 For the strong men in question 2, A completed his task in a time of 20 s, B in 25 s and C in 30s. Using the relationship $\text{power} = \frac{\text{work done}}{\text{time taken}}$, find which man exerted the most power.

4 A car's engine exerts a constant driving force F of 800 N as it increases its speed from 20 m s^{-1} to 30 m s^{-1}. Using the relationship $power = F \times v$, find the power exerted by the engine at the start and end of its motion.

5 A cyclist pedals his bike to create a constant driving force of 50 N as he climbs a hill, starting with a speed of 14 m s^{-1} and finishing with a speed of 3 m s^{-1}

What power is he exerting at the start and end of his climb?

6 Two cranes, P and Q, each lift loads of 2000 N to a height of 25 m. P takes 10 s and Q takes 15 s. How much work does each crane do and what is the rate of working (that is, the power) of each crane?

7 A car engine works at a rate of 20 kW. Find its driving force if it travelling at

a 20 m s^{-1} b 120 km h^{-1}

8 Two equally powerful athletes start a race at the same time. They finish the race with speeds of 10 m s^{-1} and 15 m s^{-1}. If their final power output is 300 W, what tractive force is each then producing?

9 Find the work done and power required when

a A crate is moved steadily for 6 m horizontally in 4 s by a horizontal rope under a tension of 120 N

b A spring balance which reads 20 N steadily lifts a suitcase 3 m upwards in 1.5 s.

10 A winch uses a rope to lift three boxes A, B and C vertically, one after the other, at constant speeds for distances and times as given in the table. For each box find, in terms of g,

a The constant tension T in the rope

b The work done by the winch

c The power exerted by the winch.

Box	A	B	C
Mass, kg	12	4	0.5
Distance, m	3	0.5	0.2
Time, s	4	4	2

11 A variable horizontal force F N pulls an object along a horizontal table through a distance x of 4 m. Find the work done by the force if

a $F = 8 - 2x$ b $F = 16 - x^2$

c $F = x^2 - x + 2$ d $F = \frac{1}{2}x + 1$

12 A horizontal rope pulls a trolley 6 m along a horizontal floor. A graph is drawn to show how the tension T in the rope varies with distance. Calculate the work done by the tension in each case

a
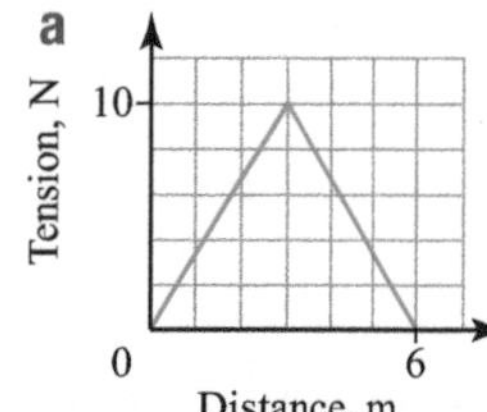

b
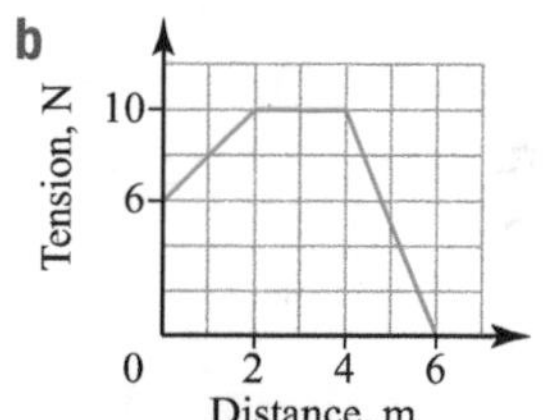

MECH

Full A Level

13 A boy pulls a block on a level track for a distance x by a string inclined at an angle θ to the horizontal. If the tension T in the string is constant, find the work done by the boy in each of these three cases:

	Tension, T	Distance, x	Angle, θ
a	10 N	8 m	60°
b	5 N	4 m	45°
c	0.2 N	0.6 m	30°

14 The three blocks in question 13 were initially at rest when the boy began to pull them. Find the power exerted by the boy in each case at the instant when the speed is

a 2 m s^{-1} b 5 m s^{-1} c 4 m s^{-1}

15 A mass at the point $(x, 0)$ moves along the x-axis of a horizontal x-y plane under a force $F = (2x + 1)$ N which acts in the plane at 30° to the x-axis. Find the power supplied by force F if the speed of the mass is 5 m s^{-1} when $x = 4$ m.

Reasoning and problem-solving

Strategy

To solve problems involving work, energy and power

1. Draw a clearly labelled diagram to show the information you are given.
2. When writing an energy equation, ensure you have the gains and losses balanced correctly.

The **energy** of a body is its ability to do work. There are different kinds of energy such as electrical, nuclear and heat. In this chapter you will study two kinds: **kinetic energy** (due to a body's motion) and **potential energy** (due to a body's position).

When you apply a force F to a body of mass m which is initially at rest, it moves with acceleration a. After travelling a distance s, it has velocity v

Newton's second law gives $F = ma$

The kinematic equation $v^2 = u^2 + 2as$ gives $v^2 = 2as$ because $u = 0$

So the work done by the force F is $F \times s = ma \times \frac{v^2}{2a} = \frac{1}{2}mv^2$

The energy of the body due to the work done to give it this velocity v is called its kinetic energy.

Key point

The **kinetic energy** (KE) of a body with mass m and velocity v is equal to $\frac{1}{2}mv^2$

When you release a mass m from height h, it falls downwards freely as a result of the gravitational force F which is equal to its weight mg

Newton's second law gives $F = ma$, so $mg = ma$, and a is the acceleration due to gravity, g

So the work done by F is $F \times h = mgh$

Before being released, the mass had energy with the potential to do this work. The energy it possessed due to its height is called its gravitational potential energy.

Key point

The **gravitational potential energy** (GPE) of a body with mass m at height h is equal to mgh

The **principle of conservation of mechanical energy** says that the total mechanical energy (the sum of KE and GPE) of a system remains constant, provided no work or energy is lost to friction or impacts.

For example, when an object falls freely under gravity, the total energy is constant because the loss in GPE when falling equals its gain in KE, provided no energy is lost to air resistance. However, if energy is lost, you would have, for this example, the energy equation:

Loss of GPE = Gain in KE + Work done against air resistance

Example 4

An 4 kg object, which is initially at rest, falls 12 m freely under gravity to hit the ground. Take $g = 9.81 \text{ m s}^{-2}$

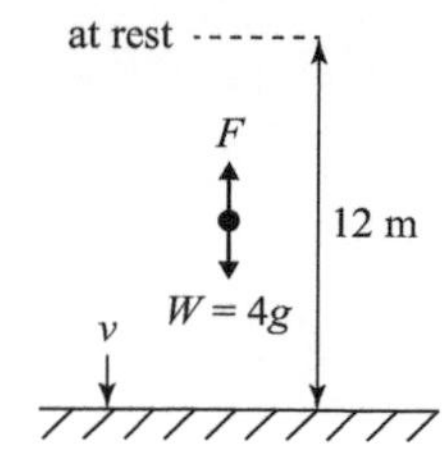

a Find its final speed v if

i Air resistance is neglected

ii Air resistance is modelled by a constant force, F of 5 N.

b Describe a limitation to this model and explain how it might be made more realistic.

a i The energy equation is GPE lost = KE gained

$$4 \times g \times 12 = \frac{1}{2} \times 4 \times v^2$$

Final velocity, $v = \sqrt{24g} = 15.3 \text{ m s}^{-1}$

ii The energy equation is GPE lost = KE gained + Work done against resistance

$$4 \times g \times 12 = \left(\frac{1}{2} \times 4 \times v^2\right) + (5 \times 12)$$

$$2v^2 = 48g - 60 = 410.9$$

Final velocity, $v = \sqrt{205.4} = 14.3 \text{ m s}^{-1}$

b It is common experience that air resistance increases with speed. The model is limited by having a constant resistance, F
A more realistic model would have F dependant on speed.

Note that part **a i** of Example 4 could also be easily solved using the kinematic equation $v^2 = u^2 + 2as$ with $a = g$. In part **b**, the acceleration, a could be found first using Newton's 2nd law. (To better understand how to use Newton's 2nd law in this way, see Example 6.) In problems where acceleration is not constant, the energy equation is likely to be the easier method.

Example 5

A car of mass 800 kg starts from rest at the foot of a slope with a constant tractive force, T N. It climbs the slope, travelling 500 m in 50 s to reach a speed of 20 m s^{-1} as it rises a vertical height of 4 m.

Stating your assumptions and taking $g = 10 \text{ m s}^{-2}$, calculate

a The gain in its GPE and KE

b The tractive force, T

c The average power of the engine over the whole journey

d The power of the engine at the end of the journey.

(*Continued on next page*)

MECH

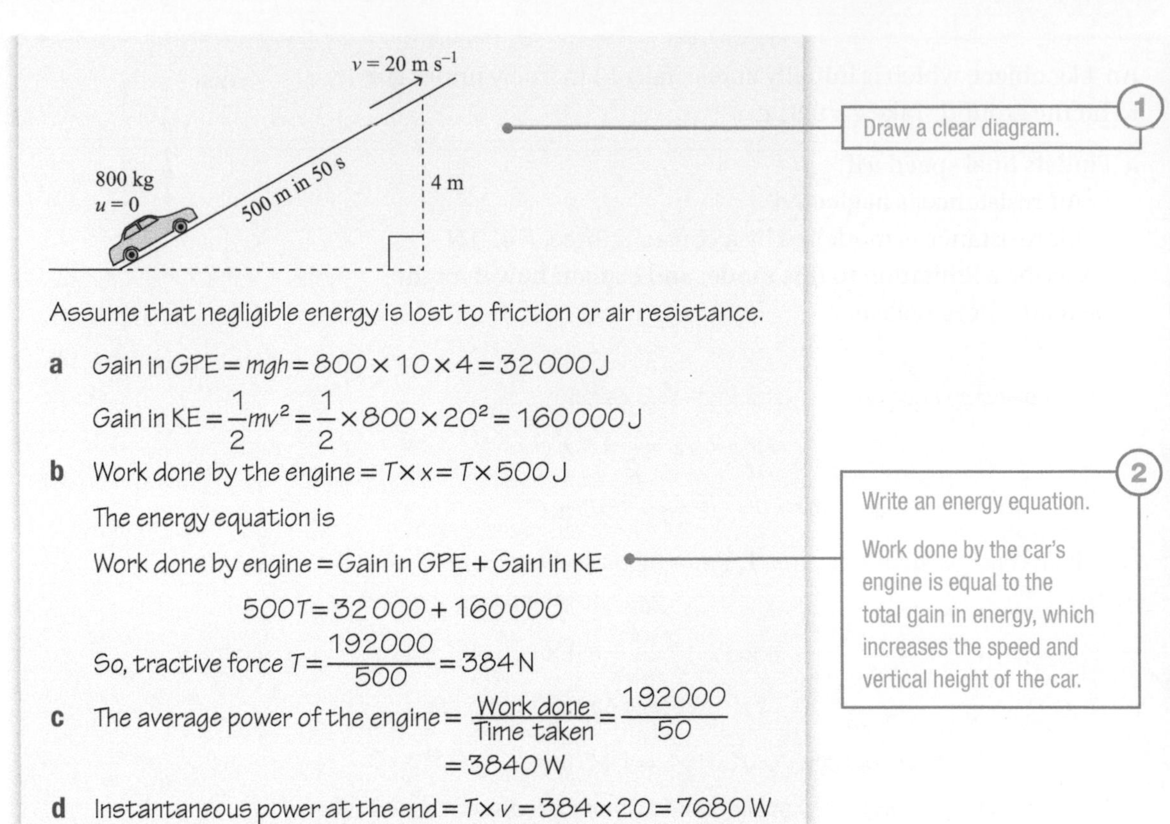

Assume that negligible energy is lost to friction or air resistance.

a Gain in GPE $= mgh = 800 \times 10 \times 4 = 32\,000$ J

Gain in KE $= \frac{1}{2}mv^2 = \frac{1}{2} \times 800 \times 20^2 = 160\,000$ J

b Work done by the engine $= T \times x = T \times 500$ J

The energy equation is

Work done by engine = Gain in GPE + Gain in KE

$$500T = 32\,000 + 160\,000$$

So, tractive force $T = \frac{192000}{500} = 384$ N

c The average power of the engine $= \frac{\text{Work done}}{\text{Time taken}} = \frac{192000}{50}$

$= 3840$ W

d Instantaneous power at the end $= T \times v = 384 \times 20 = 7680$ W

1 Draw a clear diagram.

2 Write an energy equation.

Work done by the car's engine is equal to the total gain in energy, which increases the speed and vertical height of the car.

Example 6

A car of mass 1 tonne travels from rest up a slope at 30° to the horizontal with a constant acceleration against a constant resistance R of 400 N. On reaching the top of the slope, it has a speed of 10 m s^{-1} and its engine is working at a rate of 58 kW. Calculate the length of the slope. Take g = 9.8 m s^{-1}

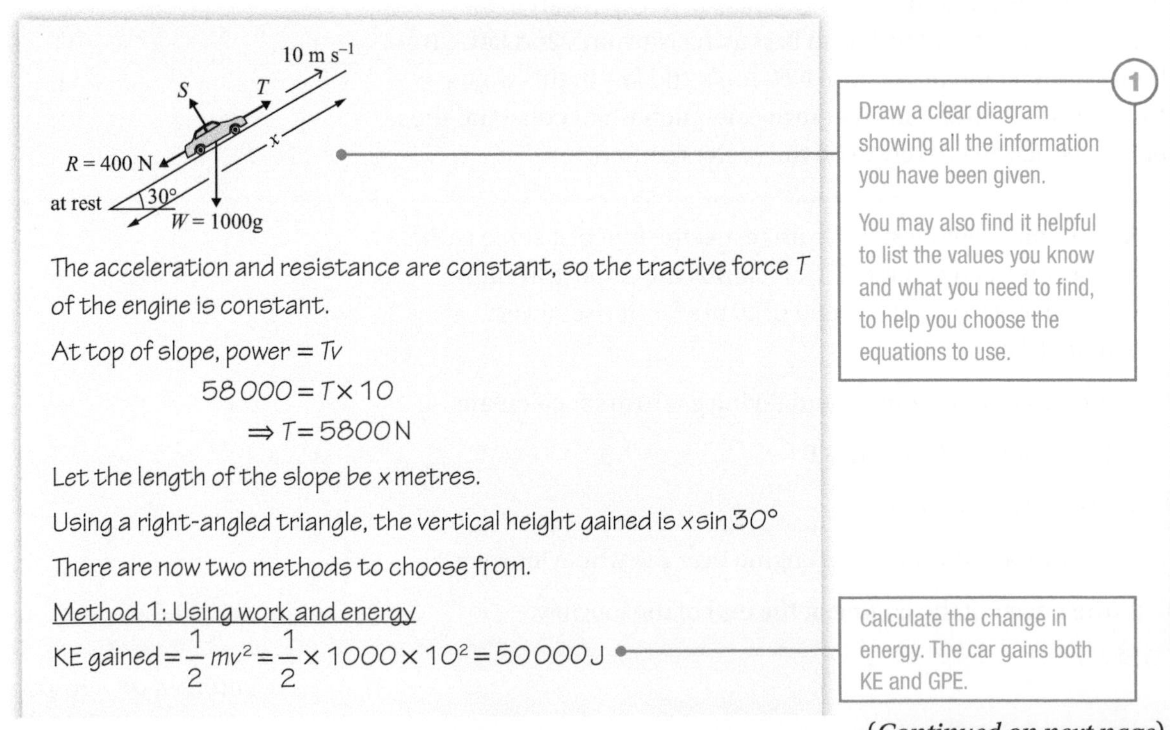

The acceleration and resistance are constant, so the tractive force T of the engine is constant.

At top of slope, power $= Tv$

$$58\,000 = T \times 10$$

$$\Rightarrow T = 5800 \text{ N}$$

Let the length of the slope be x metres.

Using a right-angled triangle, the vertical height gained is $x \sin 30°$

There are now two methods to choose from.

Method 1: Using work and energy

KE gained $= \frac{1}{2}mv^2 = \frac{1}{2} \times 1000 \times 10^2 = 50\,000$ J

1 Draw a clear diagram showing all the information you have been given.

You may also find it helpful to list the values you know and what you need to find, to help you choose the equations to use.

Calculate the change in energy. The car gains both KE and GPE.

(*Continued on next page*)

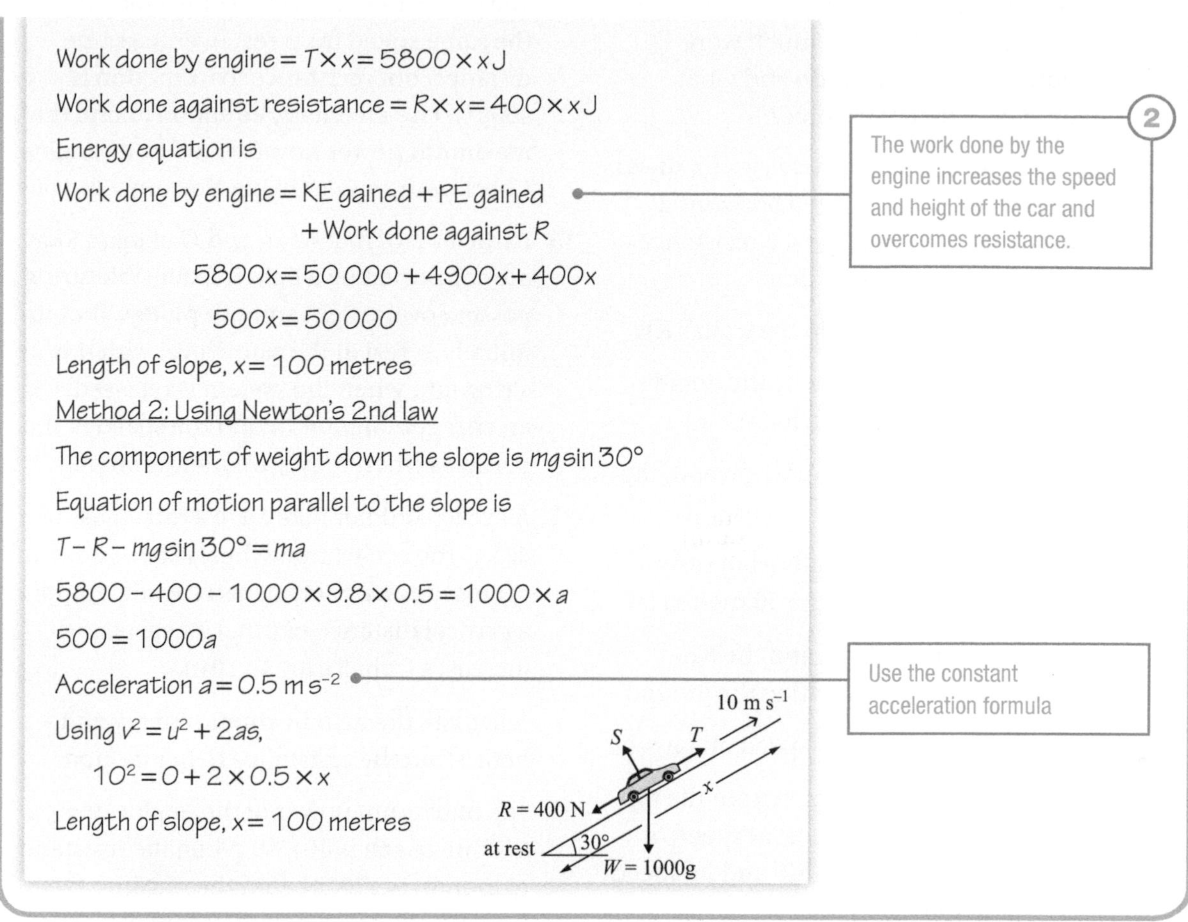

Work done by engine $= T \times x = 5800 \times x$ J

Work done against resistance $= R \times x = 400 \times x$ J

Energy equation is

Work done by engine = KE gained + PE gained + Work done against R

$$5800x = 50\,000 + 4900x + 400x$$

$$500x = 50\,000$$

Length of slope, $x = 100$ metres

Method 2: Using Newton's 2nd law

The component of weight down the slope is $mg\sin 30°$

Equation of motion parallel to the slope is

$$T - R - mg\sin 30° = ma$$

$$5800 - 400 - 1000 \times 9.8 \times 0.5 = 1000 \times a$$

$$500 = 1000a$$

Acceleration $a = 0.5$ m s^{-2}

Using $v^2 = u^2 + 2as$,

$$10^2 = 0 + 2 \times 0.5 \times x$$

Length of slope, $x = 100$ metres

2 The work done by the engine increases the speed and height of the car and overcomes resistance.

Use the constant acceleration formula

MECH

Exercise 7.1B Reasoning and problem-solving

1 A ball of mass 0.5 kg drops 12 m vertically from rest to hit the ground. Use an energy equation to find its speed on impact if

   a Air resistance is negligible,

   b Air resistance is a constant force of magnitude 4 N.

2 A rocket of mass 5 kg is fired vertically upwards with an initial speed of 24 m s^{-1}

   Use an energy equation to find its maximum height if

   a Air resistance is negligible,

   b Air resistance is a constant 4 N.

3 A 5 kg box with an initial speed of 10 m s^{-1} moves 4 m across a rough horizontal floor under a constant frictional force F and comes to rest. Find the value of F

4 A person pulls a crate for 6 m on a horizontal floor using a horizontal rope with a constant tension of 12 N. The motion is resisted by a constant frictional force F of 4 N. The crate's final velocity is 2 m s^{-1}

   Find

   a The work done by the tension,

   b The work done against friction,

   c The final power exerted by the person at the end of the motion.

5 a A lorry of mass 2 tonnes accelerates from rest to a speed of 10 m s^{-1} over 400 m on a horizontal road with negligible resistance. Find

   i The gain in its KE,

   ii The work done by the engine,

   iii The engine's tractive force.

b If the lorry had experienced a constant resistance of 50 N, how much work would the engine now do and what would be the tractive force?

6 A train of mass 50 tonnes increases its speed from $10\,\text{m s}^{-1}$ to $20\,\text{m s}^{-1}$ over a horizontal distance of 500 metres against a resistance of 2000 N due to wind and friction.

a Calculate the increase in the train's KE.

b Calculate the work done by the engine and the engine's tractive force.

c What is the initial and final power of the engine?

7 a A brick of mass 1.2 kg falls from rest from the top of a building 30 metres tall.

i What is the GPE lost and the KE gained just before hitting the ground?

ii Calculate the brick's speed on impact.

b If the brick is thrown down from the top of the building with an initial speed of $14\,\text{m s}^{-1}$, what is its total KE and its speed just before impact?

c If the brick is thrown upwards from the top of the building with a speed of $14\,\text{m s}^{-1}$, what is its speed on impact with the ground?

d If the brick experiences a constant resistance of 3 N in falling from rest, what it its velocity when striking the ground?

8 A train of mass 100 tonnes accelerates steadily from rest over 450 metres horizontally to reach a speed of $15\,\text{m s}^{-1}$

Resistance is negligible.

a Use an energy equation to find the maximum power exerted by the train's engine.

b Find the acceleration of the train and use Newton's 2nd law to check your answer.

9 The same train as in Question 8 reaches the same speed from rest over the same distance, but resistance to its motion is 4000 N. Use an energy equation to find the maximum power now exerted by its engine. Check your answer using Newton's 2nd law.

10 Particles P of mass 2 kg and Q of mass 5 kg are connected by a light inextensible string passing over a light smooth pulley. They are initially at rest at the same level with the string taut when the system is released. Use an energy equation to find the speed of the particles when they are 3 metres apart.

11 A cyclist and her bike have a total mass of 80 kg. The cyclist free-wheels 200 m downhill against a constant resistance while dropping a vertical distance of 8 m. Her speed increases from $3\,\text{m s}^{-1}$ to $10\,\text{m s}^{-1}$

Calculate the gain in kinetic energy and hence find the resistance to her motion.

12 The maximum power of the engine of a 5 tonne coach is 40 kW. When the resistance to motion is 1500 N, find the maximum speed it can reach on a horizontal road.

13 A fountain uses a pump to raise 3000 litres of water through 4 metres every minute and expels it at a speed of $8\,\text{m s}^{-1}$

Find the power of the pump.

14 Water is discharged at $6\,\text{m s}^{-1}$ through a circular nozzle of radius 4 cm, having been raised by 2 m. Find the power of the pump to the nearest watt.

15 A train of mass 1200 kg with velocity $v\,\text{ms}^{-1}$ is resisted by a force of $600 + v^2$ N. The power of its engine is $1.5v$ kW. When on level ground, calculate

a The train's maximum speed,

b Its acceleration when moving at $10\,\text{m s}^{-1}$

16 The total mass of an engine and its train is 450 000 kg. The resistance to their motion is $\frac{v^2}{4}$ N per 1000 kg at a speed of v m s^{-1}

If the maximum power of the engine is 1125 kW, find its greatest speed on horizontal ground.

17 A cyclist on a level road has a power output of 75 W when travelling at a constant speed v m s^{-1} against a resistive force $R = kv^2$ N. Calculate the value of the constant k if the cyclist is traveling at 24 km h^{-1}

18 The resistance R to the motion of a car is directly proportional to its speed v, such that $R = kv$

a A car of mass 1500 kg has a maximum speed of 45 m s^{-1} on level ground when its engine has a power output of 8 kW. Find the value of k

b Calculate the car's acceleration when it moves at 20 m s^{-1} on a level road if the power of its engine is then 6 kW.

19 An object of mass 8 kg is dragged along a rough horizontal floor from rest for a distance s m by a variable force T where $T = 36 - s^2$ N. Find its speed and the power exerted by the force T after the object has moved 4 m, given that the magnitude of the frictional force is a quarter that of the normal reaction to the floor.

20 A car of mass 1.2 tonnes travels up a slope at 30° to the horizontal at a constant speed of 25 m s^{-1} for 300 metres.

a If the slope is smooth, calculate

i The increase in the car's energy,

ii The tractive force of its engine,

iii Its power output

b If the slope is rough and the car experiences a constant frictional force down the slope of 3000 N, calculate the tractive force and power of its engine.

21

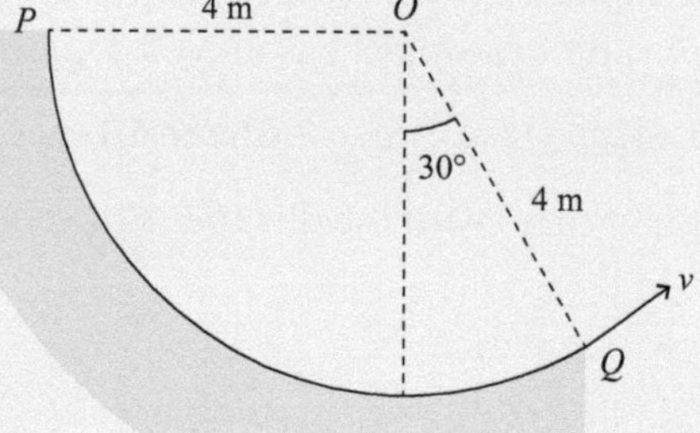

A skateboarder of mass 60 kg begins at rest at point P. He travels down a ramp shaped as an arc of a circle, centre O, with radius 4 m. Calculate his speed v when he leaves the ramp at Q when there is

a No resistance to motion,

b A constant resistance to motion of 48 N.

22 For the coach described in question 12, what is its maximum speed up a slope which makes an angle of $\sin^{-1}\frac{1}{98}$ to the horizontal?

23 A car of mass 900 kg pulls a trailer of mass 200 kg and the resistances to their motion are constant at 200 N and 80 N respectively. When travelling on a horizontal road at a maximum speed of $28\,\text{m s}^{-1}$, the car's engine is working at maximum power. The car and trailer travel with speed $8\,\text{m s}^{-1}$ at full power up a hill which makes an angle of $\arcsin\frac{1}{40}$ to the horizontal.

Calculate

a The acceleration,

b The tension in the tow-bar.

24 A smooth ring of mass m is threaded on to a fixed vertical circular hoop with centre O and radius r

The ring is projected with velocity u from its lowest point.

a Show that if the ring just reaches the highest point Q, then $u^2 = 4gr$

b If the ring passes a point P (where the acute angle $QOP = \theta$) with velocity $\frac{1}{2}u$, prove that $\cos\theta = \frac{3u^2}{8gr} - 1$ and that $\frac{8gr}{3} < u^2 < \frac{16gr}{3}$

25 For the cyclist in question 17 above, calculate the steepest gradient she can climb at $12\,\text{km h}^{-1}$ when working at the same rate, given that her total mass is 110 kg.

26 A racing car of mass 1500 kg, travelling at speed $v\,\text{m s}^{-1}$, experiences a resistance $R = 36v\,\text{N}$. The car travels down a slope inclined at 2° to the horizontal. If the maximum power of its engine is 120 kW, find the maximum speed of the car down the slope.

27 The object in question 19 is now dragged along a floor that is tilted upwards at 5° to the horizontal. All other given data stays the same. Calculate the speed of the object up the slope and the power exerted by the dragging force after the object has moved 4 metres.

28 A train of mass 240 000 kg is pulled from rest up a slope of $\arcsin\frac{1}{250}$. The resistance to motion is $2000g\,\text{N}$. The tractive force T of its engine varies with the distance s travelled, as in this table. Take $g = 9.8\,\text{m s}^{-2}$

s (m)	0	25	50	75	100	125	150
T ($\times 10^4$ N)	10	13	12	11	9	7	6

a Use a numerical method to calculate an approximate value for the work done in pulling the train 150 m up the slope.

b Calculate the speed of the train after it has travelled 100 m and the power of its engine at this point to the nearest watt.

Full A Level

7.2 Hooke's law

Fluency and skills

In many situations, it is not sensible to model **strings** as inextensible. You are now going to adopt a new model which assumes that strings are **light** and **elastic** and that they recover their **natural length** after being stretched.

A **spring** is also modelled as light and elastic. As well as recovering its natural length after being stretched, it also recovers it after being compressed. A string cannot be compressed; it simply goes slack.

In reality, if the stretch goes beyond the **elastic limit**, the string or spring is permanently deformed and Hooke's law does not apply.

For a string or stretched spring, the difference between the stretched length l' and the natural length l is the **extension** x
For a compressed spring, the difference in length is called a **compression**, x

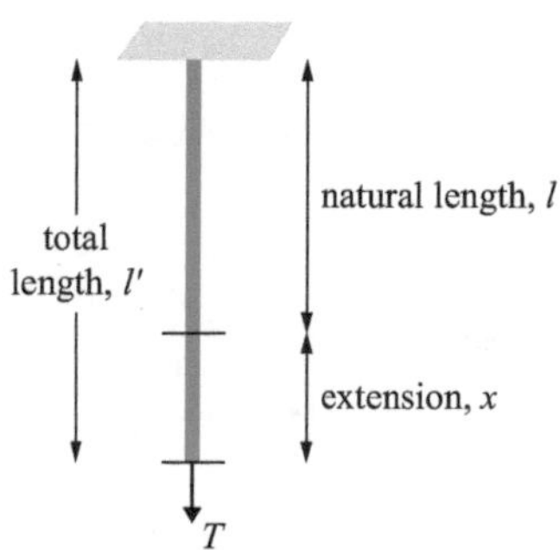

Hooke's law states that the **tension** T in an elastic string or spring is proportional to its extension x

$T \propto x$ where $x = l' - l$

So, $T = k \times x$ where the constant k is the **stiffness** of the string or spring.

Stiffness depends on the material from which the string or string is made and also on its natural length. Stiffness k is equal to $\frac{\lambda}{l}$ where the constant λ is called the **modulus of elasticity**.

The units of λ are newtons.

Key point

Hooke's law gives $T = \frac{\lambda}{l}x$ where $x = l' - l$

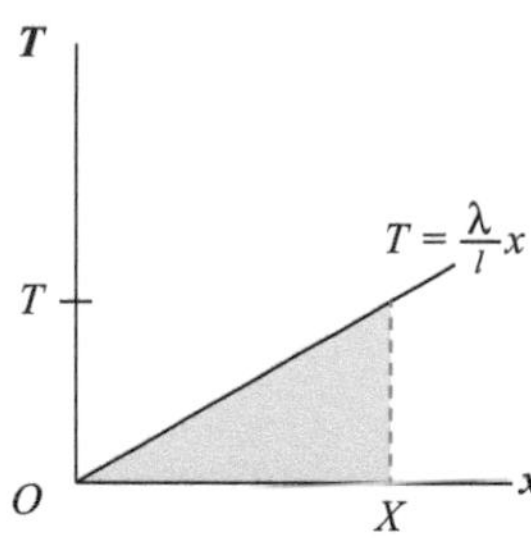

To calculate the work required to create an extension X from its natural length, imagine a small extension δx when the tension is T

The small amount of work required is force × distance = $T \times \delta x$
To calculate the total work needed, integrate for the whole extension.

There are two methods.

Using calculus

$$\text{Work required} = \int_0^X T\,dx$$
$$= \frac{\lambda}{l}\int_0^X x\,dx = \frac{\lambda}{l}\left[\frac{x^2}{2}\right]_0^X$$
$$= \frac{1}{2} \times \frac{\lambda}{l} \times X^2 = \frac{\lambda}{2l}X^2$$

Using a graph

$$\text{Work required} = \int_0^X T\,dx$$
$$= \text{Area of shaded triangle}$$
$$= \frac{1}{2} \times X \times T = \frac{1}{2} \times X \times \frac{\lambda}{l} \times X$$
$$= \frac{\lambda}{2l}X^2$$

The work done to stretch a string or spring is stored as **elastic potential energy** (EPE) in the string or spring.

Key point

The elastic potential energy, EPE, stored in a string or spring when its natural length l is extended by a distance x is

$$\text{EPE} = \frac{\lambda}{2l}x^2$$

This formula has the same shape as $\frac{1}{2}mv^2$ for kinetic energy.

To increase the extension from x_1 to x_2, the energy required is

$$\frac{\lambda}{2l}x_2^2 - \frac{\lambda}{2l}x_1^2 = \frac{\lambda}{2l}(x_2^2 - x_1^2) = \frac{\lambda}{l}(x_2 + x_1)(x_2 - x_1)$$

Example 1

A string with natural length 5 m and modulus of elasticity 10 N is extended 0.4 m.

a What is the stiffness of the string?

b Calculate the tension in the string and the EPE stored in it.

c If the string is extended a further 0.3 m, how much extra EPE is stored in it?

a Stiffness $= \frac{\lambda}{L} = \frac{10}{5} = 2\text{Nm}^{-1}$

b From Hooke's Law, tension $T = \frac{\lambda}{l}x = \frac{10}{5} \times 0.4 = 0.8\text{ N}$

EPE stored $= \frac{\lambda}{2l}x^2 = \frac{10}{2\times 5} \times 0.4^2 = 0.16\text{ J}$

c Increase in EPE $= \frac{\lambda}{2l}(x_2^2 - x_1^2) = \frac{10}{2\times 5} \times (0.7^2 - 0.4^2) = 0.33\text{ J}$

Example 2

Two strings OA and OB are tied to an object O which is held on a smooth table between two fixed points A and B which are 1.0 m apart.

OA has natural length 0.4 m and modulus of elasticity 12 N.
OB has natural length 0.3 m and modulus of elasticity 18 N.

Find the extensions of the strings and their total EPE.

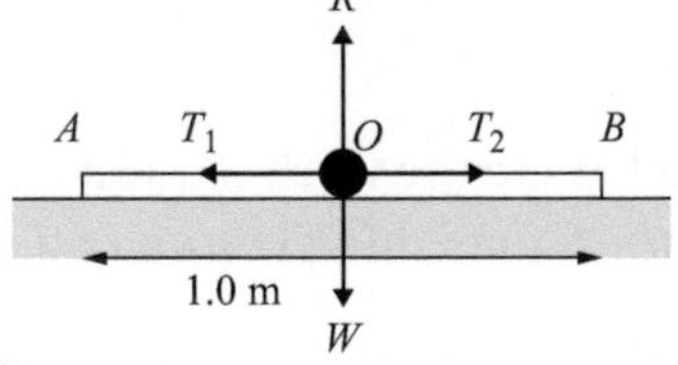

For vertical equilibrium, reaction R = weight W

For horizontal equilibrium, tension T_1 = tension T_2

The object is held in equilibrium, so the two horizontal forces balance.

Hooke's law gives:

for OA, $T_1 = \frac{\lambda_1}{l_1}x_1 = \frac{12}{0.4}x_1 = 30x_1$

for OB, $T_2 = \frac{\lambda_2}{l_2}x_2 = \frac{18}{0.3}x_2 = 60x_2$

Use Hooke's law for OA and OB

(*Continued on next page*)

So $30x_1 = 60x_2$ or $x_1 = 2x_2$ (1)

Use $T_1 = T_2$

Distance $AB = 1.0 = 0.4 + 0.3 + x_1 + x_2$

$0.3 = 2x_2 + x_2$ (2)

The extensions are $x_2 = 0.1$ m and $x_1 = 0.2$ m

Solve the simultaneous equations (1) and (2) to calculate x_1 and x_2

Total EPE $= \frac{\lambda_1}{2l_1}x_1^2 + \frac{\lambda_2}{2l_2}x_2^2$

$= \frac{12}{2\times0.4}\times0.2^2 + \frac{18}{2\times0.3}0.1^2 = 0.6 + 0.3 = 0.9$ J

MECH

Exercise 7.2A Fluency and skills

1 Five different strings N to E of natural length l m and modulus of elasticity λN are extended by x m

a Copy and complete this table, where T is the tension in each string and EPE is the elastic potential energy stored in the string.

	l, m	x, m	λ, N	T, N	EPE, J
A	2	0.5	8		
B	3	1	6		
C	1.5		15	4	
D	1.2	0.6		4.5	
E	2.5		10		1.28

b Find the stiffness of each string A to E.

2 A force of 15 N extends an elastic string by 0.2 metres. What will be the extension when the force is 30 N?

3 A force of 20 N compresses a spring by 5 cm. How far will the spring be compressed under a force of 10 N?

4 An light elastic string of natural length 2 m with a modulus of elasticity of 50 N is extended by a force of 20 N.

Calculate

a Its extension,

b The work done by the force,

c The elastic potential energy stored in the extended string.

5 A spring of natural length 1.2 m with $\lambda = 25$ N, where λ is the modulus of elasticity, is compressed by a force of 15 N.

Calculate

a The length of the compressed string,

b The spring's elastic potential energy when compressed.

6 An elastic spring of natural length 1.5 m hangs from a ceiling with an object of mass 8 kg attached at its lower end. If the modulus of elasticity, λ, is 120 N and the object hangs in equilibrium, calculate

a The tension in the spring,

b The extension of the spring.

7 Two fixed points P and Q are 3 m apart on a smooth horizontal table. Two horizontal strings PO and QO are attached to an object O which sits between P and Q

The natural length and value of λ are 0.9 m and 24 N respectively for PO, and are 0.6 m and 32 N respectively for QO Calculate the extensions of the two strings and the tensions in them when O is in equilibrium.

8 Two springs OX and OY are attached to an object O which is in equilibrium on a smooth horizontal table between two fixed points X and Y which are on the table and lie 1 m apart. OX has natural length 0.8 m and $\lambda = 15$ N. OY has natural length 1.2 m and $\lambda = 18$ N. Calculate the compressed lengths of the two springs and their total EPE.

9 Two light vertical springs AB and BC are joined at B with end A fixed to a horizontal table, and with C directly above B

A mass of 5 kg, which is attached to C, compresses both springs, and rests in equilibrium. The natural length and modulus of elasticity of AB are 0.4 m and 90 N, and those of BC are 0.2 m and 80 N. Find the distance between A and B and the total EPE stored in the springs.

10 Two vertical strings are each attached to a ceiling by one end so that they hang down. Together they hold a mass m in equilibrium at their two lower ends. They have the same natural length l and their moduli of elasticity are λ_1 and λ_2 respectively. Show that the extension of each string is $\frac{mgl}{\lambda_1 + \lambda_2}$ and calculate the tension in each string.

Reasoning and problem-solving

Strategy

To solve problems involving extensions in strings and springs

1. Sketch a clear diagram to show all information.
2. Use Hooke's law to calculate the energy stored in the string or spring.
3. Write an energy equation which balances the changes in energy and any work done.

The principle of conservation of mechanical energy says that the total mechanical energy of a system remains constant, provided there is no external work done by resisting forces such as friction.

This implies that any increase in one form of energy is balanced by a decrease in another form. When you write an energy equation, you may find it easier to think in terms of increases and decreases of KE, GPE and EPE rather than the total energy.

Example 3

A vertical elastic string, of natural length $AB = 1\,\text{m}$ with $\lambda = 49\,\text{N}$, has A fixed to a ceiling and B fixed to a mass of 2 kg. Take $g = 9.8\,\text{m s}^{-2}$

a The mass is lowered gently until in equilibrium at point O. Calculate the extension when at O

It is then pulled down a further 20 cm to point P and released.

b Calculate its velocity v when passing O and the greatest height that it reaches. Describe its subsequent motion.

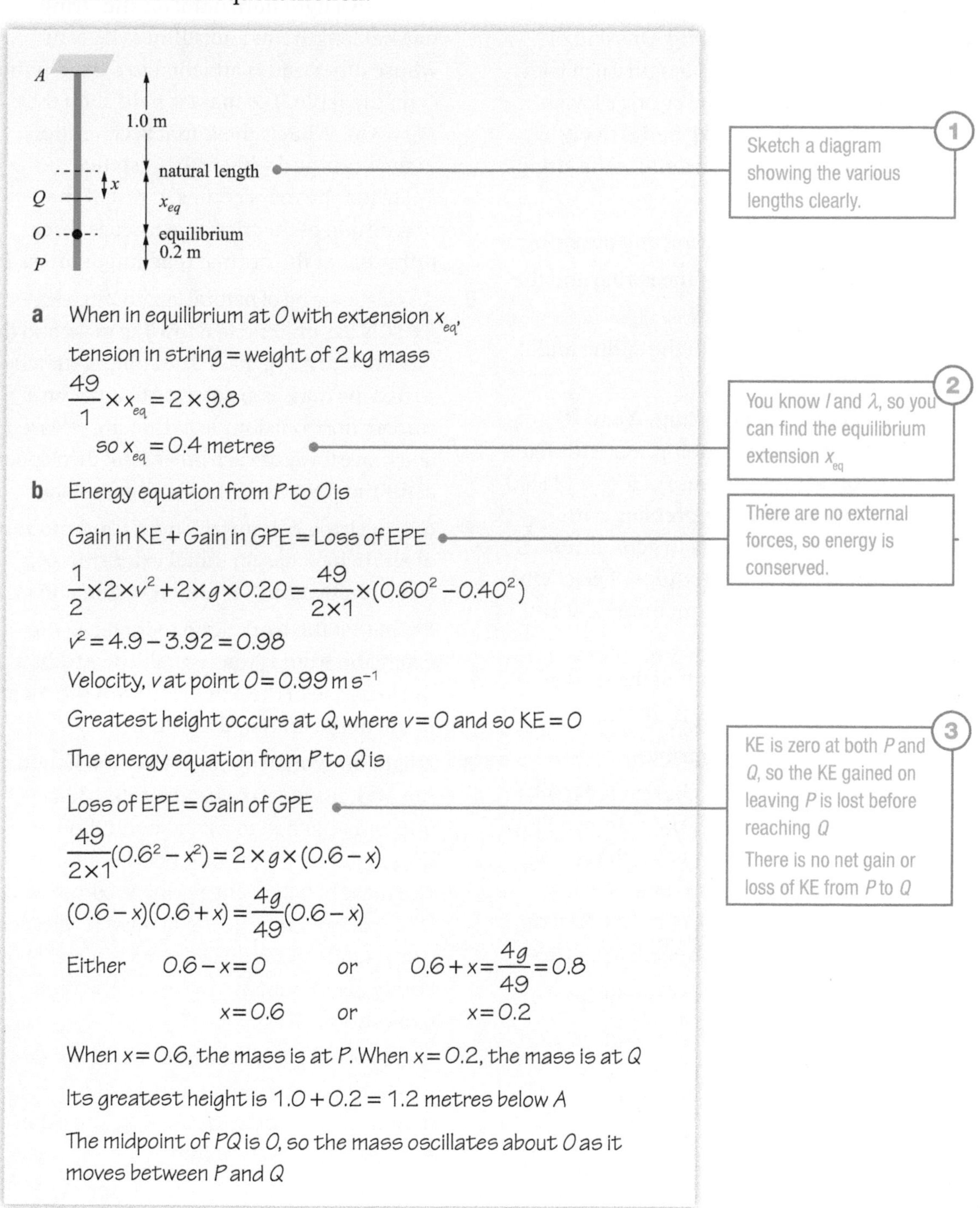

a When in equilibrium at O with extension x_{eq},

tension in string = weight of 2 kg mass

$$\frac{49}{1} \times x_{eq} = 2 \times 9.8$$

so $x_{eq} = 0.4$ metres

b Energy equation from P to O is

Gain in KE + Gain in GPE = Loss of EPE

$$\frac{1}{2} \times 2 \times v^2 + 2 \times g \times 0.20 = \frac{49}{2 \times 1} \times (0.60^2 - 0.40^2)$$

$$v^2 = 4.9 - 3.92 = 0.98$$

Velocity, v at point $O = 0.99\,\text{m s}^{-1}$

Greatest height occurs at Q, where $v = 0$ and so KE = 0

The energy equation from P to Q is

Loss of EPE = Gain of GPE

$$\frac{49}{2 \times 1}(0.6^2 - x^2) = 2 \times g \times (0.6 - x)$$

$$(0.6 - x)(0.6 + x) = \frac{4g}{49}(0.6 - x)$$

Either $0.6 - x = 0$ or $0.6 + x = \frac{4g}{49} = 0.8$

$x = 0.6$ or $x = 0.2$

When $x = 0.6$, the mass is at P. When $x = 0.2$, the mass is at Q

Its greatest height is $1.0 + 0.2 = 1.2$ metres below A

The midpoint of PQ is O, so the mass oscillates about O as it moves between P and Q

1 Sketch a diagram showing the various lengths clearly.

2 You know l and λ, so you can find the equilibrium extension x_{eq}

There are no external forces, so energy is conserved.

3 KE is zero at both P and Q, so the KE gained on leaving P is lost before reaching Q

There is no net gain or loss of KE from P to Q

Exercise 7.2B Reasoning and problem-solving

In all questions, take the value of g to be $9.8\,m\,s^{-2}$ unless told otherwise. λ is the modulus of elasticity.

1 How much work is done to stretch a light spring from a length of 2 m to a length of 3 m, if its natural length is 1.5 m and its modulus of elasticity is 25 N?

2 An elastic spring of natural length 0.5 m with $\lambda = 80$ N hangs from a ceiling. One end of an elastic string of natural length 0.6 m with $\lambda = 120$ N is attached to the spring's lower end. An object of mass 4 kg hangs freely in equilibrium from the lower end of the string.

Calculate

a The tension in the spring and the string,

b The total extension of the spring and the string,

c The total EPE stored in the spring and the string.

3 Two thin light vertical springs, A and B, stand side by side with their lower ends fixed to a horizontal table. A mass of 5 kg is placed on their upper ends, compressing both springs so that the mass is in equilibrium. If the natural length and modulus of elasticity of A are 0.8 m and 80 N, and those of B are 1.5 m and 60 N, find

a The compressed length of the springs,

b The tension in each spring,

c The total EPE in the springs.

4 a A model railway has a horizontal track which ends at a buffer made from a light, horizontal spring. In testing the spring, a mass of 2 kg compresses it by 2 cm. Subsequently, a truck of mass 0.5 kg runs into the buffer with a speed of $0.60\,m\,s^{-1}$

Calculate the stiffness of the spring, $\frac{\lambda}{l}$

How much does the buffer compress in bringing the truck to rest?

b A light, vertical spring is fixed to a floor. It is compressed by a distance a when a mass m rests on it. If the mass falls from rest onto the spring from a height of $\frac{3a}{2}$ above the top of the spring, calculate the greatest compression of the spring in the subsequent motion.

5 A mass P of 2 kg on a rough horizontal table is tied to a light horizontal elastic string, of natural length 2 m and stiffness $25\,N\,m^{-1}$, whose other end is attached to a fixed point Q on the table. The mass is held such that $PQ = 3$ m. What is the initial acceleration of the mass and what is the distance PQ when the mass comes to rest, if the magnitude of the frictional force is three-fifths that of the normal reaction on mass P?

6 An elastic string of natural length 2 m with $\lambda = 80$ N has one end tied to a 3 kg mass and the other end to a 5 kg mass. The string is stretched so that the masses are at rest 4 m apart on a smooth horizontal surface. They are released and move towards each other. Find their speeds at the moment when the string goes slack.

7 A light string of natural length l and modulus of elasticity λ has an initial extension of x_1 It is stretched further until its extension is x_2

Prove that the work done to increase the extension from x_1 to x_2 equals the product of the increase in the extension with the mean of the initial and final tensions.

8 A light horizontal spring, of natural length 2 m and modulus of elasticity 50 N, has one end attached to a fixed point A on a horizontal table. The other end is attached to a mass of 0.2 kg. The spring is compressed by 0.4 metres and the 0.2 kg mass is released from rest. How far does the mass travel before coming to instantaneous rest, if the table is smooth?

9 A mass of 1 kg is attached to the lower end of a light vertical elastic spring, of natural length $AB = 1.5$ m with $\lambda = 49$ N. The other end, A, is fixed to a ceiling.

a When the mass is in equilibrium at point P, what is the extension in the string?

b The mass is now pulled down to point Q which is 30 cm below P, and released from rest. Find its velocity v when it reaches P and the greatest height to which it rises above P. Describe its subsequent motion.

c The mass is pulled down further to point R so that it is 50 cm below P. When released from rest, how close does it get to point A?

d If the string was replaced by a light spring with the same natural length AB and modulus of elasticity λ, how would your answer to part **b** be different? How close does the mass now get to A in this case?

10 A light elastic rope of natural length 25 m with $\lambda = 1200$ N has one end tied to a bungee jumper of mass 60 kg and the other end tied to a bridge. The jumper falls from rest off the point of attachment to the bridge and reaches a speed of $v\,\text{m s}^{-1}$ after dropping a distance x m.

a Show that $30v^2 = 1788x - 24x^2 - 15000$ provided $x > k$, for constant k

State the value of k and calculate the maximum value of v

b Calculate the maximum value of x

11 A light elastic string of natural length a and modulus of elasticity λ has two masses m_1 and m_2 attached to its ends. They rest on a smooth horizontal table a distance a apart. The mass m_1 is struck in the direction of the string so that it has an initial velocity u away from m_2

Show that the maximum extension of the string is $\sqrt{\dfrac{m_1 m_2 u^2 a}{(m_1 + m_2)\lambda}}$

Full A Level

12 An object of 4 kg lies on a plane which slopes at 30° to the horizontal. A spring, of natural length 1.5 m with $\lambda = 40$ N, is attached at one end to the top of the plane with the other end attached to the object.

a If the plane is smooth, calculate the length of the spring when the object rests in equilibrium.

b If the plane is so rough that the frictional force is a fifth of the normal reaction on the object, calculate the distance along the plane in which the object can be in equilibrium.

13 A light elastic string of natural length 3 m hangs vertically from a fixed point and is stretched to a length of 4.5 m by a mass of 15 kg. The same mass is now attached to the midpoint of the same string which is stretched horizontally between two points X and Y which are 4 metres apart.

a On release, the mass falls vertically from rest. Calculate its velocity after it falls 1.5 m

b Verify that, if the mass is gently lowered until it hangs in equilibrium, its distance below the horizontal through X and Y is 1.1 m

14 A light rod AB of length 2 m is smoothly hinged to a vertical wall at A with B higher than A. A light elastic string, of natural length 1 m with $\lambda = 100$ N, is stretched from B to point C on the wall 2 m above A

A mass of m kg is fixed to B

a If angle $CAB = 60°$, calculate the value of m when the mass is in equilibrium.

b If B is placed to coincide with C and the mass m is released from rest in this position with a negligibly small horizontal velocity so that AB rotates about A, what is the maximum value of angle CAB in the subsequent motion?

7.3 Dimensional analysis

Fluency and skills

Using dimensional analysis to examine formulae can serve two purposes.

Firstly, you can use it to check whether a formula is valid or not. This process looks at the dimensional consistency of the terms in the formula. For example, it is not possible to add together a term with units in metres to a term with units in kilograms, because the dimensions of the terms are not consistent within the formula.

Some quantities, such as pure numbers and angles, have no dimensions and do not appear in dimensional equations.

Secondly, dimensional analysis can be used to help build a possible formula between variables which are suspected to be related. For example, the density of seawater and the speed of waves might affect the height of the waves and a formula might exist to connect them.

Most quantities can be expressed in terms of three basic dimensions: mass M, length L and time T. Dimensions are not the same as units, but units can sometimes help you work out dimensions. For example, speed = distance ÷ time, the units of speed are m s^{-1}, and the dimensions of speed are $\frac{\text{L}}{\text{T}}$ or LT^{-1}

You write 'dimensions of' using square brackets, so $[\text{speed}] = \text{LT}^{-1}$

Example 1

a Given that kinetic energy $\text{KE} = \frac{1}{2}mv^2$, work out the dimensions of energy.

b Show that the equation GPE = mgh is dimensionally consistent.

a $[\text{KE}] = [m] \times [v]^2 = \text{M} \times \left(\frac{\text{L}}{\text{T}}\right)^2 = \text{M L}^2\text{T}^{-2}$

b On the left-hand side, $[\text{GPE}] = [\text{KE}] = \text{M L}^2\text{T}^{-2}$

All forms of energy have the same dimensions.

On the right-hand side, $[m] \times [g] \times [h] = \text{M} \times \text{LT}^{-2} \times \text{L} = \text{M L}^2\text{T}^{-2}$

Both sides have the same dimensions, so the equation is dimensionally consistent.

Exercise 7.3A Fluency and skills

1 Which of these quantities are dimensionless?

a Speed of a car

b The ratio of height to length for a bus

c sin 60°

d Speed of rotation in radians per second

e Angle of rotation in radians

f The ratio of circumference to diameter for a circle

g Temperature

2 a Write the dimensions of velocity and acceleration.

b Use Newton's 2nd Law to write the dimensions of force and so write the dimensions of work.

c Write the dimensions of momentum.

3 Find the dimensions of

a volume b density c power

d energy e impulse f pressure

4 Show that the dimensions of the term $2as$ in the kinematic equation $v^2 = u^2 + 2as$ are the same as the dimensions v^2 and u^2

5 Show that both of the terms ut and $\frac{1}{2}at^2$ in the equation $s = ut + \frac{1}{2}at^2$ have the dimension L

6 Boyle's law gives the equation $pv = k$, where p and v are the pressure and volume of a gas. What are the dimensions of k?

7 Einstein said energy E and mass m are equivalent. If $E = k \times m$, work out the dimensions of k and show that they are the same as those of (velocity)^2

8 Given that A is an area, a is acceleration, t is time, x is distance, u, v are velocities, F is force and m is mass

i Make k the subject of each formula

ii Find the dimensions of k

a $A = kt$ b $F = kv^2$

c $F = kxt$ d $a = kmu$

e $v = \frac{kxF}{ma}$ f $\frac{u}{t} = k \times at$

9 The tension T in a string over a pulley which connects two masses m_1 and m_2 is given by $T = \frac{2mM}{m+M}g$

Show that the dimensions of $\frac{2mM}{m+M}g$ are the same as the dimensions of force.

10 The resisting force F on a ball of radius r falling vertically with velocity v through a fluid of viscosity η is given by $F = 6\pi r v \eta$
Make η the subject of the formula and so find the dimensions of viscosity η

11 Newton's law of gravitation is $F = \frac{Gm_1m_2}{r^2}$
Work out the dimensions of the constant G

12

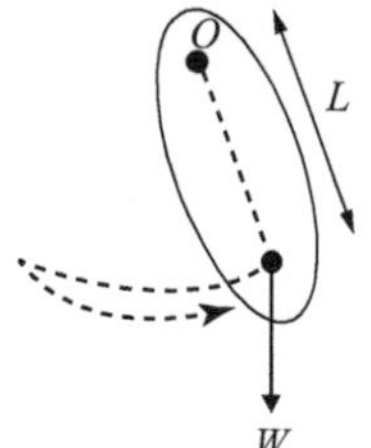

A compound pendulum of weight W and pivot O has a length L and a time period t, where $t = 2\pi\sqrt{\frac{I}{mgL}}$

Find the dimensions of the variable I and state its SI units.

13 Which of these formulae are dimensionally inconsistent?

s, h are lengths; u, v are velocities; a, g are accelerations; F is force; and t is time.

a $s = ut + \frac{1}{2}at^3$ b $Ft = m\sqrt{a}$

c $v^2 = \frac{2as}{t}$ d $\frac{1}{2}mv^2 = mgh$

e $F = \frac{m_1m_2g}{m_1+m_2}$ f $s = \frac{1}{2}t(u+v)$

MECH

Reasoning and problem-solving

Strategy

When using dimensional analysis

1. Write the dimensions of all variables in terms of M, L and T.
2. Equate powers of M, L and T and solve to work out the unknown indices.

Example 2

The time period t of a simple pendulum may depend on its mass m, its length l and the gravitational acceleration g.

Find a formula for t in terms of m, l and g

l

mg

Assume the formula takes the form

$t = k \times m^x l^y g^z$ where k is a numerical constant.

The formula's dimensional equation is

$[\text{Time}, t] = M^x \times L^y \times \left(\frac{L}{T^2}\right)^z = M^x L^{y+z} T^{-2z}$ — **1** Write all variables in terms of their dimensions.

$x = 0, \quad y + z = 0 \quad \text{and} \quad -2z = 1$

giving $x = 0, \quad y = \frac{1}{2} \quad \text{and} \quad z = -\frac{1}{2}$ — **2** Equate powers of M, L and T and solve to work out x, y and z

The formula is $t = k \times m^0 l^{\frac{1}{2}} g^{-\frac{1}{2}}$ or $t = k\sqrt{\frac{l}{g}}$ — $m^0 = 1$

Exercise 7.3B Reasoning and problem-solving

1 Given that x, y are distances; m and M are masses; F is a force; and λ is a coefficient of elasticity, find the dimensions of P and Q in the following equations, and suggest what each of P and Q might represent.

$$P = \sqrt{\frac{4Fmx}{3M(m+M)}} \qquad Q = \sqrt{\frac{\lambda(x-y)}{2mxy}}$$

2 A mass M accelerates from rest over a distance x before colliding and coalescing with a mass m. After the collision, they are brought to rest over a distance y by a constant resisting force R. It is suggested that $R = \pi\left[(m+M)g + \frac{M^2 y}{(m+M)x}g\right]$

Determine whether this formula is dimensionally consistent.

3 A small mass m is attached to a fixed point O on a table by a string of length r. It travels on a circular path about O with a speed v. Find a formula for tension T in the string of the form

$$T = k \times m^a \times v^b \times r^c$$

where k is a numerical constant, clearly stating the values of a, b and c

4 The velocity v of sound in a gas with density ρ and pressure P has the form $v = k \times \rho^x \times P^y$ where k is a numerical constant. Work out the values of x and y and rewrite the formula using your answers.

5 The lifting force F on an aeroplane's wing depends on its speed v, its surface area A and the density of the air r. If $F = k \times v^x A^y r^z$ where k is a numerical constant, calculate the values of x, y and z. Write the formula using your answers.

6 "There's a hole in my bucket, dear Liza", said Henry, as he sees liquid leaking through the hole at the bottom with a speed v. Henry says that v will depend on the depth of the liquid, d and the gravitational acceleration, g. Liza agrees with him but says that it might also depend on the density of the liquid, ρ. So, their two models are

$$v \propto d^a g^b \rho^c \quad \text{and} \quad v \propto d^x g^y$$

where a, b, c, x and y are constants.

Which model would you accept? Give your reasons, stating the values of the constants to give dimensional consistency.

7 This formula is used in the design of jet engines: $F = mgI - pA$ where F, m, p and A are force, mass, pressure and area, and g is acceleration due to gravity. What are the dimensions of the symbol I?

8 The velocity v of the wave on a violin string may depend on the string's tension T, mass m and length L. Construct a formula for v which uses these variables.

9 The speed v at which waves travel across the ocean depends on the density ρ of the water, the distance λ between the wave crests and the gravitational constant g. Construct a formula for v in terms of ρ, λ and g

There is a dimensionless quantity in the equation.

10 The flow F, in m^3 per second, of liquid down a tube depends on the tube's radius r m, its length L m, its viscosity η kg m^{-1} s^{-1} and the drop in pressure P N per m^2 along the tube.

a Construct a formula for the flow in the form $F = k \times r^{4x} L^{-x} \eta^y P^z$

b For a particular tube with $r = 0.05$, $L = 6.0$, $\eta = 0.01$, $P = 3600$ and $F = 0.15$, show that $k = 0.4$

MECH

7 Summary and review

Chapter summary

- Work done by a constant force F moving a distance x in its direction is $F \times x$ and, for a variable force, work done $= \int F\,dx$
- Energy is gained by an object when work is done on it. Mechanical energy is either kinetic energy or potential energy. The units of both energy and work are joules (J).
- Kinetic energy, KE $= \frac{1}{2}mv^2$, relates to a body's motion.
- Gravitational potential energy, GPE $= mgh$, relates to a body's vertical height in Earth's gravity.
- Elastic potential energy, EPE $= \frac{\lambda}{2l}x^2$ relates to the elasticity of a string or spring.
- Power is the rate of doing work. For a constant force F moving with a speed v, power $= F \times v$ The unit of power is watts.
- When there is no external input of work or energy, the total energy of a system is conserved. So, gains in one form of energy are balanced by losses in another form of energy.
- Hooke's law states that the tension T in a string or spring is proportional to its extension (or, for a spring, its compression), x, where $T = \frac{\lambda}{l}x$
- The stiffness of a string or spring is equal to $\frac{\lambda}{L}$
- Dimensional analysis can help you to construct a formula involving various quantities or to check the consistency of a formula.

Check and review

You should now be able to...	Try Questions
✔ Analyse problems using formulae for work, energy and power.	1–9, 12–13
✔ Use Hooke's law to solve problems involving strings and springs.	8, 9, 13
✔ Construct equations and check them using dimensional analysis.	10, 11, 14

In all questions, take the value of g to be $9.8\,\text{m s}^{-2}$ unless told otherwise.

1 A car has a mass of 1200 kg. It starts from rest and reaches a speed of $24\,\text{m s}^{-1}$ after travelling 320 metres. Assuming that resistances to motion are negligible, use an energy equation to calculate the tractive force of its engine.

2 Reindeer pull a sleigh of mass 300 kg horizontally over 100 m on ice from rest with a constant acceleration of $0.5\,\text{m s}^{-2}$ Assuming negligible resistance to motion, calculate

a The horizontal force pulling the sleigh,

b The work done by the reindeer,

c The gain in KE of the sleigh,

d The final power exerted by the reindeer.

3 A box of mass 200 kg is winched, with constant acceleration, 12 metres vertically for 10 seconds from rest by a rope. Calculate

a Its acceleration, b Its final velocity,

c The gain in its

i KE, ii GPE,

d The work done by the winch,

e The tension in the rope,

f The final power of the winch.

4 A fountain pumps 1500 litres of water 2 metres vertically every minute and expels it at a speed of $5\,\text{m s}^{-1}$

Assuming no loss of energy, find the power of the pump.

5 A cyclist of total mass 90 kg has a speed of $10\,\text{m s}^{-1}$ at point P at the top of a hill. She free-wheels downhill to point Q which is a vertical distance of 50 m below P, reaching Q with a speed of $24\,\text{m s}^{-1}$

The distance PQ by road is 360 m. There is a constant resistance to motion. Calculate its value.

6 A 2 kg mass is dragged along a rough horizontal floor from rest for a distance s m by a variable force F where $F = 24 + 2s - s^2$ N. If the magnitude of the frictional force is two-sevenths that of the normal reaction, what is the speed of the mass after moving 3 m? What is the power of the force at this point?

7 The resistance to a car's motion when moving with speed v is kv, where k is a constant and the maximum power of its engine is P

Its maximum speed down a slope is V

Its maximum speed up the same slope is $\frac{1}{2}V$

Prove that $V = \sqrt{\frac{2P}{k}}$

8 A mass of 12 kg is attached to one end of a light elastic string of natural length 1.5 m whose upper end is fixed to a ceiling. If the modulus of elasticity, λ, is 120 N and the mass is in equilibrium, calculate

a The tension in the string,

b Its extension,

c The EPE stored in it.

9 Two fixed points X and Y are 2 m apart on a smooth table. Two horizontal strings OX and OY are attached to an object O which lies between X and Y. The natural length and modulus of elasticity for OX are 0.6 m and 16 N respectively, and for OY are 0.4 m and 20 N respectively. Find the extensions of the two strings and the tensions in them when O is in equilibrium.

10 Work out whether these formulae are dimensionally consistent given that m is mass; s and l are lengths; u and v are velocities; a and g are accelerations; F is force and t is time.

a $s = vt - \frac{1}{2}at^2$ b $Ft = m(v^2 - u^2)$

c $t^2 = \pi\frac{l}{g}$ d $Fs = m \times a \times t$

11 Check the dimensional consistency of these equations.

a $Fs = \frac{1}{2}mv^2$ b Power $= mgv$

12 The forward thrust T on the blade of a turbine depends on the area A swept by the blade, the speed v of the air and density ρ of the air. Given that $T = kA^x v^y \rho^z$ where k is a dimensionless constant and that $T = 40$ when $A = 7$, $v = 13$ and $\rho = 1.2$, work out the values of x, y, z and k and write the formula for T

13 A lorry of mass 3 tonnes travels up a slope where there is no resistance to motion at an angle of $\arcsin\frac{1}{140}$ to the horizontal. Its speed increases steadily from $15\,\text{m s}^{-1}$ to $25\,\text{m s}^{-1}$ over a distance of 500 m. Find the total increase in its energy, the tractive force of its engine and its final power output.

14 An elastic string has a natural length 4 m and a modulus of elasticity of $20g$ N. The string is stretched horizontally between two points A and B that are 6 m apart, and a mass of 15 kg is attached to its midpoint. The mass drops vertically after being released from rest. Calculate its velocity after it falls 3 m.

7 Exploration

Going beyond the exams

History

Robert Hooke was a remarkable scientist who lived in the 17th century. He lends his name to Hooke's law of elasticity, devised in 1660, but he carried out research in a wide range of fields. Perhaps most notably, he illustrated his observations through a microscope in the book *Micrographia: or Some Physiological Descriptions of Minute Bodies Made by Magnifying Glasses*. His amazing drawings were the first major publication of the Royal Society in 1665. Far from the world of microscopes, he also explored the stars and planets with telescopes, drawing some very detailed sketches of Mars.

Investigation

A section of a rollercoaster can be modelled as a ball, with mass 100 kg and negligible radius, rolling around the inside of a vertical circle with radius 10 m. The ball needs to be travelling at 10 m s^{-1} at the top of the circle for it to make a complete loop-the-loop without leaving the track. Investigate what speed the ball has to be travelling as it enters the circle. How would the situation change if the ball were fixed to the circle, so that it couldn't fall off?

Note

The SI unit for energy is the same as the unit of work – the joule (J). This is named in honour of James Prescott Joule. 1 joule is the work done when lifting an object with a weight of 1 newton by 1 metre.

Research

Look up different landing vehicles that have visited other planets or moons in the Solar System. How much do they weigh, how much energy do they need to leave the surface, and in what way is the energy affected by the size of the moon or planet?

Investigation

In a bungee jump, people are tied to an elastic rope of a carefully calculated length, so that when they fall their head just dips into the water below. Investigate how the length of the rope is varied, based on the weight of the jumper as a fraction of the typical male adult.

7 Assessment

In all questions, take the value of g to be $9.8\,\text{m s}^{-2}$ unless told otherwise. λ is the modulus of elasticity.

1. A man raises a load of 20 kg through a height of 6 m using a rope and simple pulley. His maximum power output is 180 W. Assuming the rope and pulley are light and smooth

   a. How much work does he do in completing the task? **[2 marks]**

   b. What is the shortest time in which he could complete it? **[2]**

2. A child of mass 25 kg moves from rest down a slide. The total drop in height is 4 m.

   a. Assuming that there is no resistance to motion, calculate the speed of the child at the bottom of the slide. **[3]**

   b. In fact, the child reaches the bottom travelling at $6\,\text{m s}^{-1}$. The length of the slide is 6 m. Calculate

      i. The work done against friction, **[2]**

      ii. The frictional force (assuming it to be constant). **[2]**

3. A horizontal force is applied to a 6 kg body so that it accelerates uniformly from rest and moves across a horizontal plane in a straight line against a total resistance to motion of 30 N. After it has travelled 16 m, it has a speed of $4\,\text{m s}^{-1}$

   Calculate

   a. The applied force, **[2]**

   b. The total work done by the force. **[2]**

4. Two elastic strings, AB and BC, are joined at B, and the other ends are fixed to points D and E on a smooth horizontal table. AB has natural length 85 cm, and modulus of elasticity 45 N. BC has natural length 45 cm and modulus of elasticity 65 N. The distance DE is 2.4 m. Calculate the stretched lengths of the strings. **[4]**

5. Particles A and B, of mass 0.5 kg and 1.5 kg respectively, are connected by a light inextensible string of length 2.5 m. Particle A rests on a smooth horizontal table at a distance of 1.5 m from its edge. The string passes directly over the smooth edge of the table and particle B hangs suspended. The system is held at rest with the string taut and then released. Calculate the speed of A when it reaches the edge of the table. **[3]**

6. A pump, working at 3 kW, raises water from a tank at $1.2\,\text{m}^3\,\text{min}^{-1}$ and emits it through a nozzle at $15\,\text{m s}^{-1}$

   Calculate the height through which the water is raised.
   (Density of water = $1000\,\text{kg m}^{-3}$) **[4]**

7. A particle is moved along a straight line from O to A on a horizontal plane by a force F N. The length $OA = 5$ m. Find the work done by the force if

   a. $F = 12$ N **[2]**

   b. $F = (12 - 2x)$ N where x m is the distance of the particle from O **[3]**

8 The resistance to motion of a car is proportional to its speed. A car of mass 1000 kg has a maximum speed of $45\,\text{m}\,\text{s}^{-1}$ on the level when its power output is 8 kW. Calculate its acceleration when it is travelling on the level at $20\,\text{m}\,\text{s}^{-1}$ and its engine is working at 6 kW. **[6]**

9 The time, T, taken for a satellite to complete a circular orbit of radius r around the Earth (radius R) is given by

$$T^2 = \frac{4\pi^2 r^3}{gR^2}$$

Show that this formula is dimensionally consistent. **[3]**

10 A cyclist has a maximum power P

The resistance to motion is kv and in addition there is a wind blowing which exerts a constant force W

The cyclist's maximum speed against the wind is V and their maximum speed with the wind is 2 V

Show that $V = \sqrt{\frac{P}{2k}}$ **[5]**

11 A ball of mass 500 grams is fastened to one end of a light, elastic rope, whose natural length is 3 m and whose modulus of elasticity is 90 N. The other end of the rope is fastened to a bridge. The ball is held level with the fixed end, and is released from rest.

a Calculate the speed of the ball when the rope just becomes taut. **[2]**

b How far below the bridge is the lowest point reached by the ball? **[4]**

12 A particle of mass 2 kg is suspended from a point A on the end of a light spring of natural length 1 m and modulus of elasticity 196 N.

a Calculate the length of the spring when the particle hangs in equilibrium. **[3]**

The particle is now pulled down a further distance of 0.5 m and released from rest.

b Calculate the distance below A at which the particle next comes instantaneously to rest, assuming that the spring can compress to that point without the coils touching. **[5]**

c Calculate the highest position reached by the particle if, instead of the spring, an elastic string of the same natural length and modulus of elasticity had been used. **[2]**

13 An elastic string OA has natural length 2 m and modulus of elasticity 40 N. O is a fixed point on a rough horizontal plane, and a particle of mass 2 kg is attached to the string at A. The particle is projected horizontally from O with speed $v\,\text{ms}^{-1}$. It is subject to a constant resistance force of 10 N. It reaches a maximum distance x m from O, then rebounds and comes to rest exactly at O. Calculate

a The value of x **[7]**

b The value of v **[2]**

14 The frequency, f vibrations per second, with which a guitar string vibrates is thought to depend on the length, l, of the string, the density, ρ, of the steel and the tension, T, applied to it. Assuming this to be the case, find a possible formula for f in terms of l, ρ, T and a dimensionless constant k **[7]**

15 Two particles of equal mass are connected by an elastic string of length l and modulus of elasticity λ. They lie at rest on a smooth horizontal surface. One particle is set in motion with speed u

The maximum extension, x, reached by the string is given by

$$x = ku\sqrt{\frac{l}{\lambda}}$$

Use dimensional analysis to find the dimensions of k **[4]**

16 A catapult is made by attaching a light elastic string of natural length 10 cm to points A and B, which are a horizontal distance of 6 cm apart. The modulus of elasticity of the string is 5 N. A stone of mass 10 grams is placed at the centre of the string, which is then pulled back until the stone is 25 cm below the centre of AB. Calculate the greatest speed reached by the stone when it is released. State any assumptions you make in your solution. **[6]**

17 An object of mass 4 kg is released from rest at a point O and falls under gravity. It is subject to air resistance of magnitude $2.5x$ N, where x m is the distance it has fallen from O. It hits to ground at A, where $OA = 5$ m. Calculate

a The work done against resistance in travelling from O to A **[3]**

b The speed with which it hits the ground. **[3]**

18 A particle is placed at a distance r from the centre of a rough horizontal disc. The disc rotates at n revolutions per second. The maximum value of n for the particle to remain in place is given by $n = \frac{\mu}{2\pi}\sqrt{\frac{g}{r}}$, where μ is a dimensionless constant.

Show that this formula is dimensionally consistent. **[3]**

19 A car working at a rate P watts has a maximum speed V m s^{-1} when travelling on a horizontal road and experiences a resistance proportional to the square of its speed. At what rate, in terms of P, would the car have to work to double its maximum speed? **[4]**

20 A car of mass 1 tonne is towing a trailer of mass 400 kg on a level road. The resistance to motion of the car is 400 N and of the trailer is 300 N. At a certain instant, they are travelling at 10 m s^{-1} and the power output of the engine is 10.5 kW. Calculate the tension in the coupling between the car and the trailer. **[4]**

21 An object of mass m kg is accelerated from rest at a constant rate up a vertical tube of height h m. When it emerges, it rises a further $3h$ m before coming instantaneously to rest. Show that the average rate of working while the projectile is in the tube is $2m\sqrt{6g^3h}$ watts. **[8]**

22 A particle is projected vertically into the air. Assuming that the greatest height, h, reached by the particle depends only on its initial velocity, v, and the acceleration due to gravity, g, use dimensional analysis to predict the formula for h in terms of v, g and a dimensionless constant k **[5]**

23 Liquid flows through a pipe because there is a pressure difference between the ends of the pipe. The rate of flow, V m^3 s^{-1}, depends upon this pressure difference, p N m^{-2}, the viscosity of the liquid, η kg m^{-1} s^{-1}, the length of the pipe, l m, and the radius of the pipe, r m.

a Explain why it would be impossible to use dimensional analysis to determine the four unknowns in the formula $V = K\eta^{\alpha} r^{\beta} p^{\gamma} l^{\delta}$ **[2]**

b It is decided to use the pressure gradient $\frac{p}{l}$ instead of p and l separately. Use dimensional analysis to obtain a possible formula relating the rate of flow, V, to the viscosity of the liquid, the radius of the pipe and the pressure gradient. **[7]**

24 A car of mass 900 kg moves on a level road against a resistance proportional to its speed. Its power output is 6 kW and its maximum speed is 40 m s^{-1}. Calculate its acceleration when its speed is 20 ms^{-1}. **[5]**

25 A train of mass 40 tonnes accelerates uniformly. At point A it has speed 18 kmh^{-1} and has doubled its speed by the time it reaches B, where $AB = 600$ m. It is subject to a constant resistance of 3000 N. Calculate

a The work done by the engine in travelling from A to B **[3]**

b The power output of the engine when it is at the midpoint of AB **[5]**

26 A block of mass 4 kg is placed on a rough slope inclined at 30° to the horizontal. It is projected up the slope with initial speed 15 ms^{-1} and comes to rest 12 m further up the slope. Calculate the magnitude of the friction force, assuming it to be constant. **[4]**

27 A ship can travel at a constant speed of 30 kmh^{-1} when its engines are working at 15 MW. The resistance to its motion varies as the square of its speed. What power would the engines need to exert if it had to travel at a constant speed of 36 kmh^{-1}? **[6]**

28 A light elastic rope OA has natural length 10 m and modulus of elasticity 80 N. The end O is fixed to the ground and an object of mass 2 kg is attached to A

The object is projected vertically upwards from O with initial speed 40 ms^{-1}.

a Assuming that there is no air resistance, what is the maximum height reached by the object? **[4]**

b If, in fact, there is a constant air resistance of 15 N, find the maximum height reached by the object. **[4]**

c In each of the cases **a** and **b** find the speed of the object when it returns to O **[3]**

29 An elastic string of natural length 2 m and modulus of elasticity kmg is stretched between two points A and B, where A is 4 m vertically above B. A particle of mass m is attached to the midpoint of the string and is gently lowered a distance d until it is in equilibrium.

a Show that $d = \frac{1}{2k}$ provided $k \geq 0.5$ **[4]**

b Explain what happens if $k < 0.5$ and find the value of d in this case. **[3]**

30 In resisted sprint training athletes attach themselves to a wall by means of a bungee cord and then run as far as possible away from the wall until the cord pulls them back. Ashley has mass 80 kg and he is attached to the wall at O by a light elastic rope of natural length 10 m and modulus of elasticity 1600 N. He can exert a constant forward force of 400 N.

a What is the furthest that Ashley can stand from the wall? **[3]**

Ashley starts from rest at O and uses his maximum force to accelerate away from the wall.

b What is the greatest speed he achieves? **[4]**

c How far does he get from the wall before the rope pulls him back? **[4]**

8 Momentum

Forensic scientists use the principles of conservation of momentum, and momentum change during impacts, to construct models of collisions between two vehicles. Data from the accident scene, such as measurements of tyre skid marks and vehicle compression, are used to perform calculations that can reconstruct collisions and make an expert judgement of what really happened. These calculations are used in court cases to determine who was really at fault during a collision, and so it's vital that the calculations involved are accurate and give a true reflection of the incident.

Vehicle collisions are just one area in which principles relating to momentum are vitally important. Other areas include space engineering and rocket science, the design of theme park rides, and railway engineering.

Orientation

What you need to know

Maths Ch7
- Equations of motion.

Maths Ch8
- Resolving forces.
- Newton's laws of motion.

Ch7 Work energy and power
- Kinetic energy.

What you will learn

- To use the momentum equation and Newton's experimental law properly.
- To find the size of an impulse for constant and variable forces.
- To apply these concepts to problems in two dimensions using vectors.

What this leads to

Careers
- Forensic science.
- Automotive testing.
- Mechanical engineering.

8.1 Conservation of momentum

Fluency and skills

When a particle of mass m moves with velocity $\mathbf{v}$, its **momentum** is equal to the mass multiplied by the velocity. If it is acted on by a constant force $\mathbf{F}$ for a time t, then the **impulse** acting on the particle is equal to the force multiplied by the time during which the force acts.

> **Key point**
> Momentum $= m \times \mathbf{v}$ Impulse $= \mathbf{F} \times t$

The units of momentum and impulse are the same. They are newton seconds (N s).

$\mathbf{F}$ and $\mathbf{v}$ are vectors, so momentum and impulse are also vectors.

If the force $\mathbf{F}$ is constant and the particle's velocity increases from $\mathbf{u}$ to $\mathbf{v}$ then its acceleration is given by $\mathbf{a} = \frac{\mathbf{v}-\mathbf{u}}{\mathbf{t}}$

Newton's 2nd law gives $\mathbf{F} = m\mathbf{a} = \frac{m(\mathbf{v}-\mathbf{u})}{t} = \frac{m\mathbf{v}-m\mathbf{u}}{t}$

Before impact

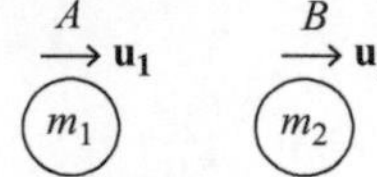

During impact

After impact

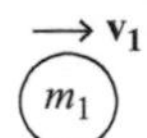

> **Key point**
> $\mathbf{F}t = m\mathbf{v} - m\mathbf{u}$
> Impulse = Change in momentum

Bodies A and B with masses m_1 and m_2 and initial velocities $\mathbf{u}_1$ and $\mathbf{u}_2$ collide directly. During the collision, forces act on the bodies for a short time t

If the force acting on B is $\mathbf{F}$, then Newton's 3rd law states that an equal and opposite force acts on A

After the collision, A and B separate with velocities $\mathbf{v}_1$ and $\mathbf{v}_2$

For A, $\quad -\mathbf{F}t = m_1\mathbf{v}_1 - m_1\mathbf{u}_1$

For B, $\quad \mathbf{F}t = m_2\mathbf{v}_2 - m_2\mathbf{u}_2$

So $\quad -(m_1\mathbf{v}_1 - m_1\mathbf{u}_1) = m_2\mathbf{v}_2 - m_2\mathbf{u}_2$

Rearrange

$$m_1\mathbf{u}_1 + m_2\mathbf{u}_2 = m_1\mathbf{v}_1 + m_2\mathbf{v}_2$$

Total initial momentum = Total final momentum

A **direct** collision means that bodies collide along their common line of travel. If they are moving towards each other, the direct collision is head-on. Most collisions in this section are direct.

> **Key point**
> The **principle of conservation of linear momentum** states that, when no *external* forces are present, the total momentum of a system of particles is unchanged by collisions between them.

'**Linear**' indicates that motion is considered as acting in a straight line.

MECH

Example 1

A body P of mass 1 kg moving with velocity 5 m s^{-1} collides directly with another body Q of mass 2 kg moving towards P with velocity 4 m s^{-1}

After the collision, Q is at rest. Find the final velocity of P

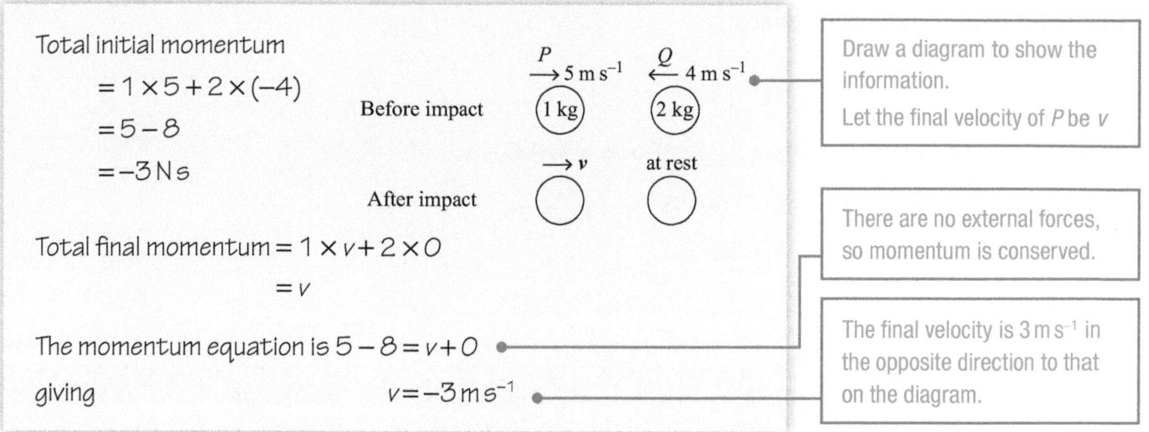

Total initial momentum

$= 1 \times 5 + 2 \times (-4)$

$= 5 - 8$

$= -3\text{ N s}$

Total final momentum $= 1 \times v + 2 \times 0$

$= v$

The momentum equation is $5 - 8 = v + 0$

giving $v = -3\text{ m s}^{-1}$

Draw a diagram to show the information.

Let the final velocity of P be v

There are no external forces, so momentum is conserved.

The final velocity is 3 m s^{-1} in the opposite direction to that on the diagram.

Example 2

When two particles **coalesce**, they combine into one particle.

A particle P of mass 1 kg with a velocity $2\mathbf{i} - 3\mathbf{j}\text{ m s}^{-1}$ collides and coalesces with a particle Q of mass 4 kg and velocity $\mathbf{i} + 2\mathbf{j}\text{ m s}^{-1}$

They move off together with a common velocity $\mathbf{v}$

Calculate the speed $|\mathbf{v}|$ and the angle which $\mathbf{v}$ makes with the $\mathbf{i}$-direction.

Final mass $= 1 + 4 = 5\text{ kg}$

The momentum equation is

$1 \times (2\mathbf{i} - 3\mathbf{j}) + 4 \times (\mathbf{i} + 2\mathbf{j}) = 5 \times \mathbf{v}$

$6\mathbf{i} + 5\mathbf{j} = 5\mathbf{v}$

Final velocity, $\mathbf{v} = 1.2\mathbf{i} + 1\mathbf{j}\text{ m s}^{-1}$

Speed, $|\mathbf{v}| = \sqrt{1.2^2 + 1^2}$

$= \sqrt{2.44}$

$= 1.56\text{ m s}^{-1}$

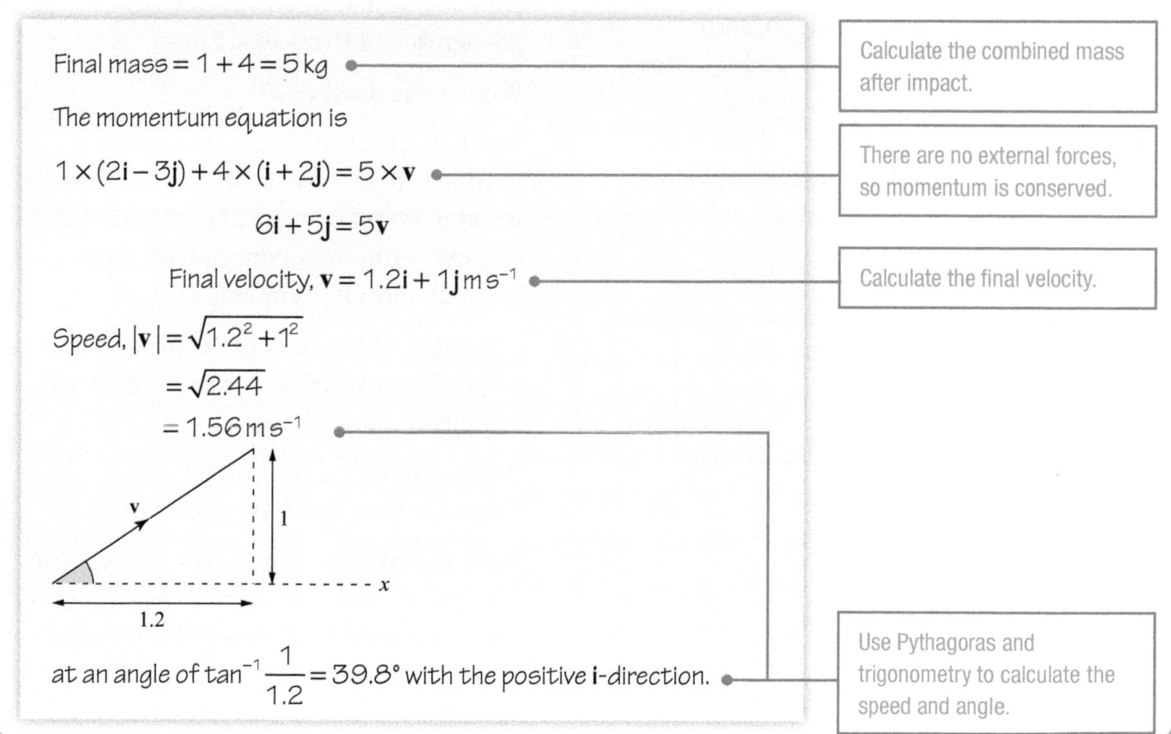

at an angle of $\tan^{-1}\dfrac{1}{1.2} = 39.8°$ with the positive $\mathbf{i}$-direction.

Calculate the combined mass after impact.

There are no external forces, so momentum is conserved.

Calculate the final velocity.

Use Pythagoras and trigonometry to calculate the speed and angle.

Example 3

A child of mass 30 kg travels in a horizontal straight line on a skateboard of mass 5 kg with a velocity of 1.4 ms^{-1}

The child jumps off the skateboard with an initial horizontal backwards velocity of 0.2 ms^{-1} while the skateboard continues forwards. All motion is in the same straight line. Calculate the final velocity, v, of the skateboard.

Total initial momentum $= (30 + 5) \times 1.4 = 49$ Ns — Total mass = 30 + 5 = 35 kg

Total final momentum $= (30 \times -0.2) + (5 \times v) = 5v - 6$ Ns

The momentum equation is $49 = 5v - 6$ — There are no external forces, so momentum is conserved.

Final velocity of skateboard, $v = \frac{55}{5} = 11 \text{ m s}^{-1}$

Exercise 8.1A Fluency and skills

1 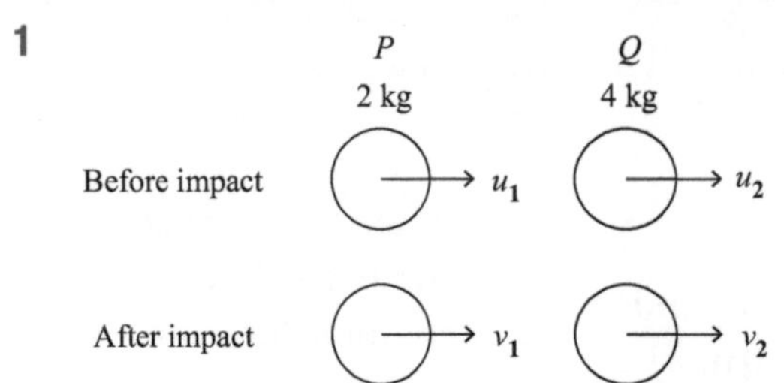

Two particles P and Q, with masses 2 kg and 4 kg respectively, move along a straight line. They collide directly, travelling with velocities u_1 and u_2 before impact and with velocities v_1 and v_2 after impact. Calculate the values, x, of the unknown velocities in this table. All velocities are given in m s^{-1}

	u_1	u_2	v_1	v_2
a	3	2	2	x
b	6	1	x	3.5
c	5	4	4	x
d	10	–2	x	4
e	1	–4	–4	x

2 Before impact: Y (m_1) moving with u_1, Z (m_2) moving with u_2. After impact: $Y + Z$ moving with v.

Two particles Y and Z, with masses m_1 kg and m_2 kg respectively, move along a straight line with velocities u_1 and u_2 m s^{-1}

They impact directly, coalesce, and then travel with velocity v m s^{-1}

Calculate v in each case.

	m_1	m_2	u_1	u_2
a	4	6	3	4
b	7	5	4	–2
c	2	8	5	–3

3 Particle A of mass 4 kg moves at 5 m s^{-1} towards particle B which is at rest. After a direct impact, A is at rest and B moves at 2 m s^{-1}

What is the mass of B?

4 A bullet of mass 40 grams travelling at 80 m s^{-1} hits and coalesces with a stationary wooden target of mass 2 kg which is free to move. Find their common speed v immediately after impact.

5 A particle with a velocity $4\mathbf{i} - 6\mathbf{j}$ m s^{-1} collides and coalesces with a similar particle with velocity $\mathbf{i} + 2\mathbf{j}$ m s^{-1}

Both particles have mass 500 grams. Calculate their velocity, $\mathbf{v}$, after impact.

6 Particle P of mass 2 kg moves with velocity $3\mathbf{i} + 3\mathbf{j}$ m s^{-1} and collides with particle Q of mass 3 kg moving with velocity $4\mathbf{i} - \mathbf{j}$ m s^{-1} After impact, P is at rest. Calculate the velocity of Q

7 Two particles A and B have masses 3 kg and 4 kg and velocities $\begin{pmatrix} 4 \\ -2 \end{pmatrix}$ m s^{-1} and $\begin{pmatrix} 3 \\ 4 \end{pmatrix}$ m s^{-1}

respectively. They collide and, after the collision, the velocity of A is $\begin{pmatrix} 0 \\ -4 \end{pmatrix}$ m s^{-1}

Calculate the velocity of B after the collision.

8 Particle P has mass 2 kg and velocity $5\mathbf{i} - \mathbf{j}$ m s^{-1}

Particle Q has mass 4 kg and velocity $-\mathbf{i} - 4\mathbf{j}$ m s^{-1}

They collide and coalesce. Calculate their velocity after impact.

9 Particle S of mass 4 kg travels with velocity $5\mathbf{i} + 4\mathbf{j}$ m s^{-1} and collides with particle T of mass 2 kg travelling with velocity $-2\mathbf{i} + \mathbf{j}$ m s^{-1}

After impact, the velocity of T is double the velocity of S. Calculate their final velocities.

10 An unexploded mine that is at rest suddenly explodes into two pieces. The pieces travel in opposite directions at speeds of 10 ms^{-1} and 16 ms^{-1}

If the first piece has a mass of 2 kg, find the mass of the other piece.

11 An object of mass 3 kg is moving horizontally at 120 ms^{-1} when it explodes into two pieces which travel in the same straight horizontal line. One piece continues forwards at 200 ms^{-1} and the other travels backwards at 80 ms^{-1} Find the masses of the two pieces.

MECH

Reasoning and problem-solving

Strategy

To solve collision problems involving the conservation of momentum

1. Draw and label a diagram showing the situation before and after the collision.
2. Check that there are no external forces involved, so that momentum is conserved.
3. Write momentum equations, ensuring correct signs for velocities, and solve them.

Example 4

The * in the diagram indicates a collision.

Three particles A, B and C, with masses 1 kg, 4 kg and 12 kg respectively, are positioned in a straight line.

Particles B and C are at rest and particle A is moving towards B with a speed of 10 m s^{-1}

After A and B collide, particle A rebounds backwards and B moves towards C with twice the speed of A

After B and C collide, they move in opposite directions with the same speed.

Show that there are no more collisions between the particles, assuming no other forces act on them.

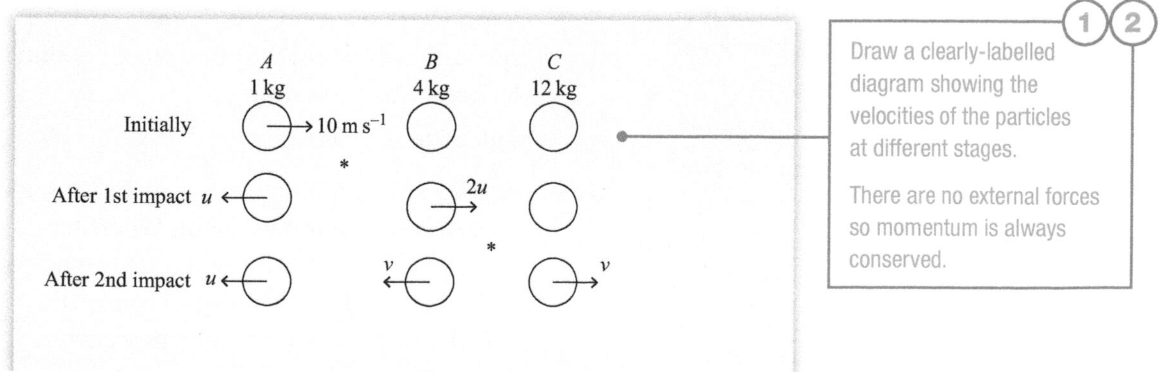

(1)(2) Draw a clearly-labelled diagram showing the velocities of the particles at different stages.

There are no external forces so momentum is always conserved.

(*Continued on the next page*)

Momentum equation when A strikes B is

$1 \times 10 + 0 = 1 \times (-u) + 4 \times 2u$

$\Rightarrow \quad 7u = 10$

$\Rightarrow \quad u = 1\frac{3}{7}\ \text{m s}^{-1}$

Momentum equation when B strikes C is

$4 \times 2u + 0 = 4 \times (-v) + 12 \times v$

$\Rightarrow \quad 8v = 8u$

$\Rightarrow \quad v = u = 1\frac{3}{7}\ \text{m s}^{-1}$

After these two collisions, A and B are moving in the same direction with the same speed and in the opposite direction to C

Hence, there are no further collisions.

3 Write a momentum equation for each collision. Note the negative signs.

Exercise 8.1B Reasoning and problem-solving

1 A hammer of mass 10 kg travelling vertically downwards with speed 24 m s^{-1} strikes the top of a vertical post of mass 2 kg and does not rebound. Find the speed of the hammer and post immediately after impact.

2 An arrow, of mass 250 grams, travelling at 25 m s^{-1}, hits directly and coalesces with a stationary target of mass 9.0 kg which can move freely. Calculate their velocity immediately after impact.

3 Two cars of mass 900 kg and 1200 kg are travelling in opposite directions on a straight road with speeds of 135 km h^{-1} and 90 km h^{-1} when they collide head-on and lock together. What is their speed in km h^{-1} immediately after impact?

4 A bullet of mass 40 grams is fired horizontally with a speed of 600 m s^{-1} into a target of mass 2 kg suspended from a vertical string. The bullet becomes embedded in the target. Find their common speed immediately after the impact.

5 Two railway trucks with masses 2 tonnes and 3 tonnes are travelling on the same line at speeds of 5 m s^{-1} and 2 m s^{-1} respectively. They collide and become coupled together. Find their common velocity after the collision if they were initially travelling

a In the same direction,

b In opposite directions.

6 Masses of 3 kg and 5 kg are joined by a stretched elastic string and held apart on a smooth horizontal table. At a certain time after they are released from rest, the lighter mass has speed 12 m s^{-1}

What is the speed of the other mass at this time?

7 When a gun is fired, the bullet shoots forwards and the gun is free to recoil backwards. The bullet has mass 50 grams and the gun has mass 2 kg. If the speed of the bullet is 250 m s^{-1} find the speed at which the gun recoils.

Given that the explosion creates an impulse, explain why the principle of conservation of momentum can still be used.

8 A soldier fires a shell of mass 10 kg. When travelling horizontally at 120 m s^{-1}, it explodes into two pieces which also initially travel horizontally. Immediately after the explosion, one piece of mass 4 kg travels backwards with speed 40 m s^{-1} whilst the other piece continues forward with a speed v m s^{-1}

Calculate the value of v

9 At a November bonfire party, a rocket of mass 0.5 kg travels in a vertical plane. When its velocity is $6\mathbf{i} + 20\mathbf{j}$ m s^{-1}, it splits into two pieces with masses 0.3 kg and 0.2 kg with velocities $-2\mathbf{i} + 18\mathbf{j}$ m s^{-1} and $a\mathbf{i} + b\mathbf{j}$ m s^{-1} respectively. Calculate a and b

10 Two particles, P of mass 3 kg and Q of mass 2 kg, lie on a smooth horizontal table connected by a light inextensible string which is slack. P is at rest and Q is moving away from P with a speed of $8\,\text{m s}^{-1}$

Calculate their common velocity immediately after the string becomes taut.

11 Three spheres A, B and C, with masses 5 kg, 4 kg and 3 kg respectively, lie in a straight line on a smooth horizontal surface with B and C at rest. Sphere A moves towards B with a speed of $18\,\text{m s}^{-1}$

On impact, A and B coalesce and move towards C. On impact, they coalesce with C and move with velocity $v\,\text{m s}^{-1}$

Calculate v

12 Three particles X, Y and Z, with masses 4 kg, 2 kg and 6 kg respectively, lie in a straight line on a smooth horizontal surface, with X and Z at rest. Y moves towards Z at $10\,\text{m s}^{-1}$ and collides with Z. After impact, Y and Z both move with speed $u\,\text{m s}^{-1}$ in opposite directions. Y now collides and coalesces with X and they continue with speed $v\,\text{m s}^{-1}$

Calculate u and v and explain why there is or is not another collision.

13 Three particles R, S and T, of mass 6 kg, 3 kg and 5 kg respectively, lie in a straight line. S is between R and T and they are on a smooth horizontal surface. R and S are joined by a light, inextensible, slack string. R and T are at rest. S moves away from R with a speed of $10\,\text{m s}^{-1}$ and the string becomes taut before S reaches T

a Calculate the speed of R after the string becomes taut.

b S now collides with T and comes to rest. Calculate the speed of T after this collision. Explain why there is a further collision between R and S

14 Three particles A, B and C, of mass 2 kg, 3 kg and 1 kg respectively, are in line with B and C at rest, B between A and C, and A moving on a smooth horizontal surface towards B with velocity u

The collision between A and B brings A to rest. B now collides with C

After this collision, B and C move in the same direction, with C moving twice as fast as B

Find the final velocities of B and C in terms of u

15 Particle A of mass 10 kg has speed $5\,\text{m s}^{-1}$
It collides directly on a smooth horizontal surface with particle B of mass m kg moving in the opposite direction at $2\,\text{m s}^{-1}$
After the collision, A continues in the same direction with a speed of $3\,\text{m s}^{-1}$
Show that, if there are no further collisions, $m \le 4$

16 A particle of mass 4 kg has a velocity of $\mathbf{i}-\mathbf{j}\,\text{m s}^{-1}$
It collides with a second particle which has a mass of m kg and a velocity of $2\mathbf{i}-3\mathbf{j}\,\text{m s}^{-1}$
After the impact, the first particle is at rest and the second particle has a velocity of $4\mathbf{i}+a\mathbf{j}\,\text{m s}^{-1}$
Find the values of m and a

17 Particles A and B have masses 2 kg and m kg and velocities $4\mathbf{i}+3\mathbf{j}$ and $p\mathbf{i}+q\mathbf{j}\,\text{m s}^{-1}$ respectively. They collide and coalese. If, after impact,

a They move parallel to the x-axis, calculate the value of m given that $q=-1$

b They move at 45° to the x-axis, calculate the final speed given that $q=3$
Use the value of m you found in part **a**.

18 A particle of mass 3 kg has a velocity of $4\,\text{m s}^{-1}$ in the x-direction. A second particle has mass m kg and velocity $6\,\text{m s}^{-1}$ in the y-direction. They collide, coalesce and begin to move together at 45° to their initial directions. Calculate the magnitude of their final velocity.

19 A smooth sphere A of mass 4 kg and centre C_1 has a velocity of $6\,\text{m s}^{-1}$ at 60° to the x-axis. A smooth sphere B of mass 2 kg is at rest with centre C_2 on the x-axis. A strikes B so that the line of centres C_1C_2 lies on the the x-axis. After impact, A moves at right angles to the x-axis. Explain why B moves along the x-axis and calculate the final velocities of A and B

Full A Level

8.2 Collisions

Fluency and skills

When two bodies with known masses and velocities collide and rebound, their velocities after impact are unknown. You need two equations to find them. One is the momentum equation. The other is given by **Newton's experimental law**. It involves the speed at which the gap between two bodies changes.

The **speed of approach** is the component of the speed along the line of impact at which the gap decreases before impact.

The **speed of separation** is the component of the speed along the line of impact at which the gap increases after impact.

Key point

$$\frac{\text{Speed of separation}}{\text{Speed of approach}} = e$$

The constant e is the **coefficient of restitution.** It depends on the **elasticity** of the bodies and has a value in the range $0 \le e \le 1$

$e = 0$ if the impact is **inelastic** (that is, with no rebound)

$e = 1$ if the impact is **perfectly elastic.**

A golf ball has $e \approx 0.8$ for an impact with a hard object, like a golf club.

Key point

If you say the velocities of the objects before impact are u_1 and u_2 and the velocities after impact are v_1 and v_2 then you can use the following equation.

$$e = \frac{v_2 - v_1}{u_1 - u_2}$$

Example 1

A particle P with mass 4 kg and velocity 20 m s^{-1} collides directly with particle Q with mass 12 kg travelling with velocity 4 m s^{-1} in the opposite direction.

Given $e = \frac{1}{2}$, calculate the velocities v_1 and v_2 after impact.

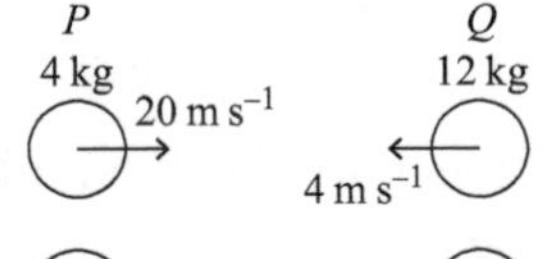

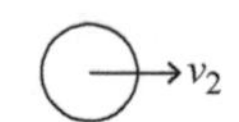

The momentum equation is

$4 \times 20 + 12 \times (-4) = 4v_1 + 12v_2$

$v_1 + 3v_2 = 8$ **(1)**

The speed of approach $= 20 + 4 = 24$ m s^{-1}

The speed of separation $= v_2 - v_1$

Work out the speed of approach and the speed of separation. Take care with the signs.

Newton's equation is

$\frac{v_2 - v_1}{24} = \frac{1}{2} \Rightarrow v_2 - v_1 = 12$ **(2)**

(1) + **(2)** gives $4v_2 = 20$

$v_2 = 5$

Solve **(1)** and **(2)** simultaneously to find v_1 and v_2

Substituting in **(1)** gives $v_1 = 8 - 15 = -7$

So, P moves at 7 m s^{-1} and Q moves at 5 m s^{-1} and both change direction after impact.

The particle in this diagram strikes a smooth, fixed plane with a velocity **u** and rebounds with velocity **v**

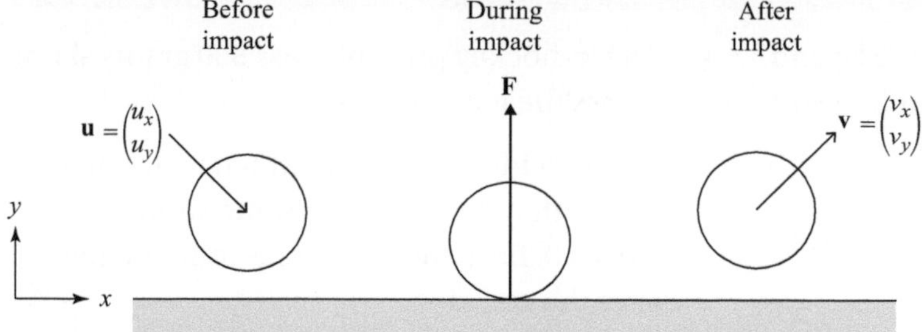

As the plane is smooth, there is no force parallel to the plane and therefore there is no change in the particle's momentum parallel to the plane. So, $u_x = v_x$

Upon impact, the force **F** acts entirely perpendicular to the plane. Momentum is not conserved in this direction (because an external force acts), but Newton's experimental law still applies perpendicular to the plane, so $e = \frac{v_y}{u_y}$

MECH

Example 2

A ball strikes a smooth fixed plane with a velocity of $\mathbf{u} = 20\sqrt{3}\mathbf{i} - 20\mathbf{j}$

The coefficient of restitution, e, is 0.8

Find the final velocity $\mathbf{v} = v_x\mathbf{i} + v_y\mathbf{j}$ after impact and the angle θ it makes with the plane.

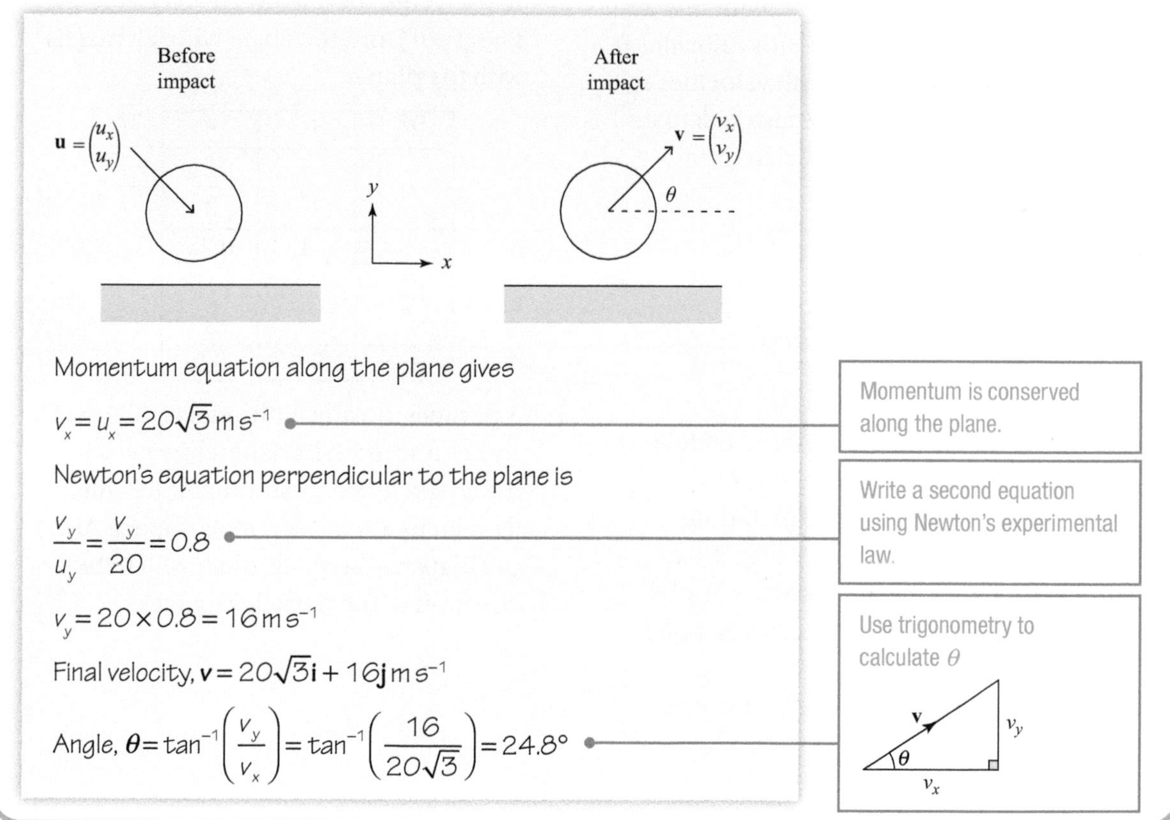

Momentum equation along the plane gives

$v_x = u_x = 20\sqrt{3}\ \text{m s}^{-1}$

Momentum is conserved along the plane.

Newton's equation perpendicular to the plane is

$\frac{v_y}{u_y} = \frac{v_y}{20} = 0.8$

Write a second equation using Newton's experimental law.

$v_y = 20 \times 0.8 = 16\ \text{m s}^{-1}$

Final velocity, $\mathbf{v} = 20\sqrt{3}\mathbf{i} + 16\mathbf{j}\ \text{m s}^{-1}$

Angle, $\theta = \tan^{-1}\left(\frac{v_y}{v_x}\right) = \tan^{-1}\left(\frac{16}{20\sqrt{3}}\right) = 24.8°$

Use trigonometry to calculate θ

Exercise 8.2A Fluency and skills

1 Two particles P and Q, with masses 2 kg and 4 kg respectively, move in a straight line and collide directly. They travel with velocities u_1 and u_2 before impact and with velocities v_1 and v_2 after impact. In each case, calculate their velocities after impact.

	u_1	u_2	e
a	5	3	$\frac{1}{2}$
b	6	−2	$\frac{3}{4}$
c	−1	−10	$\frac{2}{3}$
d	5	−4	$\frac{1}{3}$
e	8	−2	$\frac{1}{4}$

2 Two particles R and S, with masses 3 kg and 6 kg respectively, move in a straight line and collide directly. They travel with velocities u_1 and u_2 before impact and with velocities v_1 and v_2 after impact. In each case, calculate the value of v_2 and the coefficient of restitution e

	u_1	u_2	v_1
a	16	4	2
b	5	2	2.5
c	−6	4	2
d	4	−1	−2

3 Two dodgem cars at a fairground collide head-on from opposite directions and rebound. Their masses, including passengers, are 125 kg and 150 kg. Their initial speeds are 0.6 m s^{-1} and 0.5 m s^{-1} respectively. The coefficient of restitution is 0.3

Calculate their speeds just after impact.

4 An ice-hockey puck of mass 800 grams skims across the ice at 15 m s^{-1}

It hits an identical puck, which is stationary, and both move on in the same straight line. If $e = 0.6$, find their velocities immediately after the collision.

5 Two toy trains travel on the same track in the same direction and collide. The front and rear trains have masses of 0.15 kg and 0.24 kg and initial speeds of 5.0 cm s^{-1} and 8.0 cm s^{-1} respectively. After impact, the rear train has a speed of 6.0 cm s^{-1} in the same direction. Find the value of e

6 A sphere strikes a smooth horizontal plane with a velocity $\mathbf{u} = u_x\mathbf{i} - u_y\mathbf{j}$

It rebounds with a velocity $\mathbf{v} = v_x\mathbf{i} + v_y\mathbf{j}$

The coefficient of restitution is e

In each case, find the final velocity $\mathbf{v} = v_x\mathbf{i} + v_y\mathbf{j}$ and the angle which it makes with the plane.

	u_x	u_y	e
a	4	6	$\frac{1}{2}$
b	3	4	0.25
c	2	6	$\frac{1}{3}$
d	1.2	9.0	0.8

7 A particle moving with velocity $3\mathbf{i} - 4\mathbf{j}$ m s^{-1} strikes a smooth horizontal surface. If the coefficient of restitution is 0.5, find its velocity $\mathbf{v} = v_x\mathbf{i} + v_y\mathbf{j}$ after impact. Also calculate the angle through which the direction of the particle changes.

Reasoning and problem-solving

Strategy

To solve a collision problem involving elasticity

1. Draw and label a diagram showing the situation before and after the collision.
2. Check whether there are any external forces and whether momentum is conserved.
3. Write the momentum equation and Newton's equation, taking care with signs, and solve.

MECH

Example 3

Two spheres P of mass 1 kg and Q of mass 3 kg lie on a smooth horizontal plane with the line PQ at right-angles to a vertical wall. P moves at 10 m s^{-1} and collides directly with Q, which is initially at rest.

Q then hits the wall and rebounds. If the coefficient of restitution between the spheres is 0.4 and between Q and the wall is e, show that P and Q collide again if $e > \frac{1}{7}$

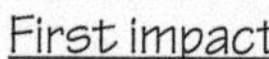

First impact

Before 1st impact

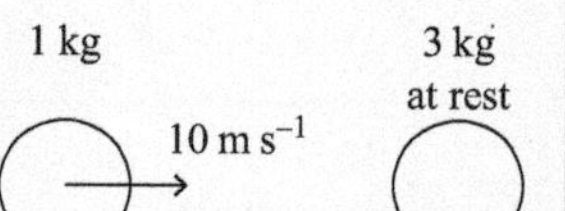

After 1st impact

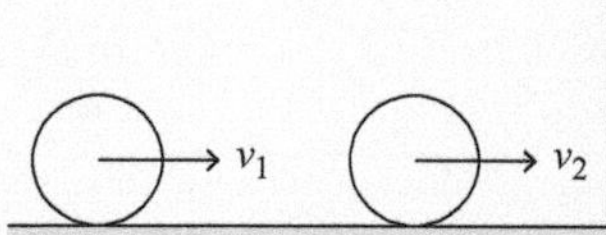

(1)(2) Draw a clearly labelled diagram showing the velocities. There is no external force at first impact so momentum is conserved.

Momentum equation is:

$$10 + 0 = v_1 + 3v_2 \quad (1)$$

Newton's equation is:

$$\frac{v_2 - v_1}{10} = 0.4 \quad \Rightarrow \quad v_2 - v_1 = 4 \quad (2)$$

(3) Write the two equations and solve simultaneously.

(1) and (2) give $v_1 = -0.5$ and $v_2 = 3.5$

P is now moving away from the wall and Q towards the wall.

Second impact

Before 2nd impact

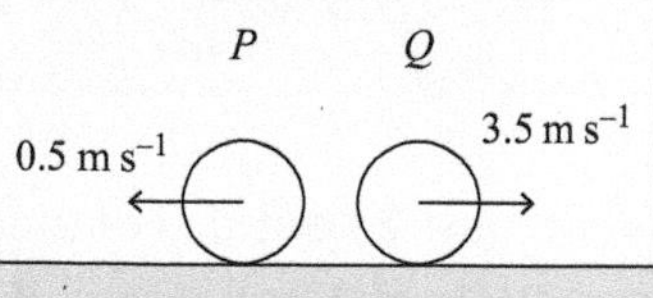

After 2nd impact

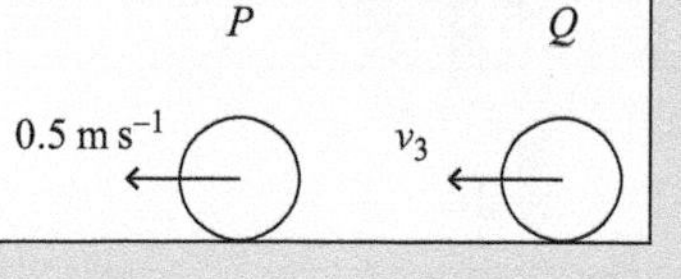

(1) Draw another diagram showing the velocities. The wall produces an external force on Q so momentum is not conserved.

Newton's equation is:

$$\frac{v_3}{3.5} = e \quad \text{so } v_3 = 3.5e$$

(3) Write only one equation since momentum is not conserved for Q

P and Q are now both moving away from the wall.

Q collides again with P if $3.5e > 0.5$ or $e > \frac{1}{7}$

Example 4

> The * in the diagram indicates a collision.

Particles P, Q and R have masses 3 kg, 2 kg, 1 kg and velocities 6 m s^{-1}, 4 m s^{-1}, 2 m s^{-1} respectively. They move in a straight line in this order and P strikes Q with perfect elasticity. A second collision between Q and R has a coefficient of restitution, e

Find the possible values of e if there are no more collisions.

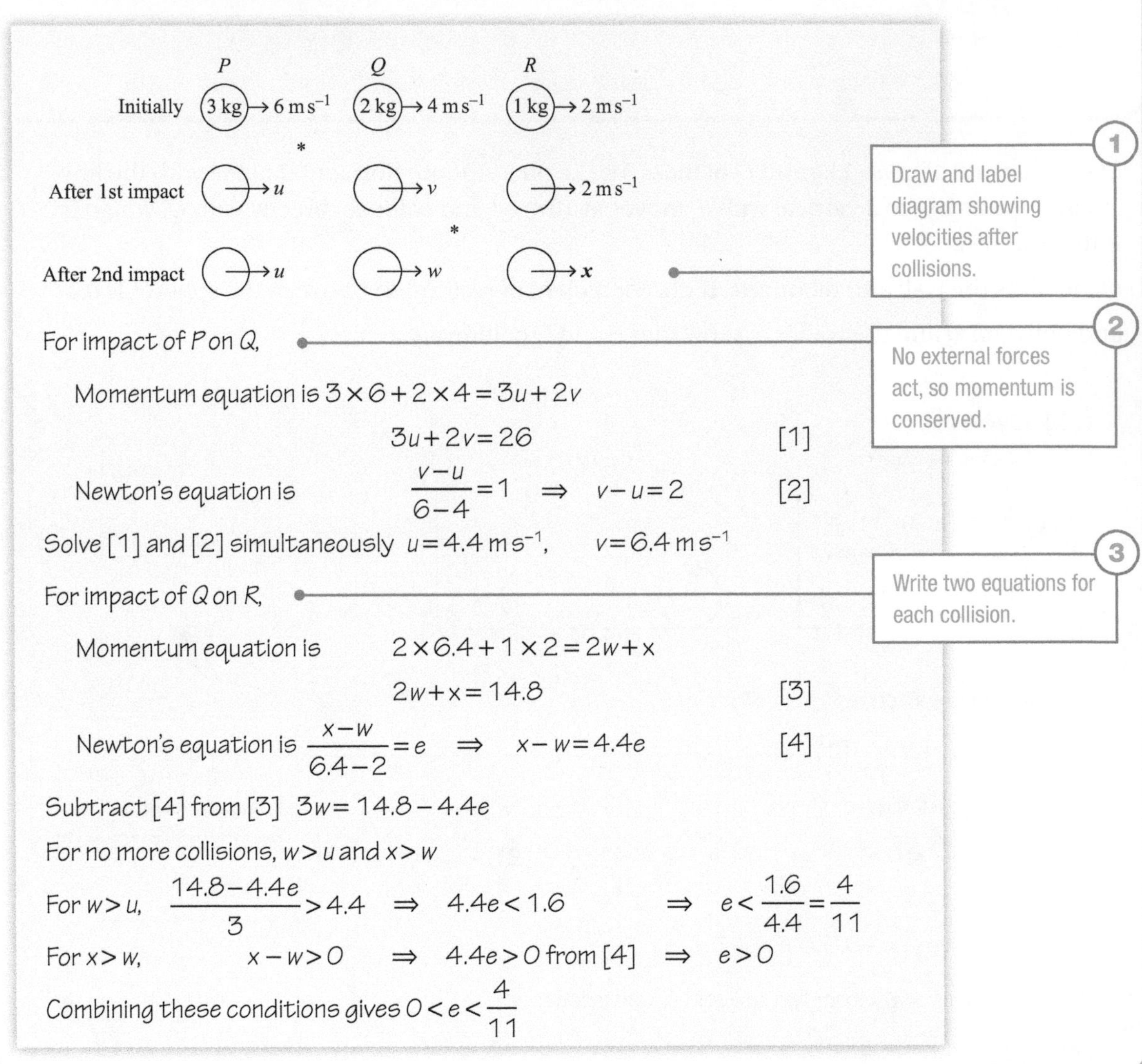

For impact of P on Q,

Momentum equation is $3\times6+2\times4=3u+2v$

$$3u+2v=26 \qquad [1]$$

Newton's equation is $\frac{v-u}{6-4}=1 \Rightarrow v-u=2 \qquad [2]$

Solve [1] and [2] simultaneously $u=4.4$ m s^{-1}, $v=6.4$ m s^{-1}

For impact of Q on R,

Momentum equation is $2\times6.4+1\times2=2w+x$

$$2w+x=14.8 \qquad [3]$$

Newton's equation is $\frac{x-w}{6.4-2}=e \Rightarrow x-w=4.4e \qquad [4]$

Subtract [4] from [3] $3w=14.8-4.4e$

For no more collisions, $w>u$ and $x>w$

For $w>u$, $\frac{14.8-4.4e}{3}>4.4 \Rightarrow 4.4e<1.6 \Rightarrow e<\frac{1.6}{4.4}=\frac{4}{11}$

For $x>w$, $x-w>0 \Rightarrow 4.4e>0$ from [4] $\Rightarrow e>0$

Combining these conditions gives $0<e<\frac{4}{11}$

1. Draw and label diagram showing velocities after collisions.

2. No external forces act, so momentum is conserved.

3. Write two equations for each collision.

Exercise 8.2B Reasoning and problem-solving

1 A ball of mass m kg is dropped vertically onto a stone floor from a height of 4 metres and rebounds to a height of 3 metres. Using $g=9.8$ m s^{-1}, calculate

a The speed at which it hits the floor and the speed at which it leaves the floor,

b The coefficient of restitution, e, between the ball and the floor.

2 Two spheres P of mass 2 kg and Q of mass 4 kg lie on a smooth horizontal plane. P is at rest. Q moves away from P at 10 m s^{-1} towards a vertical wall where $e=\frac{1}{2}$

Q hits the wall and rebounds to hit and coalesce with P

Calculate the final speed of P and Q

3 Two balls Y and Z, of mass 4 kg and m kg respectively, lie on a smooth horizontal surface between two parallel walls on a line at right angles to them. Y moves at $2\,\mathrm{m\,s^{-1}}$ towards one wall and Z moves at $4\,\mathrm{m\,s^{-1}}$ towards the other wall. They bounce off the walls, where $e=\frac{1}{3}$ and collide with each other on their return. After this collision, both balls are at rest. Calculate m

4 A particle with velocity $6\mathbf{i}-4\mathbf{j}\,\mathrm{m\,s^{-1}}$ collides with a smooth horizontal plane which contains the x-axis. Its velocity after impact is $a\mathbf{i}+2\mathbf{j}\,\mathrm{m\,s^{-1}}$

Write the value of a and find the coefficient of restitution, e

5 A ball moves with velocity $5\mathbf{i}-3\mathbf{j}$ towards a smooth horizontal plane which contains the x-axis. The coefficient of restitution e is 0.8 Calculate the speed of the ball immediately after impact and the angle it makes with the plane.

6 Two spheres R and S, of mass 2 kg and m kg respectively, lie on a smooth horizontal plane. S moves with speed u away from R (which is at rest) towards a wall fixed at right-angles to its path, rebounds from the wall and returns towards R. The coefficient of restitution for all impacts is 0.75. Calculate the value of m such that, after S impacts on R, it is at rest.

7 Two spheres, A of mass 1 kg and B of mass 3 kg, are joined by a light slack string on a smooth horizontal plane. A is at rest. B moves away from A with speed u to strike and rebound from a wall at right angles to its path, with the string still slack. Given that $e=0.5$ for all collisions, find in terms of u

a The velocities of A and B after they collide,

b The common speed of A and B after the string becomes taut.

8 Sphere B of mass 2 kg lies on a smooth horizontal plane between sphere A of mass 30 kg and a fixed vertical wall. A is at rest and B moves towards A with speed u

The value of e is 0.6 for all collisions. Show that B is at rest after its second collision with A

9 Three particles P, Q and R, with masses 6 kg, 4 kg and 8 kg respectively, lie in a straight line on a smooth horizontal surface with Q and R at rest. P moves towards Q with velocity $15\,\mathrm{m\,s^{-1}}$

If $e=\frac{3}{5}$ for all collisions, find their velocities after the third collision. Explain why there are no more collisions.

10 A rubber ball of mass m is dropped 5 metres from rest onto a horizontal floor. Taking $g=10\,\mathrm{m\,s^{-2}}$, show that the ball's speed just before the first impact is $10\,\mathrm{m\,s^{-1}}$

Given $e=0.5$, calculate the time between the first and second impacts. Use a geometric progression to find the total time for the ball to come to rest after being released.

11 Three particles X, Y and Z, of masses 2 kg, 4 kg and 4 kg respectively, lie in a straight line on a smooth horizontal plane. X and Y are joined by a light slack string. With X and Z at rest, Y moves towards Z at $6\,\mathrm{m\,s^{-1}}$ and the string becomes taut before Y reaches Z

If $e=\frac{1}{2}$ for all impacts, calculate the common speed of X and Y

a Before Y strikes Z,

b After the string becomes taut for a second time.

Explain why there are no more collisions.

Full A Level

12 A ball drops vertically and strikes a smooth slope with a speed of $10\,\mathrm{m\,s^{-1}}$

After impact, the ball moves horizontally. If the slope makes an angle of 30° with the horizontal, calculate the coefficient of restitution.

13 A ball falls freely under gravity for 10 metres until it strikes a smooth plane that is inclined at 45° to the horizontal. If $e=0.2$, calculate the speed and direction of the ball immediately after impact. Use $g=10\,\mathrm{m\,s^{-1}}$

8.3 Impulses

Fluency and skills

You already know that the impulse **I** of a constant force **F** acting on a body for a time t is equal to the change in momentum of the body.

I, F, v and **u** are all vectors.

Key point

Impulse, $\mathbf{I} = \mathbf{F} \times t$

$= m\mathbf{v} - m\mathbf{u}$

F can be very large and t very small, as in a collision.

F can be constant over time, as when a jet of water hits a fixed surface.

F can also vary with time, in which case you need to use calculus to find the impulse.

Key point

$$\mathbf{I} = \int_0^t \mathbf{F}\mathrm{d}t = \int_0^t m\frac{\mathrm{d}\mathbf{v}}{\mathrm{d}t}\,\mathrm{d}t = [m\mathbf{v}]_u^v = m\mathbf{v} - m\mathbf{u}$$

$\mathbf{F} = m\frac{\mathrm{d}\mathbf{v}}{\mathrm{d}t}$ comes from Newton's 2nd law of motion.

In each case, impulse = change in momentum.

Example 1

A ball of mass 0.2 kg moves with velocity -8 m s^{-1} horizontally towards a bat. It is struck by the bat, reverses its direction and moves away at 12 m s^{-1}

Find the impulse of the bat on the ball. What is the impulse of the ball on the bat?

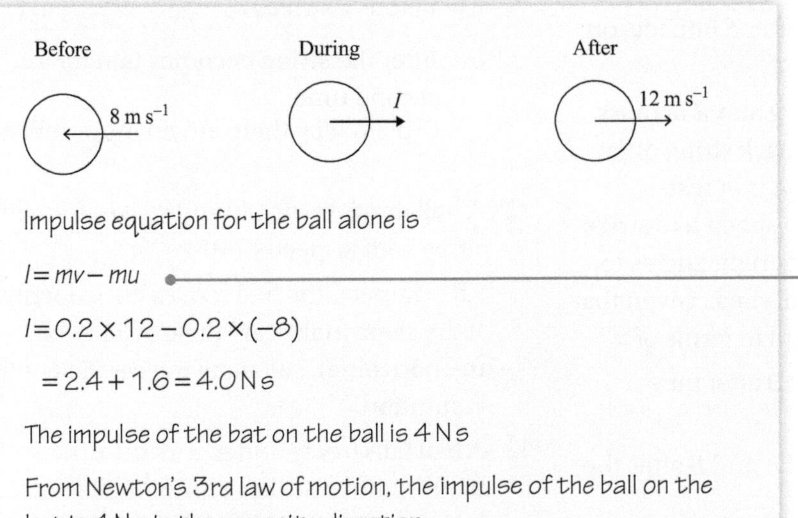

Impulse equation for the ball alone is

$I = mv - mu$

$I = 0.2 \times 12 - 0.2 \times (-8)$

$= 2.4 + 1.6 = 4.0\text{ N s}$

The impulse of the bat on the ball is 4 N s

From Newton's 3rd law of motion, the impulse of the ball on the bat is 4 N s in the opposite direction.

Apply the impulse equation for ball alone. The positive direction is to the right in the diagram.

Example 2

Water flows horizontally from a pipe of cross-section $0.02\,\text{m}^2$ at a speed of $15\,\text{m s}^{-1}$
It strikes a fixed vertical wall.

Find the force F on the wall. (Density of water $= 1000\,\text{kg m}^{-3}$)

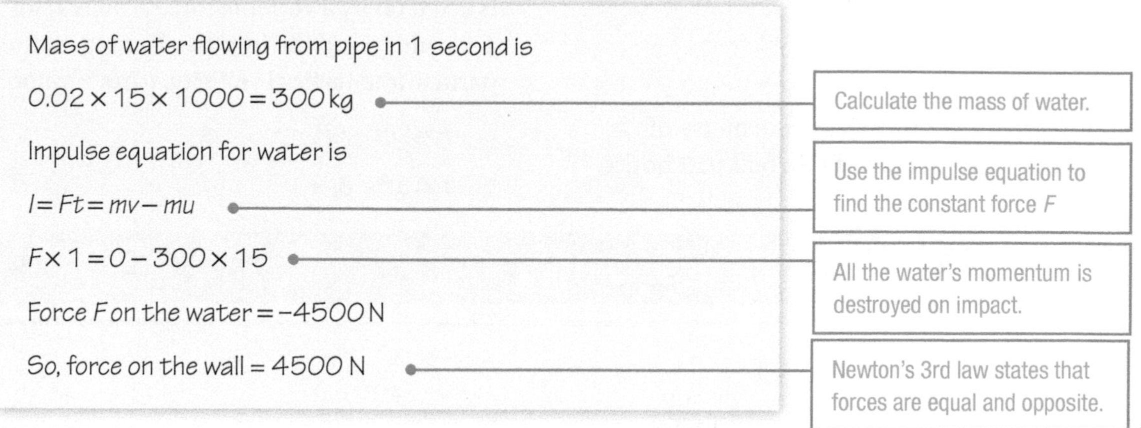

Mass of water flowing from pipe in 1 second is
0.02×15×1000=300kg
Calculate the mass of water.
Impulse equation for water is
I=Ft=mv−mu
Use the impulse equation to find the constant force F
F×1=0−300×15
All the water's momentum is destroyed on impact.
Force F on the water=−4500N
So, force on the wall = 4500 N
Newton's 3rd law states that forces are equal and opposite.

Example 3

A particle of mass 2 kg has an initial velocity u of $3\,\text{m s}^{-1}$

It is subject to a force $F = 8t - 3t^2$ newtons for 3 seconds where $0 \le t \le 3$

Find its speed v after 3 seconds.

Impulse $I = \int_0^3 F\,dt = \int_0^3 8t - 3t^2\,dt$ — Use calculus.

$= \left[4t^2 - t^3\right]_0^3 = 36 - 27 - 0 = 9\,\text{N s}$

Impulse equation is

$I = mv - mu$ — Solve to find v

$9 = 2v - 2 \times 3$

Final velocity, $v = 7.5\,\text{m s}^{-1}$

Exercise 8.3A Fluency and skills

1 A ball of mass m kg falls vertically and hits a horizontal floor with a velocity $u\,\text{m s}^{-1}$
It rebounds with velocity $v\,\text{m s}^{-1}$

Find the impulse of the floor on the ball when

a $m = 2$, $u = 12$, $v = -8$

b $m = 6$, $u = 25$, $v = -10$

2 Two identical particles A and B of mass m kg collide head-on. Particle A has an initial velocity $u\,\text{m s}^{-1}$ and a final velocity $v\,\text{m s}^{-1}$

Find the impulse that A exerts on B when

a $m = 3$, $u = 8$, $v = 2$

b $m = 5$, $u = 6$, $v = -3$

3 **a** A golf club hits a ball at rest with a horizontal impulse of 48 N s. If the ball has a mass of 50 grams, find its initial velocity.

b A cricket bat strikes a ball of mass 0.15 kg horizontally with an impulse of 36 N s

If the ball is moving horizontally towards the bat with a speed $8\,\text{m s}^{-1}$, what is the ball's final velocity?

4 Water flows from a circular pipe of radius r with a speed $v\,\mathrm{m\,s^{-1}}$ and strikes a fixed wall at right angles without any rebound. Given the density of water is $1000\,\mathrm{kg\,m^{-3}}$, find the force that the water exerts on the wall, when

a $r = 2\,\mathrm{cm}$, $v = 12\,\mathrm{m\,s^{-1}}$

b $r = 15\,\mathrm{mm}$, $v = 20\,\mathrm{m\,s^{-1}}$

5 Find the force on each square metre of ground when 1.2 cm of rain falls in 2 hours, striking the ground vertically with a speed of $72\,\mathrm{m\,s^{-1}}$

($1\,\mathrm{m^3}$ of water has a mass of 1000 kg.)

6 A 5 kg particle with an initial velocity of $2\,\mathrm{m\,s^{-1}}$ is acted on by a variable force F (in N s) for 4 seconds. Find the impulse I acting on the particle and its final velocity, $v\,\mathrm{m\,s^{-1}}$, when

a $F = 20t - 6t^2$

b $F = 3t^2 - 8t + 1$

Reasoning and problem-solving

Strategy

To solve problems involving impulses

1. Draw and label a diagram to show the information.
2. Consider the situation and briefly explain your strategy and method.
3. Write and solve equations to calculate the required values.

Example 4

A 2 kg ball with velocity $\mathbf{u} = 3\mathbf{i} - 4\mathbf{j}$ impacts a smooth plane, which is parallel to the x-axis. Given $e = 0.5$, calculate the final velocity $\mathbf{v} = v_x\mathbf{i} + v_y\mathbf{j}$ of the ball and the impulse $\mathbf{I}$ of the plane on the ball.

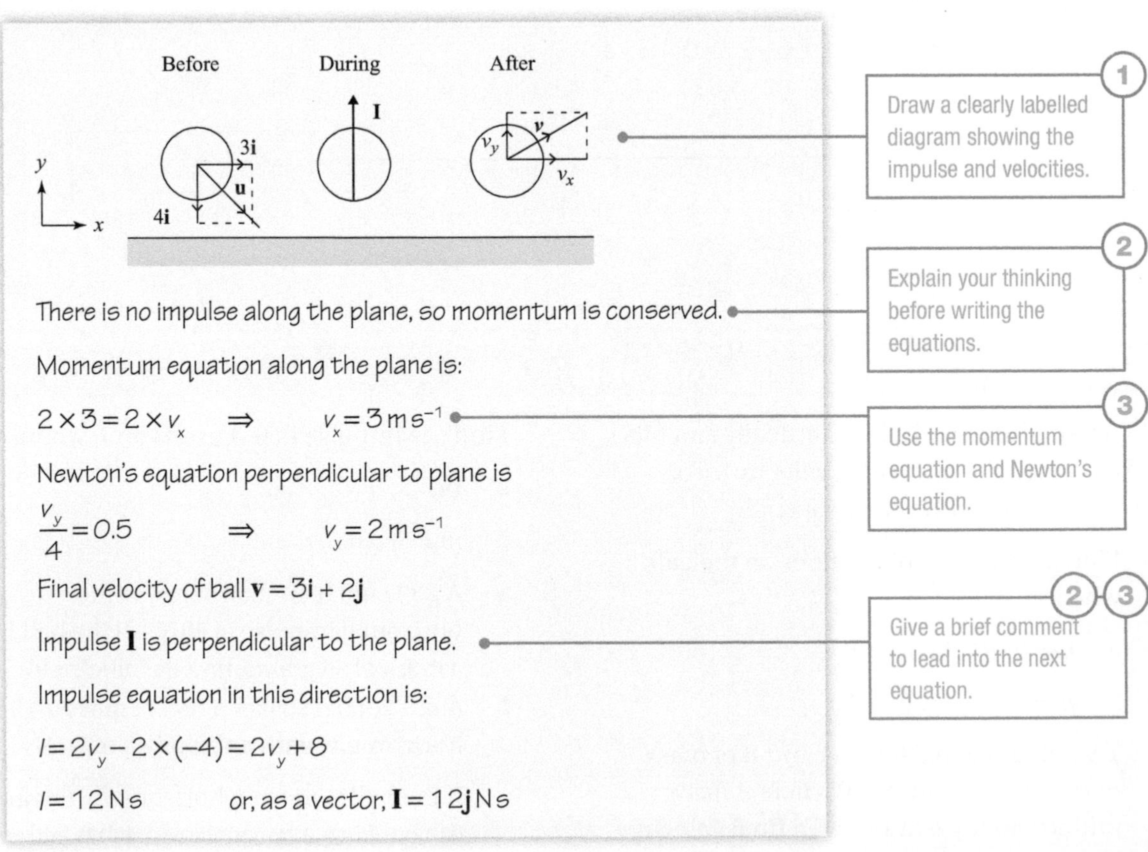

Before
During
After
I
3i
4i
u
v
v_y
v_x
y
x
There is no impulse along the plane, so momentum is conserved.
Momentum equation along the plane is:
$2 \times 3 = 2 \times v_x \quad \Rightarrow \quad v_x = 3\,\mathrm{m\,s^{-1}}$
Newton's equation perpendicular to plane is
$\frac{v_y}{4} = 0.5 \quad \Rightarrow \quad v_y = 2\,\mathrm{m\,s^{-1}}$
Final velocity of ball $\mathbf{v} = 3\mathbf{i} + 2\mathbf{j}$
Impulse $\mathbf{I}$ is perpendicular to the plane.
Impulse equation in this direction is:
$I = 2v_y - 2 \times (-4) = 2v_y + 8$
$I = 12\,\mathrm{N\,s}$ or, as a vector, $\mathbf{I} = 12\mathbf{j}\,\mathrm{N\,s}$
1
Draw a clearly labelled diagram showing the impulse and velocities.
2
Explain your thinking before writing the equations.
3
Use the momentum equation and Newton's equation.
2 3
Give a brief comment to lead into the next equation.

Example 5

Two spheres A and B of masses 4 kg and 2 kg respectively are connected by an inelastic string on a smooth horizontal plane. B is struck by an external impulse $\mathbf{J} = 24\mathbf{i} + 30\mathbf{j}$ N s

a Show that B begins to move at 75° to the line AB

b Calculate the impulsive tension in the string, $\mathbf{I}$

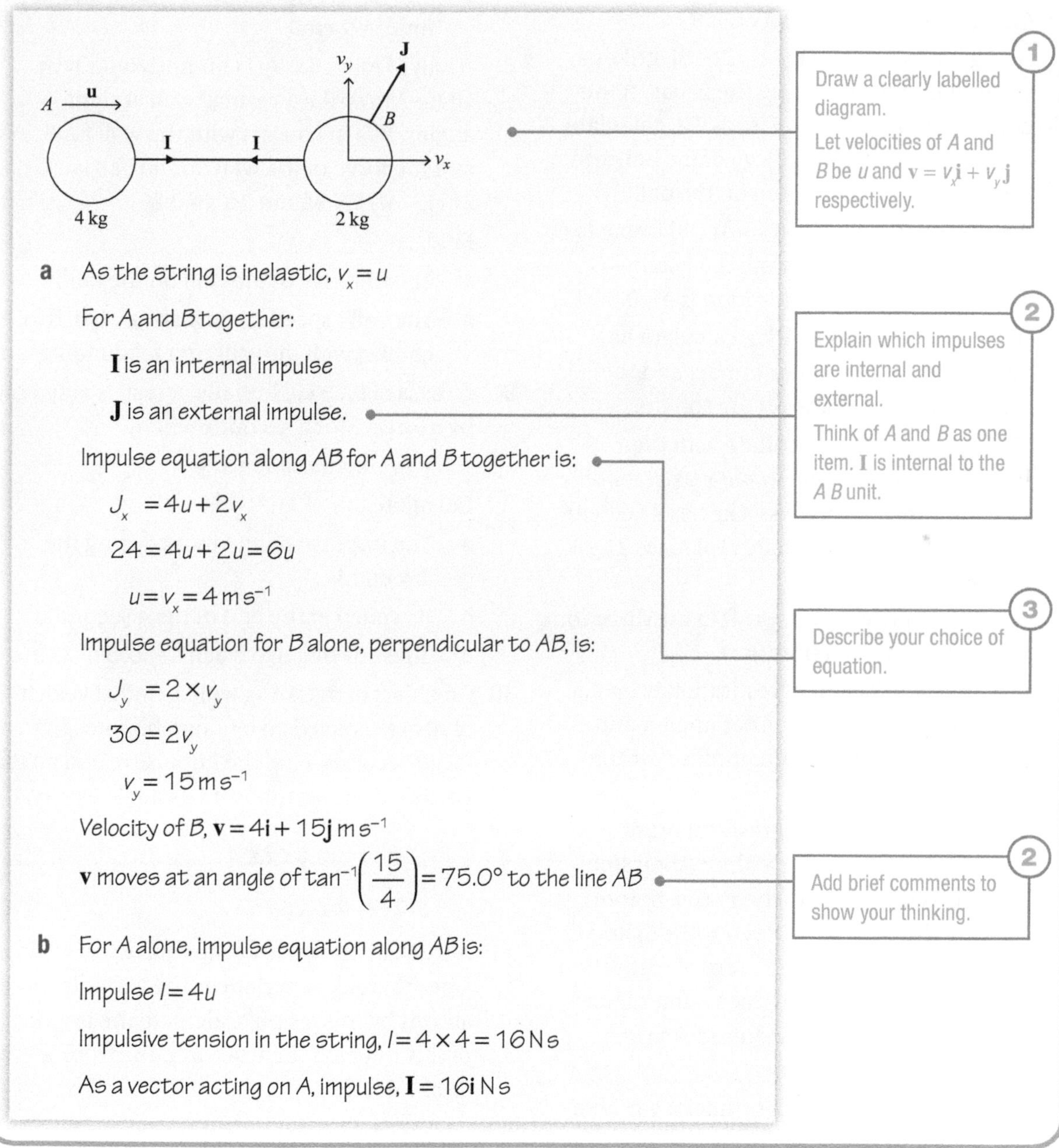

a As the string is inelastic, $v_x = u$

For A and B together:

$\mathbf{I}$ is an internal impulse

$\mathbf{J}$ is an external impulse.

Impulse equation along AB for A and B together is:

$J_x = 4u + 2v_x$

$24 = 4u + 2u = 6u$

$u = v_x = 4\text{ m s}^{-1}$

Impulse equation for B alone, perpendicular to AB, is:

$J_y = 2 \times v_y$

$30 = 2v_y$

$v_y = 15\text{ m s}^{-1}$

Velocity of B, $\mathbf{v} = 4\mathbf{i} + 15\mathbf{j}\text{ m s}^{-1}$

$\mathbf{v}$ moves at an angle of $\tan^{-1}\left(\frac{15}{4}\right) = 75.0°$ to the line AB

b For A alone, impulse equation along AB is:

Impulse $I = 4u$

Impulsive tension in the string, $I = 4 \times 4 = 16$ N s

As a vector acting on A, impulse, $\mathbf{I} = 16\mathbf{i}$ N s

1 Draw a clearly labelled diagram. Let velocities of A and B be u and $\mathbf{v} = v_x\mathbf{i} + v_y\mathbf{j}$ respectively.

2 Explain which impulses are internal and external. Think of A and B as one item. $\mathbf{I}$ is internal to the AB unit.

3 Describe your choice of equation.

2 Add brief comments to show your thinking.

You could check the answer to Example 5 part **b** by considering the impulse equation for B alone in the direction AB

$J_x - I = 2 \times v_x$ giving $I = 24 - 8 = 16$ N s in the direction towards A as in the diagram.

As a vector acting on B, impulse, $\mathbf{I} = -16\mathbf{i}$

MECH

Exercise 8.3B Reasoning and problem-solving

1 Particle A of mass 2 kg is moving at $12\,m\,s^{-1}$ and collides directly with particle B of mass 3 kg which is moving at $6\,m\,s^{-1}$ in the same direction. After impact, A and B coalesce. Calculate the impulse of A on B

2 A 5 kg ball with velocity $\mathbf{u} = 2\mathbf{i} - 3\mathbf{j}$ strikes a smooth plane containing the x-axis. If the coefficient of restitution is $e = 0.2$, calculate the final velocity $\mathbf{v} = v_x\mathbf{i} + v_y\mathbf{j}$ of the ball and the impulse $\mathbf{I}$ of the plane on the ball.

3 A sphere with velocity $\mathbf{u} = -4\mathbf{i} + 3\mathbf{j}$ impacts a smooth plane containing the y-axis. The coefficient is of restitution is $e = 0.6$

If the sphere's mass is 6 kg, calculate its velocity $\mathbf{v} = v_x\mathbf{i} + v_y\mathbf{j}$ after impact and the impulse $\mathbf{I}$ of the plane on the sphere.

4 Two smooth spheres collide with their line of centres parallel to the x-axis. Before impact, sphere A of mass 4 kg has a velocity $\mathbf{u} = 2\mathbf{i} + 3\mathbf{j}\,m\,s^{-1}$ and sphere B of mass 2 kg is at rest.

a Explain why B moves off in the direction of the x-axis after impact.

b If the coefficient of restitution is $e = 0.5$, find their velocities after impact and the magnitude of the impulse during impact.

5 When two smooth spheres, P of mass 2 kg and Q of mass 4 kg, collide, their line of centres is parallel to the y-axis. Before impact, the velocity of P is $\mathbf{u} = 4\mathbf{i} + 6\mathbf{j}$ and Q is at rest. Given the coefficient of restitution is $e = 0.4$, find their velocities immediately after impact and the magnitude of the impulse during impact.

6 A 3 kg mass travels for 5 s under a variable force F where, after t seconds, $F = 3t^2 + 2$ N

Calculate

a The impulse on the mass during the 5 s,

b The final velocity of the mass if its initial velocity is

i zero ii $4\,m\,s^{-1}$

7 A variable force F N acts for 10 s on a 5 kg mass with an initial speed of $2\,m\,s^{-1}$

If $F = 2t + 3$ at a time t s, find

a The impulse on the mass during the 10 seconds,

b The final velocity of the mass,

c The time for the mass to reach half its final velocity.

8 A ball of mass 0.2 kg is hit horizontally at $10\,ms^{-1}$ to strike a vertical wall at right angles. It is in contact with the wall for 0.25 s and the force of the wall on the ball is $2t^2(1 - 4t)$ kN where $0 \le t \le 0.25$

Find

a The impulse of the wall on the ball,

b The ball's speed at the instant when it loses contact with the wall after rebounding.

9 A mass of 0.75 kg, initially at rest, is acted on by a force F for 4 seconds where

$F = t + t^3$ N at a time t s

Calculate

a The impulse on the mass during the 4 seconds,

b Its speed at the end of the 4 seconds,

c The time taken to reach a speed of $33\,m\,s^{-1}$

10 An object of mass 4 kg with an initial velocity of $10\,m\,s^{-1}$ is acted on by a variable force F N for 12 seconds. Find the impulse acting on the particle over this time and its final velocity, if

$$F = \begin{cases} \frac{1}{2}t^2 & \text{for } 0 \le t \le 4 \\ 12 - t & \text{for } 4 \le t \le 12 \end{cases}$$

11 A truck of mass 400 kg moving at $5\,m\,s^{-1}$ on a smooth track experiences the variable force shown by this graph. Calculate the impulse of this force on the truck and the truck's final velocity.

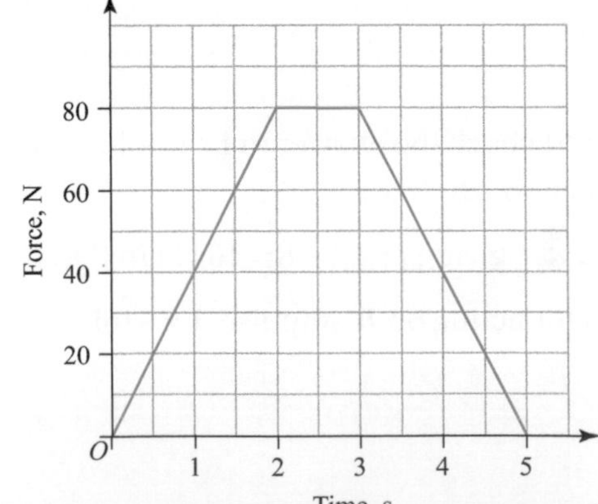

12 A box of mass 20 kg is pushed from rest over a smooth horizontal surface by the variable force shown in the diagram.

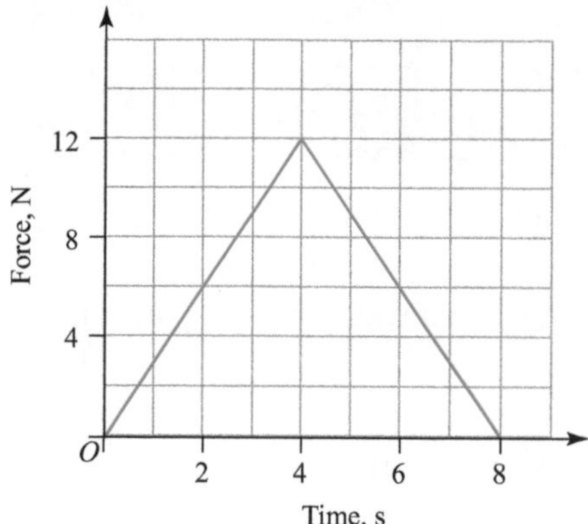

Calculate

a The impulse on the box,

b The box's final velocity,

c The time when its velocity is $0.6\,\text{m s}^{-1}$

13 A bullet of mass m and speed u strikes and embeds in a wooden disc of mass M moving in the same direction with speed v

Find the impulse of the bullet on the disc.

14 Two spheres P and Q, of mass 3 kg and 4 kg, on a smooth table are joined by a slack inelastic string. Q moves at $21\,\text{m s}^{-1}$ directly away from P which is at rest. At the moment the string becomes taut, find the velocity of both spheres and the magnitude of the impulsive tension in the string.

15 A shell of mass 1.25 kg is fired horizontally with a speed of $380\,\text{m s}^{-1}$ from a gun of mass 50 kg. The gun is brought to rest by a constant horizontal force acting over a distance of 1 metre. Calculate

a The impulse of the explosion and the initial speed of recoil of the gun,

b The magnitude of the force.

16 A smooth ball P of mass 6 kg moving at $10\,\text{m s}^{-1}$ collides with a smooth ball Q of mass 12 kg that is at rest. The direction of motion of P makes an angle of 30° with their line of centres. Given the coefficient of restitution is $e = 0.75$, find the velocities of the spheres after impact and the magnitude of the impulse during their contact.

17 A smooth ball of mass 2 kg with speed $4\sqrt{3}\mathbf{i} + 4\mathbf{j}\,\text{m s}^{-1}$ strikes a second smooth ball of mass 4 kg with speed $\sqrt{3}\mathbf{i} + \mathbf{j}\,\text{m s}^{-1}$

Just before impact, they are moving in parallel directions and the direction of motion makes an angle of 30° with the line of their centres. Given the coefficient of restitution is $e = \frac{1}{3}$, calculate their velocities after impact and the magnitude of the impulse between them during impact.

18 Two identical smooth spheres of mass m move with the same speed u

They collide when their centres are on the positive x-axis and their directions of motion are at 30° and 60°, respectively, to the x-axis. Find their speeds after impact and the magnitude of the impulse between them during impact if the spheres are perfectly elastic.

19 Two particles P and Q of masses m and M are at rest and connected by a light rod that lies in the direction $\mathbf{i}$

P is struck a blow of impulse $\boldsymbol{I} = x(\cos\theta\,\mathbf{i} + x\sin\theta\,\mathbf{j})$ which creates an impulsive thrust in the rod. Immediately after the impulse, the velocity of P is $v(\cos\phi\,\mathbf{i} + v\sin\phi\,\mathbf{j})$.

a Show that $\tan\phi = \frac{m+M}{M}\tan\theta$

b Comment on what happens to the motion of P when

i M is much larger than m,

ii M is zero.

20 Two equal mass particles, A and B, are joined by an inextensible string. Particle A is at rest. Particle B moves with velocity $u\left(\frac{\sqrt{3}}{2}\mathbf{i} + \frac{1}{2}\mathbf{j}\right)$. At the instant the string becomes taut it lies in the direction $\mathbf{j}$

Find the magnitude of the impulse in the string and the magnitude of the velocity of B at the instant the string becomes taut if

a A is held fixed, **b** A is free to move.

8 Summary and review

Chapter summary

- The momentum of a body is its mass multiplied by its velocity;
  momentum = $m \times \mathbf{v}$; the units of momentum are N s
- The principle of conservation of momentum applies when no external forces are present. It says that the total momentum is constant.
  So, total initial momentum = total final momentum
- When two bodies collide, Newton's experimental law gives the equation

$$\frac{\text{Speed of separation}}{\text{Speed of approach}} = e$$

$$e = \frac{v_2 - v_1}{u_1 - u_2}$$

- The constant e is known as the coefficient of restitution.
  If $e = 0$, the impact is inelastic.
  If $e = 1$, the bodies are perfectly elastic.
  In general, $0 < e < 1$
- When a force $\mathbf{F}$ acts on a body for a time t, the impulse $\mathbf{I} = \mathbf{F} \times t$ when $\mathbf{F}$ is constant or $\mathbf{I} = \int \mathbf{F}\,\mathrm{d}t$ when $\mathbf{F}$ is variable.
- An impulse creates a change in momentum. The impulse equation is
  Impulse of force = Change in momentum produced
- Velocity, momentum, force and impulse are all vectors, so equations involving them must take account of their directions.

Check and review

You should now be able to...	Try Questions
✔ Use the conservation of momentum equation and Newton's equation appropriately.	1–14
✔ Find the size of an impulse for constant and variable forces.	6–9, 12–15
✔ Apply these concepts to problems in two-dimensions using vectors.	2–3, 8, 12

1 Particle A of mass 2 kg moves with velocity $10\,\text{m s}^{-1}$

It collides directly with another particle B of mass 4 kg moving towards A with velocity $8\,\text{m s}^{-1}$

After the collision, B is at rest. Find the final velocity of A

2 A sphere of mass 2 kg has a velocity $5\mathbf{i} + 3\mathbf{j}\,\text{m s}^{-1}$

It collides and coalesces with a sphere of mass 4 kg with a velocity $2\mathbf{i} - 3\mathbf{j}\,\text{m s}^{-1}$
Their common velocity after the collision is $\mathbf{v}$
Calculate the speed $|\mathbf{v}|$ and the angle which $\mathbf{v}$ makes with the x-axis.

3 A particle of mass 5 kg has a velocity of $4\,\text{m s}^{-1}$ in the x-direction. A second particle has a mass m kg and a velocity of $2\,\text{m s}^{-1}$ in the y-direction. They collide, coalesce and begin to move together at 45° to their initial directions. Find their final velocity and the value of m

4 A particle with mass 2 kg and speed $20\,\text{m s}^{-1}$ collides head-on with a particle with mass 6 kg and speed $6\,\text{m s}^{-1}$

If the coefficient of restitution is $e = \frac{1}{2}$, calculate their speeds after impact.

5 A ball, which is initially at rest, drops from a height of 9 m onto a horizontal plane and rebounds to a height of 1 m. The ball then continues to bounce. Calculate the value of e and the total distance it has travelled when it hits the ground for the third time.
(Use $g = 10\,\text{m s}^{-2}$)

6 A railway truck of mass 10 tonnes and speed $3\,\text{m s}^{-1}$ strikes and couples with a lighter truck of mass 5 tonnes which is at rest on the same line. Find their common speed after impact and the impulse in the coupling between them during impact.

7 Sphere A of mass 3 kg is moving at $12\,\text{m s}^{-1}$ and collides directly with sphere B of mass m kg, which is moving at $2\,\text{m s}^{-1}$ in the same direction. After the impact, A moves at $2\,\text{m s}^{-1}$ in the opposite direction and B continues at $6\,\text{m s}^{-1}$

Calculate the values of m and e and the size of the impulse during the collision.

8 A smooth wall lies along the x-axis. A 4 kg ball with velocity $\mathbf{u} = 5\mathbf{i} - 2\mathbf{j}\,\text{m s}^{-1}$ strikes the wall where the coefficient of restitution, $e = 0.5$. Calculate the final velocity $\mathbf{v} = v_1\mathbf{i} + v_2\mathbf{j}\,\text{m s}^{-1}$ of the ball and the impulse $\mathbf{I}$ of the wall on the ball.

9 Water flows horizontally from a pipe of cross-section $0.03\,\text{m}^2$ at a speed of $25\,\text{m s}^{-1}$ It strikes a fixed vertical wall and does not rebound. Take the density of water as $1000\,\text{kg m}^{-3}$ and find the force F on the wall.

10 A particle of mass 2 kg has an initial velocity of $9\,\text{m s}^{-1}$

A force $F = 3t^2 - 6t + 2$ N acts on the particle for 3 s in the same direction as the velocity. Find the impulse on the particle and its speed after 3 seconds.

11 Three particles A, B and C with masses of 1 kg, 2 kg and 4 kg respectively are in the same line on a smooth horizontal surface in this order with A at rest. B and C collide head-on with velocities of $10\,\text{m s}^{-1}$ and $8\,\text{m s}^{-1}$ respectively. Given that $e = 0.5$ for this and any subsequent impacts, show that B then collides with A

Find all three velocities after this second collision and explain why there are no more collisions.

12 Two spheres, P of mass 2 kg and Q of mass 4 kg, lie at rest on the x-axis with P nearer the origin. They are connected by a straight string. Q is struck by an impulse $\mathbf{I} = 20\mathbf{i} + 24\mathbf{j}$ N s. Find the impulsive tension in the string and the spheres' velocities immediately after the blow.

13 A ball of mass 1 kg strikes a smooth fixed plane with a velocity of $20\,\text{m s}^{-1}$ at 30° to the plane where the coefficient of restitution is $e = 0.4$. Find the angle that the final velocity makes with the wall and also the impulse on the ball during the collision.

14 Four identical balls of mass m lie at rest on a smooth horizontal surface at the points $(0, 0)$, $(0, a)$, $(a, 0)$ and (a, a). The balls are joined together by four inelastic strings of length a
The ball at (a, a) is struck by an impulse $I\left(\frac{1}{\sqrt{2}}\mathbf{i} + \frac{1}{\sqrt{2}}\mathbf{j}\right)$
Find the initial velocities of the four balls in terms of m and I
Clearly state any assumptions that you have made.

15 A mass of 0.5 kg, initially at rest, is acted on by a variable force F for 4 s where $F = 2t + 4$ N and t is the time in seconds. Find

a The impulse acting on the mass,

b The final speed of the mass,

c The time for the mass to reach half its final speed.

8 Exploration

Going beyond the exams

Information

The term momentum refers to a 'quantity of motion of a moving body'. It is borrowed from Latin, where it means 'movement' or 'moving power'. The concept of momentum was first introduced by the French mathematician Descartes, who is perhaps most famous for his development of Cartesian graphs.

Investigation

A Newton's cradle is a device that is named after Sir Isaac Newton, and can be used to demonstrate conservation of momentum and energy. The device consists of a number of spheres (usually 5 or 6) suspended so that, when at rest, their centres are all at the same height. When a sphere at the end is lifted and released, it strikes the stationary spheres – a force is transmitted through the stationary spheres and pushes the last sphere out and upward.

Investigate, using ideas of conservation of momentum and energy, what will happen if first one ball is lifted at the end of the cradle and released. Then think about the case for two balls, three balls, and four balls.

History

Sir Isaac Newton first presented his three laws of motion in the "Principia Mathematica Philosophiae Naturalis" in 1686. His second law defines a **force** to be equal to the rate of change of momentum with respect to time, with momentum being defined to be the product of the mass, m, and its velocity, v. An alternative form is $F = ma$

Research

In physics, **angular momentum** is the rotational equivalent of linear momentum. It is an important quantity in mechanics because it is a conserved quantity, and so the total angular momentum of a system remains constant unless acted on by an external **torque**. A torque is the rotational equivalent of a force and provides a twist to an object.

Ideas of angular momentum and torque are important in understanding the motion of a gyroscope. Research gyroscopes and write a report on what they are, how they work, and how they can be used.

8 Assessment

1 An object of mass 7 kg, travelling at a speed of $4\,\text{m s}^{-1}$, is acted on by a constant force in its direction of travel which increases its speed to $10\,\text{m s}^{-1}$ in the same direction. Calculate

a The impulse exerted on the object, **[2 marks]**

b The force involved if the process took 0.35 seconds. **[2]**

2 A particle of mass 2 kg is travelling at $8\,\text{m s}^{-1}$ in a straight line. A variable braking force, acting along the same straight line, is applied so that, after t s, the magnitude of the force is $2t$ N. Calculate the time taken for the particle to come to rest. **[3]**

3 A particle of mass 5 kg is travelling with a velocity of $(4\mathbf{i} + \mathbf{j})\,\text{m s}^{-1}$ when it is subjected to an impulse of $(2\mathbf{i} - 7\mathbf{j})$ N s. Calculate the new velocity of the particle. **[3]**

4 A particle of mass 3 kg has velocity $(2\mathbf{i} - 3\mathbf{j})\,\text{m s}^{-1}$

It is acted on by a constant force of $(-\mathbf{i} + 2\mathbf{j})$ N, which changes its velocity to $0.5\mathbf{i}\,\text{m s}^{-1}$

For how long does the force act? **[4]**

5 A bullet of mass 0.1 kg is fired horizontally at a block of wood of mass 2 kg, which is stationary and free to move. The bullet enters the block travelling at $100\,\text{m s}^{-1}$
Calculate the subsequent speed of the block if

a The bullet passes through the block and emerges travelling at $40\,\text{m s}^{-1}$, **[2]**

b The bullet becomes embedded in the block. **[2]**

6 A railway truck of mass $3m$, travelling at a speed of $2v$, collides with another truck of mass $4m$, which is travelling with a speed of v

The trucks become coupled together. Find, in terms of v, the common speed of the trucks if, before impact, they were travelling

a In the same direction, **[2]**

b In opposite directions. **[2]**

7 A particle A, of mass 10 kg, is moving at $5\,\text{m s}^{-1}$ when it collides with a particle B, of mass m kg, travelling in the opposite direction at $2\,\text{m s}^{-1}$

After the collision, A travels in the same direction as before but with its speed reduced to $3\,\text{m s}^{-1}$

a If $m = 3$, find the velocity of B after the collision. **[3]**

b Show that the value of m cannot be greater than 4 **[4]**

8 Smooth uniform spheres A and B, of equal size, have masses of 3 kg and 2 kg respectively and move on a smooth horizontal surface. A has velocity $(4\mathbf{i} + 3\mathbf{j})\,\text{m s}^{-1}$ and B has velocity $(-\mathbf{i} - 5\mathbf{j})\,\text{m s}^{-1}$

They collide when the line joining their centres is parallel with $\mathbf{j}$

The coefficient of restitution between the spheres is 0.5

a Calculate the velocities of A and B after the collision. **[5]**

b Calculate the impulse received by B **[2]**

c Calculate the angle through which the motion of A has been deflected by the collision. **[2]**

9 A sledgehammer of mass 6 kg, travelling at $20\,\text{m}\,\text{s}^{-1}$, strikes the top of a post of mass 2 kg and stays in contact with the post.

a Calculate the common speed of the hammer and post immediately after impact. **[2]**

b The post and hammer are brought to rest in 0.02 s by the action of a resistive force R from the ground. By modelling R as constant, find its magnitude. **[2]**

c If, in fact, the force is given by $R = k(1 + 2t)$ N, where t s is the time from the moment of impact, find the value of the constant k given that post is again brought to rest in 0.02 s **[3]**

10 An object of mass 3 kg has velocity $(3\mathbf{i} + 2\mathbf{j})\,\text{m}\,\text{s}^{-1}$

It collides with another object, which has a mass of 2 kg and a velocity of $(\mathbf{i} - \mathbf{j})\,\text{m}\,\text{s}^{-1}$

After the impact, the first object has a velocity of $(2\mathbf{i} + \mathbf{j})\,\text{m}\,\text{s}^{-1}$

Calculate the velocity of the second object after the collision. **[2]**

11 Particles A and B have masses of 3 kg and 2 kg respectively. They are connected by a light inextensible string. The particles lie at rest on a smooth horizontal surface. The coefficient of restitution between the particles is 0.5. A is projected towards B with velocity $10\,\text{m}\,\text{s}^{-1}$

a Calculate

i The velocities of the particles after the collision, **[4]**

ii The common velocity of the particles after the string becomes taut. **[2]**

b i Explain why the answer to **a ii** is independent of e, provided $e > 0$ **[1]**

ii Describe what would happen if $e = 0$ **[1]**

12 A particle A, with mass 2 kg and velocity $10\,\text{m}\,\text{s}^{-1}$, is moving on a smooth horizontal surface. It catches up and collides with a second particle B, with mass 1 kg which is traveling with velocity $5\,\text{m}\,\text{s}^{-1}$ along the same straight line. After the impact, B then goes on to collide head-on with a vertical wall. The coefficient of restitution between A and B is 0.5, and between B and the wall is 0.75

a Calculate the velocities of A and B after they collide for the first time. **[4]**

b Calculate the velocities of the particles after they collide for a second time. **[4]**

c Describe what happens after the second impact. **[4]**

13 Three particles A, B and C have masses 3 kg, 2 kg and 1 kg respectively. They are moving, in that order, along a straight line with velocities of $3\,\text{m}\,\text{s}^{-1}$, $2\,\text{m}\,\text{s}^{-1}$ and $1\,\text{m}\,\text{s}^{-1}$ respectively. The collision between A and B, which happens first, is perfectly elastic. The second collision, between B and C, has coefficient of restitution e

a Find the velocities of A and B after the first collision. **[4]**

b Find, in terms of e, the velocity of B after the second collision. **[3]**

c Hence show that there will be no more collisions if $e \le \frac{4}{11}$ **[3]**

14 The diagram shows a snooker table *ABCD* from above. The surface is horizontal. The sides *AB* and *AD* are parallel to the **i**- and **j**-directions respectively, as shown.

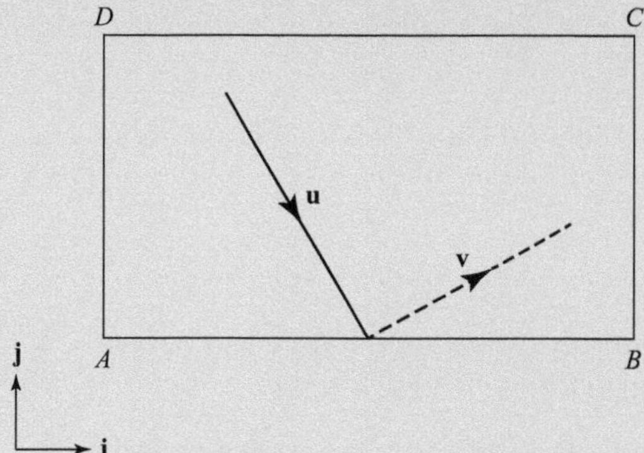

A ball of mass 0.16 kg, travelling with velocity $\mathbf{u} = (5\mathbf{i} - 7\mathbf{j})\,\text{m s}^{-1}$, strikes the cushion *AB* (which is assumed to be smooth) and rebounds with velocity $\mathbf{v}\,\text{m s}^{-1}$

The coefficient of restitution between the ball and the cushion is 0.7

a **i** Explain why the x-component of $\mathbf{v}$ is $5\,\text{m s}^{-1}$ **[1]**

ii Calculate the y-component of $\mathbf{v}$ **[2]**

iii Find the angle through which the direction of the ball has been changed. **[4]**

b The ball is in contact with the cushion for 0.1 s

The force exerted by the cushion has magnitude $kt^2(3 - 8t)$ N, where t s is the time since first contact. Find the value of the constant k **[4]**

15 The diagram shows particles *A*, *B* and *C* of masses 1 kg, 2 kg and 2 kg respectively, which are initially at rest in a straight line.

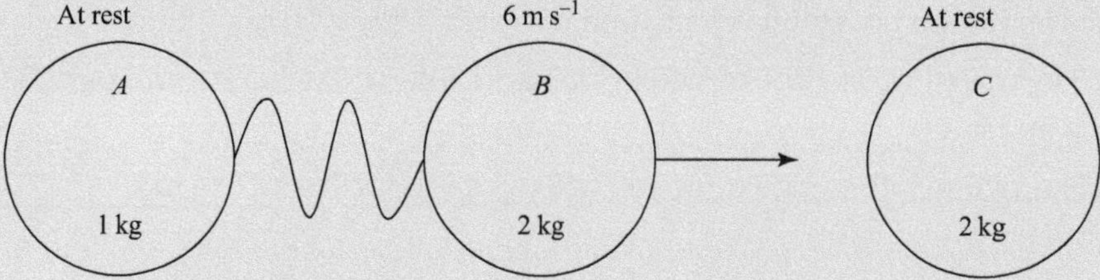

A and *B* are connected by a light inextensible string which is slack. *B* is then propelled towards *C* at $6\,\text{m s}^{-1}$

The coefficient of restitution between *B* and *C* is 0.5. The string becomes taut before *B* reaches *C*

a Find the common speed of *A* and *B* at the instant after the string becomes taut. **[2]**

b Find the impulse on *C* when *B* collides with it. **[4]**

c Show that, after the string becomes taut for a second time, there are no more collisions. **[4]**

16 Particles of mass m and $2m$, both travelling with speed u, collide head on. After the collision, one particle is travelling at twice the speed of the other. Find their speeds after impact in terms of u **[3]**

17 Particle A of mass m, moving at $7\,\text{m}\,\text{s}^{-1}$, catches up and collides with particle B, of mass km, moving at $1\,\text{m}\,\text{s}^{-1}$

After the collision, the speed of B is twice the speed of A. The coefficient of restitution is 0.75

Find the two possible values of k **[5]**

18 A particle A, of mass 4 kg, traveling at $6\,\text{m}\,\text{s}^{-1}$ catches up and collides with a particle B, of mass m kg, travelling at $1\,\text{m}\,\text{s}^{-1}$

The particles coalesce and move with speed $3\,\text{m}\,\text{s}^{-1}$

a Calculate the value of m **[2]**

b What would their common velocity after the collision have been if they had been travelling in opposite directions before the collision? **[2]**

19 A tennis player strikes a ball so that its path is exactly reversed. The ball approaches the racket at $35\,\text{m}\,\text{s}^{-1}$ and leaves at $45\,\text{m}\,\text{s}^{-1}$

The mass of the ball is 90 g. Find the magnitude of the impulse exerted on the ball. **[2]**

20 A gun of mass 500 kg, which is free to move, fires a shell of mass 5 kg horizontally at a speed of $200\,\text{m}\,\text{s}^{-1}$

Find the initial speed of recoil of the gun. **[3]**

21 Two particles A and B, of mass 4 kg and 2 kg respectively, are moving with respective speeds of $1\,\text{m}\,\text{s}^{-1}$ and $10\,\text{m}\,\text{s}^{-1}$ directly towards a fixed vertical wall. B hits the wall, rebounds and then collides with A. The coefficient of restitution between B and the wall is 0.4, and between the particles is 0.2

a Show that B is brought to rest when it collides with A **[3]**

b Calculate the speed at which A is travelling after its collision with B **[2]**

22 An object of mass 4 kg, travelling with velocity $(5\mathbf{i} + 2\mathbf{j})\,\text{m}\,\text{s}^{-1}$, is struck by a second object of mass 6 kg and velocity $\mathbf{v}$

The two objects coalesce. If their common velocity after impact is $(2\mathbf{i} - 4\mathbf{j})\,\text{m}\,\text{s}^{-1}$, find $\mathbf{v}$ **[3]**

23 A particle A, of mass 3 kg travelling at $10\,\text{m}\,\text{s}^{-1}$, catches up and collides with a second particle, B, of mass 5 kg, travelling at $2\,\text{m}\,\text{s}^{-1}$

After the collision, A is moving in the same direction with its speed reduced to $4\,\text{m}\,\text{s}^{-1}$

Calculate

a The new speed of B **[2]**

b The coefficient of restitution. **[2]**

24 A particle of mass 6 kg, travelling at $8\,\text{m}\,\text{s}^{-1}$, is brought to rest in a head-on collision with a second particle of mass 4 kg

If $e = 0.3$, calculate the initial and final velocities of the second particle. **[4]**

25 A body of mass 5 kg is travelling with velocity $(2\mathbf{i} + 3\mathbf{j})\,\text{m}\,\text{s}^{-1}$ when it receives an impulse $\mathbf{I}$ N s. This changes its velocity to $(4\mathbf{i} + 7\mathbf{j})\,\text{m}\,\text{s}^{-1}$

Find $\mathbf{I}$ and show that it has a magnitude of $10\sqrt{5}$ N s **[4]**

26 A particle of mass 8 kg is travelling with velocity $2\,\text{m}\,\text{s}^{-1}$ when it is acted upon for a period of 4 s by a force $F = (3t^2 + 1)$ N in the same direction as the velocity. Find

a The impulse received by the particle, **[2]**

b Its final velocity. **[2]**

9 Circular motion 1

The analysis of circular motion has been particularly significant in making advances in medical sciences because of the design of clinical centrifuges for use in medical laboratories. These allow technicians to separate serum, urea, blood samples and so on into their constituent parts. Measurements can then be made that support quick and accurate diagnoses of patients' medical problems. This ensures that they can be treated as soon as possible, which increases the likelihood of a quick return to good health.

Understanding of circular motion has applications in other areas of life. For example, a good understanding of circular motion is significant in the design of theme park rides, and in improving athletic performance.

Orientation

What you need to know

Maths Ch7

- Equations of motion.

What you will learn

- To use equations describing circular motion.
- To analyse horizontal circular motion.

What this leads to

Chapter 24 Circular motion 2

9.1 Kinematics of circular motion

Fluency and skills

When a point mass P moves in a circle with centre O and radius r, the distance s it travels is the arc length and θ is the angle through which the radius OP rotates.

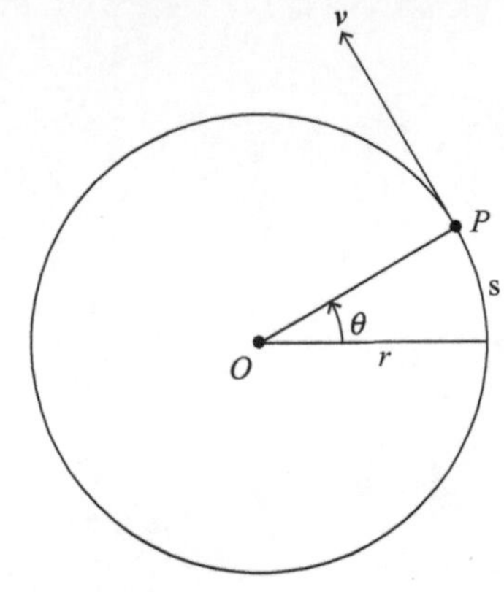

The arc length, $s = \frac{\theta}{2\pi} \times 2\pi r = r\theta$, with θ in radians.

The **linear velocity**, v of P along the tangent is the rate of change of s and the **angular velocity**, ω about O is the rate of change of θ

So: $\quad \omega = \frac{\mathrm{d}\theta}{\mathrm{d}t}$ and $v = \frac{\mathrm{d}s}{\mathrm{d}t} = r\frac{\mathrm{d}\theta}{\mathrm{d}t} = r\omega$

In this chapter, P moves at constant speed, so the magnitude of v is constant but its direction changes. Angular velocity ω is also constant and is measured in **radians per second** (rad s^{-1}), although revolutions per minute (rpm) is sometimes used.

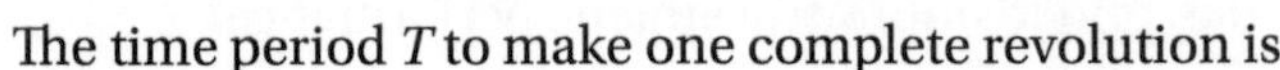

The time period T to make one complete revolution is

$$\frac{\text{Angular displacement, radians}}{\text{Angular speed, rad/sec}} = \frac{2\pi}{\omega} \text{ seconds}$$

> **Key point**
>
> For circular motion at constant speed with θ in radians
>
> $s = r\theta \qquad \omega = \frac{\mathrm{d}\theta}{\mathrm{d}t} \qquad v = r\omega \qquad T = \frac{2\pi}{\omega}$

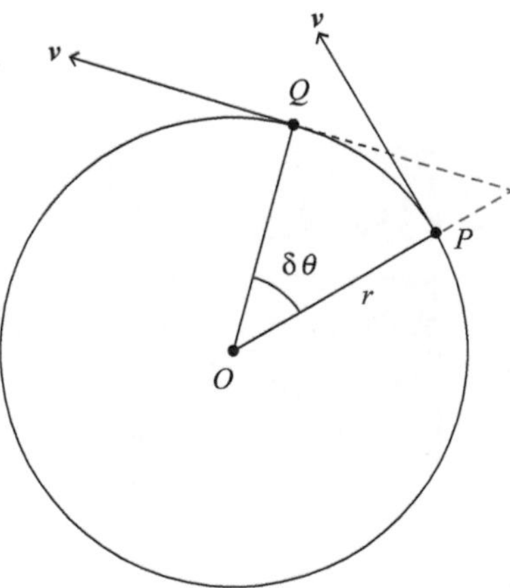

As the mass travels around the circle, it moves from P to Q through a small angle $\delta\theta$ in a time δt

At P and Q, the velocity along the tangent is v

The magnitude of this tangential velocity is constant.

However, the change in velocity along $PO = v\sin\delta\theta - 0 \approx v\delta\theta$

So the acceleration towards $O = \frac{\text{change in velocity}}{\text{time taken}} = \frac{v\delta\theta}{\delta t}$

As $\delta\theta \to 0$, the acceleration along PO tends to $v\frac{\mathrm{d}\theta}{\mathrm{d}t}$

> For a small angle $\delta\theta$ in radians, $\sin\delta\theta = \delta\theta$

where angular velocity $\frac{\mathrm{d}\theta}{\mathrm{d}t} = \omega = \frac{v}{r}$ Hence $a = v\frac{\mathrm{d}\theta}{\mathrm{d}t} = \frac{v^2}{r} = r\omega^2$

In summary, there is no linear acceleration along the tangent, but there is a linear acceleration of $r\omega^2 = \frac{v^2}{r}$ towards the centre of the circle. You can think of this result as an acceleration that changes the direction of the particle but not its speed.

Key point

A body moving on a circular path of radius r with constant angular velocity ω about the centre has

- a constant speed $v = r\omega$ along the tangent
- an acceleration $a = r\omega^2 = \dfrac{v^2}{r}$ towards the centre.

Example 1

A cyclist is mending his bike. He places it upside down with the wheel spinning 200 times every minute. A piece of grit is stuck in the tyre.

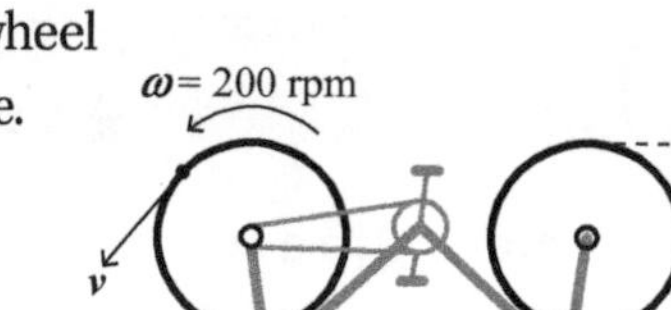

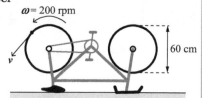

If the diameter of the tyre is 60 cm, calculate

a The angular speed of the wheel in radians per second,

b The linear tangential speed of the grit in metres per second,

c The acceleration of the grit towards the centre of the wheel.

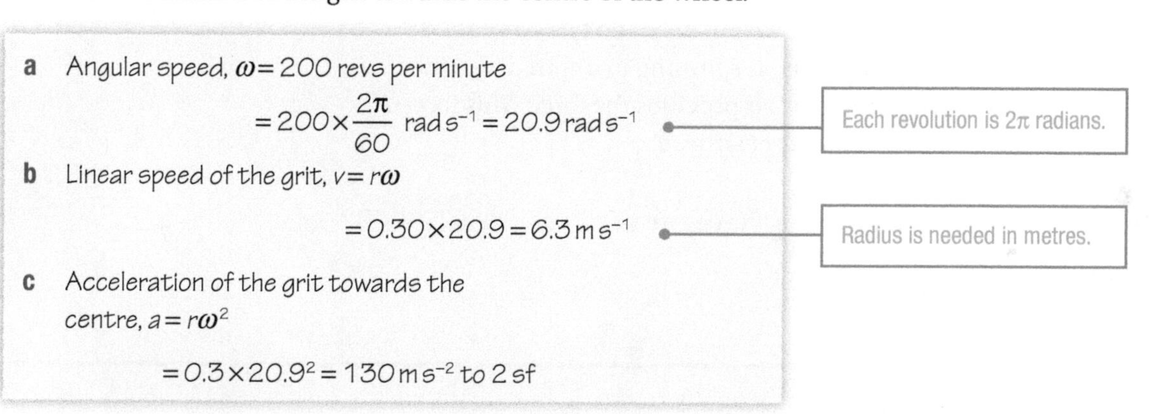

a Angular speed, $\omega = 200$ revs per minute

$$= 200 \times \frac{2\pi}{60} \text{ rad s}^{-1} = 20.9 \text{ rad s}^{-1}$$

Each revolution is 2π radians.

b Linear speed of the grit, $v = r\omega$

$$= 0.30 \times 20.9 = 6.3 \text{ m s}^{-1}$$

Radius is needed in metres.

c Acceleration of the grit towards the centre, $a = r\omega^2$

$$= 0.3 \times 20.9^2 = 130 \text{ m s}^{-2} \text{ to 2 sf}$$

Exercise 9.1A Fluency and skills

1 This table shows various speeds measured in different units. Copy and complete the table.

		rpm	rad/min	rad s^{-1}
a	Vinyl record	33		
b	Roundabout	9		
c	Wind turbine			10
d	Car engine			260

2 The London Eye, with a diameter of 120 m, makes a full revolution in 27 minutes. Calculate the magnitude of its angular velocity (in rad s^{-1}) and its linear velocity (in m s^{-1}) of a point on its circumference.

3 The hands of the Big Ben clock in London are 2.7 m and 4.2 m long. For each hand, calculate the angular velocity (in rad s^{-1}) and the speed of the tip of the hand (in m s^{-1}).

4 Given that the Earth's radius is 6371 km, calculate its angular speed in rad s^{-1} and work out the speed, in km h^{-1}, relative to the centre of the Earth, of a point which lies on the equator.

5 A mass is attached to a fixed point, P, on a smooth horizontal plane by a string 1.5 m long. The mass moves in a circular path about P, with the string taut, at a speed of 6 m s^{-1}

Calculate the angular velocity of the mass and its acceleration towards P

6 A fairground carousel rotates 10 times every minute. A child is 3 metres from the centre. Calculate the linear speed of the child and her acceleration towards the centre of the carousel.

7 The crankshaft of a car engine has a radius of 3 cm and rotates at a steady speed of 2000 rpm. Calculate the linear speed (in $m\,s^{-1}$) of a point on its circumference and the acceleration (in $m\,s^{-2}$) of that same point towards the centre of the shaft.

8 A spin dryer has a maximum angular speed of 1200 rpm and a drum with a diameter of 48 cm. What is the linear speed of a sock on the circumference of the drum when the drum is spinning at maximum speed? What is the sock's acceleration towards the centre of the drum?

Reasoning and problem-solving

Since a body moving in a circle has an acceleration of $r\omega^2$ (or $\frac{v^2}{r}$) towards the centre, Newton's 2nd law requires that there is a force equal to $m \times r\omega^2$ (or $m\frac{v^2}{r}$) towards the centre.

The force can be provided in different ways. For example, when a car rounds a horizontal bend, friction between tyres and road provides the force; when clothing is spinning in a spin dryer, the reaction between drum and clothes provides the force. This force F to the centre is called the **centripetal force.**

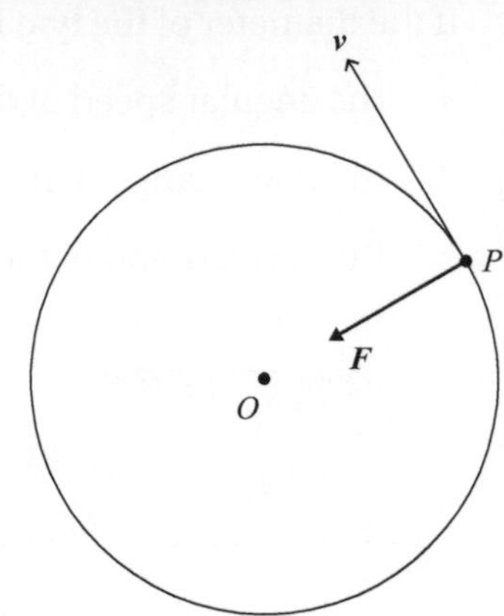

Key point

The equation of motion towards the centre is:

Centripetal force $= mr\omega^2 = m\frac{v^2}{r}$

Strategy

To solve problems involving motion in a circle

1. Draw a clear diagram showing all the forces acting on the moving object.
2. Write and solve an equation for motion towards the centre of the circle using Newton's 2nd law

Example 2

The ends of a 2 m length of string are attached to a fixed point O and a 3 kg mass. The mass rotates about O on a smooth table.

If the string breaks when the tension is greater than 530 N, find the maximum angular velocity in rpm.

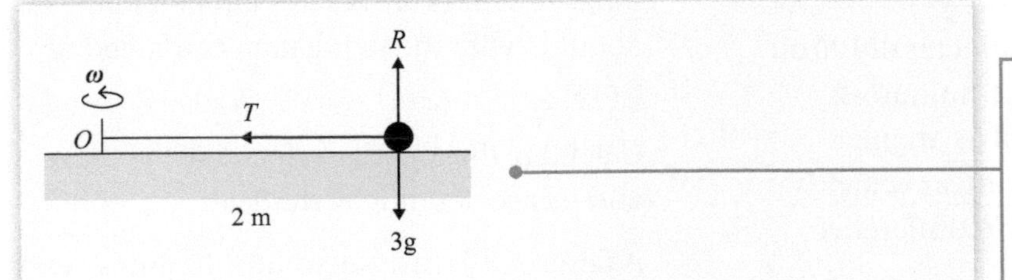

1. Draw a diagram showing all the forces. The vertical forces R and $3g$ do not contribute to the centripetal force.

(*Continued on the next page*)

Let ω be the maximum angular velocity.

On the point of breaking, the equation of motion is
$530 = m \times r\omega^2 = 3 \times 2 \times \omega^2$

Use Newton's second law for motion towards the centre of the circle O

Maximum angular velocity, $\omega = \sqrt{\frac{530}{3 \times 2}} = 9.4\text{ rad s}^{-1}$

$= \frac{9.4 \times 60}{2\pi}$

$= 90$ rpm to 2 significant figures

Exercise 9.1B Reasoning and problem-solving

1 A 2 kg mass is attached to a light, inextensible string of length 1.5 m, the other end of which is fixed to a point on a smooth horizontal table. The mass moves on the table in a circle at a speed of 3 m s^{-1} with the string taut. Calculate the tension in the string.

2 An inextensible string 1.5 m long connects a particle P of mass 0.5 kg to a fixed point O on a smooth horizontal table. If the particle moves on the table in a circle about O with the string taut and completes 90 revolutions every minute, calculate the tension in the string.

3 The engine of a toy train has a mass of 0.8 kg

It runs on a circular horizontal track of radius 1.2 m

a What produces the force that enables the engine to travel in a circle?

b If the engine has an angular speed of 10 rpm, calculate the force towards the centre of the circle, in newtons.

4 A spring balance is used to measure force. One end of a spring balance is attached to a fixed point O on a horizontal smooth surface. The other end is attached to a 5 kg mass which rotates at a steady speed in a circle of radius 20 cm about O fifty times each minute. What is the reading on the spring balance?

5 A car of mass 900 kg is driven round a circular bend on an icy road of radius 20 m The frictional force between the tyres and road cannot exceed 250 N. What is the maximum angular speed of the car if no skidding occurs? What is the reading on the car's speedometer (which shows kilometres per hour) in this case?

6 A particle of mass 100 grams rests on a rough horizontal turntable at a distance of 20 cm from the centre. The turntable rotates at a constant angular velocity. The particle is on the point of moving when the frictional force is 0.5 N. What is its maximum angular velocity, in rpm, for which the particle will not slip?

7 A railway engine of mass 50 tonnes travels on a horizontal track at 18 km h^{-1} round a circular arc of radius 0.3 km. Calculate the total sideways force exerted on its wheels by the track.

8 A mass m on a smooth horizontal table is attached by an inextensible string through a small smooth hole in the centre of the table to an equal mass which hangs freely under the table. The mass on the table moves on a circular path, of radius r, around the hole. Its speed is sufficient to hold the hanging particle at rest. Show that the required linear speed is $\sqrt{rg}$

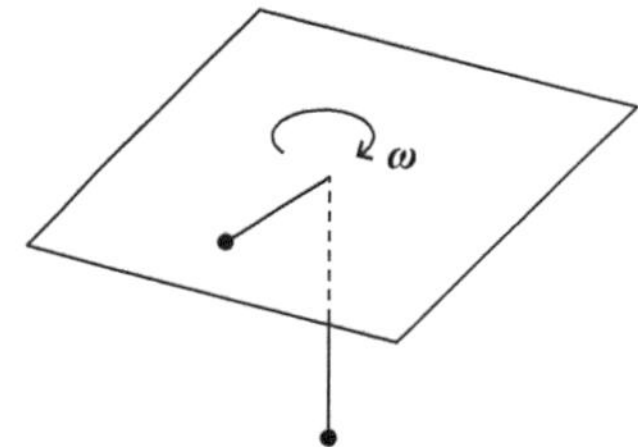

9.2 Horizontal circular motion

Fluency and skills

When circular motion takes place in a horizontal plane, the centripetal force is horizontal.

The force of gravity has no effect on horizontal motion and so does not contribute to the centripetal force.

Key point

A body moving on a circular path of radius r with constant angular velocity ω about the centre has

- a constant speed $v = r\omega$ along the tangent
- an acceleration $a = r\omega^2 = \frac{v^2}{r}$ towards the centre
- a centripetal force $F = mr\omega^2 = m\frac{v^2}{r}$

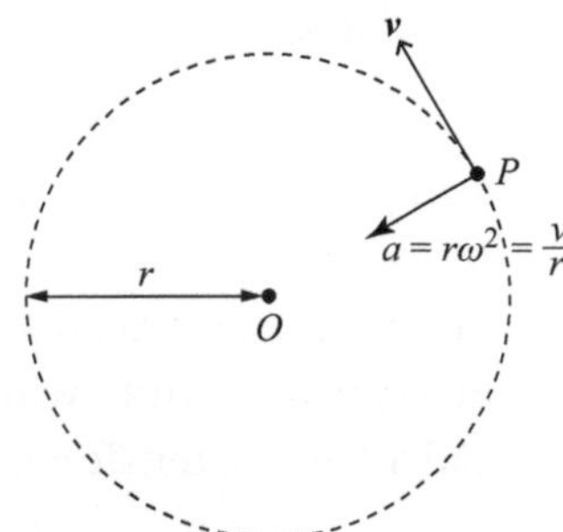

Example 1

Calculate the maximum speed (in km h^{-1}) for a car of 900 kg to round a bend of radius 60 m on a level road if

a The maximum frictional force between the tyres and road is 1440 N

b The coefficient of friction between the car's tyres and the road is 0.5 **Full A Level**

Take $g = 9.8\ \text{m s}^{-2}$

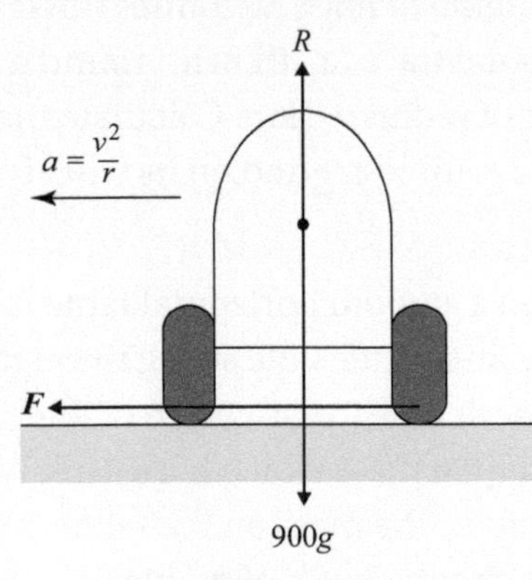

Draw a diagram showing all the forces.

The total reaction of the ground on the car through its tyres has a vertical component, R, and a frictional component, F

To prevent side-slipping, F acts towards the centre of the bend.

a Let v be the maximum speed of the car.

At maximum speed, the horizontal equation of motion is:

$$F = m \times \frac{v^2}{r}$$

R does not contribute to the centripetal force

$$1440 = 900 \times \frac{v^2}{60}$$

$$v^2 = \frac{1440 \times 60}{900} = 96$$

Maximum speed $v = \sqrt{96}\ \text{m s}^{-1} = \frac{\sqrt{96} \times 60 \times 60}{1000} = 35.3\ \text{km h}^{-1}$

(*Continued on the next page*)

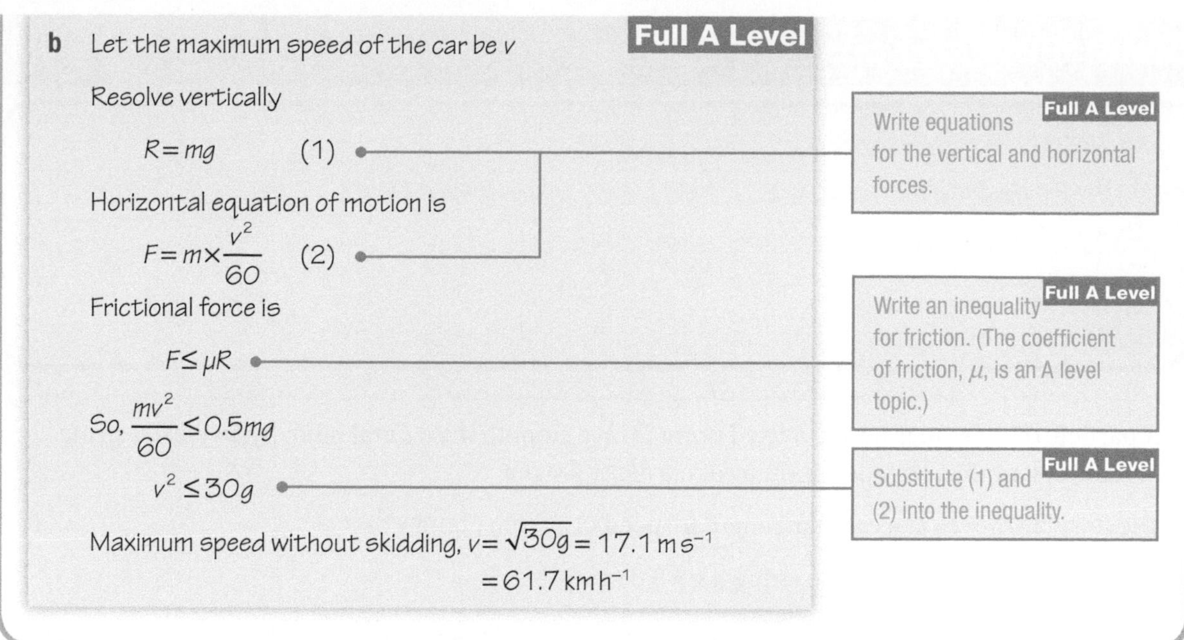

b Let the maximum speed of the car be v
Full A Level
Resolve vertically
R = mg (1)
Horizontal equation of motion is
F = m × v²/60 (2)
Frictional force is
F ≤ μR
So, mv²/60 ≤ 0.5mg
v² ≤ 30g
Maximum speed without skidding, v = √30g = 17.1 m s⁻¹
= 61.7 km h⁻¹
Full A Level
Write equations for the vertical and horizontal forces.
Full A Level
Write an inequality for friction. (The coefficient of friction, μ, is an A level topic.)
Full A Level
Substitute (1) and (2) into the inequality.

Exercise 9.2A Fluency and skills

1 A particle of mass 4 kg is attached to a fixed point O on a smooth horizontal table by a light inextensible string 0.6 m long. Calculate the tension in the string when the particle travels in a circle once every second about point O

2 A particle of mass 200 grams on a smooth horizontal table is attached by a string of length 80 cm to a fixed point on the table. What is the linear speed at which it rotates if the tension in the string is 0.5 N?

3 A string 0.6 m long can just support a mass of 10 kg without breaking. With one end fixed to a point on a smooth horizontal table, a mass of 2 kg rotates about the fixed point. Calculate the greatest number of revolutions the 2 kg mass can complete each minute without breaking the string. Use $g = 9.8\text{ m s}^{-1}$

4 A smooth circular table of radius 80 cm has a raised vertical rim. A small ball travels round the table in contact with the rim at a speed of 3 m s^{-1}

If the mass of the ball is 40 grams, calculate the force exerted by the rim on the ball.

5 A light rod OMA of length $2a$ has two equal masses m attached to it, one at its midpoint M and the other at its end A. The rod's other end O is fixed to a point on a smooth horizontal table. The rod and masses are in contact with the table and rotate about O with angular velocity ω

Calculate the tensions in the two halves of the rod in terms of m, a and ω

6 Three equal masses m are attached to points P, Q and R on a light rod OR, revolving round O in a horizontal plane, such that OP, PQ and QR are equal. Prove that the stresses in the three parts of the rod are in the ratio 3 : 5 : 6

7 A particle of mass m on a rough turntable is on the point of moving when it is a distance r from the centre and the turntable is rotating at $\omega\text{ rad s}^{-1}$

If the particle were at a distance of $2r$ from the centre, what angular speed would cause it to be on the point of moving?

Reasoning and problem-solving

Strategy

To solve problems involving motion in a horizontal circle

1. Draw a clear diagram showing all the forces acting on the moving object.
2. Decide which forces contribute to the centripetal force and write an equation for motion towards the centre of the circle.
3. Write other equations as necessary and solve them simultaneously.

Example 2

A particle of mass m is tied to a fixed point O on a smooth horizontal table by an elastic string of natural length 0.3 m and modulus of elasticity $\lambda = mg$

The particle moves at a constant angular speed of 20 rpm about point O

Calculate the extension of the string if $g = 9.8$ m s^{-2}

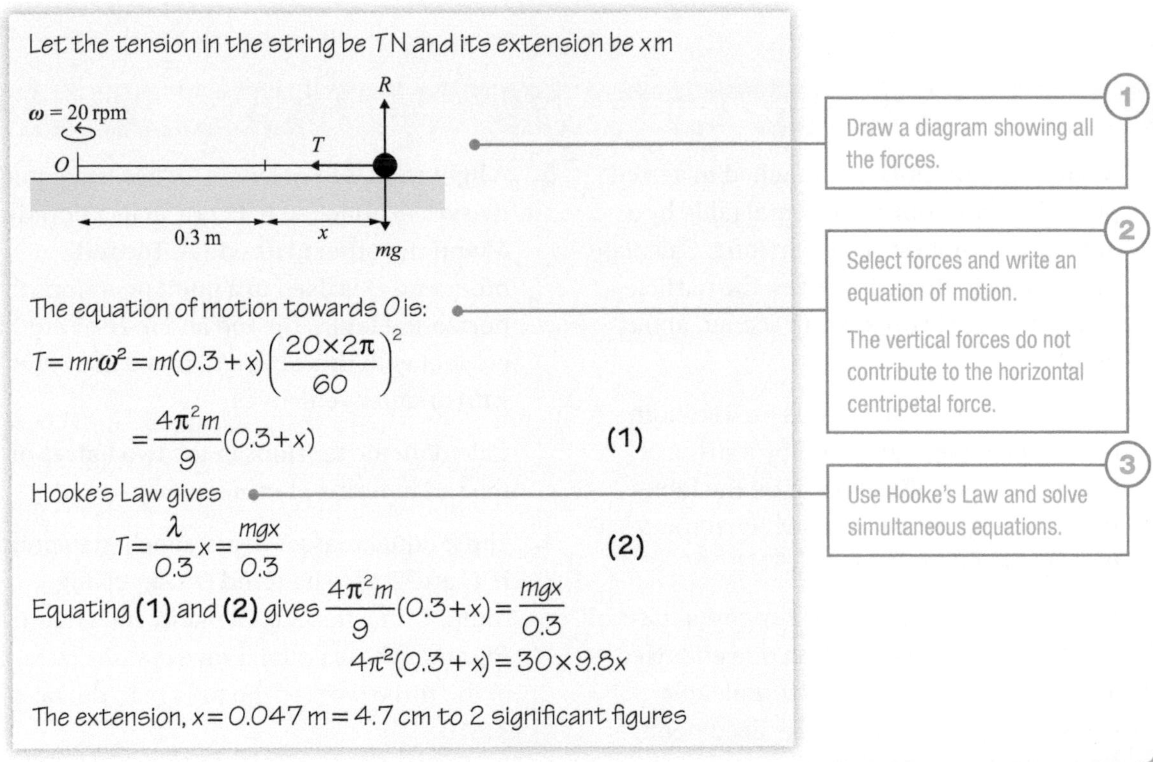

Let the tension in the string be T N and its extension be x m

(1) Draw a diagram showing all the forces.

The equation of motion towards O is:

$$T = mr\omega^2 = m(0.3 + x)\left(\frac{20 \times 2\pi}{60}\right)^2$$

$$= \frac{4\pi^2 m}{9}(0.3 + x) \qquad (1)$$

(2) Select forces and write an equation of motion.

The vertical forces do not contribute to the horizontal centripetal force.

Hooke's Law gives

$$T = \frac{\lambda}{0.3}x = \frac{mgx}{0.3} \qquad (2)$$

(3) Use Hooke's Law and solve simultaneous equations.

Equating **(1)** and **(2)** gives $\frac{4\pi^2 m}{9}(0.3 + x) = \frac{mgx}{0.3}$

$$4\pi^2(0.3 + x) = 30 \times 9.8x$$

The extension, $x = 0.047$ m $= 4.7$ cm to 2 significant figures

Exercise 9.2B Reasoning and problem-solving

Throughout this exercise use $g = 9.8$ m s^{-2} unless told otherwise.

1. An elastic string with a natural length of 0.6 m and modulus of elasticity λ of 120 N has one end attached to a fixed point O on a smooth horizontal table and the other end attached to a 2 kg mass. Calculate the extension in the string when the mass rotates at 40 rpm about point O

2. A spring balance with a natural length of 50 cm has one end attached to a fixed point O and the other end to a mass of 5 kg

   The mass rotates in circles about O at a speed of 8 rpm on a smooth horizontal surface when the spring balance reads 4 N. Calculate the modulus of elasticity of the spring balance.

Full A Level

3 Two masses m_1 and m_2 are connected by a taut string of length x

The two masses lie along a radius of a rough horizontal disc at distances x and $2x$ respectively from the centre O and rotate with uniform angular velocity ω about O. If the coefficient of friction between the disc and the particles is μ, show that they are both on the point of slipping if

$$\omega = \sqrt{\frac{\mu g(m_1 + m_2)}{x(m_1 + 2m_2)}}$$

4 Ten light spokes, each 40 cm long, connect the axle of a wheel to its light rim. A mass of 800 grams is attached to the end of each spoke. The wheel rotates at 100 rpm in a horizontal plane. Find the tension in each spoke and the total kinetic energy of the arrangement. Why does the tension in the rim not affect the tension in the spokes?

5 An elastic string of natural length 1.0 m and modulus of elasticity of 80 N has one end fixed to point O on a smooth horizontal table. The other end is attached to a mass of 3 kg which rotates about O on a circular path of radius 1.2 m. Calculate the linear speed of the mass and the total energy stored as kinetic and elastic energy.

6 A smooth ring is fixed to a point O on a smooth horizontal surface so it can rotate and act as a pivot. A light rod of length 80 cm is free to slide through the ring. The rod has masses of 10 kg and 25 kg fixed to its two ends. The rod rotates about O at a constant speed of 15 revs per second. How far is the rod's centre from O?

Full A Level

7 A car of mass m rounds a corner of radius 60 m with a coefficient of friction of 0.4 between its tyres and the horizontal road. Find the maximum speed of the car if there is no slipping. Does this speed depend on the mass of the car?

Full A Level

8 A car of mass 800 kg is travelling at 36 km h^{-1} and rounds a bend of radius 250 m on a horizontal road without skidding. Calculate the smallest value of the coefficient of friction between the tyres and the road.

9 A level turntable of radius 12 cm rotates at 33 rpm

Calculate the least coefficient of friction if a small mass stays on the turntable wherever it is placed.

10 A rough horizontal circular disc rotates about a vertical axis through its centre O

It makes two full revolutions every second. Prove that the greatest distance that a particle can be placed from O so that it is stationary relative to the disc is 6.2μ cm where μ is the coefficient of friction between the disc and particle.

11 Before its launch, a satellite on the Earth's surface has a weight of mg. When orbiting the Earth at a height h, it experiences a gravitational attractive force F which is inversely proportional to the square of its distance from the centre of the Earth.

a If the Earth's radius is R, show that the constant of proportionality equals mgR^2

b If the satellite takes 90 minutes to complete each circuit of the Earth's circumference, take R as 6370 km and calculate its angular velocity and height above the Earth's surface.

c A 'geostationary' satellite appears to be stationary in the sky above a fixed point on the equator. Calculate the height of a geostationary satellite above the Earth's surface.

9 Summary and review

Chapter summary

- When an object moves along the arc of a circle of radius r, its linear displacement, s, and the angle through which it moves, θ, respectively, are related by $s = r\theta$
- An object's linear velocity v along the tangent and its angular velocity ω about the centre of the circle are related by $v = r\omega$
- When ω is constant, the time period T for one complete revolution is $\dfrac{2\pi}{\omega}$
- The SI units for angular velocity ω are radians per second (rad s^{-1}), but revolutions per minute (rpm) are also common units.
- When v and ω are constant, there is no linear acceleration along the tangent and no angular acceleration. There is, however, an acceleration a towards the centre of the circle where $a = r\omega^2 = \dfrac{v^2}{r}$
- The acceleration towards the centre is the result of a force in this direction called the centripetal force. You can use Newton's 2nd law to write an equation of motion towards the centre involving the centripetal force.

Check and review

You should now be able to...	Try Questions
✔ Change units and convert linear to angular motion and vice versa.	1–5
✔ Write equations of motion involving a centripetal force.	6–16
✔ Solve problems involving horizontal circular motion.	8–15

1 A washing machine has two speed settings: 800 rpm and 1200 rpm. What are these speeds in rad s^{-1}?

2 A satellite travels round the Earth twenty times each day. Calculate its angular velocity in rad s^{-1}

3 A model train runs on a circular track of radius 0.8 metres. If the train's linear speed is $0.6\,\text{m s}^{-1}$, what is its angular speed in rad s^{-1} and its acceleration towards the centre of the track in m s^{-2}?

4 A car rounds a bend at a steady speed of $36\,\text{km h}^{-1}$

If the bend has a radius of 300 m and turns the car through an angle of 60°, calculate the car's angular velocity and the time it takes to round the bend.

5 A fairground ride has ten 'pods' equally spaced around the edges of a light horizontal wheel of radius 5 m

Each pod has a mass of 600 kg

What is the total kinetic energy stored in the pods when they are empty and rotating at 12 rpm?

6 A spin-dryer rotates at 1000 rpm and has a radius of 15 cm. When an item of clothing with a mass of 0.5 kg is spinning horizontally on the circumference of the drum, what is the reactive force of the drum on the item?

7 The Earth is 1.5×10^8 km from the Sun. Calculate the angular speed of the Earth around the Sun in rad s^{-1}

Given that the mass of the Earth is 6.0×10^{24} kg, calculate the centripetal force on the Earth towards the Sun.

8 One end of a light, inextensible string of length 2 m is attached to a fixed point on a smooth horizontal table. The other end is tied to a mass of 5 kg which rests on the table. The mass moves in a horizontal circle at a speed of 3 m s^{-1} with the string taut. Calculate the tension in the string.

9 A railway engine of mass 50 tonnes travels on a horizontal track round a circular arc of radius 400 m. If the maximum lateral force towards the centre of the arc that the rails can withstand is 10^5 N, calculate the maximum speed that the engine can travel round the bend.

10 A car of mass 1500 kg travels around a bend, which can be modelled as a quarter of the circumference of a circle with radius 15 m. The car will skid if the tyres have to provide a frictional force greater than 200 N. What is the shortest time in which the car can go around the bend without skidding?

11 An elastic spring, fixed at one end O, is attached to a mass of 0.5 kg at its other end. The mass rotates about O on a smooth horizontal surface. The spring has a natural length l of 40 cm and a modulus of elasticity λ of 80 N. Calculate the linear speed at which the mass travels if the spring is extended by 5 cm.

12 An elastic string has a natural length of 0.8 m and modulus of elasticity $\lambda = 100$ N. Its ends are attached to a fixed point O and to a 2 kg mass. Calculate the extension in the string when the mass rotates at 4 rad s^{-1} about O on a smooth horizontal table. What is the total of the kinetic and elastic energy in this arrangement?

13 A smooth curved wire forms a fixed horizontal hoop of radius r on which a bead of mass m is threaded. If the bead describes circles at a constant angular speed ω, calculate the magnitude and direction of the total reaction between the bead and the hoop.

14 The force of gravity F between an object of mass m and the Earth of mass M is given by $F = \frac{GmM}{d^2}$ where

$G = 6.67 \times 10^{-11}$ N m^2 kg^{-2}, $M = 5.98 \times 10^{24}$ kg and d is the distance of the object from the centre of the Earth. Taking the radius of the Earth as 6370 km, calculate the time for a satellite to orbit the Earth if its height above the Earth's surface is 250 km.

15 A horizontal turntable of radius 20 cm rotates at 45 rpm. When a particle is placed on the turntable 15 cm from the centre, it is on the point of slipping. What is the coefficient of friction between the particle and the turntable?

16 A car of mass 900 kg travelling at 54 km h^{-1} rounds a bend of radius 40 m on a horizontal road. Calculate the smallest value of the coefficient of friction if there is no sideways skidding.

Full A Level

Exploration

Going beyond the exams

Did you know?

Gradians are a measure of angle. The gradian or grad is a unit of measurement equivalent to 1/400 of a complete turn. Gradians are commonly used as the measure of angle in surveying, mining and geology.

Investigation

On a fairground ride, the rider sits in a seat which is rotated at a speed so that, eventually, it swings outwards. As the speed increases, the angle that the chains make with the vertical also increases. The situation can be modelled with the passenger and the seat modelled as a particle, and the chains holding the seat modelled as an inextensible string in tension, so that the passenger is rotating about a vertical axis.

Investigate the situation, answering questions such as:

- Is the passenger's mass significant? (What happens to empty seats?)
- Is the length of the supporting chains significant?
- How far from the vertical do passengers swing out?

Investigation

Imagine that you place several coins of the same type on a horizontal turntable at different distances from the centre of rotation and that you can vary the speed of the turntable. Which coin will slip first? The one nearest or furthest away from the centre? What changes if the coins are of different mass?

Explain your answers!

Take $g = 9.8\ \text{ms}^{-2}$ unless otherwise stated.

1 A particle of mass 2 kg is attached by a light, inextensible string of length 1.2 m to a fixed point on the surface of a horizontal, smooth table. The particle travels in a circle on the table at a speed of $2.5\ \text{m s}^{-1}$

Calculate the tension in the string. **[2 marks]**

2 A string of length 80 cm can just support a suspended mass of 40 kg without breaking. A 2 kg mass is attached to the string and the other end of the string is fastened to a point on the surface of a smooth, horizontal table. The mass is made to move in a circle on the table. Calculate, in revolutions per minute, the maximum angular speed at which the mass can revolve without breaking the string. **[4]**

3 A particle of mass 2 kg travels in a circle about a point O on a smooth horizontal plane. It is attached to O by means of a light elastic string of natural length 50 cm and modulus of elasticity 200 N. The speed of the particle is $2\sqrt{3}\ \text{m s}^{-1}$

Calculate the radius of the circle. **[5]**

4 Particles A and B, of mass 0.5 kg and 0.2 kg respectively, are connected together by means of a light inextensible string. The string is threaded through a small smooth hole, O, in a smooth horizontal table. Particle B hangs at rest 20 cm below the table surface, while particle A describes a circle about O at an angular speed of $2.8\ \text{rad s}^{-1}$

Calculate the length of the string. **[5]**

5 A car of mass m travels over a hump-backed bridge which has a cross-section in the form of an arc of a circle of radius 25 m. The speed of the car at the highest point of the bridge is $V\ \text{m s}^{-1}$

Calculate the greatest value that V can take if the car is not to leave the surface of the road at this point. **[4]**

6 A smooth wire has a bead of mass 0.005 kg threaded onto it. It is then bent round to form a circular hoop of radius 0.2 m. The hoop is fastened in a horizontal position, whilst the bead travels round it at a constant speed of $1\ \text{m s}^{-1}$

Work out the magnitude and direction of the reaction between the hoop and the bead. **[7]**

7 The gravitational attraction between objects of mass M and m that lie a distance r apart has magnitude $\frac{GMm}{r^2}$, where the universal gravitational constant $G = 6.67 \times 10^{-11}\ \text{N m}^2\text{kg}^{-2}$
Make the modelling assumptions that the Earth is fixed and the Moon travels in a circular orbit around it. The distance of Moon from the Earth is 384 000 km and it orbits once every 27.3 days. Use this information to estimate the mass of the Earth. **[4]**

8 A particle of mass 2 kg travels in a circle on a smooth horizontal plane about a point O

The particle is attached to O by means of a light elastic string of natural length 1 m and modulus of elasticity 300 N. The total energy of the system is 42 J. Calculate

a The radius of the circle, **b** The speed of the particle. **[8]**

9 Particles A and B each have mass m

Particle A lies on a smooth horizontal table and is connected to B by a light inextensible string of length $4a$ which passes through a small hole O in the table. Particle A travels in a circle of radius r about O, and B hangs at rest. Calculate the difference in total energy (potential and kinetic) between the situations $r = a$ and $r = 3a$ **[8]**

10 A car of mass m is travelling on a horizontal road round a bend of radius 50 m
The coefficient of friction between the car and the road is 0.9

a If the speed of the car is $45\,\text{km}\,\text{h}^{-1}$ find, in terms of m, the frictional force acting on the car. **[2]**

b Calculate the maximum speed at which the car could travel without skidding. **[4]**

11 A particle P, of mass 3 kg, is placed on a rough, horizontal turntable at a distance of 0.8 m from its centre O

The coefficient of friction between the particle and the turntable is 0.4

a P is connected to O by a light, inextensible string of length 0.8 m. The turntable is set in motion and its speed gradually increased until the tension in the string is 50 N. Calculate the angular speed of the turntable at this time. **[2]**

b The process is then repeated with P connected to O by a light, elastic string of natural length 0.8 m and modulus of elasticity 200 N. The speed of the turntable is gradually increased from zero to $4\,\text{rad}\,\text{s}^{-1}$

Calculate the length of the string at this time. **[3]**

12 A fairground ride is made from a large hollow cylindrical chamber placed with its axis vertical. The riders stand against the inside wall of the cylinder, and the chamber rotates about its axis. When it reaches top speed, the floor is lowered, leaving the riders supported only by the friction between themselves and the wall. The radius of the cylinder is 3 m and the coefficient of friction between a rider and the wall is $\frac{1}{3}$

What is the minimum angular speed at which the floor can be lowered so that riders do not slide down the wall? **[4]**

13 A particle is placed on a horizontal turntable at a distance r from its centre. The turntable is set in motion and the particle is on the point of slipping when the turntable is rotating with angular speed ω

The coefficient of friction between the particle and the turntable is μ
Find an expression for ω in terms of μ, r and g **[3]**

14 A rough, horizontal disc rotates at 2 revolutions per second. A particle is to be placed on the disc. The coefficient of friction between the particle and the disc is μ

Find, in terms of μ, the maximum distance from the centre at which the particle can be placed without it slipping. **[3]**

15 Two particles, each of mass m, are attached to the ends of a light, inextensible string. The string passes through a hole in the centre of a rough, horizontal turntable. One particle is placed on the turntable at a distance a from its centre, and the other hangs freely below the turntable. The coefficient of friction between the particle and the turntable is μ

The contact between the string and the hole is smooth. The turntable is rotating with angular speed ω and the suspended particle is stationary. Show that the ratio between the maximum and minimum values of ω is $\sqrt{\frac{1+\mu}{1-\mu}}$ **[5]**

10 Discrete and continuous random variables

Drivers are required by law to be insured against accidents because the outcome can be expensive for themselves and for other people involved. To calculate the insurance payments to be made by an individual driver, many factors that relate to their circumstances are taken into account. For example, their age, how many years they have been driving, where they live and the age and type of car they will be driving are considered. These factors all have an estimated probability of having an accident associated with them and these probabilities are used to calculate the cost of premiums to drivers.

This is one area where actuaries use expected values to inform the likelihood of events occurring and to calculate insurance payments for travel, homes against theft, property against fire and health of both people and animals. The associated mathematics is used extensively throughout industry and business.

Orientation

What you need to know	What you will learn	What this leads to
Maths Ch9 • Calculate the mean. • Calculate the variance. **Maths Ch10** • To model a probability distribution using the binomial distribution.	• To calculate expectation values for given random variables. • To use Poisson probability models. • To derive probability generating functions for given random variables and use them to calculate summary statistics.	**Ch 3 Continuous random variables** • Calculating the mean of a continuous random variable. • Calculating the variance of a continuous random variable.

10.1 Discrete distributions and expectation

Fluency and skills

See Maths Ch10.1 For a reminder about discrete random variables.

A probability distribution for a random experiment shows how the total probability of 1 is distributed between all the possible outcomes. If a fair coin is thrown three times, the probability distribution for the number of heads is shown in this table.

Number of heads	0	1	2	3
Probability	$\frac{1}{8}$	$\frac{3}{8}$	$\frac{3}{8}$	$\frac{1}{8}$

The number of heads can take four values only so the variable is discrete. Since its value is determined by the outcome of a random experiment, it is an example of a **discrete random variable**. Sometimes the distribution is defined using a probability distribution function, for example,

$\mathrm{P}(X=x)=\frac{x^2}{30}; x=1, 2, 3, 4$

As with all distributions, you can define measures of location and spread.

The median and mode of a discrete random variable can be calculated from tables of probabilities or distribution functions.

The median value of a random variable divides the probability distribution into two equal parts.

Key point

The median of a discrete random variable is given by M where

$\mathrm{P}(X \le M) \ge \frac{1}{2}$ and $\mathrm{P}(X \ge M) \ge \frac{1}{2}$

Solving each of these inequalities for M may give two distinct values. If this is the case, to find the median you should calculate the arithmetic mean of these values.

The median value of a random variable divides the probability distribution into two equal parts. For a discrete random variable, to find the median value you find the arithmetic mean of two values either side of the median.

Key point

The median of a discrete random variable is given by

$M=\frac{x_1+x_2}{2}$, where x_1 and x_2 are given by $\mathrm{P}\{X \ge x_1\} \ge \frac{1}{2}$ and $\mathrm{P}\{X \le x_2\} \ge \frac{1}{2}$

For sample data, the mode has been defined as the most commonly occurring value. An equivalent definition is used for random variables.

Key point

For a discrete random variable, the mode is defined as the value with the greatest probability.

Some random variables may have more than one modal value; others may have none.

Example 1

A random variable, X, has the probability distribution shown in the table. Write down the mode and median of X

x	−1	2	3	5
$P\{X=x\}$	0.05	0.55	0.3	0.1

Mode = 2

Median = 2

The mode is the x-value with the greatest probability.

$P\{X \le 2\} \ge \frac{1}{2}$ and $P\{X \ge 2\} \ge \frac{1}{2} \rightarrow M = 2$

Example 2

The number of raisins in a cake is modelled by a random variable, X, with distribution function

$$P\{X = x\} = \begin{cases} \frac{e^{-2}2^x}{x!}; & x = 1, 2, 3, 4 \\ a; & x = 4 \\ 0; & \text{otherwise} \end{cases}$$

Find the value of a and write down the median value of X

If $p_i = P(X = i)$, then $p_0 = 0.1353$, $p_1 = 0.2707$

$p_2 = 0.2707$, $p_3 = 0.1804$

$a = 1 - (0.1353 + \ldots + 0.1804) = 0.143$ (3dp)

Median = 2

Using $P(X = x) = \frac{e^{-2}2^x}{x!}$

When you throw a fair dice, the probability of getting a six is $\frac{1}{6}$

If you throw the dice 600 times this probability would predict that you would get a six one hundred times. The expected value of the number of sixes obtained if you throw a fair dice 600 times is 100

The expected value of a random variable and the arithmetic mean are closely related. If you throw a fair dice N times and the random variable, X, denotes the score on a randomly chosen throw, then the distribution of X can be written as

x	1	2	3	4	5	6
$P\{X=x\}$	$\frac{1}{6}$	$\frac{1}{6}$	$\frac{1}{6}$	$\frac{1}{6}$	$\frac{1}{6}$	$\frac{1}{6}$

The **expected value** of a random variable is the number of occurrences of the particular event that you would expect to get if the relative frequency obtained in an experiment equals the number of occurrences predicted by probability.

The expected number of each outcome is $\frac{N}{6}$ and therefore the expected total score is $1\times\frac{N}{6}+2\times\frac{N}{6}+3\times\frac{N}{6}+4\times\frac{N}{6}+5\times\frac{N}{6}+6\times\frac{N}{6}$

The expected (or mean) value of X, the score *per throw*, is therefore

$$\frac{1\times\frac{N}{6}+2\times\frac{N}{6}+3\times\frac{N}{6}+4\times\frac{N}{6}+5\times\frac{N}{6}+6\times\frac{N}{6}}{N} = 1\times\frac{1}{6}+2\times\frac{1}{6}+3\times\frac{1}{6}+4\times\frac{1}{6}+5\times\frac{1}{6}+6\times\frac{1}{6}$$

$$= 3\frac{1}{2}$$

The $\forall$ symbol means 'for all', so $\forall x$ means 'for all x'

This gives a general result for discrete random variables.

Key point

The **expected value** or **mean value** of a discrete random variable is given by
$E[X]=\sum_{\forall x} x P(X=x)$

Since this is a mean, it is sometimes given the symbol μ, a Greek m

Example 3

A random variable, S, has the probability distribution shown in the table. Find the expected value of S

s	−1	0	1	2
$P\{S=s\}$	$\frac{1}{2}$	$\frac{1}{4}$	a	a

$\frac{1}{2}+\frac{1}{4}+2a=1 \Rightarrow a=\frac{1}{8}$

Total probability on sample space equals 1

$E[S]=\sum_{\forall s} sP(S=s)=-1\times\frac{1}{2}+0\times\frac{1}{4}+1\times\frac{1}{8}+2\times\frac{1}{8}$

$=-\frac{1}{8}$

You will sometimes need to find the expected (mean) value of a function, g(X), of a random variable. This can be done by replacing the x by g(x) in the expression above.

Key point

The **expected value** of a function, g(X), of a discrete random variable is given by

$E[g(X)]=\sum_{\forall x} g(x)P(X=x)$

For example, for a random variable X, you could work out the expected value of X^2 by multiplying the values of X^2 by the corresponding probabilities and summing the results.

Example 4

A random variable, X, has the probability distribution shown in the table.

Find the expected value of

a X **b** X^2

x	−1	2	3	5
$P\{X=x\}$	0.2	0.3	0.4	0.1

a $E[X]=\sum_{\forall x} xP(X=x)$

$=-1\times0.2+2\times0.3+3\times0.4+5\times0.1$

$=2.1$

b

x	−1	2	3	5
x^2	1	4	9	25
$P\{X=x\}$	0.2	0.3	0.4	0.1

Work out the values of X^2

$E[X^2]=\sum_{\forall x} x^2P(X=x)$

$=1\times0.2+4\times0.3+9\times0.4+25\times0.1$

$=7.5$

To measure the spread of a distribution you calculate the **variance**, which is the 'expected value of the squared deviations from the mean.'

You can write the variance of X as $\text{Var}(X)$ or σ^2

Key point

The variance of a random variable X is given by

$$\text{Var}(X) = \text{E}[(X-\mu)^2] = \sum_{\forall x}(x-\mu)^2 \text{P}(X=x)$$

To make calculations simpler, you usually use an alternative form for variance.

$$\text{Var}(X) = \text{E}[X^2] - \mu^2 = \sum_{\forall x} x^2 \text{P}(X=x) - \mu^2$$

Key point

The square root of the variance, σ, is the **standard deviation** of the distribution.

STATISTICS

Example 5

A random variable, X, has the probability distribution shown in the table.

Find the expected value and variance of X

x	1	2	3	4
$\text{P}\{X=x\}$	0.3	0.2	0.4	0.1

$\mu = \text{E}[X]$

$= 1\times0.3 + 2\times0.2 + 3\times0.4 + 4\times0.1$

$= 2.3$

x	1	2	3	4
x^2	1	4	9	16
$\text{P}\{X=x\}$	0.3	0.2	0.4	0.1

Work out the values of X^2

$\text{Var}[X] = \text{E}[X^2] - \mu^2$

$= 1\times0.3 + 4\times0.2 + 9\times0.4 + 16\times0.1 - 2.3^2$

$= 1.01$

As with mean values, the variance of a function, $g(X)$, of a random variable can be found.

The variance of a function, $g(X)$, of a random variable X is given by $\text{Var}(g(X)) = \text{E}(\{g(X)\}^2) - \{\text{E}(g(X))\}^2$

Consider the classic random experiment where a fair dice is rolled once. Since the dice is fair, each of the six outcomes, integers 1–6, have an equal probability of occurring. This probability is $\frac{1}{6}$

Therefore, if X is a random variable for the score obtained when the dice is rolled once, then its probability distribution is given by

$$\text{P}\{X=x\} = \frac{1}{6};\ x=1, 2, \ldots, 6$$

This is a special case of a discrete uniform distribution. In general, the x-values don't have to be consecutive integers.

If a random variable, U, has a uniform distribution taking values 1, 2,..., n, it follows that $\text{P}\{U=u\} = k$; $u = 1, 2, \ldots, n$, where k is some constant. But $\text{P}\{S\} = 1$, where S is the set of all outcomes.

Therefore $nk = 1$ and so $k = \frac{1}{n}$

Key point

A discrete uniform random variable, U, taking values $u=1,2,\ldots,n$, has a probability distribution given by $P\{U=u\}=\frac{1}{n};\ u=1, 2,\ldots, n$

You can use this rule to calculate the mean and variance of a discrete uniform distribution.

Key point

A discrete uniform random variable, U, with distribution function $P\{U=u\}=\frac{1}{n};\ u=1, 2,\ldots, n$ has a mean and variance given by $\mu=\frac{n+1}{2}$ and $\sigma^2=\frac{n^2-1}{12}$

Example 6

An ordinary fair dice is rolled once. Calculate the mean, μ_X, and variance, σ_X^2, of X, the number scored.

$$P\{X=x\}=\frac{1}{6};\ x=1,2,\ldots,6$$

$$\mu_X=\frac{n+1}{2}=\frac{6+1}{2}=\frac{7}{2}$$

$$\sigma_X^2=\frac{n^2-1}{12}=\frac{6^2-1}{12}=\frac{35}{12}$$

Exercise 10.1A Fluency and skills

1 For each of the following probability distributions, calculate the expectation and variance of the random variables.

x	1	4	9	16	25	36
$P\{X=x\}$	$\frac{1}{6}$	$\frac{1}{6}$	$\frac{1}{6}$	$\frac{1}{6}$	$\frac{1}{6}$	$\frac{1}{6}$

y	−1	0	1	2
$P\{Y=y\}$	$2a$	$2a$	a	a

2 For each of the following probability distributions, find the mean and the variance.

a $P\{X=a\}=\frac{1}{4};\ a=1, 2, 3, 4$

b $P\{X=b\}=\frac{b}{3};\ b=1, 2$

c $P\{X=c\}=\begin{cases}\frac{1}{c};\ c=2,3\\ \frac{1}{6};\ c=4\end{cases}$

3 A fair ten-sided dice with faces numbered 1–10 is thrown once. Prove that the mean and variance of the score obtained are $\frac{11}{2}$ and $\frac{33}{4}$

4 Five cards are numbered from 1 to 5 Two cards are drawn at random with replacement and the total, T, noted. Find the expected value of T

5 A four-sided spinner with sides numbered 1–4 is biased so that odd numbered outcomes are each twice as likely as even numbered outcomes. Find the expected value and variance of the score on a randomly chosen spin.

6 A spinner can land on numbers 1–9 with equal probability. Show that the expected score for one throw is 5 and find the variance of this score.

7 A fair dice is thrown once. Find the mean value of the **square** of the score.

8 A six-sided dice with faces numbered 1–6 is biased so that the probability of any score is proportional to the square of the score. Find the expected value of S, the score on a randomly chosen throw.

9 A fair dice is thrown once. If R is the reciprocal of the score, find the mean value of R

10 X is a random variable for the score on a biased tetrahedral dice. Its probability distribution is

x	1	2	3	4
$\mathrm{P}\{X=x\}$	0.4	0.2	0.2	0.2

a Find the mean and variance.

b Find the mean and variance of the random variable Y, where $Y=2X^2$

11 A number is chosen at random from the integers 1 to 30

a Find the expected number of prime numbers.

b Explain what this means for a set of repeated trials of this experiment.

12 A fair coin is tossed until a head or four tails appear.

a Give the probability distribution of X, the number of throws required.

b What is the expected number of throws?

13 The number of made-to-measure suits ordered per month from a tailor has the following discrete probability distribution.

x	0	1	2	3
$\mathrm{P}\{X=x\}$	0.18	0.39	0.31	0.12

Find the mode and the median of X

14 A discrete random variable X has probability distribution function given by

$$P\{X=x\}=\begin{cases}\dfrac{x}{12};\ x=1,2,3\\ k;\ x=4,5,6\\ 0;\ \text{otherwise}\end{cases}$$

a Find the value of k

b Write down the modal value of X

15 Prove that a discrete uniform random variable, U, with distribution function $\mathrm{P}\{U=u\}=\frac{1}{n};\ u=1,2,\ldots,n$ has a mean and variance given by $\mu=\frac{n+1}{2}$ and $\sigma^2=\frac{n^2-1}{12}$

STATISTICS

Reasoning and problem-solving

Strategy

To solve a problem involving expectation

1. Find the probability distribution of the random variable, either in the form of a function or as a table.
2. Use standard formulae to find the mean and variance of the random variable.
3. Where necessary, find the expectation value of a function of X, g(X).
4. Give your conclusion.

You will sometimes need to calculate the expectation of linear functions of random variables.

Example 7

An ordinary six-sided dice is thrown once. X is a random variable for the score obtained. The random variable Y is defined by the equation $Y=2X-3$. By finding the possible values of Y, show that

a $\mathrm{E}[Y]=2\mathrm{E}[X]-3$ **b** $\mathrm{Var}[Y]=4\mathrm{Var}[X]$

(*Continued on the next page*)

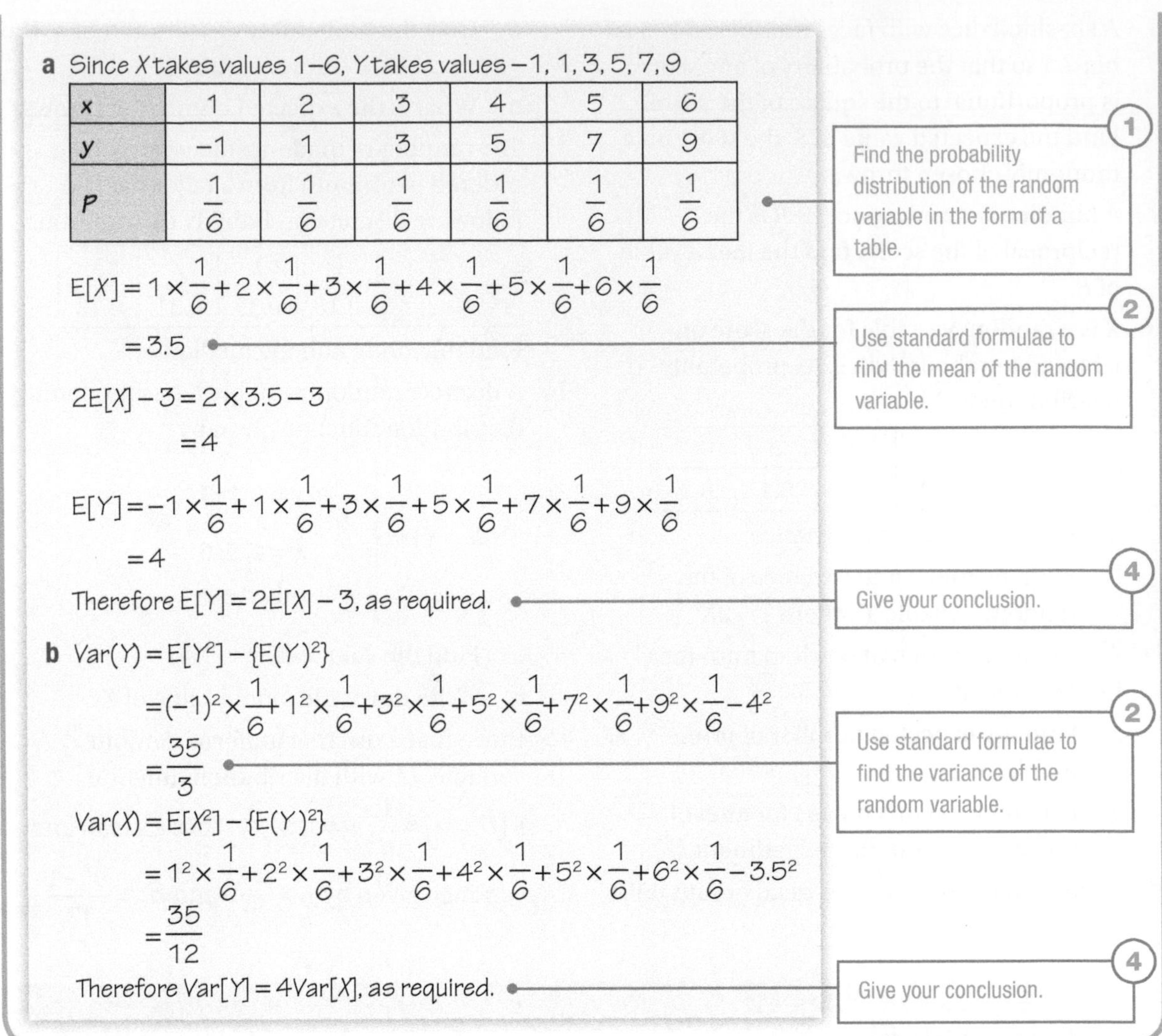

a Since X takes values 1–6, Y takes values −1, 1, 3, 5, 7, 9

x	1	2	3	4	5	6
y	−1	1	3	5	7	9
P	$\frac{1}{6}$	$\frac{1}{6}$	$\frac{1}{6}$	$\frac{1}{6}$	$\frac{1}{6}$	$\frac{1}{6}$

① Find the probability distribution of the random variable in the form of a table.

$$E[X] = 1\times\frac{1}{6}+2\times\frac{1}{6}+3\times\frac{1}{6}+4\times\frac{1}{6}+5\times\frac{1}{6}+6\times\frac{1}{6}$$

$$= 3.5$$

② Use standard formulae to find the mean of the random variable.

$$2E[X]-3 = 2\times3.5-3$$

$$= 4$$

$$E[Y] = -1\times\frac{1}{6}+1\times\frac{1}{6}+3\times\frac{1}{6}+5\times\frac{1}{6}+7\times\frac{1}{6}+9\times\frac{1}{6}$$

$$= 4$$

Therefore $E[Y] = 2E[X]-3$, as required.

④ Give your conclusion.

b $\text{Var}(Y) = E[Y^2]-\{E(Y)^2\}$

$$= (-1)^2\times\frac{1}{6}+1^2\times\frac{1}{6}+3^2\times\frac{1}{6}+5^2\times\frac{1}{6}+7^2\times\frac{1}{6}+9^2\times\frac{1}{6}-4^2$$

$$= \frac{35}{3}$$

② Use standard formulae to find the variance of the random variable.

$\text{Var}(X) = E[X^2]-\{E(Y)^2\}$

$$= 1^2\times\frac{1}{6}+2^2\times\frac{1}{6}+3^2\times\frac{1}{6}+4^2\times\frac{1}{6}+5^2\times\frac{1}{6}+6^2\times\frac{1}{6}-3.5^2$$

$$= \frac{35}{12}$$

Therefore $\text{Var}[Y] = 4\text{Var}[X]$, as required.

④ Give your conclusion.

This example illustrates a general result.

Key point

If X and Y are two discrete random variables and $Y = aX + b$, where a and b are constants, then

$E(Y) = aE(X) + b$ and $\text{Var}(Y) = a^2\text{Var}(X)$

You can also write this as $E(aX + b) = aE(X) + b$ and $\text{Var}(aX + b) = a^2\text{Var}(X)$

Example 8

A random variable, S, has probability distribution

s	1	2	3
$P\{S = s\}$	$\frac{1}{2}$	a	$2a$

where a is a positive constant.

a Find the value of a

The variable T is related to S by the equation $T = -3S + 6$

b Find the mean and variance of S and hence the mean and variance of T

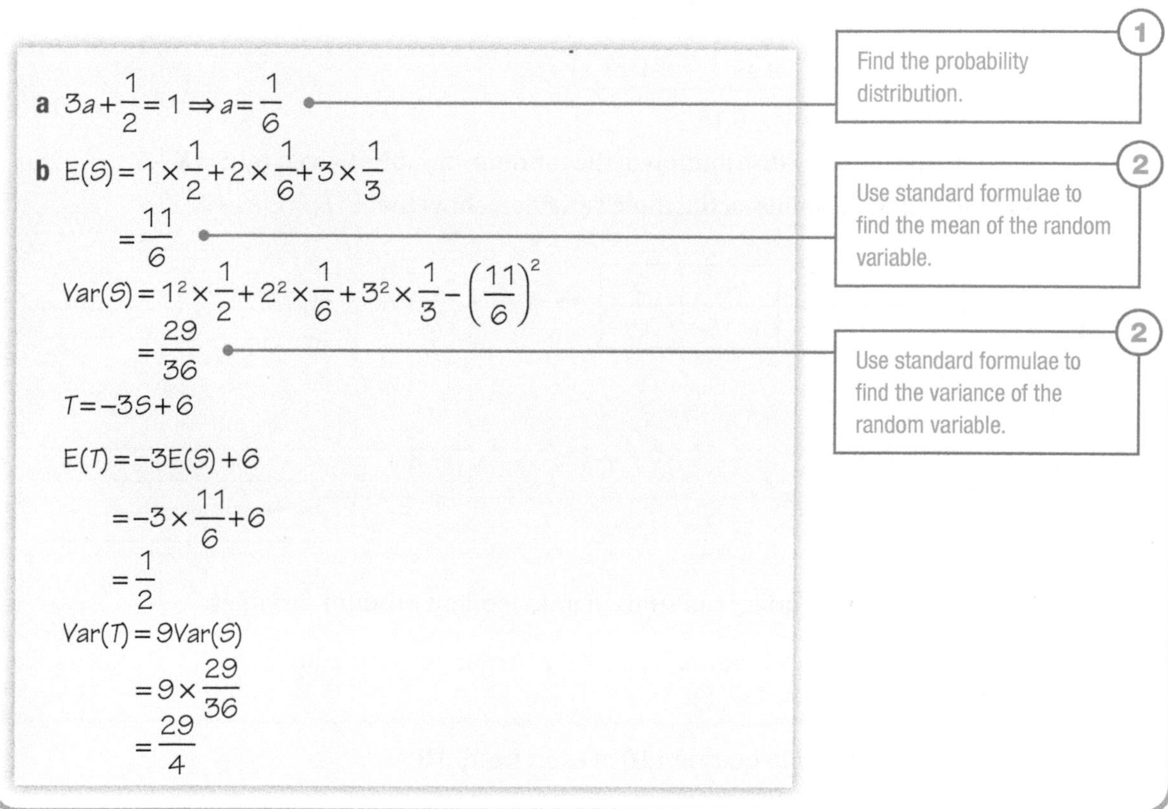

a $3a + \frac{1}{2} = 1 \Rightarrow a = \frac{1}{6}$

b $E(S) = 1 \times \frac{1}{2} + 2 \times \frac{1}{6} + 3 \times \frac{1}{3}$

$= \frac{11}{6}$

$Var(S) = 1^2 \times \frac{1}{2} + 2^2 \times \frac{1}{6} + 3^2 \times \frac{1}{3} - \left(\frac{11}{6}\right)^2$

$= \frac{29}{36}$

$T = -3S + 6$

$E(T) = -3E(S) + 6$

$= -3 \times \frac{11}{6} + 6$

$= \frac{1}{2}$

$Var(T) = 9Var(S)$

$= 9 \times \frac{29}{36}$

$= \frac{29}{4}$

1 Find the probability distribution.

2 Use standard formulae to find the mean of the random variable.

2 Use standard formulae to find the variance of the random variable.

STATISTICS

It is possible to find the mean and variance of sums of independent random variables.

Key point

For two discrete random variables, X and Y, the expected value of the sum, $X+Y$, is given by $\mathrm{E}(X+Y)=\mathrm{E}(X)+\mathrm{E}(Y)$

This simple result is surprisingly difficult to prove but the following example makes it plausible.

Example 9

The following table gives the bivariate probability distribution for the variables X and Y

		Y			
		1	**2**	**3**	**P{X = x}**
X	**1**	0.15	0.1	0.05	0.3
	2	0.2	0.1	0.4	0.7
	P{Y = y}	0.35	0.2	0.45	1

For example, $\mathrm{P}\{X=1 \text{ and } Y=1\}=0.15$

a Write down the probability distribution of the random variable T given by $T=X+Y$

b By finding the expected values of the three variables, show that $\mathrm{E}[T]=\mathrm{E}[X]+\mathrm{E}[Y]$

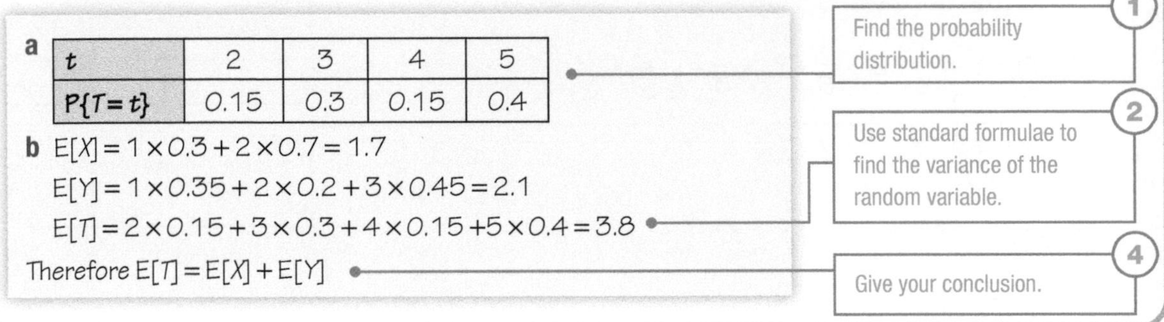

a

t	2	3	4	5
$\mathrm{P}\{T=t\}$	0.15	0.3	0.15	0.4

b $\mathrm{E}[X]=1\times0.3+2\times0.7=1.7$

$\mathrm{E}[Y]=1\times0.35+2\times0.2+3\times0.45=2.1$

$\mathrm{E}[T]=2\times0.15+3\times0.3+4\times0.15+5\times0.4=3.8$

Therefore $\mathrm{E}[T]=\mathrm{E}[X]+\mathrm{E}[Y]$

A similar result applies to the variance of sums of independent random variables.

Key point

For two independent random variables, X and Y, the variance of the sum, $X+Y$, is given by $\mathrm{Var}(X+Y)=\mathrm{Var}(X)+\mathrm{Var}(Y)$

The proof of this result is found in question **10** of Exercise 10.1B

Example 10

West End FC Under 11s football team played a set of home and away games. During home games, the number of goals scored had a mean of 3.1 and a standard deviation of 1.2

During away games, these statistics were 2.7 and 1.4 respectively. Find the mean and standard deviation of the total number of goals scored. State any necessary assumptions made, indicating to which part of the solution the assumption relates.

Let X and Y be random variables for the numbers of goals scored during home and away games respectively.

$\mathrm{E}(X+Y)=\mathrm{E}(X)+\mathrm{E}(Y)=3.1+2.7=5.8$

$\mathrm{Var}(X+Y)=\mathrm{Var}(X)+\mathrm{Var}(Y)=1.2^2+1.4^2=3.4$

Standard deviation of $X+Y=1.84$ (2dp).

For the variance result, assume that X and Y are independent.

Exercise 10.1B Reasoning and problem-solving

1 A random variable X has this probability distribution.

x	1	2	3	4	5	6
P{X = x}	$\frac{1}{6}$	$\frac{1}{6}$	$\frac{1}{6}$	$\frac{1}{6}$	a	b

Given that the mean of X is $\frac{52}{15}$, find the values of a and b

2 R is a random variable with this distribution.

r	−3	−2	−1	0	1	2	3
P{R = r}	$\frac{1}{7}$	$\frac{1}{7}$	$\frac{1}{7}$	$\frac{1}{7}$	$\frac{1}{7}$	$\frac{1}{7}$	$\frac{1}{7}$

A single value of R is chosen at random.

a Find the mean and standard deviation of R

b What is the probability that it will be one standard deviation or closer to the mean?

3 The proportion of boys born in the UK is 0.51 of all live births. Four UK births are selected at random.

a Write down all possible permutations of gender and hence calculate the probability distribution for X, a random variable for the number of girls born in the four births.

b Find the mean and variance of X

4 Jenny and John play a dice game. John says that he will get at least one six in four throws of a fair dice and will pay Jenny £1 if he fails.

a What is the probability of John winning any given game?

b How much should Jenny pay to enter the game so that in the long run neither player can expect to win anything? Give your answer to the nearest 1p.

5 A bag contains 1 red and 2 black balls. A ball is drawn from the bag and then, without the first ball being replaced, a second ball is drawn. By first finding the probability distribution, find the expected value of B, the number of black balls drawn.

6 a Find the probability of getting a double six in 24 throws of a pair of fair dice.

b In a dice game, you throw the pair of dice 3 times and receive 10 counters for every double six you throw. How many counters should you pay to enter the game if you want to break even in the long run? Give your answer to the nearest integer.

7 A random variable X has the probability distribution function

$$P\{X=x\}=\begin{cases}\frac{1}{5n}; & x=1,2,3...,5n\\ 0; & \text{otherwise}\end{cases}$$

where n is a positive integer.

a Prove that $E[X]=\frac{(1+5n)}{2}$

b Find in terms of n an expression for the variance of X

You may assume that

$$\sum_{i=1}^{n} i=\frac{n(n+1)}{2}$$

and that

$$\sum_{i=1}^{n} i^2=\frac{n}{6}(n+1)(2n+1)$$

8 A circle of radius r is inscribed in a square of side $2r$

Three darts are randomly dropped onto the square.

D is a random variable for the number of darts that land within the circle.

a Find $E[D]$

(Assume that the probability of a dart landing in the circle is equal to $\frac{\text{area of circle}}{\text{area of square}}$)

b One dart is dropped randomly onto the square 100 times and on 71 occasions the dart lands within the circle. Estimate the value of π

9 X and Y are two independent random variables. Given that $E[X]=3.2$, $Var[X]=12.1$, $E[Y]=22$ and $Var[Y]=9$, find the mean and standard deviation of $X+Y$

10 Prove that, for two independent random variables, $Var(X+Y)=Var(X)+Var(Y)$. You may assume that for independent random variables, $E[XY]=E[X]E[Y]$

11 a A random variable X has mean μ and variance σ^2

Write down the mean and variance of

i aX

ii $bX-c$

where a, b and c are constants.

b In a dice game, you throw an ordinary, fair dice once and receive in tokens five times the score obtained. You pay 20 tokens to enter the game. Find the mean and variance of your net winnings (after allowing for the entry cost).

12 A random variable, S, has probability distribution

s	1	2	3
$P\{S=s\}$	$\frac{3}{8}$	$2k$	k

where k is a positive constant.

a Find the value of k

The variable R is related to S by the equation $R=4S-3$

b Find the mean and variance of S and hence the mean and variance of R

13 A random variable X has mean 4, variance 8 and an independent variable Y has mean 5, variance 12

Write down the mean and variance of the variables S and T, where $S=X+Y$ and $T=X-Y$

14 A fair dice with faces numbered 2, 4, 6, 8, 10 and 12 is thrown once and a fair eight-sided spinner with sides numbered 1–8 is spun once. If X and Y are random variables for the scores on the dice and the spinner respectively, find the expected value and variance of $X+Y$

15 An ordinary pack of cards has the jacks, queens and kings removed. Two cards are chosen at random, with replacement, from the remaining cards. Find the mean and variance of the sum of the scores.

10.2 The Poisson distribution

Fluency and skills

The Poisson distribution can be used to model the occurrence of certain types of random events in time or space. Here, the word 'random' is used in a very particular way. In this context, events are random if the occurrence of an event at a particular point in time or space is independent of occurrences elsewhere. Further, the probability of an event happening in a small interval of a given size is the same at all times or positions in space and events can't occur simultaneously.

Key point

If X is the number of random events that occur in a given interval of time or space and λ is the mean number of occurrences in that interval, then its probability distribution function is given by

$P(X=x)=\frac{e^{-\lambda}\lambda^x}{x!}$; $x=0, 1, 2,...$

X is said to follow a Poisson distribution, parameter λ. For a Poisson distribution, the mean = variance.

Some examples of random variables that may follow a Poisson distribution are the number of radioactive disintegrations in five-second intervals, the number of stars of a given type in a given volume of space, the number of telephone calls you receive in a one-hour interval of time. It is worth noting that although the Poisson distribution function gives non-zero probabilities for outcomes 0, 1, 2..., (that is, there is no upper limit on the value of a Poisson variable), in practice, there will almost always be an upper limit on the variable.

Random events in continuous time or space constitute a **Poisson process.**

Example 1

A Geiger counter detects radioactive disintegrations at a mean rate of 23 per minute. Use the Poisson distribution function with $\lambda=23$ to find the probability that, in a randomly chosen minute, there will be

a 20 disintegrations,

b Between 21 and 23 inclusive disintegrations.

Check your answers using the Poisson function of your calculator.

a $P(X=20)=\frac{e^{-23}\lambda^{20}}{20!}$

$=0.072$ (3 dp)

Let X be a random variable for the number of disintegrations in a randomly chosen minute.

Substitute $\lambda=23$ and $x=20$ into $P(X=x)=\frac{e^{-\lambda}\lambda^x}{x!}$

b $P(21\le X\le 23)=\sum_{x=21}^{23}\frac{e^{-23}23^x}{x!}$

$=\frac{e^{-23}23^{21}}{21!}+\frac{e^{-23}23^{22}}{22!}+\frac{e^{-23}23^{23}}{23!}$

$=0.079+0.083+0.083$

$=0.245$ (3 dp)

Substitute $x=21$, $x=22$ and $x=23$, into $P(X=x)=\frac{e^{-23}23^x}{x!}$

Use your calculator to check your answers.

Example 2

A random variable X has a Poisson distribution with parameter $\lambda = 2$

a Copy the following table and use the probability distribution function of X to complete it.

x	0	1	2	3	4	5	6	>6
$P\{X = x\}$	0.135	0.271					0.012	0.005

b Use the values $x = 0, 1, 2, \ldots, 6$ to estimate the mean and variance of X

a

x	0	1	2	3	4	5	6	>6
$P\{X = x\}$	0.135	0.271	0.271	0.180	0.090	0.036	0.012	0.005

b $E(X) \approx \sum_{x=0}^{6} xP\{X = x\}$

$= 0 \times 0.135 + 1 \times 0.271 + 2 \times 0.271 + 3 \times 0.18 + 4 \times 0.09 + 5 \times 0.036$
$+ 6 \times 0.012$

$= 1.965$

$Var(X) \approx \sum_{x=0}^{6} x^2 P\{X = x\} - \mu^2$

$= 0^2 \times 0.135 + 1^2 \times 0.271 + 2^2 \times 0.271 + 3^2 \times 0.18 + 4^2 \times 0.09$
$+ 5^2 \times 0.036 + 6^2 \times 0.012 - 1.965^2$

$= 1.886$

You can use the Poisson distribution function to calculate exact values of the mean and variance of the distribution.

Key point

If X has a probability distribution function given by

$$P(X = x) = \frac{e^{-\lambda}\lambda^x}{x!};\ x = 0, 1, 2, \ldots$$

then the mean and variance of X are, respectively, given by

$$E\{X\} = \sum_{x=0}^{\infty} xP\{X = x\} = \lambda$$

$$Var\{X\} = \sum_{x=0}^{\infty} x^2 P\{X = x\} - \lambda^2 = \lambda$$

Example 3

a Write down the mean and variance of a random variable X with a Poisson distribution, $\lambda = 2$

b Find the probability that a randomly chosen value of X will be within 1 standard deviation of the mean.

a $X \sim Po(\lambda = 2)$

mean = 2, variance = 2

b Standard deviation $\sqrt{2} = 1.414$ (3 dp).

Standard deviation is the square root of the variance.

$P(2 - 1.414 < X < 2 + 1.414) = P(0.586 < X < 3.414)$

$= P(X = 1 \text{ or } 2 \text{ or } 3)$

$= 0.271 + 0.271 + 0.180$

$= 0.722$

Compare your answers to part **a** with the estimates found in Example 2 above.

The key point above shows that for a Poisson random variable, mean equals variance. This result can be used to investigate whether or not data is likely to come from a Poisson distributed population.

Example 4

The following famous data gives information about the random variable X, the number of deaths from horse kicks for 10 units of the Prussian Army over 20 years (that is, 200 observations) in the last decades of the nineteenth century.

Number of deaths, X	0	1	2	3	4
Frequency	109	65	22	3	1

By finding the mean and variance of X, decide whether this data is likely to follow a Poisson distribution.

Mean = 0.610 Variance = 0.608 (3dp)

Mean approximately equal to variance → data likely to be Poisson distributed.

Use a calculator to find these values.

Exercise 10.2A Fluency and skills

1 A random variable X has a Poisson distribution, mean 4. Find the probability that X takes values 0, 1, 2, 3, >3

2 Random events occur at a rate of three per ten seconds. Find the probability that there will be at least two events in any randomly chosen ten-second interval.

3 A random variable R is Poisson distributed, parameter 4

Find the probability that R takes values 0 or greater than 4

4 A random variable X has a Poisson distribution with parameter $\lambda = 1.1$

a Copy this table and use the probability distribution function of X to complete it.

x	0	1	2	3	4	>4
P{$X = x$}	0.333	0.366				

b Sketch the probability distribution function of X for $x = 0, 1, 2, 3, 4$

5 Random events occur at an average rate of 10 per 12 minutes. Find the probability that there will be less than three events in any randomly chosen 12-minute interval.

6 Radioactive decays occur at a rate of 34 per 12-second interval. Find the probability that, in two randomly chosen 12-second intervals, there will be at least 30 disintegrations in exactly one of the intervals.

7 A household receives on average four telephone calls per day. What is the probability that, in a randomly chosen week (of seven days), there is at least one call every day? State any assumptions you make about the appropriate probability distribution.

8 Emails arrive at your account randomly at a rate of four per two hours. Find the probability that, in a randomly chosen two-hour period, you will receive at least three emails.

9 There are on average four hazelnuts in a 200 g chocolate bar. What is the probability that, in five randomly chosen bars, there is at least one hazelnut in each bar? State any distributional or other assumptions you make.

10 A fabric-weaving machine produces cloth with flaws, which occur at a rate of 3.4 per square metre. A customer orders a 2 m^2 piece of cloth and insists that it must be flawless. Find the probability that the first piece of cloth produced for the customer meets her requirements.

11 Cars pass a traffic checkpoint at a rate of two per hour. Find the probability that, in two consecutive hours, there is a total of four cars passing. You should treat the two one-hour intervals separately.

12 The number of cars passing a rural traffic survey station in 200 four-minute intervals is shown in the table below.

Number of cars, X	0	1	2	3	4	5	6	7	8
Frequency	24	49	52	34	19	11	7	3	1

By finding the mean and variance of X, decide whether the cars are likely to be passing the station randomly.

STATISTICS

Reasoning and problem-solving

Strategy

To solve a problem involving the Poisson distribution

1. Ensure that the conditions for a Poisson distribution apply to the problem.
2. Use the probability distribution function to define the required probabilities.
3. Use a calculator, statistical tables or the Poisson distribution function to find probabilities.
4. Where required, use standard results to find the mean and variance of the distribution.
5. Give your conclusion.

Example 5

Events occur randomly at a mean rate of μ per hour. Find, in terms of μ, the probability that three events occur in a randomly chosen **two** hours.

$X_T = X_1 + X_2$

$P(X_T = 3) = P(X_1 = 3, X_2 = 0 \text{ or } X_1 = 2, X_2 = 1 \text{ or } X_1 = 1, X_2 = 2 \text{ or } X_1 = 0, X_2 = 3)$

$$P(X_T = 3) = \frac{e^{-\mu}\mu^3}{3!} \times \frac{e^{-\mu}\mu^0}{0!} + \frac{e^{-\mu}\mu^2}{2!} \times \frac{e^{-\mu}\mu^1}{1!} + \frac{e^{-\mu}\mu^1}{1!} \times \frac{e^{-\mu}\mu^2}{2!} + \frac{e^{-\mu}\mu^0}{0!} \times \frac{e^{-\mu}\mu^3}{3!}$$

$$= \frac{e^{-2\mu}\mu^3}{3!} + \frac{e^{-2\mu}\mu^3}{2!} + \frac{e^{-2\mu}\mu^3}{2!} + \frac{e^{-2\mu}\mu^3}{3!}$$

$$= e^{-2\mu}\mu^3\left(\frac{1}{3!} + \frac{1}{2!} + \frac{1}{2!} + \frac{1}{3!}\right)$$

$$= e^{-2\mu}\mu^3 \times \frac{4}{3}$$

$$= \frac{e^{-2\mu}(2\mu)^3}{3!}$$

1 Conditions for a Poisson distribution apply to the problem.

X_1 is the number of events in the first hour, X_2 is the number of events in the second hour and X_T is the total number of events in the 2 hours.

A total of 3 events in 2 hours means 3 in the first and 0 in the second or 2 in the first and 1 in the second...

2 Use the probability distribution function to define the required probabilities.

Note that this is the probability distribution function of a Poisson distribution, mean 2μ

This suggests the following result:

Key point

X_i; $i = 1, 2, \ldots, n$ is a set of n independent and identically distributed random variables each with a Poisson distribution, mean λ. If $X_T = \sum_{i=1}^{n} X_i$, then X_T has a Poisson distribution with mean $n\lambda$

In n equal intervals of time or space, the total number of random events occurring equals the sum of the number of events in each of the separate intervals. Hence, you can apply the above result.

Key point

If events occur randomly at a rate of λ per given interval of time or space, then the number of events in n such intervals is a Poisson random variable with mean $n\lambda$

Example 6

Cars pass a traffic checkpoint at a rate of two per hour.

a Explain why it is necessary to assume that the checkpoint is, for example, on a quiet rural road or a motorway with free overtaking, in order to use a Poisson distribution.

b Assuming that the conditions in part **a** are met, find the probability, that in a two-hour period, a total of more than two cars pass.

a A Poisson distribution requires random arrivals, that is, free flowing traffic/no queues.

b Let X be a random variable for the number of cars passing in 2 hours.

$\lambda = 4$ — A mean of 2 per hour is equivalent to 4 (2×2) per 2 hours.

$X \sim \text{Po}(4)$

Assume that cars arrive randomly. — **1** Conditions for a Poisson distribution apply to the problem.

$P\{X = x\} = \dfrac{e^{-\lambda}\lambda^x}{x!}$ — **2** Use the probability distribution function to define the required probabilities.

$P\{X > 2\} = 1 - P\{X \le 2\}$

$= 1 - \left(\dfrac{e^{-4}4^0}{0!} + \dfrac{e^{-4}4^1}{1!} + \dfrac{e^{-4}4^2}{2!}\right)$

$= 0.762$ — **3** Use a calculator, statistical tables or the Poisson distribution function to find probabilities.

Suppose that a number of identical red objects are scattered randomly over a large area A with an average density of r objects per unit area. Let X be a random variable for the number of these objects in a randomly chosen unit area. Suppose, further, that green objects are scattered in a similar way over the same area with an average density of s objects per unit area and with Y objects in a randomly chosen unit area. From the theory in the previous section, X and Y will each have Poisson distributions with parameters r and s respectively.

Now, considering all objects as a single collection, these will be randomly scattered over the area with a mean number per unit area of $r+s$

That is, the total number of objects in a randomly chosen unit area, $X+Y$, will follow a Poisson distribution, parameter $r+s$

Key point

If X and Y are two independent random variables that can be modelled using the Poisson distribution with distributions given by $X \sim \text{Po}(\lambda_x)$ and $Y \sim \text{Po}(\lambda_y)$, then $X+Y \sim \text{Po}(\lambda_x + \lambda_x)$

Example 7

The number of cars passing a point on a quiet rural road between 7am and 8am is given by the random variable X

The random variable Y gives the same statistic for the interval between 8am and 9am.

a Explain why the random variables could each be modelled by the Poisson distribution. The means of X and Y are 21 and 34 respectively.

b Assuming that the distribution in part **a** provides a good model, find the probability that the total number of cars passing the point between 7am and 9am is between 50 and 60 inclusive.

a On a quiet road, cars are likely to arrive randomly.

b $X \sim Po(21)$ $Y \sim Po(32)$

T is the total number of cars passing the point between 7 am and 9am

$T = X + Y$

$T \sim Po(53)$

$P\{50 \le T \le 60\} = P\{T \le 60\} - P\{T < 50\} = 0.848 - 0.322 = 0.53$ (2dp)

Exercise 10.2B Reasoning and problem-solving

1 Random events occur at a rate of three per ten seconds. Find the probability that there will be at most 33 events in any randomly chosen two-minute interval.

2 Cars pass a traffic checkpoint at a rate of two per minute. Find the probability that, in two consecutive minutes, a total of four cars pass.

3 Shoppers arrive at a supermarket at an average rate of seven every minute. Find the probability that, in a randomly chosen one-hour interval, at least 400 shoppers will arrive.

4 At a certain set of traffic lights there is on average one road traffic accident (RTA) every nine days. Find the probability that, in a randomly chosen month (of 31 days), there is at most one accident.

5 In the radioactive decay of Americium-241, alpha particles are emitted at a rate of 20 per ten-second interval. Find the probability that, in a randomly chosen one-minute interval, there will be at least 100 particles emitted.

6 Strontium-90 emits on average 100 beta particles in 10 seconds. Find the probability that, in a randomly chosen 90-second interval, there will be no fewer than 910 particles emitted.

7 I receive on average 14 emails per day. Assuming that their arrival constitutes a random process, what is the probability that in a randomly chosen seven-day period

a I receive at least twelve emails every day,

b I receive a total of at least 84 emails?

8 At a weather station during light rain, raindrops land on a one square metre detector plate at a rate of 25 per minute. Assuming that their arrival constitutes a random process, what is the probability that, in a randomly chosen ten-minute period, there will be no rain detected on a $10\,\text{cm} \times 20\,\text{cm}$ plate. State any assumptions you make.

9 Silver chain is made by linking circles of silver. Quality management rules require that the first three 5 m lengths of chain produced in a day should have at most one flaw each, otherwise the machine must be adjusted. Find the probability that no adjustment is necessary if imperfections in the chain occur at a rate of 3.2 per ten metres.

10 Suppose that there exist two types of star, A and B. The number of each type in a given volume of space follows a Poisson distribution with means 1.8 and 2.1 respectively. Assuming that these random variables are independent, find the probability that the volume contains a total of three stars.

11 Cars at a given point on the M62 motorway early in the morning travel either towards or away from Leeds with mean rates of 123 and 111 per 10 minutes respectively. Assuming that traffic is flowing freely, find the probability that the total number of cars passing the point in ten minutes is over 250

10.3 Continuous distributions 1

Fluency and skills

You know that, for discrete random variables, you give a probability distribution function by stating the probabilities of each value the variable can take. This is not possible for continuous variables because you cannot list all the individual values. In the same way as you use histograms instead of bar charts for continuous variables, probability density functions replace probability distribution functions for continuous variables.

You use a **probability density function** (pdf) of a random variable, X, to find the probability that a randomly chosen x-value will be between two values. To do this, you find the area under the function between these values. You can do this by integrating.

Key point

If X is a random variable with probability density function $f(x)$, then $P\{a<X<b\} = \int_{x=a}^{x=b} f(x)\,dx$

This is similar to histograms where area gives frequency.

Since the area under the probability density function gives a probability, it follows that the probability of X taking any exact value is zero; that is, $P\{X=x\}=0$

Probability density functions must obey two properties.

- $\int_{-\infty}^{+\infty} f(x)\,dx = 1$ (total probability is represented by total area).
- $f(x) \geq 0$ (all probabilities are greater than or equal to zero).

Example 1

A random variable, X, has a probability density function given by $f(x) = \frac{3}{8}x^2$ for $0<x<2$

a Sketch the curve $y=f(x)$ and verify that $f(x)$ is a valid pdf. On your diagram, shade the area representing $P\{X>1\}$

b Find the value of $P\{X>1\}$

a $f(x) = \frac{3}{8}x^2$ for $0<x<2$

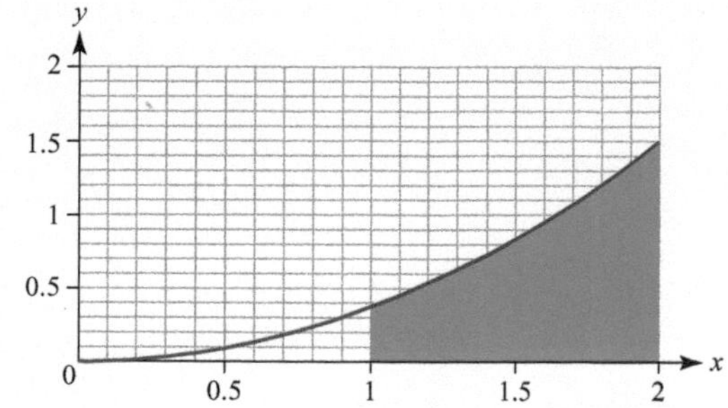

y
2
1.5
1
0.5
0
0.5
1
1.5
2
x

(*Continued on the next page*)

$f(x) > 0$ for $0 < x < 2$

All probabilities are greater than zero.

$$\int_0^2 \frac{3}{8}x^2\,dx = \left[\frac{x^3}{8}\right]_0^2$$
$$= 1 - 0$$
$$= 1$$

Total probability is represented by total area.

So $f(x)$ is a valid pdf.

b $$P\{X > 1\} = \int_1^2 \frac{3}{8}x^2\,dx$$
$$= \left[\frac{x^3}{8}\right]_1^2$$
$$= 1 - \frac{1}{8}$$
$$= \frac{7}{8}$$

Example 2

A random variable, X, has probability density function

$$f(x) = \begin{cases} -kx(x-4); & 0 < x < 4 \\ 0; & \text{otherwise} \end{cases}$$

where k is some positive constant.

Find the value of k

$$\int_0^4 -kx(x-4)\,dx = 1$$

For $f(x)$ to be a pdf, the total probability rule must apply.

$$\int_0^4 (-kx^2 + 4kx)\,dx = 1$$

Integrate.

$$\left[-k\frac{x^3}{3} + 2kx^2\right]_0^4 = 1$$
$$-\frac{64}{3}k + 32k = 1$$
$$\frac{32}{3}k = 1$$
$$k = \frac{3}{32}$$

For a continuous random variable, the median is the value that divides the total probability, and therefore the area under the corresponding pdf, into two halves.

$\int_{-\infty}^{M} f(x)dx$ gives the probability up to the median value and is therefore, by definition, equal to $\frac{1}{2}$

In practice, the lower limit of the integral will be the smallest value of x

Key point

If X is a continuous random variable with probability density function f(x), then the median value of X, denoted by M, say, is given by the equation

$$\int_{-\infty}^{M} f(x)dx = \frac{1}{2}$$

Example 3

A random variable, X, has a probability density function given by

$$f(x) = \begin{cases} \frac{1}{2}x; & 1 < x < \sqrt{5} \\ 0; & \text{otherwise} \end{cases}$$

Find the median of X

$$\int_{-\infty}^{M} f(x)dx = \frac{1}{2}$$

$$\int_{1}^{M} \frac{x}{2}dx = \frac{1}{2}$$

$$\left[\frac{x^2}{4}\right]_1^M = \frac{1}{2}$$

$$\frac{M^2}{4} - \frac{1}{4} = \frac{1}{2}$$

$$M = \sqrt{3}$$

You can also calculate the lower and upper quartiles from the probability density function. They are defined by the equations $P(X < Q_1) = \frac{1}{4}$ and $P(X < Q_3) = \frac{3}{4}$

Key point

The lower quartile (Q_1) and upper quartile (Q_3) are given by

$$\int_{-\infty}^{Q_1} f(x)dx = \frac{1}{4}$$

$$\int_{-\infty}^{Q_3} f(x)dx = \frac{3}{4}$$

where f(x) is the pdf.

STATISTICS

The mode of a continuous random variable is the value of the variable at a local maximum. This can be found by considering a sketch of the probability density function or by differentiation.

Key point

For a continuous random variable with probability density function f(x), the mode can be found from the equation

$$\frac{\mathrm{df}(x)}{\mathrm{d}x}=0$$

Example 4

A random variable, X, has probability density function

$$\mathrm{f}(x)=a(x^2-x-2);\ -1<x<1$$

where a is a constant.

a Find the value of a

b By differentiating f(x), or otherwise, find the mode of X

$\mathrm{f}(x)=a(x^2-x-2);\ -1<x<1$

a $\int_{-1}^{1} a(x^2-x-2)\mathrm{d}x=1$

$\left\{a\left(\frac{x^3}{3}-\frac{x^2}{2}-2x\right)\right\}_{-1}^{1}=1 \rightarrow -\frac{10a}{3}=1$

$a=-\frac{3}{10}$

b $\frac{\mathrm{df}(x)}{\mathrm{d}x}=0$ at modal value

$\left(-\frac{3}{10}\right)(2x-1)=0$

$x=\frac{1}{2}$

The modal value can also be found from the graph of f(x); the graph cuts the x-axis at −1 and 2 and the modal value is midway, $x=\frac{1}{2}$

You can often find results for continuous random variables by considering the discrete equivalents. So the formulae for the expected (mean) value and variance of continuous variables are found by first considering their discrete versions. Changing Σ to $\int$ and P{$X=x$} to f(x)dx give the following results.

Key point

If X is a random variable with mean μ and variance σ^2, then

$$\mu = E(X) = \int_{-\infty}^{+\infty} x f(x)\,dx$$

$$\sigma^2 = Var[X] = E[(X-\mu)^2] = \int_{-\infty}^{\infty} (x-\mu)^2 f(x)\,dx$$

For ease of calculation, the variance formula can be written

$$Var[X] = E[X^2] - \mu^2 = \int_{-\infty}^{\infty} x^2 f(x)\,dx - \mu^2$$

The continuous versions of E[X] and Var[X] are found from their discrete equivalents by changing the Σ-operators into integrals because integration is the continuous equivalent of the discrete summation, Σ

STATISTICS

Example 5

A random variable, X, has a probability density function given by

$$f(x) = \begin{cases} \frac{1}{2}x; & 1 < x < \sqrt{5} \\ 0; & \text{otherwise} \end{cases}$$

Find the mean, μ, and variance of X

$$\mu = \int_{-\infty}^{+\infty} x f(x)\,dx$$

$$= \int_{x=1}^{\sqrt{5}} \frac{x^2}{2}\,dx$$ — Substitute in f(x) and the limits.

$$= \left[\frac{x^3}{6}\right]_1^{\sqrt{5}}$$ — Integrate.

$$= \frac{(\sqrt{5})^3}{6} - \frac{1^3}{6}$$

$$= 1.70$$

$$\sigma^2 = \int_{-\infty}^{\infty} x^2 f(x)\,dx - \mu^2$$

$$= \int_1^{\sqrt{5}} \frac{x^3}{2}\,dx - 1.70^2$$ — Substitute in f(x), μ and the limits.

$$= \left[\frac{x^4}{8}\right]_1^{\sqrt{5}} - 1.70^2$$ — Integrate.

$$= \frac{(\sqrt{5})^4}{8} - \frac{1^4}{8} - 1.70^2$$

$$= 0.12$$

The mean value and variance of a function, g(X), of X can be found by replacing x by g(x)

Key point

If X is a random variable with probability density function f(x), then the mean value and variance of g(X) are given by

$$E(g(X)) = \int_{-\infty}^{\infty} g(x)f(x)\,dx$$

$$\text{Var}(g(X)) = E(\{g(X)\}^2) - \{E(g(X))\}^2$$
$$= \int_{-\infty}^{\infty} \{g(x)^2\}f(x)\,dx - \{E(g(X))\}^2$$

The above result can be used with the following for the expectation of sums of functions of random variables.

Key point

For a random variable X, the mean value of the function $g(X) + h(X)$ is given by $E(g(X) + h(X)) = E(g(X)) + E(h(X))$

Example 6

A continuous random variable X has probability density function $f(x) = \frac{x+1}{6}$ for x between 1 and 3
Find the expected value of

a $2X^2$

b $X + \frac{15}{X^3}$

a $E(2X^2) = \int_1^3 2x^2\left(\frac{x+1}{6}\right)dx$

$$= \frac{1}{3}\int_1^3 (x^3 + x^2)\,dx$$

$$= \frac{1}{3}\left[\frac{x^4}{4} + \frac{x^3}{3}\right]_1^3$$

$$= 9.56 \text{ (2 dp)}$$

b $E\left(X + \frac{15}{X^3}\right) = \int_1^3 \left(x + \frac{15}{x^3}\right)\left(\frac{x+1}{6}\right)dx$

$$= \frac{1}{6}\int_1^3 (x^2 + x + 15x^{-2} + 15x^{-3})\,dx$$

$$= \frac{1}{6}\left[\frac{x^3}{3} + \frac{x^2}{2} - 15x^{-1} - \frac{15}{2}x^{-2}\right]_1^3$$

$$= \frac{44}{9}$$

Or use $E\left(X + \frac{15}{X^3}\right) = E(X) + E\left(\frac{15}{X^3}\right)$

Exercise 10.3A Fluency and skills

1 A random variable, X, has a probability density function given by

$$f(x)=\begin{cases}\frac{1}{2}x; & 1<x<a\\ 0; & \text{otherwise}\end{cases}$$

a Sketch a graph of the probability density function.

b Prove that $a=\sqrt{5}$ and hence calculate $P(X>1.5)$

2 A random variable, X, has a pdf given by

$$f(x)=\begin{cases}kx^2; & 0<x<2\\ 0; & \text{otherwise}\end{cases}$$

Find the value of k and $P\{X<1\}$

3 A random variable, X, has probability density function

$$f(x)=\begin{cases}-ax(x-6); & 0<x<6\\ 0; & \text{otherwise}\end{cases}$$

where a is some positive constant.

Find the value of a

4 A random variable, X, has probability density function

$$f(x)=\begin{cases}ax(2x-1); & 0<x<\frac{1}{2}\\ 0; & \text{otherwise}\end{cases}$$

where a is a constant.

a Find the value of a

b Calculate the values of

i The median,

ii The lower and upper quartiles,

iii The interquartile range of X

5 a Sketch the probability density function of X, given by

$$f(x)=\begin{cases}a-|x|; & -a<x<a\\ 0; & \text{otherwise.}\end{cases}$$

b Show that $a=1$

c Find the exact values of the lower and upper quartiles and the interquartile range.

Full A Level

6 The random variable Y has a pdf given by $f(y)=\frac{2y+1}{k}$ for $0<y<4$

a Find the value of k

b Show that the first quartile is 1.79 (3 sf) and find the second quartile.

c Find $E(Y)$, $E(Y^2)$ and $Var(Y)$

7 A continuous random variable X has pdf given by

$$f(x)=\begin{cases}\frac{a(2x+1)}{4}; & 0<x<5\\ 0; & \text{otherwise.}\end{cases}$$

a Show that the value of a is $\frac{2}{15}$ and find the mean and variance of X

b Show that in three randomly chosen x-values, the probability of only one being above 4 is $\frac{4}{9}$

8 A continuous random variable X has probability density function $f(x)=\frac{2x+1}{18}$ for x between 1 and 4 Find the expected value of

a X^2

b $X+\frac{1}{X^3}$

9 A random variable, X, has probability density function

$$f(x)=\begin{cases}ax(x-5); & 0<x<5\\ 0; & \text{otherwise}\end{cases}$$

where a is some positive constant.

a Find the value of a and $P\{1<X<3\}$

b Write down the value of $P\{X=2\}$ and explain your answer.

c By differentiation, show that the mode of X is 2.5

STATISTICS

10 A random variable, X, has probability density function

$$f(x)=\begin{cases} x(x-4); & 4<x<a \\ 0; & \text{otherwise} \end{cases}$$

where a is a constant.

a Find the value of a

b Calculate the median value of X

c Find the interquartile range of X

11 A random variable, X, has a probability density function given by

$$f(x)=\begin{cases} \frac{1}{4}x; & 0<x<a \\ 0; & \text{otherwise} \end{cases}$$

a Sketch a graph of the probability density function.

b Prove that $a=2\sqrt{2}$ and hence show that the median of X is 2

12 For every parcel handled by a delivery company, an estimated time of delivery is given. Among parcels which are delivered late, the number of hours late can be modelled by a random variable T with probability density function

$$f(t)=\begin{cases} \frac{t}{5}; & 0<t<1 \\ -\frac{1}{45}t+\frac{2}{9}; & 1<t<10 \\ 0; & \text{otherwise} \end{cases}$$

a Sketch the graph of $f(t)$

b Verify that $f(t)$ is a valid probability density function.

c Write down the mode of T

d Find the median of T

Reasoning and problem-solving

Strategy

To solve a problem involving continuous random variables

1. Use the 'area equals 1' property to find any unknown constants.
2. Use integration to find probabilities and the cumulative distribution function.
3. Calculate summary statistics for location and spread.
4. Find mean and variance of functions of random variables.

You can find the expectation value and variance of a linear function of a random variable X from the probability density function. Suppose Y is a linear function of X, that is, $Y=aX+b$, where a and b are constants.

$$\begin{aligned} E(Y) &= E(aX+b) \\ &= \int_{-\infty}^{\infty}(ax+b)f(x)\,dx \\ &= \int_{-\infty}^{\infty}(axf(x)+bf(x))\,dx \\ &= a\int_{-\infty}^{\infty}xf(x)\,dx+b\int_{-\infty}^{\infty}f(x)\,dx \\ &= aE(X)+b \end{aligned}$$

since $E(X)=\int_{-\infty}^{\infty}xf(x)\,dx$ and $\int_{-\infty}^{\infty}f(x)\,dx=1$

$$\begin{aligned}
\text{Var}[Y] &= \text{Var}(aX+b)\\
&= \text{E}\left((aX+b)^2\right)-\left(\text{E}(aX+b)\right)^2 \text{ (Definition of variance)}\\
&= \text{E}\left((aX+b)^2\right)-\left(a\text{E}(X)+b\right)^2 \text{ (Using the result above)}\\
&= \text{E}\left(a^2X^2+2abX+b^2\right)-\left(a\text{E}(X)\right)^2+2ab\text{E}(X)+b^2)\\
&= a^2\text{E}\left(X^2\right)+2ab\text{E}(X)+b^2-\left(a^2\left(\text{E}(X)\right)^2+2ab\text{E}(X)+b^2\right)
\end{aligned}$$

(Using the result for sums of functions)

$$\begin{aligned}
&= a^2\left(\text{E}\left(X^2\right)-\left(\text{E}(X)\right)^2\right)\\
&= a^2\text{Var}(X)
\end{aligned}$$

Key point

If X and Y are two continuous random variables and $Y = aX + b$, where a and b are constants, then

$\text{E}(Y) = a\text{E}(X) + b$

$\text{Var}(Y) = a^2\text{Var}(X)$

These results can also be written as $\text{E}(aX + b) = a\text{E}(X) + b$ and $\text{Var}(aX + b) = a^2\text{Var}(X)$.

STATISTICS

Example 7

A random variable, X, has a pdf given by $f(x)=\begin{cases} a(x-4); & 4<x<6 \\ 0; & \text{otherwise}\end{cases}$

a Find the value of a and the mean and variance of X

The random variable Y is related to X by the equation $Y = -2X + 1$

b Find the mean and variance of Y

a $f(x) = a(x-4)$

$\int_4^6 a(x-4)\,dx = 1$

$\left[a\left(\frac{x^2}{2}-4x\right)\right]_4^6 = 1$

$a[(18-24)-(8-16)] = 1$

$2a = 1$

$a = \frac{1}{2}$

1 Use the 'area equals 1' property to find any unknown constants.

(*Continued on the next page*)

Therefore $f(x)=\frac{1}{2}(x-4)$

$$E(X)=\int_4^6 \frac{1}{2}x(x-4)dx$$
$$=\left[\frac{1}{2}\left(\frac{x^3}{3}-2x^2\right)\right]_4^6$$
$$=\left[\frac{x^3}{6}-x^2\right]_4^6$$
$$=(36-36)-\left(\frac{64}{6}-16\right)$$
$$=\frac{16}{3}$$

$$Var(X)=\int_4^6 \frac{1}{2}x^2(x-4)dx-\left(\frac{16}{3}\right)^2$$
$$=\left[\frac{1}{2}\left(\frac{x^4}{4}-\frac{4x^3}{3}\right)\right]_4^6-\left(\frac{16}{3}\right)^2$$
$$=\left[\frac{x^4}{8}-\frac{2x^3}{3}\right]_4^6-\left(\frac{16}{3}\right)^2$$
$$=(162-144)-\left(32-\frac{128}{3}\right)-\frac{256}{9}$$
$$=18+\frac{32}{3}-\frac{256}{9}$$
$$=\frac{2}{9}$$

3 Calculate summary statistics for spread.

$Y=-2X+1$

$$E(Y)=-2E(X)+1$$
$$=-2\times\frac{16}{3}+1$$
$$=-\frac{29}{3}$$

$$Var(Y)=4Var(X)$$
$$=4\times\frac{2}{9}$$
$$=\frac{8}{9}$$

4 Find mean and variance of linear functions of random variables.

As with discrete variables, the expected (mean) value and variance of the sum of two independent continuous random variables can be found.

Key point

For two independent continuous random variables, X and Y, the expected value and variance of the sum, $X+Y$, is given by

$E(X+Y)=E(X)+E(Y)$

$Var(X+Y)=Var(X)+Var(Y)$

Example 8

Two independent random variables, R and S, have distributions as follows.

$$f(r)=\begin{cases} k(r+1); & 0<r<1 \\ 0; & \text{otherwise}\end{cases}$$

$$f(s)=\begin{cases} ms^2; & 1<s<2 \\ 0; & \text{otherwise}\end{cases}$$

where k and m are constants.

a Find the values of k and m

b Find the expected value and variance of the function $2R+3S$

(*Continued on the next page*)

a $k\int_0^1 (r+1)dr = 1 \rightarrow k = \frac{2}{3};\ m\int_1^2 s^2 ds = 1 \rightarrow m = \frac{3}{7}$

b $E(2R+3S) = E(2R) + E(3S) = 2E(R) + 3E(S)$

$= 2\times\frac{2}{3}\times\int_0^1 r(r+1)dr + 3\times\frac{3}{7}\times\int_1^2 s^3 ds$

$= \frac{10}{9} + \frac{135}{28} = 5.93\ (2\ dp)$

$Var(2R+3S) = Var(2R) + Var(3S) = 4Var(R) + 9Var(S) =$

$= 4\times\left(\frac{2}{3}\times\int_0^1 r(r+1)dr - \left(\frac{5}{9}\right)^2\right) + 9\times\left(\frac{3}{7}\times\int_1^2 s^4 ds - \left(\frac{45}{28}\right)^2\right)$

$= \frac{26}{81} + \frac{2619}{3920} = 0.99\ (2\ dp)$

1 Use the 'area equals 1' property to find any unknown constants.

2R and 3S are independent random variables.

4 Find mean and variance of the functions of random variables.

STATISTICS

Exercise 10.3B Reasoning and problem-solving

1 A random variable, X, has a pdf given by

$$f(x) = \begin{cases} kx^2; & -1 < x < 3 \\ 0; & \text{otherwise} \end{cases}$$

a Find the value of k and the mean and variance of X

The random variable Y is related to X by the equation $Y = -2X + 1$

b Find the mean and variance of Y

2 In a maths exam, marks over several years were found to be modelled by a probability density function given by $f(x) = kx(60 - x)$ for $0 < x < 60$, where k is a positive constant.

a Find the value of k and $P\{X > 50\}$

One year the exam board decided to scale the marks according to the equation $y = 0.9x + 3$, where y is the scaled mark.

b What values do marks of 50 and 10 become after scaling?

c Calculate the change in the mean mark as a result of the scaling.

3 A random variable, X, has probability density function

$$f(x) = \begin{cases} ax(x-b); & 0 < x < 3 \\ 0; & \text{otherwise} \end{cases}$$

where a and b are constants.

a Given that the mean of X is $\frac{7}{4}$, find the constants a and b

b Show that the variance of X equals $\frac{43}{80}$

c If μ and σ are the mean and standard deviation of X, respectively, find the probability that a randomly chosen value of X will lie between $\mu + \sigma$ and $\mu - \sigma$. Give your answer to 2dp.

4 Demand for coal, in tonnes per week, from a wholesaler can be modelled by a probability density function

$$f(x) = \frac{3x(100-x)}{500000} \text{ for } 0 < x < 100$$

Show that the median demand is 50 tonnes per week.

5 The median of a random variable with probability density function $f(x) = ax(2 - x)$; $0 < x < 2$, $a > 0$ is M

a Find the value of a

b Show that M satisfies the equation $M^3 - 3M^2 + 2 = 0$

c Hence show that $M = 1$

6 Two independent random variables, X and Y, have distributions as follows.

$$f(x) = \begin{cases} 2x; \ 0 < x < 1 \\ 0; \ \text{otherwise} \end{cases}$$

$$f(y) = \begin{cases} my^2; \ 1 < y < 3 \\ 0; \ \text{otherwise} \end{cases}$$

where m is a constant.

a Find the values of m

b Find the expected value and variance of the function $3X + 2Y$

7 A continuous random variable X has probability density function

$$f(x) = \begin{cases} \frac{1}{6}x^2 + 1; \ 0 < x < a \\ 0; \ \text{otherwise} \end{cases}$$

a Find the value of a

b Find the values of

i $E(3X + 4)$, ii $\text{Var}(2X - 1)$

8 A random variable, X, has a probability density function given by

$$f(x) = \begin{cases} 2x^3; \ 0 < x < a \\ 0; \ \text{otherwise} \end{cases}$$

a Show that $a = 2^{\frac{1}{4}}$

b Find the mean and variance of X

c Find the mean and variance of the function $2X - 3$

Give your answers to 2 decimal places.

9 Two continuous random variables, X and Y, have means and standard deviations given by μ_X and σ_X, μ_Y and σ_Y. Write down expressions for the mean and variance of $X + Y$, stating any assumptions you make.

10 Anouk travels to work by bicycle and train. The probability density function of X, the time spent on the train in minutes, is given by

$$f(x) = \begin{cases} a(x^2 - 12x); \ 0 < x < 12 \\ 0; \ \text{otherwise} \end{cases}$$

a Find the value of a

b Find the mean and variance of X

The mean and standard deviation of the cycle ride are 7 minutes and 2 minutes respectively. The total journey time, T, is the sum of her train and cycle journey times plus 2 minutes spent walking.

c Find the mean and variance of T

10 Summary and review

Chapter summary

- The expected value or mean value and variance of a discrete random variable are given by

$$\mu = \mathrm{E}[X] = \sum_{\text{all } x} x\mathrm{P}\{X = x\}$$

$$\sigma^2 = \mathrm{Var}(X) = \mathrm{E}[(X-\mu)^2] = \sum_{\text{all } x}(x-\mu)^2\,\mathrm{P}\{X = x\} = \sum_{\text{all } x} x^2\mathrm{P}\{X = x\} - \{\mathrm{E}(X)\}^2$$

- The expected value and variance of a function, g(X), of a discrete random variable are given by

$$\mathrm{E}[\mathrm{g}(X)] = \sum_{\text{all } x}\mathrm{g}(x)\mathrm{P}\{X = x\}$$

$$\mathrm{Var}(\mathrm{g}(X)) = \mathrm{E}(\{\mathrm{g}(X)\}^2) - \{\mathrm{E}(\mathrm{g}(X))\}^2 = \sum_{\text{all } x}\{\mathrm{g}(x)^2\}\mathrm{P}\{X = x\} - \{\mathrm{E}(\mathrm{g}(X))\}^2$$

- If X and Y are two random variables and $Y = aX + b$, where a and b are constants, then $\mathrm{E}(Y) = a\mathrm{E}(X) + b$ and $\mathrm{Var}(Y) = a^2\mathrm{Var}(X)$
- If X and Y are two independent random variables, the expected value and variance of the sum, $X + Y$, are given by $\mathrm{E}(X + Y) = \mathrm{E}(X) + \mathrm{E}(Y)$ and $\mathrm{Var}(X + Y) = \mathrm{Var}(X) + \mathrm{Var}(Y)$ respectively.
- A discrete uniform random variable, U, taking values $u = 1, 2, \ldots, a$, has probability distribution given by $P\{U = u\} = \frac{1}{n};\ u = 1, 2, \ldots, n$
- A discrete uniform random variable, U, with distribution function $P\{U = u\} = \frac{1}{n};\ u = 1, 2, \ldots, n$ has a mean and variance given by $\mu = \frac{n+1}{2}$ and $\sigma^2 = \frac{n^2-1}{12}$
- If X is the number of random events that occur in a given interval of time or space and λ is the mean number of occurrences in that interval, then its probability distribution function is given by

$$\mathrm{P}\{X = x\} = \frac{\mathrm{e}^{-\lambda}\lambda^x}{x!};\ x = 0, 1, 2\ldots$$

and X is said to follow a Poisson distribution, parameter λ.

- If X has a probability distribution function given by $\mathrm{P}\{X = x\} = \frac{\mathrm{e}^{-\lambda}\lambda^x}{x!};\ x = 0, 1, 2\ldots,$ then the mean and variance of X are, respectively, given by $\mathrm{E}(X) = \lambda$ and $\mathrm{Var}(X) = \lambda$
- If $X_i;\ i = 1, 2, \ldots, n$ is a set of n independent and identically distributed random variables each with a Poisson distribution, mean λ and if $X_T = \sum_{i=1}^{n} X_i$, then X_T has a Poisson distribution with mean $n\lambda$
- If events occur randomly at a rate of λ per given interval of time or space, then the number of events in n such intervals is a Poisson random variable with mean $n\lambda$
- If X is a random variable with density function f(x), then $\mathrm{P}(a < X < b) = \int_{x=a}^{x=b}\mathrm{f}(x)\mathrm{d}x$

- Probability density functions must obey two properties.

  $\int_{-\infty}^{+\infty} f(x)dx = 1$ and $f(x) \geq 0$ for all x

- If X is a continuous random variable with probability density function $f(x)$, then the median value of X is given by the equation $\int_{-\infty}^{M} f(x)dx = \frac{1}{2}$

- The lower and upper quartiles, Q_1 and Q_3, are given by the formulae $\int_{-\infty}^{Q_1} f(x)dx = \frac{1}{4}$ and $\int_{-\infty}^{Q_3} f(x)dx = \frac{3}{4}$, where $f(x)$ is the probability density function.

- If X is a random variable with mean μ and variance σ^2, then

  $$\mu = E(X) = \int_{-\infty}^{+\infty} xf(x)dx$$

  $$\sigma^2 = Var[X] = \int_{-\infty}^{\infty} (x-\mu)^2 f(x)dx = \int_{-\infty}^{\infty} x^2 f(x)dx - \mu^2$$

- If X is a random variable with density function $f(x)$, then the mean value and variance of $g(X)$ are given by

  $$E(g(X)) = \int_{-\infty}^{\infty} g(x)f(x)dx$$

  $$Var(g(X) = E(\{g(X)\}^2) - \{E(g(X))\}^2 = \int_{-\infty}^{\infty} \{g(x)\}^2 f(x)dx - \{E(g(X))\}^2$$

Check and review

You should now be able to...	Review Questions
✔ Find the expected value and variance of random variables and their functions.	1, 2, 3
✔ Apply the Poisson distribution to suitable random experiments.	4
✔ Find probabilities and the mean and variance of continuous random variables and their functions.	5
✔ Find the mean and variance of sums of independent random variables and their functions.	6

1 Calculate the expected value and variance for this probability distribution.

x	1	3	5	7	9	11
$P(X=x)$	$\frac{1}{6}$	$\frac{1}{6}$	$\frac{1}{6}$	$\frac{1}{6}$	$\frac{1}{6}$	$\frac{1}{6}$

2 A six-sided dice with faces numbered 1–6 is biased so that the probability of any score is proportional to the cube of the score. Find the expected value of C, the score on a randomly chosen throw.

3 Random events occur at an average rate of five per four minutes. Find the probability that there will be fewer than two events in any randomly chosen four-minute interval.

4 A firm receives, on average, 18 letters per day. Assuming that their arrival constitutes a random process, what is the probability that in five randomly chosen days

a There are at least twelve letters every day,

b There is a total of at least 90 letters?

5 A continuous random variable X has probability density function given by

$$f(x)=\begin{cases}\dfrac{a(x+3)}{5}; & 0<x<2\\ 0; & \text{otherwise}\end{cases}$$

a Show that the value of a is $\frac{5}{8}$ and find the mean and variance of X

b A sample of four x-values is taken at random. Find the probability that exactly one value will be greater than one.

6 Two independent random variables, S and T, have probability density functions as follows:

$$f(s)=\begin{cases}as; & 1<s<2\\ 0; & \text{otherwise}\end{cases}$$

$$f(t)=\begin{cases}6t(1-t); & 0<t<1\\ 0; & \text{otherwise}\end{cases}$$

where a is a constant.

a Find the values of a

b Find the expected value and variance of the function $S+2T$

7 X is a random variable for the score on a randomly chosen throw of a fair, 10-sided dice. Find the expected values and variances of

a $2X$

b $X-1$

STATISTICS

10 Exploration

Going beyond the exams

Did you know?

The Poisson distribution is named after the French Mathematician, Geometer and Physicist, Simeon Poisson. Poisson is also well known for his partial differential equation that is frequently written as $\nabla^2 \phi = f$, and widely used in mechanical engineering and theoretical physics.

> **Life is good for only two things: doing mathematics and teaching it.**
> - Simeon Poisson

ICT

Many discrete distributions can be simulated in spreadsheet packages. For instance, the distribution of 100 rolls of a fair dice can be simulated in Excel.

Use the function =RANDBETWEEN(1,6) to generate a random number between 1 and 6. Then copy this formula into 100 (say) cells in the A column. Use the COUNTIF function to count how many of each number appears in the list. (An example formula can be seen in cell D3.) Then draw a graph of the results.

Try to simulate the following situations using a similar process:

- 100 rolls on a twelve-sided dice;
- 100 rolls on two six-sided dice;
- 100 rolls on a biased six-sided dice (for example, even numbers are twice as likely as odd numbers).

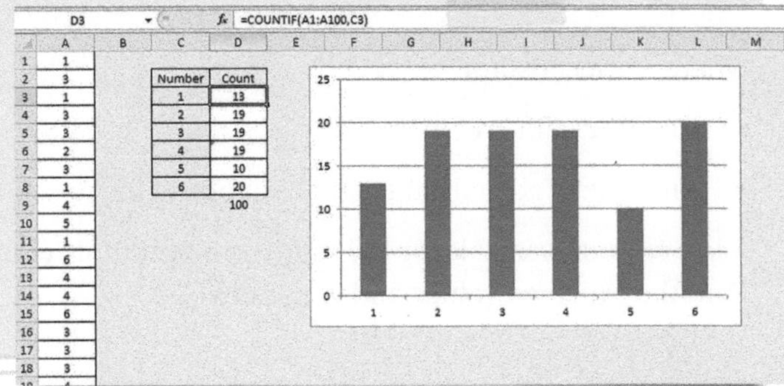

ICT

The Excel function =POISSON(x, mean, cumulative) can be used to calculate the probability of x events, with the mean average number of events per interval.

For example, by 2017 the mean average number of goals Ipswich Town scored against Norwich City per match was 1.455. Find the probability of

(a) Ipswich Town scoring 5 goals in one game against Norwich City,

(b) Ipswich Town scoring 4 or fewer goals in one game against Norwich City.

Using Excel,

(a) = POISSON(5, 1.455, false) = 0.0127 (to 3 significant figures).

(b) = POISSON(4, 1.455, true) = 0.983 (to 3 significant figures).

10 Assessment

1 The probability distribution of X and Y are as follows.

$$P(X=x)=\begin{cases}\dfrac{x}{k};\ x=1,2,3,4\\ 0;\ \text{otherwise}\end{cases}$$

y	1	2	3	4
$P\{Y=y\}$	0.3	a	a	b

Given that $P\{X=4\}=P\{Y=4\}$

a Find the value of k, a and b **[5 marks]**

b Which discrete random variable has the highest value on average? Explain how you know. **[4]**

2 This table shows the probability distribution of N, the number of people living in a household as estimated from a sample of 200 households.

n	1	2	3	4	5
$P\{N=n\}$	$5k$	$6k$	$7k$	k	k

a Find the value of k **[2]**

b Calculate the median and mode of N **[2]**

c Calculate the expected number of people per household. **[2]**

d Calculate the variance. **[3]**

3 The continuous random variable X is such that $X \sim \text{Po}(3)$

a Calculate each of these probabilities.

i $P\{X \le 5\}$ **[1]**

ii $P\{X=3\}$ **[2]**

iii $P\{4 < X \le 7\}$ **[2]**

iv $P\{X < \lambda\}$ **[2]**

b State the variance of X **[1]**

4 A shop sells apples at the rate of 2.5 per minute on average.

a Explain what assumptions you would have to make in order to model the sale of apples as a Poisson distribution. **[2]**

b Use the Poisson distribution to find the probability that in a one-minute period

i No apples are sold, **[2]**

ii Fewer than three apples are sold, **[2]**

iii More than four apples are sold. **[2]**

c Calculate the probability that, over a three-minute period, more than ten apples are sold. **[3]**

5 A random variable X has mean 4 and standard deviation 3

a Calculate $E(X^2)$ **[2]**

The random variable M is defined as $M = X + 2Y$, where X and Y are independent random variables.

b Use the information that $E(M) = 11$ and $Var(M) = 19$ to find the mean and standard deviation of the random variable Y **[4]**

6 The continuous random variable X has probability density function given by

$$f(x) = \begin{cases} kx; & 0 \le x \le 6 \\ 0; & \text{otherwise.} \end{cases}$$

a Calculate the value of k **[4]**

b Find each of these probabilities.

i $P\{X < 3\}$ **[2]** ii $P\{X > 1\}$ **[2]** iii $P\{1.5 < X < 5\}$ **[2]**

c Calculate the expectation of X **[3]**

d Calculate the variance of X **[4]**

7 The continuous random variable X has pdf given by

$$f(x) = \begin{cases} \frac{1}{9}(4x - x^2); & 0 \le x \le 3 \\ 0; & \text{otherwise} \end{cases}$$

a Sketch the probability density function of X **[2]**

b Write down the mode of X **[1]**

c Calculate the median of X **[4]**

8 The table gives the bivariate probability distribution for the variables X and Y

		Y		
		10	**20**	**P{X=x}**
X	**1**	0.2	0.1	0.3
	2	0.15	0.05	0.2
	3	0.3	0.2	0.5
	P{Y=y}	0.65	0.35	1

a Write down the probability distribution of the random variable T, where $T = X + Y$ **[3]**

b Find $E(T)$ **[1]**

c Without doing any further calculations, explain why $E(T) = E(X) + E(Y)$ but $Var(T) \neq Var(X) + Var(Y)$ **[3]**

9 Given that $X \sim Po(7)$ and $Y \sim Po(3)$, where X and Y are independent random variables,

a Write down the value of

i $E(X + Y)$ ii $Var(X + Y)$ **[2]**

b Calculate $P\{X + Y > 8\}$ **[3]**

c Find the probability that $X = 5$ and $Y = 3$ **[3]**

10 The discrete uniform variable U is defined over the interval $[2, 7]$

a Calculate $\text{E}(U)$ and $\text{Var}(U)$ **[3]**

b Find $\text{P}\{X > 3.5\}$ **[2]**

11 The discrete random variable V has probability density function $f(v) = \begin{cases} \dfrac{v^2}{30}; & v = 1, 2, 3, 4 \\ 0; & \text{otherwise} \end{cases}$

a Calculate the median of V **[2]**

b Calculate $\text{E}(5V+3)$ and $\text{Var}(5V+3)$ **[7]**

12 Given that the random variable X is such that $X \sim \text{Po}(5.5)$

a Calculate the range of values of a such that $\text{P}\{X < a\} > 0.7$ **[2]**

b Calculate the range of values of b such that $\text{P}\{X > b\} < 0.1$ **[2]**

c Calculate the probability that the value of X is within one standard deviation of the mean. **[5]**

A second, independent, random variable Y is such that $Y \sim \text{Po}(2)$

d Calculate each of these probabilities.

i $\text{P}\{X+Y < 9\}$ **[2]** ii $\text{P}\{X-Y \geq 2\}$ **[3]** iii $\text{P}\{(2X-Y)\} > E(2X-Y)$ **[4]**

13 The probability density function of Y is given by $f(y) = \begin{cases} ky^3; & 0 \leq y \leq 2 \\ 0; & \text{otherwise} \end{cases}$

a Calculate the variance of Y **[8]**

b Calculate the interquartile range of Y **[7]**

The continuous random variable W is defined by $W = 4Y - 3$

c Find the expectation and variance of W **[4]**

14 The discrete random variable T has probability density function $f(t) = \begin{cases} \dfrac{1}{A}; & t = 1, 2, 3 \\ \dfrac{t}{A}; & t = 4, 5 \\ 0; & \text{otherwise} \end{cases}$

a Calculate $\text{P}\{T > E(T)\}$ **[5]**

b Find $\text{Var}(7-3T)$. **[4]**

15 The probability density function of X is given by $f(x) = \begin{cases} \dfrac{1}{n}; & x = 1, 2, \ldots n \\ 0; & \text{otherwise} \end{cases}$

a Write down an expression for $\text{E}(X)$ **[1]**

b Prove that the variance of X is given by the formula $\text{Var}(X) = \dfrac{1}{12}(n^2 - 1)$ **[7]**

16 The number of emails a person receives during the hours of 9 am to 5 pm is modelled by a Poisson distribution. Emails are received at a rate of three every ten minutes.

a Calculate the probability of receiving more than seven emails in a ten-minute period. **[3]**

b Calculate the probability of receiving between five and eight emails (inclusive) over half an hour. **[3]**

c Calculate the probability of receiving at least three emails every ten minutes for an hour. **[5]**

17 A continuous random variable has probability density function $f(x)=\begin{cases} ax+b; & 0\le x\le 2 \\ 0; & \text{otherwise} \end{cases}$

Given that $E(X)=\frac{13}{12}$

a Calculate the values of a and b **[6]**

b Calculate the median of X **[4]**

The continuous random variable Y is defined by $Y = AX+2$

c Write down an expression for the mean of Y **[1]**

Given that the variance of Y is 47

d Calculate the values of the constant A **[4]**

18 X and Y are two independent random variables. $E(X)=12$ and $E(X+2Y)=15$

a Find the expectation of Y **[2]**

The random variables X and Y have standard deviations of 1.7 and 0.5 respectively.

b Calculate $\text{Var}(X+2Y)$ **[4]**

19 A continuous random variable $f(x)$ is defined by $f(x)=\begin{cases} k(9-x^2); & 0\le x\le 3 \\ 0; & \text{otherwise} \end{cases}$

a Show that $k=\frac{1}{18}$ **[4]**

b Calculate the expected value of X^2+X **[4]**

20 A continuous random variable $f(x)$ is defined by $f(x)=\frac{1}{9}(2+3x^2)$ for x between 1 and 2

Find $E\left(X+\frac{2}{X^3}\right)$ **[5]**

21 The continuous random variable R has probability density function

$$f(r)=\begin{cases} k\left(\frac{1}{2}+r\sin r\right); & 0\le r\le \frac{\pi}{3} \\ 0; & \text{otherwise} \end{cases}$$

a Calculate the value of k **[5]**

b Find $E\left(\sqrt{3}R+2\right)$, giving your answer in terms of π **[7]**

22 Two independent random variables, R and S, have distributions

$$P(R=r)=\begin{cases} Ar^2; & r=1, 2, 3, 4 \\ 0; & \text{otherwise} \end{cases}$$

and $f(s)=\begin{cases} \frac{1}{8}\left(s^2-\frac{1}{s^2}\right); & 1<s<k \\ 0; & \text{otherwise} \end{cases}$

a Show that $k=3$ and find the value of A **[3]**

b Calculate $E(4R+S^3)$ **[7]**

Full A Level

11 Hypothesis testing and contingency tables

The principles that underpin hypothesis testing were adopted centuries ago in response to concerns that the contractors responsible for minting coins might be tempted to syphon some of the precious metals used in the coins for their own use. The process used to prevent this became known as the *Trial of the Pyx,* which was held at regular intervals. This involved careful inspection of a sample of coins from every batch of a certain weight from the mint. The sample was compared with control coins that were stored in Pyx Chapel in Westminster Abbey in a wooden box called the 'pyx'.

This comparison of data sets underpins modern day hypothesis testing in a wide range of different areas. Such methods are used throughout manufacturing industries to control quality; they are used in medical trials of new drugs or surgical techniques and in social psychology to better understand human behaviour.

Orientation

What you need to know

Maths Ch11

- Conduct and interpret a hypothesis test.

What you will learn

- To test the parameter of a Poisson distribution.
- To find the probability of making type I and type II errors. A Level
- To perform a χ^2 test to evaluate a distribution model.
- To use a contingency table to calculate expected frequencies and χ^2 contributions.
- To calculate the number of degrees of freedom and the critical value or the p-value of the test statistic.
- To perform a test for association.

What this leads to

- Hypothesis testing for correlation.

11.1 Hypothesis testing and errors

Fluency and skills

Look back at Section 10.2 for a reminder on the Poisson distribution

You can use the Poisson distribution to model the probability of a number of independent, random events that occur over a fixed time if you know the value of λ (lambda), the parameter that gives the average number of events.

Key point

In the absence of any other information, the best estimate of λ is the average rate taken from a random sample.

If you have an estimate and then take a new sample, a hypothesis test will tell you if the sample could reasonably fit your estimate for λ

The **null hypothesis** is that the parameter takes a fixed value $H_0 : \lambda = \lambda_0$

The **alternative hypothesis** is either simply that λ_0 is wrong, $H_1 : \lambda \neq \lambda_0$ (a 2-tailed test), or that λ is higher or lower than λ_0, $H_1 : \lambda < \lambda_0$ or $H_1 : \lambda > \lambda_0$ (a 1-tailed test).

A single observation is a sample of size 1

Example 1

A Poisson distribution is believed to have a parameter of 13.2

A sample is taken to test whether or not this is the case. State the null and alternative hypotheses for this test.

The null hypothesis is $H_0 : \lambda = 13.2$ and the alternative hypothesis is $H_1 : \lambda \neq 13.2$

As with any hypothesis test you need a **test statistic**.

Key point

The **test statistic** when testing a Poisson distribution is the total number of times the event occurs in a sample of size n

If this test statistic is sufficiently different from $n\lambda$ then you have sufficient evidence to reject the null hypothesis. You can compare the test result to a critical value, or you can compare the p-value of the test statistic to the significance level.

The total number of events in the sample is compared with the number you would expect on average in that sample. The Poisson distribution only allows integer numbers of outcomes, which is why you can't compare the average number of events to λ

Find out how to use your calculator to work out cumulative probabilities.

Look back at Section 10.3 for a reminder on how to calculate individual probabilities using $P(X=n)=\frac{\lambda^n e^{-\lambda}}{n!}$

Example 2

A Poisson distribution is believed to have parameter 8.6, $X \sim Po(8.6)$. The hypotheses $H_0: \lambda = 8.6$ and $H_1: \lambda \neq 8.6$ are tested at the 10% level.

a Calculate the critical values for a sample of size 5

A sample of size 5 is taken which has an average rate of 10.4

b State, with a reason, whether the null hypothesis is accepted or rejected.

a $Y \sim Po(43)$ — As $43 = 8.6 \times 5$ (sample size × average rate)

$P(Y \le 32) = 4.97\%$ and $P(Y \le 33) = 6.93\%$

$P(Y \ge 54) = 5.86\%$ and $P(Y \ge 55) = 4.39\%$

So the critical values are 32 and 55

b $5 \times 10.4 = 52.5$ lies between the critical values so the null hypothesis is accepted. — Recall that the critical region is bounded by the critical values.

Example 3

A Poisson distribution is believed to have parameter 6.1, $X \sim Po(6.1)$

The hypotheses $H_0: \lambda = 6.1$ and $H_1: \lambda > 6.1$ are tested at the 5% level.

A sample is taken with results 3, 4, 4, 6, 7, 8, 9, 10, 13, and 14

a Calculate the total number of occurrences, n

b Consider $Y \sim Po(61)$, the Poisson distribution for the number of successes over 10 samples. Calculate the probability of obtaining a result at least this extreme if the parameter is truly 6.1 (i.e., find $P(Y \ge n)$, the p-value for the test statistic).

c State, with a reason whether the null hypothesis is accepted or rejected.

a $3+4+4+6+7+8+9+10+13+14=78$

b $P(Y \ge 78) = 0.02032$ — Find the p-value.

c Reject the null hypothesis since the p-value is less than the significance level.

There are two types of possible error you should consider when hypothesis testing.

Key point

A **type I error** is when a null hypothesis which is true is rejected.

A **type II error** is when a null hypothesis which is false is accepted.

Calculating the probability of a type II error is covered at A Level.

The probability of making a type I error is equal to the significance of the hypothesis test if the probability distribution is continuous. For discrete distributions, the probability of making a type I error may be less than the significance of the hypothesis test.

Example 4

A Poisson distribution is believed to have parameter 1.3, $X \sim \text{Po}(1.3)$

The hypotheses $H_0: \lambda = 1.3$ and $H_1: \lambda > 1.3$ are tested at the 5% level.

A sample of size 8 is taken.

a Find the critical value.

b Find the probability of making a type I error.

A type I error occurs when a result at least as extreme as the critical value is found but the true value of λ is 1.3

a $Y \sim \text{Po}(10.4)$ so $P(Y \geq 16) = 0.06405$ and $P(Y \geq 17) = 0.03681$

Therefore, the critical value is 17

Since $0.06405 > 0.05$ and $0.05 > 0.03681$

b $P(\text{type I error}) = 0.03681$

$P(\text{type I error}) = P(Y \geq 17)$

A type I error is often called a **false positive**, where a result that should be ignored attracts attention. A type II error is often called a **false negative**, where a result is overlooked which could have been important. Both types of error can have serious consequences, depending on the context.

Example 5

A kitchen in a nut-free school looks at the ingredients it is purchasing to ensure that they comply with regulations. The null hypothesis for each ingredient is that it contains no trace of nuts.

a Explain what a type I error represents and why the kitchen might not worry about making one.

b Explain what a type II error represents and why the kitchen should worry a lot about making one.

a A false positive here is that the kitchen suspects an ingredient contains nuts when it actually doesn't. This might cost a bit of money, but at least the students with allergies are safe.

b A false negative would be the kitchen believing an ingredient doesn't contain nuts when it actually does. This has potentially fatal consequences and should be avoided at all reasonable costs.

Exercise 11.1A Fluency and skills

1 A Poisson distribution with assumed parameter 5.31 is being tested to see if this is an overestimate.

   a State the null and alternative hypotheses.

   A sample of size 12 is taken to test at the 10% level.

   b Find any critical values.

2 A Poisson distribution is believed to have parameter 3.4, $X \sim \text{Po}(3.4)$. The hypotheses $H_0: \lambda = 3.4$ and $H_1: \lambda \neq 3.4$ are tested at the 10% level.

   a Calculate the critical values for a sample of size 9

   A sample of size 9 is taken which has an average rate of 3

   b State, with a reason whether the null hypothesis is accepted or rejected.

3 A Poisson distribution models a situation where the parameter is usually found to be 72

A new sample is taken to see if this has changed.

a Explain why the hypotheses are $H_0: \lambda = 72$ and $H_1: \lambda \neq 72$

The sample is of size 6 and a total of 482 outcomes occur. The critical values at the 5% significance level are 391 and 474

b Determine the conclusion of the test.

c Actually a type I error has been made. What does this mean?

4 A Poisson distribution models a situation where the parameter is usually found to be 12.5

A new sample is taken to see if this has been reduced.

a Explain why the hypotheses are $H_0: \lambda = 12.5$ and $H_1: \lambda < 12.5$

The sample is of size 14 and a total of 155 outcomes occur. The *p*-value of this statistic is 6.81%

b Determine the conclusion of the test at the 5% level.

c Actually a type II error has been made. What does this mean?

5 A Poisson distribution with assumed parameter 8.3 is being tested to see if this is a good estimate or not.

a State the null and alternative hypotheses.

A sample of size 9 is taken to test at the 10% level. A total of 87 events occur in the whole sample.

b Calculate the *p*-value of this result.

c State, with a reason whether the null hypothesis is accepted or rejected.

6 A Poisson distribution is believed to have parameter 14.2, $X \sim \text{Po}(14.2)$. The hypotheses $H_0: \lambda = 14.2$ and $H_1: \lambda > 14.2$ are tested at the 5% level. A sample is taken with results 7, 8, 10, 11, 14, 14, 15, 23, 27, 34

a Calculate the total number of occurrences, n

b Consider $Y \sim \text{Po}(142)$, the Poisson distribution for the number of successes over 10 samples. Calculate the probability of obtaining a result at least this extreme if the parameter is truly 14.2 (i.e., find $P\{Y \geq n\}$, the *p*-value for the test statistic).

c State, with a reason whether the null hypothesis is accepted or rejected.

7 A Poisson distribution with assumed parameter 3.9 is being considered. The hypotheses $H_0: \lambda = 3.9$ and $H_1: \lambda > 3.9$ are tested at the 5% level using a sample of size 15

a Calculate any critical values.

b State the probability of making a type I error.

8 A Poisson distribution is being considered. The hypotheses $H_0: \lambda = 6.3$ and $H_1: \lambda < 6.3$ are tested at the 5% level using a sample of size 20

A total of 108 events occur in the whole sample.

a Calculate the *p*-value of this result.

b State, with a reason whether the null hypothesis is accepted or rejected.

c Calculate the probability of making a type I error.

9 A medical test to detect the early presence of a disease is being designed.

a Explain why a type I error would be tolerated.

b Explain why a type II error would preferably be avoided.

10 A dice manufacturer is designing new 12-sided dice and wants to see if they are fair. Each redesign can be a costly process.

a Explain why they would want to reduce the probability of a type I error.

b Why might they not be too bothered about a type II error?

11 Describe your own situations in which

a A type I error is preferably avoided,

b A type II error is preferably avoided.

Reasoning and problem-solving

Strategy

To solve problems involving hypothesis testing

1. Conduct a test.
2. Find any critical values.
3. Accept or reject the null hypothesis.
4. Interpret the result in context.

Example 6

The Poisson distribution with a parameter of 2.8 is thought to be a good model for the number of goals scored per football match. Due to recent developments in defensive tactics it is believed that the average number of goals conceded has dropped.

The hypotheses $H_0: \lambda = 2.8$ and $H_1: \lambda < 2.8$ are tested at the 5% level.

A random sample of 24 matches is taken.

a Find any critical values.

The sample contains 49 goals in total.

b State, with a reason, whether the null hypothesis is accepted or rejected. Determine the conclusion to the hypothesis test in context.

a The critical value is 53 as $P(X \le 53) = 4.34\%$ and $P(X \le 54) = 5.69\%$ — (1)(2) Conduct a test and find the critical values.

b Since 49 is less than the critical value, there is sufficient evidence to reject the null hypothesis. — (3) Reject the null hypothesis.

There are, on average fewer goals conceded per match now. — (4) Interpret the result in context.

Type I and II errors are not exclusive to Poisson distributions. The definitions apply to any hypothesis test.

Example 7

This is a binomial distribution.

A dice is rolled 27 times to test the hypotheses $H_0: p = \frac{1}{6}$ and $H_1: p < \frac{1}{6}$ at the 5% level, where p is the probability of rolling a six. The critical value is found to be one six. Calculate the probability of making a type I error.

$X \sim B\left(27, \frac{1}{6}\right)$ and $P(X \le 1) = 0.04659$ — (2) Find the critical value.

The probability of making a type I error is the probability of the critical region. — (4) Interpret the result in context.

Exercise 11.1B Reasoning and problem-solving

1 The number of phone calls a call centre receives in a 10-minute period can be modelled by a Poisson distribution with parameter $\lambda = 6.8$

The hypotheses $H_0 : \lambda = 6.8$ and $H_1 : \lambda \neq 6.8$ are being tested at the 5% level and a sample of seven 10-minute periods are taken.

a Find any critical values.

The seven periods contain a total of 59 calls.

b State, with a reason, whether the null hypothesis is accepted or rejected. Determine the conclusion to the hypothesis test in context.

c Define a type I error and when one occurs in a hypothesis test.

d Determine the probability of making a type I error.

2 The number of points scored per game by a basketball team is believed to be well modelled by a Poisson distribution with parameter 102.3

Due to recent training, it is believed that the average number of points scored has increased. The hypotheses $H_0 : \lambda = 102.3$ and $H_1 : \lambda > 102.3$ are being tested at the 5% level. A random sample of 20 matches is taken.

a Find any critical values.

The sample contains a total of 2098 goals.

b State, with a reason whether the null hypothesis is accepted or rejected. Determine the conclusion of the hypothesis test in context.

3 The number of accidents occurring in a nursery is measured over one month by the owner. After 41 accidents are measured, they take steps to reduce the number of accidents. During the next month only 38 occur.

a Assuming the occurrence of accidents follows a Poisson distribution, state the null and alternative hypotheses to test whether the number of accidents has been reduced.

b Complete the hypothesis test at the 10% level and state the conclusion.

c Describe in words what it would mean in context if a type II error had been made.

4 Over a single season, 1064 goals were scored by all the teams collectively in a football league during the 380 matches played. On the final day of the season, 37 goals were scored in the 10 matches played that day.

a The number of goals per match is modelled by a Poisson distribution. Calculate the expected number of goals scored in 10 matches that season.

b State the null and alternative hypotheses if it is believed that the number of goals on the last day was exceptionally large.

c Determine the conclusion to the test at the 10% significance level.

d Describe in words what it would mean in context if a type I error had been made.

5 The number of accidents per month at a sports centre can be modelled by a Poisson distribution with parameter $\lambda = 0.8$ but new safety procedures have been put in place. The hypotheses $H_0 : \lambda = 0.8$ and $H_1 : \lambda < 0.8$ are being tested at the 5% level. A sample of one year is used and there were five accidents.

a Determine the p-value of the result.

b State, with a reason whether the null hypothesis is accepted or rejected. Determine the conclusion of the hypothesis test in context.

6 The number of customers coming into a town book shop in a half-hour period can be modelled by a Poisson distribution with parameter $\lambda = 15$

The hypotheses $H_0 : \lambda = 15$ and $H_1 : \lambda > 15$ are being tested at the 5% level and a sample of twelve 30-minute periods are taken. Over those periods a total of 204 customers came in.

a Find the p-value of this outcome.

b State, with a reason, whether the null hypothesis is accepted or rejected. Determine the conclusion of the hypothesis test in context.

7 The number of light bulbs in a shopping centre which burn out each month can be well modelled by a Poisson distribution with parameter $\lambda = 28$

The hypotheses $H_0 : \lambda = 28$ and $H_1 : \lambda \neq 28$ are being tested at the 10% level. A sample of four random months in a period of several years is taken and a total of 99 light bulbs were replaced in that time.

a Find the p-value of this outcome.

b State, with a reason, whether the null hypothesis is accepted or rejected. Determine the conclusion of the hypothesis test in context.

8 A coin is suspected of being biased. The hypotheses are $H_0 : p = \frac{1}{2}$ and $H_1 : p \neq \frac{1}{2}$ at the 5% level, where p is the probability of flipping a tail. The test is performed at the 5% level and the coin is flipped 16 times.

a Find any critical values.

b Determine the probability of making a type I error.

9 The probability of success for a binomial distribution is tested. It is hypothesised that the true value is 0.45 and a sample of size 12 is taken.

a Determine the probability of making a type I error if the significance level is

i 10% ii 5% iii 1%

b Find also the probability of making a type II error in each case if the true value is actually 0.6

Full A Level

11.2 Contingency tables

Fluency and skills

Correlation describes linear relationships between variables, but correlation cannot describe relationships between variables that do not have a scale. There can be many different relationships between variables. A relationship is known as an **association**.

Below is data about regional preferences for types of meat. A sample of 266 people from the north, east and south of a city were asked which meat they liked most out of beef, chicken, or pork. The **observed frequencies** are shown in the **contingency table** below. Contingency tables are used in lots of research when looking for association between two variables.

Observed frequencies	North	East	South	Totals
Beef	32	9	44	84
Chicken	41	53	20	114
Pork	8	32	28	68
Totals	81	94	88	266

A cell tells you how many of the total have both of the qualities that its row and column indicate. For example, there are 41 people from the north of the city who like chicken most.

It looks as though there might be regional preferences, but how can you be sure? The first stage is to work out what the table would look like if the variables were independent.

You would expect $266\times\frac{81}{266}\times\frac{84}{266}$ or 25.6 (to 3 sf) people to be from the north of the city with beef as their favourite.

If the variables are independent then the probability of liking beef and coming from the north of the city is the product of two probabilities. This gives the proportion of people who like beef and are from the north.

Using this method, the table below shows all the expected frequencies (values given to 3sf).

Expected frequencies	North	East	South
Beef	25.6	29.7	28.7
Chicken	34.7	40.3	39.0
Pork	20.7	24.0	23.3

The next stage is to set up a hypothesis to test. The **null hypothesis** being considered is H_0: there is no association between the variables. The **alternative hypothesis** is H_1: there is some association. This example is to be tested at the 5% significance level.

Note that you don't test for positive or negative association, just the presence of any association.

The X^2 **(chi squared) statistic** gives a way of measuring how different the observed values are from the expected values.

The X^2 statistic is a calculation involving the observed frequencies $\{O_i\}$ and the expected frequencies $\{E_i\}$.

Key point

$$X^2 = \sum_i \frac{(O_i - E_i)^2}{E_i}$$

The individual X^2 contributions to this calculation are shown in the table below.

X^2 contributions	North	East	South
Beef	1.60	14.43	8.16
Chicken	1.14	4.00	9.26
Pork	7.79	2.67	0.948

This sum of the individual contributions gives a value of $X^2 = 50.0$ (all values to 3sf).

The final stage is to decide whether you have sufficient evidence to reject the null hypothesis. To do this, you compare $X^2 = 50.0$ to the χ^2 distribution with four degrees of freedom.

Key point

The χ^2 distribution is a continous distribution but the observed values are discrete, so it is an approximation.
For this test to work accurately, it is required that every expected frequency is greater than 5

Note that observed frequencies may be 5 or lower.

The χ^2 distribution (also chi squared, but this one is lower case) with k degrees of freedom is the distribution of a sum of the squares of k independent standard normal random variables.

If the X^2 statistic is larger than the critical value found using the χ^2 distribution, or if the p-value is less than the significance level, then you have sufficient evidence to reject the null hypothesis.

You find the critical value by calculating the value of x for which $P\{X \geq x\}$ is equal to the significance level.

For $X \sim \chi_4^2$, $P\{X \geq 0.95\}$ the critical value is 9.49

Since the calculated value of $X^2 = 50.0$ is larger than the critical value, you have sufficient evidence to reject the null hypothesis.

For an $m \times n$ contingency table you should use the χ^2 distribution with $(m-1)\times(n-1)$ degrees of freedom.

Example 1

A test is performed to see if there is any association between age and movie genre preferences. The contingency table shows the recorded data.

a Draw a table of expected frequencies.

b Draw a table of X^2 contributions and calculate the X^2 test statistic.

c Calculate the number of degrees of freedom and hence the p-value of the test statistic.

d Determine the conclusion of the test.

O_i	Action	Comedy	Romance	Totals
< 25 years	41	34	27	102
25–50 years	35	38	32	105
> 50 years	7	12	19	38
Totals	83	84	78	245

(Continued on the next page)

a

E_i	Action	Comedy	Romance
< 25 years	34.6	35.0	32.5
25–50 years	35.6	36.0	33.4
> 50 years	12.9	13.0	12.1

The expected frequencies are calculated and given in this table to 3 sf.

For example, the cell that corresponds to Action and < 25 years is $245 \times \frac{83}{245} \times \frac{102}{245} = 34.6$ (3 sf)

b

	Action	Comedy	Romance
< 25 years	1.20	0.03	0.92
25–50 years	0.01	0.11	0.06
> 50 years	2.68	0.08	3.938

The X^2 contributions are provided in this table to 3 sf.

The total X^2 statistic is 9.03 (3 sf).

For example, $\frac{(41-34.6)^2}{34.6} = 1.2$ is the X^2 for Action and < 25 years.

c The 3 × 3 table has $(3-1)\times(3-1) = 4$ degrees of freedom. The *p*-value is 6.04%.

The *p*-value is $P(X \geq 9.03)$ where $X \sim \chi^2_4$. 9.03 is used as it is the test statistic.

d The *p*-value is larger than the significance level, so there is insufficient evidence to reject the null hypothesis.

STATISTICS

When expected frequencies are below five but some of the options are alike, the similar options can be grouped and their frequencies pooled. After doing so, the expected frequencies should be recalculated to ensure that there are now at least five and the number of degrees of freedom should be calculated to match the grouped table. Even if only a single expected frequency is below 5, there must be complete rows and columns after grouping.

Example 2

Voters in three constituencies are polled about their voting intentions in an upcoming election. Their options are the Red and Green parties, who have similar policies to one another but are both very different to those of the Blue party. The intentions are produced as follows to see if there is any association at the 5% level between location and intention.

O_i	Red	Green	Blue	Totals
Uppersville	9	8	85	102
Middleton	51	7	42	100
Workby	40	6	13	59
Totals	100	21	140	261

a Produce an observed frequency table using a relevant grouping, and explain why you have made that grouping.

b Calculate the number of degrees of freedom.

After grouping, the χ^2 statistic is 65.3

The critical value is 5.99

c Determine the conclusion of the test.

(*Continued on the next page*)

a

O_i	Red and Green	Blue	Totals
Uppersville	47.3	54.7	102
Middleton	46.4	53.6	100
Workby	27.4	31.6	59
Totals	121	140	261

The Red and Green columns have been grouped because, with the initial data, the Green share in Workby had an expected frequency below 5. The Red and Green columns were combined because they were said to represent similar options.

Note that all values are individually correct to 1 dp. The exact values, rather than these rounded ones, should be used in further calculations.

b Now there are $(3-1)\times(2-1)=2$ degrees of freedom.

c There is sufficient evidence to suggest that there is association between the location the voters live in and their voting intention.

Having grouped the data, any conclusion we draw must treat Red and Green intentions together.

Full A Level

When you have a 2×2 contingency table, the method explained above changes slightly and you must calculate the X^2 statistic using **Yates' correction**.

Key point

$$X^2_{\text{Yates}} = \sum\frac{(|O_i - E_i| - 0.5)^2}{E_i}$$

Example 3

Full A Level

The lunch choices of two year groups at a school are tested for association. A sample of 63 students in year 12 and year 13 are asked whether they prefer fish 'n' chips or salad.

a Draw a table of expected frequencies (correct to 3 sf).

b Draw a table of X^2 contributions and calculate the X^2 test statistic.

c At the 10% level the critical value is 2.71. Determine the conclusion of the test.

O_i	Fish 'n' chips	Salad	Totals
Year 12	19	12	31
Year 13	14	18	32
Totals	33	30	63

a

E_i	Fish 'n' chips	Salad
Year 12	16.2	14.8
Year 13	16.8	15.2

b

E_i	Fish 'n' chips	Salad
Year 12	0.33	0.36
Year 13	0.31	0.35

The X^2 contributions (3 sf) have been calculated using Yates' correction.

The test statistic is 1.35 (3 sf)

The sum of the individual contributions.

c The test statistic is less than the critical value. There is insufficient evidence to reject the null hypothesis.

Exercise 11.2A Fluency and skills

For each of the contingency tables in questions 1 to 4

a Draw a table of expected frequencies.

b Draw a table of X^2 contributions and calculate the X^2 test statistic.

c Calculate the number of degrees of freedom and hence the critical value at the 5% level.

d Calculate the p-value of the X^2 statistic.

e Determine the conclusion of the test.

1

	A	*B*	*C*	Totals
X	57	92	5	154
Y	7	32	13	52
Z	75	100	23	198
Totals	139	224	41	404

2

	A	*B*	*C*	*D*	Totals
W	16	13	16	10	55
X	8	13	8	17	46
Y	17	7	17	17	58
Z	8	15	18	18	59
Totals	49	48	59	62	218

3

	A	*B*	*C*	Totals
X	79	57	85	221
Y	61	72	62	195
Totals	140	129	147	416

Full A Level

4 (Remember to use Yates' correction.)

	A	*B*	Totals
X	70	124	194
Y	50	45	95
Totals	120	169	289

5 A test is performed to see if there is any association between the colour of a hen and how many eggs it lays per week. The contingency table shows the recorded data.

O_i	0–1	2–3	4+	Totals
Brown	6	13	16	35
White	17	20	19	56
Totals	23	33	35	91

a Draw a table of expected frequencies.

b Draw a table of X^2 contributions and calculate the X^2 test statistic at the 10% significance level.

c Calculate the number of degrees of freedom and hence the p-value of the test statistic.

d Determine the conclusion of the test.

6 A snack shop tests meal deal choices for association. A sample of sales is checked for sandwich and drink choices.

	Chicken	BLT	Ham and cheese	Egg mayo	Totals
Orange	31	28	19	6	84
Apple	19	29	17	8	73
Water	24	21	20	9	74
Totals	74	78	56	23	231

a Draw a table of expected frequencies.

b Draw a table of X^2 contributions and calculate the X^2 test statistic.

c At the 10% level the critical value is 10.6 Determine the conclusion of the test.

7 Three classes of students are polled about their preferences for an upcoming year group lunch trip. A teacher wants to determine whether there is any association between class and food preference.

O_i	Hamburger	Cheeseburger	Pizza	Hot Dog	Totals
Class A	4	5	8	6	23
Class B	2	7	6	9	24
Class C	8	2	9	4	23
Totals	14	14	23	19	70

a Calculate a table of expected frequencies.

b State which groupings should be made, if any, and calculate the number of degrees of freedom.

STATISTICS

c With grouping, the test statistic is 2.63
Determine whether the result is significant at the 5% level and state the conclusion in context.

8 Two identical twins, Frieda and Helga, and their friend Hans keep records of the snacks they eat over a month to see if there's any association between person and preference at the 5% level.

O_i	Chocolate Bars	Apples	Packets of Crisps	Totals
Frieda	15	2	11	28
Helga	12	3	14	29
Hans	10	9	21	40
Totals	37	14	46	97

a Draw a table of expected frequencies.

b Explain why Frieda and Helga's values should be grouped together.

c The X^2 statistic is 6.52
Find the critical value and determine the conclusion of the test.

Reasoning and problem-solving

Strategy

To solve problems involving possible association

1. Perform a test for association.
2. Identify positive or negative association by comparing observed and expected values for large X^2 contributions.
3. Identify where the observed and expected frequencies closely match for small X^2 contributions.

Example 4

An airline takes a survey of passengers of various heights to gauge views on the amount of legroom available.

	>1.80 m	<1.50 m	Between 1.50 m and 1.80 m	Totals
Too little	67	32	304	403
Too much	19	48	415	482
About right	48	51	379	478
Totals	134	131	1098	1363

a Perform a test for association at the 5% level.

b Comment on the association.

a

	>1.80m	<1.50m	Between 1.50m and 1.80m
Too little	39.6	38.7	324.6
Too much	47.4	46.3	388.3
About right	47.0	45.9	385.1

This is the table of expected frequencies.

	>1.80m	<1.50m	Between 1.50m and 1.80m
Too little	18.92	1.17	1.31
Too much	17.00	0.06	1.84
About right	0.02	0.56	0.096

This is the table of X^2 contributions.

The X^2 statistic is 41.0 (3 sf).

The critical value at the 5% level is $\chi_4^2 = 9.49$

There is sufficient evidence to reject the null hypothesis.

(1) Perform a test for association.

(*Continued on the next page*)

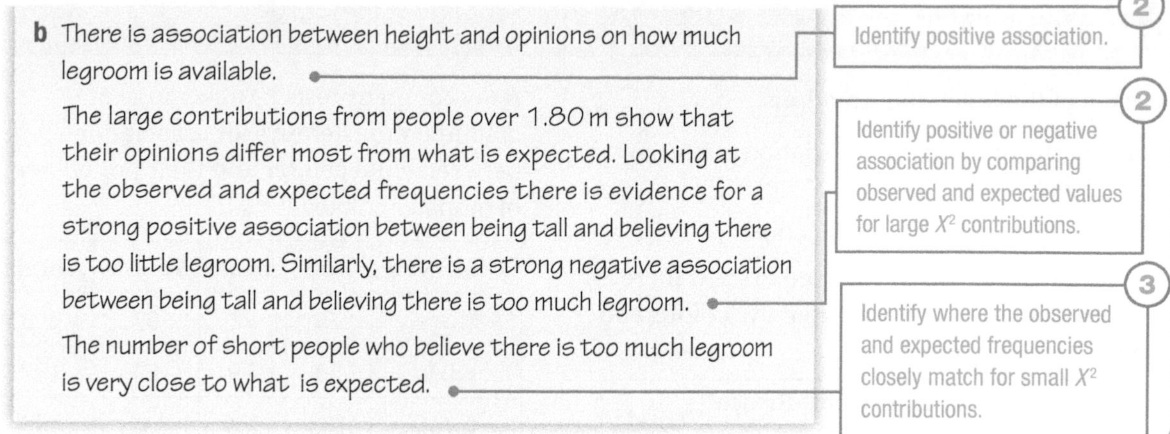

b There is association between height and opinions on how much legroom is available.

The large contributions from people over 1.80 m show that their opinions differ most from what is expected. Looking at the observed and expected frequencies there is evidence for a strong positive association between being tall and believing there is too little legroom. Similarly, there is a strong negative association between being tall and believing there is too much legroom.

The number of short people who believe there is too much legroom is very close to what is expected.

The contingency tables will not always be provided explicitly. You must be able to construct them from the question. Observed values must be absolute frequencies, but information can also be given as proportions or percentages.

Example 5

A climbing centre tracks its customers' usage to better tailor its facilities. Of the 98 who climb using ropes, $\frac{1}{7}$ are under 18, $\frac{4}{7}$ are over 65 and the rest are in between (middle). The 9 under 18s form 15% of the boulderers (non-rope climbing), 25% of the boulderers are over 65 and the other 60% are in between.

a Create a contingency table containing this information.

b Perform a hypothesis test to determine whether there is any association between age and climbing style preference at the 5% level.

c Comment on any association.

a

O_i	Under-18s	Middle	Over-65s	Totals
Ropes	14	28	56	98
Bouldering	9	36	15	60
Totals	23	64	71	158

b The expected frequencies are

Expected	Under-18s	Middle	Over-65s
Ropes	14	39.7	44
Bouldering	8.7	24.3	27

The X^2 contributions are

X^2	Under-18s	Middle	Over-65s
Ropes	0.005	3.446	3.249
Bouldering	0.01	5.629	5.31

which give a total X^2 statistic of 17.6

There are 2 degrees of freedom so the critical value is 5.99

The result is significant so there is sufficient evidence to suggest that there is association between age and preference of climbing style.

c The under-18s are in line with what you'd expect. The over 65s have a relative preference for roped climbing and the middle ages have a relative preference for bouldering.

Exercise 11.2B Reasoning and problem-solving

For each of the following situations

a Perform a test for association at the 5% level,

b Comment on the association.

1 A group of people is surveyed to see if there is any association between their eye colour and their favourite colour.

Observed	Blue eyes	Green eyes	Brown eyes	Totals
Blue	99	37	38	174
Red	34	29	33	96
Yellow	32	34	27	93
Totals	165	100	98	363

2 Live-action role players are asked which ancient tribes they most enjoy playing and which weapon they most like to carry.

Observed	Axe	Sword	Bow	Spear	Totals
Celts	10	29	20	34	93
Vikings	12	24	10	31	77
Huns	0	14	31	24	69
Totals	22	67	61	89	239

Full A Level

3 In a coastal town, the colours of rocks are compared with whether they were found on the beach or by the river.

Observed	Grey	Brown	Totals
Beach	76	64	140
River	107	59	166
Totals	183	123	306

4 A restaurant is looking for association between choices of starter and main course from its menu.

Observed	Pizza	Pasta	Totals
Soup	87	121	208
Salad	104	96	200
Totals	191	217	408

a Perform a test for association at the 10% level.

b Comment on the association.

5 A toy company tests its new range of coloured monster toys for association between child gender and their preference of monster colour.

Observed	Red	Blue	Yellow	Totals
Boys	75	71	57	203
Girls	58	79	60	197
Totals	133	150	117	400

a Perform a test for association at the 1% level.

b Comment on the association.

6 A crowd is observed at a concert. People are categorised by the colour of their hair and the colour of their clothes. The person in charge of measurements tells you the following:

'318 people attended in total. There were 96 people with black hair, of whom one third wore colourful clothes, one quarter wore light-coloured clothes and the rest wore dark-coloured clothes. Of the 120 people with brown hair, 55% wore dark-coloured clothes, 35% wore colourful clothes and the remaining 10% wore light-coloured clothes. The other people had blonde hair and were equally-divided among the different colours of clothing.'

a Construct a contingency table using the given information.

b Perform a hypothesis test at the 5% level to determine if there is any association between hair colour and choice of clothing colour.

c Comment on the association.

7 The number of yellow cards acquired per season by football teams from different regions of the country is collected over three seasons to identify whether there is any association between the two. You are told the following:

- The most yellow cards a southern team earned in a season was 68 with the second highest being 60. No team from the South equalled that amount.
- One third of the entries for the Midlands are in the 60–69 range.
- The North-West teams earned one third of the entries in the 0–59 range.
- A total of 55 entries come from London teams.

a Copy and complete the following table using the information provided.

O_i	0–59	60–69	70–79	80+	Totals
London		14		1	
Midlands			13	2	48
North-East	5	12	7		
North-West		21	10		61
South					10
Totals	90	65	40	5	200

b Explain why the table should be redrawn with columns showing the number of teams who earned 70 + cards.

c State which region should be excluded from further analysis and explain why.

d State the number of degrees of freedom. The test statistic is 12.78 and the critical value at the 5% level is 12.59, making the result significant.

e The largest contribution (3.45) is from the North-East in the 0–59 range and the smallest (0.00) is from the Midlands in the 60–69 range. Interpret this in context.

8 For a 2×2 contingency table Yates' correction can be written as

$$\chi^2_{\text{Yates}} = \frac{N\left(|ad - bc| - \frac{N}{2}\right)^2}{T_X T_Y T_A T_B}$$

for tables in the form

Observed	A	B	Totals
X	a	b	T_X
Y	c	d	T_Y
Totals	T_A	T_B	N

For every 2×2 contingency table in this section, check that this formula agrees with your value.

STATISTICS

Full A Level

11.3 Confidence intervals

Fluency and skills

Suppose you want to know the mean and variance of the heights of all the adults in a country. The most accurate values possible would be obtained by measuring every adult, but that is completely impractical. Instead, you would find a sample and calculate the mean and variance of this sample.

For a set of n data values $\{x_i\}$ the **sample mean is**

Key point

$$\bar{x} = \frac{\sum x_i}{n}$$

and the **sample variance is**

Key point

$$s^2 = \frac{1}{n-1}\sum (x_i - \bar{x})^2$$

Notice the $n-1$ in the denominator of the sample variance. Using n gives a value which is expected to underestimate the population variance, so the estimator is biased.

An unbiased estimator for a population parameter is a random variable whose expected value is equal to the parameter.

These are known to be unbiased estimators of the population mean μ and population variance σ^2. While the expected values of these random variables are equal to the population parameters, that doesn't mean that they are necessarily likely to have values close to them.

It can be shown that, for a large sample from a population, the distribution of the sample mean is distributed as

$$\bar{X} \approx \mathrm{N}\left(\mu, \frac{\sigma^2}{n}\right)$$

This is due to the central limit theorem, knowledge of which is not required for this course.

So, if you find a number of sample means, they will themselves follow the normal distribution and their mean is expected to be equal to the population mean. So the sample mean is increasingly likely to be close to the population mean as the size of the sample increases.

When the sample size equals the population size the values will obviously match exactly.

Example 1

A population follows a normal distribution with mean and variance 6.2
A sample of size 50 is taken. Give a model of the distribution of the sample mean.

The sample mean is distributed as

$$\mathrm{N}\left(6.2, \frac{6.2}{50}\right) = \mathrm{N}(6.2, 0.124)$$

Quite how close the sample mean is to the population mean depends on the population variance, which is likely to be unknown.

Although the sample variance is an unbiased estimator of the population variance, it is unlikely to be very accurate for small samples. A good rule of thumb is that a sample of at least size 30 is 'large enough' to obtain a reasonable estimate of the population variance. Then the sample variance can be used in place of σ^2

The sample data can be used to generate a **confidence interval**. This is a range of values in which it is likely that the true population mean appears.

Key point

A ***p*%-confidence interval** is generated from a sample. It is expected before generation that the population mean μ will fall into this interval with probability p%.

The p%-confidence interval generated by a sample of size n is

$$\bar{x} - z \times \frac{s}{\sqrt{n}} < \mu < \bar{x} + z \times \frac{s}{\sqrt{n}}$$

where z is calculated from p, $\bar{x}$ is the sample mean and s^2 is the sample variance.

You can calculate these yourself. $z = \Phi^{-1}\left(\frac{1+p}{2}\right)$, where Φ^{-1} is the inverse of the standard normal distribution.

p	z
99%	2.576
98%	2.326
95%	1.96
90%	1.645

Key point

The standard deviation of the sample mean, $\frac{s}{\sqrt{n}}$, is called the **standard error**.

You can never be certain that your interval contains the population mean and the more confident you want to be, the larger the interval becomes (z increases as p increases).

It is not possible to give the probability that the true mean lies in an interval once it has been generated. The p% probability only applies before a sample is taken.

STATISTICS

Example 2

A sample of size 36 is taken from a population whose standard deviation is 20.4 in order to generate a 95% confidence interval for the mean of the population.

a What is the probability that the confidence interval will contain the population mean?

The sample mean is 13.6

b Find the confidence interval.

a 95%

b The number of standard errors is $|\phi^{-1}(0.975)| = 1.96$

The interval is given by

$$13.6 - 1.96 \times \frac{20.4}{\sqrt{36}} < \mu < 13.6 + 1.96 \times \frac{20.4}{\sqrt{36}}$$

which simplifies to $6.94 < \mu < 20.3$ (3 sf)

Find out how to use your calculator to calculate the interval.

When the population variance is known it can be used without needing to be estimated from the sample. In this case, the $p\%$-confidence interval is

$$\bar{x}-z\times\frac{\sigma}{\sqrt{n}}<\mu<\bar{x}+z\times\frac{\sigma}{\sqrt{n}}$$

where z is calculated from p as before.

Example 3

A 90% confidence is to be created for a normal distribution whose variance is known to be 19.5

A sample of size 72 is taken and the sample mean is 19.3

a Calculate the standard error.

b Determine the value of z

Use tables or your calculator.

c Find the confidence interval.

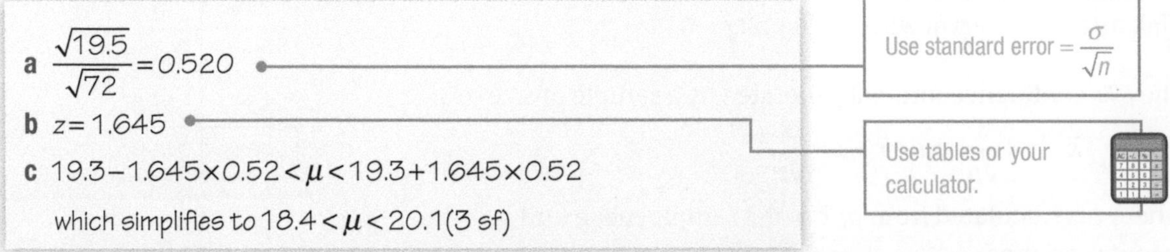

a $\frac{\sqrt{19.5}}{\sqrt{72}}=0.520$ — Use standard error $=\frac{\sigma}{\sqrt{n}}$

b $z=1.645$ — Use tables or your calculator.

c $19.3-1.645\times0.52<\mu<19.3+1.645\times0.52$

which simplifies to $18.4<\mu<20.1$ (3 sf)

Exercise 11.3A Fluency and skills

1 Find 95% confidence intervals for samples with these summary statistics.

a $n=58,\ \bar{x}=91.4,\ s^2=14.2$

b $n=41,\ \bar{x}=0.122,\ s^2=0.00563$

c $n=103,\ \bar{x}=12.7,\ s^2=126.1$

2 Find 99% confidence intervals for samples with these summary statistics.

a $n=14,\ \bar{x}=41.5,\ \sigma^2=192.5$

b $n=31,\ \bar{x}=-5.76,\ \sigma^2=91.5$

3 A sample of size 19 is drawn from a population whose variance is $\sigma^2=217.4$ The sample mean is 62.5 and the sample variance is $s^2=205.1$

a Explain why the population variance is used in finding a confidence interval.

b Find a 95% confidence interval and a 99% confidence interval using the sample data.

4 The heights of ten pole vaulters are measured to determine if they are taller than the general population. The following heights are recorded.

182.0	191.5	190.4	195.3	171.7
183.9	185.8	181.5	200.1	190.3

a Determine the mean height of the sample.

b If the population standard deviation is 6.8 cm, find a 95% confidence interval for the heights of pole vaulters.

5 The reaction times, in milliseconds, of 30 people to a stimulus are measured. Construct a 95% confidence interval for the mean reaction time.

209	207	216	207	196	205
211	206	203	207	228	204
195	217	206	220	216	217
223	228	225	217	210	218
207	211	233	226	205	192

6 The number of bees in a swarm is well approximated by a normal distribution with variance 6 250 000

a Find the probability that a sample of size 22 has a mean that lies within 2000 of the true mean

b Find the confidence interval for a sample whose mean value is 3700

7 For $X \sim N(317, 53^2)$ find

a $P\{X \le 213.12\}$

b $P\{X \ge 420.88\}$

Hence give a 95% confidence interval for a sample of size 36 with mean 317 and standard deviation 318

8 A statistician is planning a survey to find the true mean value of something believed to be well modelled by a normal distribution with standard deviation 98

Find the minimum size of the sample required for the following intervals to have a range of at most size 50

a 99% interval

b 95% interval

c 90% interval

STATISTICS

Reasoning and problem-solving

Strategy

When estimating population parameters

1. Make sure your sample is representative to avoid bias.
2. Use the sample mean and sample variance as unbiased estimators.
3. Generate a confidence interval for the population mean.
4. Answer the question in context.

Confidence intervals are a useful tool in hypothesis testing. When the true mean is unknown, a sample mean and a measure of population variance provide a test statistic for deciding whether a proposed true mean is a reasonable estimate.

Example 4

It is thought that the true mean of a population is 12.9
This is tested at the 10% significance level and 20 samples are taken to generate 20 corresponding confidence intervals. All 20 of them contain 12.9
What does this suggest?

It is expected that 18 of the intervals contain the true mean. All 20 of them contain the number 12.9, which does not automatically suggest that 12.9 is the true mean but it does not rule it out. It would be reasonable to say that 12.9 is likely to be close to the true mean.

(4) Answer the question in context.

It is important to remember that a p%-confidence interval can only be said to be p% likely to contain the true mean before it is generated. If you take repeated samples and form confidence intervals, then you would expect p% of them to contain μ

Example 5

A 90% confidence interval is generated for the average height of a human adult. The sample gives the interval 187cm to 204cm. Explain why it is not reasonable to say that this interval is 90% likely to contain the true average human adult height.

Experience suggests that 187cm is a very tall height.

It does not seem likely that the average human adult height is even larger than this.

1 Make sure your sample is representative to avoid bias.

4 Answer the question in context. Perhaps the sample was taken from a basketball team or a town with very tall people.

Increasing the sample size decreases the width of the confidence interval because the size of the standard error is decreased. A larger sample gives a more accurate estimate of the mean.

As the standard error is based on $\sqrt{n}$ though, a sample four times as large will halve the width of the interval. This can be costly in real life.

Exercise 11.3B Reasoning and problem-solving

1 A statistician wishes to know the mean of a normal distribution whose variance is unknown. They take two samples. The first, of size 51, has a mean of 61.3 and a variance of 91.6

The second, of size 39, has a mean of 60.5 and a variance of 105.1

a Generate 95% confidence intervals for the two separate samples.

The combined samples have a variance of 97.6

b Calculate the size and sample mean of the combined samples and hence generate a 95% confidence interval.

c Explain why the statistician is more likely to use the interval just generated over the previous two.

2 A population is normally distributed and it is suspected, but not known, that its variance is 432

A sample of size 32 is taken whose mean is 1071 and variance is 1905

a Generate confidence intervals from the sample using the sample variance and suspected population variance.

b Explain which interval should be used in estimating the true mean.

3 The lengths of wood cut by a machine are thought to be well modelled by a normal distribution. The differences from the intended mean length of 61 pieces of wood are given in this table.

Value	<−0.3	−0.3	−0.2	−0.1	−0.05	0	0.05	0.1	0.15	>0.15
f	0	4	8	12	12	9	5	7	4	0

a Estimate the mean length difference of the sample.

b Estimate the standard deviation of the sample.

c Construct a 95% confidence interval for the mean difference.

4 The weights (in kg) of the 73 members of a bridge club are taken to estimate the mean weight of a bridge player. The values are grouped and given in this table.

Weight (kg)	50 to 65	65 to 75	75 to 85	85 to 95	95 to 105	105 to 110
f	5	13	20	26	8	1

a Estimate the mean and standard deviation of the weights of the bridge players.

b Calculate the expected frequencies and comment on the assumption that the weights are normally distributed with your estimated mean and standard deviation.

c Construct a 95% confidence interval using the sample.

5 The masses of 10-year old boys are thought to be modelled by a normal distribution. Forty boys are weighed by their doctors and the results are given here.

32.2	29.8	36.4	39.6	26.9	28.2	41.9	40.5
32.0	32.6	32.8	34.0	17.4	33.8	29.6	24.1
28.9	34.9	31.5	27.7	27.3	38.1	33.7	36.4
16.1	32.4	33.8	29.9	25.0	46.7	30.0	33.3
28.9	29.8	37.8	22.6	28.6	33.0	32.5	28.8

a Estimate the population variance using these values.

b The 95%-confidence interval generated by this sample is $29.4 \le \mu \le 33.1$

Someone claims that there is a 95% chance that the true mean of a 10-year old boy lies in this range. Explain why they are incorrect.

11 Summary and review

Chapter summary

- A Poisson distribution has a parameter, λ, which is the average rate that the event occurs.
- When you don't know the value of λ you can estimate it as $\lambda = \lambda_0$ and then perform a hypothesis test:
  - The null hypothesis is $H_0: \lambda = \lambda_0$
  - The alternative hypothesis is either:
    - $H_1: \lambda \neq \lambda_0$, a 2-tailed test
    - $H_1: \lambda < \lambda_0$, a 1-tailed test, or
    - $H_1: \lambda > \lambda_0$, a 1-tailed test.
- If the null hypothesis is true, then the distribution of the number of events in a sample of size n is a Poisson distribution with parameter λn. If the number of events in the sample is sufficiently unlikely, then you reject the null hypothesis.
- Sometimes a null hypothesis which is true is rejected due to an unlikely result. This is called a type-1 error.

  The test gives evidence for or against λ_0 being a reasonable estimate of the parameter of the distribution.
- A contingency table shows the observed frequencies, $\{O_i\}$, from a sample which splits data according to two discrete variables.
- In a test for association the null hypothesis is that there is no association between the variables and the alternative hypothesis is that there is some association.
- You should calculate a table of expected frequencies, $\{E_i\}$, to show what the sample would look like if the two variables were independent.
- The X^2 test statistic can be calculated from the observed and expected frequencies.

$$X^2 = \sum_i \frac{(O_i - E_i)^2}{E_i}$$

- For a 2×2 contingency table, or when an observed frequency is less than 5, the test statistic is calculated using Yates' correction.

Full A Level

$$X^2_{\text{Yates}} = \sum \frac{\left(|O_i - E_i| - 0.5\right)^2}{E_i}$$

- The χ^2 distribution with k degrees of freedom is the distribution of a sum of the squares of k independent standard normal random variables.
- The X^2 test statistic can be compared with the appropriate χ^2 distribution to test for significance. For an $m \times n$ contingency table you should use the χ^2 distribution with $(m-1)\times(n-1)$ degrees of freedom.
- If the statistic is larger than the critical value, or if the p-value is less than the significance level, then the null hypothesis H_0 is rejected.
- Large X^2 contributions highlight association and you should compare observed and expected frequencies so that you can comment on the association.

- A p%-confidence interval is generated from a sample. It is expected that p% of intervals generated in that way will contain the population mean. For a population with unknown mean μ and known variance σ, that has a sample with mean $\bar{x}$, the confidence interval is

$$\bar{x}-\left|\phi^{-1}\left(\frac{1+p}{2}\right)\right|\times\frac{\sigma}{\sqrt{n}}<\mu<\bar{x}+\left|\phi^{-1}\left(\frac{1+p}{2}\right)\right|\times\frac{\sigma}{\sqrt{n}}$$

- The value $\frac{\sigma}{\sqrt{n}}$ is known as the **standard error**.

Check and review

You should now be able to...	Try Questions
✔ Test the parameter of a Poisson distribution.	1, 2
✔ Calculate the probability of making a type I error.	2
✔ Interpret the result of the test in context.	3
✔ Use a contingency table to calculate expected frequencies and X^2 contributions.	4, 7, 10
✔ Calculate the number of degrees of freedom, and the critical value or the p-value of the test statistic.	5, 6
✔ Perform a test for association.	6, 7
✔ Interpret the result of a test for association in context and describe the association.	7
✔ Generate confidence intervals.	8, 9
✔ Use Yates' correction appropriately. Full A Level	10

1 A Poisson distribution is believed to have parameter 1.9 and a test is performed to see if it is higher than this.

a State the null and alternative hypotheses.

A sample of size 19 is taken to test at the 5% level. A total of 48 events occur in the whole sample.

b Calculate the p-value of this result.

c State, with a reason whether the null hypothesis is accepted or rejected.

2 It is thought that $X \sim \text{Po}(0.75)$. The hypotheses $H_0: \lambda = 0.75$ and $H_1: \lambda < 0.75$ are tested at the 5% level. A sample of size 29 is taken.

a Find the critical value.

b Find the probability of making a type I error.

3 The number of road accidents on a country road per month is well modelled by a Poisson distribution with parameter 1.3

It is hoped that new traffic calming measures will reduce this. The hypotheses $H_0: \lambda = 1.3$ and $H_1: \lambda < 1.3$ are being tested at the 10% level. The number of accidents in next 8 months form a sample.

a Find any critical values.

There are 7 accidents in those 8 months.

b State, with a reason whether the null hypothesis is accepted or rejected. Determine the conclusion of the hypothesis test in context.

4 For this contingency table

a Draw a table of expected frequencies,

b Draw a table of X^2 contributions and calculate the X^2 test statistic.

O_i	A	B	C	Totals
X	103	71	35	209
Y	96	84	41	221
Z	111	58	49	218
Totals	310	213	125	648

5 For this contingency table, calculate

a The number of degrees of freedom,

b The critical value at the 5% significance level.

O_i	A	B	C	D	E	Totals
X	29	37	25	87	41	219
Y	14	39	93	20	53	219
Z	10	5	2	7	28	52
Totals	53	81	120	114	122	490

6 The heights and birthplaces of UK adults are tested for association at the 5% level. A large sample is taken with 24 degrees of freedom. The X^2 statistic is found to be 34.91

Calculate the p-value of this statistic and determine the conclusion of the test in context.

7 A car designer is testing public opinion on four new models (A, B, C, D) for association with preference for colour. Using the contingency table

O_i	A	B	C	D	Totals
Red	100	84	60	80	324
Blue	55	64	57	78	254
Silver	30	36	27	23	116
Totals	185	184	144	181	694

a Draw a table of expected frequencies,

b Draw a table of X^2 contributions and calculate the X^2 test statistic.

There are 6 degrees of freedom so the critical value at the 5% level is 11.07

c Determine the conclusion of the test and comment on the association.

8 A sample of size 36 is taken from a normal distribution with unknown variance. The sample mean is 14.7 and the sample variance is 29.5

a Generate a 95% confidence interval for the true mean of the distribution.

The hypotheses $H_0: \mu = 16.5$ and $H_1: \mu \neq 16.5$ are tested at the 5% level.

b Determine the conclusion of the test given the sample of size 36

c Determine the probability of making a type I error.

9 Find 95% confidence intervals for samples with the following summary statistics.

a $n = 25, \Sigma x = 100, \Sigma x^2 = 1600$

b $n = 61, \Sigma x = 301, \Sigma x^2 = 1486$

10 Given this contingency table and its table of expected values, calculate the X^2 test statistic using Yates' correction.

O_i	A	B	Totals
X	24	36	60
Y	60	24	84
Totals	84	60	144

E_i	A	B
X	35	25
Y	49	35

Full A Level

11 Exploration
Going beyond the exams

Did you know?

The theory behind confidence intervals is based around dealing with uncertainty in results that are derived from data that is taken from only a randomly selected subset of a population. Confidence intervals treat their bounds as random variables and the parameter as a fixed value.

Bayesian inference provides an alternative explanation in the form of the theory of **credible intervals**. In contrast to confidence intervals, Bayesian intervals treat their bounds as fixed and the estimated parameter as a random variable.

Life is complicated, but not uninteresting.
- Jerzy Neyman

Research

A confusion matrix is a contingency table that reports on the number of false positives, false negatives, true positives and true negatives. The F1 score is used to measure the accuracy of a test.

For example, imagine a piece of software that was being used to identify whether 16 photographs were of males or females.

It could achieve these results:

		Actual class	
		Male	Female
Predicted class	Male	5	6
	Female	2	3

The software achieved 5 true-positives, 6 false-positives, 2 false-negatives and 3 true-negatives. Research how this information can be used to calculate the F1 score.

11 Assessment

1 A Geiger counter indicates radioactive decay by a series of clicks, one click per decay. For one sample of radioactive material, the number of decays over a randomly chosen 10-second interval is observed to be 6

The sample is thought to be of a material with a mean number of decays of 2.7 per 10-second period. A scientist wishes to test the hypothesis that the source is actually a material with a greater mean.

a State the null and alternative hypotheses for this test. **[1 mark]**

b Perform the test for a significance level of 5%. **[4]**

2 Using traditional thread, a machine produces parachute material with an average of 3.2 flaws per $5\,\text{m}^2$. It is claimed that a new type of thread will reduce the number of flaws. In a random sample of $10\,\text{m}^2$ of the new material, two flaws are found.

a State the null and alternative hypotheses to test this claim. **[1]**

b Perform the test for a significance level of 5%. **[4]**

c Explain what is meant by a type 1 error and give the probability that a type 1 error is not made in the test in part **b**. **[2]**

3 A business receives on average 12 telephone calls per hour throughout the day. The day after a newspaper advertising campaign, it receives 90 calls between 10am and 4pm.

a Using a significance level of 5%, test the hypothesis that the campaign has increased the rate of telephone calls. **[4]**

b State the probability of a type 1 error for your test and explain what this means. **[2]**

(You may assume that the number of calls in a given period follows a Poisson distribution.)

4 At an accident blackspot, the rate of road traffic accidents is known to be 3.2 per month. During the month after traffic calming measures are put in place, only 2 accidents are recorded.

a Explain why a Poisson distribution may be appropriate to test whether the accident rate has been reduced. **[1]**

b Carry out this test using a significance level of 5%. You should state clearly your null and alternative hypotheses. **[4]**

c What is the largest number of accidents in a month that would result in a conclusion that the rate of accidents had been reduced? **[2]**

5 A school is recording how students spend their free time. A group of 100 students were asked to name their main pastime outside school. The results by gender, are as follows.

		Pastime				
		Reading	**Sport**	**Online**	**Television**	**Other**
Gender	**Male**	6	12	20	7	14
	Female	7	17	10	6	1

a Find the expected frequencies based upon the hypothesis that there is no association between pastime and gender. **[2]**

b Test at a significance level of 10% whether females and males differ in their choice of pastime. **[5]**

6 A teacher wishes to assess her students. She decides to use two assessment methods, a traditional problem-solving exam (A) and a multiple choice exam (B).

		Result		
		Fail	**Pass**	**Distinction**
Exam	**A**	7	18	8
	B	7	14	6

The teacher believes that there is no significant difference in exam performance between the two exams.

a Complete the following table of expected frequencies on the assumption that she is correct. **[2]**

		Result			
		Fail	**Pass**	**Distinction**	**Total**
Exam	**A**				
	B				
	Total				

b Test the teacher's belief at a significance level of 5%. **[5]**

7 A researcher is asked to assess whether patient safety is affected by weekend staffing levels. She obtains data for nursing staff drug errors and the day of the week on which they occurred. The drug errors are classified as type A: errors with no significant consequences to the patient, type B: errors with minor consequences and type C: errors with serious consequences.

		Weekday	**Weekend**	**Total**
Error	**Type A**	45	18	63
	Type B	24	8	32
	Type C	14	5	19
	Total	83	31	114

Test at a significance level of 5% whether the timing (weekday or weekend) of a drug error affects its seriousness. **[7]**

8 A batch of 37 brands of olive oil are assessed for quality by a panel of experts and rated either outstanding, good or unsatisfactory. This judgement is recorded with the price per litre of the oil.

		Quality			
		Outstanding	**Good**	**Unsatisfactory**	**Total**
Price per litre	< £7	5	5	7	17
	≥ £7	10	7	3	20
	Total	15	12	10	37

Test at a significance level of 5% whether there is an association between the price of olive oil and its quality. **[7]**

9 A random variable X has a binomial distribution, parameters $n = 12$ and p, a constant. The value of p was known to be 0.3 but is now thought to have increased.

a Write down the null and alternative hypotheses in a test of this belief. **[2]**

b Complete the following table given that the null hypothesis in part **a** is correct. Give your answers to 4 dp. **[2]**

x	6	7	8	9
$\mathrm{P}\{X \geq x\}$				

c What values of X would suggest that the belief is incorrect? Use a significance level of 5%. **[1]**

d Describe the meaning of the term 'type I error' and work out its probability for the test above. **[2]**

10 Fasting blood sugar levels of 15 healthy adult females had a mean of 4.4 mmol/L.

a Assuming these to be a random sample from a normal population with standard deviation 0.6 mmol/L, find a 95% confidence interval for the population mean. **[2]**

It is proposed that the population from which this sample was taken had a mean of 4.6 mmol/L.

b Use your answer to part **a** to determine whether or not to reject this hypothesis. You should give a reason for your answer. **[2]**

11 A sample of size 60 is taken from a normal population with unknown mean and variance.

a Explain why it is acceptable in this situation to use the **sample** variance when using normal distribution theory to find a confidence interval for the population mean. **[2]**

b Given that the sample mean and variance are 14 and 36, respectively, find an approximate 90% confidence interval for the population mean. **[2]**

12 Graphs and networks 1

To ensure fast internet connectivity in towns and cities, and throughout buildings like schools and airports, cable networks are used. Often during installation some existing conduits for services such as lighting can be used, whereas in other areas entirely new conduits will have to be put in place, and costs along certain routes may be greater than others. It is important that the cabling for such networks is done efficiently and at minimum cost. To ensure that costs are kept to a minimum, the maths of minimum spanning trees is used.

Similar calculations are used in other areas of design of networks, for example, in planning transportation networks, electricity distribution and water supply.

Orientation

What you need to know

- The construction and language of graphs.

What you will learn

- To use Kruskal's algorithm to find the minimum spanning tree for a network given as a diagram.
- To use Prim's algorithm to find the minimum spanning tree for a network given as a diagram or as a table.
- To find the shortest path between vertices in a network using Dijkstra's algorithm.
- To solve the route inspection problem for a network with odd order vertices.
- To find upper and lower bounds for the travelling salesman problem.

What this leads to

Ch13 Critical path analysis

- Activity networks.

Ch28 Graphs and networks 2

- Optimising network flows.

12.1 Graphs and networks

Fluency and skills

You use the term **graph** for a diagram involving a set of points and interconnecting lines. Each point is called a **vertex** (plural **vertices**) and a line joining two points is called an **edge**.

Key point

A graph consists of a number of points (vertices) connected by a number of lines (edges). Note that it is possible to have a vertex that does not have an edge connected to it.

What matters is which vertex is connected to which. The shape or layout of the diagram is irrelevant.

Full A Level

These three diagrams show the same set of vertices and connections – they are effectively the same graph (check that you can see this).

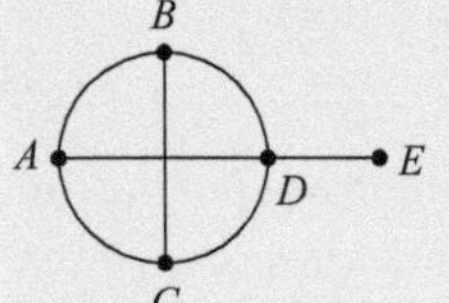

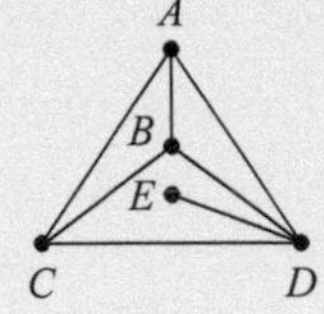

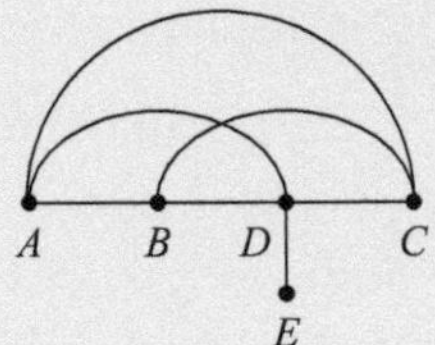

Graphs like these are called **isomorphic** (of the same form).

A graph is **connected** if you can travel from any vertex to any other vertex (perhaps passing through others on the way).

Connected

Not connected

Not connected

Two or more edges may connect the same pair of vertices. These are **multiple edges**. There may be a **loop** connecting a vertex to itself.

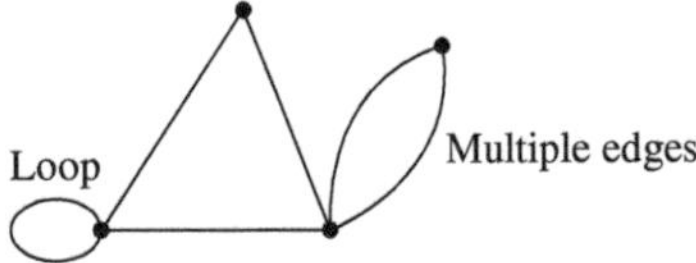

A graph with no loops or multiple edges is a **simple** graph.

The graph formed by using only some of the vertices and edges of a graph is a **subgraph** of the original graph.

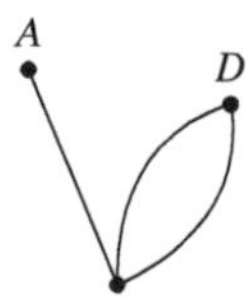

is a subgraph of

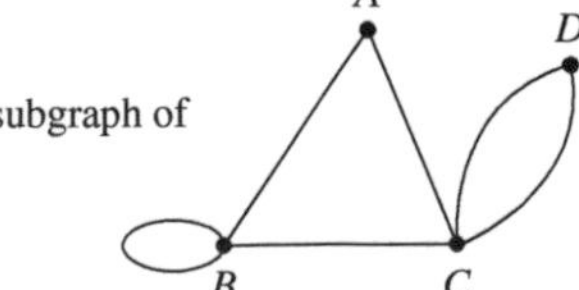

If you add extra vertices along the edges of a graph the result is called a **sub-division** of the graph.

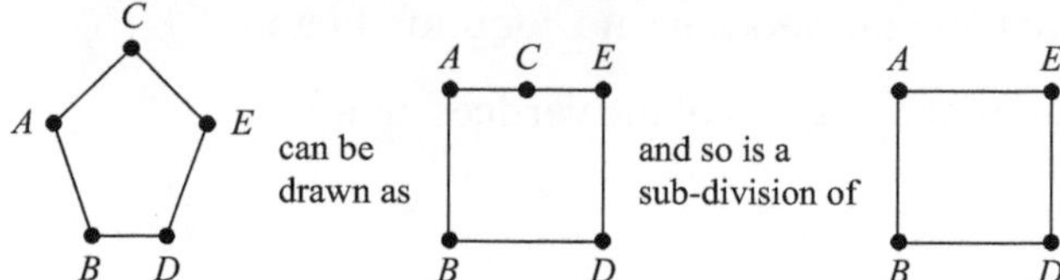

If a simple graph has an edge connecting all possible pairs of vertices, it is a **complete** graph. Complete graphs have their own notation.

> **Key point**
>
> The complete graph with n vertices is called $\boldsymbol{K_n}$

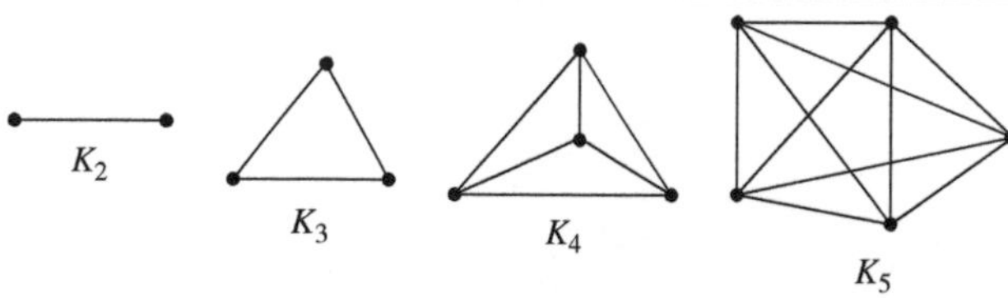

In some graphs the vertices belong to two distinct sets, and each edge joins a vertex in one set to a vertex in the other. Such a graph is called a **bipartite graph**. The graph on the right graph is a bipartite graph. The two sets of vertices are {A, B, C} and {D, E}.

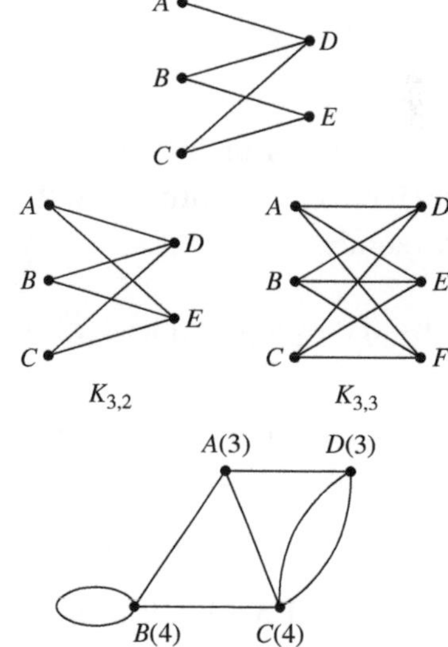

If every possible edge in a bipartite graph is present, it is a **complete bipartite graph**. Again there is a special notation.

> **Key point**
>
> The complete bipartite graph connecting m vertices to n vertices is called $\boldsymbol{K_{m,n}}$

The number of edges meeting at a vertex is called the **degree** or **order** of the vertex. The degrees of the vertices in this graph are shown in brackets.

Example 1

Draw a graph with four vertices – one with degree 4, one with degree 2 and two with degree 1

The solution is not unique. Here are two possible graphs which fit the requirements.

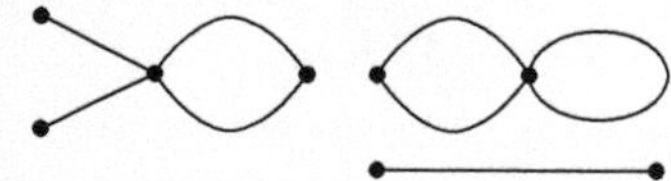

There is at least one other connected solution and one other disconnected solution. You might like to try to find these.

Example 2

a Draw a graph with vertices labelled with the integers 2, 3, ..., 9

Draw an edge between two vertices only if the two numbers have no common factor.

b Write down the number of edges and the total of the degrees of the vertices. State the relationship between these quantities.

a

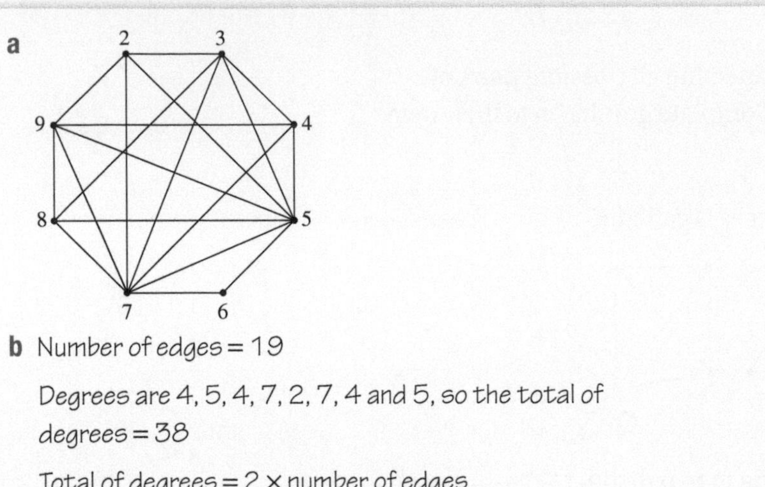

b Number of edges = 19

Degrees are 4, 5, 4, 7, 2, 7, 4 and 5, so the total of degrees = 38

Total of degrees = 2 × number of edges

The relationship in the above example is true for all graphs, because each edge contributes 1 to the degree of the vertex at each of its two ends.

Key point

Total number of degrees = 2 × number of edges

This is called the **handshaking lemma** because two hands (vertices) are involved in one handshake (edge).

A lemma is a fact which can be taken for granted in subsequent work (you can quote it without proof).

Here are two drawings of the complete graph K_4

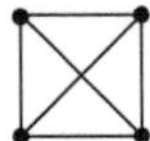

The second version is preferable because none of the edges cross. If you can draw a graph in this way is said to be **planar**, and drawing it with no crossing edges is called drawing it 'in the plane.'

Key point

A graph is planar if it can be drawn in a plane in such a way that no two edges meet each other except at a vertex.

When a graph is drawn in the plane, the areas created by its edges are called **faces**. The area outside the graph is also a face, sometimes called the **infinite face**.

If a connected graph drawn in the plane has V vertices, E edges and F faces, these values are related by **Euler's formula**:

Key point

Euler's formula

$V + F - E = 2$

Example 3

Show that Euler's formula holds for this graph.

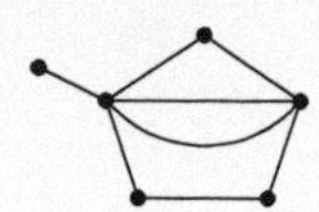

$V=6$ — The graph has 6 vertices.

$E=8$ — The graph has 8 edges.

$F=4$ — There are 4 faces, including the infinite face.

$V+F-E=6+4-8=2$, so Euler's formula holds true.

To prove Euler's formula consider any connected plane graph such as the one shown.

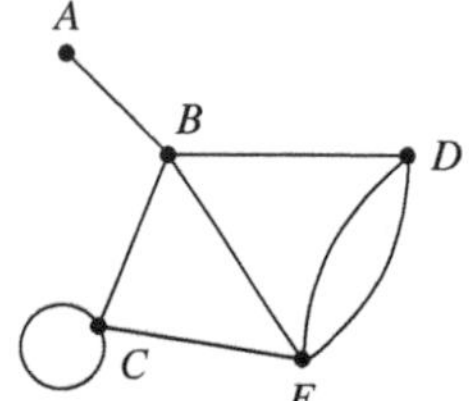

You remove edges one by one without making it disconnected.

If you remove an edge such as AB, then $E \to (E-1)$ and $V \to (V-1)$

If you remove an edge such as BE, one of the two edges DE or the loop at C, then $E \to (E-1)$ and $F \to (F-1)$

(In the diagram AB and DE have been removed.)

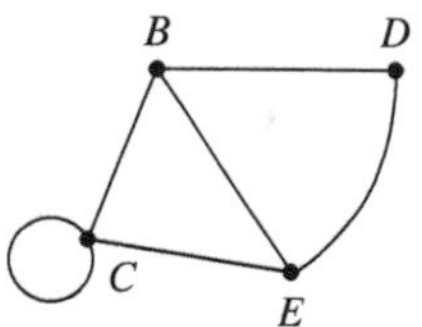

In both cases, the value of $V+F-E$ is unchanged.

Eventually, you reduce the graph to a single vertex, so $V=1$, $F=1$ and $E=0$

This gives $V+F-E=2$

This value did not change as the graph reduced, so for the original graph $V+F-E=2$

In addition to drawing a diagram, there are two ways in which you can record a graph. One way is to list its vertices and edges (the *vertex set* and *edge set*), but, more usually, you construct an **adjacency matrix** showing the vertices and the number of connections between them.

Example 4

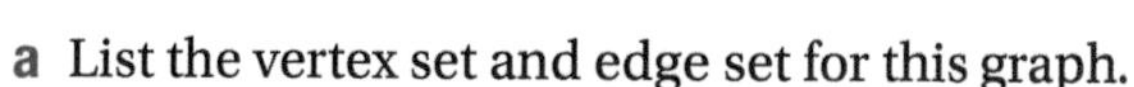

a List the vertex set and edge set for this graph.

b Draw an adjacency matrix for this graph.

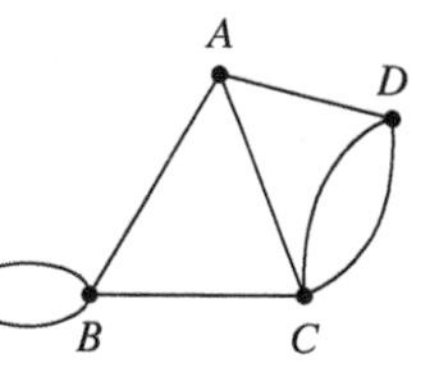

a $\{A, B, C, D\}$, $\{AB, AC, AD, BB, BC, CD\}$

b

	A	B	C	D
A	0	1	1	1
B	1	2	1	0
C	1	1	0	2
D	1	0	2	0

Count the number of direct routes between each pair of vertices.

There are 2 routes from B to itself because you can go round the loop in either direction.

DISCRETE

The adjacency matrix for a bipartite graph will contain two square blocks of zeros provided the two sets of vertices are kept separate in the table. You can then replace it by a more convenient reduced form.

Example 5

Draw an adjacency matrix for this bipartite graph.

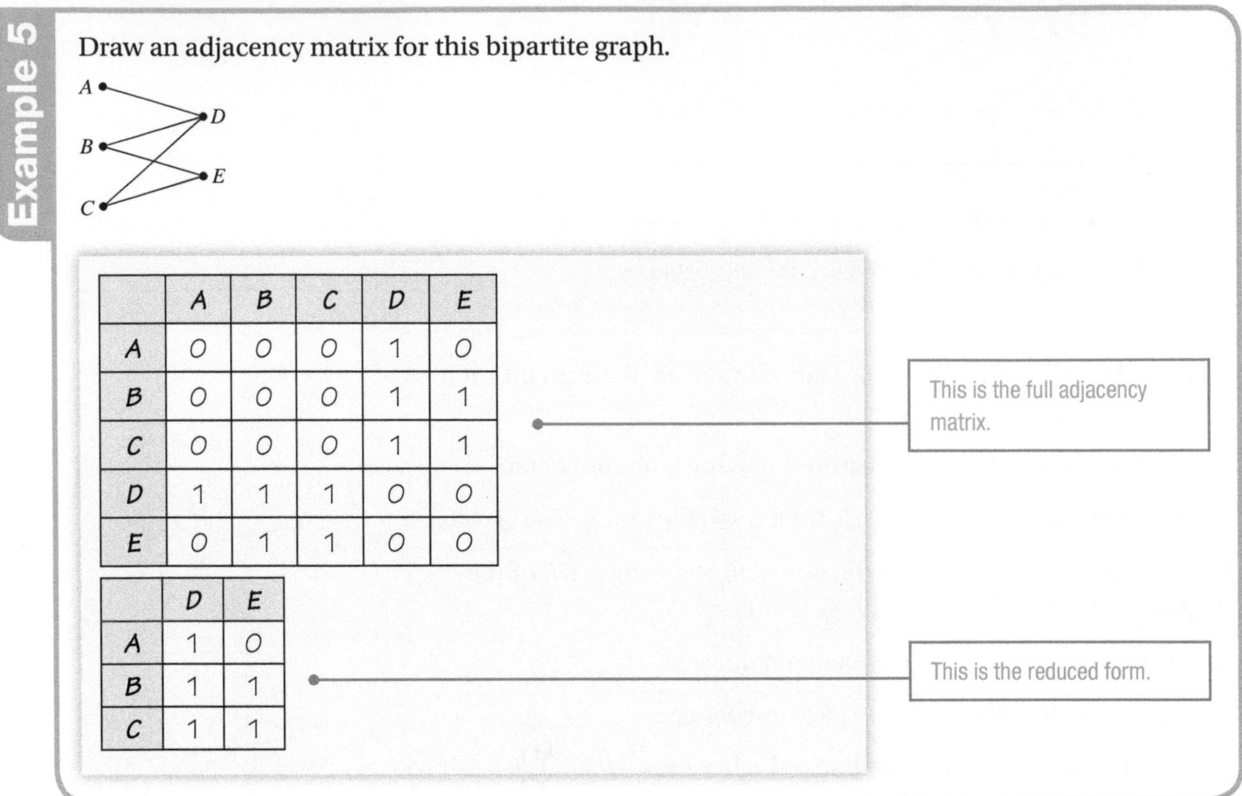

	A	B	C	D	E
A	0	0	0	1	0
B	0	0	0	1	1
C	0	0	0	1	1
D	1	1	1	0	0
E	0	1	1	0	0

	D	E
A	1	0
B	1	1
C	1	1

Sometimes one or more edges of a graph are 'one-way streets'. These edges are called **directed edges** and the graph is a **directed graph** (or **digraph**). You can still use an adjacency matrix, but this will no longer be symmetrical and you need to include 'from' and 'to' to show the direction.

Example 6

Draw an adjacency matrix for this directed graph.

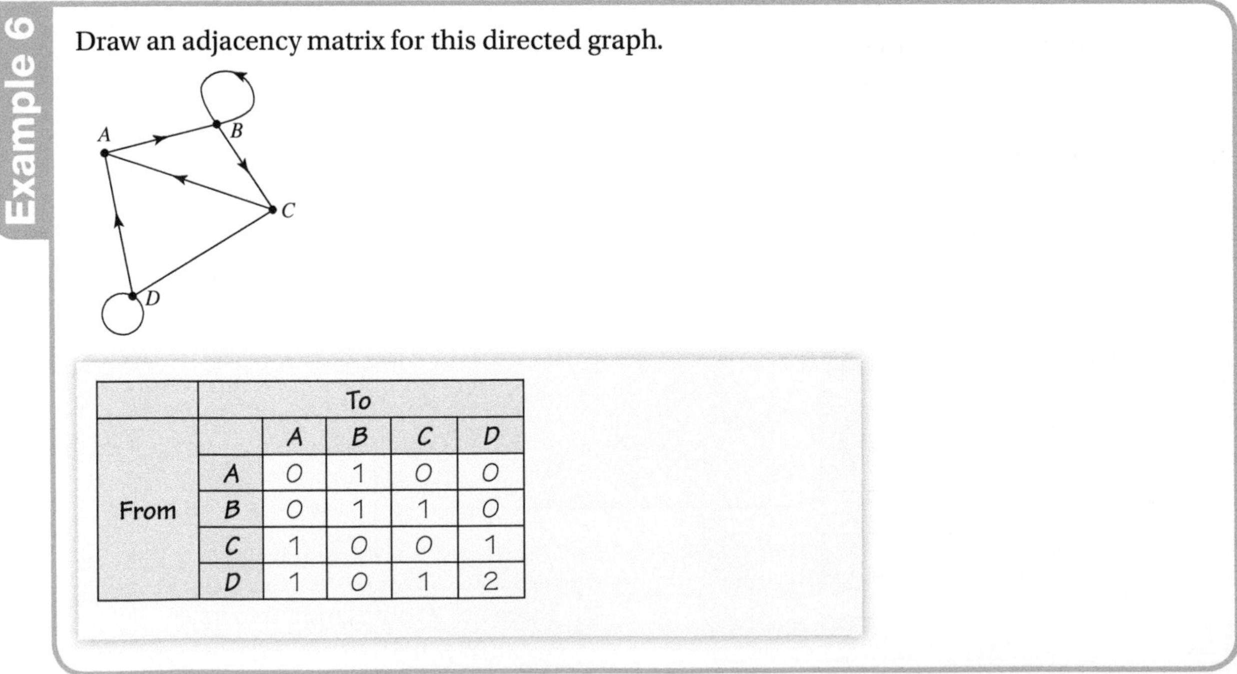

		To			
		A	B	C	D
From	A	0	1	0	0
	B	0	1	1	0
	C	1	0	0	1
	D	1	0	1	2

For every simple graph G there is a corresponding graph G', called the **complement of G**. This is formed by drawing those edges which do not exist in G and removing the edges which do. If you combine G and G' you get a complete graph.

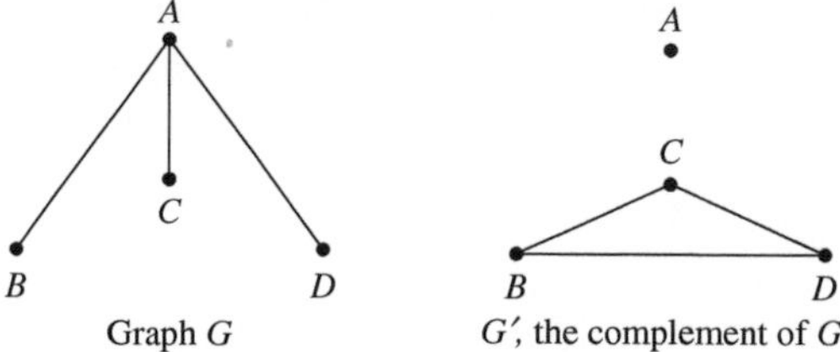

Graph G $\quad$ G', the complement of G

You find the adjacency matrix for G' by changing zeros to ones and ones to zeros in the adjacency matrix of G (ignoring the leading diagonal because no loops are added).

Example 7

Draw the adjacency matrix for this graph, G, and hence find the adjacency matrix for the complement, G'. Draw G'.

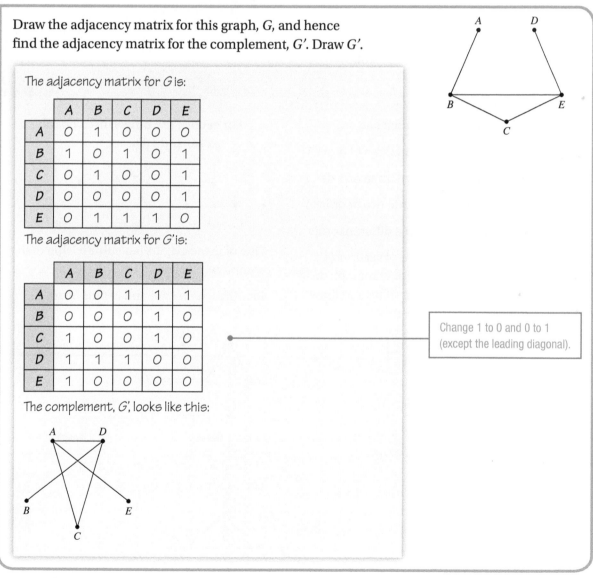

The adjacency matrix for G is:

	A	B	C	D	E
A	0	1	0	0	0
B	1	0	1	0	1
C	0	1	0	0	1
D	0	0	0	0	1
E	0	1	1	1	0

The adjacency matrix for G' is:

	A	B	C	D	E
A	0	0	1	1	1
B	0	0	0	1	0
C	1	0	0	1	0
D	1	1	1	0	0
E	1	0	0	0	0

Change 1 to 0 and 0 to 1 (except the leading diagonal).

The complement, G', looks like this:

A **network**, or **weighted graph**, is a graph with a number associated with each arc. This number is called the **weight** of the arc. Weights may correspond to distances, times, costs or many other things. A network may be **directed** (some or all the arcs are 'one-way streets'). You can represent a network using a **distance matrix**, with entries showing the weights of the arcs.

When talking about networks, use the terms 'node' and 'arc'.

Example 8

This network shows the travel times (in minutes) between five Somerset towns.

Draw a distance matrix to represent this network.

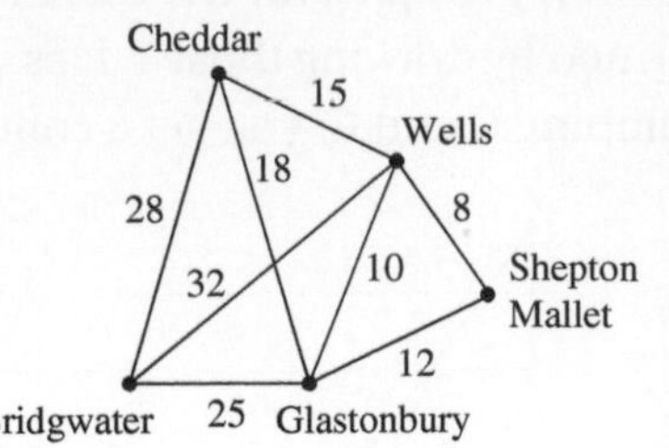

Abbreviate the towns as B, C, G, S and W.

	B	C	G	S	W
B	–	28	25	–	32
C	28	–	18	–	15
G	25	18	–	12	10
S	–	–	12	–	8
W	32	15	10	8	–

Exercise 12.1A Fluency and skills

1 Draw a connected graph which has

 a One vertex of order 3 and three of order 1

 b One vertex of order 1 and three of order 3

 c Two vertices of order 3 and two of order 1

 d Four vertices, each with a different order

2 The table shows the number of vertices (V), edges (E) and faces (F) in a planar graph. In each case, fill in the missing value and draw a possible layout for the graph.

	V	E	F
a	5	7	
b	8		6
c		9	7

3 Two of the three graphs A, B and C are sub-graphs of the following graph.

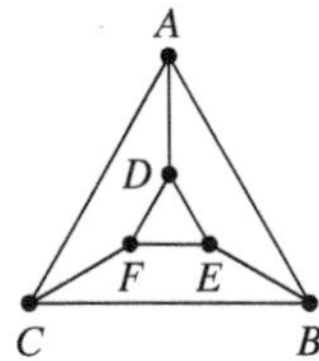

 a Identify the odd one out.

 b For each of the others, list a possible vertex set and edge set.

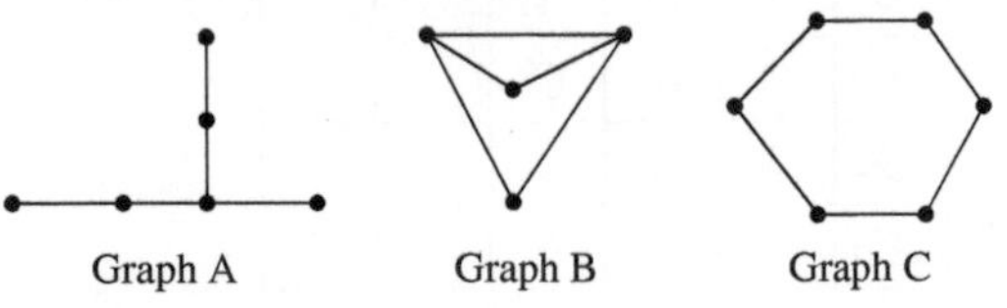

4 One of these graphs is a sub-division of the complete graph K_4

Identify which one, and list the 'extra' vertices.

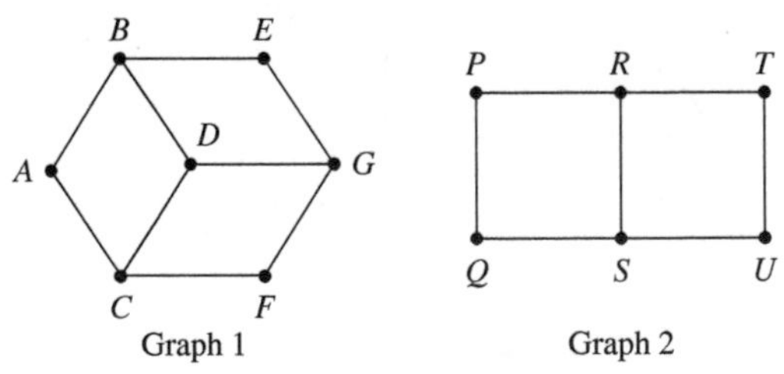

5 Write down the adjacency matrix corresponding to this graph.

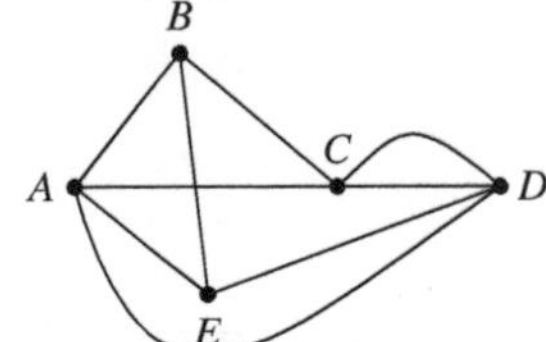

6 Draw graphs corresponding to the following adjacency matrices.

a

	A	*B*	*C*	*D*
A	0	1	1	1
B	1	0	0	1
C	1	0	0	0
D	1	1	0	0

b

	A	*B*	*C*	*D*
A	0	1	2	0
B	1	0	0	1
C	2	0	0	0
D	0	1	0	2

7 a Write down the adjacency matrix corresponding to this graph, *G*.

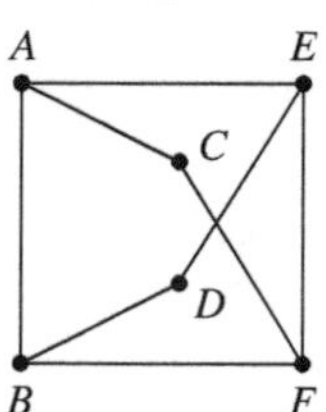

b Construct the adjacency matrix for *G'*, the complement of *G*. Hence draw *G'*.

8 Draw the bipartite graphs with these adjacency matrices.

a

	D	*E*	*F*
A	0	1	1
B	1	1	0
C	1	0	1

b

	S	*T*	*U*	*V*
P	1	1	0	1
Q	1	0	1	0
R	1	0	1	1

9 Draw the directed graph corresponding to this adjacency matrix.

		To			
		A	***B***	***C***	***D***
From	***A***	0	1	2	0
	B	0	1	1	1
	C	1	0	0	1
	D	1	0	1	0

10 Draw the directed network corresponding to this distance matrix.

		To				
		A	***B***	***C***	***D***	***E***
From	***A***	–	8	5	–	9
	B	8	–	–	–	–
	C	–	3	–	6	–
	D	2	4	11	–	–
	E	–	–	5	–	–

DISCRETE

Reasoning and problem-solving

Strategy

To solve a problem involving graphs

1. If necessary model the situation by means of a graph.
2. Use a diagram or a table, as appropriate.
3. Ensure that you use the correct graph terms.
4. Answer the question in context.

One reason for studying graph theory is that graphs and networks give a simplified representation, a mathematical model of a wide variety of situations. A familiar example of a graph used as a model is the map of the London Underground. The geography of the system and the distances between stations are not accurately represented, because the user only needs to know which line to travel on and which station follows which.

Example 9

Ice cream is available in four flavours – strawberry, vanilla, chocolate and mint. George likes strawberry and chocolate, Harriet likes vanilla, chocolate and mint, Isla likes strawberry and mint while Juan only likes chocolate.

a Use a graph to model this situation.

b If only one ice cream of each flavour is available, use your graph to find how they should be distributed so that everyone is happy.

b Juan must have chocolate, so George must have the strawberry.

As George has strawberry, Isla must have the mint.

As Isla has the mint, Harriet must have the vanilla.

4 Answer the question in context.

If the weights in a network represent direct, straight line distances, then they must obey the **triangle inequality**.

> **Key point**
>
> The triangle inequality states that the sum of the lengths of any two sides of a triangle cannot be less than the length of the third side.
>
> $AB + BC \geq AC$

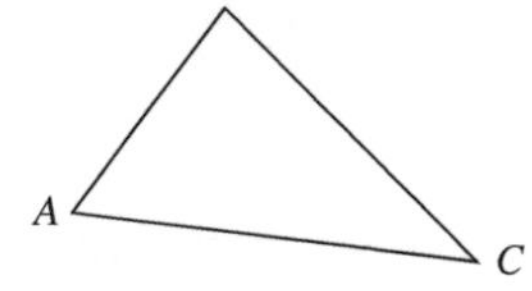

Of course, if the distances are for 'wiggly' routes or the weights represent other things, such as travel times or costs, the network may not satisfy the triangle inequality.

Example 10

This network shows the travel times, in minutes, between four towns.

a Show that this network does not satisfy the triangle inequality.

b Construct a table of shortest travel times.

c Draw a K_4 graph for the network corresponding to your answer to **b**.

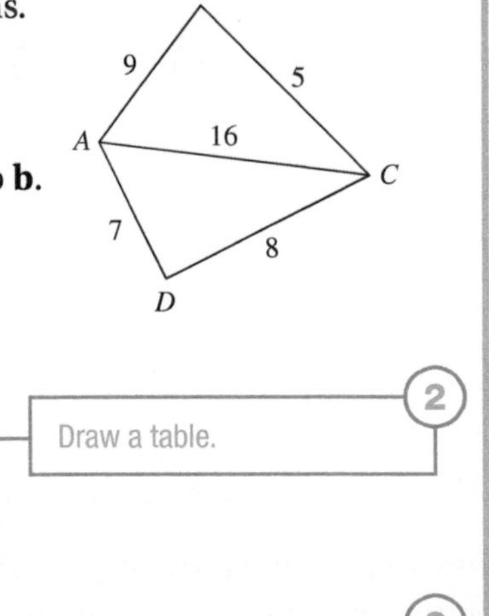

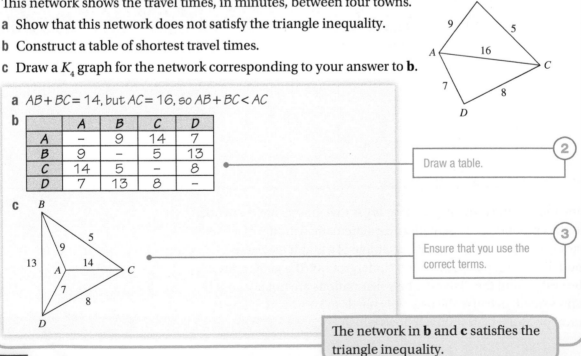

a $AB + BC = 14$, but $AC = 16$, so $AB + BC < AC$

b

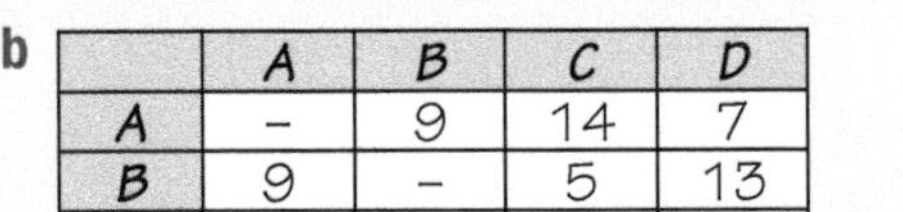

	A	B	C	D
A	–	9	14	7
B	9	–	5	13
C	14	5	–	8
D	7	13	8	–

2 Draw a table.

c

B, 5, 9, 13, 14, A, C, 7, 8, D

3 Ensure that you use the correct terms.

The network in **b** and **c** satisfies the triangle inequality.

Exercise 12.1B Reasoning and problem-solving

1 A team of four runners is to run a 4×100 m relay race. Dwayne likes to run either the first or second leg. Anton will run any leg other than the first. Kris prefers the third or fourth leg. Jason likes either the first or the last leg.

a Draw a bipartite graph to show this information.

b List the three possible ways in which the legs could be allocated so that everyone is happy.

2 The diagram shows the direct bus routes offered by an operator, together with the cost, in pounds, of a ticket on each route.

a Draw a distance matrix corresponding to this network.

b Show that the network does not satisfy the triangle inequality.

c Construct a table of cheapest costs between these five towns.

d Draw the K_5 graph corresponding to the network in **c**.

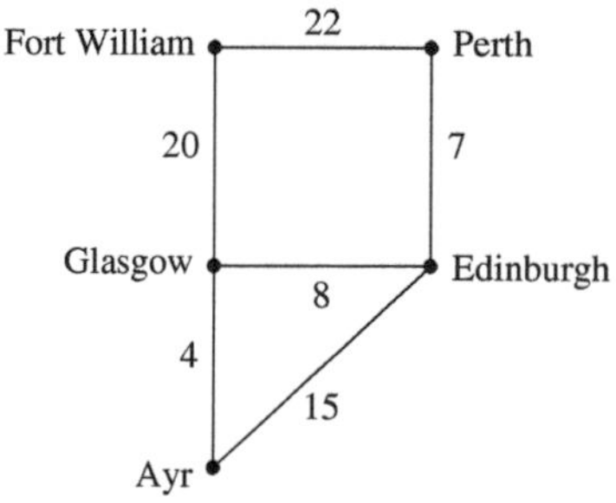

3 a State in terms of n the degree of each vertex of K_n

b Use the handshaking lemma to find the number of edges in K_n in terms of n

4 Alan has four friends – Ben, Candy, Dee and Eli. Ben is friends with Candy and Eli but doesn't know Dee. Candy is friends with Dee and Eli. Dee does not know Eli.

	Alan	Ben	Candy	Dee	Eli
Alan	0	1	1	1	1
Ben	1	0	1	0	1
Candy	1				
Dee	1				
Eli	1				

a Copy and complete the adjacency matrix showing these friendships.

b Draw the corresponding graph.

c Construct the adjacency matrix for the complement of this graph and draw the corresponding graph. What information does this graph provide?

d Is the graph K_4 a sub-graph of the graph you drew in **b**? If it is, list the vertices corresponding to this sub-graph and state what it means about the people involved.

5 The diagram shows a road junction at which traffic lights are to be installed.

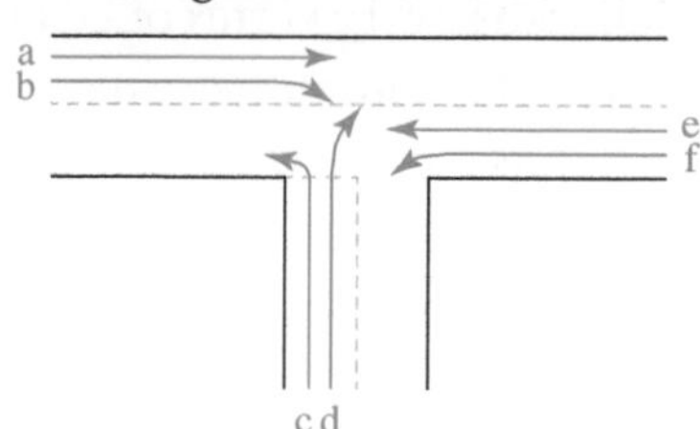

a Draw a graph with a vertex for each traffic stream to model which streams can safely flow at the same time.

b What is the maximum number of streams which can coexist?

c If the lights are to have three phases, which streams should be allowed to go in each phase?

12.2 Traversing a graph

Fluency and skills

Most problems modelled by graphs and networks involve moving around the graph. Any continuous journey around a graph with no restrictions on repeating edges or vertices is sometimes called a walk, but different types of walk have special names.

- A walk with no repeated edges is called a **trail**. A trail can visit vertices more than once. If a trail returns to its starting vertex, it is a **closed** trail.
- A walk with no repeated edges or vertices (except if you return to the start) is called a **path**. A closed path is called a **cycle**.

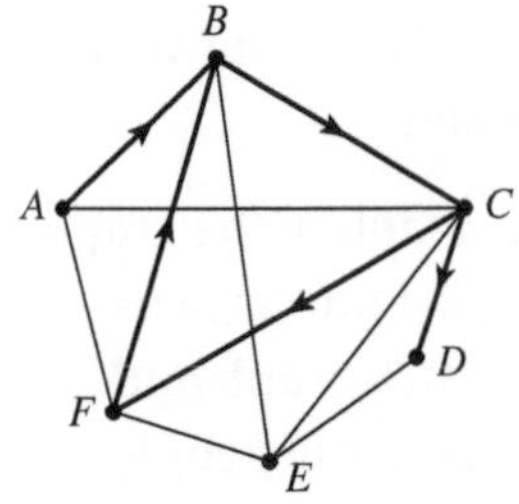

ABCFBCD is a walk

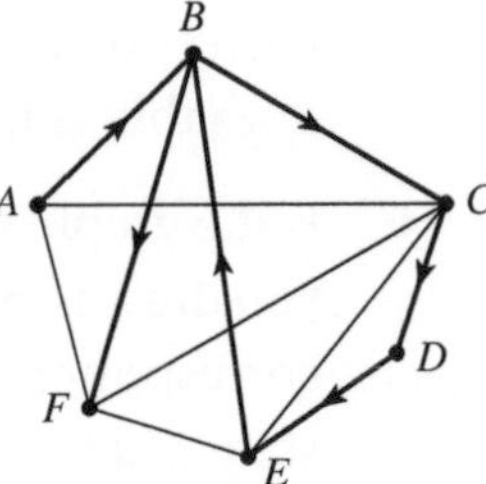

ABCDEBF is a trail

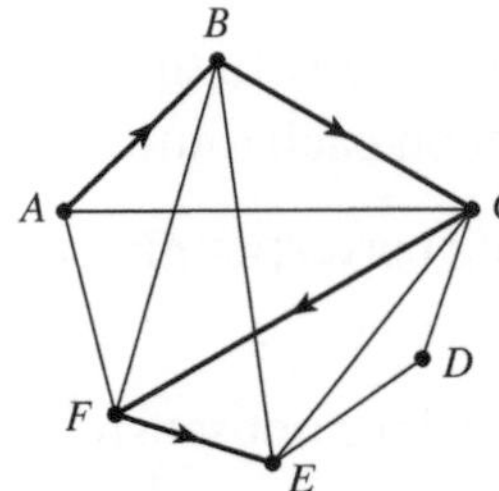

ABCFE is a path

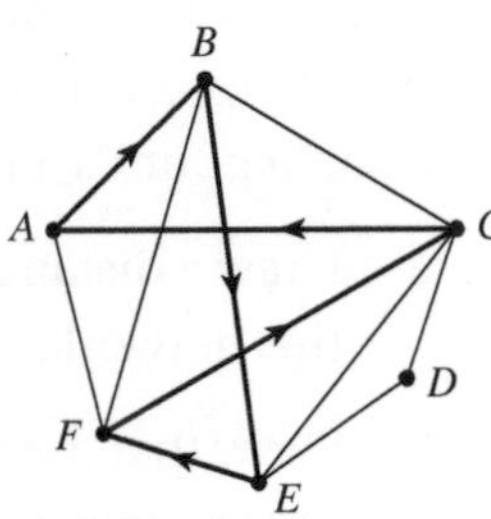

ABEFCA is a cycle

- A connected graph with no cycles is a **tree**.
- A cycle which visits every vertex once and once only is a **Hamiltonian cycle** (sometimes called a **Hamiltonian tour**). A graph which has such a cycle is a **Hamiltonian graph**.

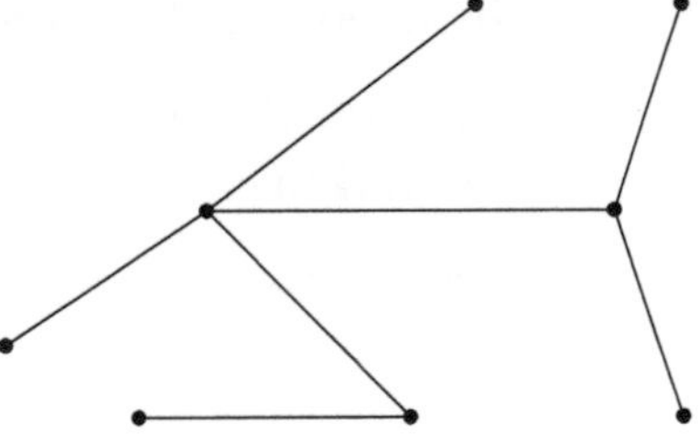
This is a tree

You may have seen puzzles asking you to draw a graph without lifting your pencil or going over a line twice. For example:

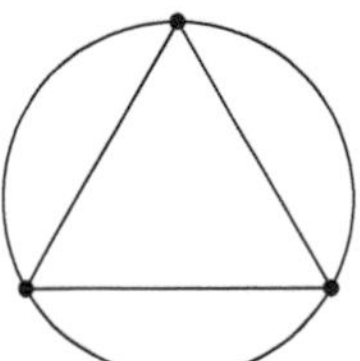

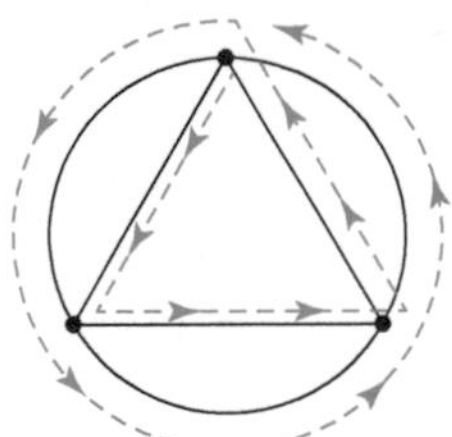

The trail starts and finishes in the same place (it is closed). A graph for which this is possible is called **traversable** or **Eulerian** (pronounced *oi-leerie-ann*).

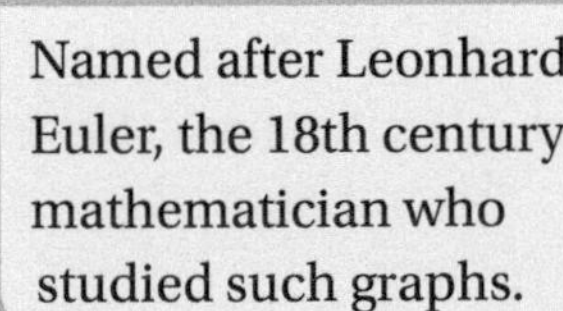
Named after Leonhard Euler, the 18th century mathematician who studied such graphs.

If a graph can only be drawn by starting and finishing at different points, it is **semi-traversable** or **semi-Eulerian**.

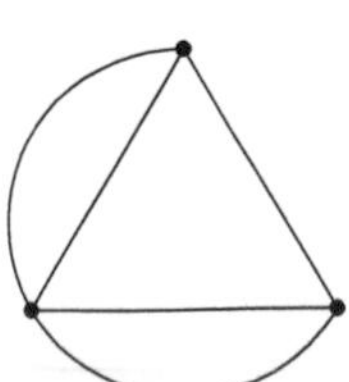

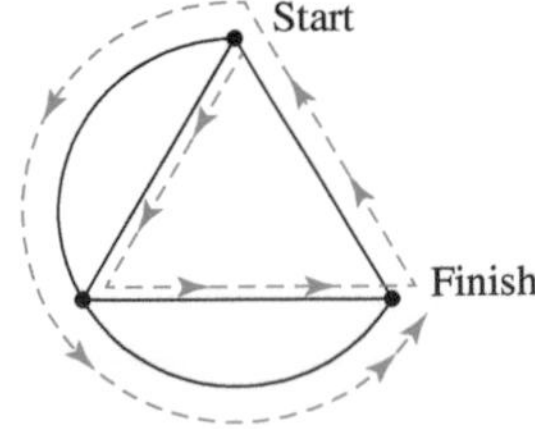

Some graphs are non-traversable (non-Eulerian). For example, try to draw this graph.

You can tell if a graph is traversable by examining whether the degrees of the vertices are odd or even.

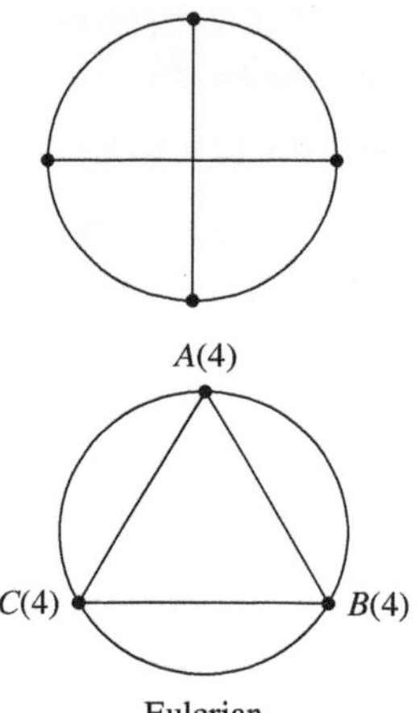

Eulerian

> **Key point**
>
> A graph is traversable (Eulerian) if all vertices are of **even** degree.

A route which traverses the graph is called an **Eulerian trail**. For this graph, *ABCABCA* is an Eulerian trail.

> **Key point**
>
> A graph is semi-traversable (semi-Eulerian) if it has two odd degree vertices. The two odd vertices have to be the start and finish vertices.

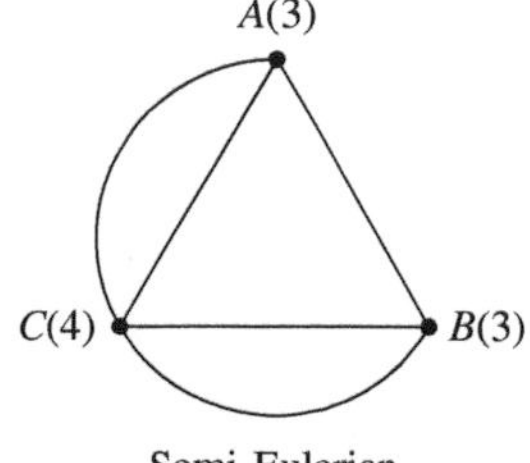

Semi-Eulerian

> **Key point**
>
> A graph with more than two odd degree vertices is non-traversable (non-Eulerian).

The handshaking lemma states that the total of degrees is twice the number of edges. The total of the degrees is therefore an even number, which means the list of degrees must have an even number of odd values.

> **Key point**
>
> The number of odd vertices in a graph is even.

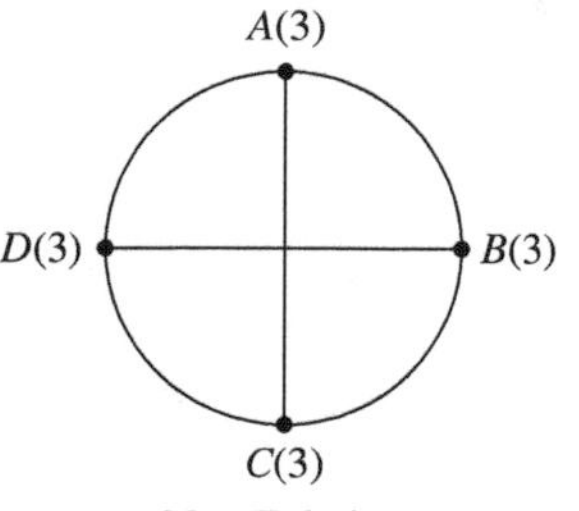

Non-Euleria n

Example 1

a Show that this graph is Hamiltonian.

b Show that the graph is semi-Eulerian. Where would a trail start and finish if it is to traverse the graph?

c Modify the graph to make it Eulerian.

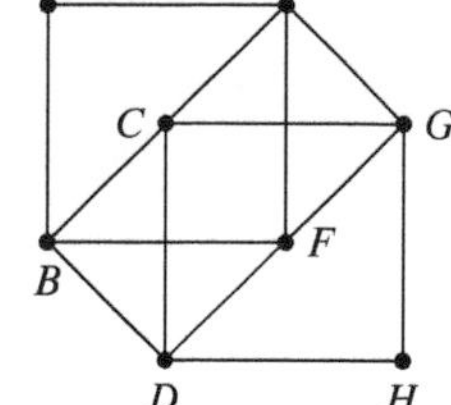

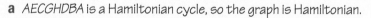

a *AECGHDBA* is a Hamiltonian cycle, so the graph is Hamiltonian.

b The degrees of the vertices are *A*(2), *B*(3), *C*(4), *D*(3), *E*(4), *F*(4), *G*(4), *H*(2)

There are two vertices with odd degree, so the graph is semi-Eulerian.

The trail would have to start at *B* and end at *D*, or vice versa.

c The graph will be Eulerian if the edge *BD* is added, making the degrees of *B* and *D* both 4

DISCRETE

Exercise 12.2A Fluency and skills

1 For the graph shown

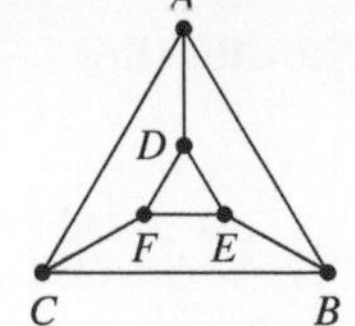

a Write a path from A to C which uses

i 3 edges ii 5 edges.

b Write a cycle, starting and ending at B, which involves (including B)

i 3 vertices ii 5 vertices.

c Write a Hamiltonian cycle.

d Explain why $EFCABC$ is not a path.

2 Which of these graphs are trees?

a

b

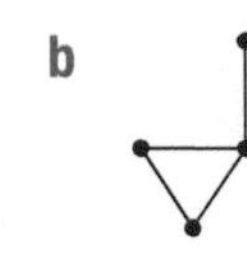

c

d

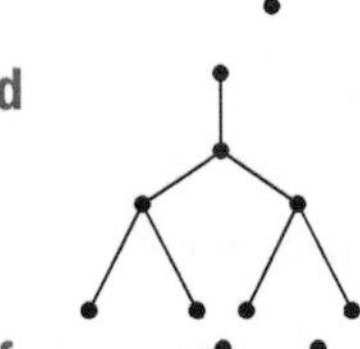

e

f

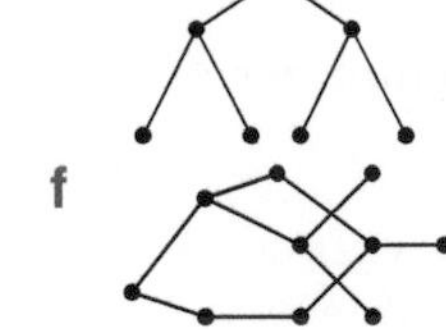

3 State whether these graphs are Eulerian, semi-Eulerian or neither.

a

b

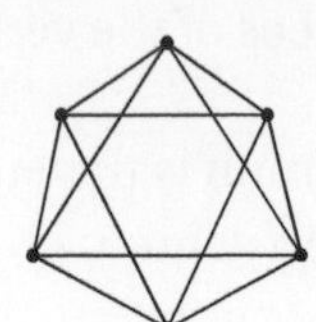

c

d

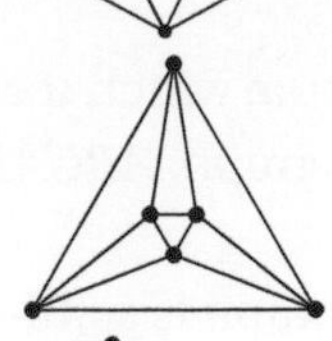

e

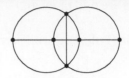

f

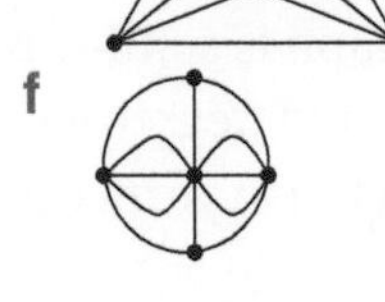

4 Draw a simple, connected graph with five vertices which is

a Eulerian but not Hamiltonian

b Hamiltonian but not Eulerian

c Neither Hamiltonian nor Eulerian

d Both Hamiltonian and Eulerian.

5 A simple connected graph has 5 vertices and 5 edges.

a State the sum of the degrees of its vertices.

b Draw the graph if exactly one of the vertices is of degree 4. State whether your graph is Eulerian, semi-Eulerian or neither.

c Draw the graph given that it is Hamiltonian.

Reasoning and problem-solving

Strategy

To solve a problem involving graphs

1. If necessary, model the situation by means of a graph.
2. Use a diagram or a table as appropriate.
3. Ensure that you use the correct graph terms.
4. Answer the question in context.

Example 2

The plan shows a building with five rooms and two doors to the outside. A prize is offered to anyone who can enter the building, pass through every door once only and return to the outside.

A B C D E

By modelling the situation as a graph, investigate whether this is possible.

(*Continued on the next page*)

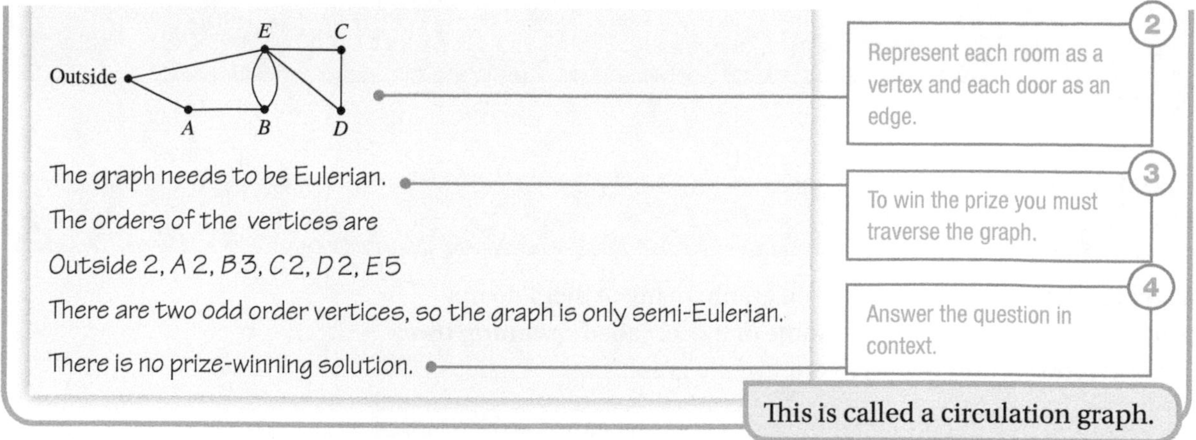

Outside
E
C
A
B
D
Represent each room as a vertex and each door as an edge.
2
The graph needs to be Eulerian.
The orders of the vertices are
Outside 2, A 2, B 3, C 2, D 2, E 5
To win the prize you must traverse the graph.
3
There are two odd order vertices, so the graph is only semi-Eulerian.
There is no prize-winning solution.
Answer the question in context.
4
This is called a circulation graph.

Exercise 12.2B Reasoning and problem-solving

1 The diagram shows the layout of a building with seven rooms. There are three doors to the outside.

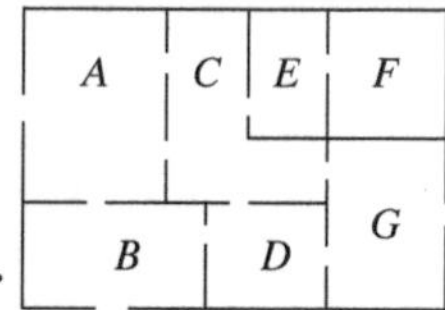

a Draw a circulation graph, with vertices for the rooms and the outside, and edges showing possible movement between them.

b Is your graph Eulerian, semi-Eulerian or neither? Give your reason.

2 The structure of a saturated hydrocarbon molecule is a tree with vertices of order 4 (the carbon atoms) and vertices of order 1 (the hydrogen atoms). Draw two distinct molecules with four carbon atoms. How many hydrogen atoms are required in each case?

3 **a** Explain why the graph shown is semi-Eulerian.

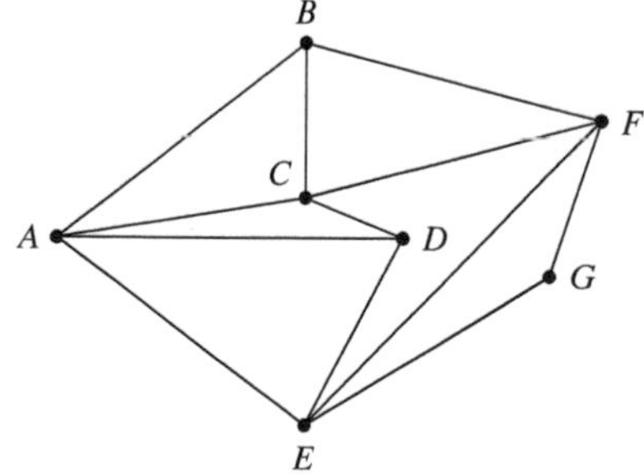

b Which edge would you need to add to the graph to make it Eulerian?

c Give an example of an Eulerian trail for your modified graph.

4 The diagram shows the town of Königsberg.

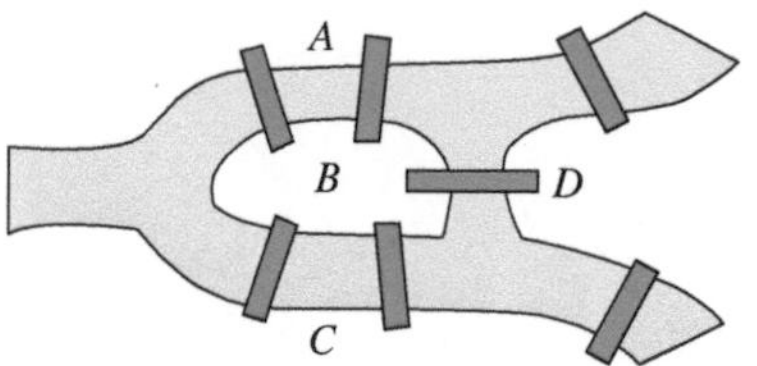

There is an island just downstream from where two rivers meet, and in the eighteenth century there were seven bridges, as shown. The citizens of Königsberg used to spend their Sunday afternoons trying to walk across each bridge once only, ending up at their starting point. By using a graph to model the situation, decide whether the citizens of Königsberg were wasting their time trying to solve this problem. (The Königsberg bridge problem was part of Euler's original study of traversability, published in 1736.)

5 There are only two distinct trees with 4 vertices.

All other such trees are isomorphic to one of these two.

How many distinct trees can you draw with

a 5 vertices **b** 6 vertices?

6 Investigate the conditions under which the complete graph K_n is traversable.

7 Investigate the conditions under which the complete bipartite graph $K_{m,n}$ is Hamiltonian.

12.3 Minimum spanning trees and Kruskal's algorithm

Fluency and skills

If you list the sub-graphs of a given graph, many of them do not contain a cycle; they are trees. Some of these, called **spanning trees** or **connectors**, contain every vertex of the graph.

> **Key point**
> A spanning tree or connector of a graph is a sub-graph that is a tree containing every vertex of the graph.

Here are two possible spanning trees of a graph.

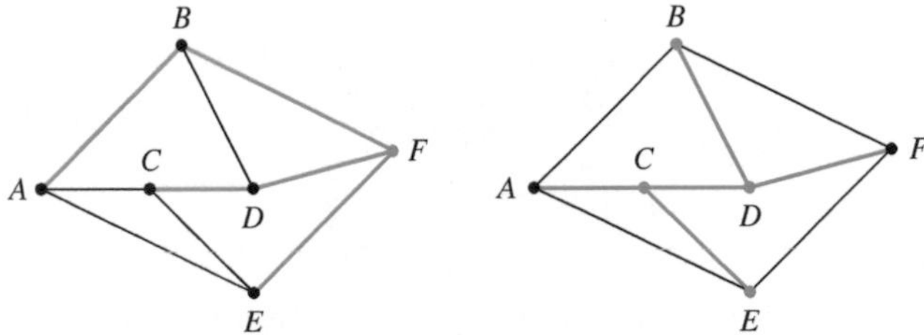

In this example, the graph has six vertices, and each spanning tree has five edges.

A graph with two vertices is spanned by one edge, and each extra vertex requires one extra edge, so, in general, the number of edges in a spanning tree is one fewer than the number of vertices.

> **Key point**
> A spanning tree for a graph with n vertices has $(n - 1)$ edges.

For a network (weighted graph), you can find the spanning tree with the lowest total weight.

> **Key point**
> For a network, the spanning tree with the lowest total weight is the minimum spanning tree (MST) or minimum connector.

This is the minimum spanning tree for this network.

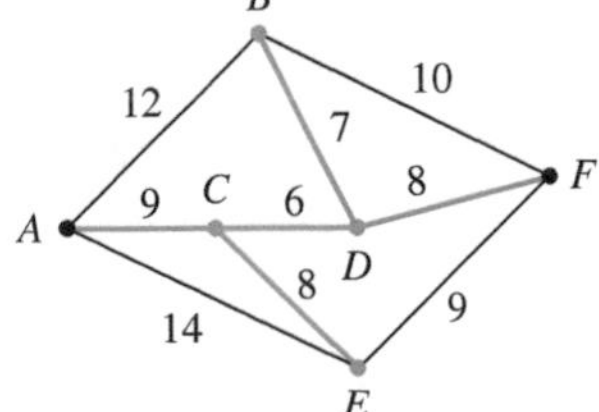

For a small network, the minimum spanning tree is obvious by inspection, but the number of possible spanning trees increases rapidly as the size of the network increases (for K_n there are n^{n-2} possible spanning trees), so, in general, you need an algorithm to find the minimum.

One such algorithm is **Kruskal's algorithm**.

Key point

Kruskal's algorithm

Step 1 Choose the arc with minimum weight.

Step 2 Choose the arc with least weight from the remaining arcs, avoiding those that form a cycle with those already chosen.

Step 3 If any nodes remain unconnected, go to **Step 2**

If at any stage there is a choice of arcs, choose at random.

Kruskal's algorithm is an example of a greedy algorithm. At each stage, you make the most advantageous choice without thinking ahead.

Example 1

Use Kruskal's algorithm to find a minimum spanning tree for the network shown. List the order in which arcs are chosen.

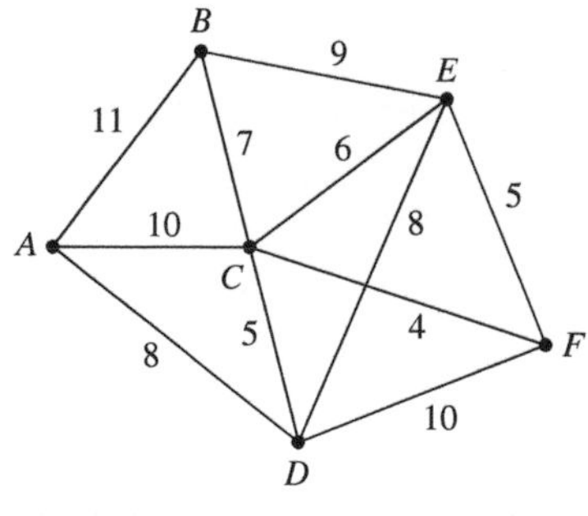

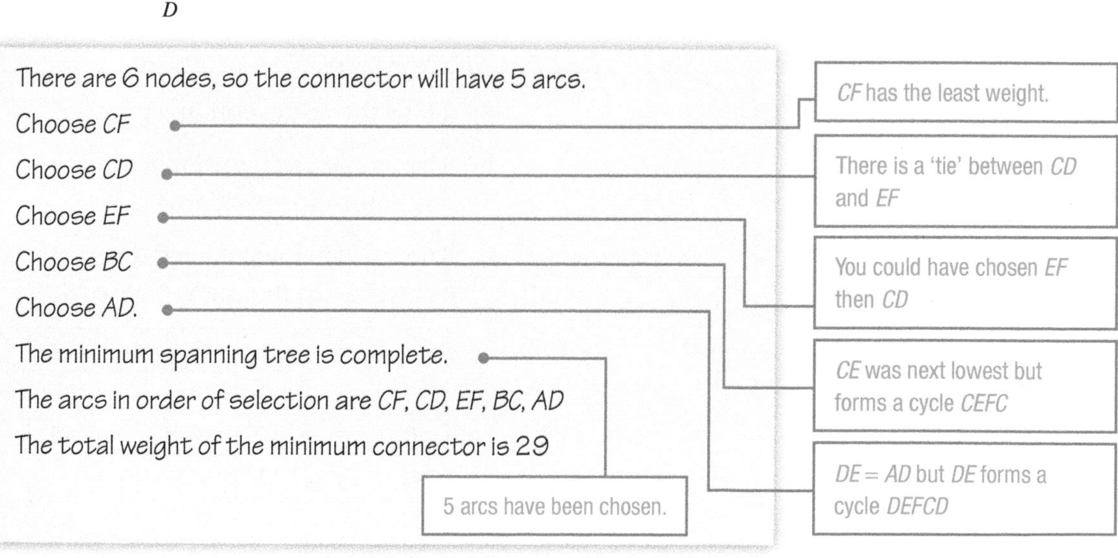

There are 6 nodes, so the connector will have 5 arcs.

Choose *CF* — *CF* has the least weight.

Choose *CD* — There is a 'tie' between *CD* and *EF*

Choose *EF* — You could have chosen *EF* then *CD*

Choose *BC* — *CE* was next lowest but forms a cycle *CEFC*

Choose *AD*. — *DE* = *AD* but *DE* forms a cycle *DEFCD*

The minimum spanning tree is complete. — 5 arcs have been chosen.

The arcs in order of selection are *CF*, *CD*, *EF*, *BC*, *AD*

The total weight of the minimum connector is 29

DISCRETE

Exercise 12.3A Fluency and skills

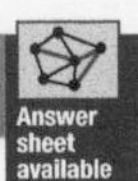

1 Use Kruskal's algorithm to find the minimum spanning tree for each of the networks shown. List the order in which you choose the arcs, and find the total weight of the connector.

a

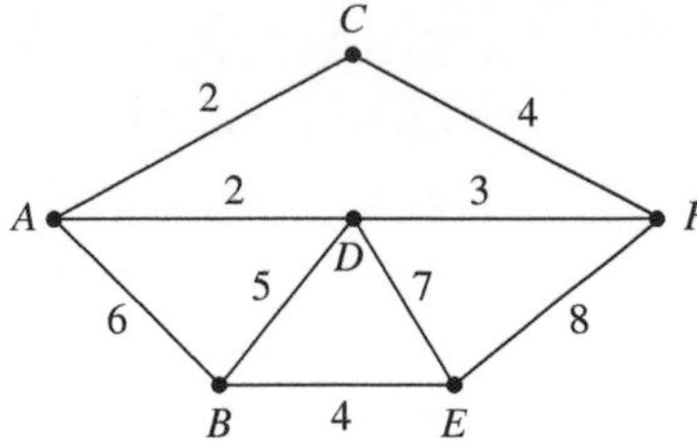

b

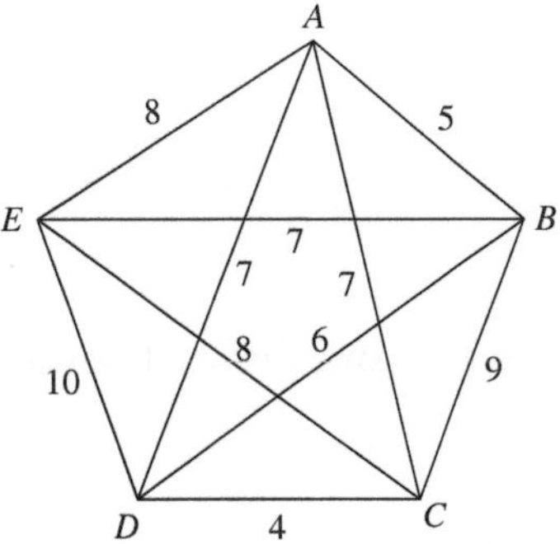

c

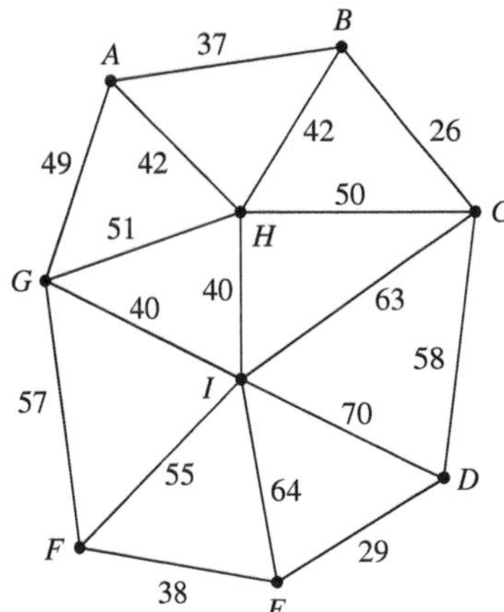

d

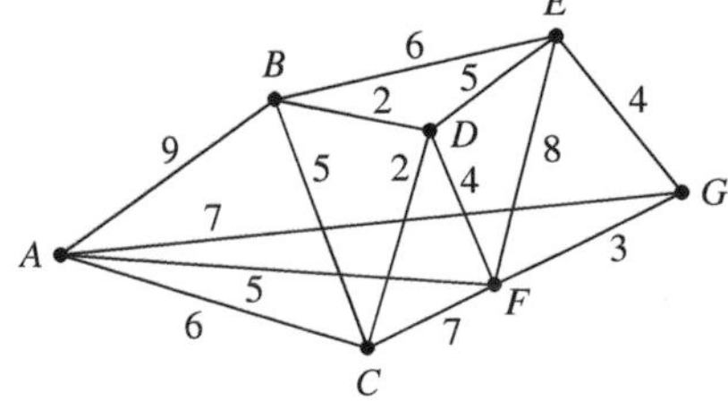

e

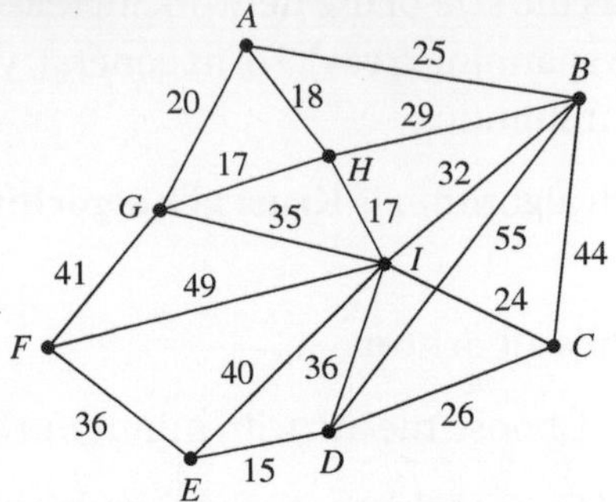

2 Use Kruskal's algorithm to find the two possible minimum spanning trees for the network shown. List the order in which you choose the arcs, and find the total weight of the connector.

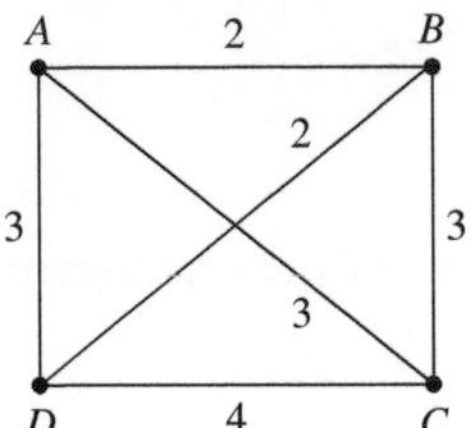

3 Find all possible minimum spanning trees for this network.

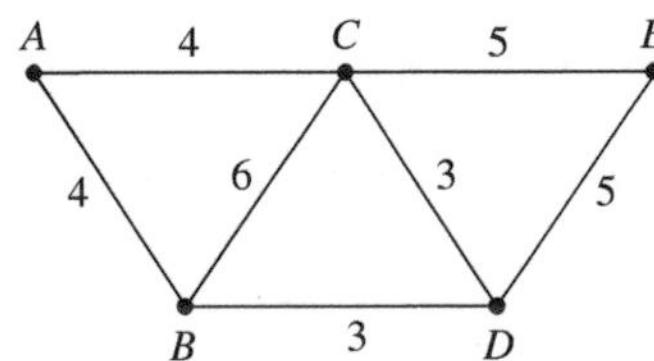

4 For the distance matrix shown,

a Draw the corresponding graph,

b Use Kruskal's algorithm to find the minimum spanning tree,

c Draw a separate labelled diagram showing just the spanning tree.

	A	B	C	D	E	F
A	–	20	42	–	35	–
B	20	–	30	–	40	–
C	42	30	–	45	–	38
D	–	–	45	–	25	55
E	35	40	–	25	–	–
F	–	–	38	55	–	–

Reasoning and problem-solving

Strategy

To solve a problem using Kruskal's algorithm

1. If necessary, draw a clear graph.
2. Apply Kruskal's algorithm, showing clearly the order in which you choose the arcs.
3. Answer the question in context.

Example 2

The network represents a country park, with seven picnic sites and a car park joined by rough paths. Distances are in metres.

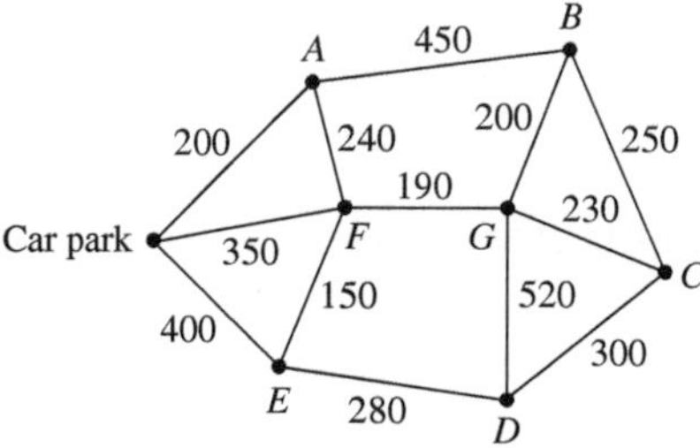

The council plans to upgrade some paths, so that all the picnic sites are accessible by wheelchair. The cost of the upgrade is £55 per metre of path. Use Kruskal's algorithm to decide which paths should be upgraded, and state how much money the project will cost.

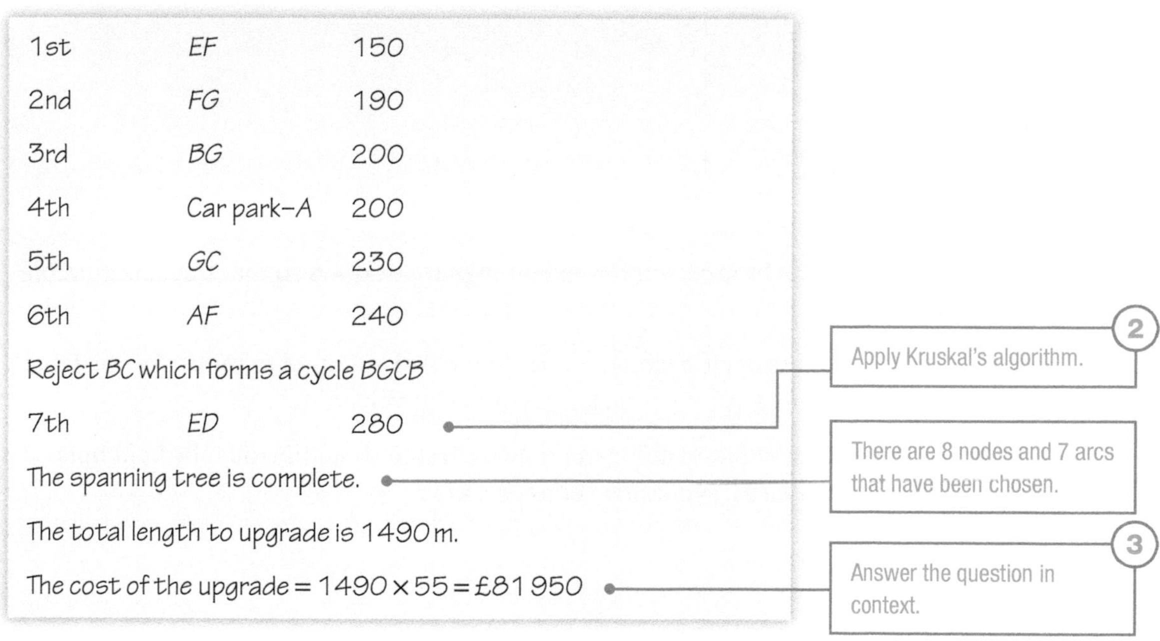

1st	EF	150
2nd	FG	190
3rd	BG	200
4th	Car park–A	200
5th	GC	230
6th	AF	240

Reject BC which forms a cycle $BGCB$

7th	ED	280

The spanning tree is complete.

The total length to upgrade is 1490 m.

The cost of the upgrade = 1490 × 55 = £81 950

(2) Apply Kruskal's algorithm.

There are 8 nodes and 7 arcs that have been chosen.

(3) Answer the question in context.

DISCRETE

Exercise 12.3B Reasoning and problem-solving

1 A farmer has five animal shelters on his land and wishes to connect them all to the water supply at the farmhouse. The table shows the distances (in metres) between the farmhouse, F, and the shelters A, B, C, D, and E (some direct connections are not possible).

	A	B	C	D	E	F
A	–	70	100	120	80	50
B	70	–	–	–	70	–
C	100	–	–	60	–	–
D	120	–	60	–	–	80
E	80	70	–	–	–	–
F	50	–	–	80	–	–

Use Kruskal's algorithm to find which connections the farmer should make to achieve the water supply as efficiently as possible. Find the total length of water piping he would need.

2 The diagram shows the roads connecting seven towns. The lengths of the roads are given in miles.

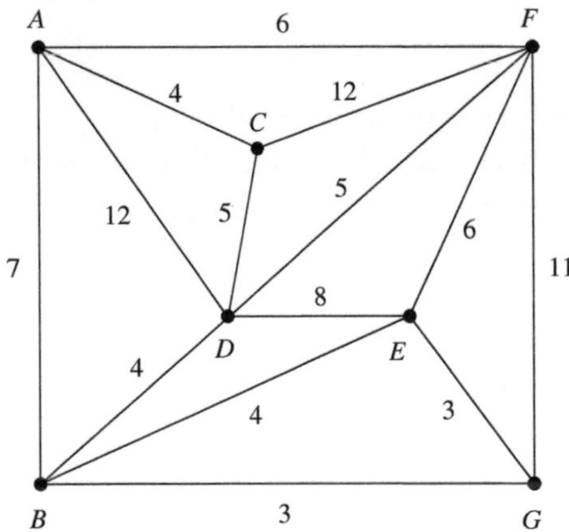

During the winter, snowploughs are used to keep enough roads open so that travel is possible between any two towns.

a Use Kruskal's algorithm to decide which roads should be kept open to minimise the length of road needing attention. Find the total length of road involved.

b The ambulance station is situated along the road from A to B, so this must be kept open. Which road would it replace in your answer to part **a**?

3 The table shows the cost, in pounds per thousand words, of translating between languages.

	English	French	German	Italian	Portuguese	Spanish
English	–	25	30	27	38	22
French	25	–	35	22	36	28
German	30	35	–	35	40	32
Italian	27	22	35	–	26	20
Portuguese	38	36	40	26	–	23
Spanish	22	28	32	20	23	–

a Use Kruskal's algorithm to find the minimum spanning tree for this network.

b From your solution to part **a**, state the cheapest sequence of translations to make a document of 10 000 words, originally in German, available in all the languages, and calculate the cost of the operation.

c Calculate the cost of translating the document directly from German into each of the other languages. Suggest reasons why, despite the extra cost, this might be a preferable course of action.

4 The diagram shows an inlet with five islands.

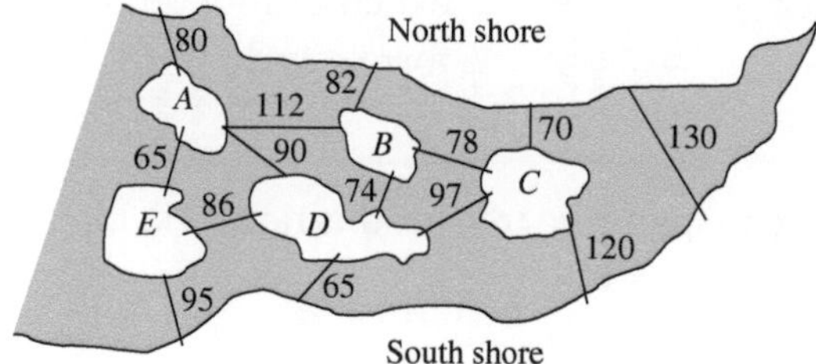

The islands and the mainland are connected by ferries or toll bridges, and the figures shown give the prices, in pounds, of a three month pass for each crossing.

A delivery firm needs to deliver to all areas.

a Draw a network to model the situation.

b Use Kruskal's algorithm to decide which passes the firm should purchase to give them the desired access as cheaply as possible.

DISCRETE

12.4 Minimum spanning trees and Prim's algorithm

Fluency and skills

Kruskal's algorithm cannot easily be adapted to a network stored as a matrix because it is difficult to check whether an arc will form a cycle with those already chosen. This means that it is not well suited to computerisation. The alternative is **Prim's algorithm**.

With Prim's algorithm, the minimum spanning tree is built up from a chosen starting node. Nodes and arcs are added one at a time to a connected sub-graph until the spanning tree is complete.

Prim's algorithm Key point

Step 1 Choose any node to be the first in the connected set.

Step 2 Choose the arc of minimum weight joining a connected node to an unconnected node. Add this arc to the spanning tree and the node to the connected set.

Step 3 If any unconnected nodes remain, go to **Step 2**

If at any stage there is a choice of arcs, choose at random.

This is also a **greedy algorithm**. It avoids the problem of forming cycles because it only considers arcs to unconnected nodes, which cannot possibly form a cycle.

Example 1

Starting from A, use Prim's algorithm to find a minimum spanning tree for this network. List the order in which arcs are chosen.

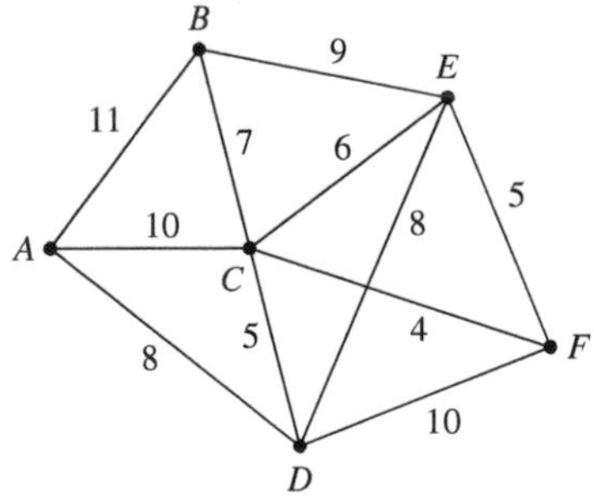

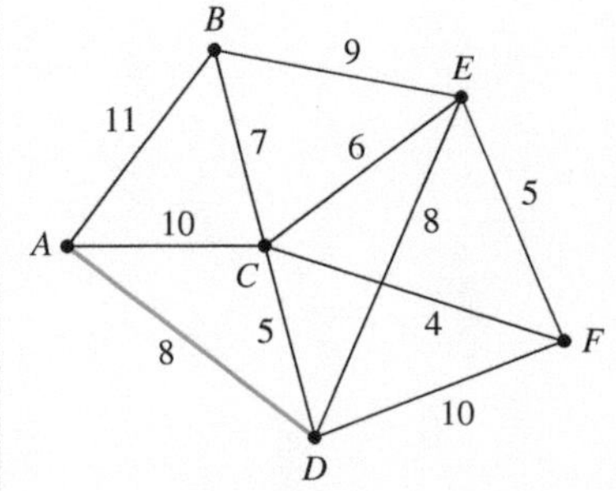

Take A as the first node in the connected set.

The arcs to unconnected nodes are AB, AC and AD

The least weight is AD, so you add this arc to the connector.

(*Continued on the next page*)

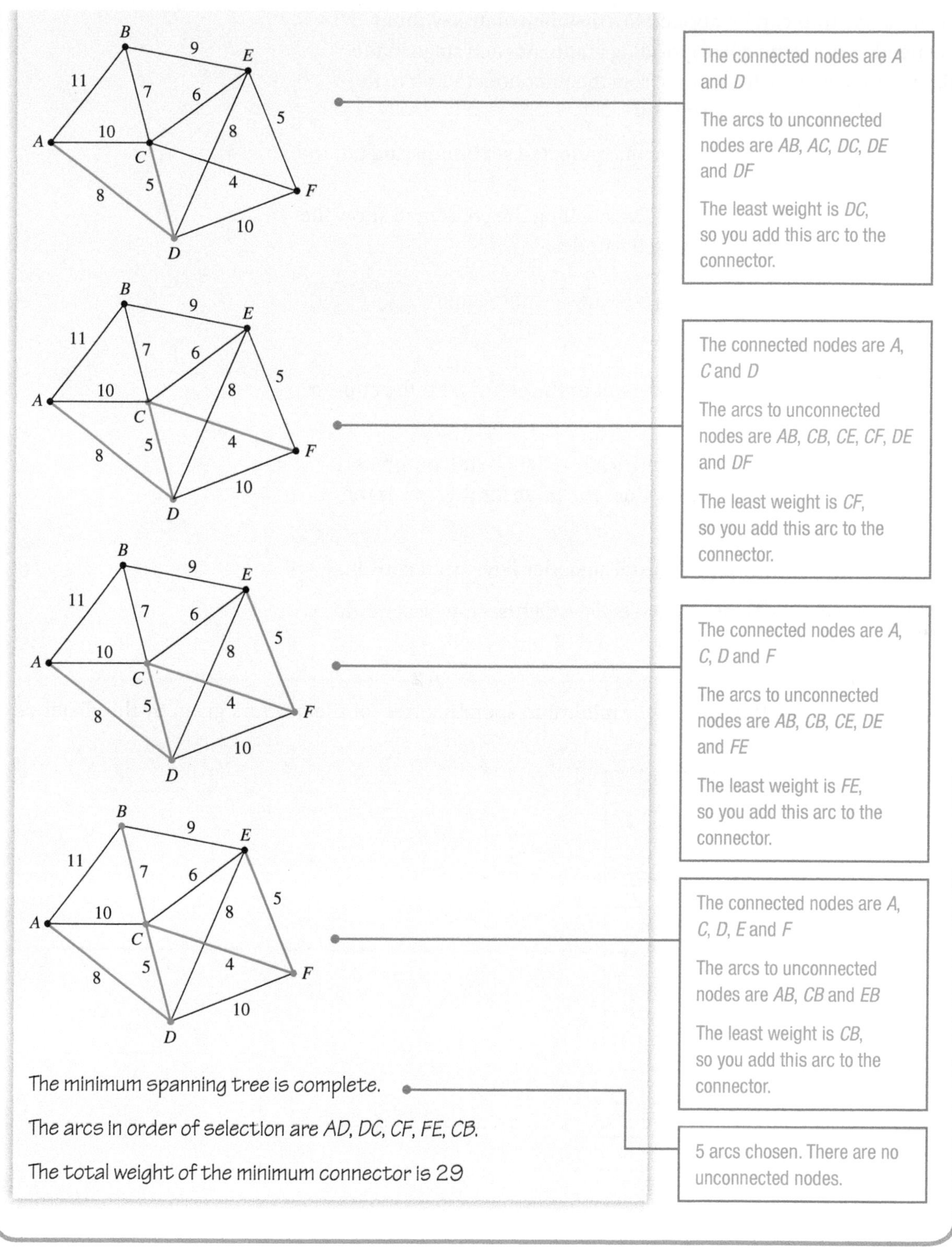
The connected nodes are A and D
The arcs to unconnected nodes are AB, AC, DC, DE and DF
The least weight is DC, so you add this arc to the connector.
The connected nodes are A, C and D
The arcs to unconnected nodes are AB, CB, CE, CF, DE and DF
The least weight is CF, so you add this arc to the connector.
The connected nodes are A, C, D and F
The arcs to unconnected nodes are AB, CB, CE, DE and FE
The least weight is FE, so you add this arc to the connector.
The connected nodes are A, C, D, E and F
The arcs to unconnected nodes are AB, CB and EB
The least weight is CB, so you add this arc to the connector.
The minimum spanning tree is complete.
The arcs in order of selection are AD, DC, CF, FE, CB.
The total weight of the minimum connector is 29
5 arcs chosen. There are no unconnected nodes.

Prim's algorithm can be applied to a distance matrix without the need to draw the corresponding graph. At each stage in the algorithm, you transfer a node from the unconnected set to the connected set. On the matrix you:

- Remove the node from the unconnected set by crossing out the row for that node.
- Add it to the connected set by labelling the column to show the order in which you choose the nodes.

Key point

Prim's algorithm on a matrix

Step 1 Select the first node.

Step 2 Cross out the row and number the column for the chosen node.

Step 3 Find the minimum undeleted weight in the numbered columns. Circle this value. The node for this row is the next chosen node.

Step 4 Repeat steps 2 and 3 until all nodes have been chosen.

If there is a choice between two equal weights, choose at random.

Example 2

Use Prim's algorithm to find the minimum spanning tree for the network given by this distance matrix.

	1					
	A	***B***	***C***	***D***	***E***	***F***
A	–	12	9	–	14	–
B	12	–	–	7	–	10
C	9	–	–	6	8	–
D	–	7	6	–	–	8
E	14	–	8	–	–	9
F	–	10	–	8	9	–

	1					
	A	B	C	D	E	F
~~A~~	~~–~~	~~12~~	~~9~~	~~–~~	~~14~~	~~–~~
B	12	–	–	7	–	10
C	9	–	–	6	8	–
D	–	7	6	–	–	8
E	14	–	8	–	–	9
F	–	10	–	8	9	–

Cross through row *A* (removes *A* from the unconnected set).

Label column *A* with 1. (*A* is the first connected node.)

(*Continued on the next page*)

	1		2			
	A	B	C	D	E	F
A		12	9		14	
B	12	–	–	7	–	10
C	9			6	8	
D	–	7	6	–	–	8
E	14	–	8	–	–	9
F	–	10	–	8	9	–

In column *A* the least weight is *AC* = 9

Circle the '9' to show *AC* has been chosen.

Cross through row *C* and label column *C* with 2

	1		2	3		
	A	B	C	D	E	F
A		12	9		14	
B	12	–	–	7	–	10
C	9	–	–	6	8	
D		7	6			8
E	14	–	8	–	–	9
F	–	10	–	8	9	–

In columns *A* and *C* the least weight is *CD* = 6

Circle the '6' to show *CD* has been chosen.

Cross through row *D* and label column *D* with 3

	1	4	2	3		
	A	B	C	D	E	F
A	–	12	9	–	14	–
B	12			7		10
C	9	–	–	6	8	
D	–	7	6	–	–	8
E	14	–	8	–	–	9
F	–	10	–	8	9	–

In columns *A*, *C* and *D* least weight is *DB* = 7

Circle the '7' to show arc *DB* has been chosen.

Cross through row *B* and label column *B* with 4

	1	4	2	3	5	
	A	B	C	D	E	F
A		12	9		14	
B	12	–	–	7	–	10
C	9			6	8	
D	–	7	6	–	–	8
E	14		8		–	9
F	–	10	–	8	9	–

In columns *A*, *B*, *C* and *D* the least weight is

CE = 8 or *DF* = 8

Choose *CE* at random.

Circle the '8' to show arc *CE* has been chosen.

Cross through row *E* and label column *E* with 5

	1	4	2	3	5	6
	A	B	C	D	E	F
A	–	12	9	–	14	–
B	12	–		7	–	10
C	9	–	–	6	8	
D	–	7	6	–	–	8
E	14		8		–	9
F	–	10	–	8	9	–

In columns *A*, *B*, *C*, *D* and *E* the least weight is *DF* = 8

Circle thc '8' to show arc *DF* has been chosen.

Cross through row *F* and label column *F* with 6

The minimum spanning tree is complete.

All the nodes are now in the connected set.

The minimum spanning tree is *AC*, *CD*, *DB*, *CE* and *DF*, with total weight 38

DISCRETE

Exercise 12.4A Fluency and skills

1 Use Prim's algorithm starting from node A to find the minimum spanning tree for the networks shown. List the order in which you choose the arcs, and find the total weight of the connector.

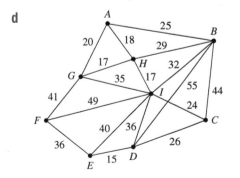

a

b

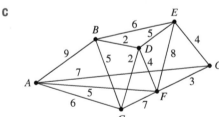

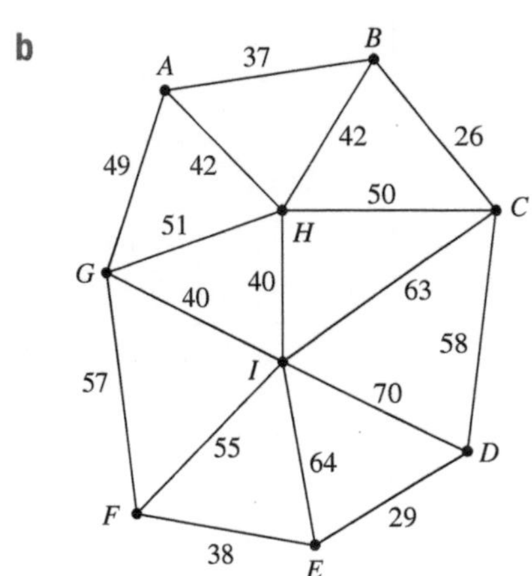

c

d

2 Use Prim's algorithm to find the minimum spanning tree for the network shown in this distance matrix.

Use A as the starting node. List the order in which the arcs are chosen, and find the total weight of the spanning tree.

	A	B	C	D	E
A	–	5	9	6	3
B	5	–	11	4	7
C	9	11	–	5	8
D	6	4	5	–	12
E	3	7	8	12	–

3 a By applying Prim's algorithm to this distance matrix, show that the network has two possible minimum spanning trees, and state the total weight.

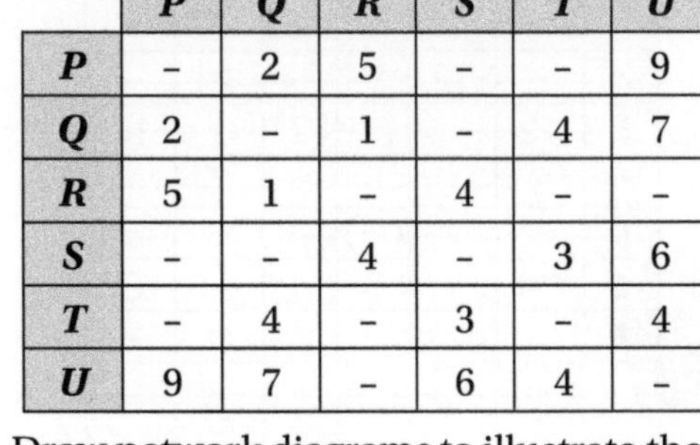

	P	Q	R	S	T	U
P	–	2	5	–	–	9
Q	2	–	1	–	4	7
R	5	1	–	4	–	–
S	–	–	4	–	3	6
T	–	4	–	3	–	4
U	9	7	–	6	4	–

b Draw network diagrams to illustrate the two possible trees.

4 Use Prim's algorithm to find the minimum spanning tree for the network shown in this distance matrix.

	A	B	C	D	E	F	G
A	–	9	17	8	22	16	12
B	9	–	6	15	–	20	14
C	17	6	–	–	12	8	19
D	8	15	–	–	10	–	6
E	22	–	12	10	–	11	–
F	16	20	8	–	11	–	13
G	12	14	19	6	–	13	–

Use A as the starting node. List the order in which you choose the arcs and calculate the total length of the spanning tree.

Reasoning and problem-solving

Strategy

To solve a problem using Prim's algorithm

1. When working with a graph, list the order in which you choose the arcs.
2. When working with a table, label the columns to show clearly the order in which you choose the arcs.
3. Answer the question in context.

Example 3

The network represents a country park, with seven picnic sites and a car park joined by rough paths. Distances are in metres. The council plans to upgrade some paths, so that all the picnic sites are accessible by wheelchair. The cost of the upgrade is £55 per metre of path.

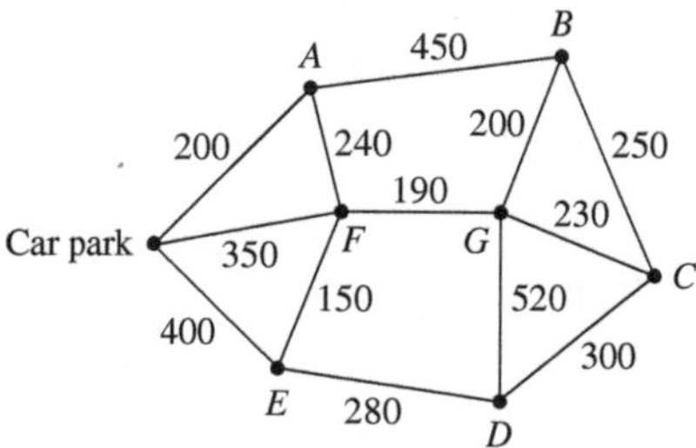

Use Prim's algorithm, starting from the car park, to decide which paths should be upgraded, and state how much money the council should earmark for the project.

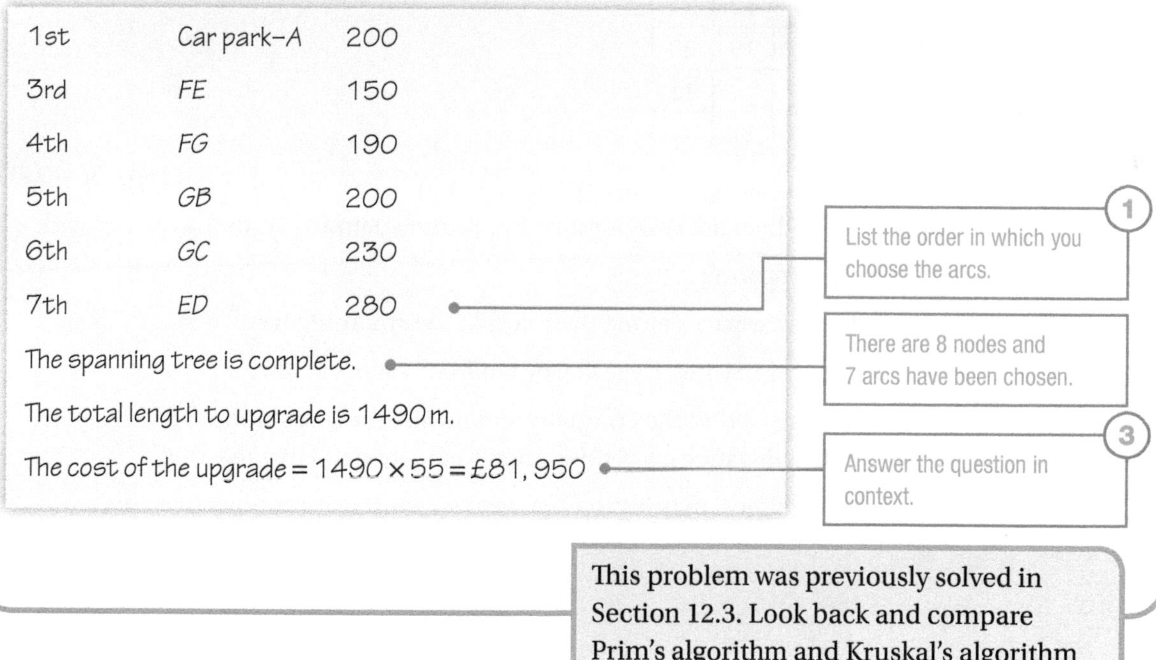

1st	Car park–A	200
3rd	FE	150
4th	FG	190
5th	GB	200
6th	GC	230
7th	ED	280

The spanning tree is complete.

The total length to upgrade is 1490 m.

The cost of the upgrade $= 1490 \times 55 = £81,950$

1
List the order in which you choose the arcs.
There are 8 nodes and 7 arcs have been chosen.
3
Answer the question in context.

This problem was previously solved in Section 12.3. Look back and compare Prim's algorithm and Kruskal's algorithm to see how they differ.

DISCRETE

Exercise 12.4B Reasoning and problem-solving

1 The table shows the distance by direct rail link between eight towns.

	A	*B*	*C*	*D*	*E*	*F*	*G*	*H*
A	–	56	20	–	–	–	–	70
B	56	–	–	15	65	–	75	88
C	20	–	–	87	95	–	120	30
D	–	15	87	–	60	–	25	112
E	–	65	95	60	–	30	40	70
F	–	–	–	–	30	–	45	–
G	–	75	120	25	40	45	–	115
H	70	88	30	112	70	–	115	–

It is decided to close some of the links, leaving just enough connections so that it is possible to travel from any town to any other by rail. Use Prim's algorithm to decide which links must be kept so that the amount of track is a minimum. Use town *A* as the starting node and record the order in which you make the links.

2 **a** Explain why Prim's algorithm is preferred to Kruskal's algorithm when the network is given in the form of a distance matrix.

	A	*B*	*C*	*D*	*E*	*F*
A	–	4	6	12	15	20
B	4	–	5	8	14	25
C	6	5	–	8	12	24
D	12	8	8	–	19	20
E	15	14	12	19	–	11
F	20	25	24	20	11	–

b **i** The distance matrix shown gives the distances, in miles, between six towns. Using Prim's algorithm, show that there are two possible minimum spanning trees for the network.

ii Draw both minimum spanning trees.

c A bus company only runs services along the roads in the spanning tree.

i How far would a person go when travelling by bus from town *A* to town *F*?

ii Where would you suggest that the company should site the bus station if they wish to minimise the greatest distance from the station to the towns? Give reasons.

3 The table shows the walking times, in minutes, between six tourist attractions in a city.

	A	*B*	*C*	*D*	*E*	*F*
A	–	4	6	4	3	2
B	4	–	7	6	5	5
C	6	7	–	6	6	5
D	4	6	6	–	3	5
E	3	5	6	3	–	2
F	2	5	5	5	2	–

The information bureau wants to install signposts along a minimum number of routes so that every attraction is accessible and the total walking time is a minimum.

a Use Prim's algorithm to find which routes they should signpost.

b Draw a tree to represent the minimum connector found in **a**.

c What is the greatest time for which a tourist would walk between any two attractions?

4 The diagram shows the roads connecting five villages, with distances in kilometres.

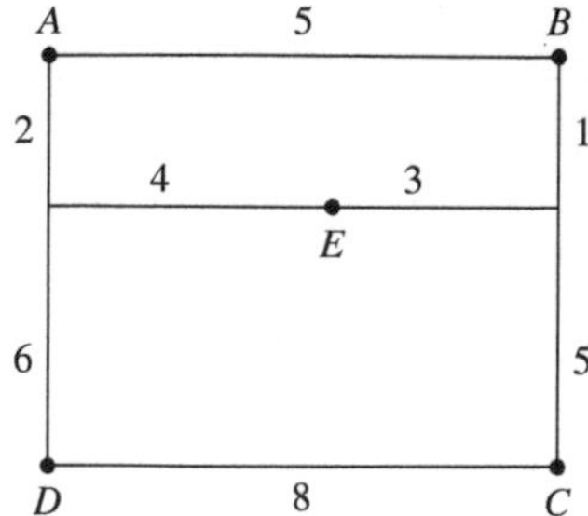

a Construct a table showing the minimum distances by road between the five villages (for example, the distance from A to E is 6 km).

b Draw a network using the distance matrix from part **a**.

c Use Prim's algorithm, starting from node A, to find the minimum connector for the network in part **b**.

As an environmental measure, the local council plans to make some stretches of road 'pedestrians, horses and cycles only,' leaving just the minimum unrestricted road so that cars can travel between the five villages.

d Explain why the result you obtained in part **c** does not give the number of kilometres of road which must stay open to cars.

e Using the original diagram, find, by inspection, the roads which should be kept open to cars.

5

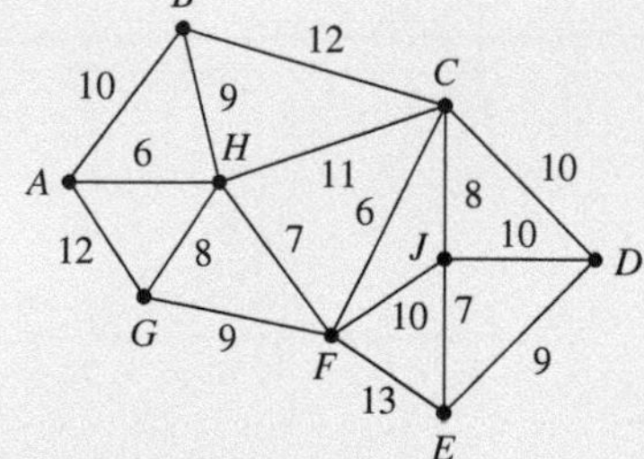

The diagram shows the weight limits (in tonnes) for lorry traffic on roads between nine towns. The council plans to ban lorries on as many roads as possible without stopping lorries from reaching all the towns. By a suitable modification to Prim's algorithm, choose the roads which should remain open to lorries, and state the heaviest lorry which would then have access to all the towns.

DISCRETE

Full A Level

12.5 The route inspection problem

Fluency and skills

The route inspection problem is sometimes called the Chinese postman problem (it is the problem rather than the postman that is Chinese. It was originally discussed by the Chinese mathematician Mei-ko Kwan).

The problem is:

For a given network, find the shortest route that travels at least once along every arc and returns to the starting point.

Ideally, you want to travel just once along each arc. You can do this if the network is Eulerian (traversable).

> **Key point**
>
> If all nodes in the network are even, the network is traversable. The distance travelled is the sum of the weights.

If there are nodes of odd degree, you must repeat some arcs to complete the route. The route inspection problem then becomes:

Which arcs must be repeated to complete the route as efficiently as possible?

Repeating an arc is the same as adding an extra arc to the network. You must choose which arcs to add so that all nodes are made even and the network becomes traversable.

Suppose you must solve the route inspection problem for the network shown on the right.

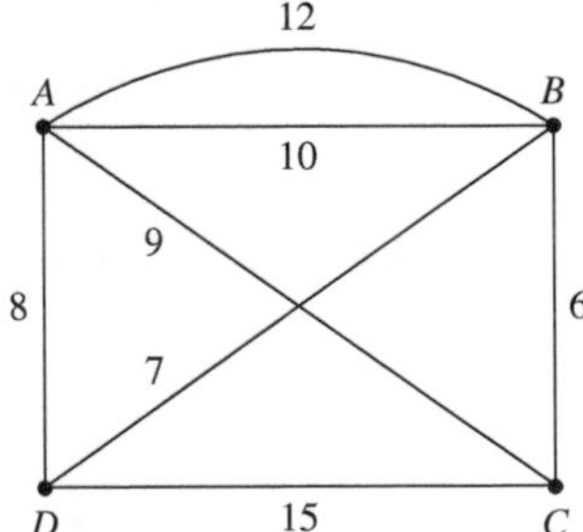

The sum of the weights is 67. The network has two nodes of odd degree, C and D, so the route must travel twice between these nodes. The shortest route from C to D is $CBD = 13$, so you double up on the arcs CB and BD, as shown.

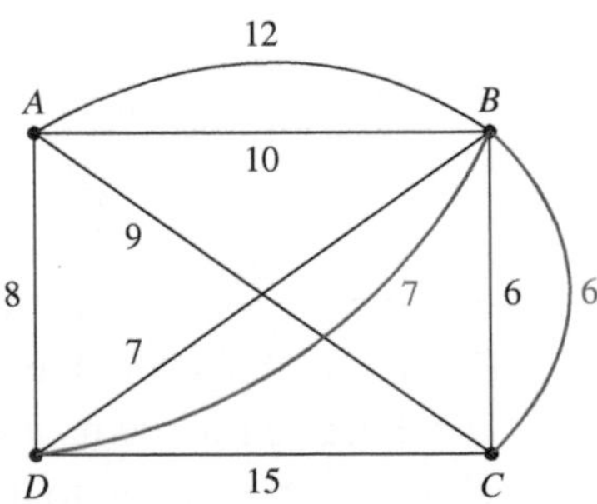

All the nodes now have even degree, so the network is Eulerian. A possible route starting from A is $ABCDABDBCA$

The total distance is $67 + (6 + 7) = 80$

There will always be an even number of odd nodes. If there are more than two, you can pair them together in various ways. You look for the best way to pair them so that the extra arcs have the lowest possible total weight. This gives the **route inspection algorithm**:

Route inspection algorithm — Key point

Step 1 Identify the odd nodes.

Step 2 List all possible pairings of the odd nodes.

Step 3 For each pairing find the shortest routes between paired nodes and the total weight of those routes.

Step 4 Choose the pairing with the smallest total.

Step 5 Repeat the shortest route arcs (equivalent to adding extra arcs to the network). The network is now traversable.

The total length of the route = (sum of original weights) + (sum of extra arc weights).

Example 1

Solve the route inspection problem for the network shown. State a possible route.

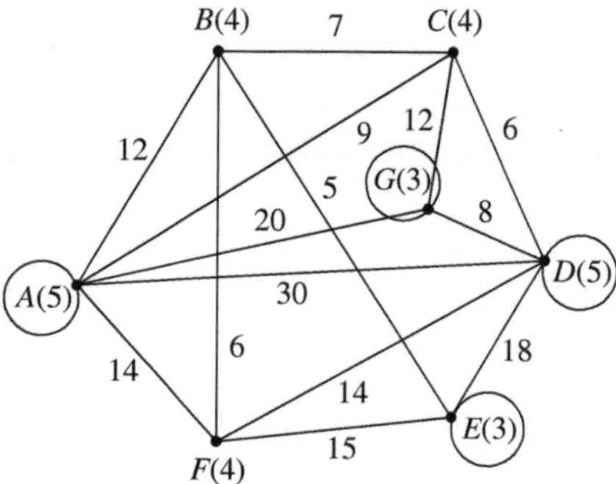

The odd nodes (circled) are A, D, E and G

You can pair these in three ways.

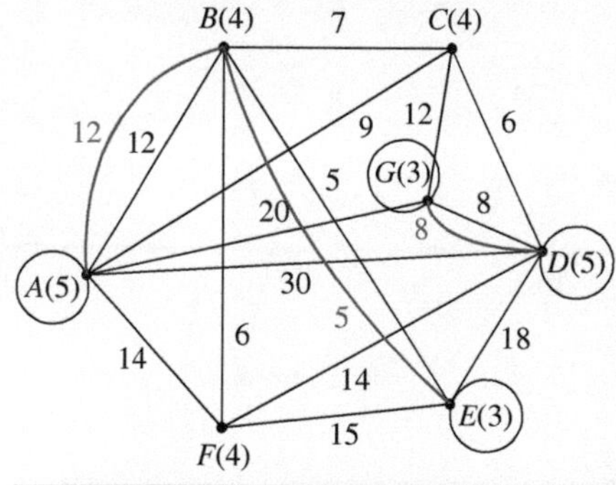

Pairing	Shortest route	Length	Total
AD	AC+CD	15	39
EG	EB+BC+CG	24	
AE	AB+BE	17	25
DG	DG	8	
AG	AG	20	38
DE	DE	18	

The best pairing is AE and DG

The extra arcs are AB, BE and DG

The total distance = 176 + 25 = 201

(sum of weights) + (weight of extra arcs).

A possible route, starting from A, is $ABCDEFABEBFDGCAGDA$

DISCRETE

The main difficulty with the route inspection algorithm is that the number of possible pairings increases very rapidly as the number of odd nodes increases.

2 odd nodes 1 pairing
4 odd nodes 3 pairings
6 odd nodes 15 pairings
8 odd nodes 105 pairings

For a network with n odd nodes, there are $(n-1)\times(n-3)\times(n-5)\times\ldots\times3\times1$ possible pairings.

You will not be expected to deal with more than 4 odd nodes unless there is extra information given to reduce the number of pairs you need to examine.

Exercise 12.5A Fluency and skills

Answer sheet available

1 This network has a total weight of 41

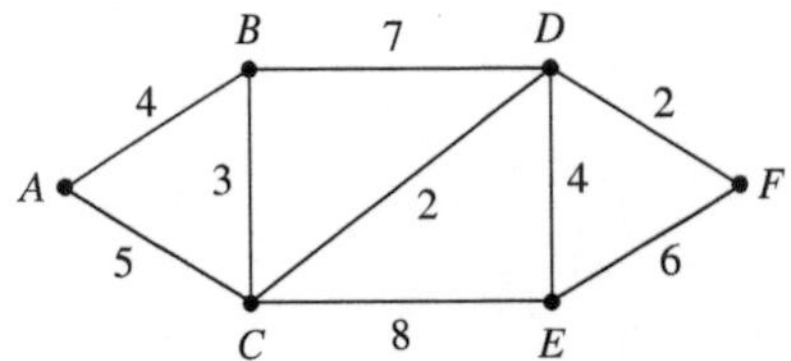

a Explain why a route inspection route starting and finishing at A will need to travel along some arcs more than once.

b Find the arcs that need to be repeated in the route.

c Find the total length of the route.

2 For the network shown,

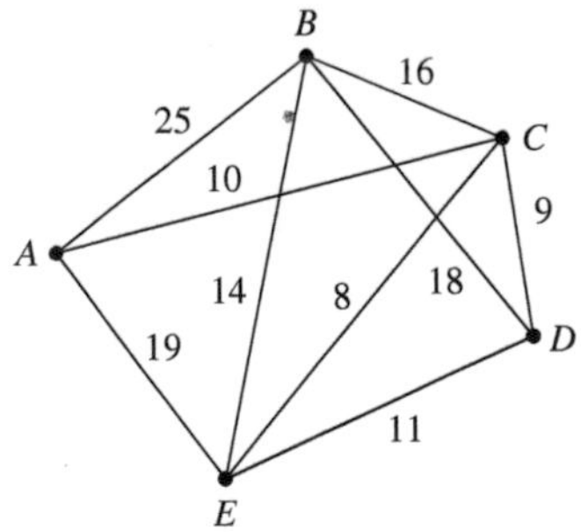

a Find which arcs which should be repeated to solve the route inspection problem,

b State the length of the route,

c State a possible route starting from node A

3 Solve the route inspection problem for each of the networks shown. In each case, show your working clearly, state which arcs will be repeated and give the total weight of the route.

a

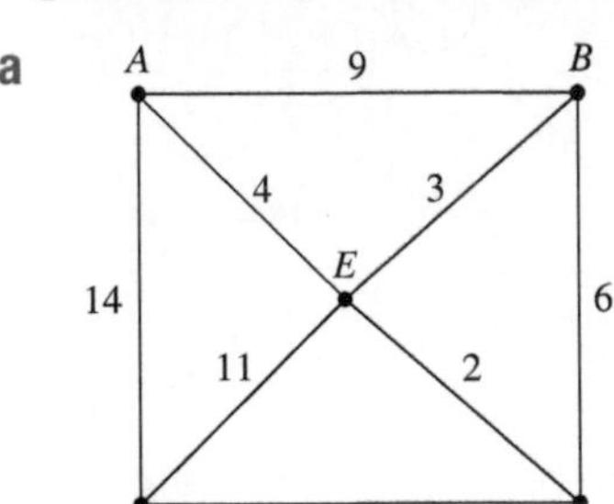

b

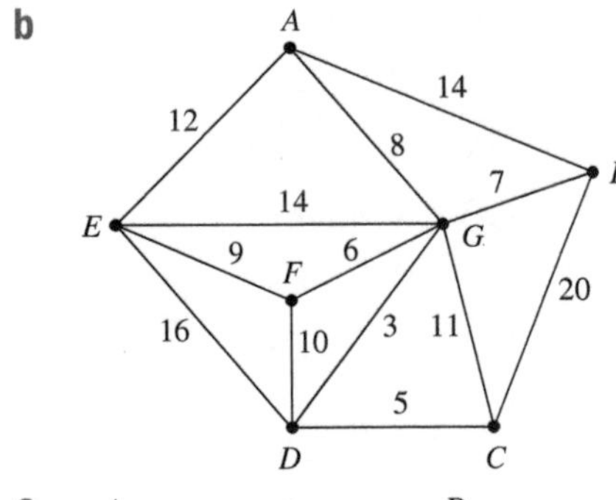

c

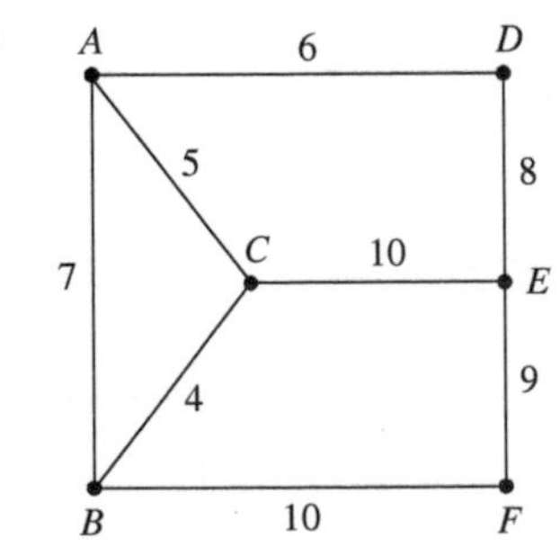

d

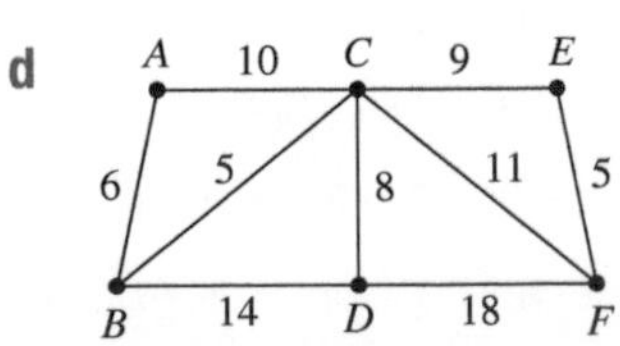

4 This network has a total weight of 82

a Show that there are two solutions to the route inspection problem. List the arcs that must be repeated in each case.

b Find the total length of the route.

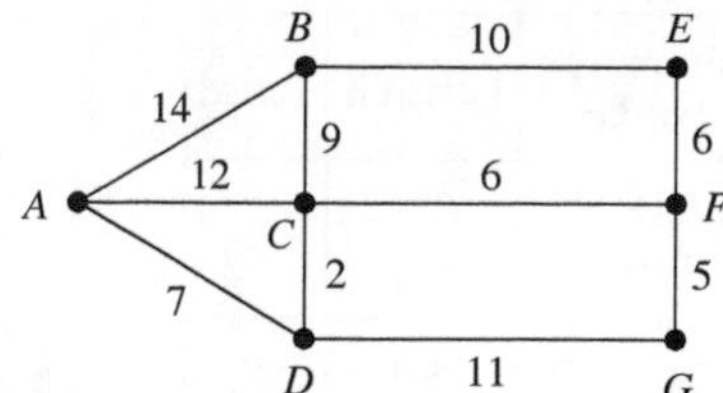

Reasoning and problem-solving

Strategy

To solve a route inspection problem

1. Identify nodes with odd degree.
2. Find the shortest routes between all pairs of odd nodes.
3. Find which arcs must be added to solve the problem.
4. Answer the question in context.

The inspection route may pass through a particular node several times. Each time, it uses one arc into and one arc out of the node, so, for example, it will pass 3 times through a node of degree 6

Key point

The inspection route passes $\frac{1}{2}n$ times through a node of degree n

Some problems may require different start and finish points. In this case, the network must be made semi-Eulerian.

Example 2

The diagram shows the times, in minutes, that a cleaning vehicle takes to sweep the paths in a park. The total time is 80 minutes. The vehicle is kept at A and must sweep every path and return to A

a Find the minimum time needed to complete the task.

b How many times will the vehicle pass through D?

A change of policy means that the vehicle can start and finish at different points.

c Find the minimum time now needed to complete the task and state the start and finish points.

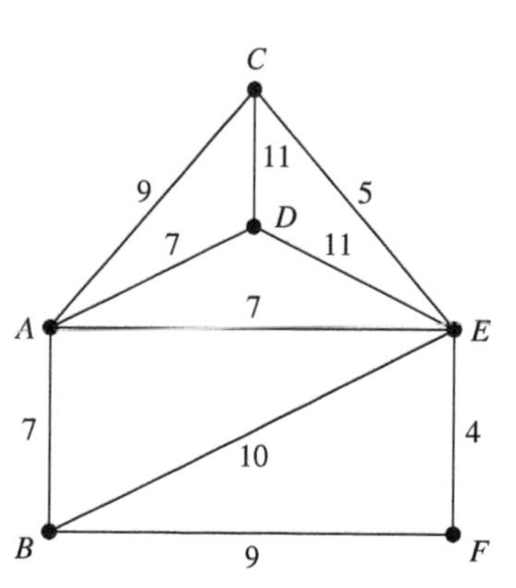

(*Continued on the next page*)

DISCRETE

a The odd nodes are B, C, D and E

Pairing	Shortest route	Length	Total
BC DE	$BE + EC$ DE	15 11	26
BD CE	$BA + AD$ CE	14 5	19
BE CD	BE CD	10 11	21

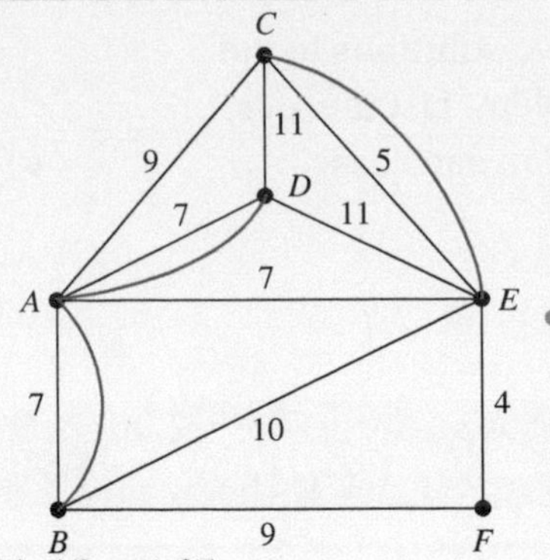

Best pairing is BD and CE, repeating BA, AD and CE

Time needed $= 80 + 19 = 99$ minutes.

b On the modified graph, D had degree 4

Route passes through D twice.

c Route will need to repeat one pair.

The shortest pair is $CE = 5$, so repeat CE

Time needed $= 80 + 5 = 85$ minutes

B and D are still odd nodes. The network is semi-Eulerian.

The route must start at B and finish at D, or vice versa.

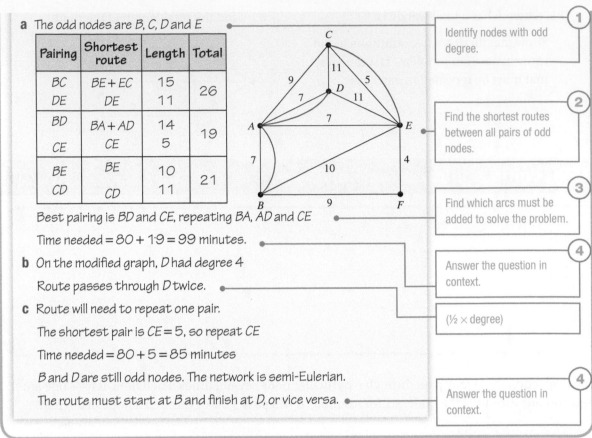

1 Identify nodes with odd degree.

2 Find the shortest routes between all pairs of odd nodes.

3 Find which arcs must be added to solve the problem.

4 Answer the question in context.

($\frac{1}{2} \times$ degree)

4 Answer the question in context.

Exercise 12.5B Reasoning and problem-solving

1 The diagram shows the map of the roads on an estate, with distances in metres. All junctions are right angles and all roads are straight apart from the crescent.

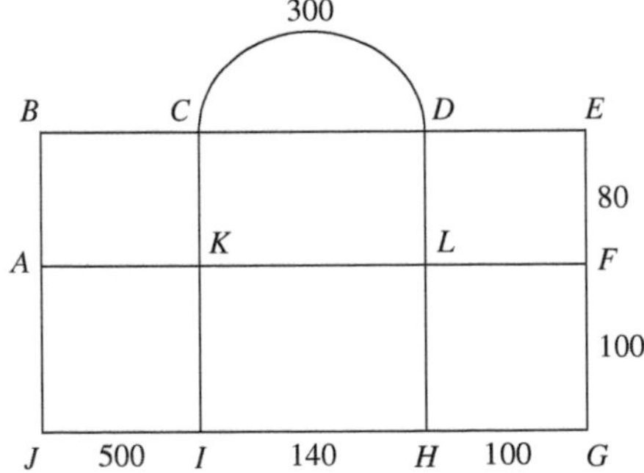

A team of workers is to paint a white line along the centre of each road.

a Assuming that the team enter and leave the estate at A and wish to travel the least possible distance on the estate, which sections of road will they will need to travel along twice? Find the distance they will travel.

b Instead, the team enters the estate at H and leaves at I to go on to another job. How will this affect your results from part **a**?

c What would be the best start and end points for the team if they wish to travel the least possible distance? How far would they go in this case?

2 The table shows the direct road links between six towns, with distances in kilometres.

	A	B	C	D	E	F
A	–	15	–	8	20	–
B	15	–	10	–	6	5
C	–	10	–	9	–	14
D	8	–	9	–	9	4
E	20	6	–	9	–	–
F	–	5	14	4	–	–

Abigail is to do a sponsored cycle ride, travelling at least once along each of these roads and starting and finishing at her home in town A

a Draw the network diagram corresponding to this table.

b Find which roads she should repeat to minimise her journey. State how long her journey will take.

3 The diagram shows the roads connecting five towns, labelled with the distances in kilometres. The total of the distances is 86 km.

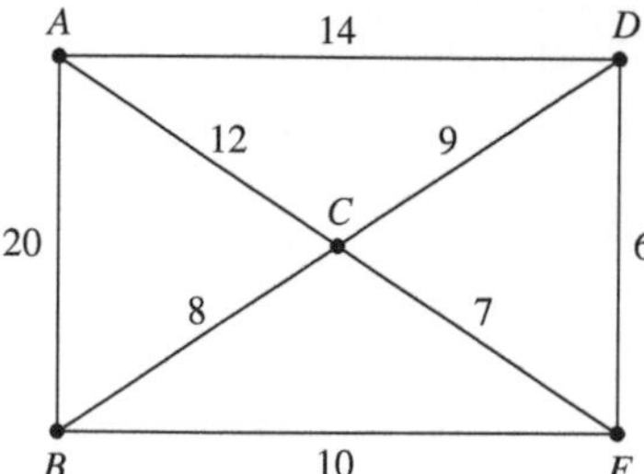

A vehicle starts from A to inspect all the roads for potholes. The vehicle must return to A when the inspection is complete.

a If the vehicle must travel both ways along each road to inspect both carriageways, explain why it can do the job in a minimum distance of 172 km.

b If the vehicle only needs to travel once along a road to inspect it, find the minimum distance it needs to travel, and state which roads must be used twice.

4 The diagram shows the paths in a nature reserve, with lengths in metres. The total length of the paths is 3350 m.

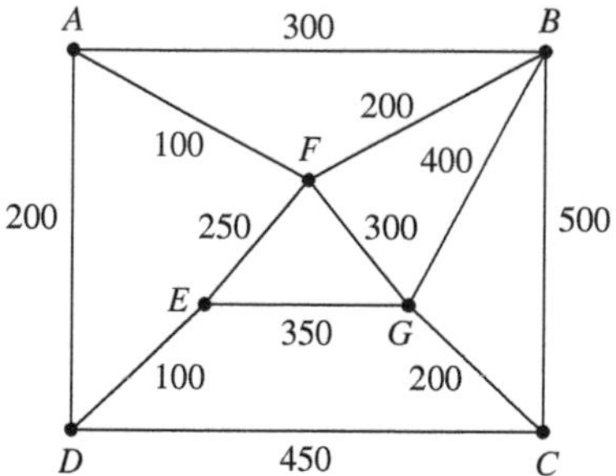

Barry lost a ring during a visit. He went back the next day to look for it.

a What is the minimum distance he needs to walk to search every path, starting and ending at A? Which paths will he travel along twice?

b Barry is sure he didn't go along BG Explain why, nevertheless, he might choose to walk along it.

c Barry's friend agrees to drop him at A and pick him up at another point. Where will he be picked up and how far might he need to walk if he wants to check every path and keep the walking distance to a minimum?

5 The diagram shows the layout of paths in a small pedestrian shopping precinct. Distances are in metres.

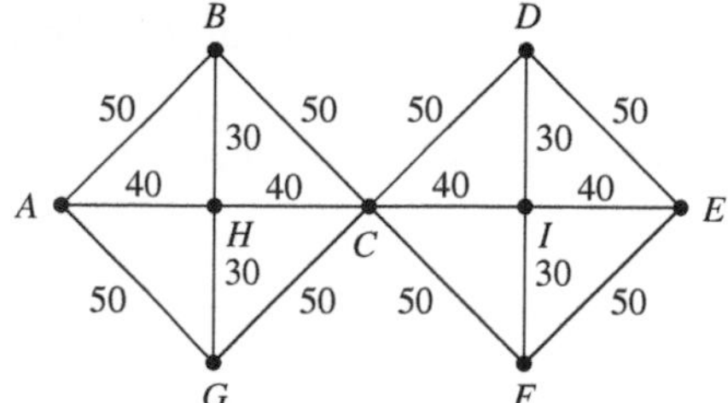

A cleaning cart must drive around the precinct, starting and ending at A

a Find the sections which should be travelled twice to minimise the distance travelled. (Hint: Make use of symmetry.)

b Find the total distance travelled.

c How many times will the cart pass through C?

6 The diagram shows a logo displayed outside a shop. The lengths shown are in cm and the logo is symmetrical, with all angles right angles.

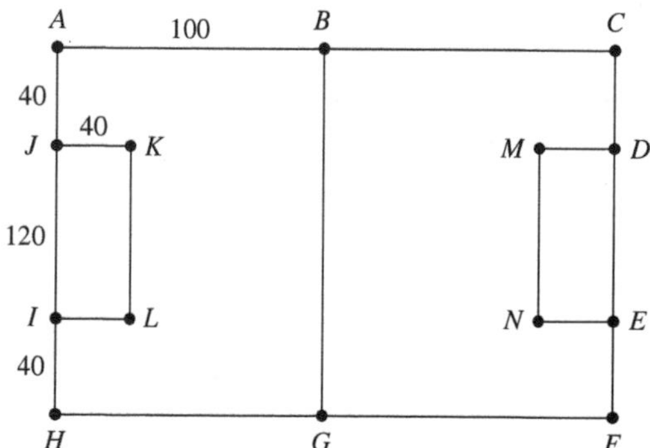

a The logo is outlined with a single rope of LEDs, starting and ending at A

Calculate the minimum length of rope needed for this, and state which sections of the logo will have a double run of rope.

b If the rope could start and end at different points, how much rope would be needed? State the start and end points.

DISCRETE

12.6 The travelling salesman problem

Fluency and skills

A travelling sales person needs to make a number of visits and return to base as efficiently a possible. The travelling salesman problem is named after this situation and can be stated as:

The travelling salesman problem (TSP) Key point

Find a route that visits every node of a network and returns to the start in the shortest possible distance.

A Hamiltonian cycle (tour) is a closed path that visits every node of a network. The **classical TSP** is to visit every node of a network *once only* and return to the start in the least possible distance: that is, to find a minimum weight Hamiltonian cycle.

However, in a practical situation, the problem may not fit this classical pattern. Firstly, there may not be a Hamiltonian cycle. For example, any complete tour of this graph would need to visit B twice. The graph is not Hamiltonian.

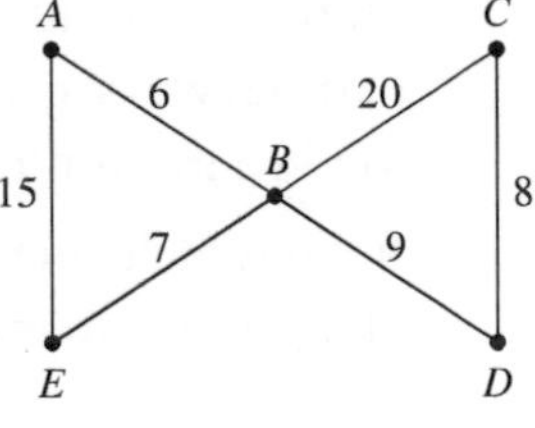

Secondly, the network may not satisfy the triangle inequality, especially if the weights are, for example, travel times.

This network has a Hamiltonian cycle $ABCA$ with total weight 36, but the best practical solution to the problem is $ACBCA$ with total weight 32

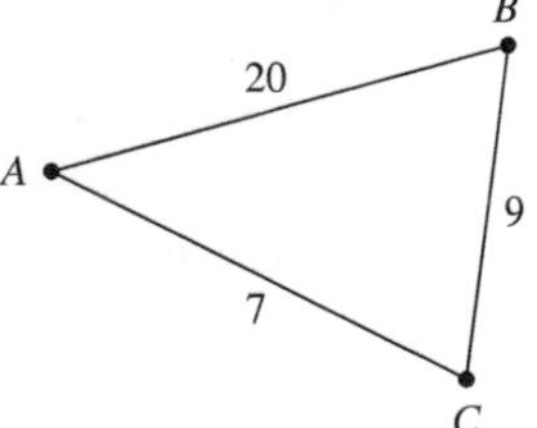

To solve a practical TSP with n nodes, you convert it to an equivalent classical TSP.

Solving a practical TSP with n nodes Key point

- Create a complete network, K_n, of shortest distances between pairs of nodes.
- Find a minimum weight Hamiltonian cycle for the complete network (this is the classical problem).
- Interpret the solution in the practical situation.

Example 1

Convert this network to a complete network of shortest distances. Find, by inspection, the solution to the classical TSP for the complete network and hence find a solution to the practical problem.

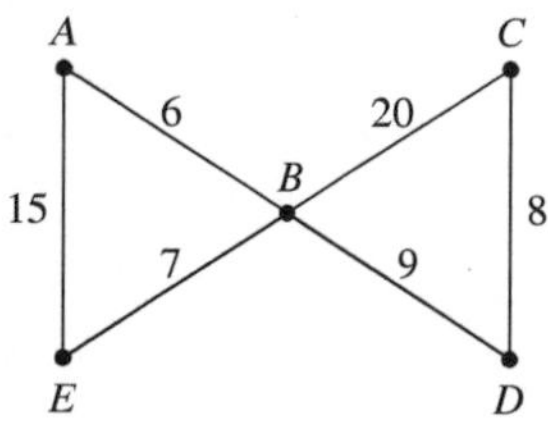

The diagram shows K_5

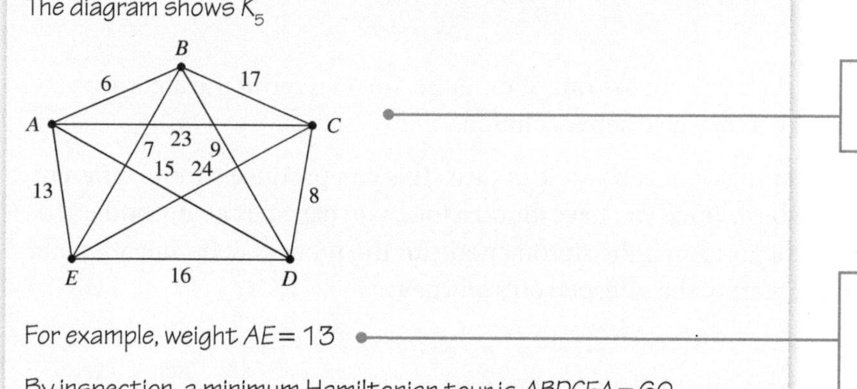

There are 5 nodes on the graph.

For example, weight $AE = 13$

$ABE = 13$ is the shortest route on the original network.

By inspection, a minimum Hamiltonian tour is $ABDCEA = 60$

In the original network:

shortest distance $CE = CDBE = 24$

shortest distance $EA = EBA = 13$

The solution to the practical problem is $ABDCDBEBA = 60$

The only known sure-fire way to find the best tour is to check every Hamiltonian cycle.

From a chosen starting node in K_n, there are $(n - 1)$ possible arcs to travel, then $(n - 2)$ second arcs and so on. Each tour gets listed twice (forwards and backward) so

Key point

A network with n nodes has $\frac{1}{2}(n-1)!$ possible tours.

This method has exponential order, so even if the network is quite small you have to check a huge number of cycles. Because of this, you need a way to:

- Find a good (not necessarily optimal) solution in a reasonable length of time.
- Decide whether the solution you have found is good enough.

An algorithm which finds a good, but perhaps not perfect solution is called a **heuristic algorithm**. For the TSP a simple heuristic method is the **nearest neighbour algorithm**, which is another example of a greedy algorithm.

Key point

The nearest neighbour algorithm

Step 1 Choose a starting node V

Step 2 From your current position, choose the arc with minimum weight leading to an unvisited node. Travel to that node.

Step 3 If there are unvisited nodes, go to **Step 2**

Step 4 Travel back to V

If at Step 2 there are equal arcs, then you choose one at random. You can rerun the algorithm making the other choice(s) to see if they give better solutions.

You can repeat the algorithm starting from each node in turn. This can produce several different tours from which you select the best. Once you have found a tour, you can start at any node. The 'starting node' referred to in the algorithm is the starting node for the process of finding the tour. It does not have to be the actual start of the salesperson's journey.

Example 2

This network shows the distances, in km, between five towns. Use the nearest neighbour algorithm, starting from each node in turn, to find a route for a travelling salesman, who is based at A and needs to visit every town.

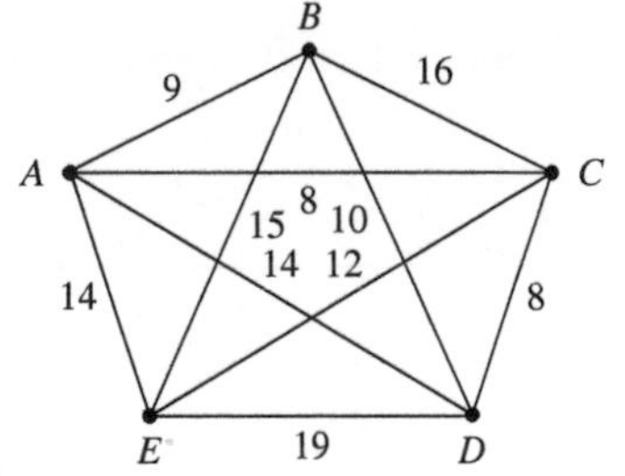

Starting from A, 'nearest' node is C, weight $AC = 8$

From C, 'nearest' unvisited node is D, $CD = 8$

From D, 'nearest' unvisited node is B, $DB = 10$

From B, 'nearest' unvisited node is E, $BE = 15$

No more unvisited nodes, so travel $EA = 14$

The total weight for $ACDBEA$ is $8 + 8 + 10 + 15 + 14 = 55$

Repeat starting from the other nodes:

Start node	Tour(s)	Total weight
A	ACDBEA	55
B	BACDEB	59
C	CABDEC	58
	CDBAEC	53
D	DCABED	59
E	ECABDE	58
	ECDBAE	53

There are, in fact, four different tours here.

For example, $BACDEB$ and $DCABED$ are the same tour.

For a salesman based at A, the best tour found is $ABDCEA = 53$ km.

For a practical situation, you convert the problem to the classical TSP and then use the nearest neighbour algorithm.

Example 3

The diagram shows the travel times, in minutes, between six locations.

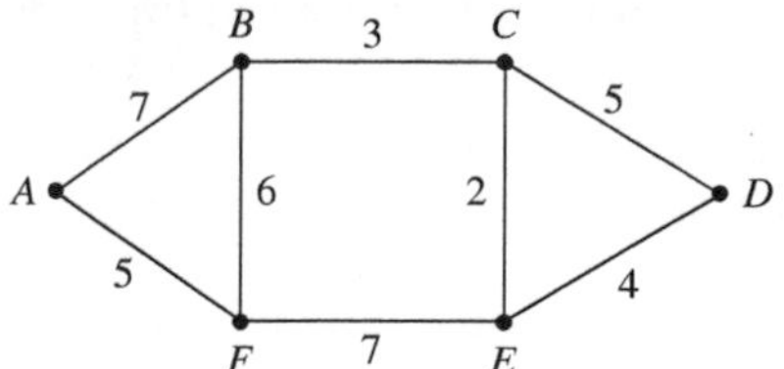

a Draw up a table of shortest travel times.

b Use the nearest neighbour algorithm to find a solution to the travelling salesman problem. List the route the salesperson will take.

c Show, by inspection, that the solution found in **b** is not optimal.

a These are the minimum travel times, for example A to D is $ABCD = 15$

	A	B	C	D	E	F
A	–	7	10	15	12	5
B	7	–	3	8	5	6
C	10	3	–	5	2	9
D	15	8	5	–	4	11
E	12	5	2	4	–	7
F	5	6	9	11	7	–

b Starting from A, 'nearest' node is F, $AF = 5$

From F, 'nearest' unvisited node is B, $FB = 6$

From B, 'nearest' unvisited node is C, $BC = 3$

From C, 'nearest' unvisited node is E, $CE = 2$

From E, 'nearest' unvisited node is D, $ED = 4$

All nodes visited, so return to A, $DA = 15$

The route is $AFBCEDA$, with total weight $= 35$

Repeat starting from the other nodes.

Start node	Tour(s)	Total weight
A	AFBCEDA	35
B	BCEDFAB	32
C	CEDBFAC	35
D	DECBFAD	35
E	ECBFADE	35
F	FABCEDF	32

The two 32 minute tours are the same tour. With A as the start node, these give $ABCEDFA$

The salesperson travels $ABCEDEFA$, because the shortest route from D to F goes through E

c On the original network, the tour $ABCDEFA = 31$, so the solution found in **b** is not optimal.

DISCRETE

Because there is no efficient algorithm for finding the length, T, of the optimal solution, you need a way of knowing what a 'good' solution is like. To decide if a solution you have found is good enough, you find two values, an **upper bound** and a **lower bound**, such that

> **Key point**
>
> lower bound $\leq T \leq$ upper bound.

The closer together the two bounds are, the more accurately you know the length of the optimal tour. You therefore try to find the smallest upper bound and the largest lower bound that you can.

If you have found a tour, it is either optimal or longer. It follows that:

> **Key point**
>
> The length of any known tour is an upper bound.

Using the nearest neighbour algorithm to find a tour automatically provides an upper bound, but not necessarily a good one. As an alternative method, you use a minimum spanning tree for the network. You could visit every node by travelling once in each direction along every arc of the minimum spanning tree, so

> **Key point**
>
> ($2 \times$ length of minimum spanning tree) is an upper bound.

This is not usually a good upper bound, but you can improve it by using short cuts.

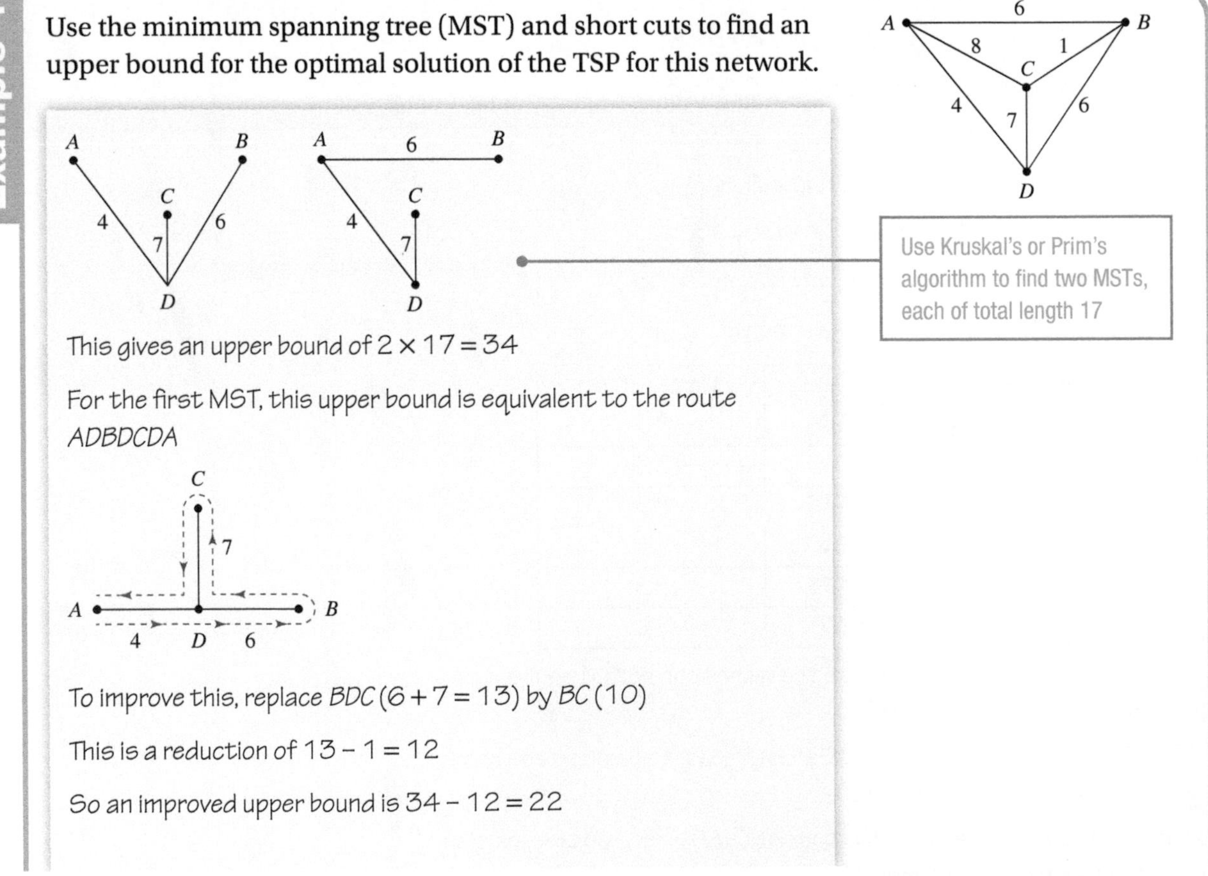

Example 4
Use the minimum spanning tree (MST) and short cuts to find an upper bound for the optimal solution of the TSP for this network.
A
B
C
D
6
8
1
4
7
6
Use Kruskal's or Prim's algorithm to find two MSTs, each of total length 17
This gives an upper bound of 2 × 17 = 34
For the first MST, this upper bound is equivalent to the route ADBDCDA
To improve this, replace BDC (6 + 7 = 13) by BC (10)
This is a reduction of 13 − 1 = 12
So an improved upper bound is 34 − 12 = 22

(*Continued on the next page*)

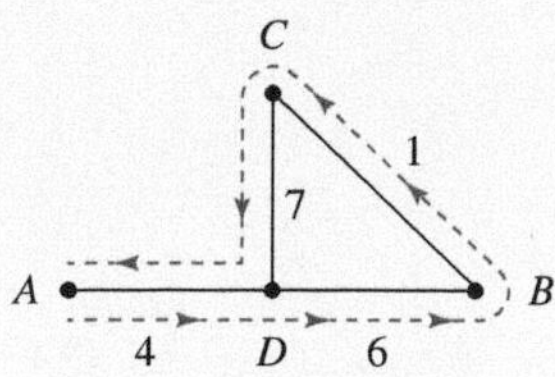

To improve this, replace CDA $(7+4=11)$ by CA (8)

This is a reduction of $11-8=3$

So an improved upper bound is $22-3=19$

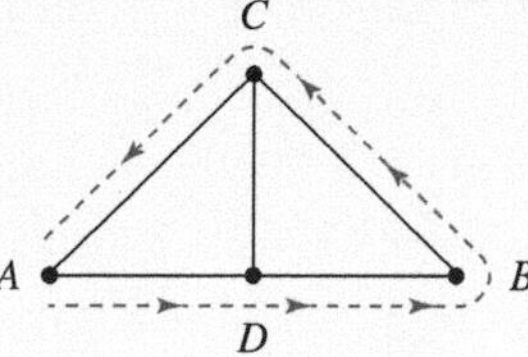

This is a tour, $ADBCA$, so there are no more short cuts.

The best upper bound using this MST is 19

For the second MST, the initial upper bound of 34 is equivalent to the route $BADCDAB$

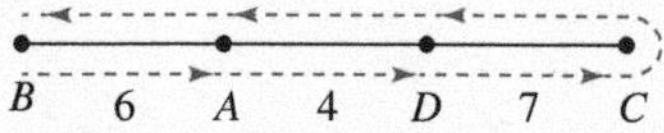

To improve this, return directly from C to B, replacing $CDAB$ $(7+4+6=17)$ by $CB(10)$

This is a reduction of $17-1=16$

The improved upper bound is $34-16=18$

This is a tour $BADCB$, so there are no more short cuts.

The best upper bound using this MST is 18

Of the two values found, the best upper bound $=18$

You always want the upper bound to be as small as possible.

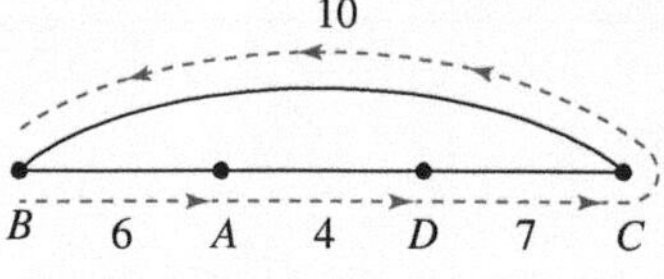

In addition to an upper bound, you need to find a lower bound. Suppose you have the network shown.

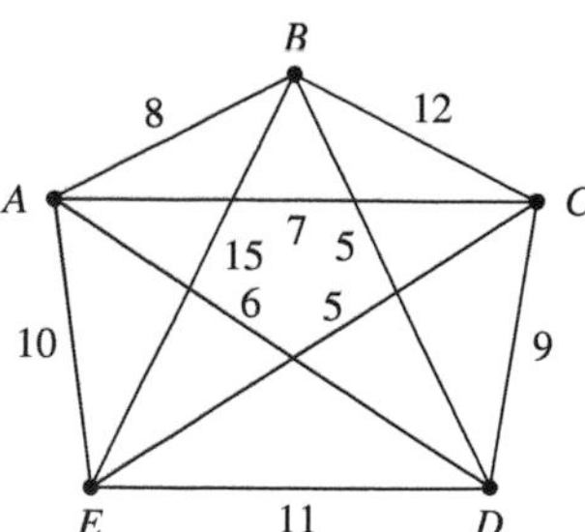

The optimal tour enters and leaves A along two of AB, AC, AD and AE

You separate these arcs from the rest of the network. The two arcs involved in the tour must total at least $6+7=13$

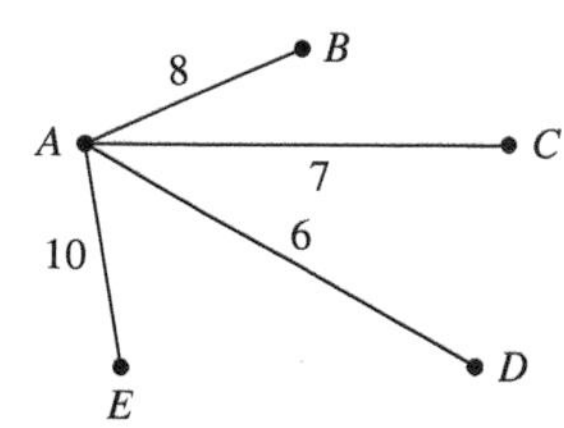

DISCRETE

The remaining three arcs of the tour join B, C, D and E

They form a spanning tree for the subgraph shown.

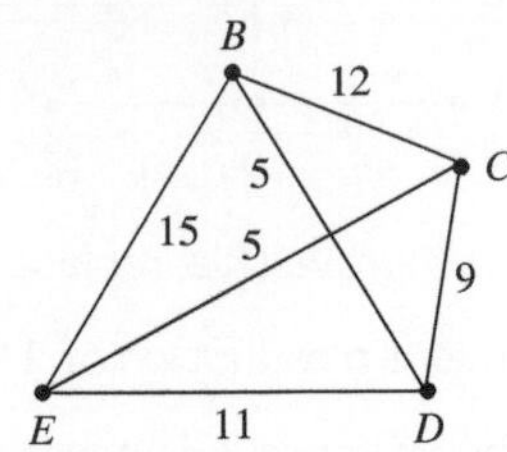

The minimum spanning tree for this subgraph is $5+5+9=19$, so the three arcs in the tour must have a total weight of at least 19

So the complete tour has a total weight of at least $13+19=32$

This is a lower bound for the tour.

To get the best lower bound available, you could repeat this process, removing each node in turn. Removing B gives a lower bound of 31, removing C gives 33, removing D gives 31 and removing E gives 33 (you should check that you can find these values). You want the lower bound to be as large as possible, so the best lower bound is 33

Key point

To find a lower bound

Step 1 Choose a node V

Step 2 Identify the two lowest weights, p and q, of the arcs connected to V

Step 3 Remove V and its connecting arcs from the network. Find the total weight, m, of the minimum spanning tree of the remaining subgraph.

Step 4 Calculate lower bound $=p+q+m$

Step 5 If possible, choose another node V and go to Step 2

Choose the largest result obtained as the best lower bound.

Example 5

For the network shown

	A	*B*	*C*	*D*	*E*
A	–	12	6	7	2
B	12	–	9	7	10
C	6	9	–	4	8
D	7	7	4	–	5
E	2	10	8	5	–

a Find a lower bound for the optimal tour by deleting node A

b Find a second lower bound by deleting node B

c State which of your values is the better lower bound,

d Given that there is an upper bound of 32, write an inequality for T, the optimal tour.

a The shortest arcs from A are $AC=6$ and $AE=2$

	1	3	2	4
	B	C	D	E
B	–	9	7	10
C	9	–	(4)	8
D	(7)	4	–	5
E	10	8	(5)	–

The table shows the graph with A removed.

(*Continued on the next page*)

The minimum spanning tree is $BD=7$, $CD=4$ and $DE=5$

A lower bound is $(6+2)+(7+4+5)=24$

Use Prim's algorithm.

b The shortest arcs from B are $BC=9$ and $BD=7$

	1	4	3	2
	A	C	D	E
~~A~~	–	6	7	2
~~C~~	6	–	(4)	8
D	7	4	–	(5)
~~E~~	(2)	8	5	

The table shows the graph with B removed.

The minimum spanning tree is $AE=2$, $CD=4$ and $DE=5$

A lower bound is $(9+7)+(2+4+5)=27$

c $27>24$, so the better lower bound is 27

d T lies between the lower and upper bounds, so $27 \le T \le 32$

The arcs of the better lower bound, BC, BD, AE, CD and DE, do not form a tour. If they did, you would know that $T=27$

Key point

If you find a lower bound which forms a tour, then that tour is the optimal solution. If none of the possible lower bounds form a tour, the optimal tour must be greater than the best lower bound.

Exercise 12.6A Fluency and skills

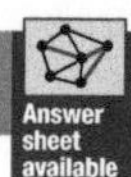

1 **a** Draw up a distance matrix for the complete network of shortest distances corresponding to the network shown.

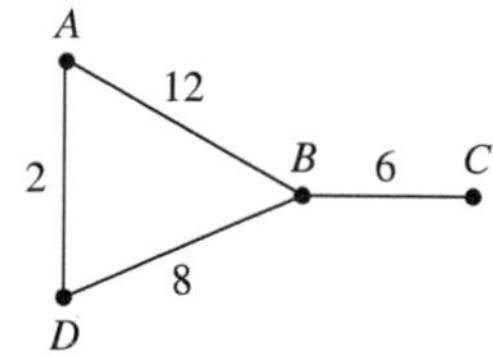

b Show that the nearest neighbour algorithm with the four possible starting nodes leads to two distinct Hamiltonian tours, each with the same total weight.

c Starting from A, list the order in which the nodes would be visited on the original network.

2 **a** Use the nearest neighbour algorithm with A as the starting node to obtain a Hamiltonian tour of the network shown.

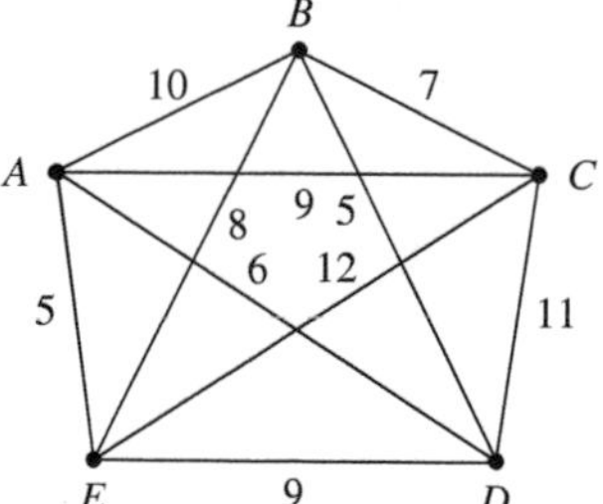

State the length of the tour.

b Show that by using B as the starting node you obtain a shorter tour.

DISCRETE

3 The network shown represents the travelling times, in minutes, between four towns.

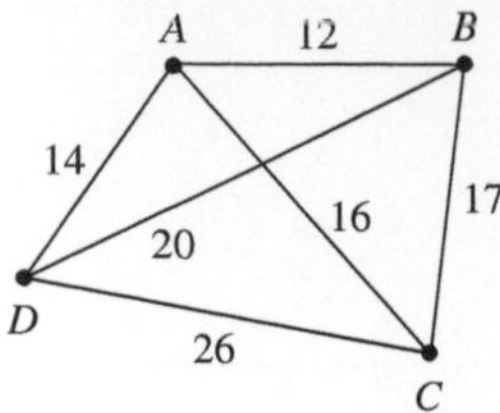

a Find the minimum spanning tree for the network and hence state an upper bound for the optimal solution to the travelling salesman problem for the network.

b By introducing two short cuts, obtain an improved upper bound.

4 The network shows the distances, in km, between six towns. A van based at A needs to deliver to the other five towns and return to A

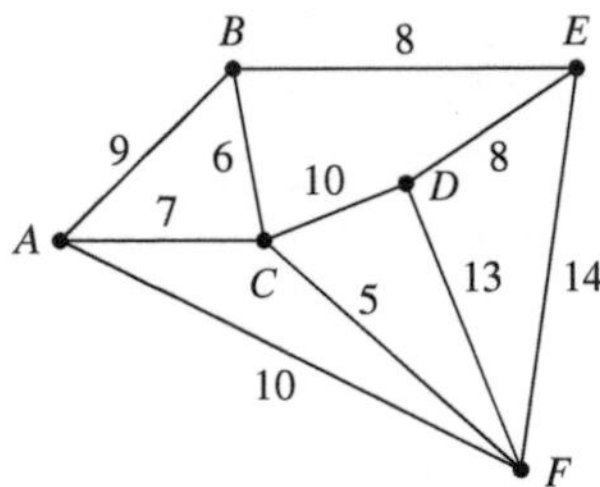

a By finding the minimum spanning tree for the network, obtain an upper bound for the total distance that the van will need to travel.

b Use short cuts to show that the optimal tour is at most 50 km.

5 a Find the two minimum spanning trees for the network shown.

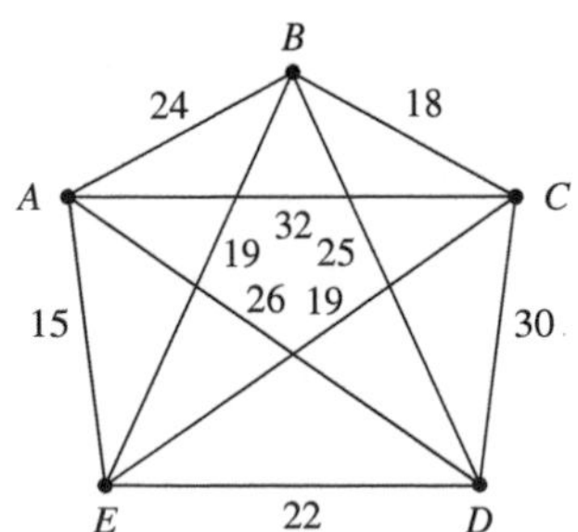

b Using the total weight of the minimum spanning tree, state an upper bound for the optimal solution to the TSP.

c Using short cuts, obtain the best upper bound you can using your minimum spanning trees.

6 a Use Prim's algorithm to find the minimum spanning tree for this network.

	A	B	C	D	E	F
A	–	9	4	7	11	8
B	9	–	12	6	8	14
C	4	12	–	6	10	7
D	7	6	6	–	5	8
E	11	8	10	5	–	13
F	8	14	7	8	13	–

b State an upper bound for the solution to the TSP for this network.

c Using short cuts, obtain an upper bound less than 45

7 a By deleting node A from the network shown, obtain a lower bound for the optimal tour. Explain, by reference to the network, why a tour of this length is not possible.

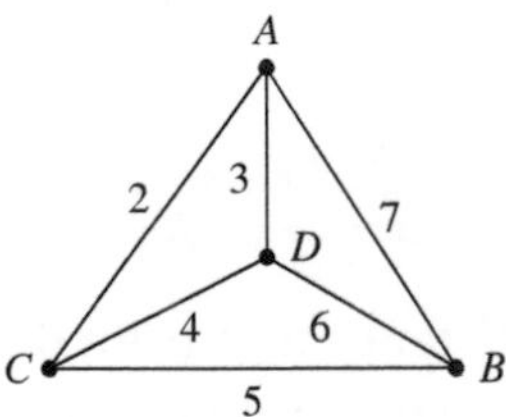

b By deleting node B, obtain another lower bound for the optimal tour and explain why this is, in fact, the length of the optimal tour.

8 By deleting each node in turn from the network shown, obtain the best lower bound available for the solution to the travelling salesman problem.

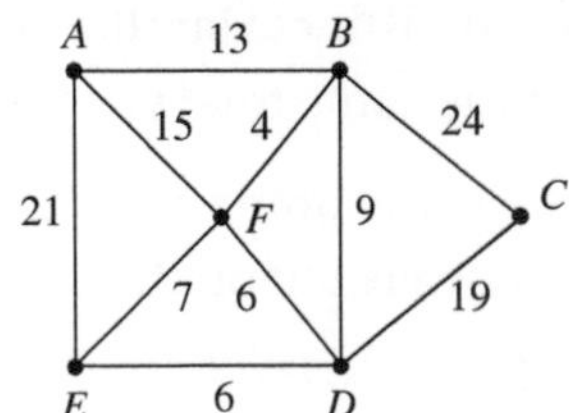

9 The table shows the distances, in km, between five locations.

	A	B	C	D	E
A	–	14	22	18	12
B	14	–	15	32	25
C	22	15	–	20	13
D	18	32	20	–	28
E	12	25	13	28	–

By deleting each node in turn, find the best lower bound for the solution to the travelling salesman problem for this network.

Reasoning and problem-solving

Strategy

To solve travelling salesman problems

1. Convert a practical problem to the classical problem by constructing a complete network of shortest distances.
2. Find possible tours using the nearest neighbour algorithm. Any tour is an upper bound for the optimal tour.
3. Find upper bounds using short cuts on the minimum spanning tree. The smallest upper bound is the best.
4. Find lower bounds by deleting a node and using the MST of the remainder. The largest lower bound is the best.
5. Answer the question in context.

Example 6

The network shows the cost of bus travel, in £, between four towns.

A tourist staying in town C wants to visit all the other towns and return to C

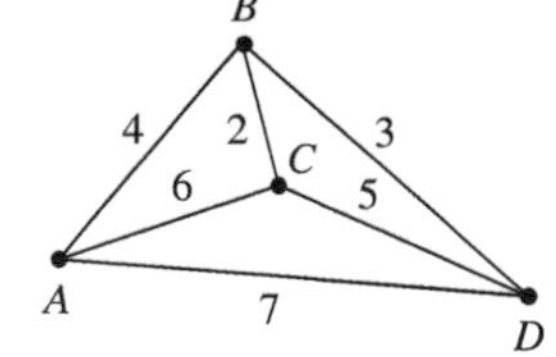

a Use the nearest neighbour algorithm starting from A to find an upper bound for the cost, T, of the trip. State the route that the tourist would take using this solution.

b By deleting first A and then B, find two possible lower bounds for the trip.

c State the best inequality you have for T

a The cheapest routes are shown in this table:

	A	B	C	D
A	–	4	6	7
B	4	–	2	3
C	6	2	–	5
D	7	3	5	–

For example the cheapest from A to D is $ABD = 7$

From A, the nearest neighbour route is $ABCDA = 4 + 2 + 5 + 7$
$= 18$

An upper bound for T is £18. The route from C is $CBDBABC$

(*Continued on the next page*)

DISCRETE

b The two lowest weight arcs from A are $AB(4)$ and $AC(6)$

The minimum spanning tree of BCD is $BC(2)$ and $BD(3)$

This gives, lower bound $= 4 + 6 + 2 + 3 = 15$

The two lowest weight arcs from B are $BC(2)$ and $BD(3)$

The minimum spanning tree of ACD is $AC(6)$ and $CD(5)$

This gives, lower bound $= 2 + 3 + 6 + 5 = 16$

c The better of the two lower bounds is £16

So the best available inequality for T is $16 \le T \le 18$

Always look for the largest lower bound.

In fact, 16 is the largest of the four possible lower bounds and the four arcs involved do not form a tour, so you can say that $16 < T \le 18$

Exercise 12.6B Reasoning and problem-solving

1 A guide is taking a party of tourists from a hotel (H) to four attractions in a city – the museum (M), the art gallery (A), the cathedral (C) and the Guild Hall (G). She estimates the walking times (in minutes) between the various locations as shown in this network.

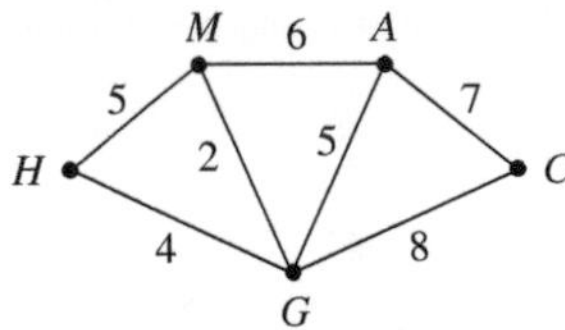

a Draw a complete network of shortest journey times.

b Use the nearest neighbour algorithm starting from A to plan a route for the party. State the total walking time and the order in which their route from the hotel would pass through the nodes of the original network.

c Show, by inspection, that the route found in **b** is not optimal.

2 The table shows the direct distances (in miles) between four locations. There are no other direct links between them.

	A	B	C	D
A	–	5		6
B	5	–		7
C			–	3
D	6	7	3	–

a Fill in the blank cells in the table with the shortest indirect routes available.

b Use the nearest neighbour algorithm with A as the starting node to find a Hamiltonian tour of your completed network. State the total length of the tour and list the order in which it would actually pass through the locations.

3 An orchestra and chorus is available as a whole unit (option A) or as four subgroups (B, C, D and E) offering smaller scale performances. A six-day music festival wants to start and finish with option A, but to have the other four options on the remaining four days. The changeover costs (in £) between the options vary, as shown in this table.

	A	B	C	D	E
A	–	250	300	150	400
B	250	–	120	360	220
C	300	120	–	260	170
D	150	360	260	–	290
E	400	220	170	290	–

a By deleting A, find a lower bound for the overall cost.

b Use the minimum spanning tree and a single short cut to find an upper bound for the cost.

c If the optimal cost is C, use your results to write inequalities satisfied by C

4 The diagram shows the locations of five nesting boxes in a bird reserve. The values shown are the lengths, in metres, of the various sections of path.

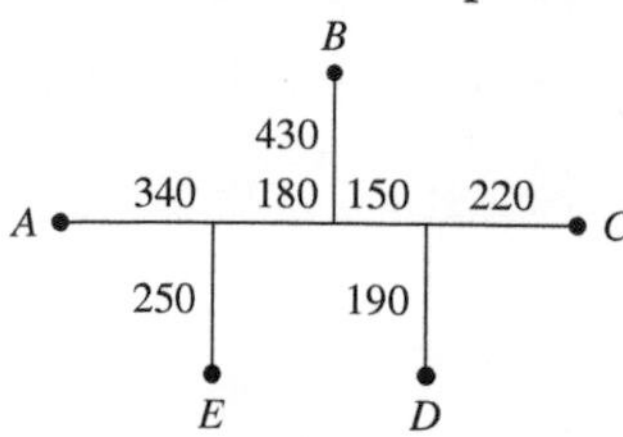

A volunteer wants to visit every nest box, starting and finishing at A

a Draw a complete network of shortest distances.

b The volunteer visits the nest boxes in alphabetical order. By deleting node A, find a lower bound for the TSP for this network and hence show that she could save no more than 120 m by finding the optimal route.

5 The table shows the distances, in km, between seven locations. A distributor based at A needs to deliver bundles of newspapers to shops in each of the other locations.

	A	B	C	D	E	F	G
A	–	5	6	7	4	9	8
B	5	–	4	5	4	8	6
C	6	4	–	9	7	8	9
D	7	5	9	–	9	7	9
E	4	4	7	9	–	6	8
F	9	8	8	7	6	–	9
G	8	6	9	9	8	9	–

a Use Prim's algorithm to find the minimum spanning tree for this network. Hence state an upper bound for the distance the distributor will need to travel.

b Using short cuts, reduce your upper bound to below 50 km.

c By deleting node A, obtain a lower bound for the distributor's journey.

d Using this and your result from part **b**, write inequalities satisfied by his optimal route.

6 The network shows the distances, in miles, between five locations.

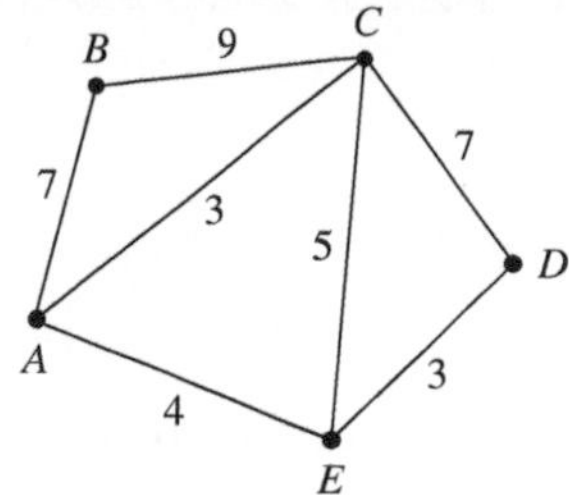

a Draw a complete network showing the shortest distances between the five locations.

b Use the nearest neighbour algorithm starting from A to obtain a possible solution to the travelling salesman problem. State the length of the route and the order in which the nodes would be visited on the original network.

7 A knight's move in chess consists of moving two squares parallel to one side of the board and then one square at right angles to this.

Belinda and Manuel play a game in which one of them places counters on six squares of the board and the other must make a 'knight's tour,' starting and ending on one of the marked squares and capturing' each of the other counters en-route, making as few moves as possible. The diagram shows the positions of the counters that Belinda has set up.

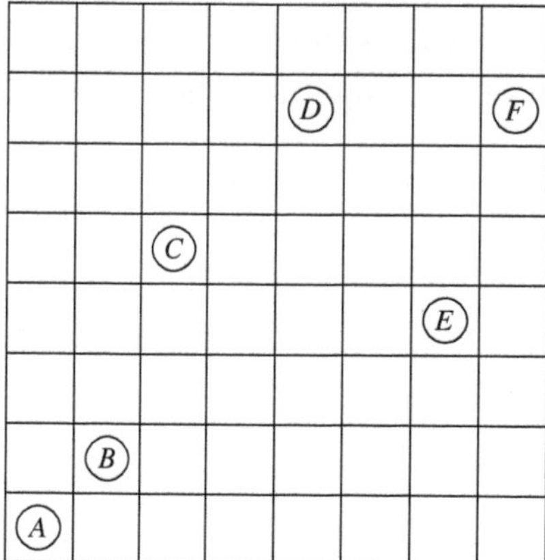

a Draw up a table to show the numbers of moves required to travel between each of these positions.

b Find upper and lower bounds for the number of moves that Manuel will have to make.

12.7 Network flows 1

Fluency and skills

A directed network can represent the routes along which a commodity flows. The commodity could be electricity, oil, freight, data or many other things. The weight of an arc is its **capacity**, which is the maximum possible flow along that arc. The network is sometimes called a **capacitated network**. Here is a simple example.

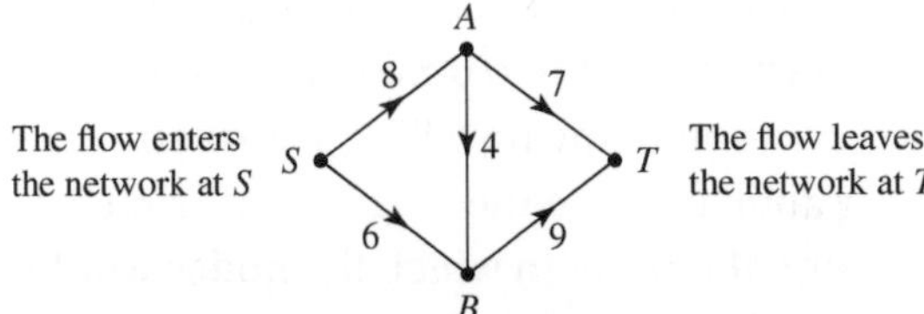

At *S*, all the arcs are directed away from the node. A node like this is called a **source.** It is usual to label a source *S*

At *T*, all the arcs are directed towards the node. A node like this is called a **sink**. It is usual to label a sink *T*

When the commodity flows through the network, there is a non-negative number - the **flow in the arc** - assigned to each arc. Together these form the **flow in the network**. The flow must satisfy certain conditions.

Firstly, the capacity of an arc is the maximum possible flow in that arc.

> **Key point**
> The **feasibility condition** states that the flow in an arc cannot be greater than its capacity.

Secondly, the commodity cannot be 'stored' at a node, so you have the following condition.

> **Key point**
> The **conservation condition** states that, at every node apart from *S* and *T*, total inflow = total outflow.

It follows from this second condition that:

> **Key point**
> Total outflow from *S* = total inflow to *T*

This total outflow/inflow is called the **value of the flow.**

A feasible flow for the network above is:

At *A* inflow = outflow = 7

At *B* inflow = outflow = 9

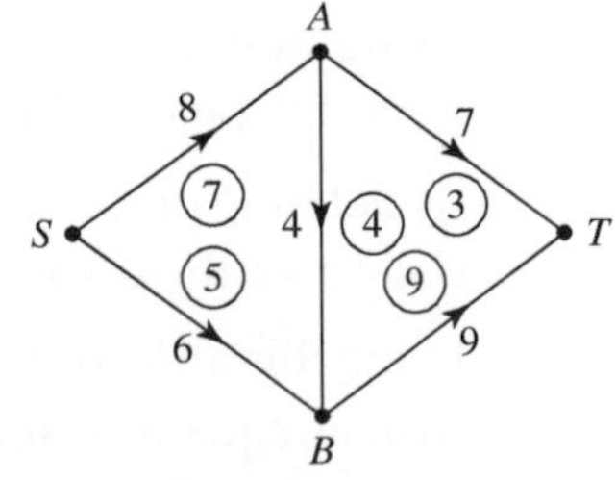

Outflow at S = inflow at T = 12

The value of the flow = 12

You circle the actual flows in the arcs. In the diagram at the bottom of the previous page, in SA, for example, the flow is 7 and the capacity (maximum possible flow) is 8. Notice that in AB and BT you have flow = capacity. These arcs are **saturated**. The other arcs are **unsaturated**.

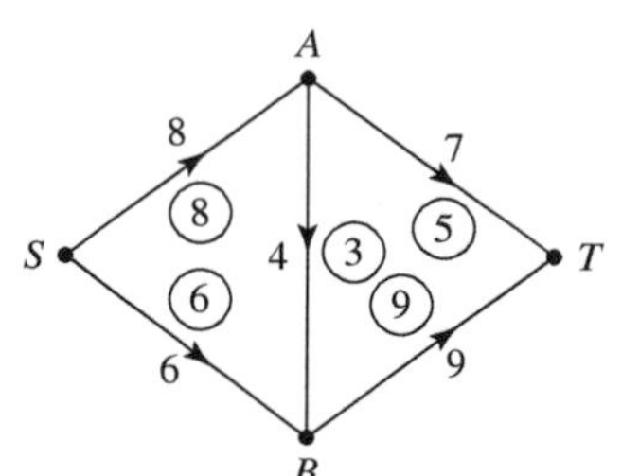

You usually need to find the maximum flow through the network. The network shown has a maximum flow of 14

The diagram on the right is one way in which this flow could take place.

Example 1

The diagram shows a capacitated network. The circled values represent a feasible flow through the network.

a Find the values of x and y

b State the value of the flow.

c State which arcs are saturated.

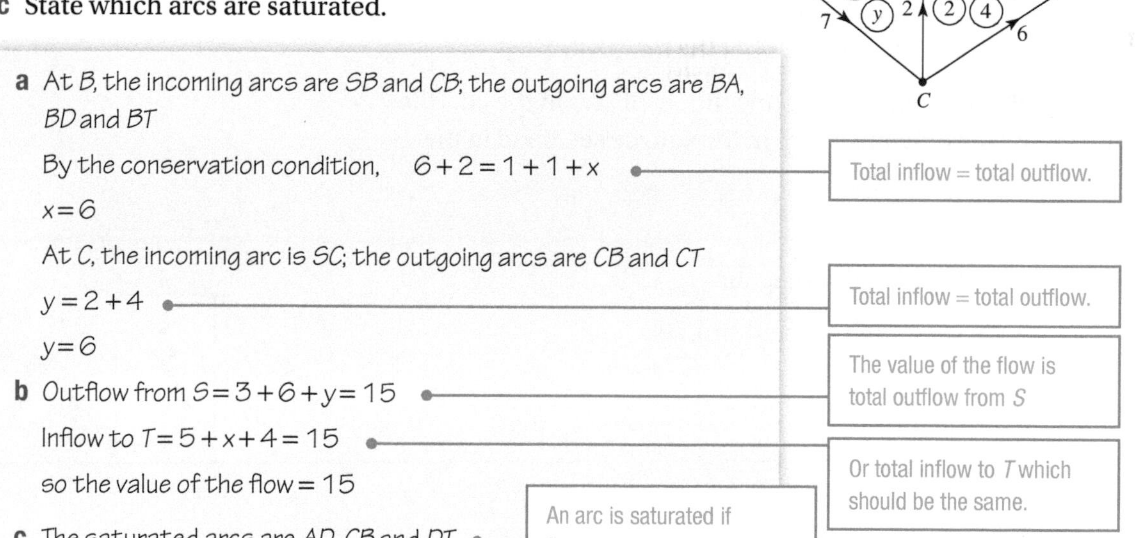

a At B, the incoming arcs are SB and CB; the outgoing arcs are BA, BD and BT

By the conservation condition, $6 + 2 = 1 + 1 + x$ — Total inflow = total outflow.

$x = 6$

At C, the incoming arc is SC; the outgoing arcs are CB and CT

$y = 2 + 4$ — Total inflow = total outflow.

$y = 6$

b Outflow from $S = 3 + 6 + y = 15$ — The value of the flow is total outflow from S

Inflow to $T = 5 + x + 4 = 15$ — Or total inflow to T which should be the same.

so the value of the flow = 15

c The saturated arcs are AD, CB and DT — An arc is saturated if flow = capacity.

You have met the idea of a 'bottleneck' in a road system, where traffic flow is restricted and affects the overall flow through the system.

In the network shown below, a flow of 9 is possible from S to C (3 along SAC, 6 along SBC) and 11 from D to T (4 along DET, 7 along DFT). However, the maximum value of the flow in the network is only 6, which is the most that can pass through the bottleneck CD

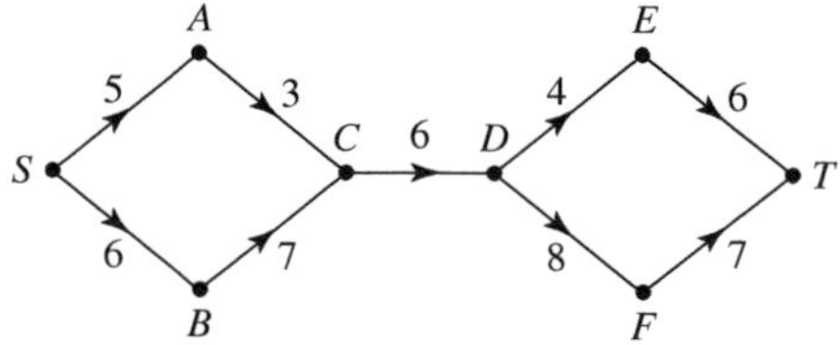

DISCRETE

A bottleneck may involve more than one arc. Here the bottleneck is formed by the arcs DE and DF. The maximum value of the flow in the network is 4

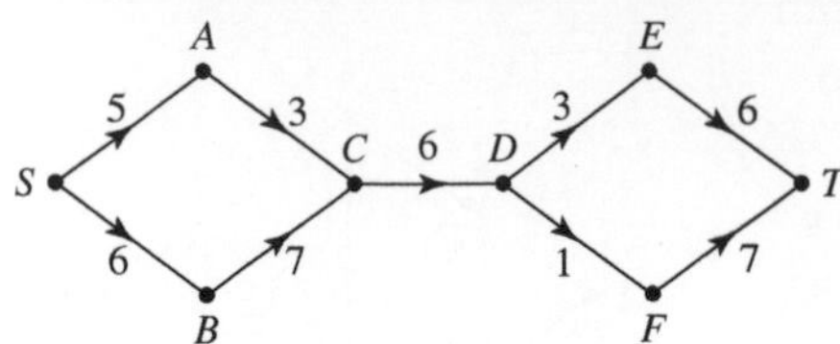

A bottleneck separates the network into two parts, one containing S and the other containing T

The value of the network flow cannot be more than the flow through a bottleneck. The maximum value of the network flow is therefore given by the flow through the 'worst' bottleneck.

This idea is formalised by the concept of a **cut**.

A **cut** is a set of arcs whose removal disconnects the network into two parts, X and Y, with X containing S and Y containing T

Key point

The **capacity of a cut** is the sum of the capacities of those arcs of the cut which are directed from X to Y

You describe a cut either by listing the set of arcs in the cut (the **cut set**), or by listing the nodes in the **source set** X and in the **sink set** Y

Example 2

For each cut shown in this diagram

a List the cut set, source set and sink set,

b Find the capacity of the cut.

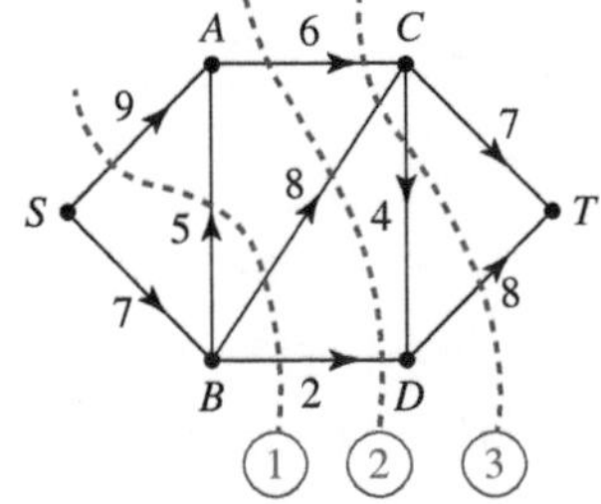

Cut 1 **a** Cut set = {SA, BA, BC, BD}
Source set X = {S, B}, sink set Y = {A, C, D, T}

b The capacity of the cut = 9 + 5 + 8 + 3 = 25

Cut 2 **a** Cut set = {AC, BC, BD}
Source set X = {S, A, B}, sink set Y = {C, D, T}

b The capacity of the cut = 6 + 8 + 2 = 16

Cut 3 **a** Cut set = {AC, BC, CD, DT}
Source set X = {S, A, B, D}, sink set Y = {C, T}

b The capacity of the cut = 6 + 8 + 8 = 22

The arc CD does not contribute to the capacity of cut 3 because it is directed from Y to X

Any flow must cross from set X to set Y. It follows that:

> **Key point**
>
> The value of any flow is less than or equal to the capacity of any cut.

The maximum flow corresponds to the 'worst bottleneck'. This is the **maximum flow–minimum cut theorem.**

> **Key point**
>
> **The maximum flow–minimum cut theorem** states that the value of the maximal flow is equal to the capacity of a minimum cut.

Example 3

List all possible cuts for the network shown, and find their capacities. Hence state the maximal flow for the network.

The diagram shows the four possible cuts.

Cut 1 $\{SA, SB\}$ Capacity $= 6 + 8 = 14$

Cut 2 $\{SA, AB, BT\}$ Capacity $= 6 + 7 = 13$

Cut 3 $\{SB, AB, BT\}$ Capacity $= 8 + 5 + 9 = 22$

Cut 4 $\{AT, BT\}$ Capacity $= 9 + 7 = 16$

The maximal flow $= 13$

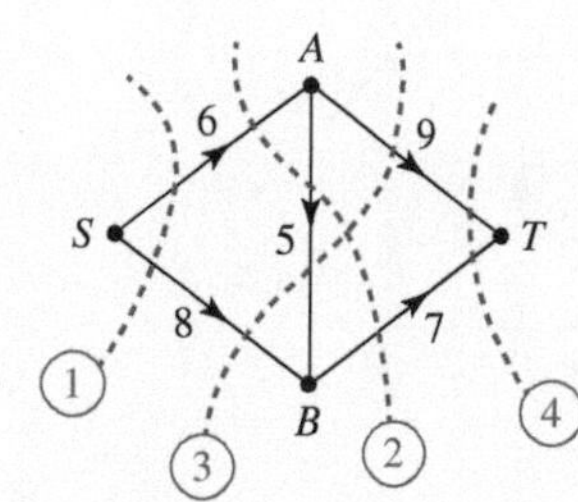

The arc AB does not contribute to the capacity of cut 2 because it is directed from Y to X

A useful consequence of the maximum flow–minimum cut theorem is:

> **Key point**
>
> If you have a flow and a cut such that the value of flow is equal to capacity of cut, then the flow is a maximum and the cut is a minimum.

Example 4

Find, by inspection, a minimum cut for the network shown. Confirm that it is a minimum cut by finding a flow with that value.

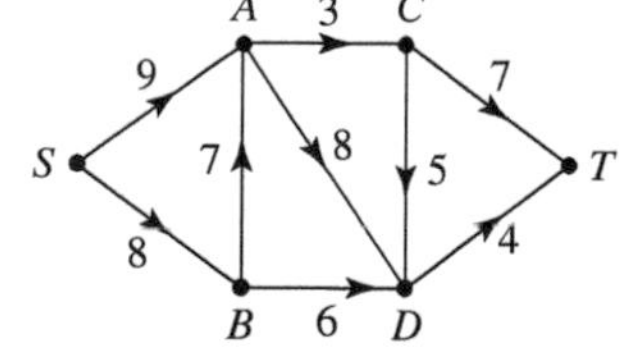

The cut $\{AC, CD, DT\}$, shown, appears to be a minimum.

The capacity of this cut $= 3 + 4 = 7$

There is a flow of value 7, consisting of 3 along $SACT$ and 4 along $SBDT$

So this flow is maximal, and the cut is a minimum.

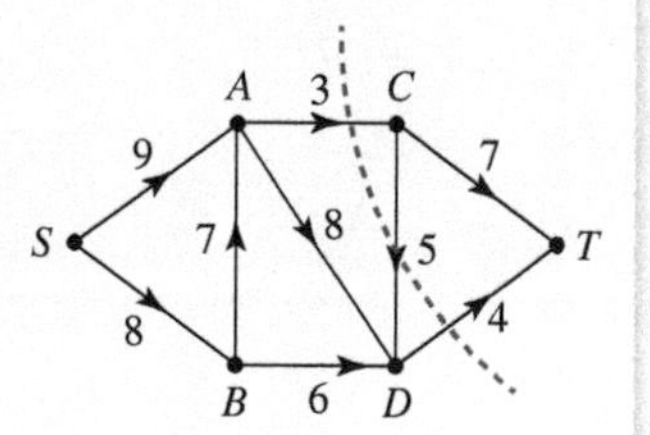

Look for arcs with the lowest weights.

DISCRETE

Exercise 12.7A Fluency and skills

1 For each of the cuts on each of these networks, state

i The set of arcs (the cut set),

ii The source set X and the sink set Y,

iii The capacity.

a

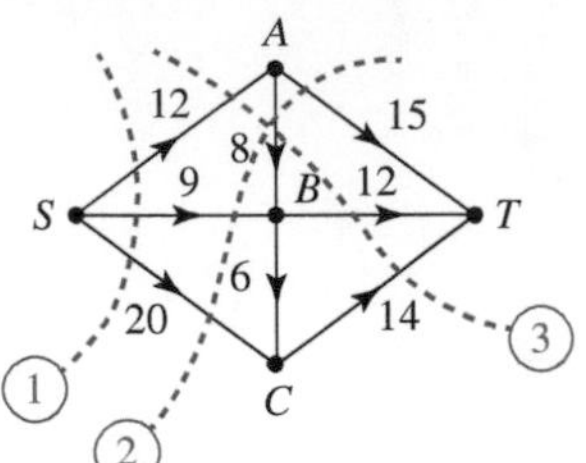

b

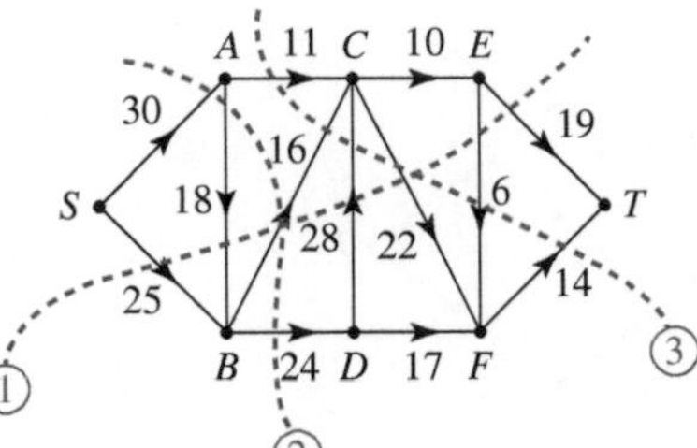

c

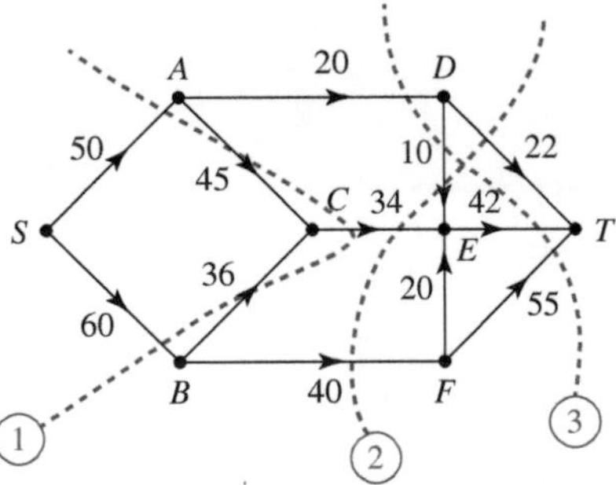

d

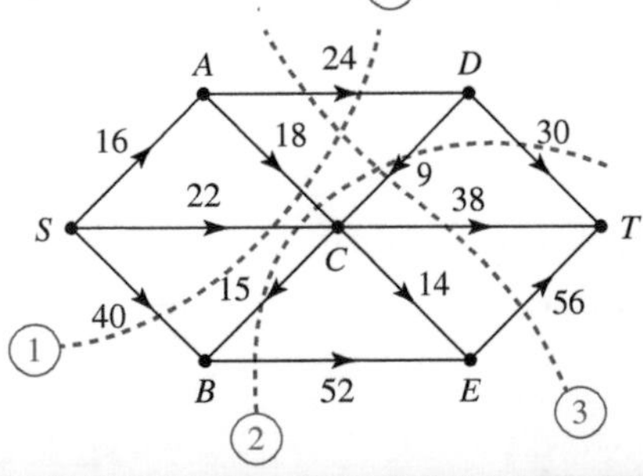

2 For the network in question **1a**

a Find a cut with a capacity of 35

b Find a flow with a value of 35

c What can you deduce from your results in **a** and **b**?

3 Find a minimum cut for the network in question **1b**, and confirm that it is a minimum by finding a flow with that value.

4 For the network shown, find, by inspection, the maximum flow and a minimum cut.

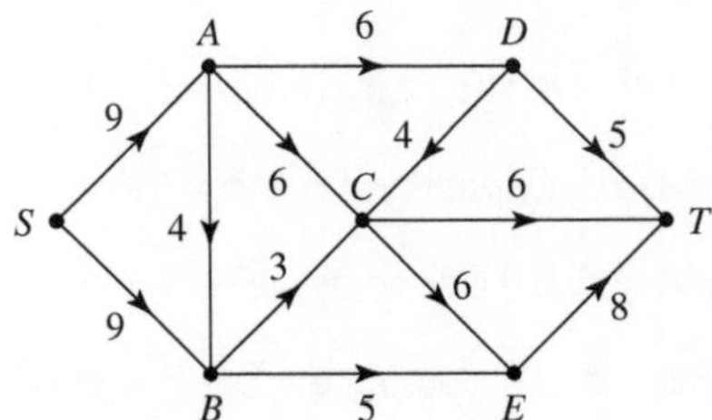

5 Find, by inspection, a minimum cut for the network shown and confirm that it is a minimum by finding a flow with that value.

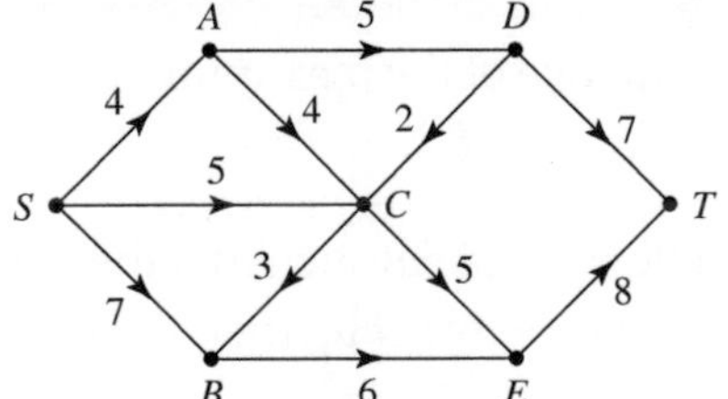

Reasoning and problem-solving

Strategy

To solve problems involving network flow

1. Draw clear diagrams, as necessary.
2. Use the standard notation. For example, circle the actual flow in an arc.
3. Use the maximum flow–minimum cut theorem.
4. Answer the question in context.

In some networks, you will have more than one source and/or sink. In this network, there are no incoming arcs at A or B. A and B are both sources. Similarly, there are no outgoing arcs at G or H. G and H are both sinks.

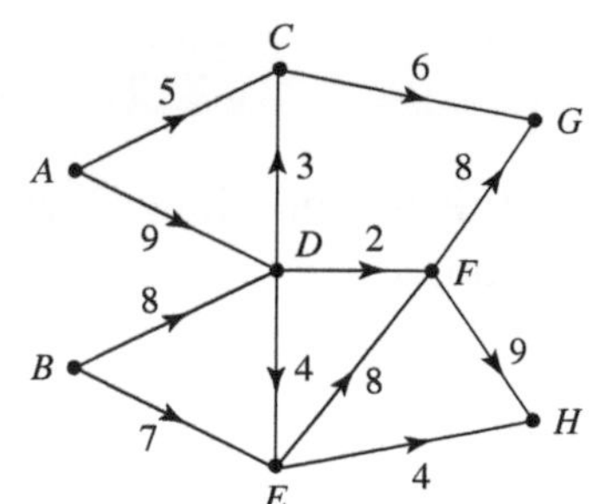

You deal with multiple sources by connecting them to a dummy **supersource**, S

Each actual source can receive from S as much flow as it needs to supply the network.

Similarly, you deal with multiple sinks by connecting them to a dummy **supersink**, T

Each sink can send to T all the flow it receives from the network.

The possible outflows from A and B are 14 and 15, so the capacities of the dummy arcs SA and SB must be at least 14 and 15

Similarly, the possible inflows to G and H are 14 and 13, so the capacities of the dummy arcs GT and HT must be at least 14 and 13

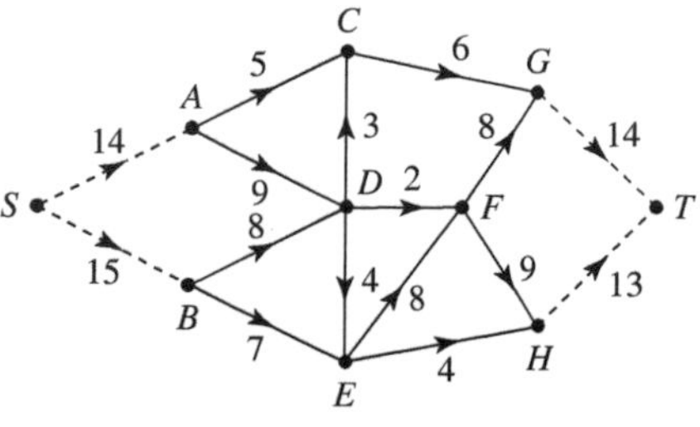

You now find the maximal flow for the modified network by the usual methods. Once this is found, you can remove the dummy arcs and nodes to leave the solution for the original network.

Example 5

a The diagram shows the capacities of the arcs in a network. Show, on the diagram, a flow of 2 along AD and 3 along ABE

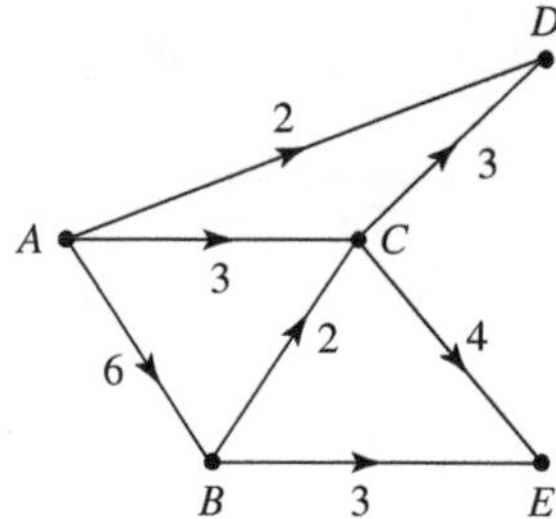

b Identify two additional flows to bring the total flow to 10

c Show that this is the maximal flow.

a D and E are sinks, so introduce a supersink T. Show the given flows on the diagram.

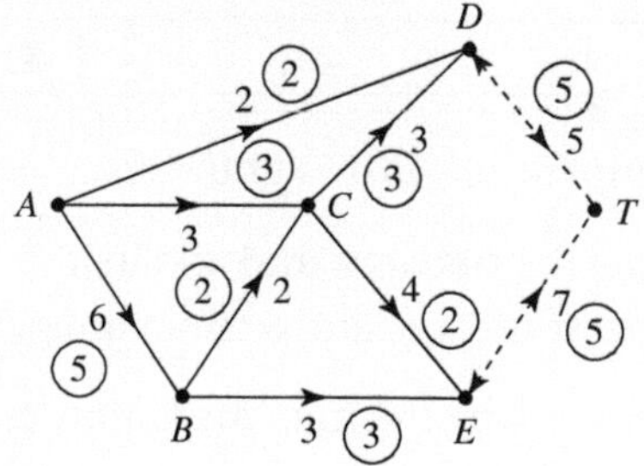

b For example, a flow of 3 along $ACDT$ and 2 along $ABCET$

c The cut {AD, AC, BC, BE} has capacity 10

Hence there is a flow and a cut of capacity 10, so this is maximal by the maximum flow–minimum cut theorem.

DISCRETE

Exercise 12.7B Reasoning and problem-solving

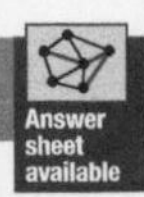

1 The network shown has a sink, G, and two sources.

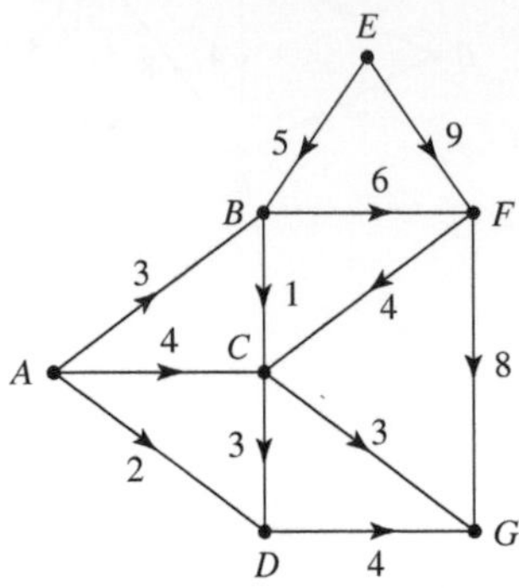

a Identify the sources.

b Introduce a supersource, S

c In a certain flow pattern, the arcs AC, AD and BC are saturated. Find the maximum value this flow can have.

d By finding a suitable cut, show that the flow found in **c** is the maximum possible for the network.

2 The network shown represents a system of one-way streets. Traffic enters the system at A and B and leaves at I and H. The weights are the maximum traffic flows, in hundreds of cars per hour, which can safely pass along the streets.

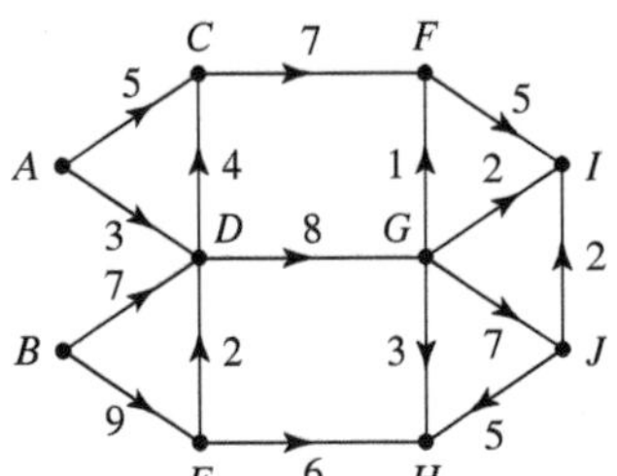

a Draw a diagram with a supersource, S, and a supersink, T

b State the maximum possible flows along $ACFI$, $BDGJH$ and BEH

c Find, by inspection, a network flow of value 19

d By finding a cut of capacity 19, show that the flow you found in **c** is maximal.

3 For the network shown,

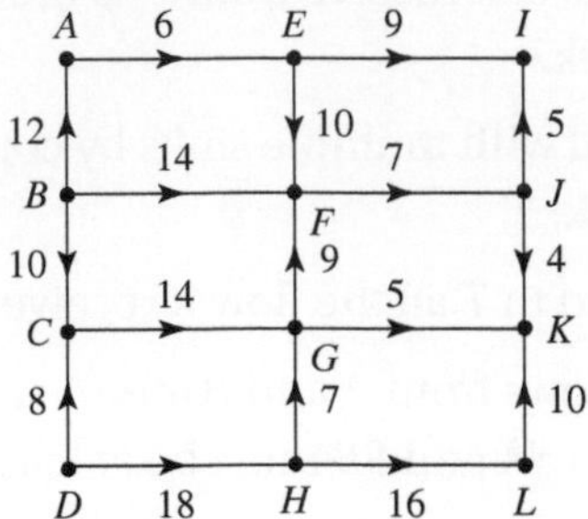

a Identify the source(s) and sink(s),

b Draw the diagram with a supersource and supersink, as necessary,

c Show that the maximum flow cannot be more than 28

d Find, by inspection, a flow with this value.

4 The table shows the flows in the arcs of a network.

		To							
		A	**B**	**C**	**D**	**E**	**F**	**G**	**H**
From	**A**	-	12	8	-	-	-	-	-
	B	-	-	-	10	-	14	-	-
	C	-	11	-	-	-	-	12	-
	D	-	-	-	-	-	-	-	-
	E	-	-	9	-	-	-	15	-
	F	-	-	8	15	-	-	-	7
	G	-	-	-	-	-	15	-	18
	H	-	-	-	-	-	-	-	-

a Identify the sources and sinks.

b Draw the network and add in a supersource, S, and a supersink, T

c Explain why the maximum flow cannot be more than 44

d Find, by inspection, a flow with this value.

12 Summary and review

Chapter summary

- A graph consists of points (vertices or nodes) connected by lines (edges or arcs).
- Graphs are isomorphic if they have the same vertices connected in the same way but drawn with a different layout.
- A graph is connected if there is a continuous route between every pair of vertices.
- A simple graph has no loops or multiple edges.
- A graph having some of the vertices and edges of a given graph, G, is a sub-graph of G
- A graph formed by adding vertices of degree 2 along edges of a given graph, G, is a sub-division of G
- A complete graph is a simple graph with all possible pairs of vertices connected. The complete graph with n vertices is called K_n
- In a bipartite graph, the vertices belong to two distinct sets, and each edge joins a vertex in one set to a vertex in the other. If every possible edge is present, it is a complete bipartite graph. The complete bipartite graph connecting m vertices to n vertices is called $K_{m,n}$
- The degree (or order) of a vertex is the number of edges connecting to it.
- Total of orders $= 2 \times$ number of edges, so the number of odd vertices is even.
- An adjacency matrix lists vertices and the number of edges connecting them.
- A network is a graph with a number, called a weight, associated with each arc (you should use node and arc when talking about networks). You can represent it using a distance matrix, with entries showing the weights of the arcs.
- If a graph or network has 'one-way streets,' then it is a directed graph (digraph) or directed network.
- Any continuous journey around a graph is a walk. A walk with no repeated edges is a trail. If a trail returns to its starting vertex, it is a closed trail.
- A walk with no repeated edges or vertices (except maybe the start) is a path. A closed path is a cycle.
- A connected graph with no cycles is a tree.
- A Hamiltonian cycle (or tour) visits every vertex of the graph once only. A graph with such a cycle is a Hamiltonian graph.
- A graph is traversable, or Eulerian, if there is a closed trail including every edge once only. The trail is an Eulerian trail. If the trail is not closed, then the graph is semi-traversable or semi-Eulerian.
- A graph is Eulerian if all vertices have even degree. It is semi-Eulerian if it has two odd degree vertices. The two odd vertices are the start and finish vertices. A graph with more than two odd degree vertices is non-traversable.
- A graph is planar if it can be drawn in a plane so that no two edges meet each other except at a vertex.
- When a graph is drawn in the plane, the areas created by its edges are called faces. The area outside the graph is the infinite face.
- A graph drawn in the plane has V vertices, E edges and F faces related by Euler's formula:

  $$V + F - E = 2$$

- A spanning tree or connector for a graph with n vertices is a tree with $(n-1)$ edges containing every vertex of the graph.
- For a network, the minimum spanning tree (MST), or minimum connector, is the spanning tree with the lowest total weight.
- Kruskal's algorithm:

  Step 1 Choose the arc with minimum weight.

  Step 2 Choose the arc with least weight from the remaining arcs, avoiding those that form a cycle with those already chosen.

  Step 3 If any nodes remain unconnected, go to Step 2
- Prim's algorithm:

  Step 1 Choose any node to be the first in the connected set.

  Step 2 Choose the arc of minimum weight joining a connected node to an unconnected node. Add this arc to the spanning tree and the node to the connected set.

  Step 3 If any unconnected nodes remain, go to Step 2
- Prim's algorithm on a matrix:

  Step 1 Select the first node.

  Step 2 Cross out the row and number the column for the chosen node.

  Step 3 Find the minimum undeleted weight in the numbered columns. Circle this value. The node for this row is the next chosen node.

  Step 4 Repeat steps 2 and 3 until all nodes have been chosen.
- The route inspection (Chinese postman) problem:

  Find the shortest route that travels at least once along every arc of a network and returns to the start. Which arcs must be repeated?
- Route inspection algorithm:

  Step 1 Identify the odd nodes.

  Step 2 List all possible pairings of the odd nodes.

  Step 3 For each pairing, find the shortest routes between paired nodes and the total weight of those routes.

  Step 4 Choose the pairing with the smallest total.

  Step 5 Repeat the shortest route arcs (equivalent to adding extra arcs to the network). The network is now traversable.

  The total length of the route = (sum of original weights) + (sum of extra arc weights).
- The travelling salesman problem (TSP):

  Find a route which visits every node of a network and returns to the start in the shortest possible distance.
- To solve a practical TSP with n nodes:

  Create a complete network, K_n, of shortest distances between pairs of nodes.

  Find a minimum weight Hamiltonian cycle for the complete network (this is the classical problem).

  Interpret the solution in the practical situation.
- The nearest neighbour algorithm:

  Step 1 Choose a starting node V

  Step 2 From your current position, choose the arc with minimum weight leading to an unvisited node. Travel to that node.

  Step 3 If there are unvisited nodes, go to Step 2

  Step 4 Travel back to V

- For the optimal tour, T, lower bound $\leq T \leq$ upper bound.
- The length of any known tour is an upper bound.
- ($2 \times$ length of minimum spanning tree) is an upper bound which may be improved by using short cuts.
- To find a lower bound:

  Step 1 Choose a node V

  Step 2 Identify the two lowest weights, p and q, of the arcs connected to V

  Step 3 Remove V and its connecting arcs from the network. Find the total weight, m, of the minimum spanning tree of the remaining subgraph.

  Step 4 Calculate lower bound $= p + q + m$

  Step 5 If possible, choose another node V and go to Step 2
- Choose the largest result obtained as the best lower bound.
- The weights in a directed network can represent flows. The weight of an arc is its capacity.
- A node is a source if all arcs are directed away from it.
- A node is a sink if all arcs are directed towards it.
- In a given situation, there is a flow in each arc. These combine to form the flow in the network.
- If the flow in an arc equals its capacity, the arc is saturated.
- Flows satisfy these conditions.
  - The feasibility condition: The flow in an arc cannot be greater than its capacity.
  - The conservation condition: At every node apart from S and T, total inflow $=$ total outflow
- Total outflow at $S =$ total inflow at T

  This total is the value of the flow.
- A cut is a set of arcs whose removal disconnects the network into two parts, X and Y, with X containing S and Y containing T
- The capacity of a cut is the sum of the capacities of those arcs of the cut that are directed from X to Y
- You describe a cut either by listing the set of arcs in the cut (the cut set) or by listing the nodes in the source set X and in the sink set Y
- The maximum flow–minimum cut theorem:

  The value of the maximal flow $=$ the capacity of a minimum cut
- It follows that if a flow and a cut are such that (value of flow) $=$ (capacity of cut), then the flow is maximum and the cut is minimum.
- If a network has multiple sources or sinks, you connect them to a dummy source, S, or a dummy sink, T

Check and review

You should now be able to...	Review Questions
✔ Draw a graph/network from an adjacency/distance matrix and vice versa.	1
✔ Understand and use the terms related to graphs and networks.	2
✔ Use Euler's formula for planar graphs.	3
✔ Use Kruskal's algorithm to find the minimum spanning tree for a network given as a diagram.	4
✔ Use Prim's algorithm to find the minimum spanning tree for a network given as a diagram or as a table.	5, 6
✔ Solve the route inspection problem for a network with nodes of odd degree.	7
✔ Find an upper bound for the travelling salesman problem using the nearest neighbour algorithm or short cuts applied to the minimum spanning tree.	8
✔ Find a lower bound for the travelling salesman problem by deleting a node and using the minimum spanning tree.	8
✔ Analyse the flow through a network and use the maximum flow–minimum cut theorem to decide if it is maximal.	9

1 Draw the graph corresponding to this adjacency matrix:

	A	*B*	*C*	*D*
A	0	1	2	0
B	1	2	1	1
C	2	1	0	1
D	0	1	1	0

2 A simple connected graph is semi-Eulerian and has 5 vertices. Draw the graph if

a It has exactly one vertex of order 3

b It is a tree,

c It has 7 edges.

3 A graph drawn in the plane has 5 vertices and 5 faces.

a How many edges does the graph have?

b Draw a graph which satisfies this data.

4 Use Kruskal's algorithm to find the minimum spanning tree for the network shown. List the order in which you choose the arcs and state the total length of the tree.

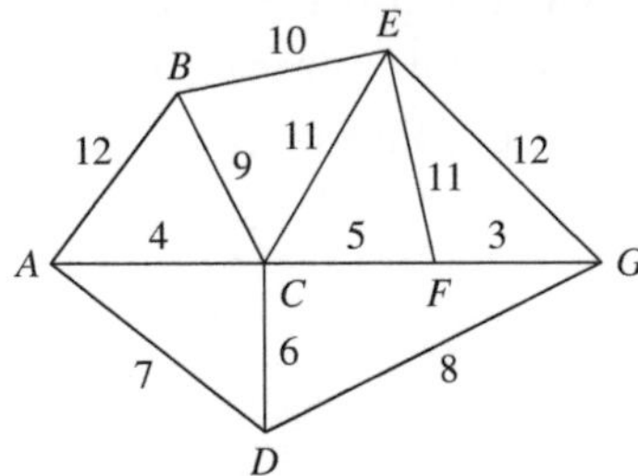

5 Use Prim's algorithm, starting from *B*, to find the minimum spanning for the network in question 4. List the order in which you choose the arcs.

6 Use Prim's algorithm with the table shown, starting from A, to find the minimum spanning tree. Show all your working. List the order in which you choose the arcs.

	A	**B**	**C**	**D**	**E**	**F**	**G**
A	–	9	12	4	6	8	14
B	9	–	10	13	8	11	3
C	12	10	–	2	9	15	7
D	4	13	2	–	13	14	4
E	6	8	9	13	–	15	8
F	8	11	15	14	15	–	13
G	14	3	7	4	8	13	-

7 The sum of the weight of the network shown is 104

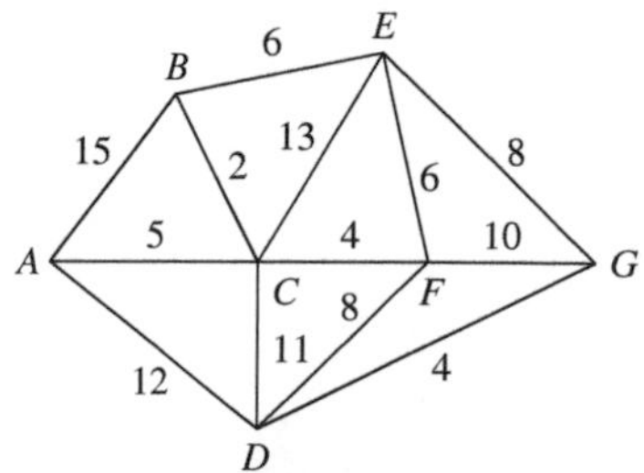

a Find the length of a Chinese postman route through this network.

b State the number of times that the route would pass through D

8 The diagram shows the distances between five locations. A delivery van starting from A has to visit all five locations and return to base.

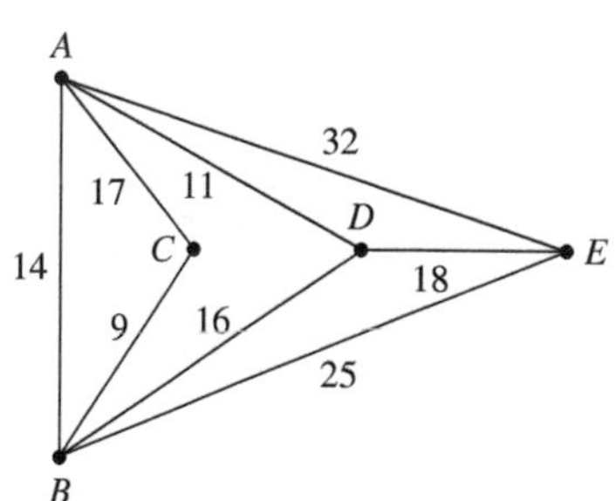

a Draw up a table of shortest distances (found by inspection),

b Use the nearest neighbour algorithm, starting from A, to find a possible route. State the length of the route and list the actual route of the van.

c The nearest neighbour algorithm starting from B gives a route of length 80 State which of the two values you have is the better upper bound.

d By deleting node A from your table, find a lower bound for the optimal tour. Hence write inequalities satisfied by T, the length of the optimal tour.

9 a Find by inspection a flow of 18 in the network shown.

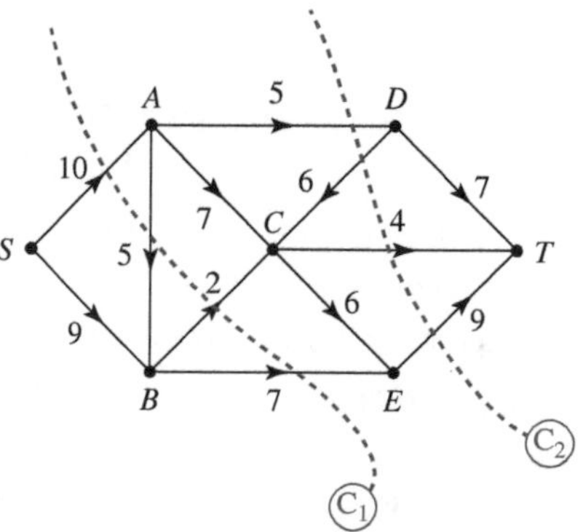

b Find the capacity of the cuts C_1 and C_2

c State, with reasons, the maximum flow in the network.

DISCRETE

Did you know?

The Chinese mathematician Meigu Guan is credited for formulating the route inspection problem. His paper was translated into English in 1962 and became known as the 'Chinese postman problem' in his honour.

Investigation

The windy postman problem is a variant of the route inspection problem. In this case, each edge may have one cost for traversing it in one direction and a different cost for traversing it in the other direction.

The cost of traveling each way is given in the brackets as (c_{ij}, c_{ji}) where $i < j$. In this example, the cost for travelling from node 3 to node 4 is 2, and the cost for travelling from node 4 to node 3 is 3.

Solve the problem starting and ending at node 3.

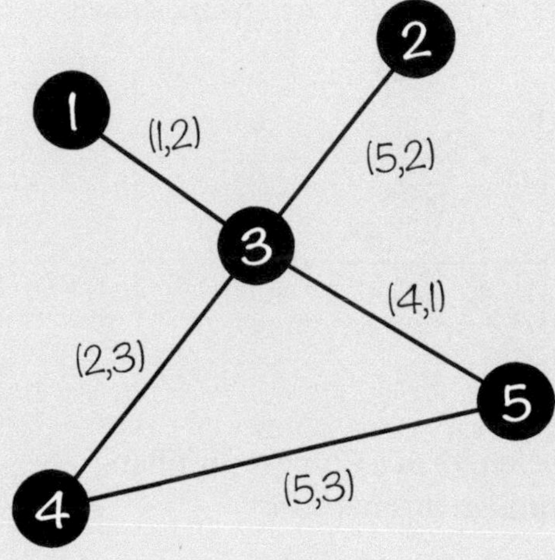

Research

Fleury's algorithm and Hierholzer's algorithm both enable you to find an Eulerian trail around a graph.

- Fleury's algorithm chooses the next edge in the path to be one whose deletion would not disconnect the graph.
- Hierholzer's algorithm requires you to find closed paths and substitute them into other closed paths.

Research how these two algorithms work.

12 Assessment

Answer sheet available

1 a Explain what is meant by a 'tree' in the context of graphs. **[2 marks]**

b Draw the network corresponding to this distance matrix. **[3]**

	A	*B*	*C*	*D*	*E*
A	–	2	7	4	–
B	2	3	–	9	–
C	7	–	–	6	–
D	4	9	6	–	5
E	–	–	–	5	2

c Give an example of a path on the network with at least four nodes. **[1]**

d Explain whether or not this network is a Hamiltonian graph. **[2]**

e How do you that know this graph is not Eulerian? **[1]**

f Which arc/s would need to be added to change this to an Eulerian graph? **[1]**

2

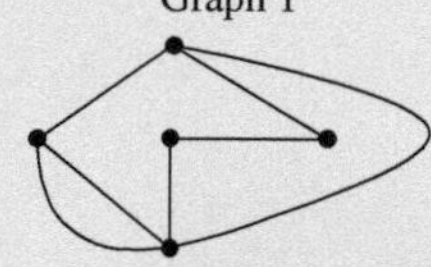

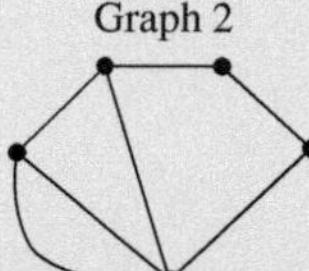

a Are graphs 1 and 2 isomorphic? Explain how you know. **[2]**

b Is graph 1 traversable, semi-traversable or non-traversable? Explain how you know. **[2]**

c Draw a subgraph of graph 2 which is also a tree. **[1]**

d Draw the graph K_4 and explain whether or not it is planar. **[3]**

3 a Draw the adjacency matrix that corresponds to this network. **[4]**

b Write down a semi-Eulerian trail for this graph and state its weight. **[2]**

c State whether or not the graph is planar and explain how you know. **[2]**

4 a Find a minimum spanning tree for the graph in question 3 using

i Kruskal's algorithm, ii Prim's algorithm starting at *B* **[6]**

In each case, list the order in which the arcs are added or rejected.

b Use your minimum spanning tree to state an upper bound for the optimal solution to the travelling salesman problem. **[3]**

5 For the network in question 3 find the length of the shortest route that travels along each arc at least once and then returns to the start point. **[6]**

6 The graph gives the distances between five points.

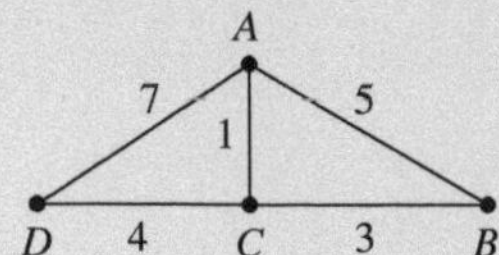

a Write down a Hamiltonian cycle for the graph shown and state its weight. [2]

b Draw a complete network of shortest distances. [2]

c Use the nearest neighbour algorithm with A as the starting node to find a Hamiltonian cycle on the network. [2]

d State the order in which the nodes would be visited in the original network. [2]

7 This diagram shows a capacitated network with a feasible flow given in circles on each edge.

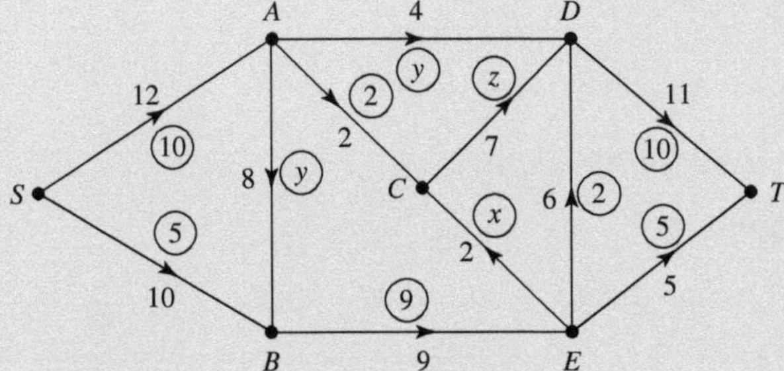

a Find the values of x, y and z [3]

b Which of the arcs are saturated? [2]

A cut is defined by source set $\{S, A, B, E\}$ and sink set $\{C, D, T\}$.

c List the arcs in the cut and state its capacity. [2]

8 a Draw the graph that corresponds to this adjacency matrix. [2]

	A	B	C
D	1	0	1
E	1	1	1
F	0	0	1

b State the name given to this type of graph. [1]

c How many vertices and how many edges must be added to this graph to create the complete graph $K_{4,3}$? [3]

d Is the graph $K_{4,3}$ Eulerian, semi-Eulerian or non-Eulerian? Explain how you know. [2]

9 **a** Use Prim's algorithm for a matrix, starting at A, to find a minimum spanning tree for the network corresponding to this distance matrix.

	A	**B**	**C**	**D**	**E**	**F**
A	–	12	9	–	7	–
B	12	–	4	5	–	1
C	9	4	–	3	8	–
D	–	5	3	–	6	–
E	7	–	8	6	–	2
F	–	1	–	–	2	–

List the order in which you select the arcs and give the weight of the tree. **[4]**

b Show that using Kruskal's algorithm leads to the same minimum spanning tree and sketch the tree. **[4]**

c Solve the route inspection problem, starting at A, for this matrix. Give a possible route and state its length. **[5]**

10 The graph shows the travel times (in minutes) between six towns. A salesperson wishes to visit each town at least once and then return to their starting point.

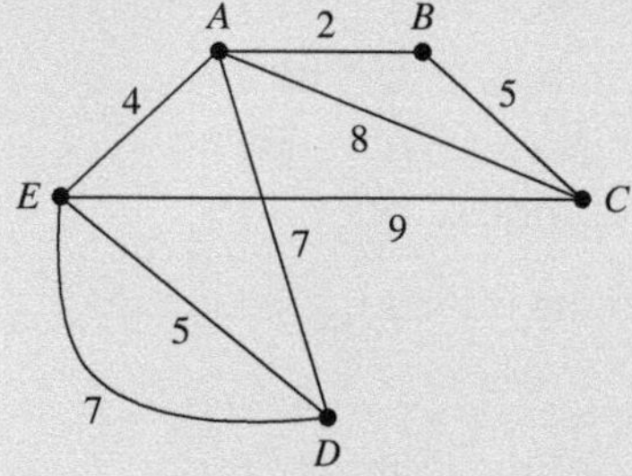

a Use a minimum spanning tree to find an upper bound for the time of the optimal tour. **[4]**

b By finding a shortcut, calculate an improved upper bound. **[2]**

c By removing each node from the network in turn, find the best lower bound for the time of the optimal tour. **[5]**

d Draw a matrix corresponding to the complete network of shortest times. **[2]**

e Use the nearest neighbour algorithm, starting and ending at C, to find a tour that visits each node at least once. **[3]**

f Show that the solution found in **d** is not optimal. **[2]**

11 Here is a capacitated network.

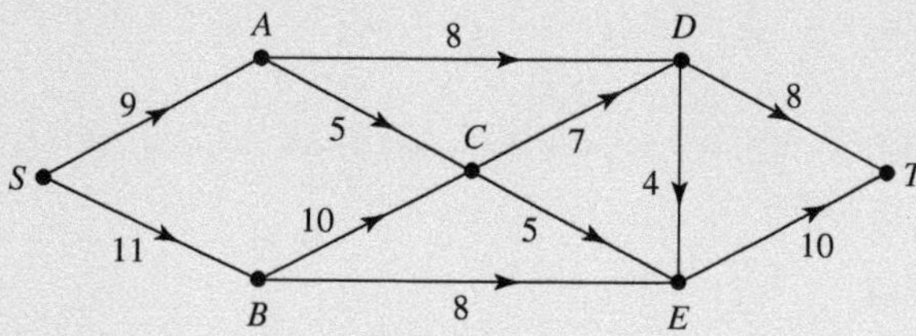

a What is the maximum possible flow along the route $SBCDET$? **[1]**

b Find a cut with a capacity of 28 and write down

i The cut set, **ii** The source set, **iii** The sink set. **[3]**

c Find, by inspection, a minimum cut and hence state the maximum flow for the network. **[3]**

12 Cables are to be laid to connect to each of the points shown in the network. The lengths (in metres) between the points are shown.

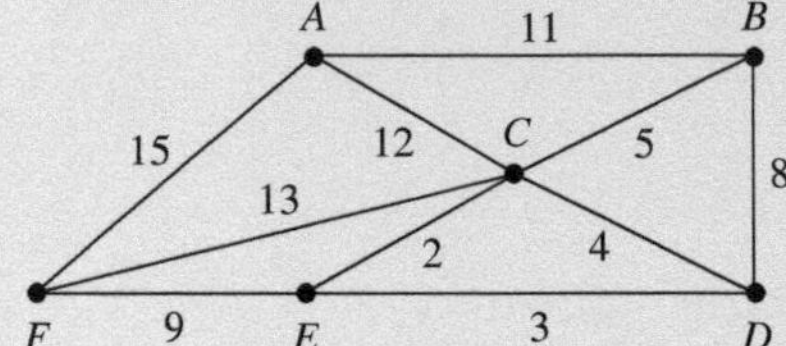

a Use a suitable algorithm to find the minimum length of cable required and state which edges are to be included. State the name of the algorithm used. **[5]**

It is discovered that there is already cable along AC and BD so this can be used and will not need to be laid.

b Explain which algorithm you chose and how you would adapt it in order to find the minimum length of cable now required. Find the additional length of cable required. **[4]**

c Find a solution to the travelling salesperson problem for this network starting and ending at A using only edges in your solution for part **a**. Then use two short cuts to show that 47 is an upper bound for the travelling salesperson problem. **[4]**

d Obtain two different lower bounds for the optimal solution to the travelling salesman problem for this network. **[4]**

e Use your solutions to part **c** and **d** to write the best inequality for the length of the optimal tour. **[2]**

f Show that the nearest neighbour algorithm does not give an optimal tour. **[5]**

13 The network represents a system of roads with the direction of travel and maximum traffic flows given on each edge.

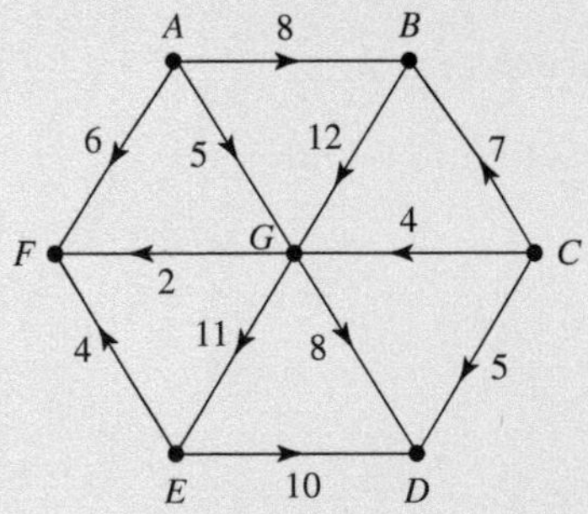

a Copy the network and add on a supersource, S, and a supersink, T. Write the minimum necessary capacity on each of the dummy edges. **[6]**

b What is the maximum flow along $SABGEFT$? **[1]**

c Given that there exists a possible flow of size 32, show that this flow is maximal. You should state the name of any theorem used. **[4]**

13 Critical path analysis 1

In large construction projects, it is essential to ensure that different contractors, equipment, building materials, and so on, are all in place at the right time. This is a massively complex undertaking. Even ensuring that facilities, such as toilets, are in place for workers at the right part of the construction site at the right time must be considered. Some projects can take many years and ensuring that access is in place to various parts of the site has to be considered. Critical path analysis is used in such situations to determine exactly what needs to be done and when.

Even projects on a much smaller scale, for example building and extension to a house, require some work of this type. Construction projects are just one example of an area in which critical path analysis is used. In almost all projects across areas as diverse as making movies to turning aircraft round at airports, such analysis is necessary to ensure smooth running and efficient use of resources.

Orientation

What you need to know	What you will learn	What this leads to
Chapter 12 Graphs and networks.	• To produce an activity network from a precedence table. • To solve problems involving activity networks. • To identify critical paths and activities.	Chapter 29 Critical path analysis 2.

13.1 Activity networks

Fluency and skills

To produce a timetable for the completion of a project, you need:

- A list of the activities involved
- Details of which activities depend on others
- How long each activity will take (its **duration**).

You record this information in a **precedence table** or **dependence table**.

For example, suppose you are planning to paint your bedroom. It might involve the following activities:

A Remove furniture

B Remove curtains

C Remove carpets

D Wash ceiling

E Wash walls

F Paint ceiling

G Paint walls

H Replace carpets

I Replace curtains

J Replace furniture

Some activities are dependent on others. For example, you can't remove the carpets until you have removed the furniture.

Here is a possible precedence table, with likely durations.

Activity	Duration (minutes)	Depends on
A	15	–
B	10	–
C	15	*A*
D	20	*A*, *B*, *C*
E	30	*A*, *B*, *C*
F	80	*D*
G	140	*E*
H	20	*F*, *G*
I	15	*F*, *G*
J	15	*H*

The timings assume that you have friends to help you, so that, for example, the ceiling can be painted at the same time as the walls.

You can now draw an **activity network**. Each activity is represented by a node, and the arcs show the order of precedence.

Activities A and B can both start straight away and A must finish before C can start. Once A, B and C are complete, D and E can start.

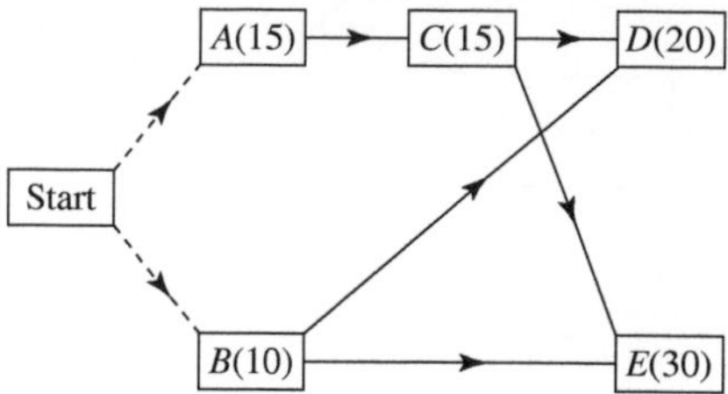

F and G must follow D and E, respectively. These must finish before H and I can start. J can start once H has finished.

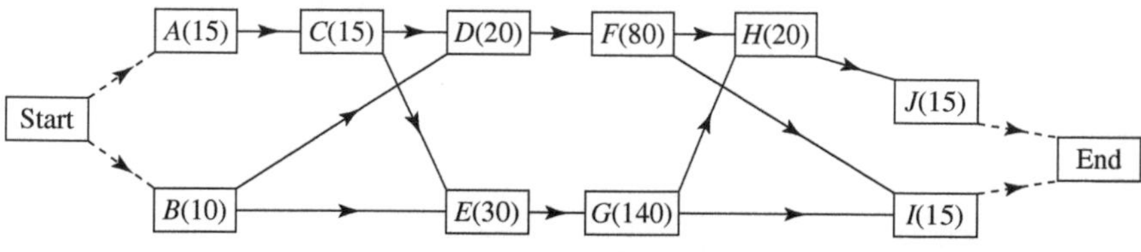

The 'Start' and 'End' boxes are not strictly needed (they are dummy activities with zero duration) but you may find them helpful.

You can now analyse the timing of the activities. For each activity you need to know:

- What is the **earliest start time**, assuming all preceding activities are completed as soon as possible?
- What is the **latest finish time**, that is, the latest you could finish the activity without increasing the overall length of the project?

It is usual to record these in a box at each node.

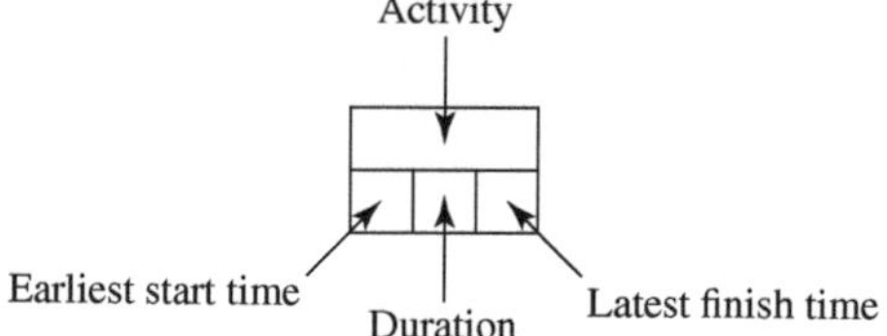

To find the earliest start times you make a **forward pass** through the network.

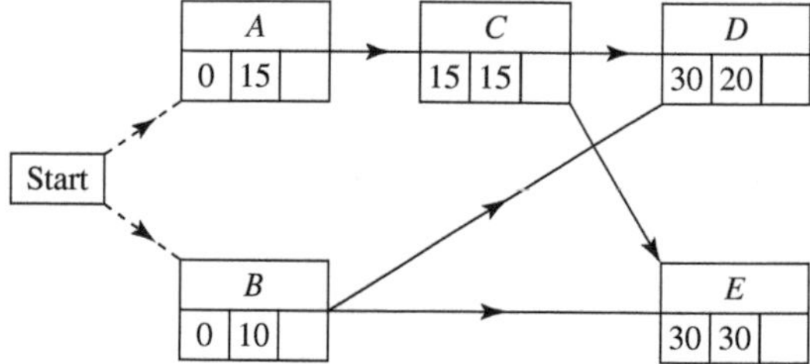

A and B can start straight away, so the earliest start time is 0
If A starts immediately, it will finish after 15 minutes. C cannot start until A finishes, so C has an earliest start time of 15 minutes.

The earliest that B can finish is 10 minutes. The earliest that C can finish is $15 + 15 = 30$ minutes. D and E cannot start until both B and C have finished, so D and E have earliest start time of 30 minutes.

Make sure that you understand how these earliest start times were found.

DISCRETE

Continuing in this way, the complete forward pass looks like this.

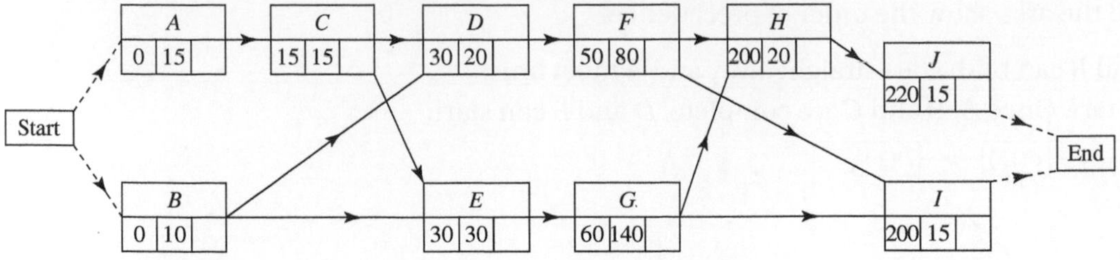

The project finishes when both *I* and *J* end. *I* can finish by 215 minutes and *J* by 235 minutes. This means that the overall project duration is 235 minutes.

You now make a **backward pass** to find the latest finish time for each activity consistent with the overall duration of 235 minutes.

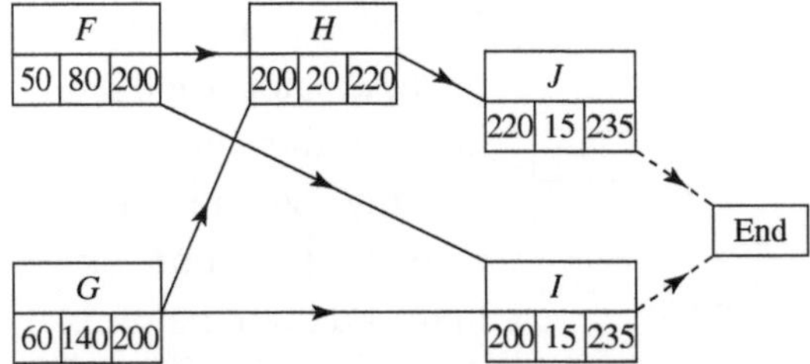

I has a duration of 15 minutes, so to finish by 235 minutes it must start by 220 minutes at the latest. Similarly, *J* must start by 220 minutes.

J follows on from *H*, so the latest *H* can finish is 220 minutes. *H* has a duration of 20 minutes, so must start by 200 minutes at the latest.

F and *G* must both finish before *H* and *I* can start. This means that the latest *F* and *G* can finish is 200 minutes.

Again, make sure that you understand how these latest finish times were found.

Continuing in this way, the complete backward pass looks like this.

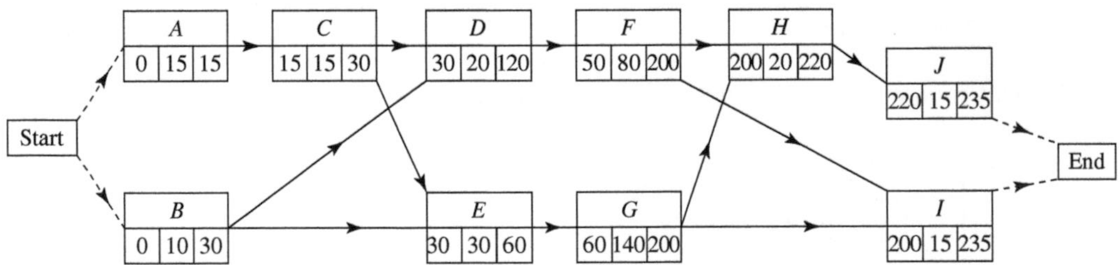

To summarise, an activity may depend on several preceding activities and be followed by several activities which depend on it.

When doing a forward pass:

Key point

The earliest start time for an activity is the maximum of the earliest times by which all preceding activities can be completed.

Completing the forward pass tells you how long the whole project will take.

Key point

The project duration is the earliest time by which all activities can be completed.

When doing a backward pass:

Key point

The latest finish time for an activity is the minimum of the latest times by which all following activities can start without affecting the project duration.

Example 1

a Draw an activity network to show the project described by this precedence table.

Activity	Duration (hours)	Depends on
A	3	–
B	6	*A*
C	8	*A*
D	10	*A*
E	7	*B*
F	7	*C*
G	9	*E, F*
H	4	*D,F*

b Find the minimum project duration and the earliest start and latest finish times for each activity.

c State the latest time at which activity *D* could start.

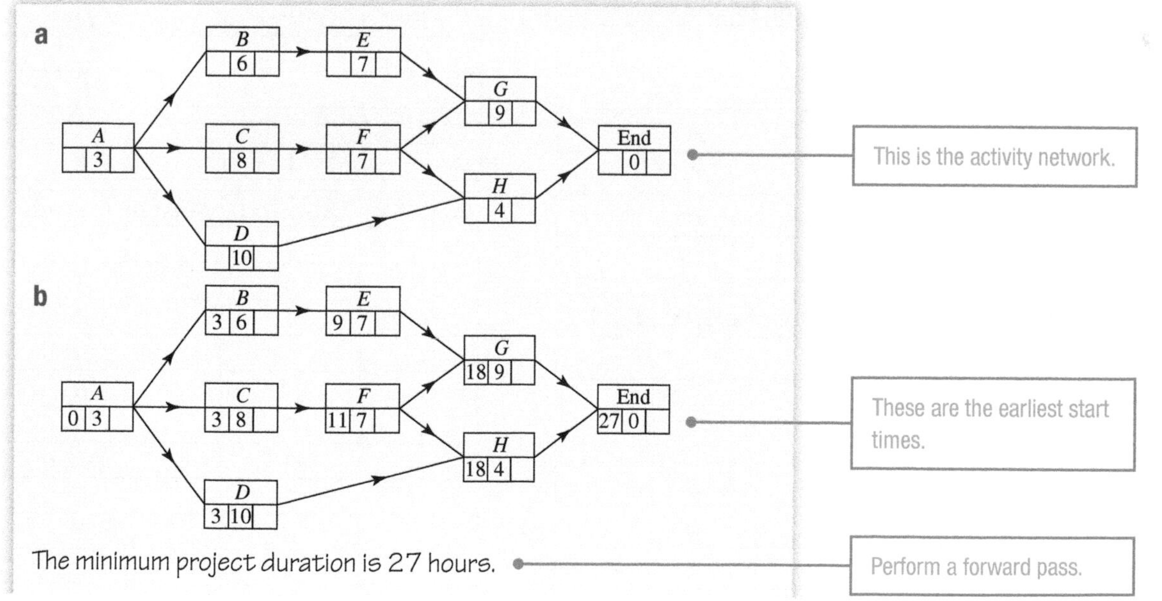

This is the activity network.

These are the earliest start times.

The minimum project duration is 27 hours.

Perform a forward pass.

(*Continued on the next page*)

DISCRETE

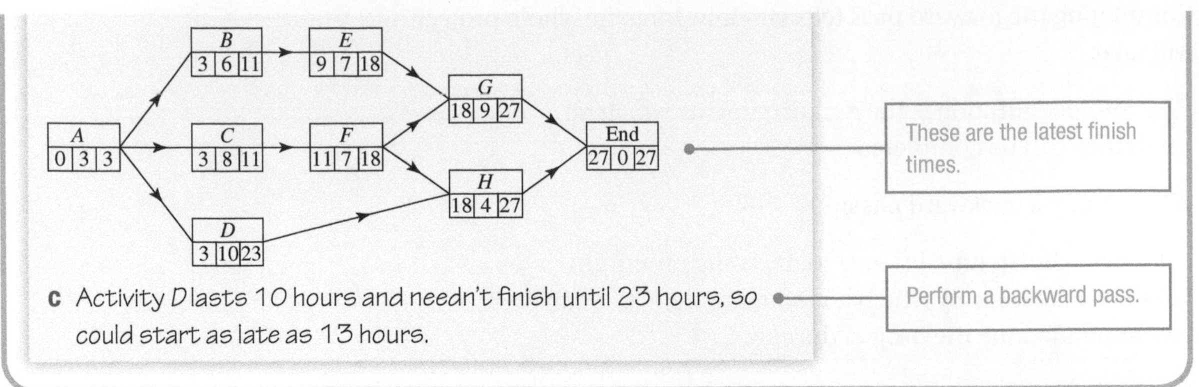

These are the latest finish times.

c Activity D lasts 10 hours and needn't finish until 23 hours, so could start as late as 13 hours.

Perform a backward pass.

Exercise 13.1A Fluency and skills

Answer sheet available

1 This diagram shows part of an activity network. Find the missing values.

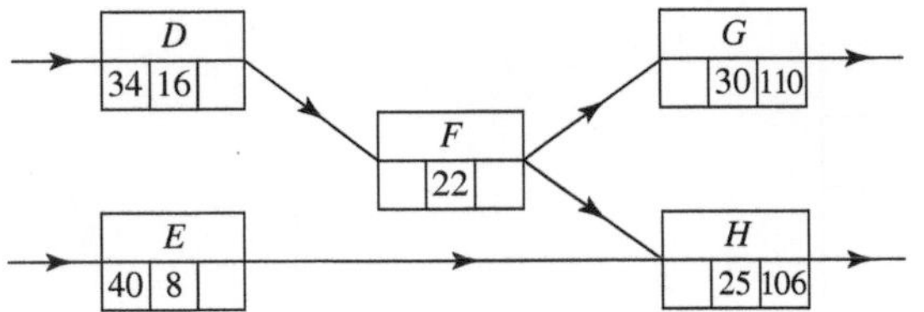

2 Draw an activity network for each of the following precedence tables. Complete a forward and a backward pass to find the project duration and the earliest start times and latest finish times.

a

Activity	Duration	Must be preceded by
A	4	–
B	2	–
C	1	A
D	5	A
E	3	B
F	2	C, E

b

Activity	Duration	Must be preceded by
A	2	–
B	5	–
C	1	A
D	3	B
E	2	B
F	2	C
G	4	D, E

c

Activity	Duration	Must be preceded by
A	3	–
B	2	A
C	3	A
D	4	A
E	5	B, C
F	3	C, D
G	1	D

d

Activity	Duration	Must be preceded by
A	6	–
B	1	–
C	12	–
D	7	A, B
E	3	D
F	4	D
G	1	E, F
H	7	F
I	2	C, G
J	3	I

e

Activity	Duration	Must be preceded by
A	3	–
B	6	A
C	4	A
D	7	A
E	11	B
F	6	B
G	3	C, D, F
H	7	D
I	8	E
J	2	E
K	3	G, I
L	2	G, H, I
M	9	K, L

Reasoning and problem-solving

Strategy

To solve problems involving activity networks

1. Before drawing the activity network, make a rough sketch to help you to produce a clear layout.
2. Draw a clear diagram (a ruler is recommended).
3. If required, do a forward and backward pass to find the project duration and the activity timings.
4. Interpret the relationship between activity durations and timings.
5. Answer the question.

Example 2

a Use an activity network to find the earliest start times and latest finish times for the activities in this project.

Activity	Duration (days)	Depends on
A	2	-
B	3	A
C	5	A
D	8	B
E	7	B
F	3	C
G	4	D
H	1	E, F
I	1	G, H

b State the minimum project duration.

c One of the activities over-runs by 5 days. What effect will this have on the project duration if the delayed activity is

i D **ii** E **iii** F (assume that the activity starts as soon as possible).

a This is the activity network:

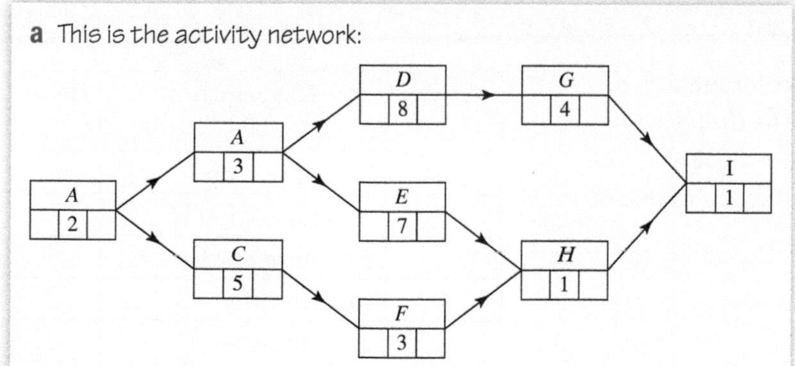

(*Continued on the next page*)

DISCRETE

These are the earliest start times:

A: 0 | 2
B: 2 | 3
C: 2 | 5
D: 5 | 8
E: 5 | 7
F: 7 | 3
G: 13 | 4
H: 12 | 1
I: 17 | 1 | 18

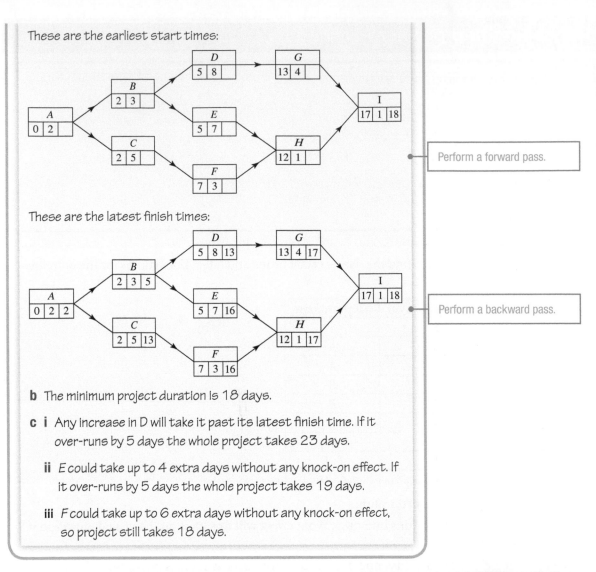

Perform a forward pass.

These are the latest finish times:

A: 0 | 2 | 2
B: 2 | 3 | 5
C: 2 | 5 | 13
D: 5 | 8 | 13
E: 5 | 7 | 16
F: 7 | 3 | 16
G: 13 | 4 | 17
H: 12 | 1 | 17
I: 17 | 1 | 18

Perform a backward pass.

b The minimum project duration is 18 days.

c **i** Any increase in D will take it past its latest finish time. If it over-runs by 5 days the whole project takes 23 days.

ii *E* could take up to 4 extra days without any knock-on effect. If it over-runs by 5 days the whole project takes 19 days.

iii *F* could take up to 6 extra days without any knock-on effect, so project still takes 18 days.

Exercise 13.1B Reasoning and problem-solving

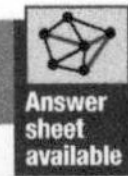

Answer sheet available

1 The following is a list of tasks for making a wooden trolley for a child. The trolley will have wheels, a square wooden base and four wooden sides with a cord attached to one of them for towing the trolley.

Activity	Description	Depends on
A	Purchase wood	
B	Purchase wheels	
C	Purchase cord	
D	Cut out sides	
E	Cut out base	
F	Attach sides to base	
G	Attach wheels to base	
H	Attach cord	

a Complete the table to show how the activities depend on each other.

b Draw the corresponding activity network.

2 Draw an activity network corresponding to this precedence table.

Activity	Preceding activity
A	–
B	–
C	–
D	A
E	A
F	B
G	B
H	C, D, E, F
I	F
J	F
K	I
L	I
M	G, H, K

3

Activity	Must be preceded by	Duration (days)
A	–	15
B	–	12
C	A	10
D	B	10
E	C, D	8
F	C, D	6
G	E	10
H	E, F	7

a Draw an activity network for this precedence table.

b Perform a forward and a backward pass on your network. State the project duration.

c Find the effect on the project duration if

i Activity C took 2 days longer than expected,

ii Activity F took 3 days longer than expected.

4

Activity	Depends on	Duration (hours)
A	–	2
B	A	5
C	A	9
D	B, C	12
E	D	3
F	C	2
G	D, F	5
H	D, F	9
I	E, G, H	5

a Draw an activity network for this precedence table.

b Perform a forward and a backward pass on your network. State the project duration.

c Activity F takes longer than expected. What is the maximum time it could take without affecting the overall project duration?

13.2 Critical paths

Fluency and skills

The time available for an activity is the gap between its earliest possible start time and its latest possible finish time. For some activities, the time available is greater than the duration of the activity. The activity has **float**.

Float = (latest finish time – earliest start time) – duration

Key point

An activity with zero float is a **critical activity**. An increase in its duration, or delay in starting it, will increase the overall project duration.

A critical activity has a float of zero.

A sequence of critical activities is a **critical path**.

Key point

Example 1

The diagram shows the result of a forward and backward pass on an activity network.

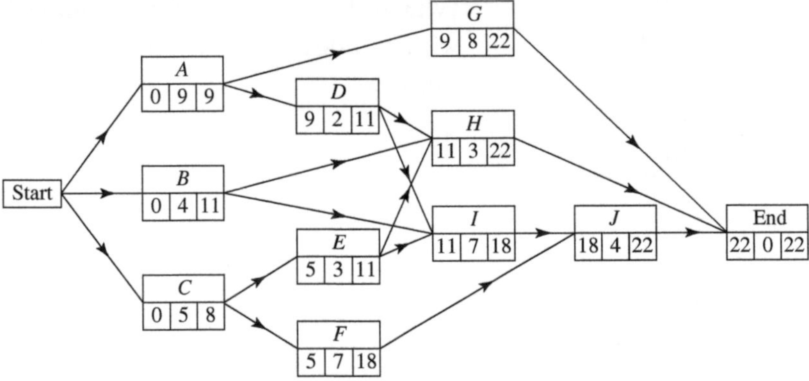

Find the float on each activity and hence identify the critical activities.

This table shows the results.

Activity	Duration	Earliest start time	Latest finish time	Total float
A	9	0	9	0
B	4	0	11	7
C	5	0	8	3
D	2	9	11	0
E	3	5	11	3
F	7	5	18	6
G	8	9	22	5
H	3	11	22	8
I	7	11	18	0
J	4	18	22	0

For example the float for $E = (11 - 3) - 5$ $= 3$

The critical activities are *A*, *D*, *I* and *J*, shown in red.

The sequence of activities $A - D - I - J$ form the critical path.

Sometimes there are two or more critical paths.

Example 2

The diagram shows the result of a forward and backward pass on an activity network.

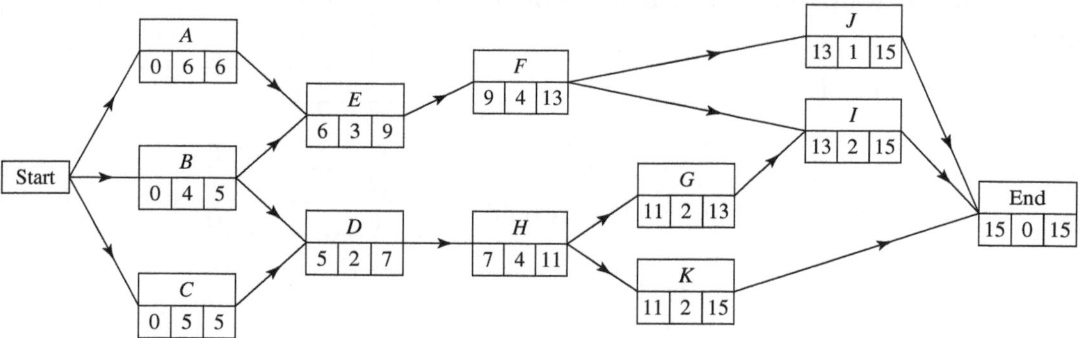

Show that there are two critical paths on this network. Illustrate them on the diagram.

Activity	Duration	Earliest start time	Latest finish time	Total float
A	6	0	6	0
B	4	0	5	1
C	5	0	5	0
D	2	5	7	0
E	3	6	9	0
F	4	9	13	0
G	2	11	13	0
H	4	7	11	0
I	2	13	15	0
J	1	13	15	1
K	2	11	15	2

The critical activities are *A, C, D, E, F, G, H* and *I*

These form two paths *A-E-F-I* and *C-D-H-G-I*, as shown in the diagram below.

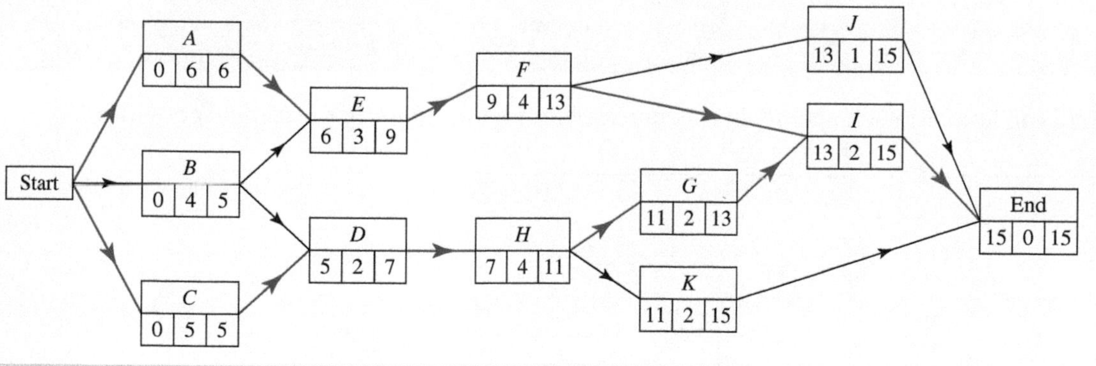

DISCRETE

Example 3

This precedence table shows a project with the duration of each activity given in days.

a Draw an activity network and find the duration of the project.

b Identify the critical path(s).

c You could reduce the duration of any of activities A–F by 1 day at the cost of £100. How much would it cost to reduce the project duration by 1 day, and how would you achieve this?

Activity	Preceded by	Duration
A	–	9
B	–	5
C	–	8
D	–	3
E	A	3
F	D	4
G	B, E	10
H	C, F	4
I	B, H	10
J	G, I	2

a This is the network with forward and backward passes completed.

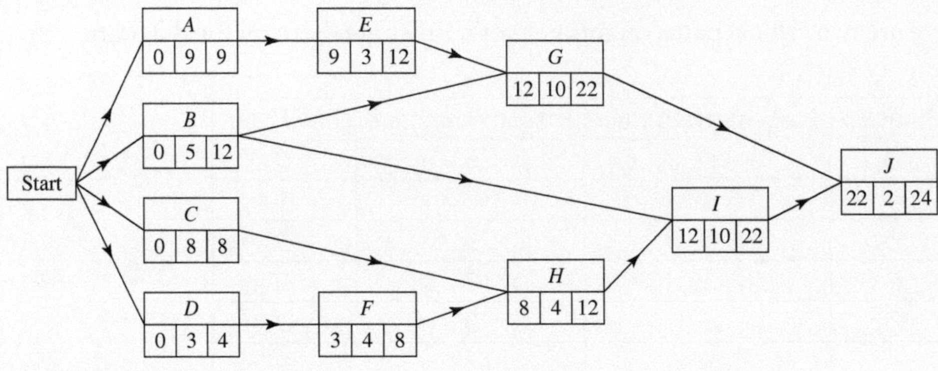

The project duration is 24 days.

b The floats of the activities are A 0, B 7, C 0, D 1, E 0, F 1, G 0, H 0, I 0, J 0. The critical activities (zero float) form two critical paths A–E–G–J and C–H–I–J.

c Both critical paths would need to be shortened by 1 day to have the desired effect on the overall duration. C must be shortened by 1 day, and either A or E by 1 day, with a total cost of £200

Exercise 13.2A Fluency and skills

1 Find the float for each activity in this network and hence identify the critical activities.

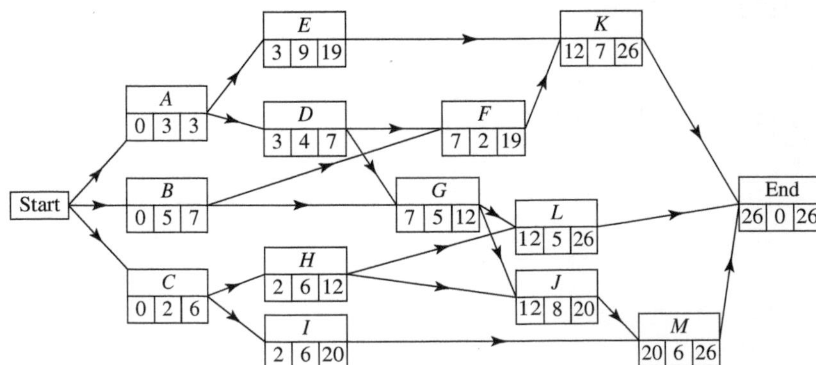

2 For each of the following networks, perform a forward and a backward pass. State the project duration. Find the float for each activity and hence identify the critical activities.

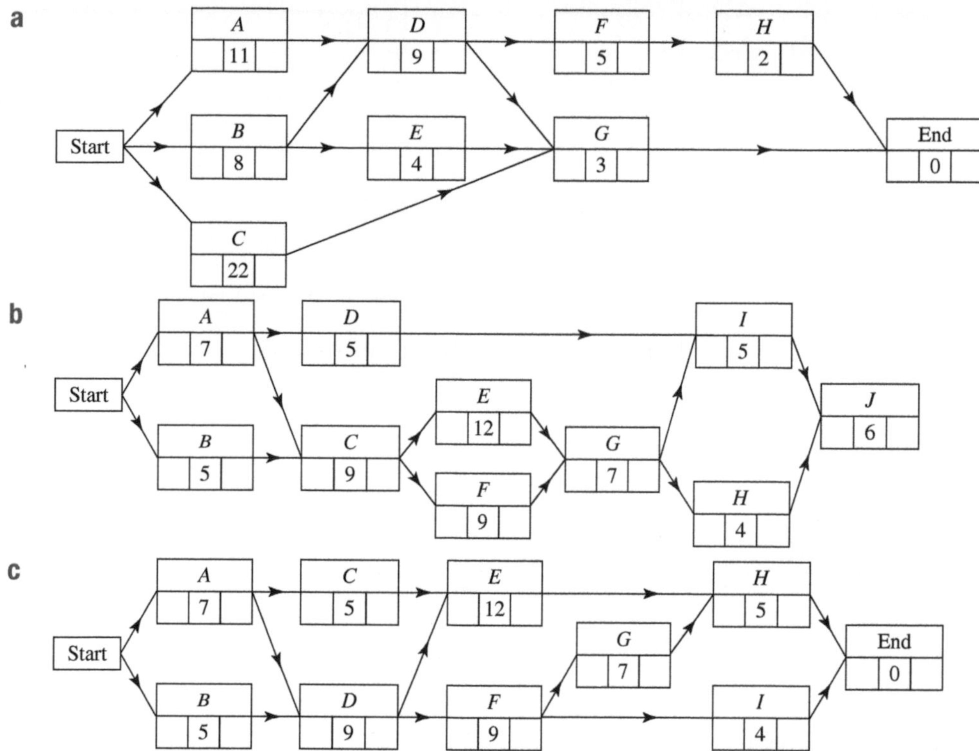

3 For this precedence table, draw an activity network and use it to find the project duration and the critical activities.

Activity	Duration	Must be preceded by
A	5	–
B	2	–
C	2	*A*
D	6	*A*, *B*
E	4	*B*
F	2	*C*, *E*

4 For this precedence table,

a Draw an activity network and find the project duration,

b Show that there are two critical paths.

Activity	Duration (hours)	Must be preceded by
A	8	–
B	6	*A*
C	7	*A*
D	11	*A*
E	12	*B*, *C*
F	8	*C*, *D*
G	5	*B*, *D*

Reasoning and problem-solving

Strategy

To solve problems involving critical activities

1. If necessary, draw a neat activity network (sketch it first to get a clear layout).
2. Perform forward and backward passes to find the earliest start times and latest finish times.
3. Calculate the float for each activity and hence identify the critical activities.
4. Answer the question.

Example 4

The network shown represents a building project, with durations in days.

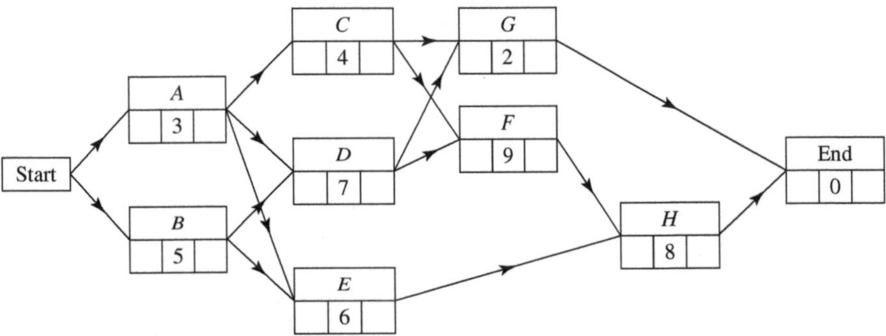

a Find the duration of the project if all goes to plan.

b Identify the critical path.

c In fact, activity C is delayed by 7 days. What effect does this have on the duration of the project?

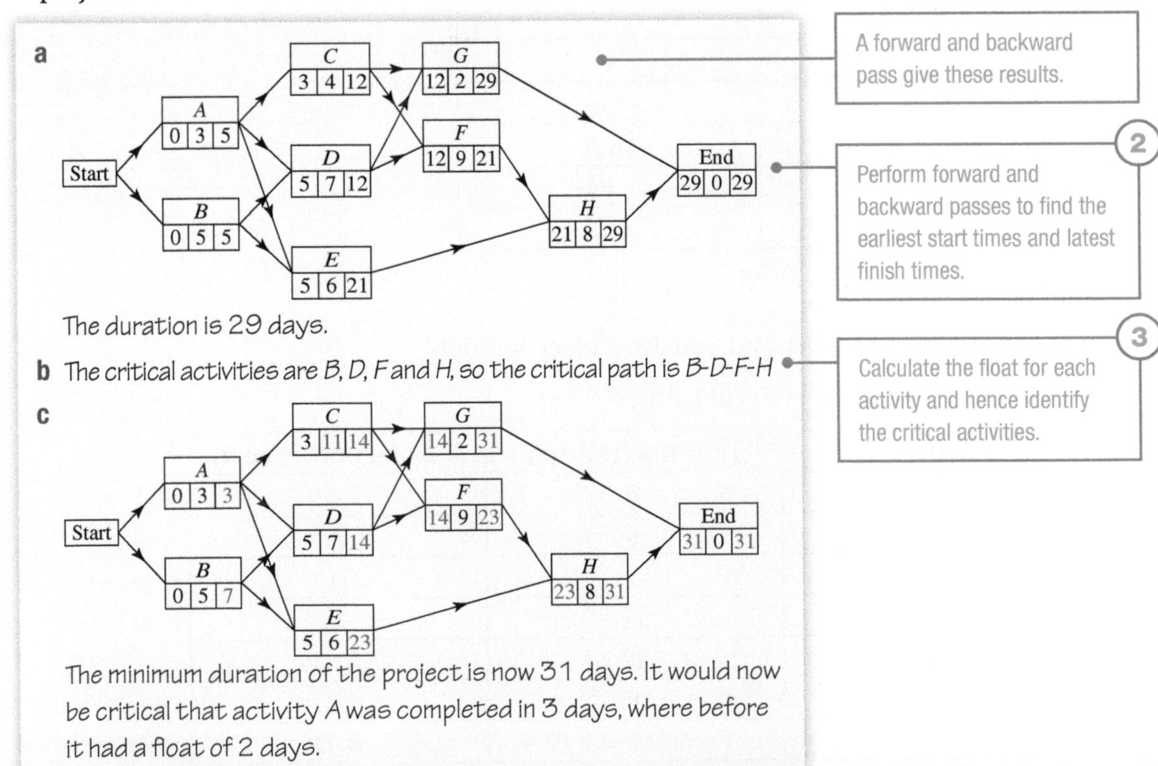

a
A forward and backward pass give these results.
Perform forward and backward passes to find the earliest start times and latest finish times.
The duration is 29 days.
b The critical activities are B, D, F and H, so the critical path is B-D-F-H
Calculate the float for each activity and hence identify the critical activities.
c
The minimum duration of the project is now 31 days. It would now be critical that activity A was completed in 3 days, where before it had a float of 2 days.

Exercise 13.2B Reasoning and problem-solving

1

Activity	Preceded by	Duration (hours)
A	-	9
B	*A*	6
C	*A*	11
D	*B*	11
E	*B*	7
F	*C*	1
G	*C*	12
H	*D*	4
I	*E. F*	10

a Find the duration of this project.

b Find the critical path(s).

c It is possible to reduce the duration of activities *C*, *D* and E at a cost of £50 per hour per activity. How much will it cost to reduce the duration of the project by 3 hours and how would this be achieved?

2 Here is a precedence table for a project.

Activity	Preceded by	Duration (days)
A	–	6
B	*A*	4
C	*A*	7
D	*B*	9
E	*B*	6
F	*B, C*	3
G	*C, E*	8
H	*D, E*	3
I	*E. F*	5

a Find the duration of the project.

b State the critical activities.

c Find the effect, if any, on the duration of the project if

i *D* was delayed by 4 days,

ii *F* and *I* were each delayed by 2 days.

3 Here is a precedence table for a project.

Activity	Must be preceded by	Duration (days)
A	–	12
B	–	9
C	*A*	6
D	*B*	8
E	*C, D*	5
F	*C, D*	6
G	*E*	7
H	*E, F*	4

a Find the duration of the project.

b Identify the critical activities and state the float of the non-critical activities.

c What is the maximum period by which activity *B* could be delayed without increasing the overall duration of the project?

4 Here is a precedence table for a project.

Activity	Depends on	Duration (hours)
A	–	10
B	*A*	10
C	–	6
D	*A, C*	5
E	*D*	9
F	*A*	10
G	*B, D*	6
H	*E, F, G*	7

a Find the duration of the project and identify the critical activities.

b Activity *B* can be split into two 5 hour tasks, one of which can be delayed until activity *F* is complete but is not required before *G* can start. Investigate whether it is possible to shorten the overall project, justifying your conclusions.

13 Summary and review

Chapter summary

- A **precedence table** or **dependence table** records activities, their duration and the order in which they can occur.
- In an **activity network**, each activity is represented by a node, and the arcs show the order of precedence.
- You analyse the project to find the **earliest start time** and **latest finish time** for each activity, and these are recorded in a box like this at each node.

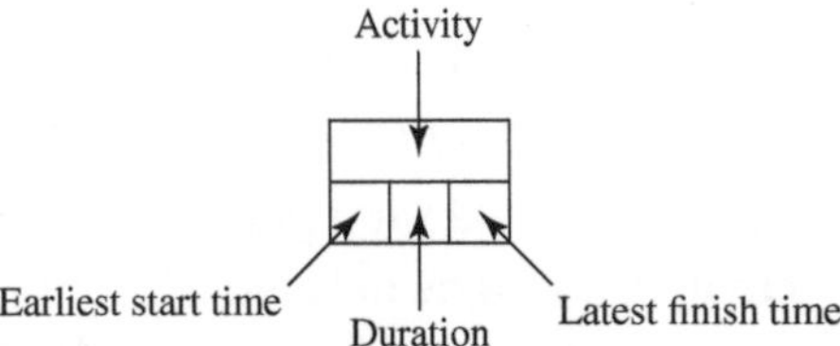

- To find the earliest start times, you make a **forward pass** through the network. The earliest start time for an activity is the maximum of the earliest times by which all preceding activities can be completed.
- The project duration is the earliest time by which all activities can be completed.
- To find the latest finish times, you make a **backward pass** through the network. The latest finish time for an activity is the minimum of the latest times by which all following activities can start without affecting the project duration.
- An activity has **float** if the time available for it (the gap between its earliest start time and its latest finish time) is greater than its duration.
- Float = (latest finish time – earliest start time) – duration
- An activity with zero float is a **critical activity**. An increase in its duration, or delay in starting it, will increase the overall project duration.
- The sequence of critical activities is the **critical path**.

Check and review

You should now be able to...	Review Questions
✔ Draw an activity network corresponding to a given precedence table.	1
✔ Perform a forward pass to find the earliest start time for each activity.	2
✔ Find the project duration.	2
✔ Perform a backward pass to find the latest finish time for each activity.	3
✔ Calculate the float for each activity.	4
✔ Identify the critical activities.	4

1

Activity	Preceded by
A	–
B	–
C	–
D	A, B
E	B, C
F	D
G	D
H	F, G
I	G
J	E, G

Draw an activity network to show this project.

2

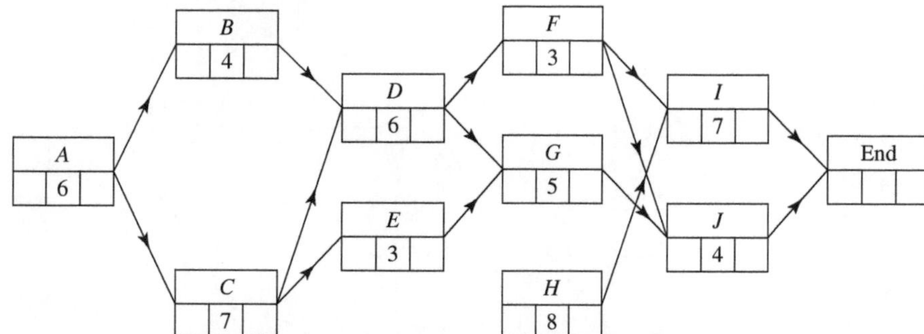

Perform a forward pass on this network to find the earliest start time for each activity (the durations are in hours). State the project duration.

3

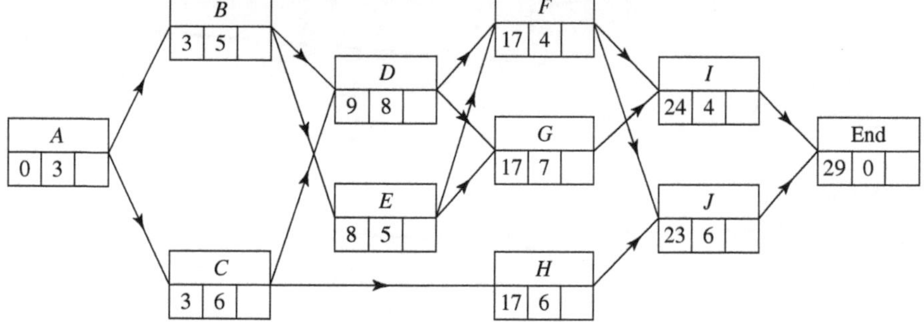

Perform a backward pass on this network to find the latest finish time for each activity (durations are in days).

4

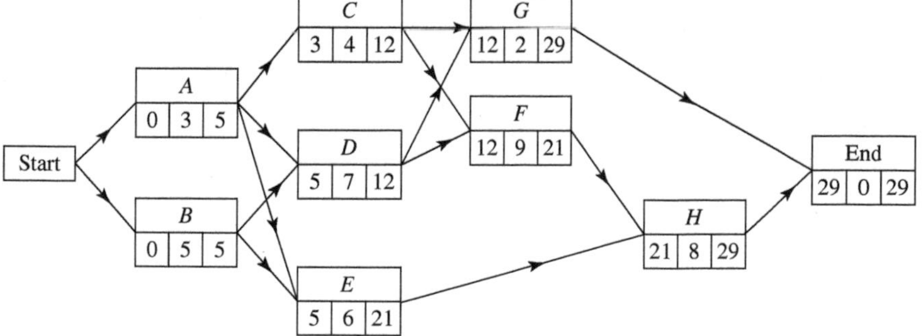

a Calculate the float for each activity in this network (durations are in days).

b State which activities are critical.

DISCRETE

History

Henry Laurence Gantt (1861–1919) was an American mechanical engineer who created and popularised the Gantt chart in the 1910s.

Gantt Charts play a key role in the management of many projects. They were used in the construction of the Hoover Dam.

The Gantt chart, because of its presentation of facts in their relation to time, is the most notable contribution to the art of management made in this generation.

- Wallace Clark

Research

A resource histogram can be used to help deal with the issue of an activity requiring multiple workers.

Look at this cascade diagram and the resources required for the activities.

Activities A and B each require 4 people and C, D, E and F each require 2 people.

The corresponding resource histogram would be as shown here.

Using the float in the cascade diagram, see if you can level out the resource histogram to minimise the number of people employed and maximise employment time of workers.

A B C D
E
F

People
F
E F
F
A B C D

Assessment

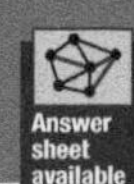

1 Draw an activity network to represent this precedence table.

Activity	Immediate predecessor(s)
A	–
B	–
C	*A*, *B*
D	*C*
E	*C*
F	*D*, *E*

[2 marks]

2 Draw an activity network to represent this precedence table.

Activity	Immediate predecessor(s)
A	–
B	–
C	–
D	–
E	*A*
F	*B*
G	*C*
H	*D*
I	*E*, *F*
J	*G*
K	*I*, *J*
L	*H*, *K*

[3]

3 Draw an activity network to represent this precedence table.

Activity	Immediate predecessor(s)
A	–
B	*A*
C	*A*
D	*A*
E	*B*
F	*B*, *C*
G	*C*
H	*D*
I	*E*, *F*
J	*F*, *G*
K	*H*
L	*I*
M	*J*, *K*

[4]

4 Find the earliest start time and latest finish time for each activity in this network.

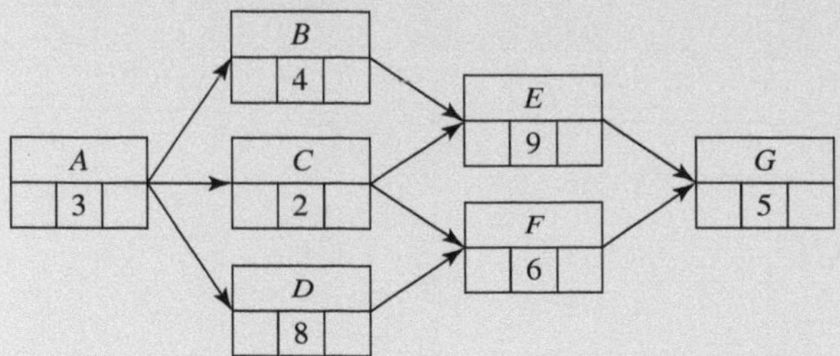

[3]

5 Find the earliest start time and latest finish time for each activity in this network.

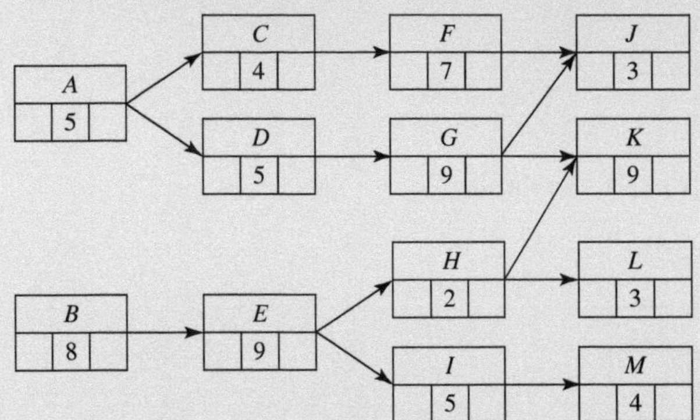

[4]

6 a Draw an activity network to represent this precedence table.

Activity	Duration	Immediate predecessor(s)
A	1	–
B	6	*A*
C	3	*A*
D	6	*A*
E	4	*B*
F	5	*B,C*
G	5	*D*

b Find the earliest start time and latest finish time for each activity in the network. [6]

7 a Draw an activity network to represent this precedence table.

Activity	Duration	Immediate predecessor(s)
A	8	–
B	8	–
C	4	–
D	9	*A, B*
E	4	*B, C*
F	8	*C*
G	9	*D, E*
H	7	*E, F*
I	7	*G, H*

b Find the earliest start time and latest finish time for each activity in the network.

c Find the critical activities. [8]

8 a Draw an activity network to represent this precedence table.

Activity	Duration	Immediate predecessor(s)
A	2	–
B	5	–
C	5	–
D	3	A, B
E	9	B
F	3	B,C
G	10	D
H	4	D, E, F
I	7	F
J	5	G, H
K	2	H, I

b Find the earliest start time and latest finish time for each activity in the network.

c Find the critical paths. **[9]**

9 a Draw an activity network to represent this precedence table.

Activity	Duration	Immediate predecessor(s)
A	10	–
B	6	–
C	2	–
D	1	–
E	9	A, B
F	15	C, D
G	8	A, E
H	3	B, C
I	6	D, F
J	4	G, H
K	3	H, I

b Find the earliest start time and latest finish time for each activity in the network.

c Find the critical path.

d The duration of task I is changed to x days ($x \geq 6$) without changing the minimum completion time of all the activities. Find a further inequality relating to the maximum value of x **[8]**

10 a Draw an activity network to represent this precedence table.

Activity	Duration	Immediate predecessor(s)
A	7	–
B	8	–
C	11	A, B
D	2	B
E	6	–
F	6	C, D
G	4	E, F
H	1	F
I	6	G, H
J	3	H
K	5	I, J

b Find the earliest start time and latest finish time for each activity in the network.

c Find the critical path.

d Given that A now takes 10 days to complete, list the critical activities. **[8]**

11 a Draw an activity network to represent this table of activities.

Activity	Duration	Immediate predecessor(s)
A	9	–
B	3	–
C	2	A
D	5	A, B
E	5	B
F	8	C, D, E
G	4	F
H	1	F
I	7	F
J	6	G, H
K	7	H, I

b Find the earliest start time and latest finish time for each activity in the network.

c Find the critical path.

d Given that G and J each now take 9 days to complete, find the new minimum completion time for the full set of activities. **[9]**

14 Linear programming and game theory 1

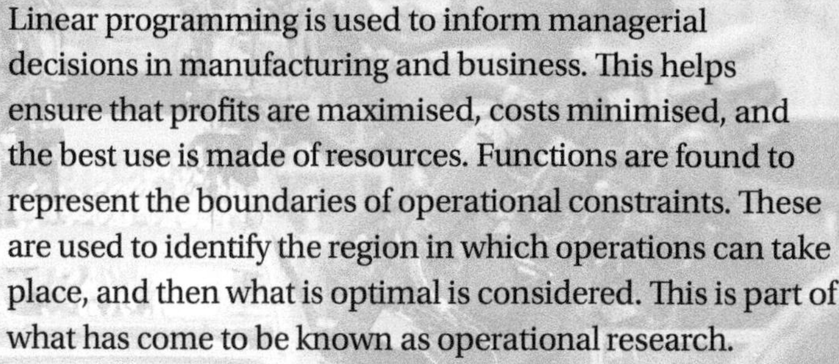

Linear programming is used to inform managerial decisions in manufacturing and business. This helps ensure that profits are maximised, costs minimised, and the best use is made of resources. Functions are found to represent the boundaries of operational constraints. These are used to identify the region in which operations can take place, and then what is optimal is considered. This is part of what has come to be known as operational research.

The development of computers has assisted with the increasingly complex calculations that are required. This builds on mathematics that was first developed during World War II to ensure that military supplies and personnel were distributed most efficiently. Nowadays, linear programming techniques have a wide range of applications. These include consideration of deploying staff, production, managing stock, marketing and financial management.

Orientation

What you need to know	What you will learn	What this leads to
KS 4 • Inequalities.	• To formulate and solve linear programming problems. • To analyse zero-sum games. • To analyse mixed-strategy games.	• Chapter 30 Applying techniques of linear programming to game theory. • The simplex method.

14.1 Constrained optimisation

Fluency and skills

Linear programming is a set of mathematical techniques which help with decision making in a variety of industrial and economic situations. The aim is to choose the best combination of a number of quantities (variables) to achieve the best (optimal) outcome. The objective is typically to maximise profit or minimise cost.

To translate the problem into mathematical terms, you first produce a **linear programming (LP) formulation**. To do this you:

- Identify the quantities you can vary. These are the **decision variables** (or **control variables**).
- Identify the limitations on the values of the decision variables. These are the **constraints**.
- Identify the quantity to be optimised. This is the **objective function**.

Example 1

A manufacturer makes two types of muesli, Standard and De Luxe.

There are two more (trivial) constraints $x \geq 0$ and $y \geq 0$

1 kg of Standard contains 800 grams of oat mix and 200 grams of fruit mix. 1 kg of De Luxe has 600 grams of oat mix and 400 grams of fruit mix. 1 kg of Standard makes 60 p profit. 1 kg of De Luxe makes 80 p.

They have 3000 kg of oat mix and 1000 kg of fruit mix in stock. They know that they can sell at most 3500 kg of Standard and 2000 kg of De Luxe.

You must decide how much of each type the manufacturer must make in order to maximise the profit. Write this problem as a linear programming formulation.

	Oat mix (kg)	Fruit mix (kg)	Max sales (kg)	Profit (£ per kg)
1 kg Standard	0.8	0.2	3500	0.60
1 kg De Luxe	0.6	0.4	2000	0.80
Availability	3000	1000		

This table summarises the information.

Suppose they make x kg of Standard and y kg of De Luxe.

These are the decision variables.

The upper limits on sales give $x \leq 3500$ and $y \leq 2000$

The constraints are the limitations on sales and the amount of raw materials available.

x kg of Standard uses $0.8x$ kg of oat mix.

y kg of De Luxe uses $0.6y$ kg of oat mix.

There is 3000 kg of oat mix, so $0.8x + 0.6y \leq 3000$

which simplifies to $4x + 3y \leq 15000$

(*Continued on the next page*)

x kg of Standard uses $0.2x$ kg of fruit mix and y kg of De Luxe uses $0.4y$ kg of fruit mix.

There is 1000 kg of oat mix, so $0.2x + 0.4y \le 1000$

which simplifies to $x + 2y \le 5000$

The profit on x kg of Standard and y kg of De Luxe is $P = 0.6x + 0.8y$

The function you want to maximise is called the objective function and is often denoted by the letter P

Maximise $P = 0.6x + 0.8y$

Subject to $4x + 3y\ 15\,000$

$x + 2y \le 5000$

$x \le 3500$

$y \le 2000$

$x \ge 0, y \ge 0$

This is the complete linear programming formulation.

It is important that you write out the problem in this form.

DISCRETE

Example 2

A paper recycler has two processing plants.

Plant A can process 4 tonnes of waste paper and 1 tonne of cardboard per hour.

Plant B can process 5 tonnes of waste paper and 2 tonnes of cardboard per hour.

It costs £400 per hour to run plant A and £600 per hour to run plant B.

There is a union agreement that each plant must get at least one third of the run time of every consignment.

A consignment of 100 tonnes of waste paper and 35 tonnes of cardboard must be shared between the plants so that costs are minimised.

Write this as a linear programming formulation.

This table summarises the information:

	Waste paper (tonnes)	Cardboard (tonnes)	Cost (£)
Plant A (1 hour)	4	1	400
Plant B (1 hour)	5	2	600
Availability	100	35	

Let x be hours for plant A and y be hours for plant B.

The decision variables are the running times.

$x \ge \frac{1}{3}(x+y)$ and $y \ge \frac{1}{3}(x+y)$

This is the union agreement written algebraically.

which simplify to

$y \le 2x$ and $x \le 2y$

$4x + 5y \ge 100$

Since the plants must process at least 100 tonnes of paper in the time available.

(*Continued on the next page*)

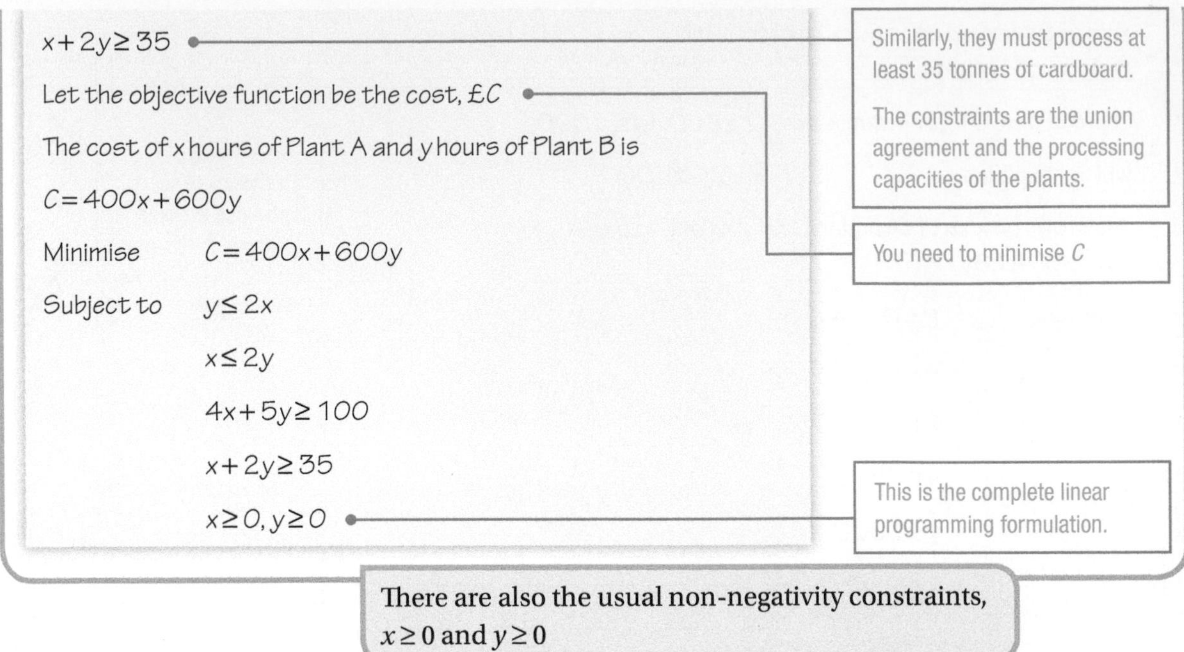

$x + 2y \geq 35$

Similarly, they must process at least 35 tonnes of cardboard.

The constraints are the union agreement and the processing capacities of the plants.

Let the objective function be the cost, £C

The cost of x hours of Plant A and y hours of Plant B is

$C = 400x + 600y$

You need to minimise C

Minimise $\quad C = 400x + 600y$

Subject to $\quad y \leq 2x$

$x \leq 2y$

$4x + 5y \geq 100$

$x + 2y \geq 35$

$x \geq 0, y \geq 0$

This is the complete linear programming formulation.

There are also the usual non-negativity constraints, $x \geq 0$ and $y \geq 0$

Once you have stated a problem as a linear programming formulation, the next stage is to solve it. If there are just two decision variables, you can use graphical methods.

Consider the muesli problem from Example 1, where x kg of Standard and y kg of De Luxe muesli were produced.

Maximise $\quad P = 0.6x + 0.8y$

Subject to $\quad x \leq 3500$

$y \leq 2000$

$4x + 3y \leq 15\,000$

$x + 2y \leq 5000$

$x \geq 0, y \geq 0$

You draw a graph to illustrate the constraints.

First, consider $x \leq 3500$ and $y \leq 2000$ by drawing the lines $x = 3500$ and $y = 2000$

Shade the regions which are **not** needed. This is called shading out.

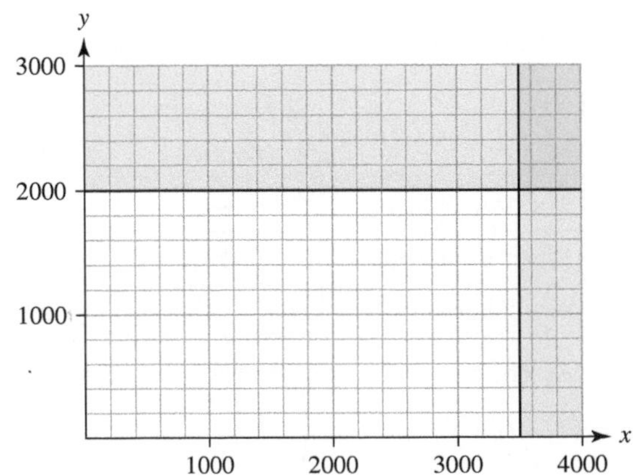

Next, consider $4x + 3y \leq 15000$ by drawing the line $4x + 3y = 15000$

As before, shade the region which does not satisfy the inequality.

Finally, consider $x + 2y \leq 5000$, $x \geq 0$ and $y \geq 0$

All allowable combinations of x and y lie in the unshaded region of the graph (including its boundary lines). This is called the **feasible region**.

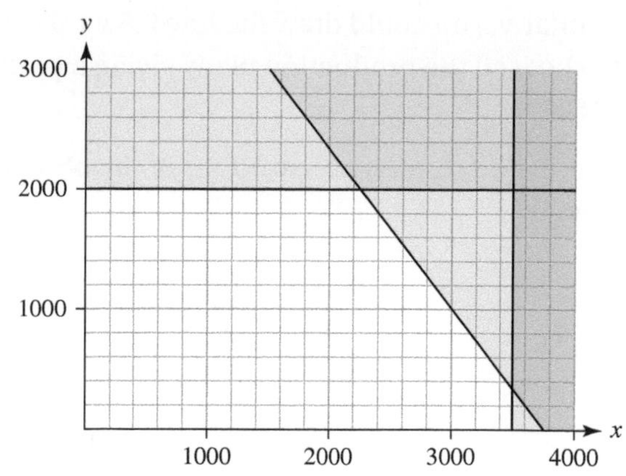

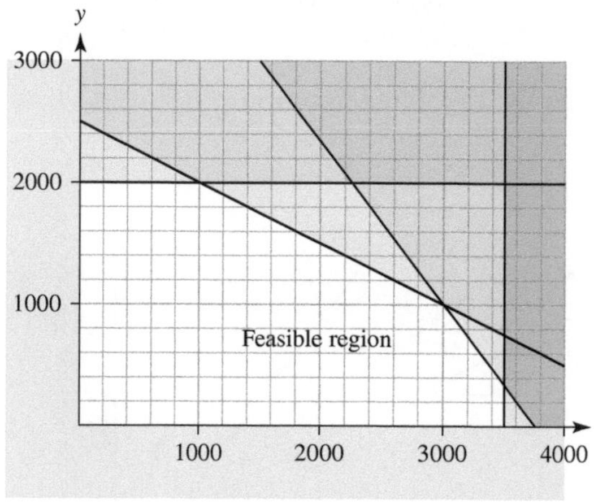

If the inequalities had been $<$ rather than $\leq$, the boundaries would not be included. The usual convention is to draw the boundary line dotted for $<$ or $>$, and continuous for $\leq$ and $\geq$

Key point

The **feasible region** is the set of (x, y) values that satisfy all the constraints. It is the unshaded region on the graph.

You now illustrate the objective function $P = 0.6x + 0.8y$. To do this, choose any values of x and y inside the feasible region. Suppose you chose $x = 1000$, $y = 1000$

This would give $P = 0.6 \times 1000 + 0.8 \times 1000 = 1400$

This plan gives £1400 profit.

There are other production plans giving $P = 1400$, for example $x = 0$, $y = 1750$ or $x = 2000$, $y = 250$
These plans all lie on the line $0.6\,x + 0.8\,y = 1400$

You draw this line on your graph.

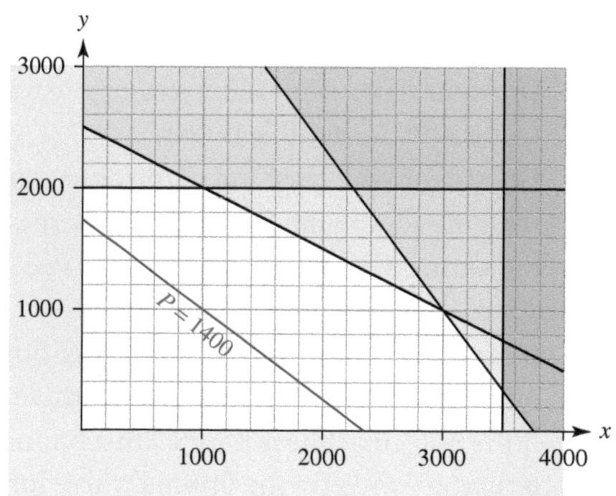

DISCRETE

Similarly, you could draw the line $0.6x + 0.8y = 1800$ to show all the production plans giving a profit of £1800

These are two possible positions of the **objective line.**

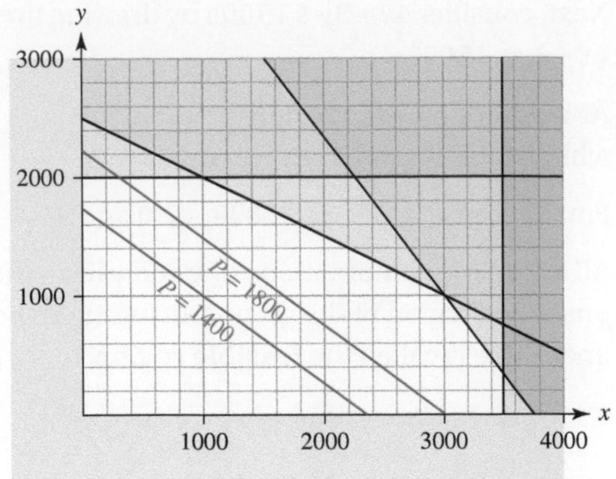

> **Key point**
>
> The **objective line** is a line joining all points (x, y) for which the objective function takes a specified value.

Profit P increases as the objective line moves to the right (always keeping the same gradient). As long it crosses the feasible region, there is a production plan giving that profit.

The maximum profit occurs at the extreme position of the line which still includes a point of the feasible region. As you move the line to the right, its last contact with the feasible region will be at a vertex.

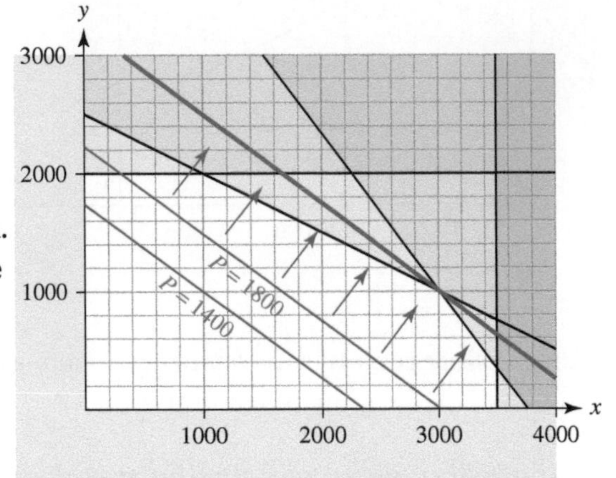

In this example, it leaves the feasible region where the lines $x + 2y = 5000$ and $4x + 3y = 15000$ intersect.

Solving these two equations simultaneously you get the solution $x = 3000$, $y = 1000$ (the point (3000, 1000))
This gives $P = 0.6 \times 3000 + 0.8 \times 1000 = 2600$

So the best production plan is 3000 kg of Standard muesli and 1000 kg of De Luxe muesli, giving a profit of £2600

In summary, to solve a linear programming problem graphically:

- Draw lines corresponding to the constraints and shade out areas to identify the feasible region.
- Draw a possible position of the objective line.
- Imagine sliding the objective line across the graph keeping it parallel to the original. Identify the vertex where it would leave the feasible region. This vertex corresponds to the optimal solution.
- If the choice of vertex is not obvious, find the coordinates of the likely vertices and test which one gives the best value of the objective function.
- If the objective line is parallel to a boundary of the feasible region, then all points on that boundary will give the optimal value for the objective function.

Example 3

This continues the waste paper problem in Example 2

Minimise $C = 400x + 600y$

Subject to $x \le 2y$

$y \le 2x$

$4x + 5y \ge 100$

$x + 2y \ge 35$

$x \ge 0, y \ge 0$

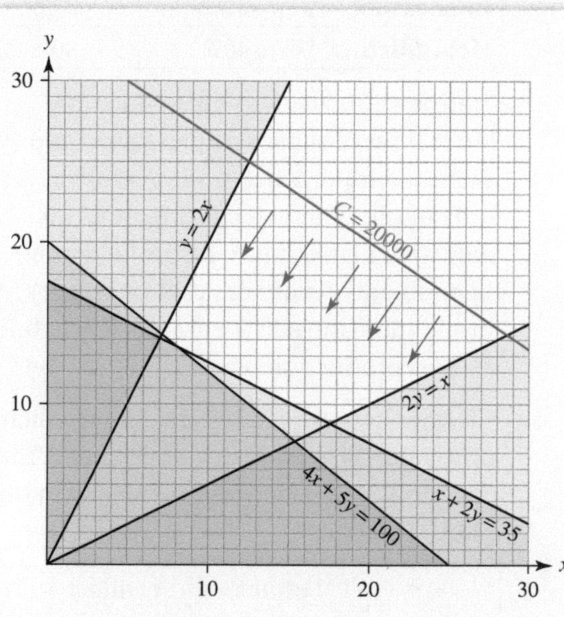

The line shown is the objective line $C = 20000$

Moving the objective line to the left parallel to itself reduces the value of C. Remember that you are trying to minimise C

Solving $y = 2x$ and $4x + 5y = 100$ gives $x = 7\frac{1}{7}, y = 14\frac{2}{7}$

This gives $C = 400 \times 7\frac{1}{7} + 600 \times 14\frac{2}{7} = 11428\frac{4}{7}$

Solving $4x + 5y = 100$ and $x + 2y = 35$ gives $x = 8\frac{1}{3}, y = 13\frac{1}{3}$

This gives $C = 400 \times 8\frac{1}{3} + 600 \times 13\frac{1}{3} = 11333\frac{1}{3}$

Solving $x + 2y = 35$ and $x = 2y$ gives $x = 17\frac{1}{2}, y = 8\frac{3}{4}$

This gives $C = 400 \times 17\frac{1}{2} + 600 \times 8\frac{3}{4} = 12250$

The minimum value of C is $11333\frac{1}{3}$, when $x = 8\frac{1}{3}, y = 13\frac{1}{3}$

The merchant should run plant A for $8\frac{1}{3}$ hours and plant B for $13\frac{1}{3}$ hours.

It is hard to see from the graph which of the three vertices gives the minimum value of C, so find C at each vertex.

DISCRETE

Exercise 14.1A Fluency and skills

1 A company makes two brands of dog food, A and B, which it sells in 500 g tins. Both brands are a mixture of meat and vegetables. The table shows the amounts of these in each brand and the amounts in stock.

	Meat (kg)	Vegetables (kg)
Brand A	0.2	0.3
Brand B	0.4	0.1
Amount in stock (kg)	8000	6000

A tin of brand A makes 10 p profit, while a tin of brand B makes 15 p. The company wants a production plan to maximise their profit. Taking the number of tins of brands A and B produced as x and y, respectively, express this problem as a linear programming formulation. You do not need to solve the problem.

2 For each of the following linear programming problems, draw a graph showing the feasible region and one possible position of the objective line. Hence find the optimal value of the objective function and the corresponding values of x and y

a Maximise $P = 3x + 2y$
Subject to $3x + 4y \leq 120$
$3x + y \leq 75$
$x \geq 10, y \geq 5$

b Maximise $P = 4x + y$
Subject to $x + y \leq 10$
$2x + y \leq 16$
$x \geq 0, y \geq 0$

c Minimise $C = 5x + 4y$
Subject to $15x + 8y \geq 90$
$y \geq x$
$y \leq 2x$

d Maximise $R = 2x + y$
Subject to $x + y \leq 14$
$x + 2y \leq 20$
$2x + 3y \leq 32$
$x \geq 0, y \geq 0$

3 Roger Teeth Ltd make fruit drinks of two types, Econofruit and Healthifruit, consisting of fruit juice, sugar syrup and water. The proportions of these in the two drinks are shown in the table.

	Fruit juice	Sugar syrup
Econofruit	20%	50%
Healthifruit	40%	30%

There are 20 000 litres of fruit juice and 30 000 litres of sugar syrup in stock (and unlimited water). The profit per litre is 30 p for Econofruit and 40 p for Healthifruit. They wish to maximise the profit. Express the problem as a linear programming formulation and solve it graphically to find the best production plan.

4 A farmer has 75 hectares of land on which to grow a mixture of wheat and potatoes. The costs and profits involved are shown in the table.

	Labour (man-hours per ha)	Fertiliser (kg per ha)	Profit (£ per ha)
Wheat	30	700	80
Potatoes	50	400	100

There are 2800 man-hours of labour and 40 tonnes of fertiliser available. The aim is to maximise the profit. Express the problem as a linear programming formulation and solve it graphically to find the best planting scheme.

Reasoning and problem-solving

Strategy

To solve a problem involving linear programming

1. Identify the decision variables and label them x, y, z…
2. Express the constraints as inequalities.
3. Identify the objective function (to be maximised or minimised).
4. If the problem has two variables, solve graphically.
5. Answer the question.

Example 4

> This example has three decision variables, but you can reduce the number of variables to two in the light of extra information. There is also a new type of constraint involved.

A manufacturer makes three types of dining chair. All chairs pass through three workshops for cutting, assembly and finishing. The cutting workshop runs for 50 hours per week, the assembly workshop for 35 hours per week and the finishing workshop for 40 hours per week. The times in the workshops for each type of chair and the profit gained from each are shown in the table.

Chair type	Cutting (hours)	Assembly (hours)	Finishing (hours)	Profit (£)
A	0.5	0.6	0.75	20
B	2	1	0.75	60
C	0.5	0.4	0.5	20

The aim is to maximise the weekly profit.

a Write this as a linear programming formulation.

b Given that half of the chairs they sell are type C, rewrite the problem with two variables and solve to find the best production plan.

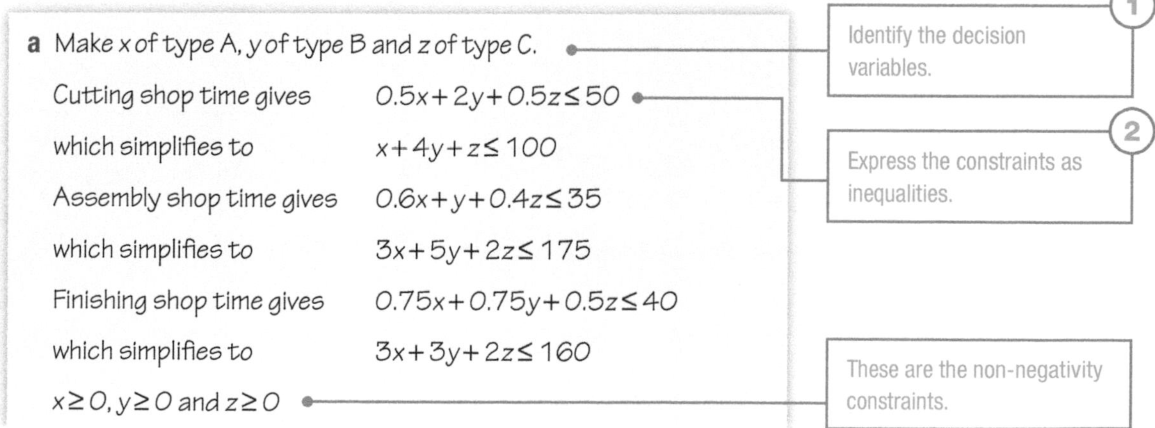

a Make x of type A, y of type B and z of type C.

Cutting shop time gives $0.5x + 2y + 0.5z \le 50$

which simplifies to $x + 4y + z \le 100$

Assembly shop time gives $0.6x + y + 0.4z \le 35$

which simplifies to $3x + 5y + 2z \le 175$

Finishing shop time gives $0.75x + 0.75y + 0.5z \le 40$

which simplifies to $3x + 3y + 2z \le 160$

$x \ge 0$, $y \ge 0$ and $z \ge 0$

(1) Identify the decision variables.

(2) Express the constraints as inequalities.

These are the non-negativity constraints.

(*Continued on the next page*)

DISCRETE

There is also the constraint that x, y and z are integers

> You can't sell fractions of a chair.

The profit is $P = 20x + 60y + 20z$

Maximise $P = 20x + 60y + 20z$

> 3 This is the objective function.
>
> You need to maximise the profit, £P

Subject to

$x + 4y + z \le 100$

$3x + 5y + 2z \le 175$

$3x + 3y + 2z \le 160$

$x \ge 0,\ y \ge 0,\ z \ge 0$

x, y and z are integers.

> This is the complete linear programming formulation.

b Given $z = x + y$

> Half of the sales are type C

Profit $P = 20x + 60y + 20(x + y) = 40x + 80y$

> Substitute $z = x + y$ into the formulation.

The constraints are

$x + 4y + (x + y) \le 100$, so $\quad 2x + 5y \le 100$

$3x + 5y + 2(x + y) \le 175$, so $\quad 5x + 7y \le 175$

$3x + 3y + 2(x + y) \le 160$, so $\quad x + y \le 32$

$x \ge 0$, $y \ge 0$, x and y are integers.

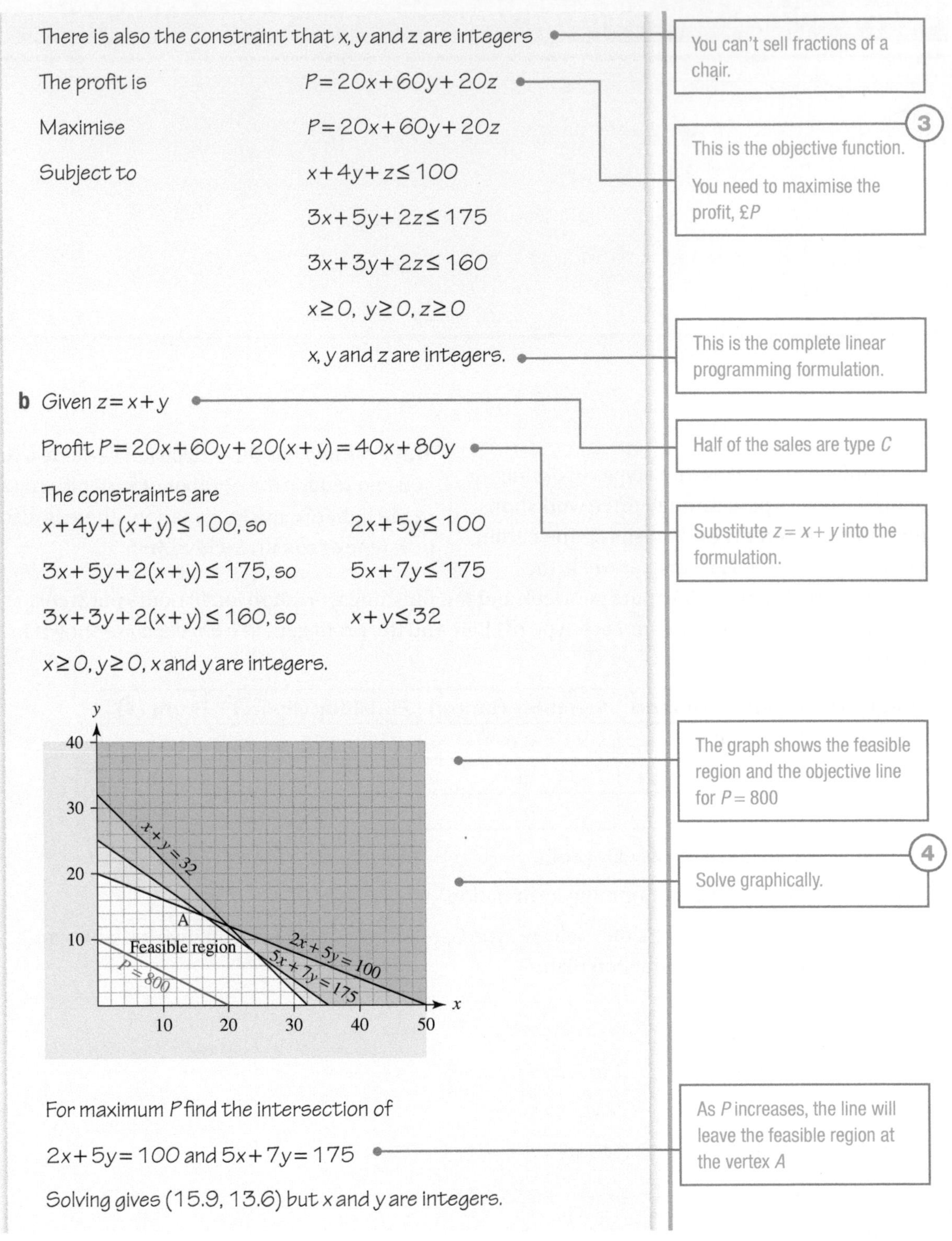

> The graph shows the feasible region and the objective line for $P = 800$

> 4 Solve graphically.

For maximum P find the intersection of

$2x + 5y = 100$ and $5x + 7y = 175$

> As P increases, the line will leave the feasible region at the vertex A

Solving gives (15.9, 13.6) but x and y are integers.

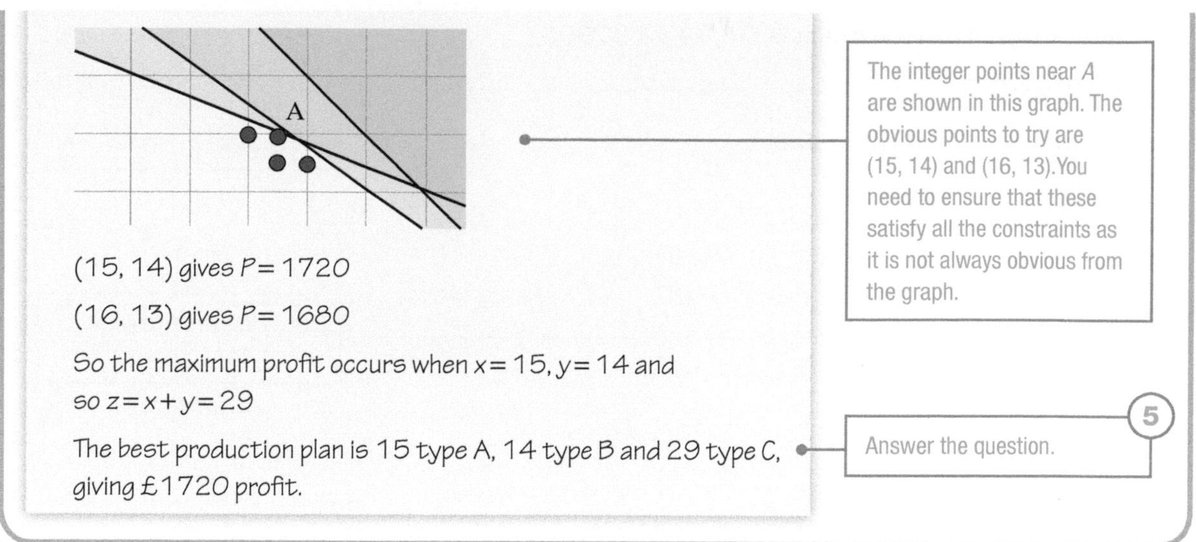

$(15, 14)$ gives $P = 1720$

$(16, 13)$ gives $P = 1680$

So the maximum profit occurs when $x = 15$, $y = 14$ and so $z = x + y = 29$

The best production plan is 15 type A, 14 type B and 29 type C, giving £1720 profit.

The integer points near A are shown in this graph. The obvious points to try are (15, 14) and (16, 13). You need to ensure that these satisfy all the constraints as it is not always obvious from the graph.

5 Answer the question.

Linear programming can also be used to decide on the most cost-effective proportions to use when blending materials.

Example 5

A company supplying vegetable oil buys from two sources, A and B. The oils are already a blend of olive oil, sunflower oil and other vegetable oils. The table shows the proportions, price and minimum weekly order of these.

	Olive oil	Sunflower oil	Other	Cost (p per litre)	Minimum order (litres)
A	50%	10%	40%	25	35000
B	20%	60%	20%	20	50000

The company wants to make a blend with at least 30% olive oil and at least 30% sunflower oil. They want to produce at least 90000 litres per week, and to minimise the cost.

Write this as a linear programming formulation and solve to find the best blend.

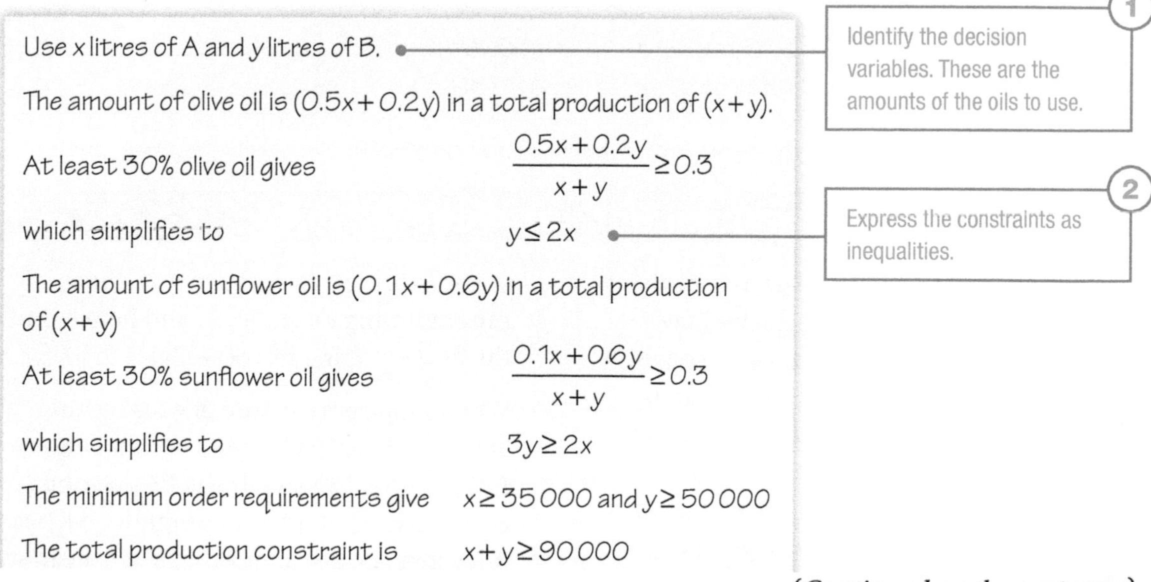

Use x litres of A and y litres of B.

The amount of olive oil is $(0.5x + 0.2y)$ in a total production of $(x + y)$.

At least 30% olive oil gives $\frac{0.5x + 0.2y}{x + y} \geq 0.3$

which simplifies to $y \leq 2x$

The amount of sunflower oil is $(0.1x + 0.6y)$ in a total production of $(x + y)$

At least 30% sunflower oil gives $\frac{0.1x + 0.6y}{x + y} \geq 0.3$

which simplifies to $3y \geq 2x$

The minimum order requirements give $x \geq 35\,000$ and $y \geq 50\,000$

The total production constraint is $x + y \geq 90\,000$

1 Identify the decision variables. These are the amounts of the oils to use.

2 Express the constraints as inequalities.

(*Continued on the next page*)

DISCRETE

The total production constraint is $x + y \geq 90\,000$

$C = 0.25x + 0.2y$

The objective function is the cost, £C

You need to minimise C

Minimise $C = 0.25x + 0.2y$

Subject to
$$y \leq 2x$$
$$3y \geq 2x$$
$$x \geq 35\,000$$
$$y \geq 50\,000$$
$$x + y \geq 90\,000$$

This is the complete linear programming formulation.

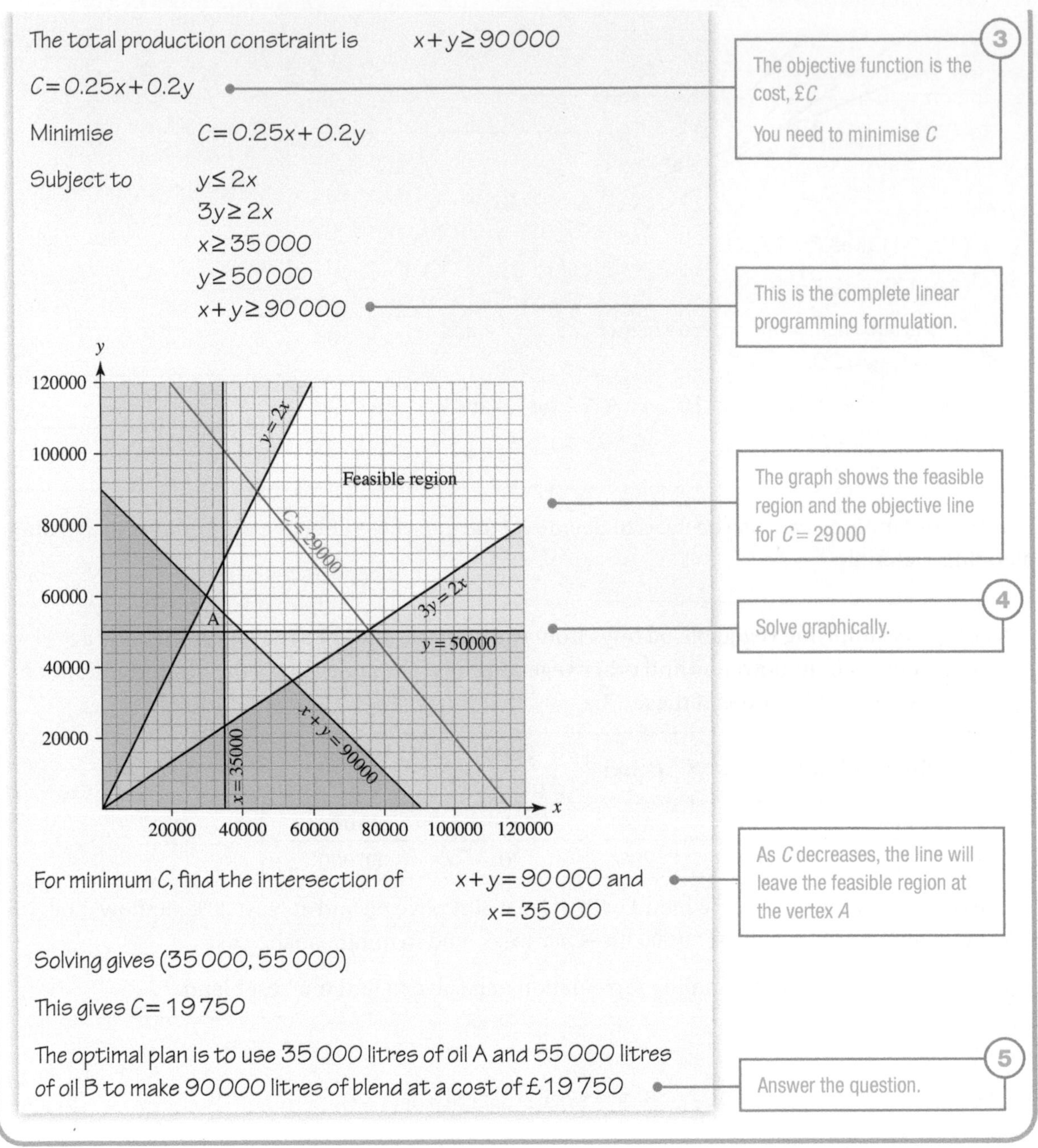

The graph shows the feasible region and the objective line for $C = 29\,000$

Solve graphically.

For minimum C, find the intersection of $x + y = 90\,000$ and $x = 35\,000$

As C decreases, the line will leave the feasible region at the vertex A

Solving gives $(35\,000, 55\,000)$

This gives $C = 19\,750$

The optimal plan is to use 35 000 litres of oil A and 55 000 litres of oil B to make 90 000 litres of blend at a cost of £19 750

Answer the question.

Exercise 14.1B Reasoning and problem-solving

1 A club which has 80 members, is organising a trip. They intend to hire vehicles they can drive themselves and travel in convoy. Only eight of the members are prepared to drive. A car, which can carry five people including the driver, costs £20 per day to hire. A minibus, which can carry 12 people including the driver, costs £60 per day to hire. They wish to minimise the hire costs. Express this problem as a linear programming formulation and hence find the best combination of vehicles to hire.

2 Whisky mac is a mixture of whisky and ginger wine. Whisky is 40% alcohol and costs £12 per litre. Ginger wine is 12% alcohol and costs £5 per litre. A bar-tender wishes to minimise the cost of making a whisky mac,

which must be at least 100 ml of liquid, at least 20% alcohol and contain at most 30 ml of alcohol. Use linear programming to find the amount of each drink in the optimal mixture.

3 A trader buys goods from a warehouse and takes them by van to his shop. He wants shampoo and washing powder. These are packed in cases of the same size. A case of shampoo weighs 6 kg and he can make a profit of £20 per case. A case of washing powder weighs 4 kg and his profit will be £14 per case. His van has room for 60 cases and can carry a maximum load of 280 kg. Find how much of each product he should buy to maximise his profit.

4 An investor has up to £20 000 to invest. She can buy 'safe' bonds yielding 5% interest or 'risky' shares yielding 10% interest. She wants to make at least 8% interest overall, and as she is a cautious investor she wants her investment in bonds to be as great as possible. How much should she invest in each?

5 A building firm has a plot of land with area 6000 m^2. The intention is to build a mixture of houses and bungalows. A house occupies 210 m^2 and a bungalow occupies 270 m^2. Planning regulations limit the total number of dwellings to 25, and insist that there must be no more than 15 of either type. A house makes £20 000 profit, and a bungalow £25 000. Formulate this situation as a linear programming problem and find graphically the best combination of dwellings.

6 There are 50 places on a trip. The group comprises x senior staff, y trainees and z children. There must be at least one adult to every two children, and at least one senior staff member to every trainee. There must be at least five trainees and at least ten senior staff. The cost of the trip is £20 for each senior staff member, £15 for each trainee and £12 for each child. It is required to minimise the total cost.

a Express this as a linear programming formulation in x, y and z

b Show that your formulation can be rewritten to minimise the objective function as

$C = 8x + 3y + 600$

and write the constraints in terms of the two variables, x and y

c Solve to find the optimum combination of people.

7 A dog food manufacturer makes three types of chew, each 10 g in weight, from two basic ingredients. The table shows the proportions of these together with the amounts of the two ingredients in stock. Ingredient 1 costs the manufacturer £2 per kg and ingredient 2 costs £1 per kg.

	Ingredient 1	Ingredient 2
Chew A	8g	2g
Chew B	6g	4g
Chew C	5g	5g
In stock	800 kg	400 kg

The manufacturer wants to make 1600 packets of mixed chews. Each must contain 60 chews, and there must be no more than 30 of each type of chew in a packet. Find the best combination of chew types in each packet to minimise the cost.

14.2 Zero-sum games

Fluency and skills

Two players A and B decide to play a game with these rules.

Each person plays a card-king (K), queen (Q) or jack (J). Depending on the cards played, either player A pays player B or player B pays player A an agreed amount.

For example, if they both play a king, player B pays player A 5 p. Record this as (5, –5), meaning that player A gains 5 p and player B loses 5 p.

Similarly, if player A plays a king and player B plays a queen, player A pays player B 4 p, Record this as (–4, 4)

The complete set of payments is shown in this table.

		B		
		K	Q	J
A	K	(5, –5)	(–4, 4)	(2, –2)
	Q	(3, –3)	(1, –1)	(4, –4)
	J	(2, –2)	(3, –3)	(–1, 1)

This is a two-person game - there are just two competitors (it could be two teams rather than two people).

On each play, one competitor's gain equals the other's loss. This is called a **zero-sum game**.

Key point

In a **zero-sum** game, the sum of the gains made by the players on each play is zero.

To record a zero-sum game, you only need to put one number in each cell, as the other is just its negative. The table becomes:

		B		
		K	Q	J
A	K	5	–4	2
	Q	3	1	4
	J	2	3	–1

This is player A's **payoff matrix**. It is conventional to record the gains of the player on the left of the table, the **row player**.

Player B's payoff matrix would look like this.

		A		
		K	**Q**	**J**
B	**K**	–5	–3	–2
	Q	4	–1	–3
	J	–2	–4	1

You analyse the game from player A's point of view.

- If she plays a king, the worst outcome is that she loses 4 p.
- If she plays a queen, the worst is that she wins 1 p.
- If she plays a jack, the worst is that she loses 1 p.

		B				
		K	**Q**	**J**	**Row minimum**	
A	**K**	5	–4	2	–4	
	Q	3	1	4	1	← max = 1
	J	2	3	–1	–1	

These are the minimum values for the rows of the table.

Her **play-safe strategy** is to play a queen, which gives the best guaranteed outcome (win 1 p). This is called a **maximin strategy**, because it maximises her minimum gain.

Key point

A play-safe strategy gives the best guaranteed outcome regardless of what the other player does.

Using the same table, you can analyse the game from player B's point of view by looking at the columns instead of the rows.

- If he plays a king, the worst outcome is that he loses 5 p.
- If he plays a queen, the worst is that he loses 3 p.
- If he plays a jack, the worst is that he loses 4 p.

These are the maximum values for the columns of the table.

		B				
		K	**Q**	**J**	**Row minimum**	
A	**K**	5	–4	2	-4	
	Q	3	1	4	1	← max = 1
	J	2	3	–1	-1	
Column maximum		5	3	4		
			↑ min = 3			

His play-safe strategy is to play a queen, which gives the best guaranteed outcome (lose 3 p). This is called a **minimax strategy**, because it minimises his maximum loss.

If both players play safe every time, it is called a **pure-strategy game**. Player A wins 1 p each time. This is the **value of the game** to player A. (The value of the game to player B is –1 p).

DISCRETE

If player A knows that player B intends to play a queen, she should play a jack instead of a queen, winning 3 p instead of 1 p. This means the solution is **unstable**.

> **Key point**
> A game has a **stable solution** if neither player can gain by changing from their play-safe strategy.

Suppose the payoff matrix is changed, like this.

		B				
		K	Q	J	Row minimum	
A	K	4	2	−4	−4	
	Q	3	4	2	2	← max = 2
	J	2	−2	1	−2	
Column maximum		4	4	2		
				↑ min = 2		

Player A's play-safe strategy is still to play a queen while player B's is now to play a jack.

Neither player can gain by changing strategy. If player A plays a queen, then player B's best option is to play a jack, and if player B plays a jack, then player A's best option is to play a queen. The game has a **stable solution**.

The **value of the game** to player A is 2 p. (The stated value of a game is always the value to the row player.)

> **Key point**
> The value of a game is the payoff to the row player if both players use their best strategy.

Notice that, in this stable solution, the maximum of the row minima is the same as the minimum of the column maxima (both = 2)

These two quantities are always equal in a stable solution.

> **Key point**
> A game has a stable solution if
> maximum of row minima = minimum of column maxima

> A stable solution can also be called a saddle point because the 2 in the table is the lowest value in its row and the highest in its column, just as the centre of a horse's saddle is its lowest point in the nose-to-tail direction and its highest point in the side-to-side direction.

Example 1

This payoff matrix describes a game between players A and B. Player A has three strategies A_1, A_2 and A_3, and player B has four strategies B_1, B_2, B_3 and B_4. Show that the game has a stable solution, and find the value of the game.

		B			
		B_1	B_2	B_3	B_4
A	A_1	8	6	9	6
	A_2	4	3	−2	2
	A_3	10	−1	1	5

(*Continued on the next page*)

		B					
		B_1	B_2	B_3	B_4	Row min	
A	A_1	8	6	9	6	6	← max = 6
	A_2	4	3	−2	2	−2	
	A_3	10	−1	1	5	−1	
Column max		10	6	9	6		
			↑ min = 6		↑ min = 6		

A's play safe strategy is A_1

B's play safe strategy is either B_2 or B_4

Max of row minima = min of row maxima, so the game has a stable solution.

The value of the game is 6

Find the row minima and choose the maximum of these.

Find the column maxima and choose the minimum of these.

In some games, one or more strategies should never be used. You can remove those rows or columns from the payoff matrix.

For example, in this game, A would never use strategy A_2, because A_1 has a better payoff in all three columns. Row 1 **dominates** row 2

		B		
		B_1	B_2	B_3
A	A_1	7	3	9
	A_2	4	1	6
	A_3	4	4	−1

Because strategy A_2 will never be used, you can eliminate row 2

		B		
		B_1	B_2	B_3
A	A_1	7	3	9
	A_3	4	4	−1

In the revised matrix, B would never use strategy B_1, because its outcomes are the same or worse than those of B_2

Column 2 dominates column 1, so you can eliminate column 1

		B	
		B_2	B_3
A	A_1	3	9
	A_3	4	−1

There is now no dominance, so the matrix cannot be simplified any further.

Key point

In a payoff matrix, row i dominates row j if, for every column, the value in row $i \geq$ value in row j

Similarly, column i dominates column j if, for every row, the value in column $i \leq$ value in column j

Example 2

Use dominance to simplify this payoff matrix as far as possible. What can be deduced from the result?

		B		
		B_1	B_2	B_3
A	A_1	5	3	3
	A_2	−2	7	−1

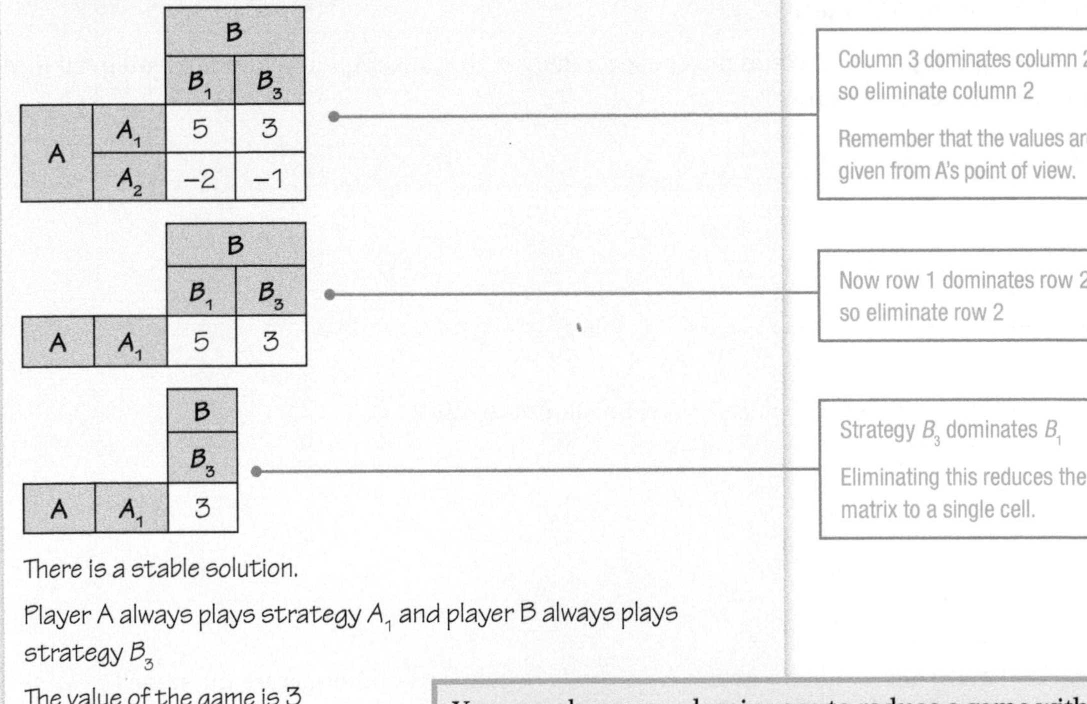

		B	
		B_1	B_3
A	A_1	5	3
	A_2	−2	−1

Column 3 dominates column 2, so eliminate column 2

Remember that the values are given from A's point of view.

		B	
		B_1	B_3
A	A_1	5	3

Now row 1 dominates row 2, so eliminate row 2

		B
		B_3
A	A_1	3

Strategy B_3 dominates B_1

Eliminating this reduces the matrix to a single cell.

There is a stable solution.

Player A always plays strategy A_1 and player B always plays strategy B_3

The value of the game is 3

You can always use dominance to reduce a game with stable solutions to a matrix with just those cells.

Exercise 14.2A Fluency and skills

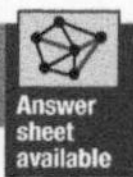

1 For each of these payoff matrices, find the play-safe strategy for each player. Determine whether the game has a stable solution and, if so, state the value of the game.

a

		B		
		B_1	B_2	B_3
A	A_1	5	3	9
	A_2	4	7	6
	A_3	2	4	5

b

		B		
		B_1	B_2	B_3
A	A_1	9	−2	3
	A_2	−1	−4	2
	A_3	4	6	3

c

		B		
		B_1	B_2	B_3
A	A_1	6	−1	4
	A_2	−1	−3	5

d

		B	
		B_1	B_2
A	A_1	−1	3
	A_2	2	7
	A_3	6	−2

e

		B			
		B_1	B_2	B_3	B_4
A	A_1	4	−2	5	9
	A_2	2	1	3	5
	A_3	3	−1	−8	4

f

		B			
		B_1	B_2	B_3	B_4
A	A_1	3	2	7	10
	A_2	5	5	7	8
	A_3	−3	2	−5	6

2 For the game shown in question **1d**, write down the payoff matrix for player B.

3 For each of these payoff matrices, use dominance to simplify the problem as far as possible. If there is a stable solution, state the strategies the players should adopt and the value of the game.

a

		B	
		B_1	B_2
A	A_1	−1	2
	A_2	2	3
	A_3	4	−2

b

		B		
		B_1	B_2	B_3
A	A_1	4	−1	4
	A_2	−1	−3	−4

c

		B		
		B_1	B_2	B_3
A	A_1	3	1	6
	A_2	2	5	3
	A_3	0	2	3

d

		B		
		B_1	B_2	B_3
A	A_1	−6	−2	1
	A_2	3	−1	−2
	A_3	−4	3	1

e

		B			
		B_1	B_2	B_3	B_4
A	A_1	−2	2	−3	−1
	A_2	−1	2	1	−1
	A_3	−2	−3	−4	−3

f

		B		
		B_1	B_2	B_3
A	A_1	3	4	2
	A_2	−1	2	1
	A_3	1	2	4
	A_4	2	1	−3

Reasoning and problem-solving

Strategy

To solve problems with zero-sum payoff matrices

1. If necessary, construct the matrix of payoffs for the row player.
2. Check whether the matrix can be reduced by using dominance.
3. Find the maximum of the row minima and the minimum of the row maxima to determine each player's play-safe strategy.
4. Decide whether the game has a stable solution (saddle point) and, if so, find its value.
5. Answer the question in context.

Example 3

Analyse the game shown in the table.

		B			
		B_1	B_2	B_3	B_4
A	A_1	0	–3	5	–9
	A_2	5	–8	–2	10
	A_3	3	10	6	9
	A_4	4	11	–3	2

		B			
		B_1	B_2	B_3	B_4
A	A_2	5	–8	–2	10
	A_3	3	10	6	9
	A_4	4	11	–3	2

Row 3 dominates row 1, so delete row 1

(2) Reduce the rows using dominance.

		B		
		B_1	B_2	B_3
A	A_2	5	–8	–2
	A_3	3	10	6
	A_4	4	11	–3

Column 3 dominates column 4, so delete column 4

(2) Reduce the columns using dominance.

		B			
		B_1	B_2	B_3	
A	A_2	5	–8	–2	–8
	A_3	3	10	6	3
	A_4	4	11	–3	–3
		5	11	6	

(3) Find row minima and column maxima.

Max of row minima = 3 (shown).

Min of column maxima = 5 (shown).

A's play-safe strategy is A_3

B's play-safe strategy is B_1

The solution is not stable.

(4) Decide whether the solution is stable.

Max of row minima ≠ min of column maxima.

Exercise 14.2B Reasoning and problem-solving

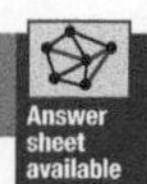

1 Analyse each of these payoff matrices and state what you can about the game and the players' strategies.

a

		B		
		B_1	B_2	B_3
A	A_1	–4	6	8
	A_2	–2	–8	10
	A_3	–5	0	3

b

		B		
		B_1	B_2	B_3
A	A_1	5	–2	6
	A_2	3	–4	–1
	A_3	–2	–2	3
	A_4	1	–5	–2

c

		B			
		B_1	B_2	B_3	B_4
A	A_1	4	0	1	–2
	A_2	3	1	–3	–1
	A_3	–1	4	3	0
	A_4	–2	2	1	–1

d

		B			
		B_1	B_2	B_3	B_4
A	A_1	3	–1	–1	0
	A_2	2	2	1	1
	A_3	1	1	0	3
	A_4	–1	–2	1	2

e

		B			
		B_1	B_2	B_3	B_4
A	A_1	–2	–5	3	–11
	A_2	13	–10	–4	8
	A_3	5	8	–2	7
	A_4	4	9	–5	0

f

		B				
		B_1	B_2	B_3	B_4	B_5
A	A_1	2	2	0	1	0
	A_2	1	0	–1	1	–2
	A_3	1	3	–1	4	–1
	A_4	4	2	0	1	0

2 X and Y play a game in which X chooses a number from the set {3, 5, 9} and Y a number from the set {2, 6, 7}. The difference between the chosen numbers, ignoring minus signs, is d. If $d > 2$, Y pays X £d; otherwise X pays Y £2d

a Construct X's payoff matrix for this game.

b Analyse the game to find the play-safe strategies for X and Y. Explain how you know that the game has no stable solution.

c Modify the game by changing one of the numbers in Y's set, so that the revised game has a stable solution.

3 A game in which there is a fixed number of points awarded on each play can be analysed as a zero-sum game by subtracting half the total score. For example, if there are 10 points available in a game between A and B, possible scores might be (10, 0) or (4, 6). These have the same effect as (5, –5) and (–1, 1), respectively, as, in the first case, A moves 10 points ahead of B and, in the second case, B moves 2 points ahead of A.

The table shows a game in which a total of 6 points is awarded for each play.

		B		
		B_1	B_2	B_3
A	A_1	(4, 2)	(3, 3)	(2, 4)
	A_2	(2, 4)	(0, 6)	(3, 3)
	A_3	(6, 0)	(4, 2)	(5, 1)

a Construct the conventional payoff matrix that is equivalent to this game.

b Find the play-safe strategy for each player and show that the game has a saddle point.

c If the game were played ten times and each player followed a pure strategy, what would be the final score?

14.3 Mixed-strategy games

Fluency and skills

Look at the game shown in this table.

		B			
		B_1	B_2	**Row minimum**	
A	A_1	−3	5	−3	
	A_2	3	−1	−1	← max = −1
Column maximum		3	5		
		↑ min = 3			

The play-safe strategies are A_2 for A and B_1 for B. If they play a pure-strategy game (always playing safe), A will win 3 every time. If the solution is stable (has a saddle point), there is no advantage to be gained in moving from the play-safe strategy, but in this case the solution is not stable. If B knows that A will play A_2, she will play B_2 instead of the play-safe strategy B_1. However, if A knows that B will do this, he will play A_1 instead of the play-safe strategy A_2

The best approach for each player is to play each strategy some of the time. This is a **mixed-strategy game.**

To analyse a mixed-strategy game, you need to use the idea of **expectation** or **expected payoff.**

Suppose you play a game where the probability of winning is $\frac{1}{4}$

If you win, you get £10

If you lose you pay £5

	Win	**Lose**
Payoff	£10	−£5
Probability	$\frac{1}{4}$	$\frac{3}{4}$

If you play the game four times, you expect to win once and lose three times.

Your expected total payoff is £10 + 3 × (−£5) = −£5

Dividing by 4, your expected average payoff per game is $\frac{1}{4} \times £10 + \frac{3}{4} \times (-£5) = -£5 \div 4 = -£1.25$

This is the sum of each payoff multiplied by its probability.

Key point

If payoffs $x_1, x_2, ..., x_n$ occur with probabilities $p_1, p_2, ..., p_n$, the **expectation** or **expected (mean) payoff** E(x) is given by

$$E(x) = x_1p_1 + x_2p_2 + ... + x_np_n = \sum_{i=1}^{n} x_ip_i$$

You can apply this to the mixed-strategy game in the initial example.

		B	
		B_1	B_2
A	A_1	−3	5
	A_2	3	−1

Suppose player A plays A_1 randomly for a proportion p of plays and A_2 for the remaining $(1 - p)$. She needs to decide the best value of p

Let the value of the game be v

If B plays B_1, the possible payoffs for A are:

	A_1	A_2
Payoff	−3	3
Probability	p	$(1-p)$

The expected payoff is $(-3) \times p + 3 \times (1 - p) = 3 - 6p$

The value of the game cannot be more than this, so $v \leq 3-6p$

If B plays B_2, the possible payoffs for A are:

	A_1	A_2
Payoff	5	−1
Probability	p	$(1-p)$

The expected payoff is $5 \times p + (-1) \times (1 - p) = 6p - 1$

The value of the game cannot be more than this, so $v \leq 6p-1$

Player A needs to maximise v subject to $v \leq 3-6p$ and $v \leq 6p-1$

This is a linear programming problem, which, in this case, can be solved graphically.

Plot lines $v = 3 - 6p$ and $v = 6p - 1$

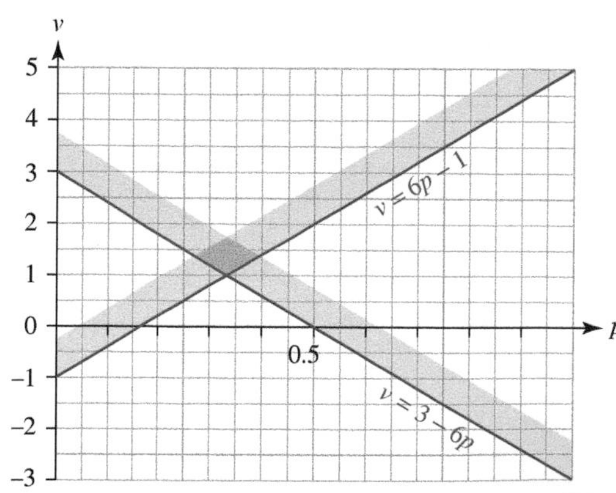

For a given value of p, the value of v occurs on the lower of the two lines.

You find the maximum value of v where the two lines intersect, so

$$3 - 6p = 6p - 1$$

$$p = \frac{1}{3}$$

Player A's best strategy is to play A_1 and A_2 randomly with probabilities $\frac{1}{3}$ and $\frac{2}{3}$

DISCRETE

The value of the game is $v = 3 - 6 \times \frac{1}{3} = 1$

You find B's best strategy in the same way.

Remember the table shows A's payoffs, and B's payoffs are the negative of these.

Suppose B plays B_1 and B_2 with probabilities q and $(1 - q)$

If A plays A_1, B's expected payoff is

$$3 \times q + (-5) \times (1 - q)) = 8q - 5$$

If A plays A_2, B's expected payoff is

$$(-3) \times q + 1 \times (1 - q)) = 1 - 4q$$

The optimal strategy occurs when these are equal:

$$8q - 5 = 1 - 4q \quad \Rightarrow \quad q = \frac{1}{2}$$

B's best strategy is to play B_1 and B_2 half of the time each, at random.

The value of the game to B is $8 \times \frac{1}{2} - 5 = -1$

This is $-v$, as you would expect.

Example 1

a Show that this game does not have a stable solution.

		B	
		B_1	B_2
A	A_1	2	−1
	A_2	−2	3

b Find the optimum mixed strategy for player A.

c Find the value of the game.

d Find the optimum mixed strategy for player B.

a

		B			
		B_1	B_2	Row minima	Max of row minima
A	A_1	2	−1	−1	−1
	A_2	−2	3	−2	
	Column maxima	2	3		
	Min of column maxima	2			

The max of the row minima ≠ min of column maxima, so there is no stable solution.

b Let the value of the game be v and let A play A_1 and A_2 with probabilities p and $(1 - p)$

(*Continued on the next page*)

If B plays B_1, then A's expected payoff is $2p + -2(1-p) = 4p - 2$, so $v \le 4p - 2$

If B plays B_2, then A's expected payoff is $-p + 3(1-p) = 3 - 4p$, so $v \le 3 - 4p$

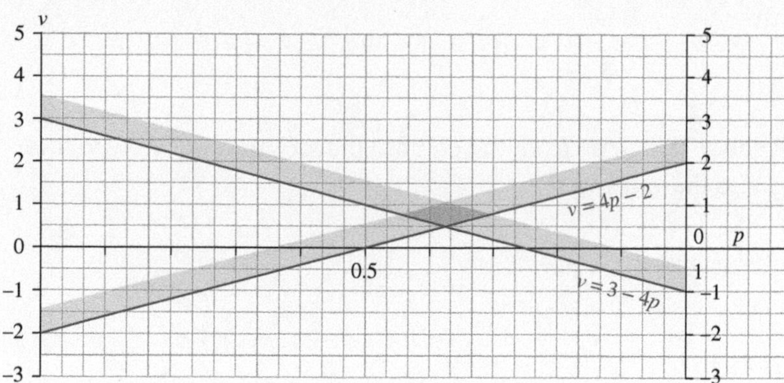

Maximum v occurs when $4p - 2 = 3 - 4p$, so $p = \frac{5}{8}$ — Equate to find p

A plays A_1 and A_2 with probabilities $\frac{5}{8}$ and $\frac{3}{8}$

c The value of the game is $v = 4 \times \frac{5}{8} - 2 = \frac{1}{2}$

d Let B play B_1 and B_2 with probabilities q and $(1-q)$

If A plays A_1, B's expected payoff is $-2q + (1-q) = 1 - 3q$

If A plays A_2, B's expected payoff is $2q - 3(1-q) = 5q - 3$

The optimal strategy occurs when these are equal:

$1 - 3q = 5q - 3$, so $q = \frac{1}{2}$ — Equate to find q

So B plays B_1 and B_2 with equal probability.

As a check, the value to B is $1 - 3 \times \frac{1}{2} = -\frac{1}{2} = -v$, as expected.

DISCRETE

Example 2

The game shown does not have a stable solution.

		B	
		B_1	B_2
A	A_1	−1	5
	A_2	4	−3

The value of the game is v. Analyse the game from the column player's point of view to find their optimal mixed strategy and the value of v

The value of the game to the column player, B, is $(-v)$

Let B play B_1 with probability q

If A plays A_1, then B's expected payoff is $q-5(1-q)=6q-5$

So $(-v)\le 6q-5$ and so $v\ge 5-6q$

If A plays A_2, then B's expected payoff is $-4q+3(1-q)=3-7q$

So $(-v)\le 3-7q$ and so $v\ge 7q-3$

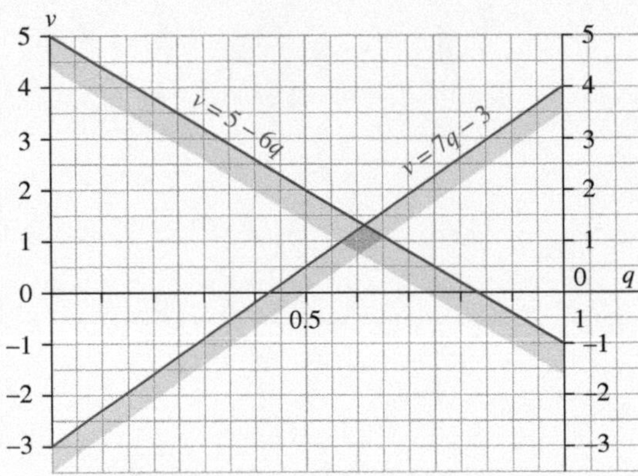

Max $(-v)$ corresponds to minimum v on the graph, that is where $7q-3=5-6q$, so $q=\frac{8}{13}$

Hence B plays B_1, B_2 with probabilities $\frac{8}{13}, \frac{5}{13}$ and $v=1\frac{4}{13}$

Alternatively, you could have used value $=-v$ to find q, by putting $1-3q=-\frac{1}{2}$ or $5q-3=-\frac{1}{2}$

Exercise 14.3A Fluency and skills

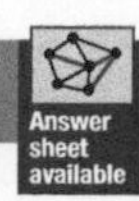

1 For the following games, decide if a mixed strategy is needed. Find the optimal strategy (pure or mixed) for both players and the value of the game.

a

		B	
		B_1	B_2
A	A_1	–2	4
	A_2	3	2

b

		B	
		B_1	B_2
A	A_1	4	2
	A_2	–1	3

c

		B	
		B_1	B_2
A	A_1	5	–3
	A_2	3	2

d

		B	
		B_1	B_2
A	A_1	1	4
	A_2	4	3

2 For each of these games, show that a mixed strategy is necessary. Find the optimal mixed strategy for each player and the value of the game.

a

		B	
		B_1	B_2
A	A_1	2	–2
	A_2	–2	3

b

		B	
		B_1	B_2
A	A_1	7	6
	A_2	5	8

c

		B	
		B_1	B_2
A	A_1	5	2
	A_2	–1	3

d

		B	
		B_1	B_2
A	A_1	–2	1
	A_2	2	0

3 Analyse each of these games from the column player's point of view to find their optimal mixed strategy and the value, v, of the game.

a

		B	
		B_1	B_2
A	A_1	–6	2
	A_2	1	-2

b

		B	
		B_1	B_2
A	A_1	9	7
	A_2	3	10

c

		B	
		B_1	B_2
A	A_1	–5	2
	A_2	1	–4

4 The table shows a game in which player A has three strategies A_1, A_2 and A_3, while player B has four strategies B_1, B_2, B_3 and B_4

		B			
		B_1	B_2	B_3	B_4
A	A_1	–1	3	7	–1
	A_2	1	5	4	2
	A_3	3	2	3	3

a Show that this game has no stable solution.

b Use dominance arguments to reduce the matrix as far as possible.

c Find the optimal mixed strategy for player A and the value of the game.

d Find the optimal mixed strategy for player B.

Reasoning and problem-solving

Strategy

To solve problems involving two-player zero-sum games

1. If necessary, construct a payoff matrix.
2. Use dominance to reduce the matrix as much as possible.
3. Find the play-safe strategies and decide whether the game has a stable solution.
4. For mixed-strategy games, find the constraints on v, the value of the game, in terms of the probability or probabilities of playing the various strategies.
5. Solve the resulting linear programming problem to find the optimal strategies and the value of the game.
6. Answer the question in context.

You can find the optimum mixed strategy for larger payoff matrices if one of the players has just two strategies, because only two of the other player's options will appear in their best strategy.

Example 3

Find the optimal strategies for the players in this 2×3 game, and find the value of the game.

		B		
		B_1	B_2	B_3
A	A_1	4	2	−1
	A_2	−3	−2	5

The game has no stable solution, so A plays A_1 and A_2 with probabilities p and $(1-p)$

(3) Decide whether the game has a stable solution.

If B plays B_1, A's expectation $= 4p - 3(1-p) = 7p - 3$

If B plays B_2, A's expectation $= 2p - 2(1-p) = 4p - 2$

If B plays B_3, A's expectation $= -p + 5(1-p) = 5 - 6p$

(4) Find the constraints on v

If the value of the game is v, you need to maximise v subject to

$v \le 7p - 3,\ v \le 4p - 2,\ v \le 5 - 6p$

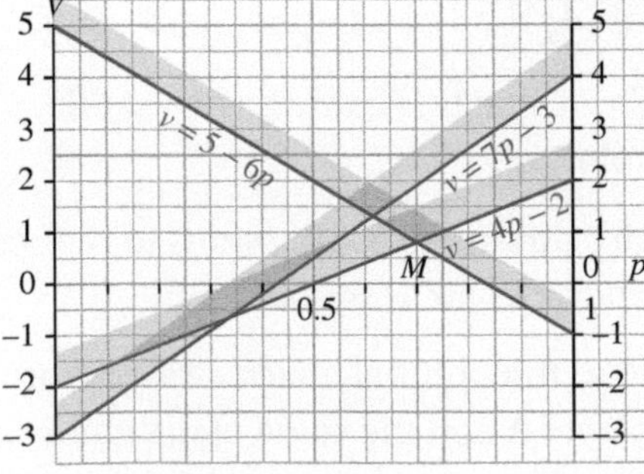

Maximum v is at vertex M on the graph.

(*Continued on the next page*)

This is the intersection of $v = 4p - 2$ and $v = 5 - 6p$

$4p - 2 = 5 - 6p \Rightarrow p = 0.7$

So A plays A_1 and A_2 randomly with probabilities 0.7 and 0.3

The value of the game is $v = 4 \times 0.7 - 2 = 0.8$

The line $v = 7p - 3$ passes above M, so B should not play B_1 because A's payoff would be greater than 0.8

B plays B_2 and B_3 with probabilities q and $(1 - q)$

If A plays A_1, B's expectation is $-2q + (1 - q) = 1 - 3q$

If A plays A_2, B's expectation is $2q - 5(1 - q) = 7q - 5$

The optimal strategy occurs where these lines intersect.

$1 - 3q = 7q - 5 \Rightarrow q = 0.6$

So B plays B_2 and B_3 with probabilities 0.6 and 0.4

The value of the game to B is $1 - 3 \times 0.6 = -0.8$, as expected.

In general, in a $2 \times n$ game the column player will only use two of the n available strategies, as the others will give the row player a greater payoff.

DISCRETE

In an $n \times 2$ game, you analyse the situation from the point of view of the column player.

Example 4

Find the optimal strategies for the players in this 3×2 game, and find the value of the game.

		B	
		B_1	B_2
A	A_1	−4	1
	A_2	0	−2
	A_3	1	−4

The game has no stable solution, so B plays B_1 and B_2 with probabilities q and $(1 - q)$

If A plays A_1, B's expectation $= 4q - (1 - q) = 5q - 1$

If A plays A_2, B's expectation $= 0 \times q + 2(1 - q) = 2 - 2q$

If A plays A_3, B's expectation $= -q + 3(1 - q) = 4 - 5q$

If the value of the game (to A) is v, then the value of the game to B is $V = -v$. You need to maximise V subject to $V \le 5q - 1$, $V \le 2 - 2q$, $V \le 4 - 5q$

(*Continued on the next page*)

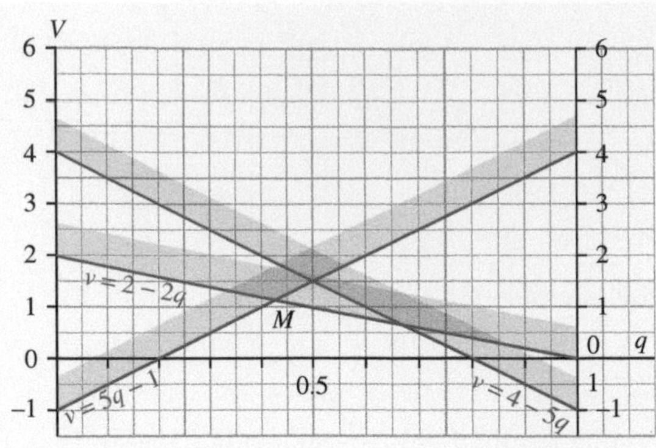

Maximum V is at vertex M on the graph.

This is the intersection of $V = 2 - 2q$ and $V = 5q - 1$

$$2 - 2q = 5q - 1 \quad \Rightarrow \quad q = \frac{3}{7}$$

So B plays B_1 and B_2 with probabilities $\frac{3}{7}$ and $\frac{4}{7}$

The value of the game to B is $V = 2 - 2 \times \frac{3}{7} = 1\frac{1}{7}$

The value of the game is $v = -V = -1\frac{1}{7}$

Line $V = 4 - 5q$ passes above M, so A should not play A_3 because B's payoff would be greater than $1\frac{1}{7}$

In general, in an $n \times 2$ game the row player will only use two of the n available strategies, as the others will give the column player a greater payoff.

A plays A_1 and A_2 with probabilities p and $(1 - p)$

If B plays B_1, A's expectation is $-4p + 0 \times (1 - p) = -4p$

If B plays B_2, A's expectation is $p - 2(1 - p) = 3p - 2$

The optimal strategy occurs where these lines intersect.

$$-4p = 3p - 2 \quad \Rightarrow \quad p = \frac{2}{7}$$

So A plays A_1 $\frac{2}{7}$ and A_2 $\frac{5}{7}$ of the time.

Exercise 14.3B Reasoning and problem-solving

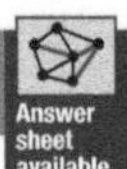

1 Analyse the games shown in these tables to find the optimal strategy for each player and the value of the game.

a

		B		
		B_1	B_2	B_3
A	A_1	−8	−2	1
	A_2	−3	3	4
	A_3	0	−4	−2

b

		B		
		B_1	B_2	B_3
A	A_1	2	1	5
	A_2	1	−4	4
	A_3	5	2	1

c

	B_1	B_2	B_3
A_1	5	2	3
A_2	1	4	2

d

		B		
		B_1	B_2	B_3
A	A_1	3	−1	1
	A_2	−2	4	2

2 The table shows a game between players X and Y, each with four possible strategies.

		Y			
		I	II	III	IV
X	I	1	4	3	2
	II	−1	3	0	3
	III	4	2	5	1
	IV	−2	6	−3	1

a Show that there is no stable solution.

b Find the optimal mixed strategy for each player and the value of the game.

3 The table shows a game with value V.

		B		
		B_1	B_2	B_3
A	A_1	−1	0	−2
	A_2	1	−2	−1
	A_3	0	−1	1

Player A has an optimal mixed strategy in which she plays A_1, A_2 with probabilities p_1 and p_2. Find and simplify three inequalities connecting p_1, p_2 and V. (You do not need to solve the problem.)

DISCRETE

Full A Level

4 The table shows a game between players A and B. A has two strategies, A_1 and A_2. B has three strategies, B_1, B_2 and B_3

		B		
		B_1	B_2	B_3
A	A_1	2	5	6
	A_2	5	2	1

a Analyse the game graphically and show that player A's optimal strategy is to play her two strategies at random with equal probability.

b Find the value of the game.

c Explain, with reference to your graph, why, in this game, player B could reasonably make use of all three available strategies.

Let B play his strategies with probabilities q_1, q_2 and q_3, respectively.

d Making use of the known value of the game, write down three equations connecting q_1, q_2 and q_3

e Putting $q_1 = q$, show that $q_2 = 2\frac{1}{2} - 4q$ and find q_3 in terms of q

f Show that $\frac{1}{2} \le q \le \frac{5}{8}$ and find the range of possible values of q_2 and q_3

14 Summary and review

Chapter summary

- In a linear programming (LP) formulation:
  - The quantities you can vary are the decision variables (control variables).
  - The limitations on the values of the decision variables are the constraints.
  - The quantity to be optimised is the objective function.
- On the graph of the constraints, the feasible region is the set of points satisfying all the constraints.
- The objective line joins all points for which the objective function takes a specified value.
- The optimal solution corresponds to a vertex of the feasible region.
- In a zero-sum game between two players, A and B, the sum of their gains and losses on each play is zero. The gains for the row player, A, are recorded in a pay-off matrix. B's gains are the negative of these entries.
- A play-safe strategy gives the best guaranteed outcome regardless of what the other player does. For A, you find the minimum value in each row. The play-safe strategy corresponds to the maximum of these row minima. For B, you find the maximum value in each column. The play-safe strategy corresponds to the minimum of these column maxima. In a pure-strategy game, both players always play safe.
- The value of a game is the payoff to the row player (A) if both players use their best strategy.
- The game has a stable solution (saddle point) if neither player gains by changing from their play-safe strategy. In this case, max of row minima = min of column maxima.
- In a payoff matrix:

  row i dominates row j if, for every column, value in row $i \geq$ value in row j
  column i dominates column j if, for every row, value in column $i \leq$ value in column j
- A row or column which is dominated can be deleted.
- If a matrix can be reduced by domination to one containing a single value, the game has a stable solution. For an unstable situation, each player should play a mixed-strategy game, that is, play each strategy some of the time.
- In a mixed-strategy game, you aim to maximise the expected payoff, where for payoffs $x_1, x_2, ..., x_n$ that occur with probabilities $p_1, p_2, ..., p_n$, the expected (mean) payoff E(x) is given by

  $$E(x) = x_1p_1 + x_2p_2 + ... + x_np_n = \sum_{i=1}^{n} x_i p_i$$
- For a 2×2 game, you assume that the row player plays their first strategy with probability p You write an inequality connecting v, their expected value, with p for each of the column player's strategies. You can solve the resulting linear programming problem graphically.
- For a $2 \times n$ or an $n \times 2$ game, you derive the linear programming formulation working from the point of view of the player with two options. From the resulting graph, you can reject all but two of the lines.

Check and review

You should now be able to...	Try Questions
✔ Use graphical methods to solve a two-variable linear programming problem.	1, 2
✔ Express a worded problem as a linear programming formulation.	2

✔ Find the play-safe strategy for each player in a two-person zero-sum game and state the value of the game.	3, 4
✔ Decide whether the game has a stable solution.	3, 4, 5, 6
✔ Use dominance to simplify a play-off matrix.	5, 6
✔ Use graphical methods to find the optimal mixed strategy for players in a game with no stable solution.	6

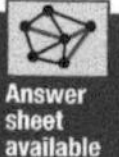

1 A linear programming problem is:

Minimise $C = 2x + 5y$

Subject to $x + y \leq 20$,

$y \geq 7$

$y \leq 2x$

$y + 3x \geq 25$

a Draw a graph of these constraints. Label the feasible region.

b Draw an objective line.

c Find the values of x and y which give the optimal solution.

2 A manufacturer makes two brands of blackcurrant jam – 'Value' and 'Luxury'. It takes 0.4 kg of blackcurrants and 0.8 kg of sugar to make 1 kg of Value jam, while 1 kg of Luxury jam requires 0.6 kg of blackcurrants and 0.6 kg of sugar. The profit is 10p per kg on Value jam and 12p per kg on Luxury jam. There are 480 kg of blackcurrants and 810 kg of sugar available. The aim is to maximise the profit.

a Express this situation as a linear programming formulation.

b By drawing a suitable graph, find the maximum profit and the corresponding quantities of the two brands of jam.

3 The table shows the payoff matrix for player A in a two-person zero-sum game.

		B	
		B_1	B_2
A	A_1	−1	3
	A_2	4	2

a Find the play-safe strategy for A and the value of the game to her.

b Find the play-safe strategy for B.

c Write down the payoff matrix for B.

4 A two-person zero-sum game is represented by this payoff matrix for player A.

		B		
		I	II	III
A	I	1	0	−2
	II	−1	4	0
	III	2	3	5

a By finding the play-safe strategy for each player, show that the game has a stable solution. Explain the implications of this.

b State the value of the game.

5 The table shows the payoff matrix for player A in a two-person zero-sum game.

		B		
		I	II	III
A	I	5	3	9
	II	4	7	6
	III	2	4	5

a Use dominance arguments to reduce the matrix as far as possible.

b State, with reasons, whether the game has a stable solution.

6 The table shows the payoffs in a game between A and B.

		B		
		B_1	B_2	B_3
A	A_1	3	−1	1
	A_2	−2	4	4

a Show that the game does not have a stable solution.

b Show that the table can be reduced to a 2×2 game.

c A plays strategy A_1 with probability p. Find two inequalities connecting v and p, where v is the value of the game to player A.

d Use a graph to find the optimal value v and the corresponding value of p

e Find B's optimal mixed strategy.

Exploration

14 Going beyond the exams

History

The American mathematical scientist, George Dantzig, is credited for developing the simplex algorithm during World War II. It was kept a secret until 1947. When it published, many other industries began to use it. Dantzig is also known for solving two other statistical problems that were unsolved at the time. He had mistaken them for homework after arriving late to a lecture!

ICT

Excel has an add-in called the Solver which can be used to solve linear programming problems. (You may need to select it from the Add-Ins.)

First formulate the following as a linear programming problem.

> *A company plans on building a maximum of 12 new shops in a large city. They will build these shops in one of three sizes for each location – a convenience store, a standard store, or a mega store. The convenience store requires £2.5 million to build and 20 employees to operate. The standard store requires £7.125 million to build and 16 employees to operate. The mega store requires £8.275 million to build and 40 employees to operate. The corporation can dedicate £80 million in construction capital, and 280 employees to staff the shops. On average, the convenience store nets £1.6 million annually, the standard store nets £2 million annually, and the mega store nets £2.6 million annually. How many of each shop should they build to maximise revenue?*

Once formulated, enter the problem in to Excel. A suggested format is shown on the left. You enter the objective function and constraints in the white cells. (The formula for the objective function in cell D9 will give you a hint of what to do.)

D9 f_x =1.6*D2+2*D3+2.6*D4

	A	B	C	D	E	F
2		Number of convenience	c			
3		Number of standard	s			
4		Number of mega	m			
9		Objective Function		0		
11		Constraint 1		0	<=	12
12		Constraint 2		0	<=	80
13		Constraint 3		0	<=	280

Click on the Solver button (on the Data tab) and complete the pop up box as shown.

Note that the solving method selected must be 'Simplex LP'.

Click on the Solve button and remember to interpret your answers in cells D2, D3 and D4 appropriately.

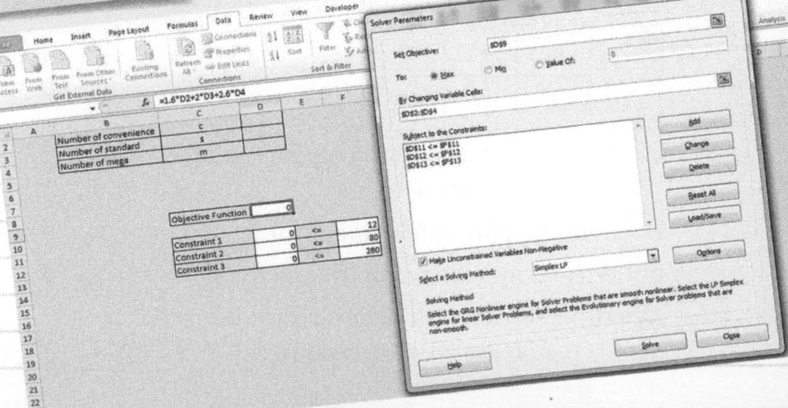

14 Assessment

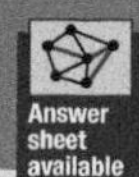

1 This payoff matrix represents a zero-sum game between two players, P and Q.

		Q		
		Q_1	Q_2	Q_3
P	P_1	6	4	9
	P_2	5	8	7
	P_3	3	5	6

a Find the play-safe strategy for each player. **[4 marks]**

b Determine whether the game has a stable solution **[1]**

c State the value of the game. **[1]**

2 This payoff matrix represents a zero-sum game between two players, M and N.

		N		
		N_1	N_2	N_3
M	M_1	6	6	8
	M_2	4	1	6
	M_3	4	2	−1

a Use dominance to simplify the problem as far as possible. **[4]**

b State what you can deduce from your answer to part **a** and give the value of the game. **[3]**

3 A linear programming problem is represented by the following constraints:

Minimise $C = 2x + 3y$

Subject to $x + y \leq 30 \quad y \leq 2x \quad 2y \geq 9 \quad y + 2x \geq 28$

a Draw a graph of these constraints and label the feasible region. **[7]**

b Draw an example of the objective function on your graph. **[2]**

4 This payoff matrix represents a zero-sum game between two players A and B.

		B			
		B_1	B_2	B_3	B_4
A	A_1	14	2	2	7
	A_2	6	2	2	9
	A_3	15	−1	1	0

a Decide whether this game has a stable solution. **[6]**

b Find the value of the game. **[1]**

5 Connor and Isla are playing a zero-sum game. They are both trying to gain the maximum number of points possible. The table below shows the number of points that Connor scores for each possible strategy that each player can use.

	I_1	I_2	I_3
C_1	2	−5	3
C_2	−1	−3	4
C_3	3	−5	2
C_4	3	−2	−1

a How many points does Connor score if he uses strategy C_3 and Isla uses strategy I_2? **[1]**

b Explain why Isla would never use strategy I_3 **[2]**

c Find the play-safe strategy for each player and explain whether the game is stable or not. You must explain your answer. **[4]**

6 A linear programming problem is represented by the following constraints.

Minimise $P = 3x + 6y + 4z$
Subject to $4x + 2y + 5z \leq 75$
$x + y + z = 24$
$x \geq 8$
$y \geq 9$
$z \geq 2$

a Rewrite these constraints, eliminating the variable z **[6]**

b Represent these constraints graphically, clearly identifying the feasible region. **[5]**

7 Sanjay and Danny are playing a card game. They each have three coloured cards and choose a card to show simultaneously so that they do not know what the other player will choose. They each start with 20 points and they gain or lose points according to the table.

		Danny		
		Red	**Blue**	**Purple**
Sanjay	**Yellow**	2	−4	−1
	Orange	−4	1	3
	Green	1	−1	1

a Find the play-safe strategy for each player. You must show your working. **[6]**

b Sanjay thinks that Danny will play safe. Which colour should he choose? **[1]**

8 A farmer has 20 hectares of land in which he grows cauliflowers and sprouts. The cost per hectare for cauliflowers is £300, whereas the cost for sprouts is £200. The farmer allocates £4800 for these products. The care of the crops requires £10 per hectare for cauliflower and £20 per hectare for sprouts. The farmer has £360 available for crop care. The profit on cauliflowers is £100 per hectare and the profit on sprouts is £125 per hectare.

a Formulate this as a linear programming problem. **[5]**

b By plotting a suitable graph, calculate the number of hectares of each crop the farmer should plant in order to maximise his profits. State the profit. **[7]**

9 A company produces two different types of vitamin tablet. The contents of the tablets are shown below.

	Vitamin C	**Niacin**	**Vitamin E**
Supervit	200 mg	30 mg	20 mg
Extravit	135 mg	20 mg	40 mg

Supervit tablets cost 15p each, whereas Extravit tablets cost 20p each. In a 30 day month, Jenny wants to spend no more than £6 on tablets. She requires no more than 5400 mg of Vitamin C, at least 700 mg of Niacin and at least 800 mg of Vitamin E.

a Formulate this as a linear programming problem. **[5]**

b By drawing a suitable graph, find the combination of tablets that Jenny should take to ensure that she has the required amount of vitamins at the minimum cost. **[9]**

15 Abstract algebra

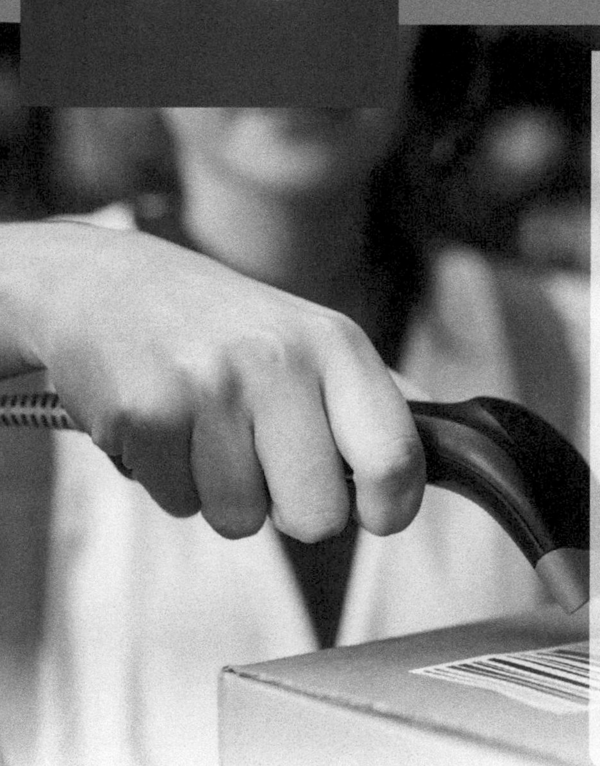

What might, at first, seem quite abstract algebraic methods that have been developed for their own sake by mathematicians can often come to have important applications. For example, modular arithmetic is used extensively around the world wherever barcodes or universal products codes are used. The bar codes you see being scanned when you are shopping involve 12-digit numbers that are used by the computer system to allocate the current price to charge the customer at the checkout. The computer will also inform other parts of the system about stock levels and automatically re-order stock if necessary to ensure the retailer is always able to cope with customers' needs.

These checking methods that draw on modular arithmetic are used by other everyday systems. For example, they provide one part of online security when credit cards are used, as well as providing a unique code (ISBN) for all books published throughout the world.

Orientation

What you need to know	What you will learn	What this leads to
KS 4 • Sets.	• To use binary operations. • To use Cayley tables. • To use modular arithmetic.	**Careers** • Computer programming.

15.1 Binary operations

Fluency and skills

A fundamental concept in mathematics is the **set**.

A set is a collection of individual objects called **elements**. A set can be finite, for example the set {1, 2, 3, 4, 5} has five elements, or infinite, for example the set of natural numbers, ℕ, is the infinite set {1, 2, 3, 4, ...}

A **binary operation** is a function with two inputs. For example, + on the set ℕ is a binary operation. The function could be more complicated; for example, the binary function $*$ on the set ℕ could be defined as $a*b=a^b+1$

In both these examples, the binary function always produces a member of the set ℕ

ℕ represents the set of natural numbers; these are the 'counting numbers': {1, 2, 3, 4, ...}

ℤ represents the set of integers (positive or negative) including 0: {0, ±1, ±2, ±3, ...}

$\mathbb{Z}^+$ is the set of positive integers: {1, 2, 3, 4 ...}

$\mathbb{Z}^-$ is the set of negative integers: {−1, −2, −3, −4, ...}

Key point

A function is a binary operation if it can be applied to any two elements of a set so that the result is also a member of the set.

Example 1

Which of these functions are binary operations on the set ℤ?

a $a*b=a-b$ **b** $a \blacksquare b=\frac{a}{b}$

a $a-b$ will always produce a member of ℤ so $*$ is a binary operation on ℤ

Any two integers can be subtracted to give another integer.

b $\frac{a}{b}$ is not, in general, a member of ℤ therefore ■ is not a binary operation on ℤ

For example, if $a=2$ and $b=3$ then $a \blacksquare b=\frac{2}{3}$ which is not an integer.

Also, $\frac{a}{b}$ is not defined when $b=0$

Key point

A binary operation, $*$, is **commutative** if $a*b=b*a$ for all a and b

Example 2

Which of these binary operations on the set ℕ are commutative?

a $a*b=a^b$ **b** $a \circ b=a+b$

a In general, $a^b \neq b^a$

For example, $2*5=2^5=32$ but $5*2=5^2=25$

Give a counter-example to prove $*$ is not commutative.

Therefore $*$ is not commutative.

b $a \circ b=b \circ a$ since addition of natural numbers is commutative.

Therefore $a \circ b$ is commutative.

Key point

A binary operation, $*$, is **associative** if $(a*b)*c = a*(b*c)$

Example 3

Which of these binary operations on $\mathbb{R}$ is associative?

a $a \quad b = 2a+b$ **b** $a*b = a \times b$

$\mathbb{R}$ is the set of real numbers.

a $(a \triangleright b) \triangleright c = (2a+b) \triangleright c$

$= 2(2a+b)+c$

$= 4a+2b+c$

$a \triangleright (b \triangleright c) = a \triangleright (2b+c)$

$= 2a+(2b+c)$

$= 2a+2b+c$

So, in general, $(a \triangleright b) \triangleright c \neq a \triangleright (b \triangleright c)$, therefore $\triangleright$ is not associative.

Apply the function to a and b first.

Now apply to b and c first.

This is not the same as the value of $(a \quad b) \quad c$

b $(a*b)*c = (a \times b) \times c$

$= abc$

$a*(b*c) = a \times (b \times c)$

$= abc$

Therefore $*$ is associative.

Since multiplication of real numbers is associative.

Sets sometimes have an **identity** element under a particular binary operation. For example, the identity element of $\mathbb{R}$ under multiplication is 1 since any member of $\mathbb{R}$ multiplied by 1 will be unchanged.

Key point

The identity element, e, of a set under an operation $*$ is such that $a*e = e*a = a$ for all values of a in the set.

The identity element for a binary operation is always unique. To prove this, assume that the binary operation $*$ has two identity elements, e and f, in the set S

Then, since f is an identity element, $e*f = e$

Also, since e is an identity element, $e*f = f$

Which implies that $e = f$

Therefore the identity element is unique.

Elements in a set sometimes have **inverses** in the set. For example in the set $\mathbb{R}$ under the operation of multiplication, the inverse of 4 is $\frac{1}{4}$ since $4 \times \frac{1}{4} = 1$

The inverse of an element combines with it to give the identity. In this case every element of the set except 0 has an inverse.

Key point

The inverse, a^{-1}, of an element a under an operation $*$ is such that $a*a^{-1} = a^{-1}*a = e$

An element, a, is a **self-inverse** if $a^{-1} = a$

Example 4

$\mathbb{Q}$ is the set of rational numbers: those that can be written as $\frac{p}{q}$ for some integers p and q
Irrational numbers cannot be written as $\frac{p}{q}$ e.g. $\sqrt{2}$

The binary operation $*$ is defined by $a*b=a+b$

a Find the identity element of the set $\mathbb{Q}$ under $*$

b Which element of $\mathbb{Q}$ is self-inverse under $*$?

a The identity element is $e=0$ since $a+0=0+a=a$

Need to show that $a*e=e*a=a$

b Need to find a such that $a*a=0$

Since the identity element is $e=0$

$a+a=0 \Rightarrow a=0$ so 0 is the only self-inverse element of $\mathbb{Q}$ under $*$

Exercise 15.1A Fluency and skills

1 Which of these are binary operations on the set $\mathbb{N}$? Explain your answers.

a $a \circ b=a-b$ **b** $a*b=a\times b$

2 Which of these are binary operations on the set $\mathbb{Z}$? Explain your answers.

a $a*b=a-b$ **b** $a \underline{\diamond}\, b=\frac{a}{b^2+1}$

3 Which of these are binary operations on the set $\mathbb{Z}^+$? Explain your answers.

a $a \diamond b=\max(a,b)$ [where $\max(a, b)$ means the maximum of a and b]

b $a*b=2a+b$

4 Which of these are binary operations on the set $\mathbb{Q}$? Explain your answers.

a $a \circ b=a^b$ **b** $a \star b=\frac{a}{b}$

5 Which of these are binary operations on the set of even integers?

a $a\star b=\frac{a}{2}+b$ **b** $a*b=3a-7b$

6 Explain whether or not each of these binary operations are associative and commutative.

a $a*b=a-b$ on the set $\mathbb{Q}$

b $a*b=a+b$ on the set $\mathbb{R}$

c $a*b=\max(a, b)$ on the set $\mathbb{N}$

d $a*b=a^2+b$ on the set $\mathbb{Z}$

e $a*b=a\times b$ on the set $\mathbb{Z}^+$

f $a*b=\min(a,b)$ on the set $\mathbb{Z}^-$

g $a*b=a\times b$ on the set $\mathbb{R}$

7 For each of the sets and binary operations in question **6**

i Identify the identity element (if it exists),

ii If the identity exists, explain which of the elements have inverses.

8 For the set of 2×2 matrices with real coefficients

a Prove that the binary operation defined as matrix multiplication is not commutative.

b Prove that the binary operation defined as matrix addition is associative and commutative.

c For each of the binary operations in parts **a** and **b**

i State the identity element.

ii Write down the general form of the inverse element.

9 For each of these sets and binary operations, identify the identity element (if it exists) and explain which of the elements have inverses.

a $a*b=a\times b$ on the set $\{1, 0, -1\}$

b $a*b=\max(a,b)$ on the set $\{2, 3, 4, 5\}$

c $a*b=a+b$ on the set of all even numbers

d $a*b=a\times b$ on the set of all odd numbers.

Reasoning and problem-solving

For a finite set, a table can be drawn showing the result of certain types of binary operation on all possible pairs of elements. This is called a **Cayley table**.

Cayley tables are named after the nineteenth-century mathematician Arthur Cayley.

For example, this is the Cayley table for the binary operation $*$ on the set $\{a, b, c\}$. Notice how the convention is to take the first element from those on the left of the table and the second from the elements on the top of the table.

$*$	$\boldsymbol{a}$	$\boldsymbol{b}$	$\boldsymbol{c}$
$\boldsymbol{a}$	$a*a$	$a*b$	$a*c$
$\boldsymbol{b}$	$b*a$	$b*b$	$b*c$
$\boldsymbol{c}$	$c*a$	$c*b$	$c*c$

Here is the Cayley table for the set $\{1, -1\}$ under multiplication.

$\times$	**1**	**−1**
1	1	−1
−1	−1	1

In this Cayley table, you can see that the identity element, e, is 1
This is because all the elements in the column and in the row corresponding to the element 1, are unchanged.

$\times$	**1**	**−1**
1	1	−1
−1	−1	1

You can also see that both 1 and −1 are self-inverses since both 1×1 and -1×-1 give the identity element, 1

$\times$	**1**	**−1**
1	1	−1
−1	−1	1

Strategy

To find the identity and inverse elements from a Cayley table

1. Look along the columns to find one the same as the initial column of elements; the element at the top of the column is the identity element, e. Also check along the relevant row to make sure it equals the initial row of elements.
2. Find all the instances of e in the table.
3. Work down the rows. For each element, a, at the start of a row the element at the top of the column containing e is the inverse of a

Example 5

For the Cayley table given,

a Find the identity element,

b Work out which of the elements are self-inverses.

Notice how the table is symmetrical along its leading diagonal. This indicates that the operation is commutative.

$*$	***a***	***b***	***c***	***d***
a	b	c	d	a
b	c	d	a	b
c	d	a	b	c
d	a	b	c	d

a $a*d=a$, $b*d=b$, $c*d=d$ and $d*d=d$ so d is the identity element.

(1) The fourth column is a, b, c, d so all of the elements are unchanged, you can also see this in the fourth row of the table.

b

$*$	a	b	c	d
a	b	c	d	a
b	c	d	a	b
c	d	a	b	c
d	a	b	c	d

(2) Find all the times the identity element, d, appears.

b and d are self-inverses since $b*b=d$ and $d*d=d$

(3) You can also see that a is the inverse of c and vice versa.

Exercise 15.1B Reasoning and problem-solving

1 For each of these Cayley tables

a Work out the identity element,

b Find the inverse of each of the elements.

i

	a	***b***	***c***
a	b	c	a
b	c	a	b
c	a	b	c

ii

	a	***b***	***c***	***d***
a	a	b	c	d
b	b	c	d	a
c	c	d	a	b
d	d	a	b	c

iii

	A	***B***	***C***	***D***
A	B	A	D	C
B	A	B	C	D
C	D	C	B	A
D	C	D	A	B

iv

	A	***B***	***C***	***D***	***E***
A	A	B	C	D	E
B	B	C	E	A	D
C	C	E	D	B	A
D	D	A	B	E	C
E	E	D	A	C	B

2 For parts **i** and **ii** of question **1**, verify that the operation is associative.

3 a Copy and complete the Cayley table for the binary operation $a\times b$ on the set $\{1, -1, i, -i\}$, where $i=\sqrt{-1}$

$\times$	**1**	**−1**	$\boldsymbol{i}$	$\boldsymbol{-i}$
1				
−1				
$\boldsymbol{i}$				
$\boldsymbol{-i}$				

b What is the identity element?

c Find the inverse of each element.

d Is the operation associative? Explain your answer.

4 The set S is the set of matrices of the form

$$\begin{pmatrix} \cos\left(\frac{n\pi}{2}\right) & -\sin\left(\frac{n\pi}{2}\right) \\ \sin\left(\frac{n\pi}{2}\right) & \cos\left(\frac{n\pi}{2}\right) \end{pmatrix} \text{ where } n = 0, 1, 2, 3$$

a Write down the members of S.

b Explain whether matrix addition is a binary operation on S.

c Copy and complete the Cayley table for matrix multiplication where the elements represent the matrices generated by those values of n

	0	**1**	**2**	**3**
0				
1		2		
2			0	
3	3			

d Write down

i The identity element,

ii The inverse of each of the elements.

e Explain the geometrical significance of this set of matrices.

5 The set S is the set of matrices of the form

$$\begin{pmatrix} \cos\left(\frac{2n\pi}{5}\right) & \sin\left(\frac{2n\pi}{5}\right) \\ -\sin\left(\frac{2n\pi}{5}\right) & \cos\left(\frac{2n\pi}{5}\right) \end{pmatrix}$$

where $n = 1, 2, \ldots k$.

a What is the smallest value of k for which matrix multiplication is a binary operation on S?

b Construct a Cayley table for matrix multiplication on S using this value of k

c Write down the inverse of each of the matrices.

6 The set S consists of the following transformations in 2D:

A = reflection in x-axis

B = reflection in y-axis

C = reflection in line $y = x$

D = reflection in line $y = -x$

E = rotation 90° anticlockwise about origin

F = rotation 180° about origin

G = rotation 90° clockwise about origin

H = no transformation

a Write down the inverse of each element.

b Write down the 2×2 matrix that represents each transformation.

c Construct a Cayley table for the binary operation $A * B$ = transformation A followed by transformation B

d Prove by counter-example that $*$ is not commutative.

7 The set S is

$$\left\{ \begin{pmatrix} -1 & 0 & 0 \\ 0 & 1 & 0 \\ 0 & 0 & 1 \end{pmatrix}, \begin{pmatrix} 1 & 0 & 0 \\ 0 & -1 & 0 \\ 0 & 0 & 1 \end{pmatrix}, \begin{pmatrix} 1 & 0 & 0 \\ 0 & 1 & 0 \\ 0 & 0 & -1 \end{pmatrix}, \begin{pmatrix} -1 & 0 & 0 \\ 0 & -1 & 0 \\ 0 & 0 & 1 \end{pmatrix}, \begin{pmatrix} -1 & 0 & 0 \\ 0 & 1 & 0 \\ 0 & 0 & -1 \end{pmatrix}, \begin{pmatrix} 1 & 0 & 0 \\ 0 & -1 & 0 \\ 0 & 0 & -1 \end{pmatrix}, \begin{pmatrix} -1 & 0 & 0 \\ 0 & -1 & 0 \\ 0 & 0 & -1 \end{pmatrix}, \begin{pmatrix} 1 & 0 & 0 \\ 0 & 1 & 0 \\ 0 & 0 & 1 \end{pmatrix} \right\}$$

a Explain how you know that matrix multiplication is a binary operation on S.

b State the identity element.

c Explain which of the elements of S are self-inverse.

d Show that matrix multiplication is commutative on S.

15.2 Modular arithmetic

Fluency and skills

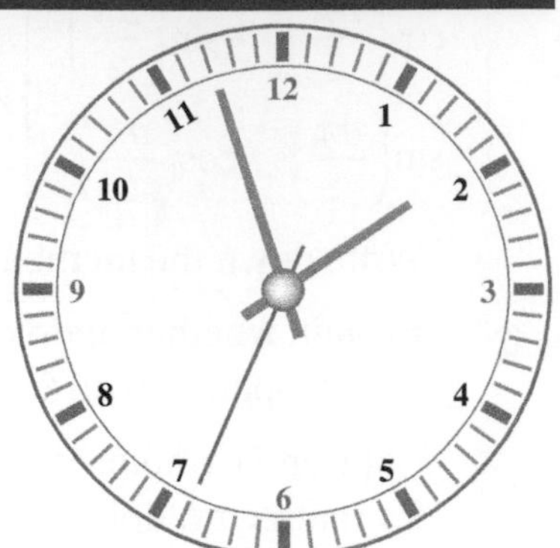

When you are using the 12-hour clock, five hours after 11:00 is 4:00 and three hours before 1:00 is 10:00. This is an example of **modular arithmetic**. The modulo in this case is 12

So you could write $11+5=4 \pmod{12}$ and $1-3=10 \pmod{12}$

Every time you reach 12 you start again from 0; 12 is said to be **congruent** to 0 in mod 12

Any modulo can be used for modular arithmetic. For example, using modulo 5, whenever you reach a multiple of 5 you start again from 0

Therefore, 5 is congruent to 0, 6 is congruent to 1 and so on.

0	**1**	**2**	**3**	**4**
5	6	7	8	9
10	11	12	13	14
15	16	17	18	19

So, for example, $3+4 \equiv 2 \pmod{5}$ and $4 \times 4 \equiv 1 \pmod{5}$

Another way of thinking about this is that $a \pmod{n}$ is the remainder when a is divided by n

You can use rules to add, subtract, multiply and divide in modular arithmetic.

If $a \equiv b \pmod{n}$, then a can be written $a = b + kn$ where k is an integer. Similarly, if $c \equiv d \pmod{n}$ then $c = d + mn$ where m is an integer.

So $a+c=(b+kn)+(d+mn)$

$=(b+d)+(k+m)n$

Therefore the solution can be written as $b+d \pmod{n}$ since $k+m$ is an integer.

You have previously used ≡ to mean 'is identical to' but it can also be used to mean 'is congruent to', as in this example.

Key point

If $a \equiv b \pmod{n}$ and $c \equiv d \pmod{n}$ then:

- $a+c=b+d \pmod{n}$
- $a-c=b-d \pmod{n}$
- $ac = bd \pmod{n}$ as long as a, b, c, d, n are integers.

These rules can be used to simplify calculations.

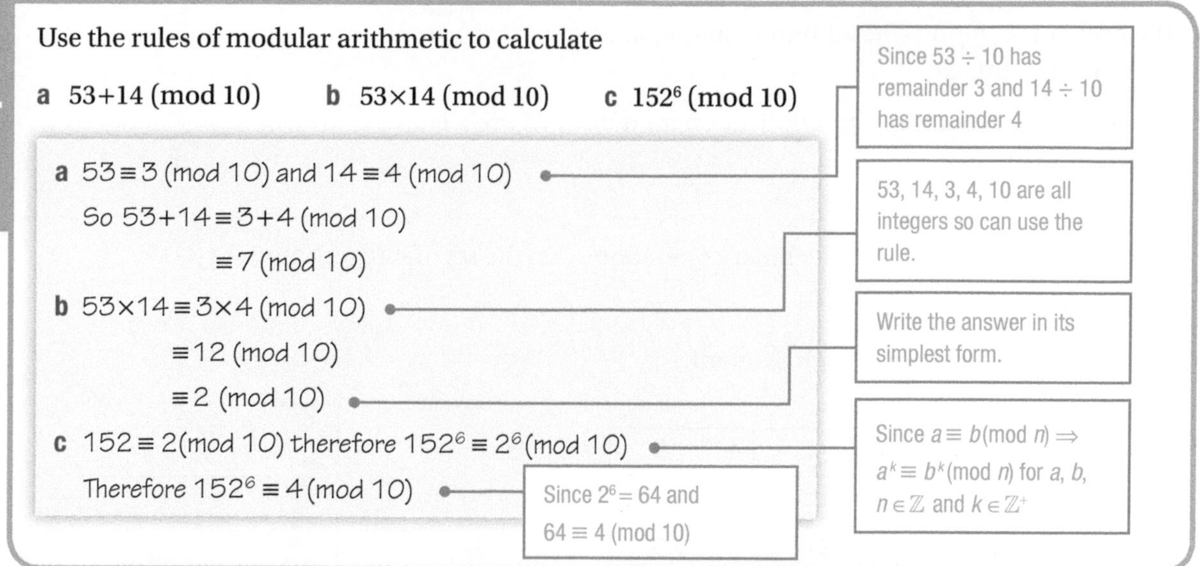

Example 1

Use the rules of modular arithmetic to calculate

a 53+14 (mod 10) **b** 53×14 (mod 10) **c** 152^6 (mod 10)

a $53 \equiv 3 \pmod{10}$ and $14 \equiv 4 \pmod{10}$

So $53+14 \equiv 3+4 \pmod{10}$

$\equiv 7 \pmod{10}$

b $53 \times 14 \equiv 3 \times 4 \pmod{10}$

$\equiv 12 \pmod{10}$

$\equiv 2 \pmod{10}$

c $152 \equiv 2 \pmod{10}$ therefore $152^6 \equiv 2^6 \pmod{10}$

Therefore $152^6 \equiv 4 \pmod{10}$

Since 53 ÷ 10 has remainder 3 and 14 ÷ 10 has remainder 4

53, 14, 3, 4, 10 are all integers so can use the rule.

Write the answer in its simplest form.

Since $a \equiv b \pmod{n} \Rightarrow a^k \equiv b^k \pmod{n}$ for $a, b, n \in \mathbb{Z}$ and $k \in \mathbb{Z}^+$

Since $2^6 = 64$ and $64 \equiv 4 \pmod{10}$

Exercise 15.2A Fluency and skills

1 Write each number in the form $a \pmod{n}$ where $0 \le a < n$, using the modulo given.

a 7 (mod 3) **b** 12 (mod 7)

c 19 (mod 4) **d** 27 (mod 9)

e 25 (mod 6) **f** –5 (mod 8)

g –55 (mod 11) **h** 48 (mod 13)

2 Use modular arithmetic to work out these calculations. Show your method clearly. Write each answer in the form $a \pmod{n}$ where $0 \le a < n$, using the modulo given.

a $21 + 19 \pmod{3}$ **b** $4 + 25 \pmod{7}$

c $68 - 34 \pmod{4}$ **d** $79 - 94 \pmod{9}$

e $5 \times 14 \pmod{6}$ **f** $25 \times 37 \pmod{8}$

g $82 \times -16 \pmod{11}$ **h** $29 \times 75 \pmod{13}$

3 Use the exponential rule of modular arithmetic to simplify these powers. Give your answers in the form $a \pmod{n}$ where $0 \le a < n$, using the modulo given.

a $37951^{37} \pmod{10}$

b $1736^{26} \pmod{5}$

c $982^{9} \pmod{5}$

d $3219^{48} \pmod{3}$

e $3268^{84} \pmod{3}$

f $12653^{11} \pmod{6}$

Reasoning and problem-solving

You can define binary operations involving modular arithmetic and draw Cayley tables of finite sets under binary operations mod n

Strategy

To construct a Cayley table for the binary operation * on the set of integers modulo n

1. Draw the table for the set {0, 1, 2,..., $(n - 1)$}
2. Calculate $a * b$ for each pair of integers and write the answers (mod n) in the tables.
3. Use the fact that $a * e = e * a = a$ for the identity element e

Key point

The notation $+_n$ represents addition modulo n, where n is a positive integer

The notation $\times_n$ represents multiplication modulo n, where n is a positive integer

Example 2

a Draw the Cayley table of the binary operation $+_6$ on the set of integers modulo 6

b State the identity element.

c Work out the inverse of each element.

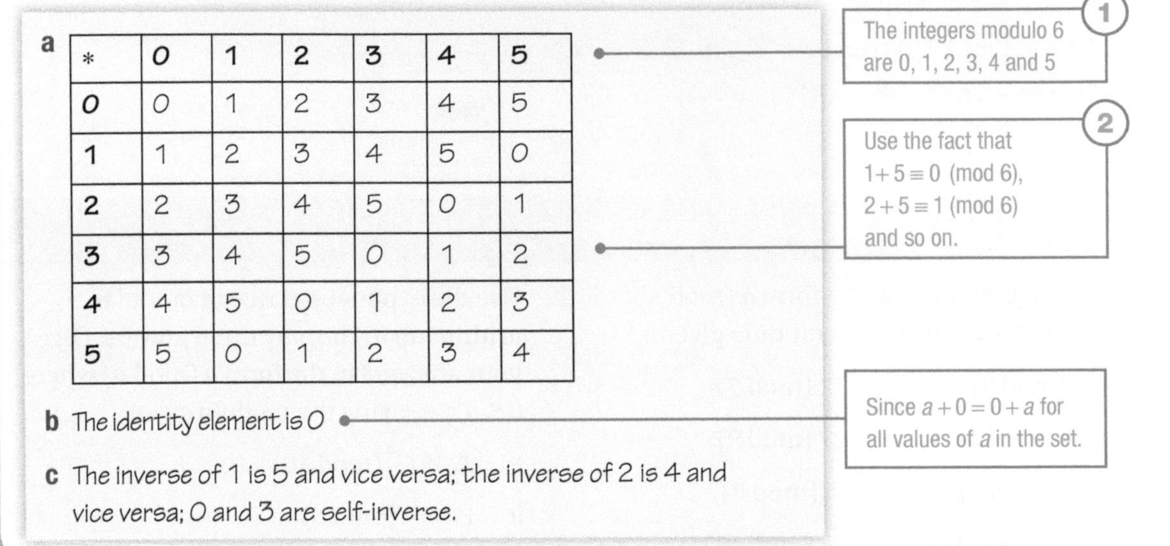

a

*	0	1	2	3	4	5
0	0	1	2	3	4	5
1	1	2	3	4	5	0
2	2	3	4	5	0	1
3	3	4	5	0	1	2
4	4	5	0	1	2	3
5	5	0	1	2	3	4

b The identity element is 0

c The inverse of 1 is 5 and vice versa; the inverse of 2 is 4 and vice versa; 0 and 3 are self-inverse.

1. The integers modulo 6 are 0, 1, 2, 3, 4 and 5

2. Use the fact that $1+5 \equiv 0 \pmod 6$, $2+5 \equiv 1 \pmod 6$ and so on.

Since $a+0=0+a$ for all values of a in the set.

Exercise 15.2B Reasoning and problem-solving

1 Draw the Cayley table for the set of integers modulo 2 under the binary operation $*$ where

$a * b = a - b \pmod 2$

2 Draw the Cayley table for the binary operation

- **a** $+_3$ on the set $\{0, 1, 2\}$
- **b** $+_4$ on the set $\{0, 1, 2, 3\}$
- **c** $\times_5$ on the set $\{1, 2, 3, 4\}$
- **d** $\times_7$ on the set $\{1, 2, 3, 4, 5, 6\}$

3 For each part of questions **1** and **2**

- **a** State the identity element,
- **b** Explain which elements have an inverse.

4 The operation $*$ is defined as $a * b = a + b + a \times b \pmod 5$ on the set $\{0, 1, 2, 3\}$

- **a** Draw the Cayley table for $*$
- **b** **i** State the identity element.
  - **ii** Write down the inverse of each of the elements.

5 The operation ♦ is defined as $a \blacklozenge b = a+b+4 \pmod 6$ on the set $\{0, 1, 2, 3, 4, 5\}$

- **a** Draw the Cayley table for ♦
- **b** **i** State the identity element.
  - **ii** Write down the inverse of each of the elements.

15 Summary and review

Chapter summary

- A binary operation is a function that can be applied to any two elements of a set so that the result is also a member of the set.
- A binary operation, $*$, is commutative if $a*b=b*a$
- A binary operation, $*$, is associative if $(a*b)*c=a*(b*c)$
- The identity element, e, of a set under an operation $*$ is such that $a*e=e*a=a$ for all values of a in the set.
- The inverse, a^{-1}, of an element a under an operation $*$ is such that $a*a^{-1}=a^{-1}*a=e$
- An element, a, is self-inverse if $a^{-1}=a$
- A Cayley table shows the results of applying a binary operation to all possible pairs of elements from a finite set.
- The notation $+_n$ represents addition modulo n, where n is a positive integer
- The notation $\times_n$ represents multiplication modulo n, where n is a positive integer

Check and review

You should now be able to...	Review Questions
✔ Decide if a function on a set is a binary operation.	1
✔ Work out if a binary operation is associative.	2
✔ Work out if a binary operation is commutative.	3, 5
✔ Work out the identity element of a binary operation.	4, 5
✔ Work out the inverses of the elements of a set under a binary operation.	4, 5
✔ Add, subtract and multiply numbers using modular arithmetic.	6
✔ Draw Cayley tables for sets under binary operations.	7, 8

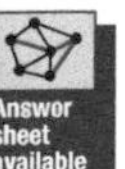

1 Decide whether or not each of these functions is a binary operation. Explain your answers.

a $a+3b$ on the set $\mathbb{R}$

b $(a+1)(b-1)$ on the set $\mathbb{Z}^+$

c $\dfrac{b}{a^2+1}$ on the set $\mathbb{R}$

d $\sqrt{ab}$ on the set $\mathbb{R}$

2 Explain whether or not each of these binary operations is associative.

a $\dfrac{a}{b}$ on the set $\mathbb{R}$

b a^2+b^2 on the set $\mathbb{Z}$

c $\dfrac{ab}{a+b}$ on the set $\mathbb{R}^+$

3 For each of the binary operations in question **2**, explain whether or not it is commutative.

DISCRETE

4 For each of the binary operations in question **2**

a Find the identity element (if it exists),

b Write down the inverse of each element where possible.

5 The Cayley table shows the binary operation $*$ over the set {A, B, C, D, E}

*	**A**	**B**	**C**	**D**	**E**
A	B	C	D	E	A
B	C	A	E	A	B
C	D	E	A	B	C
D	E	A	B	C	D
E	A	B	C	D	E

a State the identity element.

b Give the inverse of each element.

c How can you tell from the table that $*$ is commutative?

6 Use modular arithmetic to work out

a $379+612 \pmod 5$

b $1079-351 \pmod 3$

c $326\times249 \pmod 8$

d $-532\times249 \pmod{10}$

7 a Copy and complete the Cayley table for the set {0, 2, 4, 6, 8} under $+_{10}$

*	**0**	**2**	**4**	**6**	**8**
0					
2				8	
4			8		
6					
8					

b What is the identity element?

c State the inverse of each of the elements.

8 The set S consists of the matrices

$$A=\begin{pmatrix} -\frac{1}{2} & \frac{\sqrt{3}}{2} \\ -\frac{\sqrt{3}}{2} & -\frac{1}{2} \end{pmatrix}$$

$$B=\begin{pmatrix} -\frac{1}{2} & -\frac{\sqrt{3}}{2} \\ \frac{\sqrt{3}}{2} & -\frac{1}{2} \end{pmatrix}$$

$$C=\begin{pmatrix} 1 & 0 \\ 0 & 1 \end{pmatrix}$$

a Copy and complete the Cayley table for matrix multiplication on the set S.

	A	**B**	**C**
A			
B			
C			

b Write down

i The identity element,

ii The inverse of each of the elements.

c Explain the geometrical significance of this set of matrices.

History

Examples of modular arithmetic have been around for a long time. For example, in the third century Chinese book, Master Sun's Mathematical Manual, the following problem appears:

We have a number of things, but we do not know exactly how many. If we count them by threes we have two left over. If we count by fives we have three left over. If we count by sevens there are two left over. How many things are there?

Find out how this problem relates to the Chinese Remainder Theorem.

Did you know?

The check digit in an ISBN number (found on all books) is calculated using modulo 10 division.

Research

The Cayley table for the Klein four-group gives the group of permutations in twelve-note music composition. This is a method of composing music devised by Arnold Schoenberg and ensures that all twelve notes in a chromatic scale are given equal importance and hence played the same number of times.

The initial prime (P) ordering of the twelve notes of a chromatic scale can be subjected to interval-preserving transformations of retrograde (R), inversion (I) and retrograde-inversion(RI) form.

The Cayley table representing these transformations would be:

	p	R	I	RI
p	p	R	I	RI
R	R	p	RI	I
I	I	RI	p	R
RI	RI	I	R	p

Many pieces of music have been composed using this technique as a basis. Find out more by searching for Schoenberg Opus 23 movement 5.

15 Assessment

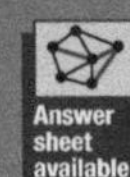

1 The binary operation ■ is defined by $a \blacksquare b = a + b^2$ on the set of natural numbers, $\mathbb{N}$

a Calculate $3 \blacksquare 7$ **[2 marks]**

b Prove by counter-example that ■ is neither associative nor commutative. **[4]**

2 Use modular arithmetic to write each of these in the form a (mod 10) where $0 \le a < n$

a $3 + 7$ b $125 - 621$

c 358×6914 d 2563^4 **[6]**

3 The operation $*$ is defined as $a * b = a^b$

a Explain whether or not $*$ is a binary operation on each of these sets.

i $\mathbb{Q}$ ii $\mathbb{R}$ iii $\mathbb{N}$ **[6]**

b Explain why the operation $*$ on the set $\mathbb{N}$ does not have an identity element. **[2]**

c Is $*$ associative on $\mathbb{N}$? Explain your answer. **[2]**

4 The binary operation $*$ is defined as $a * b = \frac{ab}{2}$ on the set $\{-2, 2\}$

a Copy and complete the Cayley table for $*$ **[2]**

*	**−2**	**2**
−2		
2		

b Write down the identity element for $*$ **[1]**

c What is the inverse of each element? **[1]**

5 The table defines the binary operation ⊗ on the set {S, T, U, V, W, X}

⊗	**S**	**T**	**U**	**V**	**W**	**X**
S	T	U	S	X	V	W
T	U	S	T	W	X	V
U	S	T	U	V	W	X
V	X	W	V	T	S	U
W	V	X	W	S	U	T
X	W	V	X	U	T	S

a Use the table to find the identity element. **[1]**

b Write down the inverse of V. **[1]**

c Which of the elements are self-inverses? **[1]**

6 a The binary operation $*$ is defined on the set $\{0, 1, 2, 3, 4\}$ as $a * b = a +_5 b$

i Copy and complete the Cayley table for $*$ [2]

*	0	1	2	3	4
0					
1					
2					
3					
4					

ii What is the identity element? [1]

iii Write down the inverse of each of the elements. [2]

b Use modular arithmetic to calculate the remainder when 18^{12} is divided by 5 [4]

7 The binary operation ♦ is defined on the set {A, B, C, D, E} by the table.

♦	A	B	C	D	E
A	D	B	E	A	C
B	E	D	C	B	A
C	B	A	D	C	E
D	C	E	A	D	B
E	A	C	B	E	D

a What is A ♦ E? [1]

b Explain why there is no identity element. [3]

c Prove that ♦ is not associative. [3]

8 The group S consists of the matrices

$$A=\begin{pmatrix} 0 & -1 \\ 1 & 0 \end{pmatrix}, B=\begin{pmatrix} -1 & 0 \\ 0 & -1 \end{pmatrix}, C=\begin{pmatrix} 0 & 1 \\ -1 & 0 \end{pmatrix}, D=\begin{pmatrix} 1 & 0 \\ 0 & 1 \end{pmatrix}$$

a Construct a Cayley table for S under matrix multiplication. [3]

b State the inverse of each of the elements. [2]

9 The operation $*$ is defined by $a*b=2a-b$

Prove that $*$ is a binary operation on the set of odd integers $\{\pm1, \pm3, \pm5, \pm7, \pm9, \ldots\}$ [4]

10 The set S consists of matrices of the form $m_n = \begin{pmatrix} 1 & 0 & 0 \\ 0 & \cos\left(\frac{n\pi}{3}\right) & -\sin\left(\frac{n\pi}{3}\right) \\ 0 & \sin\left(\frac{n\pi}{3}\right) & \cos\left(\frac{n\pi}{3}\right) \end{pmatrix}$ for $n = 1, 2, 3, 4, 5, 6$

a Copy and complete the Cayley table for the binary operation of matrix multiplication on S. **[6]**

$\times$	M_1	M_2	M_3	M_4	M_5	M_6
M_1						
M_2						
M_3						
M_4						
M_5						
M_6						

b State the identity element. **[1]**

c What is the inverse of M_2? **[1]**

d Which element is self-inverse? **[1]**

Full A Level

Mathematical formulae to learn

For A Level Further Maths

Pure Mathematics

Quadratic equations

$ax^2 + bx + c = 0$ has roots $\dfrac{-b \pm \sqrt{b^2 - 4ac}}{2a}$

Laws of indices

$a^x a^y \equiv a^{x+y}$

$a^x \div a^y \equiv a^{x-y}$

$(a^x)^y \equiv a^{xy}$

Laws of logarithms

$x = a^n \Leftrightarrow n = \log_a x$ for $a > 0$ and $x > 0$

$\log_a x + \log_a y \equiv \log_a xy$

$\log_a x - \log_a y \equiv \log_a \left(\dfrac{x}{y}\right)$

$k \log_a x \equiv \log_a (x)^k$

Coordinate geometry

A straight-line graph, gradient m passing through (x_1, y_1) has equation $y - y_1 = m(x - x_1)$

Straight lines with gradients m_1 and m_2 are perpendicular when $m_1 m_2 = -1$

Sequences

General term of an arithmetic progression: $u_n = a + (n - 1)d$

General term of a geometric progression: $u_n = ar^{n-1}$

Trigonometry

In the triangle ABC

Sine rule: $\dfrac{a}{\sin A} = \dfrac{b}{\sin B} = \dfrac{c}{\sin C}$

Cosine rule: $a^2 = b^2 + c^2 - 2bc \cos A$

Area $= \dfrac{1}{2} ab \sin C$

$\cos^2 A + \sin^2 A \equiv 1$

$\sec^2 A \equiv 1 + \tan^2 A$

$\operatorname{cosec}^2 A \equiv 1 + \cot^2 A$

$\sin 2A \equiv 2 \sin A \cos A$

$\cos 2A \equiv \cos^2 A - \sin^2 A$

$\tan 2A \equiv \dfrac{2 \tan A}{1 - \tan^2 A}$

Mensuration

Circumference and Area of circle, radius r and diameter d:

$C = 2\pi r = \pi d \qquad A = \pi r^2$

Pythagoras' Theorem:

In any right-angled triangle where a, b and c are the lengths of the sides and c is the hypotenus: $c^2 = a^2 + b^2$

Area of a trapezium $= \frac{1}{2}(a+b)h$, where a and b are the lengths of the parallel sides and h is their perpendicular separation.

Volume of a prism = area of cross section × length

For a circle of radius r, where an angle at the centre of θ radians subtends an arc of length s and encloses an associated sector of area A:

$s = r\theta \qquad A = \frac{1}{2}r^2\theta$

Complex numbers

For two complex numbers $z_1 = r_1 e^{i\theta_1}$ and $z_2 = r_2 e^{i\theta_2}$:

$$z_1 z_2 = r_1 r_2 \, e^{i(\theta_1 + \theta_2)}$$

$$\frac{z_1}{z_2} = \frac{r_1}{r_2} e^{i(\theta_1 - \theta_2)}$$

Loci in the Argand diagram:

$|z - a| = r$ is a circle radius r centred at a

$\arg(z - a) = \theta$ is a half line drawn from a at angle θ to a line parallel to the positive real axis.

Exponential form: $e^{i\theta} = \cos\theta + i\sin\theta$

Matrices

For a 2 by 2 matrix $\begin{pmatrix} a & b \\ c & d \end{pmatrix}$ the determinant $\Delta = \begin{vmatrix} a & b \\ c & d \end{vmatrix} = ad - bc$

the inverse is $\frac{1}{\Delta}\begin{pmatrix} d & -b \\ -c & a \end{pmatrix}$

The transformation represented by matrix **AB** is the transformation represented by matrix **B** followed by the transformation represented by matrix **A**.

For matrices **A**, **B**:

$(\mathbf{AB})^{-1} = \mathbf{B}^{-1}\mathbf{A}^{-1}$

Algebra

$$\sum_{r=1}^{n} r = \frac{1}{2}n(n+1)$$

For $ax^2 + bx + c = 0$ with roots α and β:

$$\alpha + \beta = \frac{-b}{a} \qquad \alpha\beta = \frac{c}{a}$$

For $ax^3 + bx^2 + cx + d = 0$ with roots α, β and γ:

$$\sum \alpha = \frac{-b}{a} \qquad \sum \alpha\beta = \frac{c}{a} \qquad \alpha\beta\gamma = \frac{-d}{a}$$

Hyperbolic functions

$$\cosh x \equiv \frac{1}{2}(e^x + e^{-x}) \qquad \sinh x \equiv \frac{1}{2}(e^x - e^{-x}) \qquad \tanh x \equiv \frac{\sinh x}{\cosh x}$$

Calculus and differential equations

Differentiation

Function	Derivative	Function	Derivative
x^n	nx^{n-1}	e^{kx}	ke^{kx}
$\sin kx$	$k\cos kx$	$\ln x$	$\frac{1}{x}$
$\cos kx$	$-k\sin kx$	$f(x) + g(x)$	$f'(x) + g'(x)$
		$f(x)g(x)$	$f'(x)g(x) + f(x)g'(x)$
		$f(g(x))$	$f'(g(x))g'(x)$

Integration

Function	Integral
x^n	$\frac{1}{n+1}x^{n+1}+c,\ n\neq -1$
$\cos kx$	$\frac{1}{k}\sin kx + c$
$\sin kx$	$-\frac{1}{k}\cos kx + c$

Function	Integral
e^{kx}	$\frac{1}{k}e^{kx}+c$
$\frac{1}{x}$	$\ln\lvert x\rvert + c,\ x\neq 0$
$f'(x)+g'(x)$	$f(x)+g(x)+c$
$f'(g(x))g'(x)$	$f(g(x))+c$

Area under a curve $= \int_a^b y\,dx\ (y \geq 0)$

Volumes of revolution about the x and y axes:

$$V_x = \pi\int_a^b y^2\,dx \qquad V_y = \pi\int_c^d x^2\,dy$$

Simple Harmonic Motion: $\ddot{x} = -\omega^2 x$

Vectors

$$\lvert x\mathbf{i}+y\mathbf{j}+z\mathbf{k}\rvert = \sqrt{(x^2+y^2+z^2)}$$

Scalar product of two vectors $\mathbf{a}=\begin{pmatrix}a_1\\a_2\\a_3\end{pmatrix}$ and $\mathbf{b}=\begin{pmatrix}b_1\\b_2\\b_3\end{pmatrix}$ is

$$\begin{pmatrix}a_1\\a_2\\a_3\end{pmatrix}\cdot\begin{pmatrix}b_1\\b_2\\b_3\end{pmatrix} = a_1b_1+a_2b_2+a_3b_3 = \lvert\mathbf{a}\rvert\,\lvert\mathbf{b}\rvert\cos\theta$$

where θ is the acute angle between the vectors **a** and **b**.

The equation of the line through the point with position vector **a** parallel to vector **b** is:

$$\mathbf{r} = \mathbf{a} + t\mathbf{b}$$

The equation of the plane containing the point with position vector **a** and perpendicular to vector **n** is:

$$(\mathbf{r}-\mathbf{a})\cdot\mathbf{n} = 0$$

Mechanics

Forces and equilibrium

Weight = mass × g

Friction: $F \leq \mu R$

Newton's second law in the form: $F = ma$

Kinematics

For motion in a straight line with variable acceleration:

$$v = \frac{\mathrm{d}r}{\mathrm{d}t} \qquad a = \frac{\mathrm{d}v}{\mathrm{d}t} = \frac{\mathrm{d}^2 r}{\mathrm{d}t^2}$$

$$r = \int v \, \mathrm{d}t \qquad v = \int a \, \mathrm{d}t$$

Statistics

The mean of a set of data: $\bar{x} = \frac{\sum x}{n} = \frac{\sum fx}{\sum f}$

The standard Normal variable: $Z = \frac{X - \mu}{\sigma}$ where $X \sim \mathrm{N}(\mu, \sigma^2)$

Mathematical notation
For A Level Further Maths

Set Notation

Set Notation	Meaning
$\in$	is an element of
$\notin$	is not an element of
$\subseteq$	is a subset of
$\subset$	is a proper subset of
$\{x_1, x_2, \ldots\}$	the set with elements $x_1, x_2, \ldots$
$\{x: \ldots\}$	the set of all x such that ...
$n(A)$	the number of elements in set A
$\varnothing$	the empty set
ε	the universal set
A'	the complement of the set A
$\mathbb{N}$	the set of natural numbers, $\{1, 2, 3, \ldots\}$
$\mathbb{Z}$	the set of integers, $\{0, \pm1, \pm2, \pm3, \ldots\}$
$\mathbb{Z}^+$	the set of positive integers, $\{1, 2, 3, \ldots\}$
$\mathbb{Z}_0^+$	the set of non-negative integers, $\{0, 1, 2, 3, \ldots\}$
$\mathbb{R}$	the set of real numbers
$\mathbb{Q}$	the set of rational numbers, $\left\{\frac{p}{q} : p \in \mathbb{Z},\ q \in \mathbb{Z}^+\right\}$
$\cup$	union
$\cap$	intersection
(x, y)	the ordered pair x, y
$[a, b]$	the closed interval $\{x \in \mathbb{R} : a \le x \le b\}$
$[a, b)$	the interval $\{x \in \mathbb{R} : a \le x < b\}$
$(a, b]$	the interval $\{x \in \mathbb{R} : a < x \le b\}$
(a, b)	the open interval $\{x \in \mathbb{R} : a < x < b\}$
$\mathbb{C}$	the set of complex numbers

Miscellaneous Symbols

Miscellaneous Symbols	Meaning
$=$	is equal to
$\neq$	is not equal to
$\equiv$	is identical to or is congruent to
$\approx$	is approximately equal to
∞	infinity
$\propto$	is proportional to
$\therefore$	therefore
$\because$	because
$<$	is less than
$\leqslant, \le$	is less than or equal to; is not greater than
$>$	is greater than
$\geqslant, \ge$	is greater than or equal to; is not less than
$p \Rightarrow q$	p implies q (if p then q)
$p \Leftarrow q$	p is implied by q (if q then p)
$p \Leftrightarrow q$	p implies and is implied by q (p is equivalent to q)

Mathematical notation – for A Level Further Maths

Miscellaneous Symbols	Meaning
a	first term for an arithmetic or geometric sequence
l	last term for an arithmetic sequence
d	common difference for an arithmetic sequence
r	common ratio for a geometric sequence
S_n	sum to n terms of a sequence
S_∞	sum to infinity of a sequence
$\cong$	is isomorphic to

Operations

Operations	Meaning
$a+b$	a plus b
$a-b$	a minus b
$a \times b$, ab, $a \cdot b$	a multiplied by b
$a \div b$, $\frac{a}{b}$	a divided by b
$\sum_{i=1}^{n} a_i$	$a_1 + a_2 + \ldots + a_n$
$\prod_{i=1}^{n} a_i$	$a_1 \times a_2 \times \ldots \times a_n$
$\sqrt{a}$	the non-negative square root of a
$\lvert a \rvert$	the modulus of a
$n!$	n factorial: $n! = n \times (n-1) \times \ldots \times 2 \times 1$, $n \in \mathbb{N}$; $0! = 1$
$\binom{n}{r}$, ${}^nC_r$, ${}_nC_r$	the binomial coefficient $\frac{n!}{r!(n-r)!}$ for $n, r \in \mathbb{Z}_0^+$, $r \leq n$ or $\frac{n(n-1)\ldots(n-r+1)}{r!}$ for $n \in \mathbb{Q}$, $r \in \mathbb{Z}_0^+$
$a \times_n b$	multiplication modulo n of a by b
$a +_n b$	addition modulo n of a and b
$G = (\langle n \rangle, *)$	n is the generator of a given group G under the operation*

Functions

Functions	Meaning
$\mathrm{f}(x)$	the value of the function f at x
$\mathrm{f}: x \mapsto y$	the function f maps the element x to the element y
f^{-1}	the inverse function of the function f
gf	the composite function of f and g which is defined by $\mathrm{gf}(x) = \mathrm{g}(\mathrm{f}(x))$
$\lim_{x \to a} \mathrm{f}(x)$	the limit of $\mathrm{f}(x)$ as x tends to a
Δx, δx	an increment of x
$\frac{\mathrm{d}y}{\mathrm{d}x}$	the derivative of y with respect to x
$\frac{\mathrm{d}^n y}{\mathrm{d}x^n}$	the nth derivative of y with respect to x
$\mathrm{f}'(x), \mathrm{f}''(x), \ldots, \mathrm{f}^{(n)}(x)$	the first, second, ..., nth derivatives of $\mathrm{f}(x)$ with respect to x
$\dot{x}, \ddot{x}, \ldots$	the first, second, ... derivatives of x with respect to t
$\int y \, \mathrm{d}x$	the indefinite integral of y with respect to x

Functions	Meaning
$\int_a^b y\,\mathrm{d}x$	the definite integral of y with respect to x between the limits $x = a$ and $x = b$

Exponential and Logarithmic Functions

Exponential and Logarithmic Functions	Meaning
e	base of natural logarithms
e^x, $\exp x$	exponential function of x
$\log_a x$	logarithm to the base a of x
$\ln x$, $\log_e x$	natural logarithm of x

Trigonometric Functions

Trigonometric Functions	Meaning
sin, cos, tan, cosec, sec, cot	the trigonometric functions
$\sin^{-1}$, $\cos^{-1}$, $\tan^{-1}$ arcsin, arccos, arctan	the inverse trigonometric functions
°	degrees
rad	radians
cosec^{-1}, $\sec^{-1}$, $\cot^{-1}$, arccosec, arcsec, arccot	the inverse trigonometric functions
sinh, cosh, tanh, cosech, sech, coth	the hyperbolic functions
$\sinh^{-1}$, $\cosh^{-1}$, $\tanh^{-1}$ cosech^{-1}, sech^{-1}, $\coth^{-1}$ arcsinh, arccosh, arctanh, arccosech, arcsech, arccoth arsinh, arcosh, artanh arcosech, arsech, arcoth	the inverse hyperbolic functions

Complex numbers

i, j	square root of -1
$x + \mathrm{i}y$	complex number with real part x and imaginary part y
$r(\cos\theta + \mathrm{i}\sin\theta)$	modulus argument form of a complex number with modulus r and argument θ
z	a complex number, $z = x + \mathrm{i}y = r(\cos\theta + \mathrm{i}\sin\theta)$
$\mathrm{Re}(z)$	the real part of z, $\mathrm{Re}(z) = x$
$\mathrm{Im}(z)$	the imaginary part of z, $\mathrm{Im}(z) = y$
$\lvert z\rvert$	the modulus of z, $\lvert z\rvert = r = \sqrt{x^2 + y^2}$
$\arg(z)$	the argument of z, $\arg(z) = \theta$, $-\pi < \theta \le \pi$
z^*	the complex conjugate of z, $x - \mathrm{i}y$

Matrices

Matrices	Meaning
$\mathbf{M}$	a matrix $\mathbf{M}$
$\mathbf{0}$	zero matrix
$\mathbf{I}$	identity matrix
$\mathbf{M}^{-1}$	the inverse of the matrix $\mathbf{M}$
$\mathbf{M}^{\mathrm{T}}$	the transpose of the matrix $\mathbf{M}$
Δ, det $\mathbf{M}$ or $\lvert\mathbf{M}\rvert$	the determinant of the square matrix $\mathbf{M}$
$\mathbf{Mr}$	Image of the column vector $\mathbf{r}$ under the transformation associated with the matrix $\mathbf{M}$

Vectors

Vectors	Meaning
$\mathbf{a}$, $\underline{a}$, $\underset{\sim}{a}$	the vector $\mathbf{a}$, $\underline{a}$, $\underset{\sim}{a}$ these alternatives apply throughout section 9
$\overrightarrow{AB}$	the vector represented in magnitude and direction by the directed line segment AB
$\hat{\mathbf{a}}$	a unit vector in the direction of $\mathbf{a}$
$\mathbf{i}$, $\mathbf{j}$, $\mathbf{k}$	unit vectors in the directions of the Cartesian coordinate axes
$\lvert\mathbf{a}\rvert$, a	the magnitude of $\mathbf{a}$
$\lvert\overrightarrow{AB}\rvert$, AB	the magnitude of $\overrightarrow{AB}$
$\begin{pmatrix} a \\ b \end{pmatrix}$, $a\mathbf{i} + b\mathbf{j}$	column vector and corresponding unit vector notation
$\mathbf{r}$	position vector
$\mathbf{s}$	displacement vector
$\mathbf{v}$	velocity vector
$\mathbf{a}$	acceleration vector
$\mathbf{a}\cdot\mathbf{b}$	the scalar product of $\mathbf{a}$ and $\mathbf{b}$

Differential equations

Differential equations	Meaning
ω	angular speed

Probability and statistics

Probability and statistics	Meaning
A, B, C, etc.	events
$A \cup B$	union of the events A and B
$A \cap B$	intersection of the events A and B
$\mathrm{P}(A)$	probability of the event A
A'	complement of the event A
$\mathrm{P}(A \mid B)$	probability of the event A conditional on the event B
X, Y, R etc.	random variables
x, y, r etc.	values of the random variables X, Y, R etc.
$x_1, x_2, \ldots$	values of observations

$f_1, f_2, \ldots$	frequencies with which the observations $x_1, x_2, \ldots$ occur
$\mathrm{p}(x)$, $\mathrm{P}(X=x)$	probability function of the discrete random variable X
$p_1, p_2, \ldots$	probabilities of the values $x_1, x_2, \ldots$ of the discrete random variable X
$\mathrm{E}(X)$	expectation of the random variable X
$\mathrm{Var}(X)$	variance of the random variable X
$\sim$	has the distribution
$\mathrm{B}(n, p)$	binomial distribution with parameters n and p, where n is the number of trials and p is the probability of success in a trial
q	$q=1-p$ for binomial distribution
$\mathrm{N}(\mu, \sigma^2)$	Normal distribution with mean μ and variance σ^2
$Z \sim \mathrm{N}(0, 1)$	standard Normal distribution
ϕ	probability density function of the standardised Normal variable with distribution $\mathrm{N}(0, 1)$
Φ	corresponding cumulative distribution function
μ	population mean
σ^2	population variance
σ	population standard deviation
$\overline{x}$	sample mean
s^2	sample variance
s	sample standard deviation
H_0	null hypothesis
H_1	alternative hypothesis
r	product moment correlation coefficient for a sample
ρ	product moment correlation coefficient for a population

Mechanics

Mechanics	**Meaning**
kg	kilograms
m	metres
km	kilometres
m/s, $\mathrm{m\,s^{-1}}$	metre(s) per second (velocity)
$\mathrm{m/s^2}$, $\mathrm{m\,s^{-2}}$	metre(s) per second square (acceleration)
F	Force or resultant force
N	newton
N m	newton metre (moment of a force)
t	time
s	displacement
u	initial velocity
v	velocity or final velocity
a	acceleration
g	acceleration due to gravity

Answers

Chapter 1

Exercise 1.1A

1 a $x=\pm 5i$ b $x=\pm 11i$
c $x=\pm 2\sqrt{5}i$ d $x=\pm 2\sqrt{2}i$
e $z=\pm 3i$ f $z=\pm 2\sqrt{3}i$

2 a $7-6i$ b $-7-10i$
c $18-27i$ d $26+25i$
e $-41-63i$ f $27-10i$

3 a $7+17i$ b $39-27i$
c $3+8i$ d $65-72i$

4 a $-i$ b 1
c i d $-8i$
e 81 f $-112+180i$

5 a $\frac{6}{5}-\frac{3}{5}i$ b $-\frac{5}{13}+\frac{1}{13}i$
c $-\frac{2}{5}+\frac{11}{5}i$ d $-\frac{1}{5}-\frac{7}{5}i$
e $(1-2\sqrt{2})+(-2-\sqrt{2})i$ f $-\sqrt{2}$

6 a $2+4i$ b $-11+10i$
c $-\frac{5}{17}+\frac{14}{17}i$ d $-\frac{5}{13}-\frac{14}{13}i$

Exercise 1.1B

1 $a=\pm 3, b=\pm 12$
2 $b=7,\ a=34$
3 $z=4+i$
4 $x=2-i$
5 $z_2=8+2i, z_1=3-5i$
6 $w=-2-\frac{1}{2}i, z=4+7i$
7 $z_2=-3-2i, z_1=\frac{1}{2}+i$
8 a $w=3+5i$ or $w=-3-5i$
b $w=1-2i$ or $w=-1+2i$
c $w=5-2i$ or $w=-5+2i$
9 $-2+\sqrt{2}i$ and $2-\sqrt{2}i$
10 $w=-3\pm i, z=1\pm 3i$
11 a $z=1+7i$ or $z=1-7i$
b $z=2-3i$ or $z=-2+3i$
12 a $-7-24i$
b $4282+1475i$
c $44+8i$
13 $a=\pm 3, b=\pm 96$

Exercise 1.2A

1 a $5+2i$ b $8-i$
c $-5i-6$ d $\sqrt{2}+i\sqrt{3}$
e $\frac{1}{3}-4i$ f $-\frac{2}{3}i-5$

2 a 85 b 18
c $-4i$ d $\frac{77}{85}-\frac{36}{85}i$
e $9-2i$ f $\frac{77}{85}+\frac{36}{85}i$

3 a 8 b $-2\sqrt{6}$
c $2\sqrt{2}i$ d $\frac{1}{2}-\frac{\sqrt{3}}{2}i$
e 8 f 24

4 a $x=-\frac{5}{2}\pm\frac{\sqrt{3}}{2}i$ b $x=\frac{3}{2}\pm\frac{\sqrt{11}}{2}i$
c $x=-\frac{7}{4}\pm\frac{\sqrt{7}}{4}i$ d $x=\frac{5}{3}\pm\frac{\sqrt{2}}{3}i$

5 a $x^2-3x-28=0$ b $x^2-6x+34=0$
c $x^2+2x+82=0$ d $x^2+10x+41=0$

6 a $\sqrt{3}-i$ b $z^2-2\sqrt{3}z+4=0$

7 a $x^2-4x+5=0$ b $x^2-8x+25=0$
c $x^2+2x+50=0$ d $x^2+10x+29=0$
e $x^2-2ax+(9+a^2)=0$ f $x^2-10x+(25+b^2)=0$

8 a $x^3+x^2-7x+65=0$
b $x^3-2x-4=0$
c $x^3-2\sqrt{3}x^2+7x=0$
d $x^3+(2\sqrt{2}-3)x^2+(3-6\sqrt{2})x-9=0$

9 a $6+2i$
b $2z^3-23z^2+68z+40=0$

10 a $x^3+9x^2+25x+25=(x+5)(x^2+4x+5)$
$f(x)=0\Rightarrow x^2+4x+5=0$
b $x=-5, -2\pm i$

11 a $1-8i$ is also a root giving quadratic factor $x^2-2x+65$
$x^4+4x^3+66x^2+364x+845$
$=(x^2-2x+65)(x^2+6x+13)=0$
$\Rightarrow x^2+6x+13=0$
b $x=1\pm 8i, -3\pm 2i$

12 a $a=1, b=4, c=30$
b $a=1, b=-3, c=-185$

13 a $a=-2, b=1, c=-20$
b $a=27, b=-38, c=26$

14 $(x-7i)(x+7i)=x^2+49$
$x^4-x^3+43x^2-49x-294=(x^2+49)(x^2-x-6)=0$
So can be written $(x^2+49)(x-3)(x+2)=0$
$(A=49, B=-3, C=2)$

Exercise 1.2B

1 a Let $z=a+bi, w=c+di, a, b, c, d\in\mathbb{R}$

$$
\begin{aligned}
(zw)^* &= ((a+bi)(c+di))^* \\
&= ((ac-bd)+(bc+ad)i)^* \\
&= (ac-bd)-(bc+ad)i \\
&= (a-bi)(c-di) \\
&= z^*w^*
\end{aligned}
$$

b

$$
\begin{aligned}
(z^*)^* &= ((a+bi)^*)^* \\
&= (a-bi)^* \\
&= a+bi \\
&= z
\end{aligned}
$$

c $\left(\frac{z}{w}\right)^* = \left(\frac{a+bi}{c+di}\right)^*$

$= \left(\frac{(a+bi)(c-di)}{(c+di)(c-di)}\right)^*$

$= \left(\frac{(ac+bd)+(bc-ad)i}{c^2-d^2i^2}\right)^*$

$= \frac{ac+bd-(bc-ad)i}{c^2+d^2}$

$\frac{z^*}{w^*} = \frac{a-bi}{c-di}$

$= \frac{(a-bi)(c+di)}{(c-di)(c+di)}$

$= \frac{ac+bd-(bc-ad)i}{c^2+d^2}$

$= \left(\frac{z}{w}\right)^*$

2 a Let $z = a+bi$, $a, b \in \mathbb{R}$
$z+z^* = (a+bi)+(a-bi)$
$= 2a$ so a real number

b $z-z^* = (a+bi)-(a-bi)$
$= 2bi$ so an imaginary number

c $zz^* = (a+bi)(a-bi)$
$= a^2 - abi + abi - b^2i^2$
$= a^2 + b^2$ so a real number

3 $z = 3 \pm 7i$

4 $w = 2+9i$ or $w = -2+9i$

5 $z = \sqrt{3} - i$

6 $w = 4+2i$

7 a $k = -119$ **b** $x = (7), 1 \pm 4i$

8 $k = -8$
$x = -\frac{3}{2} \pm \frac{\sqrt{23}}{2}i, 1$

9 $x = 1, 3, \pm i$

10 $x = -3, 7, 1+\sqrt{2}i$ and $1-\sqrt{2}i$

11 a $k = 10$

b $x = -6 \pm i, 3, -1$

12 $A = -10$
$x = 5 \pm i, \pm\sqrt{2}i$

13 a e.g. $x^4 - 4x^3 + 24x^2 - 40x + 100 = 0$

b e.g. $x^4 + 12x^3 + 62x^2 + 156x + 169 = 0$

14 a $(x+1)(x^2 - 20x + 109)$

b

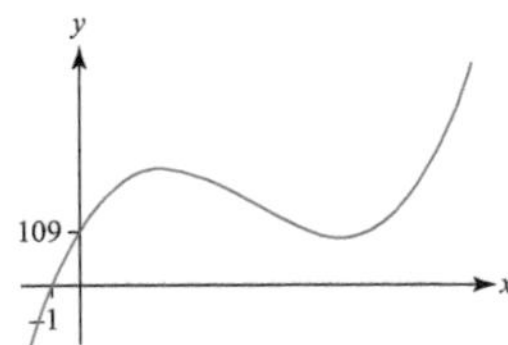

15 a $4x^4 + 12x^3 - 35x^2 - 300x + 625 = (4x^2 - 20x + 25)(x^2 + 8x + 25)$

b

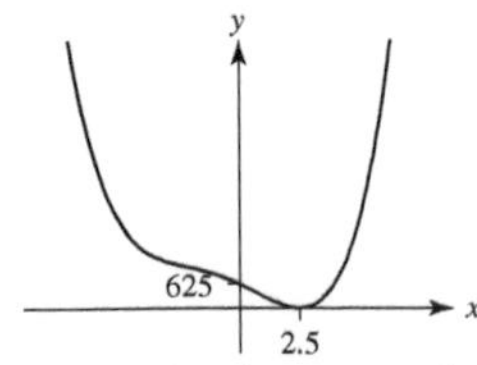

16 a Two real and two complex roots

b $x = -5$ (repeated), $2+i$, $2-i$

17 $x^3 - (a+b+c)x^2 + (ab+ac+bc)x - abc = 0$

18 a -5 **b** -12

19 a $-\alpha-\beta-\gamma-\delta$ **b** $\alpha\beta\gamma\delta$

20 $a = 2, b = -41, c = 336, d = -1318, e = 2262$

21 a Order 5 (quintic)

b $x^5 - 6x^4 + 10x^3 - 20x^2 + 9x + 306 = 0$

Exercise 1.3A

1 $\overrightarrow{OA} = 5+3i$
$\overrightarrow{OB} = -3-6i$
$\overrightarrow{OC} = 6$
$\overrightarrow{OD} = -1+2i$
$\overrightarrow{OE} = -4i$
$\overrightarrow{OF} = -6+5i$

2 $u = -3+5i, v = 2-7i, w = 4i, z = -4+i$

3

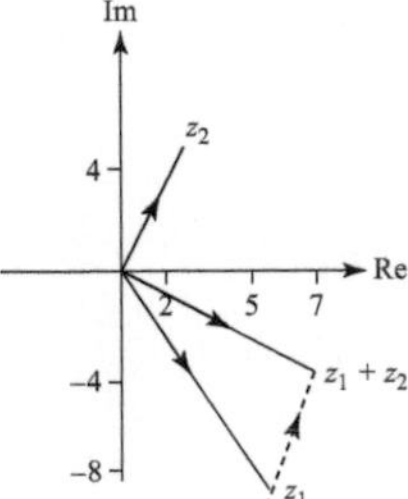

4

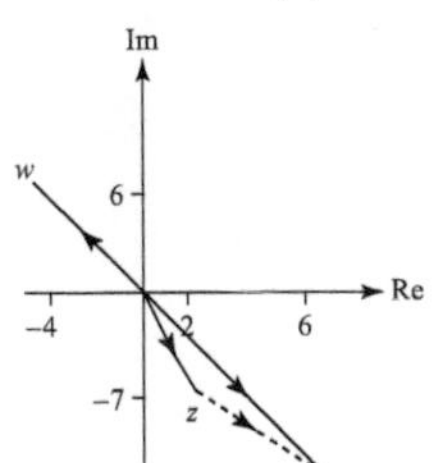

5

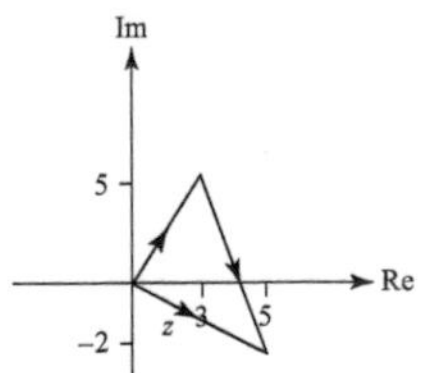

6

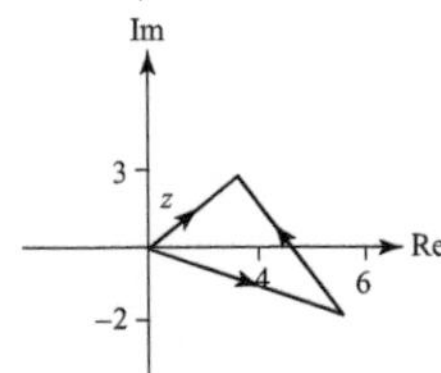

7

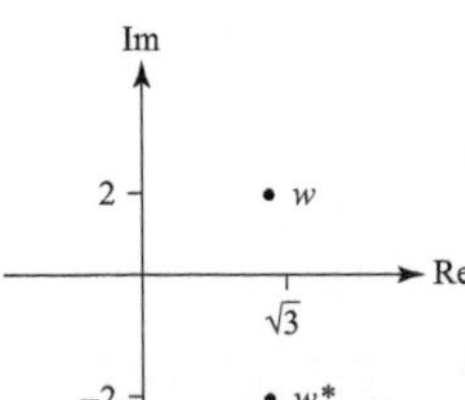

w^* is a reflection of w in the real axis.

8

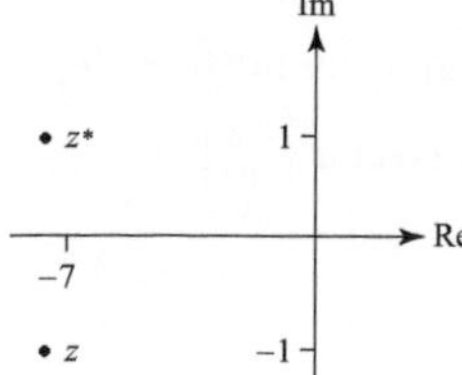

z is a reflection of z^* in the real axis.

9 a $x = \pm 4i$

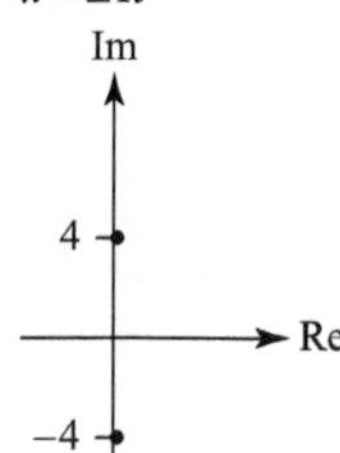

b $x = \pm 4\sqrt{5}i$

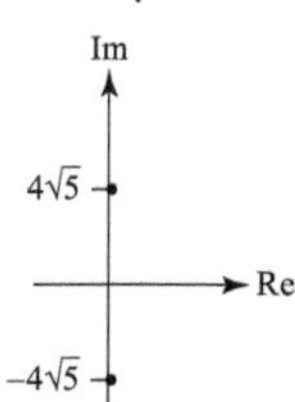

In both cases the points are reflections of each other in the real axis

10 a $z = -1 \pm \sqrt{3}i$

b

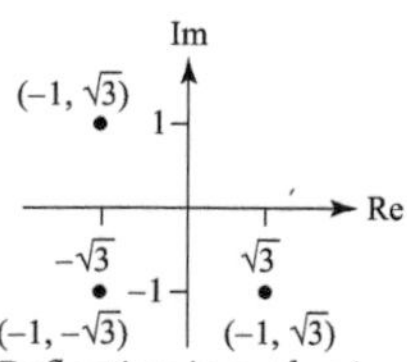

c Reflection in real axis.

11 a $z = 1 \mp 5i$

b

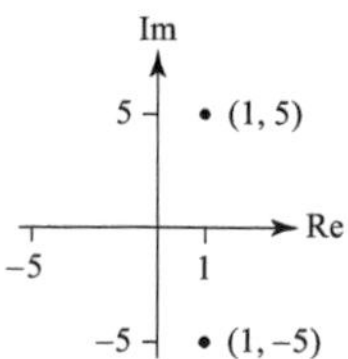

c Reflection in real axis.

12 a

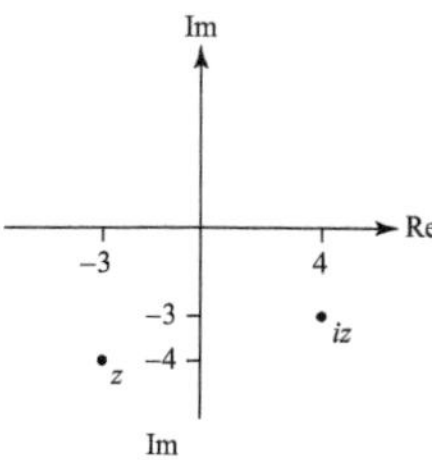

b

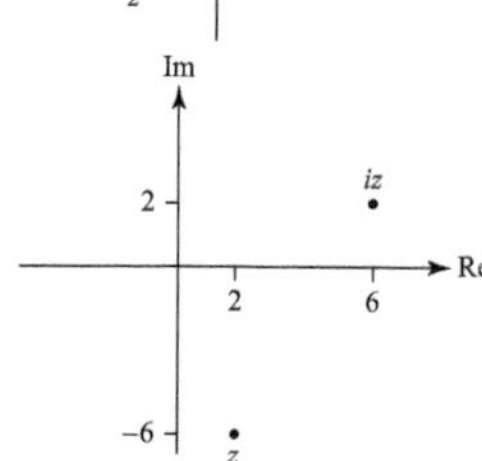

In both cases iz is the image of z rotated $\frac{\pi}{2}$ radians (anti–clockwise) about the origin.

13 a

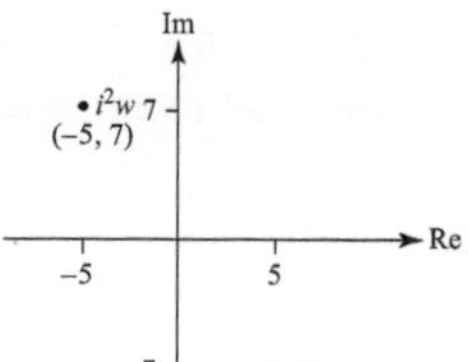

b

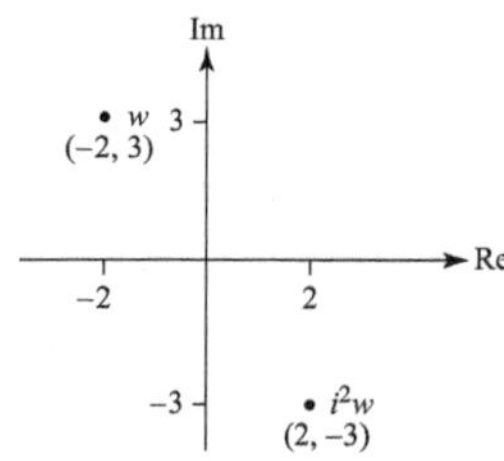

In both cases, i^2w is w rotated π radians around the origin.

Exercise 1.3B

1 a $z_2 = 2 + 5i$

b

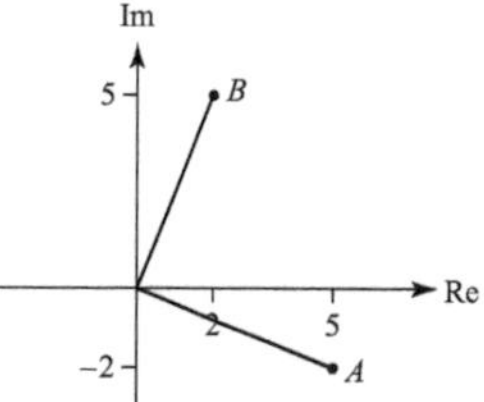

c Gradient of OA is $-\frac{2}{5}$, gradient of OB is $\frac{5}{2}$

$-\frac{2}{5} \times \frac{5}{2} = -1$ so OA and OB are perpendicular

$\therefore$ AOB is a right angle

2 a $x = -1 \pm \sqrt{7}i, 1$

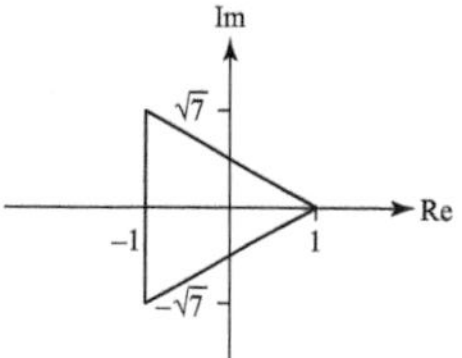

b Isosceles triangle

c $2\sqrt{7}$

3 6 square units

4 a $A = -2, B = 3$

b $x = -\frac{3}{2} \pm \frac{\sqrt{3}}{2}i, 1 \pm \sqrt{2}i$

c

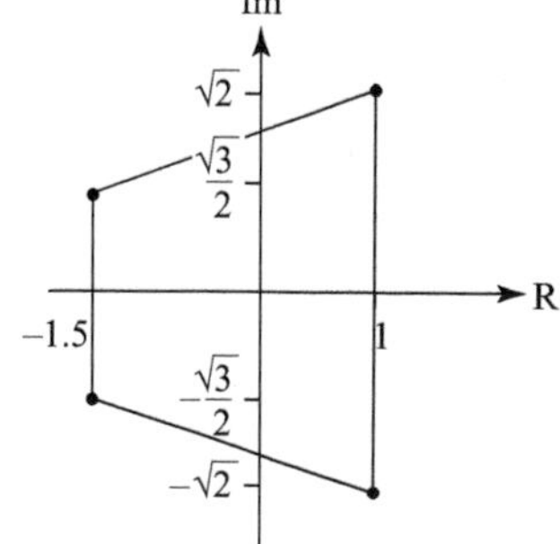

d A trapezium

e 5.70 square units

5 a $(-2)^4+8(-2)^3+40(-2)^2+96(-2)+80=0$

$x^4+8x^3+40x^2+96x+80=(x+2)^2(x^2+4x+20)=0$, so $x=-2$ is repeated

b $x=-2\pm 4i$

c

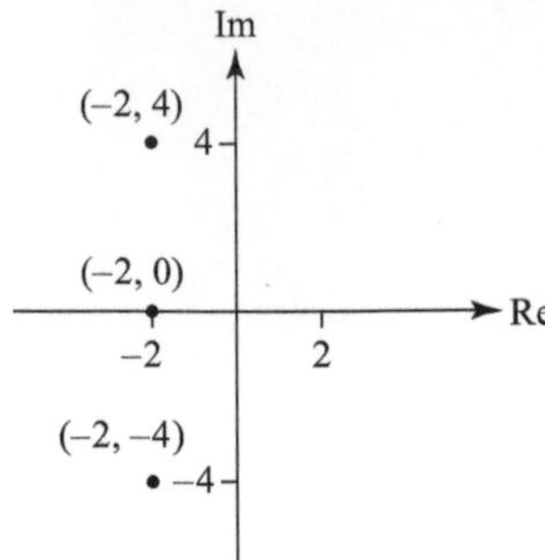

6 a $x=-1$ and 2

b 9 square units

7 a $x^3-x^2+9x-9=(x-1)(x^2+9)=0$

$\Rightarrow x=1, \pm 3i$

$A(1, 0), B(0, 3), C(0, -3)$

$|AB|=\sqrt{1^2+3^2}=\sqrt{10}$

$|AC|=\sqrt{1^2+3^2}=\sqrt{10}$

$|BC|=\sqrt{0^2+6^2}=6$

So ABC is isosceles

b 3 square units

8 a $A=(5,-2)$

$B=(7,3)$

$C=(2,5)$

Length $OA=\sqrt{5^2+2^2}=\sqrt{29}$

Length $AB=\sqrt{(7-5)^2+(3+2)^2}=\sqrt{29}$

Length $BC=\sqrt{(7-2)^2+(5-3)^2}=\sqrt{29}$

Length $OC=\sqrt{29}$

Gradient $OA=-\frac{2}{5}$

Gradient $AB=\frac{5}{2}$ So OA and AB are perpendicular

Gradient $BC=-\frac{2}{5}$ so AB and BC are perpendicular

Gradient $OC=\frac{5}{2}$ so OC and OA are perpendicular.

Therefore it is a square.

b 29 square units

9 a Q represents $2+4i$, R represents $6-4i$,

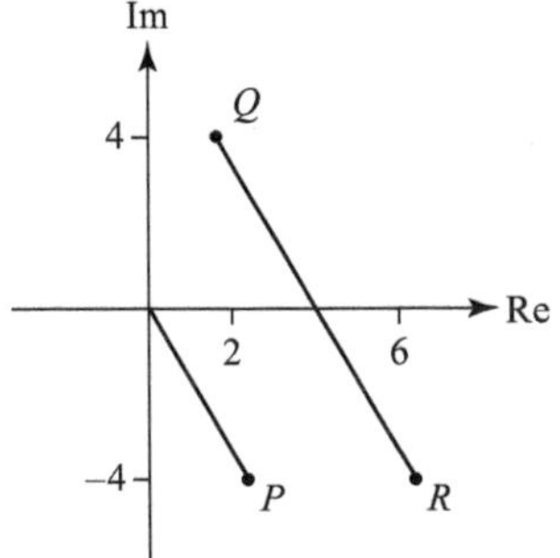

b OP has been enlarged by scale factor 2 centre the origin then translated by the vector $\begin{pmatrix} 2 \\ 4 \end{pmatrix}$.

Or alternatively: OP has been enlarged by scale factor 2 centre $(-2, -4)$.

10 a $-2+9i$

b CB is an enlargement of OA centre the origin, scale factor 2, then translated by the vector $\begin{pmatrix} -2 \\ 9 \end{pmatrix}$

11 $|a-c|(b+d)$

12

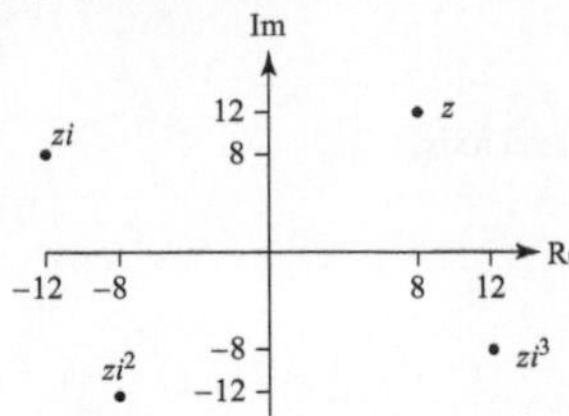

z is rotated around the origin by $\frac{\pi}{2}$ radians (anticlockwise) for zi, by π radians for zi^2 and by $\frac{3\pi}{2}$ radians (anticlockwise) for zi^3

13 Let $z=a+bi$

Then $iz=ai+bi^2=-b+ai$

Gradient of $OA=\frac{b}{a}$

Gradient of $OB=\frac{a}{-b}$

$\frac{b}{a}\times\frac{a}{-b}=-1$ so OA is perpendicular to OB

14 17 square units

15 3 square units

16 10

17 a $\frac{z}{w}=\frac{12-5i}{3+2i}=\frac{(12-5i)(3-2i)}{(3+2i)(3-2i)}$

$=\frac{36-15i-24i-10}{9+4}$

$=\frac{26-39i}{13}=2-3i$

b

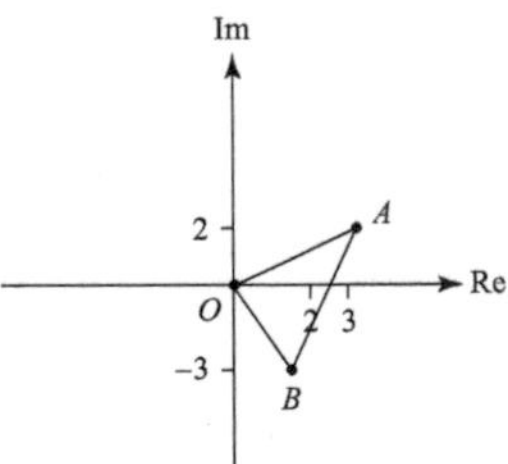

c Gradient of $OB=-\frac{3}{2}$

Gradient of $OA=\frac{2}{3}$

$-\frac{3}{2}\times\frac{2}{3}=-1$ therefore OB and OA are perpendicular so OAB is a right–angled triangle.

d $\frac{13}{2}$ square units

18

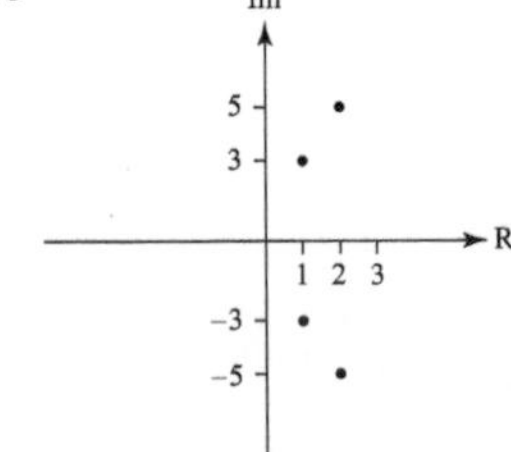

19 a i A kite

ii An isosceles triangle

iii A trapezium

b i $a=1, b=0, c=0, d=0, e=-16$

ii $a=1, b=-4, c=6, d=-4, e=-80$

Exercise 1.4A

1 a Modulus $=13$
Argument $=0.395^c$

b Modulus $=5$
Argument $=-0.644^c$

c Modulus $=3$
Argument $=-\frac{\pi}{2}$

d Modulus $=10$
Argument $=-2.21^c$.

e Modulus $=5\sqrt{2}$
Argument $=1.71$

f Modulus $=\sqrt{2^2+1^2}=\sqrt{5}$
Argument $=-2.68^c$

g Modulus $=2$
Argument $=-\frac{3\pi}{4}$

h Modulus $=3$
Argument $=0.955^c$

2 a $zw=(1+3i)(-5+2i)=-11-13i$

$|zw|=\sqrt{11^2+13^2}=\sqrt{290}$

$|z|=\sqrt{1^2+3^2}=\sqrt{10}$

$|w|=\sqrt{5^2+2^2}=\sqrt{29}$

$|z||w|=\sqrt{10}\sqrt{29}=\sqrt{290}=|zw|$ as required.

$\frac{z}{w}=\frac{1+3i}{-5+2i}$

$=\frac{(1+3i)(5+2i)}{(-5+2i)(5+2i)}=\frac{1}{29}-\frac{17}{29}i$

$\left|\frac{z}{w}\right|=\sqrt{\left(\frac{1}{29}\right)^2+\left(\frac{17}{29}\right)^2}=\sqrt{\frac{10}{29}}$

$\frac{|z|}{|w|}=\frac{\sqrt{10}}{\sqrt{29}}=\sqrt{\frac{10}{29}}=\left|\frac{z}{w}\right|$ as required

b $zw=(-2-i)(\sqrt{5}i)=\sqrt{5}-2\sqrt{5}i$

$|zw|=\sqrt{(\sqrt{5})^2+(2\sqrt{5})^2}=5$

$|z|=\sqrt{2^2+1^2}=\sqrt{5}$

$|w|=\sqrt{5}$

$|z||w|=\sqrt{5}\sqrt{5}=5=|zw|$ as required.

$\frac{z}{w}=\frac{-2-i}{\sqrt{5}i}$

$=\frac{(-2-i)i}{\sqrt{5}i\cdot i}=-\frac{1}{\sqrt{5}}+\frac{2}{\sqrt{5}}i$

$\left|\frac{z}{w}\right|=\sqrt{\left(\frac{1}{\sqrt{5}}\right)^2+\left(\frac{2}{\sqrt{5}}\right)^2}=1$

$\frac{|z|}{|w|}=\frac{\sqrt{5}}{\sqrt{5}}=1=\left|\frac{z}{w}\right|$ as required

c $zw=(-\sqrt{3}+6i)(1-\sqrt{3}i)=5\sqrt{3}+9i$

$|zw|=\sqrt{(5\sqrt{3})^2+(9)^2}=2\sqrt{39}$

$|z|=\sqrt{(\sqrt{3})^2+6^2}=\sqrt{39}$

$|w|=\sqrt{1^2+(\sqrt{3})^2}=2$

$|z||w|=2\sqrt{39}=|zw|$ as required.

$\frac{z}{w}=\frac{-\sqrt{3}+6i}{1-\sqrt{3}i}$

$=\frac{(-\sqrt{3}+6i)(1+\sqrt{3}i)}{(1-\sqrt{3}i)(1+\sqrt{3}i)}=-\frac{7\sqrt{3}}{4}+\frac{3}{4}i$

$\left|\frac{z}{w}\right|=\sqrt{\left(\frac{7\sqrt{3}}{4}\right)^2+\left(\frac{3}{4}\right)^2}=\frac{\sqrt{39}}{2}$

$\frac{|z|}{|w|}=\frac{\sqrt{39}}{2}=\left|\frac{z}{w}\right|$ as required

3 a $zw=(1+i)(3+\sqrt{3}i)=(3-\sqrt{3})+(3+\sqrt{3})i$

$\arg(zw)=\tan^{-1}\left(\frac{3+\sqrt{3}}{3-\sqrt{3}}\right)=\frac{5\pi}{12}$

$\arg z=\tan^{-1}1=\frac{\pi}{4}$

$\arg w=\tan^{-1}\frac{\sqrt{3}}{3}=\frac{\pi}{6}$

$\arg z+\arg w=\frac{\pi}{4}+\frac{\pi}{6}=\frac{5\pi}{12}$ as required

$\frac{z}{w}=\frac{1+i}{3+\sqrt{3}i}$

$=\frac{(1+i)(3-\sqrt{3}i)}{(3+\sqrt{3}i)(3-\sqrt{3}i)}=\frac{3+\sqrt{3}}{12}+\frac{3-\sqrt{3}}{12}i$

$\arg\left(\frac{z}{w}\right)=\tan^{-1}\frac{3-\sqrt{3}}{3+\sqrt{3}}=\frac{\pi}{12}$

$\arg z-\arg w=\frac{\pi}{4}-\frac{\pi}{6}=\frac{\pi}{12}$ as required

b $zw=i(2-2i)=2+2i$

$\arg(zw)=\tan^{-1}\left(\frac{2}{2}\right)=\frac{\pi}{4}$

$\arg z=\frac{\pi}{2}$

$\arg w=-\tan^{-1}\frac{2}{2}=-\frac{\pi}{4}$

$\arg z+\arg w=\frac{\pi}{2}+\frac{-\pi}{4}=\frac{\pi}{4}$ as required

$\frac{z}{w}=\frac{i}{2-2i}$

$=\frac{i(2+2i)}{(2-2i)(2+2i)}=-\frac{1}{4}+\frac{1}{4}i$

$\arg\left(\frac{z}{w}\right)=\pi-\tan^{-1}\frac{0.25}{0.25}=\frac{3\pi}{4}$

$\arg z-\arg w=\frac{\pi}{2}-\frac{-\pi}{4}=\frac{3\pi}{4}$ as required

c $zw=-2i(\sqrt{3}-3i)=-6-2\sqrt{3}i$

$\arg(zw)=-\pi+\tan^{-1}\left(\frac{2\sqrt{3}}{6}\right)=-\frac{5\pi}{6}$

$\arg z=-\tan^{-1}\frac{3}{\sqrt{3}}=-\frac{\pi}{3}$

$\arg w=-\frac{\pi}{2}$

$\arg z+\arg w=-\frac{\pi}{3}+\frac{-\pi}{2}=-\frac{5\pi}{6}$ as required

$\frac{z}{w}=\frac{\sqrt{3}-3i}{-2i}$

$=\frac{i(\sqrt{3}-3i)}{-2i\cdot i}=\frac{3}{2}+\frac{\sqrt{3}}{2}i$

$\arg\left(\frac{z}{w}\right)=\tan^{-1}\frac{\sqrt{3}}{3}=\frac{\pi}{6}$

$\arg z-\arg w=-\frac{\pi}{3}-\frac{-\pi}{2}=\frac{\pi}{6}$ as required

4 $|w| = \sqrt{3}$

$\arg(w) = \frac{2\pi}{3}$

5 $|z_2| = \frac{\sqrt{3}}{6}$

$\arg(z_2) = -\frac{11}{12}\pi$

6 $|w| = 8\sqrt{3}$

$\arg(w) = \frac{\pi}{2}$

7 **a** $3i$ **b** -5 **c** $-5\sqrt{3}+5i$ **d** $-\frac{\sqrt{3}}{2}-\frac{3}{2}i$

8 **a** $z = 3\sqrt{2}\left(\cos\frac{\pi}{4}+i\sin\frac{\pi}{4}\right)$

b $z = 2\left(\cos\left(-\frac{\pi}{3}\right)+i\sin\left(-\frac{\pi}{3}\right)\right)$

c $z = 4\left(\cos\left(-\frac{5\pi}{6}\right)+i\sin\left(-\frac{5\pi}{6}\right)\right)$

d $z = \sqrt{97}(\cos(1.99)+i\sin(1.99))$

9 **a**

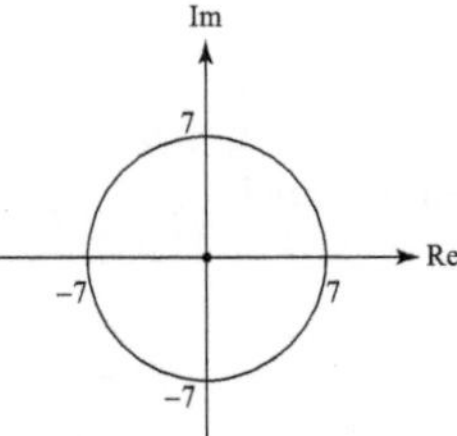

b

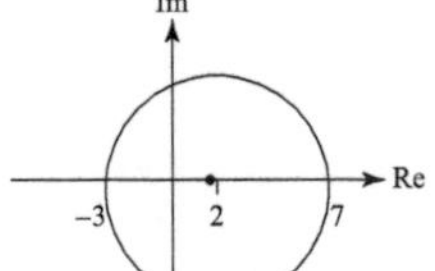

c

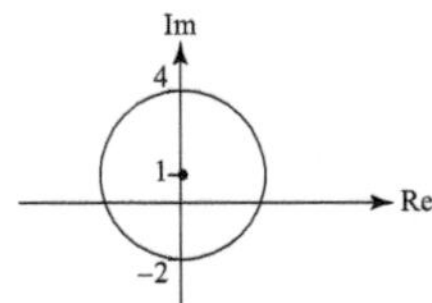

d

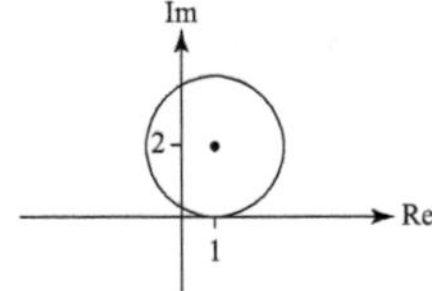

e

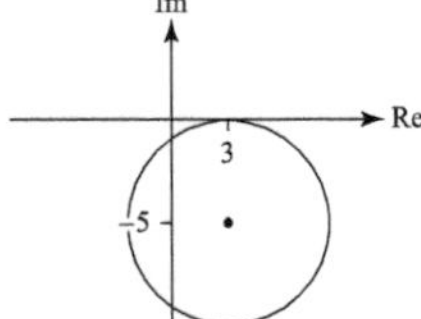

f

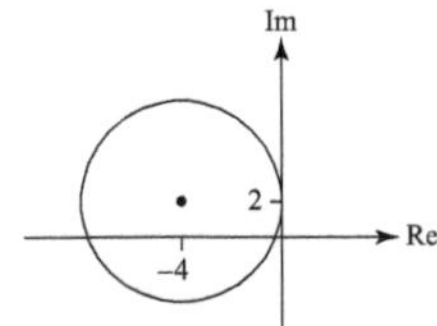

10 **a**

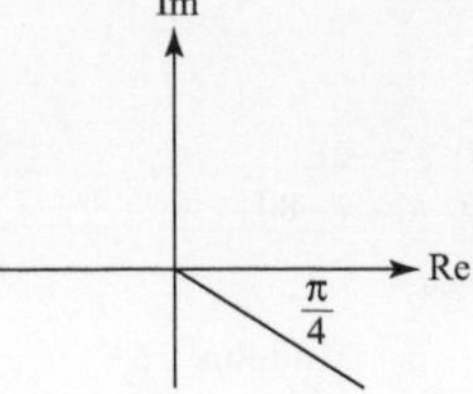

b

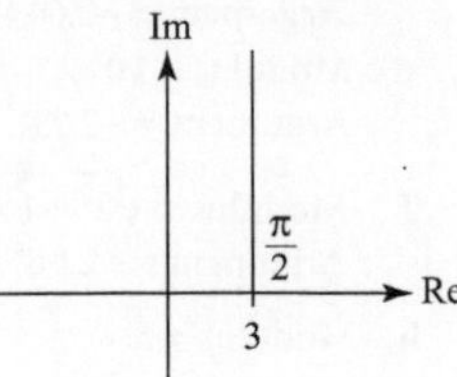

c

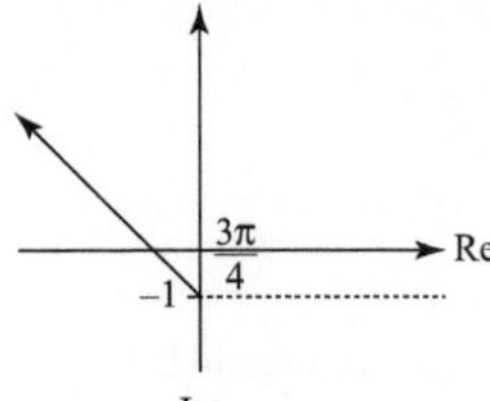

d

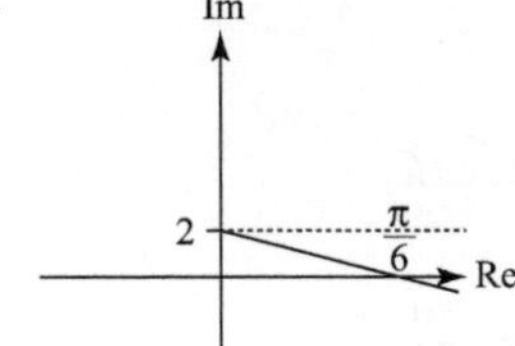

e

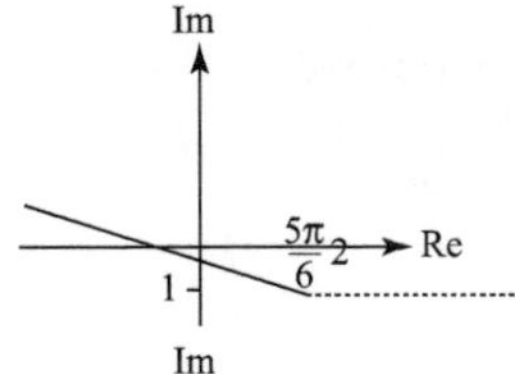

f

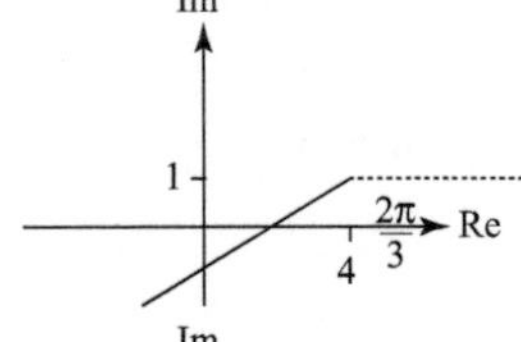

g

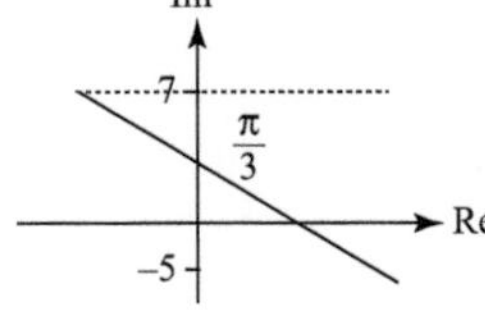

11 **a**

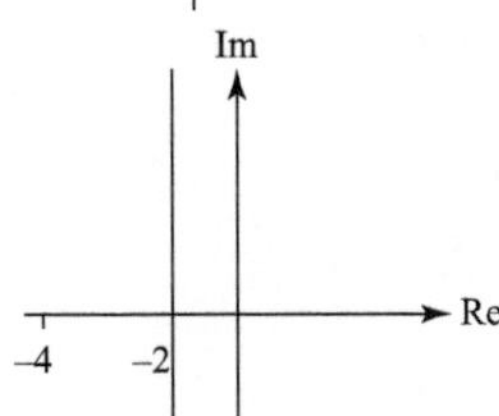

b

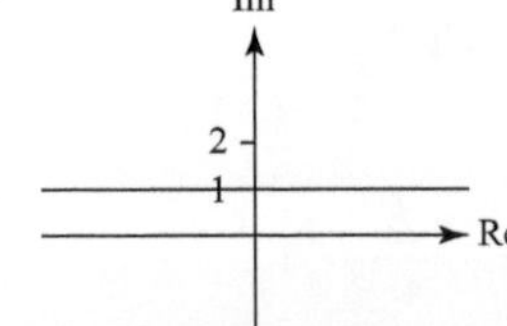

c

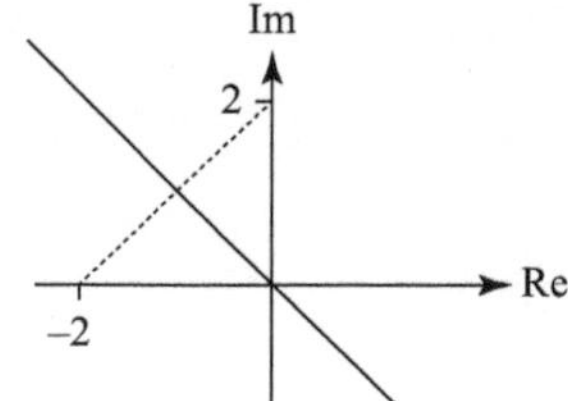

d

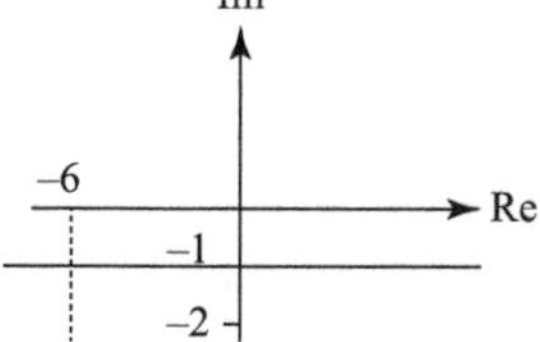

e

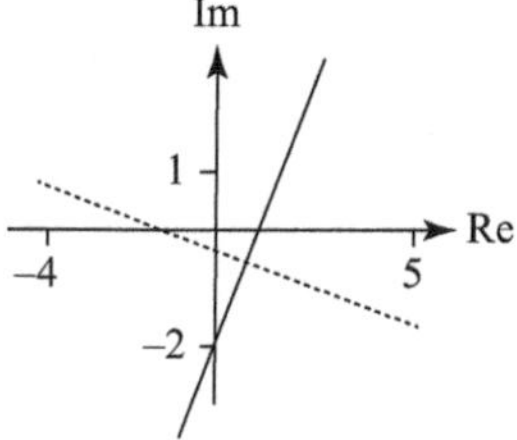

Exercise 1.4B

1 Let $z = |z|(\cos A + i\sin A)$ and $w = |w|(\cos B + i\sin B)$

Then $zw = |z|(\cos A + i\sin A)|w|(\cos B + i\sin B)$

$= |z||w|(\cos A\cos B + i\sin A\cos B + i\sin B\cos A + i^2 \sin A\sin B)$

$= |z||w|(\cos A\cos B - \sin A\sin B + i(\sin A\cos B + \sin B\cos A))$

$= |z||w|(\cos(A+B) + i\sin(A+B))$

So $|zw| = |z||w|$

And $\arg(zw) = A + B = \arg z + \arg w$ as required

2 Let $z = x + iy$

Then $|x + iy + 3 - 2i| = 4$

$(x+3)^2 + (y-2)^2 = 16$

Circle centre (−3, 2) and radius 4

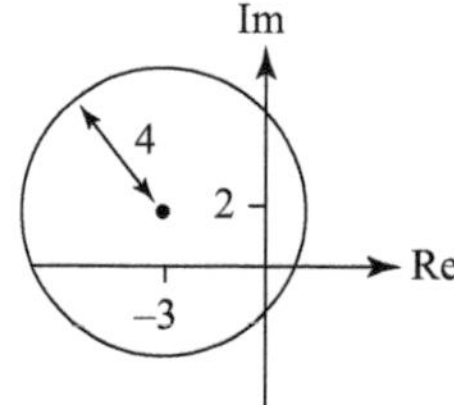

3 a Let $z = x + iy$

$|x + iy - 2 - i| = 2$

$(x-2)^2 + (y-1)^2 = 4$

Circle centre (2, 1) and radius 2

b $\sqrt{3}$ square units

4 a $x = 2.5$ b $y = -x$

c $x + 2y = 3$ d $8x - 2y = 7$

e $2x = 7$ f $13x - 7y = 10$

5 a $y = x + 3$ b $x = -5$

c $y = \sqrt{3}x + 1 + 2\sqrt{3}$ d $y = -\sqrt{3}x - 1 + 4\sqrt{3}$

6 $y = 5x + 3$

7 $y = \sqrt{2} - x$

8 $x - 2y + 3 = 0$

9 a

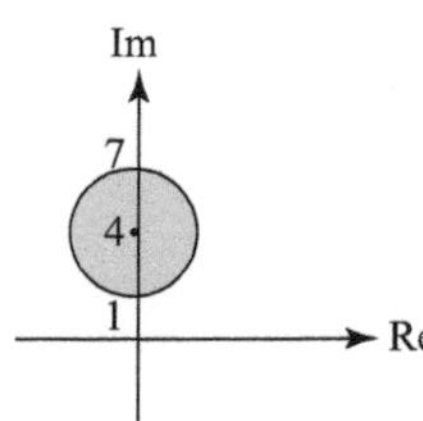

b

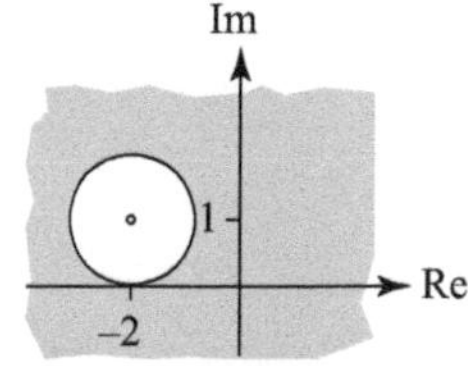

c

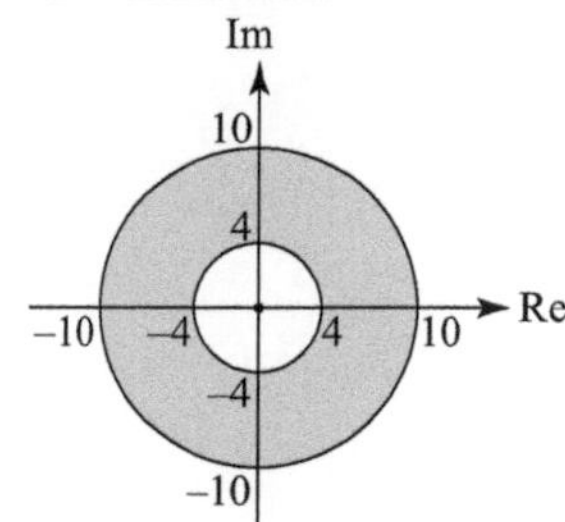

d

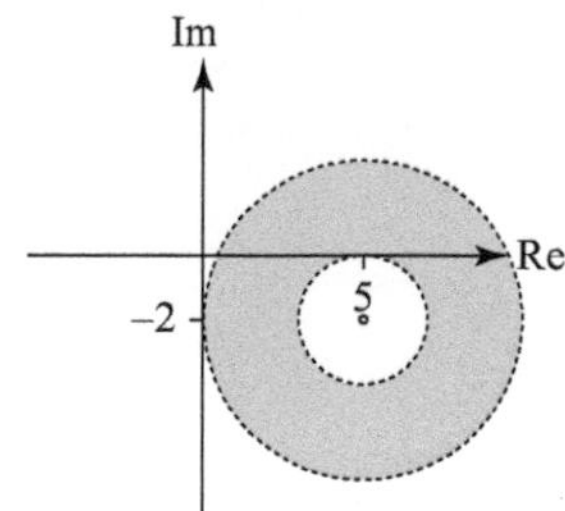

10 42π square units

11 a

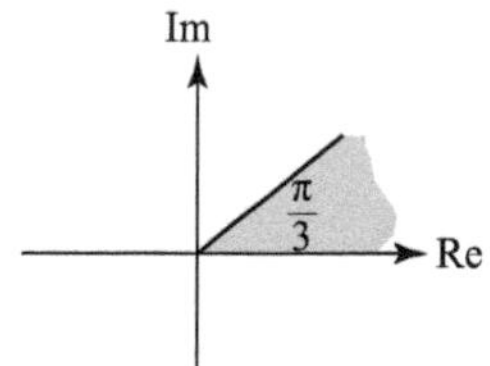

b

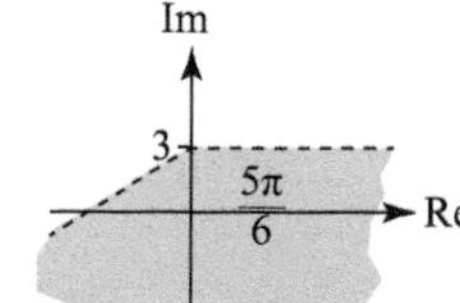

c

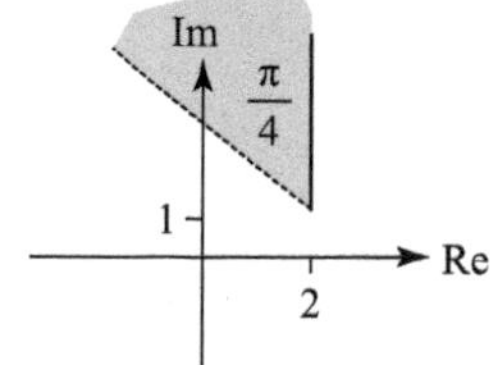

12 16π square units

13 a

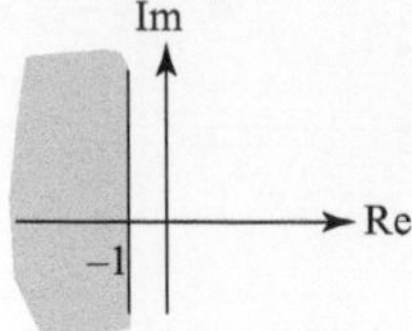

b

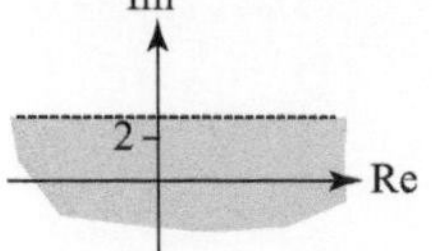

c

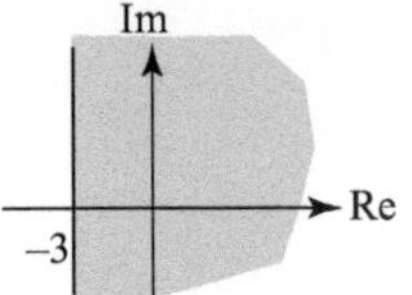

d

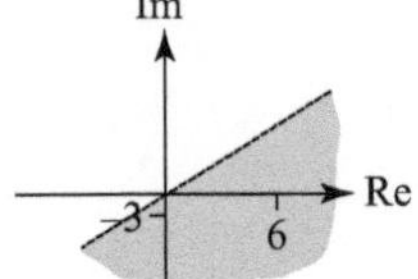

e

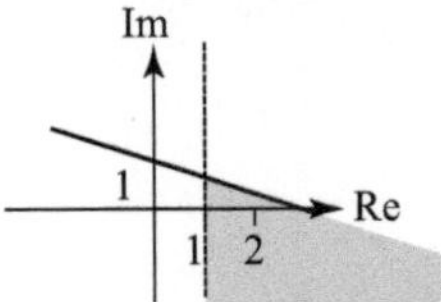

14

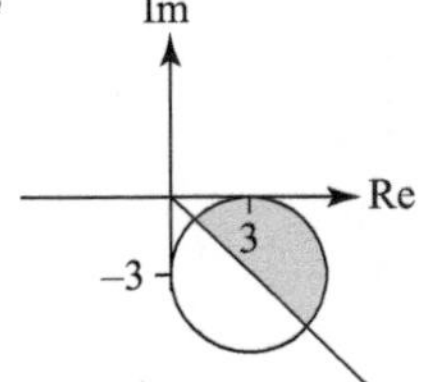

15

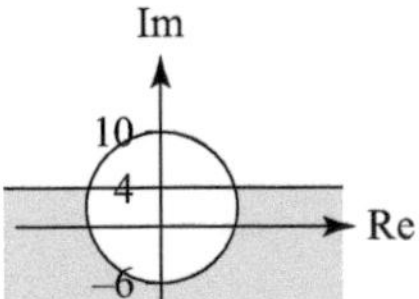

16

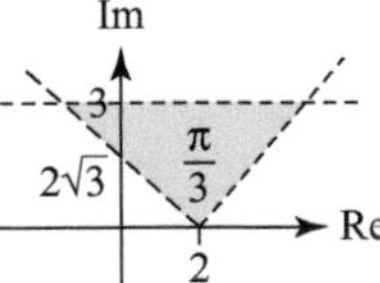

17

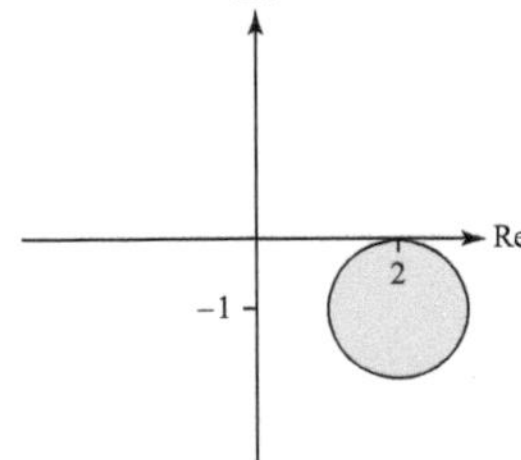

18

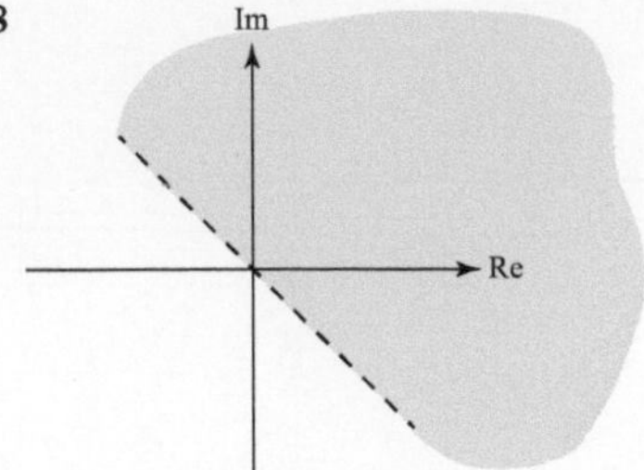

19

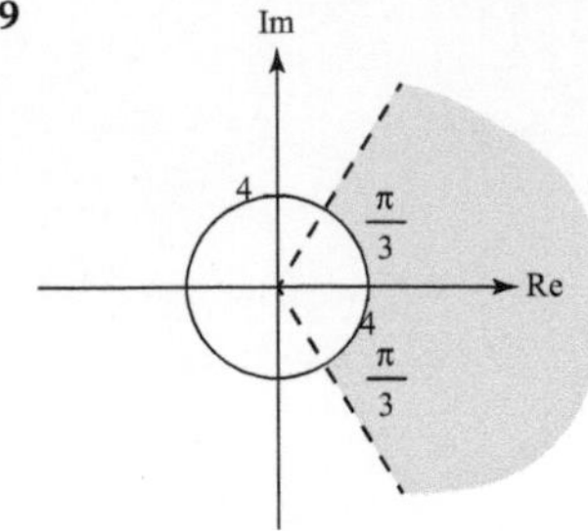

20

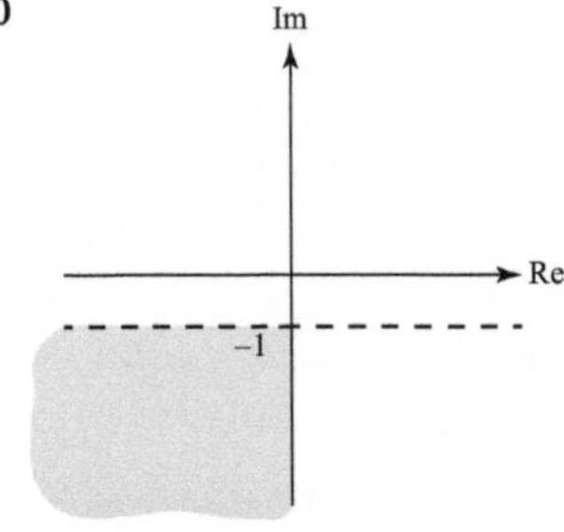

21 a Let $z = x + iy$

$|x+iy+3| = 2|x+iy-6i|$

$|x+iy+3|^2 = 4|x+iy-6i|^2$

$(x+3)^2 + y^2 = 4x^2 + 4(y-6)^2$

$x^2 + 6x + 9 + y^2 = 4x^2 + 4y^2 - 48y + 144$

$3x^2 - 6x + 3y^2 - 48y + 135 = 0$

$x^2 - 2x + y^2 - 16y + 45 = 0$

$(x-1)^2 - 1 + (y-8)^2 - 64 + 45 = 0$

$(x-1)^2 + (y-8)^2 = 20$

Circle centre (1, 8) radius $2\sqrt{5}$

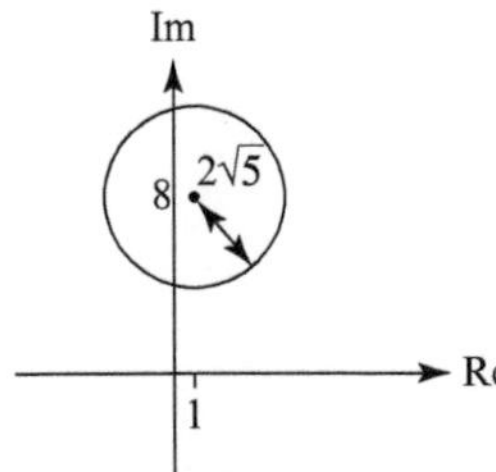

b

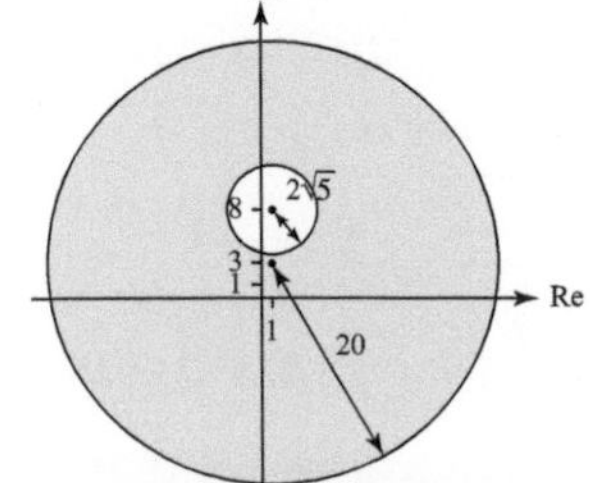

22 **a** $(x-3)^2+(y+1)^2=9$

b $x=y$

c $z=\frac{2+\sqrt{2}}{2}+\frac{2+\sqrt{2}}{2}i$ or $z=\frac{2-\sqrt{2}}{2}+\frac{2-\sqrt{2}}{2}i$

23 $z=2\sqrt{2}+(3-2\sqrt{2})i$

Review exercise 1

1 **a** $3-7i$ **b** $-6-9i$ **c** $12-5i$

d $-22-7i$ **e** $4-28i$ **f** $-40-42i$

g $5+4i$ **h** $-2+3i$ **i** $\frac{2}{13}+\frac{23}{13}i$

j $\frac{3}{41}-\frac{69}{82}i$ **k** $\frac{10}{41}+\frac{8}{41}i$ **l** $1+\frac{2}{3}i$

2 $a=1, b=14, c=53$

3 **a** -16 **b** $z=7\pm4i, 2$

4 $x=3\pm2i$ and $x=-2\pm3i$

5 **a** Modulus $=\sqrt{85}$
Argument $=1.35^c$

b Modulus $=3\sqrt{2}$
Argument $=-\frac{\pi}{4}$

c Modulus $=7$
Argument $=\frac{\pi}{2}$

d Modulus $=2$
Argument $=-\frac{\pi}{2}$

e Modulus $=\sqrt{17}$
Argument $=1.82^c$

f Modulus $=5$
Argument $=-2.21^c$

6 **a** $10(\cos(0.644)+i\sin(0.644))$

b $13(\cos(2.75)+i\sin(2.75))$

c $2\sqrt{2}\left(\cos\left(-\frac{3\pi}{4}\right)+i\sin\left(-\frac{3\pi}{4}\right)\right)$

d $2\left(\cos\left(-\frac{\pi}{6}\right)+i\sin\left(-\frac{\pi}{6}\right)\right)$

e $\sqrt{5}\left(\cos\left(-\frac{\pi}{3}\right)+i\sin\left(-\frac{\pi}{3}\right)\right)$

7 **a** $\sqrt{3}+i$ **b** $\frac{\sqrt{6}}{2}-\frac{\sqrt{6}}{2}i$

8 **a** $3\sqrt{2}\left(\cos\left(-\frac{\pi}{6}\right)+i\sin\left(-\frac{\pi}{6}\right)\right)$

b $\sqrt{2}\left(\cos\left(-\frac{\pi}{2}\right)+i\sin\left(-\frac{\pi}{2}\right)\right)$

c $\frac{\sqrt{2}}{2}\left(\cos\left(\frac{\pi}{2}\right)+i\sin\left(\frac{\pi}{2}\right)\right)$

9

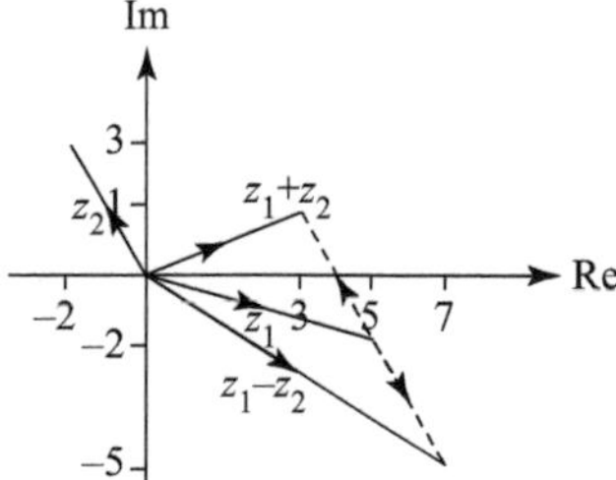

10 **a**

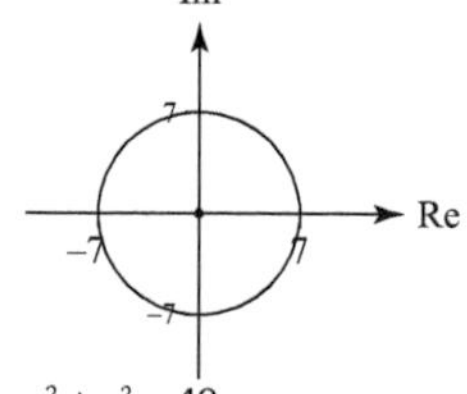

$x^2+y^2=49$

b

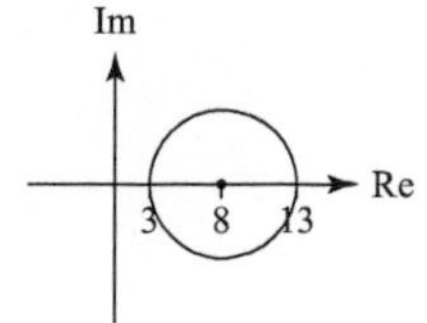

$(x-8)^2+y^2=25$

c

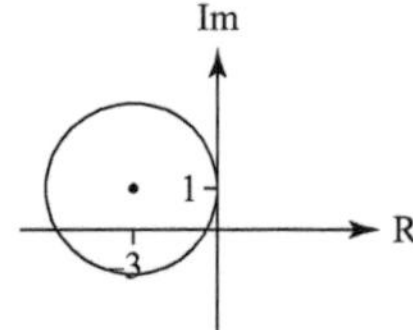

$(x+3)^2+(y-1)^2=9$

d

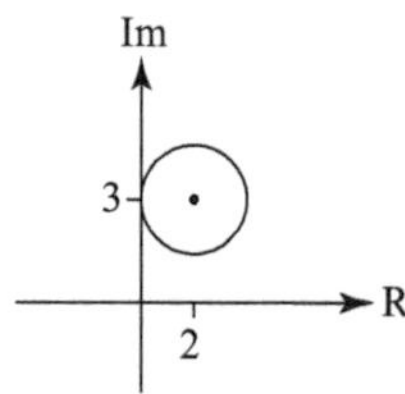

$(x-2)^2+(y-3)^2=4$

11 **a**

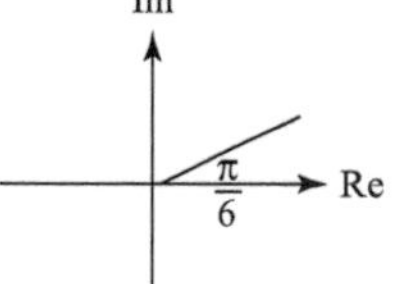

b

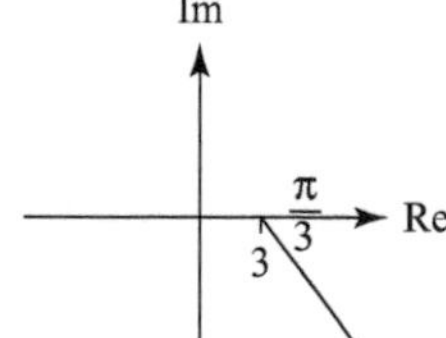

c

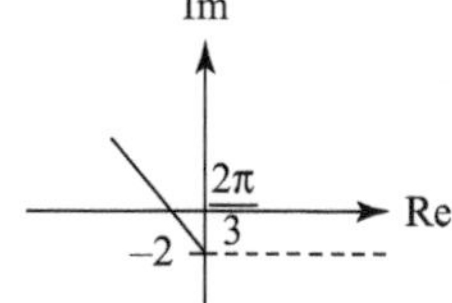

d

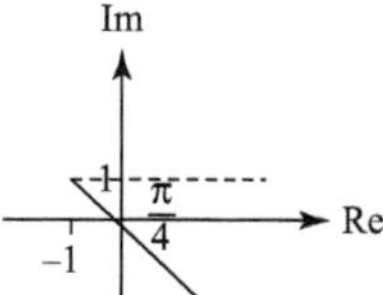

12 **a**

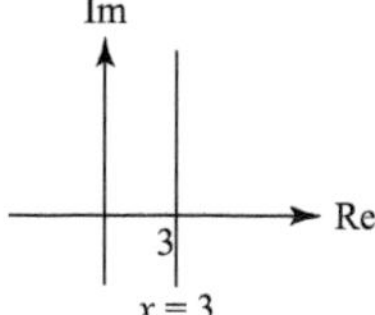

$x=3$

b

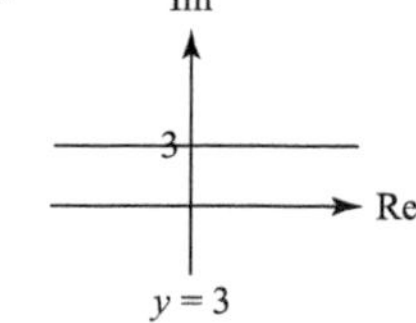

$y=3$

c

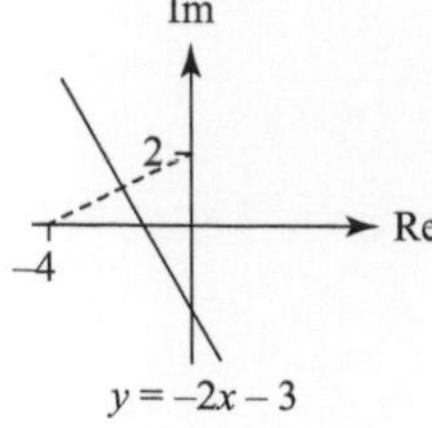

d

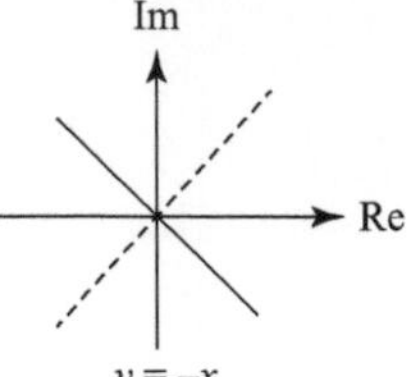

13 a

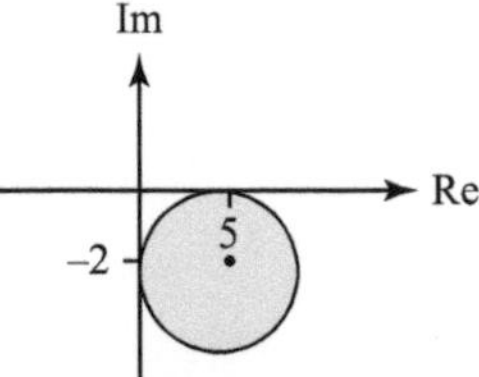

b

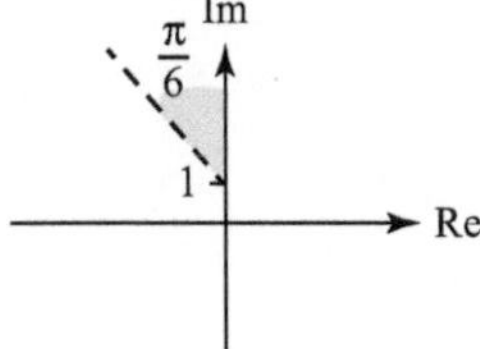

c

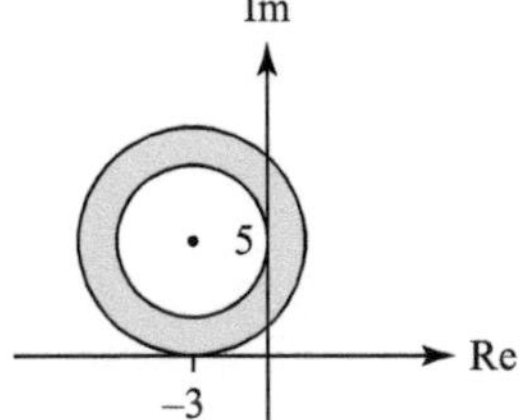

d

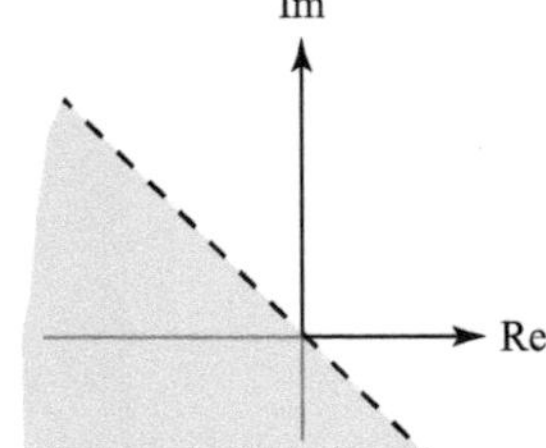

14

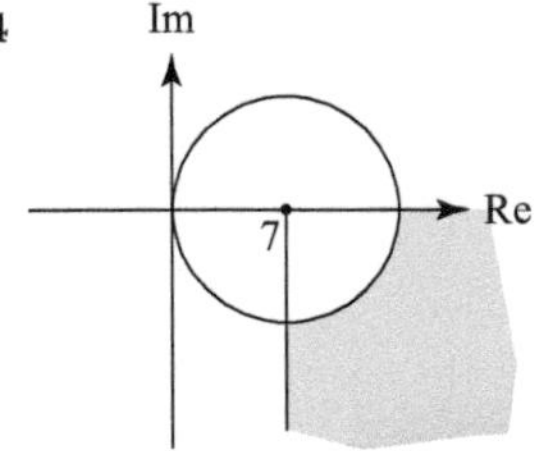

Assessment 1

1 a i $-2+14i$ **ii** $96-50i$ **iii** $\frac{15}{101}+\frac{52}{101}i$

b $|z| = 14.2, \arg z = -0.885^c$

2 a $-1+2i, -1-2i$

b

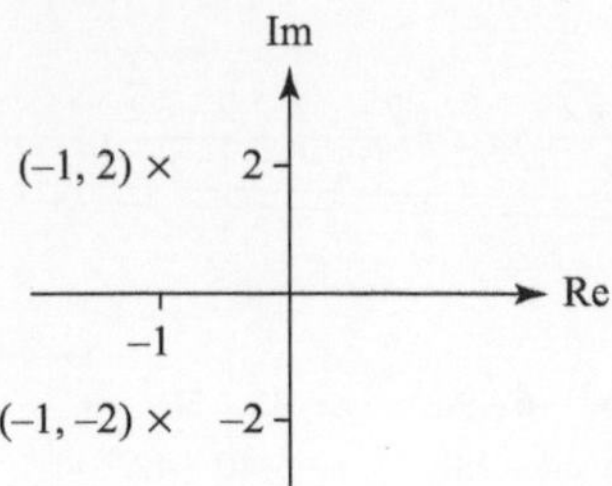

c Reflection in real axis

3 a $z = \frac{(5+2i)(3-i)}{(3+i)(3-i)}$

$= \frac{15-5i+6i-2i^2}{9-i^2}$

$= \frac{17+i}{10}$

$= \frac{17}{10}+\frac{1}{10}i$

$\left(a = \frac{17}{10}, b = \frac{1}{10}\right)$

b i $\frac{17}{5}$ **ii** $\frac{1}{5}i$ **iii** $\frac{29}{10}$

c i

Im

$\left(-\frac{1}{10}, \frac{17}{10}\right)$ ×

$\left(\frac{17}{10}, \frac{1}{10}\right)$ ×

Re

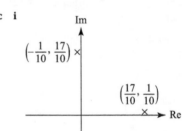

ii iz is a rotation of z 90° anticlockwise around origin.

4 $b = \pm4, a = \pm1$

5 $b = \pm1, a = \pm\sqrt{2}$

6 a $2\pm3i, -2$

b 12 square units

7 a $5+i, 7$

b $a = -17, b = 96$

8 $z = 6+9i$

9 $w = -3-7i, z = 2-9i$

10 a $|z| = 6, \arg z = \frac{\pi}{3}$

b $|w| = 2, \arg z = \frac{3\pi}{4}$

c i $|zw| = 12$

ii $\left|\frac{z}{w}\right| = 3$

iii $\arg(zw) = -\frac{11\pi}{12}$

iv $\arg\left(\frac{z}{w}\right) = -\frac{5\pi}{12}$

11 a $w = \sqrt{3}\left(\cos(-0.615)+i\sin(-0.615)\right)$

b $\sqrt{33}$

12 a $z = 2\left(\cos\left(-\frac{\pi}{6}\right)+i\sin\left(-\frac{\pi}{6}\right)\right)$

b i $\arg(zw) = -\frac{\pi}{4}$ **ii** $\arg\left(\frac{z}{w}\right) = -\frac{\pi}{12}$

c $|w| = 5$

13 a i

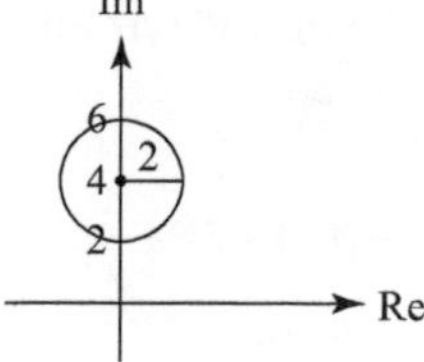

ii

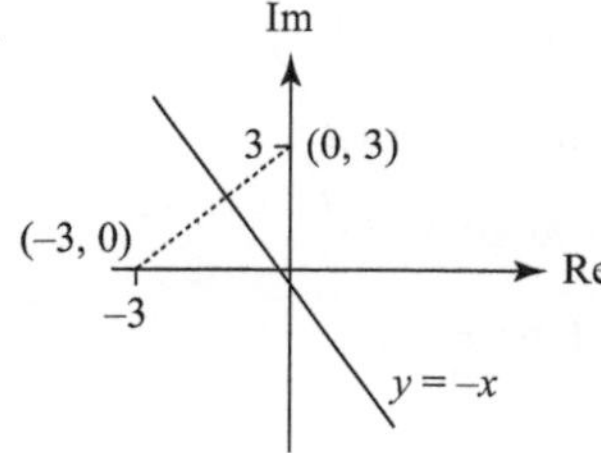

b i $x^2+(y-4)^2=4$

ii $y=-x$

14 a $|x+iy+3-4i|=4$

$(x+3)^2+(y-4)^2=16$

Therefore a circle (with centre (–3, 4) and radius 4)

b, c

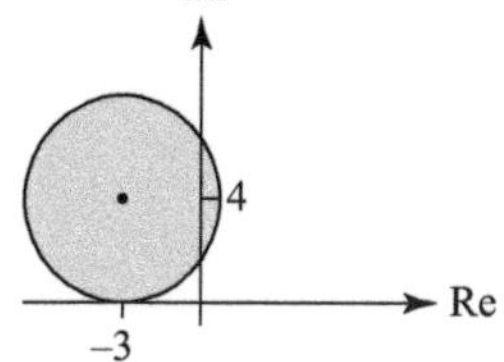

15 a

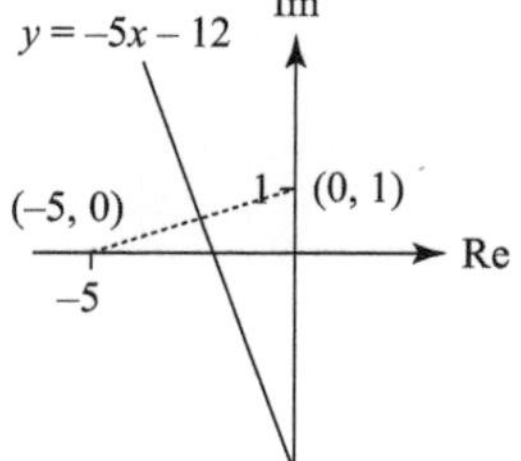

b $y=-5x-12$

16 $-8+5i$ and $8-5i$

17 $z=\sqrt{5}-i$ or $-\sqrt{5}+i$

18 28 square units

19 a $-i-4$

b $x^3+11x^2+41x+51=0$

20 a $z=-2-i$ and $z=3+5i$

b $z^4-2z^3+15x^2+106z+170=0$

21 So $a=1$, $b=-4$, $c=56$, $d=-104$, $\mathrm{e}=676$

22 $w=2\pm i$, $z=-1\mp 2i$

23 $w=6\pm i$, $z=2\mp 3i$

24 a $2(3i)^5+(3i)^4+36(3i)^3+18(3i)^2+162(3i)+81$

$=486i+81-972i-162+486i+81$

$=0$ so $3i$ is a solution

b $(x^2+9)^2(2x+1)$

25 $x=-1\pm i$, 3 (repeated)

26 a $\beta=3\left(\cos\left(-\frac{5\pi}{6}\right)+i\sin\left(-\frac{5\pi}{6}\right)\right)$

b i $|\alpha\beta|=9$ **ii** $\left|\frac{\alpha}{\beta}\right|=1$ **iii** $\arg(\alpha\beta)=0$

c $x^2+3\sqrt{3}x+9=0$

27 a $\frac{z_1}{z_2}=\frac{(6-a)}{5}+\frac{(3+2a)}{5}i$

b $a=\pm 9$

28 a $w=\frac{1}{2}\left(\cos\left(-\frac{\pi}{3}\right)+i\sin\left(-\frac{\pi}{3}\right)\right)$

b $z=\frac{\sqrt{3}}{24}+\frac{1}{24}i$

29 Circle, centre (3, –1), radius 1

30 a

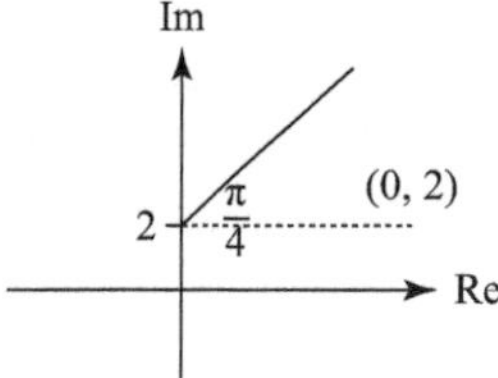

b $y=x+2$

31

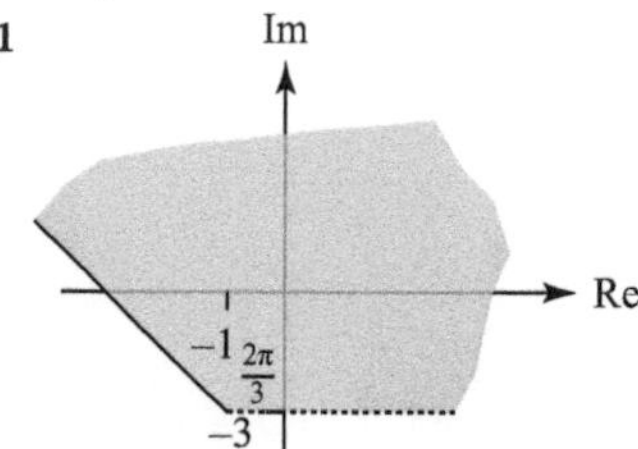

32

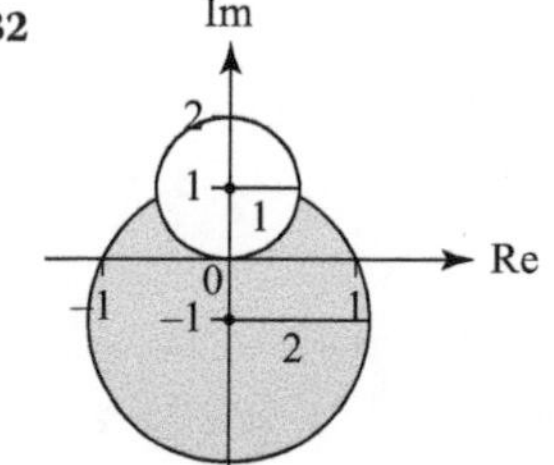

33 5 square units

34 $\arg(z-(\sqrt{3}+i))=-\frac{5\pi}{6}$

35 $|x+yi+2|=3|x+yi-2i|$

$(x+2)^2+y^2=9(x^2+(y-2)^2)$

$x^2+4x+4+y^2=9x^2+9y^2-36y+36$

$8x^2-4x+8y^2-36y+32=0$

$x^2-\frac{x}{2}+y^2-\frac{9}{2}y+4=0$

$\left(x-\frac{1}{4}\right)^2+\left(y-\frac{9}{4}\right)^2=\frac{9}{8}$

So a circle, centre $\left(\frac{1}{4},\frac{9}{4}\right)$ and radius $\frac{3}{4}\sqrt{2}$

36 a

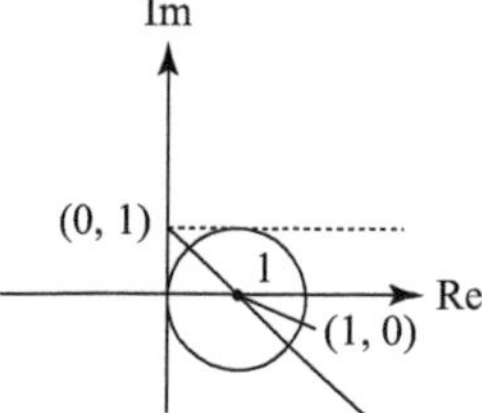

b $z=\left(\frac{2-\sqrt{2}}{2}\right)+\frac{\sqrt{2}}{2}i$ or $z=\left(\frac{2+\sqrt{2}}{2}\right)-\frac{\sqrt{2}}{2}i$

37 a $z=6-i$ or $z=-2-i$

b

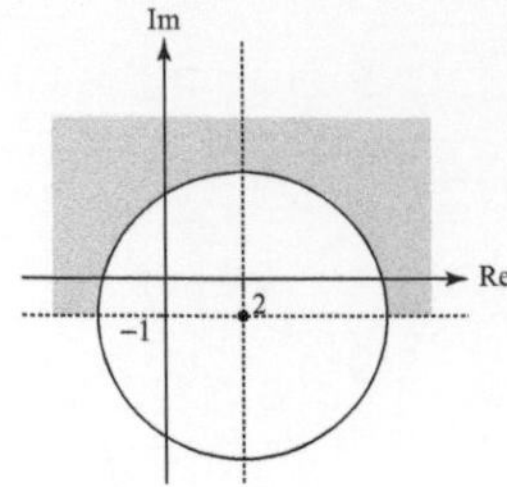

38 a

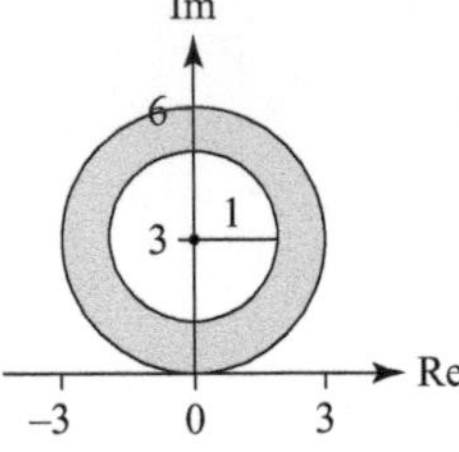

b 8π

39 a

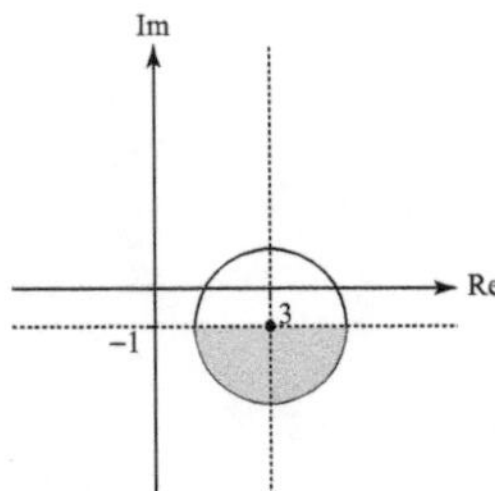

b 2π

Chapter 2

Exercise 2.1A

1 The sums and products are:

a 5 and 9 **b** −6 and 7 **c** 8 and −12

d −10 and −5 **e** 4 and $\frac{8}{3}$ **f** $-\frac{1}{4}$ and $\frac{3}{2}$

2 The sum, sum of the products in pairs and the product of the roots are:

a −4, −9 and 14 **b** 7, −11 and −12

c 13, 22 and 26 **d** $\frac{-5}{2}, \frac{17}{2}$ and $\frac{21}{2}$

e $\frac{1}{4}, \frac{3}{4}$ and −2 **f** $\frac{-3}{4}, \frac{-3}{2}$ and $\frac{-5}{8}$

3 a $-\frac{k}{3}$ **b** $-\frac{k}{3}-\frac{1}{3}$ **c** $\frac{k^2}{9}+\frac{8}{3}$

d $\frac{44}{3}+\frac{4k}{3}$ **e** $-\frac{k^3}{27}-\frac{4k}{3}$ **f** $\frac{k}{4}$

4 $x^3-5x^2+4x+2=0$

5 a i 26 **ii** 0

b i 30 **ii** $-\frac{11}{12}$

6 a $-\frac{a}{2}$ **b** $\frac{b}{2}$ **c** $-\frac{c}{2}$ **d** $\frac{a^2}{4}-b$

e $\frac{a^3}{8}+\frac{3ab}{4}-\frac{3c}{2}$ **f** $1-\frac{a}{2}+\frac{b}{2}-\frac{c}{2}$ **g** $-\frac{b}{c}$ **h** $\frac{b^2}{4}-\frac{ac}{2}$

7 $ax^4+bx^3+cx^2+dx+e=a(x-\alpha)(x-\beta)(x-\gamma)(x-\delta)$

$=a(x^4-\alpha x^3-\beta x^3-\gamma x^3-\delta x^3+\alpha\beta x^2+\alpha\gamma x^2+\alpha\delta x^2$
$+\beta\gamma x^2+\beta\delta x^2+\gamma\delta x^2-\alpha\beta\gamma x-\alpha\beta\delta x-\alpha\gamma\delta x-\beta\gamma\delta x+\alpha\beta\gamma\delta)$

$=a(x^4-(\alpha+\beta+\gamma+\delta)x^3+(\alpha\beta+\alpha\gamma+\alpha\delta+\beta\gamma+\beta\delta+\gamma\delta)x^2$
$-(\alpha\beta\gamma+\alpha\beta\delta+\alpha\gamma\delta+\beta\gamma\delta)x+\alpha\beta\gamma\delta)$

Therefore,

a $-a(\alpha+\beta+\gamma+\delta)=b\Rightarrow\alpha+\beta+\gamma+\delta=-\frac{b}{a}$

b $a(\alpha\beta+\alpha\gamma+\alpha\delta+\beta\gamma+\beta\delta+\gamma\delta)=c$
$\Rightarrow\alpha\beta+\alpha\gamma+\alpha\delta+\beta\gamma+\beta\delta+\gamma\delta=\frac{c}{a}$

c $-a(\alpha\beta\gamma+\alpha\beta\delta+\alpha\gamma\delta+\beta\gamma\delta)=d$
$\Rightarrow\alpha\beta\gamma+\alpha\beta\delta+\alpha\gamma\delta+\beta\gamma\delta=-\frac{d}{a}$

d $a(\alpha\beta\gamma\delta)=e\Rightarrow\alpha\beta\gamma\delta=\frac{e}{a}$

8 a $\frac{k^2}{9}-\frac{14}{3}$

b $\frac{2}{k}$

Exercise 2.1B

1 a $y^3+5y^2+3y+1=0$
b $y^3+9y^2+45y+27=0$

2 a $y^3+4y^2+3y=0$
b $y=0, -1$ or -3
c $x=2, 1$ or -1

3 $8y^3-4y^2-14y+9=0$

4 $y^4+16y^3+42y^2+96y+197=0$

5 a $y^3+3y^2-18y-56+k=0$
b $x=4, -2, 7$ (since $x=y+4$)

6 a $x=3\pm i, 5$
b $m=-11$
$n=40$

7 a 24
b $4+24y-11y^2+119y^3=0$

8 a $x=-\frac{7}{2}, 9\pm 2i$
b $p=-115$
$q=1498$

9 $b=5$

10 $\alpha^3=4\alpha+3$; $\beta^3=4\beta+3$ and $\gamma^3=4\gamma+3$
Hence $\alpha^3+\beta^3+\gamma^3=4(\alpha+\beta+\gamma)+9$
But from $x^3=4x+3$ or $x^3-4x-3=0$, $\alpha+\beta+\gamma=0$
Hence $\alpha^3+\beta^3+\gamma^3=4(0)+9=9$

11 $2y^3-4y^2-4y+9=0$

12 a 4
b $-y^3+8y^2-15y+3=0$ or $y^3-8y^2+15y-3=0$
c 3

Exercise 2.2A

1 a If $a=-3$ and $b=4$ then
$-\frac{1}{3}<\frac{1}{4}$
but if $a=3$ and $b=4$ then $\frac{1}{3}>\frac{1}{4}$

b If $a=2$, $b=4$, $x=5$ and $y=6$ then
$2-4<5-6$ $(-2<-1)$
but if $a=2$, $b=3$, $x=5$ and $y=7$
then $2-3>5-7$ $(-1>-2)$

c If $a=2$, $b=4$, $x=5$ and $y=6$ then $2\times4<5\times6$ $(8<30)$,
but if $a=-2$, $b=-4$, $x=-1$ and $y=6$ then
$-2\times-4>-1\times6$ $(8>-6)$

2 a $x>0$ **b** $x>0$ or $x\le-3$

c $x<-2$ or $x\ge-\frac{1}{2}$ **d** $2<x<11$

3 a

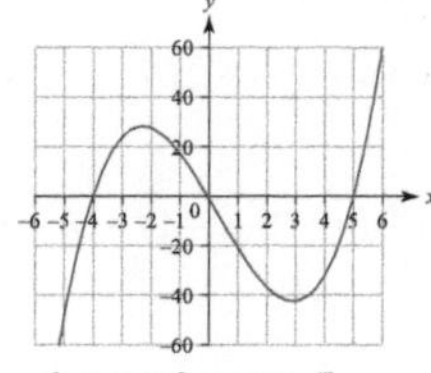

$-4 < x < 0$ or $x > 5$

b

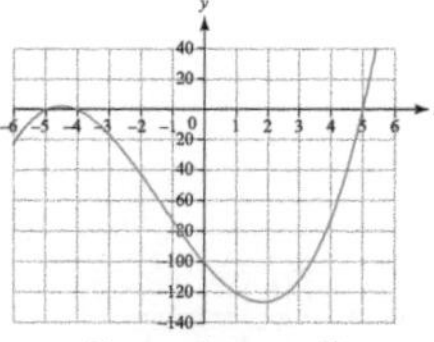

$x < -5$ or $-4 < x < 5$

c

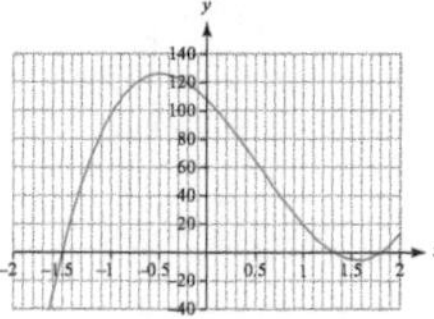

$x \leq -\frac{3}{2}$ or $\frac{4}{3} \leq x \leq \frac{9}{5}$

d

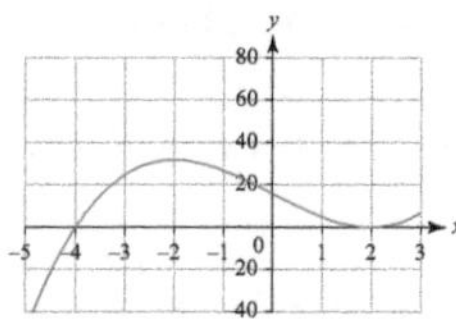

$x \geq -4$

e

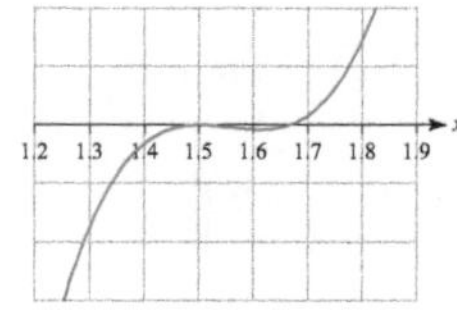

$x > \frac{5}{3}$

f

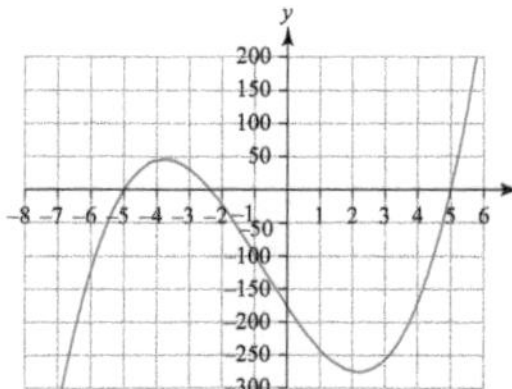

$x < -5$ or $-\frac{7}{3} < x < 5$

g

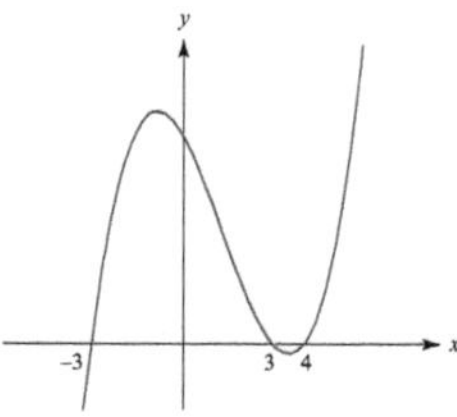

$x \leq -3$ or $3 \leq x \leq 4$

4 a $x < -1$ or $0 < x < 2$ or $x > 9$

b $-2 < x < -1$ or $3 < x < 7$

c $-\frac{1}{2} \leq x \leq \frac{5}{4}$ or $\frac{4}{3} \leq x \leq 11$

d $(x-1)^2(x-3)^2 \geq 0 \to \{x: x \in \mathbb{R}\}$

e $(2x-7)^2(3x+2)^2 > 0 \to \{x: x \in \mathbb{R}\}$

f $-7 < x < -\sqrt{5}$ or $\sqrt{5} < x < 7$

g $-3 < x < -2$ or $4 < x < 5$

5 a $x > 2$ or $x < 0$

b $x > 2$ or $x < -3$

c $x > 5$ or $x < 2$

d $-2 < x < -1$ or $x > 0$

e $-5 < x < -3$ or $x > 4$

f $-\frac{4}{3} < x < \frac{3}{2}$ or $x > 7$

g $-\frac{3}{2} < x < 4$ or $x < -\frac{5}{2}$

h $-2 < x < \frac{2}{3}$ or $x > \frac{3}{2}$

6 a $-1 < x < 0$

b $-4 < x < -3$

c $-2 < x < 3$

d $x < -3$ or $0 < x < 5$

e $x < -7$ or $-6 < x < -4$

f $x < -6$ or $-\frac{5}{2} < x < \frac{3}{4}$

g $-\frac{6}{5} < x < \frac{1}{2}$ or $x > \frac{9}{2}$

h $x < -\frac{5}{2}$ or $-1 < x < 2$

Exercise 2.2B

1 a $(-3,0)$, $(2.5,0)$, $(7,0)$ and $(0,105)$

b $(x+3)(2x-5)(x-7) = 0$

$\to 2x^3 - 13x^2 - 22x + 105 = 0$

c $A = 2, B = -13, C = -22$ and $D = 105$

d $(-0.7, 115)$ and $(5.1, -80)$

e $x < -3$ or $2.5 < x < 7$

2 a $(-3,0)$, $(2,0)$, $(7,0)$ and $(0,-84)$

The turning point at $(2,0)$ means there is a repeated root.

b The equation of the function is $(x+3)(x-7)(x-2)^2 = 0$

$x^4 - 8x^3 - x^2 + 68x - 84 = 0$

c $A = 1, B = -8, C = -1, D = 68$ and $E = -84$

d $(-1.5, -156)$, $(2,0)$ and $(5.5, -156)$

e $x < -3$ or $x > 7$

3 $0 \leq x \leq 2$ and $6 \leq x$

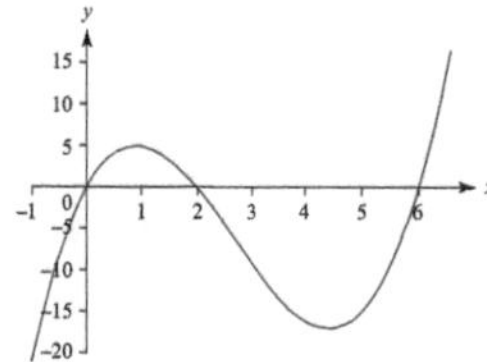

4 $-4 \leq x \leq -2$; $2 \leq x \leq 4$

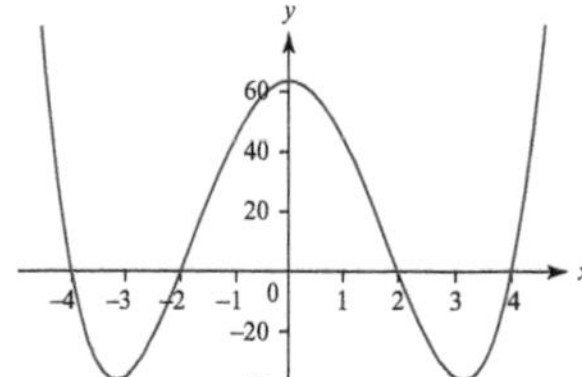

5 a $x > 3$ or $-3 < x < 2$

b $-\frac{11}{4} < x < 0$; $x < -4$

c $x \geq 5$ or $3 \leq x < 4$

d $-6 \leq x \leq -2$; $x > -1$

6 $-3 < x < 2$ or $x > 7$

7 $-5 \leq x \leq -3$; $3 \leq x \leq 5$

8 $0 < x < 1$; $x > 4$

9 a $y \geq -3+\sqrt{8}; y \leq -3-\sqrt{8}$

b

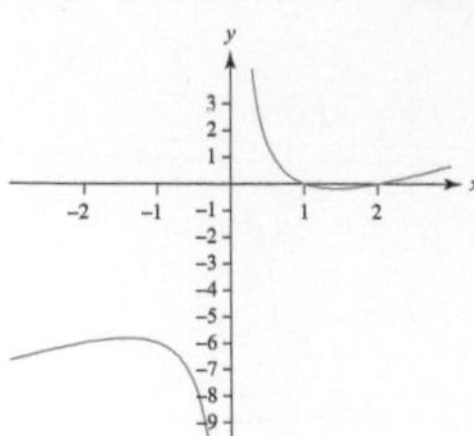

10 a $-2 < x < -1.5$ or $x > 4$

b

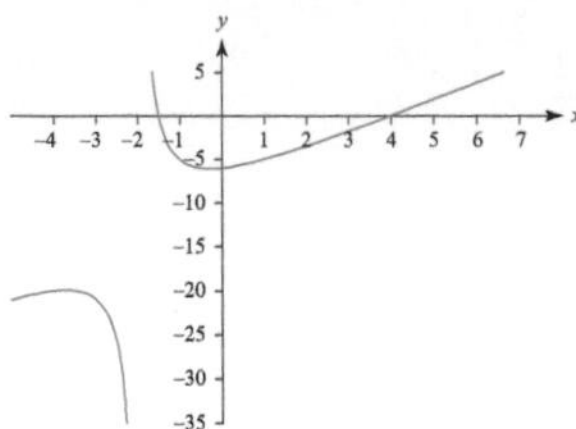

11 a $x \geq 0$

b

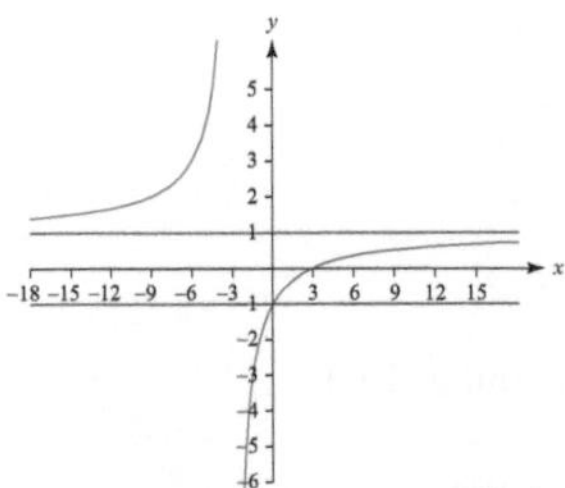

12 Let the minimum value of $x+\frac{1}{x}$ be k. Thus $x+\frac{1}{x}=k$, so $x^2-kx+1=0$

If x is real then $k^2-4\times1\times1\geq0$, i.e. $k^2\geq4$ or $k\leq-2$ or $k\geq2$

Since x is positive, $k\geq2$

Alternatively:

Find x such that $x+\frac{1}{x}<2$

$x+\frac{1}{x}-2<0$

$\frac{x^2+1-2x}{x}<0$

$\frac{(x-1)^2}{x}<0$

The numerator is squared so is positive for all x.

So, the denominator must be negative, but this is not possible since $x>0$

So there is no value of x for which $x+\frac{1}{x}$ is less than 2

Exercise 2.3A

1 a $\sum_1^5 r^3 \equiv 1^3+2^3+3^3+4^3+5^3$

b $\sum_4^n r^2 \equiv 4^2+5^2+.....+n^2$

c $\sum_1^n (r^2-2r) \equiv (1^2-2\times1)+(2^2-2\times2)+(3^2-2\times3)$
$+(4^2-2\times4)+....+(n^2-2\times n)$
$=-1+0+3+8+...+(n^2-2\times n)$

d $\sum_1^n \frac{1}{m+2} \equiv \frac{1}{3}+\frac{1}{4}+\frac{1}{5}+\frac{1}{6}+.....+\frac{1}{n+2}$

e $\sum_1^6 (-1)^r r^3 \equiv -1+8-27+64-125+216$

f $\sum_{n-3}^n r(r+1) \equiv (n-3)(n-2)+(n-2)(n-1)+(n-1)(n)$
$+(n)(n+1)$

2 a $\sum_1^{16}(2r-1)$ **b** $\sum_1^n r^5$

c $\sum_1^{n+1}\frac{1}{r}$ **d** $\sum_1^{14}(-1)^{r+1}3r$

e $\sum_1^n r(r+2)$ **f** $\sum_1^n (-1)^{r+1}\frac{(2r-1)(r+3)}{(r+2)}$

3 a $\sum_{r=1}^n (r^2+6r-3)=\frac{1}{6}n(n+1)(2n+1)+6\times\frac{1}{2}n(n+1)-3n$

$=n\left[\frac{1}{6}(n+1)(2n+1)+3(n+1)-3\right]$

$=\frac{1}{6}n\left[(n+1)(2n+1)+18(n+1)-18\right]$

$=\frac{1}{6}n(2n^2+3n+1+18n+18-18)$

$=\frac{1}{6}n(2n^2+21n+1)$ as required

b $\sum_{r=1}^n (2r^2-7r)=2\times\frac{1}{6}n(n+1)(2n+1)-7\times\frac{1}{2}n(n+1)$

$=\frac{1}{3}n(n+1)(2n+1)-\frac{7}{2}n(n+1)$

$=n(n+1)\left[\frac{1}{3}(2n+1)-\frac{7}{2}\right]$

$=\frac{1}{6}n(n+1)\left[2(2n+1)-21\right]$

$=\frac{1}{6}n(n+1)(4n-19)$ as required

c $\sum_{r=1}^n (8-5r^2)=8n-5\times\frac{1}{6}n(n+1)(2n+1)$

$=n\left[8-\frac{5}{6}(n+1)(2n+1)\right]$

$=\frac{1}{6}n\left[48-5(n+1)(2n+1)\right]$

$=\frac{1}{6}n(48-10n^2-15n-5)$

$=-\frac{1}{6}n(10n^2+15n-43)$ as required

d $\sum_{r=1}^n (r+1)^2=\sum_{r=1}^n (r^2+2r+1)$

$=\frac{1}{6}n(n+1)(2n+1)+2\times\frac{1}{2}n(n+1)+n$

$$=n\left[\frac{1}{6}(n+1)(2n+1)+(n+1)+1\right]$$

$$=\frac{1}{6}n\left[(n+1)(2n+1)+6(n+1)+6\right]$$

$$=\frac{1}{6}n(2n^2+3n+1+6n+6+6)$$

$$=\frac{1}{6}n(2n^2+9n+13) \text{ as required}$$

e $$\sum_{r=1}^{n}(r^3+2r+3)=\frac{1}{4}n^2(n+1)^2+2\times\frac{1}{2}n(n+1)+3n$$

$$=n\left[\frac{1}{4}n(n+1)^2+(n+1)+3\right]$$

$$=\frac{1}{4}n\left[n(n+1)^2+4(n+1)+12\right]$$

$$=\frac{1}{4}n(n^3+2n^2+n+4n+4+12)$$

$$=\frac{1}{4}n(n^3+2n^2+5n+16) \text{ as required}$$

f $$\sum_{r=1}^{n}(2r^3+3r^2)=2\times\frac{1}{4}n^2(n+1)^2+3\times\frac{1}{6}n(n+1)(2n+1)$$

$$=\frac{1}{2}n^2(n+1)^2+\frac{1}{2}n(n+1)(2n+1)$$

$$=\frac{1}{2}n(n+1)\left[n(n+1)+(2n+1)\right]$$

$$=\frac{1}{2}n(n+1)(n^2+n+2n+1)$$

$$=\frac{1}{2}n(n+1)(n^2+3n+1) \text{ as required}$$

g $$\sum_{r=1}^{n}(r^3-2r^2+3r)$$

$$=\frac{1}{4}n^2(n+1)^2-2\times\frac{1}{6}n(n+1)(2n+1)+3\times\frac{1}{2}n(n+1)$$

$$=n(n+1)\left[\frac{1}{4}n(n+1)-\frac{1}{3}(2n+1)+\frac{3}{2}\right]$$

$$=\frac{1}{12}n(n+1)\left[3n(n+1)-4(2n+1)+18\right]$$

$$=\frac{1}{12}n(n+1)(3n^2+3n-8n-4+18)$$

$$=\frac{1}{12}n(n+1)(3n^2-5n+14) \text{ as required}$$

h $$\sum_{r=1}^{n}(r+2)^3=\sum_{r=1}^{n}r^3+6r^2+12r+8$$

$$=\frac{1}{4}n^2(n+1)^2+6\times\frac{1}{6}n(n+1)(2n+1)+12\times\frac{1}{2}n(n+1)+8n$$

$$=n\left[\frac{1}{4}n(n+1)^2+(n+1)(2n+1)+6(n+1)+8\right]$$

$$=\frac{1}{4}n\left[n(n+1)^2+4(n+1)(2n+1)+24(n+1)+32\right]$$

$$=\frac{1}{4}n(n^3+2n^2+n+8n^2+12n+4+24n+24+32)$$

$$=\frac{1}{4}n(n^3+10n^2+37n+60) \text{ as required}$$

4 **a** There are $3n$ terms. **b** The $(2n+1)$th term is $(2n+1)^2-(2n+1)+3$ $=4n^2+2n+3$

5 **a** $\frac{3n(3n+1)}{2}$ **b** $\frac{n(2n-1)(4n-1)}{3}$

c $n^2(2n-1)^2$ **d** n^2

e $\frac{(n+1)(n+2)}{3}$ **f** $\frac{n(n+1)(n-1)}{3}$

6 **a** $\frac{n}{3}(n^2-1)$ **b** 330

7 **a** rth term $=r(r+1)(r+2)$

Sum to n terms $=\frac{n(n+1)(n+2)(n+3)}{4}$

b rth term $=(2r-1)(r+3)$

Sum to n terms $=\frac{n}{6}(4n^2+21n-1)$

c rth term $=r(r+2)(r+4)$

Sum to n terms $=\frac{n(n+1)(n+4)(n+5)}{4}$

d rth term $=4r-2$

Sum to n terms $=2n^2$

e rth term $=r^2+1$

Sum to n terms $=\frac{n(2n^2+3n+7)}{6}$

8 **a** $n^2+4n-32$

b $\frac{(n-2)(n-1)(n-3)}{3}-2$

c $(n+1)(2n+1)(2n^2+3n+4)-\frac{n(n-1)(n^2-n+6)}{4}$

9 **a** $\frac{n(1-3n)}{2}$

b $1-\frac{1}{n+1}\equiv\frac{n}{n+1}$

c $\frac{n(3n+5)}{2(n+1)(n+2)}$

d $\frac{11}{6}-\frac{3n^2+12n+11}{(n+1)(n+2)(n+3)}$

e $\frac{2n}{2n+1}$

f $\frac{n(n+7)}{12(n+3)(n+4)}$

Exercise 2.3B

1 **a** First term is $2(1)^2+4(1)=6$
Sum of first two terms is $2(2)^2+4(2)=16$, so second term is $16-6=10$
Sum of first three terms is $2(3)^2+4(3)=30$, so third term is $30-16=14$

b The nth term is $4n+2$

2 **a** $\frac{n(n+1)}{2}$

b Altogether there are $\frac{n(n+1)(n+2)}{6}$

c 599 kg (3sf)

3 Number of cannon balls $= \dfrac{n(2n+1)(7n+1)}{6}$

4 a $\dfrac{1}{1^2} - \dfrac{1}{2^2} = \dfrac{3}{4}$

$\dfrac{1}{2^2} - \dfrac{1}{3^2} = \dfrac{5}{36}$

$\dfrac{1}{3^2} - \dfrac{1}{4^2} = \dfrac{7}{144}$

$\dfrac{1}{4^2} - \dfrac{1}{5^2} = \dfrac{9}{400}$

$\dfrac{1}{n^2} - \dfrac{1}{(n+1)^2}$

b $\displaystyle\sum_1^n \left(\frac{1}{r^2} - \frac{1}{(r+1)^2}\right) \equiv \frac{n(n+2)}{(n+1)^2}$

The sum to infinity $\displaystyle\lim_{n\to\infty}\left(1 - \frac{1}{(n+1)^2}\right) = 1$

5 $\dfrac{m(m+1)(m+2)}{6}$

6 a $\dfrac{1}{(2r-1)^2} - \dfrac{1}{(2r+1)^2} \equiv \dfrac{(2r+1)^2 - (2r-1)^2}{(2r-1)^2(2r+1)^2}$

$\equiv \dfrac{8r}{[(2r-1)(2r+1)]^2}$

$\equiv \dfrac{8r}{((2r)^2-1)^2}$

b $\displaystyle\sum_1^n \frac{8r}{((2r)^2-1)^2} \equiv \sum_1^n \frac{1}{(2r-1)^2} - \frac{1}{(2r+1)^2}$

$\equiv \left(\dfrac{1}{1^2} - \dfrac{1}{3^2}\right) + \left(\dfrac{1}{3^2} - \dfrac{1}{5^2}\right) + \left(\dfrac{1}{5^2} - \dfrac{1}{7^2}\right) + \ldots +$

$\left(\dfrac{1}{(2n-3)^2} - \dfrac{1}{(2n-1)^2}\right) + \left(\dfrac{1}{(2n-1)^2} - \dfrac{1}{(2n+1)^2}\right)$

$\equiv 1 - \dfrac{1}{(2n+1)^2}$

$\displaystyle\sum_1^n \frac{r}{((2r)^2-1)^2} = \frac{n(n+1)}{2n(2n+1)^2}$

7 a $r(r+1) - r(r-1) \equiv r^2 + r - r^2 + r \equiv 2r$

b $\Sigma r(r+1) - r(r-1) \equiv$

1×2	−	1×0
2×3	−	2×1
3×4	−	3×2
⇓		
$(n-2) \times (n-1)$	−	$(n-3) \times (n-2)$
$(n-1) \times n$	−	$(n-2) \times (n-1)$
$n \times (n+1)$	−	$(n-1) \times n$

Hence $2\Sigma r \equiv \Sigma r(r+1) - r(r-1) \equiv 1 \times 0 + n \times (n+1)$

$2\Sigma r \equiv n(n+1)$

$\displaystyle\sum_1^n r \equiv \frac{n(n+1)}{2}$

8 a $(2r+1)^3 - (2r-1)^3 \equiv 8r^3 + 12r^2 + 6r + 1 - (8r^3 - 12r^2 + 6r - 1)$

$\equiv 24r^2 + 2$

b $\Sigma[(2r+1)^3 - (2r-1)^3] = 24\Sigma r^2 + 2n$

$\Sigma[(2r+1)^3 - (2r-1)^3]$

3^3	−	1^3
5^3	−	3^3
7^3	−	5^3
↓		↓
$[(2n-3)^3$	−	$(2n-5)^3]$
$[(2n-1)^3$	−	$(2n-3)^3]$
$[(2n+1)^3$	−	$(2n-1)^3]$

$\equiv (2n+1)^3 \quad - \quad 1$

Hence $24\displaystyle\sum_1^n r^2 + 2n \equiv (2n+1)^3 - 1$

$24\displaystyle\sum_1^n r^2 \equiv (2n+1)^3 - 2n - 1$

$\equiv 8n^3 + 12n^2 + 4n$

$\equiv 4n(n+1)(2n+1)$

Hence $\displaystyle\sum_1^n r \equiv \frac{n(n+1)(2n+1)}{6}$

9 a $\Sigma(2r-1)^2 = 4\Sigma r^2 - 4\Sigma r + \Sigma 1$

$= \dfrac{4n(n+1)(2n+1)}{6} - \dfrac{4n(n+1)}{2} + n \rightarrow$

$\dfrac{n[4(n+1)(2n+1) - 12(n+1) + 6]}{6}$

$= \dfrac{n(8n^2-2)}{6} \rightarrow \dfrac{n(2n-1)(2n+1)}{3}$

b $1^2 + 3^2 + 5^2 + \ldots + (2r-1)^2 \equiv \displaystyle\sum_1^{2n} r^2 - 4\sum_1^n r^2$

$= \dfrac{2n(2n+1)(4n+1)}{6} - \dfrac{4n(n+1)(2n+1)}{6}$

$= \dfrac{n(2n+1)[(4n+1) - 2(n+1)]}{3}$

$= \dfrac{n(2n+1)[(2n-1)}{3}$

10 a $-\dfrac{1}{x} + \dfrac{3}{x+2} - \dfrac{2}{x+4} \equiv$

$\dfrac{-(x+2)(x+4) + 3x(x+4) - 2x(x+2)}{x(x+2)(x+4)}$

$\equiv \dfrac{-x^2 - 6x - 8 + 3x^2 + 12x - 2x^2 - 4x}{x(x+2)(x+4)}$

$\equiv \dfrac{2x-8}{x(x+2)(x+4)} \equiv \dfrac{2(x-4)}{x(x+2)(x+4)}$

11 $\left(2r^2 + 3r + \dfrac{1}{r} - \dfrac{1}{r+1}\right) \equiv \dfrac{2r^3(r+1) + 3r^2(r+1) + (r+1) - r}{r(r+1)}$

$\equiv \dfrac{2r^4 + 2r^3 + 3r^3 + 3r^2 + r + 1 - r}{r(r+1)}$

$\equiv \dfrac{2r^4 + 5r^3 + 3r^2 + 1}{r(r+1)}$

$\displaystyle\sum \frac{2r^4 + 5r^3 + 3r^2 + 1}{r(r-1)} \equiv \sum_1^n \left(2r^2 + 3r + \frac{1}{r} - \frac{1}{r+1}\right)$

$\equiv 2\displaystyle\sum_1^n r^2 + 3\sum_1^n r + \sum_1^n \left(\frac{1}{r} - \frac{1}{r+1}\right)$

$\equiv 2\dfrac{n(n+1)(2n+1)}{6} + 3\dfrac{n(n+1)}{2} + \left(\dfrac{1}{1} - \dfrac{1}{2}\right) + \left(\dfrac{1}{2} - \dfrac{1}{3}\right)$

$+ \left(\dfrac{1}{3} - \dfrac{1}{4}\right) + \left(\dfrac{1}{4} - \dfrac{1}{5}\right) + \ldots + \left(\dfrac{1}{n-1} - \dfrac{1}{n}\right) + \left(\dfrac{1}{n} - \dfrac{1}{n+1}\right)$

$\equiv \dfrac{n(n+1)(2n+1)}{3} + \dfrac{3n(n+1)}{2} + 1 - \dfrac{1}{n+1}$

$\equiv \dfrac{2n(n+1)(2n+1)(n+1) + 9n(n+1)(n+1) + 6(n+1) - 6}{6(n+1)}$

$\equiv \dfrac{n(4n^3 + 19n^2 + 26n + 17)}{6(n+1)}$

Exercise 2.4A

1 a When $n=1$, $\sum_{r=1}^{n} 1 = 1$

and $n=1$ so true for $n=1$

Assume true for $n=k$ and consider $n=k+1$:

$$\sum_{r=1}^{k+1} 1 = \sum_{r=1}^{k} 1 + 1$$
$$= k+1$$

So true for $n=k+1$

The statement is true for $n=1$ and by assuming it is true for $n=k$ it is shown to be true for $n=k+1$, therefore, by mathematical induction, it is true for all $n \in \mathbb{N}$

b When $n=1$, $\sum_{r=1}^{n} r = 1$

and $\frac{1}{2}n(n+1) = \frac{1}{2} \times 1(1+1)$
$$= \frac{1}{2}(2)$$
$$= 1$$

so true for $n=1$

Assume true for $n=k$ and consider $n=k+1$:

$$\sum_{r=1}^{k+1} r = \sum_{r=1}^{k} r + (k+1)$$
$$= \frac{1}{2}k(k+1) + k + 1$$
$$= \frac{1}{2}(k+1)(k+2)$$
$$= \frac{1}{2}(k+1)(k+1+1)$$

So true for $n=k+1$

The statement is true for $n=1$ and by assuming it is true for $n=k$ it is shown to be true for $n=k+1$, therefore, by mathematical induction, it is true for all $n \in \mathbb{N}$

c When $n=1$, $\sum_{r=1}^{n} (2r+3) = 2\times1+3$
$$= 5$$

and $n(n+4) = 1(1+4)$
$$= 5$$

so true for $n=1$

Assume true for $n=k$ and consider $n=k+1$:

$$\sum_{r=1}^{k+1} (2r+3) = \sum_{r=1}^{k} (2r+3) + 2(k+1) + 3$$
$$= k(k+4) + 2k + 5$$
$$= k^2 + 6k + 5$$
$$= (k+1)(k+5)$$
$$= (k+1)(k+1+4)$$

So true for $n=k+1$

The statement is true for $n=1$ and by assuming it is true for $n=k$ it is shown to be true for $n=k+1$, therefore, by mathematical induction, it is true for all $n \in \mathbb{N}$

d When $n=1$, $\sum_{r=1}^{n} r(r+1) = 1(1+1)$
$$= 2$$

and $\frac{1}{3}n(n+1)(n+2) = \frac{1}{3} \times 1(1+1)(1+2)$
$$= \frac{1}{3}(2)(3)$$
$$= 2$$

so true for $n=1$

Assume true for $n=k$ and consider $n=k+1$:

$$\sum_{r=1}^{k+1} r(r+1) = \sum_{r=1}^{k} r(r+1) + (k+1)(k+2)$$
$$= \frac{1}{3}k(k+1)(k+2) + (k+1)(k+2)$$
$$= \frac{1}{3}(k+1)(k+2)(k+3)$$
$$= \frac{1}{3}(k+1)(k+1+1)(k+1+2)$$

So true for $n=k+1$

The statement is true for $n=1$ and by assuming it is true for $n=k$ it is shown to be true for $n=k+1$, therefore, by mathematical induction, it is true for all $n \in \mathbb{N}$

e When $n=1$, $\sum_{r=1}^{n} (r-1)^2 = (1-1)^2$
$$= 0$$

and $\frac{1}{6}n(n-1)(2n-1) = \frac{1}{6} \times 1(1-1)(2\times1-1)$
$$= \frac{1}{6}(0)(1)$$
$$= 0$$

so true for $n=1$

Assume true for $n=k$ and consider $n=k+1$:

$$\sum_{r=1}^{k+1} (r-1)^2 = \sum_{r=1}^{k} (r-1)^2 + (k+1-1)^2$$
$$= \frac{1}{6}k(k-1)(2k-1) + k^2$$
$$= \frac{1}{6}k[(k-1)(2k-1) + 6k]$$
$$= \frac{1}{6}k[2k^2 - 2k - k + 1 + 6k]$$
$$= \frac{1}{6}k(2k^2 + 3k + 1)$$
$$= \frac{1}{6}k(2k+1)(k+1)$$
$$= \frac{1}{6}(k+1)(k+1-1)(2(k+1)-1)$$

So true for $n=k+1$

The statement is true for $n=1$ and by assuming it is true for $n=k$ it is shown to be true for $n=k+1$, therefore, by mathematical induction, it is true for all $n \in \mathbb{N}$

f When $n=1$, $\sum_{r=1}^{n} (r+1)(r-1) = (1+1)(1-1)$
$$= 0$$

and $\frac{1}{6}n(2n+5)(n-1) = \frac{1}{6} \times 1(2\times1+5)(1-1)$
$$= \frac{1}{6}(7)(0)$$
$$= 0$$

so true for $n=1$

Assume true for $n=k$ and consider $n=k+1$:

$$\sum_{r=1}^{k+1} (r+1)(r-1) = \sum_{r=1}^{k} (r+1)(r-1) + (k+1+1)(k+1-1)$$
$$= \frac{1}{6}k(2k+5)(k-1) + k(k+2)$$
$$= \frac{1}{6}k[(2k+5)(k-1) + 6(k+2)]$$

$$=\frac{1}{6}k(2k^2-2k+5k-5+6k+12)$$
$$=\frac{1}{6}k(2k^2+9k+7)$$
$$=\frac{1}{6}k(2k+7)(k+1)$$
$$=\frac{1}{6}(k+1)(2(k+1)+5)(k+1-1)$$

So true for $n=k+1$

The statement is true for $n=1$ and by assuming it is true for $n=k$ it is shown to be true for $n=k+1$, therefore, by mathematical induction, it is true for all $n\in\mathbb{N}$

2 a When $n=1, \sum_{r=1}^{n}(r+1)^2=2^2=4$

and $\frac{1}{6}n(2n^2+9n+13)=\frac{1}{6}(1)(24)=4$

So the statement is true when $n=1$

Assume statement is true for $n=k$ and substitute $n=k+1$ into the formula:

$$\sum_{r=1}^{k+1}(r+1)^2=\sum_{r=1}^{k}(r+1)^2+(k+1+1)^2$$
$$=\frac{1}{6}k(2k^2+9k+13)+(k+2)^2$$
$$=\frac{1}{6}(2k^3+9k^2+13k+6k^2+24k+24)$$
$$=\frac{1}{6}(2k^3+15k^2+37k+24)$$
$$=\frac{1}{6}(k+1)(2k^2+13k+24)$$
$$=\frac{1}{6}(k+1)(2(k+1)^2+9(k+1)+13)$$

So the statement is true when $n=k+1$

The statement is true for $n=1$ and by assuming it is true for $n=k$ it is shown to be true for $n=k+1$, therefore, by mathematical induction, it is true for all $n\in\mathbb{N}$

b When $n=1, \sum_{r=1}^{n}5^{n-1}=5^0=1$

and $\frac{1}{4}(5^n-1)=\frac{1}{4}(4)=1$

So the statement is true when $n=1$

Assume statement is true for $n=k$ and substitute $n=k+1$ into the formula:

$$\sum_{r=1}^{k+1}5^{r-1}=\sum_{r=1}^{k}5^{r-1}+5^{k+1-1}$$
$$=\frac{1}{4}(5^k-1)+5^k$$
$$=\frac{1}{4}(5^k-1+4(5^k))$$
$$=\frac{1}{4}(5(5^k)-1)$$
$$=\frac{1}{4}(5^{k+1}-1)$$

So the statement is true when $n=k+1$

The statement is true for $n=1$ and by assuming it is true for $n=k$ it is shown to be true for $n=k+1$, therefore, by mathematical induction, it is true for all $n\in\mathbb{N}$

3 When $n=1, \sum_{r=1}^{n}r^3=1^3$

$$=1$$

and $\frac{1}{4}n^2(n+1)^2=\frac{1}{4}\times1^2(1+1)^2$

$$=\frac{1}{4}(1)(2)^2$$
$$=1$$

so true for $n=1$

Assume true for $n=k$ and consider $n=k+1$:

$$\sum_{r=1}^{k+1}r^3=\sum_{r=1}^{k}r^3+(k+1)^3$$
$$=\frac{1}{4}k^2(k+1)^2+(k+1)^3$$
$$=\frac{1}{4}(k+1)^2[k^2+4(k+1)]$$
$$=\frac{1}{4}(k+1)^2[k^2+4k+4]$$
$$=\frac{1}{4}(k+1)^2(k+2)^2$$
$$=\frac{1}{4}(k+1)^2(k+1+1)^2$$

So true for $n=k+1$

The statement is true for $n=1$ and by assuming it is true for $n=k$ it is shown to be true for $n=k+1$, therefore, by mathematical induction, it is true for all $n\in\mathbb{N}$

4 When $n=1, \sum_{r=1}^{2n}r=1+2$

$$=3$$

and $n(2n+1)=1(2\times1+1)$

$$=3$$

so true for $n=1$

Assume true for $n=k$ and consider $n=k+1$:

$$\sum_{r=1}^{2(k+1)}r=\sum_{r=1}^{2k}r+(2k+1)+(2k+2)$$
$$=k(2k+1)+4k+3$$
$$=2k^2+k+4k+3$$
$$=2k^2+5k+3$$
$$=(2k+3)(k+1)$$
$$=(k+1)(2(k+1)+1)$$

So true for $n=k+1$

The statement is true for $n=1$ and by assuming it is true for $n=k$ it is shown to be true for $n=k+1$, therefore, by mathematical induction, it is true for all $n\in\mathbb{N}$

5 When $n=1, \sum_{r=1}^{2n}r^2=1^2+2^2=5$

and $\frac{1}{3}n(2n+1)(4n+1)=\frac{1}{3}(1)(3)(5)=5$

So the statement is true when $n=1$

Assume statement is true for $n=k$ and substitute $n=k+1$ into the formula.

$$\sum_{r=1}^{2(k+1)}r^2=\sum_{r=1}^{2k}r^2+(2k+1)^2+(2k+2)^2$$

$$= \frac{1}{3}k(2k+1)(4k+1) + (4k^2+4k+1) + (4k^2+8k+4)$$

$$= \frac{1}{3}(8k^3+6k^2+k+12k^2+12k+3+12k^2+24k+12)$$

$$= \frac{1}{3}(8k^3+30k^2+37k+15)$$

$$= \frac{1}{3}(k+1)(8k^2+22k+15)$$

$$= \frac{1}{3}(k+1)(2k+3)(4k+5)$$

$$= \frac{1}{3}(k+1)(2(k+1)+1)(4(k+1)+1)$$

So statement is true when $n = k+1$

The statement is true for $n = 1$ and by assuming it is true for $n = k$ it is shown to be true for $n = k+1$, therefore, by mathematical induction, it is true for all $n \in \mathbb{N}$

6 a When $n = 1$, $\sum_{r=1}^{n} 2^r = 2^1$

$$= 2$$

and $2(2^n - 1) = 2(2^1 - 1)$

$$= 2$$

so true for $n = 1$

Assume true for $n = k$ and consider $n = k + 1$:

$$\sum_{r=1}^{k+1} 2^r = \sum_{r=1}^{k} 2^r + 2^{k+1}$$

$$= 2(2^k - 1) + 2^{k+1}$$

$$= 2(2^k) - 2 + 2(2^k)$$

$$= 4(2^k) - 2$$

$$= 2(2^{k+1} - 1)$$

So true for $n = k + 1$

The statement is true for $n = 1$ and by assuming it is true for $n = k$ it is shown to be true for $n = k + 1$, therefore, by mathematical induction, it is true for all $n \in \mathbb{N}$

b When $n = 1$, $\sum_{r=1}^{n} 3^r = 3^1$

$$= 3$$

and $\frac{3}{2}(3^n - 1) = \frac{3}{2}(3^1 - 1)$

$$= \frac{3}{2}(2)$$

$$= 3$$

so true for $n = 1$

Assume true for $n = k$ and consider $n = k + 1$:

$$\sum_{r=1}^{k+1} 3^r = \sum_{r=1}^{k} 3^r + 3^{k+1}$$

$$= \frac{3}{2}(3^k - 1) + 3^{k+1}$$

$$= \frac{3}{2}\left(3^k - 1 + \frac{2}{3}(3^{k+1})\right)$$

$$= \frac{3}{2}\left(3^k - 1 + \frac{2}{3}(3)3^k\right)$$

$$= \frac{3}{2}(3^k - 1 + 2(3^k))$$

$$= \frac{3}{2}(3(3^k) - 1)$$

$$= \frac{3}{2}(3^{k+1} - 1)$$

So true for $n = k + 1$

The statement is true for $n = 1$ and by assuming it is true for $n = k$ it is shown to be true for $n = k + 1$, therefore, by mathematical induction, it is true for all $n \in \mathbb{N}$

c When $n = 1$, $\sum_{r=1}^{n} 4^r = 4^1$

$$= 4$$

and $\frac{4}{3}(4^n - 1) = \frac{4}{3}(4^1 - 1)$

$$= \frac{4}{3}(3)$$

$$= 4$$

so true for $n = 1$

Assume true for $n = k$ and consider $n = k + 1$:

$$\sum_{r=1}^{k+1} 4^r = \sum_{r=1}^{k} 4^r + 4^{k+1}$$

$$= \frac{4}{3}(4^k - 1) + 4^{k+1}$$

$$= \frac{4}{3}\left(4^k - 1 + \frac{3}{4}(4^{k+1})\right)$$

$$= \frac{4}{3}\left(4^k - 1 + \frac{3}{4}(4)4^k\right)$$

$$= \frac{4}{3}(4^k - 1 + 3(4^k))$$

$$= \frac{4}{3}(4(4^k) - 1)$$

$$= \frac{4}{3}(4^{k+1} - 1)$$

So true for $n = k + 1$

The statement is true for $n = 1$ and by assuming it is true for $n = k$ it is shown to be true for $n = k + 1$, therefore, by mathematical induction, it is true for all $n \in \mathbb{N}$

d When $n = 1$, $\sum_{r=1}^{n} 2^{r-1} = 2^{1-1}$

$$= 1$$

and $2^n - 1 = 2^1 - 1$

$$= 1$$

so true for $n = 1$

Assume true for $n = k$ and consider $n = k + 1$:

$$\sum_{r=1}^{k+1} 2^{r-1} = \sum_{r=1}^{k} 2^{r-1} + 2^{k+1-1}$$

$$= 2^k - 1 + 2^k$$

$$= 2(2^k) - 1$$

$$= 2^{k+1} - 1$$

So true for $n = k + 1$

The statement is true for $n = 1$ and by assuming it is true for $n = k$ it is shown to be true for $n = k + 1$, therefore, by mathematical induction, it is true for all $n \in \mathbb{N}$

e When $n = 1$, $\sum_{r=1}^{n} 3^{r-1} = 3^{1-1}$

$$= 1$$

and $\frac{1}{2}(3^n - 1) = \frac{1}{2}(3^1 - 1)$

$$= 1$$

so true for $n = 1$

Assume true for $n = k$ and consider $n = k + 1$:

$$\sum_{r=1}^{k+1} 3^{r-1} = \sum_{r=1}^{k} 3^{r-1} + 3^{k+1-1}$$

$$= \frac{1}{2}(3^k - 1) + 3^k$$

$$= \frac{1}{2}(3^k - 1 + 2(3^k))$$

$$= \frac{1}{2}(3(3^k) - 1)$$

$$= \frac{1}{2}(3^{k+1} - 1)$$

So true for $n = k + 1$

The statement is true for $n = 1$ and by assuming it is true for $n = k$ it is shown to be true for $n = k + 1$, therefore, by mathematical induction, it is true for all $n \in \mathbb{N}$

f When $n = 1$, $\sum_{r=1}^{n} \left(\frac{1}{2}\right)^r = \left(\frac{1}{2}\right)^1$

$$= \frac{1}{2}$$

and $1 - \left(\frac{1}{2}\right)^n = 1 - \left(\frac{1}{2}\right)^1$

$$= \frac{1}{2}$$

so true for $n = 1$

Assume true for $n = k$ and consider $n = k + 1$:

$$\sum_{r=1}^{k+1} \left(\frac{1}{2}\right)^r = \sum_{r=1}^{k} \left(\frac{1}{2}\right)^r + \left(\frac{1}{2}\right)^{k+1}$$

$$= 1 - \left(\frac{1}{2}\right)^k + \left(\frac{1}{2}\right)^{k+1}$$

$$= 1 - \left(\frac{1}{2}\right)^k \left(1 - \frac{1}{2}\right)$$

$$= 1 - \left(\frac{1}{2}\right)^k \left(\frac{1}{2}\right)$$

$$= 1 - \left(\frac{1}{2}\right)^{k+1}$$

So true for $n = k + 1$

The statement is true for $n = 1$ and by assuming it is true for $n = k$ it is shown to be true for $n = k + 1$, therefore, by mathematical induction, it is true for all $n \in \mathbb{N}$

7 When $n = 1$, $\sum_{r=1}^{n} \frac{1}{r(r+1)} = \frac{1}{1(1+1)}$

$$= \frac{1}{2}$$

and $\frac{n}{n+1} = \frac{1}{1+1}$

$$= \frac{1}{2}$$

so true for $n = 1$

Assume true for $n = k$ and consider $n = k + 1$:

$$\sum_{r=1}^{k+1} \frac{1}{r(r+1)} = \sum_{r=1}^{k} \frac{1}{r(r+1)} + \frac{1}{(k+1)(k+1+1)}$$

$$= \frac{k}{k+1} + \frac{1}{(k+1)(k+2)}$$

$$= \frac{k(k+2)+1}{(k+1)(k+2)}$$

$$= \frac{k^2+2k+1}{(k+1)(k+2)}$$

$$= \frac{(k+1)^2}{(k+1)(k+2)}$$

$$= \frac{k+1}{k+2}$$

$$= \frac{k+1}{k+1+1}$$

So true for $n = k + 1$

The statement is true for $n = 1$ and by assuming it is true for $n = k$ it is shown to be true for $n = k + 1$, therefore, by mathematical induction, it is true for all $n \in \mathbb{N}$

8 When $n = 2$, $\sum_{r=2}^{n} \frac{1}{r(r-1)} = \frac{1}{2(2-1)}$

$$= \frac{1}{2}$$

and $\frac{n-1}{n} = \frac{2-1}{2}$

$$= \frac{1}{2}$$

so true for $n = 2$

Assume true for $n = k$ and consider $n = k + 1$:

$$\sum_{r=2}^{k+1} \frac{1}{r(r-1)} = \sum_{r=2}^{k} \frac{1}{r(r-1)} + \frac{1}{(k+1)(k+1-1)}$$

$$= \frac{k-1}{k} + \frac{1}{k(k+1)}$$

$$= \frac{(k-1)(k+1)+1}{k(k+1)}$$

$$= \frac{k^2-1+1}{k(k+1)}$$

$$= \frac{k^2}{k(k+1)}$$

$$= \frac{k}{k+1}$$

$$= \frac{(k+1)-1}{k+1}$$

So true for $n = k + 1$

The statement is true for $n = 2$ and by assuming it is true for $n = k$ it is shown to be true for $n = k + 1$, therefore, by mathematical induction, it is true for all $n \in \mathbb{N}$

9 When $n = 1$, $\sum_{r=1}^{n} \frac{1}{r^2+2r} = \frac{1}{1^2+2}$

$$= \frac{1}{3}$$

and $\frac{n(3n+5)}{4(n+1)(n+2)} = \frac{1(3+5)}{4(1+1)(1+2)}$

$$= \frac{8}{24}$$

$$= \frac{1}{3}$$

so true for $n = 1$

Assume true for $n = k$ and consider $n = k + 1$:

$$\sum_{r=1}^{k+1} \frac{1}{r^2+2r} = \sum_{r=1}^{k} \frac{1}{r^2+2r} + \frac{1}{(k+1)^2+2(k+1)}$$

$$= \frac{k(3k+5)}{4(k+1)(k+2)} + \frac{1}{k^2+2k+1+2k+2}$$

$$= \frac{k(3k+5)}{4(k+1)(k+2)} + \frac{1}{k^2+4k+3}$$

$$= \frac{k(3k+5)}{4(k+1)(k+2)} + \frac{1}{(k+3)(k+1)}$$
$$= \frac{k(3k+5)(k+3)+4(k+2)}{4(k+1)(k+2)(k+3)}$$
$$= \frac{k(3k^2+14k+15)+4(k+2)}{4(k+1)(k+2)(k+3)}$$
$$= \frac{3k^3+14k^2+15k+4k+8}{4(k+1)(k+2)(k+3)}$$
$$= \frac{(k+1)(3k^2+11k+8)}{4(k+1)(k+2)(k+3)}$$
$$= \frac{(3k+8)(k+1)^2}{4(k+1)(k+2)(k+3)}$$
$$= \frac{(k+1)(3k+8)}{4(k+2)(k+3)}$$
$$= \frac{(k+1)(3(k+1)+5)}{4(k+1+1)(k+1+2)}$$

So true for $n = k + 1$

The statement is true for $n = 1$ and by assuming it is true for $n = k$ it is shown to be true for $n = k + 1$, therefore, by mathematical induction, it is true for all $n \in \mathbb{N}$

Exercise 2.4B

1 a When $n = 1$, $n^2 + 3n = 1 + 3$
$= 4$
$= 2 \times 2$

so true for $n = 1$

Assume true for $n = k$ and consider $n = k + 1$:

$(k+1)^2 + 3(k+1) = k^2 + 2k + 1 + 3k + 3$
$= k^2 + 5k + 4$
$= (k^2 + 3k) + 2k + 4$
$= 2A + 2k + 4$ since $k^2 + 3k$ divisible by 2
$= 2(A + k + 2)$

So true for $n = k + 1$

The statement is true for $n = 1$ and by assuming it is true for $n = k$ it is shown to be true for $n = k + 1$, therefore, by mathematical induction, it is true for all $n \in \mathbb{N}$

b When $n = 1$, $5n^2 - n = 5 - 1$
$= 4$
$= 2 \times 2$

so true for $n = 1$

Assume true for $n = k$ and consider $n = k + 1$:

$5(k+1)^2 - (k+1) = 5k^2 + 10k + 5 - k - 1$
$= 5k^2 + 9k + 4$
$= (5k^2 - k) + 10k + 4$
$= 2A + 10k + 4$ since $5k^2 - k$ divisible by 2
$= 2(A + 5k + 2)$

So true for $n = k + 1$

The statement is true for $n = 1$ and by assuming it is true for $n = k$ it is shown to be true for $n = k + 1$, therefore, by mathematical induction, it is true for all $n \in \mathbb{N}$

c When $n = 1$, $8n^3 + 4n = 8 + 4$
$= 12$
$= 12 \times 1$

so true for $n = 1$

Assume true for $n = k$ and consider $n = k + 1$:

$8(k+1)^3 + 4(k+1) = 8(k^3 + 3k^2 + 3k + 1) + 4k + 4$
$= 8k^3 + 24k^2 + 24k + 8 + 4k + 4$
$= 8k^3 + 24k^2 + 28k + 12$
$= (8k^3 + 4k) + 24k^2 + 24k + 12$
$= 12A + 12(2k^2 + 2k + 1)$ since $8k^3 + 4k$ divisible by 12
$= 12(A + 2k^2 + 2k + 1)$

So true for $n = k + 1$

The statement is true for $n = 1$ and by assuming it is true for $n = k$ it is shown to be true for $n = k + 1$, therefore, by mathematical induction, it is true for all $n \in \mathbb{N}$

d When $n = 1$, $11n^3 + 4n = 11 + 4$
$= 15$
$= 3 \times 5$

so true for $n = 1$

Assume true for $n = k$ and consider $n = k + 1$:

$11(k+1)^3 + 4(k+1) = 11(k^3 + 3k^2 + 3k + 1) + 4k + 4$
$= 11k^3 + 33k^2 + 33k + 11 + 4k + 4$
$= 11k^3 + 33k^2 + 37k + 15$
$= (11k^3 + 4k) + 33k^2 + 33k + 15$
$= 3A + 3(11k^2 + 11k + 5)$ since $11k^3 + 4k$ divisible by 3
$= 3(A + 11k^2 + 11k + 5)$

So true for $n = k + 1$

The statement is true for $n = 1$ and by assuming it is true for $n = k$ it is shown to be true for $n = k + 1$, therefore, by mathematical induction, it is true for all $n \in \mathbb{N}$

2 When $n = 1$, $7n^2 + 25n - 4$
$= 7 + 25 - 4$
$= 28$
$= 2 \times 14$

so true for $n = 1$

Assume true for $n = k$ and consider $n = k + 1$:

$7(k+1)^2 + 25(k+1) - 4 = 7(k^2 + 2k + 1) + 25k + 25 - 4$
$= 7k^2 + 14k + 7 + 25k + 25 - 4$
$= 7k^2 + 39k + 28$
$= (7k^2 + 25k - 4) + 14k + 32$
$= 2A + 2(7k + 16)$ since $7k^2 + 25k - 4$ divisible by 2
$= 2(A + 7k + 16)$

So true for $n = k + 1$

The statement is true for $n = 1$ and by assuming it is true for $n = k$ it is shown to be true for $n = k + 1$, therefore, by mathematical induction, it is true for all $n \in \mathbb{N}$

3 When $n = 2$, $n^3 - n = 2^3 - 2$
$= 6$
$= 3 \times 2$

so true for $n = 2$

Assume true for $n = k$ and consider $n = k + 1$:

$(k+1)^3 - (k+1) = (k^3 + 3k^2 + 3k + 1) - k - 1$
$= k^3 + 3k^2 + 2k$
$= (k^3 - k) + 3k^2 + 3k$
$= 3A + 3(k^2 + k)$ since $k^3 - k$ divisible by 3
$= 3(A + k^2 + k)$

So true for $n = k + 1$

The statement is true for $n = 2$ and by assuming it is true for $n = k$ it is shown to be true for $n = k + 1$, therefore, by mathematical induction, it is true for all $n \in \mathbb{N}$, $n \geq 2$

4 When $n = 1$, $10n^3 + 3n^2 + 5n - 6$
$= 10 + 3 + 5 - 6$
$= 12$
$= 6 \times 2$

so true for $n = 1$

Assume true for $n = k$ and consider $n = k + 1$:

$10(k+1)^3 + 3(k+1)^2 + 5(k+1) - 6$
$= 10(k^3 + 3k^2 + 3k + 1) + 3(k^2 + 2k + 1) + 5(k+1) - 6$
$= 10k^3 + 30k^2 + 30k + 10 + 3k^2 + 6k + 3 + 5k + 5 - 6$
$= 10k^3 + 33k^2 + 41k + 12$
$= (10k^3 + 3k^2 + 5k - 6) + 30k^2 + 36k + 18$
$= 6A + 30k^2 + 36k + 18$ since $10k^3 + 3k^2 + 5k - 6$ divisible by 6
$= 6(A + 5k^2 + 6k + 3)$

So true for $n = k + 1$

The statement is true for $n = 1$ and by assuming it is true for $n = k$ it is shown to be true for $n = k+1$, therefore, by mathematical induction, it is true for all $n \in \mathbb{N}$

5 a When $n = 1$, $6^n + 9 = 6^1 + 9$
$= 15$
$= 5 \times 3$

so true for $n = 1$

Assume true for $n = k$ and consider $n = k+1$:

$6^{k+1} + 9 = 6(6^k) + 9$
$= (6^k + 9) + 5(6^k)$
$= 5A + 5(6^k)$ since $6^k + 9$ divisible by 5
$= 5(A + 6^k)$

So true for $n = k+1$

The statement is true for $n = 1$ and by assuming it is true for $n = k$ it is shown to be true for $n = k+1$, therefore, by mathematical induction, it is true for all $n \in \mathbb{N}$

b When $n = 1$, $3^{2n} - 1 = 3^2 - 1$
$= 8$
$= 8 \times 1$

so true for $n = 1$

Assume true for $n = k$ and consider $n = k+1$:

$3^{2(k+1)} - 1 = 3^{2k}3^2 - 1$
$= 9(3^{2k}) - 1$
$= (3^{2k} - 1) + 8(3^{2k})$
$= 8A + 8(3^{2k})$ since $3^{2k} - 1$ divisible by 8
$= 8(A + 3^{2k})$

So true for $n = k+1$

The statement is true for $n = 1$ and by assuming it is true for $n = k$ it is shown to be true for $n = k+1$, therefore, by mathematical induction, it is true for all $n \in \mathbb{N}$

c When $n = 1$, $2^{3n+1} - 2 = 2^4 - 2$
$= 14$
$= 7 \times 2$

so true for $n = 1$

Assume true for $n = k$ and consider $n = k+1$:

$2^{3(k+1)+1} - 2 = 2^{3k+4} - 2$
$= 2^3(2^{3k+1}) - 2$
$= 8(2^{3k+1}) - 2$
$= (2^{3k+1} - 2) + 7(2^{3k+1})$
$= 7A + 7(2^{3k+1})$ since $3^{2k} - 1$ divisible by 7
$= 7(A + 2^{3k+1})$

So true for $n = k+1$

The statement is true for $n = 1$ and by assuming it is true for $n = k$ it is shown to be true for $n = k+1$, therefore, by mathematical induction, it is true for all $n \in \mathbb{N}$

6 When $n = 1$, $5^n - 4n + 3 = 5^1 - 4 + 3$
$= 4$
$= 4 \times 1$

so true for $n = 1$

Assume true for $n = k$ and consider $n = k+1$:

$5^{k+1} - 4(k+1) + 3 = 5(5^k) - 4k - 4 + 3$
$= (5^k - 4k + 3) + 4(5^k) - 4$
$= 4A + 4(5^k - 1)$ since $5^k - 4k + 3$ divisible by 4
$= 4(A + 5^k - 1)$

So true for $n = k+1$

The statement is true for $n = 1$ and by assuming it is true for $n = k$ it is shown to be true for $n = k+1$, therefore, by mathematical induction, it is true for all $n \in \mathbb{N}$

7 When $n = 1$, $3^n + 2n + 7 = 3^1 + 2 + 7$
$= 12$
$= 4 \times 3$

so true for $n = 1$

Assume true for $n = k$ and consider $n = k+1$:

$3^{k+1} + 2(k+1) + 7 = 3(3^k) + 2k + 9$
$= 3(3^k + 2k + 7) - 4k - 12$
$= 3(4A) - 4(k+3)$ since $3^{k+1} + 2(k+1) + 7$ divisible by 4
$= 4(3A - k - 3)$

So true for $n = k+1$

The statement is true for $n = 1$ and by assuming it is true for $n = k$ it is shown to be true for $n = k+1$, therefore, by mathematical induction, it is true for all $n \in \mathbb{N}$

8 When $n = 1$, $7^n - 3n + 5 = 7 - 3 + 5$
$= 9$
$= 3 \times 3$

so true for $n = 1$

Assume true for $n = k$ and consider $n = k+1$:

$7^{k+1} - 3(k+1) + 5 = 7(7^k) - 3k + 2$
$= (7^k - 3k + 5) + 6(7^k) - 3$
$= 3A + 3(2(7^k) - 1)$ since $7^k - 3k + 5$ divisible by 3
$= 3(A + 2(7^k) - 1)$

So true for $n = k+1$

The statement is true for $n = 1$ and by assuming it is true for $n = k$ it is shown to be true for $n = k+1$, therefore, by mathematical induction, it is true for all $n \in \mathbb{N}$

9 a When $n = 1$, $\sum_{r=n+1}^{2n} r^2 = \sum_{r=2}^{2} r^2$
$= 2^2$
$= 4$

and $\frac{1}{6}n(2n+1)(7n+1) = \frac{1}{6}(3)(8)$
$= 4$

so true for $n = 1$

Assume true for $n = k$ and consider $n = k+1$:

$$\sum_{r=(k+1)+1}^{2(k+1)} r^2 = \sum_{r=k+1}^{2k} r^2 + (2k+1)^2 + (2k+2)^2 - (k+1)^2$$
$$= \frac{1}{6}k(2k+1)(7k+1) + 4k^2 + 4k + 1 + 4k^2 + 8k + 4 - k^2 - 2k - 1$$
$$= \frac{1}{6}(14k^3 + 9k^2 + k + 42k^2 + 60k + 24)$$
$$= \frac{1}{6}(14k^3 + 51k^2 + 61k + 24)$$
$$= \frac{1}{6}(k+1)(2k+3)(7k+8)$$
$$= \frac{1}{6}(k+1)(2(k+1)+1)(7(k+1)+1)$$

So true for $n = k+1$

The statement is true for $n = 1$ and by assuming it is true for $n = k$ it is shown to be true for $n = k+1$, therefore, by mathematical induction, it is true for all $n \in \mathbb{N}$

b When $n = 1$, $\sum_{r=n}^{2n} r^3 = \sum_{r=1}^{2} r^3$
$= 1 + 8$
$= 9$

and $\frac{3}{4}n^2(5n+1)(n+1) = \frac{3}{4}(1)(6)(2)$
$= 9$

so true for $n = 1$

Assume true for $n = k$ and consider $n = k+1$:

$$\sum_{r=(k+1)}^{2(k+1)} r^3 = \sum_{r=k}^{2k} r^3 + (2k+1)^3 + (2k+2)^3 - k^3$$

$$= \frac{3}{4}k^2(5k+1)(k+1) + 8k^3 + 12k^2 + 6k + 1$$

$$+ 8k^3 + 24k^2 + 24k + 8 - k^3$$

$$= \frac{3}{4}(5k^4 + 6k^3 + k^2 + 20k^3 + 48k^2 + 40k + 12)$$

$$= \frac{3}{4}(5k^4 + 26k^3 + 49k^2 + 40k + 12)$$

$$= \frac{3}{4}(k+1)(5k^3 + 21k^2 + 28k + 12)$$

$$= \frac{3}{4}(k+1)(k+1)(k+2)(5k+6)$$

$$= \frac{3}{4}(k+1)^2(k+1+1)(5(k+1)+1)$$

So true for $n = k+1$

The statement is true for $n = 1$ and by assuming it is true for $n = k$ it is shown to be true for $n = k+1$, therefore, by mathematical induction, it is true for all $n \in \mathbb{N}$

c When $n = 1$, $8^n - 5^n = 8^1 - 5^1$

$= 3$

$= 3 \times 1$

so true for $n = 1$

Assume true for $n = k$ and consider $n = k+1$:

$8^{k+1} - 5^{k+1} = 8(8^k) - 5(5^k)$

$= 5(8^k - 5^k) + 3(8^k)$

$= 5(3A) + 3(8^k)$ since $8^k - 5^k$ divisible by 3

$= 3(5A + 8^k)$

So true for $n = k+1$

The statement is true for $n = 1$ and by assuming it is true for $n = k$ it is shown to be true for $n = k+1$, therefore, by mathematical induction, it is true for all $n \in \mathbb{N}$

Exercise 2.5A

1 a $1 + 2x + 2x^2 + \frac{4x^3}{3} + \frac{2x^4}{3} + \ldots$

b $1 - x + \frac{1}{2}x^2 - \frac{x^3}{6} + \frac{x^4}{24} + \ldots$

c $1 - 3x + \frac{9}{2}x^2 - \frac{9x^3}{2} + \frac{27x^4}{8} + \ldots$

d $1 + \frac{x}{2} + \frac{1}{8}x^2 + \frac{x^3}{48} + \frac{x^4}{384} + \ldots$

e $1 - \frac{x}{3} + \frac{1}{18}x^2 + \frac{x^3}{162} + \frac{x^4}{1944} + \ldots$

2 a i $-x - \frac{x^2}{2} - \frac{x^3}{3} - \frac{x^4}{4} - \frac{x^5}{5} \ldots - \frac{x^r}{r} + \ldots$

ii $-1 \le x < 1$

b i $2x - 2x^2 + \frac{8x^3}{3} - 4x^4 + \frac{32x^5}{5} \ldots + \frac{(2x)^r}{r} + \ldots$

ii $-\frac{1}{2} < x \le \frac{1}{2}$

c i $-3x - \frac{9x^2}{2} - 9x^3 - \frac{81x^4}{4} - \frac{243x^5}{5} \ldots + \frac{(-3x)^r}{r} + \ldots$

ii $-\frac{1}{3} \le x < \frac{1}{3}$

d i $\frac{x}{2} - \frac{x^2}{8} + \frac{x^3}{24} - \frac{x^4}{64} + \frac{x^5}{180} \ldots + \frac{x^r}{(2^r)r} + \ldots$

ii $-2 < x \le 2$

e i $\frac{-x}{3} - \frac{x^2}{18} - \frac{x^3}{81} - \frac{x^4}{324} - \frac{x^5}{1215} \ldots - \frac{x^r}{(3)r} + \ldots$

ii $-3 \le x < 3$

3 a $2x - \frac{4x^3}{3} + \frac{4x^5}{15} - \frac{8x^7}{315} + \ldots + \frac{(-1)^{r+1}(2x)^{2r-1}}{(2r-1)!} + \ldots$

b $\frac{x}{2} - \frac{x^3}{48} + \frac{x^5}{3840} - \frac{x^7}{645\,120} + \ldots + \frac{(-1)^{r+1}(x)^{2r-1}}{2^{2r-1}(2r-1)!} + \ldots$

c $-x + \frac{x^3}{3!} - \frac{x^5}{5!} + \frac{x^7}{7!} + \ldots + \frac{(-1)^r(x)^{2r-1}}{(2r-1)!} + \ldots$

d $-3x + \frac{9x^3}{2} - \frac{81x^5}{40} + \frac{243x^7}{560} + \ldots + \frac{(-1)^r(x)^{2r-1}}{3^{2r-1}(2r-1)!} + \ldots$

e $\frac{3x}{2} - \frac{9x^3}{16} + \frac{81x^5}{1280} - \frac{243x^7}{71\,680} + \ldots + \frac{(-1)^{r+1}(3x)^{2r-1}}{2^{2r-1}(2r-1)!} + \ldots$

4 a $1 - 8x^2 + \frac{32x^4}{3} - \frac{256x^6}{45} + \ldots + \frac{(-1)^{r+1}(x)^{2r}}{4^{2r}(2r)!} + \ldots$

b $1 - \frac{x^2}{18} + \frac{x^4}{1944} - \frac{x^6}{524\,880} + \ldots + \frac{(-1)^{r+1}(x)^{2r}}{3^{2r}(2r)!} + \ldots$

c $1 - \frac{x^2}{2!} + \frac{x^4}{4!} - \frac{x^6}{6!} + \ldots + \frac{(-1)^{r+1}(x)^{2r}}{(2r)!} + \ldots$

d $1 - 2x^2 + \frac{2x^4}{3} - \frac{4x^6}{45} + \ldots + \frac{(-1)^{r+1}(x)^{2r}}{2^{2r}(2r)!} + \ldots$

e $1 - \frac{x^2}{8} + \frac{x^4}{384} - \frac{x^6}{46\,080} + \ldots + \frac{(-1)^{r+1}(x)^{2r}}{2^{2r}(2r)!} + \ldots$

5 a i $1 + \frac{x}{2} - \frac{x^2}{8} + \frac{x^3}{16} + \ldots$

ii $-1 < x < 1$

b i $1 - \frac{x}{2} - \frac{x^2}{8} - \frac{x^3}{16} + \ldots$

ii $-1 < x < 1$

c i $1 + \frac{x}{3} - \frac{x^2}{9} + \frac{5x^3}{81} + \ldots$

ii $-1 < x < 1$

d i $1 + 3x + \frac{3x^2}{2} - \frac{x^3}{2} + \ldots$

ii $-\frac{1}{2} < x < \frac{1}{2}$

e i $1 - \frac{15x}{2} + \frac{135x^2}{8} - \frac{135x^3}{16} + \ldots$

ii $-\frac{1}{3} < x < \frac{1}{3}$

f i $1 + x + x^2 + x^3 + \ldots$

ii $-1 < x < 1$

g i $3 - 3x + 3x^2 - 3x^3 + \ldots$

ii $-1 < x < 1$

6 a $x^2 - \frac{x^6}{3!} + \frac{x^{10}}{5!} - \frac{x^{14}}{7!} + \ldots$

b $x^{\frac{1}{2}} - \frac{x^{\frac{3}{2}}}{3!} + \frac{x^{\frac{5}{2}}}{5!} - \frac{x^{\frac{7}{2}}}{7!} + \ldots$

c $1 - \frac{x^4}{2!} + \frac{x^8}{4!} - \frac{x^{12}}{6!} + \ldots$

d $1 - \frac{x}{2!} + \frac{x^2}{4!} - \frac{x^3}{6!} + \ldots$

Exercise 2.5B

1 a $x + x^2 + \frac{x^3}{3} + \ldots$

b $1 + x + \frac{x^2}{2} - \ldots$

c $1 + x - \frac{x^2}{2} + \ldots$

d $1 - x^2 + \frac{x^4}{3} + ...$

2 $x - \frac{x^3}{2!} + \frac{x^5}{4!} - ...$ Valid for all x.

3 a $1 + \frac{3x}{2} - \frac{9x^2}{8} + ...$ Valid when $-\frac{1}{3} < x < \frac{1}{3}$

b $-2x - \frac{2x^3}{3} - \frac{2x^5}{5} + ...$ Valid when $-1 < x < 1$

c $x + x^2 + x^3 + ...$ Valid when $-1 < x < 1$

d $1 - x - x^2 + ...$ Valid when $-1 < x < 1$

4 a i $1 + \frac{1}{2x} - \frac{1}{8x^2} + \frac{1}{16x^3} + ...$

ii $\frac{1}{x} - \frac{1}{2x^2} + \frac{1}{3x^3} - \frac{1}{4x^4} ...$

b $x \ge 1$ or $x < -1$

5 a i $x + \frac{x^3}{3!} + \frac{x^5}{5!} + ...$

ii $1 - \frac{x^2}{2} + \frac{x^4}{24}$

iii $\frac{5}{6} - \frac{x^2}{2} + \frac{x^4}{6} - \frac{x^6}{16} + \frac{13x^4}{1152} - \frac{x^{10}}{1152} + \frac{x^{12}}{41472}$

b The range of values of x for which $\ln(1 + \cos x)$ is valid is $-1 < \cos x \le 1$
or $\cos^{-1}(-1) < x \le \cos^{-1}(1)$
or $-\pi < x \le 0$

6 a 0.5488 (4sf)

b i 1.2528 (5 sf)

ii The series is only valid for $-1 < x \le 1$
$x = 2.5$ is outside this range, so you have to make x in this range.

7 a $f(x) = e^{2x}$ so $f(0) = e^0 = 1$
$f'(x) = 2e^{2x}$ so $f'(0) = 2e^0 = 2$
$f''(x) = 4e^{2x}$ so $f''(0) = 4e^0 = 4$
$f'''(x) = 8e^{2x}$ so $f'''(0) = 8e^0 = 8$
$f^{iv}(x) = 16e^{2x}$ so $f^{iv}(0) = 16e^0 = 16$

Hence $e^{2x} \equiv 1 + x(2) + \frac{x^2}{2!}(4) + \frac{x^3}{3!}(8) + \frac{x^4}{4!}(16) + ...$

$\equiv 1 + 2x + 2x^2 + \frac{4x^3}{3} + \frac{2x^4}{3} + ...$

b $e^x = 1 + x + \frac{x^2}{2!} + \frac{x^3}{3!} + ...$

so $e^{2x} \equiv 1 + 2x + \frac{4x^2}{2!} + \frac{8x^3}{3!} + \frac{16x^3}{4!} + ...$

$\equiv 1 + 2x + 2x^2 + \frac{4x^3}{3} + \frac{2x^4}{3} + ...$

c $\sqrt{e} \equiv 1 + 2\left(\frac{1}{4}\right) + 2\left(\frac{1}{4}\right) + \frac{4\left(\frac{1}{4}\right)^3}{3} + \frac{2\left(\frac{1}{4}\right)^4}{3} + ...$

$\approx 1 + 0.5 + 0.125 + 0.02083333333... + 0.002604166666...$
$\approx 1.6484375...$

8 a $f(x) = \sin x$ so $f(0) = \sin 0 = 0$
$f'(x) = \cos x$ so $f'(0) = \cos 0 = 1$
$f''(x) = -\sin x$ so $f''(0) = -\sin 0 = 0$
$f'''(x) = -\cos x$ so $f'''(0) = -\cos 0 = 1$
$f^{iv}(x) = \sin x$ so $f^{iv}(0) = \sin 0 = 0$

Hence $\sin x \equiv 0 + x(1) + \frac{x^2}{2!}(0) + \frac{x^3}{3!}(-1) + \frac{x^4}{4!}(0)$

$+ \frac{x^5}{5!}(-1) + \frac{x^6}{6!}(0) + \frac{x^7}{7!}(-1) + ...$

$\equiv x - \frac{x^3}{3!} + \frac{x^5}{5!} - \frac{x^7}{7!} + ...$

b Hence $\sin\frac{\pi}{2} \approx \left(\frac{\pi}{2}\right) - \frac{\left(\frac{\pi}{2}\right)^3}{3!} + \frac{\left(\frac{\pi}{2}\right)^5}{5!} - \frac{\left(\frac{\pi}{2}\right)^7}{7!} + ...$

$\approx 0.99984...$

Hence percentage error between 0.99984 and the exact value of $1 = \frac{(0.99984 - 1) \times 100}{1}$

$\approx -0.016\%$

Review exercise 2

1 a Sum of roots = −5
Sum of the products in pairs = 2
Product of the roots = −1

b Sum of roots = −9
Sum of the products in pairs = −11
Product of the roots = 8

2 $\left(\frac{y}{2}\right)^3 - 4\left(\frac{y}{2}\right)^2 + 6\left(\frac{y}{2}\right) - 2 = \frac{y^3}{8} - y^2 + 3y - 2$

3 a 102

b $510y^3 - 1428y^2 + 1291y - 371 = 0$

4 a $\pm i, -2$

b $a = 2$
$b = 1$

5 a −1

b i $y^3 + 2y^2 - 2y - 2 = 0$ **ii** 2

6 a −3 **b** $-\frac{3}{2}$ **c** 6

7 a $x \le 0$ or $x \ge 5$

b $0 < x < 4$

c $-3 < x < -2$

d $x \le \frac{5}{3}$ or $x > \frac{5}{2}$

e $2 < x < 3$ or $x < 1$

f $-15 \le x < -4$

8 a $x < -\frac{3}{2}$ or $0 < x < \frac{2}{3}$

b $-6 < x < -1$ or $x > 8$

c $-\frac{8}{3} \le x \le 1$ or $\frac{5}{2} \le x \le 8$

d $-5 \le x \le -1$ or $3 \le x \le 7$

9 a $x \le 1$ or $x \ge 3$

b $-3 < x \le 7$

10 a $27 + 64 + 125 + 216 + 343 + 512 + 729 + 1000 = 3016$

b $n^2 + (n+1)^2 + (n+2)^2 + (n+3)^2 + (n+4)^2$

c $6 + 14 + 24 + ... + ((n-1)^2 + 5 \times (n-1)) + (n^2 + 5n)$

d $-\frac{1}{2} - 1 + \infty + ... + . \frac{1}{(n-4)} + \frac{1}{n-3}$

e 56

f $2(n-1)(3(n-1) - 2) + 2(n)(3n-2) + 2(n+1)(3(n+1) - 2)$
$+ 2(n+2)(3(n+2) - 2) + 2(n+3)(3(n+3) - 2)$
$= 2(n-1)(3n-5) + 2n(3n-2) + 2(n+1)(3n+1)$
$+ 2(n+2)(3n+4) + 2(n+3)(3n+7)$

11 a $n(2-n)$

b $\frac{n(n+1)(2n-5)}{3}$

c $\frac{n(n+1)(n^2+n+10)}{4}$

12 a $(2n+1)^2(n+1)^2 - \frac{n^2(n-1)^2}{4} + 2n + 4$

b When $n = 6$, $\sum_{n}^{2n+1} (r^3 + 2) = 8072$

13 a $\frac{1}{2} - \frac{1}{n+2}$

b $\frac{1}{3} - \frac{1}{n+3}$

14 1

15 $t^2(t+1)^2 - t^2(t-1)^2 \equiv t^2(t^2+2t+1) - t^2(t^2-2t+1)$
$\equiv t^4 + 2t^3 + t^2 - t^4 + 2t^3 - t^2$
$\equiv 4t^3$
$\Sigma\, t^2(t+1)^2 - t^2(t-1)^2$
$\equiv [1^2(2)^2 - 1^2(0)^2] + [2^2(3)^2 - 2^2(1)^2]$
$+ [3^2(4)^2 - 3^2(2)^2] + \ldots + (n-1)^2(n)^2 - (n-1)^2(n-2)^2$
$+ n^2(n+1)^2 - n^2(n-1)^2$
$\equiv (4+0) + (36-4) + (144-36) + \ldots\ldots + n^2(n+1)^2$

Hence $\Sigma 4t^3 \equiv n^2(n+1)^2$

So $\sum_1^n t^3 = \frac{n^2(n+1)^2}{4}$

16 $\sum_1^k \left(\frac{1}{n^2} - \frac{1}{(n+1)^2}\right) = \frac{1}{1} - \frac{1}{4}$
$+ \frac{1}{4} - \frac{1}{9}$
$+ \frac{1}{9} - \frac{1}{16}$
$\downarrow$
$\downarrow$
$+ \frac{1}{(k-1)^2} - \frac{1}{(k)^2}$
$+ \frac{1}{(k)^2} - \frac{1}{(k+1)^2}$

Thus $\sum_1^k \left(\frac{1}{n^2} - \frac{1}{(n+1)^2}\right) = \frac{1}{1} - \frac{1}{(k+1)^2} =$
$\frac{(k+1)^2 - 1}{(k+1)^2} = \frac{k^2+2k}{(k+1)^2} = \frac{k(k+2)}{(k+1)^2}$

17 When $n = 1$, $3^{2n+1} + 1 = 3^{2+1} + 1$
$= 28$
$= 4 \times 7$
so true for $n = 1$
Assume true for $n = k$ and consider $n = k + 1$:
$3^{2(k+1)+1} + 1 = 3^{2k+3} + 1$
$= 3^2 3^{2k+1} + 1$
$= 9(3^{2k+1}) + 1$
$= (3^{2k+1} + 1) + 8(3^{2k+1})$
$= 4A + 8(3^{2k+1})$ since true for $n = k$
$= 4(A + 2(3^{2k+1}))$
So true for $n = k + 1$
The statement is true for $n = 1$ and by assuming it is true for $n = k$ it is shown to be true for $n = k + 1$, therefore, by mathematical induction, it is true for all $n \in \mathbb{N}$

18 When $n = 1$, $2^{2n} - 3n + 2 = 2^2 - 3 + 2$
$= 3$
so true for $n = 1$
Assume true for $n = k$ and consider $n = k + 1$:
$2^{2(k+1)} - 3(k+1) + 2 = 2^{2k+2} - 3k - 3 + 2$
$= 2^2 2^{2k} - 3k + 2 - 3$
$= (2^{2k} - 3k + 2) + 3(2^{2k}) - 3$
$= 3A + 3(2^{2k} - 1)$ since true for $n = k$
$= 3(A + 2^{2k} - 1)$
So true for $n = k + 1$
The statement is true for $n = 1$ and by assuming it is true for $n = k$ it is shown to be true for $n = k + 1$, therefore, by mathematical induction, it is true for all $n \in \mathbb{N}$

19 Let $n = 1$, then $6^n - 1 = 5$ which is divisible by 5
So true for $n = 1$
Assume true for $n = k$ and let $n = k + 1$
$6^{k+1} - 1 = 6(6^k) - 1$
$= 5(6^k) + 6^k - 1$
$= 5(6^k) + 5A$ for some integer A since $6^k - 1$ is divisible by 5
$= 5(6^k + A)$
So divisible by 5
True for $n = 1$ and true for $n = k$ implies true for $n = k + 2$ therefore true for all $n \in \mathbb{N}$

20 a The validity of $(1+x^2)^{-\frac{1}{2}}$ is $-1 < x^2 < 1$;
Since x^2 cannot be negative the validity becomes $-1 < x < 1$

b The validity of $\ln\left(1 - \frac{x}{2}\right)$ is $-1 < -\frac{x}{2} \le 1$ and hence $-2 < -x < 2$
Or, multiplying through by -1, $2 > x > -2$ or, as is more commonly written $-2 < x < 2$

21 a $x^2 - \frac{x^4}{2} + \frac{x^6}{3} - \ldots$

b $2x - \frac{2x^2}{3} - \frac{8x^3}{9} + \ldots$

22 $1 + \frac{x}{3} + \frac{x^2}{18} + \frac{x^3}{162} + \ldots$

Hence $\sqrt[3]{e} \equiv e^{\frac{1}{3}}$ so, substituting $x = 1$,

$\sqrt[3]{e} \approx 1 + \frac{1}{3} + \frac{1}{18} + \frac{1}{162} + \ldots \approx \frac{113}{81} \approx 1.3951$ (4 dp)

Assessment 2

1 $\alpha + \beta = -\frac{4}{3}$

2 $\sum \alpha\beta\delta = -1$

3 $x < -3$ or $x > 1$

4 a $1 - 15x + 150x^2 - 1250x^3 + \ldots$

b $|x| < \frac{1}{5}$

5 a $= 1 - \frac{x}{2} + \frac{x^2}{8} - \frac{x^3}{48} + \ldots$

b 1.05123

6 $2n(n+1)^2$

7 a $u^3 = 1$ **b** $u = 1$ **c** $x = 1$

8 $x < -\frac{1}{13}$

9 a $\frac{n}{2}(2n+3)(n-1)$ **b** 4200

10 a $\frac{n}{6}(4n^2 + 21n - 1)$ **b** 3035 **c** 2012.5

11 a $f(x) = \frac{1}{2}(x-2)(x-3)(x+4)$

b $g(x) = \frac{1}{2}\left(\frac{1}{2}x + 1 - 2\right)\left(\frac{1}{2}x + 1 - 3\right)\left(\frac{1}{2}x + 1 + 4\right)$
$= \frac{1}{16}(x-2)(x-4)(x+10)$

12 a $\alpha\beta\gamma = -1$
$\sum \alpha\beta = -\frac{3}{2}$

$$\sum \alpha = \frac{1}{2}$$

b $u^3 + 2u^2 - 5u + 2 = 0$

13 a $-2x^2 - 2x^4 - \frac{8}{3}x^6 - 4x^8 - \ldots$

b $-\frac{1}{\sqrt{2}} \le x < \frac{1}{\sqrt{2}}$

14 $\frac{x}{3} - \frac{x^2}{3} + \frac{13}{81}x^3 - \frac{4}{81}x^4$

15 a $f(1) = 0$

Therefore, by factor theorem, $x - 1$, is a factor.

b $2x^3 + 3x^2 - 8x + 3 = (x-1)(2x^2 + 5x - 3)$
$= (x-1)(2x-1)(x+3)$

c $x < -3$ or $\frac{1}{2} < x < 1$

16 When $n = 1$, $\sum_{r=1}^{n} 4^{r-1} = 1$

$\frac{1}{3}(4^n - 1) = \frac{1}{3}(4-1) = 1$

So true for $n = 1$

Assume true for $n = k$

$$\sum_{r=1}^{k+1} 4^{r-1} = 4^{(k+1)-1} + \sum_{r=1}^{k} 4^{r-1}$$
$$= 4^k + \frac{1}{3}(4^k - 1)$$
$$= \frac{4}{3}(4^k) - \frac{1}{3}$$
$$= \frac{1}{3}(4^{k+1} - 1)$$

True for $n = 1$ and assuming true for $n = k$ implies true for $n = k + 1$, hence true for all $n \in \mathbb{N}$

17 When $n = 1$, $4n^3 + 8n = 12$ which is divisible by 12 so true for $n = 1$

Assume true for $n = k$

$$4(k+1)^3 + 8(k+1)$$
$$= 4(k^3 + 3k^2 + 3k + 1) + 8(k+1)$$
$$= 4k^3 + 8k + 12k^2 + 12k + 4 + 8$$
$$= 4k^3 + 8k + 12k^2 + 12k + 12$$
$$= 4k^3 + 8k + 12(k^2 + k + 1)$$

So divisible by 12 since $4k^3 + 8k$ and $12(k^2 + k + 1)$ are divisible by 12

True for $n = 1$ and assuming true for $n = k$ implies true for $n = k + 1$, hence true for all $n \in \mathbb{N}$

18 When $n = 1$, $4^{2n+1} - 1 = 4^3 - 1 = 63$

$63 = 21 \times 3$ hence a multiple of 3

Assume true for $n = k$

$$4^{2(k+1)+1} - 1 = 4^{2k+3} - 1$$
$$= 4^2 4^{2k+1} - 1$$
$$= 16(4^{2k+1} - 1) + 15$$

So a multiple of 3 since $4^{2k+1} - 1$ assumed to be a multiple of 3 and $15 = 5 \times 3$ so a multiple of 3

True for $n = 1$ and assuming true for $n = k$ implies true for $n = k + 1$, hence true for all $n \in \mathbb{N}$

19 a

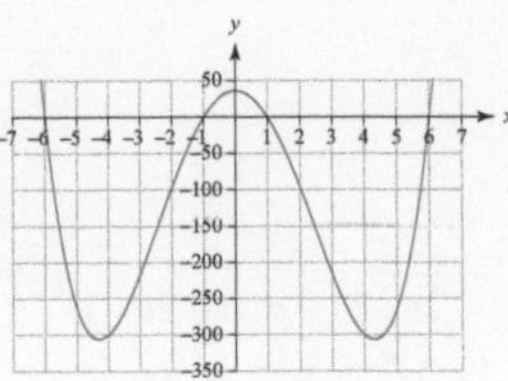

$-4 < x < 3$ or $x > 5$

b $x \le -13$ or $4 < x < 5$

20 a $\frac{1}{x} - \frac{1}{x+1} = \frac{x+1-x}{x(x+1)}$
$= \frac{1}{x(x+1)}$

b $1 - \frac{1}{n+1}$

c $\frac{99}{100}$

d $\frac{1}{100}$

21 a $\frac{1}{2}\left(\frac{1}{x-1} - \frac{1}{x+1}\right) = \frac{1}{2}\left(\frac{(x+1)-(x-1)}{(x-1)(x+1)}\right)$

$\frac{1}{2}\left(\frac{2}{x^2-1}\right) = \frac{1}{x^2-1}$

b $\frac{1}{2} + \frac{1}{4} - \frac{1}{2n} - \frac{1}{2n+2} = \frac{3}{4} - \frac{1}{2n} - \frac{1}{2n+2}$

c $\frac{(3 \times 20 + 2)(20 - 1)}{4 \times 20 \times (20+1)} = \frac{589}{840}$

d $\sum_{r=2}^{\infty} \frac{1}{r^2 - 1} = \frac{3}{4}$

22 $\sum_{r=1}^{n} r^2 = \frac{1}{6}(n+1)(2n^2 + n) = \frac{n}{6}(n+1)(2n+1)$

23 a $u^4 + 27u + 81 = 0$ **b** $u^4 + 2u^2 - u + 1 = 0$

24 $1 < x < 1.25$ and $x \le -1$

25 a $k \le -3, k \ge 5$

b $f(x) \le -3$ or $f(x) \ge 5$

$-1 \le \sin\theta \le 1$

Which is not in $f(x) \ge 5$ or $f(x) \le -3$

26 a $5x - 7x^2 + \frac{68}{3}x^3$ **b** $-\frac{1}{4} < x \le \frac{1}{4}$

27 a $\frac{1}{2}\left(\frac{(x+1)(x+2) - 2x(x+2) + x(x+1)}{x(x+1)(x+2)}\right)$
$= \frac{x^2 + 3x + 2 - 2x^2 - 4x + x^2 + x}{2x(x+1)(x+2)}$
$= \frac{1}{x(x+1)(x+2)}$

b $= \frac{n^2 + 3n + 2 - 2n - 4 + 2n + 2}{4(n+1)(n+2)} = \frac{n(n+3)}{4(n+1)(n+2)}$

c $\frac{5}{1848}$

28 a $\frac{2x(x(x+2)) + 3(x+2) - 3x}{x(x+2)}$
$= \frac{2x^3 + 4x^2 + 3x + 6 - 3x}{x(x+2)}$
$= \frac{2x^3 + 4x^2 + 6}{x(x+2)}$

b $\dfrac{n(2n^3+8n^2+19n+19)}{2(n+1)(n+2)}$

29 When $n=2$, $\sum_{r=1}^{n}\dfrac{1}{r^2-1}=\dfrac{1}{3}$

$$\frac{3n^2-n-2}{4n(n+1)}=\frac{12-2-2}{4(2)(3)}=\frac{8}{24}=\frac{1}{3}$$

So true for $n=2$

Assume true for $n=k$

$$\sum_{r=1}^{k+1}\frac{1}{r^2-1}=\frac{1}{(k+1)^2-1}+\sum_{r=1}^{k}\frac{1}{r^2-1}$$

$$=\frac{1}{(k+1)^2-1}+\frac{3k^2-k-2}{4k(k+1)}$$

$$=\frac{1}{k^2+2k}+\frac{3k^2-k-2}{4k(k+1)}$$

$$=\frac{4(k+1)}{4k(k+2)(k+1)}+\frac{(3k^2-k-2)(k+2)}{4k(k+1)(k+2)}$$

$$=\frac{4(k+1)+(3k^2-k-2)(k+2)}{4k(k+1)(k+2)}$$

$$=\frac{4k+4+3k^3+6k^2-k^2-2k-2k-4}{4k(k+1)(k+2)}$$

$$=\frac{3k^3+5k^2}{4k(k+1)(k+2)}$$

$$=\frac{k^2(3k+5)}{4k(k+1)(k+2)}$$

$$=\frac{k(3k+5)}{4(k+1)(k+2)}$$

$$\frac{3(k+1)^2-(k+1)-2}{4(k+1)(k+1+1)}=\frac{3k^2+6k+3-k-1-2}{4(k+1)(k+2)}$$

$=\dfrac{3k^2+5k}{4(k+1)(k+2)}$ as required]

True for $n=2$ and assuming true for $n=k$ implies true for $n=k+1$, hence true for all $n\in\mathbb{N}$. Since $\dfrac{1}{r^2-1}$ not defined at $r=1$

Chapter 3

Exercise 3.1A

1 **a** **i** $\left(0,-\dfrac{1}{2}\right)$ $(-1,0)$

ii $x=2$ $y=1$

iii

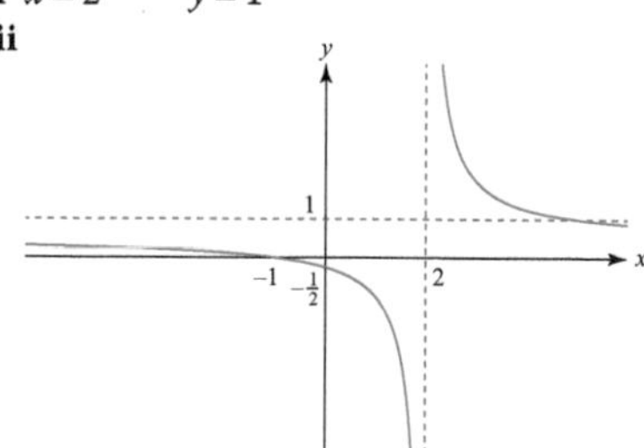

b **i** $\left(0,-\dfrac{1}{2}\right)$ $(1,0)$

ii $x=-2$ $y=1$

iii

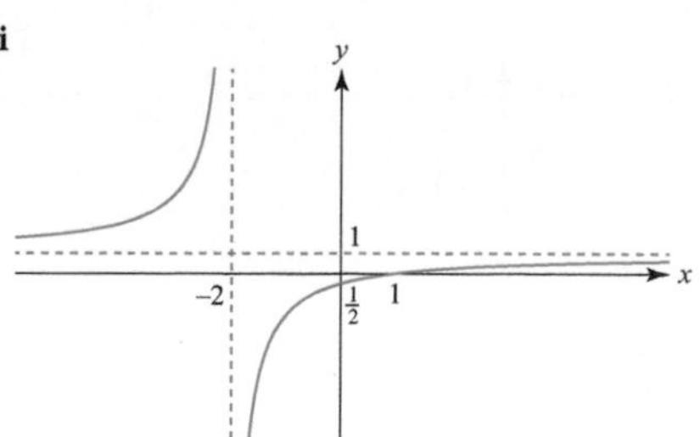

c **i** $(-5,0)$ $\left(0,-\dfrac{5}{3}\right)$

ii $x=3$ $y=1$

iii

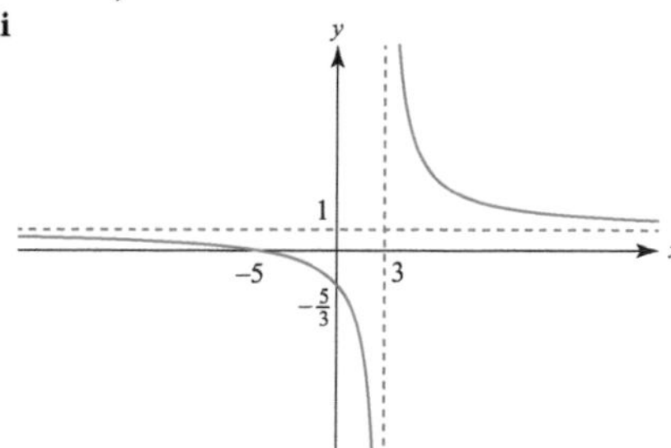

d **i** $\left(0,-\dfrac{5}{3}\right)$ $\left(\dfrac{5}{6},0\right)$

ii $x=-\dfrac{3}{8}$ $y=\dfrac{3}{4}$

iii

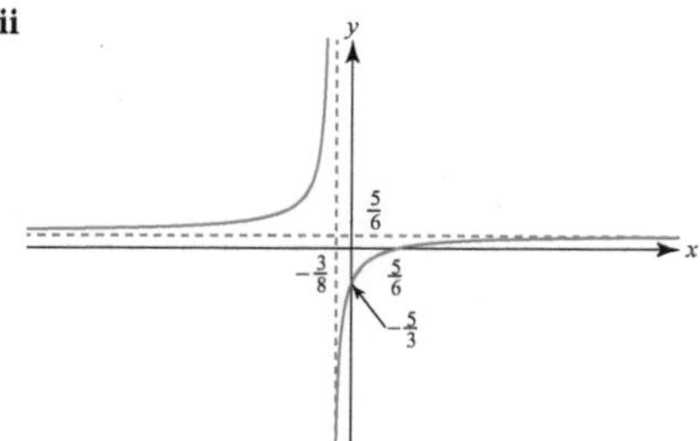

e **i** $(0,-1)$ $(-12,0)$

ii $x=\dfrac{12}{5}$ $y=\dfrac{1}{5}$

iii

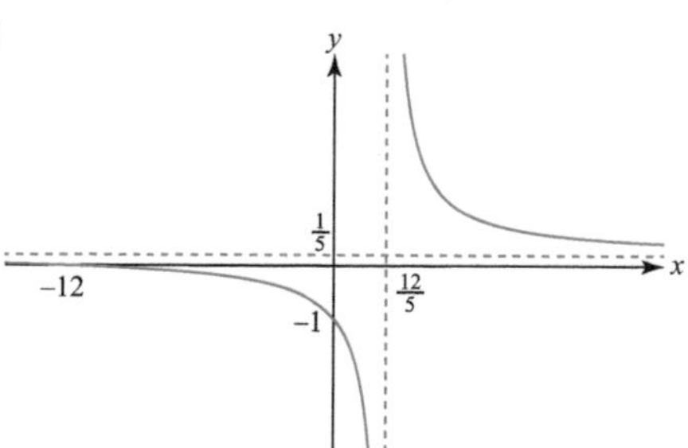

f **i** $\left(0,\dfrac{1}{2}\right)$ $\left(\dfrac{10}{7},0\right)$

ii $x=\dfrac{20}{3}$ $y=\dfrac{7}{3}$

iii

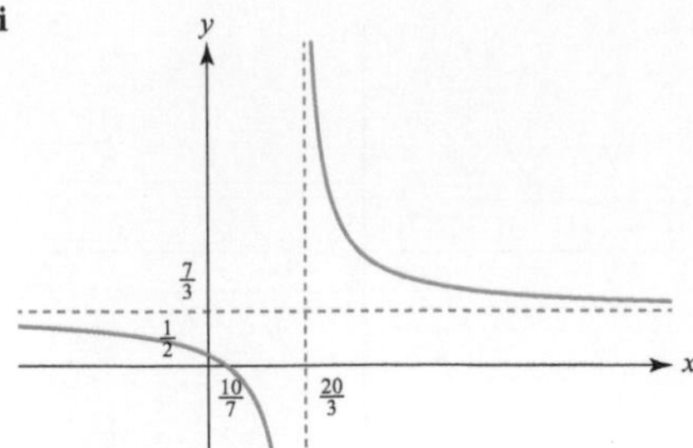

2 a intercepts: $(0,-a)$; $(-a,0)$

asymptotes: $x=\frac{1}{2}$; $y=\frac{1}{2}$

b intercepts: $\left(0,\frac{4}{b}\right)$; $\left(-\frac{4}{3},0\right)$

asymptotes: $x=-b$; $y=3$

c intercepts: $\left(0,\frac{a}{5}\right)$; $\left(\frac{a}{2},0\right)$

asymptotes: $x=\frac{5}{3}$; $y=\frac{2}{3}$

d intercepts: $\left(0,\frac{-a}{b}\right)$; $(a,0)$

asymptotes: $x=-\frac{b}{2}$; $y=\frac{1}{2}$

3 a intercepts: $(-3,0)$; $\left(0,-\frac{3}{4}\right)$

asymptotes: $x=4$; $y=1$

points of intersection: $(11,2)$

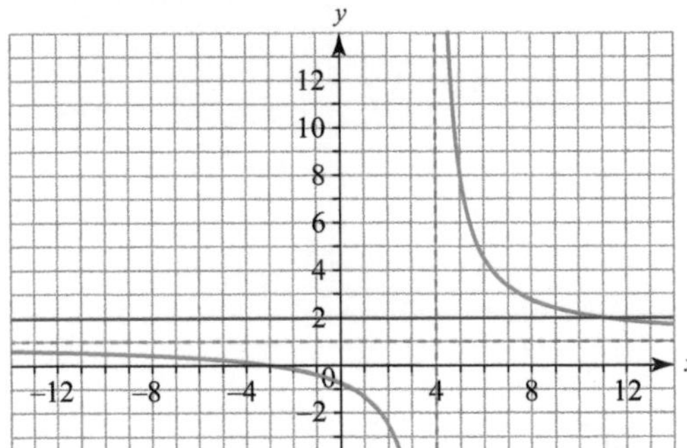

b intercepts: $\left(0,-\frac{3}{2}\right)$; $(-6,0)$

asymptotes: $x=4$; $y=1$

points of intersection: $(6,6)$

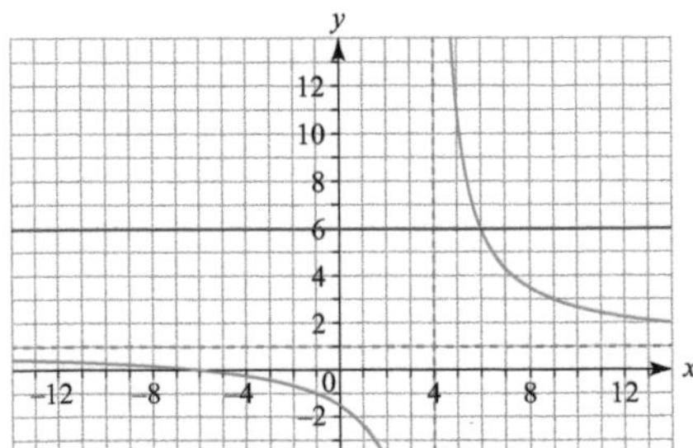

c intercepts: $(0,-1)$; $(-12,0)$

asymptotes: $x=\frac{12}{5}$; $y=\frac{1}{5}$

points of intersection: $(-12, 0)$

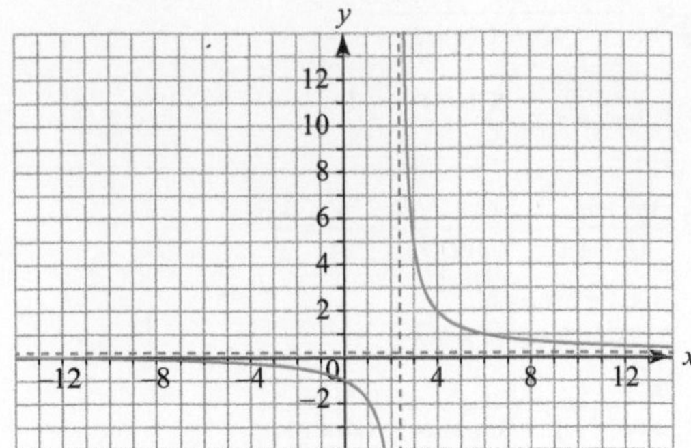

d intercepts: $\left(0,-\frac{4}{3}\right)$; $\left(\frac{4}{3},0\right)$

asymptotes: $x=\frac{3}{2}$; $y=-\frac{3}{2}$

points of intersection: $\left(\frac{20}{13}, -8\right)$

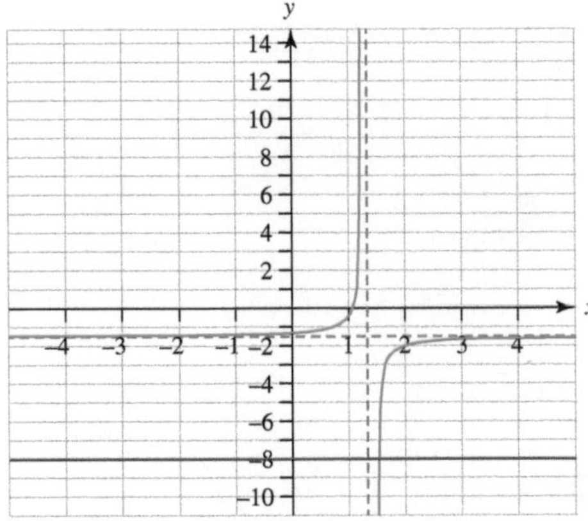

e intercepts: $(-5,0)$; $(0,5)$; $\left(-\frac{1}{2},0\right)$; $(0,1)$

asymptotes: $x=-1$; $y=1$

points of intersection: $(-2,-3)$; $(1,3)$

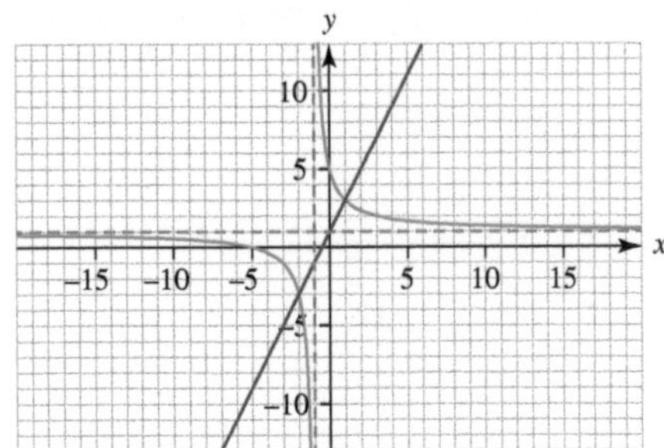

f intercepts: $(3,0)$; $\left(0,-\frac{6}{5}\right)$; $(0,3)$

asymptotes: $x=-5$; $y=2$

points of intersection: $(-7,10)$; $(3,0)$

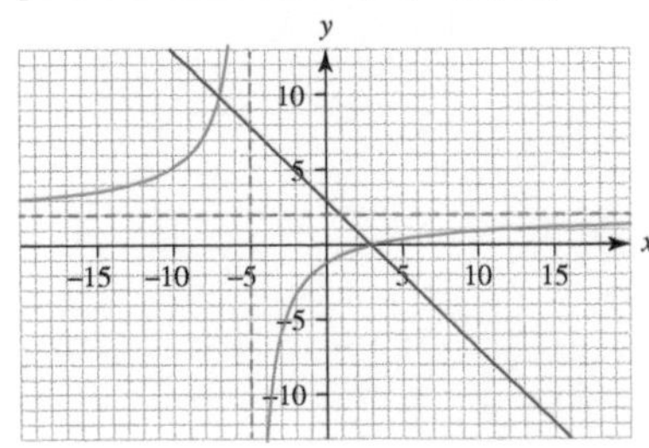

g intercepts: $\left(-\frac{15}{7},0\right)$; $\left(\frac{5}{3},0\right)$; $\left(0,\ \frac{10}{3}\right)$; $(0,15)$
asymptotes: $x=-3$; $y=-6$
points of intersection: $(-5,-20)$; $(-1,8)$

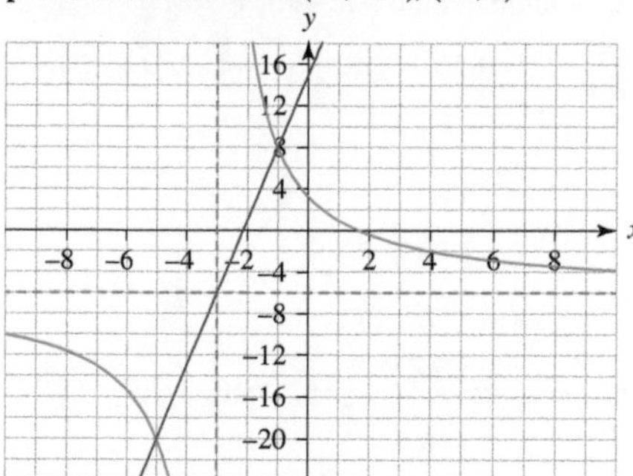

h intercepts: $(-6,0)$; $(24,0)$; $(0,24)$; $(0,-6)$
asymptotes: $x=\frac{8}{3}$; $y=\frac{8}{3}$
points of intersection: $(4,20)$; $(20,4)$

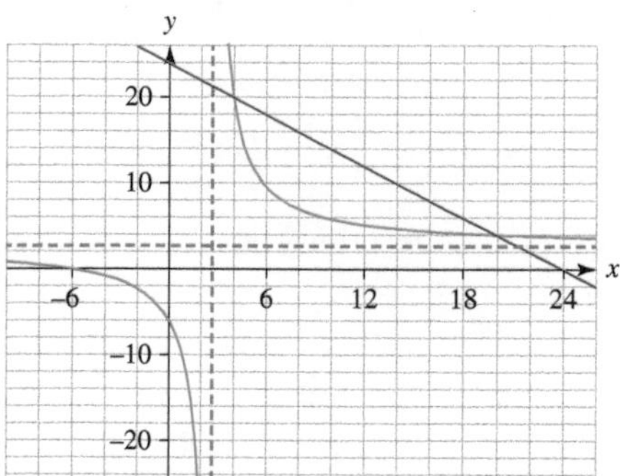

4 a intercepts: $(2,0)$; $\left(0,-\frac{2}{5}\right)$
asymptotes: $x=-5$; $y=1$
points of intersection: $(-6,8)$
$\frac{x-2}{x+5}<8$ when $x<-6$ or $x>-5$

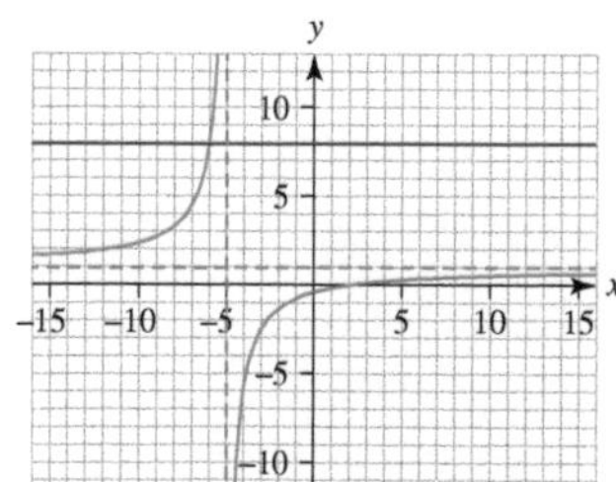

b intercepts: $\left(0,-\frac{5}{2}\right)$; $(15,0)$
asymptotes: $x=-6$; $y=1$
points of intersection: $(-13,4)$
$\frac{x-15}{x+6}\le 4$ when $x\le -13$ or $x>-6$

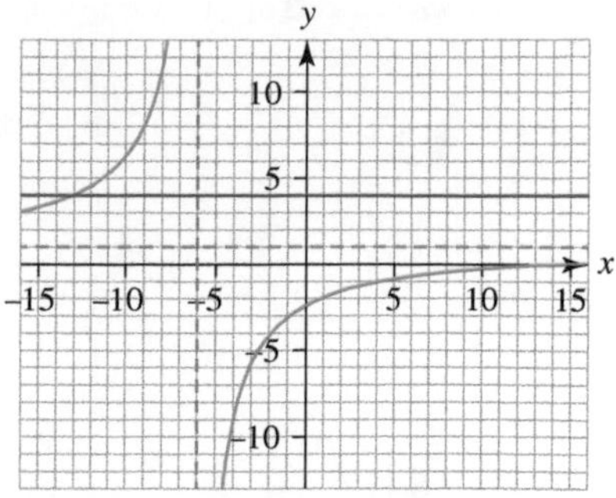

c intercepts: $\left(0,-\frac{1}{4}\right)$; $\left(\frac{1}{4},0\right)$
asymptotes: $x=-\frac{4}{3}$; $y=\frac{4}{3}$
points of intersection: $\left(-\frac{9}{2},2\right)$

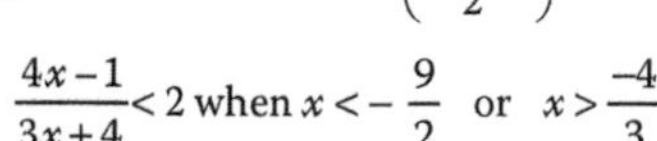
$\frac{4x-1}{3x+4}<2$ when $x<-\frac{9}{2}$ or $x>\frac{-4}{3}$

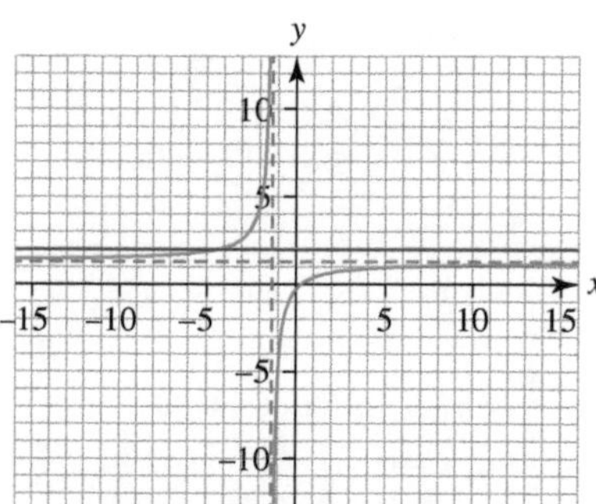

d intercepts: $\left(0,-\frac{5}{3}\right)$; $\left(-\frac{5}{3},0\right)$
asymptotes: $x=\frac{3}{2}$; $y=\frac{3}{2}$
points of intersection: $(2,11)$
$\frac{3x+5}{2x-3}\ge 11$ when $\frac{3}{2}<x\le 2$

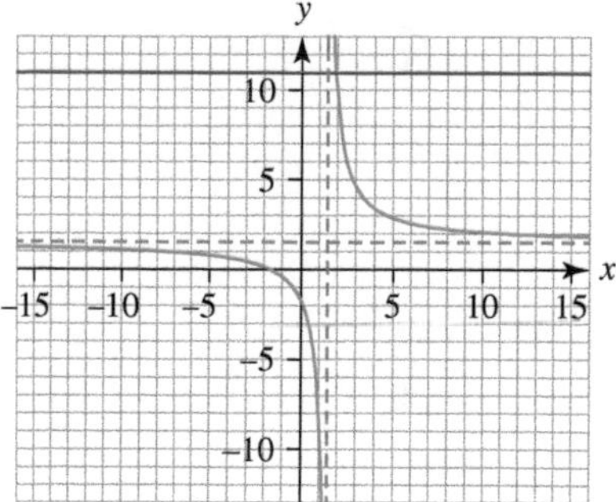

Exercise 3.1B

1 a intercepts: $(-5,0)$; $(0,5)$; $(0,1)$; $\left(\frac{-1}{2},0\right)$
asymptotes: $x=-1$; $y=1$
points of intersection: $(-2,-3)$; $(1,3)$

$\frac{x + 5}{x + 1} > 2x + 1$ when $x < -2$ or $-1 < x < 1$

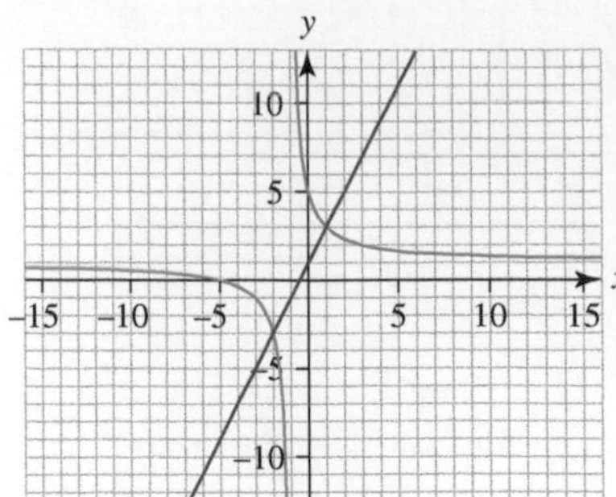

b intercepts: $(3,0)$; $\left(0,-\frac{6}{5}\right)$; $(0,3)$

asymptotes: $x = -5$; $y = 2$

points of intersection: $(-7, 10)$; $(3, 0)$

$\frac{2x - 6}{x + 5} < 3 - x$ when $x < -7$ or $-5 < x < 3$

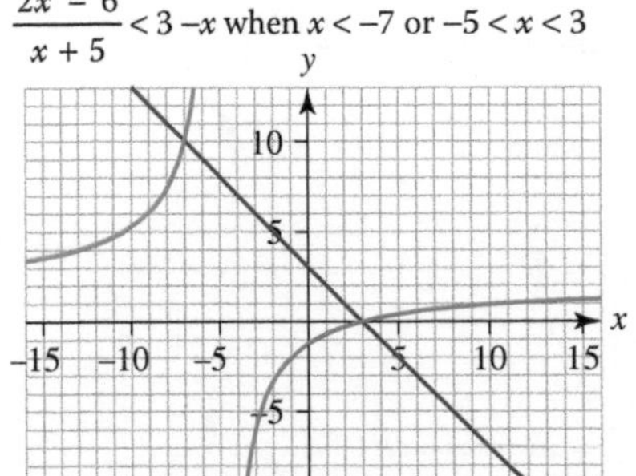

c intercepts: $\left(\frac{9}{2},0\right)$; $(-8,0)$; $(0,-2)$; $(0,-9)$

asymptotes: $x = 4$; $y = 1$

points of intersection: $(2,-5)$; $(7,5)$

$\frac{x + 8}{x - 4} \geq 2x - 9$ when $x \leq 2$, $4 < x \leq 7$

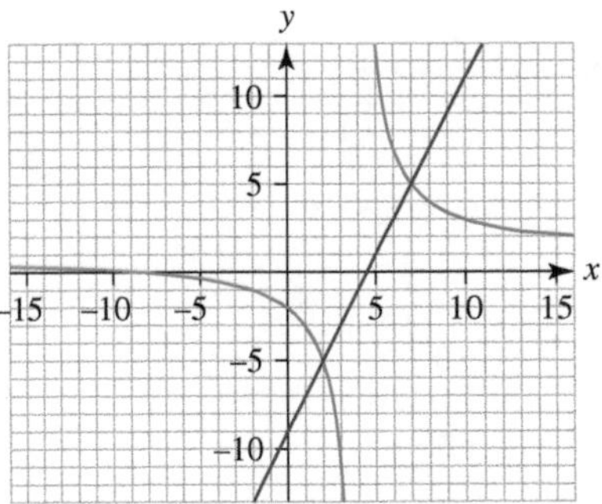

d intercepts: $(-3,0)$; $(7,0)$; $(0,-2)$; $(0,14)$

asymptotes: $x = \frac{3}{2}$; $y = 1$

points of intersection: $(2,10)$; $(6,2)$

$\frac{2x + 6}{2x - 3} \geq 14 - 2x$ when $\frac{3}{2} < x \leq 2$, $x \geq 6$

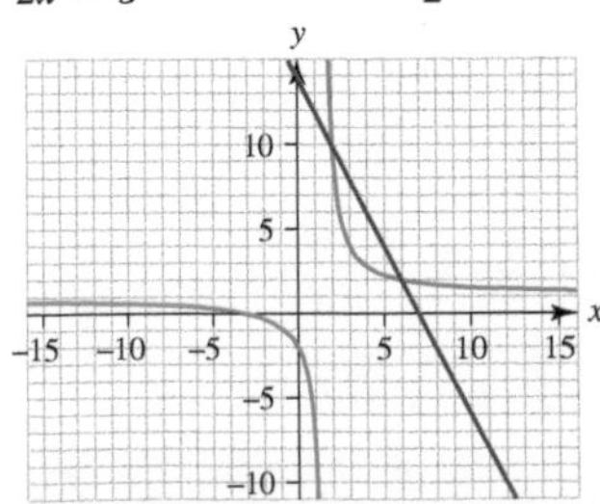

e intercepts: $\left(-\frac{15}{7},0\right)$; $\left(\frac{5}{3},0\right)$; $\left(0, \frac{10}{3}\right)$; $(0,15)$

asymptotes: $x = -3$; $y = -6$

points of intersection: $(-5,-20)$; $(-1,8)$

$\frac{10 - 6x}{x + 3} \leq 7x + 15$ when $-5 \leq x < -3$, $x \geq -1$

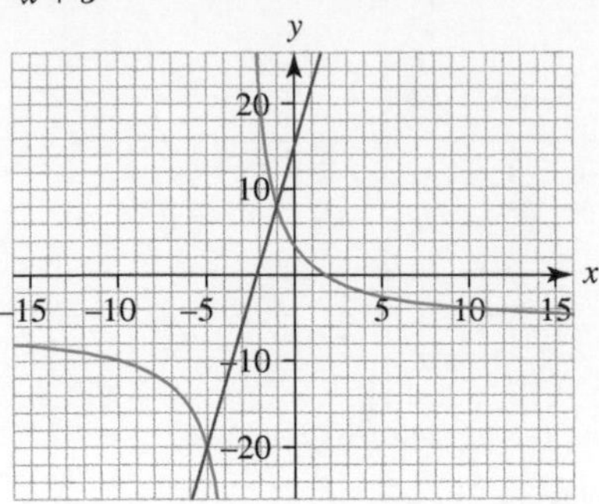

f intercepts: $(-6,0)$; $(24,0)$; $(0,24)$; $(0,-6)$

asymptotes: $x = \frac{8}{3}$; $y = \frac{8}{3}$

points of intersection: $(4, 20)$; $(20,4)$

$\frac{8x + 48}{3x - 8} > 24 - x$ when $\frac{8}{3} < x < 4$ or $x > 20$

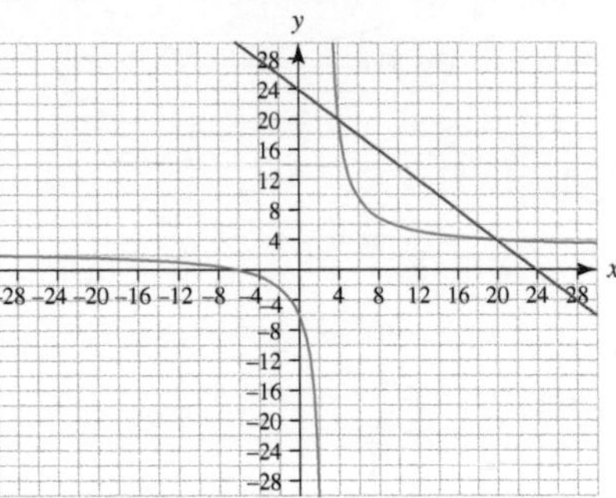

2 a intercepts: $\left(0,-\frac{1}{2}\right)$; $(-2,0)$

asymptotes: $x = 4$; $y = 1$

points of intersection: $(2, -2)$; $(7,3)$

$-2 < \frac{x+2}{x-4} < 3$ when $x < 2$ or $x > 7$

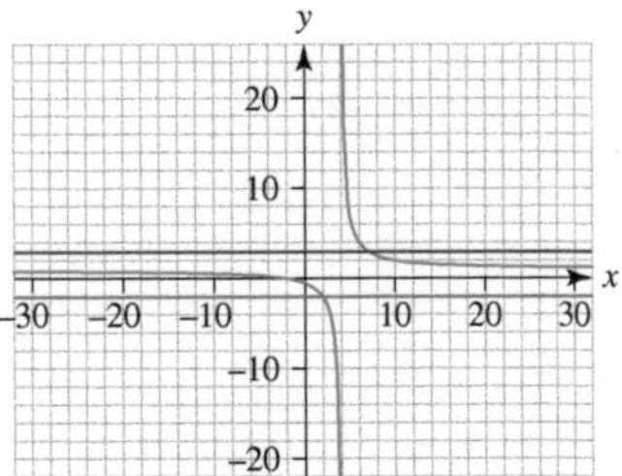

b intercepts: $(0,-1)$; $\left(-\frac{5}{2},0\right)$

asymptotes: $x = \frac{5}{2}$; $y = 1$

points of intersection: $\left(\frac{-5}{2}, 0\right)$; $(2, -9)$

$0 \geq \frac{2x+5}{2x-5} \geq -9$ when $\frac{-5}{2} \leq x \leq 2$

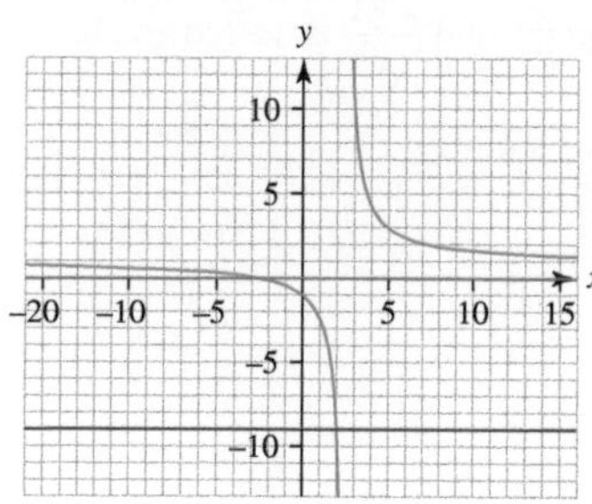

c intercepts: $(0, 1)$; $\left(\frac{5}{3}, 0\right)$

asymptotes: $x = \frac{5}{2}$; $y = \frac{3}{2}$

points of intersection: $(0, 1)$; $(2, -1)$

$-1 \le \frac{3x-5}{2x-5} \le 1$ when $0 \le x \le 2$

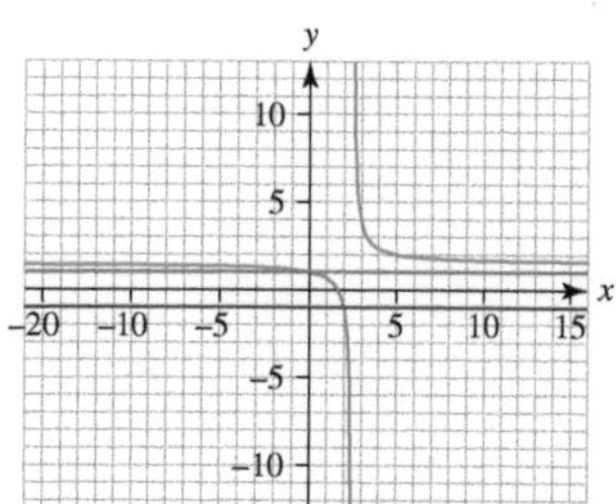

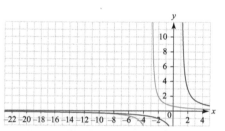

d intercepts: $(12, 0)$; $(2, 0)$; $(0, 2)$; $(0, 6)$

asymptotes: $x = 6$; $y = 1$

points of intersection between curve and lines:

$(0, 2)$; $(4, 4)$; $(7, -5)$; $(12, 0)$

$2 - x < \frac{x-12}{x-6} \le 6 - \frac{x}{2}$ when $0 < x \le 4$ or $7 < x \le 12$

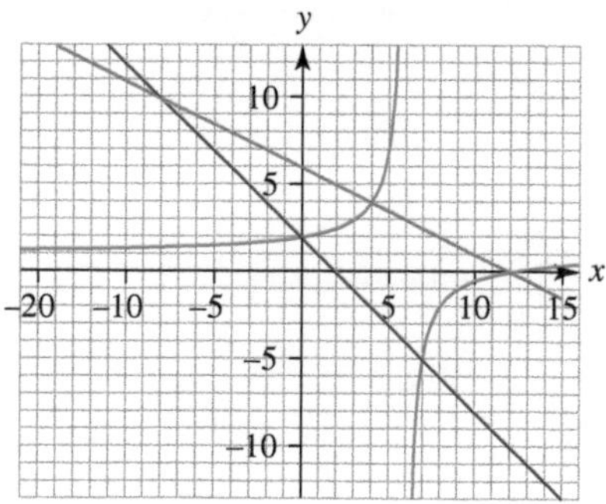

e intercepts: $(9, 0)$; $\left(\frac{11}{3}, 0\right)$; $(0, 11)$; $(0, 6)$; $(0, 3)$

asymptotes: $x = 3$; $y = 2$

points of intersection between curve and lines:

$(-3, 4)$; $(1, 8)$; $(5, -4)$; $(9, 0)$

$11 - 3x \le \frac{2x-18}{x-3} \le \frac{9-x}{3}$ when $5 \le x \le 9$

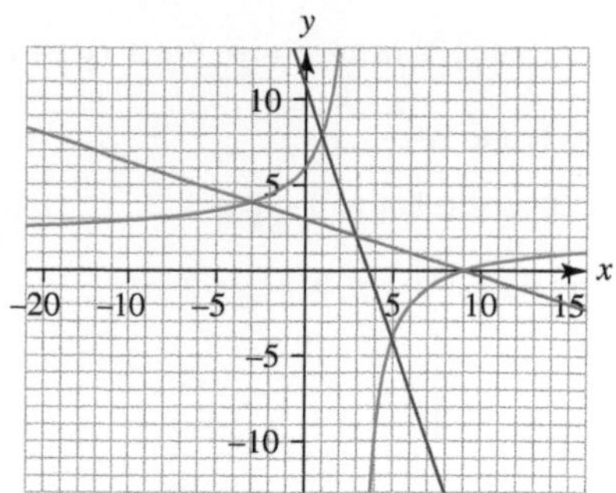

f intercepts: $\left(0, \frac{2}{3}\right)$; $(0, -3)$;

asymptotes: $x = 1$; $x = -3$; $y = 0$

points of intersection: $\left(-11, -\frac{1}{4}\right)$

$\frac{2}{x+3} > \frac{3}{x-1}$ when $x < -11$ or $-3 < x < 1$

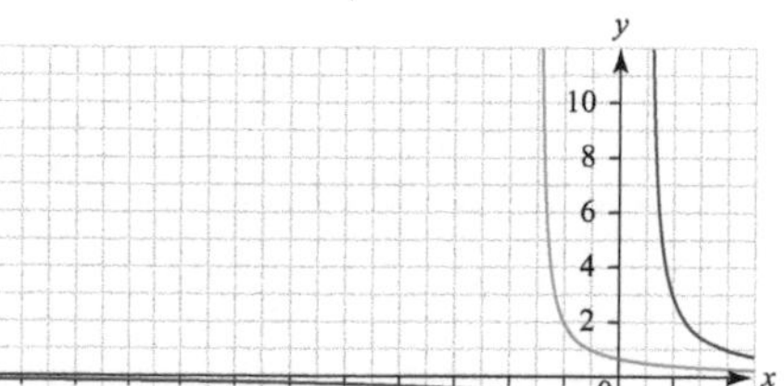

g intercepts: $\left(0, \frac{5}{4}\right)$; $\left(0, -\frac{3}{2}\right)$

asymptotes: $x = -4$; $x = 2$; $y = 0$

points of intersection: $\left(11, \frac{1}{3}\right)$

$\frac{5}{x+4} \le \frac{3}{x-2}$ when $x < -4$, $2 < x \le 11$

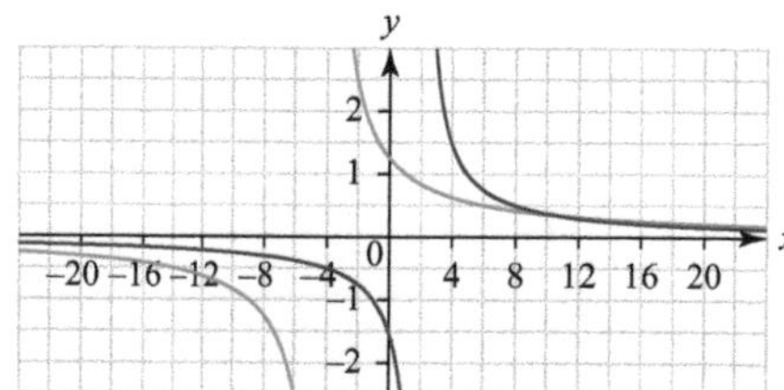

3 The graph shows that $\frac{2x}{x+2}$ is only less than 1 in the range $-2 < x < 2$

At all other times the function is greater than or equal to 1

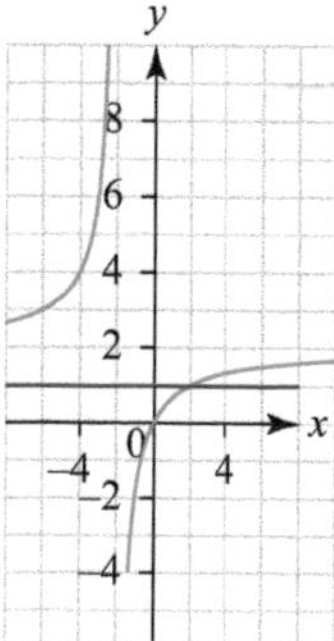

4 The two appropriate graphs are $y = \dfrac{3}{x-1}$ and $y = x+1$

The inequality is thus $\dfrac{3}{x-1} \geq x+1$

The curve is 'above' the line in the ranges $x \leq -2$ and $1 < x \leq 2$

Therefore the solution is $x \leq -2,\ 1 < x \leq 2$

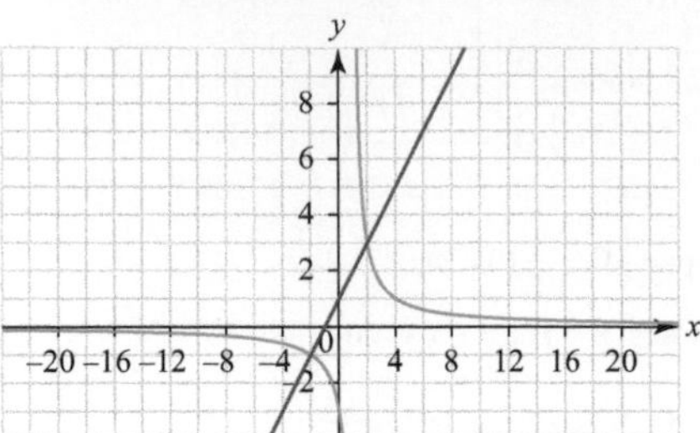

5 The inequality is $-1 \leq \dfrac{x-5}{x+5} \leq 1$

The curve lies between the straight lines only when $x \geq 0$

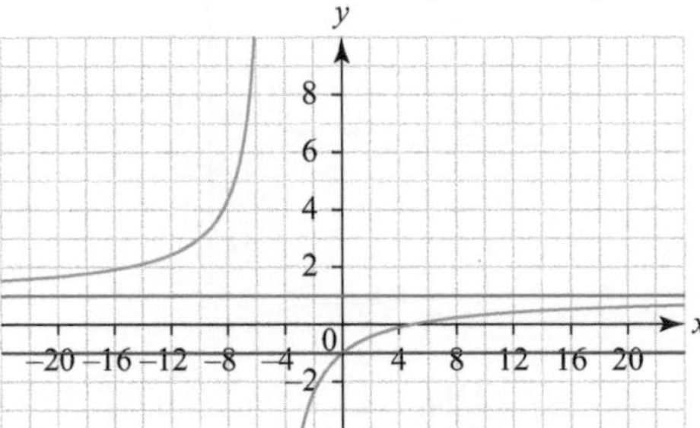

6 The inequality is $-1 \leq \dfrac{x+1}{1-x} \leq 1$

The curve lies between the straight lines only when $x \leq 0$

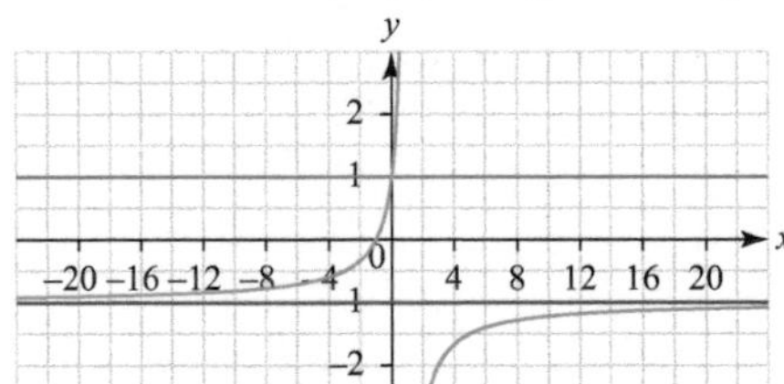

7 **a** $y = \dfrac{2x}{x+1} - \dfrac{x-3}{x-5}$

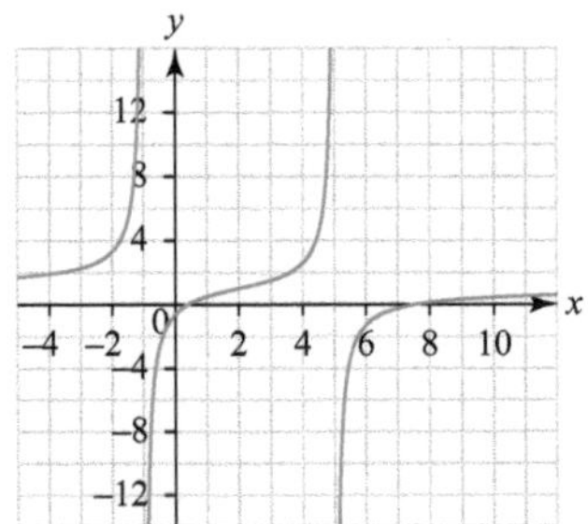

b The graph crosses the y-axis when $x = 0$

At this point, $y = 0 - \dfrac{-3}{-5} = \dfrac{3}{5}$

c The equations of its vertical asymptotes are where denominators of the fractions are 0 i.e. where $x = -1$ and $x = 5$

Its horizontal asymptote occurs when x gets very large, and y approaches $\dfrac{2}{1} - \dfrac{1}{1} = 1$

8 $\dfrac{x}{x+1}$ is the blue graph and $\dfrac{x+1}{x}$ is the red graph.

$\dfrac{x}{x+1} \geq \dfrac{x+1}{x}$ when $x < -1$,

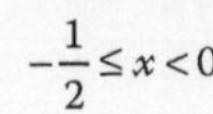

$-\dfrac{1}{2} \leq x < 0$

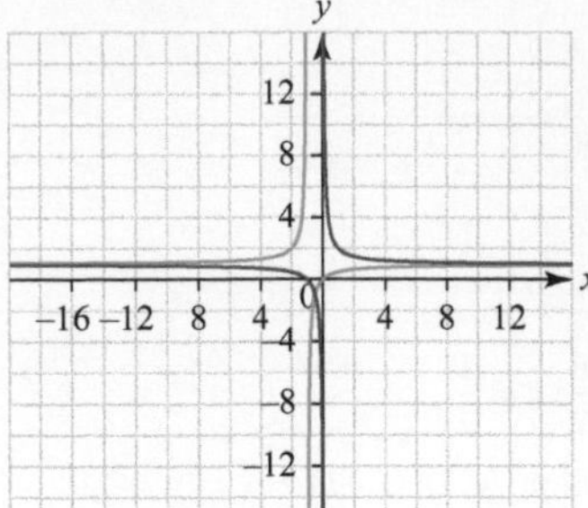

9 **a** $x^2 - 2x - 15 = 0 \rightarrow x^2 - 2x = 15 \rightarrow x - 2 = \dfrac{15}{x}$

Thus $x^2 - 2x - 15 = 0$ is equivalent to $x - 2 = \dfrac{15}{x}$

$x - 2 = \dfrac{15}{x}$ when $x = -3$ or 5

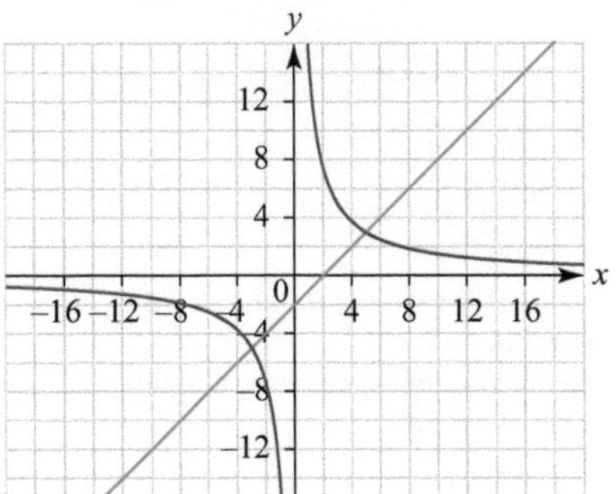

Alternatively, $x^2 - 2x - 15 = 0 \rightarrow x - 2 + x^2 - 3x - 13 = 0$
$\rightarrow x - 2 = -x^2 + 3x + 13$ when $x = -3$ or 5

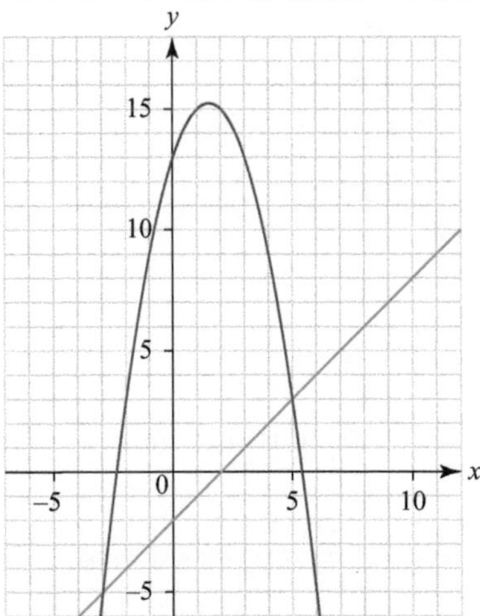

A similar alternative is $x^2 - x - 17 = x - 2$

b $0 = x^2 - 2x - 15 = (x-5)(x+3) \Rightarrow x = 5, -3$

10 $x^3 - x^2 - 6x = 0 \rightarrow x^2(x-1) = 6x \rightarrow x^2 = \dfrac{6x}{x-1}$

Thus $A = 6$ and $B = -1$

x^2 is the blue graph and $\frac{6x}{x-1}$ is the red graph.

$x^2 = \frac{6x}{x-1}$ when $x = -2$ or $x = 0$ or $x = 3$

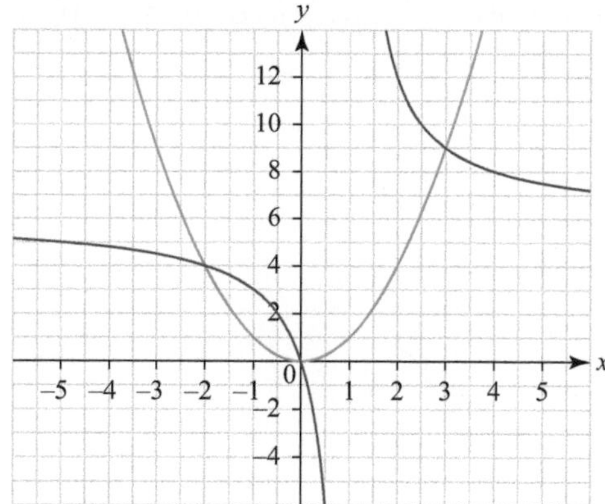

11 $\frac{1}{x}$ is the blue graph, $\frac{x}{2}$ is the red graph and $\frac{3}{x+2}$ is the green graph.

Blue meets green when $\frac{1}{x} = \frac{3}{x+2} \to x = 1$

Green has an asymptote at $x = -2$, and blue has an asymptote at $x = 0$

Hence green is the largest in the range $-2 < x < 0$

Blue meets red when $x^2 = 2$, i.e. $x = \pm\sqrt{2}$

Thus $\frac{1}{x}$ is the largest when $x < -2$ or $0 < x < 1$;

Red meets green when $\frac{x}{2} = \frac{3}{x+2} \to x^2 + 2x - 6 = 0$

$\to (x+1)^2 - 1 - 6 = 0 \to x = -1 + \sqrt{7}$

Thus $\frac{x}{2}$ is the largest when $x > -1 + \sqrt{7}$

$\frac{3}{x+2}$ is the largest when $-2 < x < 0$ or $1 < x < -1 + \sqrt{7}$

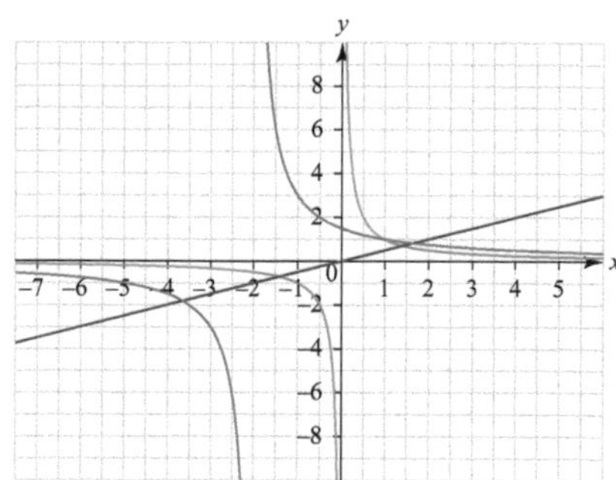

Exercise 3.2A

1 a Intercepts are $(0, -1)$, $(3, 0)$ and $(-3, 0)$
There are no vertical asymptotes as denominator never zero.
As $x \to \infty$, $y \to 1$ so $y = 1$ is a horizontal asymptote.

b Intercept is $(0, -1)$
Denominator $= 0$ when $x = \pm 3$ so vertical asymptotes at $x = 3$ and $x = -3$
As $x \to \infty$, $y \to 1$ so $y = 1$ is a horizontal asymptote.

c Intercept is $(0, 0)$
There are no vertical asymptotes as denominator never zero.
As $x \to \infty$, $y \to 1$ so $y = 1$ is a horizontal asymptote.

d Intercepts are $(0, 0)$ and $(-5, 0)$
There are no vertical asymptotes as denominator never zero.
As $x \to \infty$, $y \to 1$ so $y = 1$ is a horizontal asymptote.

e When $x = 0$ denominator is zero so $x = 0$ is vertical asymptote.
Denominator also zero when $x = -4$ so this is another vertical asymptote.
Numerator never equal to zero line doesn't cross x axis.
As $x \to \infty$, $y \to 1$ so $y = 1$ is a horizontal asymptote.

f When $x = 0$ denominator is zero so $x = 0$ is vertical asymptote.
Denominator also zero when $x = -4$ so this is another vertical asymptote.
When $y = 0$, $x^2 - 3x - 4 = 0$
$(x-4)(x+1) = 0$ so $x = 4$ or $x = -1$
So intercepts are $(4, 0)$ and $(-1, 0)$
As $x \to \infty$, $y \to 1$ so $y = 1$ is a horizontal asymptote.

2 a The denominator $= 0$ when $x = -2 \pm \sqrt{7}$ so the graph has asymptotes and does not exist for these values of x. Otherwise the graph is defined. Hence it has range
$x \neq -2 \pm \sqrt{7}$, $y \geq \frac{3}{7}$ or $y \leq 0$

b The denominator $= 0$ when $x = 1$ or $x = 2$ so the graph has asymptotes and does not exist for these values of x. Otherwise the graph is defined, and hence has range
$x \neq 1, 2$
$y < 1$ or $y \geq 9$.

c The denominator $= 0$ when $x = \frac{3 \pm \sqrt{5}}{2}$ so the graph has asymptotes and does not exist for these values of x

The graph exists when $y \neq 1$ and $y < 1$ or $y \geq 5$

When $y = 1$, the equation $1 = \frac{x^2 - 3x - 4}{x^2 - 3x + 1}$ has no real solutions, so $y = 1$ is excluded from the range.

d The denominator is never zero so the graph exists for all values of x.
$-0.112 \leq y \leq 1.11$ (3sf)
The function takes the value $y = 0$ at $x = 2$, so this is included in the range.

3 a $(-2, -4)$ is a local maximum point.
$(0, 0)$ is a local minimum point.

b $(2, 3)$ is a local maximum point.
$(-2, -3)$ is a local minimum point.

c $\left(0, \frac{3}{2}\right)$ is a local maximum point.

d $\left(-2, -\frac{5}{4}\right)$ is a local maximum point.

4 a

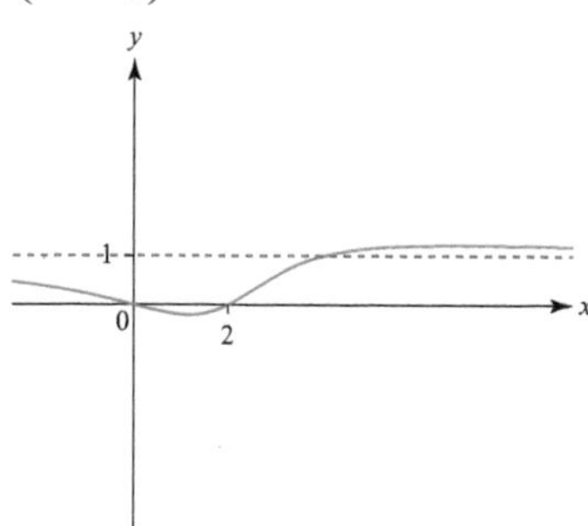

b

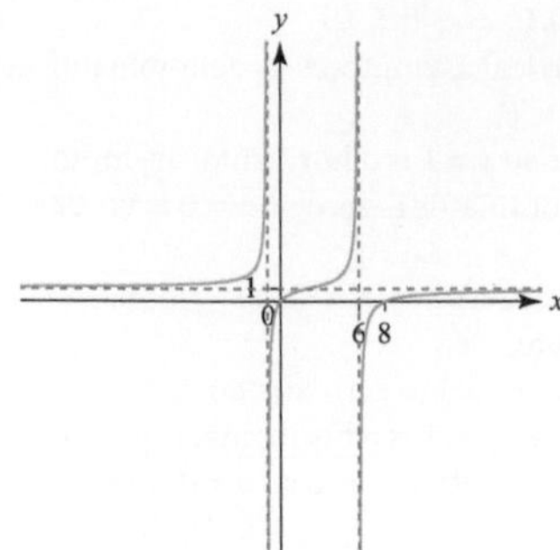

c

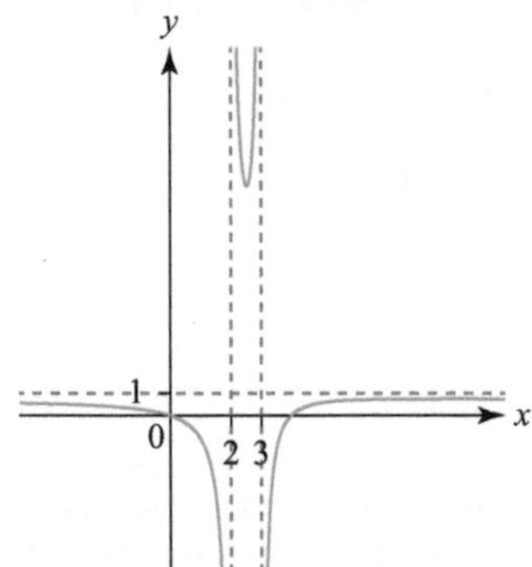

d

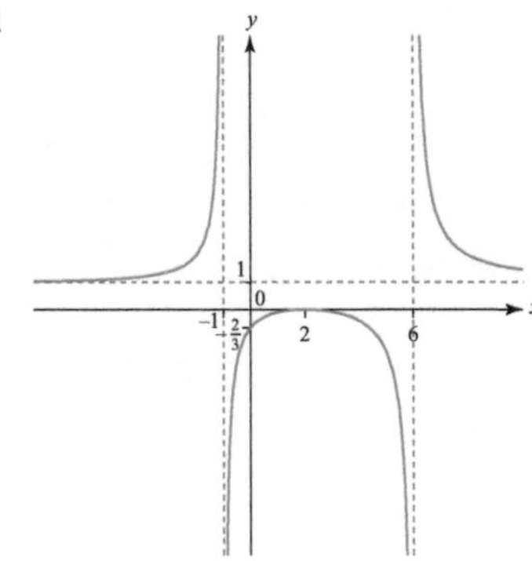

Exercise 3.2B

1 $\frac{x}{x^2+1} = y$

$y(x^2+1) = x$

$yx^2 - x + y = 0$

$y = 0$ at $x = 0$; otherwise y is real if $1 - 4y(y) \geq 0$

$1 - 4y^2 \geq 0$

$4y^2 - 1 \leq 0$

$y^2 \leq \frac{1}{4}$

$-\frac{1}{2} \leq y \leq \frac{1}{2}$

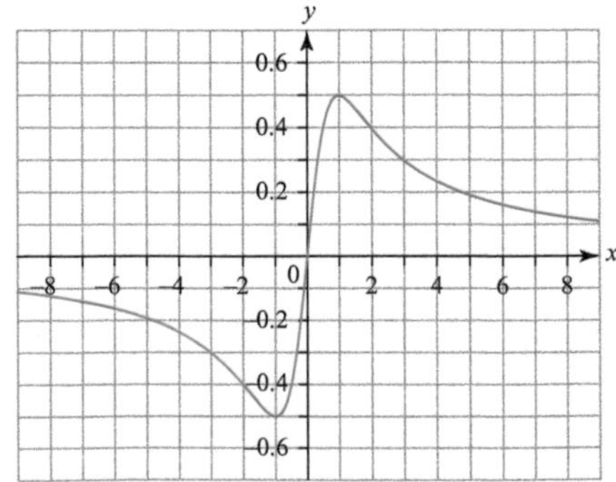

2 $\frac{x^2}{x^2+1} = y$

$y(x^2+1) = x^2$

$(y-1)x^2 + y = 0$

$y = 0$ at $x = 0$; otherwise y has real roots if $0 - 4y(y-1) \geq 0$

$4y(y-1) \leq 0$

$0 \leq y \leq 1$

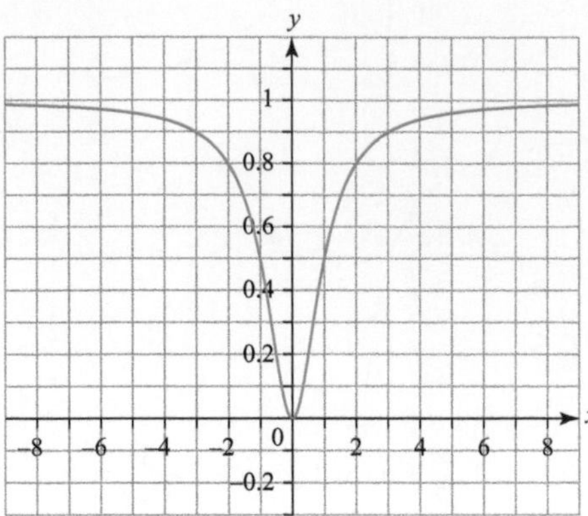

3 $\frac{x^2}{x^2-1} = y$

$y(x^2-1) = x^2$

$(y-1)x^2 - y = 0$

There are no values of x for which $y = 1$; otherwise y has no real roots if $0 + 4y(y-1) < 0$

$4y(y-1) < 0$

$0 < y < 1$

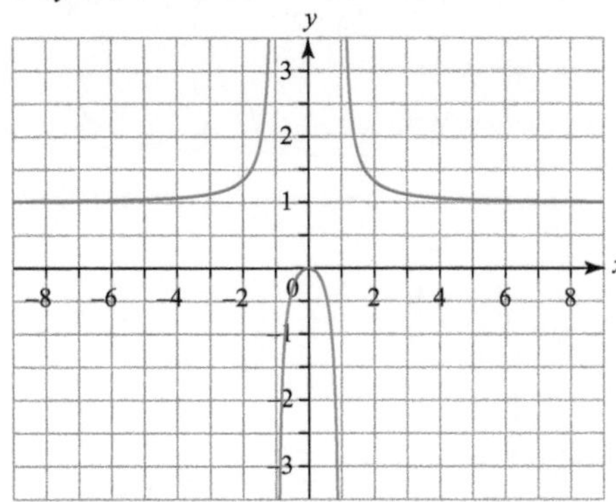

4 $y = \frac{x^2+5}{x-2}$

$y(x-2) = x^2 + 5$

$x^2 - yx + (2y+5) = 0$

y has real roots if $y^2 - 4(2y+5) \geq 0$

$y^2 - 8y - 20 \geq 0$

$(y-10)(y+2) \geq 0$

$y < -2$ or $y > 10$ At $y = -2$, we have the equation

$0 = x^2 + 2x + 1 = (x+1)^2$

hence we have a local maximum at $(-1, -2)$

At $y = 10$, we have the equation

$0 = x^2 - 10x + 25 = (x-5)^2$

hence we have a local minimum at $(5, 10)$

This apparent contradiction is explained by the fact that these are local maxima and minima.

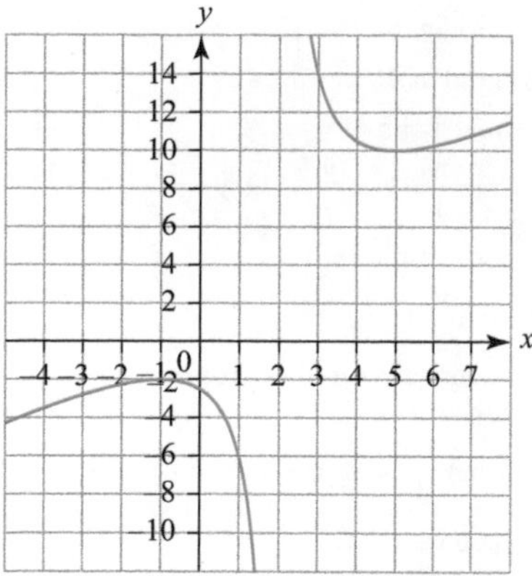

5

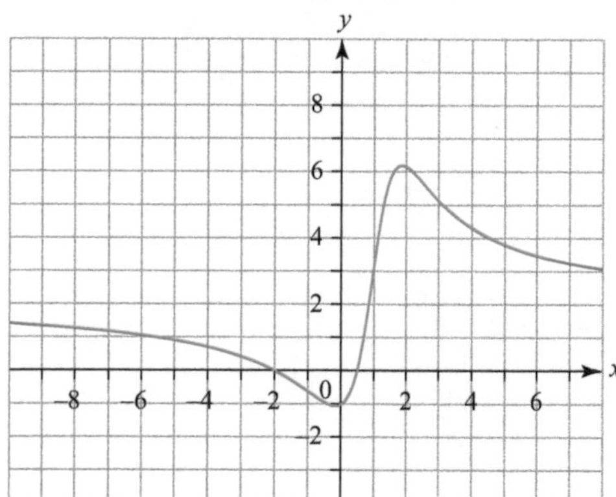

6 a $c \le 1$

Hence greatest possible value of c is 1

b When $c = 1$

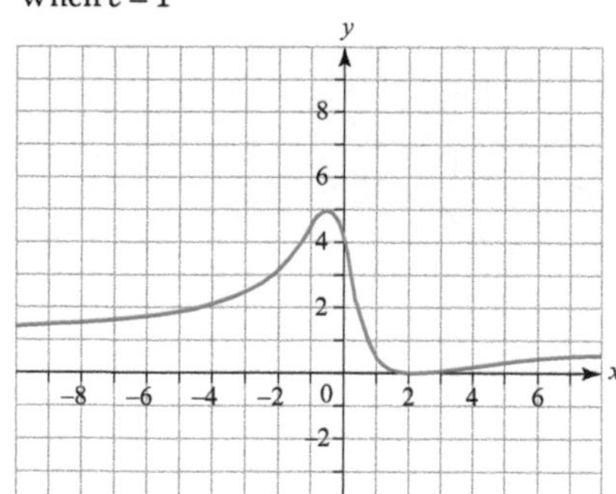

7 $C \le -9, C \ge -1$

local maximum at $\left(-\frac{1}{3}, -9\right)$

local minimum at $(1, -1)$

8 $y = \frac{x^2 + A}{x^2 - B^2}$

$y(x^2 - B^2) = x^2 + A$

$(y-1)x^2 - (B^2y + A) = 0$

The value $y = 1$ cannot be attained for any value of x

Otherwise, y has no real roots if $0 + 4(y-1)(B^2y + A) < 0$

$4(y-1)(B^2y + A) < 0$

$(y-1)(B^2y + A) < 0$

$-\frac{A}{B^2} < y < 1$

For $y = \frac{x^2 + 4}{x^2 - 3^2}$, $A = 4$ and $B = 3$, and so y has no real roots in the range

$-\frac{4}{9} < y < 1$

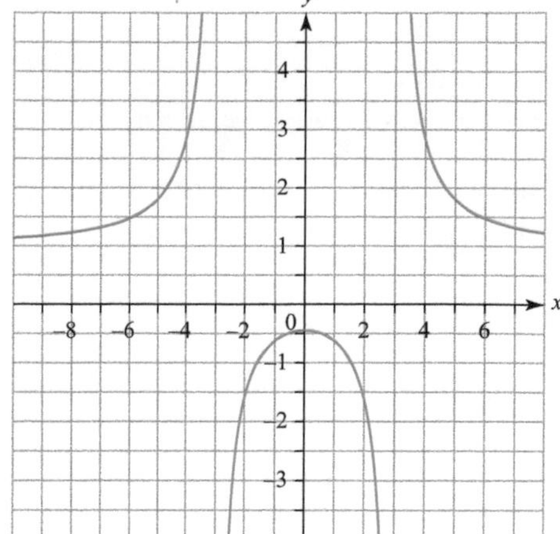

Max turning point has coordinates $\left(0, \frac{-4}{9}\right)$

9 a $y = \frac{4}{x^2 - x}$ is the blue graph and $y^2 = \frac{4}{x^2 - x}$ is the red graph.

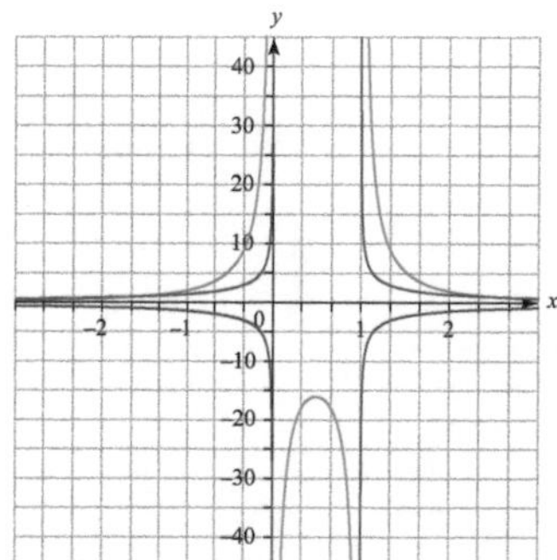

b The graphs intersect at $x = \frac{1}{2}(1 \pm \sqrt{17})$ Hence the inequality holds for

$\frac{1}{2}(1 - \sqrt{17}) < x < 0$ and $1 < x < \frac{1}{2}(1 + \sqrt{17})$

c $y^2 = \frac{4}{x^2 - x}$ is negative between $x = 0$ and $x = 1$ and the square root of a negative number does not have any real values.

10 a $y = \frac{x^2 - A}{x^2 - 4x - A}$

$y(x^2 - 4x - A) = x^2 - A$

$(y-1)x^2 - 4yx - A(y-1) = 0$

$y = 1$ is attained at $x = 0$ otherwise y has real roots if

$16y^2 + 4A(y-1)^2 \ge 0$

$16y^2$ and $4(y-1)^2$ are both ≥ 0, this holds when $A \ge 0$

b i The graph shows there is no turning point.

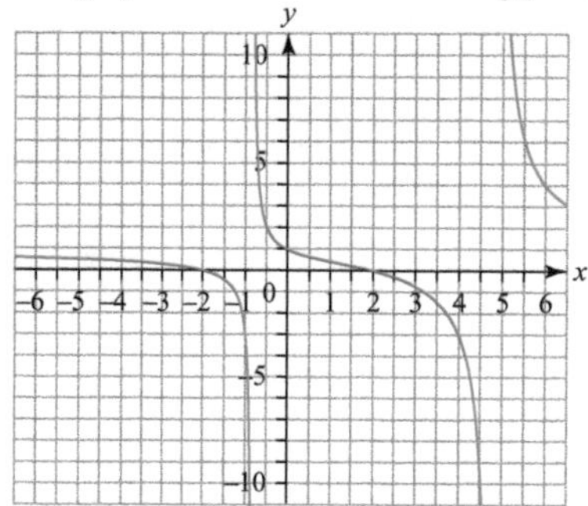

ii $x = \pm 2$

11 If $y = \dfrac{kx}{x^2 - x + 1}$

$yx^2 - (k + y)x + y = 0$

$y = 0$ is attained at $x = 0$ otherwise y has no real roots if

$(k + y)^2 - 4y^2 < 0$

$k^2 + 2ky - 3y^2 < 0$

$y^2 - \dfrac{2k}{3}y - \dfrac{k^2}{3} > 0$

$\left(y - \dfrac{k}{3}\right)^2 - \dfrac{k^2}{3} - \dfrac{k^2}{9} > 0$

$\left(y - \dfrac{k}{3}\right)^2 - \dfrac{4k^2}{9} > 0$

$\Rightarrow$ y has no real roots if $y < \dfrac{-k}{3}$, $y > k$ for no real roots

Hence the minimum and maximum values of y are $-\dfrac{k}{3}$ and k

When $k = 6$, the minimum and maximum values of y are -2 and 6

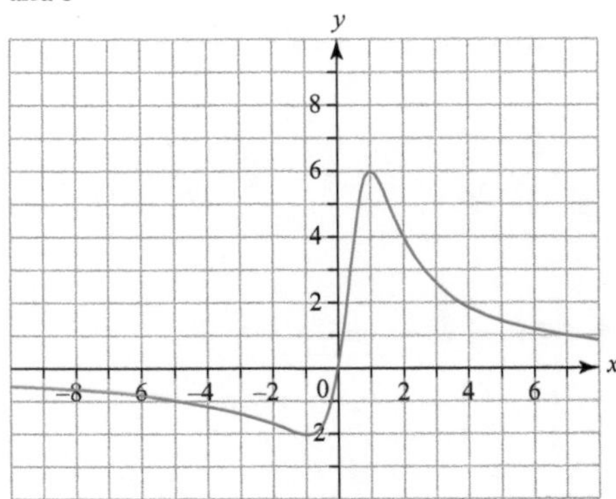

Exercise 3.3A

1

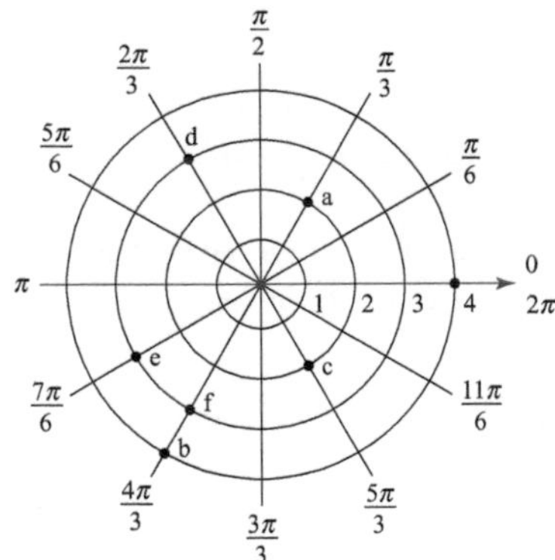

a $(1, \sqrt{3})$ **b** $(-2, -2\sqrt{3})$

c $(1, -\sqrt{3})$ **d** $\left(\dfrac{-3}{2}, \dfrac{3\sqrt{3}}{2}\right)$

e $\left(-\dfrac{3\sqrt{3}}{2}, -\dfrac{3}{2}\right)$ **f** $\left(-\dfrac{3}{2}, -\dfrac{3\sqrt{3}}{2}\right)$

2 a $\left(4, \dfrac{\pi}{6}\right)$ **b** $\left(6, \dfrac{2\pi}{3}\right)$

c $\left(8, \dfrac{5\pi}{3}\right)$ **d** $(13, 4.32)$

3 a $\left(4, -\dfrac{5\pi}{6}\right)$ **b** $\left(10, -\dfrac{\pi}{3}\right)$

c $(5, 2.21)$ **d** $(\sqrt{37}, 1.41)$

4 $r = \sqrt{5}$, circle centre origin, radius $\sqrt{5}$

5 Circle, centre origin, radius 4; $x^2 + y^2 = 16$

6 a Sketch should show graph of $y = x$

$\theta = \dfrac{\pi}{4} \Rightarrow \tan\theta = \tan\dfrac{\pi}{4} = 1 \Rightarrow \dfrac{y}{x} = 1 \Rightarrow y = x$

b Sketch should show graph $x = 0$

$\theta = \dfrac{\pi}{2} \Rightarrow \tan\theta = \tan\dfrac{\pi}{2}$

so gradient is undefined, hence lies on $x = 0$

c Sketch should show $y = -x$

$\theta = \dfrac{3\pi}{4} \Rightarrow \tan\theta = \tan\dfrac{3\pi}{4} = -1 \Rightarrow \dfrac{y}{x} = -1 \Rightarrow y = -x$

d Sketch should show $y = x\dfrac{\sqrt{3}}{3}$

$\theta = -\dfrac{\pi}{6} \Rightarrow \dfrac{y}{x} = \tan\left(-\dfrac{\pi}{6}\right)$ so $y = -\dfrac{\sqrt{3}}{3}x$

7 a

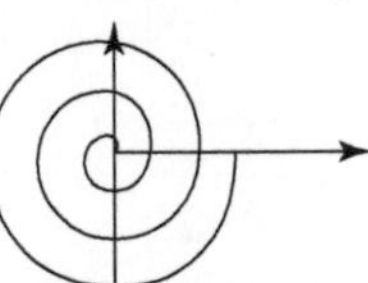

b

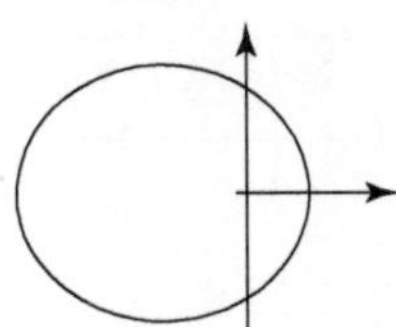

8 a i maximum = 1, minimum = 0

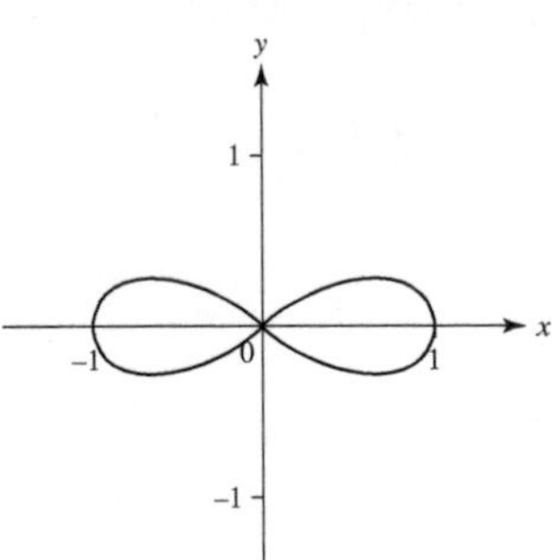

ii maximum = 1, minimum = −1

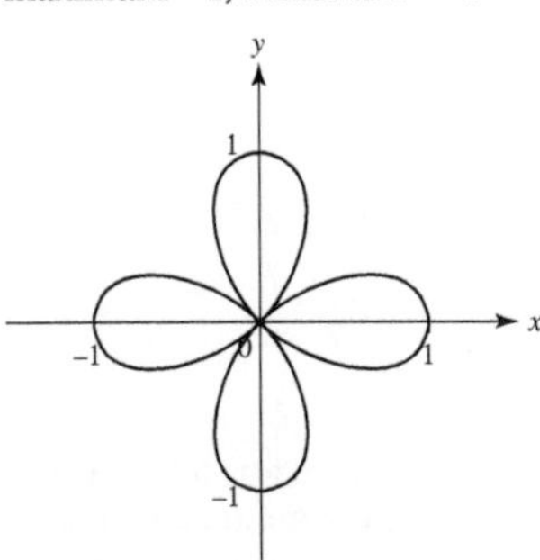

b maximum = 1, minimum = 0

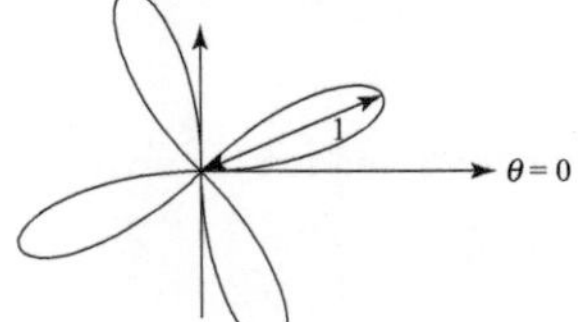

c **i** maximum = 2, minimum = 0

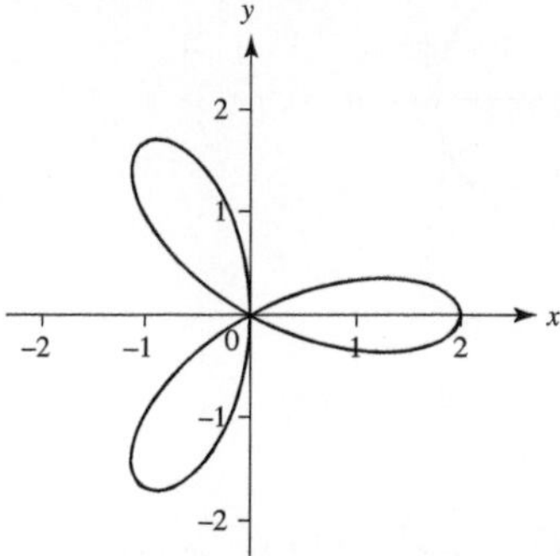

ii maximum = 2, minimum = −2

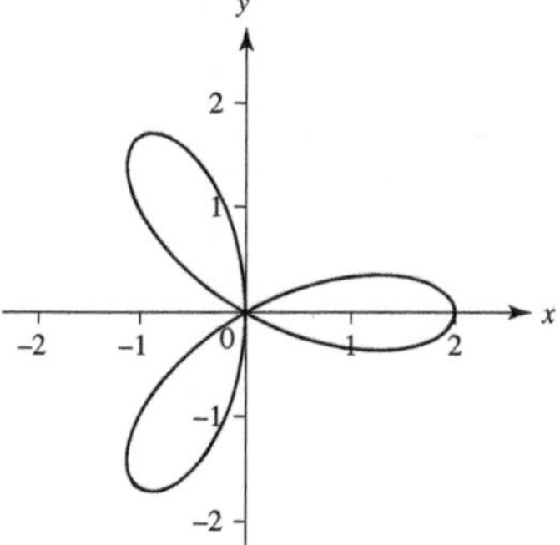

d **i** maximum = 4, minimum = 0

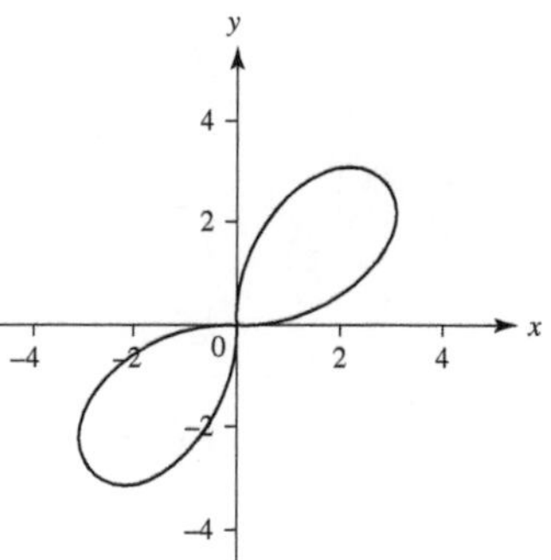

ii maximum = 4, minimum = −4

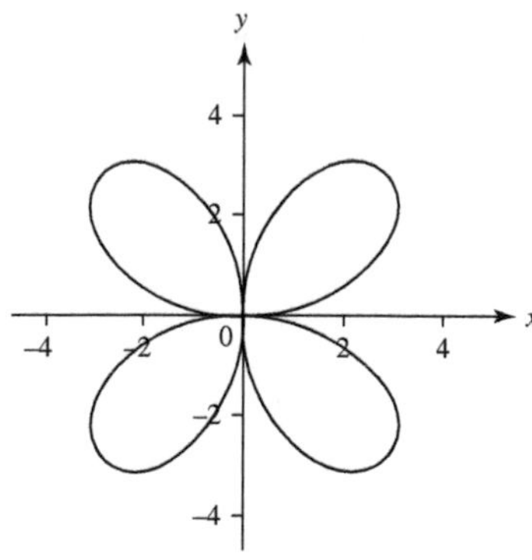

e Max = 2, min = 0

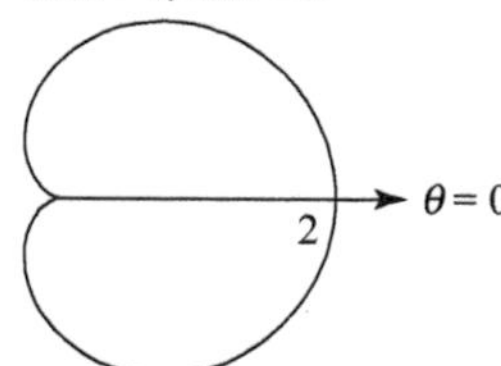

f Max = 5, min = 3

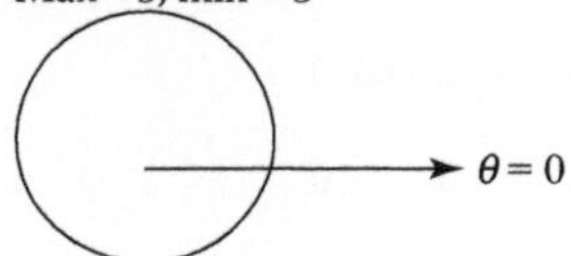

g Max = 5, min = 1

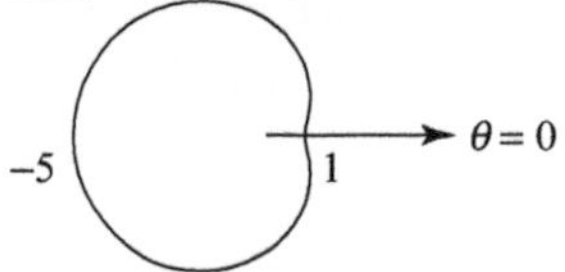

h **i** maximum = 8, minimum = 0

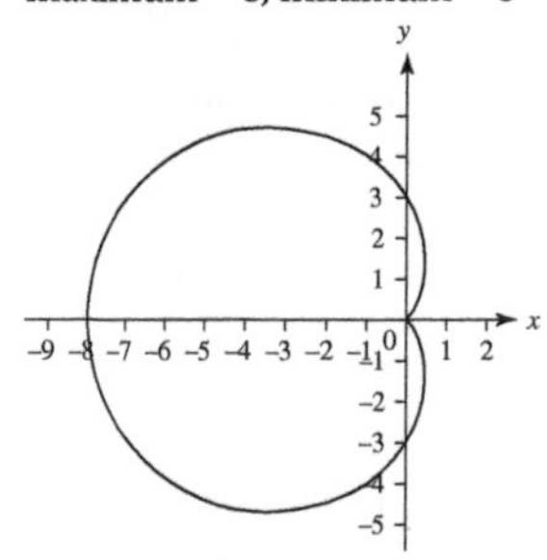

ii maximum = 8, minimum = −2

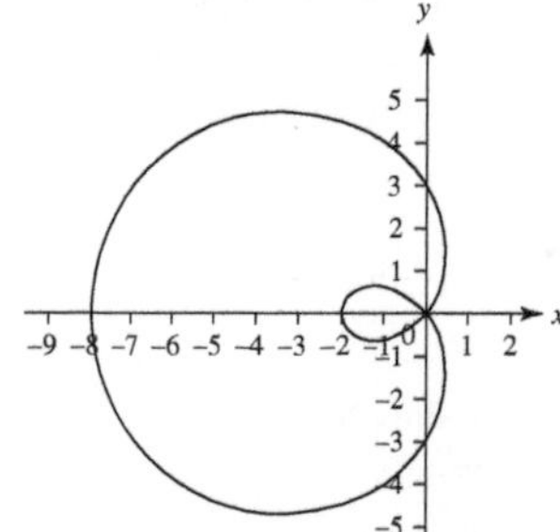

i Max = 4π, min = 0

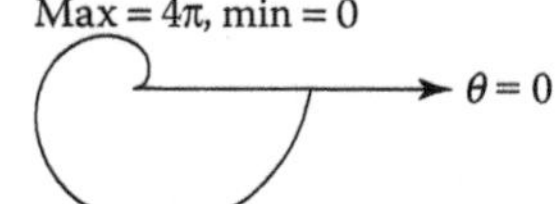

j **i** maximum = 2, minimum = 0

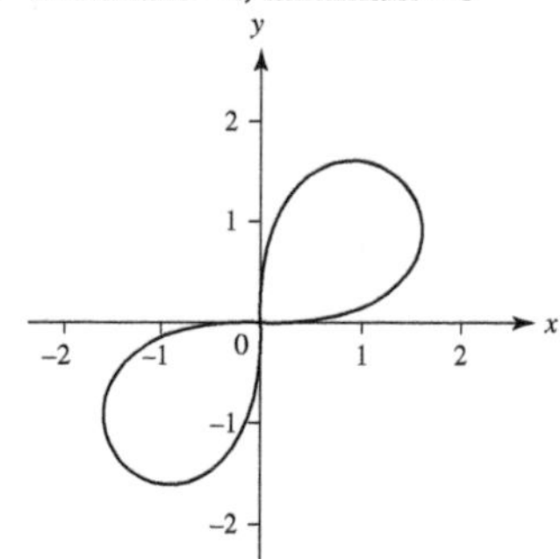

ii maximum = 2, minimum = −2

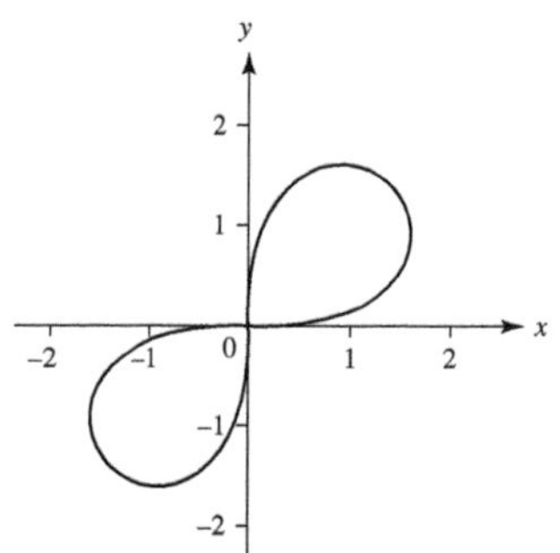

Exercise 3.3B

1 a $\theta=\dfrac{\pi}{4}$ **b** $\theta=\arctan 2$

c $\theta=\dfrac{\pi}{2}, r\geq 0$ **d** $\theta=0 \quad r\geq 0$

e $r=\dfrac{5}{\sin\theta}$ **f** $r=\dfrac{3}{\cos\theta}$

g $r(\sin\theta-\cos\theta)=1\Rightarrow r$
$=\dfrac{1}{\sin\theta-\cos\theta} \quad \theta\neq n\pi+\dfrac{\pi}{4}$

h $r=\dfrac{c}{\sin\theta-m\cos\theta} \quad \theta\neq n\pi+\arctan(m)$

i $r=5$

j $r=10\cos\theta+24\sin\theta$ **k** $r=6\cos\theta$

l $r=8\sin\theta$ **m** $r=10(\cos\theta+\sin\theta)$

n $r=\dfrac{\sin\theta}{\cos^2\theta}$

o $r^2=\dfrac{5}{1+\sin 2\theta} \quad \theta\neq n\pi-\dfrac{\pi}{4}$

2 a $y=1-x$
a straight line, gradient -1, y-intercept 1.

b $y=2x+5$
a straight line, gradient 2, y-intercept 5.

c $x=2$
a line parallel to y-axis cutting x-axis at 2.

d $y=-\dfrac{2}{3}$
a line parallel to x-axis cutting y-axis at $-\dfrac{2}{3}$

e $x^2+(y-2)^2=2^2$
a circle centre $(0,2)$ radius 2.

f $(x-4)^2+(y-3)^2=25=5^2$
a circle centre $(4,3)$ radius 5

3 a $\left(2,\dfrac{\pi}{6}\right),\left(2,\dfrac{5\pi}{6}\right)$

b $\left(\dfrac{1}{2},\pm\dfrac{\pi}{12}\right),\left(\dfrac{1}{2},\pm\dfrac{5\pi}{12}\right),\left(\dfrac{1}{2},\pm\dfrac{7\pi}{12}\right),\left(\dfrac{1}{2},\pm\dfrac{11\pi}{12}\right)$

c $\left(\dfrac{\sqrt{2}}{2},\dfrac{\pi}{8}\right),\left(\dfrac{\sqrt{2}}{2},-\dfrac{7\pi}{8}\right),$

4 Spiral has equation $r=a\theta$ and circle has equation $r=b$
Intersect when $a\theta=b\Rightarrow\theta=\dfrac{b}{a}$
So single point of intersection at $\left(b,\dfrac{b}{a}\right)$

5 a i

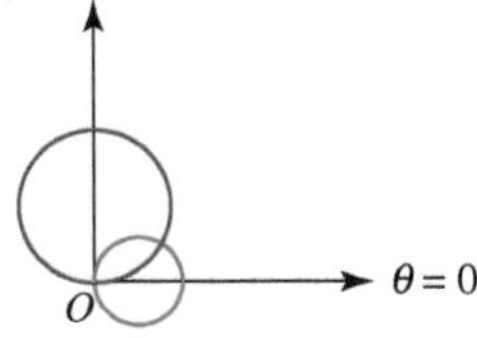

ii $\left(\dfrac{\sqrt{3}}{2},\dfrac{\pi}{6}\right)$

b i

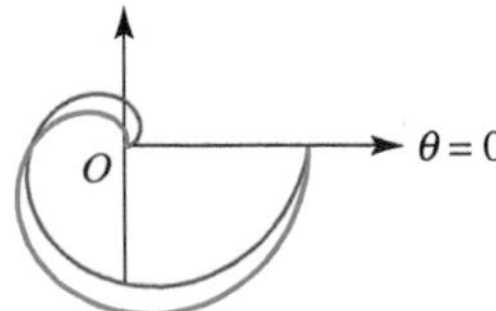

ii $(0,\ 0),(\pi,\pi),(2\pi,2\pi)$

c i

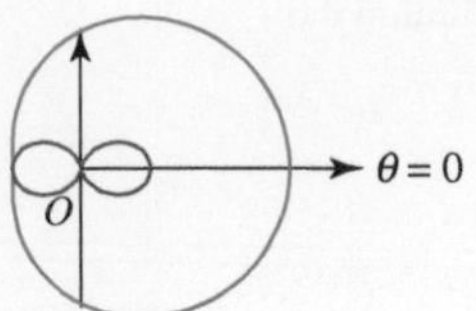

ii $(1,\pi)$

d i

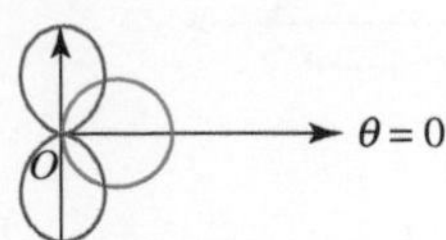

ii $\left(\dfrac{-1+\sqrt{5}}{2},\ 0.905\right),\left(\dfrac{-1+\sqrt{5}}{2},\ 5.38\right)$

e

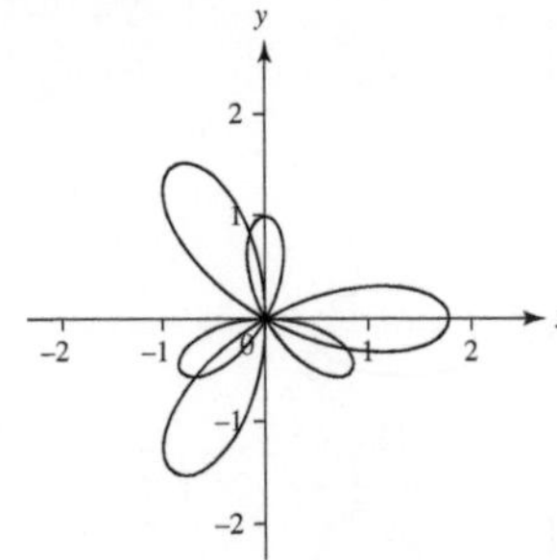

$\left(\dfrac{\sqrt{3}}{2},\dfrac{5\pi}{9}\right),\left(\dfrac{\sqrt{3}}{2},\dfrac{11\pi}{9}\right),\left(\dfrac{\sqrt{3}}{2},\dfrac{17\pi}{9}\right)$

f $\left(\dfrac{\sqrt{3}}{2},\dfrac{2\pi}{9}\right),\left(\dfrac{\sqrt{3}}{2},\dfrac{5\pi}{9}\right),\left(\dfrac{\sqrt{3}}{2},\dfrac{8\pi}{9}\right),\left(\dfrac{\sqrt{3}}{2},\dfrac{11\pi}{9}\right),$
$\left(\dfrac{\sqrt{3}}{2},\dfrac{14\pi}{9}\right),\left(\dfrac{\sqrt{3}}{2},\dfrac{17\pi}{9}\right)$

6 a $-1\leq\cos\theta\leq 1$
$-4\leq 4\cos\theta\leq 4$
$a-4\leq a+4\cos\theta\leq a+4$
So if $a>4$ then $a+4\cos\theta>0$ for all θ. So $r>0$. To pass through the pole $r=0$.
So curve never passes through pole.

b $r(\theta)=a+4\cos\theta$
$r(-\theta)=a+4\cos(-\theta)=a+4\cos\theta=r(\theta)$
Since $r(-\theta)=r(\theta)$ curve is symmetrical about the initial line.

c i

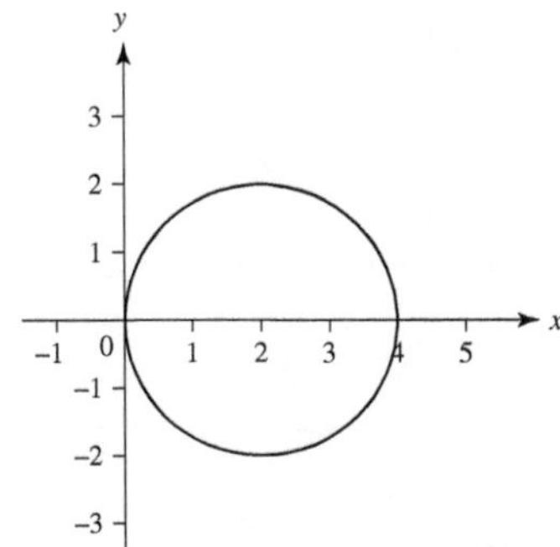

ii

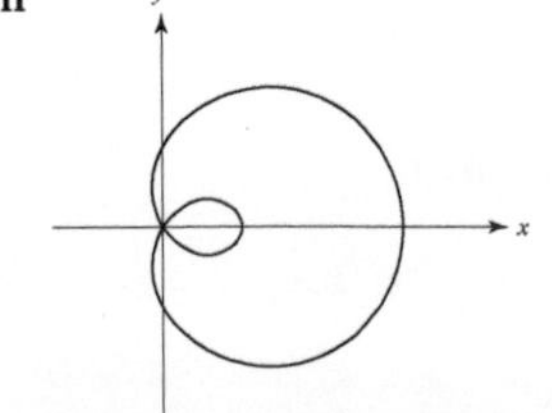

iii

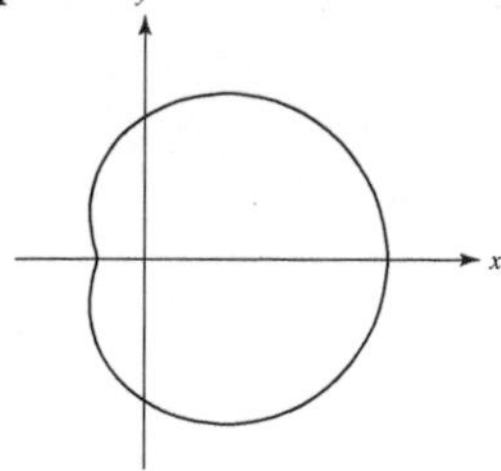

iv

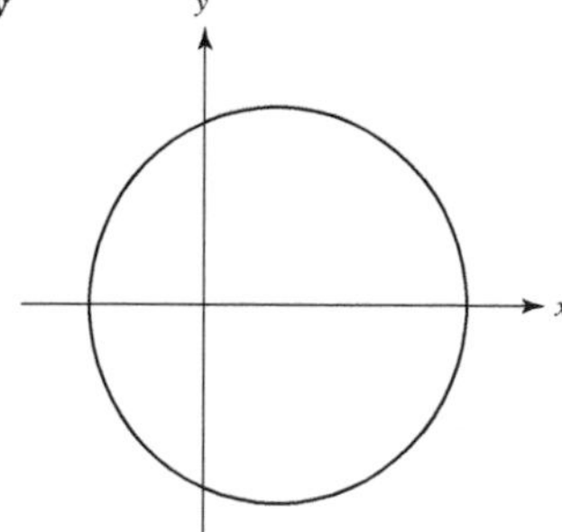

d $-1 \le \cos\theta \le 1$

$\Rightarrow a-4 \le r \le a+4$

7 $\sec^2\theta = 2\tan\theta$

$\Rightarrow 1 = 2\sin\theta\cos\theta = \sin 2\theta$

$\Rightarrow 2\theta = \frac{\pi}{2}, \frac{5\pi}{2} \ldots$

$\Rightarrow \theta = \frac{\pi}{4}, \frac{5\pi}{4} \ldots$

$\theta = \frac{\pi}{4}, r = \sqrt{2}$

$\theta = \frac{5\pi}{4}, r = -\sqrt{2}$

So they intersect at $\left(\sqrt{2}, \frac{\pi}{4}\right)$t and $\left(-\sqrt{2}, \frac{5\pi}{4}\right)$.

But these are the same point, so they only intersect at one point.

8 a

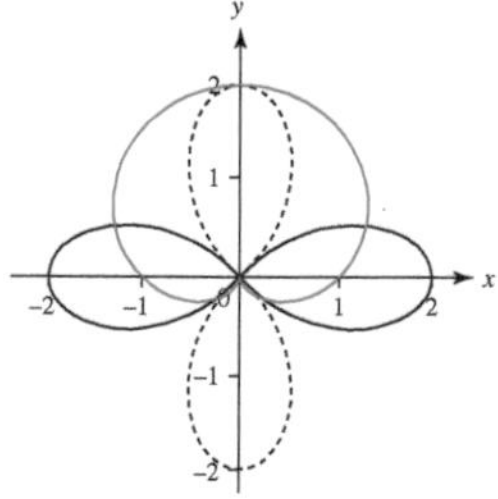

(1.390, 0.401), (1.390, 2.741), (0.360, –0.695), (0.360, –2.45)

Since $1 + \sin\theta$ is always non-negative, extending r to permit negative values does not add any additional solutions.

b

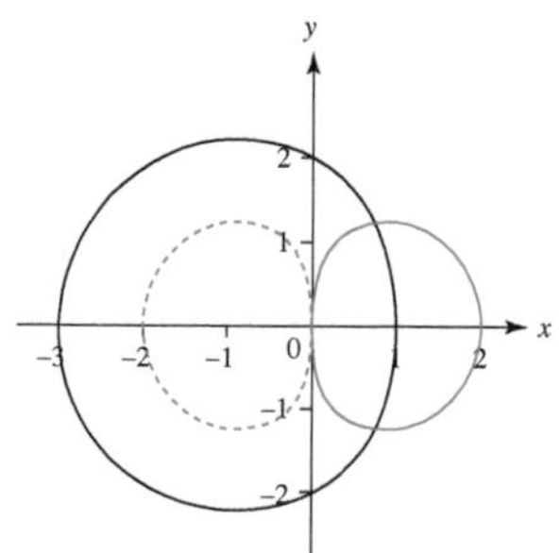

Points of intersection are (1.464, 1.005), (1.464, –1.005)
Since $2 - \cos\theta$ is always non-negative, extending r to permit negative values does not add any additional solutions.

9 $2\sin^2\theta = \cos 2\theta = 1 - 2\sin^2\theta$

$4\sin^2\theta = 1$

$\sin^2\theta = \frac{1}{4}$

$\sin\theta = \pm\frac{1}{2}$

$\theta = \pm\frac{\pi}{6}, \pm\frac{5\pi}{6}$

$r = 2\sin^2\theta = \frac{1}{2}$ so points of intersection are

$\left(\frac{1}{2}, \pm\frac{\pi}{6}\right), \left(\frac{1}{2}, \pm\frac{5\pi}{6}\right)$

The second pair of points are reflections of the first pair through the y-axis, so these form a rectangle, with Cartesian coordinates $\left(\pm\frac{\sqrt{3}}{4}, \pm\frac{1}{4}\right)$,

Area $= 2 \times \frac{1}{4} \times \frac{\sqrt{3}}{4} = \frac{\sqrt{3}}{4}$

Exercise 3.4A

1 a

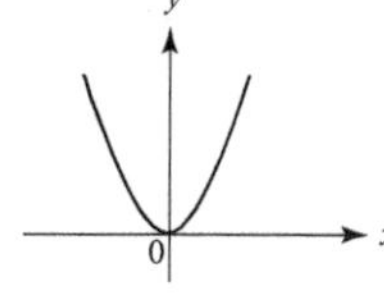

b

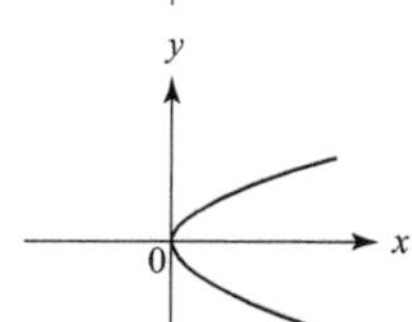

c

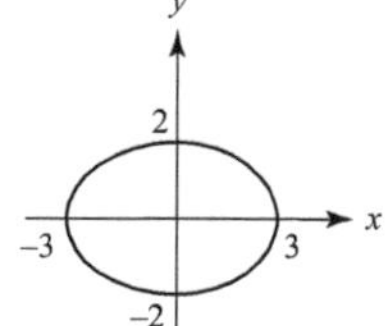

d

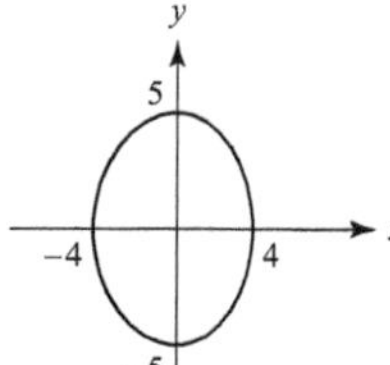

e

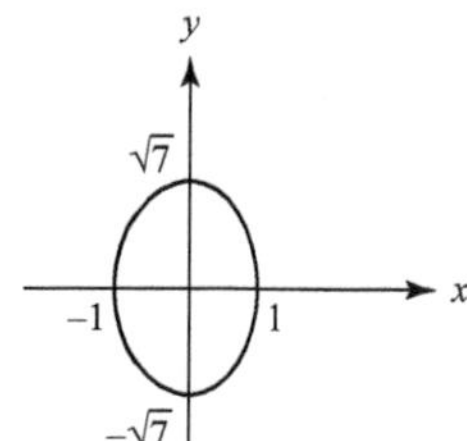

f

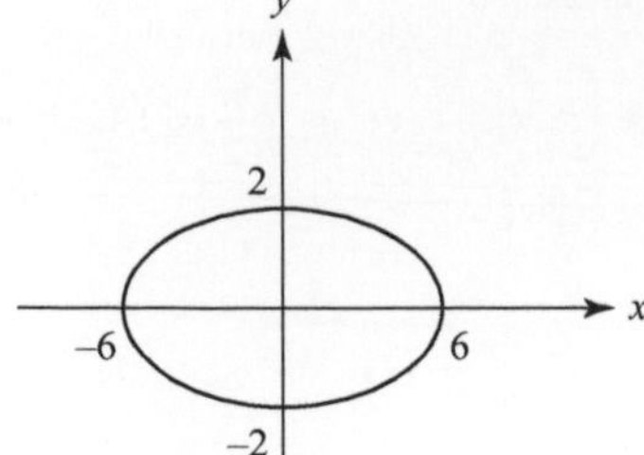

g

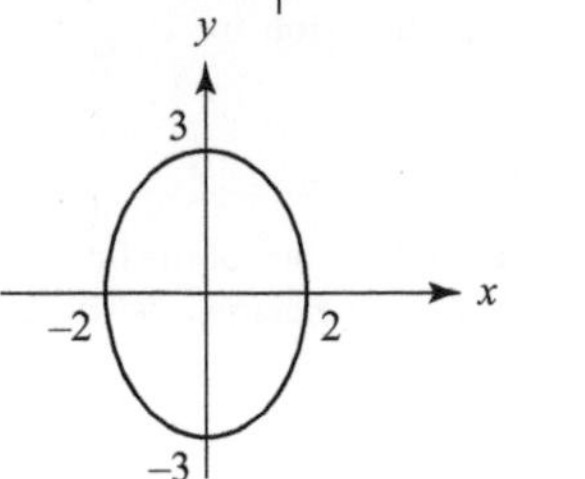

h

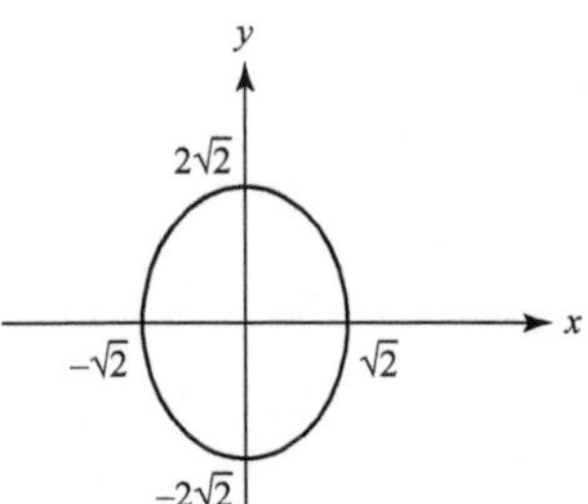

2 a

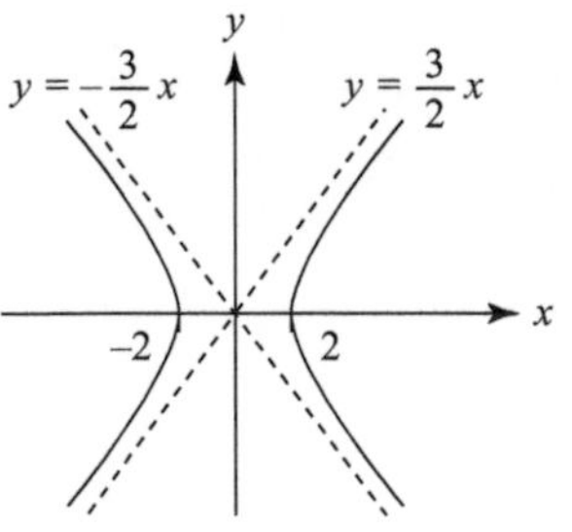

Asymptotes: $y = \pm\frac{3}{2}x$

b

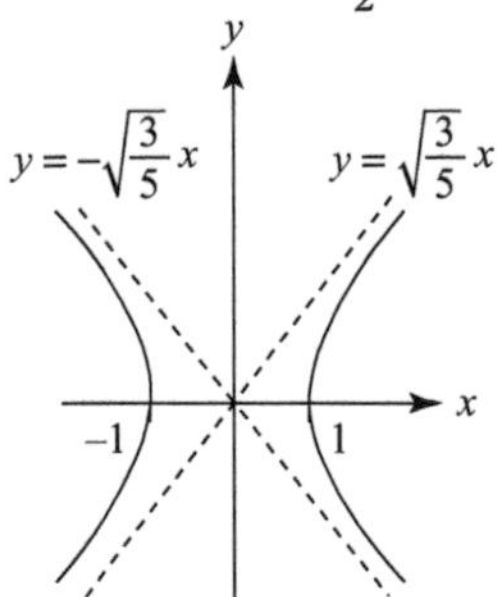

Asymptotes: $y = \pm\sqrt{\frac{3}{5}}x$

c

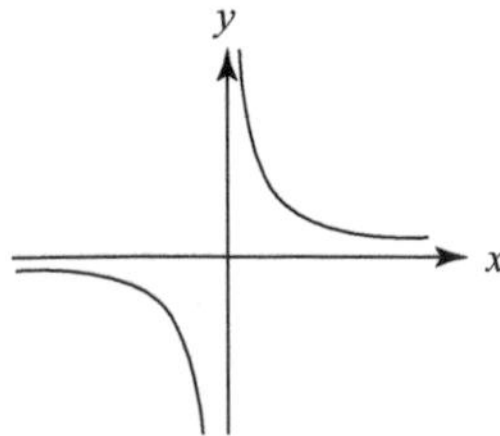

Asymptotes $x = 0,\ y = 0$

d

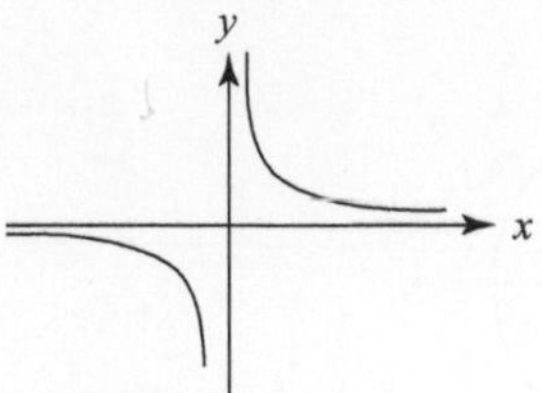

Asymptotes $x = 0,\ y = 0$

e

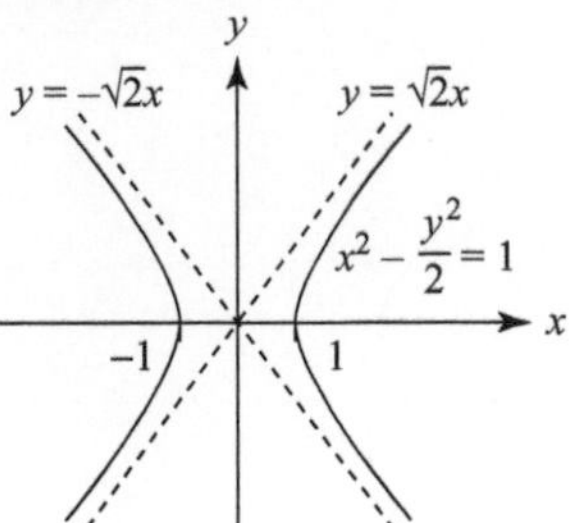

Asymptotes: $y = \pm\sqrt{2}x$

f

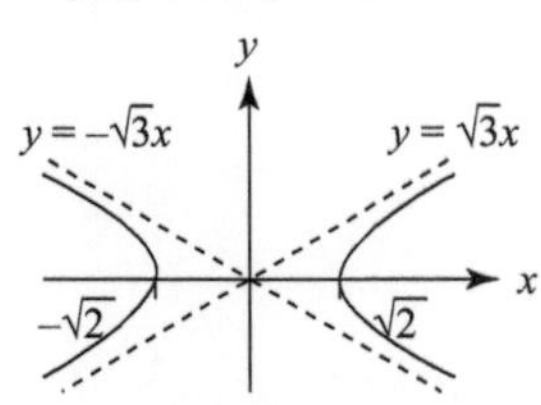

Asymptotes: $y = \pm\sqrt{3}x$

g

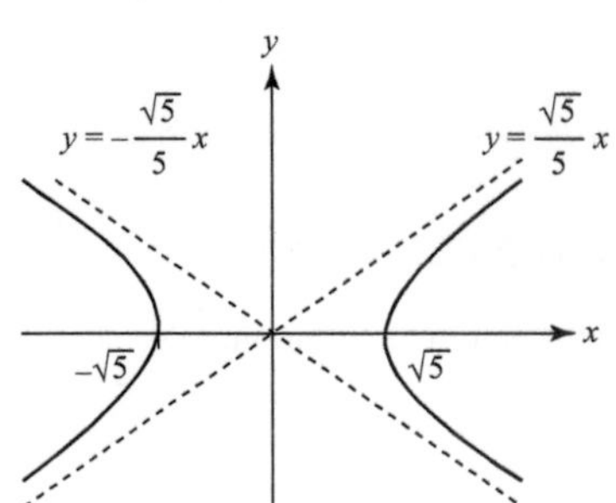

Asymptotes: $y = \pm\frac{\sqrt{5}}{5}x$

h

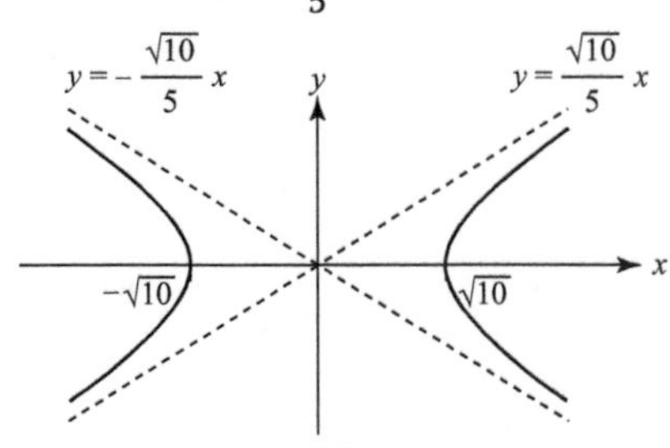

Asymptotes: $y = \pm\frac{\sqrt{10}}{5}x$

3 a

y

x

0

b (0, 0) and (5, 10)

4 a

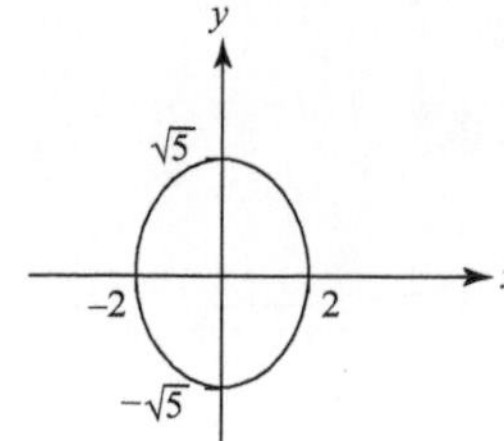

b (−2, 0) and $\left(\frac{2}{9}, \frac{20}{9}\right)$

5 a

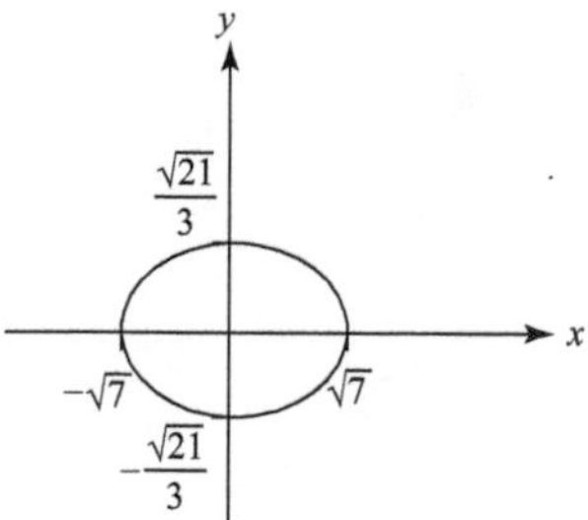

b $x = \frac{7-3y}{2}$

Substitute to give $\left(\frac{7-3y}{2}\right)^2 + 3y^2 = 7$

$\Rightarrow 49 - 42y + 9y^2 + 12y^2 = 28$

$\Rightarrow 21y^2 - 42y + 21 = 0$

$\Rightarrow y^2 - 2y + 1 = 0$

$\Rightarrow (y-1)^2 = 0$

So only one solution: $y = 1, x = 2$

Therefore must be a tangent

6 a Ellipse $\frac{x^2}{27} + \frac{y^2}{8} = 1$

b Hyperbola $\frac{x^2}{256} - \frac{y^2}{128} = 1$

7 $\frac{9x^2}{8} - \frac{y^2}{8} = 1$

8 a Rectangular hyperbola

b

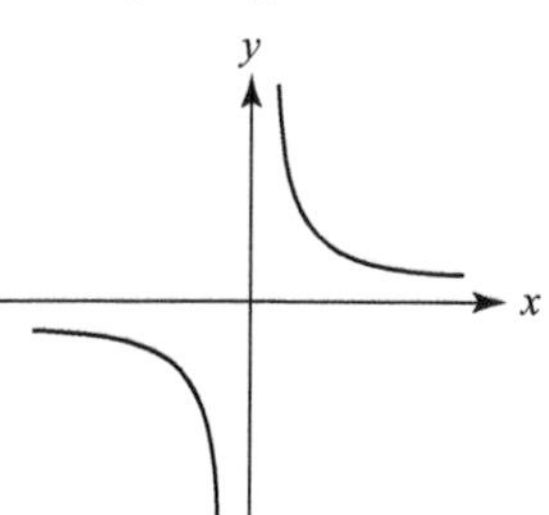

Asymptotes at $x = 0,\ y = 0$

c $\frac{14}{3}\sqrt{10}$ (or 14.8)

9 a

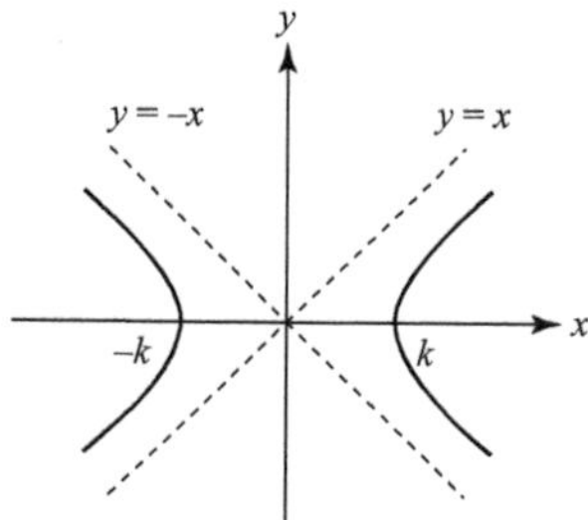

b The tangent lines $y = \pm x$ both make an angle $\frac{\pi}{4}$ with the x and y axes, and thus make an angle $\frac{\pi}{2}$ with each other, which is a right angle.

Exercise 3.4B

1 a

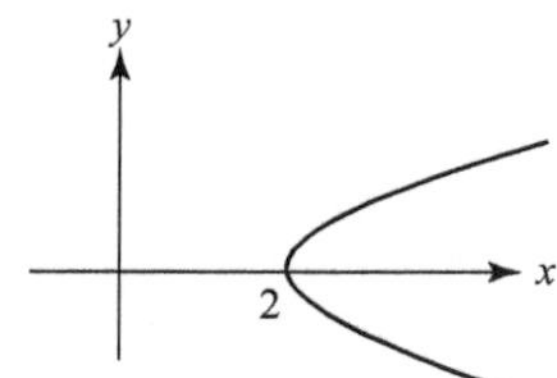

b

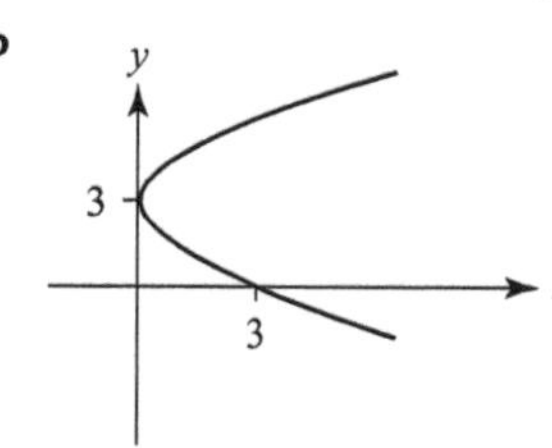

c

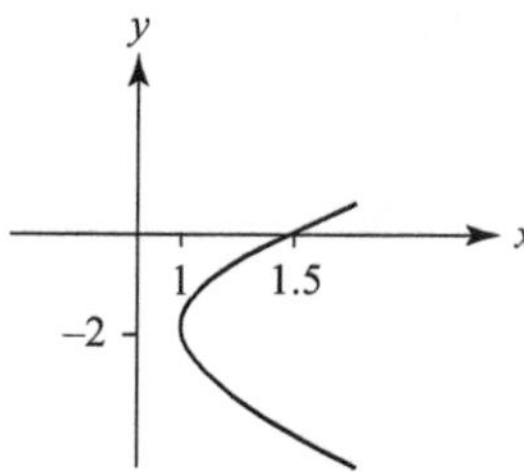

d

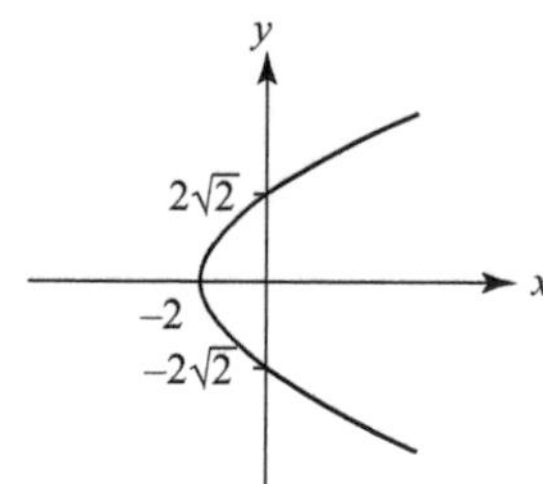

e

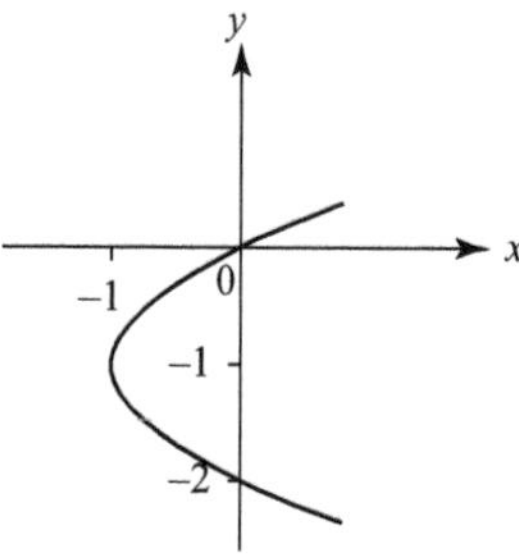

f

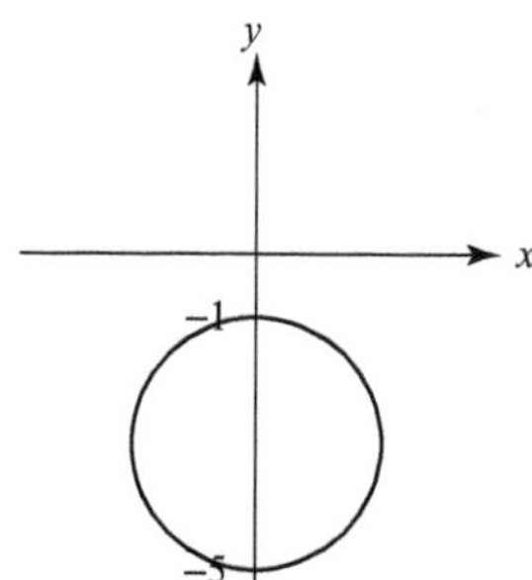

g

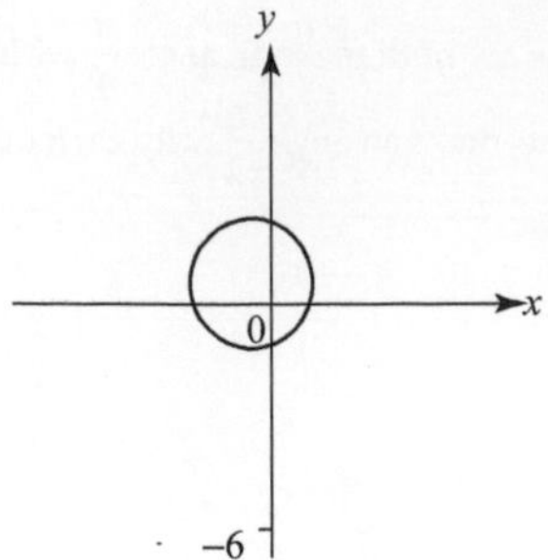

h

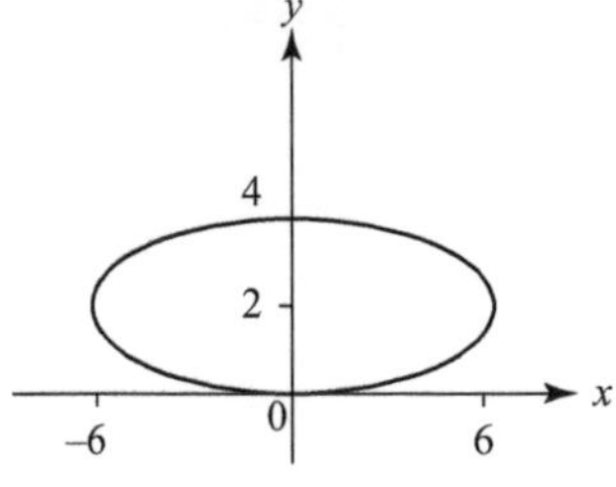

i

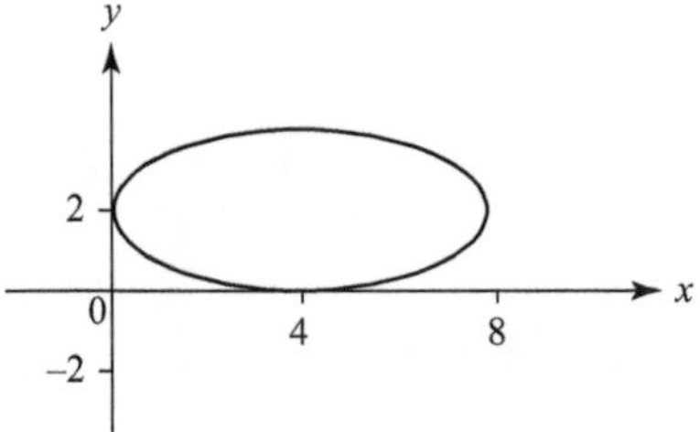

j

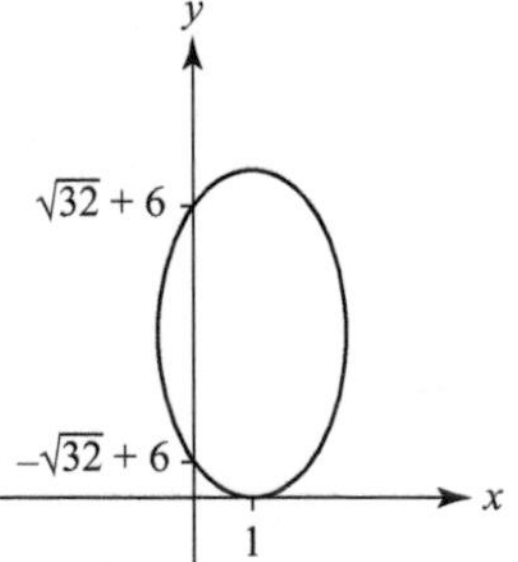

2 a

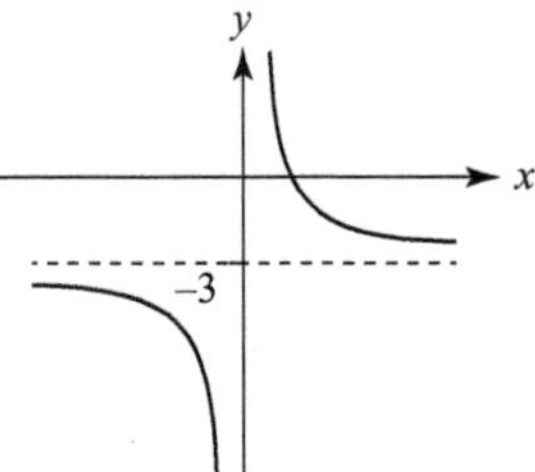

Asymptotes $x = 0$, $y = -3$

b

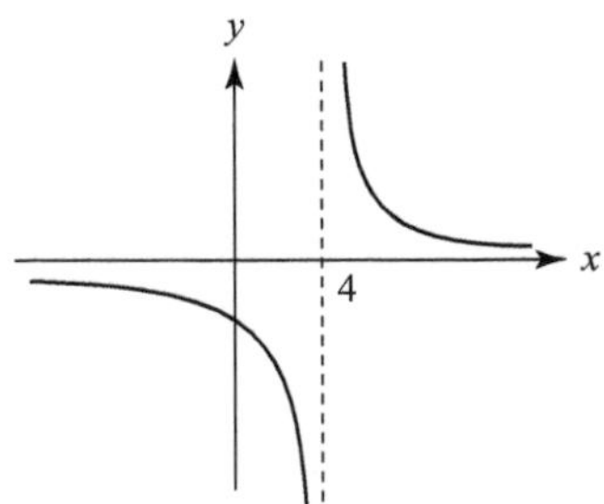

Asymptotes $x = 4$, $y = 0$

c

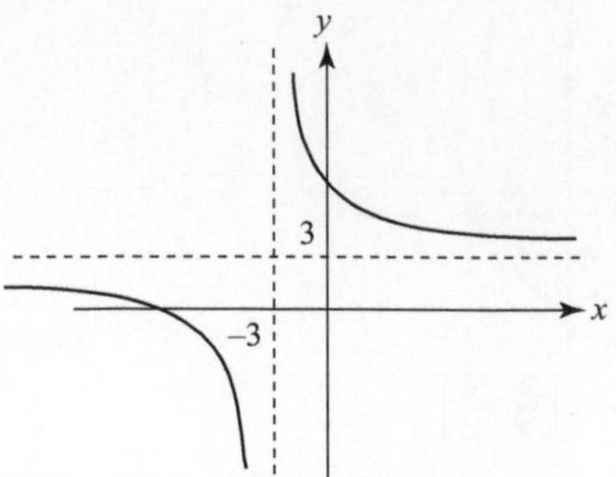

Asymptotes $x = -3$, $y = 3$

d $x(y-2) = 4$

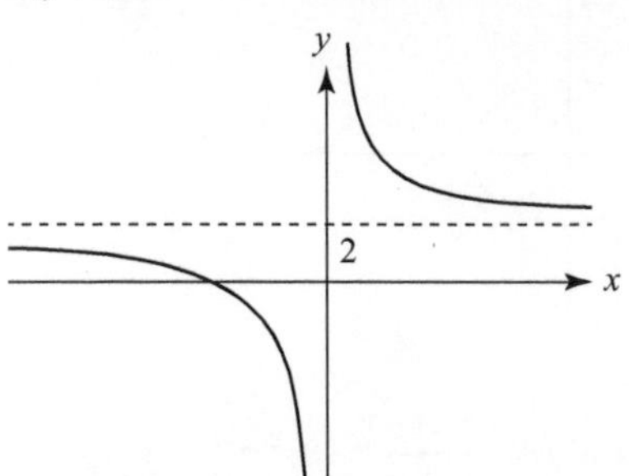

Asymptotes $x = 0$, $y = 2$

e $y(x-5) = 2$

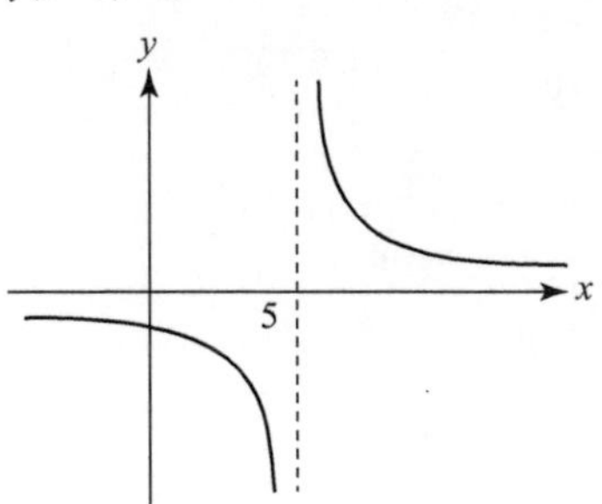

Asymptotes $x = 5$, $y = 0$

f

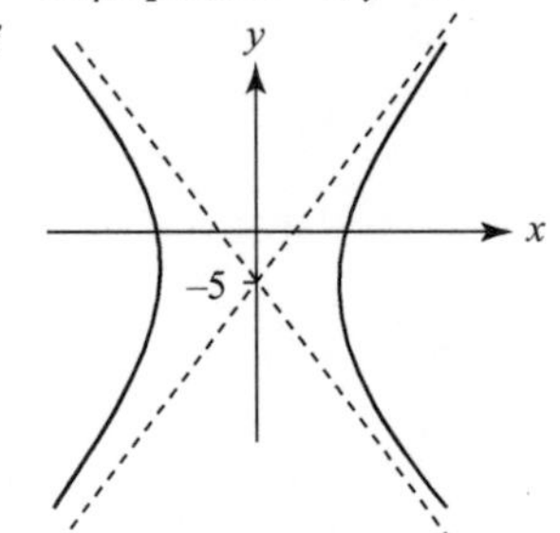

Asymptotes $y + 5 = \pm\dfrac{3}{\sqrt{2}}x$

g

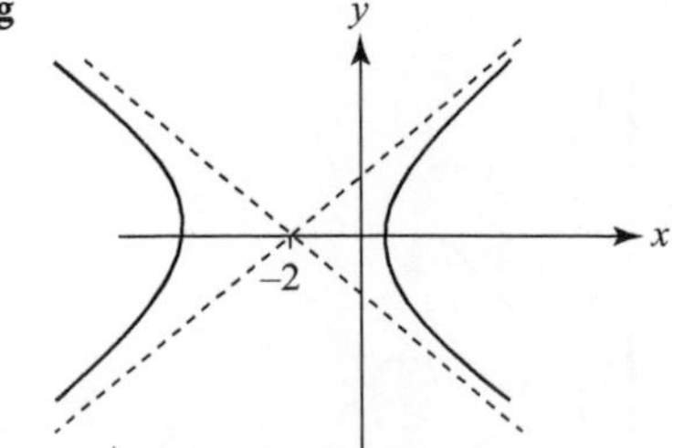

Asymptotes $y = \pm\dfrac{5}{4}(x + 2)$

h

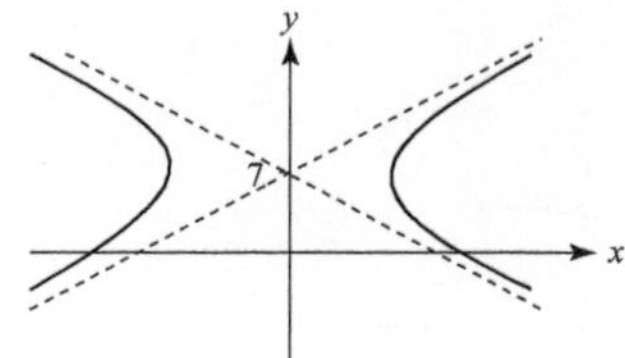

Asymptotes $y-7=\pm\frac{1}{2}x$

i

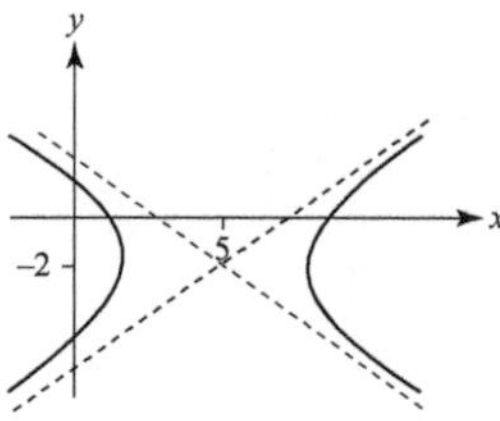

Asymptotes $y+2=\pm\frac{\sqrt{2}}{2}(x-5)$

j

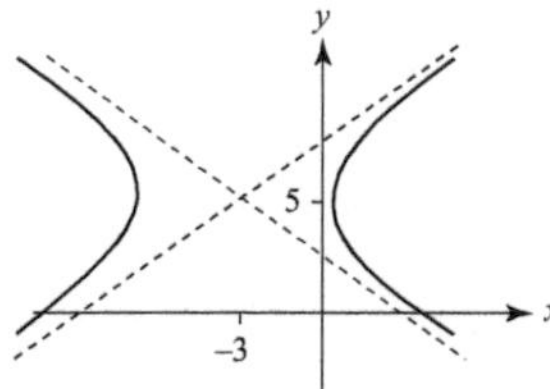

Asymptotes $y-5=\pm\frac{\sqrt{3}}{3}(x+3)$

3 a

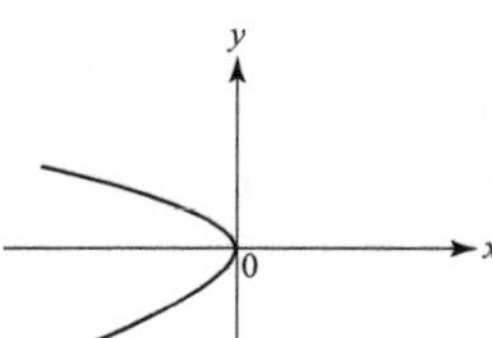

b

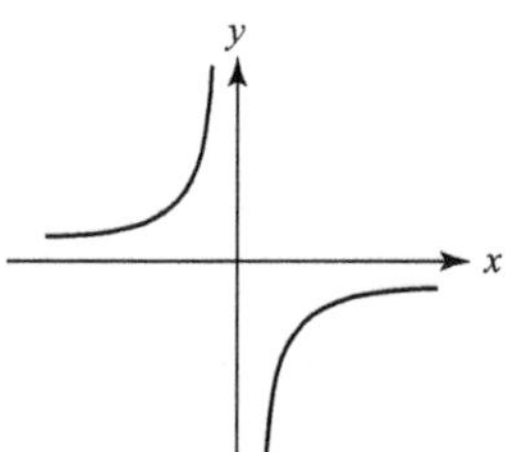

4 a $y^2=8\left(\frac{x}{3}\right)\Rightarrow y^2=\frac{8}{3}x$

b $(4y)^2=8x\Rightarrow y^2=\frac{x}{2}$

c $(-y)^2=8x\Rightarrow y^2=8x$

d $y^2=8(-x)\Rightarrow y^2=-8x$

5 a $(y+3)^2=-4(x-2)$ **b** $x^2=-4y$

6 a $\frac{x^2}{12}+\frac{y^2}{16}=1$ **b** $\frac{x^2}{300}+\frac{y^2}{4}=1$

c $\frac{x^2}{12}+\frac{y^2}{4}=1$ **d** $\frac{x^2}{4}+\frac{y^2}{12}=1$

7 $\frac{(x-3)^2}{8}+\frac{(y+1)^2}{4}=1$

8 a i $-x(y-1)=4$ **ii** $y=1, x=0$

b i $(x-5)(y-3)=4$ **ii** $y=3, x=5$

c i $y(x-1)=4$ **ii** $x=1, y=0$

d i $y(x+1)=4$ **ii** $y=0, x=-1$

9 a i $\frac{x^2}{9}-\frac{y^2}{729}=1$ **ii** $y=\pm 9x$

b i $\frac{x^2}{9}-\frac{y^2}{81}=1$ **ii** $y=\pm 3x$

c i $\frac{(x+1)^2}{9}-\frac{(y-4)^2}{81}=1$ **ii** $y-4=\pm 3(x+1)$

d i $\frac{y^2}{9}-\frac{x^2}{81}=1$ **ii** $x=\pm 3y\Rightarrow y=\pm\frac{1}{3}x$

10 a $y+k=\pm\frac{1}{a}(x-k)$

b Asymptotes $x-k=\pm\frac{1}{a}(y+k)$

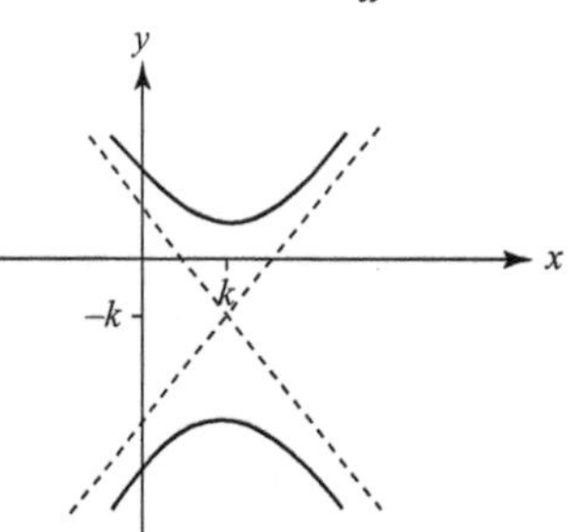

Exercise 3.5A

1 a 0 **b** 1.54 **c** $\frac{3}{4}$

d $\frac{5}{3}$ **e** 1 **f** $\frac{3}{5}$

2 a

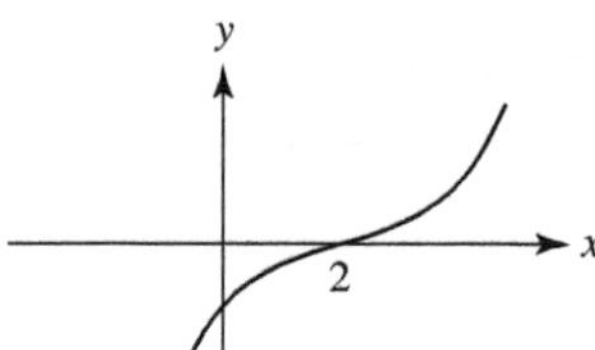

b

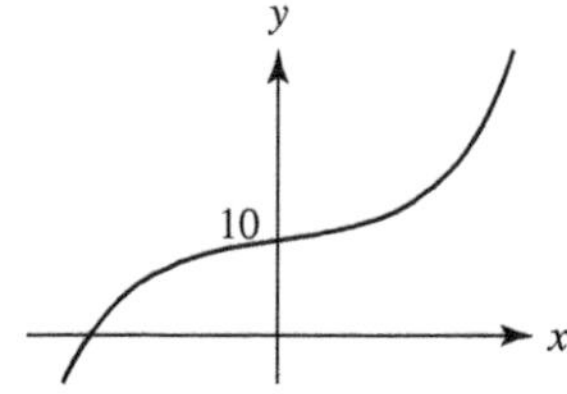

c

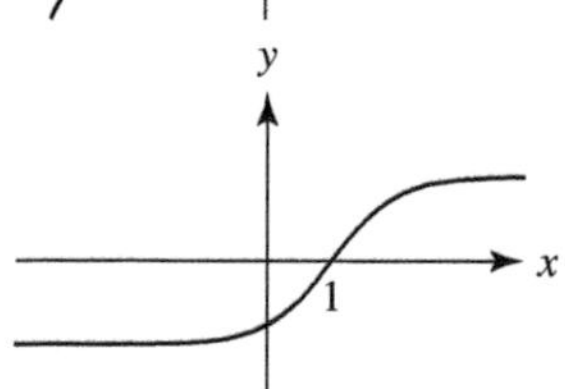

d

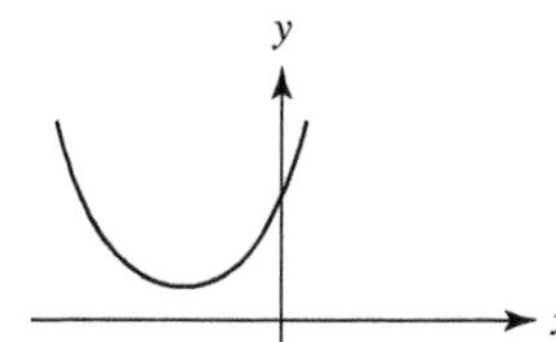

e

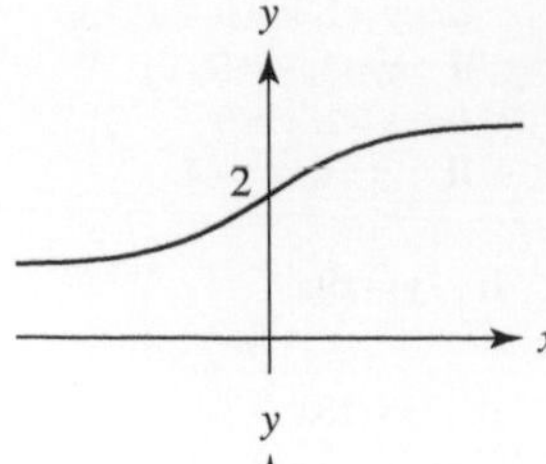

f

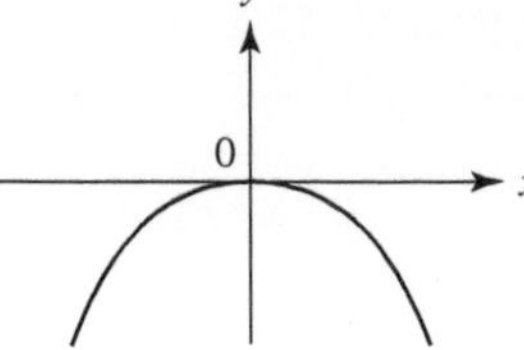

3 a $y = f(x)$

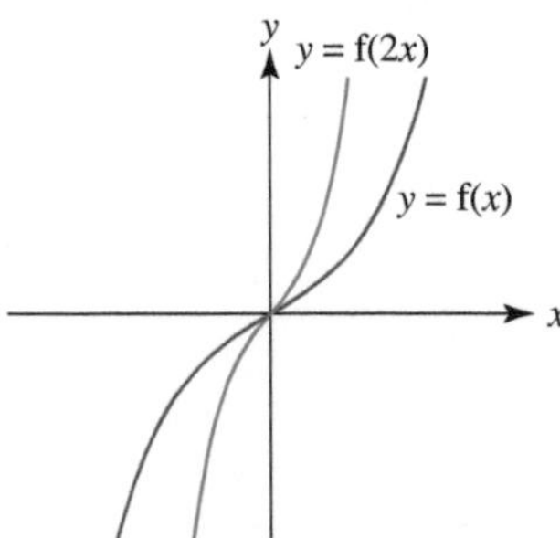

b $\sinh(2x) = 2 \Rightarrow x = 0.722$

4

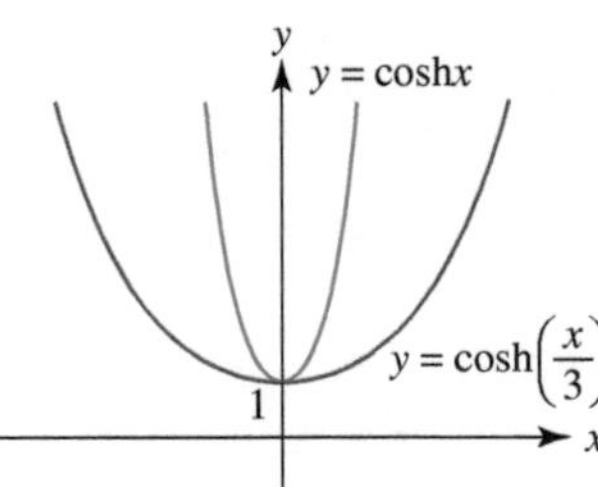

5 a

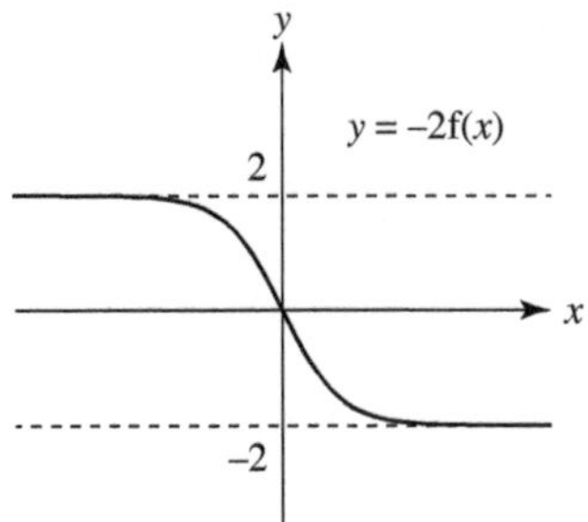

b $-2\tanh(x) = 1 \Rightarrow \tanh(x) = -\frac{1}{2}$
$\Rightarrow x = -0.549$

6 a

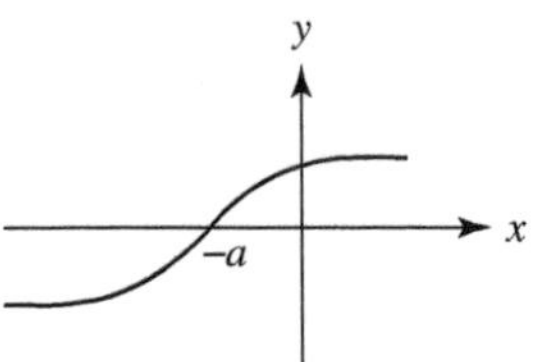

b $y = \pm 1$

7 a

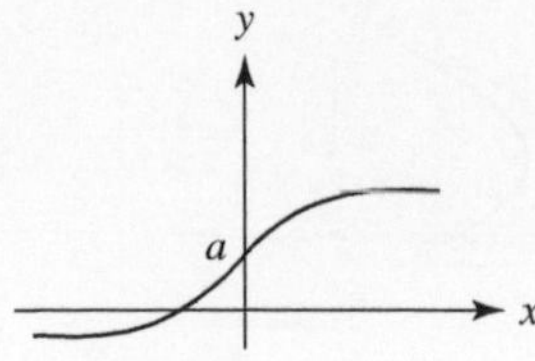

b $y = a+1; y = a-1$

8 a

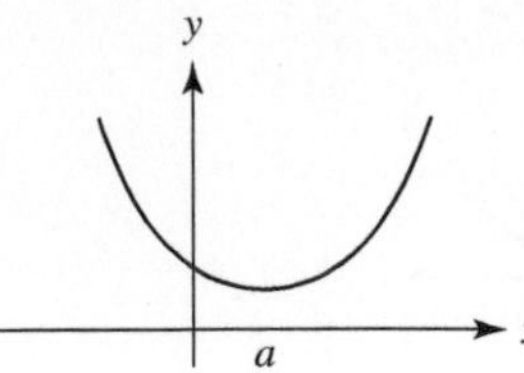

b

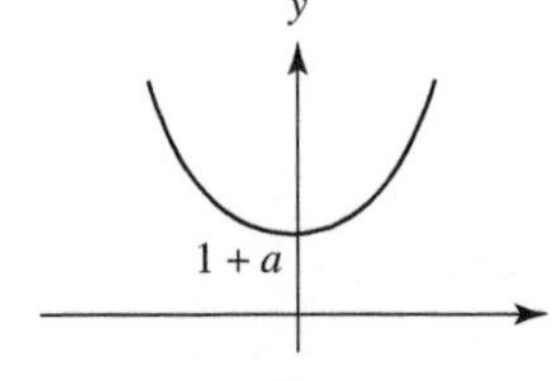

c

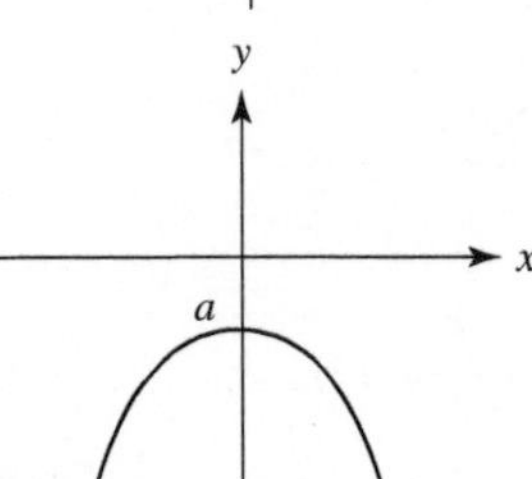

9 a 2.31 b 1.32, −1.32 c −0.549
d 0.763, −2.76 e 0.698 f 0.549

10 a 0 b 0 c ln(2)
d $\ln(4), \ln\left(\frac{1}{4}\right)$ e ln(2) f ln(9)

11 a $\ln\frac{1}{2}$ b $0, \ln\left(\frac{1}{5}\right)$
c $\ln\left(\frac{3 \pm \sqrt{6}}{3}\right)$ d $\frac{1}{4}\ln 2$

Exercise 3.5B

1 a $\ln(2+\sqrt{5}), -\ln(3+\sqrt{10})$ b $0, \pm\ln\left(\frac{3+\sqrt{5}}{2}\right)$
c $\pm\frac{1}{2}\ln(5+2\sqrt{6})$ d $\pm\frac{1}{2}\ln(3+2\sqrt{2})$

2 a 0 b $\ln(-4+\sqrt{17}), \ln(1+\sqrt{2})$
c $0, \ln\left(\frac{-2+\sqrt{13}}{3}\right)$ d $\frac{1}{2}\ln 3$
e $\frac{1}{2}\ln(3+\sqrt{10})$ f $\pm\frac{1}{3}(\ln(4+\sqrt{15}))$
g $\frac{1}{6}\ln(3)$

3 $LHS = \left(\frac{e^x + e^{-x}}{2}\right)^2 - \left(\frac{e^x - e^{-x}}{2}\right)^2$
$= \frac{1}{4}(e^{2x} + 2 + e^{-2x}) - \frac{1}{4}(e^{2x} - 2 + e^{-2x})$
$= \frac{1}{4}(2+2)$
$= 1 = RHS$

4 a $\sinh(A)\cosh(B)+\sinh(B)\cosh(A)$

$$\equiv\left(\frac{e^A-e^{-A}}{2}\right)\left(\frac{e^B+e^{-B}}{2}\right)+\left(\frac{e^B-e^{-B}}{2}\right)\left(\frac{e^A+e^{-A}}{2}\right)$$

$$\equiv\frac{e^{A+B}+e^{A-B}-e^{-(A-B)}-e^{-(A+B)}}{4}$$

$$+\frac{e^{A+B}+e^{-(A-B)}-e^{A-B}-e^{-(A+B)}}{4}$$

$$\equiv\frac{2e^{A+B}-2e^{-(A+B)}}{4}$$

$$\equiv\frac{e^{A+B}-e^{-(A+B)}}{2}$$

$\equiv\sinh(A+B)$ as required

b $\sinh(A)\cosh(B)-\sinh(B)\cosh(A)$

$$=\left(\frac{e^A-e^{-A}}{2}\right)\left(\frac{e^B+e^{-B}}{2}\right)-\left(\frac{e^B-e^{-B}}{2}\right)\left(\frac{e^A+e^{-A}}{2}\right)$$

$$=\frac{e^{A+B}+e^{A-B}-e^{-(A-B)}-e^{-(A+B)}}{4}$$

$$-\frac{e^{A+B}-e^{A-B}+e^{-(A-B)}-e^{-(A+B)}}{4}$$

$$=\frac{2e^{A-B}-2e^{-(A-B)}}{4}$$

$$=\frac{e^{A-B}-e^{-(A-B)}}{2}$$

$=\sinh(A-B)$ as required

c $2\sinh(x)\cosh(x)=2\left(\frac{e^x+e^{-x}}{2}\right)\left(\frac{e^x-e^{-x}}{2}\right)$

$$=\frac{(e^{2x}+1-1-e^{2x})}{2}$$

$$=\frac{(e^{2x}-e^{-2x})}{2}$$

$=\sinh(2x)$ as required

5 a $\tanh x=\frac{\sinh x}{\cosh x}$

$$=\frac{\frac{e^x-e^{-x}}{2}}{\frac{e^x+e^{-x}}{2}}$$

$$=\frac{e^x-e^{-x}}{e^x+e^{-x}}$$

$$=\frac{e^{2x}-1}{e^{2x}+1}$$

b $\frac{2\tanh x}{1+\tanh^2 x}\equiv\frac{2\left(\frac{e^{2x}-1}{e^{2x}+1}\right)}{1+\left(\frac{e^{2x}-1}{e^{2x}+1}\right)^2}$

$$\equiv\frac{2(e^{2x}-1)(e^{2x}+1)}{(e^{2x}+1)^2+(e^{2x}-1)^2}$$

$$\equiv\frac{2(e^{4x}-1)}{(e^{4x}+2e^{2x}+1)+(e^{4x}-2e^{2x}+1)}$$

$$\equiv\frac{2(e^{4x}-1)}{2e^{4x}+2}$$

$$\equiv\frac{e^{4x}-1}{e^{4x}+1}$$

$\equiv\tanh(2x)$ as required

6 a $2\cosh^2 x-1\equiv2\left(\frac{e^x+e^{-x}}{2}\right)^2-1$

$$\equiv\frac{2(e^{2x}+2+e^{-2x})}{4}-1$$

$$\equiv\frac{e^{2x}+e^{-2x}+2}{2}-\frac{2}{2}$$

$$\equiv\frac{e^{2x}+e^{-2x}}{2}$$

$\equiv\cosh(2x)$ as required

b $2\cosh^2 x-1+\cosh x=5$

$2\cosh^2 x+\cosh x-6=0$

$\Rightarrow\cosh x=\frac{3}{2},(-2)$

$\Rightarrow\cosh x=\pm\ln\left(\frac{3}{2}+\sqrt{\left(\frac{3}{2}\right)^2-1}\right)$

$=\pm\ln\left(\frac{3+\sqrt{5}}{2}\right)$

7 Let $y=\sinh^{-1}x$ then $x=\sinh y=\frac{e^y-e^{-y}}{2}$

$2x=e^y-e^{-y}$

$e^{2y}-2xe^y-1=0$

$(e^y-x)^2=x^2+1$

$e^y=x\pm\sqrt{x^2+1}$

$e^y>0$ therefore, $x-\sqrt{x^2+1}$ not valid as $x^2+1>x^2$

so $\sqrt{x^2+1}>x\Rightarrow x-\sqrt{x^2+1}<0$

So $y=\ln(x+\sqrt{x^2+1})$ as required

8 a Let $y=\tanh^{-1}x$ then $x=\tanh y=\frac{e^y-e^{-y}}{e^y+e^{-y}}$

$xe^y+xe^{-y}=e^y-e^{-y}$

$(1+x)e^{-y}=(1-x)e^y$

$\frac{1+x}{1-x}=e^{2y}$

$2y=\ln\left(\frac{1+x}{1-x}\right)$

So $\text{artanh}\, x=\frac{1}{2}\ln\left(\frac{1+x}{1-x}\right)$ as required

b The range of $y=\tanh x$ is $-1<y<1$

9 $\pm\ln\left(\frac{5}{2}+\sqrt{\frac{21}{4}}\right)=\pm\ln\left(\frac{5+\sqrt{21}}{2}\right)$

10 $\ln\left(\frac{1+\sqrt{5}}{2}\right)$, $\ln(-1+\sqrt{2})$

11 $\pm\ln(2+\sqrt{3})$, $\pm\ln\left(\frac{3+\sqrt{5}}{2}\right)$

12 a $3\sinh\left(\frac{x}{2}\right)-\sinh(x)\equiv 3\sinh\left(\frac{x}{2}\right)-2\sinh\left(\frac{x}{2}\right)\cosh\left(\frac{x}{2}\right)$

$\equiv \sinh\left(\frac{x}{2}\right)\left(3-2\cosh\left(\frac{x}{2}\right)\right)$ as required

b $0, \pm 2\ln\left(\frac{3+\sqrt{5}}{2}\right)$

13 a $\frac{1}{1+\cosh x}+\frac{1}{1-\cosh x}\equiv\frac{(1-\cosh x)+(1+\cosh x)}{(1+\cosh x)(1-\cosh x)})$

$\equiv\frac{2}{1-\cosh^2(x)}$

$\equiv -\frac{2}{\sinh^2(x)}$ as required

b $\ln\left(\frac{1+\sqrt{5}}{2}\right), \ln\left(\frac{-1+\sqrt{5}}{2}\right)$

14 a $2\sinh x\cosh x=\cosh^2 x$

$\cosh^2 x-2\sinh x\cosh x=0$

$\cosh x(\cosh x-2\sinh x)=0$

$\cosh x\neq 0$ so $\cosh x-2\sinh x=0\Rightarrow\cosh x=2\sinh x$

$\Rightarrow\frac{1}{2}=\frac{\sinh x}{\cosh x}=\tanh x$ as required

b $x=\frac{1}{2}\ln(3)$

15 1, 1.84

16 a $\sqrt{5}$

b $\frac{2\sqrt{5}}{5}$

17 a $2\sqrt{2}$

b $\frac{2\sqrt{2}}{3}$

18 a $\frac{\sqrt{3}}{3}$

b $\frac{2\sqrt{3}}{3}$

19 a $\frac{d}{dx}(\sinh x)=\frac{d}{dx}\left(\frac{e^x-e^{-x}}{2}\right)=\frac{e^x+e^{-x}}{2}=\cosh x$

$\frac{d}{dx}(\cosh x)=\frac{d}{dx}\left(\frac{e^x+e^{-x}}{2}\right)=\frac{e^x-e^{-x}}{2}=\sinh x$

b $\frac{d}{dx}(\sinh x)_{x=0}=\frac{e^0+e^0}{2}=\cosh 0=1$

$\frac{d}{dx}(\cosh x)_{x=0}=\frac{e^0-e^0}{2}=\sinh 0=0$

c When $x>0$, $e^{-x}>0\Rightarrow e^x+e^{-x}>e^x>e^x-e^{-x}$

$\Rightarrow e^x+e^{-x}>e^x-e^{-x}\Rightarrow\frac{e^x+e^{-x}}{2}>\frac{e^x-e^{-x}}{2}\Rightarrow\cosh x>\sinh x$

So the gradient of $y=\sinh x$ is greater than the gradient of $y=\cosh x$ when $x>0$.

d 1

Review exercise 3

1

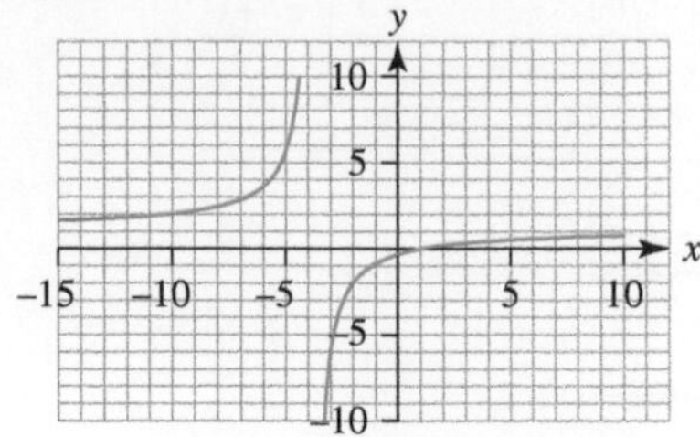

2 a

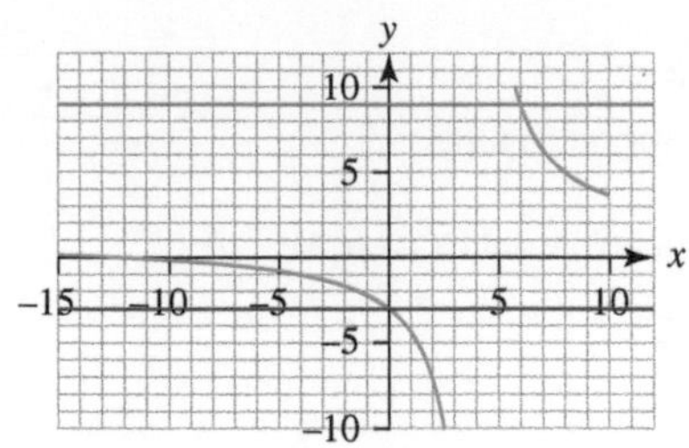

Intercepts are $(-12, 0)$ and $(0, -3)$

Asymptotes are $x=4$ and $y=1$

Solution to the inequality is $x<0$ or $x>6$

b

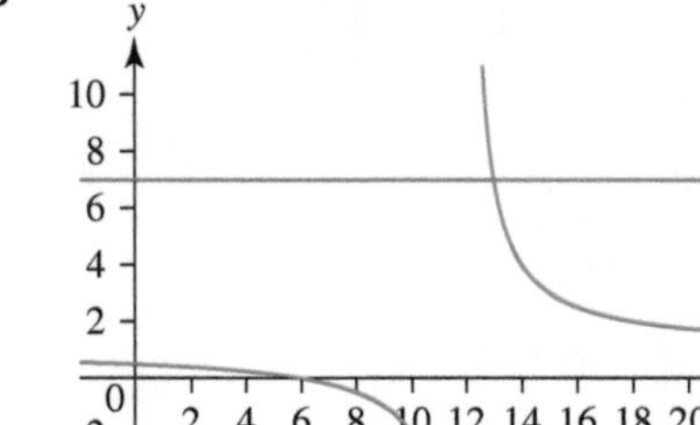
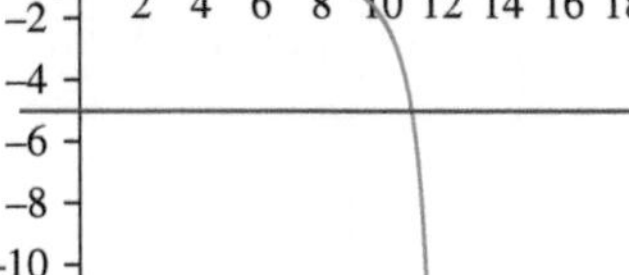

Intercepts are $(6, 0)$ and $\left(0, \frac{1}{2}\right)$

Asymptotes are $x=12$ and $y=1$

Solution to the inequality is $x\leq 11$ or $x\geq 13$

3 a

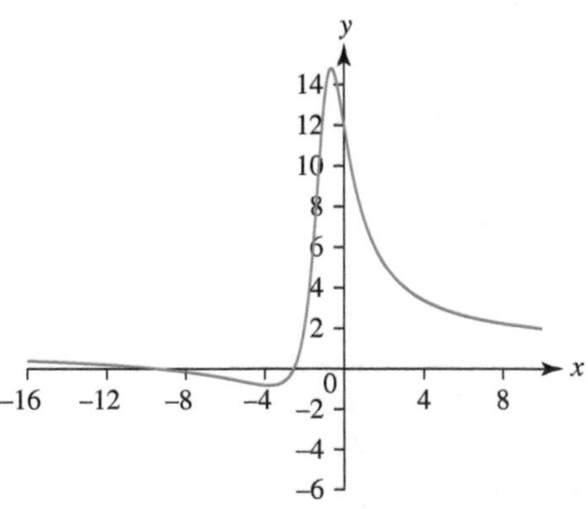

Intercepts are $(-6\pm 2\sqrt{3}, 0)$ (i.e.$\approx(-9.46, 0)$ and $(-2.54, 0)$) and $(0, 12)$

Asymptotes are $y=1$ and there are no vertical asymptotes;

b $y = \dfrac{x^2 + 12x + 24}{x^2 + 2x - 3}$

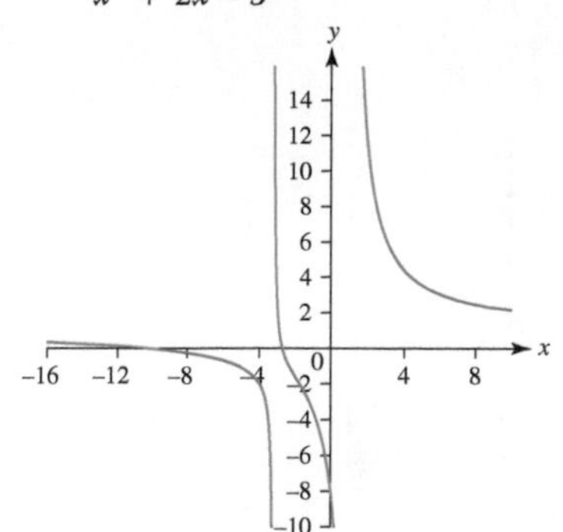

Intercepts are $(-6 \pm 2\sqrt{3}, 0)$ (i.e. $\approx (-9.46, 0)$ and $(-2.54, 0)$) and $(0, -8)$

Asymptotes are $x = -3$ and $x = 1$ and $y = 1$

4 $y < 1$ or $y \geq 5$.

5 $\left(\dfrac{5}{2}, 5\right)$.

no maximum.

6 **a** $\left(\dfrac{5}{2}, \dfrac{5\sqrt{3}}{2}\right)$ **b** $\left(\dfrac{5}{2}, -\dfrac{5\sqrt{3}}{2}\right)$

7 **a** $\left(4, \dfrac{\pi}{6}\right)$ **b** $\left(4, \dfrac{5\pi}{6}\right)$

c $\left(4, \dfrac{7\pi}{6}\right)$ **d** $\left(4, \dfrac{11\pi}{6}\right)$

8 **a** $\tan^{-1} 3$ **b** $\dfrac{1}{\sin\theta - 2\cos\theta}$

c $r = 4$ **d** $2(\cos\theta + \sin\theta)$

9 **a** centre is $(0, 2)$ radius 2

b $(x^2 + y^2 + 4y) = 4(x^2 + y^2)$

10 **a** **i**

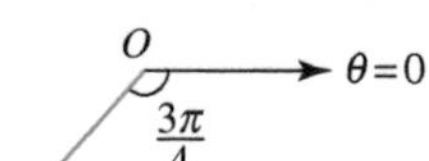

ii $y = x$

b **i**

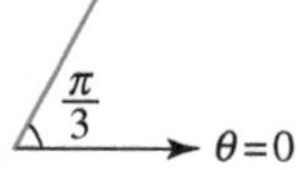

ii $y = \sqrt{3}x$

11 **a** **i**

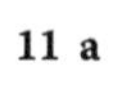

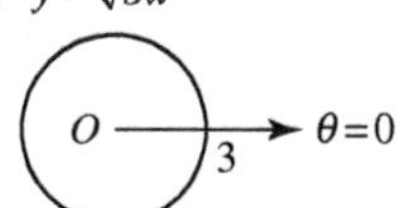

ii max = 3, min = 3

b **i**

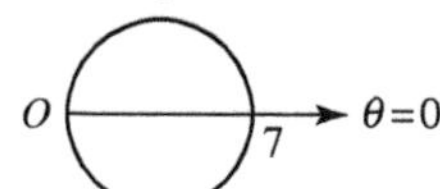

ii max = 7, min = 0

c **i**

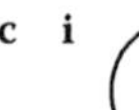

ii max = 4, min = 0

d **i**

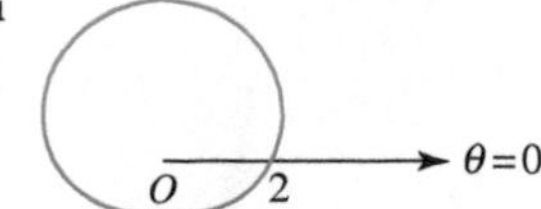

ii max = 3, min = 1

e **i**

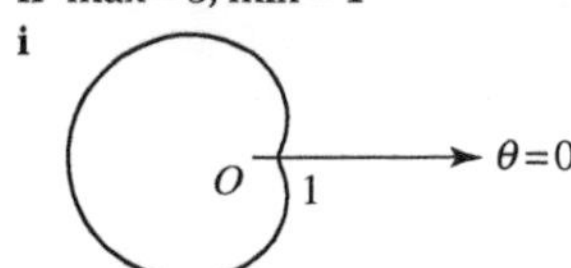

ii max = 7, min = 1

f **i**

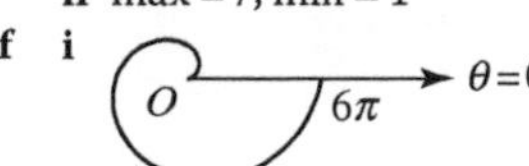

ii max = 6π , min = 0

12 $\left(2, \dfrac{\pi}{3}\right), \left(2, \dfrac{5\pi}{3}\right)$

13 If $r \geq 0$ on both curves, then solutions are

$\left(\dfrac{\sqrt{3}}{2}, \dfrac{\pi}{6}\right), \left(\dfrac{\sqrt{3}}{2}, \dfrac{7\pi}{6}\right)$

If $r < 0$ on both curves, then there are 4 additional solutions:

$\left(\dfrac{\sqrt{3}}{2}, \dfrac{\pi}{3}\right), \left(\dfrac{\sqrt{3}}{2}, \dfrac{5\pi}{6}\right), \left(\dfrac{\sqrt{3}}{2}, \dfrac{4\pi}{3}\right), \left(\dfrac{\sqrt{3}}{2}, \dfrac{11\pi}{6}\right)$

14 **a**

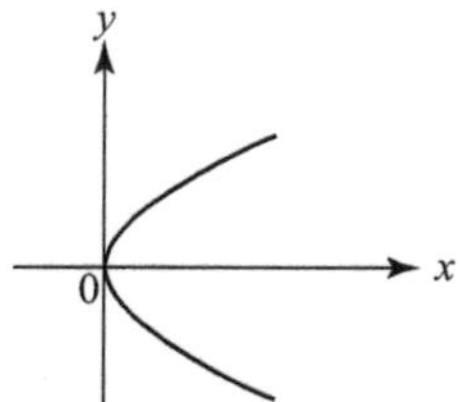

b (9, 6) and (1, −2)

15 **a**

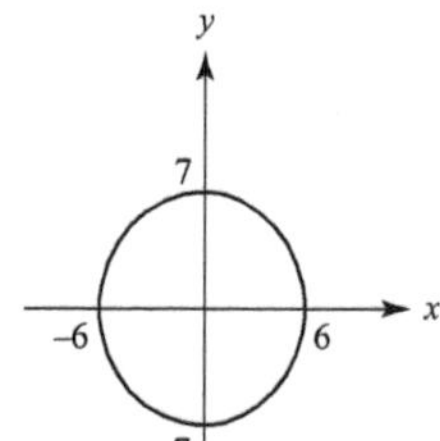

b

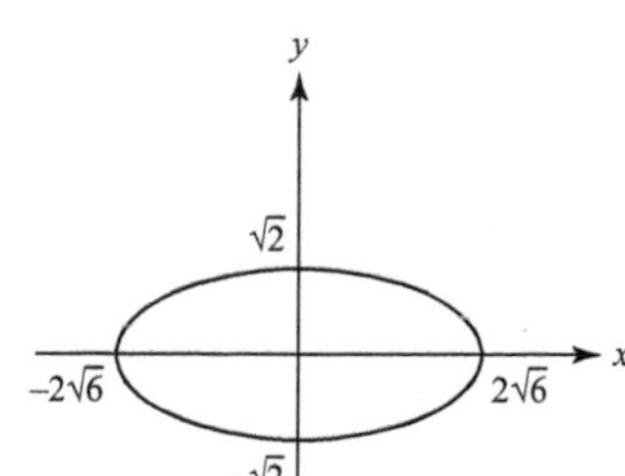

16

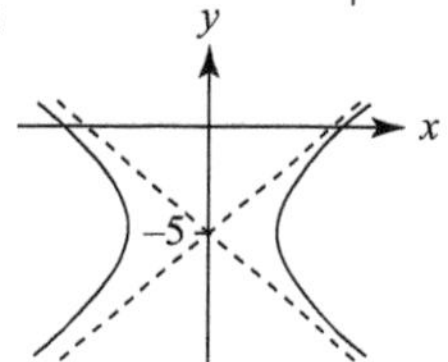

Asymptotes $y = \pm\dfrac{\sqrt{2}}{2}x$

17 a

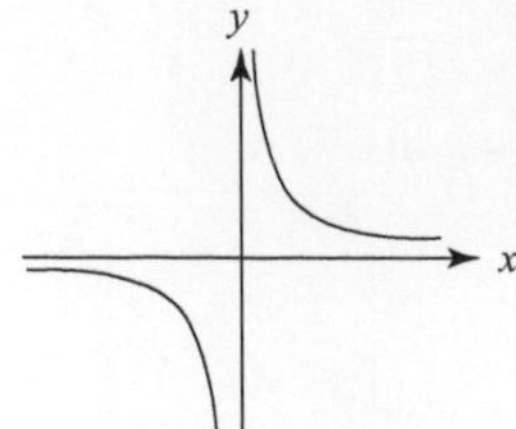

Asymptotes, $x=0$, $y=0$

b $y=1-x$

Substitute to give $x(1-x)=5 \Rightarrow -x^2+x-5=0$

$b^2-4ac=1^2-4(-1)(-5)=-19$

$b^2-4ac<0$ so no solutions

Hence they do not intersect.

18 a

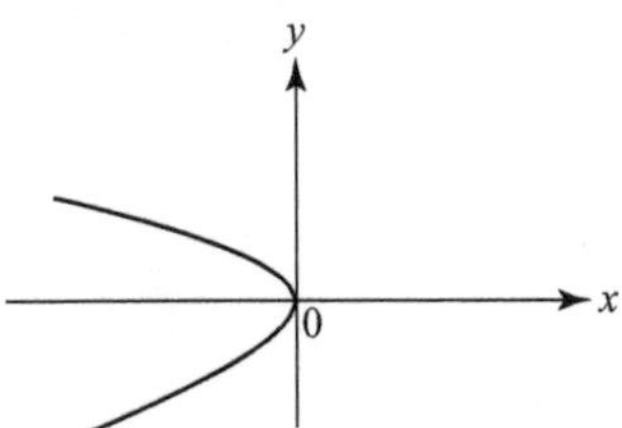

b

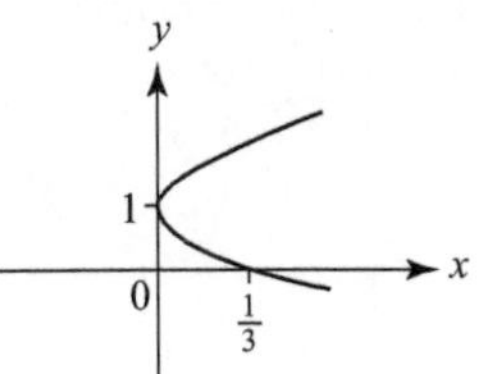

19 a

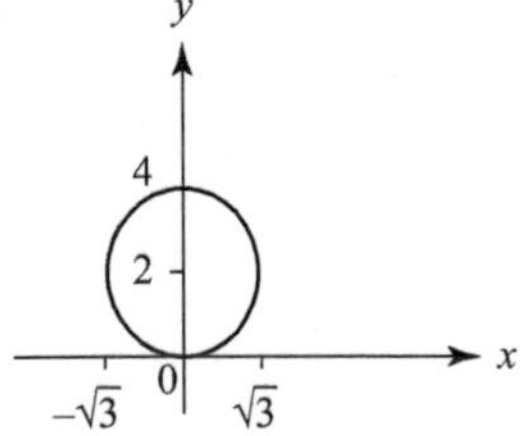

b

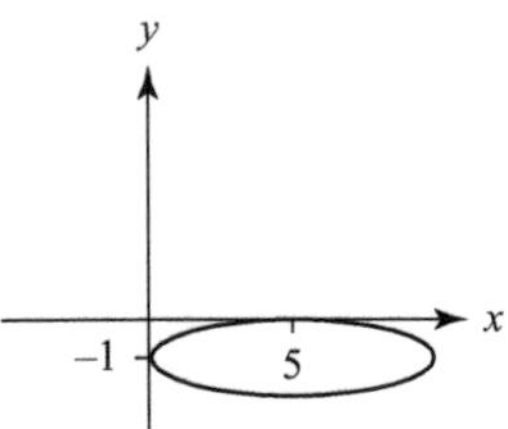

20 a Asymptotes at $y=-5\pm\frac{2}{\sqrt{3}}x$

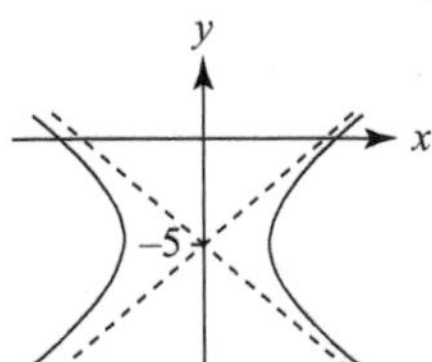

b Asymptotes at $y-2=\pm\frac{2}{3}(x+3)$

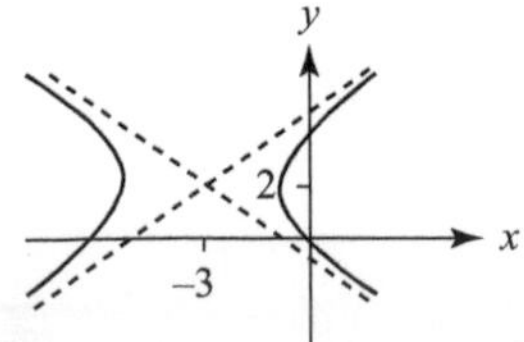

c Asymptotes at $x=0$, $y=0$

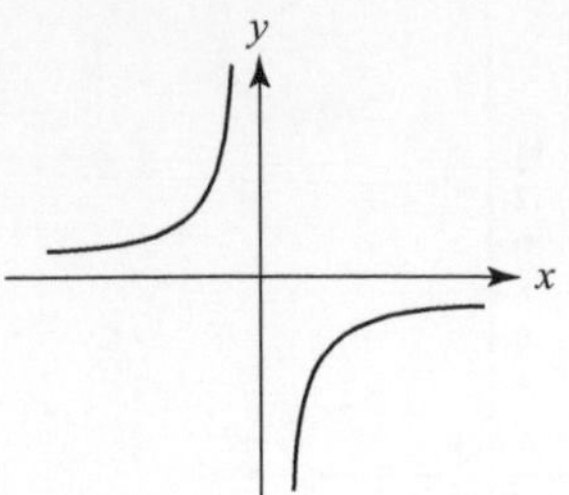

d Asymptotes at $x=-3$, $y=4$

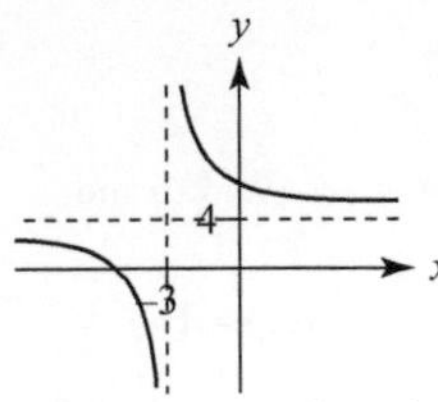

21 a $\frac{y^2}{8}+x^2=2 \Rightarrow \frac{x^2}{2}+\frac{y^2}{16}=1$

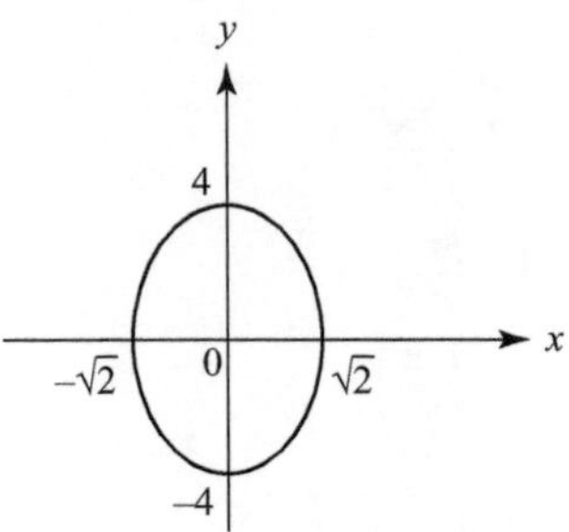

b $6(-y)^2-2(-x)^2=6 \Rightarrow y^2-\frac{x^2}{3}=1$

Asymptotes are at $y=\pm\sqrt{\frac{3}{3x}}$

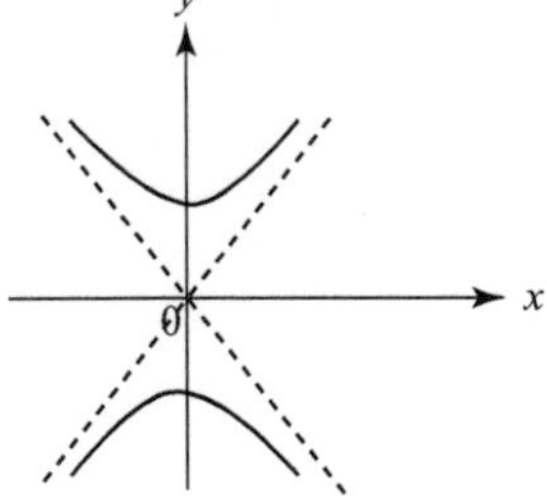

22 a $\frac{13}{5}$

b $-\frac{3}{5}$

23 a

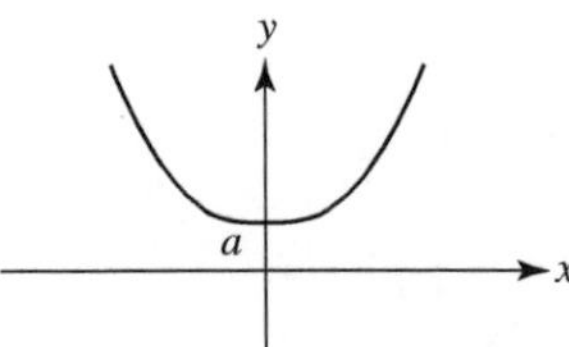

b

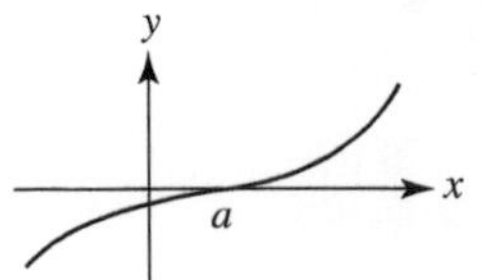

c

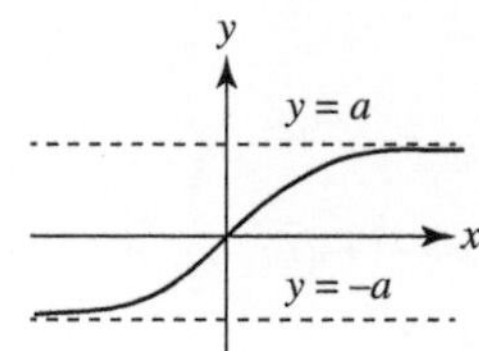

d

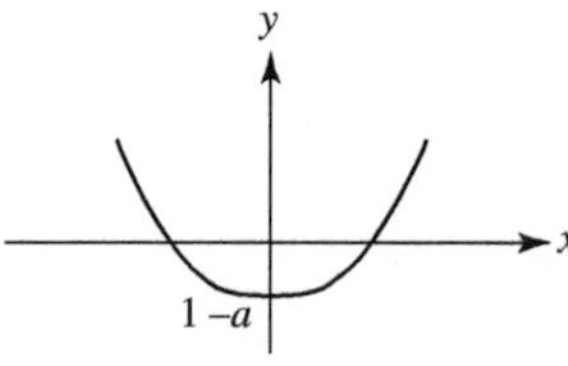

24 $x = \ln(-2+\sqrt{5}), \ln(1+\sqrt{2})$

25 a $x = \ln 3$

b $x = \ln\left(1+\sqrt{2}\right)$

26 Let $y = \cosh^{-1} 2x$

Then $2x = \cosh y = \dfrac{e^y + e^{-y}}{2}$

$4x = e^y + e^{-y} \Rightarrow e^{2y} - 4xe^y + 1 = 0$

$e^y = 2x + \sqrt{4x^2 - 1} \Rightarrow y = \ln(2x + \sqrt{4x^2 - 1})$

27 a $\sinh^{-1}(2) = \ln\left(2+\sqrt{5}\right)$

b $\ln\left(\dfrac{\sqrt{3}}{3}\right)$

Assessment 3

1 a

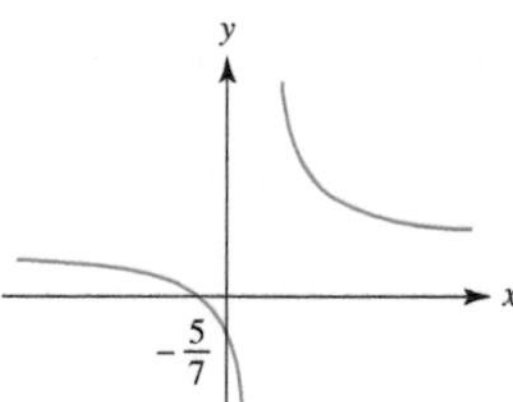

b $x = 7$

$y = 1$

c $(1,-1), (5,-5)$

2 a

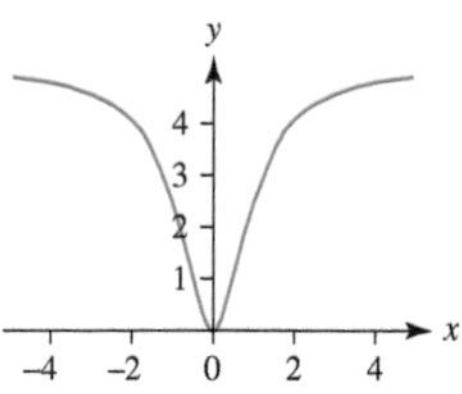

b $\left(\pm\dfrac{1}{2}, 1\right)$

3 a $\left(2\sqrt{3}, -\dfrac{\pi}{3}\right)$

b i

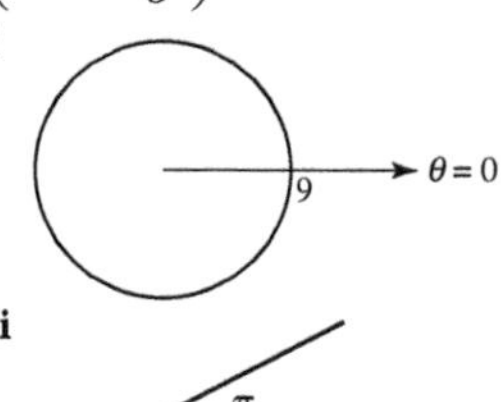

ii

c i $x^2 + y^2 = 81$ **ii** $y = \dfrac{1}{\sqrt{3}}x$

4 a $\left(-2\sqrt{3}, -2\right)$ **b** $x^2 + y^2 = 2y$

c

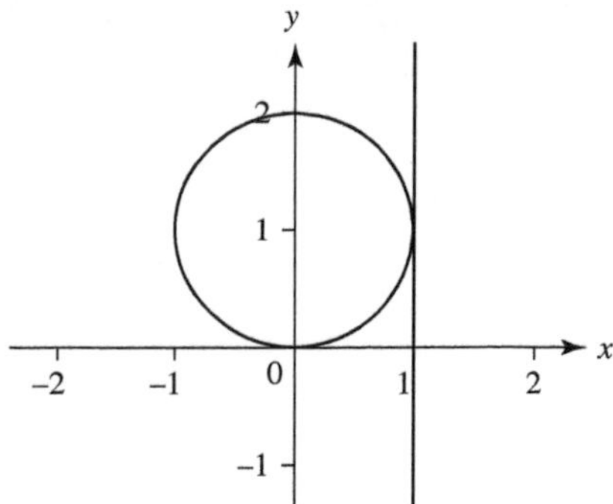

5 a

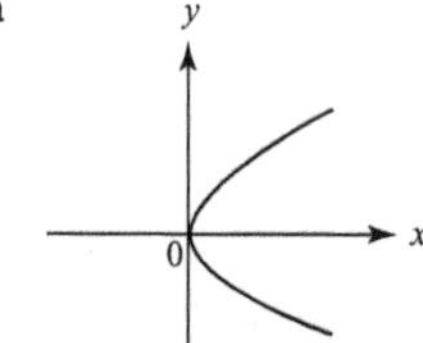

Parabola

b $\dfrac{3}{4}$

6 a

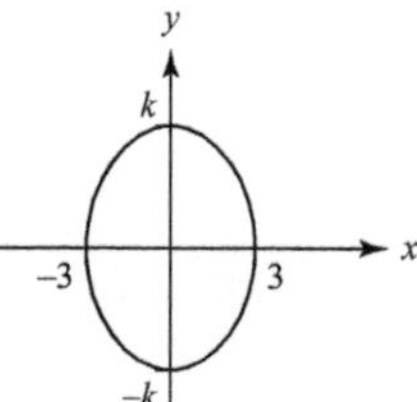

b

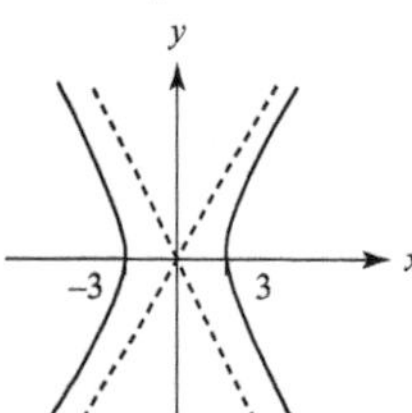

Asymptotes $y = \pm\dfrac{k}{3}x$

7 a

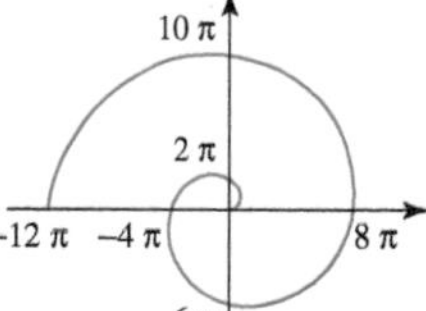

b

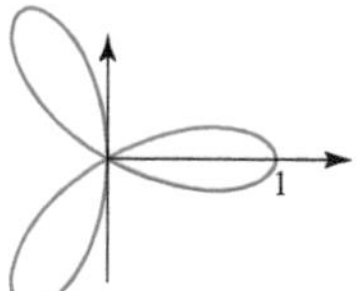

c

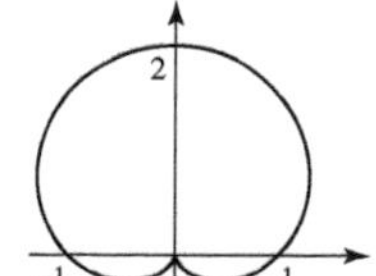

d

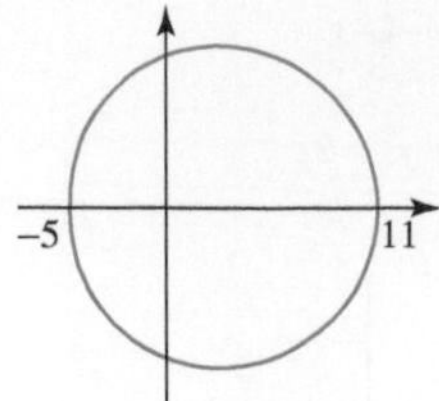

8 a $\sinh(\ln 2) = \dfrac{e^{\ln 2} - e^{-\ln 2}}{2}$

$= \dfrac{e^{\ln 2} - e^{\ln\left(\frac{1}{2}\right)}}{2}$

$= \dfrac{2 - \frac{1}{2}}{2}$

$= \dfrac{3}{4}$

b $x = 1.8$

9 a

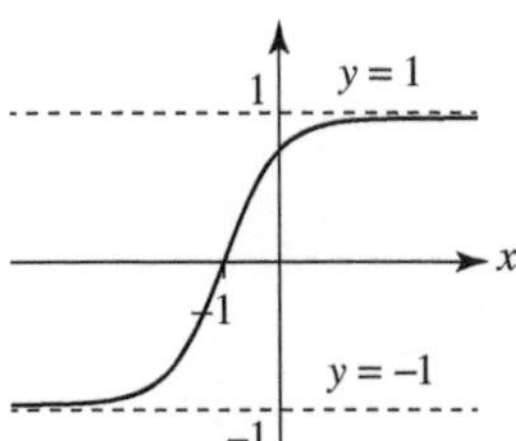

b $y = 1$

$y = -1$

c $\tanh x = \dfrac{\sinh x}{\cosh x}$

$= \dfrac{e^x - e^{-x}}{2} \div \dfrac{e^x + e^{-x}}{2}$

$= \dfrac{2(e^x - e^{-x})}{2(e^x + e^{-x})}$

$= \dfrac{e^x - e^{-x}}{e^x + e^{-x}}$

$= \dfrac{e^{2x} - 1}{e^{2x} + 1}$ as required

d $x = \dfrac{1}{2}\ln 3$

10 a

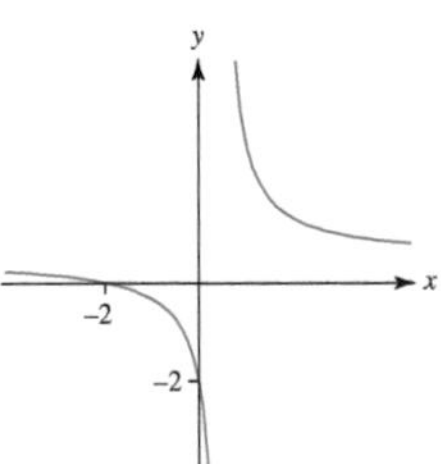

b $-4 < x \leq -1$ or $\dfrac{1}{2} < x \leq 8$

11 a

b $x > \dfrac{\sqrt{13} - 1}{2}$ or $\dfrac{-1 - \sqrt{13}}{2} < x < 1$

12 a $x = -1 \pm \sqrt{7}$

b $x = 1$ or $-\dfrac{8}{3}$, $y = \dfrac{4}{3}$

c $x > -1 + \sqrt{7}, -2 < x < \dfrac{1}{2}, x > -1 + \sqrt{7}$

13 a $(x-5)(y+1) = xy - 5y + x - 5$

So $xy - 5y + x - 9 = 0$

$\Rightarrow (x-5)(y+1) - 4 = 0$

$\Rightarrow (x-5)(y+1) = 4$

$(k = 4)$

b

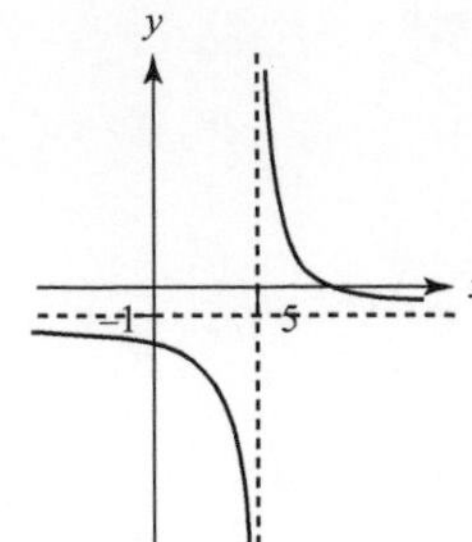

Asymptotes $x = 5$, $y = -1$

14

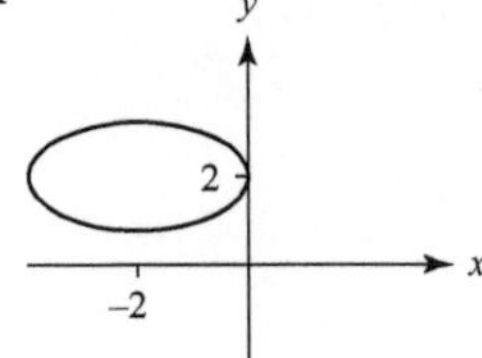

15 $r\sin\theta = \dfrac{1}{r\cos\theta}$

$r^2\sin\theta\cos\theta = 1$

$r^2\left(\dfrac{1}{2}\sin 2\theta\right) = 1$

$r^2 = \dfrac{2}{\sin 2\theta}$

$r^2 = 2\operatorname{cosec} 2\theta$

16 a $(x^2 + y^2)^3 = 36x^2y^2$

b

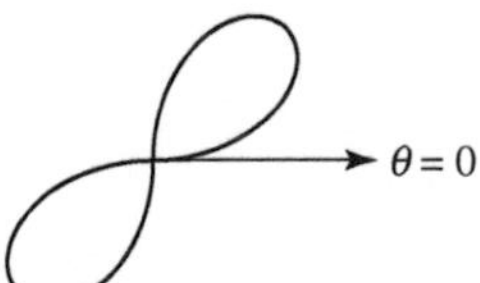

c Max value of r is 3

This occurs at $\theta = \dfrac{\pi}{4}, \dfrac{5\pi}{4}$

17 a $r = 7$ is maximum

$r = 1$ is minimum

b

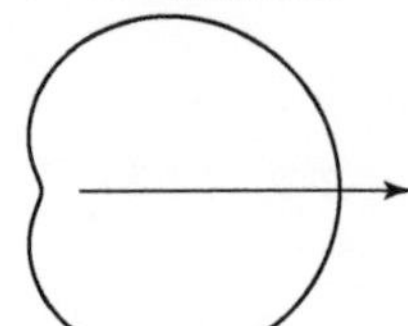

18 a

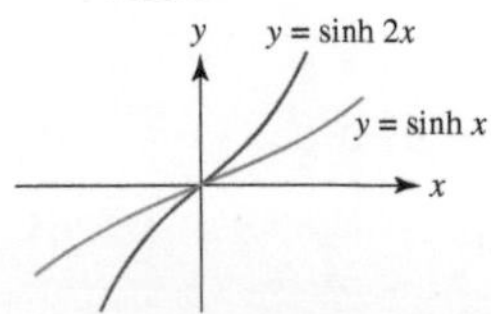

b $y = \dfrac{e^x - e^{-x}}{2}$

$2y = e^x - e^{-x}$

$e^{2x} - 2ye^x - 1 = 0$

$(e^x - y)^2 - y^2 - 1 = 0$

$(e^x - y)^2 = y^2 + 1$

$e^x - y = \pm\sqrt{y^2+1}$

$e^x = y \pm \sqrt{y^2+1}$

$e^x > 0 \therefore\ e^x = y + \sqrt{y^2+1}$ as $(y - \sqrt{y^2+1}) < 0$

$x = \ln(y + \sqrt{y^2+1})$ as required.

19 a $\dfrac{y^2}{10} - \dfrac{x^2}{5} = 1$

b

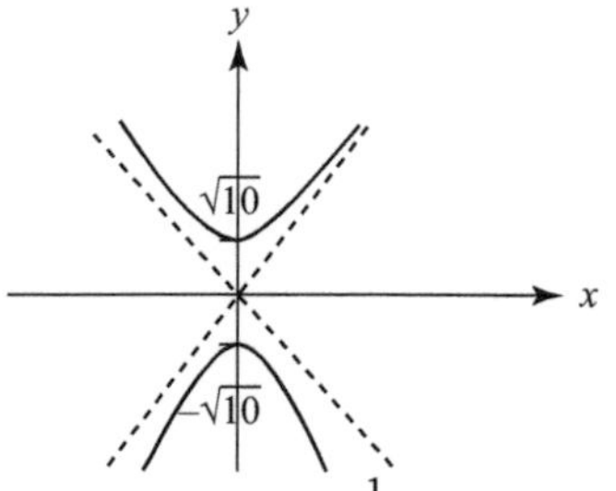

Asymptotes at $x = \pm\dfrac{1}{\sqrt{2}}y$

(or $y = \pm\sqrt{2}x$)

20 a $y^2 = 48x$

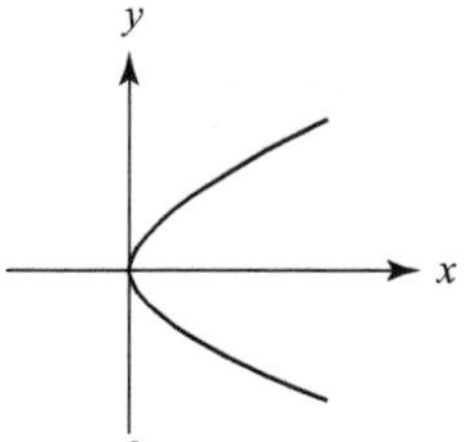

b $(y+8)^2 = 12(x-3)$

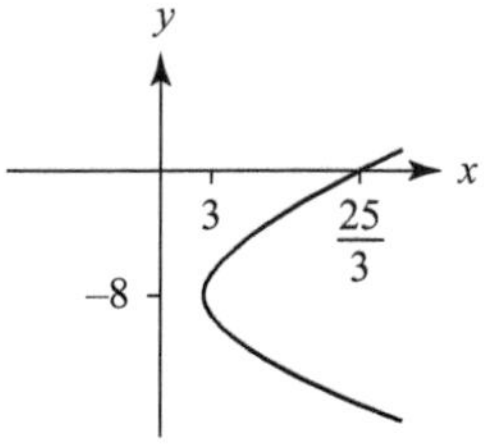

c $y = -\dfrac{x^2}{12}$

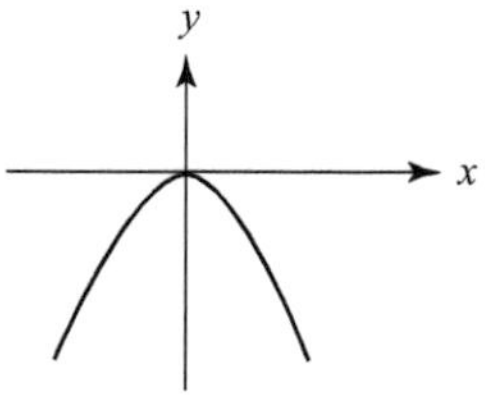

21 a i and ii

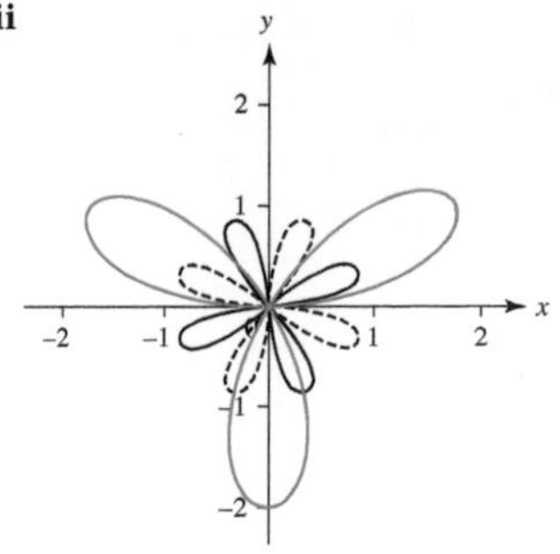

b Dotted sections only included when r can take negative values.

22 a $2\sinh x + \cosh x \equiv 2\left(\dfrac{e^x - e^{-x}}{2}\right)\left(\dfrac{e^x + e^{-x}}{2}\right)$

$\equiv 2\left(\dfrac{e^{2x} - e^{-2x}}{4}\right)$

$\equiv \dfrac{e^{2x} - e^{-2x}}{2}$

$\equiv \sinh(2x)$ as required

b i $x = \dfrac{1}{2}\ln 5, 0$ **ii** $x = \dfrac{1}{2}\ln(2+\sqrt{5}), \dfrac{1}{2}\ln\left(\dfrac{-3+\sqrt{13}}{2}\right)$ 1

Chapter 4

Exercise 4.1A

1 a $\dfrac{45}{4}$ (11.25) **b** $\dfrac{15}{4}$ (3.75)

2 a

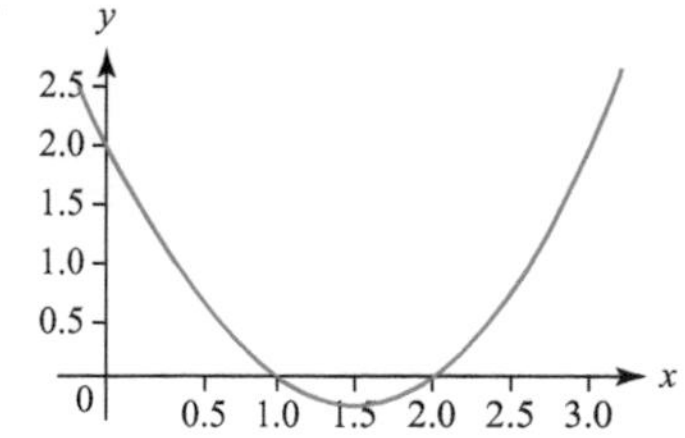

b $\dfrac{1}{6}$

c i $-\dfrac{1}{6}$ **ii** $\dfrac{5}{6}$

3 a

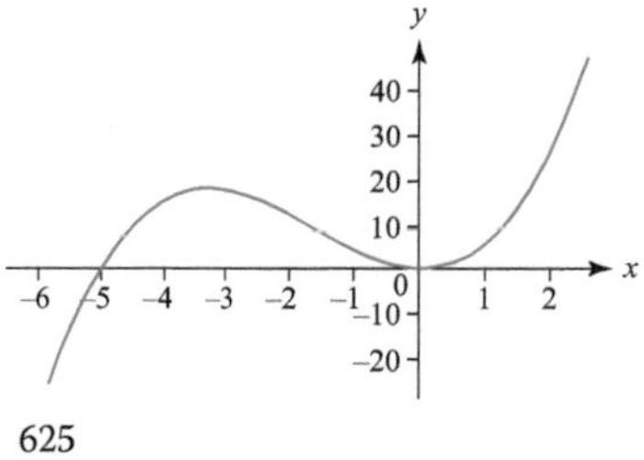

b $\dfrac{625}{12}$

c i $\dfrac{125}{12}$ **ii** $\dfrac{26}{3}$ (8.67)

4 a 256 **b** 121 **c** 11 **d** $\dfrac{1}{16}$

5 a 9 **b** $\dfrac{113}{3}$ (37.7)

c 5 **d** $\dfrac{23}{3}$ (7.67)

6 a $-\frac{325}{2}$ (–162.5) **b** $-\frac{99}{2}$ (–49.5)

c 0 **d** $\frac{245}{2}$ (122.5)

7 $\frac{1}{5-2}\int_2^5 x^{-2}dx = \frac{1}{3}\left[-x^{-1}\right]_2^5$

$= \frac{1}{3}\left(-\frac{1}{5}\right) - \frac{1}{3}\left(-\frac{1}{2}\right)$

$= \frac{1}{10}$ (0.1)

8 $\frac{1}{4-1}\int_1^4 x^{\frac{1}{2}}dx = \frac{1}{3}\left[\frac{2}{3}x^{\frac{3}{2}}\right]_1^4$

$= \frac{1}{3}\left(\frac{16}{3}\right) - \frac{1}{3}\left(\frac{2}{3}\right)$

$= \frac{14}{9}$ (1.56)

9 $\frac{1}{9-4}\int_4^9 x^{-\frac{1}{2}}dx = \frac{1}{5}\left[2x^{\frac{1}{2}}\right]_4^9$

$= \frac{1}{5}(6) - \frac{1}{5}(4)$

$= \frac{2}{5}$ (0.4)

10 a $A = 1, B = \frac{3}{2}$ and $c = -\frac{1}{2}$

b $\frac{1}{9-1}\int_1^9 x^{-\frac{1}{2}} + \frac{3}{2}\,dx = \frac{1}{8}\left[2x^{\frac{1}{2}} + \frac{3}{2}x\right]_1^9$

$= \frac{1}{8}\left(6 + \frac{27}{2}\right) - \frac{1}{8}\left(2 + \frac{3}{2}\right)$

$= 2$

11 a $A = \frac{3}{5}, B = -\frac{1}{5}, c = \frac{1}{2}, d = \frac{3}{2}$ **b** $\frac{8}{25} - \frac{6}{25}\sqrt{2}$

12 a $-\frac{13}{3}$ (–4.33) **b** $-\frac{229}{200}$ (–1.145)

c $\frac{1}{12}$ (0.0833) **d** $\frac{28}{5}$ (5.6)

13 a $1 - 2T + \frac{4}{3}T^2$ **b** $4T^2 + 4T + \frac{7}{3}$

14 $\frac{26}{9}\sqrt{2}$

15 $\frac{45}{4}a^3 - \frac{3}{8a^3}$

16 $= -\frac{1}{6}X^4$

Exercise 4.1B

1 a 3.6 ms^{-1}

b $\frac{1}{2}\left[\frac{2}{5}t^3 - \frac{1}{5}t\right]_1^3 = \frac{1}{2}\left(\frac{54}{5} - 3\right) - \frac{1}{2}\left(\frac{2}{5} - \frac{1}{5}\right)$

$= 5$ ms^{-2} as required.

2 a $\frac{1}{5-0}\int_0^5 \frac{3t^2 - 5}{5}\,dt = \frac{1}{25}[t^3 - 5t]_0^5$

$= \frac{1}{25}(125 - 25) - 0$

$= 4$ ms^{-1}

b 3 ms^{-1}

3 $s = 2t^{\frac{3}{2}}$

$v = 3t^{\frac{1}{2}}$

$\left(a = \frac{3}{2}t^{-\frac{1}{2}}\right)$

Mean acceleration $= \frac{1}{1-\frac{1}{2}}\int_1^4 \frac{3}{2}t^{-\frac{1}{2}}\,dt$

$= 2\left[3t^{\frac{1}{2}}\right]_{0.5}^1$

$= 2\left(3 - \frac{3\sqrt{2}}{2}\right)$

$= 6 - 3\sqrt{2}$ ms^{-2}

4 a $\frac{1}{3-1}\int_1^3 \frac{t}{10} - 10t^{-3}dt = \frac{1}{2}\left[\frac{t^2}{20} + 5t^{-2}\right]_1^3$

$= \frac{1}{2}\left(\frac{9}{20} + \frac{5}{9}\right) - \frac{1}{2}\left(\frac{1}{20} + 5\right)$

$= -\frac{91}{45}$ ms^{-2}

b $\frac{11}{6}$ (1.83) ms^{-1}

5 Mean value $= \frac{1}{b-a}\int_a^b mx + c\,dx$

$= \frac{1}{b-a}\left[\frac{mx^2}{2} + cx\right]_a^b$

$= \frac{1}{b-a}\left(\frac{mb^2}{2} + cb\right) - \frac{1}{b-a}\left(\frac{ma^2}{2} + ca\right)$

$= \frac{1}{b-a}\left(\frac{m}{2}(b^2 - a^2) + c(b-a)\right)$

$= \frac{m(b+a)(b-a)}{2(b-a)} + \frac{c(b-a)}{b-a}$

$= \frac{m(b+a)}{2} + c$ as required

6 Mean value $= \frac{1}{a-0}\int_0^a x^2\,dx$

$= \frac{1}{a}\left[\frac{x^3}{3}\right]_0^a$

$= \frac{1}{a}\frac{a^3}{3}$

$= \frac{a^2}{3}$ as required

7 a

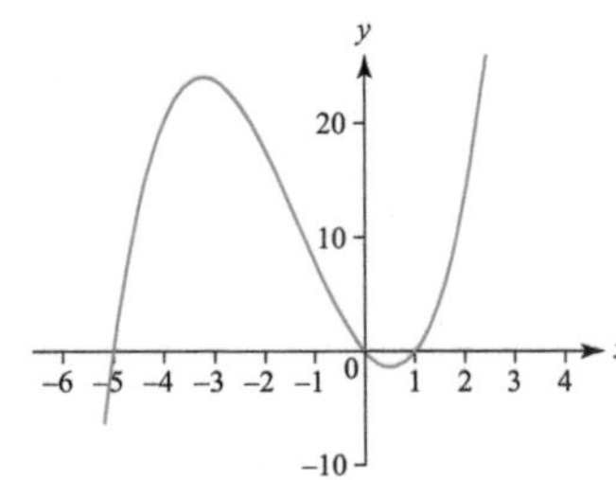

b $\frac{443}{6}$ (73.8)

c i $\frac{175}{12}$ **ii** $= -\frac{11}{12}$

8 $\frac{49}{36}$

9 -8

10 7

11 6, -9

12 $b=4$

$a=-1$

13 $\frac{1}{3-1}\int_1^3 x^4-2x^3+3x-5\,dx=\frac{1}{2}\left[\frac{x^5}{5}-\frac{x^4}{2}+\frac{3}{2}x^2-5x\right]_1^3$

$=\frac{1}{2}\left(\frac{243}{5}-\frac{81}{2}+\frac{27}{2}-15\right)-\frac{1}{2}\left(\frac{1}{5}-\frac{1}{2}+\frac{3}{2}-5\right)$

$=\frac{26}{5}$

$\frac{1}{-1--2}\left[\frac{x^5}{5}-\frac{x^4}{2}+\frac{3}{2}x^2-5x\right]_{-2}^{-1}$

$=\left(-\frac{1}{5}-\frac{1}{2}+\frac{3}{2}+5\right)-\left(-\frac{32}{5}-8+6+10\right)$

$=\frac{21}{5}$

So [1, 3] is bigger by 1.

14 Mean speed of A: $\frac{1}{\frac{1}{2}}\int_0^{1.5} t^2+t\,dt=2\left[\frac{t^3}{3}+\frac{t^2}{2}\right]_0^{0.5}$

$=\frac{1}{3}$

Mean speed of B: $\frac{1}{\frac{1}{2}}\int_0^{0.5} t^{\frac{1}{2}}dt=2\left[\frac{2}{3}t^{\frac{3}{2}}\right]_0^{0.5}$

$=\frac{\sqrt{2}}{3}$

B is $\frac{1}{3}(\sqrt{2}-1)$ faster than A

15 $a=\frac{2}{5}$, $b=27\sqrt{2}$, $c=-1$

16 $\frac{1}{8}\ln\left(\frac{4k^2+8k+1}{4k^2-3}\right)$

17 **a** $=2+\frac{1}{2}\ln 3$

b No as function not continuous in this range.

18 **a** $\frac{3}{2\pi}$ **b** 0

c $\frac{1}{3}(e^3-1)$ **d** $\frac{4}{3}e^3+\frac{2}{3}$

Exercise 4.2A

1 $\frac{9207}{5}\pi$ (5785)

2 **a** $\int_0^2 8-x^3\,dx=\left[8x-\frac{x^4}{4}\right]_0^2$

$=(16-2)-0$

$=12$ as required

b $\frac{576}{7}\pi$ (259)

3 **a** $\frac{1}{20}$ **b** $\frac{\pi}{252}$ (0.0125)

4 $\frac{59}{30}\pi$ (6.18)

5 $\frac{5261}{105}\pi$ (157)

6 $\pi\int_{0.5}^1\left(\frac{1}{x^2}\right)^2 dx=\pi\int_{0.5}^1 x^{-4}dx$

$=\pi\left[-\frac{1}{3}x^{-3}\right]_{0.5}^1$

$=\pi\left(-\frac{1}{3}--\frac{8}{3}\right)$

$=\frac{7}{3}\pi$

7 **a** $\int_1^3 \frac{2}{5}x^{-2}+\frac{1}{5}x^{-\left(\frac{3}{2}\right)}\,dx=\left[-\frac{2}{5}x^{-1}-\frac{2}{5}x^{-\frac{1}{2}}\right]_1^3$

$=-\frac{2}{15}-\frac{2}{5\sqrt{3}}-\left(-\frac{2}{5}-\frac{2}{5}\right)$

$=-\frac{2}{15}-\frac{2\sqrt{3}}{15}+\frac{4}{5}$

$=\frac{1}{15}(-2-2\sqrt{3}+12)$

$=\frac{10-2\sqrt{3}}{15}$

b **i** 0.405 **ii** 0.101

8 **a** $\frac{256}{15}\pi$

b 8π

9 $x=\left(\frac{y}{4}\right)^{\frac{1}{4}}$

$\pi\int_2^8\left(\left(\frac{y}{4}\right)^{\frac{1}{4}}\right)^2 dy=\pi\int_2^8\frac{1}{2}y^{\frac{1}{2}}\,dy$

$=\pi\left[\frac{1}{3}y^{\frac{3}{2}}\right]_2^8$

$=\pi\left(\frac{16}{3}\sqrt{2}-\frac{2}{3}\sqrt{2}\right)$

$=\frac{14}{3}\sqrt{2}\,\pi$

10 $\frac{36}{35}\sqrt{3}\pi$

Exercise 4.2B

1 **a** $\frac{19}{12}$ **b** $\frac{109}{30}\pi$

2 **a** 12 **b** $\frac{912}{7}\pi$

3 **a** $\int_0^{\sqrt{3}} 16-x^4dx=\left[16x-\frac{x^5}{5}\right]_0^{\sqrt{3}}$

$=16\sqrt{3}-\frac{9}{5}\sqrt{3}-0$

$=\frac{71}{5}\sqrt{3}$

Area of R $=\frac{71}{5}\sqrt{3}-7\times\sqrt{3}$

$=\frac{36}{5}\sqrt{3}$ as required

b $\frac{792}{5}\sqrt{3}\pi$ (862)

4

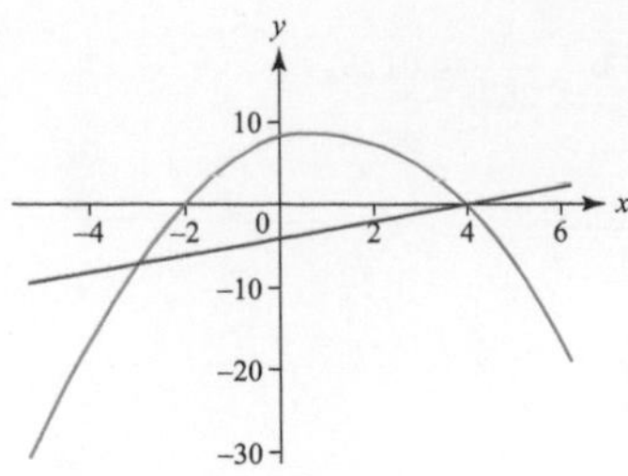

a $\int_0^4 8+2x-x^2 dx = \left[8x+x^2-\frac{x^3}{3}\right]_0^4$

$=32+16-\frac{64}{3}$

$=\frac{80}{3}$

Area required $=\frac{80}{3}+\frac{1}{2}\times 4\times 4$

$=\frac{104}{3}$ as required.

b $\frac{1696}{15}\pi$ (355)

5 a $\frac{10}{3}$ square units b $\frac{8}{3}\pi$

6 $V=\pi\int_0^b (ax)^2 dx$

$=\pi\int_0^b a^2x^2 dx$

$=\pi\left[\frac{a^2}{3}x^3\right]_0^b$

$=\frac{\pi a^2}{3}(b^3-0)$

$=\frac{\pi(ab)^2}{3}b$

$=\frac{1}{3}\pi r^2 h$ where $r=ab$ and $h=b$

7 $V=\pi\int_0^h r^2\, dx$

$=\pi\left[r^2x\right]_0^h$

$=\pi(r^2h-0)$

$=\pi r^2 h$ as required

8 $V=\pi\int_0^r \sqrt{r^2-x^2}^{\,2} dx$

$=\pi\int_0^r r^2-x^2 dx$

$=\pi\left[r^2x-\frac{x^3}{3}\right]_0^r$

$=\pi\left(r^3-\frac{r^3}{3}-0\right)$

$=\pi\left(\frac{2}{3}r^3\right)$

$=\frac{2}{3}\pi r^3$

9 a $A=\frac{\pm\sqrt{6}}{3}$

b $\frac{\pi}{15}(27\sqrt{2}-\sqrt{6})$

10 a A has coordinates (7, 14), B has coordinates (7, −14)

b 686π

11 a $\frac{\pi}{4}\ln\left(\frac{5}{2}\right)$ b $\frac{2}{3}\sqrt{3}\pi$

c $\frac{127}{14}\pi$ d $\frac{9}{4}\pi\left(1-\frac{1}{e}\right)$ or $\frac{9\pi(e-1)}{4e}$

12 $\frac{48\pi}{5}$

Review exercise 4

1 $\frac{241}{3}$

2 $\frac{1}{8-2}\int_2^8 \frac{1+x}{\sqrt{x}}\, dx = \frac{1}{6}\int_2^8 x^{-\frac{1}{2}}+x^{\frac{1}{2}}\, dx$

$=\frac{1}{6}\left[2x^{\frac{1}{2}}+\frac{2}{3}x^{\frac{3}{2}}\right]_2^8$

$=\frac{1}{6}\left(4\sqrt{2}+\frac{32}{3}\sqrt{2}\right)-\frac{1}{6}\left(2\sqrt{2}+\frac{4}{3}\sqrt{2}\right)$

$=\frac{17}{9}\sqrt{2}$

3 $-\frac{75}{8}$

4 $\frac{1}{1-a}\int_a^1 x^{-3}-\frac{1}{2}x^{-2} dx = \frac{1}{1-a}\left[-\frac{1}{2}x^{-2}+\frac{1}{2}x^{-1}\right]_a^1$

$=\frac{1}{1-a}\left(-\frac{1}{2}+\frac{1}{2}\right)-\frac{1}{1-a}\left(-\frac{1}{2a^2}+\frac{1}{2a}\right)$

$=\frac{1}{2(a-1)}\frac{(a-1)}{a^2}$

$=\frac{1}{2a^2}$ as required

5 a 12 ms^{-1}

b 15.4 ms^{-2}

6 291π (914)

7 a $\int_{\frac{1}{\sqrt{3}}}^2 3-x^{-2} dx = \left[3x+\frac{1}{x}\right]_{\frac{1}{\sqrt{3}}}^2$

$=\left(6+\frac{1}{2}\right)-\left(\frac{3}{\sqrt{3}}+\sqrt{3}\right)$

$=\frac{13}{2}-2\sqrt{3}$

$=\frac{13-4\sqrt{3}}{2}$

b $\frac{503-192\sqrt{3}}{24}\pi$ (22.3)

8 50π

9 $\pi\int_0^{\frac{1}{2}} 16y^4 dy = \pi\left[\frac{16}{5}y^5\right]_0^{\frac{1}{2}}$

$=\pi\left(\frac{1}{10}-0\right)$

$=\frac{\pi}{10}$

10 a $\int_0^2 y^4+1\,dy=\left[\frac{y^5}{5}+y\right]_0^2$

$=\frac{32}{5}+2-0$

$=\frac{42}{5}$ as required

b $\frac{3226}{45}\pi$ (225)

11 a

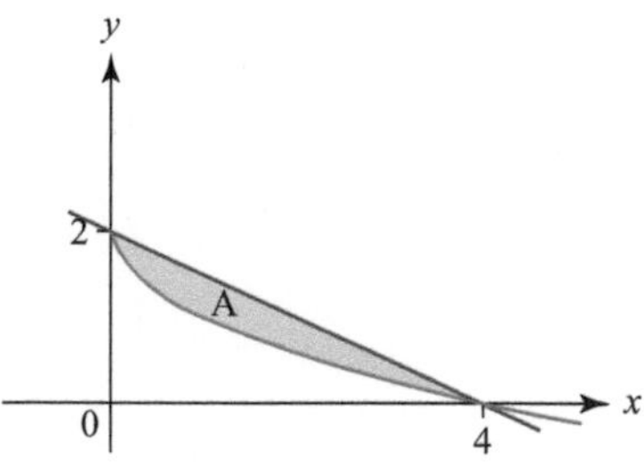

b $\frac{8}{3}\pi$

c $\frac{64}{15}\pi$ (13.4)

12 a $\frac{49}{48}$

b $\frac{901}{960}\pi$ (2.95)

13 a $\frac{32}{3}\pi$

b $\frac{128}{15}\pi$

Assessment 4

1 a $\frac{16}{3}$ **b** $-\frac{1}{7}$

2 a $\frac{1}{2}x-\frac{3}{2}x^{-\frac{3}{2}}$ **b** $\frac{3}{4}$

3 $\frac{9}{5}\sqrt{3}+\frac{4}{5}$

4 a 36 ms^{-1} **b** 18

5 6

6 a 2 **b** $\frac{20}{3}$ **c** $\frac{496}{15}\pi$ or 103.9

7 18π

8 $\frac{619}{21}\pi$ or 92.6

9 1170

10 9.44 ms^{-1}

11 $a=6, b=2$

12 $k=2$ or $-\frac{2}{5}$

13 128

14 a $a=3, b=2$

b 4π

15 a

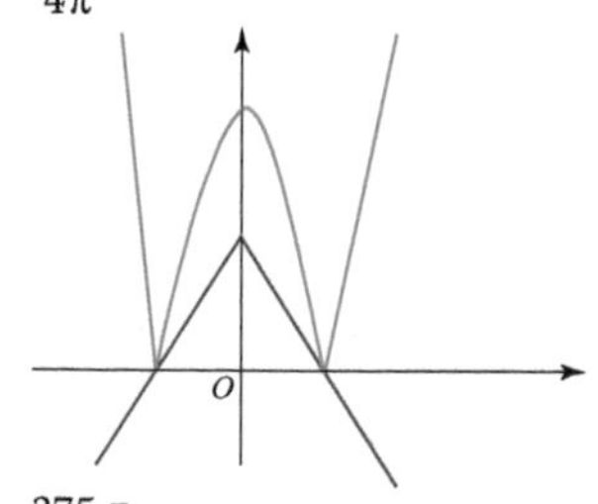

b $375\,\pi$

16 $\frac{1}{2-0}\int_0^1 e^{2x}+x^3+2dx$

$=\frac{1}{2}\left[\frac{1}{2}e^{2x}+\frac{x^4}{4}+2x\right]_0^2$

$=\frac{1}{2}\left(\frac{1}{2}e^4+4+4\right)-\frac{1}{2}\left(\frac{1}{2}+0+0\right)$

$=\frac{1}{4}e^4+\frac{15}{4}$

17 $\frac{1}{2(3-a)}\ln\left(\frac{3}{2a-3}\right)$

18 a $\frac{3}{2\pi}$ or 0.477

b $\frac{1}{2}+\frac{3\sqrt{3}}{4\pi}$ or 0.913

19 $-\frac{3}{4}+\frac{7}{2}\ln 2$

20 a 4

b $2\pi(e^2-1)$

21 177π

22 a

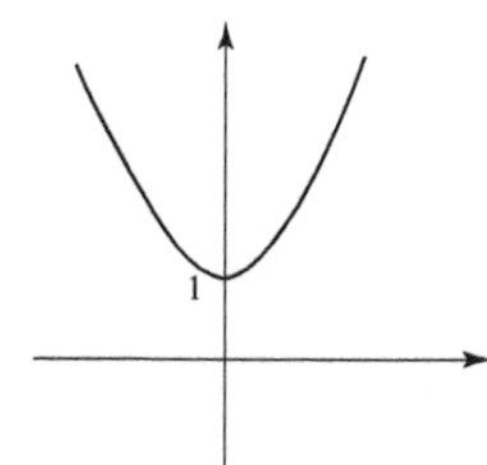

b $\frac{\pi}{32}\left(e^4-8-e^{-4}\right)$

23 $\frac{2}{3}$

24 $2\pi-\frac{\pi^2}{2}$

Chapter 5

Exercise 5.1A

1 a **i** 3×2 **ii** 3×1 **iii** 2×2 **iv** 2×4

b the matrix in part **iii** is a square matrix

2 a $\begin{pmatrix}4 & 3 & 2\\ 5 & -1 & 6\end{pmatrix}$ **b** $\begin{pmatrix}-4 & -6\\ 1 & -3\end{pmatrix}$

c $\begin{pmatrix}-4 & 0 & 16\\ 8 & 20 & -12\\ 32 & -24 & 0\end{pmatrix}$ **d** $\begin{pmatrix}-2 & 12\\ 6 & 10.5\\ 1 & -4\end{pmatrix}$

3 a i $\begin{pmatrix}9 & -3\\ -5 & 2\end{pmatrix}$ **ii** Not possible as different order matrices

iii $\begin{pmatrix}13 & 6\\ -5 & -6\\ -5 & -1\end{pmatrix}$ **iv** $\begin{pmatrix}24 & 12\\ 6 & -18\\ -6 & 0\end{pmatrix}$

b i $\begin{pmatrix}45 & -20\\ 0 & -10\end{pmatrix}-\begin{pmatrix}0 & 1\\ -5 & 4\end{pmatrix}=\begin{pmatrix}45 & -21\\ 5 & -14\end{pmatrix}$

ii $\begin{pmatrix}-10 & -4\\ 14 & 0\\ 6 & 2\end{pmatrix}+\begin{pmatrix}56 & 28\\ 14 & -42\\ -14 & 0\end{pmatrix}=\begin{pmatrix}46 & 24\\ 28 & -42\\ -8 & 2\end{pmatrix}$

$=2\begin{pmatrix}23 & 12\\ 14 & -21\\ -4 & 1\end{pmatrix}$

4 a $\begin{pmatrix} -6 & -15 & 3 & -18 \\ -10 & 9 & -1 & -10 \end{pmatrix}$ b $\begin{pmatrix} 16 & 2 \\ 36 & 72 \\ 30 & -15 \end{pmatrix}$

c $\begin{pmatrix} 0 \\ 13 \end{pmatrix}$ d $(-8 \quad -5 \quad -10)$

e $\begin{pmatrix} -4 & 8 & 20 \\ -3 & 6 & 15 \end{pmatrix}$ f $\begin{pmatrix} -5 & 6 & 4 \\ -2 & -12 & 4 \\ 6 & -3 & -5 \end{pmatrix}$

5 $\begin{pmatrix} 3 & 2 & 0 \\ -1 & 0 & -2 \end{pmatrix}\begin{pmatrix} 4 & 1 \\ 0 & 3 \\ -3 & 0 \end{pmatrix}$

$= \begin{pmatrix} (3\times4)+(2\times0)+(0\times-3) & (3\times1)+(2\times3)+(0\times0) \\ (-1\times4)+(0\times0)+(-2\times-3) & (-1\times1)+(0\times3)+(-2\times0) \end{pmatrix}$

$= \begin{pmatrix} 12+0+0 & 3+6+0 \\ -4+0+6 & -1+0+0 \end{pmatrix}$

$= \begin{pmatrix} 12 & 9 \\ 2 & -1 \end{pmatrix}$

6 a Not possible since 1 column in **A** but 2 rows in **B**

b $\begin{pmatrix} 51 \\ -7 \end{pmatrix}$

c $\begin{pmatrix} -14 \\ -72 \end{pmatrix}$

d Not possible since 3 columns in **C** but 2 rows in **B**

e $\begin{pmatrix} 36 \\ 40 \end{pmatrix}$

f Not possible as **C** not a square matrix.

7 a $\begin{pmatrix} 7a & 8a+6 & 3a \\ 2-a & 2 & 3+a \end{pmatrix}$ b $\begin{pmatrix} 3a & -a^2 \\ 6 & -2a \\ -3a & a^2 \end{pmatrix}$

c $(a \quad a)$

d $\begin{pmatrix} a^2-6 & 6a \\ -4a & a^2-6 \end{pmatrix}$

8 $\mathbf{A}^3 = \begin{pmatrix} 2 & 5 \\ -1 & 2 \end{pmatrix}\begin{pmatrix} 2 & 5 \\ -1 & 2 \end{pmatrix}\begin{pmatrix} 2 & 5 \\ -1 & 2 \end{pmatrix}$

$= \begin{pmatrix} 2 & 5 \\ -1 & 2 \end{pmatrix}\begin{pmatrix} 4-5 & 10+10 \\ -2-2 & -5+4 \end{pmatrix}$

$= \begin{pmatrix} 2 & 5 \\ -1 & 2 \end{pmatrix}\begin{pmatrix} -1 & 20 \\ -4 & -1 \end{pmatrix}$

$= \begin{pmatrix} -2-20 & 40-5 \\ 1-8 & -20-2 \end{pmatrix}$

$= \begin{pmatrix} -22 & 35 \\ -7 & -22 \end{pmatrix}$ $k = 35$

9 $\begin{pmatrix} 3k & 2k \\ 12-2k & 8 \end{pmatrix}$

10 $\begin{pmatrix} 20+2k & 4 & 8 \\ 4+k^2 & -1 & k \\ 12+2k & -6 & 0 \end{pmatrix}$

Exercise 5.1B

1 $a=-3$
$b=-5$
$c=-1$
$d=2$

2 $a=-3$
$b=4$
$c=-3$

3 a $x=-1, y=4$
b $x=-2, y=4$

4 a $x=1, -\frac{3}{5}$
b $x=-2, -5$
c $x=\pm3$
d $x=-2$

5 $a=7, b=-3, c=2$

6 $x=3, y=4, z=-1$

7 $a=-1, b=-2, c=6$

8 a $\begin{pmatrix} 2 & 5 \\ 6 & -1 \end{pmatrix}\begin{pmatrix} x \\ y \end{pmatrix} = \begin{pmatrix} 6 \\ 3 \end{pmatrix}$

b $\begin{pmatrix} 1 & 1 & -1 \\ 2 & 1 & 1 \\ 3 & 2 & 2 \end{pmatrix}\begin{pmatrix} x \\ y \\ z \end{pmatrix} = \begin{pmatrix} -4 \\ 4 \\ 10 \end{pmatrix}$

9 $\begin{pmatrix} 1 & 5 \\ 0 & 1 \end{pmatrix}^1 = \begin{pmatrix} 1 & 5 \\ 0 & 1 \end{pmatrix}$ and $\begin{pmatrix} 1 & 5\times1 \\ 0 & 1 \end{pmatrix} = \begin{pmatrix} 1 & 5 \\ 0 & 1 \end{pmatrix}$ so true for $n=1$

Assume true for $n=k$

$\begin{pmatrix} 1 & 5 \\ 0 & 1 \end{pmatrix}^{k+1} = \begin{pmatrix} 1 & 5 \\ 0 & 1 \end{pmatrix}^k \begin{pmatrix} 1 & 5 \\ 0 & 1 \end{pmatrix}$

$= \begin{pmatrix} 1 & 5k \\ 0 & 1 \end{pmatrix}\begin{pmatrix} 1 & 5 \\ 0 & 1 \end{pmatrix}$

$= \begin{pmatrix} 1 & 5+5k \\ 0 & 1 \end{pmatrix}$

$= \begin{pmatrix} 1 & 5(k+1) \\ 0 & 1 \end{pmatrix}$ so true for $n=k+1$

Since true for $n=1$ and assuming true for $n=k$ implies true for $n=k+1$, therefore true for all positive integers n

10 $\begin{pmatrix} 5 & 4 \\ 0 & 1 \end{pmatrix}^1 = \begin{pmatrix} 5 & 4 \\ 0 & 1 \end{pmatrix}$ and $\begin{pmatrix} 5^1 & 5^1-1 \\ 0 & 1 \end{pmatrix} = \begin{pmatrix} 5 & 4 \\ 0 & 1 \end{pmatrix}$ so true for $n=1$

Assume true for $n=k$

$\begin{pmatrix} 5 & 4 \\ 0 & 1 \end{pmatrix}^{k+1} = \begin{pmatrix} 5 & 4 \\ 0 & 1 \end{pmatrix}^k \begin{pmatrix} 5 & 4 \\ 0 & 1 \end{pmatrix}$

$= \begin{pmatrix} 5^k & 5^k-1 \\ 0 & 1 \end{pmatrix}\begin{pmatrix} 5 & 4 \\ 0 & 1 \end{pmatrix}$

$= \begin{pmatrix} 5(5^k) & 4(5^k)+5^k-1 \\ 0 & 1 \end{pmatrix}$

$= \begin{pmatrix} 5^{k+1} & 5(5^k)-1 \\ 0 & 1 \end{pmatrix}$

$= \begin{pmatrix} 5^{k+1} & 5^{k+1}-1 \\ 0 & 1 \end{pmatrix}$ so true for $n=k+1$

Since true for $n=1$ and assuming true for $n=k$ implies true for $n=k+1$, therefore true for all positive integers n

11 $\begin{pmatrix}-2 & -1\\ 9 & 4\end{pmatrix}^1 = \begin{pmatrix}-2 & -1\\ 9 & 4\end{pmatrix}$ and $\begin{pmatrix}1-3\times1 & -1\\ 9\times1 & 3\times1+1\end{pmatrix} = \begin{pmatrix}-2 & -1\\ 9 & 4\end{pmatrix}$

so true for $n = 1$

Assume true for $n = k$

$$\begin{pmatrix}-2 & -1\\ 9 & 4\end{pmatrix}^{k+1} = \begin{pmatrix}-2 & -1\\ 9 & 4\end{pmatrix}^{k}\begin{pmatrix}-2 & -1\\ 9 & 4\end{pmatrix}$$

$$= \begin{pmatrix}1-3k & -k\\ 9k & 3k+1\end{pmatrix}\begin{pmatrix}-2 & -1\\ 9 & 4\end{pmatrix}$$

$$= \begin{pmatrix}-2+6k-9k & -1+3k-4k\\ -18k+27k+9 & -9k+12k+4\end{pmatrix}$$

$$= \begin{pmatrix}-2-3k & -1-k\\ 9k+9 & 3k+4\end{pmatrix}$$

$$= \begin{pmatrix}1-3(k+1) & -(k+1)\\ 9(k+1) & 3(k+1)+1\end{pmatrix} \text{ so true for } n = k+1$$

Since true for $n = 1$ and assuming true for $n = k$ implies true for $n = k + 1$, therefore true for all positive integers n

12 $\begin{pmatrix}1 & 4\\ 0 & 2\end{pmatrix}^1 = \begin{pmatrix}1 & 4\\ 0 & 2\end{pmatrix}$ and $\begin{pmatrix}1 & 4(2^1-1)\\ 0 & 2^1\end{pmatrix} = \begin{pmatrix}1 & 4\\ 0 & 2\end{pmatrix}$ so true for $n = 1$

Assume true for $n = k$

$$\begin{pmatrix}1 & 4\\ 0 & 2\end{pmatrix}^{k+1} = \begin{pmatrix}1 & 4\\ 0 & 2\end{pmatrix}^{k}\begin{pmatrix}1 & 4\\ 0 & 2\end{pmatrix}$$

$$= \begin{pmatrix}1 & 4(2^k-1)\\ 0 & 2^k\end{pmatrix}\begin{pmatrix}1 & 4\\ 0 & 2\end{pmatrix}$$

$$= \begin{pmatrix}1 & 4+8.2^k-8\\ 0 & 2.2^k\end{pmatrix}$$

$$= \begin{pmatrix}1 & 4(2.2^k-1)\\ 0 & 2.2^k\end{pmatrix}$$

$$= \begin{pmatrix}1 & 4(2^{k+1}-1)\\ 0 & 2^{k+1}\end{pmatrix} \text{ so true for } n = k+1$$

Since true for $n = 1$ and assuming true for $n = k$ implies true for $n = k + 1$, therefore true for all positive integers n

13 Let $\mathbf{A} = \begin{pmatrix}a_1 & a_2\\ a_3 & a_4\\ a_5 & a_6\end{pmatrix}$, $\mathbf{B} = \begin{pmatrix}b_1 & b_2\\ b_3 & b_4\\ b_5 & b_6\end{pmatrix}$, $\mathbf{C} = \begin{pmatrix}c_1 & c_2\\ c_3 & c_4\\ c_5 & c_6\end{pmatrix}$

$$\mathbf{A}+(\mathbf{B}+\mathbf{C}) = \begin{pmatrix}a_1 & a_2\\ a_3 & a_4\\ a_5 & a_6\end{pmatrix} + \begin{pmatrix}b_1+c_1 & b_2+c_2\\ b_3+c_3 & b_4+c_4\\ b_5+c_5 & b_6+c_6\end{pmatrix}$$

$$= \begin{pmatrix}a_1+b_1+c_1 & a_2+b_2+c_2\\ a_3+b_3+c_3 & a_4+b_4+c_4\\ a_5+b_5+c_5 & a_6+b_6+c_6\end{pmatrix}$$

$$= \begin{pmatrix}(a_1+b_1)+c_1 & (a_2+b_2)+c_2\\ (a_3+b_3)+c_3 & (a_4+b_4)+c_4\\ (a_5+b_5)+c_5 & (a_6+b_6)+c_6\end{pmatrix}$$

$$= \begin{pmatrix}a_1+b_1 & a_2+b_2\\ a_3+b_3 & a_4+b_4\\ a_5+b_5 & a_6+b_6\end{pmatrix} + \begin{pmatrix}c_1 & c_2\\ c_3 & c_4\\ c_5 & c_6\end{pmatrix}$$

$= (\mathbf{A}+\mathbf{B})+\mathbf{C}$ as required

14 Let $\mathbf{A} = \begin{pmatrix}a_1 & a_2\\ a_3 & a_4\end{pmatrix}$, $\mathbf{B} = \begin{pmatrix}b_1 & b_2\\ b_3 & b_4\end{pmatrix}$

$$\mathbf{A}+\mathbf{B} = \begin{pmatrix}a_1 & a_2\\ a_3 & a_4\end{pmatrix} + \begin{pmatrix}b_1 & b_2\\ b_3 & b_4\end{pmatrix}$$

$$= \begin{pmatrix}a_1+b_1 & a_2+b_2\\ a_3+b_3 & a_4+b_4\end{pmatrix}$$

$$= \begin{pmatrix}b_1+a_1 & b_2+a_2\\ b_3+a_3 & b_4+a_4\end{pmatrix}$$

$$= \begin{pmatrix}b_1 & b_2\\ b_3 & b_4\end{pmatrix} + \begin{pmatrix}a_1 & a_2\\ a_3 & a_4\end{pmatrix}$$

$= \mathbf{B}+\mathbf{A}$ as required

15 e.g. let $\mathbf{A} = \begin{pmatrix}2 & 0\\ 1 & 1\end{pmatrix}$, $\mathbf{B} = \begin{pmatrix}0 & 1\\ 2 & 2\end{pmatrix}$

$$\mathbf{AB} = \begin{pmatrix}2 & 0\\ 1 & 1\end{pmatrix}\begin{pmatrix}0 & 1\\ 2 & 2\end{pmatrix}$$

$$= \begin{pmatrix}0 & 2\\ 2 & 3\end{pmatrix}$$

$$\mathbf{BA} = \begin{pmatrix}0 & 1\\ 2 & 2\end{pmatrix}\begin{pmatrix}2 & 0\\ 1 & 1\end{pmatrix}$$

$$= \begin{pmatrix}1 & 1\\ 6 & 2\end{pmatrix}$$

So $\mathbf{AB} \neq \mathbf{BA}$

16 Let $\mathbf{A} = \begin{pmatrix}a_1 & a_2\\ a_3 & a_4\end{pmatrix}$, $\mathbf{B} = \begin{pmatrix}b_1 & b_2\\ b_3 & b_4\end{pmatrix}$, $\mathbf{C} = \begin{pmatrix}c_1 & c_2\\ c_3 & c_4\end{pmatrix}$

$\mathbf{A}(\mathbf{B}+\mathbf{C})$

$$= \begin{pmatrix}a_1 & a_2\\ a_3 & a_4\end{pmatrix}\left[\begin{pmatrix}b_1 & b_2\\ b_3 & b_4\end{pmatrix} + \begin{pmatrix}c_1 & c_2\\ c_3 & c_4\end{pmatrix}\right]$$

$$= \begin{pmatrix}a_1 & a_2\\ a_3 & a_4\end{pmatrix}\begin{pmatrix}b_1+c_1 & b_2+c_2\\ b_3+c_3 & b_4+c_4\end{pmatrix}$$

$$= \begin{pmatrix}a_1(b_1+c_1)+a_2(b_3+c_3) & a_1(b_2+c_2)+a_2(b_4+c_4)\\ a_3(b_1+c_1)+a_4(b_3+c_3) & a_3(b_2+c_2)+a_4(b_4+c_4)\end{pmatrix}$$

$$= \begin{pmatrix}(a_1b_1+a_2b_3)+(a_1c_1+a_2c_3) & (a_1b_2+a_2b_4)+(a_1c_2+a_2c_4)\\ (a_3b_1+a_4b_3)+(a_3c_1+a_4c_3) & (a_3b_2+a_4b_4)+(a_3c_2+a_4c_4)\end{pmatrix}$$

$$= \begin{pmatrix}a_1b_1+a_2b_3 & a_1b_2+a_2b_4\\ a_3b_1+a_4b_3 & a_3b_2+a_4b_4\end{pmatrix} + \begin{pmatrix}a_1c_1+a_2c_3 & a_1c_2+a_2c_4\\ a_3c_1+a_4c_3 & a_3c_2+a_4c_4\end{pmatrix}$$

$$= \begin{pmatrix}a_1 & a_2\\ a_3 & a_4\end{pmatrix}\begin{pmatrix}b_1 & b_2\\ b_3 & b_4\end{pmatrix} + \begin{pmatrix}a_1 & a_2\\ a_3 & a_4\end{pmatrix}\begin{pmatrix}c_1 & c_2\\ c_3 & c_4\end{pmatrix}$$

$= \mathbf{AB}+\mathbf{AC}$

17 Let $\mathbf{A} = \begin{pmatrix}a_1 & a_2\\ a_3 & a_4\end{pmatrix}$

$$\mathbf{A}\begin{pmatrix}1 & 0\\ 0 & 1\end{pmatrix} = \begin{pmatrix}a_1 & a_2\\ a_3 & a_4\end{pmatrix}\begin{pmatrix}1 & 0\\ 0 & 1\end{pmatrix}$$

$$= \begin{pmatrix}a_1 & a_2\\ a_3 & a_4\end{pmatrix}$$

$$\begin{pmatrix}1&0\\0&1\end{pmatrix}\mathbf{A}=\begin{pmatrix}1&0\\0&1\end{pmatrix}\begin{pmatrix}a_1&a_2\\a_3&a_4\end{pmatrix}$$
$$=\begin{pmatrix}a_1&a_2\\a_3&a_4\end{pmatrix}$$

So $\mathbf{A}\begin{pmatrix}1&0\\0&1\end{pmatrix}=\begin{pmatrix}1&0\\0&1\end{pmatrix}\mathbf{A}$ as required

18 Let $\mathbf{B}=\begin{pmatrix}b_1&b_2&b_3\\b_4&b_5&b_6\\b_7&b_8&b_9\end{pmatrix}$

$$\mathbf{B}\begin{pmatrix}1&0&0\\0&1&0\\0&0&1\end{pmatrix}=\begin{pmatrix}b_1&b_2&b_3\\b_4&b_5&b_6\\b_7&b_8&b_9\end{pmatrix}\begin{pmatrix}1&0&0\\0&1&0\\0&0&1\end{pmatrix}$$
$$=\begin{pmatrix}b_1&b_2&b_3\\b_4&b_5&b_6\\b_7&b_8&b_9\end{pmatrix}$$

$$\begin{pmatrix}1&0&0\\0&1&0\\0&0&1\end{pmatrix}\mathbf{B}=\begin{pmatrix}1&0&0\\0&1&0\\0&0&1\end{pmatrix}\begin{pmatrix}b_1&b_2&b_3\\b_4&b_5&b_6\\b_7&b_8&b_9\end{pmatrix}$$
$$=\begin{pmatrix}b_1&b_2&b_3\\b_4&b_5&b_6\\b_7&b_8&b_9\end{pmatrix}$$

So $\mathbf{B}\begin{pmatrix}1&0&0\\0&1&0\\0&0&1\end{pmatrix}=\begin{pmatrix}1&0&0\\0&1&0\\0&0&1\end{pmatrix}\mathbf{B}$ as required

19 a $\begin{pmatrix}8&14&15&17\\6&11&7&9\\9&18&19&12\end{pmatrix}$

$\begin{pmatrix}5\\2\\3\\1\end{pmatrix}$ or $\begin{pmatrix}0.05\\0.02\\0.03\\0.01\end{pmatrix}$

b $\begin{pmatrix}8&14&15&17\\6&11&7&9\\9&18&19&12\end{pmatrix}\begin{pmatrix}0.05\\0.02\\0.03\\0.01\end{pmatrix}$

$$=\begin{pmatrix}1.3\\0.82\\1.5\end{pmatrix}$$

c £3620

20 a $\begin{pmatrix}5&20&7&6&50\\4&15&15&8&100\\12&4&2&30&50\end{pmatrix}\begin{pmatrix}12\\8\\18\\30\\1\end{pmatrix}$

$\begin{pmatrix}576\\778\\1162\end{pmatrix}$

b £25.16

21 a $ae+bg=a\left(\dfrac{d}{ad-bc}\right)+b\left(-\dfrac{c}{ad-bc}\right)$

$$=\frac{ad}{ad-bc}-\frac{bc}{ad-bc}$$
$$=\frac{ad-bc}{ad-bc}$$
$=1$ as required.

b $h=\dfrac{a}{ad-bc},\ f=-\dfrac{b}{ad-bc}$

22 When $n=1$, $\begin{pmatrix}0&1\\2&0\end{pmatrix}^{2n}=\begin{pmatrix}0&1\\2&0\end{pmatrix}^2$
$$=\begin{pmatrix}2&0\\0&2\end{pmatrix}$$

$$2^n\mathbf{I}=2\begin{pmatrix}1&0\\0&1\end{pmatrix}$$
$$=\begin{pmatrix}2&0\\0&2\end{pmatrix}$$

So true for $n=1$

Assume true for $n=k$ and consider $n=k+1$:

$$\begin{pmatrix}0&1\\2&0\end{pmatrix}^{2(k+1)}=\begin{pmatrix}0&1\\2&0\end{pmatrix}^{2k}\begin{pmatrix}0&1\\2&0\end{pmatrix}^2$$
$$=2^k\begin{pmatrix}1&0\\0&1\end{pmatrix}\begin{pmatrix}2&0\\0&2\end{pmatrix}$$
$$=2^k\begin{pmatrix}2&0\\0&2\end{pmatrix}$$
$$=2^k\cdot 2\begin{pmatrix}1&0\\0&1\end{pmatrix}$$
$$=2^{k+1}\mathbf{I}$$

So true for $n=k+1$

The statement is true for $n=1$ and by assuming it is true for $n=k$ it is shown to be true for $n=k+1$, therefore, by mathematical induction, it is true for all $n\in\mathbb{N}$

Exercise 5.2A

1 a

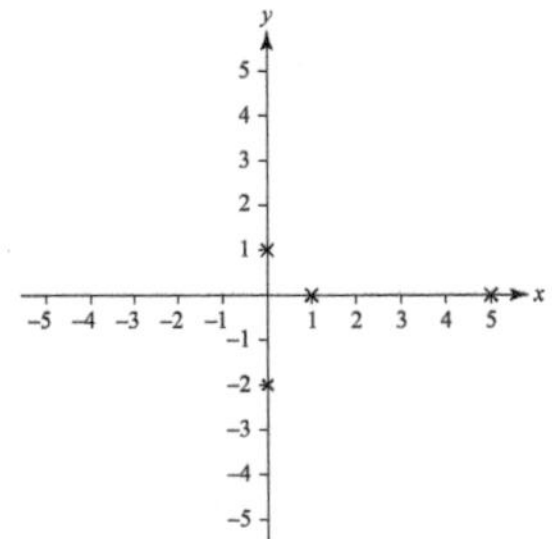

b $\begin{pmatrix}5&0\\0&-2\end{pmatrix}$

2 a

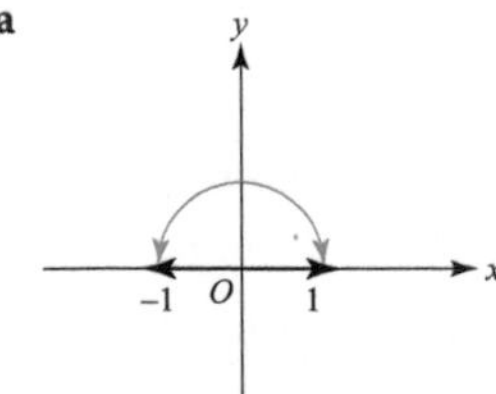

b $\begin{pmatrix} 0 \\ -1 \end{pmatrix}$

c $\begin{pmatrix} -1 & 0 \\ 0 & -1 \end{pmatrix}$

3 a $\begin{pmatrix} 0 & 1 \\ 1 & 0 \end{pmatrix}$ **b** $\begin{pmatrix} \frac{\sqrt{3}}{2} & -\frac{1}{2} \\ \frac{1}{2} & \frac{\sqrt{3}}{2} \end{pmatrix}$

c $\begin{pmatrix} 5 & 0 \\ 0 & 5 \end{pmatrix}$ **d** $\begin{pmatrix} 1 & 0 \\ 0 & 4 \end{pmatrix}$

4 a (0, −2), (4, 3) and (−8, 3)

b Stretch parallel to x-axis of scale factor 2

5 $\theta = 315°$

6 a $\begin{pmatrix} 6 \\ -1 \end{pmatrix}, \begin{pmatrix} -1 \\ 2 \end{pmatrix}$

b Stretch scale factor 0.5 parallel to y-axis

7 a Enlargement scale factor 3 centre the origin

b Stretch scale factor 2 parallel to y-axis

c Rotation by 135° anticlockwise about origin

d Rotation by 270° anticlockwise about origin

e Reflection in y-axis

f Rotation by 20° clockwise about origin or 340° anticlockwise.

8 a i $\mathbf{BA} = \begin{pmatrix} 0 & -1 \\ -1 & 0 \end{pmatrix}$ **ii** $\mathbf{AB} = \begin{pmatrix} -1 & 0 \\ -1 & 1 \end{pmatrix}$

b $\mathbf{BA} \neq \mathbf{AB}$ since matrix multiplication is not commutative.

9 a **A** is reflection in x-axis, **B** is rotation about origin of 90° clockwise or 270° anticlockwise

b i $\mathbf{BA} = \begin{pmatrix} 0 & -1 \\ -1 & 0 \end{pmatrix}$ **ii** $\mathbf{AB} = \begin{pmatrix} 0 & 1 \\ 1 & 0 \end{pmatrix}$

c i Reflection in $y = -x$ **ii** Reflection in $y = x$

10 a $\begin{pmatrix} -\frac{\sqrt{2}}{2} & \frac{\sqrt{2}}{2} \\ \frac{\sqrt{2}}{2} & \frac{\sqrt{2}}{2} \end{pmatrix}$ **b** $\begin{pmatrix} 0 & -2 \\ -2 & 0 \end{pmatrix}$ **c** $\begin{pmatrix} 0 & 1 \\ 2 & 0 \end{pmatrix}$

11 $\left(\frac{3}{2}\sqrt{2}, 7\sqrt{2}\right), \left(-\frac{\sqrt{2}}{2}, 3\sqrt{2}\right), (2\sqrt{2}, -2\sqrt{2}), (4\sqrt{2}, 2\sqrt{2})$

12 a $\begin{pmatrix} -\frac{\sqrt{2}}{2} & -\frac{\sqrt{2}}{2} \\ \frac{\sqrt{2}}{2} & -\frac{\sqrt{2}}{2} \end{pmatrix}$

b $\mathbf{A}^2 = \begin{pmatrix} -\frac{\sqrt{2}}{2} & -\frac{\sqrt{2}}{2} \\ \frac{\sqrt{2}}{2} & -\frac{\sqrt{2}}{2} \end{pmatrix}\begin{pmatrix} -\frac{\sqrt{2}}{2} & -\frac{\sqrt{2}}{2} \\ \frac{\sqrt{2}}{2} & -\frac{\sqrt{2}}{2} \end{pmatrix}$

$= \begin{pmatrix} \frac{1}{2} - \frac{1}{2} & \frac{1}{2} + \frac{1}{2} \\ -\frac{1}{2} - \frac{1}{2} & -\frac{1}{2} + \frac{1}{2} \end{pmatrix}$

$= \begin{pmatrix} 0 & 1 \\ -1 & 0 \end{pmatrix}$

c $\sin\theta = -1 \Rightarrow \theta = -90$
and $\cos(-90) = 0$
So a clockwise rotation of 90°,
or show geometrically.

d An anticlockwise rotation by 45° about the origin.

13 a $\begin{pmatrix} -1 & 0 \\ 0 & -1 \end{pmatrix}$; rotation of 180° around origin

b $\begin{pmatrix} -\frac{1}{2} & -\frac{\sqrt{3}}{2} \\ \frac{\sqrt{3}}{2} & -\frac{1}{2} \end{pmatrix}$; rotation of 120° anticlockwise around origin

14 a $\begin{pmatrix} 1 & 0 \\ 0 & -3 \end{pmatrix}$

b A stretch of scale factor −3 parallel to the y-axis.

15 a $\mathbf{M} = \begin{pmatrix} k\cos\theta & -k\sin\theta \\ k\sin\theta & k\cos\theta \end{pmatrix}$

b $\theta = (-45°), 135°, k = 2\sqrt{2}$

16 a (4, 1, 0) and (2, −5, 3)

b Reflection in $y = 0$

17 a $\begin{pmatrix} 1 & 0 & 0 \\ 0 & \frac{\sqrt{3}}{2} & -\frac{1}{2} \\ 0 & \frac{1}{2} & \frac{\sqrt{3}}{2} \end{pmatrix}$ **b** $\begin{pmatrix} 1 & 0 & 0 \\ 0 & 1 & 0 \\ 0 & 0 & -1 \end{pmatrix}$

c $\begin{pmatrix} \frac{\sqrt{2}}{2} & 0 & \frac{\sqrt{2}}{2} \\ 0 & 1 & 0 \\ -\frac{\sqrt{2}}{2} & 0 & \frac{\sqrt{2}}{2} \end{pmatrix}$

18 a Rotation around x-axis, $\theta = 300°$

b $\begin{pmatrix} 3 \\ 2 - \frac{\sqrt{3}}{2} \\ -\frac{1}{2} - 2\sqrt{3} \end{pmatrix}$

19 Anticlockwise rotation of 300° around z-axis.

20 a Reflection in plane $z = 0$

b $\mathbf{M} = \begin{pmatrix} 1 & 0 & 0 \\ 0 & 1 & 0 \\ 0 & 0 & -1 \end{pmatrix}$

21 a Rotation of 90° (anticlockwise) around the x-axis.

b Reflection in $x = 0$

c Rotation of 180° around the z-axis.

d Rotation of 120° anticlockwise around y-axis.

22 a $\begin{pmatrix} -\frac{\sqrt{2}}{2} & 0 & -\frac{\sqrt{2}}{2} \\ 0 & 1 & 0 \\ \frac{\sqrt{2}}{2} & 0 & -\frac{\sqrt{2}}{2} \end{pmatrix}$ **b** $\left(-1 - \frac{\sqrt{2}}{2}, 0, 1 - \frac{\sqrt{2}}{2}\right)$

23 a $\begin{pmatrix} \frac{\sqrt{3}}{2} & \frac{1}{2} & 0 \\ -\frac{1}{2} & \frac{\sqrt{3}}{2} & 0 \\ 0 & 0 & 1 \end{pmatrix}$ **b** $\left(\frac{3}{2}, -\frac{\sqrt{3}}{2}, 1\right)$

Exercise 5.2B

1 a $y = 0$
b $y = -2x$

2 a $\begin{pmatrix} 3 & -2 \\ 2 & 3 \end{pmatrix}\begin{pmatrix} x \\ y \end{pmatrix} = \begin{pmatrix} x \\ y \end{pmatrix}$

$3x - 2y = x \Rightarrow x = y$

$2x + 3y = y \Rightarrow x = -y$

(0, 0) is only point that satisfies both of these equations, therefore (0, 0) is the only invariant point.

b $\begin{pmatrix} 2 & 0 & 1 \\ 0 & 3 & -2 \\ 1 & 0 & -4 \end{pmatrix}\begin{pmatrix} x \\ y \\ z \end{pmatrix} = \begin{pmatrix} x \\ y \\ z \end{pmatrix}$

$2x + z = x \Rightarrow z = -x$

$3y - 2z = y \Rightarrow y = z$

$x - 4z = z \Rightarrow x = 5z$

(0, 0, 0) is only point that satisfies all of these equations, therefore (0, 0, 0) is the only invariant point.

3 a No difference, e.g. reflection in y-axis affects only the x-coordinates and stretch parallel to y-axis affects only the y-coordinates. Or use a sketch. Or show that combined transformation is $\begin{pmatrix} -1 & 0 \\ 0 & k \end{pmatrix}$ for both orders.

b No difference, e.g. use a sketch or show that combined transformation is $k\begin{pmatrix} \cos\theta & -\sin\theta \\ \sin\theta & \cos\theta \end{pmatrix}$ in both cases.

c Different. Reflection in $y = x$ followed by stretch along x-axis will be $\begin{pmatrix} 0 & k \\ 1 & 0 \end{pmatrix}$, other way around is $\begin{pmatrix} 0 & 1 \\ k & 0 \end{pmatrix}$.

4 a Stretch parallel to x-axis of scale factor 3 and stretch parallel to y-axis of scale factor -2.

b Enlargement of scale factor 4 centre the origin and reflection in $y = x$.

5 a $y = \frac{1}{5}x$ **b** $y = 3x$ and $y = -x$

6 a e.g. A stretch parallel to the y-axis and a reflection in the y-axis (or rotation by 180° or 360° about origin)

b Reflection in y-axis

7 e.g. Reflection in line $y = x$ (or $y = x + k$)

8 Enlargement, centre the origin

9 a $y = \left(\frac{-1 \pm \sqrt{5}}{2}\right)x$ **b** $y = mx$ for any m

c $y = 2x$ and $y = -\frac{1}{2}x + c$ for any c.

10 a **i**; **bi** $y = 0$; line of invariant points
b **aii**; **bii** (0, 0)
c **aiii**; **biii** $x = 0$; line of invariant points
d **aiv**; **biv** $y = x + c$ for any c; not lines of invariant points, only (0, 0) is invariant.
$y = -x$; line of invariant points.

11 a $y = x$

b $\begin{pmatrix} -2 & 3 \\ -3 & 4 \end{pmatrix}\begin{pmatrix} x \\ x+c \end{pmatrix} = \begin{pmatrix} -2x + 3(x+c) \\ -3x + 4(x+c) \end{pmatrix}$

$= \begin{pmatrix} x + 3c \\ x + 4c \end{pmatrix}$

$= \begin{pmatrix} x + 3c \\ (x + 3c) + c \end{pmatrix}$

So invariant lines are $y = x + c$

12 $\begin{pmatrix} 1 & 0 \\ 0 & -1 \end{pmatrix}$

13 a Plane $x = 0$ **b** Plane $y = z$
c z-axis **d** Plane $x = y$

Exercise 5.3A

1 $\det\begin{pmatrix} -7 & 3 \\ 5 & -4 \end{pmatrix} = -7 \times -4 - 3 \times 5$

$= 28 - 15$

$= 13$ as required

2 a -18 **b** -3
c $-5a$ **d** 2

3 $\det\begin{pmatrix} a+b & 2a \\ 2b & a+b \end{pmatrix} = (a+b)(a+b) - (2a)(2b)$

$= a^2 + 2ab + b^2 - 4ab$

$= a^2 - 2ab + b^2$

$= (a-b)^2$ as required

4 $k = 3, 4$

5 a Non-singular
b Singular
c Non-singular unless $a = 0$
d Singular

6 a $x = \frac{5}{2}$
b $x = \pm\sqrt{3}$
c $x = 0, \frac{-3}{2}$
d $x = 0, 4$

7 $y = -1, -2$

8 a i -30 **ii** 1

b i $\mathbf{AB} = \begin{pmatrix} -1 & 2 \\ 0 & 5 \end{pmatrix}\begin{pmatrix} 0 & -3 \\ 2 & 4 \end{pmatrix} = \begin{pmatrix} 4 & 11 \\ 10 & 20 \end{pmatrix}$

$\det(\mathbf{AB}) = 4 \times 20 - 11 \times 10 = -30$ as required

ii $\mathbf{A} + \mathbf{B} = \begin{pmatrix} -1 & 2 \\ 0 & 5 \end{pmatrix} + \begin{pmatrix} 0 & -3 \\ 2 & 4 \end{pmatrix} = \begin{pmatrix} -1 & -1 \\ 2 & 9 \end{pmatrix}$

$\det(\mathbf{A} + \mathbf{B}) = -1 \times 9 - -1 \times 2 = -7$

9 a -2
b $\frac{1}{9}$

10 a -8
b 16

11 $\frac{1}{7}$

12 a 5 **b** $\frac{5}{3}$

13 a $\frac{1}{27}\begin{pmatrix} 5 & -2 \\ -4 & 7 \end{pmatrix}$ **b** $\frac{1}{20}\begin{pmatrix} 1 & 3 \\ 8 & 4 \end{pmatrix}$

c $-\frac{1}{x}\begin{pmatrix} -3 & -5 \\ x & 2x \end{pmatrix}$ **d** $\frac{1}{50}\begin{pmatrix} 2\sqrt{5} & \sqrt{5} \\ -2\sqrt{5} & 4\sqrt{5} \end{pmatrix}$

14 $\begin{pmatrix} 6 & 1 \\ -9 & -1 \end{pmatrix}^{-1} = \frac{1}{6\times-1-1\times-9}\begin{pmatrix} -1 & -1 \\ 9 & 6 \end{pmatrix}$

$= \frac{1}{3}\begin{pmatrix} -1 & -1 \\ 9 & 6 \end{pmatrix}$

$= \begin{pmatrix} -\frac{1}{3} & -\frac{1}{3} \\ 3 & 2 \end{pmatrix}$ as required.

Alternatively show that

$\begin{pmatrix} 6 & 1 \\ -9 & -1 \end{pmatrix}\begin{pmatrix} -\frac{1}{3} & -\frac{1}{3} \\ 3 & 2 \end{pmatrix} = \mathbf{I}$

15 a $\begin{pmatrix} \frac{3}{b} & -\frac{2}{b} \\ -\frac{1}{a} & \frac{1}{a} \end{pmatrix}$

b $\det(\mathbf{A}^{-1}) = \frac{3}{b}\cdot\frac{1}{a} - \left(-\frac{2}{b}\right)\left(-\frac{1}{a}\right)$

$= \frac{3}{ab} - \frac{2}{ab}$

$= \frac{1}{ab}$

$= \frac{1}{\det(A)}$ as required

16 $\begin{pmatrix} 3 & -2 \\ 4 & -3 \end{pmatrix}\begin{pmatrix} 3 & -2 \\ 4 & -3 \end{pmatrix} = \begin{pmatrix} 3\times3+-2\times4 & 3\times-2+-2\times-3 \\ 4\times3+-3\times4 & 4\times-2+-3\times-3 \end{pmatrix}$

$= \begin{pmatrix} 1 & 0 \\ 0 & 1 \end{pmatrix}$ therefore it is self-inverse.

17 $a = \pm\sqrt{7}$

18 $a = 3$

19 a $x = \frac{5}{3}, -2$ b $\frac{1}{10}\begin{pmatrix} 2x & x \\ 1-x & x+1 \end{pmatrix}$

20 a $\mathbf{T}^{-1} = \frac{1}{4}\begin{pmatrix} 2 & 0 \\ 0 & 2 \end{pmatrix}$

$= \frac{1}{2}\begin{pmatrix} 1 & 0 \\ 0 & 1 \end{pmatrix}$

$= \frac{1}{2}\mathbf{I}$ as required

b Enlargements scale factor $2\left(\frac{1}{2}\right)$ centre the origin.

21 a $\frac{\sqrt{2}}{2}\begin{pmatrix} -1 & -1 \\ 1 & -1 \end{pmatrix}$

b $\mathbf{A}^{-1}\mathbf{A} = \frac{1}{\sqrt{2}}\begin{pmatrix} -1 & 1 \\ -1 & -1 \end{pmatrix}\frac{\sqrt{2}}{2}\begin{pmatrix} -1 & -1 \\ 1 & -1 \end{pmatrix}$

$= \frac{1}{2}\begin{pmatrix} 1+1 & 1-1 \\ 1-1 & 1+1 \end{pmatrix}$

$= \frac{1}{2}\begin{pmatrix} 2 & 0 \\ 0 & 2 \end{pmatrix}$

$= \begin{pmatrix} 1 & 0 \\ 0 & 1 \end{pmatrix}$ as required, therefore the inverse is as given.

22 a Always self-inverse.

b Only self-inverse if enlargement of scale factor ±1

c Only self-inverse when rotated by a multiple of 180°

d Always self-inverse.

Exercise 5.3B

1 a $x = 2, y = -3$ b $x = -5, y = -1$

c $x = 3, y = \frac{1}{2}$ d $x = -\frac{1}{4}, y = \frac{1}{3}$

2 a $x = 2, y = -5, z = -3$

b $x = \frac{1}{4}, y = -\frac{1}{2}, z = 1$

3 $x = -\frac{1}{a} - 6, y = 2 + 2a$

4 $x = \frac{3k-1}{3(k+1)}, y = \frac{4k}{3(k+1)}$

5 $\begin{pmatrix} -2 & 4 \\ -1 & -2 \end{pmatrix}$

6 $\begin{pmatrix} 4 & -2 \\ 0 & -1 \end{pmatrix}$

7 $\frac{b+2a}{b}$

8 $\begin{pmatrix} 17 & 16 \\ -9 & -8 \end{pmatrix}$

9 $\begin{pmatrix} 5 & 0 \\ 4 & -1 \end{pmatrix}$

10 e.g. $\sqrt{2}\begin{pmatrix} 1 & 0 \\ 0 & 1 \end{pmatrix}$, or $\begin{pmatrix} 0 & \sqrt{2} \\ \sqrt{2} & 0 \end{pmatrix}$

11 a $\mathbf{B}^{-1}$ b $\mathbf{A}$

12 $\mathbf{A} = \mathbf{C}^{-1}\mathbf{BC} \Rightarrow \mathbf{CA} = \mathbf{CC}^{-1}\mathbf{BC}$

$\Rightarrow \mathbf{CA} = \mathbf{BC}$

$\Rightarrow \mathbf{CAC}^{-1} = \mathbf{BCC}^{-1}$

$\Rightarrow \mathbf{CAC}^{-1} = \mathbf{B}$ as required

13 $\mathbf{ABA}^{-1} = \mathbf{I} \Rightarrow \mathbf{A}^{-1}\mathbf{ABA}^{-1} = \mathbf{A}^{-1}\mathbf{I}$

$\Rightarrow \mathbf{BA}^{-1} = \mathbf{A}^{-1}$

$\Rightarrow \mathbf{BA}^{-1}\mathbf{A} = \mathbf{A}^{-1}\mathbf{A}$

$\Rightarrow \mathbf{B} = \mathbf{I}$

14 $(\mathbf{ABC})^{-1}(\mathbf{ABC}) = \mathbf{I}$

$\Rightarrow (\mathbf{ABC})^{-1}\mathbf{ABCC}^{-1} = \mathbf{IC}^{-1}$

$\Rightarrow (\mathbf{ABC})^{-1}\mathbf{AB} = \mathbf{C}^{-1}$

$\Rightarrow (\mathbf{ABC})^{-1}\mathbf{ABB}^{-1} = \mathbf{C}^{-1}\mathbf{B}^{-1}$

$\Rightarrow (\mathbf{ABC})^{-1}\mathbf{A} = \mathbf{C}^{-1}\mathbf{B}^{-1}$

$\Rightarrow (\mathbf{ABC})^{-1}\mathbf{AA}^{-1} = \mathbf{C}^{-1}\mathbf{B}^{-1}\mathbf{A}^{-1}$

$\Rightarrow (\mathbf{ABC})^{-1} = \mathbf{C}^{-1}\mathbf{B}^{-1}\mathbf{A}^{-1}$ as required

15 $\mathbf{P}$ self-inverse $\Rightarrow \mathbf{P} = \mathbf{P}^{-1}$

$\Rightarrow \mathbf{PP} = \mathbf{P}^{-1}\mathbf{P}$

$\Rightarrow \mathbf{P}^2 = \mathbf{I}$

$\mathbf{I} = \mathbf{I}^{-1} \Rightarrow \mathbf{P}^2 = \mathbf{I}^{-1}$ as required

16 $\mathbf{PQP} = \mathbf{I} \Rightarrow \mathbf{P}^{-1}\mathbf{PQP} = \mathbf{P}^{-1}\mathbf{I}$

$\Rightarrow \mathbf{QP} = \mathbf{P}^{-1}$

$\Rightarrow \mathbf{QPP}^{-1} = \mathbf{P}^{-1}\mathbf{P}^{-1}$

$\Rightarrow \mathbf{Q} = (\mathbf{P}^{-1})^2$ as required

17 (5, –1)

18 Let $\mathbf{A} = \begin{pmatrix} a_1 & a_2 \\ a_3 & a_4 \end{pmatrix}$ and $\mathbf{B} = \begin{pmatrix} b_1 & b_2 \\ b_3 & b_4 \end{pmatrix}$

$\mathbf{AB} = \begin{pmatrix} a_1 & a_2 \\ a_3 & a_4 \end{pmatrix}\begin{pmatrix} b_1 & b_2 \\ b_3 & b_4 \end{pmatrix} = \begin{pmatrix} a_1b_1+a_2b_3 & a_1b_2+a_2b_4 \\ a_3b_1+a_4b_3 & a_3b_2+a_4b_4 \end{pmatrix}$

Then

$\det(\mathbf{AB}) = (a_1b_1+a_2b_3)(a_3b_2+a_4b_4) - (a_1b_2+a_2b_4)(a_3b_1+a_4b_3)$

$= (a_1b_1a_3b_2 + a_1b_1a_4b_4 + a_2b_3a_3b_2 + a_2b_3a_4b_4) - (a_1b_2a_3b_1 + a_1b_2a_4b_3 + a_2b_4a_3b_1 + a_2b_4a_4b_3)$

$= a_1b_1a_4b_4 - a_1b_2b_3a_4 - a_2b_1a_3b_4 + a_2b_2a_3b_3$

$= (a_1a_4 - a_2a_3)(b_1b_4 - b_2b_3)$

$= \det(\mathbf{A})\det(\mathbf{B})$ as required

19 Let $\mathbf{A} = \begin{pmatrix} a_1 & a_2 \\ a_3 & a_4 \end{pmatrix}$ then $\mathbf{A}^{-1} = \frac{1}{a_1a_4 - a_2a_3}\begin{pmatrix} a_4 & -a_2 \\ -a_3 & a_1 \end{pmatrix}$

$$\det(\mathbf{A}^{-1}) = \frac{a_4}{a_1a_4 - a_2a_3}\cdot\frac{a_1}{a_1a_4 - a_2a_3} - \frac{-a_2}{a_1a_4 - a_2a_3}\cdot\frac{-a_3}{a_1a_4 - a_2a_3}$$

$$= \frac{a_4a_1 - a_2a_3}{(a_1a_4 - a_2a_3)^2}$$

$$= \frac{1}{a_1a_4 - a_2a_3}$$

$$= \frac{1}{\det(\mathbf{A})}$$

$= [\det(\mathbf{A})]^{-1}$ as required

20 a i $\begin{pmatrix} \cos\theta & -\sin\theta \\ \sin\theta & \cos\theta \end{pmatrix}$ **ii** $\begin{pmatrix} \cos\theta & \sin\theta \\ -\sin\theta & \cos\theta \end{pmatrix}$

b $\begin{pmatrix} \cos\theta & -\sin\theta \\ \sin\theta & \cos\theta \end{pmatrix}\begin{pmatrix} \cos\theta & \sin\theta \\ -\sin\theta & \cos\theta \end{pmatrix} = \begin{pmatrix} 1 & 0 \\ 0 & 1 \end{pmatrix}$

$\Rightarrow \cos\theta\cos\theta - \sin\theta(-\sin\theta) \equiv 1$

$\Rightarrow \cos^2\theta + \sin^2\theta \equiv 1$ as required.

21 a Square with vertices at (0, 0), $(x, 0)$, $(0, y)$ and (x, y)

$\mathbf{T} = \begin{pmatrix} a & 0 \\ 0 & b \end{pmatrix}$ is a stretch of a in x-direction and b in y-direction.

$$\begin{pmatrix} a & 0 \\ 0 & b \end{pmatrix}\begin{pmatrix} 0 & x & 0 & x \\ 0 & 0 & y & y \end{pmatrix} = \begin{pmatrix} 0 & ax & 0 & ax \\ 0 & 0 & by & by \end{pmatrix}$$

So a rectangle with vertices at (0, 0), $(ax, 0)$, $(0, by)$ and (ax, by).

b Area of original square is xy

Area of image is $(ax)(by) = abxy$

$= \det(\mathbf{T}) \times xy$ as required since $\det(\mathbf{T}) = ab - 0 = ab$

Review exercise 5

1 a A has order 2×2, B has order 2×3, C has order 3×2, D has order 2×3

b i $\begin{pmatrix} 8 & 0 & 2 \\ 5 & 3 & 0 \end{pmatrix}$ **ii** $\begin{pmatrix} -3 & -15 \\ 12 & 0 \\ 0 & 6 \end{pmatrix}$

iii $\begin{pmatrix} 8 & -6 & 5 \\ -10 & -9 & 0 \end{pmatrix}$ **iv** $\begin{pmatrix} 3 & 1 \\ -2 & 6 \end{pmatrix}$

c i $\begin{pmatrix} 10 & 20 & -6 \\ 60 & 40 & 4 \end{pmatrix}$

ii **BA** not possible as 3 columns in **B** but 2 rows in **A**

iii $\begin{pmatrix} 8 & -2 \\ 11 & -25 \end{pmatrix}$ **iv** $\begin{pmatrix} 14 & -62 \\ 24 & 8 \\ -8 & 24 \end{pmatrix}$

v **AC** not possible as 2 columns in **A** but 3 rows in **C**

vi $\begin{pmatrix} -8 & 7 & -3 \\ 32 & -8 & 12 \\ 0 & -2 & 0 \end{pmatrix}$ **vii** $\begin{pmatrix} 28 & 36 \\ -72 & 136 \end{pmatrix}$

viii $\mathbf{C}^2$ not possible as **C** not a square matrix

2 a $\begin{pmatrix} a^2-a & 3a-2 \\ 3a & 6 \\ 2a^2-2a & 4a-6 \end{pmatrix}$

b $\begin{pmatrix} -a-6 & 1+9a \\ -2a & 3+2a^2 \\ 2a^2 & -2a-9 \end{pmatrix}$

c $\begin{pmatrix} a^2+3a & 3a+6 \\ a^2+2a & 3a+4 \end{pmatrix}$

d $\begin{pmatrix} 1+6a & -9 & a^2-3 \\ 2a^2 & 1-3a & a \\ -2a & 2a^2-3 & 3a \end{pmatrix}$

3 a i (2, –1) reflection in line $y = x$

ii (–7, 2) stretch scale factor 7 parallel to x-axis

iii (–2, –1) rotation of 90° anticlockwise about origin

iv $\left(\frac{1}{2}(2+\sqrt{3}), \frac{1}{2}(1-2\sqrt{3}\right)$; rotation of 210° anticlockwise about origin

b i 5 **ii** 35 **iii** 5 **iv** 5

c i determinant negative so does not preserve orientation

ii, iii, iv have positive determinants so do preserve orientation

4 a $(3, -3\sqrt{2}, \sqrt{2})$; rotation of 45° anticlockwise around x-axis.

b (3, –2, –4); reflection in plane $z = 0$

5 a $\begin{pmatrix} -1 & 0 \\ 0 & 1 \end{pmatrix}$

b $\begin{pmatrix} \frac{\sqrt{2}}{2} & \frac{\sqrt{2}}{2} \\ -\frac{\sqrt{2}}{2} & \frac{\sqrt{2}}{2} \end{pmatrix}$

c $\begin{pmatrix} -3 & 0 \\ 0 & -3 \end{pmatrix}$

6 a $\begin{pmatrix} \frac{\sqrt{2}}{2} & 0 & \frac{\sqrt{2}}{2} \\ 0 & 1 & 0 \\ -\frac{\sqrt{2}}{2} & 0 & \frac{\sqrt{2}}{2} \end{pmatrix}$

b $\begin{pmatrix} 1 & 0 & 0 \\ 0 & -1 & 0 \\ 0 & 0 & 1 \end{pmatrix}$

7 a $\begin{pmatrix} 0 & 1 \\ -2 & 0 \end{pmatrix}$

b $\begin{pmatrix} 0 & -1 \\ 1 & 0 \end{pmatrix}$

c $\begin{pmatrix} -\frac{\sqrt{3}}{2} & \frac{1}{2} \\ \frac{1}{2} & \frac{\sqrt{3}}{2} \end{pmatrix}$

8 a Invariant point is (0, 0)

b Line of invariant points is $x = 3y$

9 a $y = -x$ and $y = \frac{3}{2}x$

b $y = 3x$ and $y = -\frac{1}{3}x + c$

10 a $\det\begin{pmatrix} 2 & -1 \\ 4 & -3 \end{pmatrix} = 2\times-3--1\times4=-2$

b $\det\begin{pmatrix} 3a & b \\ -2a & b \end{pmatrix} = 3ab--2ab=5ab$

11 a $x=\frac{4}{3}$

b $x=\pm\sqrt{2}$

12 a $-\frac{1}{9}\begin{pmatrix} -2 & -3 \\ 1 & 6 \end{pmatrix}$

b $-\frac{1}{3}\begin{pmatrix} -2 & -1 \\ -3 & 0 \end{pmatrix}$

c Does not exist since $\det\begin{pmatrix} 3 & -6 \\ -4 & 8 \end{pmatrix} = 3\times8--6\times-4=0$

d $\frac{1}{ab}\begin{pmatrix} 3b & -2b \\ -7a & 5a \end{pmatrix}$

13 $\begin{pmatrix} 2 & 3 \\ 0 & 1 \end{pmatrix}$

14 $\begin{pmatrix} 2 & -7 \\ -5 & -3 \end{pmatrix}$

15 a $x=-5,\ y=2$ **b** $x=\frac{1}{5},\ y=-1$

Assessment 5

1 a $\begin{pmatrix} 15 & 6 \\ -9 & 3 \end{pmatrix}$

b $\begin{pmatrix} 8 & -6 \\ 0 & -4 \end{pmatrix}$

c $\begin{pmatrix} 9 & -1 \\ -3 & -1 \end{pmatrix}$

d $\begin{pmatrix} -20 & 19 \\ 12 & -7 \end{pmatrix}$

2 a $\begin{pmatrix} 12 & 9 \\ -7 & 25 \end{pmatrix}$

b $\begin{pmatrix} 30 & -17 \\ 9 & 7 \end{pmatrix}$

c **M** and **N** do not commute

3 a $\begin{pmatrix} 2 & 7 \\ 9 & 11 \end{pmatrix}$ **b** $\begin{pmatrix} 4 & 14 \\ 18 & 22 \end{pmatrix}$ **c** $\begin{pmatrix} 67 & 91 \\ 117 & 184 \end{pmatrix}$

4 $a=4,\ b=1,\ c=16$

5 a Enlargement, 6

b the origin, 90 degrees

c reflection, the x-axis

6 a $\mathbf{M}=\begin{pmatrix} -\frac{\sqrt{2}}{2} & -\frac{\sqrt{2}}{2} \\ \frac{\sqrt{2}}{2} & -\frac{\sqrt{2}}{2} \end{pmatrix}$

b $\mathbf{N}=\begin{pmatrix} \frac{\sqrt{2}}{2} & \frac{\sqrt{2}}{2} \\ \frac{-\sqrt{2}}{2} & \frac{\sqrt{2}}{2} \end{pmatrix}$

c $\begin{pmatrix} 0 & -1 \\ 1 & 0 \end{pmatrix}$

A rotation by 90 degrees anticlockwise about the origin. This the combined effect of the two rotations represented by **M** and **N**.

7 a The point (0, 0)

b Any point $(\lambda,-2\lambda)$

c Any point $(-3\lambda,\ \lambda)$

8 a The y-axis

b The z-axis

9 $\begin{pmatrix} x \\ y \end{pmatrix}=\begin{pmatrix} \frac{57}{25} \\ \frac{22}{25} \end{pmatrix}$

10 $\begin{pmatrix} x \\ y \end{pmatrix}=\begin{pmatrix} 4 \\ 3 \end{pmatrix}$

11 a 2 **b** (0, 0)

12 a 12 **b** 0

13 a $\begin{pmatrix} 9 & -16 \\ -5 & 9 \end{pmatrix}$ **b** $\frac{1}{22}\begin{pmatrix} 3 & 2 \\ 5 & 4 \end{pmatrix}$

14 a $\mathbf{A}^{-1}=\begin{pmatrix} 2 & -1 \\ -5 & 3 \end{pmatrix}$

b $\mathbf{B}^{-1}=\frac{1}{2}\begin{pmatrix} 1 & 4 \\ 1 & 2 \end{pmatrix}$

c $\begin{pmatrix} -9 & 5.5 \\ -4 & 2.5 \end{pmatrix}$

d $\begin{pmatrix} -5 & 11 \\ -8 & 18 \end{pmatrix}$

$(\mathbf{B}^{-1}\mathbf{A}^{-1})^{-1}=AB$

Chapter 6

Exercise 6.1A

1 $\mathbf{r}=\begin{pmatrix} 0 \\ 4 \\ 3 \end{pmatrix}+\lambda\begin{pmatrix} 1 \\ -2 \\ 0 \end{pmatrix}$

2 $\mathbf{r}=7\mathbf{i}-\mathbf{j}+2\mathbf{k}+\lambda(5\mathbf{i}-3\mathbf{j}+\mathbf{k})$

3 $\mathbf{r}=\begin{pmatrix} -7 \\ 2 \\ -5 \end{pmatrix}+\lambda\begin{pmatrix} 11 \\ -11 \\ 3 \end{pmatrix}$ or alternatively

$\mathbf{r}=\begin{pmatrix} 4 \\ -9 \\ -2 \end{pmatrix}+\lambda\begin{pmatrix} 11 \\ -11 \\ 3 \end{pmatrix}$

4 $\mathbf{r}=2\mathbf{i}+3\mathbf{k}+\lambda(\mathbf{j}-4\mathbf{k})$ or alternatively

$\mathbf{r}=2\mathbf{i}+\mathbf{j}-\mathbf{k}+\lambda(\mathbf{j}-4\mathbf{k})$

5 $x=5;\ \frac{y-2}{4}=\frac{z+7}{-1}$

6 $\frac{x-4}{8}=\frac{y+2}{-2}=\frac{z}{3}$

7 $\frac{x-9}{2}=\frac{y-4}{1}=\frac{z+3}{-2}$ or alternatively $\frac{x-1}{2}=\frac{y-0}{1}=\frac{z-5}{-2}$

8 $\frac{x-0}{-2}=\frac{y-5}{3}=\frac{z-1}{1}$ or alternatively $\frac{x-4}{-2}=\frac{y+1}{3}=\frac{z+1}{1}$

9 a $\frac{x-0}{-4}=\frac{z+2}{8}$ and $y=1$ **b** $\frac{x+3}{5}=\frac{y-0}{-1}=\frac{z-7}{3}$

10 a $\mathbf{r}=\begin{pmatrix}5\\1\\-3\end{pmatrix}+\lambda\begin{pmatrix}2\\-3\\1\end{pmatrix}$

b $\mathbf{r}=\begin{pmatrix}-2\\0\\5\end{pmatrix}+\lambda\begin{pmatrix}4\\-2\\-3\end{pmatrix}$

11 a (3, 2, 1) lies on the line.

b (3, 2, 1) does not lie on the line

12 a (−2, 0, 4) does not line on the line

b (−2, 0, 4) lies on the line.

Exercise 6.1B

1 a Not the same line.

b The equations represent the same line.

c The same line.

2 a Not the same line.

b Not the same line.

c Represent the same line.

d Represent the same line.

e Not the same line.

3 a Intersect at (−11, 16)

b Intersect at (−2, 10)

c Intersect at (1, −1)

4 a Intersect at (7, −6, −2)

b lines are parallel

c skew lines.

d intersect at (−4, 2, 0)

5 $3\sqrt{10}$

6 $\mathbf{r}=\begin{pmatrix}5\\-6\end{pmatrix}+\lambda\begin{pmatrix}-3\\-2\end{pmatrix}$

7 $\mathbf{r}=\begin{pmatrix}1\\2\end{pmatrix}+\lambda\begin{pmatrix}-1\\2\end{pmatrix}$

8 a $\mathbf{r}=\begin{pmatrix}3\\1\\0\end{pmatrix}+\lambda\begin{pmatrix}1\\2\\-1\end{pmatrix}$

b $\mathbf{r}=\begin{pmatrix}-3\\0\\1\end{pmatrix}+\lambda\begin{pmatrix}-1\\-1\\2\end{pmatrix}$

c $\mathbf{r}=\begin{pmatrix}3\\0\\1\end{pmatrix}+\lambda\begin{pmatrix}1\\-1\\2\end{pmatrix}$

9 $\mathbf{r}=\begin{pmatrix}1\\3\\5\end{pmatrix}+\lambda\begin{pmatrix}2\\1\\4\end{pmatrix}$

Exercise 6.2A

1 a 1 **b** 0 **c** 0 **d** 1

e 2 **f** 0 **g** 1 **h** 3

2 a 4 **b** −5

c $-\sqrt{2}$ **d** −3

3 $6a^2-4a$

4 k −6

5 a 46.4° **b** 26.5°

c 63.9° **d** 101.6°

6 a $\dfrac{a}{a^2+1}$ **b** $\dfrac{5-4a^2}{5a^2+5}$

7 $\begin{pmatrix}6\\-4\end{pmatrix}\cdot\begin{pmatrix}-8\\-12\end{pmatrix}=-48+48=0$ so perpendicular

8 $\begin{pmatrix}4\\-1\\5\end{pmatrix}\cdot\begin{pmatrix}-2\\-3\\1\end{pmatrix}=(4\times-2)+(-1\times-3)+(5\times1)$

$=0$ therefore they are perpendicular

9 $3\times2-5\times4-2\times-7=0$ so perpendicular

10 $2\times4+0+1\times-3=5$ so not perpendicular

11 $a=1$

12 $b=2$

13 $c=1, 2$

Exercise 6.2B

1 $\begin{pmatrix}-1\\3\end{pmatrix}\cdot\begin{pmatrix}6\\2\end{pmatrix}=-6+6=0$ so perpendicular

2 $\begin{pmatrix}6\\7\\-5\end{pmatrix}\cdot\begin{pmatrix}2\\-1\\1\end{pmatrix}=12-7-5=0$ so perpendicular

3 $\begin{pmatrix}3\\-2\\6\end{pmatrix}\cdot\begin{pmatrix}2\\6\\1\end{pmatrix}=6-12+6=0$ therefore they are perpendicular.

4 $\theta=45°$

5 $2=5+3t$ ①

$1+2s=11-4t$ ②

$-3+s=3-t$ ③

Solving ① and ② gives $s=7$ and $t=-1$

Check in ③

$-3+7=4$

$3-t=4$

So the lines intersect

$$\cos\theta=\frac{\begin{pmatrix}0\\2\\1\end{pmatrix}\cdot\begin{pmatrix}3\\-4\\-1\end{pmatrix}}{\left|\begin{pmatrix}0\\2\\1\end{pmatrix}\right|\left|\begin{pmatrix}3\\-4\\-1\end{pmatrix}\right|}\quad\frac{-8-1}{\sqrt{5}\sqrt{26}}=\frac{-9}{\sqrt{5}\sqrt{26}}$$

$\theta=180-142.1=38°$ to the nearest degree

6 93°

7 $-1+\lambda=-2\Rightarrow\lambda=-1$

$2+\lambda=2-\mu\Rightarrow\mu=1$

Check: $3-\lambda=4$

$1+3\mu=4$ hence they intersect

$\theta=43.1°$

8 $a=\pm2\sqrt{2}$

9 $a=\pm2$

10 a $\left(2, 3, \dfrac{7}{2}\right)$ **b** $\cos\theta=\dfrac{67}{9\sqrt{161}}$

11 $k=2\pm\sqrt{3}$

12 $k=-\dfrac{1}{4}$

13 $\dfrac{\sqrt{89}}{2}$

14 $\dfrac{\sqrt{11}}{2}$

15 $\mathbf{a}\cdot\mathbf{b}=(a_1\mathbf{i}+a_2\mathbf{j}+a_3\mathbf{k})\cdot(b_1\mathbf{i}+b_2\mathbf{j}+b_3\mathbf{k})$

$=a_1b_1\mathbf{i}\cdot\mathbf{i}+a_1b_2\mathbf{i}\cdot\mathbf{j}+a_1b_3\mathbf{i}\cdot\mathbf{k}+a_2b_1\mathbf{j}\cdot\mathbf{i}+a_2b_2\mathbf{j}\cdot\mathbf{j}$
$+a_2b_3\mathbf{j}\cdot\mathbf{k}+a_3b_1\mathbf{k}\cdot\mathbf{i}+a_3b_2\mathbf{k}\cdot\mathbf{j}+a_3b_3\mathbf{k}\cdot\mathbf{k}$

$=a_1b_1\mathbf{i}\cdot\mathbf{i}+a_2b_2\mathbf{j}\cdot\mathbf{j}+a_3b_3\mathbf{k}\cdot\mathbf{k}$ Using fact that **i, j** and **k** are perpendicular

$=a_1b_1+a_2b_2+a_3b_3$ as required since $\mathbf{i}\cdot\mathbf{i}=\mathbf{j}\cdot\mathbf{j}=\mathbf{k}\cdot\mathbf{k}=1$

16 Let $\mathbf{a}=a_1\mathbf{i}+a_2\mathbf{j}+a_3\mathbf{k}$, $\mathbf{b}=b_1\mathbf{i}+b_2\mathbf{j}+b_3\mathbf{k}$ and $\mathbf{c}=c_1\mathbf{i}+c_2\mathbf{j}+c_3\mathbf{k}$

$\mathbf{b}+\mathbf{c}=(b_1\mathbf{i}+b_2\mathbf{j}+b_3\mathbf{k})+(c_1\mathbf{i}+c_2\mathbf{j}+c_3\mathbf{k})$

$=(b_1+c_1)\mathbf{i}+(b_2+c_2)\mathbf{j}+(b_3+c_3)\mathbf{k}$

$\mathbf{a}\cdot(\mathbf{b}+\mathbf{c})=(a_1\mathbf{i}+a_2\mathbf{j}+a_3\mathbf{k})((b_1+c_1)\mathbf{i}+(b_2+c_2)\mathbf{j}+(b_3+c_3)\mathbf{k})$

$=a_1(b_1+c_1)\mathbf{i}\cdot\mathbf{i}+a_1(b_2+c_2)\mathbf{i}\cdot\mathbf{j}+a_1(b_3+c_3)\mathbf{i}\cdot\mathbf{k}$

$+a_2(b_1+c_1)\mathbf{j}\cdot\mathbf{i}+a_2(b_2+c_2)\mathbf{j}\cdot\mathbf{j}+a_2(b_3+c_3)\mathbf{j}\cdot\mathbf{k}$

$+a_3(b_1+c_1)\mathbf{k}\cdot\mathbf{i}+a_3(b_2+c_2)\mathbf{k}\cdot\mathbf{j}+a_3(b_3+c_3)\mathbf{k}\cdot\mathbf{k}$

$=a_1(b_1+c_1)\mathbf{i}\cdot\mathbf{i}+a_2(b_2+c_2)\mathbf{j}\cdot\mathbf{j}+a_3(b_3+c_3)\mathbf{k}\cdot\mathbf{k}$

$=a_1(b_1+c_1)+a_2(b_2+c_2)+a_3(b_3+c_3)$

$=a_1b_1+a_1c_1+a_2b_2+a_2c_2+a_3b_3+a_3c_3$

$=(a_1b_1+a_2b_2+a_3b_3)+(a_1c_1+a_2c_2+a_3c_3)$

$=\mathbf{a}\cdot\mathbf{b}+\mathbf{a}\cdot\mathbf{c}$ (using qu 8 or prove as above) as required

17 **a** Let $\mathbf{a}=\begin{pmatrix}a_1\\a_2\\a_3\end{pmatrix}$

Then $\mathbf{a}\cdot\mathbf{a}=\begin{pmatrix}a_1\\a_2\\a_3\end{pmatrix}\cdot\begin{pmatrix}a_1\\a_2\\a_3\end{pmatrix}$

$=a_1^2+a_2^2+a_3^2$

$=|\mathbf{a}|^2$

since $|\mathbf{a}|=\sqrt{a_1^2+a_2^2+a_3^2}$

b $a=b-c$ using the 'triangle rule'

$|\mathbf{a}|^2=\mathbf{a}\cdot\mathbf{a}$

$=(\mathbf{b}-\mathbf{c})\cdot(\mathbf{b}-\mathbf{c})$

$=\mathbf{b}\cdot\mathbf{b}+\mathbf{c}\cdot\mathbf{c}-2\mathbf{b}\cdot\mathbf{c}\cos\theta$

$|\mathbf{b}|^2+|\mathbf{c}|^2-2|\mathbf{b}||\mathbf{c}|\cos\theta$ as required

18 **a** When $y=0$, $\frac{y-5}{5}=-1$

So $\frac{z-3}{3}=-1\Rightarrow z=0$

Therefore it cuts through x-axis

$\frac{x+2}{-4}=-1\Rightarrow x=2$

So intercept is (2, 0, 0)

b 55.6°

19 $\frac{27}{2}\sqrt{6}$

Exercise 6.3A

1 6
2 3
3 $2\sqrt{2}$ (2.83)
4 $\frac{\sqrt{114}}{3}$ (3.56)
5 1
6 **a** $\sqrt{66}$ (8.12) **b** $3\sqrt{10}$ (9.49)
c $\frac{4}{3}\sqrt{30}$ (7.30) **d** $\sqrt{14}$ (3.74)
e $\sqrt{134}$ (11.58)
7 **a** $\frac{5}{2}\sqrt{2}$ (3.54 to 3 significant figures)
b $\frac{5}{6}\sqrt{6}$ (2.04 to 3 significant figures)
c $\frac{4\sqrt{5}}{5}$ (1.79 to 3 significant figures

Exercise 6.3B

1 $\frac{\sqrt{110}}{5}$ (2.10)
2 $\frac{\sqrt{30}}{6}$ (0.913)
3 $\frac{\sqrt{329}}{10}$ (5.74)
4 **a** $\frac{7\sqrt{3}}{3}$ (4.04 to 3 sig. fig.)
b (7, 2, –8)
c $\sqrt{14}$ (3.74 to 3 sig. fig.)
5 (–2, 5, 2)
6 $\left(\frac{5}{3},\frac{2}{3},-\frac{16}{3}\right)$
7 $\sqrt{\left(\frac{19}{7}\right)}$ (1.65)
8 For L_1:

$x=\left|\begin{pmatrix}5\\-2\\-4\end{pmatrix}+\lambda\begin{pmatrix}1\\2\\1\end{pmatrix}\right|$

$x^2=(5+\lambda)^2+(-2+2\lambda)^2+(-4+\lambda)^2$

$\frac{d(x^2)}{d\lambda}=2(5+\lambda)+4(-2+2\lambda)+2(-4+\lambda)$

$\frac{d(x^2)}{d\lambda}=0\Rightarrow\lambda=\frac{1}{2}$

Therefore, $x=\sqrt{\left(5+\frac{1}{2}\right)^2+(-2+2\left(\frac{1}{2}\right))^2+\left(-4+\frac{1}{2}\right)^2}$

$=\frac{\sqrt{174}}{2}$ (6.60)

For L_2

$x=\left|\begin{pmatrix}-1\\3\\2\end{pmatrix}+\lambda\begin{pmatrix}1\\1\\0\end{pmatrix}\right|$

$x^2=(-1+\lambda)^2+(3+\lambda)^2+2^2$

$\frac{d(x^2)}{d\lambda}=2(-1+\lambda)+2(3+\lambda)$

$\frac{d(x^2)}{d\lambda}=0\Rightarrow\lambda=-1$

Therefore, $x=\sqrt{(-1-1)^2+(3-1)^2+2^2}$

$=2\sqrt{3}$ (3.46)

Therefore L_2 comes closer to the origin (by 3.13 units) than L_1.

9 For point A:

$$x = \left| \begin{pmatrix} 4 \\ -3 \\ 7 \end{pmatrix} + \lambda \begin{pmatrix} 1 \\ 3 \\ -2 \end{pmatrix} - \begin{pmatrix} 4 \\ 1 \\ 6 \end{pmatrix} \right|$$

$$x^2 = (\lambda)^2 + (-4+3\lambda)^2 + (1-2\lambda)^2$$

$$\frac{d(x^2)}{dt} = 2\lambda + 6(-4+3\lambda) - 4(1-2\lambda)$$

$$\frac{d(x^2)}{dt} = 0 \Rightarrow \lambda = 1$$

Therefore, $x = \sqrt{(1)^2 + (-4+3)^2 + (1-2)^2}$

$= \sqrt{3}$ (1.73)

For point B:

$$x = \left| \begin{pmatrix} 4 \\ -3 \\ 7 \end{pmatrix} + \lambda \begin{pmatrix} 1 \\ 3 \\ -2 \end{pmatrix} - \begin{pmatrix} 6 \\ -2 \\ 6 \end{pmatrix} \right|$$

$$x^2 = (-2+\lambda)^2 + (-1+3\lambda)^2 + (1-2\lambda)^2$$

$$\frac{d(x^2)}{dt} = 2(-2+\lambda) + 6(-1+3\lambda) - 4(1-2\lambda)$$

$$\frac{d(x^2)}{dt} = 0 \Rightarrow \lambda = \frac{1}{2}$$

Therefore, $x = \sqrt{\left(-2+\frac{1}{2}\right)^2 + \left(-1+\frac{3}{2}\right)^2 + (1-1)^2}$

$= \frac{\sqrt{10}}{2}$ (1.58)

Therefore point B comes closer to the line (by 0.151) than line A.

10 $\frac{1}{\sqrt{2}}$ (= 0.71)

Review exercise 6

1 a $\frac{x-1}{2} = \frac{y+1}{-3} = \frac{z-4}{1}$

b $\mathbf{r} = \mathbf{i} - \mathbf{j} + 4\mathbf{k} + s(2\mathbf{i} - 3\mathbf{j} + \mathbf{k})$

2 a $\frac{x-3}{5} = \frac{y}{-4} = \frac{z-5}{-2}$

b $\mathbf{r} = 3\mathbf{i} + 5\mathbf{k} + s(5\mathbf{i} - 4\mathbf{j} - 2\mathbf{k})$

3 a Skew

b Parallel

c Intersect at (17, 9, −2)

d Intersect at (10, −7, 7)

4 37

5 a −13 b 60.6°

6 109.1°

7 $\begin{pmatrix} -2 \\ 8 \\ -4 \end{pmatrix} \cdot \begin{pmatrix} -4 \\ 2 \\ 6 \end{pmatrix} = 8 + 16 - 24$

$= 0$ therefore they are perpendicular

8 $\sqrt{3}$

9 a −2 (3**i** − 6**k**) = −6**i** + 12**k** = 3(−2**i** + 4**k**) so the lines are parallel

b 1.67 (3 sf)

10 $\frac{\sqrt{46}}{\sqrt{5}}$ (= 3.03)

11 a they do not intersect, and they are not parallel, therefore they must be skew

b $\frac{3}{11}\sqrt{11}$ (0.905 to 3 sf)

Assessment 6

1 a $\mathbf{r} = 2\mathbf{i} + 4\mathbf{j} - \mathbf{k} + \lambda(\mathbf{i} + \mathbf{j} - 2\mathbf{k})$

b $-\mathbf{i} + \mathbf{j} + 5\mathbf{k} = 2\mathbf{i} + 4\mathbf{j} - \mathbf{k} + \lambda(\mathbf{i} + \mathbf{j} - 2\mathbf{k})$

Coefficients of **i**: $-1 = 2 + \lambda \Rightarrow \lambda = -3$

Check coefficients of **j**: $4 - 3 = 1$

Check coefficients of **k**: $-1 - 3(-2) = 5$

So (−1, 1, 5) lies on the line.

2 a $\mathbf{r} = \begin{pmatrix} 3 \\ 0 \\ 1 \end{pmatrix} + \lambda \begin{pmatrix} 1 \\ -1 \\ 1 \end{pmatrix}$

b $3 + \lambda = 8 \Rightarrow \lambda = 5$

Check: $0 - 5 = -5$ as required

$1 + 5 = 6$ not 7 so (8, −5, 7) does not lie on the line.

3 $\mathbf{r} = \begin{pmatrix} 2 \\ 1 \\ 0 \end{pmatrix} + \lambda \begin{pmatrix} 8 \\ -5 \\ 3 \end{pmatrix}$

4 a −3 b $\theta = 87.1°$

5 $\frac{1}{6}\sqrt{3}$

6 $(-4\mathbf{i} + \mathbf{k}).(2\mathbf{i} - \mathbf{j} + 8\mathbf{k}) = -8 + 0 + 8 = 0$

Therefore they are perpendicular.

7 4

8 $\begin{pmatrix} -12 \\ -14 \\ 18 \end{pmatrix}$

9 $\frac{x-2}{-1} = \frac{y-0}{4} = \frac{z+3}{3}$

10 a $a = 2, b = 7, c = -3$

b $\mathbf{r} = \begin{pmatrix} 2 \\ 7 \\ -3 \end{pmatrix} + t \begin{pmatrix} 2 \\ -3 \\ 1 \end{pmatrix}$

11 $\frac{\sqrt{42}}{14}$

12 So first line closer

13 a Intersect at (1, −2, 6)

b parallel

c skew

14 a $\frac{x-2}{-1} = \frac{y=0}{4} = \frac{z+3}{3}$

b They are not parallel as the direction vectors are not equal, so it is not the same line.

15 B and C

16 $\frac{1}{3}\sqrt{137}$

17 $\frac{\sqrt{42}}{14}$

18 a 48.2°

b $\left(\frac{-1}{3}, \frac{19}{3}, \frac{-4}{3}\right)$

19 3.32 square units

20 a they do not intersect, and they are not parallel, therefore they must be skew

b $\frac{4}{3}\sqrt{3}$

(2.31 to 3 sig. fig)

c (−4, 2, −3)

d $\frac{\sqrt{165}}{2}$

Chapter 7

Exercise 7.1 A

1 a 80 J b 600 J c 150 kJ
2 C does the most work
3 A exerted the most power
4 At start, 16 000 W, at end 24 000 W
5 Initial power 700 W final power 150 W
6 Work: 50 000 J
Power P = 5000 W
Power Q = $3333\frac{1}{3}$ W
7 a 1000 N b 600 N
8 First: 30 N Second: 20 N
9 a 720J, 180 W b 60J, 40 W
10 Box A
a $12g$ N b $36g$ J
c $9g$ W
Box B
a $4g$ N b $2g$ J
c $0.5g$ W
Box C
a $0.5g$ N b $0.1g$ J
c $0.05g$ W
11 a 16 J b $42\frac{2}{3}$ J c $21\frac{1}{3}$ J
12 a 30 J b 46 J
13 a 40 J b 14.1 J c 0.104 J
14 a 10 W b 17.7 W c 0.693 W
15 39.0 W

Exercise 7.1 B

1 a 15.3 m s^{-1} b 6.57 m s^{-1}
2 a 29.4 m b 27.2 m
3 62.5 N
4 a 72 J b 24 J c 24 W
5 a i 10^5J ii 10^5J iii 250 N
b 120000 J, 300 N
6 a 7.5×10J b 8.5×10^6 J, 17000 N
c 170000 W, 340000 W
7 a i 352.8 J, 352.8 J ii 24.2 m s^{-1}
b 470.4 J, 28 m s^{-1}
c 28 m s^{-1}
d 20.9 m s^{-1}
8 a 375 kW b $\frac{1}{4}$ m s^{-2}
9 435 000 W
10 3.55 m s^{-1}
11 3640 J, 13.2 N
12 26.7 m s^{-1}
13 3560 W
14 1134 W
15 a 30 m s^{-1} b $\frac{2}{3}$ m s^{-2}
16 21.5 m s^{-1}
17 $\frac{81}{320}$
18 a 3.95 b 0.147 m s^{-2}
19 3.32 m s^{-1}, 66.5 W
20 a i 1764000 J ii 5880 N iii 147 kW
b 8880 N, 222 kW
21 a 8.24 m s^{-1} b 7.38 m s^{-1}
22 20 m s^{-1}
23 a 0.391 m s^{-2} b 207 N
24 a Energy equation from bottom to top of loop is:
GPE gained = KE lost
$m \times g \times 2r = \frac{1}{2}mu^2$
$u^2 = 4gr$
b energy equation from bottom to point P is
GPE gained = KE lost
$mgr + mgr\cos\theta = \frac{1}{2}mu^2 - \frac{1}{2}m\left(\frac{u}{2}\right)^2$
$gr(1+\cos\theta) = \frac{3u^2}{8}$
$\cos\theta = \frac{3u^2}{8gr} - 1$
θ is acute, so cos θ
$0 < \cos\theta < 1$
$0 < \frac{3u^2}{8gr} - 1 < 1$
$8gr \times 1 < 3u^2 < 2 \times 8gr$
$\frac{8gr}{3} < u^2 < \frac{16gr}{3}$
25 1.05°
26 65.3 m s^{-1}
27 2.08 m s^{-1}, 41.5 W
28 a 1500×10^4 J b 8.40 m s^{-1}, 756 312 W

Exercise 7.2 A

1 a A: 2 N, 0.5 J B: 2 N, 1 J
C: 0.4 m, 0.8 J D: 9 N, 1.35 J
E: 0.8 m, 3.2 N
b A 4 N m^{-1} B 2 N m^{-1}
C 10 N m^{-1} D 7.5 N m^{-1}
E 4 N m^{-1}
2 0.4 m
3 2.5 m
4 a 0.8 m b 8 J c 8 J
5 a 0.72 m b 5.4 J
6 a 78.4 N b 0.98 m
7 $x_1 = 1$ m, $x_2 = 0.5$ m, $26\frac{2}{3}$ N
8 $x_1 = \frac{4}{9}$ m, $x_2 = \frac{5}{9}$ m, 4.17 J
9 0.260 m, 8.33 J
10 For vertical equilibrium
$T_1 + T_2 = mg$
$\frac{\lambda_1}{l}x + \frac{\lambda_2}{l}x = mg$
$(\lambda_1 + \lambda_2)x = mgl$
Extension $x = \frac{mgl}{\lambda_1 + \lambda_2}$
$\frac{mg\lambda_1}{\lambda_1+\lambda_2}$ N, $\frac{mg\lambda_2}{\lambda_1+\lambda_2}$ N

Exercise 7.2B

1 $16\frac{2}{3}$ J
2 a 39.2 N in both b 0.441 m c 8.64 J
3 a 0.65 m b 15 N, 34 N c 15.6 J
4 a 980, 1.36 m b $3a$
5 6.62 m s^{-2}, 1.94 m
6 5.77 m s^{-1}; 3.46 m s^{-1}
7 Either using calculus

$$\text{Work done} = \int_{x_2}^{x_1} T\,dx = \frac{\lambda}{l}\left[\frac{x^2}{2}\right]_{x_1}^{x_2}$$
$$= \frac{1}{2}\times\frac{\lambda}{l}\times(x_2^2 - x_1^2)$$
$$= \frac{1}{2}\times\frac{\lambda}{l}(x_2 + x_1)(x_2 - x_1)$$
$$= \frac{T_1 + T_2}{2}(x_2 - x_1)$$

Or using the graph of $T = \frac{\lambda x}{l}$

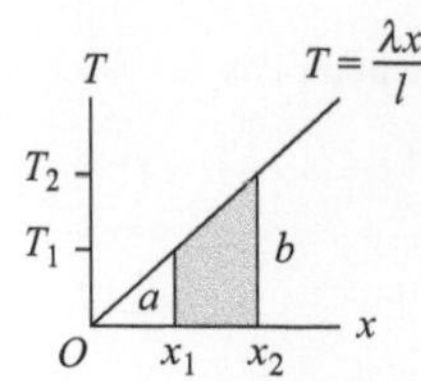

Work done = shaded area on T–x graph

$= \text{area of a trapezium}$

$= \frac{1}{2}h(a+b)$

$= \frac{1}{2}(T_1+T_2)(x_2-x_1)$

8 0.8 m

9 a 0.3 m

b 1.71 m s^{-1}, 0.3 m above P

The mass now falls back to Q and continues to oscillate about P, 0.3 m above and below.

c 1.23 m below A

d The mass will oscillate about P, 0.2 m above and below. 1.3 m below A

10 a Energy equation gives:

KE gained + EPE gained = GPE lost

$\frac{1}{2}\times 60\times v^2+\frac{1}{2}\times\frac{1200}{25}(x-25)^2=60\times g\times x$

$30v^2+24(x^2-50x+625)=588x;$

$30v^2=1788x-24x^2-15000$

25, 24.7 m s^{-1}

b 64.9 m

11 Max extension occurs when m_1 and m_2 have the same velocity v and string is no longer extending.

energy equation at this point is

Loss of KE of m_1 = Gain in KE of m_2 + Gain in EPE

$\frac{1}{2}m_1u^2-\frac{1}{2}m_1v^2=\frac{1}{2}m_1v^2+\frac{1}{2}\times\frac{\lambda}{a}\times x^2$

$\frac{\lambda}{a}x^2=m_1u^2-(m_1+m_2)v^2$ (1)

There are no external impacts or impulses so momentum is conserved.

Momentum equation gives

$m_1u=(m_1+m_2)v$ (2)

Substitute (2) in (1) to eliminate v.

$\frac{\lambda}{a}x^2=m_1u^2-(m_1+m_2)\left(\frac{m_1u}{m_1+m_2}\right)^2$

$=\frac{m_1(m_1+m_2)u^2-(m_1u)^2}{m_1+m_2}$

$=\frac{m_1m_2u^2}{m_1+m_2}$

Max extension, $x=\sqrt{\frac{axm_1m_2u^2}{\lambda(m_1+m_2)}}$

12 a 2.235 m **b** 0.509 m

13 a 3.13 m s^{-1}

b Resolving vertically for equilibrium

$2T\cos\theta=15g$

where $\theta=\tan^{-1}\left(\frac{2}{1.1}\right)=61.1892\ldots°$

Hooke's law gives $T=\frac{30g}{1.5}(\sqrt{2^2+1.1^2}-1.5)$

$=153.3783\ldots$ N

$2T\cos\theta=2\times153.3783\ldots\times\cos 61.1892\ldots°=147.8317$ N

$15g=15\times9.8=147$ N

$147\approx147.8$, mass hangs in equilibrium

14 a 10.2 kg **b** 117°

Exercise 7.3 A

1 b, c, e, f, g

2 a [velocity] = [L][T]$^{-1}$ [acceleration] = [L][T]$^{-2}$

b [work] = [M][L]^2[T]$^{-2}$

c [momentum] = [M][L][T]$^{-1}$

3 a [L]^3 **b** [M][L]$^{-3}$ **c** [M][L]^2[T]$^{-3}$

d [M][L]^2[T]$^{-2}$ **e** [M][L][T]$^{-1}$ **f** [M][L]$^{-1}$[T]$^{-2}$

4 $2as=[a]\,[s]$

$=[L][T]^{-2}[L]$

$=[L]^2\,[T]^{-2}$

$v^2=u^2=([L][T]^{-1})^2$

$=[L]^2\,[T]^{-2}$

5 $ut=[L][T]^{-1}[T]=[L]$

$\frac{1}{2}at^2=[L][T]^{-2}[T]^2=[L]$

6 ML^{2}T^{-2}

7 $E=k\times m$

$[E]=ML^2T^{-2}$

$[m]=M$

$[k]=\frac{ML^2T^{-2}}{M}=L^2T^{-2}$

velocity = distance/time

$[\text{velocity}]^2=\left(\frac{L}{T}\right)^2=L^2T^{-2}$

8 a i $k=\frac{A}{t}$ **ii** [L]^2[T]$^{-1}$

b i $k=\frac{F}{v^2}$ **ii** [M][L]$^{-1}$

c i $k=\frac{F}{xt}$ **ii** [M][T]$^{-3}$

d i $k=\frac{a}{mu}$ **ii** [M]$^{-1}$[T]$^{-1}$

e i $k=\frac{vma}{xF}$ **ii** [T]$^{-1}$

f i $k=\frac{u}{at^2}$ **ii** [T]$^{-1}$

9 $\frac{2mMg}{m+M}=\frac{[M]^2[L][T]^{-2}}{[M]}$

$=[M][L][T]^{-2}$

$=[\text{force}]$

10 [M][T]$^{-1}$

11 M^{-1}L^{3}T^{-2}

12 [M][L]^2

13 a Inconsistent

b Inconsistent

c Inconsistent

d Consistent

e Consistent

f Consistent

Exercise 7.3 B

1 P might be a velocity, Q might be an angular velocity

2 Consistent

3 $T=\frac{kmv^2}{r}$

4 $v=k\rho^{-\frac{1}{2}}P^{\frac{1}{2}}=k\sqrt{\frac{P}{\rho}}$

5 $v=kv^2Ar$

6 Henry's as v does not depend on ρ

7 I is dimensionless

8 $v = k\sqrt{\frac{TL}{m}}$

9 $V = k\sqrt{\lambda g}$

10 a $F = k \times R^4 L^{-1} \eta^{-1} P$

b Substitute

$0.15 = \frac{k \times 0.05^4 \times 3600}{6 \times 0.01}$

$k = 0.4$

Review exercise 7

1 1080 N

2 a 150 N

b 15 000 J

c 15 000 J

d 1500 Watts

3 a 0.24 m s^{-2}

b 2.4 m s^{-1}

c i 576 J ii 23 520 J

d 24 096 J

e 2008 N

f 4819 Watts

4 802.5 Watts

5 63 N

6 Speed $v = 7.43$ m s^{-1}

156 Watts

7 Newton's 2nd Law gives:

down slope

$\frac{p}{v} + mg\sin\alpha - kv = 0$

up slope

$\frac{2p}{v} - mg\sin\alpha - k\frac{v}{2} = 0$

add equations

$\frac{3p}{v} - 3\frac{kv}{2} = 0$

$v^2 = 2\frac{p}{k}$

$v = \sqrt{\frac{2p}{k}}$

8 a 118 N b 1.47 m c 86.4 J

9 $x = 0.652$ m

$y = 0.348$ m

17.4 N

10 a Consistent b Inconsistent

c Consistent d Inconsistent

11 a Dimensions are consistent

b Dimensions are consistent

12 $T = 0.0282\, Av^2\rho$

13 Total increase in energy = 705 000J to 3 sf

tractive force $T =$ 1410 N

Power at end = 35250 W

14 2.48 m s^{-1}

Assessment 7

1 a 1176 J b 6.53 s

2 a 8.85 m s^{-1}

b i 530 J ii 88.3 N

3 a 33 N b 528 J

4 $AB = 1.65$ m, $BC = 0.745$ m

5 4.70 m s^{-1}

6 3.83 m

7 a 60 J b 35 J

8 0.221 m s^{-2}

9 [LHS] $\equiv \mathrm{T}^2$

[RHS] $\equiv \frac{\mathrm{L}^3}{\mathrm{LT}^{-2} \times \mathrm{L}^2} \equiv \mathrm{T}^2$

So formula is dimensionally consistent.

10 Driving force = $\frac{\text{power}}{\text{velocity}}$

Against the wind: $\frac{P}{V} - kV - W = 0$ ....[1]

With the wind: $\frac{P}{2V} - 2kV + W = 0$ ....[2]

Add [1] and [2]: $\frac{3P}{2V} - 3kV = 0 \Rightarrow V = \sqrt{\frac{P}{2k}}$

11 a 7.67 m s^{-1} b 4.17 m

12 a 1.1 m b 0.6 m below A c 0.2 m

13 a 4 m b 8.94 m s^{-1}

14 $f = \frac{k}{l^2}\sqrt{\frac{T}{\rho}}$

15 $\mathrm{M}^{\frac{1}{2}}$

16 28.5 m s^{-1}

17 a 31.25 J b 9.08 m s^{-1}

18 formula is dimensionally consistent

19 $P_1 = 8P$ W

20 400 N

21 Let speed at top of tube be v, and zero PE be at ground level.

Energy at top of tube = $\frac{1}{2}mv^2 + mgh$

Energy at highest point = $4mgh$

These are equal, so $v^2 = 6gh$

Let time in tube be t and acceleration be a

$v = at$ and $v^2 = 2ah$

These give $t = \frac{v}{3g} = \frac{\sqrt{6gh}}{3g} = \sqrt{\frac{2h}{3g}}$

Total work done = $4mgh$, so rate of working

$= \frac{4mgh}{t} = 4mgh\sqrt{\frac{3g}{2h}} = 2m\sqrt{6g^3h}$

22 $h = \frac{kv^2}{g}$

23 a Three dimensions give three equations, so can't find four unknowns.

b $V = \frac{kr^4p}{\eta l}$

24 0.25 m s^{-2}

25 a 3.3 MJ

b 43.5 kW to 3 s.f.

26 17.9 N

27 25.9 MW to 3 s.f.

28 a 26.4 m

b 23.9 m

c 29.7 m s^{-1}

29 a $k \ge 0.5$

b $d = \frac{1-k}{k}$

30 a 12.5 m

b 10.6 m s^{-1}

c 20 m

Chapter 8

Exercise 8.1 A

1 a $2\frac{1}{2}$ m s^{-1} b 1 m s^{-1}

c $4\frac{1}{2}$ m s^{-1} d -2 m s^{-1}

e $-1\frac{1}{2}$ m s^{-1}

2 a 3.6 m s^{-1} b 1.5 m s^{-1} c $= -1.4$ m s^{-1}

3 10 kg

4 $\frac{3.2}{2.04} = 1.57\ \text{m s}^{-1}$

5 $\frac{5}{2}\mathbf{i} - 2\mathbf{j}\ \text{m s}^{-1}$

6 $\begin{pmatrix} 6 \\ 1 \end{pmatrix}\ \text{m s}^{-1}$

7 $\begin{pmatrix} 6 \\ 5.5 \end{pmatrix}\ \text{m s}^{-1}$

8 $\begin{pmatrix} 1 \\ -3 \end{pmatrix}\ \text{m s}^{-1}$

9 $\begin{pmatrix} 2 \\ 2\frac{1}{4} \end{pmatrix}\ \text{m s}^{-1}$ and $\begin{pmatrix} 4 \\ 4\frac{1}{2} \end{pmatrix}\ \text{m s}^{-1}$

$2\mathbf{i} + 2\frac{1}{4}\mathbf{j}$ and $4\mathbf{i} + 4\frac{1}{2}\mathbf{j}$

10 1.25 kg

11 Masses are $2\frac{1}{7}$ kg and $\frac{6}{7}$ kg

Exercise 8.1 B

1 20 m s^{-1}

2 0.676 m s^{-1} to 3 sf in the direction of motion of the arrow before the collision

3 6.43 km h^{-1} to 3 sf

4 $\frac{24}{2.04} = 11.8\ \text{m s}^{-1}$

5 **a** 3.2 m s^{-1} **b** 0.8 m s^{-1}

6 7.2 m s^{-1}

7 6.25 m s^{-1}
Taking gun and bullet together, the explosion is not an external impulse, so does not affect the total momentum.

8 227 m s^{-1} to 3sf

9 $a = 18, b = 23$

10 3.2 m s^{-1} in the direction Q was originally moving

11 $\frac{90}{12} = 7\frac{1}{2}\text{m s}^{-1}$

12 $u = 5\ \text{m s}^{-1}$

$v = -\frac{10}{6} = -1\frac{2}{3}\text{m s}^{-1}$ There are no more collisions as X and Y together are moving in the opposite direction to Z.

13 **a** $3\frac{1}{3}\text{ms}^{-1}$

b 2 m s^{-1}
R is now moving towards S at rest.
So there is another collision between R and S.

14 $\frac{2}{5}u$ and $\frac{4}{5}u$

15 **a** Momentum equations gives
$10 \times 5 + m \times (-2) = 10 \times 3 + mv$
$mv = 20 - 2m$

b No further collisions if
$v \geq 3$
$\frac{20 - 2m}{m} \geq 3$
$20 \geq 5m$
$m \leq 4$

16 $m = 2$
$a = -5$

17 **a** $m = 6$

b $V = 3\sqrt{2}\ \text{ms}^{-1}$

18 $v = \frac{12\sqrt{2}}{5}\ \text{ms}^{-1}$

19 $v_A = \begin{pmatrix} 0 \\ 3\sqrt{3} \end{pmatrix}, v_B = \begin{pmatrix} 6 \\ 0 \end{pmatrix}$

Exercise 8.2 A

1 **a** $v_2 = 4\ \text{m s}^{-1}$
$v_1 = 3\ \text{m s}^{-1}$

b $v_2 = \frac{8}{3}\ \text{m s}^{-1}$
$v_1 = \frac{10}{3}\ \text{m s}^{-1}$

c $v_2 = -9\ \text{m s}^{-1}$
$v_1 = -3\ \text{m s}^{-1}$

d $v_2 = 0\ \text{m s}^{-1}$
$v_1 = -3\ \text{m s}^{-1}$

e $v_2 = 2\frac{1}{6}\ \text{ms}^{-1}$
$v_1 = 2\frac{1}{6} - 2\frac{1}{2} = -\frac{1}{3}\text{ms}^{-1}$

2 **a** $11\ \text{m s}^{-1}, \frac{3}{4}$ **b** $3.25\ \text{m s}^{-1}, \frac{1}{4}$

c $0\ \text{m s}^{-1}, \frac{1}{5}$ **d** $2\ \text{m s}^{-1}, \frac{4}{5}$

3 0.18 m s^{-1} 0.15 m s^{-1}

4 $v_2 = 12\ \text{m s}^{-1}$
$v_1 = 3\ \text{m s}^{-1}$

5 0.733 to 3 sf

6 **a** $\mathbf{v} = 4\mathbf{i} + 3\mathbf{j}\ \text{m s}^{-1}$
at $\tan^{-1}\left(\frac{3}{4}\right) = 36.9°$
to x-axis

b $\mathbf{v} = 3\mathbf{i} + \mathbf{j}\ \text{m s}^{-1}$
at $\tan^{-1}\left(\frac{1}{3}\right) = 18.4°$
to x-axis

c $\mathbf{v} = 2\mathbf{i} + 2\mathbf{j}\ \text{m s}^{-1}$
at $\tan^{-1}\left(\frac{2}{2}\right) = 45°$
to x-axis

d $\mathbf{v} = 1.2\mathbf{i} + 7.2\mathbf{j}\ \text{m s}^{-1}$
at $\tan^{-1}\left(\frac{7.2}{1.2}\right) = 80.5°$

7 $\mathbf{v} = 3\mathbf{i} + 2\mathbf{j}\ \text{m s}^{-1}$
at an angle of
86.8°

Exercise 8.2 B

1 **a** 8.85 m s^{-1} (3 sf), 7.67 m s^{-1} (3 sf) **b** 0.866 (3sf)

2 $3\frac{1}{3}\ \text{m s}^{-1}$

3 2 kg

4 $\frac{1}{2}$

5 $= 5.55\ \text{m s}^{-1}$
25.6° to the plane.

6 1.5 kg

7 a $\dfrac{9}{16}u$ and $\dfrac{5}{16}u$

b $\dfrac{3}{8}u$

8 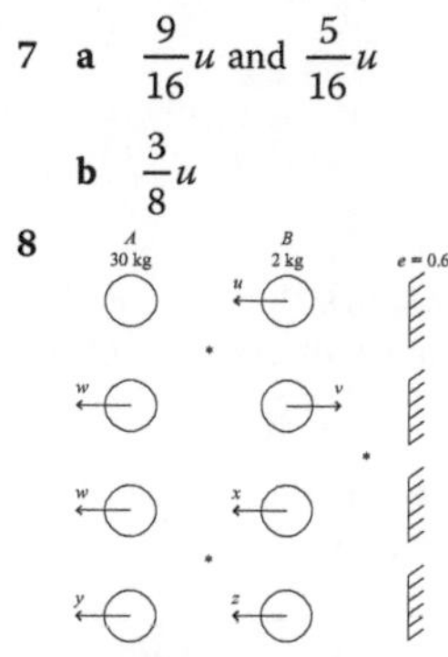

For 1st impact A and B
Momentum equation is
$2u + 0 = 30\text{w} - 2v$
$u = 15\text{w} - v$ (1)
Newton's equation is
$\dfrac{w+v}{u} = 0.6$
$0.6u = w + v$ (2)
Solve (1) + (2)
$1.6\,u = 16w$
$w = 0.1u$
$v = 0.5u$

For impact of B on wall, Newton's equation
$x = 0.6v = 0.3u$
$x > w$ so B impacts on A
Momentum equation is
$30\text{w} + 2x = 30y + 2z$
$3u + 0.6u = 30y + 2z$
$1.8\,u = 15y + z$ (3)
Newton's equation is
$\dfrac{y-z}{x-w} = 0.6$
$y - z = 0.12u$ (4)
Solve (3) + (4)
$1.8u + 0.12u = 16y$
$y = \dfrac{1.92u}{16} = 0.12u$
From (4)
$z = y - 0.12u = 0$
Hence, sphere B is at rest.

9

For P, Q impact
Momentum equation
$6 \times 15 = 6u + 4v$
$45 = 3u + 2v$ (1)
Newton's equation
$\dfrac{v-u}{15} = \dfrac{3}{5}$
$v - u = 9$ (2)
Solve (1) + 3 × (2)
$5v = 45 + 27$
$v = 14.4$
$u = v - 9 = 5.4$

For Q, R impact
Momentum equation
$4 \times 14.4 = 4w + 8x$
$14.4 = w + 2x$ (3)
Newton's equation
$\dfrac{x-w}{14.4} = \dfrac{3}{5}$
$x - w = 8.64$ (4)
Solve (3) + (4)
$3x = 23.04$
$x = 7.68$
$w = v - 2x = -0.96$
For 2nd P, Q impact
Momentum equation is
$6 \times 5.4 + 4 \times (-0.96) = 6y + 4z$
$28.56 = 6y + 4z$ (5)
Newton's equation is
$\dfrac{z-y}{5.4-(-0.96)} = \dfrac{3}{5}$
$z - y = 3.816$ (6)
Solve (5) + 6 × (6)
$10z = 51.456$
$y = z - 3.816 = 1.3296$
$z = 5.1456$
Final velocities are 1.33 ms^{-1} to 3 sf, 5.15 ms^{-1} to 3 sf and 7.68 ms^{-1} so $v_P < v_Q < v_R$ so there are no more collisions.

10 3 s

11 a 4 m s^{-1}

b 2 m s^{-1}

As z is moving faster than X and Y together and they are moving in the same direction, there are no more collisions.

12 $\dfrac{1}{3}$

13 10.2 m s^{-1}

33.7° below the horizontal.

Exercise 8.3 A

1 a −40 N s b −210 N s

2 a 18 N s b 45 N s

3 a 960 m s^{-1} b 232 m s^{-1}

4 a 181 N b 283 N

5 0.12 N

6 a 32 N s

8.4 m s^{-1}

b 4 N s

2.8 m s^{-1}

Exercise 8.3 B

1 7.2 N s

2 $\mathbf{v} = 2\mathbf{i} + 0.6\mathbf{j}$ m s^{-1}

$\mathbf{I} = 18\mathbf{j}$ N s

3 $\mathbf{v} = 2.4\mathbf{i} + 3\mathbf{j}$ m s^{-1}

$\mathbf{I} = 38.4\mathbf{i}$ N s

4 a 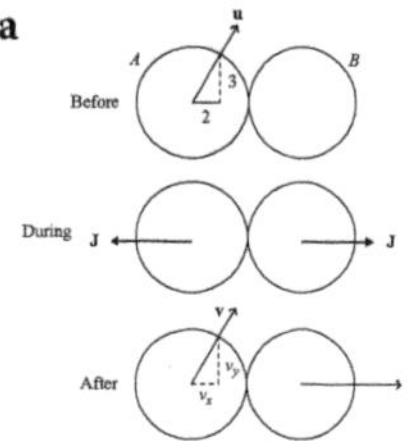

No impulse perpendicular to x-axis. Momentum conserved so B does not change its velocity in the y-direction.

b $4\mathbf{i}$ N s

5 Final velocities are
$4\mathbf{i} + 0.4\mathbf{j}$ m s^{-1} for P
and $2.8\mathbf{j}$ m s^{-1} for Q
$11.2\mathbf{j}$ N s

6 **a** 135 Ns
b **i** 45 ms^{-1} **ii** 41 ms^{-1}

7 **a** 120 Ns **b** 26 ms^{-1} **c** 6.1s

8 **a** 2.6 Ns **b** 3 ms^{-1}

9 **a** 72 Ns **b** 96 ms^{-1} **c** 3s

10 Impulse $= 42\frac{2}{3}$ N s
Final velocity $= 20\frac{2}{3}$ m s^{-1}

11 240 N s
Final velocity $v = 5.6$ m s^{-1}

12 **a** 48 N s
b 2.4 m s^{-1}
c $\sqrt{8}$ secs

13 $J = \dfrac{Mm(u - v)}{M + m}$

14 12 m s^{-1}
36 N s

15 **a** 475 N s
9.5 m s^{-1}
b 2260 N (3 sf)

16 Velocities are 5 m s^{-1} and
5.20 m s^{-1}
Impulse = 60 N s

17 Final velocities are
$\frac{4\sqrt{3}}{3}\mathbf{i} + 4\mathbf{j}$ and $\frac{7\sqrt{3}}{3}\mathbf{i} + \mathbf{j}$
Impulse $= \frac{16\sqrt{3}}{3}$ N s

18 After impact the velocities are
$$\sqrt{v_1^2 + v_2^2} = \sqrt{\frac{u^2}{4} + \frac{u^2}{4}} = \frac{u}{\sqrt{2}}$$
and
$$\sqrt{w_1^2 + w_2^2} = \sqrt{\frac{3u^2}{4} + \frac{3u^2}{4}} = \sqrt{\frac{3}{2}}\,u$$
Impulse
$$J = m\frac{u\sqrt{3}}{2} - m\frac{u}{2} = \frac{\sqrt{3} - 1}{2}mu$$

19

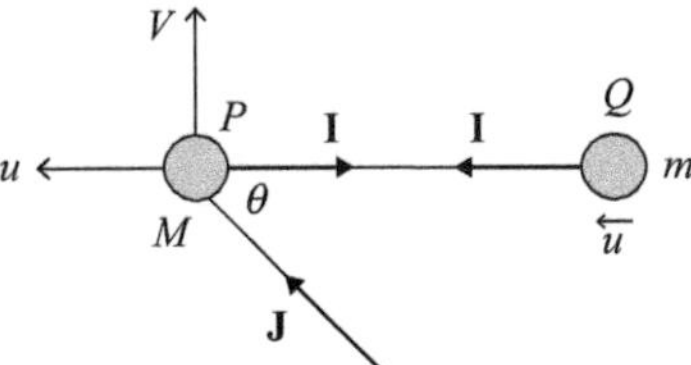

Let velocity of Q be **u** and components of velocity of P be u and v.
Impulse equation perpendicular to the rod is
$J\sin\theta = Mv$
Impulse equation along the rod for P and Q together is
$J\cos\theta = Mu + mu$

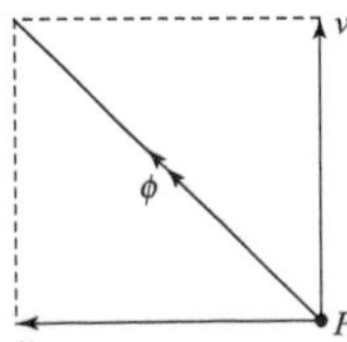

If angle of final velocity of P to rod $= \phi$
$$\tan\phi = \frac{v}{u} = \frac{J\sin\theta}{M} \times \frac{M + m}{J\cos\theta}$$
$$\tan\phi = \frac{M + m}{M}\tan\theta$$

$M >> m$ $\tan\phi$ grows very large $\Rightarrow \phi$ tends to 90°
Q effectively becomes fixed and P swings around Q in a circle.

$M = 0$ $\tan\phi = \tan\theta \Rightarrow \phi = \theta$
A massless Q has no effect on the motion of P which moves in the direction of the impulse.

20 **a** No impulse perpendicular to AB, so component $\frac{u}{2}\mathbf{j}$ is unchanged
b Magnitude of velocity of $B = \frac{u}{2}$
Magnitude of final velocity of
$$B = \sqrt{\frac{3}{16}u^2 + \frac{1}{4}u^2}$$
$$= \frac{\sqrt{7}}{4}u$$
Impulse $= \dfrac{mu\sqrt{3}}{4}$

Review exercise 8

1 -6 m s^{-1}

2 speed $= \sqrt{v_1^2 + v_2^2} = \sqrt{10}$ m s^{-1}
at $\tan^{-1}\left(\frac{-1}{3}\right) = -18.4°$ (3 sf) to the x-direction.

3 $m = 10$,
so final velocity $= \begin{pmatrix} \frac{4}{3} \\ \frac{4}{3} \end{pmatrix}$ ms^{-1}

4 $9\frac{1}{4}$ m s^{-1} and $3\frac{3}{4}$ m s^{-1}

5 $e = \dfrac{2\sqrt{5}}{6\sqrt{5}} = \dfrac{1}{3}$
$11\frac{10}{3}$ m

6 $v = 2$ m s^{-1}
10000 N s

7 $m = 10\frac{1}{2}$ kg
$\dfrac{6 + 2}{12 - 2} = e \Rightarrow e = \dfrac{4}{5}$
42 N s

8 Final velocity $= 5\mathbf{i} + \mathbf{j}$ m s^{-1}
$\mathbf{I} = 12\mathbf{j}$ N s

9 18750 N

10 6 N s
12 m s^{-1}

11 Velocities are
8 m s^{-1} for A to the left
4 m s^{-1} for B to the left
1 m s^{-1} for C to the right
As $8 > 4 > -1$ there are no more collisions.

12 Velocity of P is $\frac{10}{3}\mathbf{i}$
Velocity of Q is $1\frac{10}{3}\mathbf{i} + 6\mathbf{j}$
Impulse in string $= \frac{20}{3}$ N s

13 13.0°, $\mathbf{J} = 14\mathbf{j}$ N s

14 Assume no friction and that the balls are modelled as particles.

$0, \frac{I}{2\sqrt{2}m}\mathbf{i}, \frac{I}{2\sqrt{2}m}(\mathbf{i}+\mathbf{j}), \frac{I}{2\sqrt{2}m}\mathbf{j}$

15 a 32 Ns b 64 m s^{-1} c 2.47 s

Assessment 8

1 a 42 Ns b 120 N
2 4 s
3 $\mathbf{v} = (4.4\mathbf{i} - 0.4\mathbf{j})\text{ m s}^{-1}$
4 4.5 s
5 a 3 m s^{-1} b 4.76 m s^{-1}
6 a $V = \frac{10v}{7}$ b $V = \frac{2v}{7}$
7 a $V = \frac{14}{3}\text{ m s}^{-1}$
b $50 - 2m = 30 + mv$ gives $v = \frac{20-2m}{m}$

B must travel at least as fast as *A*

$\frac{20-2m}{m} \geq 3 \Rightarrow m \leq 4$

8 a A has velocity $(4\mathbf{i} - 1.8\mathbf{j})\text{ m s}^{-1}$, B has velocity $(-\mathbf{i} + 2.2\mathbf{j})\text{ m s}^{-1}$
b $14.4\mathbf{j}\text{ N s}$ c $\tan^{-1}\left(\frac{-1.8}{4}\right) - \tan^{-1}\left(\frac{3}{4}\right) = -61.1°$
9 a 15 m s^{-1} b 6000 N c 5882
10 $\mathbf{v} = (2.5\mathbf{i} + 0.5\mathbf{j})\text{ m s}^{-1}$
11 a i $v_A = 4\text{ m s}^{-1}, v_B = 9\text{ m s}^{-1}$ ii $v = 6\text{ m s}^{-1}$
b i No momentum is lost, so final momentum $5v$ equals initial momentum 30 provided that $e > 0$
ii $e = 0$ means the particles would not separate after collision, so the string would never become taut. The particles would move together at 6 m s^{-1}
12 a $v_A = 7.5\text{ m s}^{-1}, v_B = 10\text{ m s}^{-1}$
b $v_A = 0, v_B = 7.5\text{ m s}^{-1}$
c *B* is brought to rest, so there will be no more collisions.
13 a *A* $u_1 = 2.2\text{ m s}^{-1}$, for *B* $v_1 = 3.2\text{ m s}^{-1}$
b $v_2 = \frac{7.4 - 2.2e}{3}$
c If there are no more collisions, B must be moving faster than A

$\frac{7.4-2.2e}{3} \geq 2.2 \Rightarrow e \leq \frac{4}{11}$

14 a i The cushion is smooth, so no impulse and hence no change of momentum in the i-direction
ii 4.9 m s^{-1} iii 98.9°
b 2380
15 a 4 m s^{-1} b 6 N s
c The total momentum of *A* and *B* is now $1 \times 4 + 2 \times 1 = 6\text{ N s}$. This will be unchanged by the collision between *A* and *B* and the string again becomes taut. Their common velocity is then v, where $3v = 6$, $v = 2\text{ m s}^{-1}$. They are moving slower than *C*, so there are no more collisions.
16 $\frac{u}{4}$ and $\frac{u}{2}$
17 4.25 or 0.3125
18 a 6 b 1.8 m s^{-1}
19 7.2 N s
20 2 m s^{-1}
21 a When *A* and *B* collide the momentum equation is $4 - 8 = 4v_A + 2v_B$ .....[1] The restitution equation is $0.2 \times 5 = 1 = v_B - v_A$ ....[2] Solving [1] and [2] $v_B = 0$
b 1 m s^{-1}
22 $-8\mathbf{j}\text{ m s}^{-1}$
23 a 5.6 m s^{-1} b 0.2
24 $u = 7.38\text{ m s}^{-1}, v = 4.62\text{ m s}^{-1}$
25 $\mathbf{I} = (10\mathbf{i} + 20\mathbf{j})\text{ N s}$

Magnitude:

$|I| = 10\sqrt{5}$ Ns

26 a 68 Ns b 10.5 m s^{-1}

Chapter 9

Exercise 9.1 A

1

	rpm	rad min^{-1}	rad s^{-1}
a	33	207	3.5
b	9	57	0.94
c	95	600	10
d	2480	15600	260

2 0.0039 rad s^{-1}
0.23 m s^{-1}
3 Hour hand: 0.0004 m s^{-1}; 0.00015 rad s^{-1}
Minute hand: 0.0073 m s^{-1}; 0.0017 rad s^{-1}
4 7.3×10^{-5} rad s^{-1}, 1670 km h^{-1}
5 4 rad s^{-1}, 24 m s^{-2}
6 Linear speed $= \frac{60\pi}{60} = \pi\text{ m s}^{-1}$

Acceleration to centre $= \frac{\pi^2}{3} = 3.3\text{ m s}^{-2}$

7 Linear speed $= \frac{120\pi}{60} = 2\pi\text{ m s}^{-1}$

Acceleration to centre $= \frac{(2\pi)^2}{0.03} = 1320\text{ m s}^{-2}$

8 Linear speed $= \frac{576\pi}{60} = 30.2\text{ m s}^{-1}$

Acceleration to centre $= \frac{30.2^2}{0.24} = 3800\text{ m s}^{-2}$

Exercise 9.1 B

1 12 N
2 66.6 N
3 1.05 N
4 27.4 N
5 0.118 rad s^{-1}
8.5 km h^{-1}
6 47.7 rpm
7 4170 N
8

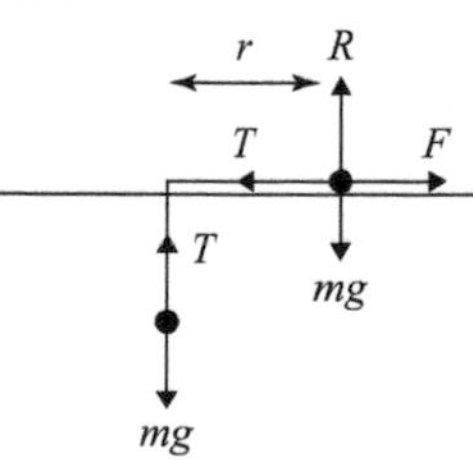

Let linear speed of mass on table = v

Equation of motion of this mass is

Tension $T = m \times \frac{v^2}{r}$

Resolve vertically for the hanging mass

Tension = mg

So $\frac{mv^2}{r} = mg$

Required velocity, $v = \sqrt{rg}$

Exercise 9.2 A

1 94.7 N

2 $v = \sqrt{\frac{0.5 \times 0.8}{0.2}} = \sqrt{2}\ \text{m s}^{-1}$

3 It completes 86 revs per minute

4 0.45 N

5 Tension $T = 3ma\omega^2$
Tension $S = 2ma\omega^2$

6
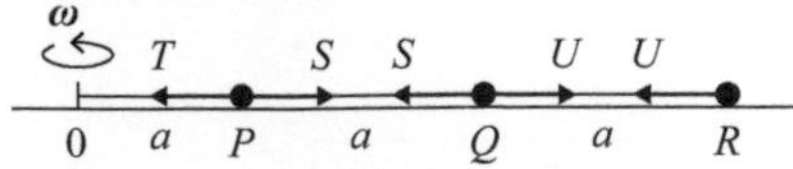

Let tensions be T, S and U as shown
Equation of motion about 0

For R is: $U = m \times 3a \times \omega^2$ (1)
For S is: $S - U = m \times 2a \times \omega^2$ (2)
For T is: $T - S = m \times a \times \omega^2$ (3)
(1) gives $U = 3ma\omega^2$
Add (1)(2) $S = 5ma\omega^2$
Add (1)(2)(3) $T = 6ma\omega^2$
Hence, ratio of tension U:S:T is 3:5:6

7 $\frac{\omega}{\sqrt{2}}$

Exercise 9.2 B

1 0.13 m

2 3.125 N

3
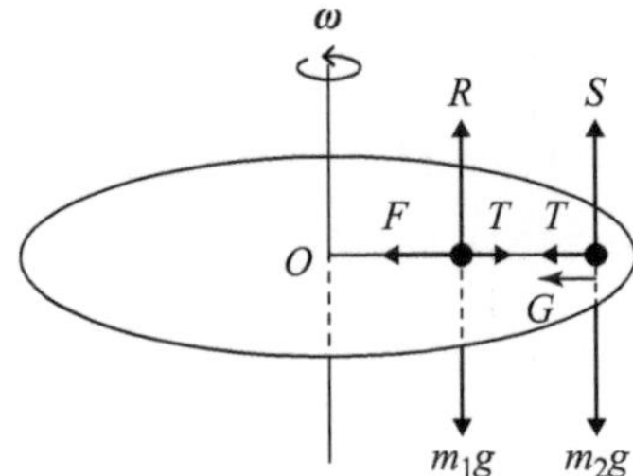

Let vertical reactions be R and S and frictional forces be F and G
Tension in string = T
Resolve vertically
$R = m_1 g$
$S = m_2 g$
On point of slipping,
$F = \mu R = \mu m_1 g$
and
$G = \mu S = \mu m_2 g$
Horizontal equations of motion to centre are
$F - T = m_1 \times x \times \omega^2$
and
$G + T = m_2 \times 2x \times \omega^2$
Add to eliminate T
$G + F = \omega^2(m_1 x + 2m_2 x)$
Substitute for F and G
$(m_1 + m_2)\mu g = x\omega^2(m_1 + 2m_2)$
Hence,

$$\omega = \sqrt{\frac{\mu g(m_1 + m_2)}{x(m_1 + 2m_2)}}$$

4 The tension in the rim is at right angle to that in the spokes and thus has no effect.

5 2.53 m s^{-1}
11.2 J

6 17.1 cm

7 Maximum speed is 15.3 m s^{-1}
The speed v does not depend on the mass of the car.

8 $\mu \geq \frac{320}{800 \times 9.8} = 0.04$

9 0.15

10
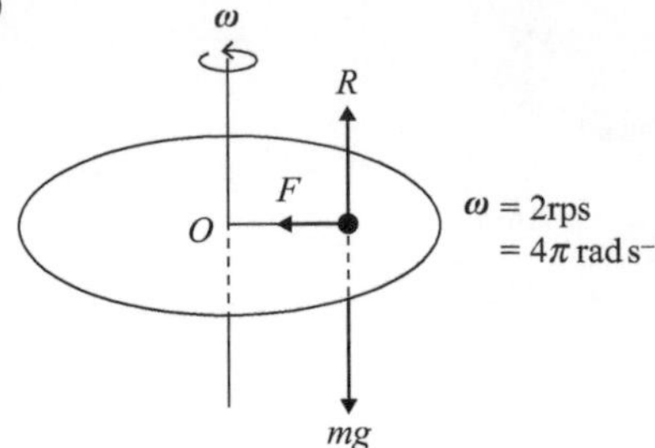

Resolve vertically
$R = mg$
Equation of motion to centre is
$F = mr(4\pi)^2$
For limiting friction,
$F = \mu R$
$= \mu mg$
So
$\mu mg = mr(4\pi)^2$
Greatest distance,

$r = \mu \times \frac{9.8}{(4\pi)^2}$ m
$= \mu \times 6.2$ cm

11 a $F \alpha \frac{1}{(R^2)}$

$F = \frac{k}{(R^2)}$

Constant
$k = mgR^2$

b Angular velocity
= 0.00116 rad s^{-1} to 3sf
Height
= 227 km

c Height of geostationary satellite
= 35800 km to 3sf.

Review exercise 9

1 800 rpm $= \frac{800 \times 2\pi}{60} = 83.8$ rad s^{-1}

1200 rpm $= \frac{1200 \times 2\pi}{60} = 125.7$ rad s^{-1}

2 0.0015 rad s^{-1}

3 $v = r\omega$ so $\omega = \frac{v}{r} = \frac{0.6}{0.8} = \frac{3}{4}$ rad s^{-1}

Acceleration $= \frac{v^2}{r} = \frac{0.6^2}{0.8} = 0.45$ rad s^{-2}

4 $\frac{1}{30}$ rad s^{-1}
31.4 sec

5 118 kJ to 3sf

6 822 N

7 Angular speed = 1.99×10^{-7} rad s^{-1}
Force = 3.56×10^{19} N

8 22.5 N

9 28.3 m s^{-1}

10 16.7 s

11 3 m s^{-1}
12 23.2 J
13 Reaction $= m\sqrt{g^2+r^2\omega^4}$

$\theta = \tan^{-1}\left(\dfrac{r\omega^2}{g}\right)$

14 89.3 minutes
15 0.34
16 0.574

Assessment 9

1 10.4 N
2 149 rev min^{-1}
3 60 cm
4 70 cm
5 15.7 m s^{-1}
6 Magnitude = 0.055 N
Direction = 63°
7 6.02×10^{24} kg
8 a 1.2 m b 6 m s^{-1}
9 3*mga*
10 a $\dfrac{12.5^2m}{50}=\dfrac{25m}{8}$ N

b 75.6 km h^{-1}
11 a 5.07 rad s^{-1} b 0.932 m
12 3.13 rad s^{-1}
13 $\omega=\sqrt{\dfrac{\mu g}{r}}$
14 $r=\dfrac{\mu g}{16\pi^2}$
15 Maximum friction is μmg.
For maximum speed $mg+\mu mg = ma\omega_1^2$

$\Rightarrow \omega_1^2=\dfrac{g(1+\mu)}{a}$

For minimum speed $mg-\mu mg = ma\omega_2^2$

$\Rightarrow \omega_2^2=\dfrac{g(1-\mu)}{a}$

Hence $\dfrac{\omega_1}{\omega_2}=\sqrt{\dfrac{1+\mu}{1-\mu}}$

Chapter 10

Exercise 10.1A

1 $E(X)=1\times\dfrac{1}{6}+4\times\dfrac{1}{6}+\ldots+36\times\dfrac{1}{6}=\dfrac{91}{6}$

$\text{Var}(X)=1^2\times\dfrac{1}{6}+4^2\times\dfrac{1}{6}+\ldots+36^2\times\dfrac{1}{6}-\left(\dfrac{91}{6}\right)^2=149.14$ (2 d.p.)

$E[Y]=-1\times\dfrac{1}{3}+1\times\dfrac{1}{6}+2\times\dfrac{1}{6}=\dfrac{1}{6}$

$\text{Var}[Y]=1\times\dfrac{1}{3}+1\times\dfrac{1}{6}+4\times\dfrac{1}{6}-\left(\dfrac{1}{6}\right)^2=\dfrac{41}{36}$

2 a $E[X]=1\times\dfrac{1}{4}+2\times\dfrac{1}{4}+\ldots+4\times\dfrac{1}{4}=\dfrac{5}{2}$

$\text{Var}(X)=1^2\times\dfrac{1}{4}+2^2\times\dfrac{1}{4}+\ldots+4^2\times\dfrac{1}{4}-\left(\dfrac{5}{2}\right)^2=\dfrac{5}{4}$

b $E[X]=1\times\dfrac{1}{3}+2\times\dfrac{2}{3}=\dfrac{5}{3}$

$\text{Var}(X)=1^2\times\dfrac{1}{3}+2^2\times\dfrac{2}{3}-\left(\dfrac{5}{3}\right)^2=\dfrac{2}{9}$

c $E[X]=2\times\dfrac{1}{2}+3\times\dfrac{1}{3}+4\times\dfrac{1}{6}=\dfrac{8}{3}$

$\text{Var}(X)=2^2\times\dfrac{1}{2}+3^2\times\dfrac{1}{3}+4^2\times\dfrac{1}{6}-\left(\dfrac{8}{3}\right)^2=\dfrac{5}{9}$

3 $E(X)=1\times\dfrac{1}{10}+2\times\dfrac{1}{10}+\ldots+10\times\dfrac{1}{10}=\dfrac{11}{2}$

$\text{Var}(X)=1^2\times\dfrac{1}{10}+2^2\times\dfrac{1}{10}+\ldots+10^2\times\dfrac{1}{10}-\left(\dfrac{11}{2}\right)^2=\dfrac{33}{4}$

4 $E[T]=\dfrac{150}{25}=6$

5 $E(\text{score})=1\times\dfrac{1}{3}+2\times\dfrac{1}{6}+3\times\dfrac{1}{3}+4\times\dfrac{1}{6}=\dfrac{7}{3}$

$\text{Var(score)}=1^2\times\dfrac{1}{3}+2^2\times\dfrac{1}{6}+3^2\times\dfrac{1}{3}+4^2\times\dfrac{1}{6}-\left(\dfrac{7}{3}\right)^2=\dfrac{11}{9}$

6 $E(X)=1\times\dfrac{1}{9}+2\times\dfrac{1}{9}+\ldots+9\times\dfrac{1}{9}=\dfrac{45}{9}=5$

$\text{Var}(X)=1^2\times\dfrac{1}{9}+2^2\times\dfrac{1}{9}+\ldots+9^2\times\dfrac{1}{9}-5^2=\dfrac{285}{9}-25=\dfrac{20}{3}$

7 $E(X^2)=1^2\times\dfrac{1}{6}+2^2\times\dfrac{1}{6}+\ldots+6^2\times\dfrac{1}{6}=\dfrac{91}{6}$

8

s	1	2	3	4	5	6
P(S = s)	*k*	4*k*	9*k*	16*k*	25*k*	36*k*

$k=\dfrac{1}{91}\Rightarrow E[S]=1\times\dfrac{1}{91}+2\times\dfrac{4}{91}+\ldots+6\times\dfrac{36}{91}=4.85$ (2 d.p.)

9 $E[R]=\dfrac{1}{1}\times\dfrac{1}{6}+\dfrac{1}{2}\times\dfrac{1}{6}+\ldots+\dfrac{1}{6}\times\dfrac{1}{6}=0.41$ (2 d.p.)

10 a

x	P(*X* = *x*)	*x*P(*X* = *x*)	*x*²P(*X* = *x*)
1	0.4	0.4	0.4
2	0.2	0.4	0.8
3	0.2	0.6	1.8
4	0.2	0.8	3.2
		2.2	6.2

If the mean and variance of X are μ and σ^2, then

$\mu=\sum_{\text{all }x}x\text{P}(X=x)=2.2$ and $\sigma^2=\sum_{\forall x}x^2\text{P}(X=x)-{}^2$

$=6.2-2.2^2=1.36$

b $E(Y)=2E(X^2)=2\times6.2=12.4$; $\text{Var}(Y)=\text{Var}(2X^2)=4\text{Var}(X^2)$
$=4E(X^4)-4\{E(X^2)\}^2$
$=4(1\times0.4+16\times0.2+81\times0.2+256\times0.2)-4\times6.22$
$=130.24$

11 a Expected value $0\times\dfrac{2}{3}+1\times\dfrac{1}{3}=\dfrac{1}{3}$

b If the experiment is performed a large number of times, approximately $\dfrac{1}{3}$ of the results will be prime and, with high probability, the proportion will get closer to $\dfrac{1}{3}$ as number of trials increases.

12 a

x	1	2	3	4
P(*X* = *x*)	$\frac{1}{2}$	$\frac{1}{4}$	$\frac{1}{8}$	$\frac{1}{8}$

b $E[X]=1\times\dfrac{1}{2}+2\times\dfrac{1}{4}+3\times\dfrac{1}{8}+4\times\dfrac{1}{8}=\dfrac{15}{8}$

13 Mode = 1 Median = 1

14 a $\dfrac{1}{6}$ b 3

15 $\mu = E(U) = \sum_{u=1}^{n}\frac{u}{n} = \frac{1}{n}\sum_{u=1}^{n}u = \frac{1}{n}\times\frac{n(n+1)}{2} = \frac{n+1}{2}$

$\sigma^2 = Var(U) = \sum_{i=1}^{n}\frac{u^2}{n} - \mu^2 = \frac{1}{n}\times\frac{1}{6}n(n+1)(2n+1) - \left(\frac{n+1}{2}\right)^2$

$= \frac{2(n+1)(2n+1)-3(n+1)^2}{12} = \frac{n^2-1}{12}$

Exercise 10.1B

1 $a = \frac{1}{5}, b = \frac{2}{15}$

2 a By symmetry, the mean of this distribution is 0

r	$P(R=r)$	$rP(R=r)$	$r^2P(R=r)$
−3	$\frac{1}{7}$	$-\frac{3}{7}$	$\frac{9}{7}$
−2	$\frac{1}{7}$	$-\frac{2}{7}$	$\frac{4}{7}$
−1	$\frac{1}{7}$	$-\frac{1}{7}$	$\frac{1}{7}$
0	$\frac{1}{7}$	0	0
1	$\frac{1}{7}$	$\frac{1}{7}$	$\frac{1}{7}$
2	$\frac{1}{7}$	$\frac{2}{7}$	$\frac{4}{7}$
3	$\frac{1}{7}$	$\frac{3}{7}$	$\frac{9}{7}$
		0	4

$\sigma^2 = \sum_{\forall r} r^2P(R=r) - \mu^2 = 4$. Therefore $\sigma = 2$.

b The probability that it will be one standard deviation or closer to the mean is therefore $\frac{5}{7}$.

3 a BBBB GBBB BGBB BBGB BBBG GGBB GBGB GBBG BGGB BGBG BBGG GGGB GGBG GBGG BGGG GGGG

x	$P(X=x)$
0	$0.51^4 = 0.07$
1	$4\times0.51^3\times0.49 = 0.26$
2	$6\times0.51^2\times0.49^2 = 0.37$
3	$4\times0.51\times0.49^3 = 0.24$
4	$0.49^4 = 0.06$

b $E[X] = 0\times0.07 + 1\times0.26 + \ldots + 4\times0.06 = 1.96 = $ (2 d.p.)
$Var[X] = 0\times0.07 + 1^2\times0.26 + \ldots + 4^2\times0.06 - 1.96^2 = 1.02$ (2 d.p.)

4 a P(John wins) $= P(X\geq1) = 1 - P\{X=0\} = 1 - \left(\frac{5}{6}\right)^4 = 0.518$ (3 d.p.)

b Jenny should pay John 48p to enter the game.

5

b	1	2
$P(B=b)$	$\frac{2}{3}\times\frac{1}{2}+\frac{1}{3}\times1=\frac{2}{3}$	$\frac{2}{3}\times\frac{1}{2}=\frac{1}{3}$

$E[B] = 1\times\frac{2}{3} + 2\times\frac{1}{3} = \frac{4}{3}$

6 a $P(X\geq1) = 1 - P(X=0) = 1 - \left(\frac{35}{36}\right)^{24} = 0.491$ (3 d.p.)

b 1 counter

7 a $E(X) = \frac{1}{5n}\sum_{x=1}^{5n}x = \frac{5n(5n+1)}{2\times5n} = \frac{1+5n}{2}$

b $Var(X) = \left(\frac{1}{5n}\sum_{x=1}^{5n}x^2\right) - \left(\frac{1+5n}{2}\right)^2 = \frac{1}{5n}\times\frac{5n}{6}\times$

$(5n+1)(10n+1) - \frac{(1+5n)^2}{4} = \frac{25n^2-1}{12}$

8 a P(one dart landing in circle) $= \frac{\pi r^2}{4r^2} = \frac{\pi}{4}$

$E[D] = 0\times\left(1-\frac{\pi}{4}\right)^3 + 1\times\frac{\pi}{4}\left(1-\frac{\pi}{4}\right)^2\times3 + 2\times\left(1-\frac{\pi}{4}\right)\left(\frac{\pi}{4}\right)^2$

$\times3 + 3\times\left(\frac{\pi}{4}\right)^3 = 2.36$ (2 d.p.)

b Estimate of $\frac{\pi}{4} = 0.71$. π estimated as 2.84

9 $E(X+Y) = E(X) + E(Y) = 3.2 + 22 = 25.2$
$Var(X+Y) = Var(X) + Var(Y) = 12.1 + 9 = 21.1$
Standard deviation of $X+Y = 4.59$ (2 d.p.)

10 $Var(X+Y) = E\left((X+Y)^2\right) - [E(X+Y)]^2$

$= E(X^2) + E(Y^2) + 2E(XY) - \left\{\{E(X)\}^2 + \{E(Y)\}^2 + 2E(X)E(Y)\right\}$

$= Var(X) + Var(Y) + 2\{E(XY) - E(X)E[Y]\}$

$= Var(X) + Var(Y)$

because $E[XY] = E(X)E(Y)$ for independent X and Y.

11 a i $a\mu, a^2\sigma^2$ ii $b\mu - c, b^2\sigma^2$

b X – score on a randomly chosen game. W – Net winnings. W = 5X – 20.

$E(X) = 3.5$ $Var(X) = \frac{105}{36}$

$E(W) = 5E(X) - 20 = 5\times3.5 - 20 = -2\frac{1}{2}$

$Var(W) = 25Var(X) = \frac{2625}{36} = 72.9$ (1 dp)

12 a $\frac{5}{24}$

b $E(S) = \frac{11}{6}$; $Var(S) = \frac{5}{9}$

$E(R) = \frac{13}{3}$; $Var(R) = \frac{80}{9}$

13 9 and 20; −1, 20.

14 $E(X) = 2\times\frac{1}{6} + 4\times\frac{1}{6} + \ldots + 12\times\frac{1}{6} = 7$

$Var(X) = 2^2\times\frac{1}{6} + 4^2\times\frac{1}{6} + \ldots + 12^2\times\frac{1}{6} - 7^2 = \frac{35}{3}$

$E(Y) = 1\times\frac{1}{8} + 2\times\frac{1}{8} + \ldots + 8\times\frac{1}{8} = \frac{9}{2}$

$Var(Y) = 1^2\times\frac{1}{8} + 2^2\times\frac{1}{8} + \ldots + 8^2\times\frac{1}{8} - \left(\frac{9}{2}\right)^2 = \frac{21}{4}$

$E(X+Y) = \frac{23}{2}$

$Var(X+Y) = \frac{203}{12}$

15 Let X be the score obtained.

$E(X) = 1\times\frac{1}{10} + 2\times\frac{1}{10} + \ldots + 10\times\frac{1}{10} = 5.5$

$Var(X) = 1^2\times\frac{1}{10} + 2^2\times\frac{1}{10} + \ldots + 10^2\times\frac{1}{10} - 5.5^2 = 8.25$

$E(X_1 + X_2) = 11$

$Var(X_1 + X_2) = 16.5$

Exercise 10.2A

1 $P(X=x)=\frac{e^{-\mu}\mu^x}{x!}$. 0.01832, 0.07326, 0.1465, 0.1954, 0.5665

2 0.8009

3 0.39 (2 d.p.)

4

x	0	1	2	3	4	>4
P(X = x)	0.333	0.366	0.201	0.074	0.020	0.006

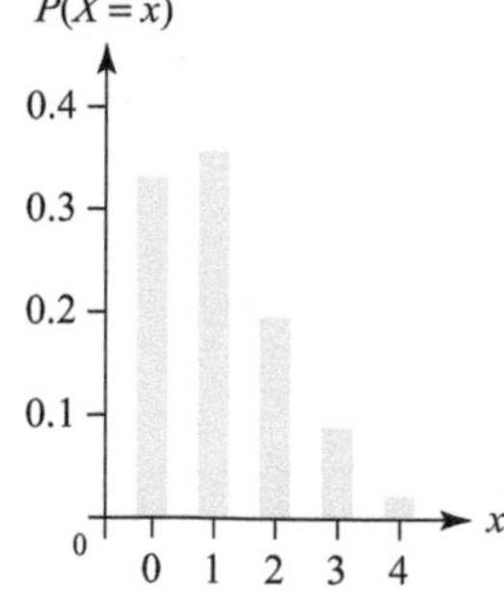

5 0.0028 (2 s.f.).

6 0.35 (2 d.p.).

7 0.98168. Assume independence from one day to the next. $0.98168^7 = 0.88$ (2 d.p.)

8 0.76

9 $P(X \geq 1) = 0.982$. Number in any bar independent of any other bar. $0.982^5 = 0.91$ (2 d.p.)

10 0.0011 (2s.f.)

11 0.20 (2 d.p.)

12 $\mu = 2.29$ $\sigma^2 = 2.80$

Data likely to be approximately Poisson distributed → cars passing randomly. In addition, the mean and variance of the data are fairly similar which is consistent with a Poisson distribution (in which the mean and variance are equal).

Exercise 10.2B

1 0.347 (3 d.p.)

2 0.195 (3 d.p.)

3 0.841 (3 d.p.)

4 0.142 (3 d.p.)

5 0.97214

6 0.374 (3 d.p.)

7 a 0.1215 (4 d.p.) b 0.9314

8 Drops arrive independently and at a constant average rate. 0.0067(2 s.f.)

9 0.145 (3 d.p.)

10 0.20 (2dp)

11 0.14 (2dp)

Exercise 10.3A

1 a

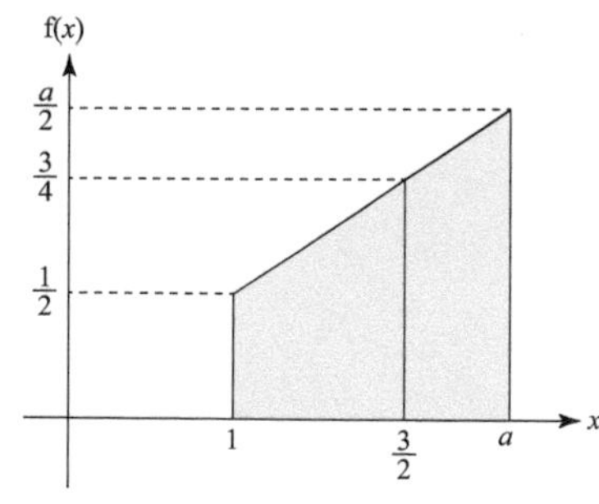

b $a=\sqrt{5}$. $P(X>1.5)=\frac{11}{16}$

2 $k=\frac{3}{8}$. $P(X<1)=\frac{3}{8}\int_0^1 x^2\,dx=\frac{1}{8}$

3 $\frac{1}{36}$

4 a $\int_0^{\frac{1}{2}} ax(2x-1)\,dx=1$, $\int_0^{\frac{1}{2}}(2ax^2-ax)\,dx=1$,

$\frac{2a}{24}-\frac{a}{8}=1$, $a=-24$

b i $-24\int_0^M (2x^2-x)\,dx=\frac{1}{2}$ $\rightarrow -32M^3+24M^2-1=0$ which is satisfied by $M=\frac{1}{4}$ or by symmetry.

ii $-24\int_0^{Q_1}(2x^2-x)\,dx=\frac{1}{4}$ $\rightarrow -64Q_1^{\,3}+48Q_1^{\,2}-1=0$ which is satisfied by $Q_1=0.16318$ (5 d.p.).

$-24\int_0^{Q_3}(2x^2-x)\,dx=\frac{3}{4}$ $\rightarrow -64Q_3^{\,3}+48Q_3^{\,2}-3=0$ which is satisfied by $Q_3=0.33682$ (5 d.p.).

iii IQR = 0.3362 − 0.16318 = 0.1736 (4 d.p.)

5 a

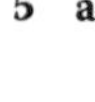

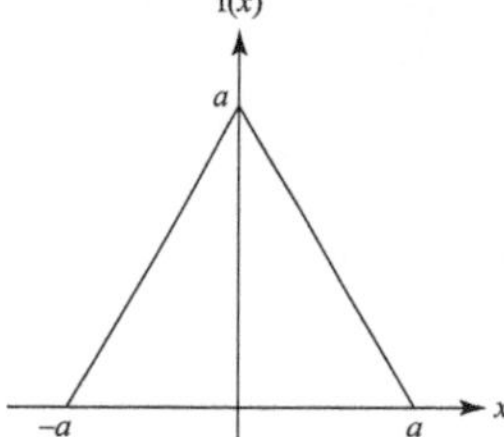

b $\frac{1}{2}\times 2a\times a=1$, $a=\pm 1$. $a\neq -1$ ($f(x)>0$ for all x). $a=1$

c $Q_1=\frac{\sqrt{2}-2}{2}$, $Q_3=\frac{-\sqrt{2}+2}{2}$ by symmetry. IQR $=2-\sqrt{2}$

6 a $k=20$

b Area under pdf between $y=0$ and $y=Q_1$ is

$\frac{1}{20}(1+2Q_1+1)\times\frac{Q_1}{2}=\frac{1}{4}$

$Q_1^{\,2}+Q_1-5=0$, $Q_1=1.79$ (3 sf). Area under pdf between $y=0$ and $y=M$ is

$\frac{1}{20}(2+2M)\times\frac{M}{2}=\frac{1}{2}$

$M^2+M-10=0$, $M=2.70$ (3 sf)

c $E(Y)=\frac{1}{20}\int_0^4 y(2y+1)\,dy=\frac{1}{20}\left(\frac{128}{3}+\frac{16}{2}\right)=\frac{38}{15}$

$E(Y^2)=\frac{1}{20}\int_0^4 y^2(2y+1)\,dy=\frac{1}{20}\left(128+\frac{64}{3}\right)=\frac{112}{15}$

$\text{Var}(Y)=\frac{112}{15}-\left(\frac{38}{15}\right)^2=\frac{236}{225}$

7 a $\frac{a}{4}\int_0^5(2x+1)\,dx=\frac{a}{4}(25+5)=\frac{15}{2}a=1\rightarrow a=\frac{2}{15}$

$E(X)=\frac{1}{30}\int_0^5 x(2x+1)\,dx=\frac{1}{30}\left(\frac{250}{3}+\frac{25}{2}\right)=\frac{115}{36}$

$\text{Var}(X)=\frac{1}{30}\int_0^5 x^2(2x+1)\,dx-\left(\frac{115}{36}\right)^2$

$=\frac{1}{30}\left(\frac{625}{2}+\frac{125}{3}\right)-\left(\frac{115}{36}\right)^2=1.60$ (2 d.p.).

b $P(X>4) = \frac{1}{30}\int_4^5 (2x+1)dx = \frac{1}{30}(25+5-16-4) = \frac{1}{3}$

$P(\text{one value above } 4) = 3\times\frac{1}{3}\times\left(\frac{2}{3}\right)^2 = \frac{4}{9}$

8 $E(X^2) = \frac{33}{4}$

$E\left(X+\frac{1}{X^3}\right) = 2.86$ (2 dp)

9 a $-\frac{6}{125}$, 0.54 (2dp)

b $P(X=2)=0$. Exact values of continuous random variables have probability zero.

c $\frac{5}{2}$

10 a 4.395 (3 dp) **b** 4.481 **c** 0.241 (3 dp)

11 a

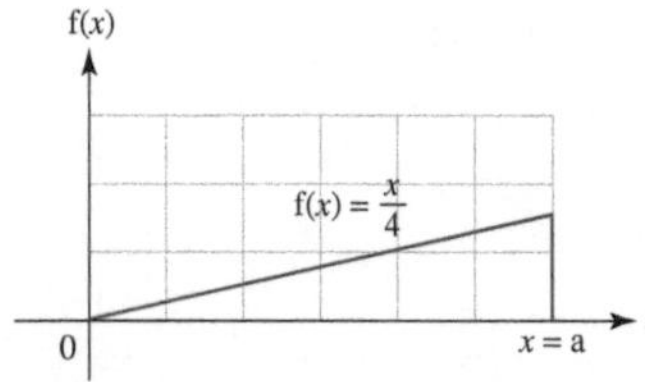

b Since the area under graph equals 1,

$\left(\frac{a}{2}\right)\frac{a}{4} = 1$

$a^2 = 8$

$a = \sqrt{8} = 2\sqrt{2}$ (since $a > 0$)

For the median M,

$\frac{M}{2}\left(\frac{M}{4}\right) = \frac{1}{2}$

$M^2 = 4$

$M = 2$ (since $M > 0$)

12 a

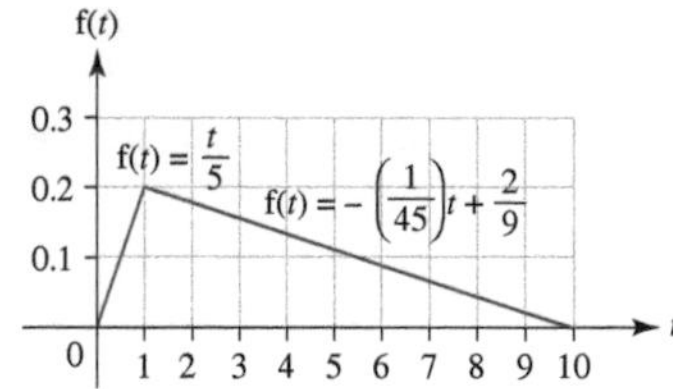

b $f(t) \geq 0$ for all t and area under graph is $\left(\frac{1}{2}\right)\times 10\times 0.2 = 1$

c 1 **d** 3.29 (2dp)

Exercise 10.3B

1 a $k = \frac{3}{28}$, $E(X) = \frac{15}{7}$, $Var(X) = \frac{156}{245}$

b $E(Y) = -\frac{23}{7}$, $Var(Y) = \frac{624}{245}$

2 a $k = \frac{1}{36000}$, $P(X > 50) = \frac{2}{27}$

b 50: $0.9\times 50 + 3 = 48$. 10: $0.9\times 10 + 3 = 12$

c Original mean 30, by symmetry. New mean: $0.9\times 30 + 3 = 30$. Zero change.

3 a $a = -\frac{1}{9}$, $b = 4$

b $-\frac{1}{9}\int_0^3 x^3(x-4)dx - \left(\frac{7}{4}\right)^2 = -\frac{1}{9}\left(\frac{3^5}{5} - 3^4\right) - \left(\frac{7}{4}\right)^2 = \frac{43}{80}$

c 0.61

4 $f(x) = \frac{3x(100-x)}{500\,000}$ for $0 < x < 100$

$\frac{3}{500\,000}\int_0^M (100x - x^2)dx = \frac{1}{2}$

$\frac{3}{500\,000}\left(50M^2 - \frac{M^3}{3}\right) = \frac{1}{2}$. $M = 50$.

5 a $a = \frac{3}{4}$

b $\frac{3}{4}\int_0^M x(2-x)dx = \frac{3}{4}\left[x^2 - \frac{x^3}{3}\right]_0^M = \frac{3}{4}\left(M^2 - \frac{M^3}{3}\right) = \frac{1}{2} \rightarrow$

$M^3 - 3M^2 + 2 = 0$

c $1^3 - 3\times 1^2 + 2 = 0$. Hence $M = 1$

6 a $\frac{3}{26}$

b $E(3X+2Y) = 6.62$ (2 dp)

$Var(3X+2Y) = 1.54$ (2 dp)

7 a 0.952 **b i** 5.461 **ii** 0.305 (3 dp)

8 a $a = 2^{\frac{1}{4}}$

b $E(X) = 0.951\cdots$

$Var(X) = 0.037\ldots$

c $E(2X-3) = -1.10$

$Var(2X-3) = 0.15$ (2dp)

9 $E(X+Y) = \mu_X + \mu_Y$

$Var(X+Y) = \sigma_X{}^2 + \sigma_Y{}^2 \rightarrow sd(X+Y) = \sqrt{(\sigma_x^2 + \sigma_y^2)}$

Assume X and Y are independent.

10 a $a = -\frac{1}{288}$

b $E(X) = 6$

$Var(X) = 7.2$

c $T = X + Y + 2$ where Y is cycle time.

$E(Y) = 7$, $Var(Y) = 4$

$E(T) = E(X+Y+2) = 6+7+2 = 15$;

$Var(X+Y+2) = 7.2 + 4 = 11.2$

Review exercise 10

1 $E(X) = 1\times\frac{1}{6} + 3\times\frac{1}{6} + \ldots + 11\times\frac{1}{6} = 6$

$Var(X) = 1^2\times\frac{1}{6} + 3^2\times\frac{1}{6} + \ldots + 11^2\times\frac{1}{6} - 6^2 = \frac{35}{3}$

2

c	1	2	3	4	5	6
P(C = c)	k	8k	27k	64k	125k	216k

$k = \frac{1}{441} \Rightarrow E[C] = 1\times\frac{1}{441} + 2\times\frac{8}{441} + \ldots + 6\times\frac{216}{441} = 5.16$ (2 d.p.)

3 $P(X=x) = \frac{e^{-\mu}\mu^x}{x!}$; $\mu = 5$. $P(X<2) = 0.04$ (2 d.p.).

4 a Let N be a random variable for the number of letters in one day. Assume random arrivals in time and therefore a Poisson distribution. $N \sim \text{Po}(18)$. $\text{P}(N \geq 12) = 0.945$. Assume independence from one day to the next. $0.945^5 = 0.75$ (2 d.p.)

b Let M be a random variable for the number of letters in five days. $M \sim \text{Po}(90)$. $\text{P}(M \geq 90) = 0.51$ (2 d.p.).

5 a $\frac{a}{5}\int_0^2 (x+3)\text{d}x = \frac{a}{5}(2+6) = \frac{8}{5}a = 1 \rightarrow a = \frac{5}{8}$

$\text{E}(X) = \frac{1}{8}\int_0^2 x(x+3)\text{d}x = \frac{1}{8}\left(\frac{8}{3}+6\right) = \frac{13}{12}$;

$\text{Var}(X) = \frac{1}{8}\int_0^2 x^2(x+3)\text{d}x - \left(\frac{13}{12}\right)^2 = \frac{1}{8}(4+8) - \left(\frac{13}{12}\right)^2$

$= 0.33$ (2 d.p.).

b $\text{P}(X>1) = \frac{1}{8}\int_1^2 (x+3)\text{d}x = \frac{1}{8}\left(2+6-\frac{1}{2}-3\right) = \frac{9}{16}$

$\text{P(one value above 1)} = 4 \times \frac{9}{16} \times \left(\frac{7}{16}\right)^3 = 0.19$ (2 d.p.)

6 a $\frac{2}{3}$

b $\text{E}(S+2T) = \frac{23}{9}$

$\text{VarE}(S+2T) = 0.28$ (2dp)

7 a 11, 33 **b** $\frac{9}{2}, \frac{33}{4}$

Assessment 10

1 a $k = 10$
$b = 0.4$
$a = 0.15$

b $\text{E}(X) = 1\times\frac{1}{10} + 2\times\frac{2}{10} + 3\times\frac{3}{10} + 4\times\frac{4}{10}$
$= 3$
$\text{E}(Y) = 1\times0.3 + 2\times0.15 + 3\times0.15 + 4\times0.4$
$= 2.65$
So X is higher on average

2 a $k = \frac{1}{20}$

b Median is 2

Mode is 3

c $\text{E}(N) = 1\times\frac{5}{20} + 2\times\frac{6}{20} + 3\times\frac{7}{20} + 4\times\frac{1}{20} + 5\times\frac{1}{20}$
$= \frac{47}{20}$

d $\text{Var}(N) = \frac{133}{20} - \left(\frac{47}{20}\right)^2$
$= \frac{451}{400}$ or 1.1275

3 a **i** 0.9161 **ii** 0.224 **iii** 0.1728 **iv** 0.4232

b 3

4 a Assume apples are sold individually and at a constant rate.

b **i** 0.0821 **ii** 0.5438 **iii** 0.1088

c 0.1378

5 a $E(X^2) = 25$ **b** $\sqrt{2.5}$

6 a $k = \frac{1}{18}$

b **i** $\text{P}(X<3) = \left[\frac{1}{36}x^2\right]_0^3$
$= \frac{1}{36}(9-0)$
$= \frac{1}{4}$

ii $\text{P}(X>1) = \left[\frac{1}{36}x^2\right]_1^6$
$= \frac{1}{36}(36-1)$
$= \frac{35}{36}$

iii $\text{P}(1.5 < X < 5) = \left[\frac{1}{36}x^2\right]_{1.5}^5$
$= \frac{91}{144}$

c 4

d 2

7 a

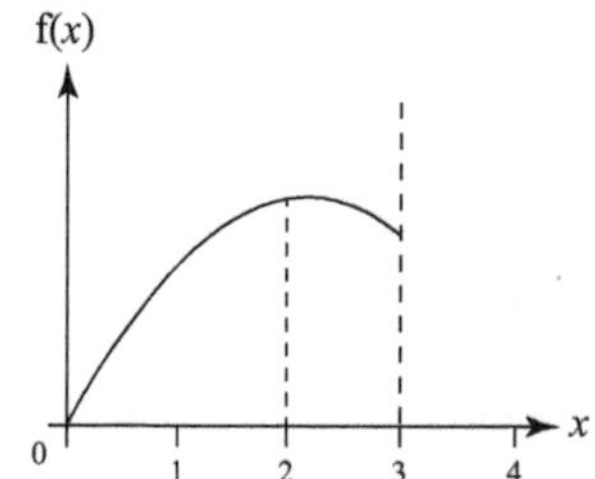

b 2

c 1.79

8 a

t	11	21	12	22	13	23
$P(T=t)$	0.2	0.1	0.15	0.05	0.3	0.2

b $E(T) = 15.7$

c $E(T) = E(X) + E(Y)$ for any discrete random variables X and Y where $T = X + Y$

but $\text{Var}(T) = \text{Var}(X) + \text{Var}(Y)$ only if X and Y are independent.

In this case, X and Y are not independent, we know this since, e.g.

$P(X=2) = 0.2$, $P(Y=10) = 0.6$,

$P(X=2 \text{ and } Y=10) = 0.15$

$P(X=2)\times P(Y=10) = 0.2\times0.6 = 0.12 \neq 0.15$

9 a **i** 10 **ii** 10

b 0.6672 **c** 0.0286

10 a $\text{E}(U) = 4.5$

$\text{Var}(U) = \frac{1}{12}(6^2-1)$

$= \frac{35}{12}$ or 2.92

b $\text{P}(X>3.5) = \text{P}(X \geq 4)$

$= \frac{2}{3}$

11 a 3

b $E(5V+3) = 5E(V)+3$

$= 5\left(\frac{10}{3}\right)+3$

$= \frac{59}{3}$ or 19.7

$Var(5V+3) = 5^2 Var(V)$

$= 25\left(\frac{31}{45}\right)$

$= \frac{155}{9}$ or 17.2

12 a $a > 7$ or $a \geq 8$

b $b > 8$ or $b \geq 9$

c 0.6078

d i 0.6620 **ii** 0.8641 **iii** 0.4126

13 a $\frac{8}{75}$ **b** 0.45

c $E(W) = 4E(Y)-3$

$= 4\left(\frac{8}{5}\right)-3$

$= \frac{17}{5}$

$Var(W) = 16Var(Y)$

$= 16\left(\frac{8}{75}\right)$

$= \frac{128}{75}$ or 1.71

14 a $P(T > 3.92) = P(T = 4)+P(T = 5)$

$= \frac{4}{12}+\frac{5}{12}$

$= \frac{3}{4}$

b $Var(7-3T) = 9Var(T)$

$= \frac{227}{16}$ or 14.2

15 a $E(X) = \frac{n+1}{2}$

b $E(X^2) = \frac{1}{n}+\frac{4}{n}+\frac{9}{n}+\ldots+\frac{n^2}{n}$

$= \frac{1}{n}\left(1+4+9+\ldots+n^2\right)$

$= \frac{1}{n}\left(\frac{n}{6}(n+1)(2n+1)\right)$

$= \frac{1}{6}(n+1)(2n+1)$

$Var(X) = \frac{1}{6}(n+1)(2n+1)-\left(\frac{n+1}{2}\right)^2$

$= \frac{1}{6}(n+1)(2n+1)-\frac{1}{4}(n+1)^2$

$= \frac{1}{24}(n+1)[4(2n+1)-6(n+1)]$

$= \frac{1}{24}(n+1)(2n-2)$

$= \frac{1}{12}(n+1)(n-1)$

$= \frac{1}{12}(n^2-1)$ as required

16 a 0.0119 **b** 0.401 **c** 0.0368

17 a $a = \frac{1}{8}$, $b = \frac{3}{8}$

b $m = -3+\sqrt{17}$ (1.12)

c $E(Y) = \frac{13}{12}A+2$

d 12

18 a 1.5 **b** 3.89

19 a $\int_0^3 k(9-x^2)dx = 1$

$k\left[9x-\frac{x^3}{3}\right]_0^3 = 1$

$k(27-9) = 1$

$18k = 1$

$k = \frac{1}{18}$

b $= \frac{117}{40}$ (2.925)

20 $= \frac{7}{4}+\frac{2}{3}\ln 2$ (2.21)

21 a $k = \frac{2}{\sqrt{3}}$

b $E(\sqrt{3}R+2) = \sqrt{3}\frac{2}{\sqrt{3}}\left(\frac{\sqrt{3}}{3}\pi-\frac{\pi^2}{36}-1\right)+2$

$= 2\left(\frac{\sqrt{3}}{3}\pi-\frac{\pi^2}{36}-1\right)+2$

$= \frac{2\sqrt{3}}{3}\pi-\frac{\pi^2}{18}$

22 a $\int_1^3 \frac{1}{8}\left(s^2-\frac{1}{s^2}\right)ds = \frac{1}{8}\left[\frac{s^3}{3}+\frac{1}{s}\right]_1^3$

$= \frac{1}{8}\left(\left(9+\frac{1}{3}\right)-\left(\frac{1}{3}+1\right)\right)$

$= \frac{1}{8}\left(\frac{28}{3}-\frac{4}{3}\right) = 1$ as required

$A(1^2+2^2+3^2+4^2) = 1$

$A = \frac{1}{30}$

b 28

Chapter 11

Exercise 11.1A

1 a $H_0: \lambda = 5.31$, $H_1: \lambda < 5.31$

b 53 as $P(X \leq 53) = 0.0976$ and $P(X \leq 54) = 0.1224$

2 a The critical values are 21 and 41

b $3 \times 9 = 27$ is between the critical values so the null hypothesis is accepted.

3 a The test is to see if the parameter has changed, but we don't know if we expect it to be higher or lower.

b The statistic is larger than the upper critical value, so the result is significant. There is sufficient evidence to suggest that the parameter is not 72.

c Actually the parameter is still 72, but an extreme result has incorrectly led us to reject that conclusion.

4 a The parameter is suspected to have decreased, so the alternative hypothesis takes the form given.

b The p-value is larger than the significance level, so the result is not significant. The null hypothesis is not rejected and you conclude that there is insufficient evidence, at the 5% level, to suggest that the parameter is less than 12.5.

c Actually the parameter has decreased, but the result was not extreme enough to lead us to that conclusion.

5 a $H_0 : \lambda = 8.3,\ H_1 : \lambda \neq 8.3$

b 0.0885

c Since the p-value is larger than half the significance level (2-tailed test) there is insufficient evidence to reject the null hypothesis.

6 a 163

b 0.04505

c Since the p-value is less than the significance level there is sufficient evidence to reject the null hypothesis.

7 a 72 as $P(X \geq 72) = 0.0482$ and $P(X \geq 71) = 0.0617$

b 0.0482

8 a 0.0568

b Since the p-value is larger than the significance level there is insufficient evidence to reject the null hypothesis.

c 0.0469 (as 107 is the critical value)

9 a A false positive can be treated by re-testing, which hopefully will avoid error and show the person as disease-free.

b A false negative means a person with the disease would go untreated for longer and potentially suffer as a result.

10 a Since re-designs are costly, they wouldn't want to repeatedly go through the process unnecessarily.

b If they are convinced that the dice are fair from their testing then probably they are close enough to being fair that people can't tell the difference even when they aren't fair.

11 a and b Students' own answers and reasons.

Exercise 11.1B

1 a 34 and 63 as $P(X \leq 34) = 0.0243$, $P(X \leq 35) = 0.0350$, $P(X \geq 63) = 0.0188$, and $P(X \geq 62) = 0.0256$.

b The result is outside of the critical region so there is insufficient evidence to reject the null hypothesis. You conclude that the average rate of phone calls to the call centre is reasonably estimated at 6.8 per 10 minutes.

c A type I error is when a true null hypothesis is falsely rejected. It occurs when a result in the critical region is obtained.

d The probability is $P(X \leq 34) + P(X \geq 62) = 0.0429$ (to 3 s.f.)

2 a The critical value is 2122 points.

b As $2098 < 2122$ there is insufficient evidence to reject the null hypothesis. You conclude that the average number of points scored per game has not increased.

3 a $H_0 : \lambda = 41,\ H_1 : \lambda < 41$

b $P(X \leq 38) = 35.6\% > 10\%$. There is insufficient evidence to suggest that the number of accidents has been reduced.

c It would mean that you believe the rate of accidents has not reduced when actually it has.

4 a 28

b $H_0 : \lambda = 28,\ H_1 : \lambda > 28$.

c $P(X \geq 37) = 5.89\% < 10\%$. There is sufficient evidence to suggest that the number of goals on the final day was exceptionally large.

d It would mean that the rate of goal scoring on the final day was not exceptionally large despite our belief that it was.

5 a $P(X \leq 5) = 8.38\%$ (3s.f.)

b Since the p-value is larger than the significance level, the result is not significant. There is insufficient evidence to suggest that there are fewer accidents per month at the sports centre.

6 a 0.0420

b The p-value is smaller than the significance level. There is sufficient evidence to reject the null hypothesis. You conclude that the average rate of customers coming into the book shop has increased.

7 a 0.117

b The p-value is larger than half the significance level (2-tailed test). There is insufficient evidence to reject the null hypothesis.

8 a 3 and 13 as $P(X \leq 3) = P(X \geq 13) = 0.0106$ and $P(X \leq 4) = P(X \geq 12) = 0.0384$

b $0.0213 = P(X \leq 3) + P(X \geq 13)$ (3s.f.)

9 a i The critical region is 0, 1, 2, 9, 10, 11, 12. $\alpha = 0.0777$

ii The critical region is 0, 1, 10, 11, 12 $\alpha = 0.0162$

iii The critical region is 0, 11, 12 $\alpha = 0.00185$

b i $\beta = 0.772$ ii $\beta = 0.916$ iii $\beta = 0.980$

Exercise 11.2A

1 a

Expected	A	B	C
X	53.0	85.4	15.6
Y	17.9	28.8	5.3
Z	68.1	109.8	20.1

b

X^2	A	B	C
X	0.30	0.51	7.23
Y	6.63	0.35	11.3
Z	0.69	0.87	0.42

X^2 Total	28.3

c There are 4 degrees of freedom, so the critical value is 9.49.

d The p-value is 0.00001

e The test statistic is larger than the critical value (or the p-value is less than the significance level), so there is sufficient evidence to reject the null hypothesis.

2 a

Expected	A	B	C	D
W	12.4	12.1	14.9	15.6
X	10.3	10.1	12.4	13.1
Y	13.0	12.8	15.7	16.5
Z	13.3	13.0	16.0	16.8

b

X^2	A	B	C	D
W	1.07	0.07	0.08	2.04
X	0.53	0.81	1.59	1.17
Y	1.20	2.61	0.11	0.02
Z	2.09	0.31	0.26	0.09

X^2 Total	14.04275

c There are 9 degrees of freedom, so the critical value is 16.9.
d The p-value is 0.120839
e The test statistic is smaller than the critical value (or the p-value is larger than the significance level), so there is insufficient evidence to reject the null hypothesis.

3 a

Expected	A	B	C
X	74.4	68.5	78.1
Y	65.6	60.5	68.9

b

X^2	A	B	C
X	0.29	1.94	0.61
Y	0.33	2.20	0.69

X^2 Total	6.055767

c There are 2 degrees of freedom so the critical value is 5.99.
d The p-value is 0.04842
e The test statistic is larger than the critical value (or the p-value is less than the significance level), so there is sufficient evidence to reject the null hypothesis.

4 a

Expected	A	B
X	80.6	113.4
Y	39.4	55.6

b

X^2	A	B
X	1.25	0.89
Y	2.56	1.82

Yates:	6.52749

c There is 1 degree of freedom so the critical value is 3.84.
d The p-value is 0.010623
e The test statistic is larger than the critical value (or the p-value is less than the significance level), so there is sufficient evidence to reject the null hypothesis.

5 a

E_i	0–1	2–3	4+
Brown	8.8	12.7	13.5
White	14.2	20.3	21.5

b

X^2	0-1	2-3	4+
Brown	0.916	0.007	0.479
White	0.57	0.005	0.30

The total X^2 statistic is 2.27802
c There are $(3 - 1) \times (2 - 1) = 2$ degrees of freedom so the p-value of the test statistic is 0.320139
d The p-value is larger than the significance level so there is insufficient evidence to reject the null hypothesis.

6 a

E_i	Chicken	BLT	Ham & Cheese	Egg Mayo
Orange	26.9	28.4	20.4	8.4
Apple	23.4	24.6	17.7	7.3
Water	23.7	25.0	17.9	7.4

b

X^2	Chicken	BLT	Ham & Cheese	Egg Mayo
Orange	0.622	0.005	0.091	0.668
Apple	0.822	0.768	0.027	0.074
Water	0.004	0.636	0.237	0.362

The total statistic X^2 is 4.31525

c The statistic is less than the critical value so there is insufficient evidence to reject the null hypothesis.

7 a

E_i	Hamburger	Cheeseburger	Pizza	Hot Dog
Class A	4.6	4.6	7.6	6.2
Class B	4.8	4.8	7.9	6.5
Class C	4.6	4.6	7.6	6.2

b Group hamburger and cheeseburger data all together. 4 degrees of freedom.
c The critical value is 9.49, so the result is not significant. There is insufficient evidence to say there is any association between which class a student is in and their food preference.

8 a

Expected	Chocolate Bars	Apples	Packets of Crisps
Frieda	10.7	4.0	13.3
Helga	11.1	4.2	13.8
Hans	15.3	5.8	19.0

b They are twins and so it isn't unreasonable to suggest that they might have similar preferences.
c There are $(3 - 1) \times (2 - 1) = 2$ degrees of freedom so the critical value is 5.99. The result is significant; there is sufficient evidence to suggest there is association between the person and the snacks they eat.

Exercise 11.2B

1 a

Expected	Blue Eyes	Green Eyes	Brown Eyes
Blue	79.1	47.9	47.0
Red	43.6	26.4	25.9
Yellow	42.3	25.6	25.1

X^2	Blue Eyes	Green Eyes	Brown Eyes
Blue	5.01	2.49	1.71
Red	2.13	0.25	1.94
Yellow	2.50	2.74	0.143

The total X^2 statistic is 18.9. For 4 degrees of freedom at the 5% significance level the critical value is 9.49. The test statistic is larger than the critical value so the result is significant at the 5% level. There is sufficient evidence to reject the null hypothesis.
b The largest contribution is from people with blue eyes who say blue is their favourite colour, significantly more than expected. The number of people with green eyes whose favourite colour is red and those with brown eyes whose favourite colour is yellow is very close to what is expected.

2 a

Expected	Axe	Sword	Bow	Spear
Celts	8.6	26.1	23.7	34.6
Vikings	7.1	21.6	19.7	28.7
Huns	6.4	19.3	17.6	25.7

X^2	Axe	Sword	Bow	Spear
Celts	0.242	0.329	0.588	0.0125
Vikings	3.40	0.270	4.741	0.189
Huns	6.351	1.476	10.179	0.112

The total X^2 statistic is 27.9. For 6 degrees of freedom at the 5% significance level the critical value is 12.6. The test statistic is larger than the critical value so the result is significant at the 5% level. There is sufficient evidence to reject the null hypothesis.

b The association suggests that people who play the different tribes choose their weapons accordingly and not independently. The largest contributions are from people who play as Huns, using far fewer Axes than expected and far more bows than expected. People who play as Vikings use more Axes than expected and fewer Bows than expected. Spears are the most popular weapon but they occur around the expected frequency for each tribe.

3 **a**

Expected	Grey	Brown
Beach	83.7	56.3
River	99.3	66.7

X^2	Grey	Brown
Beach	0.62	0.93
River	0.53	0.78

Using Yates' correction the total X^2 statistic is 2.86. At the 5% significance level with one degree of freedom the critical value is 3.84. The statistic is lower than the critical value so there is insufficient evidence to reject the null hypothesis.

b Although in the observed data there are more brown stones on the beach than by the river, the numbers are not dissimilar enough from the expected values to suggest there is any association.

4 **a**

Expected	Pizza	Pasta
Soup	97.4	110.6
Salad	93.6	106.4

X^2	Pizza	Pasta
Soup	1.001	0.881
Salad	1.041	0.916

With Yates' correction the test statistic is 3.839. The critical value with 1 degree of freedom at the 10% level is 2.706. The test statistic is larger than the critical value so there is sufficient evidence to reject the null hypothesis. You conclude that there is association between choice of starter and main meal.

b No individual contribution is very large or small but together they suggest association. Pasta + Soup and Pizza + Salad go together slightly more than you would expect, and the other pairs go together slightly less than you would expect.

5 **a**

Expected	Red	Blue	Yellow
Boys	67.5	76.1	59.4
Girls	65.5	73.9	57.6

X^2	Red	Blue	Yellow
Boys	0.834	0.345	0.095
Girls	0.859	0.356	0.098

The total X^2 test statistic is 2.59. The critical value with 2 degrees of freedom at the 1% level is 9.21. The test statistic is less than the critical value so there is insufficient evidence to reject the null hypothesis. You conclude that there is no association between child gender and monster colour preference.

b The numbers of boys and girls who prefer the yellow monster match the expected numbers well. There are no large contributions to the test statistic.

6 **a**

Observed	Dark	Colourful	Light	Totals
Black	40	32	24	96
Brown	66	42	12	120
Blonde	34	34	34	102
Totals	140	108	70	318

b

Expected	Dark	Colourful	Light
Black	42	32.6	21.1
Brown	52.8	40.8	26.4
Blonde	44.9	34.6	22.5

X^2	Dark	Colourful	Light
Black	0.121	0.011	0.389
Brown	3.28	0.038	7.87
Blonde	2.65	0.01	5.9

There are 4 degrees of freedom. The critical value at the 5% level is 9.49. The test statistic is 20.3. This is larger than the critical value so the result is significant. There is sufficient evidence to suggest that there is association between hair colour and choice of clothing colour.

c People with black hair have broadly typical choices of clothing. Light-coloured clothes are particularly favoured by those with blonde hair and not favoured by those with brown hair. The reverse is true for dark-coloured clothes.

7 **a**

Observed	0–59	60–69	70–79	80+	Totals
London	30	14	10	1	55
Midlands	17	16	13	2	48
North-East	5	12	7	2	26
North-West	30	21	10	0	61
South	8	2	0	0	10
Totals	90	65	40	5	200

b The expected frequencies for 80+ are below 5 and the most-similar grouping is with 70–79.

c The South should be excluded as the expected frequencies are below 5. It is not clear which other region is similar enough to group them with.

d 6

e The North East teams earned below 60 yellow cards much less often than expected. The Midlands teams earned between 60 and 69 yellow cards almost exactly as often as expected.

8 Check correct from Ex 11.2A Q4 and Ex 11.2B Qs 3 and 4

Exercise 11.3A

1 a $90.43 \le \mu \le 92.37$
 b $0.09903 \le \mu \le 0.144497$
 c $10.53 \le \mu \le 14.87$

2 a $31.95 \le \mu \le 51.05$
 b $-10.19 < \mu < -1.33$

3 a Since the population variance is known, the sample variance does not need to be used to estimate it.
 b 95%: $55.87 \le \mu \le 69.13$ 99%: $53.79 \le \mu \le 71.21$

4 a 187.25cm b $183.0 \le \mu \le 191.5$

5 $208.5 \le \mu \le 215.8$

6 a 0.939
 b $2655 \le \mu \le 4745$

7 a 2.5%
 b 2.5% $213.12 \le \mu \le 420.88$

8 Need $25 \ge z \times \frac{\sigma}{\sqrt{n}}$, so $n \ge \left(\frac{98z}{25}\right)^2$.
 a 102 b 60 c 42

Exercise 11.3B

1 a $58.67 \le \mu \le 63.93$, $57.28 \le \mu \le 63.72$
 b Size = 90, mean = 60.95. $58.91 \le \mu \le 62.99$
 c The sample is larger and there is no reason to believe the two samples are significantly different. The combined sample should give better estimates for the true mean.

2 a $1063.80 \le \mu \le 1078.20$ from 432, and $1055.88 \le \mu \le 1086.12$ from 1905.
 b The sample variance is very different from the population variance but the sample size is large enough that the sample variance should be used. The confidence intervals themselves aren't too dissimilar and if it was really important to tell the difference then likely a larger sample should be taken.

3 a mean estimate is –0.05
 b standard deviation estimate is 0.119
 c $-0.0799 \le \mu \le -0.020$

4 a mean estimate is 82.8, standard deviation estimate is 11.7
 b Expected frequencies are 4.5, 13.7, 23.6, 20.2, 8.7 and 1.4, which are very close to the recorded values. The assumption of a normal distribution seems justified.
 c $80.1 \le \mu \le 85.5$

5 a estimate of population variance is 35.41
 b Although 95% of all intervals generated in this way are expected to contain the true mean value, it is impossible to say for certain what the probability is that a generated interval contains the true mean.

Review exercise 11

1 a The null hypothesis is $H_0: \lambda = 1.9$ and the alternative hypothesis is $H_0: \lambda = 1.9$
 b $P(X \ge 48) = 0.0332$ (3s.f.).
 c The p-value is less than the significance level so there is $H_0: \lambda = 1.9$ sufficient evidence to reject the null hypothesis.

2 a The critical value is 13 as $P(X \le 13) = 3.11\%$ and $P(X \le 14) = 5.29\%$ (both 3s.f.).
 b The probability of making a type I error is 0.0311 (3s.f.) as that is the probability of obtaining a result in the critical region if the parameter actually is 0.75.

3 a The critical value is 5 as $P(X \le 5) = 5.34\%$ and $P(X \le 14) = 5.29\%$ (both 3s.f.).
 b 7 is outside the critical region so there is insufficient evidence to reject the null hypothesis. You conclude that there are not fewer accidents per month than previously, and that 1.3 is a reasonable estimate of the rate of accidents per month.

4 a

E_i	A	B	C
X	100.0	68.7	40.3
Y	105.7	72.6	42.6
Z	104.3	71.7	42.1

 b

X^2	A	B	C
X	0.091	0.077	0.701
Y	0.895	1.775	0.062
Z	0.432	2.603	1.148

The X^2 test statistic is 7.78 (3s.f.).

5 a For a 5×3 table there are $4 \times 2 = 8$ degrees of freedom.
 b The critical value at the 5% significance level is 15.5 (3s.f.).

6 The p-value is 0.0697 (3s.f.). This is larger than the significance level so there is insufficient evidence to reject the null hypothesis. You conclude that there is no association between height and birthplace of UK adults.

7 a

E_i	A	B	C	D
Red	86.4	85.9	67.2	84.5
Blue	67.7	67.3	52.7	66.2
Silver	30.9	30.8	24.1	30.3

 b

X^2	A	B	C	D
Red	2.151	0.042	0.777	0.240
Blue	2.385	0.166	0.350	2.086
Silver	0.028	0.894	0.357	1.739

The X^2 statistic is 11.22.
 c The test statistic is larger than the critical value so there is sufficient evidence to reject the null hypothesis. You conclude that there is association between the public's preference for the four car models and their choice of colour. For model A there is positive association with Red and negative association with Blue. For model D there is positive association with Blue and negative association with Silver. Models B and C appear to have colour preferences in roughly the expected proportion.

8 a $14.7 - \phi^{-1}(0.975) \times \frac{\sqrt{29.5}}{\sqrt{36}} \le \mu \le 14.7 + \phi^{-1}(0.975) \times \frac{\sqrt{29.5}}{\sqrt{36}}$, $12.93 \le \mu \le 16.47$
 b The test statistic is $\frac{14.7 - 16.5}{\sqrt{\frac{29.5}{36}}} = -1.988$. $\phi(-1.988) = 0.0234$
This is smaller than the significance level so the result is significant. There is sufficient evidence to suggest that the population mean is not 16.5
 c The size of the test is the probability of making a type I error, which for a continuous distribution is simply the significance level, 5%.

9 a $1.2282 < \mu < 6.7718$
 b $4.9066 < \mu < 4.9623$

10 The X^2 statistic using Yates' correction is 12.96. The contributions are shown in this table.

X^2	A	B
X	3.15	4.41
Y	2.25	3.15

Assessment 11

1 a $H_0: \lambda = 2.7$
$H_1: \lambda > 2.7$
b 0.057. No reason to reject the null hypothesis that the mean rate is 2.7

2 a $H_0: \lambda = 6.4$ $H_1: \lambda < 6.4$
b $P(X \le 2) = 0.046$ (3dp)
For a 5% one-tailed test, reject the null hypothesis; evidence suggests that the number of flaws has been reduced.
c Rejecting the null hypothesis when it is true.
P(Type I error) = 0.046 Rejecting the null hypothesis when it is true. $1 - 0.046 = 0.954$

3 a For a 5% one-tailed test, reject the null hypothesis; evidence suggests that the number of calls has increased.
b 0.047. This is the probability that the null hypothesis is rejected given that it is true.

4 a The Poisson distribution models the occurrence of random events in time; road traffic accidents can be modelled as random events.
b $H_0: \lambda = 3.2$ $H_1: \lambda < 3.2$
Under H_0, $P(X \le 2) = 0.380$ (3dp)
For a 5% one-tailed test, no reason to reject the null hypothesis; no evidence to suggest that the number of accidents has been reduced.
c Only zero accidents would result in a conclusion that the rate of accidents had been reduced.

5 a

		Pastime					
		Reading	Sport	Online	Television	Other	Total
Gender	Male	7.67	17.11	17.7	7.67	8.85	59
	Female	5.33	11.89	12.3	5.33	6.15	41
	Total	13	29	30	13	15	100

b Reject null hypothesis; evidence suggests that there is an association between pastime and gender.

6 a

		Result			
		Fail	Pass	Distinction	Total
Exam	A	7.7	17.6	7.7	33
	B	6.3	14.4	6.3	27
	Total	14	32	14	60

b No reason to reject teacher's belief.

7 There is no reason to reject the hypothesis that weekdays and weekends result in the same proportion of the different errors.

8 There is no reason to reject the hypothesis that there is no association between grade of oil and price.

9 a $H_0: p = 0.3$; $H_1: p > 0.3$
b Under H_0

x	6	7	8	9
$P(X \ge x)$	0.1179	0.0386	0.0095	0.0017

c $x = 7, 8, 9, 10, 11, 12$
d A type I error occurs when the null hypothesis is rejected given that it is true.
0.0386

10 a $4.4 \pm 1.96\frac{0.6}{\sqrt{15}} = (4.10, 4.70)$ (2 d.p.)
b No reason to reject the hypothesis as 4.6 is within the interval (4.10, 4.70).

11 a Large sample, which suggests sample variance close to population variance.
b $14 \pm 1.645\sqrt{\frac{36}{60}} = (12.73, 15.27)$ (2 d.p.)

Chapter 12

Exercise 12.1A

1 (these are examples – others may be possible)
a
b
c
d

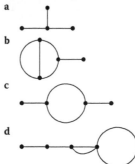

2

	V	E	F
a	5	7	4
b	8	12	6
c	4	9	7

a
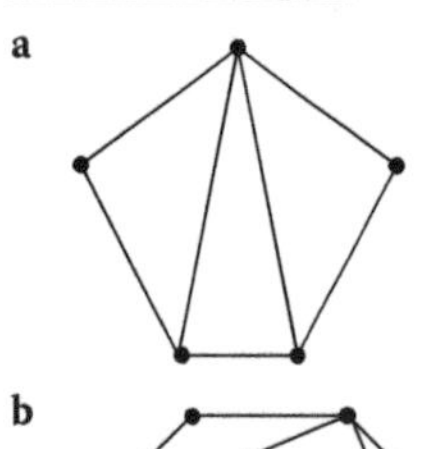
b
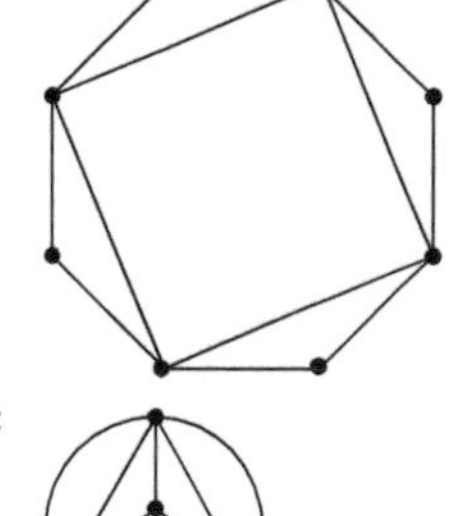
c
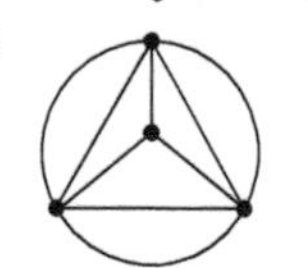

3 a Graph B is the odd one out.
b For graph A: {*A*, *B*, *C*, *D*, *E*, *F*}, {*AD*, *DF*, *FC*, *FE*, *EB*} other sets are possible
For graph C: {*A*, *B*, *C*, *D*, *E*, *F*}, {*AD*, *DF*, *FE*, *EB*, *BC*, *CA*} other sets are possible

4 Graph 1, the extra vertices are *A*, *E* and *F*.

5

	A	B	C	D	E
A	0	1	1	1	1
B	1	0	1	0	1
C	1	1	0	2	0
D	1	0	2	0	1
E	1	1	0	1	0

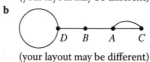

6 a
B C A D
(your layout may be different)
b
D B A C
(your layout may be different)

7 a

	A	B	C	D	E	F
A	0	1	1	0	1	0
B	1	0	0	1	0	1
C	1	0	0	0	0	1
D	0	1	0	0	1	0
E	1	0	0	1	0	1
F	0	1	1	0	1	0

b

	A	B	C	D	E	F
A	0	0	0	1	0	1
B	0	0	1	0	1	0
C	0	1	0	1	1	0
D	1	0	1	0	0	1
E	0	1	1	0	0	0
F	1	0	0	1	0	0

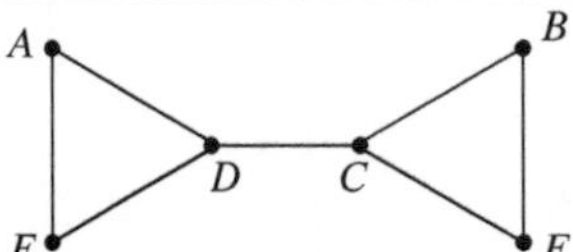

8 a

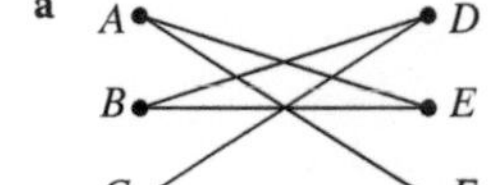

b

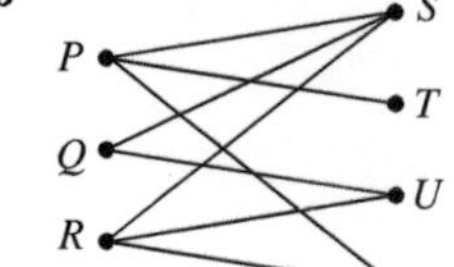

9

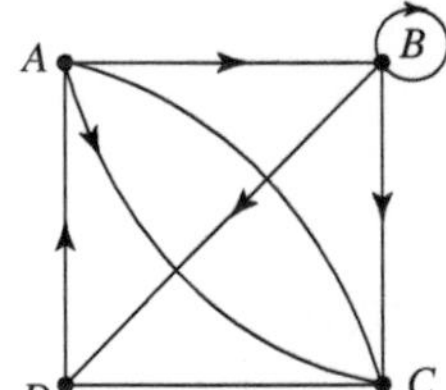

10

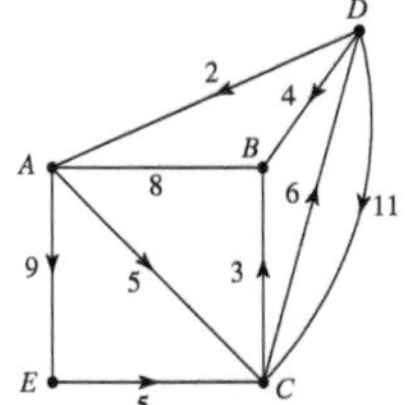

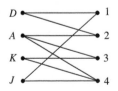

Exercise 12.1B

1 a

b $D1\ A2\ K3\ J4$ or $D2\ A3\ K4\ J1$ or $D2\ A4\ K3\ J1$

2 a

	P	E	F	G	A
P	–	7	22	–	–
E	7	–	–	8	15
F	22	–	–	20	–
G	–	8	20	–	4
A	–	15	–	4	–

b $EG + GA < EA$

c

	P	E	F	G	A
P	–	7	22	15	19
E	7	–	28	8	12
F	22	28	–	20	24
G	15	8	20	–	4
A	19	12	24	4	–

d

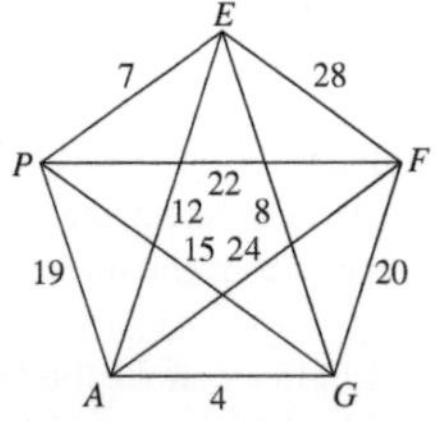

3 a $n - 1$

b Sum of degrees = twice number of edges

Number of edges $= \frac{1}{2}n(n-1)$

4 a

	A	B	C	D	E
A	0	1	1	1	1
B	1	0	1	0	1
C	1	1	0	1	1
D	1	0	1	0	0
E	1	1	1	0	0

b

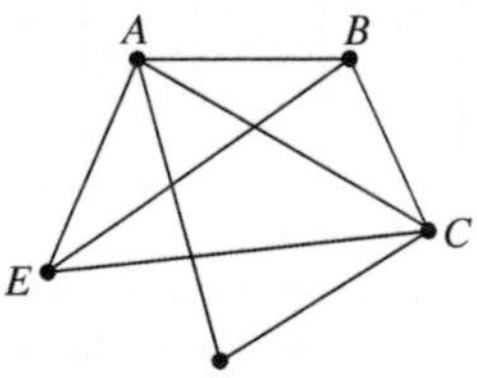

c

	A	B	C	D	E
A	0	0	0	0	0
B	0	0	0	1	0
C	0	0	0	0	0
D	0	1	0	0	1
E	0	0	0	1	0

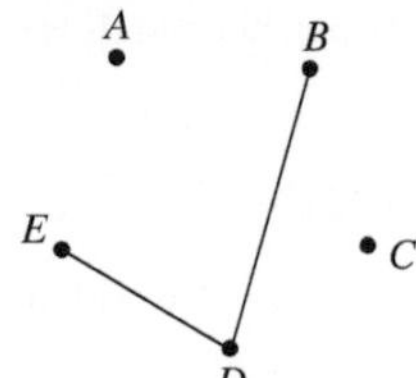

This graph shows which members of the group do not know each other.'

d Yes, {A, B, C, E}, these four all know each other.

5 a

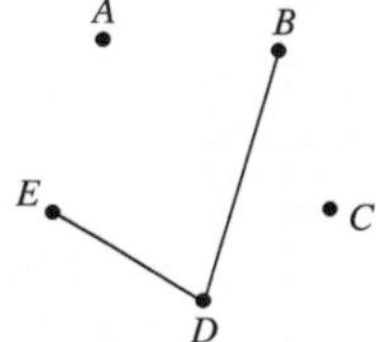

b 3

c Must have *fcd* (the only triple containing *d*), *afe* (the only one with *e*) and *abc* (the only one with *b*).

Exercise 12.2A

1 a i *ADFC*
ii *ABEDFC* (these are just examples)
b i *BACB*
ii *BADFEB* (these are just examples)
c For example *BADEFCB*
d It visits *C* twice.

2 Graphs **a**, **d** and **e**

3 a neither b Eulerian c semi–Eulerian
d Eulerian e Eulerian f neither

4 (These are just examples)

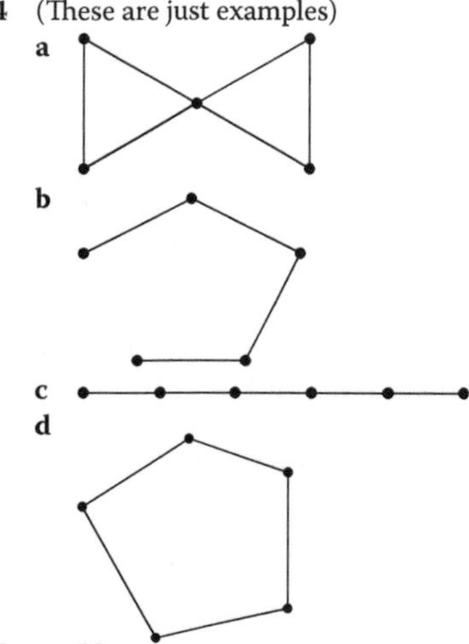

5 a 10

b

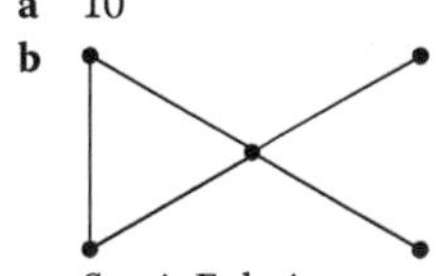

Semi–Eulerian

c

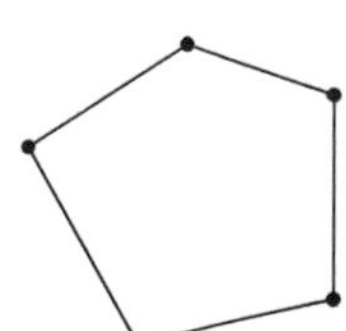

Exercise 12.2B

1 a

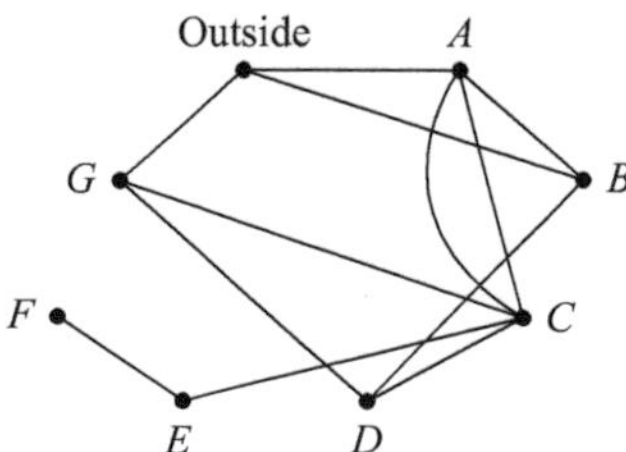

b non–Eulerian because 6 vertices of odd order

2 For example

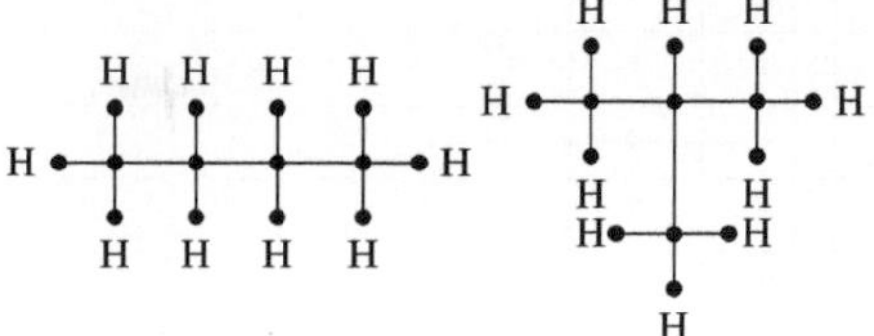

10 hydrogen in each case.

3 a Two odd order vertices – *B* and *D*
b *BD*
c For example *ABFCAEGFEDBCDA*

4

A
B D
C

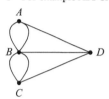

All vertices odd, so non–Eulerian.

5 a 3

b 6

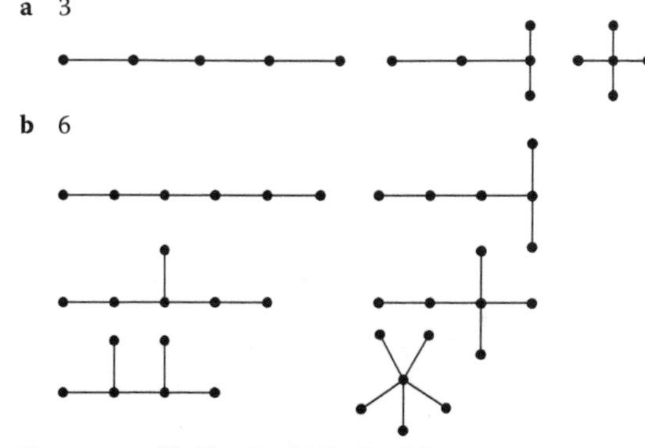

6 K_n is traversable if and only if n is odd.

7 $K_{m,n}$ is Hamiltonian if and only if $m = n$

Exercise 12.3A

1 a *AC*(2), *AD*(2), *DF*(3), *BE*(4), *BD*(5) or *AD*(2), *AC*(2), *DF*(3), *BE*(4), *BD*(5)
Total 16.
b *CD* (4), *AB* (5), *BD* (6), *BE* (7) or *CD* (4), *AB* (5), *BD* (6), *CE* (7)
Total 22
c *BC* (26), *DE* (29), *AB* (37), *EF* (38), *GI* (40), *HI* (40), *AH* or *BH* (42), *FI* (55) or *BC* (26), *DE* (29), *AB* (37), *EF* (38), *HI* (40), *GI* (40), *AH* or *BH* (42), *FI* (55)
Total 307
d *BD*/*CD* (2), *CD*/*BD* (2), *FG* (3), *EG*/*DF* (4) *DF*/*EG* (4), *AF* (5)
Total 20
e *DE* (15), *HG*/*HI* (17), *HI*/*HG* (17), *AH* (18), *CI*(24), *AB* (25), *CD* (26), *EF* (36)
Total 178

2 *AB*/*BD* (2), *BD*/*AB* (2), *AC* (3) or *AB*/*BD* (2), *BD*/*AB* (2), *BC* (3)
Total 7

3 There are 4 trees. The order of choosing them is
BD/*CD*(3), *CD*/*BD*(3), *AB*(4), *CE*(5)
BD/*CD*(3), *CD*/*BD*(3), *AB*(4), *ED*(5)
BD/*CD*(3), *CD*/*BD*(3), *AC*(4), *CE*(5)
BD/*CD*(3), *CD*/*BD*(3), *AC*(4), *ED*(5)

4 a

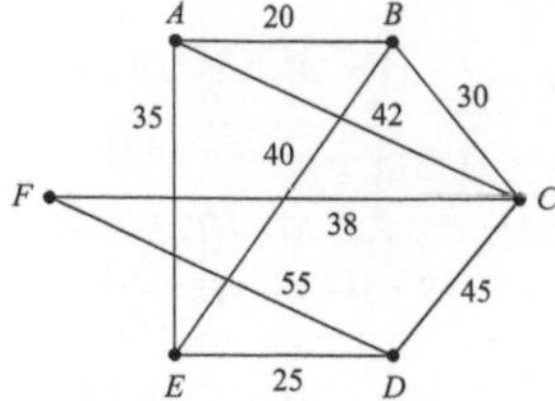

b *AB* (20), *DE* (25), *BC* (30), *AE* (35), *CF* (38) (Total 148)

c

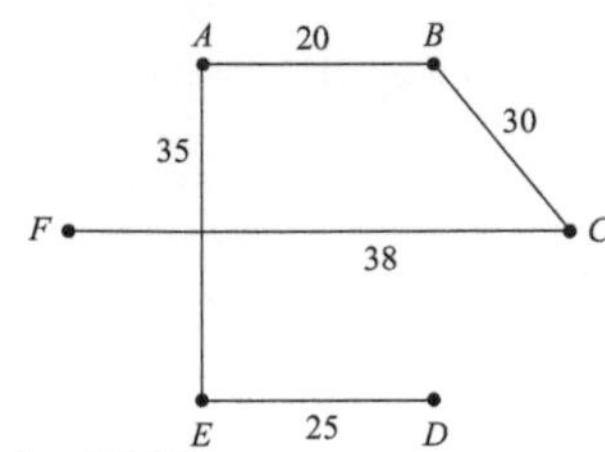

Exercise 12.3B

1

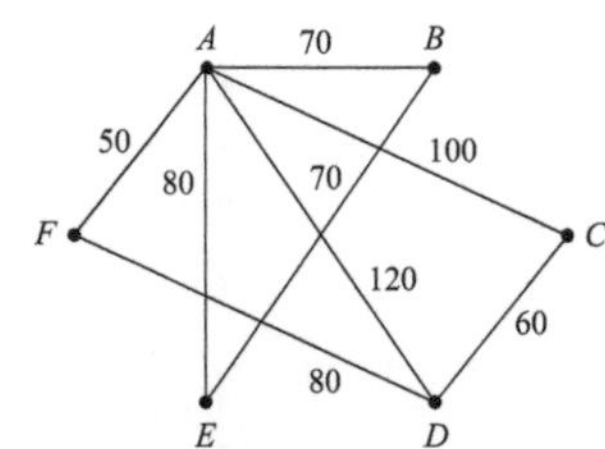

AF (50), *CD* (60), *AB*/*BE* (70), *BE*/*AB* (70), *DF* (80)
Total 330

2 a *BG*/*EG* (3), *EG*/*BG* (3), *AC*/*BD* (4), *BD*/*AC* (4), *CD*/*DF* (5), *DF*/*CD* (5)
Total 24miles

b *CD*

3 a From a sketch of the network: *IS* (20), *ES*/*IF* (22), *IF*/*ES* (22), *PS* (23), *GE* (30)

b Starting with German: *GE* (30), *ES* (22), *SP* (23), *SI* (20), *IF* (22)
Total cost £1170

c *GE* (30), *GF* (35), *GI* (35), *GP* (40), *GS* (32)
Total cost £1720
May be better because successive translations can gradually change the meaning of the text. May also be quicker because translators can work at the same time.

4 a and b

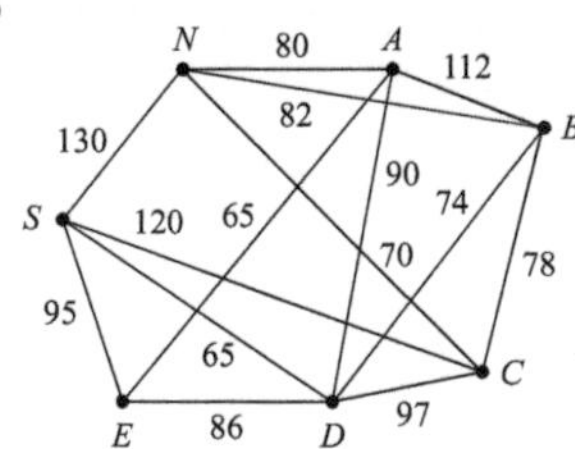

Exercise 12.4A

1 a *AB* (5), *BD* (6), *CD* (4), *BE* (7)
Total 22

b *AB* (37), *BC* (26), *AH*/*BH* (42), *HI* (40), *GI* (40), *FI* (55), *EF* (38), *DE* (29)
Total 307

c *AF* (5), *FG* (3), *DF* (4), *CD*/*BD* (2), *BD*/*CD* (2), *EG* (4)
Total 20

d *AH* (18), *HG*/*HI* (17), *HI*/*HG* (17), *CI* (24), *AB* (25), *CD* (26), *DE* (15), *EF* (36)
Total 178

2

	1	3	5	4	2
	A	*B*	*C*	*D*	*E*
A	–	5	9	6	3
B	(5)	–	11	4	7
C	9	11	–	(5)	8
D	6	(4)	5	–	12
E	(3)	7	8	12	–

AE (3), *AB* (5), *BD* (4), *CD* (5)
Total 17

3 a Assume start from *P* (but you could start from anywhere, which would change the labeling of the columns).
Circle, in order, *PQ* (2), *RQ* (1), *RS* (4), *ST* (3), *TU* (4) total 14, with columns labelled 1, 2, 3, 4, 5, 6 **or** circle *PQ* (2), *RQ* (1), *QT* (4), *ST* (3), *UT* (4) with columns labelled 1, 2, 3, 4, 5, 6

b

2 1 4 3 4
P *Q* *R* *S* *T* *U* or *P* 2 *Q* 1 *R* 4 *S* 3 *T* 4 *U*

4 Circle, in order, *AD* (8), *DG* (6), *AB* (9), *BC* (6), *CF* (8), *ED* (10)
Total 47 with columns labelled 1, 4, 5, 2, 7, 6, 3

Exercise 12.4B

1 Circle, in order, *AC* (20), *CH* (30), *AB* (56), *BD* (15), *DG* (25), *EG* (40), *EF* (30)
Total 216, with columns labelled 1, 4, 2, 5, 7, 8, 6, 3

2 a Because of the difficulty of checking for cycles

b i Circle, in order, *AB* (4), *BC* (5), *BD* (8), *CE* (12), *EF* (11) total 40 with columns labelled 1, 2, 3, 4, 5, 6
or
circle *AB* (4), *BC* (5), *CD* (8), *CE* (12), *EF* (11) with same labelling and total

ii

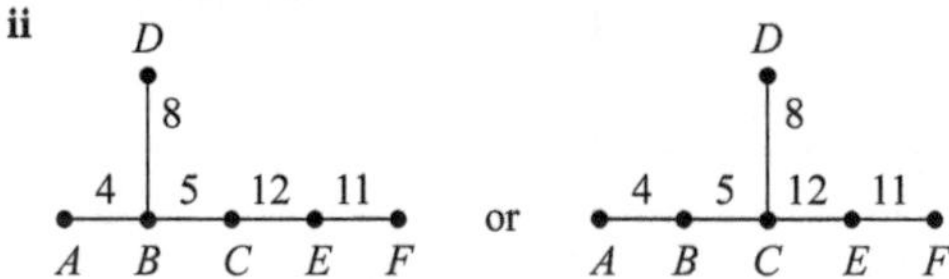

c i $4 + 5 + 12 + 11 = 32$ miles

ii *C*, gives lowest maximum distance 23 miles

3 a Circle, in order, *AF* (2), *EF* (2), *DE* (3), *AB* (4), *CF* (5), with columns labelled 1, 5, 6, 4, 3, 2

b

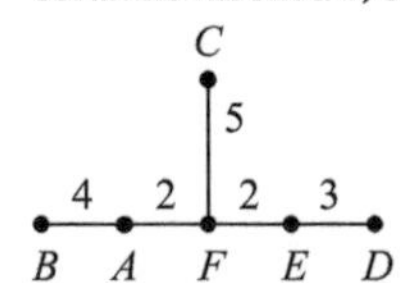

c 11 mins, between *B* and *C*, or between *B* and *D*.

4 a

	A	B	C	D	E
A	–	5	11	8	6
B	5	–	6	13	4
C	11	6	–	8	8
D	8	13	8	–	10
E	6	4	8	10	–

b

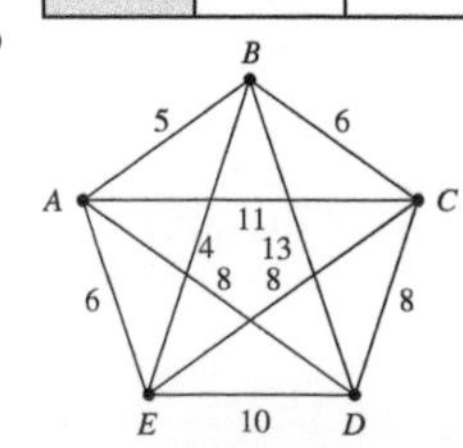

c AB (5), BE (4), BC (6), AD/CD (8)
Total 23 with columns labelled 1, 2, 4, 5, 3

d Because, for example, BE and BC might share a stretch of road

e Close AB, CD with all others open.

5 At each stage choose the road with the greatest capacity BC (12), CH (11), AB (10), CD (10), AG (12), DJ (10), FJ (10), EF (13)
A lorry of 10 tonnes can go everywhere.

Exercise 12.5A

1 a There are two nodes, B and E, of odd degree, so the route must travel twice between these.

b The shortest route between these is $BCDE = 9$, so repeat arcs BC, CD and DE.

c Total length $= 41 + 9 = 50$

2 a Odd nodes are A and D. Least distance is $ACD = 19$, so repeat AC and CD.

b Total weight $= 130$, so route $= 130 + 19 = 149$

c For example, $ABCACDCEBDEA$

3 a Odd nodes are A, B, C, D
$AB(AEB) + CD = 7 + 7 = 14$. $AC(AEC) + BD(BCD) = 6 + 13 = 19$
$AD + BC(BEC) = 13 + 5 = 18$.
Repeat AE, EB, CD. Total weight $= 56$, so route $= 56 + 14 = 70$

b Odd nodes are A, B, C, F
$AB + CF(CDGF) = 14 + 14 = 28$. $AC(AGDC) + BF(BGF) = 16 + 13 = 29$
$AF(AGF) + BC(BGDC) = 14 + 15 = 29$
Repeat AB, CD, DG, GF. Total weight $= 135$, so route $= 135 + 28 = 163$

c Odd nodes are A, B, C, E
$AB + CE = 7 + 10 = 17$. $AC + BE(BCE) = 5 + 14 = 19$.
$AE(ADE) + BC = 14 + 4 = 18$
Repeat AB, CE. Total weight $= 59$, so route $= 59 + 17 = 76$

d Odd nodes are B, C, D, F
$BC + DF = 5 + 18 = 23$. $BD(BCD) + CF = 13 + 11 = 24$
$BF(BCF) + CD = 16 + 8 = 24$
Repeat BC, DF
Total weight $= 86$, so route $= 86 + 23 = 109$

4 a The odd nodes are A, B, D and F.
The pairings are $AB + DF = 14 + 8 = 22$, $AD + BF = 7 + 15 = 22$, and $AF + BD = 15 + 11 = 26$.
Either repeat AB, CD and CF or AD, BC and CF.

b Total length $= 82 + 22 = 104$

Exercise 12.5B

1 a Odd nodes A, F, H, I.
$AF + HI = 740 + 140 = 880$. $AH + FI = 740 + 340 = 1080$
$AI + FH = 600 + 200 = 800$.
Repeat (AJ, JI or AK, KI) and (FG, GH or FL, LH)
Total of weights $= 3240$, so route $= 3240 + 800 = 4040$ m

b Need to repeat route from A to F (AK, KL, LF)
Length of route $= 3240 + 740 = 3980$ m

c Least distance between odd vertices is $HI = 140$, so repeat this. Start at A and finish at F (or vice versa).
Total route $= 3240 + 140 = 3380$ m

2 a

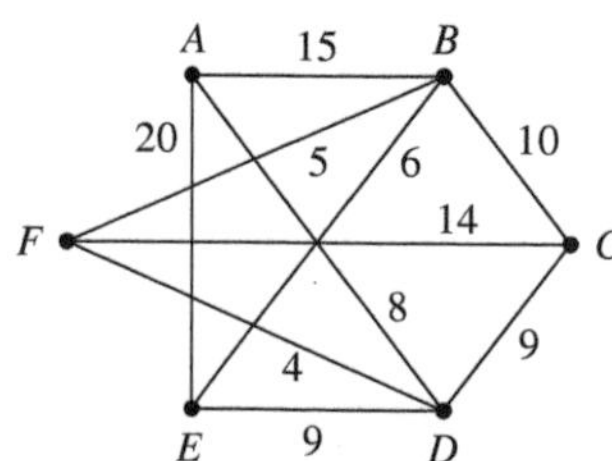

b Odd nodes A, C, E, F
$AC(ADC) + EF(EBF) = 17 + 11 = 28$. $AE(ADE) + CF(CDF) = 17 + 13 = 30$
$AF(ADF) + CE(CBE) = 12 + 16 = 28$
Repeat AD, DF, CB, BE or repeat AD, DE, EB, BF
Total weight $= 100$, so route $= 100 + 28 = 128$ km

3 a If all arcs are to be repeated then every node becomes even. The network is Eulerian and distance $= 86 \times 2 = 172$ km

b Odd nodes A, B, D, E
$AB + DE = 20 + 6 = 26$. $AD + BE = 14 + 10 = 24$
$AE(ACE) + BD(BED) = 19 + 16 = 35$
Repeat AD, BE. Distance $= 86 + 24 = 110$ km.

4 a Odd nodes A, C, D, E
$AC(AFGC) + DE = 600 + 100 = 700$. $AD + CE(CGE) = 200 + 550 = 750$
$AE(ADE) + CD = 300 + 450 = 750$
Repeat AF, FG, GC, DE
Total distance $= 3350 + 700 = 4050$ m

b Removing BG would create two more odd nodes, so there would be more repeats and he would go further.

c Pick up at C, so the only repeat needed is $DE = 100$, Total distance $= 3450$ m.

5 a There are 6 odd nodes (A, B, D, E, F, G), but it is clear the least pairing is
$AB + EF + DG = 200$ or the symmetrical combination.
Repeat AB, EF, DC, CG

b Distance $= 680 + 200 = 880$ m.

c With repeats C has order 8, so route passes through 4 times.

6 a Odd nodes B, D, E, G, I, J.
Discounting symmetries and the obvious bad pairings such as DI, consider
$BG + DE + IJ = 440$ and $BJ + GI + DE = 400$
Repeat BA, AJ, DE, GH, HI (or symmetry) gives rope length 1800 cm.

b Repeat DE and IJ. Start at B, end at G (or vice versa). Rope length 1640 cm.

Exercise 12.6A

1 a

	A	B	C	D
A	–	10	16	2
B	10	–	6	8
C	16	6	–	14
D	2	8	14	–

b From A: $ADBCA = 32$ is the same tour as from C: $CBDAC = 32$
From B: $BCDAB = 32$ is the same tour as from D: $DABCD = 32$

c Both tours would visit $ADBCBDA$.

2 a $AEBDCA = 38$

b $BDAECB = 35$

3 a MST is AB, AC, AD with total length 42. Upper bound $= 84$ min.

b For example, replace CAB by CB and BAD by BD. Route is $ACBDA = 67$ min.

4 a MST is AC, BC, BE, CF, DE with total length 34. Upper bound $= 68$ km.

b For example, replace ACB by AB and $DEBCF$ by DF. Route is $ABEDFCA = 50$ km.

5 a MST is AE, BC, BE, DE or AE, BC, CE, DE with total length 74.

b 148

c For example, from the second MST replace AED by AD and $DECB$ by DB.
Route is $ADBCEA = 103$.

6 a MST is AC, BD, CD, CF, DE
Total 28

b 56

c Start, for example, with ACDBDEDCFCA. Then replace DCF by DF, $ACDB$ by AB, BDE by BE. Upper bound = 41

7 a Arcs $(AC + AD) + (BC + CD) = 14$.
Not possible because a tour of length 14 would have to use AC, CD, DA which form a cycle ACD.

b Arcs $(BC + BD) + (AC + AD) = 16$
These form a cycle $ACBDA$, so this must be the optimal tour.

8 The bounds by deleting A, B, C, D, E, F in turn give lower bounds of 63, 59, 72, 60, 55, 57. The best is 72.

9 The bounds by deleting A, B, C, D, E in turn are 74, 72, 72, 77, 72. The best is 77.

Exercise 12.6B

1 a

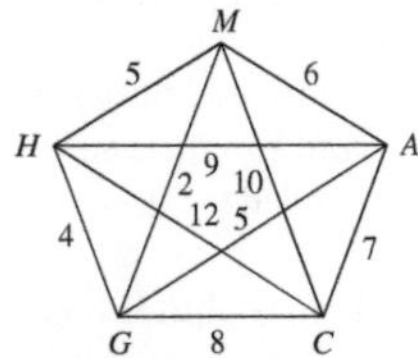

b $AGMH(G)CA = 31$ min. On original network the route is $HGMACGH$.

c $HGCA(G)MH = 30$ min, so $HGMACH$ is not optimal.

2 a

	A	B	C	D
A	–	5	9	6
B	5	–	10	7
C	9	10	–	3
D	6	7	3	

b $ABDCA$, 24 miles. Actual route $ABDCDA$.

3 a $(AB + AD) + (BC + CD + CE) = £950$

b MST is AD, AB, BC, CE, total 690. Upper bound is £1380 using $ADABCECBA$ but replacing $ECBA$ by EA gives a tour of £1240.

c $950 \le C \le 1240$

4 a

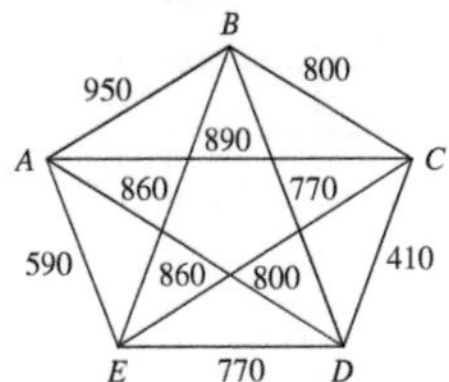

b Lower bound = $(AD + AE) + (BD + CD + DE) = 3400$ m
$ABCDEA = 3520$
so $3400 \le$ optimal tour ≤ 3520, so maximum saving is 120 m

5 a MST is AE, BC, BD, BE, BG, EF, length 29 km. Upper bound = 58 km

b For example, replace $DBEF$ by DF GBC by GC and AEF by AF.

c Route $AFDBGCBEA = 48$ km

d Lower bound is $(AB + AE) + (BC + BD + BE + BG + EF) =$ 34 km
$34 \le T \le 48$

6 a

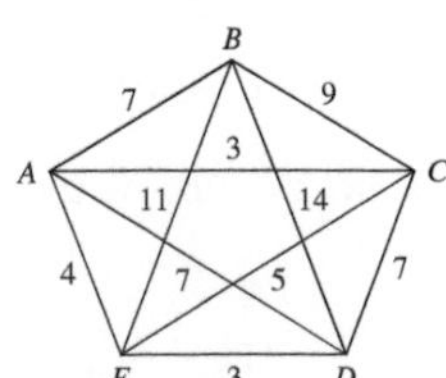

b $ACEDBA = 32$ miles. Actual order is $ACEDEABA$.

7 a

	A	B	C	D	E	F
A	–	4	2	4	3	5
B	4	–	2	4	3	5
C	2	2	–	4	3	3
D	4	4	4	–	3	3
E	3	3	3	3	–	2
F	5	5	3	3	2	–

b Upper bound 16 (for example, using nearest neighbour starting at A, $ACBDFEA$), lower bound 15 (for example, remove C, and choose $AE, BE, EF, DF + AC, BC$)

Exercise 12.7A

1 a Cut 1: **i** $\{SA, SB, SC\}$
ii $X = \{S\}, Y = \{A, B, C, T\}$
iii 41

Cut 2: **i** $\{AT, AB, SB, SC\}$
ii $X = \{S, A\}, Y = \{B, C, T\}$
iii 52

Cut 3: **i** $\{SA, AB, BT, CT\}$
ii $X = \{S, B, C\}, Y = \{A, T\}$
iii 38 (not 46, because AB is directed from the sink to the source)

b Cut 1: **i** $\{SB, AB, BC, DC, CF, EF, ET\}$
ii $X = \{S, A, C, E\}, Y = \{B, D, F, T\}$
iii 90

Cut 2: **i** $\{SA, AB, BC, BD\}$
ii $X = \{S, B\}, Y = \{A, C, D, E, F, T\}$
iii 70

Cut 3: **i** $\{AC, BC, DC, CF, EF, FT\}$
ii $X = \{S, A, B, D, F\}, Y = \{C, E, T\}$
iii 69

c Cut 1: **i** $\{SA, AC, BC, SB, CE\}$
ii $X = \{S, C\}, Y = \{A, B, D, E, F, T\}$
iii 144

Cut 2: **i** $\{BF, CE, DE, DT\}$
ii $X = \{S, A, B, C, D\}, Y = \{E, F, T\}$
iii 106

Cut 3: **i** $\{AD, DE, ET, FT\}$
ii $X = \{S, A, B, C, E, F\}, Y = \{D, T\}$
iii 117

d Cut 1: **i** $\{SB, SC, AC, AD\}$
ii $X = \{S, A\}, Y = \{B, C, D, E, T\}$
iii 104

Cut 2: **i** $\{SC, AC, BE, CB, DC, DT\}$
ii $X = \{S, A, B, D\}, Y = \{C, E, T\}$
iii 131

Cut 3: **i** $\{AD, DC, CT, ET\}$
ii $X = \{S, A, B, C, E\}, Y = \{D, T\}$
iii 118

2 a $\{SA, SB, BC, CT\}$

b SAT (12), SBT (9), SCT (14)

c Optimal flow = 35 (max flow = min cut)

3 Cut $\{CE, EF, FT\} = 24$, flow for example $SACET$ (10) and $SBDFT$ (14)

4 Cut $\{SA, AB, BC, BE\} = 17$, flow for example $SADT$ (5), $SACT$ (4), $SBCT$ (2), $SBCET$ (1), $SBET$ (5)

5 Cut $\{SA, AC, CD, ET\} = 12$
Flow 4 along $SADT$, 5 along $SCET$ and 3 along $SBET$

Exercise 12.7B

1 a Sources A, E

b SE (14) and SA (9)

c 15

d $\{DG, CG, FG\} = 15$

2 a Add the arcs SA (8), SB (16), HT (14), IT (9)
b $ACFI$ (5), $BDGJH$ (5), BEH (6)
c Those in **b** plus $ADGH$ (3)
d {EH, DG, GF, FI} = 19 so maximal (max flow = min cut}

3 a Sources B, D and sinks I, K
b Add arcs SB (36), SD (26), IT (14), KT (19)
c Cut {AE, EF, FJ, GK, KL} = 28
d For example $BAEI$ (6), $BFJI$ (5), $BFJK$ (2), $BCGK$ (5), $DHLK$ (10)

4 a Sources A, E and sinks D, H
b
c Cut {SA, SE} = 44
d For example ABD (10), $ABFD$ (2), $ACBFD$ (8), EGH (15), $ECGH$ (3), $ECGFH$ (6)

Review exercise 12

1
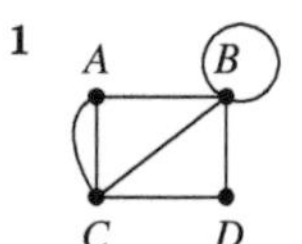

2 a
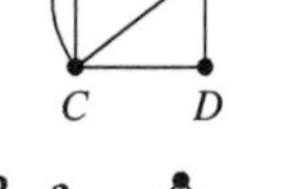

b

c
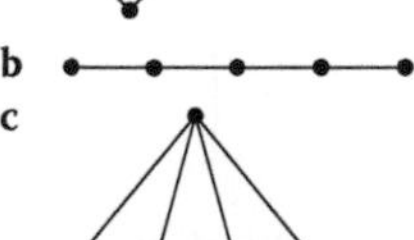

3 a 8 edges (from $V + F - E = 2$)
b
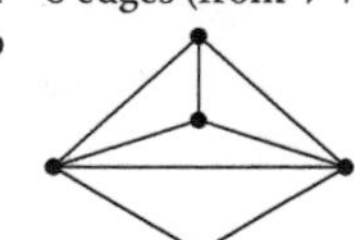

4 FG (3), AC (4), CF (5), CD (6), BC (9), BE (10) Total 37.

5 BC, AC, CF, FG, CD, BE

6

	1	5	3	2	6	7	4
	A	B	C	D	E	F	G
A	–	9	12	4	6	8	14
B	9	–	10	13	8	11	(3)
C	12	10	–	(2)	9	15	7
D	(4)	13	2	–	13	14	4
E	(6)	8	9	13	–	15	8
F	(8)	11	15	14	15	–	13
G	14	3	7	(4)	8	13	–

AD, CD, DG, BG, AE, AF

7 a Odd nodes A, B, C, G
AB (ACB) + CG (CFG) = 7 + 14 = 21. AC + BG (BEG) = 5 + 14 = 19
AG (ADG) + BC = 16 + 2 = 18
Repeat arcs AD, DG, BC. Length of route = 104 + 18 = 122
b D now has degree 6, so route passes through D 3 times.

8 a

	A	B	C	D	E
A	–	14	17	11	29
B	14	–	9	16	25
C	17	9	–	25	34
D	11	16	25	–	18
E	29	25	34	18	–

b $ADBCEA$ = 99. Actual route is $ADBCBEDA$. Upper bound because the optimal tour must be ≤ any known tour.
c and d MST is CB, BA, AD, DE, total 52. So 104 is an upper bound. Shortcut replacing $EDABC$ by EC gives upper bound 86.

9 a $SADT$ (5), $SACT$ (4), $SBCET$ (2) $SBET$ (7)
b $C_1 = 19$, $C_2 = 18$
c From max flow– min cut theorem, the max flow = 18

Assessment 12

1 a A connected graph with no cycles
b
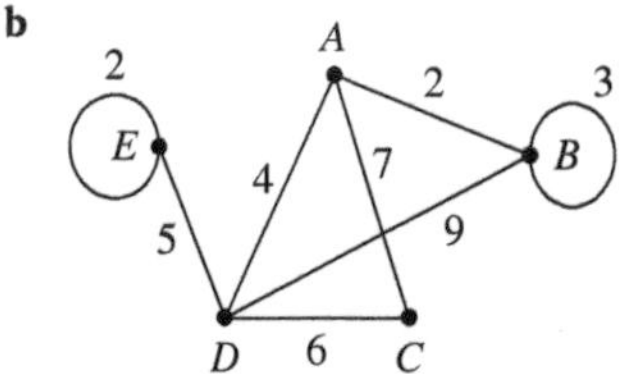

c For example, $EDAC$
d Any cycle would need to visit D twice as it is the only vertex that is directly connected to E. So there is no Hamiltonian cycle, and the graph is not Hamiltonian.
e There are two vertices with odd order (A and E) (it is semi-Eulerian).
f Add the edge AE.

2 a Yes. For example, demonstrate by labelling and listing edges, using a table, or redrawing graph 1 with the right-most arc redrawn further to the left so that it looks like graph 2.
b Semi-traversable since it has two odd order vertices.
c For example,
d
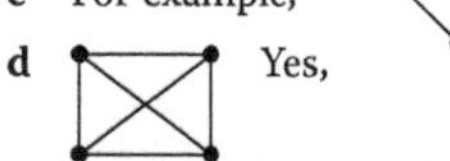
Yes,
For example, it can be drawn like this:
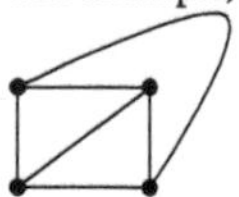

3 a

		To				
From		A	B	C	D	E
	A	0	1	1	1	0
	B	1	0	1	1	0
	C	1	2	0	1	0
	D	0	1	1	0	1
	E	0	0	0	1	2

b There is no Eulerian trail. There is, however, a semi-Eulerian trail of $EDCBCADBA$ with a total weight of 53.
c This cannot be done since one arc (either BD or AC) has no weight.
d Yes,
BD can be moved to around the outside so that none of the edges cross.

6 a For example, $ABCDA$
For example, 19

b

c $ACBDA$

d $ACBCDCA$

7 a Vertex E: $9 = x + 2 + 5 \Rightarrow x = 2$
Vertex C: $2 + 2 = z \Rightarrow z = 4$
Vertex A: $10 = y + 2 + y \Rightarrow y = 4$
Or:
Vertex B: $y + 5 = 9$ so $y = 4$
Vertex D: $4 + z + 2 = 10$ so $z = 4$
Vertex C: $2 + x = z$ so $x = 2$

b AC, BE, ET
EC, AD

c $\{AD, AC, EC, ED, ET\}$
Capacity of cut $= 4 + 2 + 2 + 6 + 5$
$= 19$

8 a

b bipartite

c AF, BD, BF to form $K_{3,3}$ then additional vertex and 3 edges to expand to $K_{4,3}$.
So 1 vertex
6 edges

d Non-Eulerian, as all four of the vertices in the set of 4 will have order 3.

9 a

	A	B	C	D	E	F
A	–	12	9	–	7	–
B	12	–	4	5	–	1
C	9	4	–	3	8	–
D	–	5	3	–	6	–
E	7	–	8	6	–	2
F	–	1	–	–	2	–

AE, EF, BF, BC, CD
Weight = 17

b

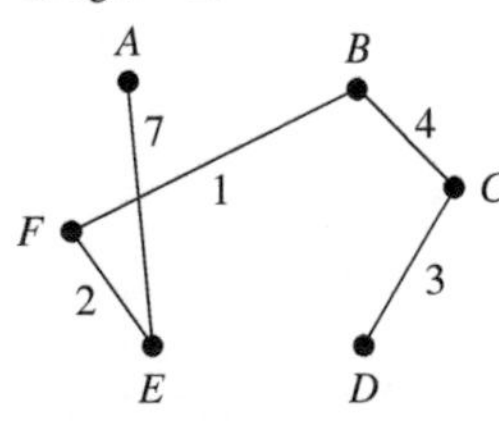

Add BF, EF, CD, BC
Reject BD, DE
Add AE

c Vertices with odd degree: A, D
Shortest route between A and D is ACD so repeat AC, CD.
Possible Route: $ABCDCEDBFEACA$
Length $= 57 + 12 = 69$

10 a a Minimum spanning tree is

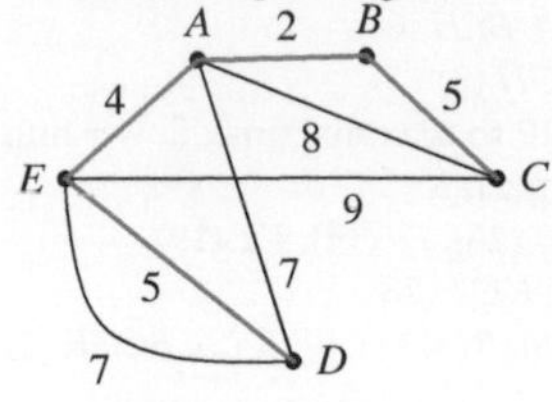

So upper bound $= 2 \times 16$
$= 32$ minutes

b For example, a short cut is AD instead of AED
So improved upper bound is 30

c Removing A gives MST of weight 19
So lower bound is $19 + 2 + 4 = 25$

Removing B gives MST of weight 16
So lower bound is $16 + 2 + 5 = 23$

Removing C gives MST of weight 11
So lower bound is $11 + 5 + 8 = 24$

Removing D gives MST of weight 11
So lower bound is $11 + 5 + 7 = 23$

Removing E gives MST of weight 14
So lower bound is $14 + 4 + 5 = 23$

So best lower bound is 25 minutes

d

	A	B	C	D	E
A	–	2	7	7	4
B	2	–	5	9	6
C	7	5	–	14	9
D	7	9	14	–	5
E	4	6	9	5	–

e $CBAEDC$
Length of tour is $5 + 2 + 4 + 5 + 14 = 30$ minutes
Tour on original network is:
$CBAEDEC$

f For example, a shorter tour is $ABCEDA$ with length 28 minutes

11 a 4

b i $\{AD, CD, CE, BE\}$
ii $\{S, A, B, C\}$
iii $\{D, E, T\}$

c Minimum cut: $\{DT, ET\}$
has capacity 18
The maximum flow is therefore 18.

12 a For example, Kruskals: Add CE, DE,
Reject CD
Add BC
Reject BD
Add EF, AB
Length $= 2 + 3 + 5 + 9 + 11 = 30$ m
Kruskal's or Prim's

b Use algorithm to find minimum spanning tree.
Redraw the network with weight 0 on AC and BD. Then use algorithm as before
Using Kruskal's,
Add AC, BD, CE, DE
Reject CD, BC
Add EF
Length reqired is $2 + 3 + 9 = 14$ m

c Upper bound $= 2 \times 30 = 60$

ABCEFEDECBA (in either direction)

Short cuts from *A* to *F* and *D* to *C* give the route *AFEDCBA* with weight 47

d removing *A* gives MST of weight 19. So lower bound is $19 + 11 + 12 = 42$

removing *B* gives MST of weight 26

So lower bound is $26 + 5 + 8 = 39$

Removing *C* gives MST of weight 31

So lower bound is $31 + 2 + 4 = 37$

Removing *D* gives MST of weight 27

So lower bound is $27 + 3 + 4 = 34$

Removing *E* gives MST of weight 33

So lower bound is $33 + 2 + 3 = 38$

Greatest lower bound is 42

e $42 \le$ length of optimal tour ≤ 47

f Matrix/Complete graph of shortest lengths:

	A	*B*	*C*	*D*	*E*	*F*
A	–	11	12	16	14	15
B	11	–	5	8	7	16
C	12	5	–	4	2	11
D	16	8	4	–	3	12
E	14	7	2	3	–	9
F	15	16	11	12	9	–

ABCEDFA = 48

BCEDFAB = 48

CEDBAFC = 50

DECBAFD = 48

ECDBAFE = 49

FECDBAF = 49

None are in the range so not optimal.

13 a

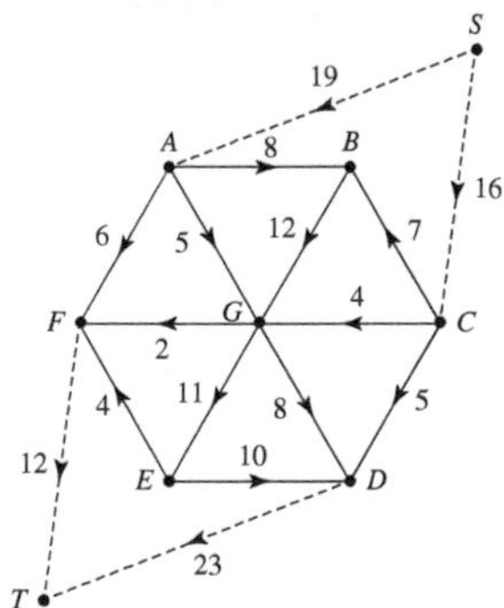

b 4

c Either cut {*AF, GF, GE, GD, CD*} or cut {*AF,AG,BG,CG,CD*} have capacity 32.

Therefore since there exists a possible flow of size 32, the max flow-min cut theorem tells us this is the maximal flow.

Chapter 13

Exercise 13.1A

1

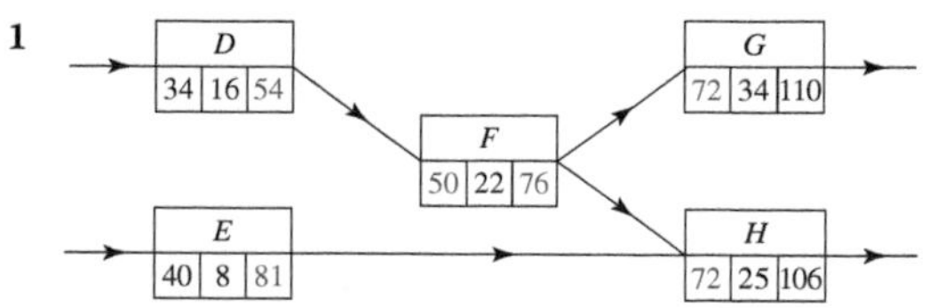

2 a

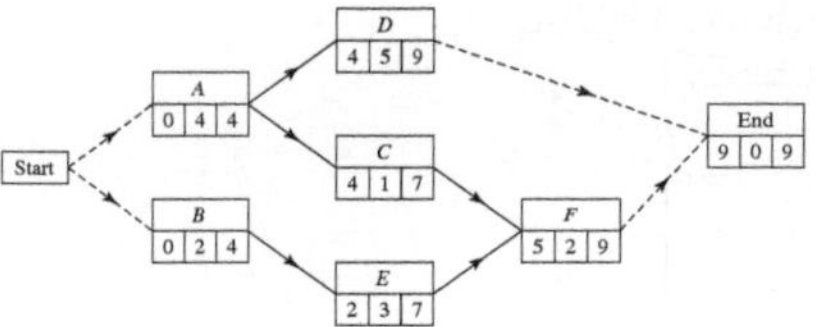

Project duration = 9

b

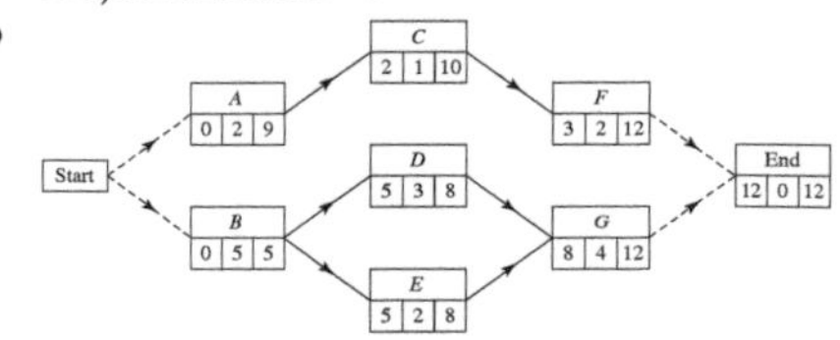

Project duration = 12

c

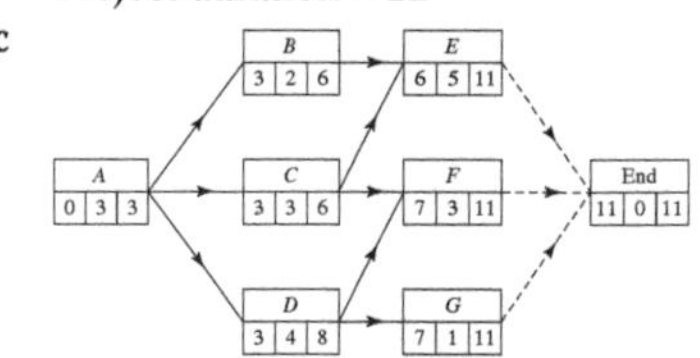

Project duration = 11

d

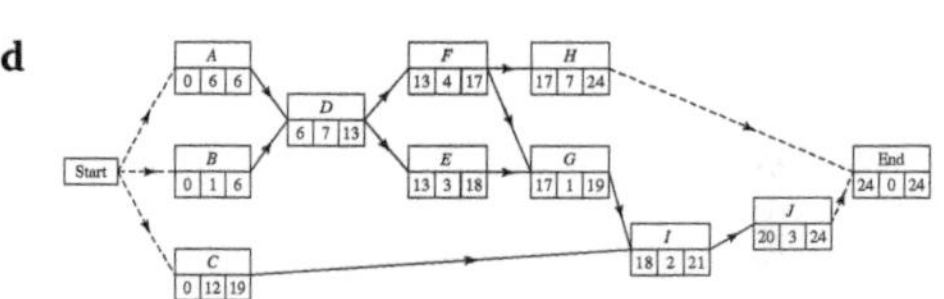

Project duration = 24

e

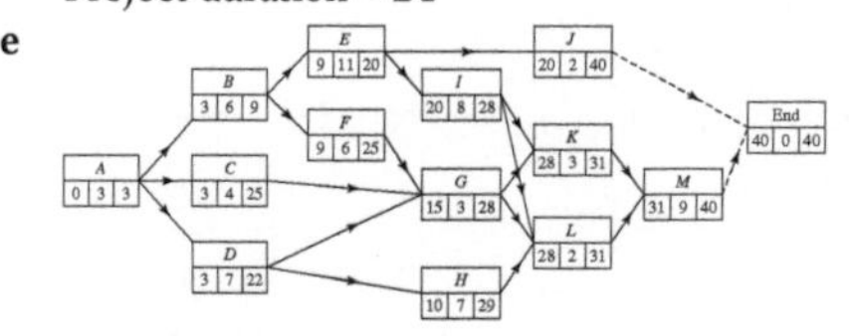

Project duration = 40

Exercise 13.1B

1 a

Activity	Description	Depends on
A	Purchase wood	–
B	Purchase wheels	–
C	Purchase cord	–
D	Cut out sides	*A*
E	Cut out base	*A*
F	Attach sides to base	*D, E*
G	Attach wheels to base	*B, E*
H	Attach cord	*C, D*

b

Start, A, B, C, D, E, F, G, H, End

(you may have decided on slightly different precedence)

2

3 a & b

Project duration = 43 days

c i Project duration would be 45 days

ii Project duration would be unchanged at 43 days

4 a & b

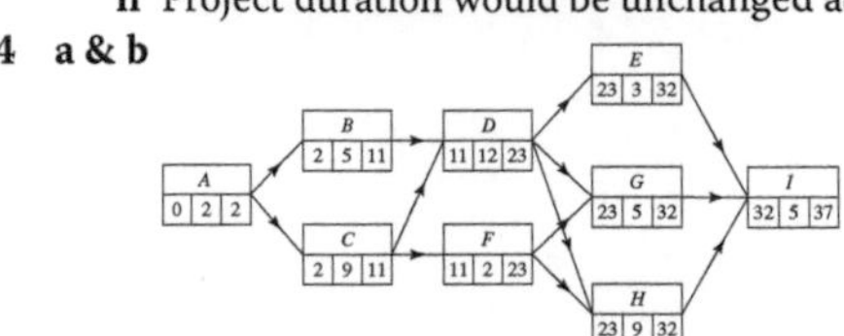

Project duration = 37 hours

c 12 hours

Exercise 13.2A

1 Floats are *A* 0, *B* 2, *C* 4, *D* 0, *E* 7, *F* 10, *G* 0, *H* 4, *I* 12, *J* 0, *K* 7, *L* 9, *M* 0

Critical activities are *A, D, G, J, M*

2 a

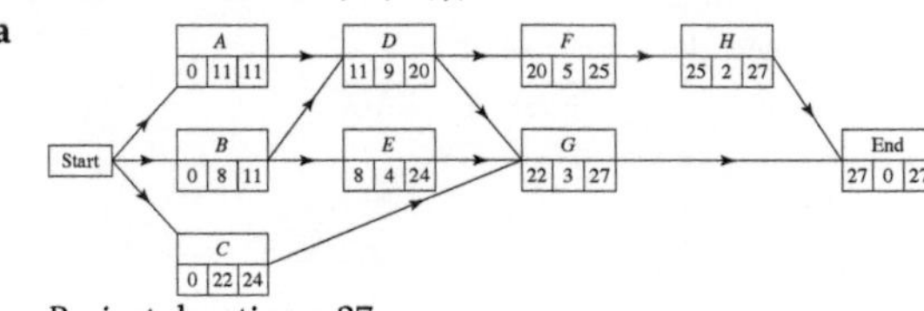

Project duration = 27

Floats are *A* 0, *B* 3, *C* 2, *D* 0, *E* 12, *F* 0, *G* 2, *H* 0

Critical activities are *A, D, F, H*

b

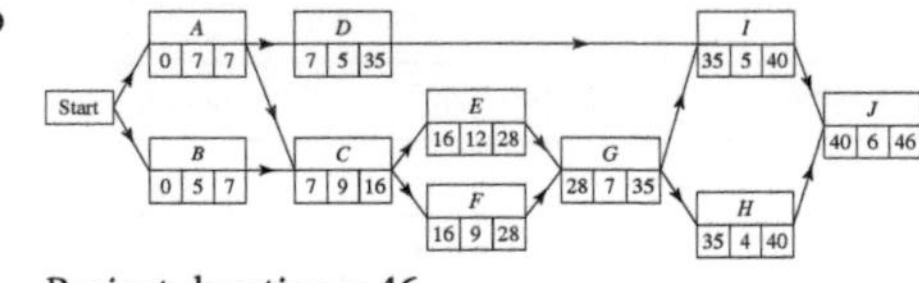

Project duration = 46

Floats are *A* 0, *B* 2, *C* 0, *D* 23, *E* 0, *F* 3, *G* 0, *H* 1, *I* 0, *J* 0

Critical activities are *A, C, E, G, I, J*

c

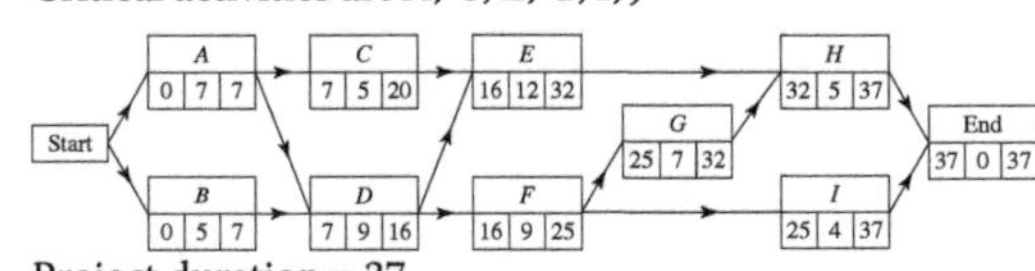

Project duration = 37

Floats are *A* 0, *B* 2, *C* 8, *D* 0, *E* 4, *F* 0, *G* 0, *H* 0, *I* 8

Critical activities are *A, D, F, G, H*

3

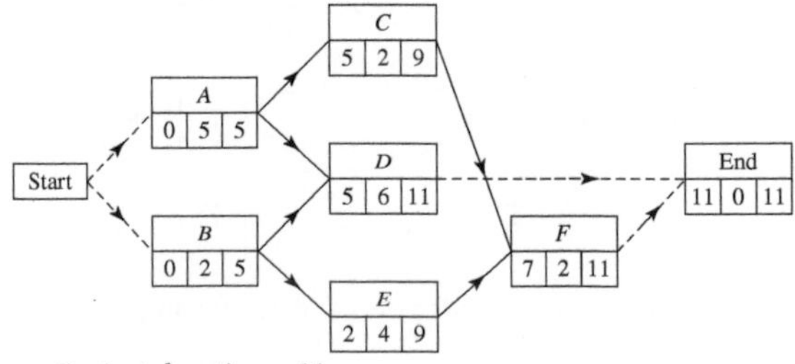

Project duration = 11

Critical activities are *A* and *D*

4 a

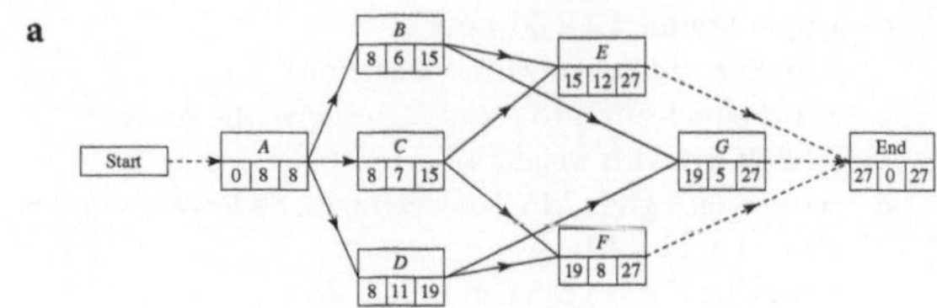

Duration = 27 hours

b Critical activities are *A, C, D, E* and *F*, forming critical paths *A-C-E* and *A-D-F*.

Exercise 13.2B

1 a

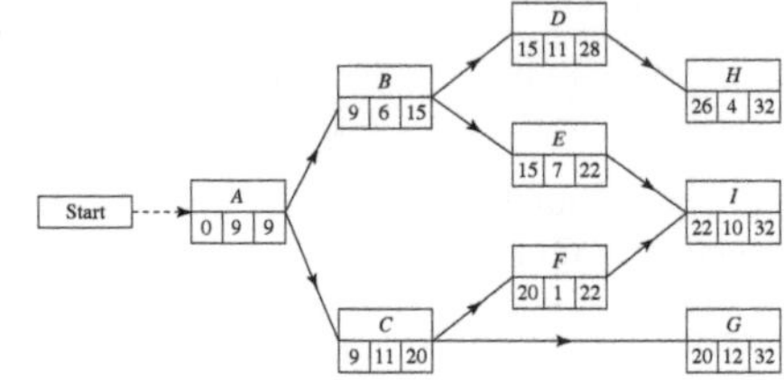

Duration = 32 hours.

b Critical paths *A-B-E-I* and *A-C-G*.

c Reduce *C* and *E* by 3 hours and *D* by 1 hour. Total cost £350.

2 a

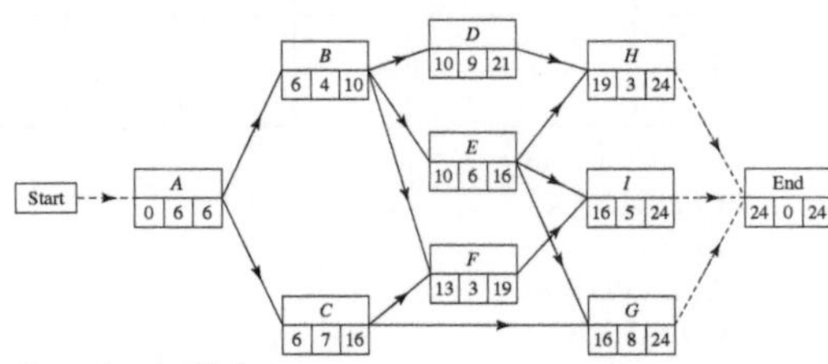

Duration = 24 days

b *A, B, E, G*

c i *D* has 2 days float, so 4 days delay extends the project by 2 days.

ii If *F* is delayed by 2 days then *I* cannot start until 18. If *I* is then delayed by 2 days it means the project is extended by 1 day.

3 a

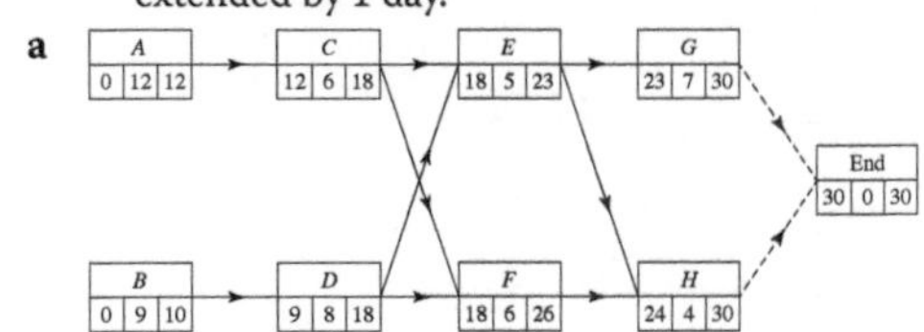

Duration = 30 days

b Critical activities *A, C, E, G*

Floats *B* 1, *D* 1, *F* 2, *H* 2

c 1 day

4 a

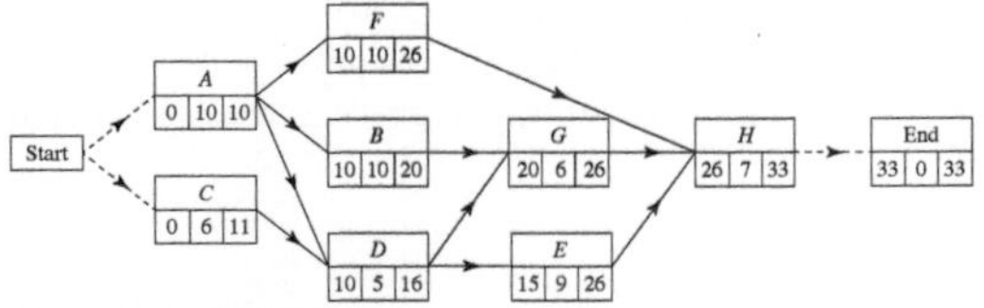

Duration = 33 hours

Critical activities *A, B, G, H*

b Split *B* into B_1 in its current place and B_2 following *F*. The duration is then 32 hours. The new critical activities are *A, F*, B_2, *H*

Review exercise 13

1

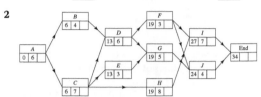

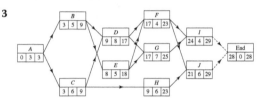

Start, A, B, C, D, E, F, G, H, I, J, End

2

Duration = 29 hours

3

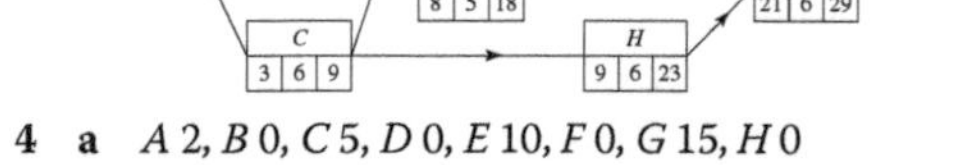

4 a A 2, B 0, C 5, D 0, E 10, F 0, G 15, H 0

b B, D, F, H

Assessment 13

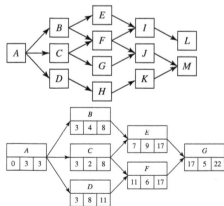

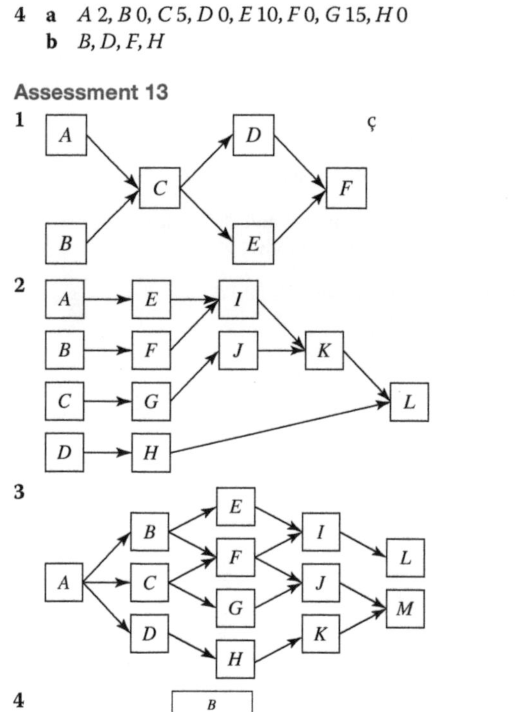

4

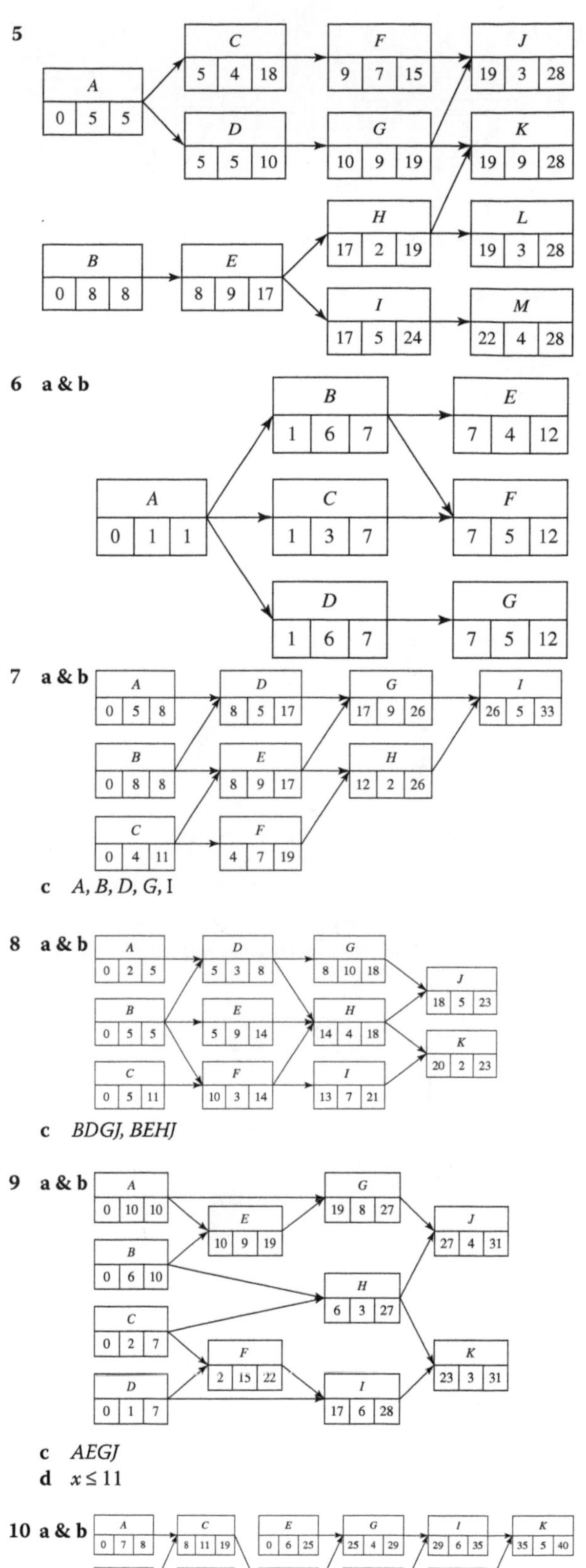

7 c $A, B, D, G,$ I

8 c $BDGJ$, $BEHJ$

9 c $AEGJ$

d $x \leq 11$

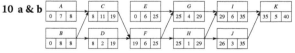

10 c $BCFGIK$

d $ACFGIK$

11 a & b

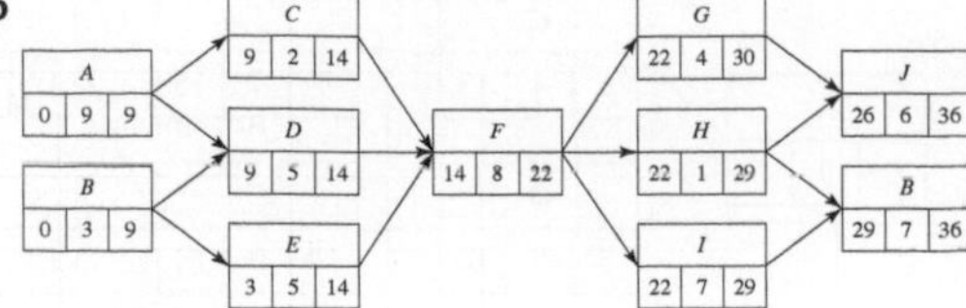

c *ADFIK*

d The earliest start time for J becomes 31 days and the earliest finish time becomes 40 days.

Chapter 14

Exercise 14.1A

1 If profit is £P, then

Maximise $P = 0.1x + 0.15y$

subject to $0.2x + 0.4y \leq 8000$ (or $x + 2y \leq 40000$)

$0.3x + 0.1y \leq 6000$ (or $3x + y \leq 60000$)

$x \geq 0, y \geq 0$

2 a The graph shows the objective line at $P = 50$

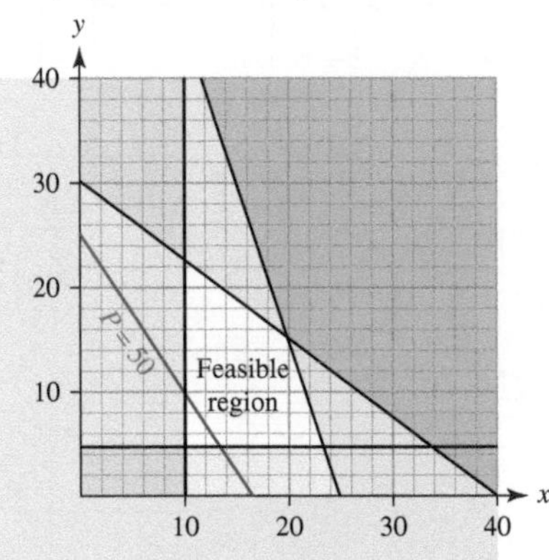

Optimal value $P = 90, x = 20, y = 15$

b The graph shows the objective line at $P = 8$.

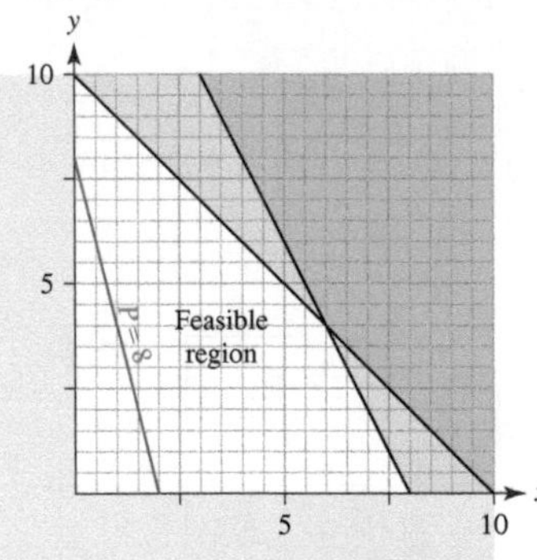

Optimal value $P = 32, x = 8, y = 0$

c The graph shows the objective line at $C = 50$.

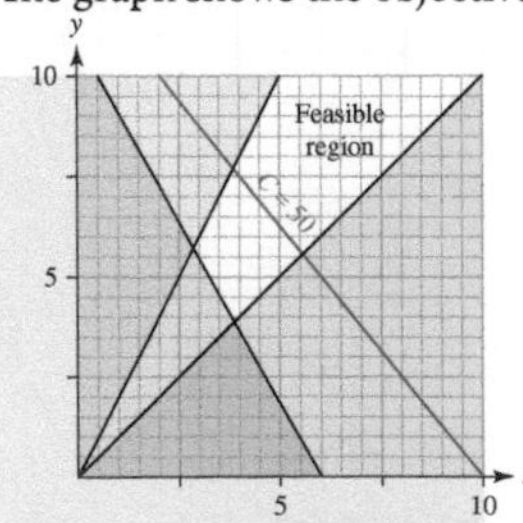

Optimal value $C = 39, x = 3, y = 6$

d The graph shows the objective line at $R = 15$.

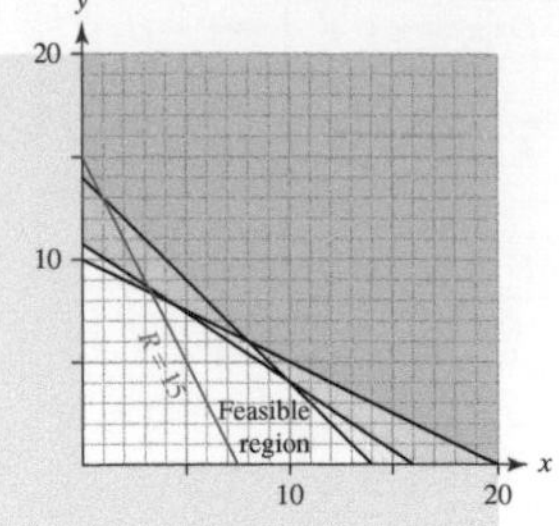

Optimal value $R = 28, x = 14, y = 0$.

3 x litres of Econofruit, y litres of Healthifruit, profit £P

Maximise $P = 0.3x + 0.4y$

subject to $0.2x + 0.4y \leq 20000$ (or $x + 2y \leq 100000$)

$0.5x + 0.3y \leq 30000$ (or $5x + 3y \leq 300000$)

$x \geq 0, y \geq 0$

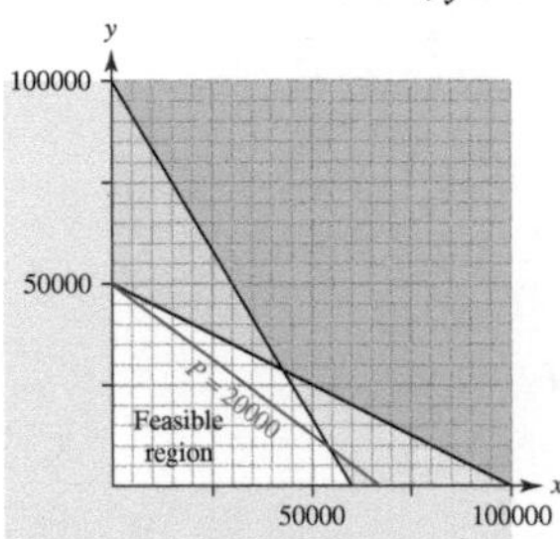

Objective line drawn at $P = 20000$

Optimal value $P = 24285\frac{5}{7}$ when $x = 42857\frac{1}{7}, y = 28571\frac{3}{7}$

Make 42857 litres of Econofruit and 28571 litres of Healthifruit, making £24285 profit.

4 x ha of wheat, y ha of potatoes. Profit £P

Maximise $P = 80x + 100y$

subject to $3x + 5y \leq 280$

$7x + 4y \leq 400$

$x + y \leq 65$

$x \geq 0, y \geq 0$

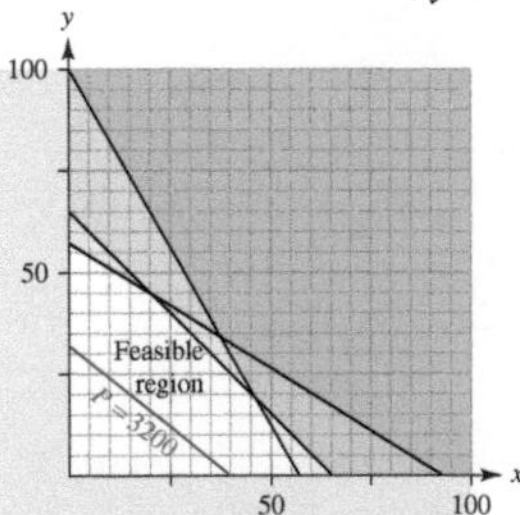

Objective line drawn at $P = 3200$

Optimal value $P = 6050$

when $x = 22.5, y = 42.5$

Plant 22.5 ha wheat and 42.5 ha of potatoes. making £6050 profit.

Exercise 14.1B

1 So hire 2 cars and 6 minibuses at a cost of £400.

2 So mix $28\frac{4}{7}$ ml of whisky with $71\frac{3}{7}$ ml ginger wine at a cost of £0.70.

3 Buy 13 cases of shampoo and 47 cases of cleaner, making a profit of £918.

4 Invest £8000 in bonds and £12000 in shares.

5 Build 13 houses and 12 bungalows, making a profit of £560000

6 x senior staff, y trainees, z children. Cost is £C.

a Minimise $C = 20x + 15y + 12z$

subject to
$x + y + z = 50$
$2x + 2y \geq z$
$x \geq y$
$x \geq 10, y \geq 5, z \geq 0$
x, y, z are integers

b Substitute $z = 50 - (x + y)$ in the objective function and constraints.

$C = 20x + 15y + 12(50 - x - y) = 8x + 3y + 600$

$2x + 2y \geq z$ becomes $3x + 3y \geq 50$

Complete LP statement is

Minimise $C = 8x + 3y + 600$

subject to
$3x + 3y \geq 50$
$x \geq y$
$x \geq 10, y \geq 5$
x, y are integers

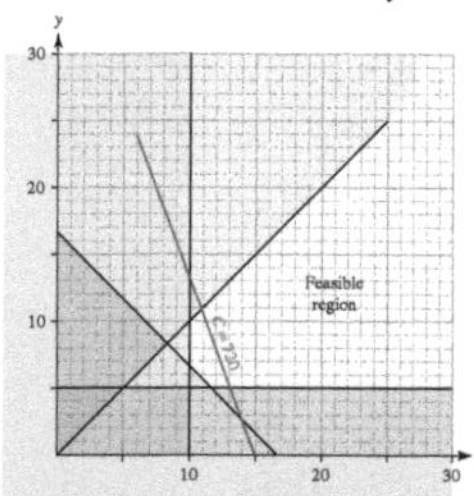

c Take 10 senior staff, 7 trainees and 33 children at a cost of £701.

7 The possible solutions are:
10 chew A, 20 chew B, 30 chew C
9 chew A, 23 chew B, 28 chew C
8 chew A, 26 chew B, 26 chew C
7 chew A, 29 chew B, 24 chew C
all of which give a cost of 95p.

Exercise 14.2A

1 a

		B				
		B_1	B_2	B_3	Row minima	Max of row minima
A	A_1	5	3	9	3	
	A_2	4	7	6	4	4
	A_3	2	4	5	2	
	Column maxima	5	7	9		
	Min of column maxima	5				

Play-safe strategies A_2 and B_1. Not a stable solution. Value = 4.

b

		B				
		B_1	B_2	B_3	Row minima	Max of row minima
A	A_1	9	−2	3	−2	
	A_2	−1	−4	2	−4	
	A_3	4	6	3	3	3
	Column maxima	9	6	3		
	Min of column maxima			3		

Play-safe strategies A_3 and B_3. Stable solution. Value = 3.

c

		B				
		B_1	B_2	B_3	Row minima	Max of row minima
A	A_1	6	−1	4	−1	−1
	A_2	−1	−3	5	−3	
	Column maxima	6	−1	5		
	Min of column maxima		−1			

Play–safe strategies A_1 and B_2. Stable solution. Value = −1.

d

		B			
		B_1	B_2	Row minima	Max of row minima
A	A_1	−1	3	−1	
	A_2	2	7	2	2
	A_3	6	−2	−2	
	Column maxima	6	7		
	Min of column maxima	6			

Play–safe strategies A_2 and B_1. Not a stable solution. Value = 2.

e

		B					
		B_1	B_2	B_3	B_4	Row minima	Max of row minima
A	A_1	4	−2	5	9	−2	
	A_2	2	1	3	5	1	1
	A_3	3	−1	−8	4	−8	
	Column maxima	4	1	5	9		
	Min of column maxima		1				

Play-safe strategies A_2 and B_2. Stable solution. Value = 1.

f

		B					
		B_1	B_2	B_3	B_4	Row minima	Max of row minima
A	A_1	3	2	7	10	2	
	A_2	5	5	7	8	5	5
	A_3	−3	2	−5	6	−5	
	Column maxima	5	5	7	10		
	Min of column maxima	5	5				

Play-safe strategies A_2 and B_1 or B_2. Stable solution. Value = 5.

2

		A		
		A_1	A_2	A_3
B	B_1	1	−2	−6
	B_2	−3	−7	2

3 a Row 2 dominates row 1.

		B	
		B_1	B_2
A	A_2	2	3
	A_3	4	−2

b Row 1 dominates row 2. After row 2 deleted, column 2 dominates columns 1 and 3, leaving this matrix.

		B
		B_2
A	A_1	−1

This is stable. Play A_1 and B_2. Value = −1

c Row 2 dominates row 3, then column 1 dominates column 3.

		B	
		B_1	B_2
A	A_1	3	1
	A_2	2	5

d Row 3 dominates row 1, then column 3 dominates column 2.

		B	
		B_1	B_3
A	A_2	3	−2
	A_3	−4	1

e Row 2 dominates rows 1 and 3, then column 1 dominates columns 2 and 3.

		B	
		B_1	B_4
A	A_2	−1	−1

This is stable. Play A_2 and B_1 or B_4. Value = −1.

f Row 1 dominates rows 2 and 4, then column 1 dominates column 2.

		B	
		B_1	B_3
A	A_1	3	2
	A_3	1	4

Exercise 14.2B

1 a Dominance eliminates row 3 then column 3.

		B			
		B_1	B_2	Row minima	Max of row minima
A	A_1	−4	6	−4	−4
	A_2	−2	−8	−8	
	Column maxima	−2	6		
	Min of column maxima	−2			

Play-safe strategies are A_1 and B_1. Value = −4. Unstable solution.

b Stable solution where A plays A_1 or A_3 and B plays B_2. Value −2.

c Dominance eliminates row 4 then column 2.

		B				
		B_1	B_2	B_3	Row minima	Max of row minima
A	A_1	4	1	−2	−2	
	A_2	3	−3	−1	−3	
	A_3	−1	3	0	−1	−1
	Column maxima	4	3	0		
	Min of column maxima			0		

Play-safe strategies are A_3 and B_4. Value = −1. Unstable solution.

d Stable solution where A plays A_2 and B plays B_3. Value 1.

e Stable solution where A plays A_3 and B plays B_3. Value −2.

f Stable solution where A plays A_4 and B plays B_3 or B_5. Value 0.

2 a

		Y		
		Choose 2	Choose 6	Choose 7
X	Choose 3	−2	3	4
	Choose 5	3	−2	−4
	Choose 9	5	3	−4

b Play-safe strategies are A_1 and B_2. The solution is unstable as the max of the row minima (−2) does not equal the min of the column maxima (3).

c For example, change Y's set to {3, 6, 7}

3 a

		B		
		B_1	B_2	B_3
A	A_1	1	0	−1
	A_2	−1	−3	0
	A_3	3	1	2

b

		B				
		B_1	B_2	B_3	Row minima	Max of row minima
A	A_1	1	0	−1	−1	
	A_2	−1	−3	0	−3	
	A_3	3	1	2	1	1
	Column maxima	3	1	2		
	Min of column maxima		1			

Play-safe strategies A_3 and B_2. Max of row minima = min of column maxima = 1, so stable solution (saddle point).

c Player A would win 10 – 0.

Exercise 14.3A

1 a Max of row mins = 2
Min of column maxs = 3
Solution is not stable
Let A play $A1$ with probability p
$v \le -2p + 3(1-p) = 3 - 5p$
$v \le 4p + 2(1-p) = 2p + 2$
Graph as shown.

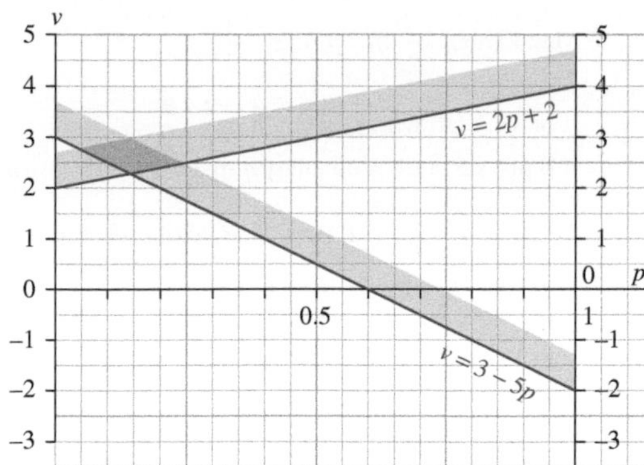

Max v when $3 - 5p = 2p + 2$, so $p = \frac{1}{7}$

This gives $v = 2\frac{2}{7}$

A plays A_1, A_2 with probabilities $\frac{1}{7}$, $\frac{6}{7}$

If B plays B_1 with probability q then

$v = -2q + 4(1 - q) = 2\frac{2}{7}$

This gives $q = \frac{2}{7}$, so B plays B_1, B_2

with probabilities $\frac{2}{7}$, $\frac{5}{7}$

b Max of row mins = 2
Min of column maxs = 3
Solution is not stable
Let A play A_1 with probability p
$v \le 4p - (1 - p) = 5p - 1$
$v \le 2p + 3(1 - p) = 3 - p$
Graph as shown.

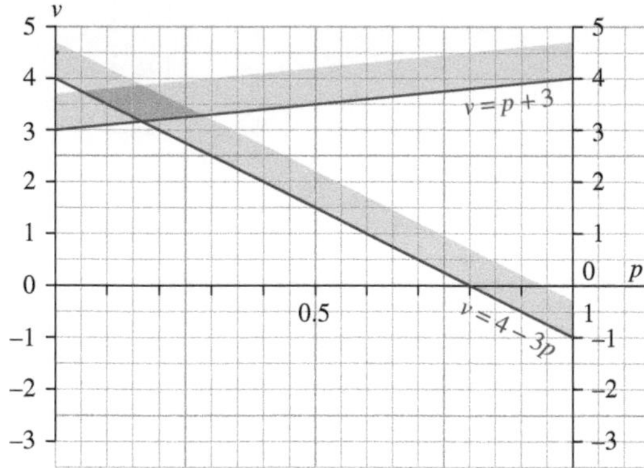

Max v when $5p - 1 = 3 - p$, so $p = \frac{2}{3}$

This gives $v = 2\frac{1}{3}$

A plays A_1, A_2 with probabilities $\frac{2}{3}$, $\frac{1}{3}$

If B plays B_1 with probability q then

$v = 4q + 2(1 - q) = 2\frac{1}{3}$

This gives $q = \frac{1}{6}$, so B plays B_1, B_2 with probabilities $\frac{1}{6}$, $\frac{5}{6}$

c Max of row mins = 2, min of column maxs = 2
Solution is stable. A plays A_2 and B plays B_2. Value = 2.

d Max of row mins = 3
Min of column maxs = 4
Solution is not stable
Let A play A_1 with probability p
$v \le p + 4(1 - p) = 4 - 3p$
$v \le 4p + 3(1 - p) = p + 3$
Graph as shown.

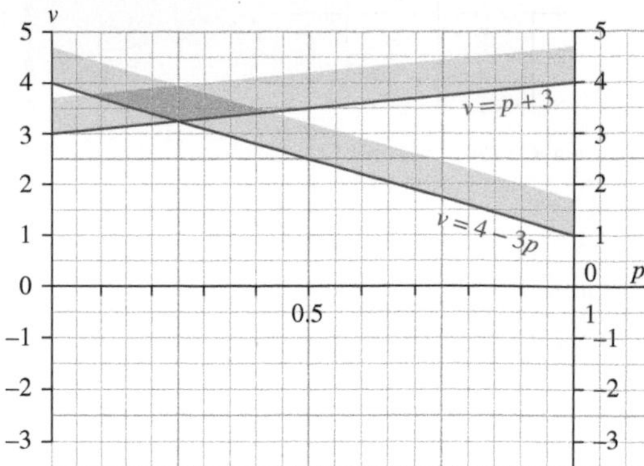

Max v when $4 - 3p = p + 3$, so $p = \frac{1}{4}$

This gives $v = 3\frac{1}{4}$

A plays A_1, A_2 with probabilities $\frac{1}{4}$, $\frac{3}{4}$

If B plays B_1 with probability q then $v = q + 4(1 - q) = 3\frac{1}{4}$

This gives $q = \frac{1}{4}$, so B plays B_1, B_2 with probabilities $\frac{1}{4}$, $\frac{3}{4}$

2 a Max of row mins = −2
Min of column maxs = 2
Solution is not stable
Let A play A_1 with probability p
$v \le 2p - 2(1 - p) = 4p - 2$
$v \le -2p + 3(1 - p) = 3 - 5p$
Graph as shown.

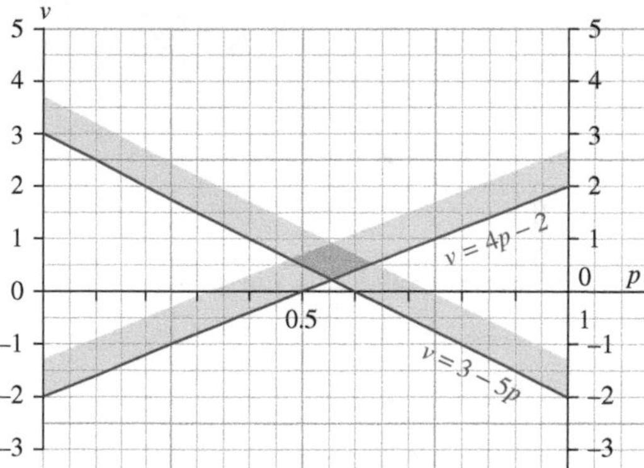

Max v when $4p - 2 = 3 - 5p$, so $p = \frac{5}{9}$

This gives $v = \frac{2}{9}$

A plays A_1, A_2 with probabilities $\frac{5}{9}$, $\frac{4}{9}$

If B plays B_1 with probability q then $v = 2q - 2(1 - q) = \frac{2}{9}$

This gives $q = \frac{5}{9}$, so B plays B_1, B_2 with probabilities $\frac{5}{9}$, $\frac{4}{9}$

b Max of row mins = 6
Min of column maxs = 7
Solution is not stable
Let A play A_1 with probability p
$v \le 7p + 5(1 - p) = 2p + 5$
$v \le 6p + 8(1 - p) = 8 - 2p$
Graph as shown.

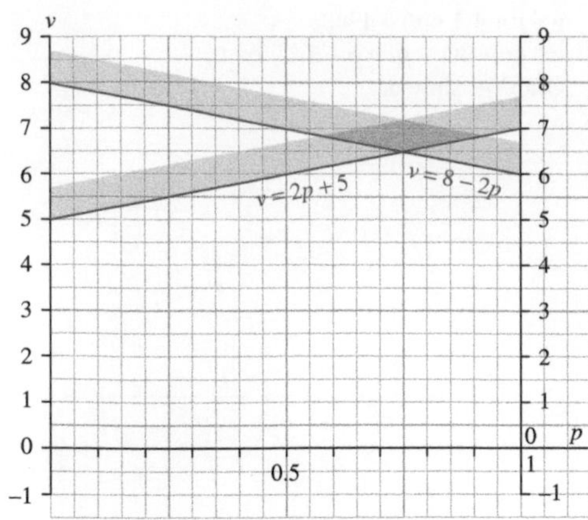

Max v when $2p + 5 = 8 - 2p$, so $p = \frac{3}{4}$

This gives $v = 6\frac{1}{2}$

A plays A_1, A_2 with probabilities $\frac{3}{4}$, $\frac{1}{4}$

If B plays B_1 with probability q then $v = 7q + 6(1 - q) = 6\frac{1}{2}$

This gives $q = \frac{1}{2}$, so B plays B_1, B_2 with probabilities $\frac{1}{2}$, $\frac{1}{2}$

c Max of row mins = 1
Min of column maxs = 3
Solution is not stable
Let A play A_1 with probability p
$v \le 5p - (1 - p) = 6p - 1$
$v \le 2p + 3(1 - p) = 3 - p$
Graph as shown.

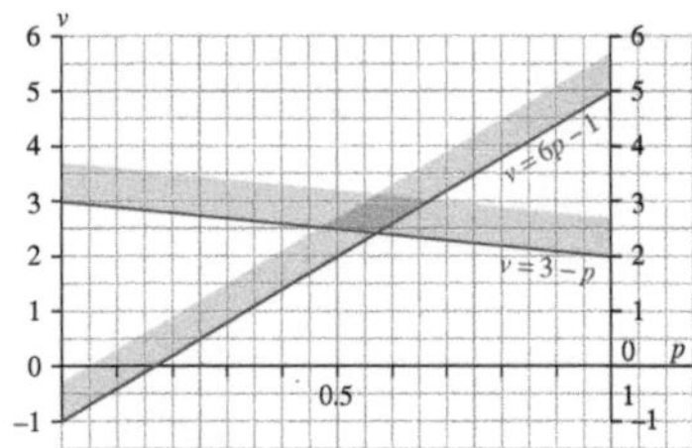

Max v when $6p - 1 = 3 - p$, so $p = \frac{4}{7}$

This gives $v = 2\frac{3}{7}$

A plays A_1, A_2 with probabilities $\frac{4}{7}$, $\frac{3}{7}$

If B plays B_1 with probability q then $v = 5q + 2(1 - q) = 2\frac{3}{7}$

This gives $q = \frac{1}{7}$, so B plays B_1, B_2 with probabilities $\frac{1}{7}$, $\frac{6}{7}$

d Max of row mins = −1
Min of column maxs = 1
Solution is not stable
Let A play A_1 with probability p
$v \le -2p + 2(1 - p) = 2 - 4p$
$v \le p$
Graph as shown.

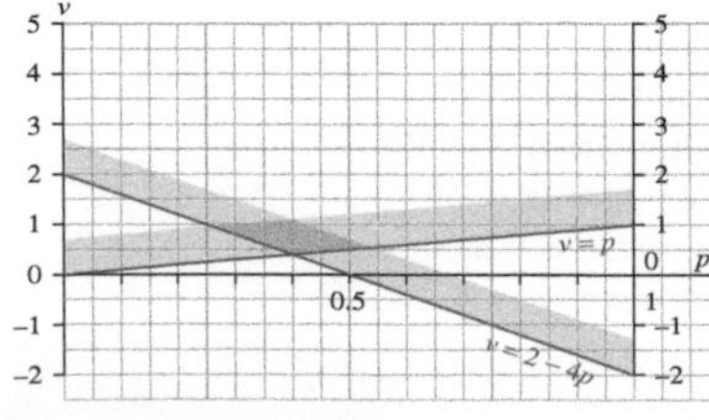

Max v when $2 - 4p = p$, so $p = \frac{2}{5}$

This gives $v = \frac{2}{5}$

A plays A_1, A_2 with probabilities $\frac{2}{5}$, $\frac{3}{5}$

If B plays B_1 with probability q then $v = -2q + (1 - q) = \frac{2}{5}$

This gives $q = \frac{1}{5}$, so B plays B_1, B_2 with probabilities $\frac{1}{5}$, $\frac{4}{5}$

3 a Let B play B_1 with probability q
$(-v) \le 6q - 2(1 - q)$ and so $v \ge 2 - 8q$
$(-v) \le -q + 2(1 - q)$ and so $v \ge 3q - 2$
Graph as shown.

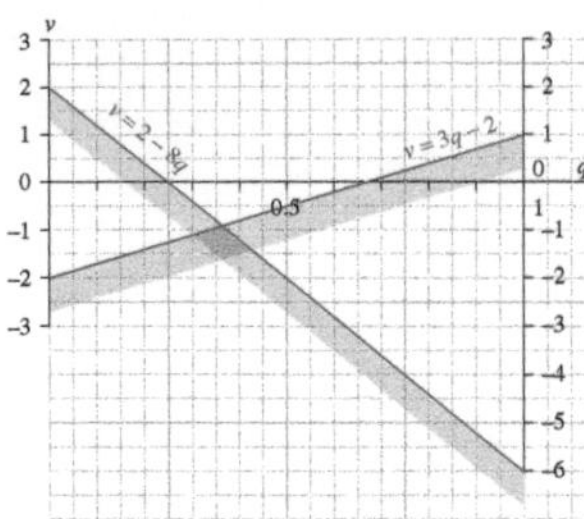

Max $(-v)$ when $2 - 8q = 3q - 2$, so $q = \frac{4}{11}$

This gives $v = -\frac{10}{11}$

B plays B_1, B_2 with probabilities $\frac{4}{11}$, $\frac{7}{11}$

b Let B play B_1 with probability q
$(-v) \le -9q - 7(1 - q)$ and so $v \ge 2q + 7$
$(-v) \le -3q - 10(1 - q)$ and so $v \ge 10 - 7q$
Graph as shown.

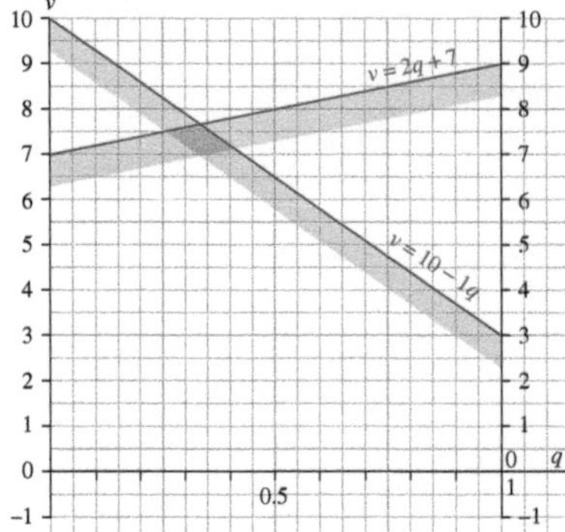

Max $(-v)$ when $2q + 7 = 10 - 7q$, so $q = \frac{1}{3}$

This gives $v = 7\frac{2}{3}$

B plays B_1, B_2 with probabilities $\frac{1}{3}$, $\frac{2}{3}$

c Let B play B_1 with probability q
$(-v) \le 5q - 2(1 - q)$ and so $v \ge 2 - 7q$
$(-v) \le -q + 4(1 - q)$ and so $v \ge 5q - 4$
Graph as shown.

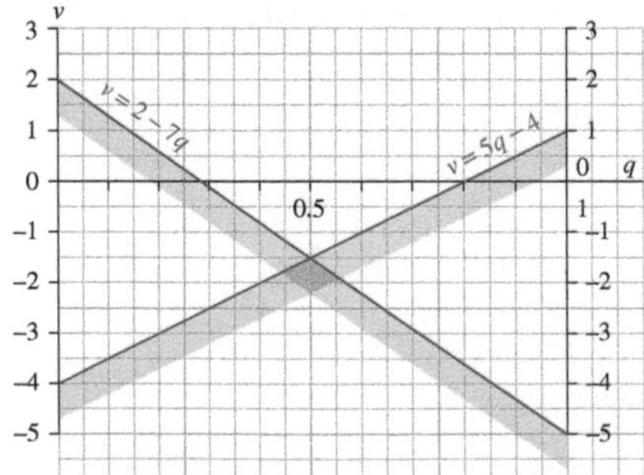

Max $(-v)$ when $2 - 7q = 5q - 4$, so $q = \frac{1}{2}$

This gives $v = -1\frac{1}{2}$

B plays B_1, B_2 with probabilities $\frac{1}{2}$, $\frac{1}{2}$

4 a

		B					
		B_1	B_2	B_3	B_4	Row minima	Max of row minima
A	A_1	−1	3	7	−1	−1	
	A_2	1	5	4	2	1	
	A_3	3	2	3	3	2	2
	Column maxima	3	5	7	3		
	Min of column maxima	3			3		

Max of row minima ≠ min of column maxima, so no stable solution.

b Column 4 dominates column 3, so remove column 3

Then row 2 dominates row 1, so remove row 1.

Then column 3 dominates column 1, so remove column 1.

The table is now:

		B	
		B_2	B_4
A	A_2	5	2
	A_3	2	3

c Let A play A_2 with probability p

$v \le 5p + 2(1 - p) = 3p + 2$

$v \le 2p + 3(1 - p) = 3 - p$

Graph as shown.

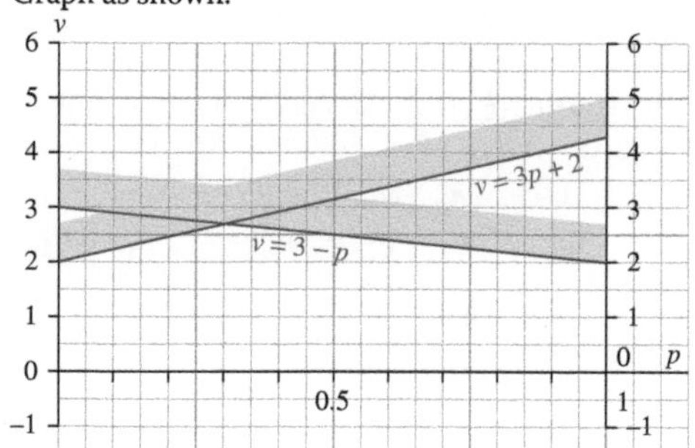

Max v when $3p + 2 = 3 - p$, so $p = \frac{1}{4}$

This gives $v = 2\frac{3}{4}$

A plays A_2, A_3 with probabilities $\frac{1}{4}$, $\frac{3}{4}$

d If B plays B_2 with probability q then $v = 5q + 2(1 - q) = 2\frac{3}{4}$

This gives $q = \frac{1}{4}$, so B plays B_2, B_4 with probabilities $\frac{1}{4}$, $\frac{3}{4}$

Exercise 14.3B

1 a Row 2 dominates row 1, so remove row 1

Column 2 dominates column 3, so removes column 3

This leaves:

		B	
		B_1	B_2
A	A_2	−3	3
	A_3	0	−4

Max of row mins = −3

Min of column maxs = 0

Solution is not stable

Let A play A_2 with probability p

$v \le -3p$

$v \le 3p - 4(1 - p) = 7p - 4$

Graph as shown.

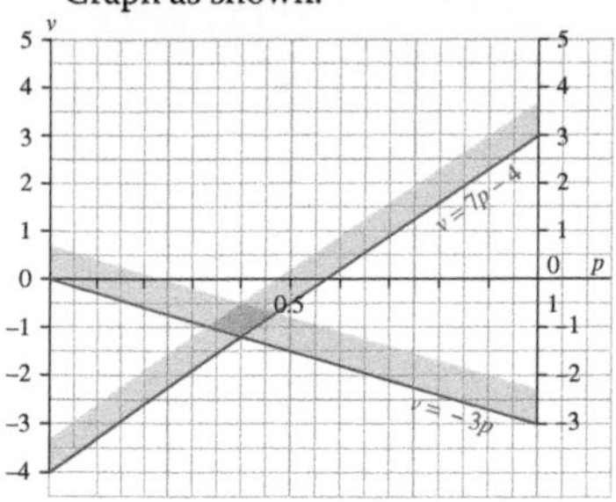

Max v when $-3p = 7p - 4$, so $p = \frac{2}{5}$

This gives $v = -1\frac{1}{5}$

A plays A_2, A_3 with probabilities $\frac{2}{5}$, $\frac{3}{5}$

If B plays B_1 with probability q then $v = -3q + 3(1 - q) = -1\frac{1}{5}$

This gives $q = \frac{7}{10}$, so B plays B_1, B_2 with probabilities $\frac{7}{10}$, $\frac{3}{10}$

b Column 2 dominates column 1, so remove column 1

Row 1 dominates row 2, so remove row 2

This leaves:

		B	
		B_2	B_3
A	A_1	1	5
	A_3	2	1

Max of row mins = 1

Min of column maxs = 2

Solution is not stable

Let A play A_1 with probability p

$v \le p + 5(1 - p)$ and so $v \le 5 - 4p$

$v \le 2p + (1 - p)$ and so $v \le p + 1$

Graph as shown.

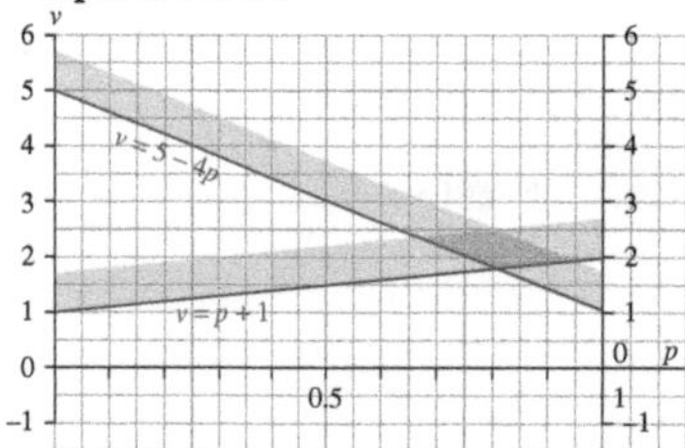

Max v when $5 - 4p = p + 1$, so $p = \frac{4}{5}$

This gives $v = 1\frac{4}{5}$

A plays A_1, A_3 with probabilities $\frac{4}{5}$, $\frac{1}{5}$

If B plays B_2 with probability q then $v = q + 2(1 - q) = 1\frac{4}{5}$

This gives $q = \frac{1}{5}$, so B plays B_2, B_3 with probabilities $\frac{1}{5}$, $\frac{4}{5}$

c Max of row mins = 2

Min of column maxs = 3

Solution is not stable

Let A play A_1 with probability p

$v \le 5p + (1 - p)$ and so $v \le 4p + 1$

$v \le 2p + 4(1 - p)$ and so $v \le 4 - 2p$

$v \le 3p + 2(1 - p)$ and so $v \le p + 2$

Graph as shown.

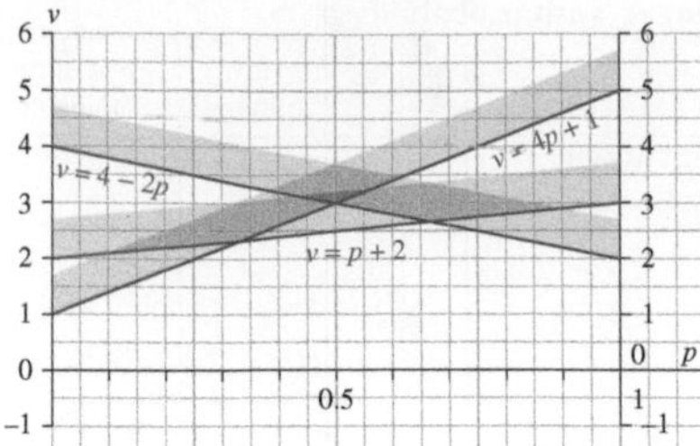

B would never play B_1

Max v when $p + 2 = 4 - 2p$, so $p = \frac{2}{3}$

This gives $v = 2\frac{2}{3}$

A plays A_1, A_2 with probabilities $\frac{2}{3}, \frac{1}{3}$

If B plays B_2 with probability q then $v = 2q + 3(1 - q) = 2\frac{2}{3}$

This gives $q = \frac{1}{3}$, so B plays B_2, B_3 with probabilities $\frac{1}{3}, \frac{2}{3}$

d Max of row mins = −1

Min of column maxs = 2

Solution is not stable

Let A play A_1 with probability p

$v \le 3p - 2(1 - p)$ and so $v \le 5p - 2$

$v \le -p + 4(1 - p)$ and so $v \le 4 - 5p$

$v \le p + 2(1 - p)$ and so $v \le 2 - p$

Graph as shown.

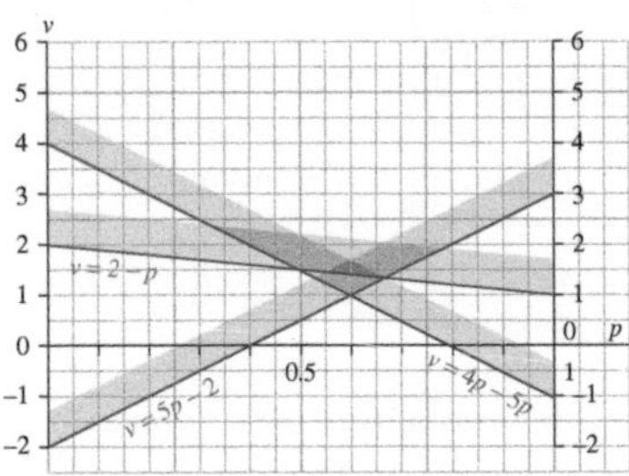

B would never play B_3

Max v when $5p - 2 = 4 - 5p$, so $p = \frac{3}{5}$

This gives $v = 1$

A plays A_1, A_2 with probabilities $\frac{3}{5}, \frac{2}{5}$

If B plays B_1 with probability q then $v = 3q - (1 - q) = 1$

This gives $q = \frac{1}{2}$, so B plays B_1, B_2 with probabilities $\frac{1}{2}, \frac{1}{2}$

2 a

		Y					
		I	II	III	IV	Row minima	Max of row minima
X	I	1	4	3	2	1	1
	II	−1	3	0	3	−1	
	III	4	2	5	1	1	1
	IV	−2	6	−3	1	−3	
	Column maxima	4	6	5	3		
	Min of column maxima				3		

Max of row minima ≠ min of column maxima, so no stable solution.

b Column 4 dominates column 2, so remove column 2

Then row 1 dominates row 4, so remove row 4

Then column 1 dominates column 3. so remove column 3.

This leaves:

		Y	
		I	IV
X	I	1	2
	II	−1	3
	III	4	1

Let Y play I with probability q

$(-v) \le -q - 2(1 - q)$ and so $v \ge 2 - q$

$(-v) \le q - 3(1 - q)$ and so $v \ge 3 - 4q$

$(-v) \le -4q - (1 - q)$ and so $v \ge 3q + 1$

Graph as shown.

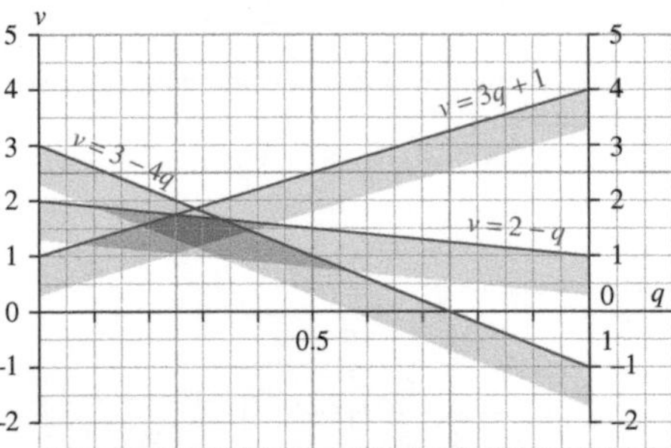

X would never play I

Max $(-v)$ when $3 - 4q = 3q + 1$, so $q = \frac{2}{7}$

This gives $v = 1\frac{6}{7}$

Y plays I, IV with probabilities $\frac{2}{7}, \frac{5}{7}$

If X plays II with probability p then $v = -p + 4(1 - p) = 1\frac{6}{7}$

This gives $p = \frac{3}{7}$, so X plays II, III with probabilities $\frac{3}{7}, \frac{4}{7}$

3 $V \le -p_1 + p_2$

$V \le -2p_2 - (1 - p_1 - p_2)$, giving $V \le p_1 - p_2 - 1$

$V \le -2p_1 - p_2 + (1 - p_1 - p_2)$, giving $V \le 1 - 3p_1 - 2p_2$

4 a Let A play A_1 with probability p

$v \le 2p + 5(1 - p) = 5 - 3p$

$v \le 5p + 2(1 - p) = 3p + 2$

$v \le 6p + (1 - p) = 5p + 1$

Graph as shown.

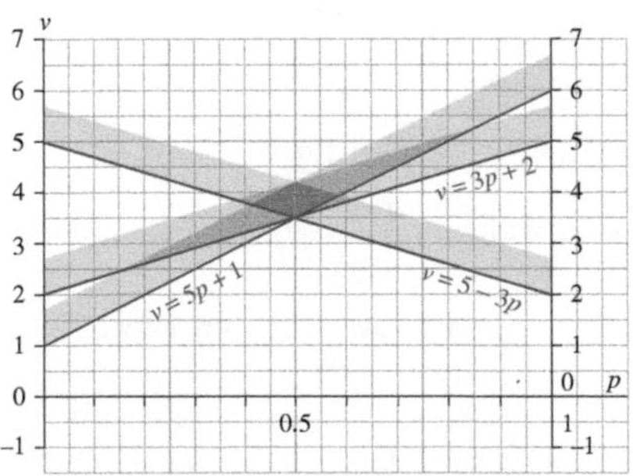

Max v when $5 - 3p = 3p + 2$, so $p = \frac{1}{2}$

A plays A_1, A_2 with probabilities $\frac{1}{2}, \frac{1}{2}$

b This gives $v = 3\frac{1}{2}$

c All three line pass through the same point, so none of the strategies is worse than the others.

d $q_1 + q_2 + q_3 = 1$ [1]

$2q_1 + 5q_2 + 6q_3 = 3\frac{1}{2}$ [2]

$5q_1 + 2q_2 + q_3 = 3\frac{1}{2}$ [3]

e From part **d**, [3] – [1], $q_2 = 2\frac{1}{2} - 4q$. Sub in [1] gives

$q_3 = 3q - 1\frac{1}{2}$

f $q_2 \geq 0$, which gives $q \leq \frac{5}{8}$

$q_3 \geq 0$, which gives $\frac{1}{2} \leq q$

Hence $\frac{1}{2} \leq q \leq \frac{5}{8}$

Max $q_2 = 2\frac{1}{2} - 4 \times \frac{1}{2} = \frac{1}{2}$. Max $q_3 = 3 \times \frac{5}{8} - 1\frac{1}{2} = \frac{3}{8}$

So $0 \leq q_2 \leq \frac{1}{2}$, $0 \leq q_3 \leq \frac{3}{8}$

Review exercise 14

1 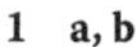 **a, b**

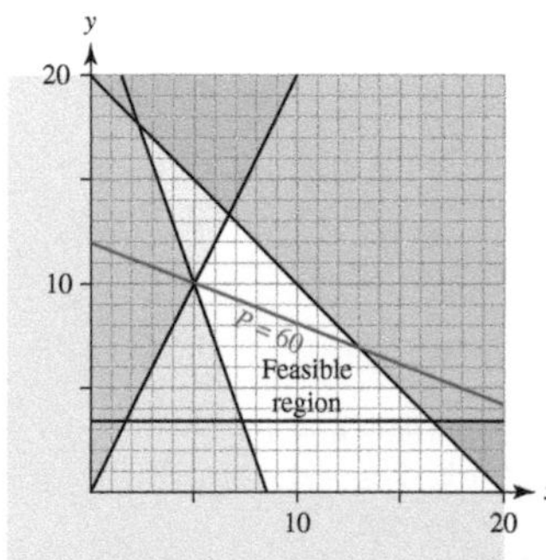

c $x = 6\frac{2}{3}$, $y = 13\frac{1}{3}$

2 **a** x = Value, y = Luxury, profit P = pence.

Maximise $P = 10x + 12y$

subject to $2x + 3y \leq 2400$

$4x + 3y \leq 4050$

$x \geq 0, y \geq 0$

b

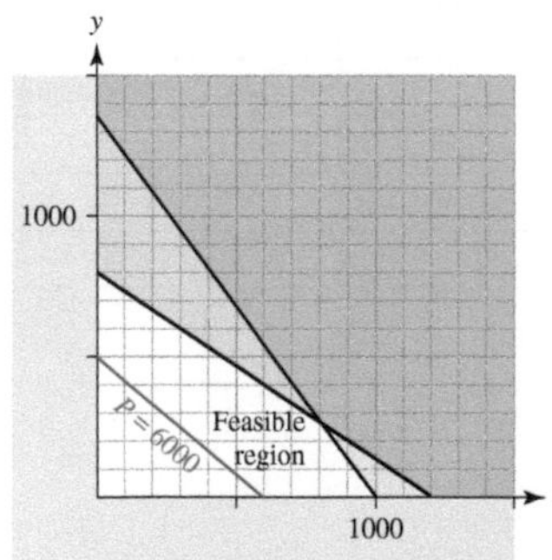

Maximum $P = 11250$, $x = 825$, $y = 250$

Make 825 kg of Value and 250 kg of Luxury

Profit = £112.50

3 **a** Play-safe strategy for A is A_2, with value 2.

b Play-safe strategy for B is B_2

c

		A	
		A_1	A_2
B	B_1	1	−4
	B_2	−3	−2

4 **a**

		B				
		I	II	III	Row minima	Max of row minima
A	I	1	0	−2	−2	
	II	−1	4	0	−1	
	III	2	3	5	2	2
	Column maxima	2	4	5		
	Min of column maxima	2				

Play-safe strategies are III for A and I for B.

Max of row minima = min of column maxima, so the solution is stable, that is a saddle point.

Neither player can gain by changing from their play-safe strategy.

b Value = 2

5 **a** Row 2 dominates row 3, so eliminate row 3

Column 3 dominates column 1 so eliminate column 1.

This leaves:

		B	
		II	III
A	I	3	9
	II	7	6

No further reduction is possible.

b The game does not have a stable solution because if it did the table could be reduced to a single value using dominance.

6 **a** Max of row mins = −1

Min of column maxs = 2

Solution is not stable

b Column 2 dominates column 3, so table becomes:

		B	
		B_1	B_2
A	A_1	3	−1
	A_2	−2	4

c $v \leq 3p - 2(1 - p) = 5p - 2$

$v \leq -p + 4(1 - p) = 4 - 5p$

d Graph as shown.

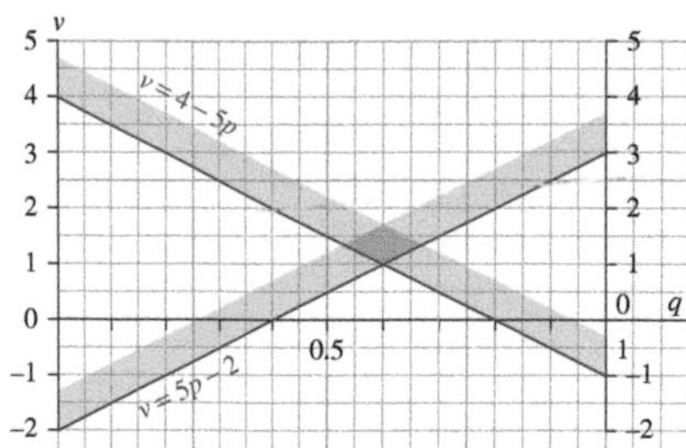

Max v when $5p - 2 = 4 - 5p$ giving $p = 0.6$.

A plays A_1, A_2 with probabilities 0.6, 0.4

When $p = 0.6$, $v = 1$

e If B plays B_1 with probability q then

$v = 3q - 2(1 - q) = 1$

This gives $q = 0.6$

B plays B_1, B_2 with probabilities 0.6, 0.4.

Assessment 14

1 a

		B				
		B_1	B_2	B_3	Row minima	Max of row minima
A	A_1	6	4	9	4	
	A_2	5	8	7	5	5
	A_3	3	5	6	3	
	Column maxima	6	8	9		
	Min of column maxima	6				

Play-safe strategy for A is A_2 and for B is B_1

b Game not stable as max and min values are different.

c Value of the game is 5

2 a Row M_1 dominates M_2 so M_2 can be removed.

		N		
		N_1	N_2	N_3
M	M_1	6	6	8
	M_3	4	2	−1

Column N_2 dominates N_1 so N_1 can be removed.

		N	
		N_2	N_3
M	M_1	6	8
	M_3	2	−1

Row M_1 dominates M_2 so M_2 can be removed.

		N	
		N_2	N_3
M	M_1	6	8

Column N_2 dominates N_3 so N_3 can be removed.

		N
		N_2
M	M_1	6

b You can deduce that there is a stable solution to the game. M should always play M_1 and N should always play N_2.
The value of the game is 6.

3 a Attempt to draw graph of $ax + by = c$
Correct plot of $x + y = 30$
Correct plot of $y = 2x$
Correct plot of $2y = 9$
Correct plot of $y + 2x = 28$
Correct shading of feasible region and identification in some suitable way.

b Attempt to draw graph of $ax + by = c$
Correct plot of $2x + 3y = m$

4 a

		B				
		B_1	B_2	B_3	B_4	Row minima
A	A_1	14	2	2	7	2
	A_2	6	2	2	9	2
	A_3	15	−1	1	0	−1
	Column maxima	15	2	2	9	

Maximum of row minima = 2
Minimum of column maxima = 2
The game is therefore stable as these values are equal.

b The value of the game is 2.

5 a 5

b Column I_3 is dominant over column I_2 (all values are lower) so Isla would never use this strategy.

c Table can be reduced to

	I_1	I_2	Row minima
C_1	2	−5	−5
C_2	−1	−3	−3
C_3	3	−5	−5
C_4	3	−2	−2
Column maxima	3	−2	

Play-safe for Connor is C_4 (= −2)
Play-safe for Isla is I_2 (= −2)
The game is stable as these values are equal.

6 a $x + y \le 22$

b Correct plot of $x = 12$
Correct plot of $y = 9$
Correct plot of $x + y = 22$
Correct plot of $x + 3y = 45$
Clear identification of feasible region.

7 a

		Danny			
		Red	Blue	Purple	Row minima
Sanjay	Yellow	2	−4	−1	−4
	Orange	−4	1	3	−4
	Green	1	−1	1	−1
	Column maxima	2	1	3	

Play-safe strategy for Sanjay is Green
Play-safe strategy for Danny is Blue

b Sanjay should play Orange.

8 a Assume x = number of hectares of cauliflowers and y = number of hectares of sprouts
Maximise $P = 100x + 125y$
Subject to
$x + y \le 20$ (land area)
$300x + 200y \le 4800$ (cost per hectare)
$10x + 20y \le 360$ (crop care)
$x \ge 0$
$y \ge 0$

b Farmer should plan 4 hectares of cauliflowers and 16 hectares of sprouts
Maximum Profit is £2400

9 a Let x = number of Supervit tablets and y = number of Extravit tablets
minimise $C = 15x + 20y$
Subject to
$15x + 20y \leq 600$ (cost)
$200x + 135y \leq 5400$ (Vit C)
$30x + 20y \geq 700$ (Niacin)
$20x + 40y \leq 800$
$x \geq 0, y \geq 0$

b Jenny should buy 16 tablets of Supervit and 12 tablets of Extravit.
Total cost is £4.80

Chapter 15

Exercise 15.1A

1 a $\circ$ is not a binary operation on $\mathbb{N}$ since when $a \leq b$, then $a - b \leq 0$ so $a \circ b$ is not a member of $\mathbb{N}$.

b $*$ is a binary operation on $\mathbb{N}$ since $a \times b$ is a natural number for all natural numbers a and b.

2 a $*$ is a binary operation on $\mathbb{Z}$ since $a - b$ is an integer for all integers a and b.

b $\lozenge$ is not a binary operation on $\mathbb{Z}$ since $\frac{a}{b^2+1}$ is not necessarily an integer, for example $\frac{3}{2^2+1} = \frac{3}{5}$ which is not an integer.

3 a $\lozenge$ is a binary operation on $\mathbb{Z}^+$ since the max of a and b will be either a or b, therefore a positive integer.

b $*$ is a binary operation on $\mathbb{Z}^+$ since $2a + b$ is a positive integer for all integers a and b.

4 a $\circ$ is not a binary operation on $\mathbb{Q}$ since a^b not always rational, for example if $a = 2$ and $b = \frac{1}{2}$ then $a \circ b = 2^{\frac{1}{2}} = \sqrt{2}$ which is irrational.

b $*$ is not a binary operation on $\mathbb{Q}$ since $\frac{a}{b}$ is not defined for $b = 0$.

5 a $*$ is not a binary operation on the set of even integers since $\frac{a}{2}$ could be odd so $\frac{a}{2} + b$ would be odd, e.g. $a = 6$, $b = 8 \Rightarrow a * b = \frac{6}{2} + 8 = 11$ which is not in the set.

b $*$ is a binary operation on the set of even integers since if a and b are even then $3a$ and $7b$ must also be even so $3a - 7b$ is even.

6 a Is not commutative, e.g. $5 - 3 \neq 3 - 5$
Is not associative since $(a - b) - c = a - b - c$ but $a - (b - c) = a - b + c$

b Is commutative and associative since addition of numbers is commutative: $a + b = b + a$ and associative: $a + (b + c) = (a + b) + c$

c Is commutative and associative and since $a * b$ is the maximum of the two values so the order they are written makes no difference.

d Is not commutative: $a^2 + b \neq b^2 + a$
or associative: $a * (b * c) = a * (b^2 + c) = a^2 + b^2 + c$ but $(a * b) * c = (a^2 + b) * c = (a^2 + b)^2 + c = a^4 + 2a^2b + b^2 + c$

e Is commutative and associative since multiplication of numbers is commutative: $a \times b = b \times a$ and associative: $a \times (b \times c) = (a \times b) \times c$

f Is commutative and associative and since $a * b$ is the minimum of the two values so the order they are written makes no difference.

g Is commutative and associative since multiplication of numbers is commutative: $a \times b = b \times a$ and associative: $a \times (b \times c) = (a \times b) \times c$

7 a i identity element does not exist: $a - 0 = a$ but $0 - a \neq a$

b i identity element is 0
ii All elements have inverses; the inverse of a is $-a$

c i Identity element is 1
ii The only element with an inverse is 1, which is self-inverse.

d i Identity element does not exist

e i Identity element is 1
ii The only element with an inverse is 1, which is self-inverse.

f i The identity element is -1
ii The only element with an inverse is -1, which is self-inverse.

g i Identity element is 1.
ii All elements except 0 have an inverse; the inverse of a is $\frac{1}{a}$

8 a Is not commutative: $\begin{pmatrix} a_1 & a_2 \\ a_3 & a_4 \end{pmatrix}\begin{pmatrix} b_1 & b_2 \\ b_3 & b_4 \end{pmatrix}$

$= \begin{pmatrix} a_1b_1 + a_2b_3 & a_1b_2 + a_2b_4 \\ a_3b_1 + a_4b_3 & a_3b_2 + a_4b_4 \end{pmatrix}$

but $\begin{pmatrix} b_1 & b_2 \\ b_3 & b_4 \end{pmatrix}\begin{pmatrix} a_1 & a_2 \\ a_3 & a_4 \end{pmatrix}$

$= \begin{pmatrix} a_1b_1 + a_3b_2 & a_2b_1 + a_4b_2 \\ a_1b_3 + a_3b_4 & a_2b_3 + a_4b_4 \end{pmatrix}$ and in general $a_1b_1 + a_3b_2$ $\neq a_1b_1 + a_2b_3$ etc.

b $\begin{pmatrix} a_1 & a_2 \\ a_3 & a_4 \end{pmatrix} + \left(\begin{pmatrix} b_1 & b_2 \\ b_3 & b_4 \end{pmatrix} + \begin{pmatrix} c_1 & c_2 \\ c_3 & c_4 \end{pmatrix}\right)$

$= \begin{pmatrix} a_1 & a_2 \\ a_3 & a_4 \end{pmatrix} + \begin{pmatrix} b_1 + c_1 & b_2 + c_2 \\ b_3 + c_3 & b_4 + c_4 \end{pmatrix}$

$= \begin{pmatrix} a_1 + b_1 + c_1 & a_2 + b_2 + c_2 \\ a_3 + b_3 + c_3 & a_4 + b_4 + c_4 \end{pmatrix}$ so associative.

$\begin{pmatrix} a_1 & a_2 \\ a_3 & a_4 \end{pmatrix} + \begin{pmatrix} b_1 & b_2 \\ b_3 & b_4 \end{pmatrix} = \begin{pmatrix} a_1 + b_1 & a_2 + b_2 \\ a_3 + b_3 & a_4 + b_4 \end{pmatrix}$

$= \begin{pmatrix} b_1 & b_2 \\ b_3 & b_4 \end{pmatrix} + \begin{pmatrix} a_1 & a_2 \\ a_3 & a_4 \end{pmatrix}$

since addition of real numbers is commutative.

c i Multiplicative identity is $\begin{pmatrix} 1 & 0 \\ 0 & 1 \end{pmatrix}$ and addition identity is $\begin{pmatrix} 0 & 0 \\ 0 & 0 \end{pmatrix}$

ii Multiplicative inverse of $\begin{pmatrix} a & b \\ c & d \end{pmatrix}$ is

$\frac{1}{ad - bc}\begin{pmatrix} d & -b \\ -c & a \end{pmatrix}$, it exists only when $ad - bc \neq 0$

Addition inverse of $\begin{pmatrix} a & b \\ c & d \end{pmatrix}$ is of $\begin{pmatrix} -a & -b \\ -c & -d \end{pmatrix}$ and always exists.

9 a The identity element is 1; the numbers 1 and -1 are both self-inverse but 0 does not have an inverse.

b The identity element is 2, which is also self-inverse but the other elements do not have an inverse

c The identity element is 0, all elements have an inverse; the inverse of a is $-a$.

d The identity element is 1, which is also self-inverse but the other elements do not have an inverse.

Exercise 15.1B

1 i a Identity element is c

b Inverse of a is b and vice-versa; c is self-inverse.

ii a Identity element is a

b Inverse of b is d and vice-versa; a and c are self-inverse.

iii a Identity element is B

b All elements are self-inverse.

iv a Identity element is A

b Inverse of B is D and vice-versa; inverse of C is E and vice-versa; A is self-inverse.

2 a $a*(b*c)=a*b=c$ and $(a*b)*c=c*c=c$

$b*(a*c)=b*a=c$ and $(b*a)*c=c*c=c$

$c*(a*b)=c*c=c$ and $(c*a)*b=a*b=c$

No need to check, e.g. $a*(c*b)$ as function is commutative so $b*c=c*b$

Therefore it is associative.

b $a*(b*c)=a*d=d$ and $(a*b)*c=b*c=d$

$a*(b*d)=a*a=a$ and $(a*b)*d=b*d=a$

$a*(d*c)=a*b=b$ and $(a*d)*c=d*c=b$

$b*(a*c)=b*c=d$ and $(b*a)*c=b*c=d$

$b*(a*d)=b*d=a$ and $(b*a)*d=b*d=a$

$b*(c*d)=b*b=c$ and $(b*c)*d=d*d=c$

$c*(a*b)=c*b=d$ and $(c*a)*b=c*b=d$

$c*(a*d)=c*d=b$ and $(c*a)*d=c*d=b$

$c*(b*d)=c*a=c$ and $(c*b)*d=d*d=c$

No need to check, e.g. $a*(c*b)$ as function is commutative so $b*c=c*b$

Therefore it is associative.

3 a

×	1	−1	i	$-i$
1	1	−1	i	$-i$
−1	−1	1	$-i$	i
i	i	$-i$	−1	1
$-i$	$-i$	i	1	−1

b The identity element is 1

c The inverse of i is $-i$ and vice-versa; 1 and −1 are self-inverse.

d The operation is associative since multiplication of complex numbers is associative.

4 a $S=\left\{\begin{pmatrix}1&0\\0&1\end{pmatrix},\begin{pmatrix}0&-1\\1&0\end{pmatrix},\begin{pmatrix}-1&0\\0&-1\end{pmatrix},\begin{pmatrix}0&1\\-1&0\end{pmatrix}\right\}$

b Not a binary operation since A + B not necessarily a member of S, for example $\begin{pmatrix}1&0\\0&1\end{pmatrix}+\begin{pmatrix}1&0\\0&1\end{pmatrix}=\begin{pmatrix}2&0\\0&2\end{pmatrix}$ which is not in S.

c

	0	1	2	3
0	0	1	2	3
1	1	2	3	0
2	2	3	0	1
3	3	0	1	2

d i Identity is $\begin{pmatrix}1&0\\0&1\end{pmatrix}$ or 0

ii 0 and 2 are self-inverses, 1 is inverse of 3 and vice-versa.

e Represent rotations of $0, \frac{\pi}{2}, \pi$ and $\frac{3\pi}{2}$ radians anti-clockwise around the origin.

5 a Minimum value of k is 5

b

	1	2	3	4	5
1	2	3	4	5	1
2	3	4	5	1	2
3	4	5	1	2	3
4	5	1	2	3	4
5	1	2	3	4	5

c 1 is inverse of 4 and vice-versa; 2 is inverse of 3 and vice-versa; 5 is a self-inverse.

6 a A, B, C, D, F, H are self-inverses; E is the inverse of G and vice-versa

b $A=\begin{pmatrix}1&0\\0&-1\end{pmatrix}, B=\begin{pmatrix}-1&0\\0&1\end{pmatrix}, C=\begin{pmatrix}0&1\\1&0\end{pmatrix},$

$D=\begin{pmatrix}0&-1\\-1&0\end{pmatrix}, E=\begin{pmatrix}0&-1\\1&0\end{pmatrix}, F=\begin{pmatrix}-1&0\\0&-1\end{pmatrix},$

$G=\begin{pmatrix}0&1\\-1&0\end{pmatrix}, H=\begin{pmatrix}1&0\\0&1\end{pmatrix}$

c

*	A	B	C	D	E	F	G	H
A	H	F	E	G	C	B	D	A
B	F	H	G	E	D	A	C	B
C	G	E	H	F	B	D	A	C
D	E	G	F	H	A	C	B	D
E	D	C	A	B	F	G	H	E
F	B	A	D	C	G	H	E	F
G	C	D	B	A	H	E	F	G
H	A	B	C	D	E	F	G	H

d e.g. A * C = E but C * A = G

7 a All the possible combinations of matrices of this form are in S so it is a binary operation

b $\begin{pmatrix}1&0&0\\0&1&0\\0&0&1\end{pmatrix}$ since this is the identity matrix, I, and

AI = IA = A for all matrices A

c $\begin{pmatrix}a&0&0\\0&b&0\\0&0&c\end{pmatrix}^2=\begin{pmatrix}a^2&0&0\\0&b^2&0\\0&0&c^2\end{pmatrix}$ if $a, b, c=\pm1$ then

$a^2, b^2, c^2=1$ therefore all the matrices are self-inverse

d $\begin{pmatrix}a&0&0\\0&b&0\\0&0&c\end{pmatrix}\begin{pmatrix}d&0&0\\0&e&0\\0&0&f\end{pmatrix}=\begin{pmatrix}ad&0&0\\0&be&0\\0&0&cf\end{pmatrix}$

and $\begin{pmatrix}d&0&0\\0&e&0\\0&0&f\end{pmatrix}\begin{pmatrix}a&0&0\\0&b&0\\0&0&c\end{pmatrix}=\begin{pmatrix}ad&0&0\\0&be&0\\0&0&cf\end{pmatrix}$

So matrix multiplication is commutative on S

Exercise 15.2A

1 e.g.

a $7 \pmod 3 = 4$

b $12 \pmod 7 = 5$

c $19 \pmod 4 = 3$

d $27 \pmod 9 = 0$

e $25 \pmod 6 = 1$

f $-5 \pmod 8 = 3$

g $-55 \pmod{11} = 0$

h $48 \pmod{13} = 9$

2 a $21 + 19 \pmod 3 = 0 + 1 \pmod 3 = 1 \pmod 3$

b $4 + 25 \pmod 7 = 4 + 4 \pmod 7 = 8 \pmod 7 = 1 \pmod 7$

c $68 - 34 \pmod 4 = 0 - 2 \pmod 4 = -2 \pmod 4 = 2 \pmod 4$

d $79 - 94 \pmod 9 = 7 - 4 \pmod 9 = 3 \pmod 9$

e $5 \times 14 \pmod 6 = 5 \times 2 \pmod 6 = 12 \pmod 6 = 4 \pmod 6$

f $25 \times 37 \pmod 8 = 1 \times 5 \pmod 8 = 5 \pmod 8$

g $82 \times -16 \pmod{11} = 5 \times 6 \pmod{11} = 30 \pmod{11}$
$= 8 \pmod{11}$

h $29 \times 75 \pmod{13} = 3 \times 10 \pmod{13} = 30 \pmod{13}$
$= 4 \pmod{13}$

3 a $37951 \equiv 1^{37} \pmod{10}$
$\equiv 1 \pmod{10}$

b $1736 \equiv 1^{26} \pmod 5$
$\equiv 1 \pmod 5$

c $982 \equiv 2^{9} \pmod 5$
$\equiv 2 \pmod 5$

d $3219 \equiv 0^{48} \pmod 3$
$\equiv 0 \pmod 3$

e $3268 \equiv 1^{84} \pmod 3$
$\equiv 1 \pmod 3$

f $12653 \equiv 5 \pmod 6$
$\equiv (-1) \pmod 6$

so $12653^{11} \equiv (-1)^{11} \pmod 6$
$\equiv (-1) \pmod 6$
$\equiv 5 \pmod 6$

Exercise 15.2B

1 a

*	0	1
0	0	1
1	1	0

2 a

*	0	1	2
0	0	1	2
1	1	2	0
2	2	0	1

b

*	0	1	2	3
0	0	1	2	3
1	1	2	3	0
2	2	3	0	1
3	3	0	1	2

c

*	1	2	3	4
1	1	2	3	4
2	2	4	1	3
3	3	1	4	2
4	4	3	2	1

d

*	1	2	3	4	5	6
1	1	2	3	4	5	6
2	2	4	6	1	3	5
3	3	6	2	5	1	4
4	4	1	5	2	6	3
5	5	3	1	6	4	2
6	6	5	4	3	2	1

3 In question 1

i identity element is 0

ii both elements are self-inverse

In question 2

a i identity element is 0

ii 1 is the inverse of 2 and vice-versa; 0 is self-inverse

b i identity element is 0

ii 1 is the inverse of 3 and vice-versa; 0 and 2 are self-inverse

c i identity element is 1

ii 2 is the inverse of 3 and vice-versa; 1 and 4 are self-inverse

d i identity element is 1

ii 2 is the inverse of 4 and vice-versa; 3 is the inverse of 5 and vice-versa; 1 and 6 are self-inverse

4 a

*	0	1	2	3
0	0	1	2	3
1	1	3	0	2
2	2	0	3	1
3	3	2	1	0

b i Identity is 0

ii 0 and 3 are self-inverses, 1 is inverse of 2 and vice-versa.

5 a

◆	0	1	2	3	4	5
0	4	5	0	1	2	3
1	5	0	1	2	3	4
2	0	1	2	3	4	5
3	1	2	3	4	5	0
4	2	3	4	5	0	1
5	3	4	5	0	1	2

b i Identity is 2

ii 2 and 5 are self-inverses, 0 is inverse of 4 and vice-versa; 1 is inverse of 3 and vice-versa.

Review exercise 15

1 a Is a binary operation since $a + 3b$ always a natural number if a and b are natural numbers.

b Is not a binary operation, if $b = 1$ then $(a+1)(b-1) = (a-1) \times 0 = 0$ which is not a positive integers.

c Is a binary operation: if $a, b \in \mathbb{Q}$ then they can be written $a = \frac{p}{q}, b = \frac{m}{n}$ for some integers p, q, m, n.

So $\frac{b}{a^2+1} = \frac{\frac{m}{n}}{\left(\frac{p}{q}\right)^2+1} = \frac{mq^2}{n(p^2+q^2)} \in \mathbb{Q}$ since p, q, m, n are integers.

d Is not a binary operation, if one of a and b is negative then ab will be negative so $\sqrt{ab}$ will not be a real number, e.g. $a = 1, b = -1, \sqrt{ab} = \sqrt{-1} = i$.

2 a Not associative since $\frac{\frac{a}{b}}{c}=\frac{ac}{b}$ but $\frac{a}{\frac{b}{c}}=\frac{a}{bc}$

b Is not associative: $(a^2+b^2)^2+c^2=a^4+2a^2b^2+b^4+c^2$
$a^2+(b^2+c^2)^2=a^2+b^4+2b^2c^2+c^4\neq a^4+2a^2b^2+b^4+c^2$

c Associative since $\frac{a\left(\frac{bc}{b+c}\right)}{a+\left(\frac{bc}{b+c}\right)}=\frac{abc}{a(b+c)+bc}=\frac{abc}{ab+ac+bc}$

and $\frac{\left(\frac{ab}{a+b}\right)c}{\left(\frac{ab}{a+b}\right)+c}=\frac{abc}{ab+c(a+b)}=\frac{abc}{ab+ac+bc}$

3 a Not commutative, in general $\frac{a}{b}\neq\frac{b}{a}$

b Commutative since $a^2+b^2=b^2+a^2$

c Commutative since $\frac{ab}{a+b}=\frac{ba}{b+a}$

4 a i Identity element does not exist (since not commutative).

ii Therefore no inverse elements.

b i Identity element is 0 since $0^2+a^2=a^2+0^2=a^2$

ii 0 is self-inverse but no other elements have an inverse in $\mathbb{Z}$ since $a^2+(a^{-1})^2=0\Rightarrow(a^{-1})^2\leq 0$ which is only possible if a^{-1} is imaginary.

c i No identity element.

ii So no inverse elements.

5 a Identity element is E

b Inverse of A is D and vice-versa; inverse of B is C and vice-versa; E is self-inverse.

c Cayley table is symmetric along leading diagonal therefore * is commutative.

6 a $379+612 \pmod 5\equiv 4+2 \pmod 5\equiv 6 \pmod 5\equiv 1 \pmod 5$

b $1079-351 \pmod 3\equiv 2-0 \pmod 3\equiv 2 \pmod 3$

c $326\times 249 \pmod 8\equiv 6\times 1 \pmod 8\equiv 6 \pmod 8$

d $-532\times 249 \pmod{10}\equiv -2\times 9 \pmod{10}\equiv 8\times 9 \pmod{10}\equiv 72 \pmod{10}\equiv 2 \pmod{10}$

7 a

*	0	2	4	6	8
0	0	2	4	6	8
2	2	4	6	8	0
4	4	6	8	0	2
6	6	8	0	2	4
8	8	0	2	4	6

b Identity element is 0

c The inverse of 2 is 8 and vice-versa; the inverse of 4 is 6 and vice-versa; 0 is self-inverse.

8 a

	A	B	C
A	B	C	A
B	C	A	B
C	A	B	C

b i identity is $\begin{pmatrix}1&0\\0&1\end{pmatrix}$ or C

ii C is self-inverse, A is inverse of B and vice-versa.

c Rotations of 0°, 120°, 240° clockwise around origin.

Assessment 15

1 a 52

b e.g. $2\blacksquare(3\blacksquare 1)=2+(3+1^2)^2=18$
$(2\blacksquare 3)\blacksquare 1=(2+3^2)+1^2=12$
So not associative
e.g. $2\blacksquare 3=2+3^2=11$
$3\blacksquare 2=3+2^2=7$ so not commutative

2 a $3+7\equiv 0 \pmod{10}$

b $125-621\equiv 5-1 \pmod{10}$
$\equiv 4 \pmod{10}$

c $358\times 6914\equiv 8\times 4 \pmod{10}$
$32\equiv 2 \pmod{10}$

d $2563^4\equiv 3^4 \pmod{10}$
$81\equiv 1 \pmod{10}$

3 a i If b is a fraction, then a^b may not be rational, e.g. if $b=\frac{1}{2}$, $a=2$ then $2^{\frac{1}{2}}=\sqrt{2}$ which is irrational.
So not a binary operation on $\mathbb{Q}$.

ii If a is negative then a^b may not be a real number, e.g. $a=-1$, $b=\frac{1}{2}$. Then $(-1)^{\frac{1}{2}}=i$ which is an imaginary number
So not a binary operation on $\mathbb{R}$

iii If a and b are natural numbers then a^b will also be a natural number so it is a binary operation on $\mathbb{N}$.

b In general $a^b\neq b^a$ so operation is not commutative, therefore no identity element

c $(a*b)*c=(a^b)^c=a^{bc}$
$a*(b*c)=a^{(b^c)}$
so not associative

4 a

*	−2	2
−2	2	−2
2	−2	2

b 2

c both elements are self-inverses

5 a the identity element is U

b $V^{-1}=X$

c U and W are self-inverses

6 a i

*	0	1	2	3	4
0	0	1	2	3	4
1	1	2	3	4	0
2	2	3	4	0	1
3	3	4	0	1	2
4	4	0	1	2	3

ii identity element is 0

iii inverse of 1 is 4 and vice-versa; inverse of 2 is 3 and vice-versa; 0 is self-inverse

b e.g. $18^{12}\equiv 3^{12} \pmod 5$
$\equiv (3^4)^3 \pmod 5$
$\equiv 81^3 \pmod 5$
$\equiv 1^3 \pmod 5$
$\equiv 1 \pmod 5$

7 a C

b The operation is not commutative
There is no element, e, such that $x\blacklozenge e=e\blacklozenge x=x$
Therefore no identity element

c e.g. $(A\blacklozenge B)\blacklozenge C=B\blacklozenge C=C$
but $A\blacklozenge(B\blacklozenge C)=A\blacklozenge C=E$
so not associative

8 a

	A	B	C	D
A	B	C	D	A
B	C	D	A	B
C	D	A	B	C
D	A	B	C	D

b D and B are self-inverses
A is inverse of C and vice-versa

9 Let $a = 2m + 1$, $b = 2n + 1$ where m and n are integers.
Then $a * b = 2(2m+1)-(2n+1)$
$= 4m+2-2n-1$
$= 2(2m-n)+1$
Which is an odd number since $2(2m-n)$ is even.
Hence it is a binary operation

10 a

×	M_1	M_2	M_3	M_4	M_5	M_6
M_1	M_2	M_3	M_4	M_5	M_6	M_1
M_2	M_3	M_4	M_5	M_6	M_1	M_2
M_3	M_4	M_5	M_6	M_1	M_2	M_3
M_4	M_5	M_6	M_1	M_2	M_3	M_4
M_5	M_6	M_1	M_2	M_3	M_4	M_5
M_6	M_1	M_2	M_3	M_4	M_5	M_6

b M_6
c M_4
d M_3

Index